Typical properties of selected materials used in engineering[1] (SI units)

Material	Density, kg/m³	Ultimate strength: Tension, MPa	Ultimate strength: Compression,[2] MPa	Ultimate strength: Shear, MPa	Yield strength[3]: Tension, MPa	Yield strength[3]: Shear, MPa	Modulus of elasticity, GPa	Modulus of rigidity, GPa	Coefficient of thermal expansion, 10^{-6}/°C	Ductility, percent elongation in 50 mm	Poisson's ratio ν
STEEL:											
Structural (ASTM-A36)	7860	400	—	—	250	145	200	79	11.7	23	0.27–0.30
High-strength–low-alloy ASTM-A242	7860	480	—	—	345	210	200	79	11.7	21	0.27–0.30
Quenched and tempered alloy ASTM-A514	7860	825	—	—	690	380	200	79	11.7	18	0.27–0.30
Stainless, (302)											
Cold-rolled	7920	860	—	—	520	—	190	73	17.3	12	0.27–0.30
Annealed	7920	620	—	—	275	152	190	73	17.3	50	0.27–0.30
CAST IRON:											
Gray, 4.5% C, ASTM A-48	7200	170	650	240	—	—	70	28	12.1	0.5	0.2–0.3
Malleable, ASTM A-47	7300	340	620	330	230	—	165	64	12.1	10	0.2–0.3
ALUMINUM:											
Alloy 2014-T6 (4.4% Cu)	2800	480	—	290	410	220	72	27	23.0	13	0.33
Alloy 6061-T6 (1% Mg)	2710	290	—	185	255	140	69	26	23.6	17	0.33
YELLOW BRASS (65% Cu, 35% Zn):											
Cold-rolled	8470	540	—	300	435	250	105	39	20	8	0.34
Annealed	8470	330	—	220	105	65	105	39	20	62	0.34
PHOSPHOR BRONZE:											
Cold-rolled (510)	8860	560	—	—	520	275	110	41	17.8	10	0.34
Spring temper (524)	8780	840	—	—	—	—	110	—	18.4	4	0.34
MAGNESIUM ALLOY (8.5% Al)	1800	380	—	165	275	—	45	—	26	7	0.35
TITANIUM ALLOY (6% Al, 4% V)	4460	900	—	—	825	—	114	—	9.5	10	0.33
MONEL ALLOY 400 (Ni-Cu):											
Cold-worked	8830	675	—	—	580	345	180	—	13.9	22	0.32
Annealed	8830	550	—	—	220	125	180	—	13.9	46	0.32
TIMBER, air dry:[4]											
Douglas fir	525	—	50	7.5	—	—	12.5	—	Varies:	—	—
Southern pine	610	—	50	9.2	—	—	11	—	3.0 to 4.5	—	—
CONCRETE:											
Medium strength	2320	—	28	—	—	—	25	—	10	—	0.1–0.2
High strength	2320	—	40	—	—	—	30	—	10	—	0.1–0.2
Nylon, molded	1100	55	—	—	—	—	2	—	81	50	0.4
Polystyrene	1050	48	90	55	—	—	3	—	72	4	—
Neoprene	1250	24	—	—	—	—	—	—	—	500	0.45–0.5
Granite	2770	20	240	35	—	—	70	—	7	—	0.2–0.3
Glass, 98% silica	2190	—	50	—	—	—	65	28	80	—	0.2–0.27

[1] Properties of metals vary widely as a result of variations in composition, heat treatment, and mechanical working.
[2] For ductile metals the compression strength is generally assumed to be equal to the tension strength.
[3] Offset of 0.2 percent.
[4] Timber properties are for loading parallel to the grain.

MECHANICS OF SOLIDS
CONCEPTS AND APPLICATIONS

MECHANICS OF SOLIDS
CONCEPTS AND APPLICATIONS

William B. Bickford
Arizona State University

IRWIN
Homewood, IL 60430
Boston, MA 02116

Senior sponsoring editor:	*Bill Stenquist*
Developmental editor:	*Kelley Butcher*
Marketing manager:	*Robb Linsky*
Project editor:	*Ethel Shiell*
Production manager:	*Irene H. Sotiroff*
Designer:	*Michael Warrell*
Cover designer:	*Trish Lawrence*
Art manager:	*Kim Meriwether*
Compositor:	*Bi-Comp, Inc.*
Typeface:	*10/12 Times Roman*
Printer:	*R. R. Donnelley & Sons*

Library of Congress Cataloging-in-Publication Data

Bickford, William (William B.)
Mechanics of solids: concepts and applications / William B. Bickford.
p. cm.
Includes index.
ISBN 0-256-11425-0 (5.25″ version)
ISBN 0-256-12344-6 (3.5″ version)
1. Strength of materials. 2. Mechanics, Applied. I. Title.
TA405.B498 1993
620.1′1292—dc20 91–38441

Printed in the United States of America

1 2 3 4 5 6 7 8 9 0 DOC 9 8 7 6 5 4 3 2

To my wife Diane—for her love and support

PREFACE

This text is intended for the first course in the study of solid mechanics, or mechanics of materials, usually taken by engineering students in the sophomore or junior year. It is assumed that the student has completed the basic courses in physics, calculus, and statics.

OBJECTIVES

The purpose of the study of solid mechanics is twofold. First, students must be introduced to the basic ideas and concepts used in the area of solid mechanics. This includes a thorough treatment of the three basic ideas of equilibrium, deformation, and material behavior properties. These three ideas are emphasized and kept in focus throughout this text, with careful study of how their combination leads to specific theories about the transmission of forces in typical structural members.

Second, students need ample practice in applying these theories to practical situations that involve the transmission of axial, torsional, and bending-force resultants. Relatively simple structures are examined in this text to analyze and design for strength and stiffness. Both of these goals require continual awareness of all the assumptions that are necessary parts of a given theory, to understand and avoid situations where application of theory is unwarranted.

A traditionally difficult aspect of developing a solid mechanics text has been in striking a balance between theory and the many practical applications that are important to students who will go on to use the knowledge in actual practice. Here there seems to be a dichotomy between the available texts. Some treat equilibrium, deformation, and material behavior in a single introductory chapter, while others devote separate chapters to each of these basic ideas.

I have chosen the second option, because a major goal is to keep the three basic ideas clearly in focus when developing a theory or applying the results of the theory to actual situations. In my experience, a careful separate treatment of each of the basic ideas provides an excellent framework for the study of elementary theories in solid mechanics.

CONTENTS

Chapter 1 presents an overview of the typical areas of application of solid mechanics. The two basic concerns of solid mechanics—strength and stiffness—are introduced and discussed. The transition from rigid-body mechanics to the mechanics of deformable bodies, with the need to describe deformations resulting in changes in the size and shape of a body, is considered. The three basic ideas of solid mechanics are introduced.

Chapter 2 contains a comprehensive treatment of equilibrium. The idea that equilibrium must always be considered when developing a specific theory is emphasized. The concept of stress is introduced and illustrated by several familiar examples. Stress transformations and Mohr's circle are treated in detail.

Chapter 3 is devoted to the discussion of deformations and to the resulting changes in size and shape of a body. Simple extensional and shear-strain situations associated with specially shaped bodies are discussed in detail. Strain-displacement relations are developed and transformation of strain is considered.

The behavior and properties of materials is covered in Chapter 4. This study is taken from the standpoint that in order to answer questions pertaining to both strength and stiffness, one must relate the force and displacement variables discussed, respectively, in Chapters 2 and 3. The qualitative and quantitative results obtained from the uniaxial tension test are described. Elastic versus inelastic, brittle versus ductile, and time-dependent behavior including fatigue are discussed. The concepts of factor of safety and theories of failure are introduced and applied to specific stress states.

Chapters 5, 6, and 7 are devoted to axial, torsional, and bending deformations, respectively. In each chapter, the three basic ideas of equilibrium, deformation, and material behavior are combined to produce the appropriate relations between the internal forces and the stresses, for the purpose, in Chapters 5, 6, and 7, of answering questions of strength. Each chapter contains a section of important and useful exceptions to the theory.

Chapter 8 deals with the stiffness question for the bending deformation problem. Integration of $EIv'' = M$ and $EIv'''' = -w$ are demonstrated as techniques for determining the transverse deflection v. Singularity functions are introduced and their use explored in the many situations where loadings are not continuous along the length of the beam. The methods of superposition and integration-moment-area are presented. Statically indeterminate problems are treated using superposition and integration.

Combined loading problems are covered in Chapter 9. Combined axial and bending loads, as well as combined axial, bending, and torsional loads, are considered. Thin-walled cylindrical and spherical vessels under internal pressure are treated, as are situations where such vessels are simultaneously subjected to external loads.

Chapter 10 is devoted to alternate formulations of the classes of problems treated in the previous chapters, using several energy methods. Strain energy and complementary energy ideas, including Castigliano's theorems, are presented and applied to the problem of axial, torsional, and bending deformations. The application of complementary principles when analyzing statically indeterminate systems is discussed. Stiffness and flexibility approaches are shown in connection with the strain energy and complementary principles, respectively.

Stability of structures is covered in Chapter 11, with emphasis on the potential loss of stability in the presence of compressive stresses. The stability of discrete systems is analyzed using the principle of stationary potential energy. Column stability for a range of boundary conditions is considered. Buckling of eccentrically loaded columns as well as initially unstraight columns is addressed.

It would be impossible to cover every topic in this text in an introductory solid mechanics course; therefore, selected topics throughout the text have been marked as optional with a red outline box. Further streamlining options are suggested here:

Sections	Approximate # of periods
2.1–2.6	5
3.1–3.2	2
4.1–4.6, 4.9–4.10	5
5.1–5.3, 5.4.1 or 5.4.2, 5.5.1	5
6.1–6.3, 6.4.1 or 6.4.2, 6.5.1	5
7.1–7.8	7
8.1–8.3, 8.4, or 8.5, or 8.6	5
9.1–9.4	4
10.3–10.4	4

or

Sections	Approximate # of periods
2.1–2.3	2
3.1–3.2	2
4.1–4.6, 4.9–4.10	5
5.1–5.3, 5.4.1 or 5.4.2, 5.5.1	5
6.1–6.3.1, 6.4.1 or 6.4.2, 6.5.1	5
7.1–7.8	7
8.1–8.3, 8.4, or 8.5, or 8.6	5
2.5–2.7	3
9.1–9.4	4
10.2–10.4	4

In a course where stability is to be covered, the appropriate number of periods must be subtracted from the above outlines, or the energy methods of Chapter 10 deleted.

PEDAGOGY

Each chapter has a brief chapter outline and introduction, an abundance of illustrations, many completely worked examples (over 200 in all), and, at the end of each chapter, a closure section that summarizes concepts learned in the chapter. In developing the theory and applying it in each example problem, attention is focused on the three basic ideas of equilibrium, deformation, and material behavior.

Homework problems generally appear at the end of text sections. These problems are intended to assist the student in understanding and applying the basic ideas. More challenging problems are indicated by red outline boxes. There are approximately 1,500 homework problems in all.

At the end of several of the chapters, the student is asked to apply computer-programming skills to the solution of solid mechanics problems. In these computer exercises, the student is asked to develop algorithms for solving specific classes of problems.

Problems that are less well defined and that may require an iterative solution are referred to as Analysis/Design problems; these are placed at the ends of several chapters. They are intended to enable the student to begin to see the real nature of the design process. The use of the computer may be appropriate in solving these.

SUPPLEMENTS

The Instructor's Manual, available to teachers from the publisher, contains complete solutions to the homework problems.

SOFTWARE

Interactive software for the IBM PC, PS/2, and Macintosh is included at the back of the book. The software, developed with the assistance of Robert L. Rankin of Arizona State University, includes approximately 100 exercises and tutorials designed to amplify the theory and applications discussed in the text and illustrated in the example problems. Instructions for loading the software can be found on the disk label.

William B. Bickford

ACKNOWLEDGMENTS

Publishing a new solid mechanics text is a challenging undertaking that requires a great investment of time, resources, and expertise. As the editors for *Mechanics of Solids: Concepts and Applications,* we are pleased to acknowledge the assistance of many reviewers in the development of this project. The following people reviewed the manuscript in part or whole:

Nicholas J. Altiero, *Michigan State University*
David J. McGill, *Georgia Technical Institute*
George G. Adams, *Northeastern University*
Hamid Saadatmanesh, *University of Arizona*
Daniel L. Garber, Jr., *University of Maryland*
Stephen R. Swanson, *University of Utah*
Donald A. Grant, *University of Maine*
William L. Bingham, *North Carolina State University*
Gordon Batson, *Clarkson University*
Richard Nielsen, *University of Idaho*
Terry L. Kohutek, *Texas A & M University*
Robert Baren, *Temple University*
David Wang, *University of Wisconsin—Platteville*
Louis C. Tartaglione, *University of Lowell*
James R. Morgan, *Texas A & M University*
Kenneth B. Oster, *University of Missouri—Rolla*
Robert A. Thompson, *University of Dayton*
David R. Budney, *University of Alberta*
David H. Sanders, *University of Nevada—Reno*
Edwin C. Rossow, *Northwestern University*
R. N. Dubey, *University of Waterloo*
E. C. Aifantis, *Michigan Technological University*

Hareesh V. Tippur, *Auburn University*
Thalia Anagnos, *San Jose State University*

The author worked hard to use each person's suggestions to improve the quality of the book; this text has been greatly enhanced as a result of their comments. We hope that you find *Mechanics of Solids: Concepts and Applications* useful for your course.

Bill Stenquist
Senior Sponsoring Editor

Kelley Butcher
Developmental Editor

TO THE STUDENT

Prior to beginning the study of solid mechanics you will have completed a course in statics. Statics deals almost exclusively with the application of Newton's laws to the equilibrium of particles and rigid bodies. The primary tool for investigating mechanics is the free-body diagram (FBD). The importance of this step in understanding, setting up, and solving mechanics problems cannot be overstated.

Draw Free-Body Diagrams

No real body behaves as a particle or rigid body; thus it is necessary to consider the deformations that accompany the forces transmitted throughout the body. Since bodies are composed of real materials that behave differently under the action of forces, it is also necessary to distinguish between the behaviors of different materials. The study of the various aspects of material behavior constitutes an essential part of solid mechanics.

New concepts such as stress and strain are necessary to adequately describe and analyze deformable solids. You are already somewhat familiar with these ideas from everyday experience. The pressure that you feel on your body due to contact with the chair or other support as you read these words is an example of a stress. During the course of studying this book you will become much more familiar with the concept of stress as it pertains to the transmission of forces in solids. The process of stretching a rubber band (a deformable body) produces stretching, an example of strain. The stretching of a rubber band, bending of a yardstick, and twisting of a yardstick are similar to the three basic types of deformations that will be considered in this course. Develop a habit of deforming (in a nondestructive manner) many of the simple objects you encounter every day, to help you understand the ideas of deformation and strain.

The study of solid mechanics becomes very rewarding as it increases awareness of your ability to use basic principles of mechanics to analyze and design an enormous variety of objects comprised of solid material. Aircraft, spacecraft, buildings, automobiles, and machines of all kinds are but a few of the items whose successful design depends strongly upon the proper knowledge of solid mechanics. May you find the study of solid mechanics interesting and rewarding!

William B. Bickford

CONTENTS

LIST OF SYMBOLS

A area, point
a, b, c dimensions, distances, constants
C centroid, constant of integration, compressive force
c distance from neutral axis to outer surface of a beam
D displacement (translation or rotation), diameter
d diameter, dimension, distance
E modulus of elasticity, Young's modulus
E_s secant modulus of elasticity
E_t tangent modulus of elasticity
e eccentricity, dimension, distance, unit volume change (dilatation, volumetric strain)
F force, flexibility
G modulus of elasticity in shear
g acceleration of gravity
H distance, force, reaction, horsepower
h height, dimension
I moment of inertia (or second moment) of a plane area
I_x, I_y, I_z moments of inertia with respect to x, y, and z axes
I_{xy} product of inertia with respect to the x and y axes
I_p polar moment of inertia
I_1, I_2 principal moments of inertia
J torsion constant
K stress-concentration factor
k spring constant, stiffness
L length, distance, span length
L_e effective length of a column
ln natural logarithm
M bending moment, couple, mass
M_p plastic moment for a beam
M_y yield moment for a beam
m moment per unit length, mass per unit length
N axial force
n factor of safety, number, ratio, integer, revolutions per minute (rpm)
O origin of coordinates
P force, concentrated load, axial force, power

P_{allow}	allowable load (or working load)
P_{cr}	critical load for a column
P_u	ultimate load
P_y	yield load
p	pressure (force per unit area)
Q	force, concentrated load, first moment (or static moment) of a plane area
q	intensity of distributed load (force per unit distance), shear flow
R	reaction, radius, force
r	radius, distance, radius of gyration ($r = \sqrt{I/A}$)
S	section modulus of the cross section of a beam, shear center, stiffness, force
s	distance, length along a curved line
T	twisting couple or torque, temperature, tensile force
T_u	ultimate torque
T_y	yield torque
t	thickness, intensity of distributed torque (torque per unit distance)
U	strain energy
u_0	strain energy density (strain energy per unit volume)
U^*	complementary energy
u_0^*	complementary energy density (complementary energy per unit volume)
V	shear force, volume
v	deflection of a beam, velocity
v', v'', etc.	dv/dx, d^2v/dx^2, etc.
W	weight, work
w	load per unit of length (force per unit length)
X	statical redundant
x, y, z	rectangular coordinates, distances
$\bar{x}, \bar{y}, \bar{z}$	coordinates of centroid
α	angle, coefficient of thermal expansion, nondimensional ratio, spring constant, stiffness
β	angle, nondimensional ratio, spring constant, stiffness
γ	shear strain, specific weight (weight per unit volume)
$\gamma_{xy}, \gamma_{yz}, \gamma_{zx}$	shear strains in the xy, yz, and zx planes
γ_θ	shear strain for inclined axes
$\gamma_{x_1y_1}$	shear strain in the x_1y_1 plane
δ, Δ	deflection, displacement, elongation
ε	normal strain
$\varepsilon_x, \varepsilon_y, \varepsilon_z$	normal strains in the x, y, and z directions
ε_θ	normal strain for inclined axes
$\varepsilon_{x_1}, \varepsilon_{y_1}$	normal strains in the x_1 and y_1 directions

$\varepsilon_1, \varepsilon_2, \varepsilon_3$ principal normal strains
ε_y yield strain
θ angle, angle of twist per unit length, angle of rotation of beam axis
θ_p angle to a principal plane or to a principal axis
κ curvature ($\kappa = 1/\rho$)
ρ radius, radius of curvature, radial distance in polar coordinates, mass density (mass per unit volume, specific mass)
ν Poisson's ratio
σ normal stress
$\sigma_x, \sigma_y, \sigma_z$ normal stresses on planes perpendicular to the x, y, and z axes
σ_θ normal stress on inclined plane
$\sigma_{x_1}, \sigma_{y_1}$ normal stresses on planes perpendicular to the rotated x_1y_1 axes
$\sigma_1, \sigma_2, \sigma_3$ principal stresses
σ_{allow} allowable stress (or working stress)
σ_{cr} critical stress for a column ($\sigma_{\text{cr}} = P_{\text{cr}}/A$)
σ_{pl} proportional-limit stress
σ_r residual stress
σ_u ultimate stress
σ_y yield stress
τ shear stress
$\tau_{xy}, \tau_{yz}, \tau_{zx}$ shear stresses on planes perpendicular to the x, y, and z axes and parallel to the y, z, and x axes
τ_θ shear stress on inclined plane
$\tau_{x_1y_1}$ shear stress on plane perpendicular to the rotated x_1 axis and parallel to the y_1 axis
τ_{allow} allowable stress (or working stress) in shear
τ_u ultimate stress in shear
τ_y yield stress in shear
ϕ angle, angle of twist
ψ nondimensional ratio
ω angular velocity, angular frequency ($\omega = 2\pi f$)

Greek Alphabet

A	α	Alpha	N	ν	Nu
B	β	Beta	Ξ	ξ	Xi
Γ	γ	Gamma	O	o	Omicron
Δ	δ	Delta	Π	π	Pi
E	ε	Epsilon	P	ρ	Rho
Z	ζ	Zeta	Σ	σ	Sigma
H	η	Eta	T	τ	Tau
Θ	θ	Theta	Υ	υ	Upsilon
I	ι	Iota	Φ	ϕ	Phi
K	κ	Kappa	X	χ	Chi
Λ	λ	Lambda	Ψ	ψ	Psi
M	μ	Mu	Ω	ω	Omega

MECHANICS OF SOLIDS
CONCEPTS AND APPLICATIONS

CHAPTER 1 INTRODUCTION

CHAPTER OUTLINE

1.1 USES OF SOLID MECHANICS

Over the past several thousand years there has been a continuing record of the achievements of the human race in constructing a wide variety of structures, machines, monuments, and so forth. Many of these were not only successful at the time they were built, but have withstood the test of time and continue to exist to the present day. The pyramids, the temples of Greece, the Roman Colliseum, the many castles scattered over the surface of the earth—these have lasted hundreds or even thousands of years, some still serving their original purpose. Over these past several thousand years, humankind has built other structures as well, such as catapults, ships, and bridges that functioned successfully over shorter spans of time.

Before the middle of the 17th century, these structures were built mainly on the basis of experience. In each generation, the "engineers" served long apprenticeships to learn the tricks of their trade from the old pros, experts who had many years of experience, perhaps involving as many failures as successes. There is no clear evidence that these engineers of old had developed any significant capability of supporting their experiments with any form of "calculating." It is an interesting fact that engineers in ancient times were employed primarily for the purpose of helping their rulers wage war.

More recent successes of note include structures such as the Eiffel Tower in Paris, the Sears Tower in Chicago (the world's tallest building), the Boeing 747 commercial aircraft, and several of the space shuttle vehicles. On a smaller, more ordinary scale, the automobile, the electric drill, the bicycle, and many other everyday items are structures that routinely perform functions in the real world. The relatively few failures of structures in the modern world attest to the success of engineers, whose work is an integral part of the design and construction of machines and structures we all use daily.

A more or less scientific approach to solid mechanics, or strength of materials as it is often called, began with Leonardo da Vinci (1452–1519). He was the first to apply the principles of statics to determine the internal forces in members, and the first to perform experiments on the strength of engineering materials. He experimented with iron wires and with beams. As a result of his testing of beams he concluded that the strength of a beam supported at both ends varies inversely as the length and directly as the width.

Galileo Galilei (1564–1642)

Italian physicist, astronomer, and mathematician; performed experiments to test the strength of bars and beams; founded the science of dynamics. (Source: Deutsches Museum, München.)

Although Galileo (1564–1642) is best known for his construction of a telescope around the year 1609 and his subsequent contributions to astronomy, he also made the first attempts to apply stress analysis logically. His results—published in *Two New Sciences* in 1638—represent the beginning of the science of strength of materials. His experiments consisted of simple tension tests, from which he concluded that the strength of a bar is proportional to its cross-sectional area and independent of its length. He also conducted bending experiments and concluded, incorrectly, that the stresses necessary to counteract the bending are distributed uniformly over the cross section. The famous sketch associated with his bending investigations appears on the cover and is repeated in Fig. 1.1. For the cantilever beam, he observed that in

Figure 1.1 Illustration of Galileo's bending test

order to keep the strength constant the cross-sectional dimensions must be increased at a greater rate than the length.

There are many other individuals, too numerous to mention, who have contributed to the development of solid mechanics or strength of materials. For very interesting and detailed accounts of the history of strength of materials the reader is referred to *A History of Elasticity and Strength of Materials* by I. Todhunter and K. Pearson, and to *History of Strength of Materials* by Stephen P. Timoshenko.

1.2 AREAS OF APPLICATION

The areas of application of solid mechanics are essentially unlimited. Automobiles, ships, air and spacecraft, buildings and other structures, machines of all sizes and shapes performing a wide variety of tasks—all need to be analyzed on the basis of whether or not they are "strong enough" and "stiff enough." Any mechanical system that must function in the presence of forces, temperature changes, and so forth, is generally designed to fulfill requirements of strength and flexibility as dictated by the principles of solid mechanics. We will use the term **structure** to indicate any mechanical system whose design is influenced by the proper application of the principles of solid mechanics. Several diverse areas of application of the discipline of solid mechanics are presented below.

1.2.1 Aerospace Vehicles

There are many applications of the mechanics of solids in the area of aerospace vehicles. These include airplanes, satellites, missiles, and space shuttles. The main inputs to consider for such vehicles include

1. Weight.
2. Aerodynamic forces (i.e., lift and drag).
3. Propulsion forces.
4. Dynamic or acceleration forces due to maneuvers.
5. Landing forces.
6. Thermal inputs due to such effects as aerodynamic and solar heating.

Any aerospace vehicle must continually perform its mission under the action of any combination of these inputs. The following are two examples of the typical considerations that must be given to aerospace vehicles.

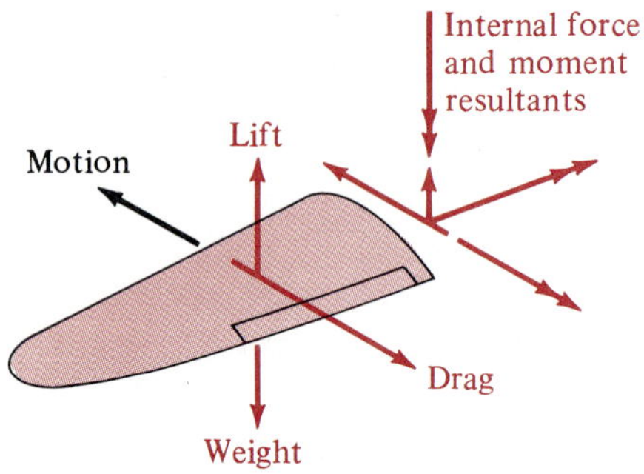

Figure 1.2 Typical aircraft wing

Example 1.1 Aircraft wing The forces affecting any aircraft wing are similar, whether it be the wing of a small private aircraft, a large commercial jetliner, a transonic fighter, or a shuttle wing. Consider a typical aircraft wing, indicated in Fig. 1.2.

Of interest are the internal forces that must exist at a typical location along the wing in order for the wing, or any portion of the wing, to be in equilibrium. Ultimately, we wish to determine whether the wing will function properly under the external forces present; that is, is the wing constructed so as to be strong enough and stiff enough?

Example 1.2 Fighter landing on an aircraft carrier When a fighter lands on the deck of an aircraft carrier as shown in Fig. 1.3, substantial forces can result. There can be large vertical deceleration forces between the landing deck of the carrier and the tires of the fighter, in addition to the horizontal force from the cable that "catches" the fighter. The structural design of the fighter must be such that the internal forces the fighter is capable of developing are sufficient for the fighter to survive many such landings.

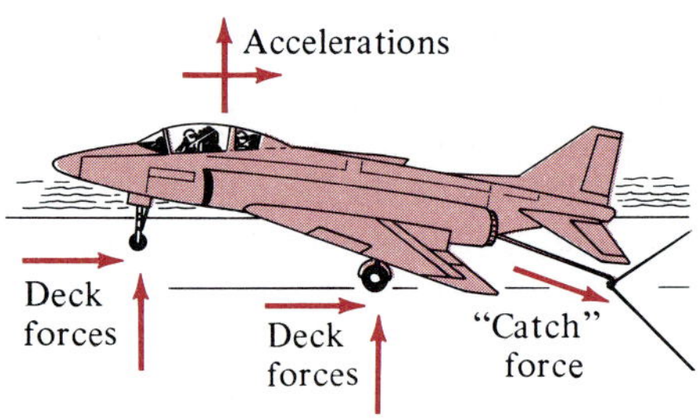

Figure 1.3 Landing forces on a typical fighter

1.2.2 Internal Combustion Engine and Auto Chassis

Many of the components of an internal combustion engine and chassis are subjected to high levels of force and temperature change. Inputs to consider for typical components include

1. Possible high temperatures and temperature gradients.
2. Combustion chamber pressures.

3. High translational and rotational acceleration loads.
4. Significant contact forces between gears.
5. Loads from rough roads.

With proper maintainence, a modern internal combustion engine and chassis are expected to operate for several tens of thousands of miles without major failures.

Example 1.3 Connecting rod A connecting rod in a typical internal combustion engine is indicated in Fig. 1.4. The connecting rod must be able to transmit forces between the crankshaft and the piston while undergoing fairly high accelerations, and while under the action of contact loads at the locations where it is connected to the piston and the crankshaft.

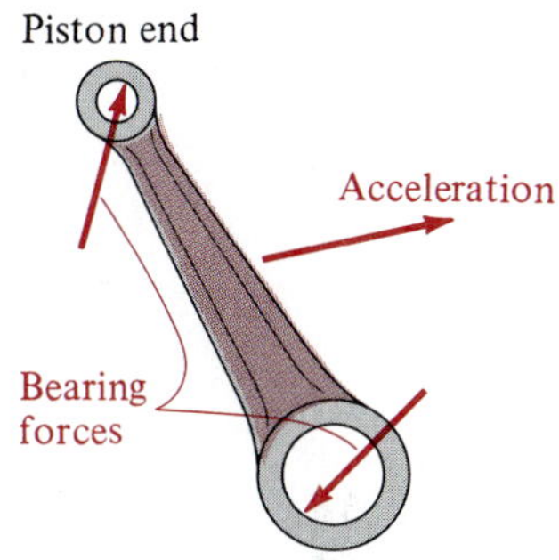

Figure 1.4 Connecting rod for an internal combustion engine

Example 1.4 Gears in a transmission During operation of an automobile, the power generated by the engine is transmitted to the back wheels by the drive train. The transmission, which is part of the drive train, functions by a system of gears. A typical gear interface is indicated in Fig. 1.5a. The torque transmitted through the drive train results in forces between the gear teeth as indicated in Fig. 1.5b. The gear teeth must be strong enough to transmit these forces.

(a) Gear system (Courtesy of Boston Gear.)

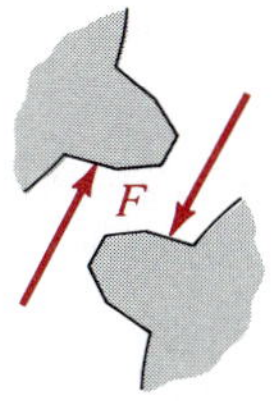

(b) Gear tooth force

Figure 1.5 A typical gear interface

1.2.3 General Structural Analyses

Principles of solid mechanics are central to the successful design and construction of any structure. Examples of a wide range of structures of varying sizes, and the forces that act on them, are presented below.

Example 1.5 Multistory building Among the inputs that must be considered for a structural design of multistory buildings are

1. Weight of the structure and its contents.
2. Wind loads.
3. Ground motions due to earthquakes.
4. Structural loads associated with solar heating.

In order to withstand these inputs, a skyscraper must be designed to be strong enough and stiff enough. A sketch of a typical skyscraper such as the World Trade Center in Chicago is given in Fig. 1.6a.

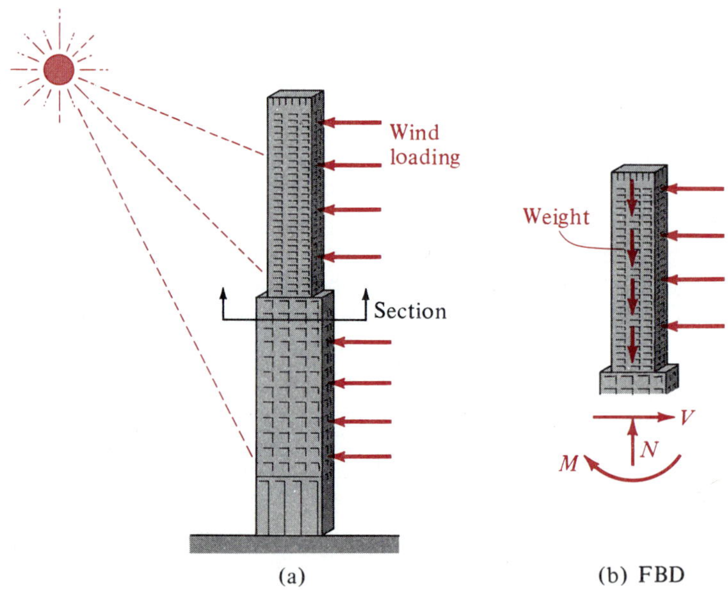

Figure 1.6 Typical skyscraper

Indicated in the figure are the weight and wind loadings. The building must be strong enough to withstand the internal forces and moments at any level due to the weight and wind loadings applied above that level, as indicated in Fig. 1.6b. The building must also be stiff enough so that occupants near the top will not experience undue motion caused by wind loads. Thermal loads resulting from solar heating can also be significant.

Example 1.6 Supertanker on the high seas A supertanker weighing several hundred thousand tons is another example of a structure. A supertanker in high seas is sketched in Fig. 1.7. Although the pressure on the hull would vary uniformly with depth in smooth seas, the pressure distribution on the hull would be much less uniform during high seas. Due to the extremely large mass of the tanker, there are also significant inertia loadings caused by pitch, roll, and yaw motions. The supertanker would have to be strong enough to withstand the water pressures necessary to balance its weight, the inertial loads, and the uneven pressure distributions associated with the high seas. Again, thermal load resulting from exposure to the sun can also be significant.

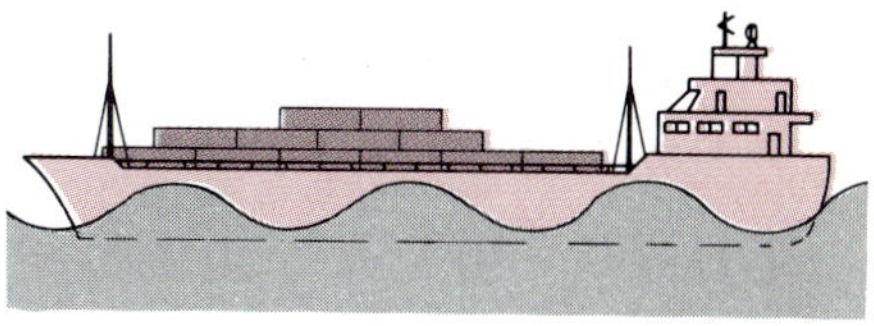

Figure 1.7 Supertanker in high seas

Example 1.7 Microprocessor lead wire Toward the other end of the size spectrum, a very thin gold lead wire, part of a small electronic device, is indicated in Fig. 1.8. The gold lead wire, perhaps a few ten-thousandths of an inch in diameter, might connect a microcomputer chip to its surroundings. The wire could be contained within a piece of equipment mounted in an aerospace vehicle that experiences very large accelerations as a result of vibrations. The wire, therefore, must be strong enough to withstand the forces arising from large vibrational accelerations. In addition, currents passing through the wire during operation of the equipment can cause cyclic thermal loads; these, too, must be considered.

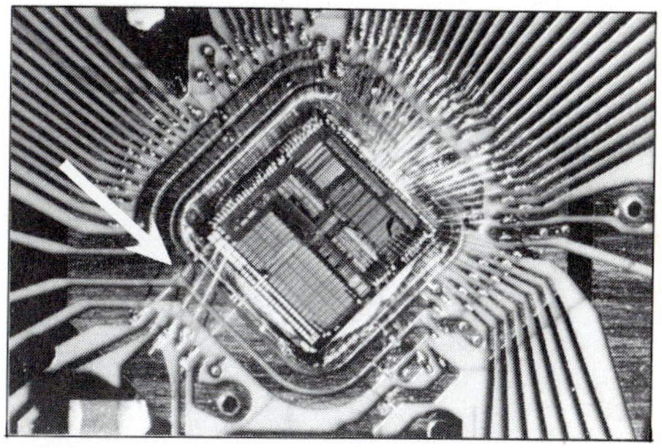

Figure 1.8 Thin gold lead wire in an electronic device

The world's first 32-bit, single-chip central processing unit packaged on its finstrate powers the HP 9000 desktop computer. (Photo courtesy of Hewlett-Packard Company.)

The phrases "strong enough" and "stiff enough," used in the examples above, are terms whose precise meanings will be discussed numerous times throughout the text. It is customary to ask whether the strength of a structure or system is adequate, rather than to ask whether it is "strong enough," although both phrases have the same meaning. Flexibility is a term often used in connection with the idea of stiffness. If the deformations of a structure are too large, the structure is termed too flexible or, alternatively, not stiff enough. Flexibility and stiffness are reciprocal terms—the greater the flexibility, the smaller the stiffness. Generally, strength requirements rather than flexibility requirements govern the design of a structure; if a structure is strong enough to withstand the loads it is subjected to, its stiffness is usually adequate. The engineer designing a structure must ensure that both the strength and stiffness are adequate to meet all requirements.

1.3 THE TWO BASIC PROBLEMS OF SOLID MECHANICS

From a careful consideration of the example applications discussed above it follows that there are two basic problems to address in order to assess whether the design of a particular structure is adequate. One problem is to determine the **strength** of the structure; the other is to determine the **stiffness** of the structure.

The strength problem can be stated roughly as follows. Given the external loads, determine the reactions and the corresponding internal forces present in the structure. Then determine if the strength of the structure is adequate to sustain the internal forces resulting from the loads. (This is the "strong enough" part of a solid mechanics problem.)

The stiffness problem is basically concerned with the deformations of the structure. By stiffness we generally mean a resistance to deformation. In order to guard against excessive deformation, that is, too much flexibility, we must be able to determine the deformations resulting from external loads. (This is the "stiff enough" part of the problem.)

In some situations it is possible to first solve the strength problem and then, using the results of the strength problem, proceed with a separate evaluation of the stiffness. This situation corresponds to the **statically determinate** case usually discussed in a statics course. In a typical **statically indeterminate** case, the strength and stiffness problems must be considered jointly in order to determine whether the structure is strong enough and stiff enough.

We will see in later chapters that in order to solve both of these problems we must have information about the properties of the material(s) from which the structure is made. Material properties generally fall into one of two categories: those properties associated with the strength characteristics and those properties associated with the stiffness characteristics. We will discuss material properties in detail in Chapter 4.

As an example of a strength consideration, a structure composed of steel may be adequate to withstand all of the loads to which it is subjected, whereas a structure of exactly the same size and shape composed of wood, rubber, or concrete may not have adequate strength. The strength type material properties mentioned above are used to assist in the evaluation of the strength of the structure.

Similarly, the stiffness of structures of the same size and shape is different when the structures are composed of different materials such as steel, wood, rubber, or concrete. The stiffness type material properties mentioned above are used to assist in the evaluation of the stiffness of the structure.

Thus it is necessary to have knowledge of material properties in order to answer questions about strength or stiffness. The following are two examples of strength and stiffness considerations for simple structures.

Example 1.8 Consider the problem of a weight W, an **external load,** suspended by a tension member of weight w as indicated in Fig. 1.9a. A free-body diagram (FBD) of the tension member in Fig. 1.9b shows that for equilibrium, the reaction at the support is $W + wL$, the total weight. The free-body diagram in Fig. 1.9c indicates that at an intermediate position the **internal force** $T(x) = W + wx$ is again equal to the weight below the section in question. Depending upon the diameter of the tension member and the magnitude of the weight $W + wL$, it is possible for the tension member to fracture; that is, it may not be strong enough. Specifically, if the weight is very large and the member is composed of a relatively weak material such as rubber, fracture is likely unless the area of the rubber is very large. On the other hand, if the tension member is a steel wire, a fairly small diameter wire could support a reasonably large weight W. Material properties would be examined to assess whether or not the internal force $T = W + wx$ is excessive. Thus it is clear that the strength of the structure is dependent on the material used. A measure of the stiffness of this structure could be obtained by determining the total elongation of the member under the action of the weights W and wL. This would require knowledge of stiffness type material properties for the particular material used.

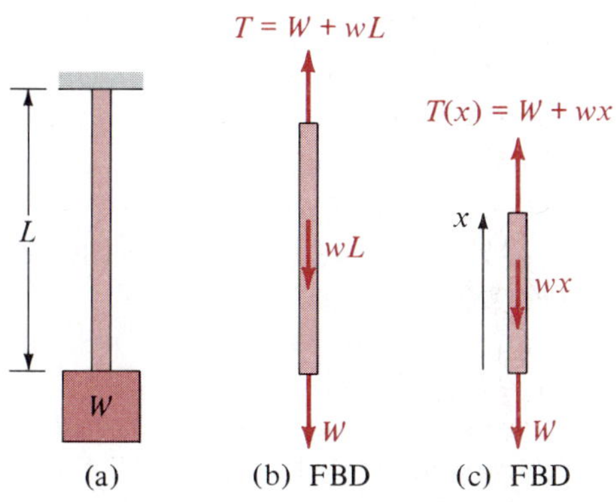

Figure 1.9 Weight W hanging from a tension member

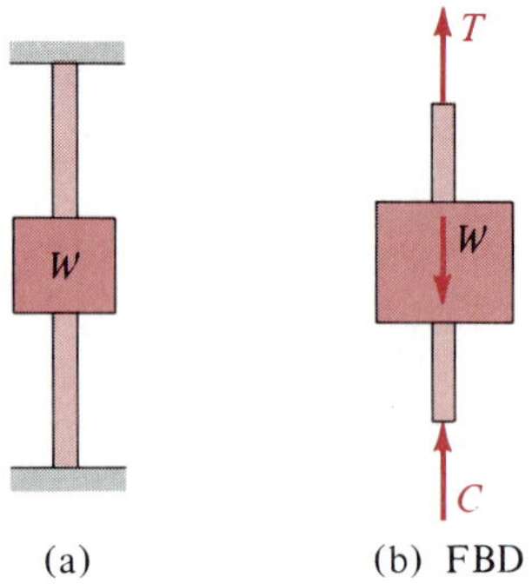

Figure 1.10 Hanging weight W supported at both the top and bottom

Example 1.9 In Example 1.8, the equilibrium problem was easily solved to yield the internal forces so that the strength of the structure could be addressed directly. Contrast this with the following problem of a similar nature. Two members, assumed for simplicity to be straight and circular in cross section, are used to support a weight W (an external load) as indicated in Fig. 1.10a.

A free-body diagram of a part of the structure that exposes the internal forces in both members is indicated in Fig. 1.10b. The weight W, an **external load**, must clearly be supported by the tension force T in the upper portion of the structure and by the compression force C in the lower portion. T and C are **internal forces**. The one equation of equilibrium resulting from the requirement that the resultant force in the vertical direction be zero indicates that $W = T + C$; that is, one equation and two unknowns (C and T). Thus, for this problem, the internal forces cannot be determined solely on the basis of statics: the structure is statically indeterminate. Without knowledge of the internal forces T and C, it is not possible to evaluate the strength of the structure.

Equations are needed involving the deformations and the stiffness type material properties of the structure, in order to form an additional equation for determining the two unknowns T and C. In other words, it is necessary to supplement the equation of equilibrium $W = T + C$ with an equation that results from considering the deformations and the material behavior of the structure. After solving the two equations for the force and displacement unknowns, it would be possible to evaluate both the strength and stiffness of the structure.

These two examples are typical of the character of solid mechanics problems in general. Throughout the text the reader will at times be asked questions regarding the strength of a structure and at other times questions regarding the stiffness. Note that for a statically determinate structure, it is possible to solve the strength problem without considering the stiffness; once the strength problem has been solved the stiffness problem can then be solved using the results of the strength problem. For a statically indeterminate structure, however, the formulation of the problem must involve both the internal forces and the force-displacement relations. Once the internal forces and deformations have been determined, it is possible to evaluate both the strength and stiffness of the structure. These ideas are indicated in the flowchart shown in Fig. 1.11.

The process is as follows: The equations of equilibrium relating the external loads to the reactions and internal forces are written. If the number of internal forces and reactions equals the number of independent equations of equilibrium, the structure is said to be statically determinate. In this case the equations are solved for the reactions and internal forces. The internal forces are used in connection with strength type material properties to assess the strength of the structure.

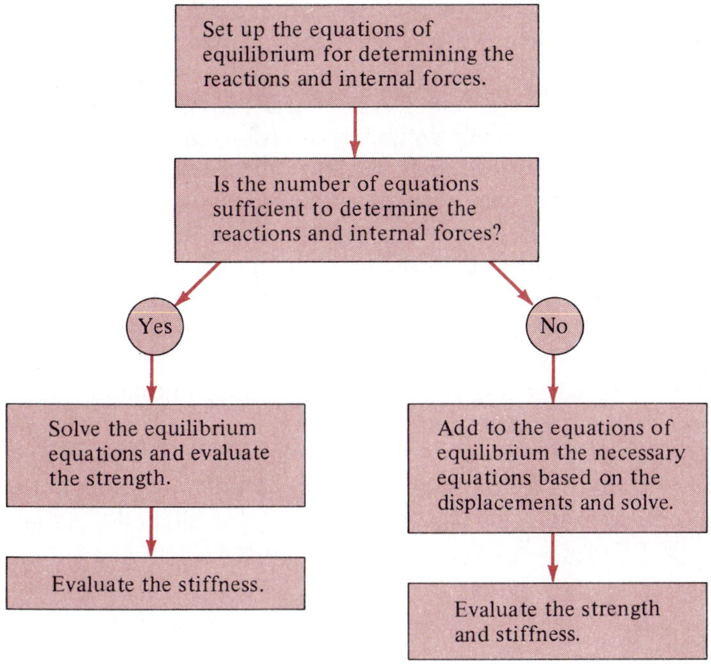

Figure 1.11 Flowchart for problems involving statically determinate and statically indeterminate structures

The stiffness can then be evaluated on the basis of an additional independent analysis involving the stiffness type material properties.

For a statically indeterminate structure, the number of internal forces and reactions exceeds the number of independent equations of equilibrium. The additional equations must be supplied by considering the displacements and the material behavior of the structure. After the equilibrium equations and the additional displacement equations are solved, it is possible to evaluate both the strength and stiffness of the structure. In either situation, the end result is that information is available for answering questions regarding the two basics problems.

1.4 THE THREE BASIC IDEAS OF SOLID MECHANICS

Recall from basic mechanics that there are essentially two components for setting up and solving a statics or dynamics problem; these are referred to as **kinetics** and **kinematics**. Basically, kinetics is concerned with Newton's second law, usually written as $\mathbf{F} = m\mathbf{a}$, or in the case of a static problem, $\mathbf{F} = \mathbf{0}$, that is, no unbalanced force. Kinematics is the description of the geometry of the motion, independent of its causes or

Sir Isaac Newton (1642–1727)

English mathematician who formulated the laws of gravity and motion and the elements of differential calculus. (Source: Deutsches Museum, München.)

effects. Recall also that for the most part in statics and dynamics, a body is assumed to be either a **particle** or a **rigid body**.

The assumption that a body is a particle is an idealization that is appropriate when the body behaves as if it is essentially a point mass. The only motion a particle can undergo is translation; any discussion of rotation or deformation is meaningless when dealing with a particle. For a problem in particle dynamics, a free-body diagram of the body in question is constructed and then used to fill in the $\mathbf{F}$ side of $\mathbf{F} = m\mathbf{a}$, where $\mathbf{a}$ is the acceleration of the mass center. $\mathbf{F} = m\mathbf{a}$ is often referred to as an equation of motion. In a situation where $\mathbf{F}$ is known, $\mathbf{a} = \mathbf{F}/m$ can be integrated to determine the motion of the mass center. On the other hand, if $\mathbf{a}$ is known, $\mathbf{F}$ is easily determined. For the special and important case when $\mathbf{a}$ is zero, we have a statics problem, with $\mathbf{F} = \mathbf{0}$ referred to as the **equation of equilibrium**.

The assumption that a body of finite extent is rigid is also an idealization, characterized by the requirement that any two points in the body remain equidistant, regardless of the motion of the body. As indicated in Fig. 1.12, a rigid body can translate and/or rotate.

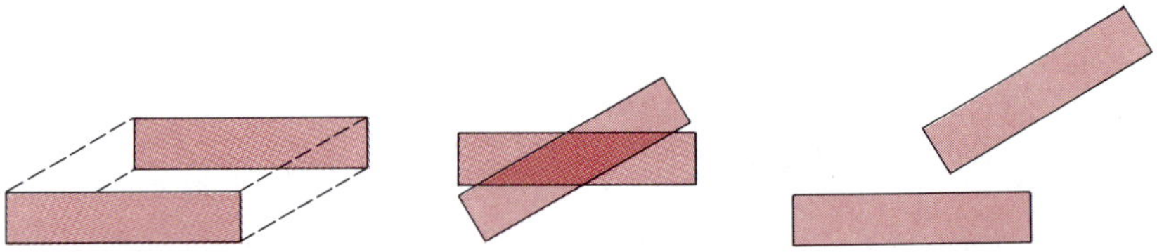

(a) Translation (b) Rotation (c) Translation and rotation

Figure 1.12 Rigid-body motions

For a problem in rigid-body dynamics, it is necessary to enforce two equations, namely, $\mathbf{F} = m\mathbf{a}$, which describes the motion $\mathbf{a}$ of the mass center of the body, and $\mathbf{M} = d\mathbf{H}/dt$, which describes the angular motion of the body relative to the mass center. $\mathbf{M}$ is the moment acting on the body and $\mathbf{H}$ is the angular momentum. If $\mathbf{a}$, the acceleration of the mass center, and all of the angular motion terms which contribute to $d\mathbf{H}/dt$ are zero, we again have a statics problem with the two equations $\mathbf{F} = \mathbf{0}$ and $\mathbf{M} = \mathbf{0}$, referred to as the **equations of equilibrium**. (Note that in order to formulate a dynamics problem for a body, we need only know its mass and its mass moments of inertia. Once having made the assumption that the body is rigid, the nature of the material(s) that compose the body is immaterial.)

No body is truly rigid. Any real body deforms, that is, changes its size and/or its shape, when subjected to mechanical or thermal loads.

Examples of changes in size and shape are given in Fig. 1.13. Fig. 1.13a shows a balloon whose size increases with an increase in internal pressure. Fig. 1.13b shows a beam changing shape under the action of a transverse load.

When it is appropriate to consider a body as deformable, it is clearly necessary to relax the assumption that the body is rigid. The relaxation of the rigid-body assumption has two important consequences. First, the rigid-body kinematics consisting only of the description of translations and rotations must be replaced by a kinematics that is capable of describing the changes in size and shape associated with the **deformation** of the body. Second, given two bodies of the same initial size and shape that are composed of different materials and that are subjected to exactly the same set of mechanical and thermal inputs, the qualitative and quantitative characters of the deformation of the two bodies will clearly depend on the materials which constitute the body. Thus the relaxation of the rigid body assumption results in a branching into two considerations:

1. The need for a type of kinematics that is capable of describing the deformations of the body.
2. The need for information about the material behavior of the body.

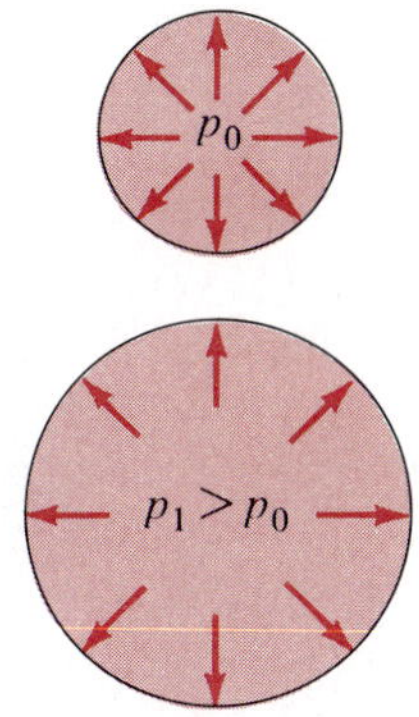

(a) Change in size

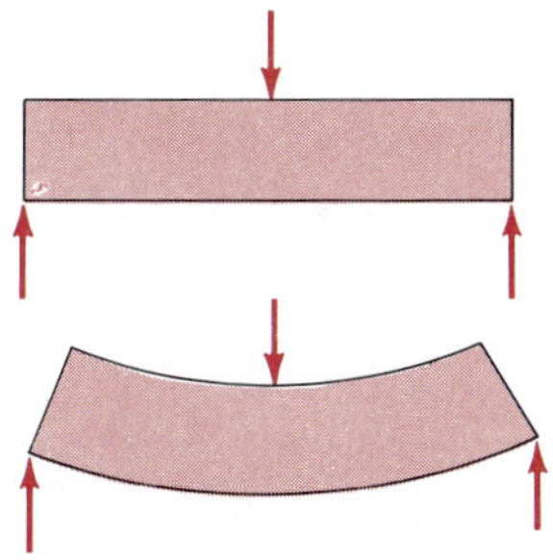

(b) Change in shape

Figure 1.13 Deformations involving changes in size and shape

The study of the mechanics of deformable bodies can be considered an extension of the study of statics and dynamics in that we will still require the body to behave in accordance with Newton's laws of motion. However, we will need to generalize the kinematics, and include the material behavior, in order to fully describe the response of a deformable body to mechanical and thermal inputs.

With these points in mind we outline the **three basic ideas of solid mechanics** as follows:

1. **Equilibrium.** In the analysis of any physics problem there are one or more physical principles or laws that are known to be true. For the statics of deformable bodies this basic physical law derives from Newton, and states that the equations of equilibrium $\Sigma\mathbf{F} = \mathbf{0}$ and $\Sigma\mathbf{M} = \mathbf{0}$ must be satisfied. We will use these equations involving forces and moments to analyze the overall equilibrium of a deformable body, as well as the manner in which the forces and moments are transmitted into the interior of the body. Occasionally we will use an equation of motion rather than an equation of equilibrium as the basic physical principle we must satisfy. In any case the first basic idea involves the satisfaction of Newton's laws. Free-body diagrams are indispensable when considering equilibrium.

2. **Description of the deformation.** In contrast to a rigid body, a deformable body is capable of responding to external stimuli such as forces, temperature changes, and accelerations by undergoing changes in size and shape. For example, a rubber band deforms by elongating under the action of a tensile force, that is, it changes in length. Similarly, a pink eraser can be deformed or bent from an original shape that is roughly rectangular into a final shape that is curved. In solid mechanics, kinematics is the study of the changes in geometry resulting from the deformation of the body, independent of the causes of the deformation.

3. **Material behavior.** The goal of the study of material behavior is to describe, both qualitatively and quantitatively, the relations between the deformations of a solid and the forces, temperature changes, and accelerations that cause them. As stated in Section 1.3, we will be concerned with those aspects of material behavior relating to strength and those relating to stiffness. It is a well-known fact that both the qualitative and quantitative aspects of material behavior must be determined experimentally.

One other consideration arises in passing from rigid-body mechanics to deformable-body mechanics, regarding the difference between the initial, or undeformed, and the final, or deformed, configurations of the body. For a statics problem, where the body is assumed to be rigid, there are no changes in the size or shape of the body that arise from any loads; the initial and deformed configurations are the same. This is not true for a deformable solid. The equilibrium configuration for a body will necessarily be the deformed configuration that results **after** the loads are applied. If the deformations are large enough to change the initial configuration significantly, the equilibrium equations will have to involve the deformations, which are among the unknowns for the problem. The analysis of bodies undergoing large deformations is very complex and in general will not be covered in this text. Most deformations are small enough that the equilibrium equation for the initial configuration can still be used.

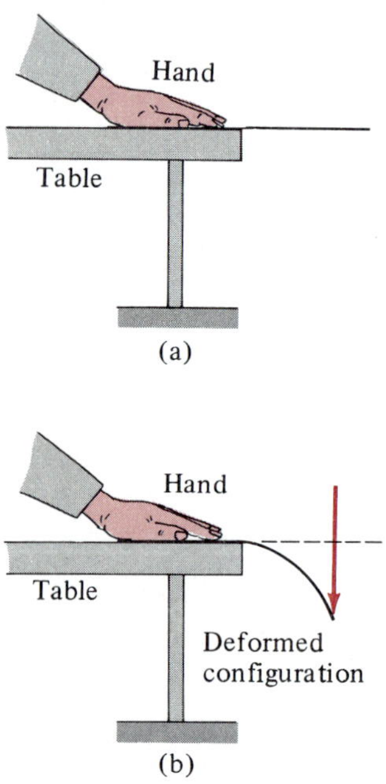

Figure 1.14 Deformation of coat hanger

As an illustration, take an ordinary piece of coat hanger wire, place it on the edge of a table as indicated in Fig. 1.14a, and apply a force as indicated in Fig. 1.14b. We know that a relatively small force can produce a substantial vertical deflection of the wire. The initial and deformed configurations are substantially different. In particular, the location of the load has changed appreciably, so that equilibrium equations written for the initial configuration would no longer apply to the deformed configuration.

Let us outline several very simple situations involving deformable bodies and indicate how the three basic ideas are used to formulate and solve for the internal forces and deformations.

Example 1.10 Consider the problem of a spring that is subjected to a force F as indicated in Fig. 1.15a.

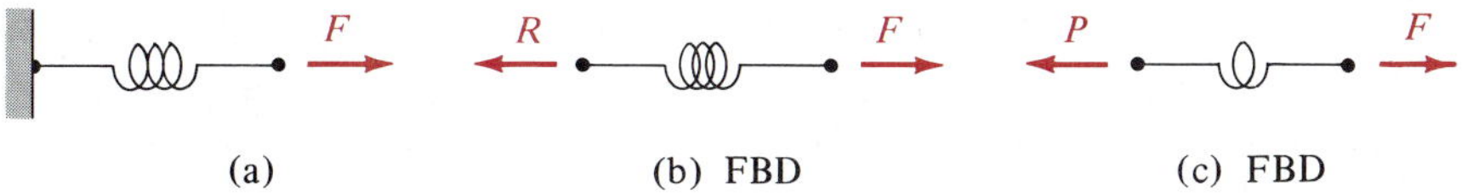

Figure 1.15 Loaded spring

Equilibrium The external force F is assumed to have been applied slowly so that the problem can be analyzed as an equilibrium problem. A free-body diagram of the entire spring is indicated in Fig. 1.15b. The reaction force R provided by the support we determine by writing the equilibrium equation for the entire spring, namely

$$\Sigma F_X = F - R = 0$$

showing that the reaction force R is equal in magnitude to the applied force F. (Note that this is an example of a statically determinate system in that all the reactions can be determined by writing and solving the equilibrium equations.) To determine the internal force at some location along the length of the spring, it is necessary to draw a free-body diagram of a part of the spring that exposes the desired internal force, as in Fig. 1.15c. Denoting the unknown internal force as P and writing the equilibrium equation for the free-body diagram of Fig. 1.15c, there results

$$\Sigma F_X = F - P = 0$$

showing that the internal force P is also equal in magnitude to the external load F. If the internal forces are all that are required, the problem is solved. The deformations and the material behavior do not enter into the problem.

Deformations and the Material Behavior If we wish to determine the elongation of the spring, we need additional information about the behavior of the spring. For a spring, the information about both the deformation and the material behavior is contained in the "force-deflection" relation for the spring. The "force" in this relation is the constant value of the internal force transmitted by the spring, whereas the "deflection" refers to the elongation of the spring. This force-deflection relation, which is determined experimentally, appears typically as in Fig. 1.16, where f is the internal force and δ the elongation. Several different possibilities for the basic behavior of the spring are indicated.

As we see in the figure, there is commonly a portion of the curve near $f = 0$, $\delta = 0$ where the shape of the curve is at least approximately linear. Not surprisingly, a spring whose behavior can be considered to be linear is called a linear spring, with the corresponding force deflection relation written as

$$f = k\,\delta \tag{1.1}$$

where k is called the spring constant. It represents the slope of the linear portion of the curve as shown in Fig. 1.16.

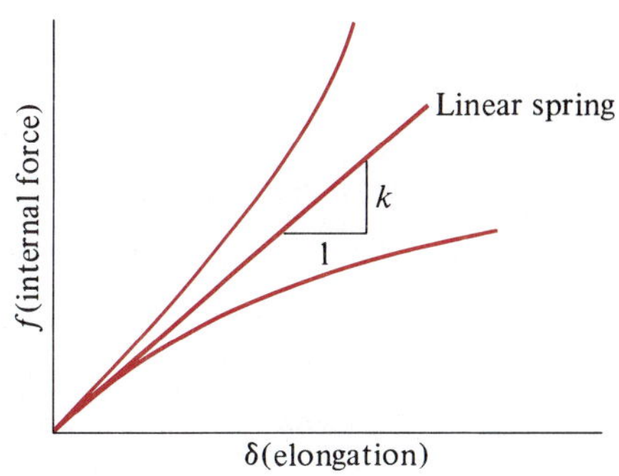

Figure 1.16 Force deflection relation for a spring

Thus Eqn. (1.1) can be used to determine the elongation δ of the spring according to

$$\delta = f/k$$

completing the problem of determining the internal forces and the deformations. For this idealized case, only the total elongation of the spring is determined from the formulation; displacements at other locations along the spring are not determined specifically. The formulations required to determine detailed information of this type are presented in Chapters 5 through 8.

Example 1.11 Consider the problem of two linear springs loaded as indicated in Fig. 1.17.

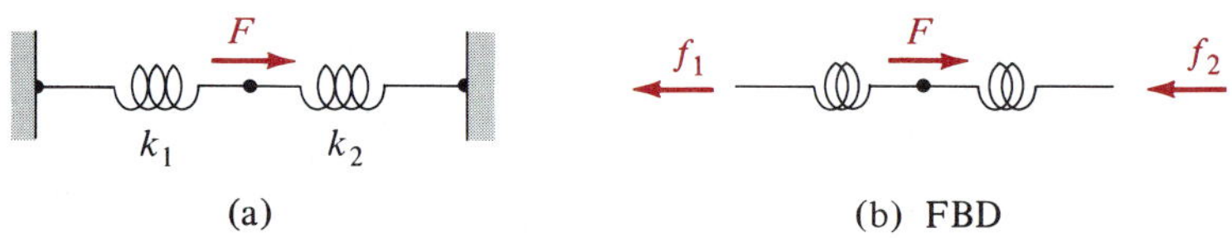

Figure 1.17 Two springs in series

Equations of Equilibrium The load F is applied slowly. A free-body diagram exposing the internal forces in the two springs is given in Fig. 1.17b. Requiring equilibrium yields

$$\Sigma F = -f_1 + F - f_2 = 0$$

or

$$f_1 + f_2 = F \qquad (a)$$

The problem is statically indeterminate since there is only one equation for determining the two unknown internal forces f_1 and f_2. The problem of determining the internal forces in the springs cannot be solved without additional information.

Deformation and Material Behavior The additional equations for determining f_1 and f_2 are supplied by statements regarding the force-displacement behavior of the springs. First, the deformation statement follows quite simply from the observation that elongation of spring 1 must be the same as the shortening of spring 2. Refer to this common value as δ. Second, assuming that each spring is linear with spring constants k_1 and k_2, respectively, the force-displacement relations for the two springs are

$$f_1 = k_1 \delta \qquad \text{and} \qquad f_2 = k_2 \delta \qquad (b)$$

Substituting these two relations into Eqn. (a) yields

$$(k_1 + k_2)\,\delta = F$$

or

$$\delta = F/(k_1 + k_2)$$

The two internal forces are determined from Eqns. (*b*) as

$$f_1 = k_1 F/(k_1 + k_2)$$

and (*c*)

$$f_2 = k_2 F/(k_1 + k_2)$$

Note that the force F is split between the two springs according to

$$f_1/f_2 = k_1/k_2$$

This result demonstrates one of the fundamental properties of linear structures: when two linear structures are combined (as in Fig. 1.17) to resist a load, the structures supply forces whose relative values are in proportion to their stiffnesses; that is, the stiffer of the two structures carries more of the load.

Thus we see that for this statically indeterminate problem, simultaneous equations involving both the internal forces f_1 and f_2 and the deformation δ must be developed in order to solve the problem. This need to simultaneously consider the equilibrium, deformation, and material behavior in the formulation and solution of a statically indeterminate problem will be demonstrated numerous times throughout the text.

1.5 CLOSURE

The purposes of this introductory chapter have been threefold. The first purpose was to introduce the student to the scope of solid mechanics and its many and varied applications, and to the idea that in the process of designing a structure, a responsible engineer must consider the effects of many types of environment to which the structure may be subjected. Mechanical and thermal conditions frequently result in the primary loads, but other environments such as electrical, magnetic, and chemical can also be important, and must not be ignored.

The second purpose was to identify the two basic questions of concern in the analysis and design of a structure: (*a*) is the structure strong enough, and (*b*) is the structure stiff enough? These questions are generally asked within the context of the possible modes of failure: Strong enough can mean that the structure is capable of sustaining the necessary internal forces without fracturing. Stiff enough can mean that the deformations are not excessive or that the structure does not buckle.

The third purpose was to encourage the student to begin to think of the formulation of problems in solid mechanics in terms of the three basic ideas:

1. Equations of equilibrium or equations of motion.
2. Geometry of deformation.
3. Material behavior and properties.

If questions as to strength and stiffness are to be answered, the formulation of any problem in the mechanics of deformable bodies **must** include, either explicitly or implicitly, each of these three basic ideas. Wherever appropriate and possible, the application of the three basic ideas in setting up and solving problems will be indicated in the text. Additionally, whenever the student is setting up and solving a problem he or she should keep these three basic ideas in focus, and check to see that each has been included.

Some insight into the relationship between the three basic ideas can be gained by referring to Fig. 1.18.

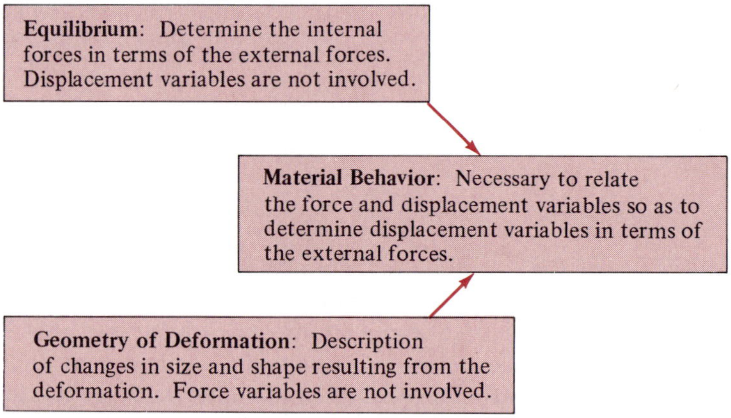

Figure 1.18 Relationship between the basic ideas of solid mechanics

The figure clearly indicates that the basic idea of material behavior provides a bridge between the force variables determined in the study of equilibrium, and the displacement variables described in the discussion of deformation. Without this bridge it would be impossible to determine the displacements in terms of the external loads, generally considered to be the inputs to the system.

Finally, we recall that by a problem in solid mechanics we mean that, in general, we wish to determine **both** the internal forces **and** the deformations that result from any loads. The formulation and solution of a statically indeterminate problem involves simultaneous consideration of the internal forces and the deformations, whereas for a statically determinate problem, the internal forces can be determined independently, followed by a separate analysis to determine the deformations.

The formulation of a problem in solid mechanics results in an equation or set of equations, which is solved to determine the force and displacement variables. As in any area of physics, the equations must be dimensionally homogeneous; that is, the dimensions of each term in an equation must be the same. The student should regularly check the dimensions of equations, both to keep in mind the dimensions of quantities in the equations, and also to ensure dimensional homogeneity. Both the SI system of units, where mass is measured in kilograms, force in newtons, length in meters, and time in seconds, and the U.S. customary system of units, where mass is measured in slugs, force in pounds force, length in feet, and time in seconds, will be employed in the examples and problems.

CHAPTER 2 EQUILIBRIUM AND STRESS

CHAPTER OUTLINE

2.1 INTRODUCTION

Equilibrium is the basic idea embodying Newton's laws in the study of the mechanics of deformable bodies. The second law, stating that $\Sigma\mathbf{F} = \mathbf{0}$ and $\Sigma\mathbf{M} = \mathbf{0}$ for a body in equilibrium, and the third law, stating that action = reaction, form the basis for our study of external and internal forces. The force inputs to a solid mechanics problem consist primarily of external forces of two basic types: action at a distance forces, such as gravity, that are distributed throughout the body, and forces applied to the surface of the body by direct contact with another body. An example of a body at rest on a horizontal surface is shown in Fig. 2.1a. The corresponding free-body diagram of Fig. 2.1b shows the contact forces and the distributed gravitational forces (action at a distance forces) adding up to the weight of the body.

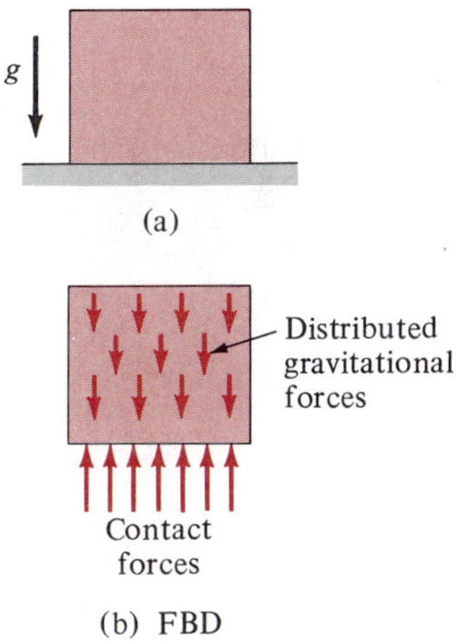

Figure 2.1 Distributed gravitational and contact forces

These external forces generally produce deformations of the body that are resisted by internal forces. The distribution and intensity of the internal forces give rise naturally to the concept of stress, a quantity we must understand to ultimately answer the question as to the strength of the body.

The purpose of this chapter is to relate the external forces to the internal forces, and then to begin to relate the internal forces to the stresses, which will presently be defined. The end result is that the stresses are related to the external loads. In this chapter and throughout the text, the student should keep firmly in focus that equilibrium is the basic idea underlying the relations between the external and internal force resultants and also between the internal force resultants and the stresses.

As in statics, the basic tool for all of the developments in this chapter is the **free-body diagram (FBD).** The free-body diagram is used to develop the equilibrium equations for a finite body, and for any portion of a solid for which equations of equilibrium or equations of motion are desired. The free-body diagram is used in the definition of stress and in the investigation of the properties of stress. The free-body diagram is an absolutely indispensable tool in studying all aspects of equilibrium.

2.2 INTERNAL FORCE RESULTANTS

In any situation involving a real body used as a structure, forces are transmitted throughout the body in accordance with the principles of force transmission discussed in statics. In the mechanics of deformable bodies we are interested in the **distribution of the internal forces** associated with the force transmission, so as to determine whether the strength of the body is sufficient to transmit these internal force distributions. As an example, one has only to take a piece of chalk and "bend" it until it breaks to see that the strength of the chalk is insufficient to transmit the internal forces associated with the applied moment. Determining the distribution of the internal forces in a body is a very important step in formulating and solving a solid mechanics problem.

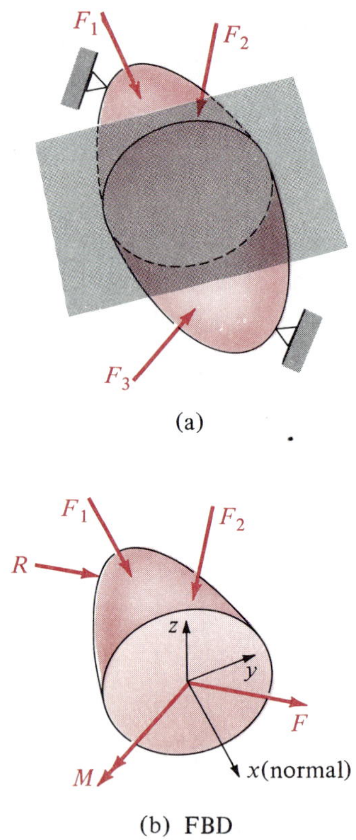

Figure 2.2 Internal force and moment resultants for a three-dimensional problem

By definition, **internal forces** are unseen forces that act in the interior of the body. These forces arise in connection with the ability of the solid to resist changes in its size or shape, and can only be "seen" by passing an imaginary plane or planes through the body at the location that exposes the resultant internal force and moment whose values are sought. This process is often referred to as **sectioning**. If the entire body is in equilibrium under the action of the external forces including the reactions, any portion of the body obtained by sectioning must also be in equilibrium. This equilibrium requirement for the section enables us to determine the resultant force and moment acting on the internal surface exposed by the section. The exposed force and moment are referred to as **internal force resultants**. Note that only the *resultants* of the internal force can be determined by drawing a free-body diagram of the section and requiring equilibrium. The precise distribution of force associated with these internal force resultants cannot be known without additional information.

To illustrate the concept specifically, consider a body that is acted on by forces and/or couples as shown in Fig. 2.2a.

The body is assumed to be in equilibrium under the action of the external forces $\mathbf{F}_1$, $\mathbf{F}_2$, . . ., and the reactions. Pass a section through the body as indicated in Fig. 2.2a and construct a free-body diagram of the portion of the body on one side of the section, exposing the internal force resultants **F** and **M** as shown in Fig. 2.2b.

After constructing the desired free-body diagram, the equations of equilibrium $\Sigma\mathbf{F} = \mathbf{0}$ and $\Sigma\mathbf{M} = \mathbf{0}$ can be written for the sectioned portion of the body in order to determine the internal forces **F** and **M**.

The importance of correctly drawing the proper free-body diagram cannot be overstated. It is the single most important step in the formulation and solution of a problem in the mechanics of deformable bodies.

To assist in illustrating this step, define a local coordinate system whose x-axis is normal to the section as shown in Fig. 2.2b. The y- and z-axes lie in the plane of the section, and with x form a right-handed coordinate system. In general the internal forces transmitted are statically equivalent to a force **F** and a couple or moment **M** that are determined by solving the equilibrium equations, namely, $\Sigma \mathbf{F} = \mathbf{0}$ and $\Sigma \mathbf{M} = \mathbf{0}$.

Note on the basis of Newton's third law that if the portion on the other side of the section in Fig. 2.2a had been considered, the internal force and moment would be $-\mathbf{F}$ and $-\mathbf{M}$, respectively. Note also that if the origin of the coordinate system were chosen as some other location on the exposed face, the resultant moment **M** with respect to the new origin would in general be different, while the resultant force **F** would remain the same.

For a structure that is assumed to be two-dimensional such as the one in Fig. 2.3a, it is sufficient to assume that the internal force consists of the two in-plane components F_x and F_y along with a moment M about the z-axis.

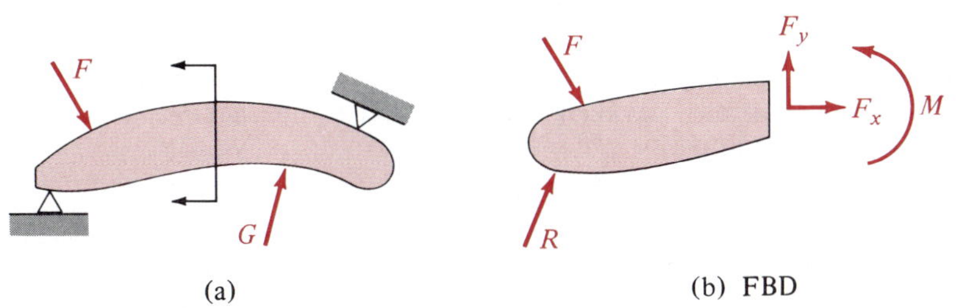

Figure 2.3 Internal force resultants for a two-dimensional problem

For the free-body diagram shown in Fig. 2.3b, the necessary equilibrium equations are $\Sigma F_x = 0$, $\Sigma F_y = 0$, and $\Sigma M_z = 0$, for determining the internal forces F_x, F_y, and M.

To determine the unknown internal forces using the technique of sectioning, it is important to keep in mind which of the forces are external and which are internal. Several examples illustrating this idea follow.

Example 2.1 Consider the problem of a curved bar loaded by the colinear external forces F as shown in Fig. 2.4a. A free-body diagram of the portion to the left of the section at the middle of the bar is shown in Fig. 2.4b.

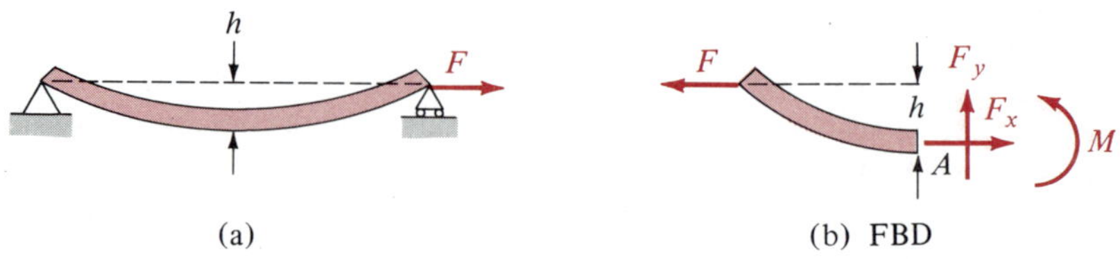

Figure 2.4 Curved bar under tension

The free-body diagram of Fig. 2.4b shows the three possible internal force resultants F_x, F_y, and M and the external force **F**. The equilibrium equations are

$$\Sigma F_x = -F + F_x = 0$$
$$\Sigma F_y = F_y = 0$$

and

$$\Sigma M_A = M + h\,F = 0$$

from which $F_x = F$, $F_y = 0$, and $M = -Fh$.

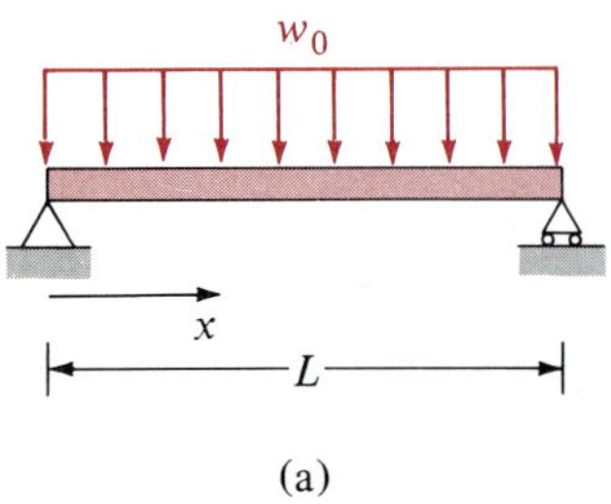

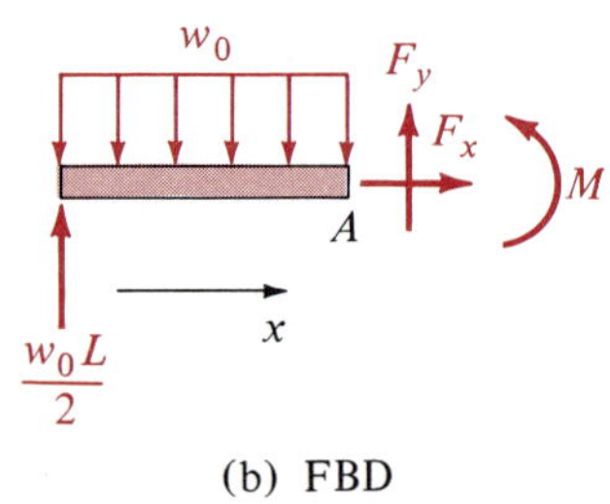

Figure 2.5 Bar acted upon by distributed loading

Example 2.2 Consider the uniform bar loaded by the uniformly distributed loading w_0 and supported as shown in Fig. 2.5a. A free-body diagram of the bar is indicated in Fig. 2.5b.

It is left to the student to verify by drawing a free-body diagram of the entire bar that the reactions are each $w_0L/2$. Enforcing the necessary equilibrium equations results in

$$\Sigma F_x = F_x = 0$$
$$\Sigma F_y = \frac{w_0L}{2} - w_0x + F_y = 0$$
$$\Sigma M = M - \frac{w_0L}{2}x + (w_0x)\frac{x}{2} = 0$$

from which $F_y = -\,w_0(L/2 - x)$ and $M = w_0x(L - x)/2$. On the free-body diagram of Fig. 2.5b, the reaction $w_0L/2$ and the total distributed load w_0x are external forces while F_x, F_y, and M are internal forces. Note from Fig. 2.5b that the distributed loading w_0 can be replaced by its resultant w_0x acting at $x/2$. This results in the free-body diagram shown in Fig. 2.6.

The student should verify that the three equilibrium equations again yield $F_y = -w_0(L/2 - x)$ and $M = w_0x(L - x)/2$. Note carefully that replacing the distributed load by its resultant on a free-body diagram of a section is allowable only *after* choosing the section and determining the amount and location of load involved.

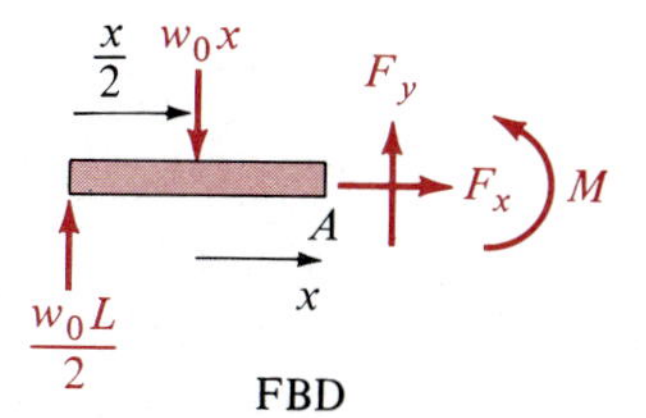

Figure 2.6 Replacement of the distributed loading by its resultant

---

Example 2.3 Consider the problem of a uniform bar of cross section A, mass density ρ, and length L acting under its own weight $W = \rho AgL$ as indicated in Fig. 2.7a.

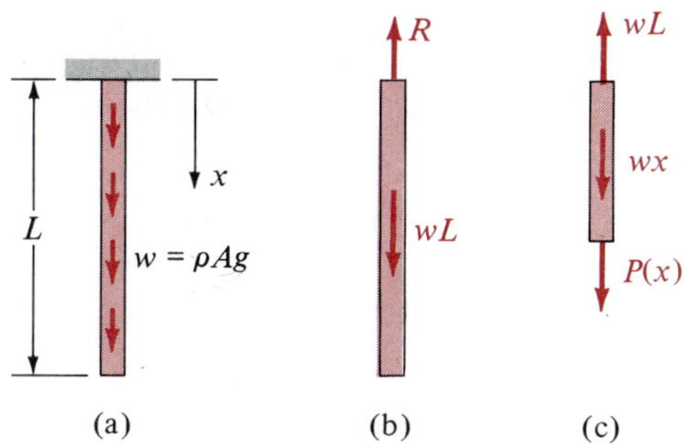

Figure 2.7 Bar hanging under its own weight

In contrast to the previous example, this distributed loading acts in the direction of the axis of the bar rather than perpendicular to the axis of the bar. The intensity of the distributed loading is given by dividing the total weight $W = wL = \rho AgL$ by the length L to obtain $W/L = w = \rho Ag$, which has dimensions F/L.

From the free-body diagram of the entire bar indicated in Fig. 2.7b, it follows that the only equilbrium equation is

$$\Sigma F_x = wL - R = 0$$

from which $R = wL = W$, the weight of the bar. To determine the internal force $P(x)$ at a location x units from the top of the bar, a free-body diagram of the section above the section at x is drawn as in Fig. 2.7c. Writing the single equilibrium equation gives

$$\Sigma F_x = -wL + wx + P(x) = 0$$

yielding $P(x) = w(L - x)$ for the internal force transmitted. Note here that when $x = 0$, $P(0) = wL = R$ as a check, and that when $x = L$, $P(L) = 0$, which is also a check in that there is no force transmitted at the bottom of the bar. The student should verify by considering a free-body diagram of the portion below the section in question that the same result is obtained.

PROBLEMS

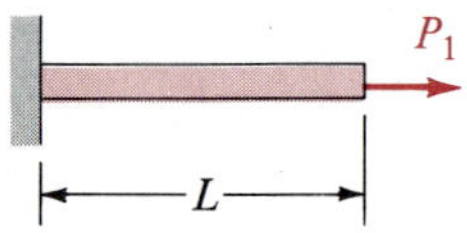

P. 2.1

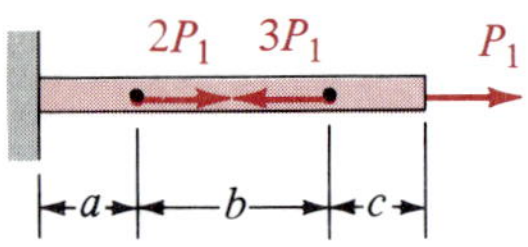

P. 2.2

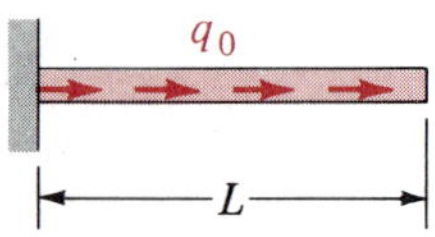

P. 2.3

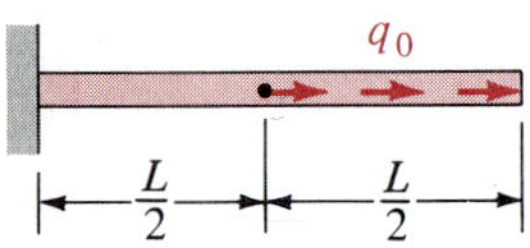

P. 2.4

2.1–2.4 For each of the following problems, let the internal axial force resultant be designated by $P(x)$, defined as positive when the bar is in tension. Choose the section(s) and then draw the free-body diagrams necessary to determine $P(x)$ at all locations along the length of the bar; hence plot $P(x)$ as a function of x. In problems 2.3 and 2.4, q_0 represents a uniform load per unit of length.

2.5 A uniform bar rotates at a constant angular rate ω about an axis passing through one end as shown. Draw a free-body diagram of the portion indicated and use Newton's second law to determine the internal force $P(x)$. Recall that the force required to maintain the circular motion of such a body is given by $F = mr\omega^2$, where r is the distance from the axis to the mass center of the body of mass m.

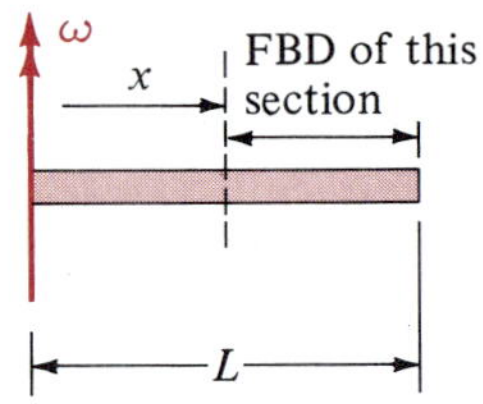

Ps. 2.5 and 2.6

2.6 Repeat problem 2.5 using d'Alembert's principle to determine the equivalent loading $q(x)$ and then integrate $dP/dx + q = 0$ to determine $P(x)$.

2.7 A variable cross-section bar hangs under its own weight as shown. The area varies linearly with position along the bar as indicated. Take the specific weight of the material to be γ and determine $P(x)$.

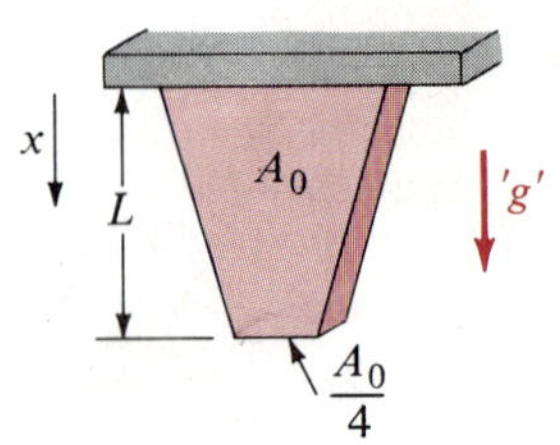

Ps. 2.7 and 2.8

2.8 Repeat problem 2.7 if the bar has a conical shape with radius r_0 at the top and $r_0/4$ at the bottom.

2.9 A concrete pier has the shape of a pyramid as shown. It is loaded by a force $P = 1000$ kN in addition to the load due to its weight. With $\gamma_{concrete} = 25.4$ kN/m^3, $B = 4$ m, $b = 1$ m, and $H = 3$ m, determine $P(x)$.

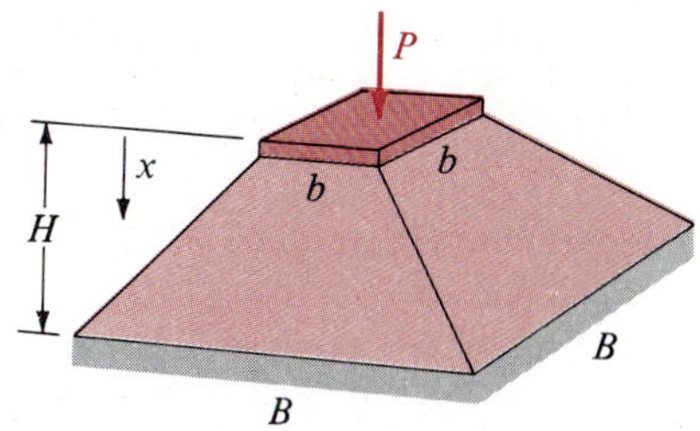

Ps. 2.9 and 2.10

2.10 Repeat problem 2.9 with $P = 300$ k, $\gamma_{concrete} = 150$ lbf/ft^3, $B = 15$ ft, $b = 4$ ft, and $H = 10$ ft.

2.11 A rocket 9 m in length, of mass 5000 kg, is modeled as a uniform rod as indicated. If the thrust (assumed to be applied at one end as shown) provided by the rocket motors produces a rigid-body acceleration of 15 g's, determine $P(x)$.

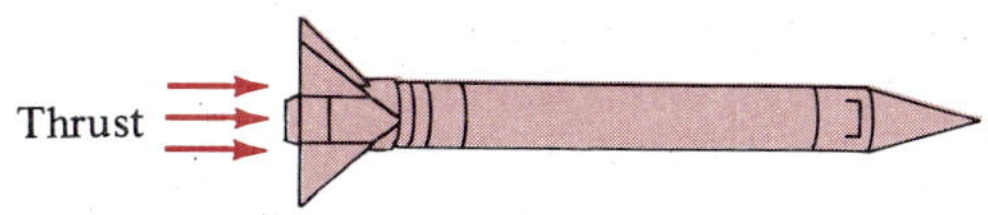

Ps. 2.11 and 2.12

2.12 Repeat problem 2.11 if the mass of the rocket is 300 lbf sec^2/ft and the length is 30 ft.

2.13 Consider the problem of a bar of length L loaded over a portion of its length ε by the constant distributed loading q_0 as shown. Plot the internal axial force resultant as a function of x for several values of ε/L. Allow the distance ε to shrink while at the same time requiring q_0 to grow such that the product $q_0\varepsilon = F$, a force with a fixed value. This shows how a concentrated load can be considered to be the limit of a distributed loading. Note that as ε approaches 0, the intensity of loading q_0 must approach infinity.

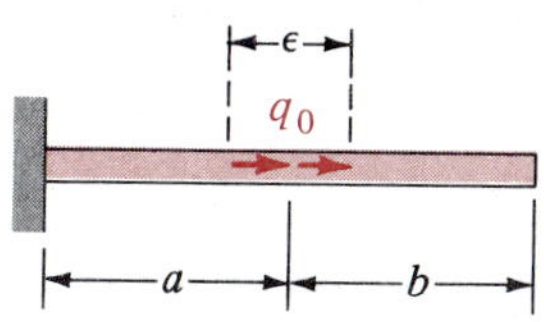

P. 2.13

2.14–2.16 For each of the following, choose sections and draw the free-body diagrams necessary to determine $P(x)$ at all locations along the length of the bar; hence, plot $P(x)$ as a function of x.

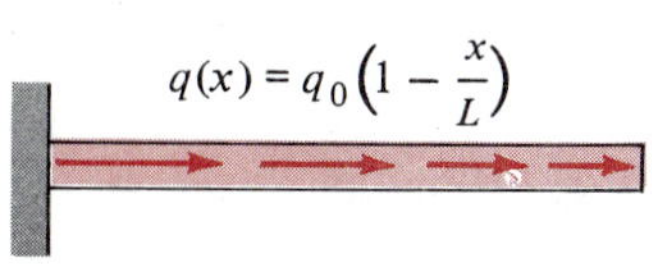

P. 2.14

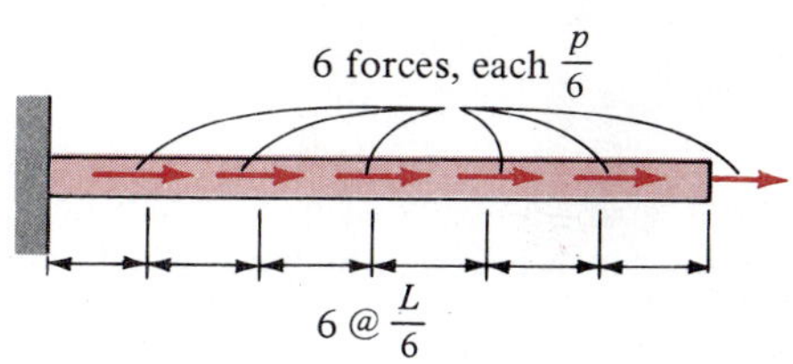

P. 2.15

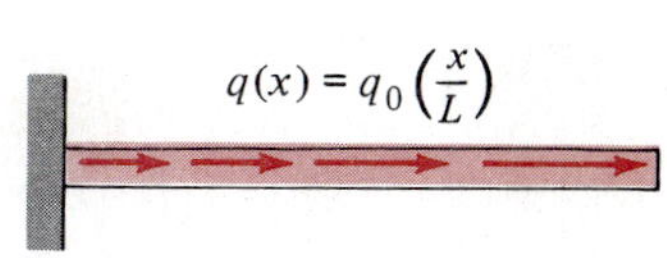

P. 2.16

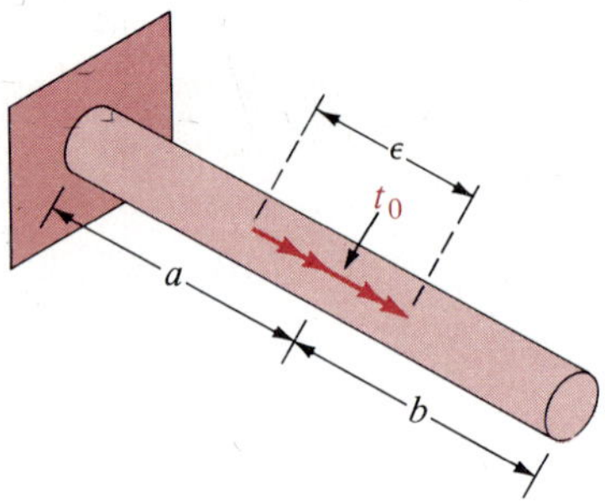

P. 2.17

2.17 Consider the problem of a bar of length L loaded over a portion of its length by the constant distributed torque loading t_0 as shown. Plot the internal axial torque resultant as a function of x for several values of ε/L. Allow the distance ε to shrink while at the same time requiring t_0 to grow such that the product $t_0\varepsilon = T_0$, i.e., a torque with a fixed value. This shows how a "concentrated torque" can be considered to be the limit of a distributed loading. Note that as ε approaches 0, the intensity of loading t_0 must approach infinity.

2.18–2.20 For each of the following problems, let the internal axial torque resultant be designated by $T(x)$, defined as positive in accordance with the right-hand rule. Choose the sections and then draw the free-body diagrams necessary to determine $T(x)$ at all locations along the length of the bar; hence, plot $T(x)$ as a function of x.

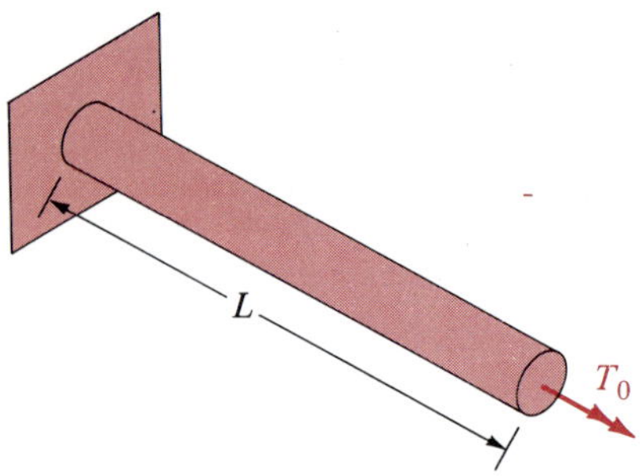

P. 2.18

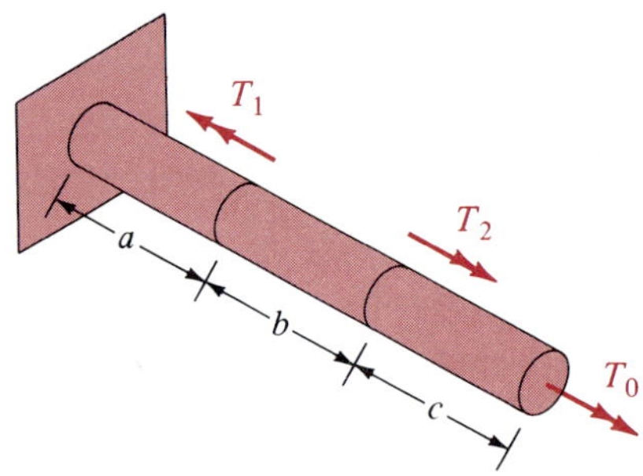

P. 2.19

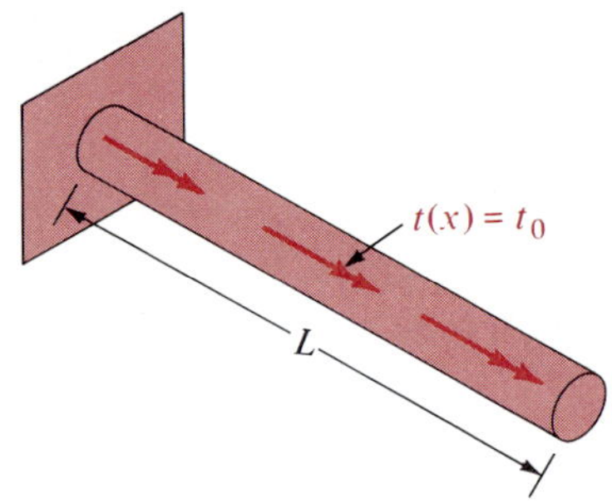

P. 2.20

2.21–2.25 For each of the following problems, let the internal force and moment be designated by $V(x)$ and $M(x)$ respectively. Take the positive directions of V and M to be as shown.

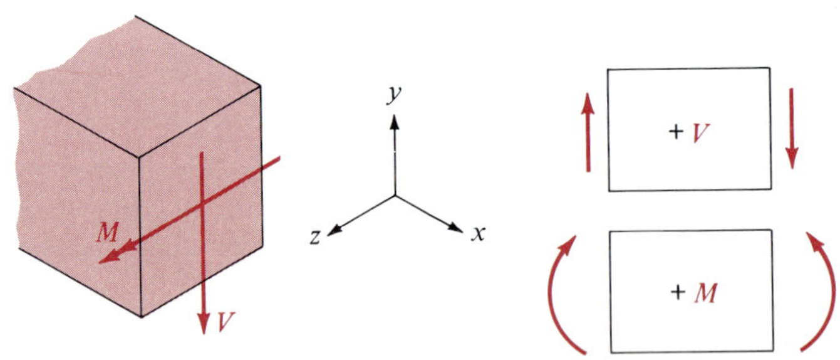

Using Figs. P. 2.21 through 2.25 on the following page, choose the section(s) and then draw the free-body diagrams necessary to determine $V(x)$ and $M(x)$ at all locations along the length of the bar; hence, plot $V(x)$ as a function of x and also $M(x)$ as a function of x.

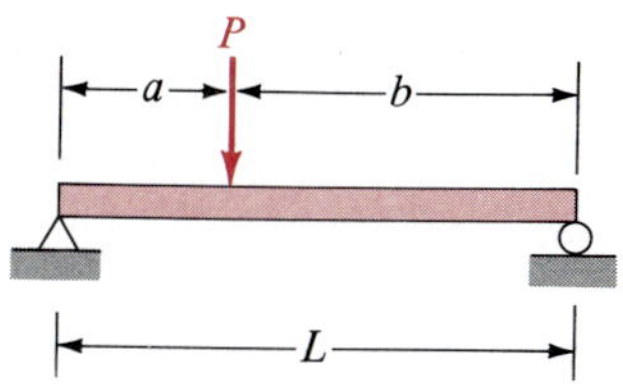

P. 2.21

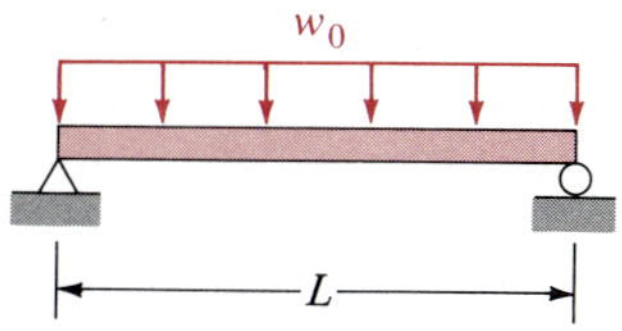

P. 2.22

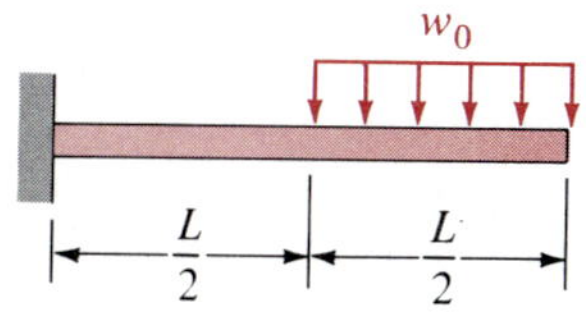

P. 2.23

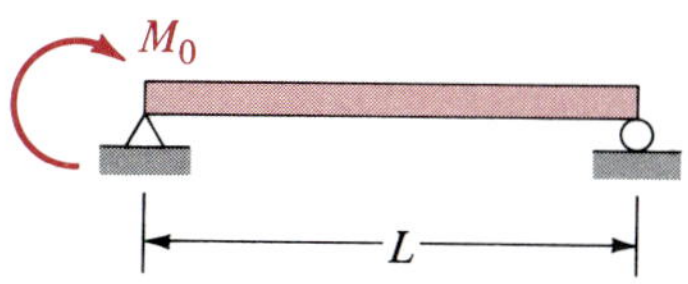

P. 2.24

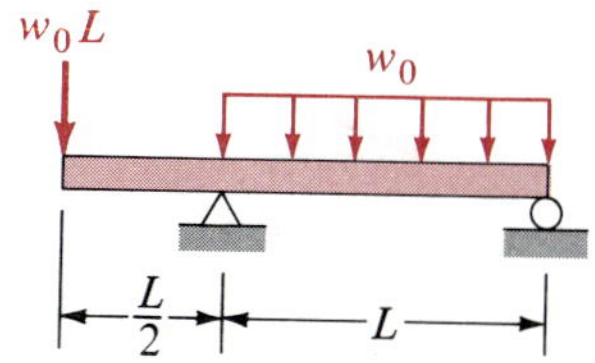

P. 2.25

2.26–2.29 In each of the following, consider that the structure is two dimensional, that is, it lies in a plane. Take the internal force resultant to consist of an axial force, a shear force, and a moment. Hence draw the free-body diagrams necessary to determine the internal force resultants at each of the indicated sections. The notation $[w] = F/L$ indicates that w, the weight per unit of length, is the loading. In problems 2.26 and 2.27, sections AA and BB are just to the left and just above, respectively, the locations of the loads and/or the middle of the portions of the bars. In problems 2.28 and 2.29, sections AA and BB are at the middle of the segments and just above the joint, respectively.

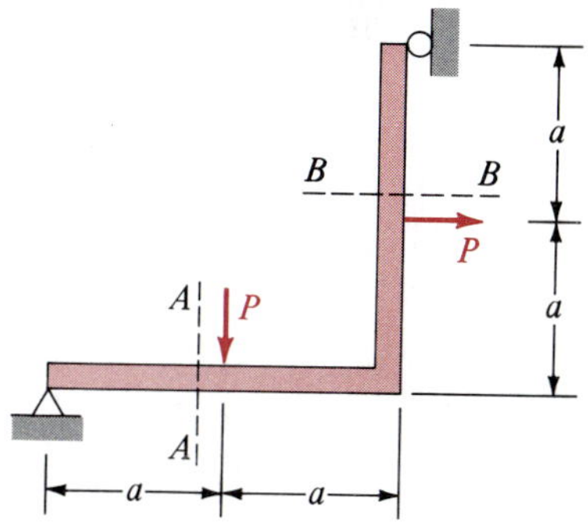

P. 2.26

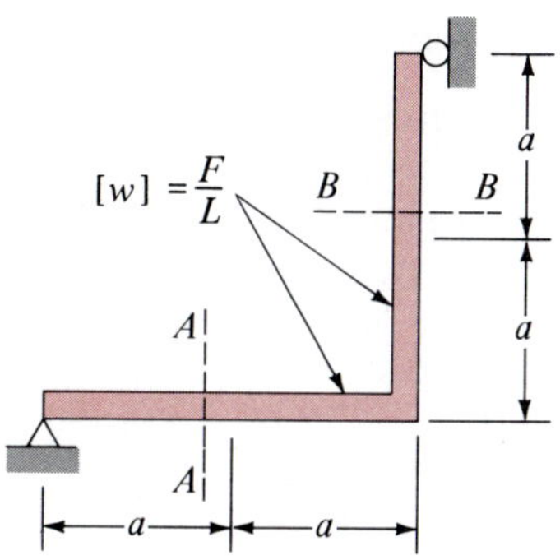

P. 2.27

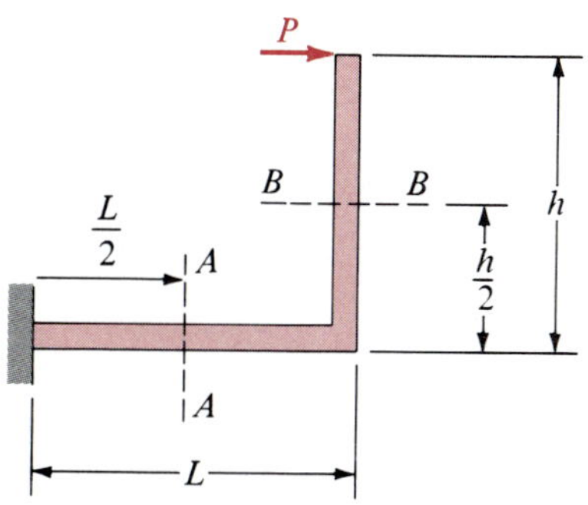

P. 2.28

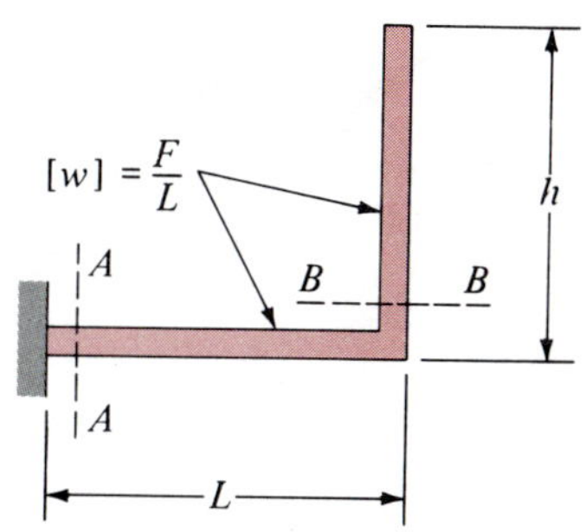

P. 2.29

2.30–2.32 In each of the following, consider the structure to be three dimensional, with the possibility of an axial and two shear forces at the section in question, as well as the possibility of all three moment components. Hence draw the free-body diagrams necessary to determine the internal force resultants at each of the indicated sections. The notation $[w] = F/L$ indicates that w, the weight per unit of length, is the loading.

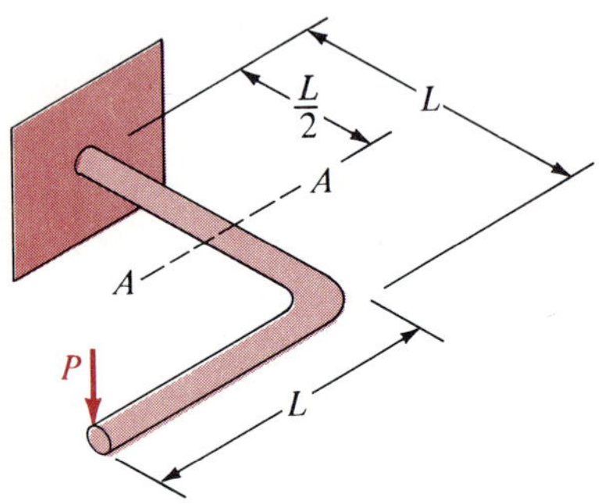

P. 2.30

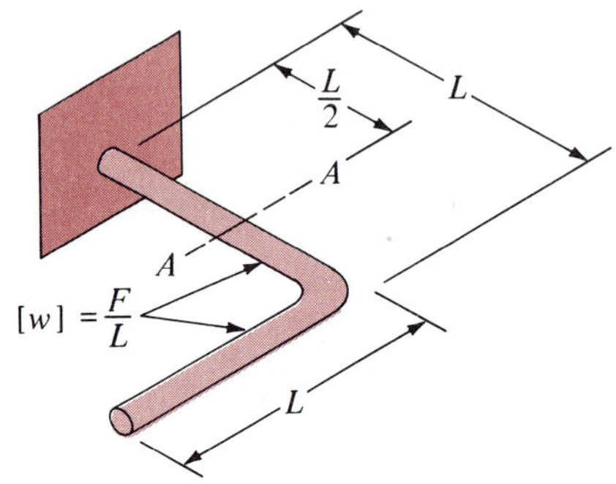

P. 2.31

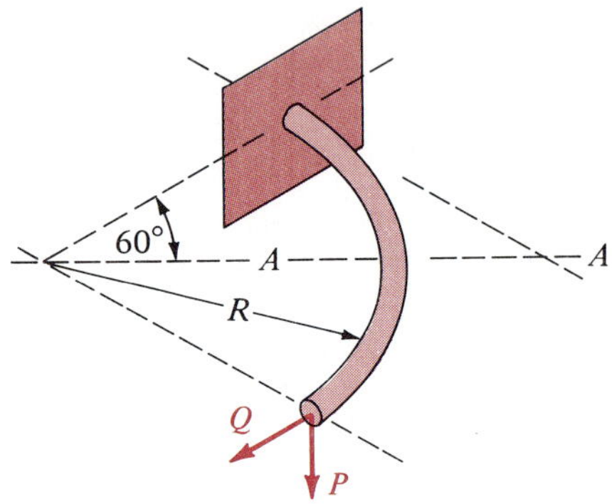

P. 2.32

2.3 STRESS—DISTRIBUTION OF INTERNAL FORCES

Stress is a defined quantity that is indispensable for formulating and solving problems in the mechanics of deformable bodies. The student is already familiar with several physical situations that involve stress, but where the quantity may not have been referred to as a stress. Physical pressure in its many forms is an example of what will be defined as normal stress, whereas a distribution of tangential forces associated with friction has some of the characteristics of what will be defined as shear stress.

2.3.1 Formal Definition of Stress and Stress Notation

To properly evaluate the strength of a structure it is necessary to view stress in a more general fashion than simply as a normal pressure. In this section we will begin to develop the properties of stress that are needed for this comprehensive evaluation of strength.

Stress is defined at a **point** on a **surface**. The point can be located on the bounding surface or boundary of a deformable body as indicated in Fig. 2.8a, or on an imaginary internal surface resulting from the process of sectioning as indicated in Fig. 2.8b.

In the first case, the surface is defined by the normal **n** to the tangent plane at P, whereas in the second case the direction of the normal **n** is perpendicular to the plane section. In both cases, the normal serves to define the orientation of the surface at the point in question. In what follows we will define stress formally and investigate several of its properties.

Consider a body subjected generally to the action of body forces **B** and contact forces $\mathbf{F}_i$ as indicated in Fig. 2.9a.

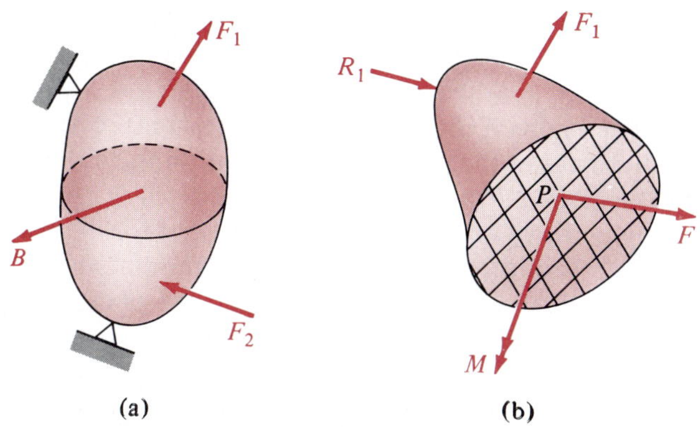

Figure 2.9 Body under the action of body and surface forces

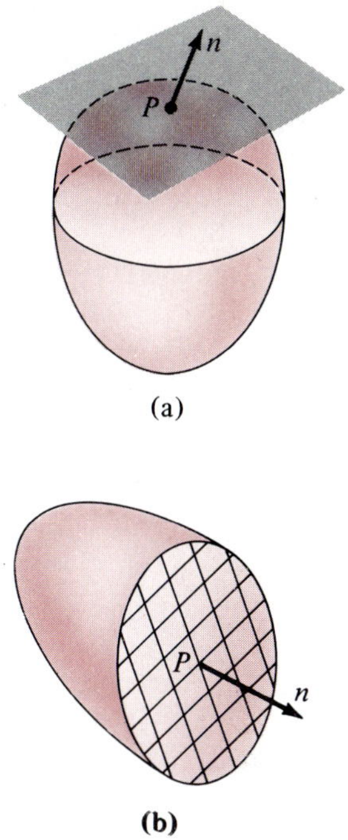

Figure 2.8 Surfaces for defining stress

Pass a plane section through the body as indicated in Fig. 2.9b, exposing internal force distributions that are statically equivalent to the resultants **F** and **M**. Consider a point P on the surface, and an infinitesimal area ΔA as shown in Fig. 2.10.

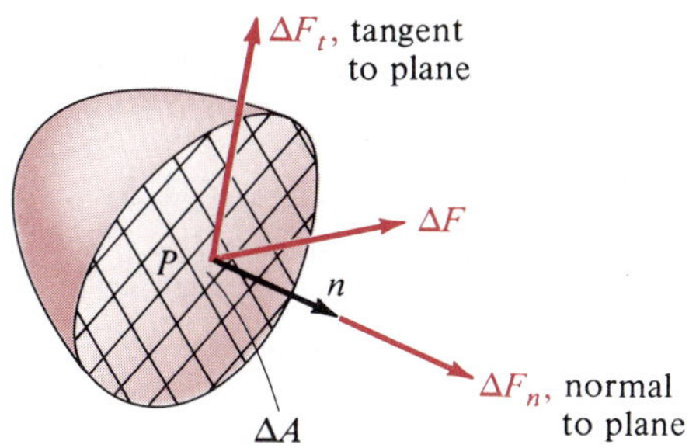

Figure 2.10 Point and plane for definition of stress

Associated with the infinitesimal area ΔA there is in general a force ΔF_n normal to the plane and a force ΔF_t lying in the plane. The **normal stress at P on the plane whose normal is n** is defined by the limit

$$\lim_{\Delta A \to 0} \frac{\Delta F_n}{\Delta A} = \sigma_n(P) \tag{2.1}$$

and has dimensions F/L^2. Note that $\sigma_n(P)$ can be either negative or positive. A negative $\sigma_n(P)$ indicates a compressive stress and a positive $\sigma_n(P)$ indicates a tensile stress.

The **shear stress at P on the plane whose normal is n** is defined by the limit

$$\lim_{\Delta A \to 0} \frac{\Delta F_t}{\Delta A} = \tau_n(P) \tag{2.2}$$

also having dimensions F/L^2. In general, there is no restriction on the direction of the shear at the point P in question; thus, at the point P it is possible for the force ΔF_t and hence the shear stress $\tau_n(P)$ to have any direction in the plane of the section. Also note that for the plane associated with a given section through a body, the normal and shear stress will in general be different at each point P on the plane.

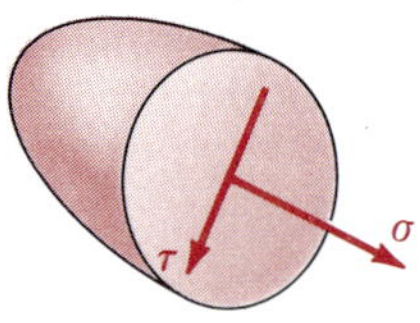

Figure 2.11 Stresses represented as arrows

It is customary to represent stresses as arrows on a sketch as indicated in Fig. 2.11. Remember, however, that what is actually being represented is a force ΔF associated with the stress acting on the area ΔA in question, and that the stress components must be multiplied by an appropriate area before they can be included in the equilibrium equations as the forces on a free-body diagram.

One of the first steps when setting up a problem in solid mechanics is to define a coordinate system for describing locations, displacements, and forces. Stresses must also be referred to a coordinate system, often a right-handed rectangular cartesian system such as shown in Fig. 2.12a.

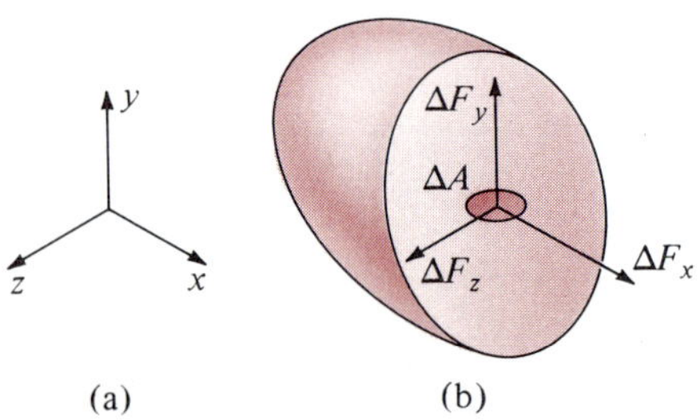

Figure 2.12 Right-handed rectangular cartesian coordinate system

The notation for stresses referred to such a system is defined as follows. First consider a section through a body that is parallel to the yz-plane indicated in Fig. 2.12b. For a differential element ΔA at a point P on the imaginary plane, there is an elemental force $\Delta \mathbf{F}$ which we represent as

$$\Delta \mathbf{F} = \mathbf{i}\, \Delta F_x + \mathbf{j}\, \Delta F_y + \mathbf{k}\, \Delta F_z$$

as indicated in Fig. 2.12b. The positive stress components on this face are defined according to

$$\sigma_x = \lim_{\Delta A \to 0} \frac{\Delta F_x}{\Delta A} \tag{2.3a}$$

$$\tau_{xy} = \lim_{\Delta A \to 0} \frac{\Delta F_y}{\Delta A} \tag{2.3b}$$

$$\tau_{xz} = \lim_{\Delta A \to 0} \frac{\Delta F_z}{\Delta A} \tag{2.3c}$$

Note carefully that when the normal is in the positive x-direction the positive shear stresses τ_{xy} and τ_{xz} are in the positive y- and z-directions, respectively.

Newton's third law as applied to stress then dictates that when the normal to the face is in the negative x-direction, the positive shear stresses τ_{xy} and τ_{xz} are in the negative y- and z-directions, respectively.

This procedure can be repeated for planes whose normals are in the y- and z-directions. For a state of stress referred to an xyz-coordinate system, the positive stress components are indicated in Fig. 2.13.

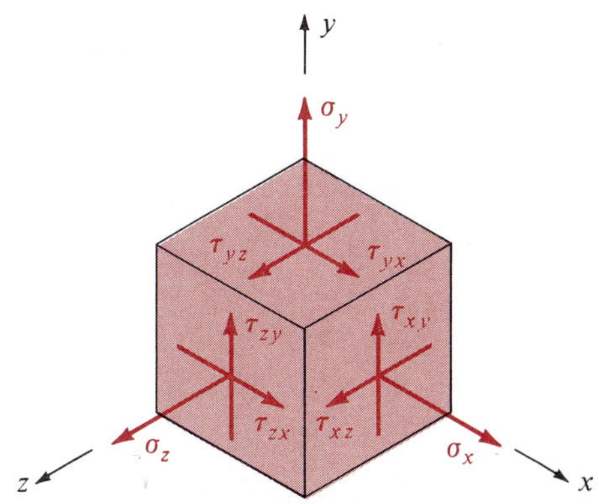

Figure 2.13 Positive stress components

Each stress in the collection of stresses

$$\begin{matrix} \sigma_x & \tau_{xy} & \tau_{xz} \\ \tau_{yx} & \sigma_y & \tau_{yz} \\ \tau_{zx} & \tau_{zy} & \sigma_z \end{matrix}$$

is often referred to as a **component** of stress. The collection of nine stresses is referred to as the **stress tensor**. The so-called tensor character of stress is demonstrated by the way it behaves under a transformation of coordinates; this is discussed later in the chapter. By summing moments about the z, y, and x axes, respectively, it can be shown that $\tau_{xy} = \tau_{yx}$, $\tau_{xz} = \tau_{zx}$, and $\tau_{yz} = \tau_{zy}$, so that there are only six independent stress components. The equality of these shear stress components is referred to as the **symmetry of the stress tensor**. The practical significance of this result is that at any point, the shear stresses on any two perpendicular planes must be numerically equal, as shown in Fig. 2.14.

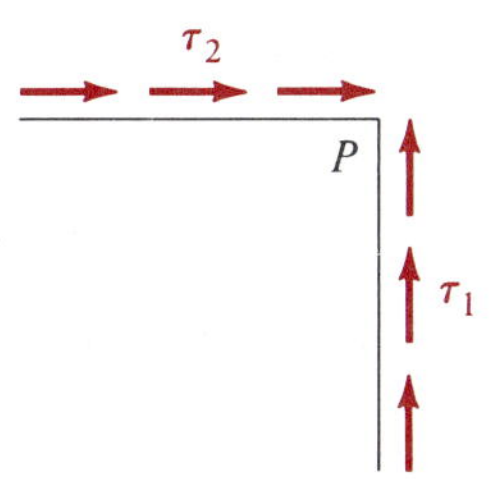

Figure 2.14 Equality of shear stresses at a point

As the corner P is approached, the value of the shear stress distribution τ_1 must be equal to the value of the shear stress distribution τ_2. In particular if either τ_1 or τ_2 is known to be zero at P, it follows that both must vanish at P.

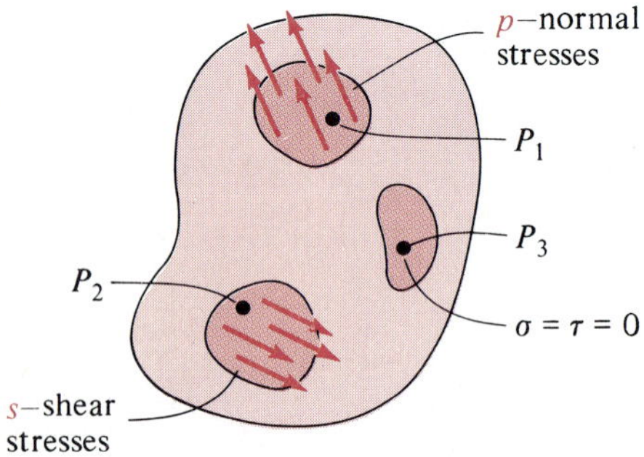

Figure 2.15 Stresses on the boundary of a body

An important aspect of stress in a solid mechanics problem concerns the forces and stresses applied to the boundary of a body. Consider in this regard Fig. 2.15, depicting the surface of a body to which a normal stress distribution p and a shear stress distribution s have been applied over portions of the surface as shown.

If we consider the stress components at points P_1, P_2, and P_3 it is clear that

$$\text{at point } P_1 \qquad \sigma(P_1) = p(P_1),\ \tau(P_1) = 0$$
$$\text{at point } P_2 \qquad \sigma(P_2) = 0,\ \tau(P_2) = s(P_2)$$
$$\text{at point } P_3 \qquad \sigma(P_3) = 0,\ \tau(P_3) = 0$$

In general, if there are no surface or contact forces applied on a portion of the surface of a body, that portion of the surface is said to be **stress free**; that is, the normal and shear stresses vanish for that portion of the surface. In this text we will use this idea primarily to conclude that there can be no resultant force or moment associated with a surface or portion of a surface that is stress free.

Example 2.4 Consider a rectangular bar loaded by a uniform pressure as shown in Fig. 2.16.

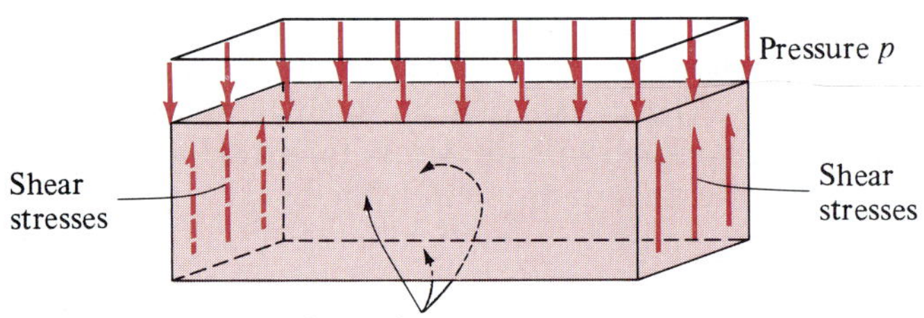

Figure 2.16 Stress-free faces for a uniformly loaded bar

The bar is supported by shear stresses $\tau \neq 0$ applied at the ends as shown. For the top face $\sigma = -p$ and $\tau = 0$. For the front, back, and bottom faces $\sigma = \tau = 0$, thus these are stress-free faces. Knowing that these three surfaces are stress free leads us to conclude that no forces or moments are transmitted across these faces. In many of the applications treated in this text it is important to be able to recognize when a face is stress free.

In summary, we make the following observations about the essential characteristics of stress.

1. Stress is defined at a point on a tangent plane at the boundary of a body, or at a point on an imaginary plane in the interior of a body.

2. The physical dimensions of stress are FL^{-2}, or force per unit of area. Common units of stress are pounds force per square inch (psi), or thousands of pounds force per square inch (ksi), newtons per square meter (N/m^2) or pascals (Pa), and millions of pascals per square meter (MPa).

3. The direction of stress is not restricted; thus the normal stress can be either negative or positive and the shear stress can act in any direction within the plane on which the shear stress is defined.

4. Stress satisfies Newton's third law. On an interior plane obtained by sectioning, a stress component is either negative on both sides of the section or positive on both sides of the section, with the direction of the shear stress also opposite on the two sides of the section. (See Fig. 2.17.) This property follows directly from the fact that the forces used to define stress satisfy Newton's third law.

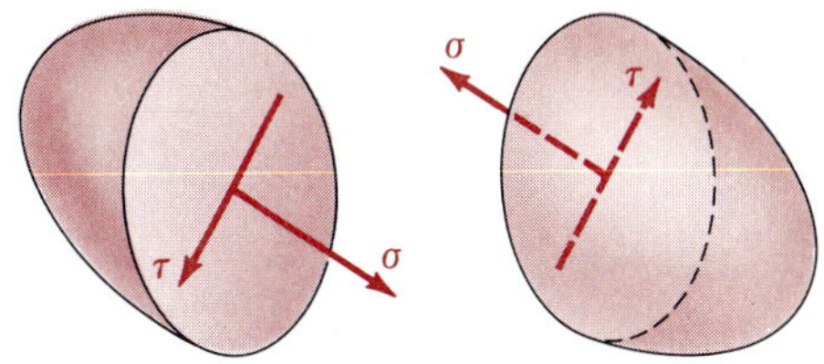

Figure 2.17 Third-law property of stress

PROBLEMS

2.33–2.36 State the values of all the stress components for each of the blocks shown.

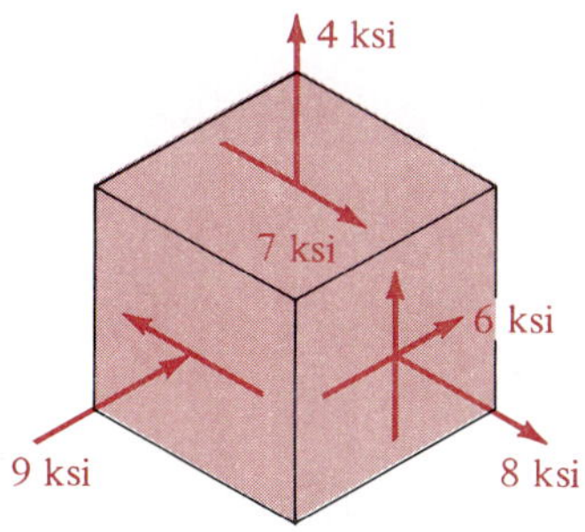

P. 2.33

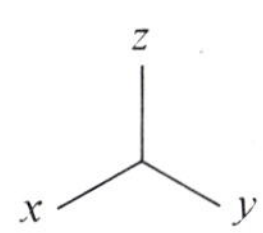

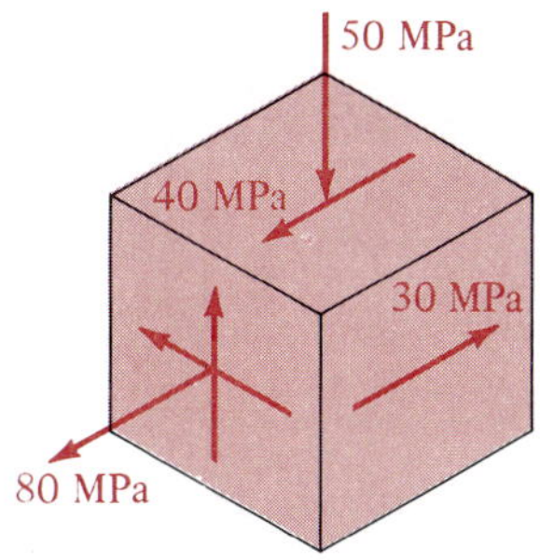

P. 2.34

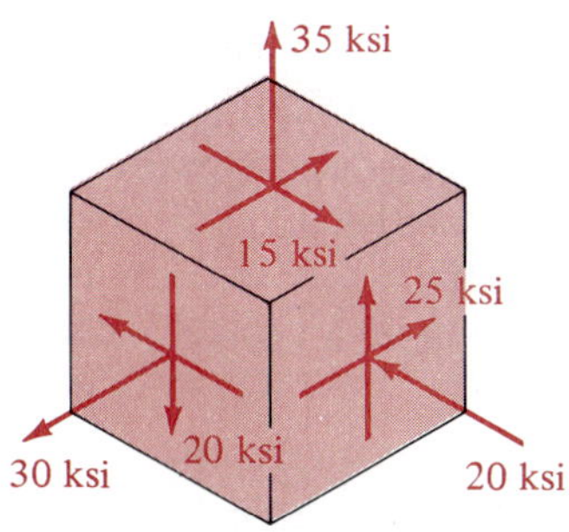

P. 2.35

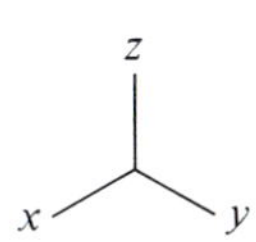

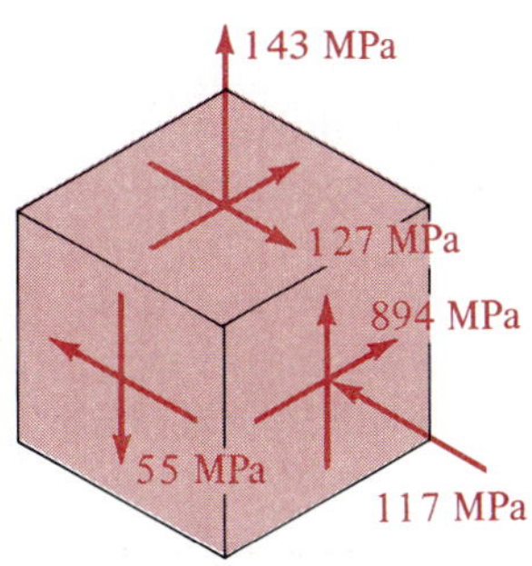

P. 2.36

2.37–2.40 State the values of all the stress components for each of the blocks shown.

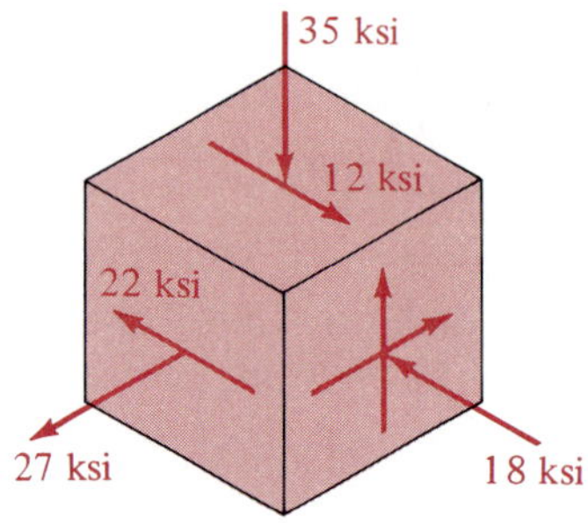

P. 2.37

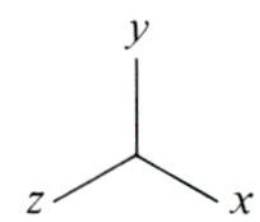

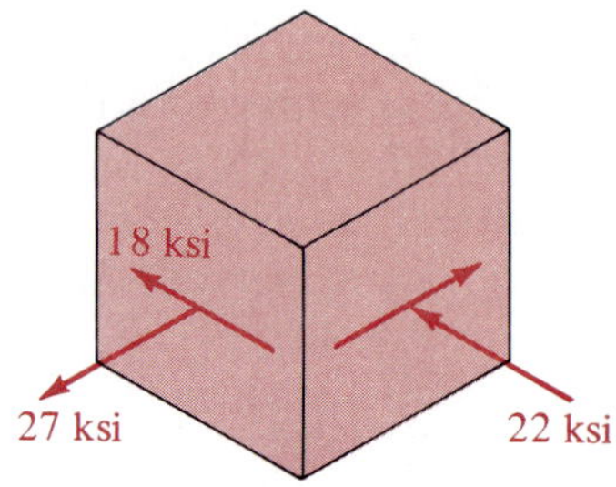

P. 2.38

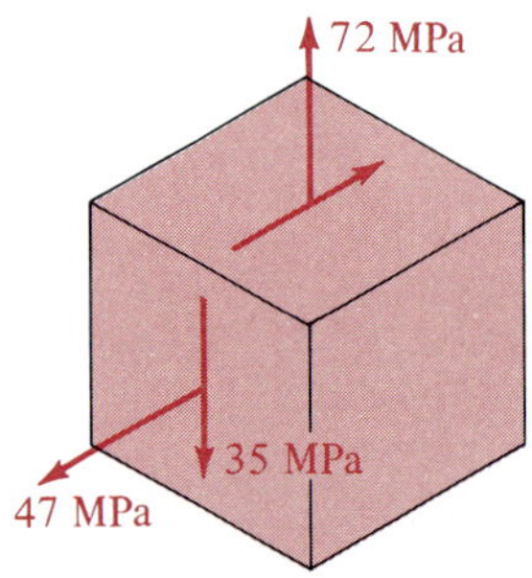

P. 2.39

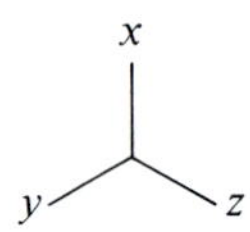

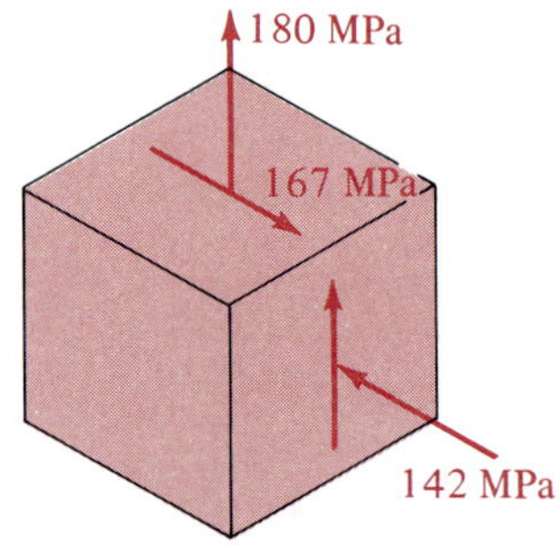

P. 2.40

(a)

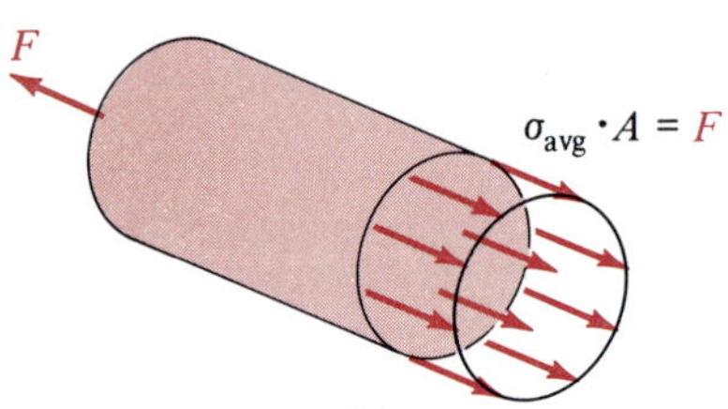

(b)

Figure 2.18 Average stress in a bar under a centrally applied axial load

2.3.2 Average Stresses

Many structures are designed so that the individual components transmit either tensile, compressive, or shear forces. The stresses are a measure of the distribution of the tensile or compressive force over the cross section, and are related to the internal forces transmitted across a section. A meaningful measure of the load-carrying capacity of a member can be obtained by computing what is termed the average tensile stress, average compressive stress, or average shear stress in the bar. Consider the problem of a straight circular bar acted upon by a centrally applied axial force F as shown in Fig. 2.18a.

We define σ_{avg}, the **average normal stress** acting on a typical section shown in Fig. 2.18b, as

$$\sigma_{avg} = \frac{F}{A}$$

or equivalently as $F = \sigma_{avg} \cdot A$. The value of σ_{avg} is determined by using F, the value of the internal force transmitted, divided by the local area A. The dimension of the average normal stress is force per unit area or $[\sigma] = F/L^2$. The average normal stress is positive or negative depending upon whether the force transmitted is tensile or compressive, respectively.

For a bar of arbitrary cross section we will determine in Chapter 5 that the line of action of the force transmitted must be through the centroid of the cross section in order to accurately assess the stresses on the basis of the equation $\sigma = P/A$. If the force does not pass through the centroid, bending results and the stresses can no longer be computed as $\sigma = P/A$. We will also discuss in detail in Chapter 5 that $\sigma = P/A$ is not valid in the neighborhood of any sudden changes in the area resulting from holes, notches, and so forth.

In the SI system, force is measured in newtons (N) and area in square meters (m^2). Then 1 N/m^2 = 1 pascal or 1 Pa, 1 kN/m^2 = 1 kPa, and 10^6 N/m^2 = 1 MPa. In the U.S. system, force is usually measured in pounds force (lbf) and area in square inches (in^2). Then 1 lbf/in^2 = 1 psi, 1 kip/in^2 = 1000 lbf/in^2 = 1 ksi. Length units of feet are also sometimes used. With 1 lbf = 4.448 N and 1 m = 39.36 in., the conversion between stress in the two systems of units is

$$1\ \text{lbf/in}^2 = 4.448\ \text{N}/(0.0254\ \text{m})^2 = 6.891 \times 10^3\ \text{Pa}$$

or

$$1\ \text{ksi} = 6.891\ \text{MPa}$$

Example 2.5 A simple structure consisting of two straight bars each with an area of 1 in^2 is shown in Fig. 2.19a.

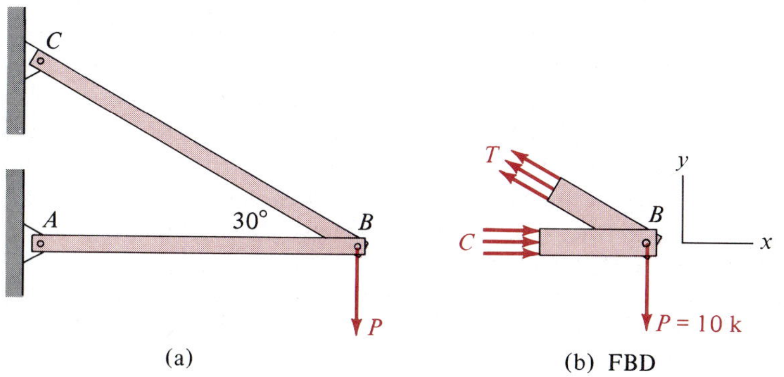

Figure 2.19 Simple truss structure

A free-body diagram of a portion of the structure near point B is shown in Fig. 2.19b. The equilibrium equations are

$$\Sigma F_x = C - T\frac{\sqrt{3}}{2} = 0$$

and

$$\Sigma F_y = \frac{T}{2} - 10{,}000 \text{ lbf} = 0$$

from which $T = 20{,}000$ lbf and $C = 17{,}320$ lbf. T is the known tensile force transmitted by bar BC so that we can compute the average normal stress $(\sigma_{avg})_{BC}$ in bar BC as

$$(\sigma_{avg})_{BC} = \frac{T}{A_{BC}} = 20{,}000 \text{ lbf}/1.0 \text{ in.}^2 = 20.0 \text{ ksi}$$

Similarly, the average normal stress $(\sigma_{avg})_{AB}$ in bar AB is

$$(\sigma_{avg})_{AB} = \frac{-C}{A_{AB}} = -17{,}320 \text{ lbf}/1.0 \text{ in.}^2 = -17.32 \text{ ksi}$$

where the negative sign indicates a compressive stress.

The use of the average normal stress value for member BC assumes that the stress is constant over the face in question as indicated in Fig. 2.20a; this represents one possible stress distribution whose resultant is T.

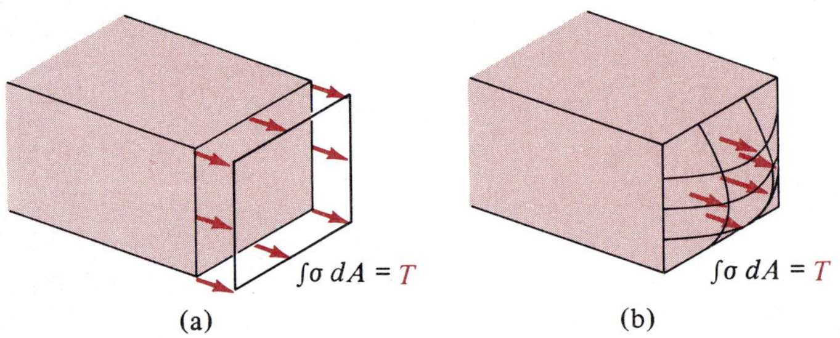

Figure 2.20 Possible stress distributions in member BC

Any other distribution of normal stress having the same resultant, such as the distribution shown in Fig. 2.20b, would have a maximum that is larger than the average. Thus the average normal stress provides a lower bound for the actual maximum normal stress. Nevertheless, average normal stress values computed according to $\sigma = F/A$ are commonly used as a measure of the normal stress in a bar that transmits a tensile or compressive force applied through the centroid.

Example 2.6 A thin-walled cylindrical pressure vessel with hemispherical caps of radius R is shown in Fig. 2.21a. If the internal pressure is p, determine the average normal stress in the walls near the center of the cylindrical portion of the container.

Solution A section is passed through the pressure vessel as shown in Fig. 2.21a resulting in the free-body diagram shown in Fig. 2.21b. The exposed gas or fluid is acted upon by the uniform pressure producing the central force $p \cdot A = p \cdot \pi R^2$. The internal force in the walls of the vessel must balance this pressure force. The area over which the internal force acts is $A \approx 2\pi Rt$, so that the average normal stress in the cylinder is given by

$$\Sigma F_x = -p\pi R^2 + \sigma_{avg} 2\pi Rt = 0$$

or

$$\sigma_{avg} = \frac{pR}{2t}$$

on a plane normal to the axis of the cylinder. Any shear stresses that possibly exist in the wall must have a zero resultant.

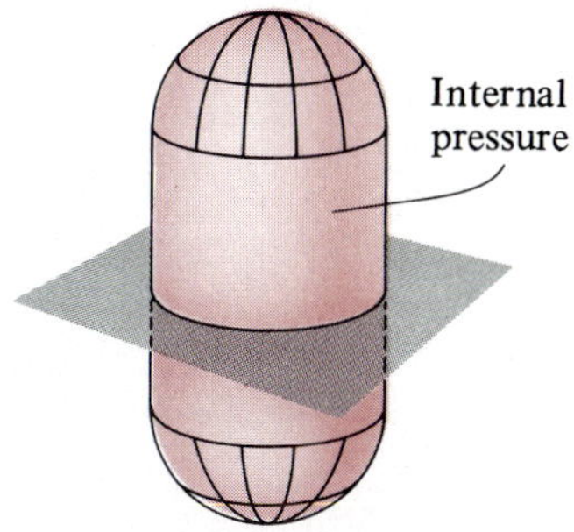

(a)

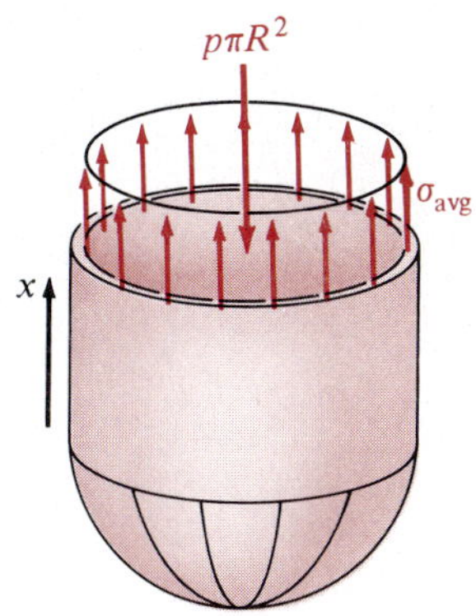

(b) FBD

Figure 2.21 Wall stresses in a cylindrical pressure vessel

Example 2.7 The shear strength of a glue joint is frequently determined on the basis of a test similar to the one shown in Fig. 2.22a. Determine the average shear stress in the glue.

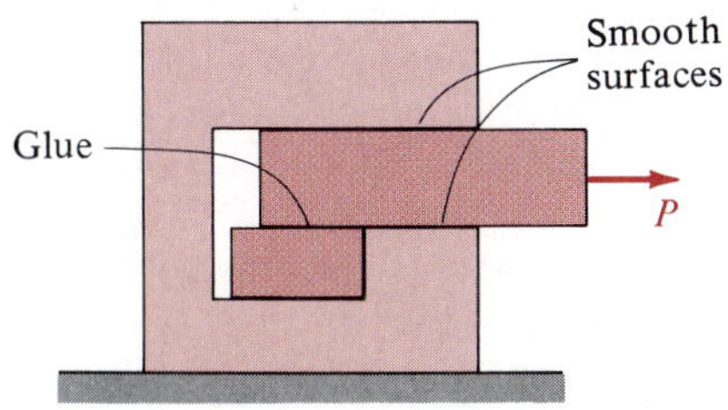

(a)

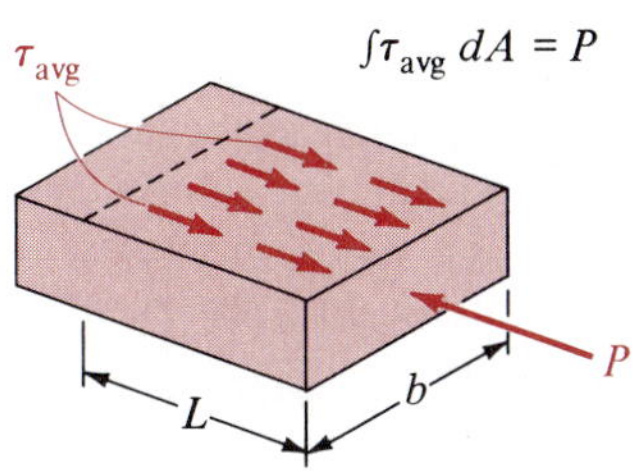

(b)

Figure 2.22 Average shear stress in a glue joint

Solution Referring to the free-body diagram of the lower part as shown in Fig. 2.22b it is seen that the total force that must be transmitted by the glue is the applied load P. Note that only the horizontal forces that act on the free-body diagram are shown. The average shear stress in the glue is easily computed to be

$$\tau_{avg} = \frac{P}{A} = \frac{P}{(b \cdot L)}$$

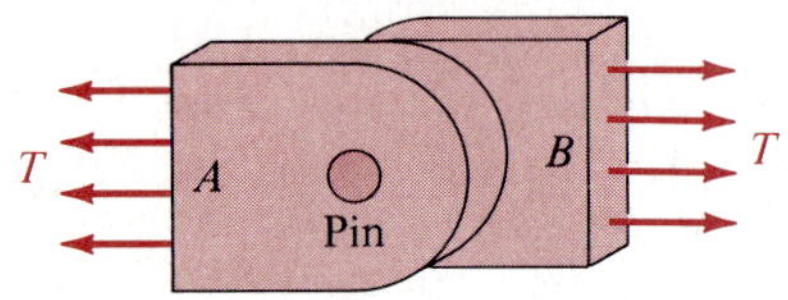

Figure 2.23 Single-shear connection

Example 2.8 One of the most important applications for the concept of average stress is in analyzing structures where individual members of the structure are connected by pins. One of the simplest types of these connections is indicated in Fig. 2.23.

This connection is called a **single shear** connection. Single shear means that the total load is transmitted in shear at one location in the pin. In analyzing the manner in which the forces are actually transmitted by the connection, there are several average stresses of interest. Consider a free-body diagram of member B, as shown in Fig. 2.24a.

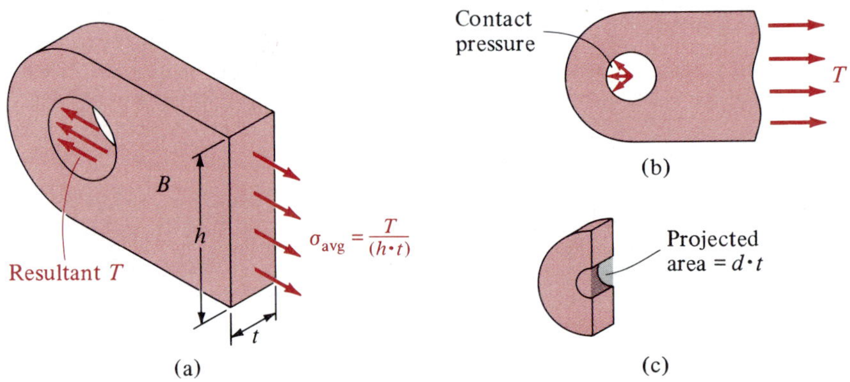

Figure 2.24 Free-body diagrams of member B

As the figure indicates, the average normal stress is given by $\sigma_{avg} = T/\text{area} = T/(h \cdot t)$ away from the hole. For equilibrium, the resultant force acting on the inside surface of the hole, caused by contact with the pin, must have a resultant of T. We *assume* that the contact force on the surface of the hole is primarily associated with contact pressures as shown in Fig. 2.24b. The actual distribution cannot be known without a detailed stress analysis of the contact between the pin and the member. As a measure of the contact stress we compute an average normal stress referred to as the **bearing stress** and defined as

$$\sigma_{bearing} = \frac{T}{(t \cdot d)}$$

where $t \cdot d$ is the projected area as indicated in Fig. 2.24c. Note that a bearing stress is always compressive. The actual maximum compressive stress between the pin and the hole can exceed the bearing stress by a significant amount, depending for instance on the mismatch between the diameter of the hole and the diameter of the pin. Nevertheless, assuming that the pin or rivet fits snugly in the hole, the bearing stress as computed above is commonly used as a measure of the compressive stress between the pin and hole.

Another important average stress is the average normal stress acting on the section shown in Fig. 2.25, the reduced area associated with the section through the hole in member B.

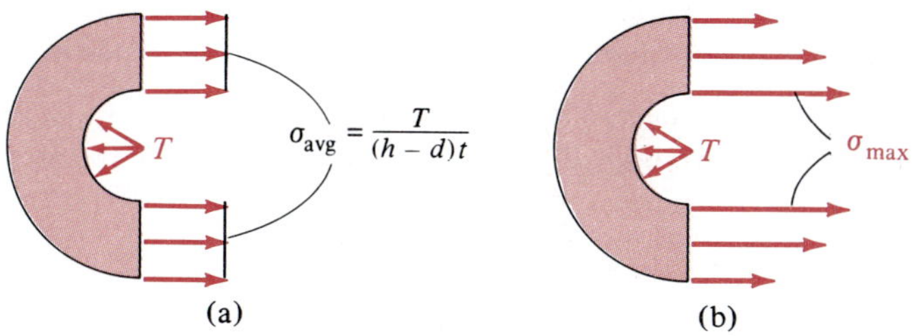

Figure 2.25 Average normal stress and actual stress distribution on the reduced area asociated with the section through the hole

With the reduced area given by $A = (h - d)t$, the average normal stress is $\sigma_{avg} = T/((h - d)t)$, clearly larger than the average normal stress on a section away from the hole. As indicated in Fig. 2.25b, the maximum normal stress acting on the section through the hole can be several times the average value, due to the sudden change in the area over which T acts. This idea will be discussed further in the section on stress concentrations in Chapter 5.

Consider also a free-body diagram of the portion of the pin that is in contact with the member B as shown in Fig. 2.26.

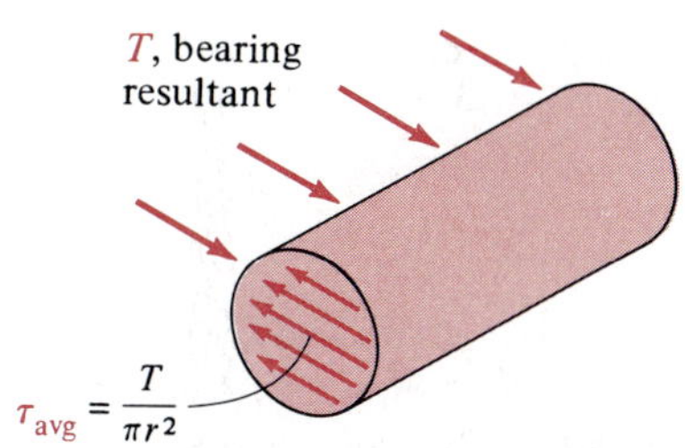

Figure 2.26 Average shear stress for the equilibrium of a portion of the pin

The resultant of the pressure between the pin and the hole as well as the internal force resultant must be equal to the load T. The internal force resultant arises from shear stresses that act on the exposed face of the pin. We compute an average shear stress as

$$\tau_{\text{avg}} = \frac{T}{A_{\text{pin}}}$$

This average shear stress serves to establish a lower bound on the actual maximum shear stress on the face of the pin.

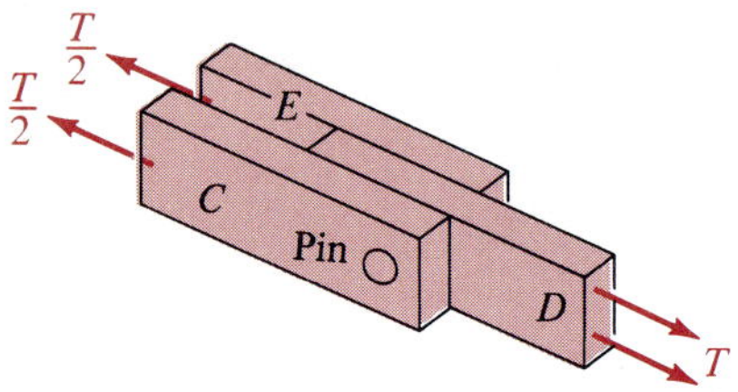

Figure 2.27 Double-shear connection

Example 2.9 Another method of connection using a pin is shown in Fig. 2.27. Determine the bearing stresses, the average shear stress in the pin, and the average normal stress at a section through the pin for all the bars. Take the dimensions of each of the bars to be h by t and the diameter of the holes to be d.

Solution This mode of using a pin to transmit the force T is commonly referred to as **double shear**. Free-body diagrams of the pin and each of the members are shown in Fig. 2.28.

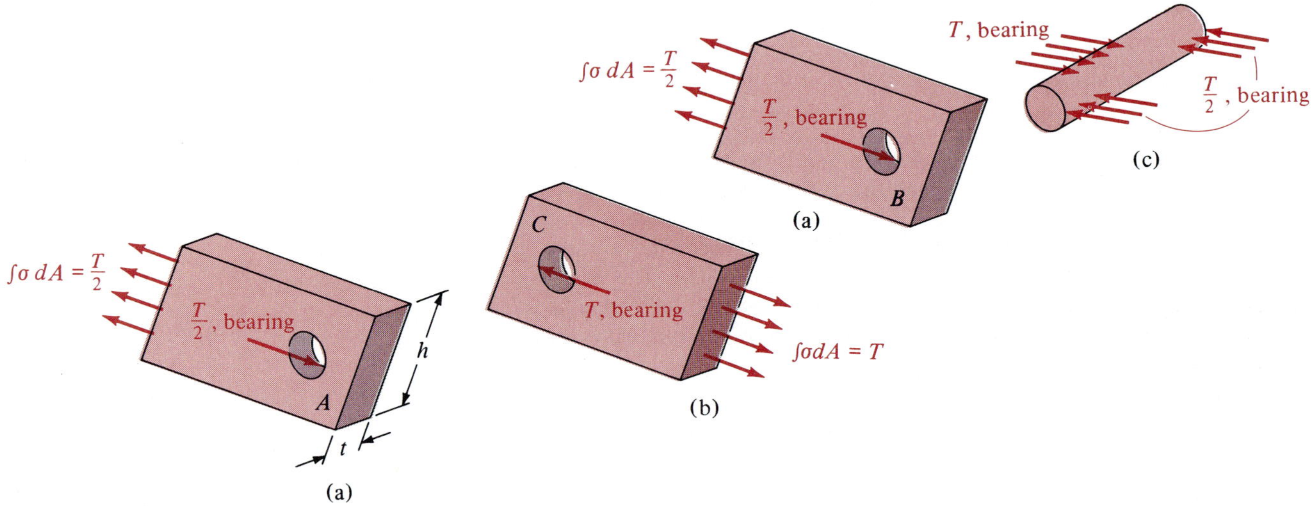

Figure 2.28 Free-body diagrams for double shear

Referring to the free-body diagrams shown in Figs. 2.28a, the average bearing stresses on the surface of the holes in members A and B can be computed in the same fashion as outlined previously for single shear, namely,

$$\sigma_{\text{bearing}} = \frac{(T/2)}{(d \cdot t)}$$

with the average normal stress on sections through the holes in members A and B given by

$$\sigma_{\text{avg}} = \frac{(T/2)}{((h - d)t)}$$

Referring to Fig. 2.28b the average bearing stress at the hole in member C is

$$\sigma_{\text{bearing}} = \frac{T}{(d \cdot t)}$$

with the average normal stress on a section through the hole in member C as

$$\sigma_{\text{avg}} = \frac{T}{((h - d)t)}$$

Consider next the free-body diagram of the pin shown in Fig. 2.28c. In order to determine the average shear stress in the pin, pass sections through the pin corresponding to the surfaces of the members A and B as indicated in Fig. 2.29.

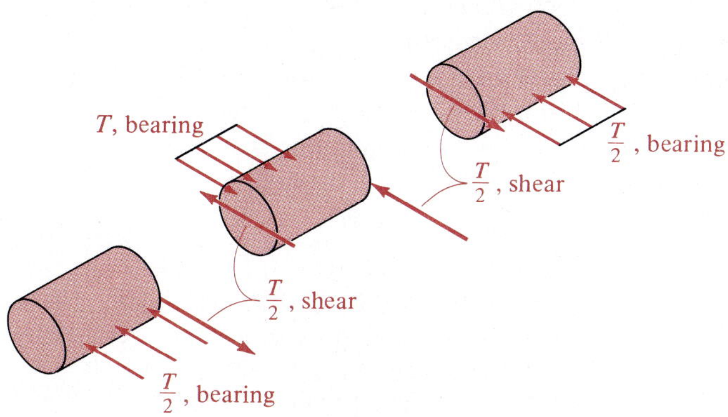

Figure 2.29 Free-body diagrams of portions of the pin

The free-body diagrams clearly indicate that the pin carries half the load in shear at each location. For this reason, the pin is said to be acting in double shear. The average shear stress acting on each of the exposed internal surfaces of the pin is

$$\tau_{\text{avg}} = \frac{(T/2)}{A_{\text{pin}}}$$

Consider again the free-body diagram of a portion of the pin in double shear, assuming a uniform bearing pressure as in Fig. 2.30a.

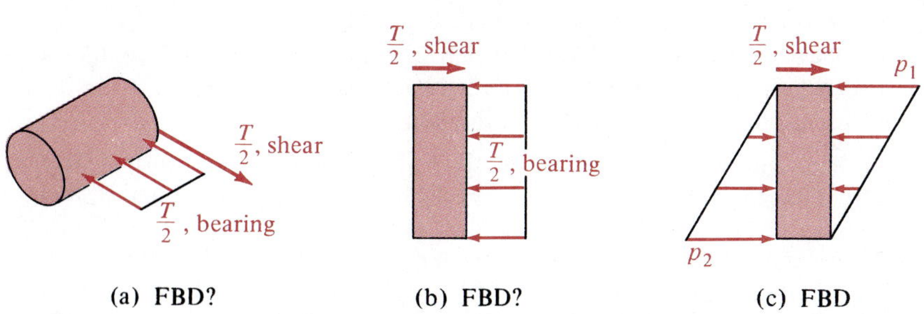

Figure 2.30 Free-body diagram of a portion of a pin in double shear

Consider also a side view of the free-body diagram of a portion of the pin as in Fig. 2.30b. Note that the free-body diagram cannot possibly be correct: a moment equation about an axis perpendicular to the plane of the paper cannot possibly be satisfied. Equilibrium is violated! A reasonable distribution of pressures on the pin which could possibly be in equilibrium is shown in Fig. 2.30c. The student is asked to analyze this possible solution, in the problems at the end of the section. Computed as in this example, average bearing stress is only an approximation of the actual stresses acting between the pin and the surface of the hole.

In summary, it is meaningful to use an average normal stress or an average shear stress in certain simple loading situations, such as a centrally applied tensile or compressive load in a straight member, as in Example 2.5, a glued joint, as in Example 2.7, or a pin transmitting a shear force, as in Examples 2.8 and 2.9. Other situations where average normal or shear stresses are meaningful will become clearer after studying the material in Chapters 5 through 8.

PROBLEMS

2.41 A woman weighing 115 lbf walks in high heels along a level floor as shown. A dynamic analysis indicates that an instant after her heel makes contact with the floor, the force of contact is approximately 1.8 times her weight. Estimate the average normal pressure between her heel and the floor assuming a square area of contact as shown.

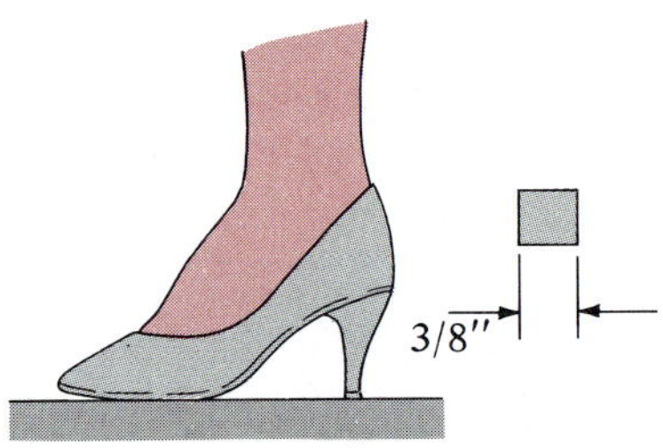

P. 2.41

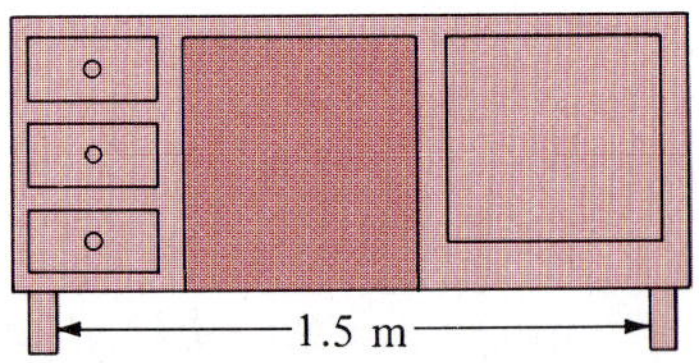

P. 2.42

2.42 The desk and its contents weigh 240 lbf. Assuming the contact area between each of the legs and the floor to be a circle of diameter 0.2 in., estimate the average normal stress between the floor and a leg.

2.43 For the truss structure shown, determine (*a*) the average normal stresses on the reduced sections at *A* and *C*, (*b*) the bearing stresses at *A* and *C*, and (*c*) the shear stresses in the pins, acting in double shear, at *A* and *C*. Take all members to be square with dimensions of $b = 2.0$ in., all pin diameters to be 0.5 in., $P = 20$ k and $\alpha = 30°$.

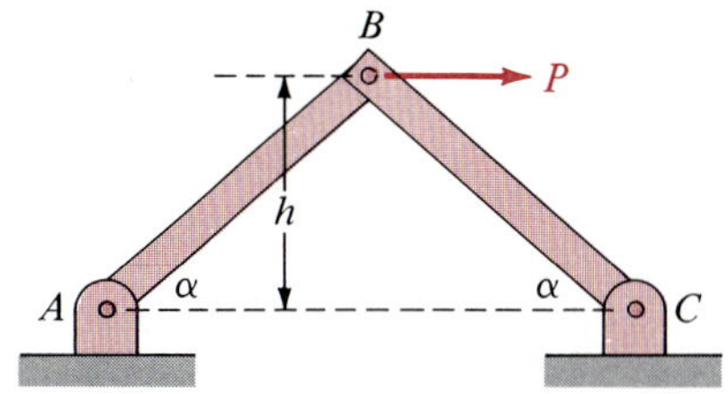

Ps. 2.43 and 2.44

2.44 Repeat problem 2.43 for $P = 80$ kN, $b = 45$ mm, the pin diameters to be 15 mm, and $\alpha = 36°$.

2.45 The segmented bar is loaded by axial forces as shown. Determine the internal force in each portion of the bar and hence plot $P(x)$ as a function of x. Determine and plot the average normal stress as a function of x. Take $d_1 = 2.0$ in., $d_2 = 1.5$ in., $d_3 = 2.75$ in., $P_1 = 12$ kips, $P_2 = 15$ kips, and $P_3 = 20$ kips.

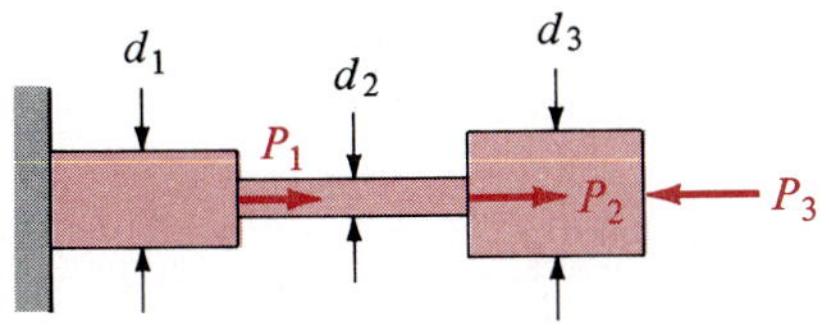

Ps. 2.45 and 2.46

2.46 Repeat problem 2.45 for $d_1 = 60$ mm, $d_2 = 40$ mm, $d_3 = 55$ mm, $P_1 =$ 50 kN, $P_2 = 70$ kN, and $P_3 = 60$ kN.

2.47 Determine (*a*) the average shearing stress in the 15 mm diameter pin, acting in single shear, and (*b*) the bearing stresses in the support and in member *ABC*. All members are 25 mm thick in the pin direction.

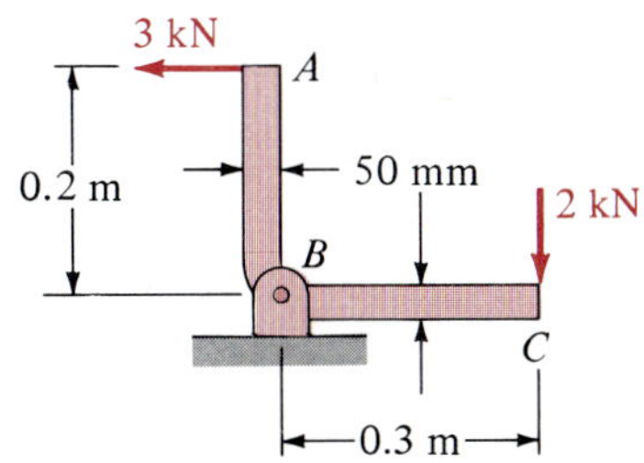

P. 2.47

2.48 The fixture shown is designed to test the strength of glue joints. If the two pieces separate when P is 1200 lbf, what is the strength of the glue in shear? Take $b = 4$ in. and the thickness of the glue joint to be 4 in.

2.49 Repeat problem 2.48 if $P = 6$ kN, $b = 90$ mm, and the thickness of the glue joint is 80 mm.

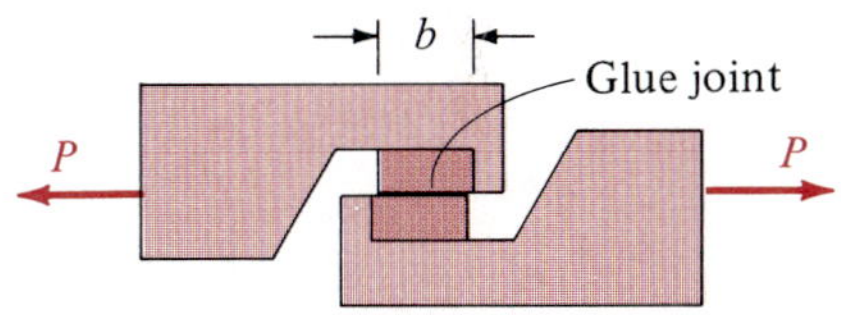

Ps. 2.48 and 2.49

2.50 A concrete pier rests on the earth as shown. Determine the average bearing pressure between the earth and the pier. Take the specific weight of the concrete to be 150 lbf/ft^3, $B = 40$ ft, $b = 10$ ft, and $h = 30$ ft.

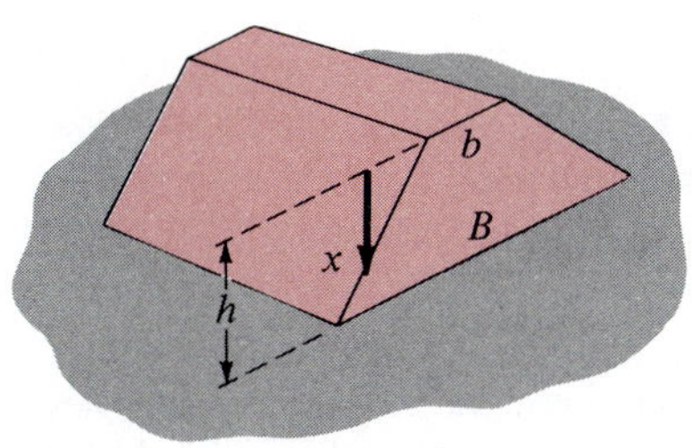

Ps. 2.50 and 2.51

2.51 For the concrete pier of problem 2.50, determine the average normal stress in the pier as a function of x.

2.52 Two plates are connected by a row of rivets as shown. Neglecting friction and assuming that each rivet carries the same shear load, determine (*a*) the maximum bearing stress in the joint, (*b*) the average shear stress in the rivets, and (*c*) the maximum average tensile stress in the plate. Take the diameter of the rivets to be 0.5 in. and $P = 3$ kips.

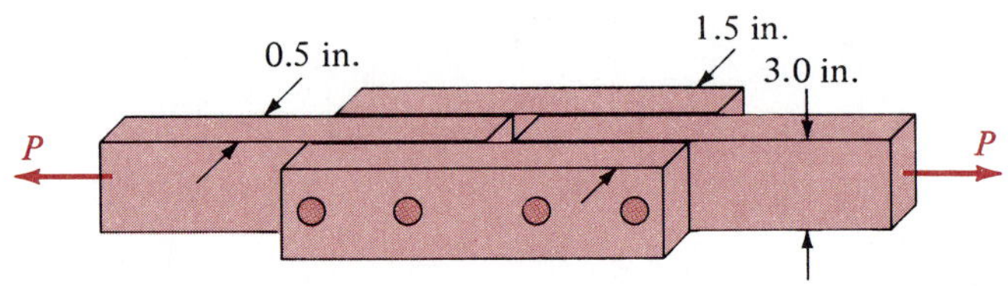

P. 2.52

2.53 It is proposed that washers be formed by punching holes through disks as shown in the figure. Develop a relationship between the punching force P and the average shear stress in the disk.

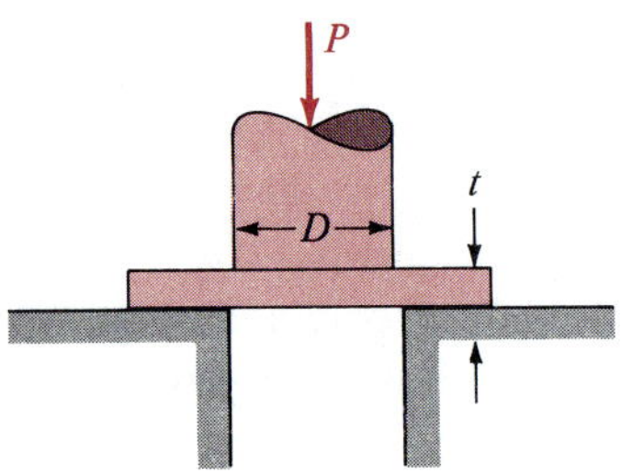

P. 2.53

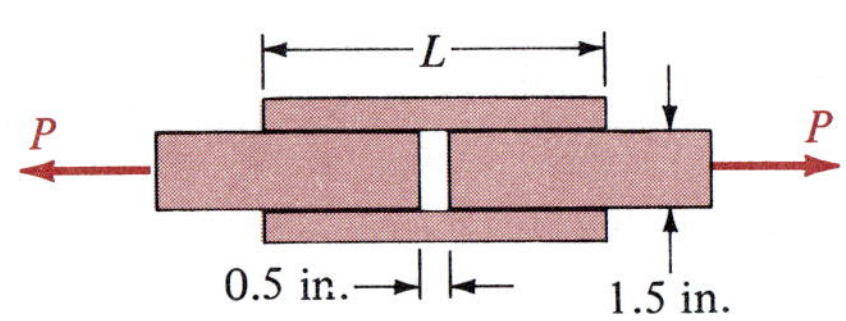

P. 2.54

2.54 A tension member is formed by gluing a splice plate to two 2 × 4s as shown. If the allowable shear stress in the glue is to be 200 psi, what must be the length of the splice plate to transmit a load of 6000 lbf? Note the actual dimensions of a 2 × 4 are 1.5 in. by 3.5 in.

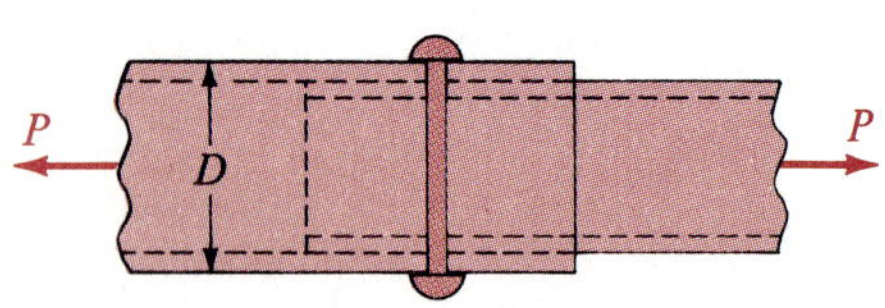

Ps. 2.55 and 2.56

2.55 The two circular pipes are connected by a pin as shown. Determined the average shear stress in the pin and the bearing stresses in the pipes. Take $P =$ 10 kN, $D = 90$ mm, the thickness of both pipes to be 8 mm, and the diameter of the rivet to be 10 mm.

2.56 Repeat problem 2.55 when a torque T of 150 N · m is applied instead of the axial force P.

2.57 Two pieces of PVC pipe are glued as shown. Determine the average shear stress in the glue. Take $P = 3$ kips, $D = 3.0$ in., $L = 4.0$ in., and $t = 0.1$ in.

2.58 Repeat problem 2.57 when a torque T of 6 in · k is applied instead of the axial force P. Also determine the average shear stress in the smaller tube.

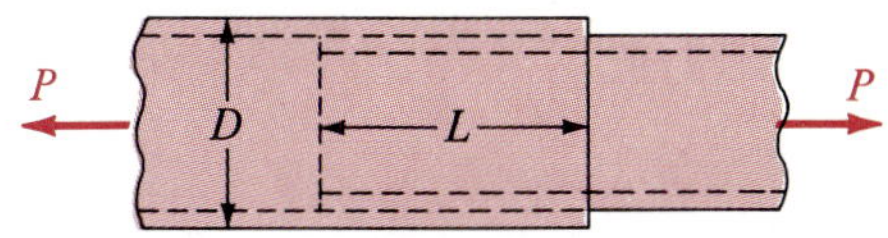

Ps. 2.57 and 2.58

2.59 The wrench shown is operated by inserting the two 0.3 in. diameter pins in the holes and applying the force P shown. If the torque developed is to be 6 in · k, what is the average shear stress in one of the pins? Take $d = 4$ in.

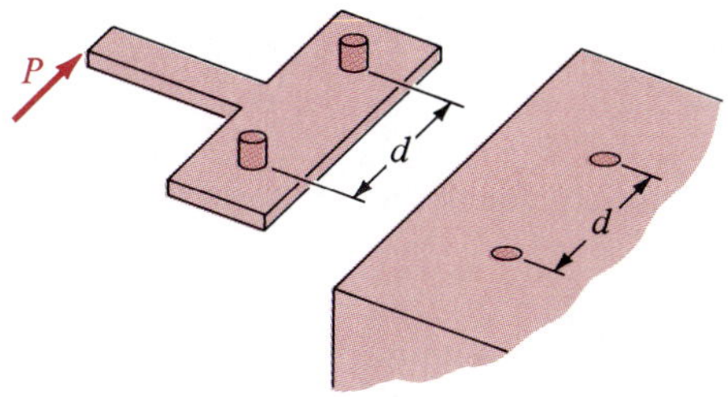

P. 2.59

2.60 The blade on a rotary lawn mower is often attached to the crankshaft of the engine by a shear key as shown. If the torque transmitted is T, what is the average shear stress in the key? Take the cross section of the key to be square of dimension b.

2.61 Consider the more realistic free-body diagram of the key in problem 2.60 as shown. Estimate the values of the bearing pressures on the key by assuming they have the distributions shown. Determine the magnitudes of the bearing pressures and the shear stress in the key in terms of the torque T.

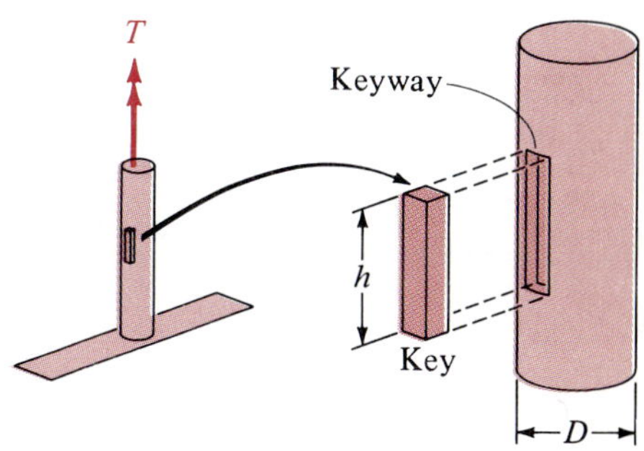

P. 2.60

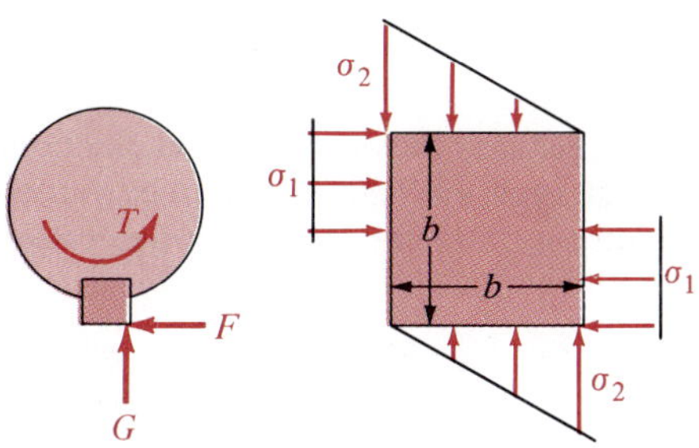

P. 2.61

2.62 A uniform bar of area A and mass density ρ rotates about an axis through its center at a rate ω as shown. Determine the average normal stress as a function of the position x.

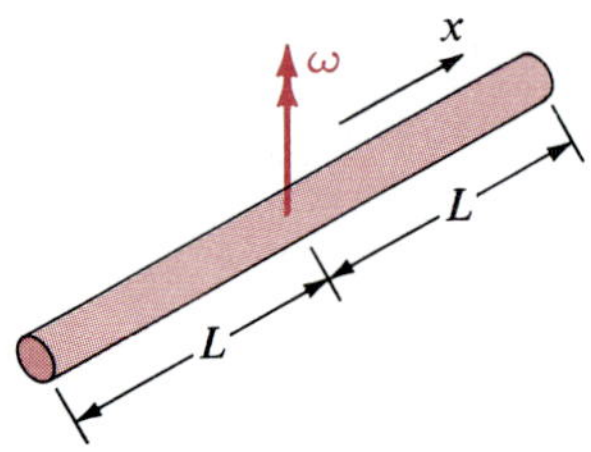

P. 2.62

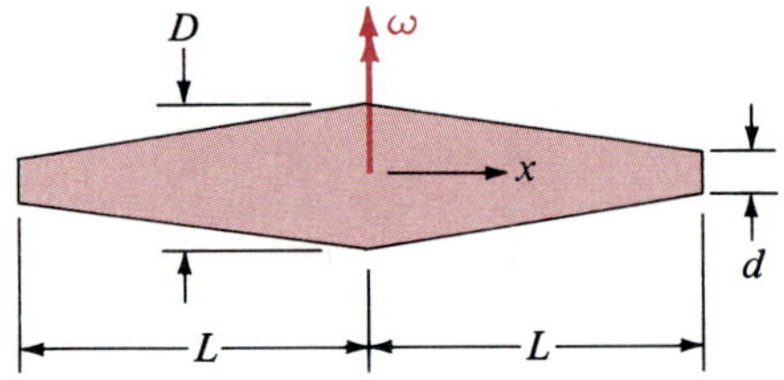

P. 2.63

2.63 Repeat problem 2.62 for the nonuniform bar of variable circular cross section shown.

2.64 The free-body diagram of a pin in single shear is assumed to appear as in the figure shown: rather than a uniform pressure acting on the projected area of one side of the pin, it is assumed that linearly varying pressures act on the projected area of both sides of the pin. Determine the values of p_1 and p_2 in order that the pin be in equilibrium. What is the ratio of p_1 to the average bearing pressure determined in the usual fashion?

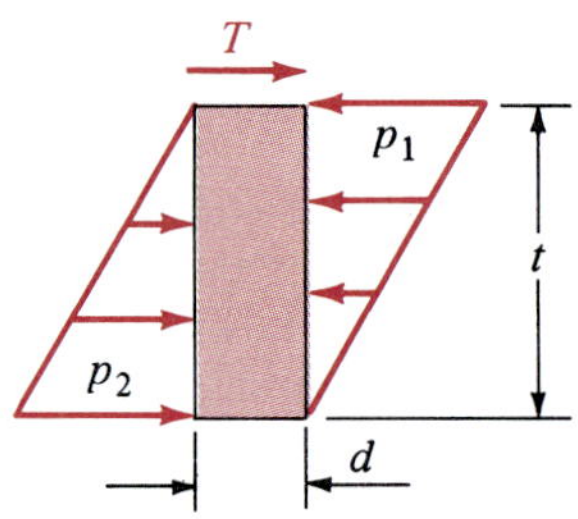

P. 2.64

2.65 Idealize the contact area between the heel and floor in problem 2.41 to be circular and the actual pressure distribution to be parabolic as shown. What is the maximum pressure p_0 in terms of the average pressure p_{avg}?

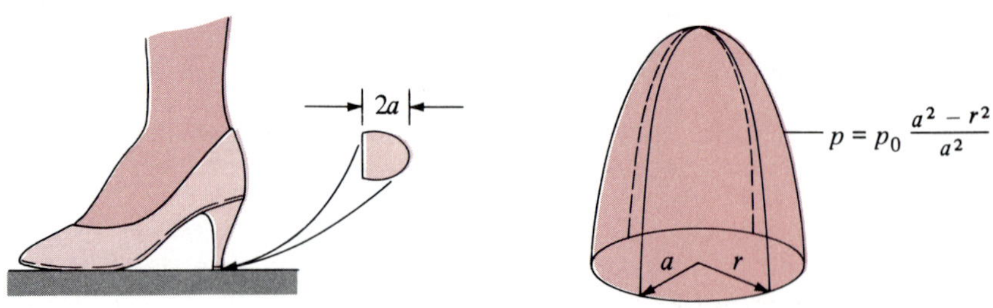

P. 2.65

2.66 The 3500 lbm rear-wheel drive automobile shown accelerates from rest at (1/3)g. Estimate the average normal and shear stresses between the back tires and the road.

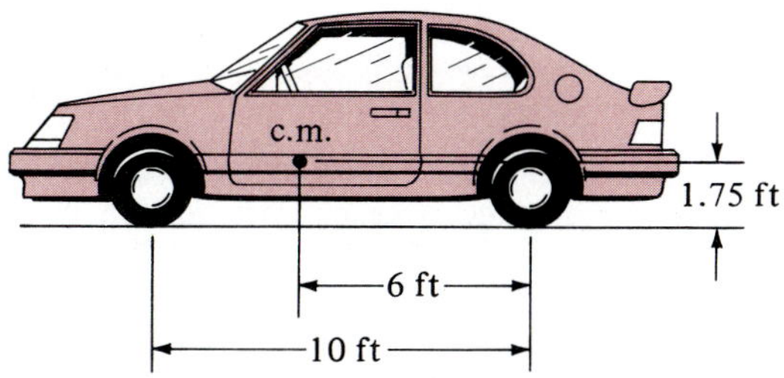

P. 2.66

2.4 STRESS TRANSFORMATIONS

For a body subjected to forces as indicated in Fig. 2.31, consider two different plane sections that contain the point P and for which the normals are **n** and **n′**, respectively, as shown in Figs. 2.31a and 2.31c. Take the stresses on the plane whose normal is **n** to be σ_x, τ_{xy}, and τ_{xz} and those on the plane whose normal is **n′** to be $\sigma_{x'}$, $\tau_{x'y'}$, and $\tau_{x'z'}$ as indicated in Figs. 2.31b and 2.31d, respectively.

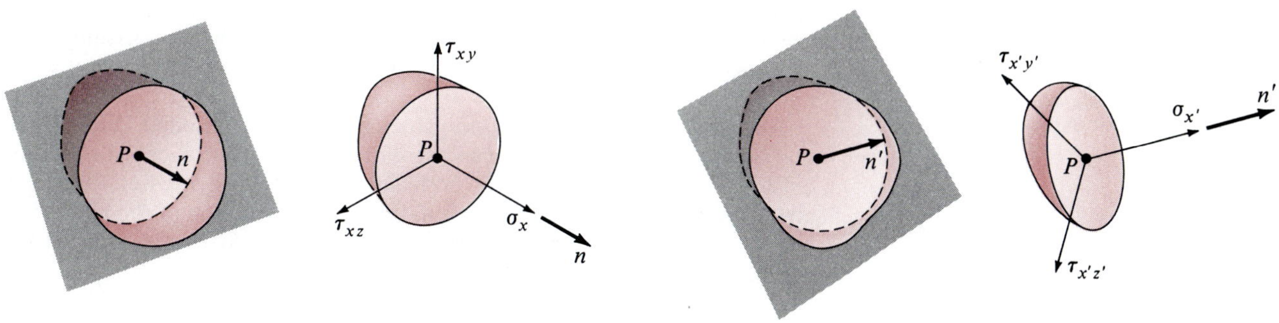

(a) Plane section with normal n (b) (c) Plane section with normal n' (d)

Figure 2.31 Different planes through a point P

At a point P common to both planes the two sets of stresses will in general be different. The character of these differences lies at the heart of what is termed a **stress transformation.** To understand stress transformation, we must ask how the stresses depend upon the orientation of the plane passing through the point P. We will use equilibrium to develop the relationships between the stresses acting on different planes passing through the point. Before looking at the question of a stress transformation in general, let us examine several specific examples.

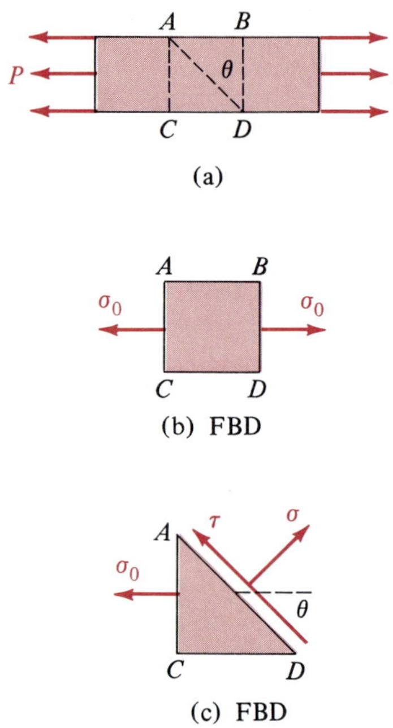

Figure 2.32 Stresses on different planes for a bar in pure tension

Example 2.10 Consider a bar acted upon by a tensile force as indicated by Fig. 2.32a. Determine the average stresses on a section through the bar inclined at an angle θ.

Solution Fig. 2.32b shows a segment of the bar on whose faces the average stress $\sigma_0 = P/A$ acts. AB and CD are stress free. Fig. 2.32c shows a triangular element one of whose faces is vertical and acted upon by the normal stress σ_0 and another face inclined as shown with the normal to the inclined face making an angle of θ to the horizontal. CD is again a stress-free face. Recognizing that the stresses σ and τ must be such that the triangular element is in equilibrium, a summation of the forces in the horizontal and vertical directions yields

$$\Sigma F_H = -\sigma_0 \cdot A + (\sigma \cdot A_\theta) \cos \theta - (\tau \cdot A_\theta) \sin \theta = 0$$

and

$$\Sigma F_V = \qquad + (\sigma \cdot A_\theta) \sin \theta + (\tau \cdot A_\theta) \cos \theta = 0$$

where $A_\theta = A/\cos \theta$ is the area of face AD. Note that each of the forces $\sigma_0 A$, σA_θ, and τA_θ represents the average stress times the corresponding area. These two equations can be solved to yield the average normal stress $\sigma = \sigma_0 \cos^2 \theta$ and the average shear stress $\tau = -\sigma_0 \cos \theta \sin \theta$, showing that the average shear stress actually acts in the direction opposite to what was assumed in the free-body diagram of Fig. 2.32c. The maximum shear stress occurs at $\theta = 45°$ and has the value $\tau = -\sigma_0/2$.

The important point is that on planes through the bar other than the plane perpendicular to the direction of the external load P, the stresses that must exist for equilibrium consist of both normal and shear stresses; that is, the stresses on a plane at a point depend upon the orientation of the plane that is passed through the point even though the external loads and internal force resultants are unchanged.

Example 2.11 Consider also the problem of a square element acted upon by shear stresses, as in Fig. 2.33a. Determine the average normal and average shear stresses on the diagonals.

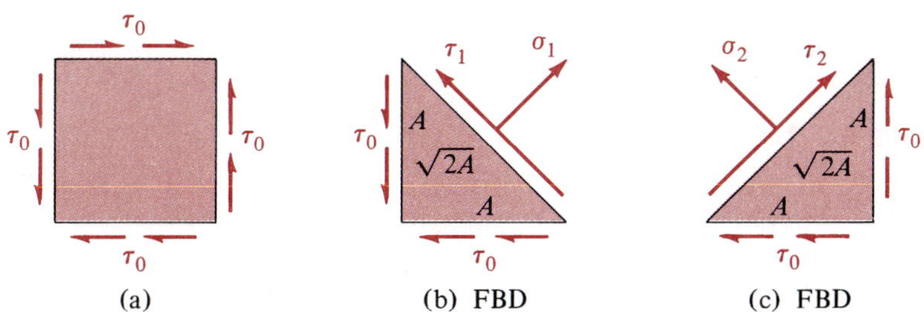

Figure 2.33 Square element in pure shear

Solution Free-body diagrams for sections taken along each of the diagonals are shown in Fig. 2.33b and 2.33c. Denoting the area of one of the faces on which the shear stress τ_0 acts by A, a sum of the forces in the directions corresponding to σ_1 and τ_1 for the free-body diagram of Fig. 2.33b yields

$$\Sigma F_{\sigma_1} = \sigma_1 \sqrt{2}\, A - \tau_0 A/\sqrt{2} - \tau_0 A/\sqrt{2} = 0$$

and

$$\Sigma F_{\tau_1} = \tau_1 \sqrt{2}\, A - \tau_0 A/\sqrt{2} + \tau_0 A/\sqrt{2} = 0$$

from which $\sigma_1 = \tau_0$ and $\tau_1 = 0$. For the free-body diagram of Fig. 2.33c the same equations are

$$\Sigma F_{\sigma_2} = \sigma_2 \sqrt{2}\, A + \tau_0 A/\sqrt{2} + \tau_0 A/\sqrt{2} = 0$$

and

$$\Sigma F_{\tau_2} = \tau_2 \sqrt{2}\, A - \tau_0 A/\sqrt{2} + \tau_0 A/\sqrt{2} = 0$$

from which $\sigma_2 = -\tau_0$ and $\tau_2 = 0$. Thus an element oriented at 45° is acted upon by the stresses shown in Fig. 2.34, that is, a tensile stress of magnitude τ_0 in one direction and a compressive stress of τ_0 in the perpendicular direction. This example indicates once again that the stresses on a plane at a point depend upon the orientation of the plane that is passed through the point.

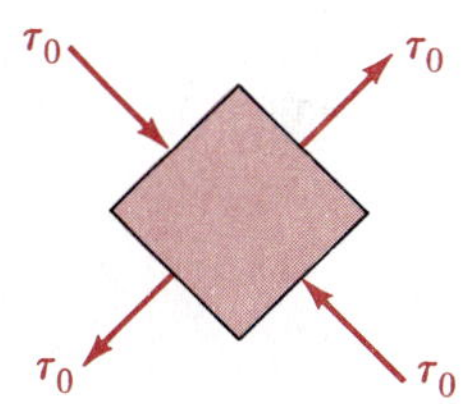

Figure 2.34 Element at 45° for pure shear

The state of stress at a point P generally consists of all six of the stress components. There are many applications, however, where the geometry of the body and the way it is loaded are such that the state of stress is essentially two dimensional. Such a state is frequently associated with a thin flat plate that is loaded in its plane, as in Fig. 2.35. The z-direction is normal to the plane of the plate.

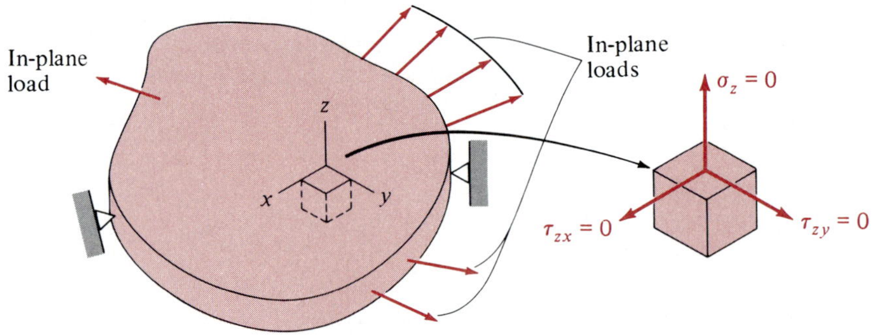

Figure 2.35 Plate loaded in its plane—plane stress

If the z = constant surfaces of the plate are stress free and the boundary of the plate is loaded by forces independent of the z coordinate in the plane of the plate as indicated, the plate is said to be in a state of **plane stress.** The basic assumption for plane stress is that the stress components σ_z, τ_{yz}, and τ_{xz}, all of which are zero on the two free surfaces of the plate, are zero everywhere in the plate. The only nonzero components of the stress are σ_x, σ_y, and $\tau_{xy} = \tau_{yx}$.

Consider then the state of stress at a point P in the interior of a plate, as in Fig. 2.36.

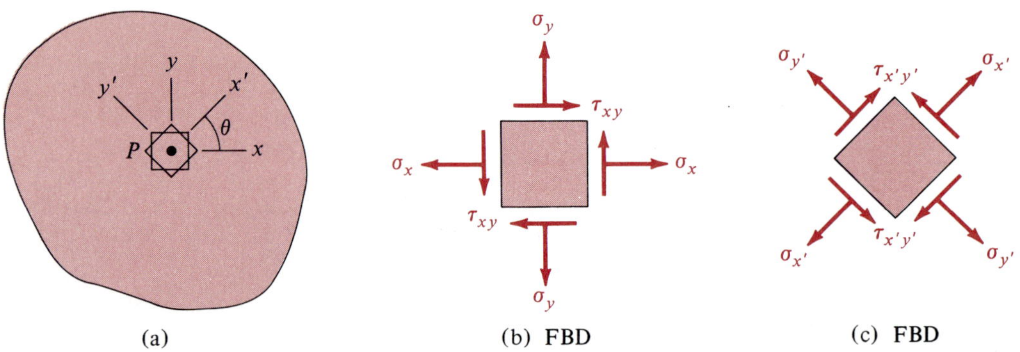

Figure 2.36 States of plane stress on elements with different orientations

A free-body diagram of an infinitesimal square element in the neighborhood of a point P in the plate is indicated in Fig. 2.36b. The sides of the element are parallel to the xy coordinate system at P and the thickness of the element in the z-direction is taken to be t. The element is taken to be small enough so that the stress values do not vary over the dimensions of the element. As shown, the stresses acting on the faces of the elements are σ_x, σ_y, and $\tau_{xy} = \tau_{yx}$. Consider also the stresses $\sigma_{x'}$, $\sigma_{y'}$, and $\tau_{x'y'} = \tau_{y'x'}$ acting on a differently oriented infinitesimal square element at P as shown in Fig. 2.36c. The sides of this element are parallel to the $x'y'$ coordinate system at P. For this plane stress situation, a stress transformation relates the stress components on the two differently oriented elements. As we saw in the two previous examples, a stress transformation is basically an equilibrium consideration.

Now let us consider the free-body diagram in Fig. 2.37. Note carefully that the orientation of the inclined face is defined by the normal to the face. The angle θ is the angle through which the x-axis has been rotated to coincide with the direction of the x'-axis.

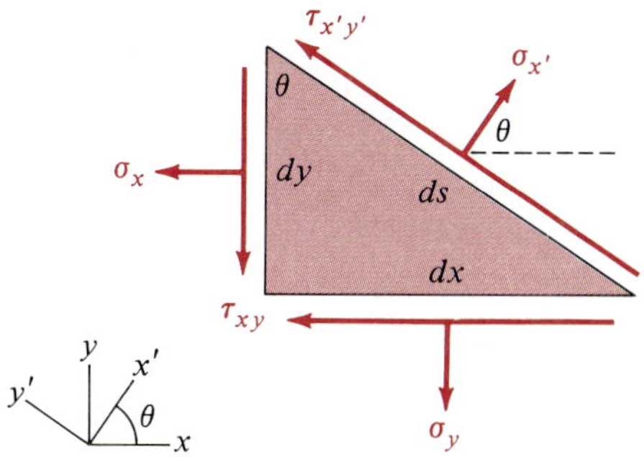

Figure 2.37 Free-body diagram for developing the plane stress transformation

With t as the thickness of the element, the equilibrium equations are

$$\begin{aligned}\Sigma F_{x'} &= \sigma_{x'} t\, ds - \sigma_x t\, dy \cos\theta - \tau_{xy} t\, dy \sin\theta \\ &\quad - \tau_{xy} t\, dx \cos\theta - \sigma_y t\, dx \sin\theta = 0\end{aligned}$$

and

$$\begin{aligned}\Sigma F_{y'} &= \tau_{x'y'} t\, ds + \sigma_x t\, dy \sin\theta - \tau_{xy} t\, dy \cos\theta \\ &\quad + \tau_{xy} t\, dx \sin\theta - \sigma_y t\, dx \cos\theta = 0\end{aligned}$$

Using the fact that $dy = ds \cos\theta$ and $dx = ds \sin\theta$, these two equilibrium equations can be solved for $\sigma_{x'}$ and $\tau_{x'y'}$ to yield

$$\sigma_{x'} = \sigma_x \cos^2\theta + \sigma_y \sin^2\theta + 2\tau_{xy} \sin\theta \cos\theta \tag{2.4}$$

and

$$\tau_{x'y'} = -(\sigma_x - \sigma_y) \cos\theta \sin\theta + \tau_{xy}(\cos^2\theta - \sin^2\theta) \tag{2.5}$$

By adding $\pi/2$ to θ, the stress component $\sigma_{y'}$ can be obtained as

$$\sigma_{y'} = \sigma_y \cos^2\theta + \sigma_x \sin^2\theta - 2\tau_{xy} \sin\theta \cos\theta \tag{2.6}$$

Equations (2.4), (2.5), and (2.6) are the transformation equations for plane stress. Given the stresses σ_x, σ_y, and τ_{xy} on two perpendicular planes passing through a point P, the stresses on any other plane passing through P can be determined using these equations.

By using the double-angle formulas $\cos^2\theta = (1 + \cos 2\theta)/2$, $\sin^2\theta = (1 - \cos 2\theta)/2$, and $2 \sin\theta \cos\theta = \sin 2\theta$, the transformation equations (2.4), (2.5), and (2.6) can be expressed as

$$\sigma_{x'} = \frac{\sigma_x + \sigma_y}{2} + \frac{\sigma_x - \sigma_y}{2} \cos 2\theta + \tau_{xy} \sin 2\theta \tag{2.7}$$

$$\sigma_{y'} = \frac{\sigma_x + \sigma_y}{2} - \frac{\sigma_x - \sigma_y}{2} \cos 2\theta - \tau_{xy} \sin 2\theta \tag{2.8}$$

$$\tau_{x'y'} = -\frac{\sigma_x - \sigma_y}{2} \sin 2\theta + \tau_{xy} \cos 2\theta \tag{2.9}$$

These double-angle forms for the transformation equations will be very useful later in the chapter.

Using either of the forms of the stress transformation equations, we note the following checks on our development:

$$
\begin{array}{llll}
\text{for } \theta = 0 & \sigma_{x'} = \sigma_x & \text{and} & \tau_{x'y'} = \tau_{xy} \\
\text{for } \theta = \pi/2 & \sigma_{x'} = \sigma_y & \text{and} & \tau_{x'y'} = -\tau_{xy} \\
\text{for } \theta = \pi & \sigma_{x'} = \sigma_x & \text{and} & \tau_{x'y'} = \tau_{xy} \\
\text{for } \theta = 3\pi/2 & \sigma_{x'} = \sigma_y & \text{and} & \tau_{x'y'} = -\tau_{xy}
\end{array}
$$

The student should verify that by taking $\tau_{xy} = \tau_0$, $\sigma_x = \sigma_y = 0$, and $\theta = 45°$, the results of Example 2.11 are obtained.

Adding equations 2.7 and 2.8 results in

$$\sigma_{x'} + \sigma_{y'} = \sigma_x + \sigma_y \tag{2.10}$$

thus, the sum of the normal stress components on any two perpendicular planes at a point is a constant. This is the so-called **first stress invariant.** Practically it can be used in a situation where σ_x and σ_y are known, $\sigma_{x'}$ has been computed, and the value of $\sigma_{y'}$ is also desired. Using the first stress invariant, $\sigma_{y'}$ can be easily obtained from Eqn. (2.10) as

$$\sigma_{y'} = \sigma_x + \sigma_y - \sigma_{x'}$$

It is left for the student to show in the problems that there is also a second stress invariant for the plane stress case given by

$$\sigma_x\sigma_y - (\tau_{xy})^2 = \sigma_{x'}\sigma_{y'} - (\tau_{x'y'})^2 \tag{2.11}$$

Example 2.12 Consider the problem of a wood member loaded in tension as shown in Fig. 2.38.

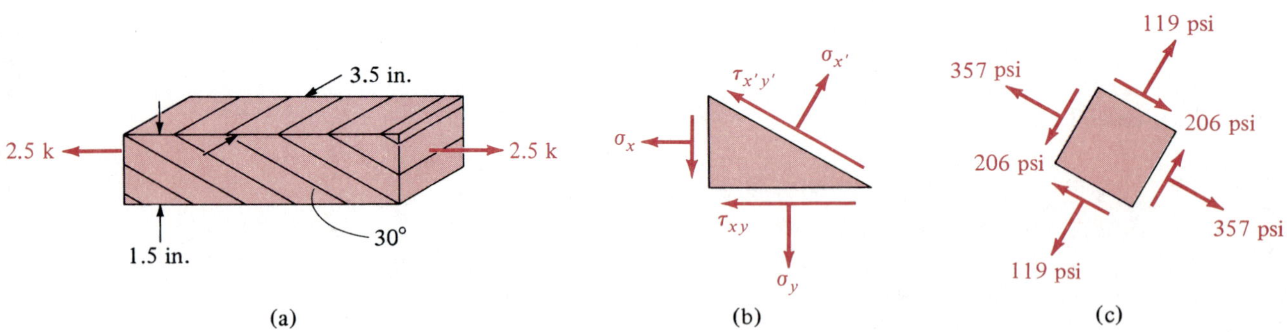

Figure 2.38 Wood member in tension

This example is meant only to illustrate the use of transformation equations; clearly it would not be wise to align the grain of the wood in this manner. One would suspect that failure of the member would occur along the grain—thus one would want to know the stress levels on faces parallel to grain. Referring to Fig. 2.38b, σ_x = 2500 lbf/(3.5 in.)(1.5 in.) = 476 psi, $\sigma_y = 0$, and $\tau_{xy} = 0$. Also, $\theta = 60°$ for the face on which $\sigma_{x'}$ and $\tau_{x'y'}$ act. Using Eqns. (2.4) and (2.5) then yields

$$\sigma_{x'} = [476(0.5)^2 + 0(0.866)^2 + 2(0)(0.5)(0.866)] \text{ psi} = 119 \text{ psi}$$

and

$$\tau_{x'y'} = [(0 - 476)(0.5)(0.866) + 0(.5^2 - 0.866^2)] \text{ psi} = -206 \text{ psi}$$

respectively, for the average normal and shear stress on the plane. The negative sign on $\tau_{x'y'}$ indicates that the shear stress actually acts down the plane as shown in Fig. 2.38c. Using the idea of the first invariant, the average normal stress σ_y on a perpendicular plane is 357 psi. The stresses on an element oriented with its sides parallel to the grain is shown in Fig. 2.38c. In all likelihood, these values would be used to assess or predict failure of the member.

This example illustrates one of the main uses of stress transformations: to determine the planes within the body on which the worst state of stress and hence failure is most likely to occur. What is meant by "worst" depends upon the characteristics of the material the body is made of; these will be discussed in Chapter 4.

PROBLEMS

2.67–2.81 For each of the following figures, draw a free-body diagram and write the equations of equilibrium to determine the normal and shear stress on the inclined face indicated.

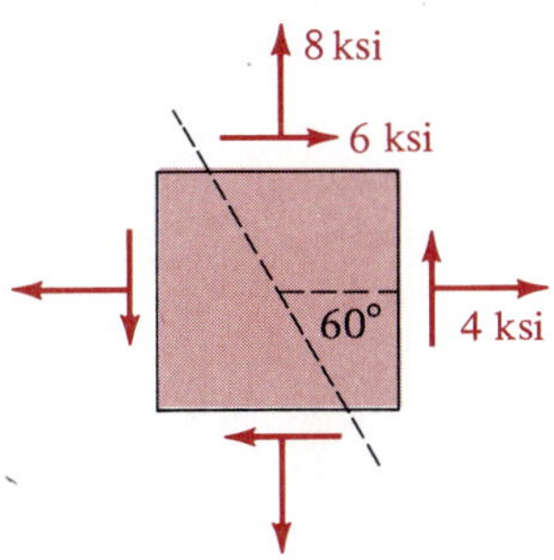

Ps. 2.67 and 2.82

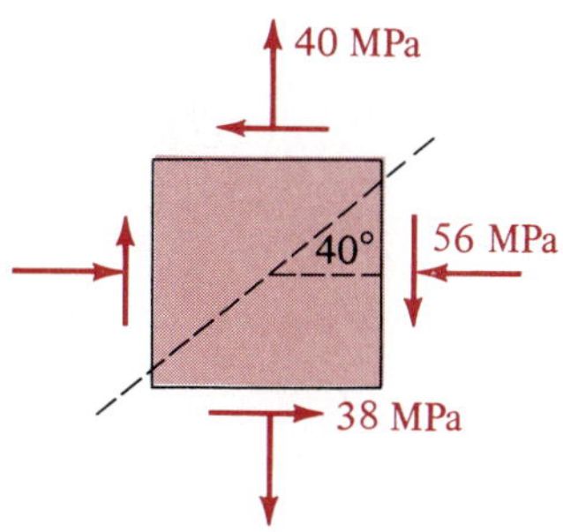

Ps. 2.68 and 2.83

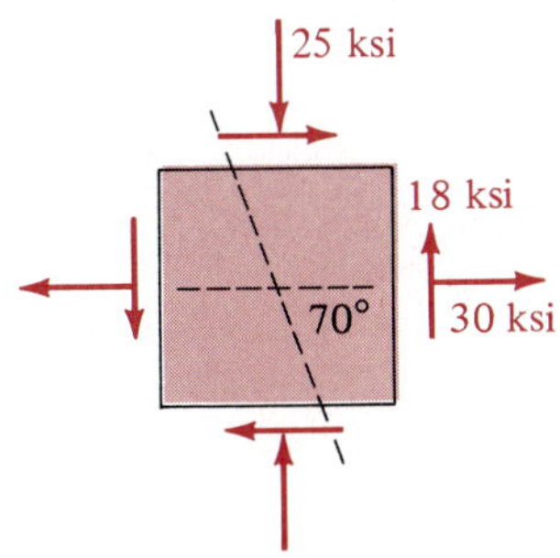

Ps. 2.69 and 2.84

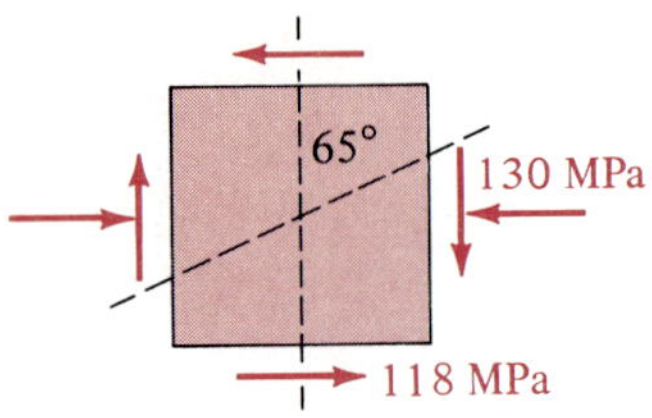

Ps. 2.70 and 2.85

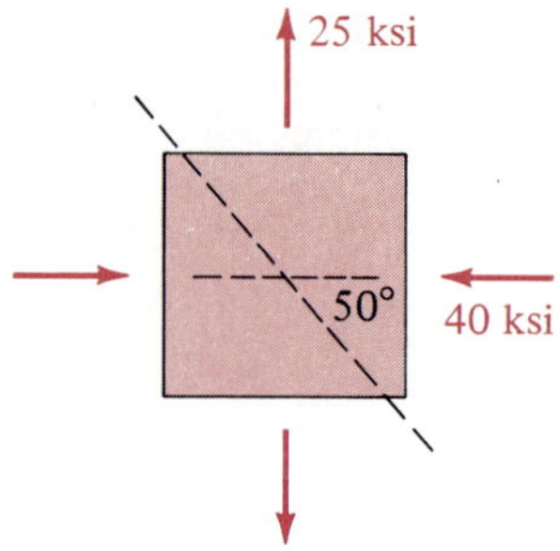

Ps. 2.71 and 2.86

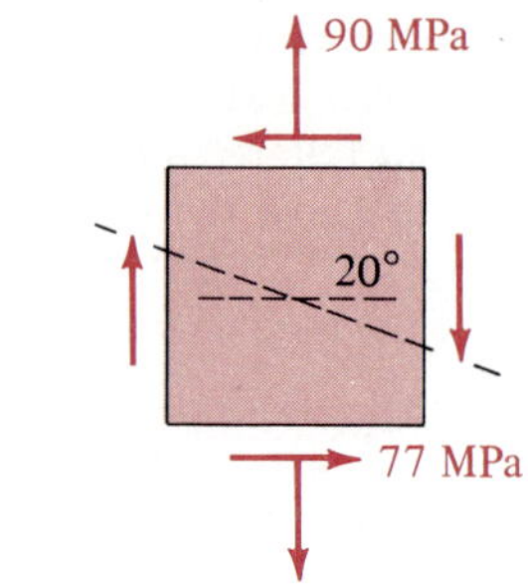

Ps. 2.72 and 2.87

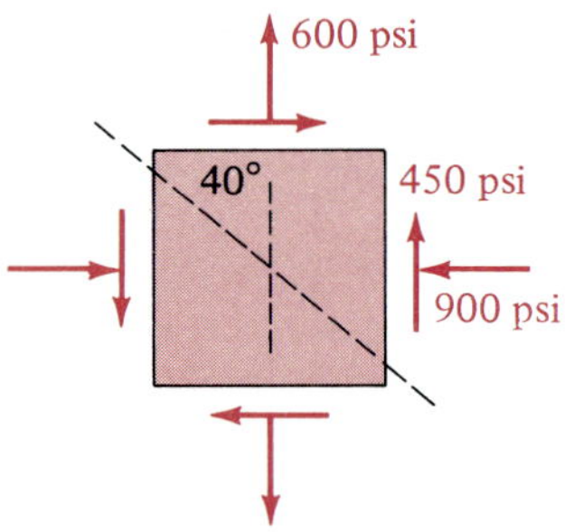

Ps. 2.73 and 2.88

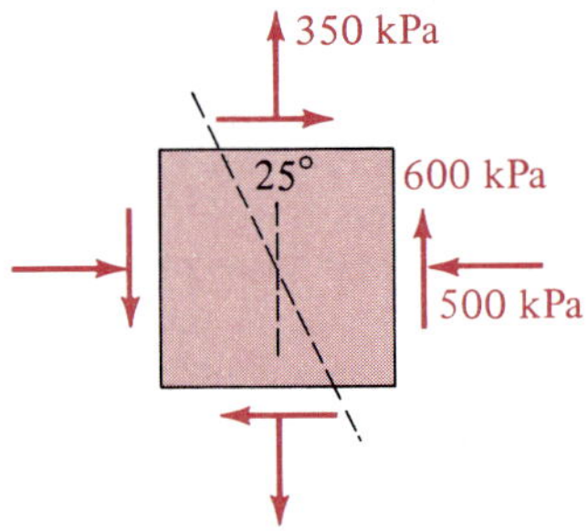

Ps. 2.74 and 2.89

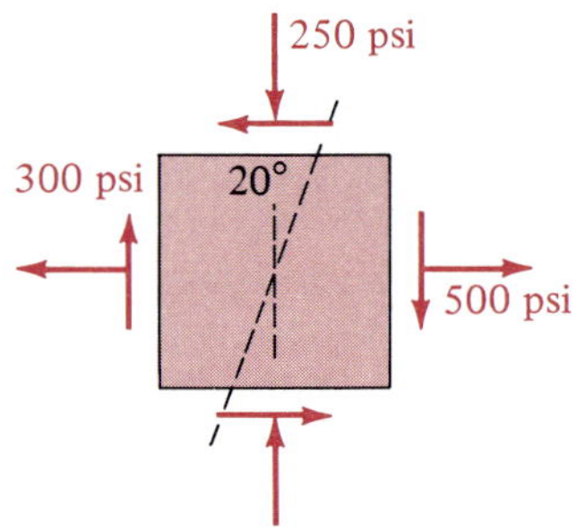

Ps. 2.75 and 2.90

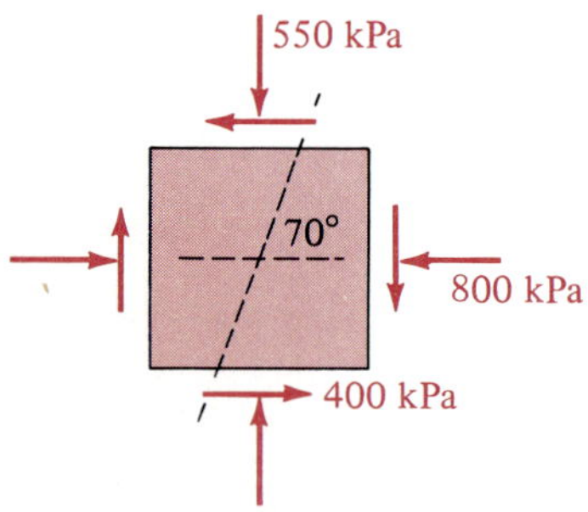

Ps. 2.76 and 2.91

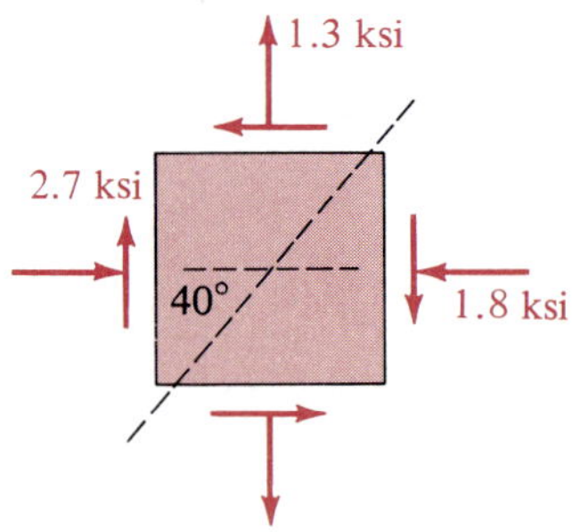

Ps. 2.77 and 2.92

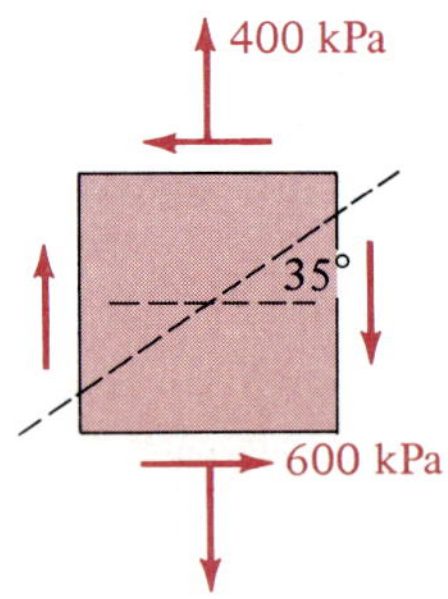

Ps. 2.78 and 2.93

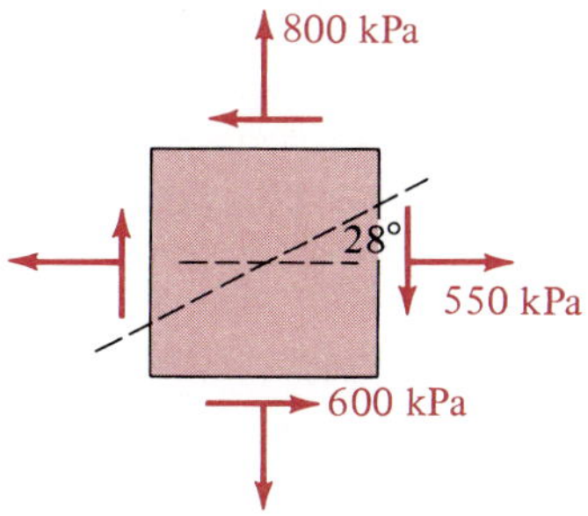

Ps. 2.79 and 2.94

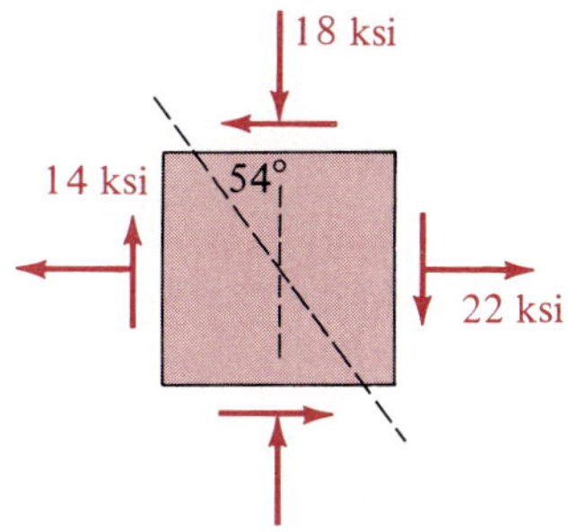

Ps. 2.80 and 2.95

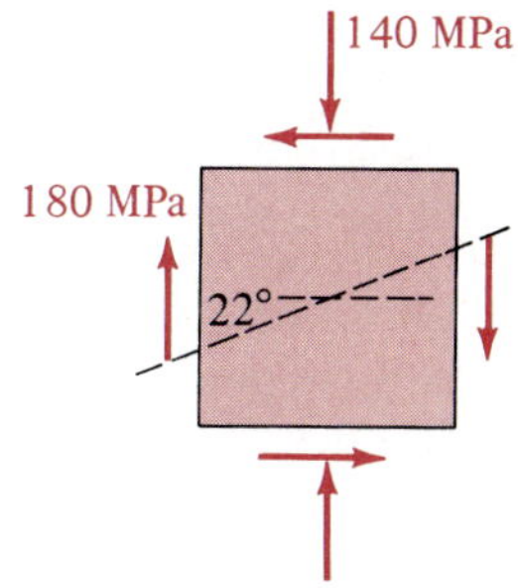

Ps. 2.81 and 2.96

2.82–2.96 Repeat problems 2.67–2.81 using the stress transformation equations.

2.97 Use the stress transformation equations to prove that $\sigma_{x'}\sigma_{y'} - (\tau_{x'y'})^2 = \sigma_x\sigma_y - (\tau_{xy})^2$. This combination of the stresses is known as the second stress invariant for plane stress.

2.5 PRINCIPAL STRESSES

Real structures are composed of real materials. Any real material fails when subjected to a large enough stress. Many of the theories of failure that will be discussed in Chapter 4 are based on experimental evidence indicating that materials fail when the normal or shear stress at a point reaches a critical value. It then becomes necessary to be able to determine maximum normal and shear stresses within a body in order to compare these with the critical values associated with the theories of failure. The maximum and minimum normal stresses at a point are called **principal stresses.** In this section we will develop the means for determining principal stresses and the maximum shear stress for a plane stress situation.

For the plane stress case discussed in the previous section, we developed the stress transformation equations

$$\sigma_{x'} = \frac{\sigma_x + \sigma_y}{2} + \frac{\sigma_x - \sigma_y}{2}\cos 2\theta + \tau_{xy}\sin 2\theta \qquad (2.12)$$

and

$$\tau_{x'y'} = -\frac{\sigma_x - \sigma_y}{2}\sin 2\theta + \tau_{xy}\cos 2\theta \qquad (2.13)$$

Given σ_x, σ_y, and τ_{xy} at a point P, $\sigma_{x'}$ and $\tau_{x'y'}$ are functions of θ. In line with the discussion above, it is desirable to be able to determine the maximum (minimum) values of these stresses and the planes on which they act. The possible maximum (minimum) of $\sigma_{x'}$ is obtained by equating to zero its derivative with respect to θ, namely

$$0 = \frac{d\sigma_{x'}}{d\theta} = -(\sigma_x - \sigma_y)\sin 2\theta + 2\tau_{xy}\cos 2\theta$$

from which

$$\frac{\sin 2\theta}{\cos 2\theta} = \tan 2\theta = \frac{2\tau_{xy}}{\sigma_x - \sigma_y} \tag{2.14}$$

with the two distinct solutions

$$\theta_1 = \frac{1}{2}\tan^{-1}\frac{2\tau_{xy}}{\sigma_x - \sigma_y}$$

and

$$\theta_2 = \frac{1}{2}\left(\tan^{-1}\frac{2\tau_{xy}}{\sigma_x - \sigma_y} + \pi\right)$$
$$= \theta_1 + \pi/2$$

The two planes defined by θ_1 and θ_2 are referred to as **principal planes.** Note that the two principal planes are perpendicular. Note also that by setting $\tau_{x'y'} = 0$ there results

$$\frac{\sin 2\theta}{\cos 2\theta} = \frac{2\tau_{xy}}{\sigma_x - \sigma_y}$$

which is precisely Eqn. (2.14), obtained by seeking the maximum and mimimum normal stresses.

This is in fact an alternate definition of principal planes, namely, those planes on which the shear stress is zero.

Substituting the solutions from Eqn. (2.14) into Eqn. (2.12), the values of the normal stresses on these two planes are

$$\sigma_1 = \frac{\sigma_x + \sigma_y}{2} + \sqrt{\left(\frac{\sigma_x - \sigma_y}{2}\right)^2 + \tau_{xy}^2} \tag{2.15a}$$

$$\sigma_2 = \frac{\sigma_x + \sigma_y}{2} - \sqrt{\left(\frac{\sigma_x - \sigma_y}{2}\right)^2 + \tau_{xy}^2} \tag{2.15b}$$

and are referred to as the principal stresses. The reader is encouraged to check that the shear stress is zero on both principal planes.

The possible maximum (minimum) values of the shear stress at P are determined by taking

$$\frac{d\tau_{x'y'}}{d\theta} = 0 = -(\sigma_x - \sigma_y)\cos 2\theta - 2\tau_{xy}\sin 2\theta$$

from which

$$\tan 2\theta = -\frac{(\sigma_x - \sigma_y)}{2\tau_{xy}} \tag{2.16}$$

Note that the right-hand sides of Eqs. (2.14) and (2.16) are negative reciprocals indicating that the 2θ values of the solutions to the two equations are separated by 90°. It follows that the angle between the directions of the principal stresses and the maximum shear stress is 45°. The values of the corresponding maximum shear stresses can be determined by substituting the solutions for θ from Eqn. (2.16) into Eqn. (2.13) and are

$$\tau_{\text{max,min}} = \pm\sqrt{\left(\frac{\sigma_x - \sigma_y}{2}\right)^2 + \tau_{xy}^2} \tag{2.17}$$

Thus the two planes on which the maximum shear stresses occur are also separated by 90° and lie midway between the principal planes. As opposed to the principal planes where there is no shear stress, the planes on which the maximum shear stress acts are not free of normal stress. By substituting the θ values for the maximum shear stress planes into Eqn. (2.12), it can be shown that the normal stress on both of the maximum shear stress planes is

$$\sigma = \frac{\sigma_x + \sigma_y}{2}$$

All this information can be displayed very nicely as indicated in Fig. 2.39. The directions of the principal planes and the corresponding principal stresses are indicated in Fig. 2.39b, with the maximum shear stresses and the planes upon which they act shown in Fig. 2.39c.

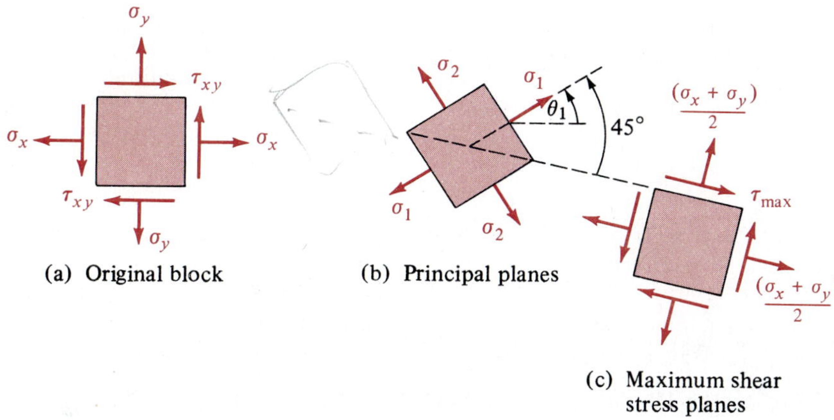

Figure 2.39 Values and locations of the principal stresses, principal planes, and maximum shear stress

Note that if the element on which the principal stresses are shown is displayed as a square, the orientation of the planes on which the maximum shear stresses act can be obtained by constructing the diagonal(s) of the principal stress element, as in Figs. 2.39b and 2.39c.

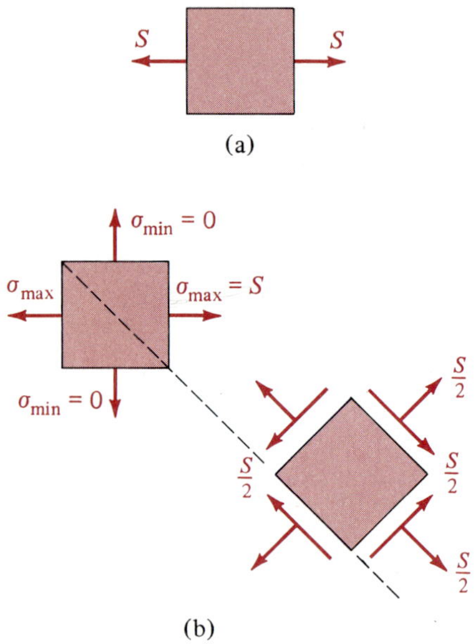

Figure 2.40 Principal stress and maximum shear stress information for pure tension

Example 2.13 The element representing the state of stress for **pure tension,** given as $\sigma_x = S$, $\sigma_y = \tau_{xy} = 0$, is indicated in Fig. 2.40a. Substituting into Eqn. (2.14) for the principal planes yields

$$\tan 2\theta = \frac{2 \cdot 0}{(S - 0)} = 0$$

so that $\theta_1 = 0$ and $\theta_2 = \pi/2$, the original planes. This result should have been anticipated since there was no shear stress on the faces of the original element. At any rate the principal stresses are $\sigma_1 = S$ and $\sigma_2 = 0$. Using Eqn. (2.16), the maximum shear stresses occur where

$$\tan 2\theta = \frac{-(S - 0)}{2 \cdot 0} = -\infty$$

from which $2\theta = -\pi/2$. θ is then $-\pi/4$ with maximum shear stress given from Eqn. (2.17) as

$$\tau_{max} = S/2$$

The normal stress on these two planes is $(S + 0)/2 = S/2$. The elements that convey all this information are shown in Fig. 2.40b.

Example 2.14 The element representing the state of stress for **pure shear** given as $\tau_{xy} = S$, $\sigma_x = \sigma_y = 0$, is indicated in Fig. 2.41a. Substituting into Eqn. (2.14) for the principal planes yields

$$\tan 2\theta = \frac{2S}{(0 - 0)} = \infty$$

so that $2\theta = \pi/2$ with $\theta_1 = \pi/4$ and $\theta_2 = 3\pi/4$. Using equation (2.15) the principal stresses are

$$\sigma_{1,2} = (0 + 0)/2 \pm \sqrt{\left(\frac{0 - 0}{2}\right)^2 + S^2} = \pm S$$

and act on the 45° planes as shown in Fig. 2.41b. The maximum shear stresses occur where

$$\tan 2\theta = \frac{2 \cdot 0}{(S - 0)}$$

from which $2\theta = 0$ and π, with $\theta = 0$ and $\pi/2$. The original directions represent the planes on which the maximum shear stresses act.

The elements that convey all this information are shown in Fig. 2.41b.

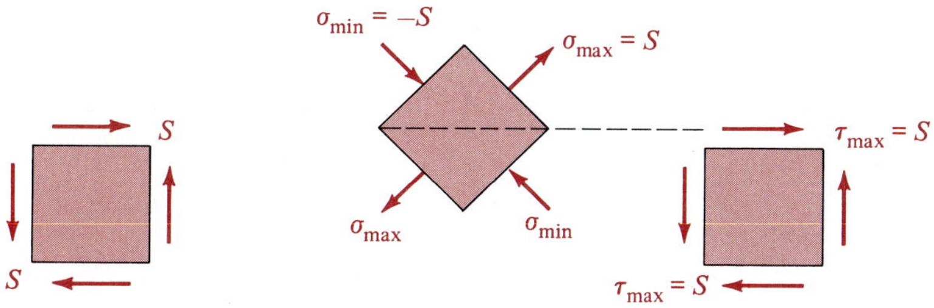

Figure 2.41 Principal stress and maximum shear stress information for pure shear

---

Example 2.15 At a point P in an element under plane stress, the stresses are represented as in Fig. 2.42a. Determine the principal stresses and planes and the maximum shear stress. Show the elements on which they act.

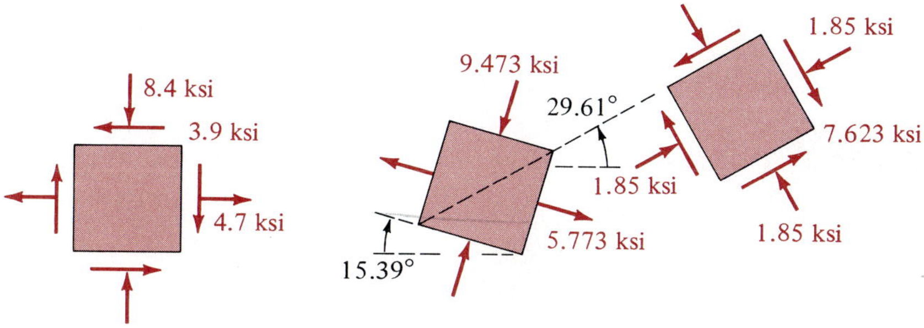

Figure 2.42 A state of plane stress

Solution Substituting into Eqn. (2.14) for the locations of the principal stresses yields

$$\tan 2\theta = 2\,\frac{-3900}{4700 - (-8400)} = -0.5954$$

from which $2\theta_1 = -30.77°$ or $\theta_1 = -15.39°$. θ_2 can be taken as either $-105.39°$ or as $74.61°$. Substituting the values of θ_1 and θ_2 into Eqn. (2.12) yields $\sigma_1 = 5773$ psi and $\sigma_2 = -9473$ psi. As determined from Eqn. (2.17), the value of the maximum shear stress is $\tau_{max} = 7623$ psi, and occurs on the planes bisecting the principal planes. The directions of the principal planes and the corresponding principal stresses, along with the maximum shear stresses and the planes upon which they act, are shown in Fig. 2.42b.

All of the results regarding the principal stresses and maximum shear stress developed in this section are based on the assumption of plane stress, which, as was indicated in Fig. 2.35, to which the student is referred, is characterized by $\sigma_z = \tau_{zx} = \tau_{zy} = 0$. From the idea that a principal plane is a plane on which there is no shear stress, it follows that the z face is a principal plane. Consider then the element on which the principal stresses σ_1 and σ_2 act, shown in Fig. 2.43.

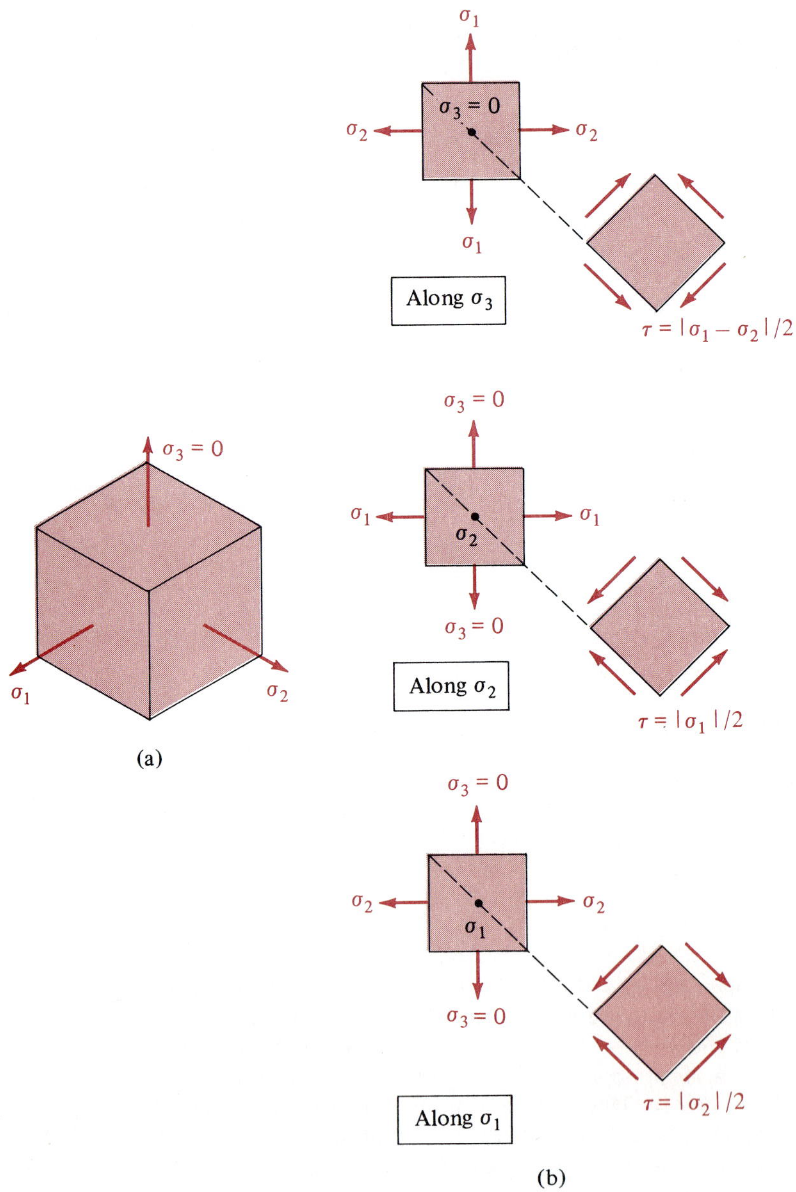

Figure 2.43 Maximum shear stresses

This same element is shown viewed along each of the principal axes in Fig. 2.43b. The value of the shear stress on the bisecting 45° plane is shown in each case. If σ_1 and σ_2 are of the same sign, either $|\sigma_1|/2$ or $|\sigma_2|/2$ will be larger than the in-plane shear stress $|\sigma_1 - \sigma_2|/2$, with the conclusion that the largest shear stress at the point is not the in-plane shear stress $|\sigma_1 - \sigma_2|/2$. The largest shear stress turns out to be

$$\tau_{abs} = \max(|\sigma_1|/2, |\sigma_2|/2, |\sigma_1 - \sigma_2|/2)$$

and will be referred to as the **absolute maximum shear stress.** The student must remember to select the largest of these three values when determining the maximum absolute shear stress at a point for a body in a state of plane stress.

Example 2.16 For the state of stress given by $\sigma_x = 10$ ksi, $\sigma_y = 6$ ksi, and $\tau_{xy} = 4$ ksi, determine the absolute maximum shear stress.

Solution The principal stresses are given by

$$\sigma_{1,2} = \{(10 + 6)/2 \pm \sqrt{\left(\frac{10 - 6}{2}\right)^2 + 4^2}\} \text{ ksi} = (8 \pm 4.47) \text{ ksi}$$

$$= 12.47 \text{ ksi and } 3.53 \text{ ksi}$$

The in-plane maximum shear stress is 4.47 ksi with the shear stresses on the other two 45° planes as $\tau = 12.47$ ksi/2 = 6.24 ksi and $\tau = 3.53$ ksi/2 = 1.77 ksi, so that

$$\tau_{abs} = 6.24 \text{ ksi}$$

that is, not the in-plane maximum.

PROBLEMS

2.98–2.112 For each of the following determine the principal stresses and show them on properly oriented elements. Determine the maximum in-plane shear stress, the corresponding normal stress, and the properly oriented element on which they act. Also determine the absolute maximum shear stress.

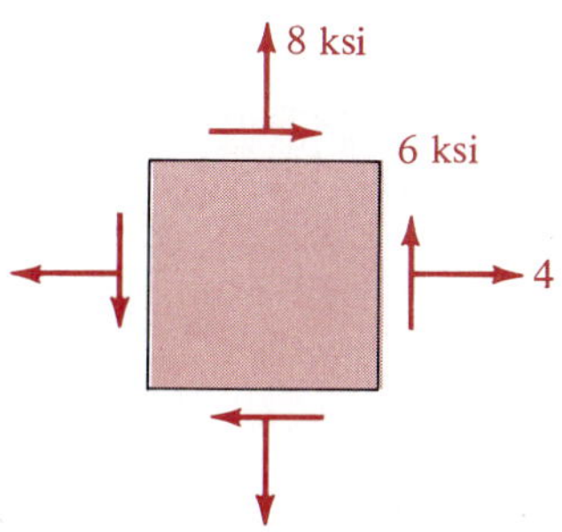

Ps. 2.98 and 2.113

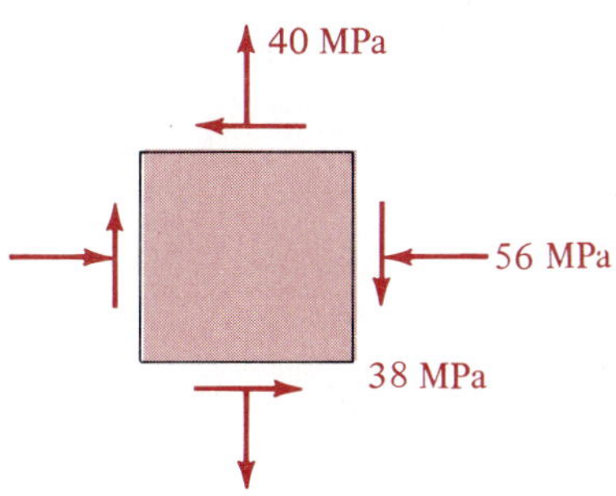

Ps. 2.99 and 2.114

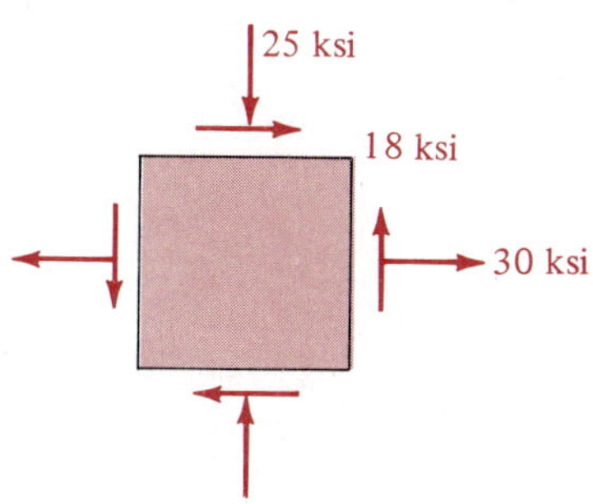

Ps. 2.100 and 2.115

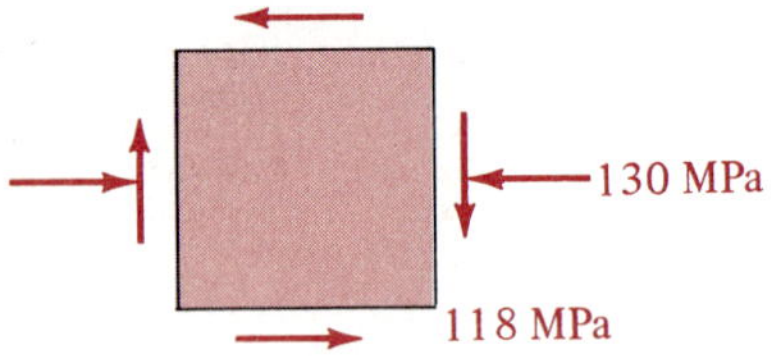

Ps. 2.101 and 2.116

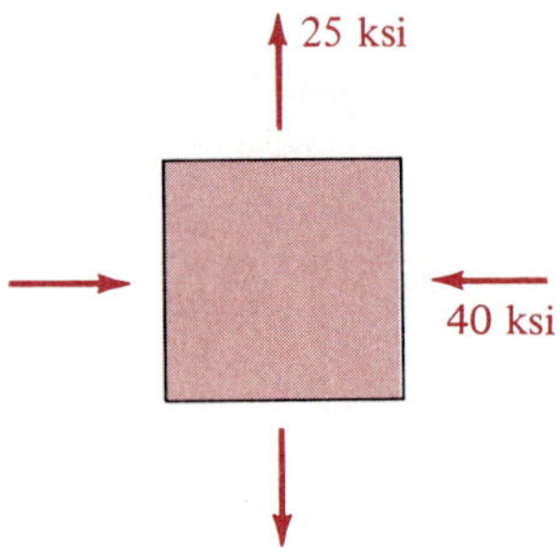

Ps. 2.102 and 2.117

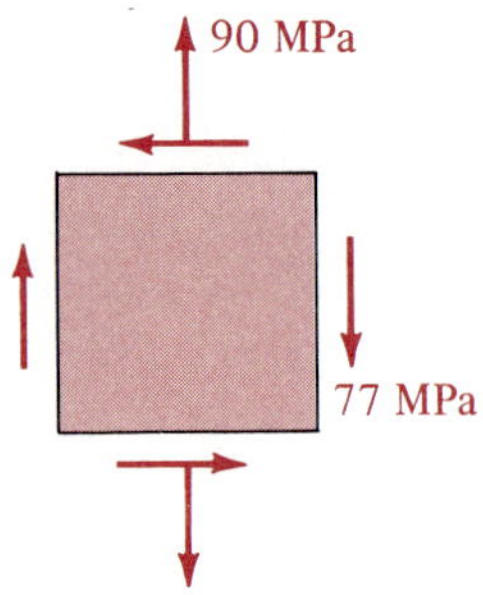

Ps. 2.103 and 2.118

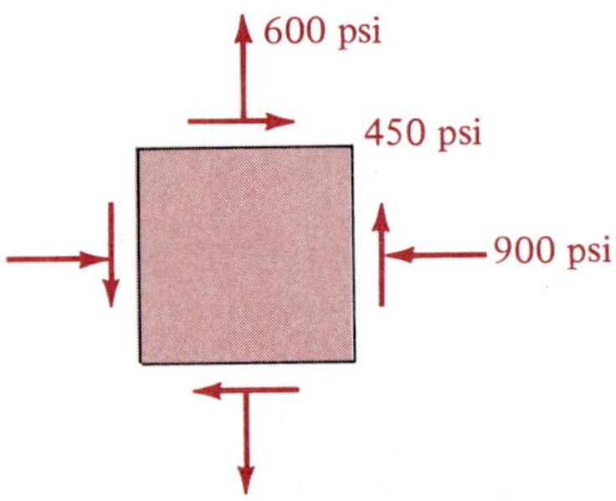

Ps. 2.104 and 2.119

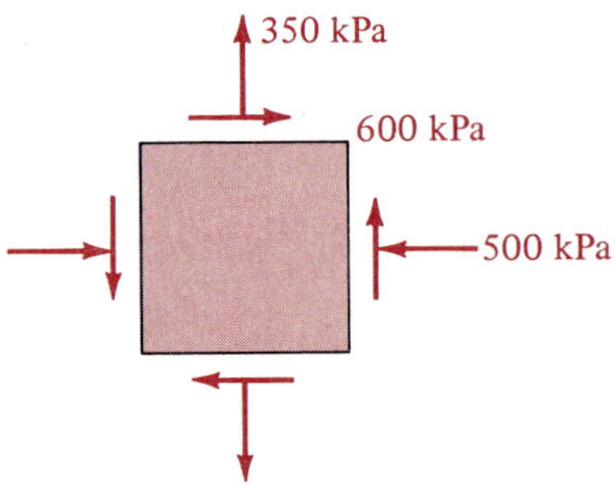

Ps. 2.105 and 2.120

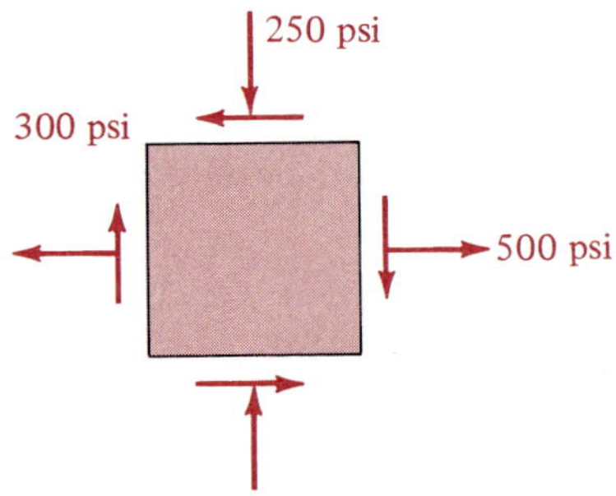

Ps. 2.106 and 2.121

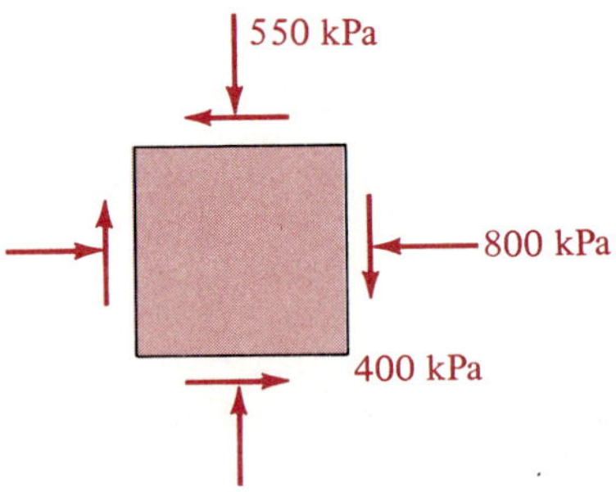

Ps. 2.107 and 2.122

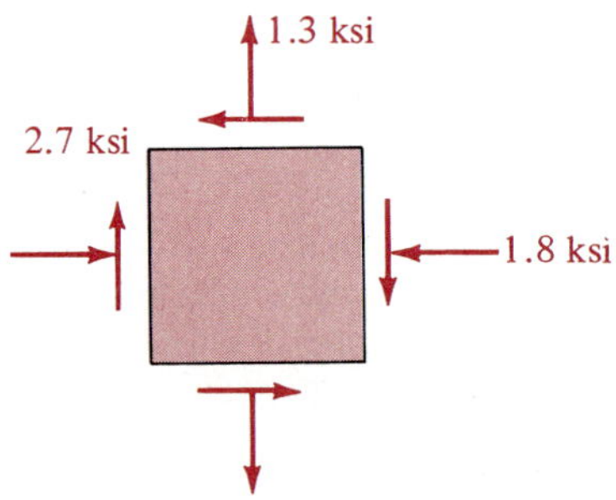

Ps. 2.108 and 2.123

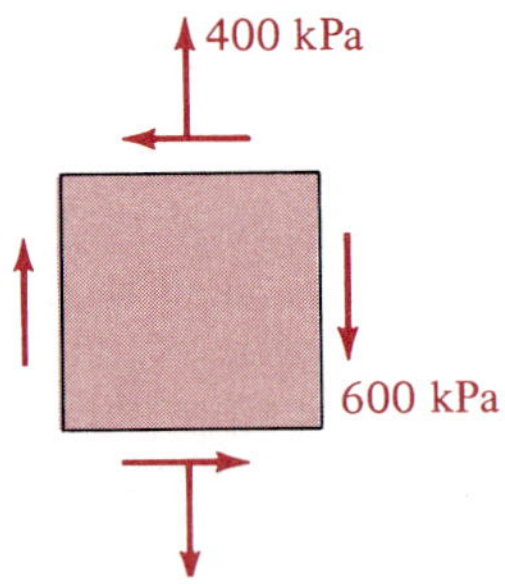

Ps. 2.109 and 2.124

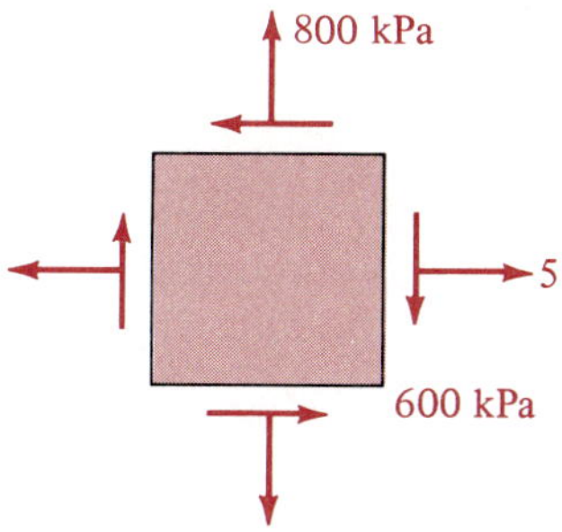

Ps. 2.110 and 2.125

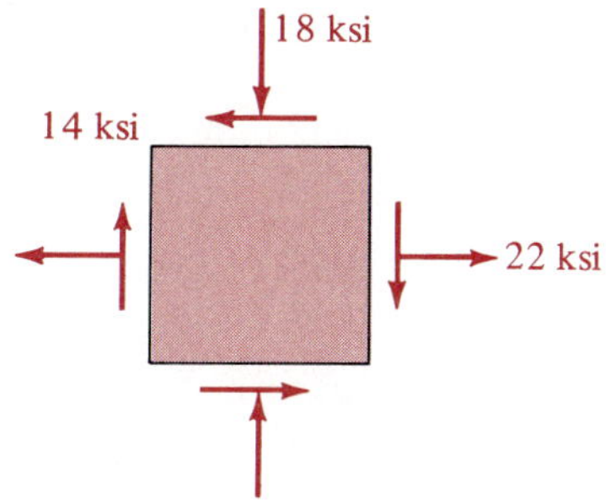

Ps. 2.111 and 2.126

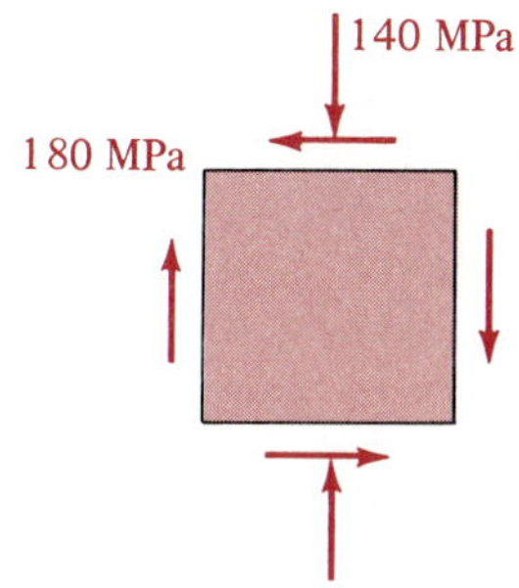

Ps. 2.112 and 2.127

2.6 MOHR'S CIRCLE FOR PLANE STRESS

In the last section we used the stress transformation equations to determine principal stresses, principal planes, and the maximum shear stress for plane states of stress. The calculations were straightforward but somewhat tedious. In this section we will present a graphical tool called **Mohr's circle**, after the German engineer Otto Mohr (1835–1918), who first used it to study plane stress. The circle is based on the stress transformation equations and provides a graphical solution to the problem of determining principal stresses, principal planes, and maximum shear stresses. We will not present a complete derivation of all the properties of the circle, but rather we will outline its use and demonstrate its application to plane states of stress.

Otto Mohr (1835–1918)

German civil engineer who developed the circle of stress called Mohr's circle.

Recall that the stress transformation equations expressed in double-angle form are

$$\sigma_{x'} = \frac{\sigma_x + \sigma_y}{2} + \frac{\sigma_x - \sigma_y}{2} \cos 2\theta + \tau_{xy} \sin 2\theta \tag{2.22}$$

and

$$\tau_{x'y'} = -\frac{\sigma_x - \sigma_y}{2} \sin 2\theta + \tau_{xy} \cos 2\theta \tag{2.23}$$

Squaring and adding these two equations yields

$$\left[\sigma_{x'} - \frac{(\sigma_x + \sigma_y)}{2}\right]^2 + \tau_{x'y}^2 = (\sigma_x - \sigma_y)^2/4 + \tau_{xy}^2$$

This equation can be interpreted as the equation of a circle in the $\sigma_{x'}$, $\tau_{x'y'}$ plane whose center C has the coordinates

$$C : [(\sigma_x + \sigma_y)/2,\ 0]$$

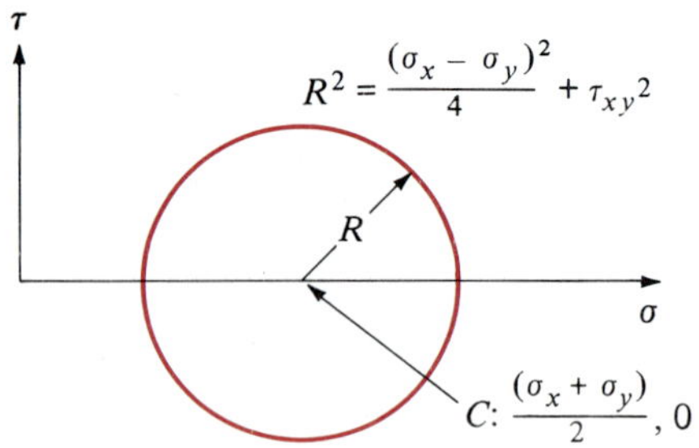

Figure 2.44 Mohr's circle for plane stress

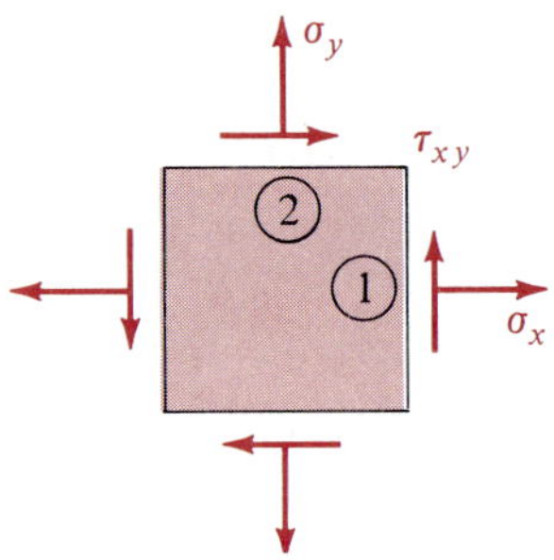

Figure 2.45 Labeling of element faces

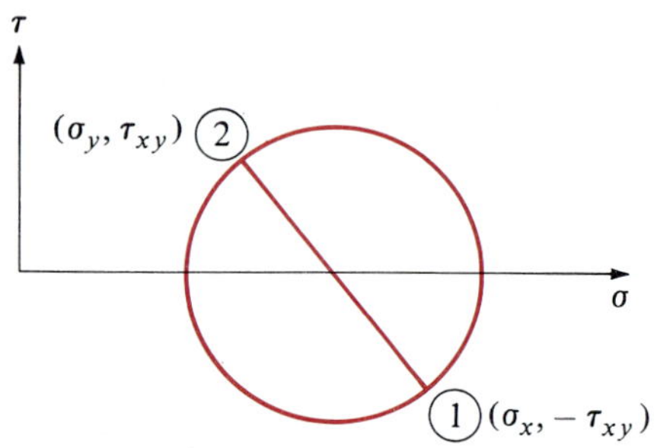

Figure 2.46 Plotting and labeling of points on Mohr's circle

and whose radius R is given by

$$R = \sqrt{\frac{(\sigma_x - \sigma_y)^2}{4} + \tau_{xy}^2}$$

as indicated in Fig. 2.44. This circle is called Mohr's circle. The construction of Mohr's circle is carried out using the following steps:

1. Sketch the element showing the stresses σ_x, σ_y, and τ_{xy}. Carefully label the faces in some definite manner, such as faces 1 and 2 in Fig. 2.45.

2. Sketch the coordinate axes for the circle, calling them σ and τ as indicated in Fig. 2.46.

3. Plot *and label* the points $(\sigma_x, -\tau_{xy})$, corresponding to face 1, and (σ_y, τ_{xy}), corresponding to face 2, as indicated in Fig. 2.46. Signs must be observed; that is, a negative stress component must be plotted accordingly. The shear stress τ_{xy} is positive when directed as shown for face 1 in Fig. 2.45 and is plotted as negative on Mohr's circle.

When the σ- and τ-axes are directed as shown in Fig. 2.46, it is necessary to plot τ_{xy} as a negative quantity on Mohr's circle in order that rotation 2θ of the diameter of Mohr's circle be in the same direction as the rotation θ of the plane stress element.

4. Draw the diameter of Mohr's circle between points 1 and 2 and sketch in or use a compass to construct the circle as shown in Fig. 2.46. This completes the construction of the circle.

The key to understanding and using Mohr's circle is to keep in mind that there is an absolute correspondence between the faces of the element and the points on Mohr's circle. For this reason it is advisable to number the faces of the element and the corresponding points on the circle.

Properties of Mohr's Circle (Refer to Fig. 2.47.) First note that the points on Mohr's circle corresponding to the principal stresses are

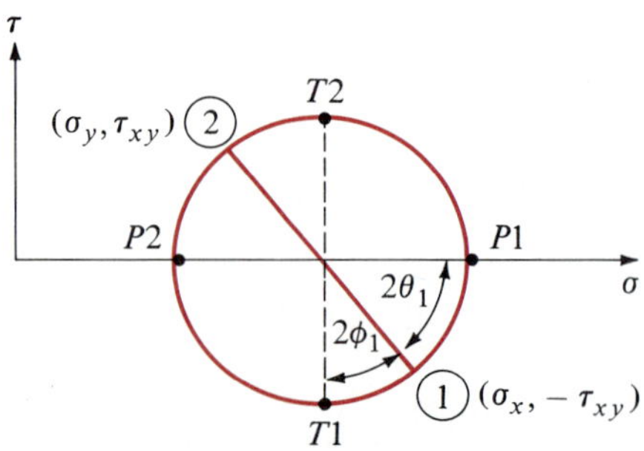

Figure 2.47 Properties of Mohr's circle

designated as $P1$ and $P2$. The points corresponding to the maximum in-plane shear stress values are designated as $T1$ and $T2$. The angles corresponding to the principal planes and the planes on which the maximum in-plane shear stresses act are shown as $2\theta_1$ and $2\phi_1$, respectively.

1. Each point on Mohr's circle represents the state of stress on a corresponding face of the element. Points 1 and 2 that were used to plot the circle are examples.
2. Rotation of the diameter of Mohr's circle through an angle 2θ corresponds to a rotation through an angle θ, in the same direction, of the corresponding normal to the face in the element.
3. The values of the principal stresses are given by the coordinates of the points on the σ-axis labeled $P1$ and $P2$.
4. The values of the maximum in-plane shear stresses are given by the coordinates of the points at the top and bottom of the circle labeled $T1$ and $T2$.
5. The locations of the principal planes in the element are located by rotating the normals to faces 1 and 2 in the element through angles θ_1 and $\theta_1 + \pi/2$ corresponding to the rotation of the diameter of Mohr's circle through angles $2\theta_1$ and $2\theta_1 + \pi$.
6. The locations of the planes on which the maximum in-plane shear stresses act in the element are located by rotating the normals to faces 1 and 2 through angles ϕ_1 and $\phi_1 + \pi/2$, corresponding to the rotation of the diameter of Mohr's circle through angles $2\phi_1$ and $2\phi_1 + \pi$. The directions on the element are determined by recalling that, as shown in Fig. 2.48, a shear stress that is positive (negative) on Mohr's circle acts in the negative (positive) direction on the element rotated through the angle ϕ_1.

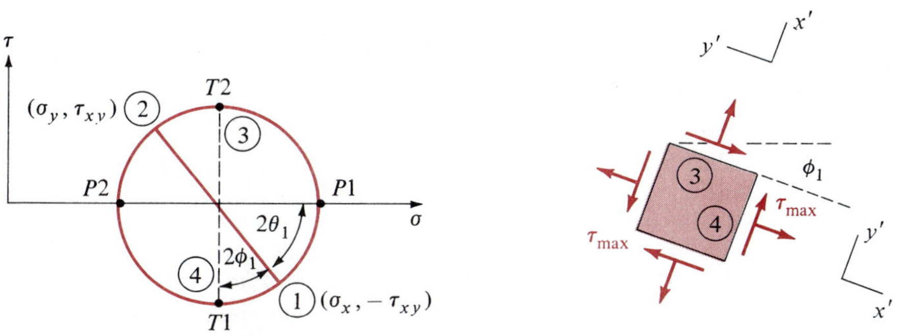

Figure 2.48 Shear stress directions

As shown, the shear stress corresponding to point 3 on Mohr's circle is positive and points in the negative y'-direction for the coordinate system associated with face 3. Similarly the shear stress corresponding to point 4 on Mohr's circle is negative and points in the positive y'-direction for the coordinate system associated with face 4.

Mohr's circle can be used in place of the transformation equations. Once Mohr's circle has been constructed, the stresses on any face can be determined by following the rules outlined above. As pointed out in the previous section, the absolute maximum shear stress at the point must be taken as

$$\tau_{abs} = \max(|\sigma_1 - \sigma_2|/2, |\sigma_1|/2, |\sigma_2|/2)$$

where σ_1 and σ_2 are the principal stresses.

Example 2.17 For the state of stress represented by the element in Fig. 2.49a, use Mohr's circle to determine the principal stresses and planes, the maximum in-plane shear stress, and the planes on which it acts. Determine also the absolute maximum shear stress.

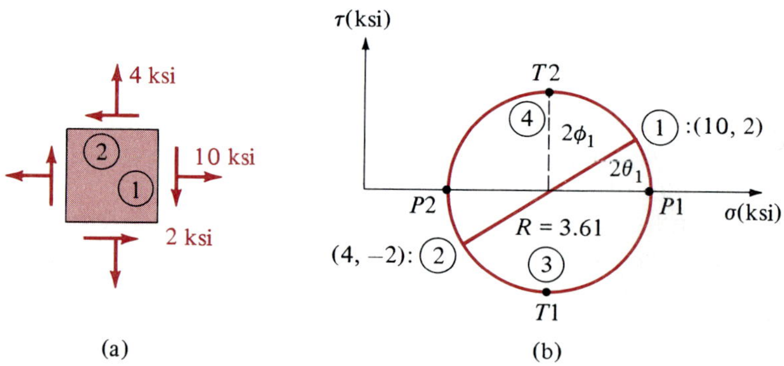

Figure 2.49 Block and Mohrs' circle

Solution The values of the stress components shown in Fig. 2.49a are $\sigma_x = 10$ ksi, $\sigma_y = 4$ ksi, and $\tau_{xy} = -2$ ksi. Point 1 with coordinates $(\sigma, \tau) = (10, -(-2))$ ksi and point 2 with coordinates $(\sigma, \tau) = (4, -2)$ ksi are plotted on Mohr's circle as shown in Fig. 2.49b. The center of the circle has the coordinates $(\sigma, \tau) = ((10 + 4)/2, 0)$ ksi $= (7, 0)$ ksi and the radius is given by $R = \sqrt{[(10 - 4)/2]^2 + 2^2}$ ksi $= 3.61$ ksi. The angle $2\theta_1$ is given by $\tan 2\theta_1 = 2/3$, from which $2\theta_1 = 33.69°$ or $\theta_1 = 16.85°$. It follows from the figure that $2\phi_1 = 90° - 2\theta_1 = 56.31°$, or $\phi_1 = 28.15°$. The values of the principal stresses correspond to points $P1$ and $P2$ and are

$$\sigma_{max} = (7.00 + 3.61) \text{ ksi} = 10.61 \text{ ksi}$$

and

$$\sigma_{min} = (7.00 - 3.61) \text{ ksi} = 3.39 \text{ ksi}$$

The maximum in-plane shear stress is numerically equal to the radius of Mohr's circle, that is, $\tau_{max} = 3.61$ ksi.

On Mohr's circle the diameter is rotated clockwise 33.70° to travel from point 1 to $P1$. Thus in the element, the normal to face 1 is rotated clockwise through an angle of $33.70°/2 = 16.85°$ to the plane on which σ_{max} acts. This is indicated in Fig. 2.50a. The principal stress σ_{min} acts on a perpendicular plane as shown.

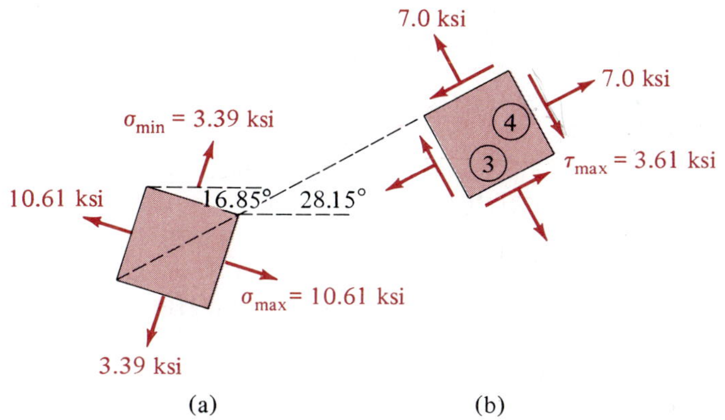

Figure 2.50 Principal planes, principal stresses, and maximum shear stress

To get from point 1 to $T2$ on Mohr's circle, rotate the diameter $2\phi_1 = 56.31°$ counterclockwise. Thus on the element rotate the normal to face 1 counterclockwise through an angle of $\phi_1 = 56.31°/2 = 28.15°$ to obtain the face on which the maximum in-plane shear stress acts. The normal stress on the face corresponding to $T2$ is 7.0 ksi. These stresses are indicated in Fig. 2.50b. Note also that the diagonals of the element on which the principal stresses act define the orientation of the element on which the maximum in-plane shear acts. For this example the absolute maximum shear stress is $\tau_{abs} = 10.61 \text{ ksi}/2 = 5.81$ ksi.

Example 2.18 For the state of stress shown in Fig. 2.51a, use Mohr's circle to determine the principal stresses, the maximum shear stress and the planes on which they act. Determine also the absolute maximum shear stress.

Solution As shown in Fig. 2.51a, $\sigma_x = 30$ MPa, $\sigma_y = -40$ MPa, and $\tau_{xy} = -10$ MPa. Point 1 with coordinates $(\sigma, \tau) = [30, -(-10)]$ MPa and point 2 with coordinates $(\sigma, \tau) = (-40, -10)$ MPa are plotted on Mohr's circle as shown in Fig. 2.51b. The center of the circle has the coordinates $(\sigma, \tau) = [(30 - 40)/2, 0]$ MPa $= (-5, 0)$ MPa. The radius is given by $R = \sqrt{\{[30 - (-40)]/2\}^2 + (-10)^2}$ MPa $= 36.4$ MPa. The angle $2\theta_1$ is given by $\tan 2\theta_1 = 10/35$, from which $2\theta_1 = 15.95°$ or $\theta_1 = 7.97°$. It follows from the figure that $2\phi_1 = 90° - 2\theta_1 = 74.05°$, or $\phi_1 = 37.03°$. The values of the principal stresses correspond to points $P1$ and

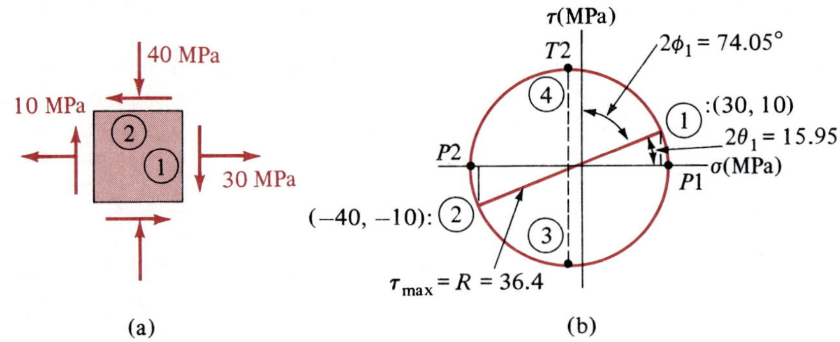

Figure 2.51 Mohr's circle for a plane stress element

$P2$ and are

$$\sigma_{max} = (-5 + 36.4) \text{ ksi} = 31.4 \text{ MPa}$$

and

$$\sigma_{min} = (-5 - 36.4) \text{ ksi} = -41.4 \text{ MPa}$$

The maximum in-plane shear stress is numerically equal to the radius of Mohr's circle, or $\tau_{max} = 36.4$ MPa.

On Mohr's circle the diameter is rotated clockwise $2\theta_1 = 15.95°$ to travel from point 1 to $P1$. Thus in the element, the normal to face 1 is rotated clockwise through an angle of $15.95°/2 = 7.97°$ to the plane on which $\sigma_{max} =$ 31.4 MPa acts. This is indicated in Fig. 2.52. The principal stress $\sigma_{min} = -41.4$ MPa acts on a perpendicular plane as shown.

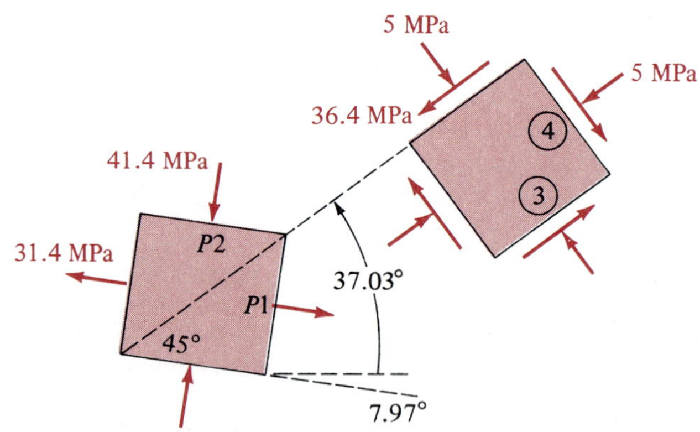

Figure 2.52 Principal planes, principal stresses, and maximum shear stress

To get from point 1 to $T2$ on Mohr's circle, rotate the diameter $2\phi_1 = 74.05°$ counterclockwise. Thus on the element rotate the normal to face 1 counterclockwise through an angle of $\phi_1 = 74.05°/2 = 37.03°$ to obtain the face on which the maximum shear stress acts. The normal stress on the face corresponding to $T2$ is -5 MPa. These stresses are indicated in Fig. 2.52. Note also that the diagonals of the square element on which the principal stresses act define the orientation of the element on which the maximum in-plane shear stress acts. For this example the absolute maximum shear stress is $\tau_{max} = 36.4$ MPa.

A careful consideration of Examples 2.17 and 2.18 yields the following very useful observation. Consider a typical element on which the principal stresses act as shown in Fig. 2.53a.

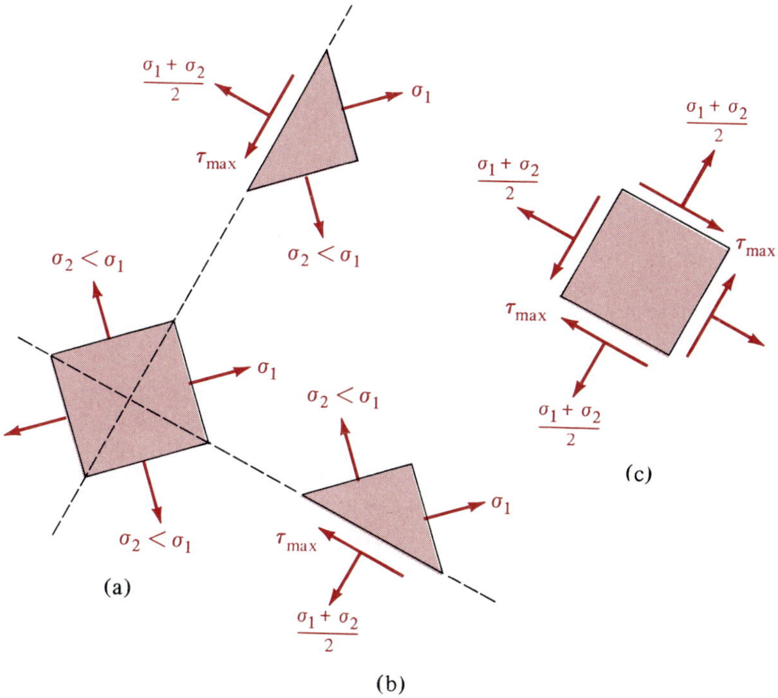

Figure 2.53 Construction for planes on which the maximum shear stresses act

Free-body diagrams of the portions of the element to one side of each of the two diagonals is shown in Figs. 2.53b. By visually summing forces in the directions of the diagonals, the directions of the in-plane shear stress τ_{max} can easily be deduced as shown in Figs 2.53b. The element on which τ_{max} and the corresponding average normal stress $\sigma_{avg} = (\sigma_x + \sigma_y)/2 = (\sigma_1 + \sigma_2)/2$ act is then shown in Fig. 2.53c. Thus, once the principal planes and directions have been determined from

Mohr's circle, these observations show a method for easily determining the directions of the maximum in-plane shear stresses on the 45° planes, using basic equilibrium concepts.

Example 2.19 For the state of stress defined in the previous example and shown in Fig. 2.54a, use Mohr's circle to determine the stresses on a plane 30° from the x-axis as shown.

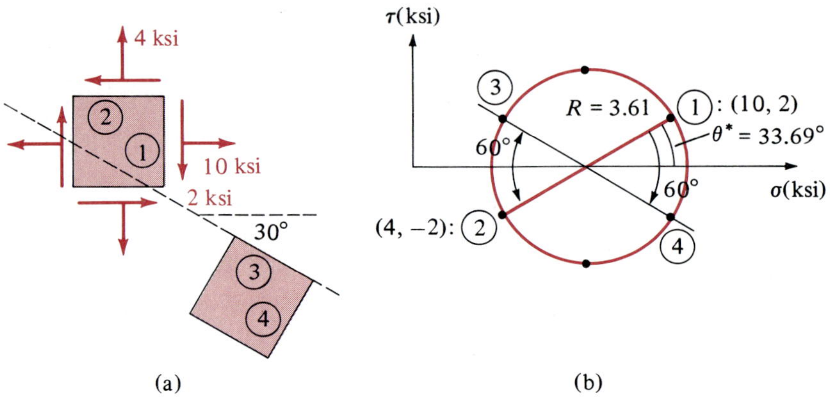

Figure 2.54 Use of Mohr's circle for the transformation of stress

Solution To determine the state of stress on the 30° plane indicated, the normal to face 1 can be rotated clockwise through an angle of 30° to obtain the normal to face 4, the face in question. This corresponds to a 60° clockwise rotation of the diameter of Mohr's circle. The angle θ^* that the diameter between points 1 and 2 makes with the σ-axis is given by

$$\theta^* = \tan^{-1}(2/3) = 33.69°$$

so that the angle that the diameter between points 3 and 4 makes with the σ-axis is 26.31°. The coordinates of point 4 are then given by

$$\sigma = (7.00 + 3.61 \cos(26.31°)) \text{ ksi} = 10.24 \text{ ksi}$$

and

$$\tau = -3.61 \sin(26.31) \text{ ksi} = -1.60 \text{ ksi}$$

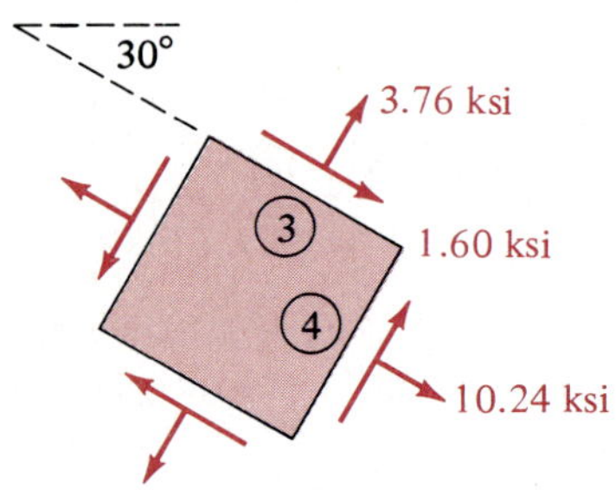

Figure 2.55 State of stress on an element rotated 30°

The corresponding state of stress is indicated in Fig. 2.55. The normal stress of 3.76 ksi on face 3 is computed using the first stress invariant.

We can also consider face 3 as having rotated the normal to face 2 through an angle of 30° clockwise. This corresponds to rotating the diameter of Mohr's circle clockwise through an angle of 60° from point 2, which would again put us at point 3 on Mohr's circle.

PROBLEMS

2.113–2.127 Resolve problems 2.98–2.112 using Mohr's circle.
2.128–2.133 For each of the following use Mohr's circle to determine the stresses on the element oriented according to the dotted line.

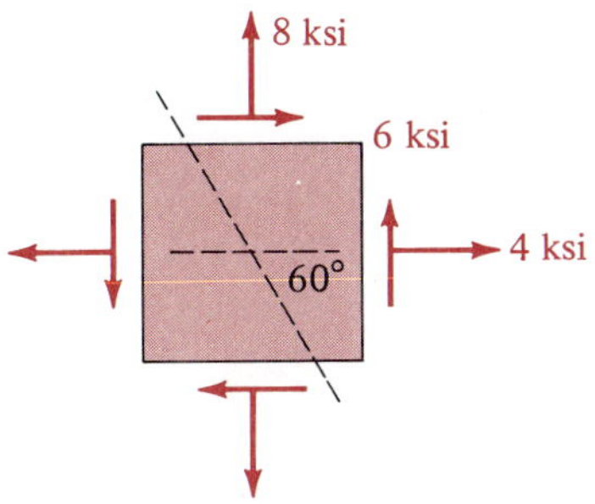

P. 2.128

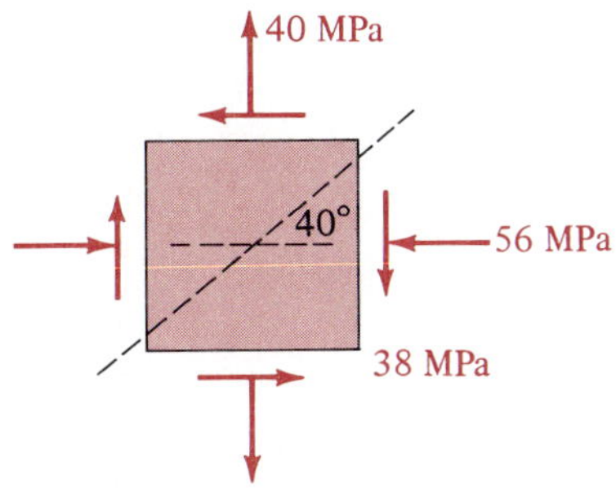

P. 2.129

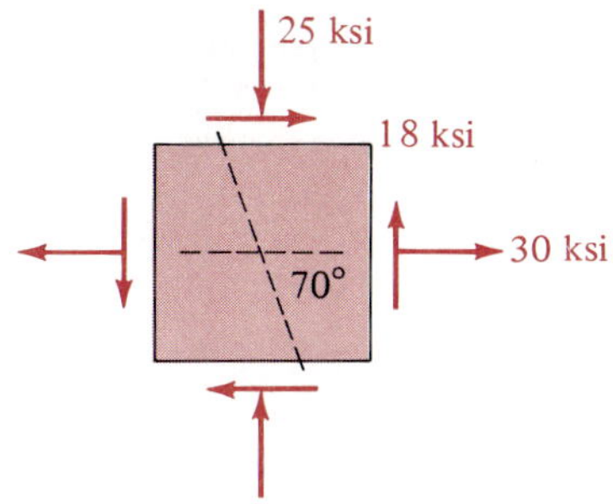

P. 2.130

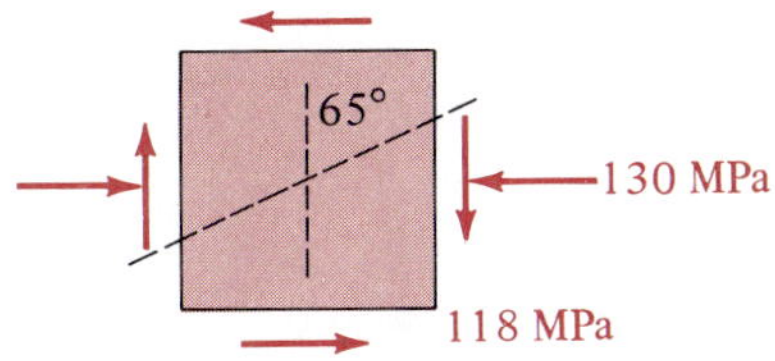

P. 2.131

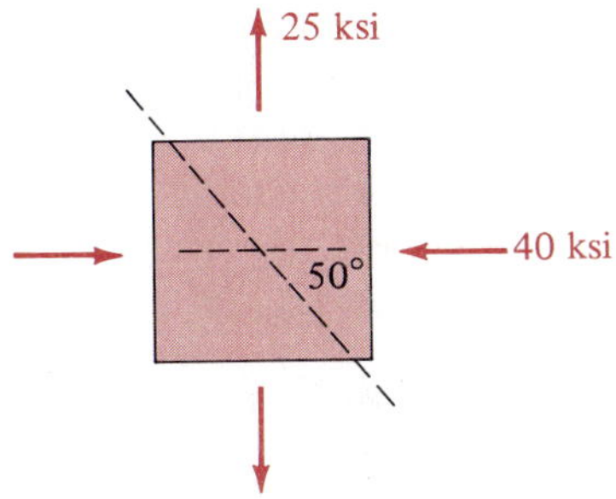

P. 2.132

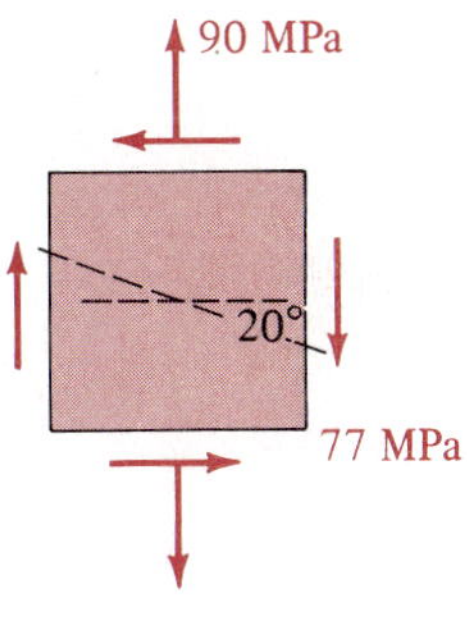

P. 2.133

2.7 CLOSURE

Equilibrium is an indispensable ingredient in the mechanics of solids. Formulating and solving a solid mechanics problem without satisfying equilibrium, at least approximately, would be something like beginning a solo journey down an uncharted river blindfolded—it makes absolutely no sense and could be very dangerous. The student should keep in mind that equilibrium must always be considered, either implicitly or explicitly, when formulating and solving a problem in solid mechanics.

The inputs to a deformable solid consist primarily of known external forces of two types: action at a distance or body forces, and contact or surface forces. In accordance with the requirements of equilibrium, these external loads are transmitted throughout the interior of the body, resulting in internal forces. These internal forces are in turn related to the stresses. There are two basic reasons for wanting to determine the internal force resultants and hence the stresses. First, in many situations the success of a structure is judged by its strength; therefore, the stresses must not exceed the values the material is capable of withstanding. Second, determining the internal force resultants, and hence the stresses, begins the process of evaluating the stiffness of the structure.

In this chapter we have considered several examples of stresses when the shape of the body is rectangular and the loading is pure tension or pure shear. In subsequent chapters we will develop elementary theories for more generally shaped and loaded bodies that will enable us once again to (*a*) determine the stresses in terms of the internal force resultants, in order to evaluate the strength of the structure, and (*b*) determine the deformations, in order to complete the evaluation of the stiffness of the structure.

CHAPTER 3 DEFORMATION AND STRAIN

CHAPTER OUTLINE

3.1 INTRODUCTION

Kinematics is defined in *Webster's* as "a branch of dynamics that deals with aspects of motion apart from consideration of force and mass." In the mechanics of solids, we are for the most part concerned with problems where any "motions" are assumed to take place slowly, so as to produce negligible accelerations and velocities; essentially, we are dealing with problems in statics. The kinematics that applies to the mechanics of solids deals with changes in the size and shape of a body as a result of its deformations. In solid mechanics, kinematics is often referred to as the "geometry of deformation."

In this chapter we will discuss rigid-body motions, displacements in general, and deformations. Displacement of the body will refer to the difference in the original and final positions of the body. A description of the displacements will lead naturally to the definition of strains as measures of the changes in the size and shape of the body, resulting from certain types of relative deformations between neighboring points in a body.

3.2 DISPLACEMENTS

In order to adequately describe deformations it is essential to understand precisely what is meant by displacement. In general, displacements are changes in position, consisting of rigid-body motions and deformations. We will illustrate these ideas by several examples before discussing displacements in a general situation.

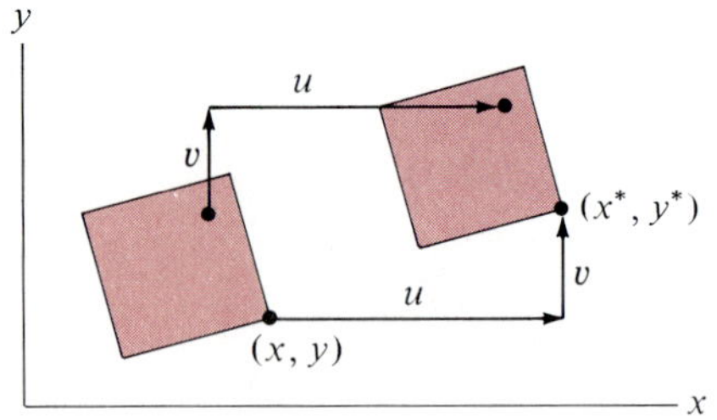

Figure 3.1 Rigid-body translation

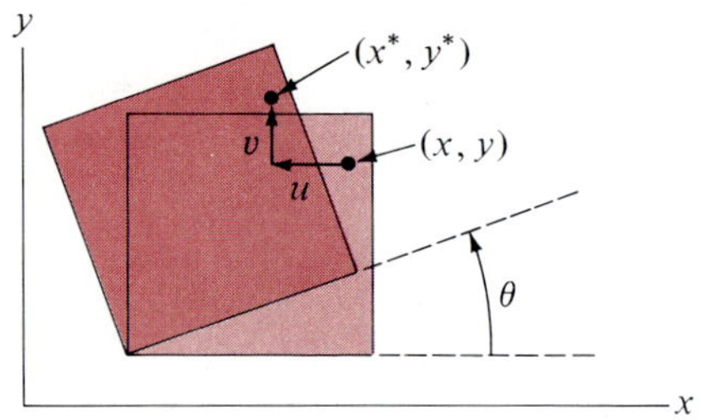

Figure 3.2 Rigid-body rotation

Example 3.1 Consider a two-dimensional rigid body, translated with no rotation to a new position as shown in Fig. 3.1. This is one example of a rigid-body displacement; it is called a **rigid-body translation**. The displacement components u and v in the x and y directions respectively are the constants given by $u = x^* - x$ and $v = y^* - y$, respectively. It is assumed that w, the component of displacement in the direction perpendicular to the plane of the paper, is identically zero. The displacement components u and v are the same for every point in the body. There is no change in the size or shape of the body, that is, no deformation.

Fig. 3.2 pictures the same body where the displacements are a result of rotation through an angle θ about one edge. The displacement components $u = x^* - x$ and $v = y^* - y$ in the x and y directions respectively are different for different points of the body. This type of rigid-body displacement is called a **rigid-body rotation**. The orientation of the body changes but there is no change in the size or shape of the body.

Both of these rigid-body displacements simply move the body to a new position. Any rigid-body displacement can be considered to be a combination of a rigid-body translation and a rigid-body rotation. The resulting displacements are such that there is no change in the distance between any two points in the body.

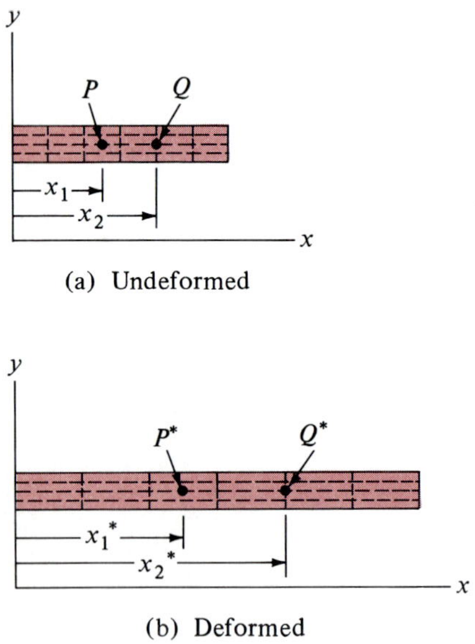

Figure 3.3 Undeformed and deformed configurations

Example 3.2 Consider the bar in Fig. 3.3, shown undeformed in Fig. 3.3a and deformed in Fig. 3.3b. This deformation is known as an **extensional deformation** of the bar. The two points P and Q displace to points P^* and Q^* as indicated.

Consider the displacement of point P, whose undeformed coordinates are given by (x_1, y). The deformed position of P is denoted by P^* with coordinates given by (x_1^*, y). Similarly, the undeformed position of point Q is given by (x_2, y) with the corresponding deformed position given by (x_2^*, y). For this extensional deformation, the change in position or displacement of P occurs only in the x-direction and is given by $u(P) = x_1{}^* - x_1$. The displacement of Q is given by $u(Q) = x_2^* - x_2$. The other two displacement components in the y- and z-directions are given by $v = 0$ and $w = 0$, respectively. It is clear that in general $u(P) \neq u(Q)$; the displacement component u depends upon which point in the bar is considered. We express this as $u = u(x)$. The fact that u does not depend upon y is a result of the way the deformation was defined. Whether a real body would deform in this manner in response to external loads remains to be seen. The displacement $u = u(x)$ results in deformations that change the size and shape of the body.

Example 3.3 A rectangular block is deformed as shown in Fig. 3.4. This type of deformation is frequently referred to as a simple **shear deformation**.

The points P: (x, y) and P^*: (x^*, y^*) represent the undeformed and deformed positions of a general point in the body. For the deformation shown, the displacement components are given by

$$u = x^* - x = u(y) = \Delta\left(\frac{y}{h}\right)$$

and

$$v = y^* - y = 0$$

$$w = 0$$

in the x-, y-, and z-directions, respectively. The representations $u = u(y) = \Delta(y/h)$, $v = w = 0$ show that the values of the displacements u depend only upon the original y coordinate: all points at a given distance above the x-axis displace the same amount in the x-direction and do not displace at all in the y-direction. These displacements indicate a deformation that results primarily in changes in the shape of the body. In particular, there is a change in the angle between lines originally parallel to the x- and y-axes.

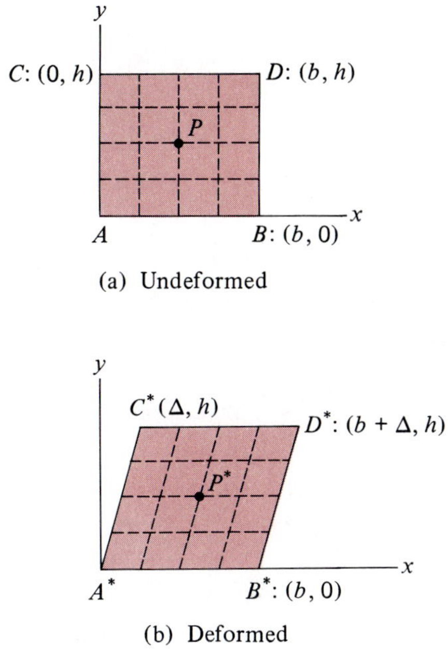

Figure 3.4 Shear deformation of a block

In a more general case we are concerned with characterizing the displacements in an irregularly shaped body deforming in the xy plane as in Fig. 3.5.

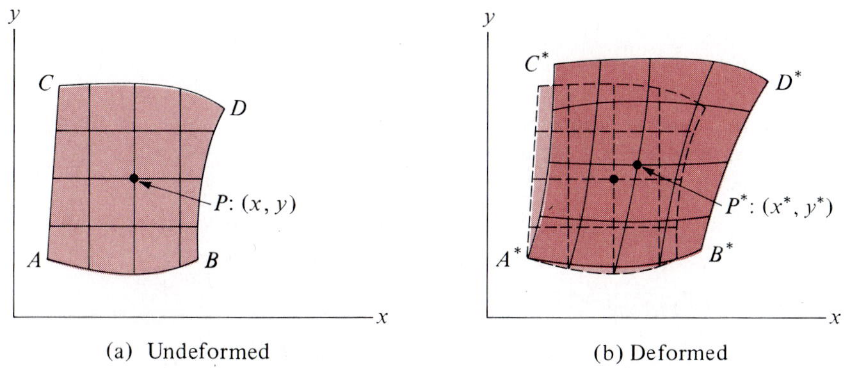

Figure 3.5 General two-dimensional deformation

The undeformed geometry is shown in Fig. 3.5a; both the undeformed and deformed geometries are indicated in Fig. 3.5b. The displacements u in the x-direction and v in the y-direction are given by the differences in the undeformed and deformed positions according to

$$u(x, y) = x^* - x$$

and

$$v(x, y) = y^* - y$$

For a deformable body undergoing an extensional deformation, the distance between two points can change, resulting generally in a change in the size of the body. For a body undergoing a shear deformation, the angle between two lines can change, generally resulting in changes in shape. Both of these types of deformation are seen in the deformation pictured in Fig. 3.5b. It is these changes in the size and shape that distinguish a deformable body from a rigid body; this we will study in the following sections.

PROBLEMS

3.1 A block is rotated through an angle θ as shown. Determine the displacements $u(x, y)$ and $v(x, y)$ of a point P in terms of the angle θ. Specialize these to the case where θ is small.

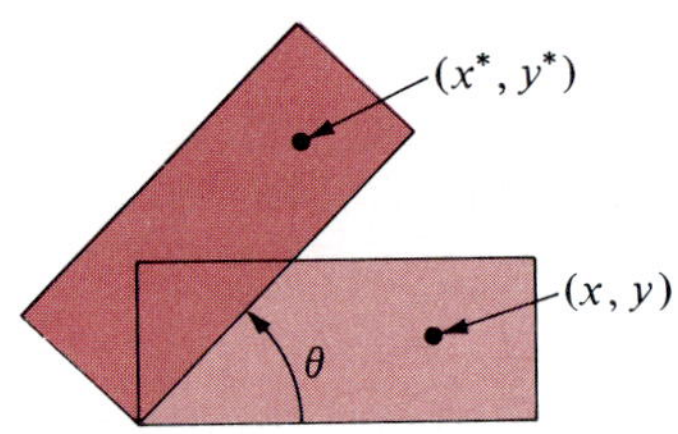

P. 3.1

3.2 A rectangle is deformed in a uniform fashion into another rectangle as shown. Assume $u = \alpha + \beta x$ and $v = \gamma + \delta y$ and thus determine α, β, γ, and δ in terms of b, h, B, and H.

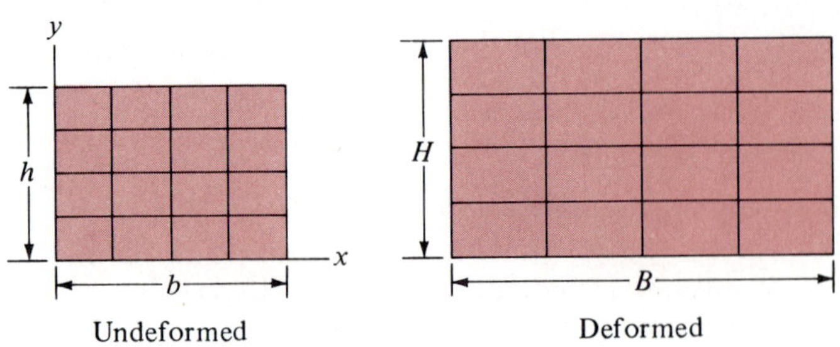

P. 3.2

3.3 A rectangle is deformed in a uniform fashion into a rhombus as shown. There is no displacement of any point in the y-direction. Assume $u = \alpha + \beta y$ and thus determine α and β in terms of b, h, and U.

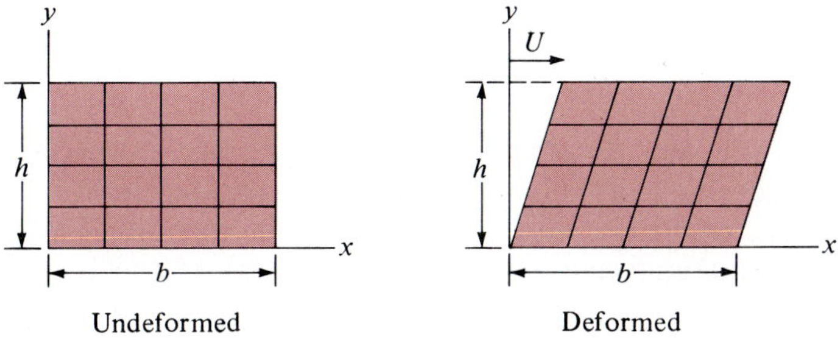

P. 3.3

3.4 A bar is deformed so that u, the x component of displacement at a point P, is proportional to the x coordinate of P. Given that the displacement at the end $x = L$ is Δ, determine u as a function of x in terms of the parameters Δ and L.

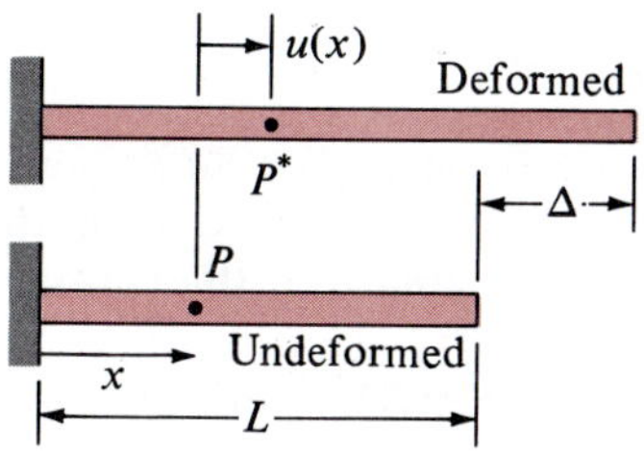

Ps. 3.4 and 3.5

3.5 Repeat problem 3.4 if the displacement of a point P located a distance x from the left end of the bar is proportional to the square of the distance x.

3.6 A rectangle is deformed in a uniform fashion into a quadrilateral as shown, that is, vertical and horizontal lines remain straight after the deformation. Assume $u = \alpha_1 + \beta_1 x + \gamma_1 y + \delta_1 xy$ and $v = \alpha_2 + \beta_2 x + \gamma_2 y + \delta_2 xy$ and thus determine α_1, β_1, γ_1, δ_1, α_2, β_2, γ_2, and δ_2 in terms of b, h, U, and V.

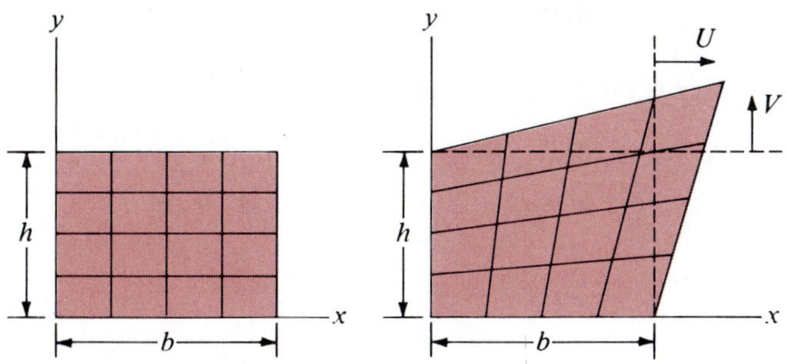

P. 3.6

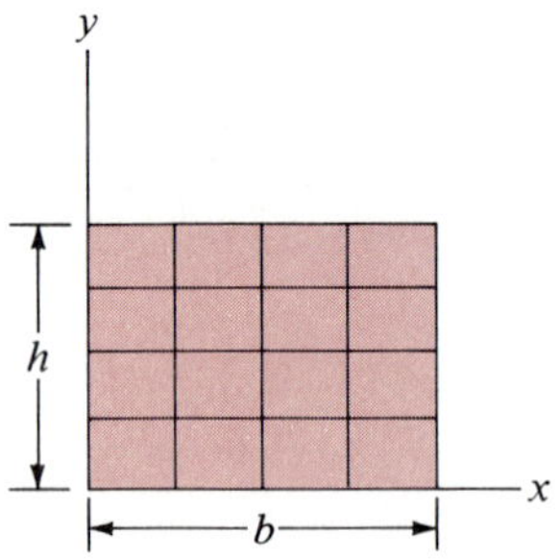

(a)

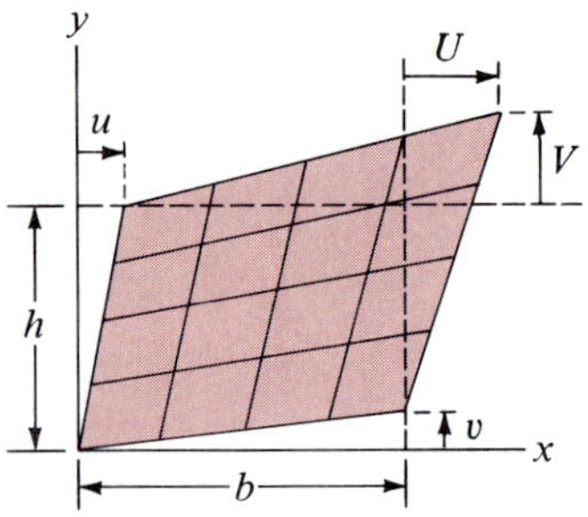

(b)

P. 3.7

3.7 A rectangle is deformed in a uniform fashion into a quadrilateral as shown, that is, vertical and horizontal lines remain straight after the deformation. Assume $u = \alpha_1 + \beta_1 x + \gamma_1 y + \delta_1 xy$ and $v = \alpha_2 + \beta_2 x + \gamma_2 y + \delta_2 xy$ and thus determine $\alpha_1, \beta_1, \gamma_1, \delta_1, \alpha_2, \beta_2, \gamma_2$, and δ_2 in terms of b, h, u, v, U, and V.

3.3 STRAINS—RELATIVE DISPLACEMENTS

As we will see shortly, a deformable body is able to support various types of loads due to its ability to generate internal forces when the size and/or shape of the body is changed. These internal forces are associated with stretching and shearing deformations of the body. These stretching and shearing deformations in turn are due to relative displacements between neighboring points in the body. Measures of these relative deformations are termed strains—these are the focus of this section. We will first investigate extensional and shearing strains in situations where the deformations are relatively simple, and then discuss the general case.

3.3.1 Extensional Strain

Consider the familiar problem of the stretching of a rubber band as indicated in Fig. 3.6. An overall measure of the stretching is given simply by the change in the length of the rubber band

$$\text{Stretch} = \text{Change in length of the rubber band} = L^* - L$$

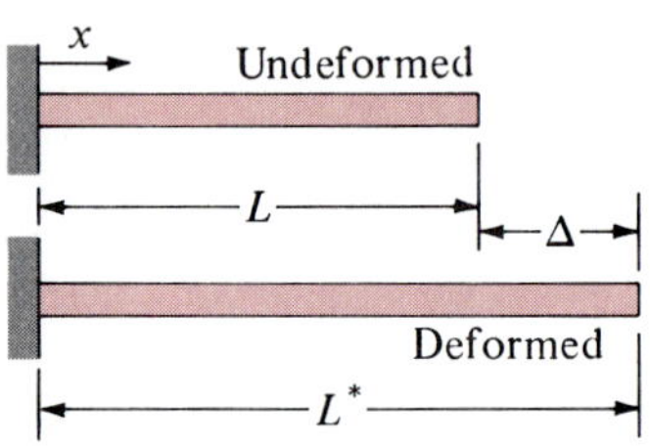

Figure 3.6 Stretching of a rubber band

Expressed in terms of the displacement Δ of the end of the rubber band this becomes

$$\text{Stretch} = \Delta$$

The **average extensional strain** or the average normal strain in the bar is computed by dividing the stretch by the original length to obtain

$$\varepsilon_{avg} = \Delta/L$$

By considering the change in length per unit of original length for an infinitesimal segment of the bar, we obtain the extensional strain at a point. Consider then a small segment PQ of the rubber band such as in Fig. 3.7.

As a result of the stretching of the rubber band, points P and Q deform to points P^* and Q^*. The extensional strain at P is defined as

$$\varepsilon_x(P) = \lim_{PQ \to 0} \frac{P^*Q^* - PQ}{PQ} \tag{3.1}$$

where PQ and P^*Q^* denote the lengths of the line segments between P and Q and between P^* and Q^*, respectively. Referring to Fig. 3.7 it can be seen that

$$PQ = \Delta x$$

and

$$P^*Q^* = OQ^* - OP^* = [x + \Delta x + u(Q)] - [x + u(P)]$$

or

$$P^*Q^* - PQ = u(Q) - u(P) = \Delta u$$

so that

$$\varepsilon_x(P) = \lim_{PQ \to 0} \frac{P^*Q^* - PQ}{PQ} = \lim_{\Delta x \to 0} \frac{\Delta u}{\Delta x}$$

$$= \frac{du}{dx}(P) = u'(P) \qquad (3.2)$$

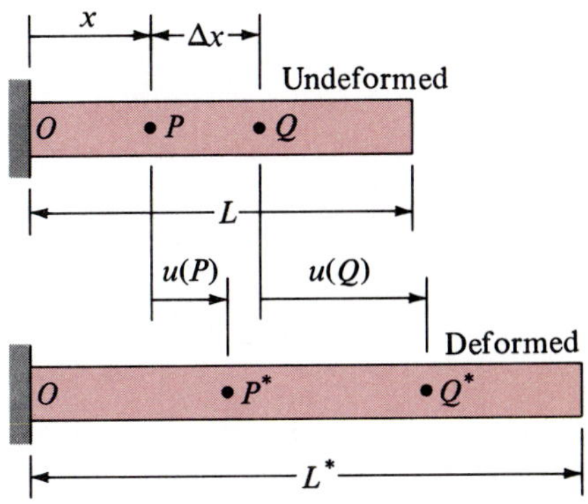

Figure 3.7 Strain at a point

Equation (3.2) is referred to as a **strain-displacement relation**. Note that as defined, $\varepsilon_x(P)$ is dimensionless since $[u] = L$ and $[x] = L$. For simple problems involving extensional deformations of straight bars we will frequently drop the subscripts and write

$$\varepsilon = u'$$

for the strain displacement relation. Generally for a bar fixed at the left end, a plot of the displacement appears as shown in Fig. 3.7.1. As indicated, the extensional strain at a point along the axis of the bar is simply the slope of the u versus x curve.

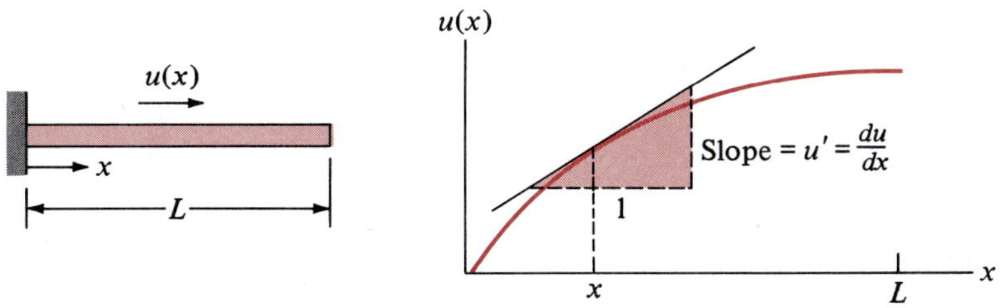

Figure 3.7.1 Strain as the slope of the displacement curve

The strain displacement relation $\varepsilon = u'$ is based on the assumption that ε is small. If ε is not small, a different definition for strain is necessary. The student is referred to problems 3.26–3.29 for further information on this topic. Extensional strains are generally a measure of the change in the lengths of lines in a body undergoing a deformation and will be used frequently throughout the text.

Example 3.4 Consider the slender deformable bar shown in Fig. 3.8.

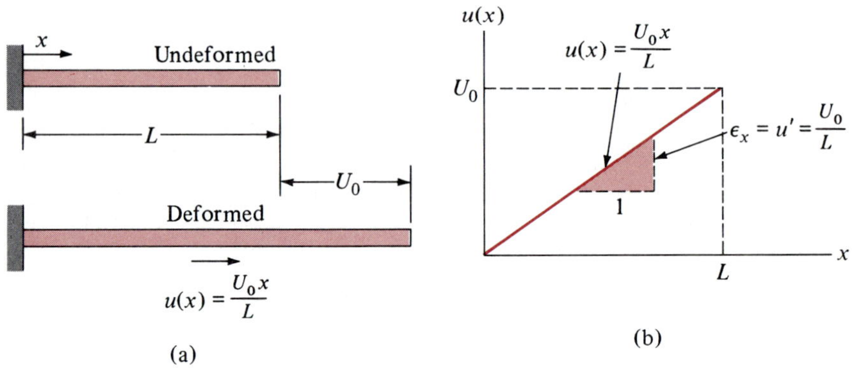

Figure 3.8 Displacements and strain in a slender bar

The displacement of a point in the direction of the axis of the bar is taken to be a linear function of position measured from the left end of the bar, that is, $u(x) = U_0(x/L)$ as indicated in Fig. 3.8b. According to the result given by Eq. (3.2), the strain at a point in the bar is

$$\varepsilon_x = u' = U_0/L$$

This constant value is the same as the average strain in the bar given by the total elongation divided by the length, or

$$\varepsilon_{\text{avg}} = \frac{u(L) - u(0)}{L} = \frac{U_0}{L}$$

This particular deformation will be referred to as **uniform extension**.

One use of the strain-displacement relation $\varepsilon = u'$ is in situations where the strains are known or have been determined. The equation $\varepsilon = u'$ can be integrated to determine the displacements according to

$$\int_{x_0}^{x} \varepsilon \, dx = \int_{x_0}^{x} u' \, dx = u(x) - u(x_0)$$

As an illustration consider Example 3.4, where it is known that the strain in the bar of length L is the uniform value U_0/L. Using the strain-displacement relation yields

$$\varepsilon = u' = \frac{U_0}{L}$$

Integration results in

$$u = \frac{U_0}{L} x + C_1$$

which, with the knowledge that $u(0) = 0$ gives the value $C_1 = 0$ and $u(x) = U_0 x/L$.

Example 3.5 The extensional strain in a uniform bar hanging under its own weight as shown in Fig. 3.9 is known to be $\varepsilon = a(L - x)$ where a is a constant depending on the physics of the problem. Using the strain-displacement relation there results

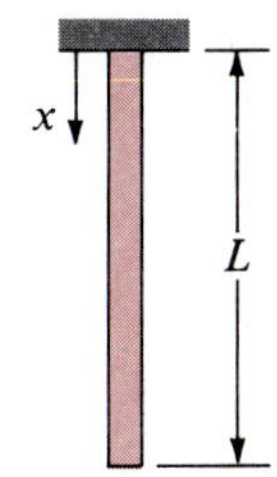

Figure 3.9 Hanging bar

$$u' = \varepsilon = a(L - x)$$

Integrating yields

$$u = \int \varepsilon \, dx = -a \frac{(L - x)^2}{2} + C_1$$

The constant of integration is evaluated by noting that at $x = 0$, the displacement u is zero; that is,

$$0 = u(0) = \frac{-aL^2}{2} + C_1$$

or

$$C_1 = \frac{aL^2}{2}$$

Eliminating C_1 the final expression for the displacement u is

$$u = a \frac{2Lx - x^2}{2}$$

The strains and displacements are shown in Fig. 3.10.

These two curves are related by a differential relationship, the strain-displacement relation $\varepsilon = u'$, which can be very useful in helping to construct the strain curve given the displacement curve or vice versa. The student should become familiar with this sort of graphical perspective of the integration.

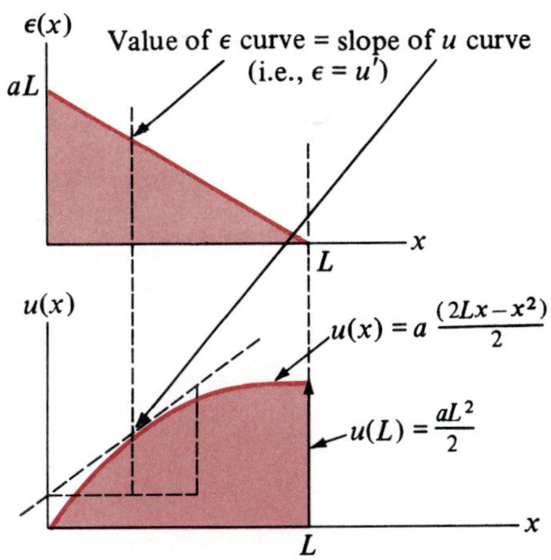

Figure 3.10 Strains and displacements in a hanging bar

Example 3.6 Consider the deformation of a straight prismatic bar into a circular arc subtending an angle θ as shown in Fig. 3.11.

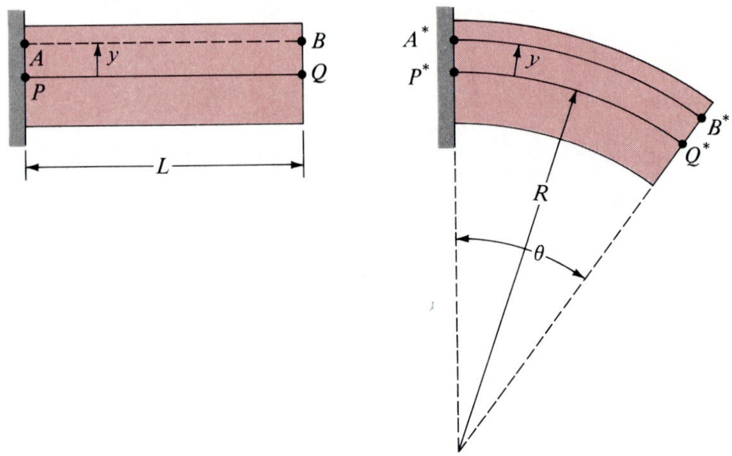

Figure 3.11 Deformation into a circular arc

It is assumed that any portion of the middle surface, given by $y = 0$, is unchanged in length as a result of the deformation, so that $P^*Q^* = PQ = L$. The radius of curvature of the middle surface is taken to be R, thus $L = R\theta$. The average extensional strain at a point a distance y above the middle surface is given by

$$\begin{aligned}\varepsilon_{\text{avg}} &= \frac{A^*B^* - AB}{AB} \\ &= \frac{(R + y)\theta - L}{L} \\ &= \frac{R\,\theta + y\,\theta - L}{L} \\ &= \frac{y\,\theta}{L} = \frac{y}{R}\end{aligned}$$

The maximum average strain occurs at the point furthest from the middle surface.

PROBLEMS

3.8 The rigid bar AB is rotated through an angle θ as indicated. What is the corresponding average extensional strain in the wire BC? Specialize the result when θ is small.

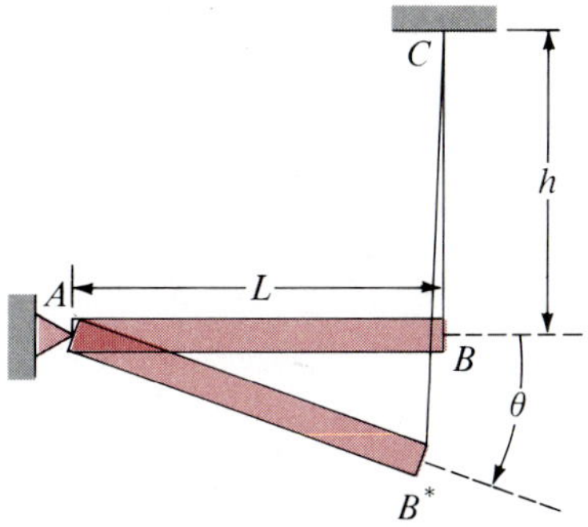

P. 3.8

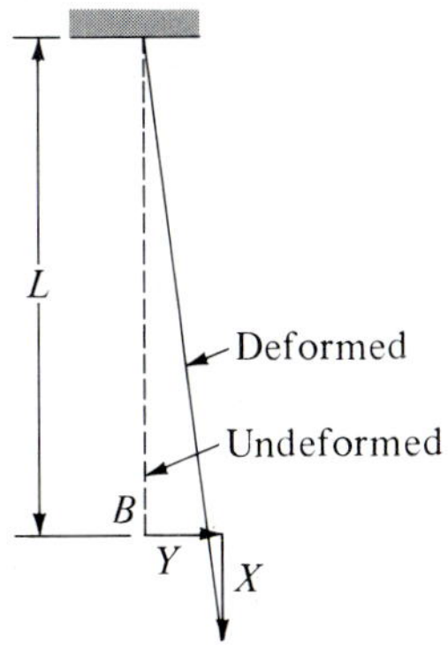

P. 3.9

3.9 The rope hangs in the vertical position as shown. If point B is moved to the new position indicated, determine the change in the average extensional strain. Specialize the result to the case where X and Y are both small compared to L.

3.10 A rubber band 6 in. long is placed around a 3-in.-diameter rigid tube. What is the average extensional strain in the rubber band?

3.11 The diameter of the hoop shown is increased by an amount Δd. Determine the corresponding average circumferential strain in the hoop.

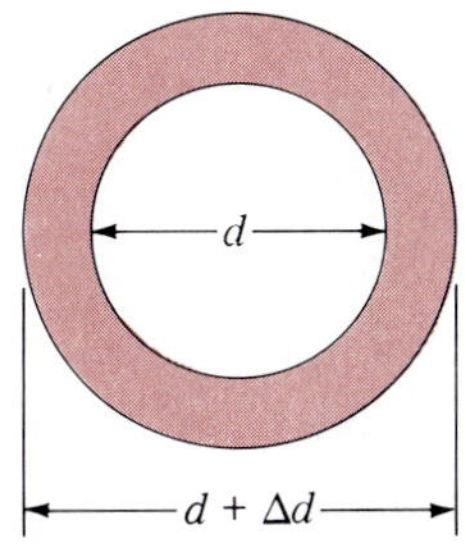

P. 3.11

3.12 A thin bar restrained against displacement at A is deformed in the direction of its axis such that point B moves 1.0 in. to the right. Determine the average extensional strains in segments AB and BC.

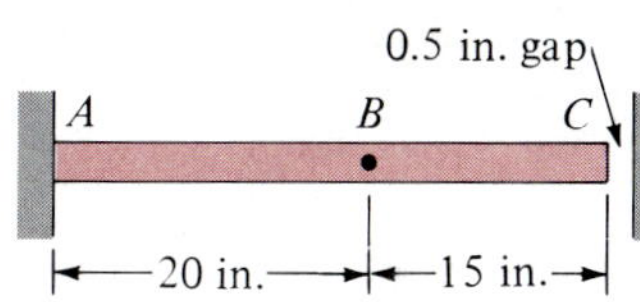

P. 3.12

3.13 In the truss structure shown, point B is displaced 2 in. downward. Determine the average extensional strains in members AB and BC.

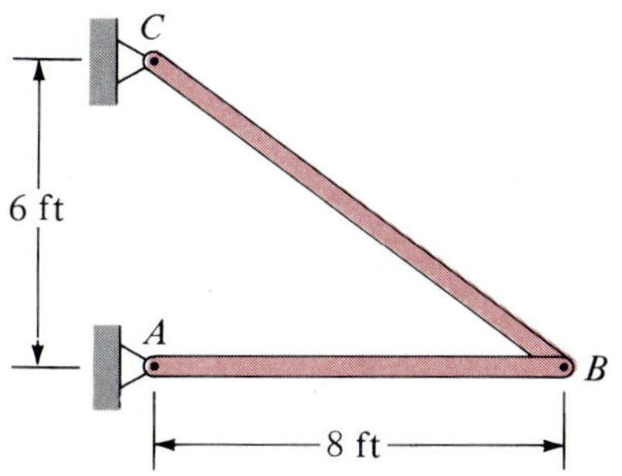

P. 3.13

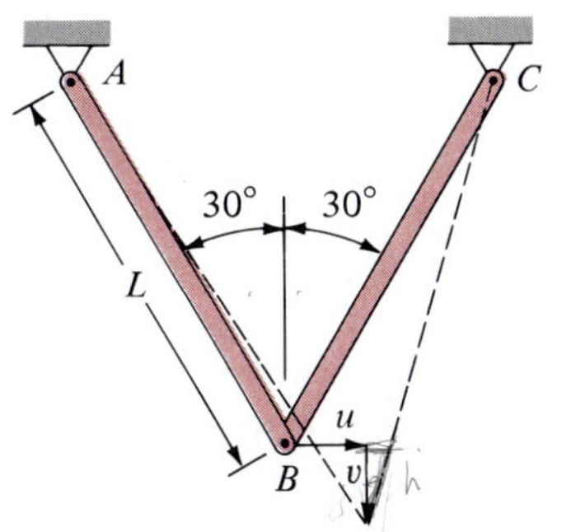

P. 3.14

3.14 Point B is displaced as shown. Determine the average extensional strains in members AB and BC. Specialize these when u and v are small compared to L.

3.15 Point D on the truss shown displaces 6 mm to the right and 4 mm down. Determine the average extensional strains in members AD and BD.

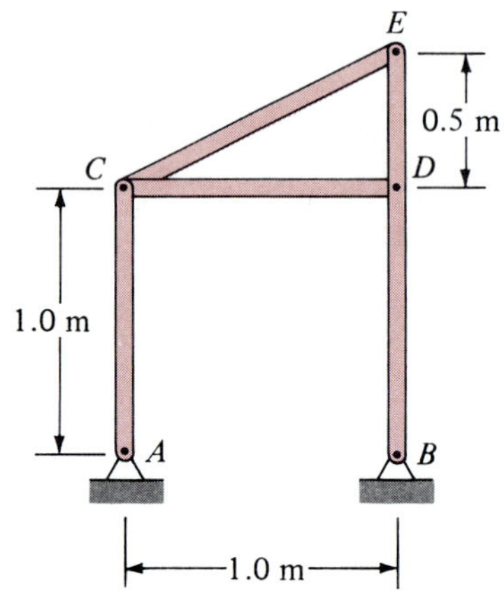

P. 3.15

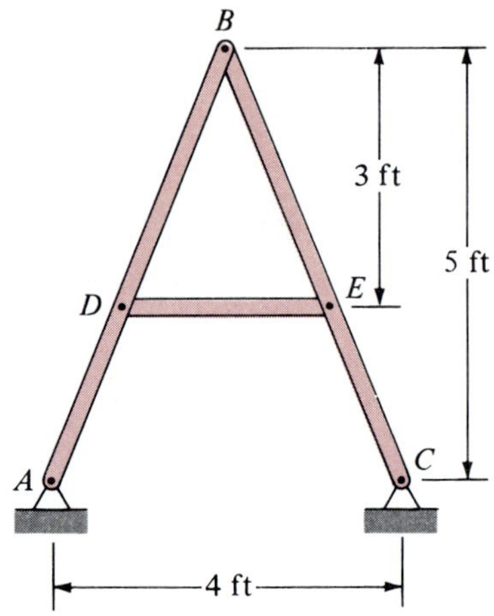

P. 3.16

3.16 A truss consists of two relatively rigid members AB and BC connected by a wire DE as shown. If point C is moved 1.2 in. to the right, what is the average extensional strain in the wire?

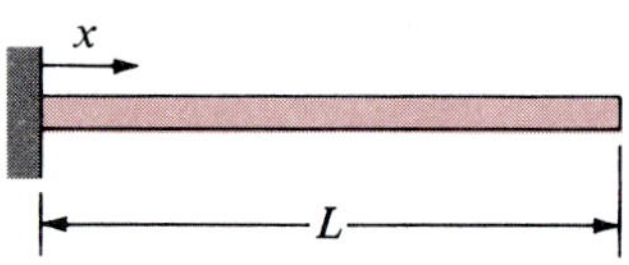

Ps. 3.17 and 3.18

3.17 The extensional strain in the rod shown is given by $\varepsilon = 0.001 + 0.002\,(x/L)^2$. Given that the displacement is zero at $x = 0$, determine the displacement $u(x)$ in the rod and hence the elongation of the rod.

3.18 A wire is stretched so that the positive extensional strain is proportional to the square of the distance from one end. If the extensional strain has the value of 0.02 at the midpoint of the bar of original length 1 meter, determine the final length.

3.19 A toy balloon, approximately spherical, is 6 in. in diameter. It is further inflated so that it is still approximately spherical with a diameter of 9 in. What is the change in the average circumferential extensional strain in the balloon?

3.20 A thick-walled tube is deformed such that the outer diameter increases by Δb and the inner diameter by Δa. Determine the average radial extensional strain and an average circumferential extensional strain.

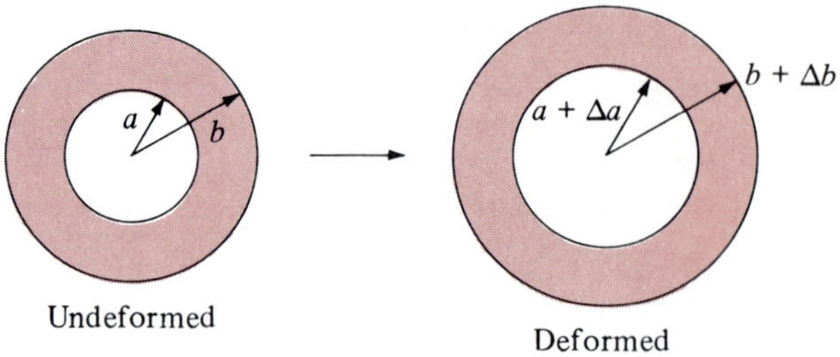

P. 3.20

3.21 A rectangle is deformed into another rectangle as shown. Determine the average extensional strains of (*a*) the line AB, (*b*) the line AC, and (*c*) the line AD. Take u/L to be small.

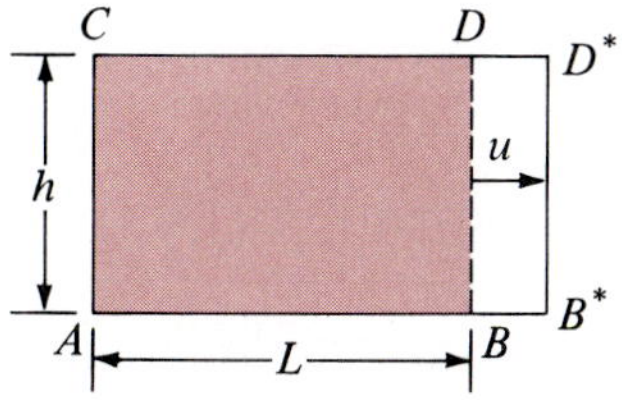

P. 3.21

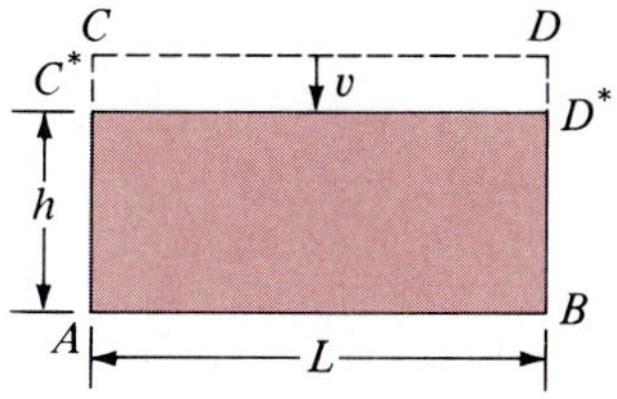

P. 3.22

3.22 A rectangle is deformed into another rectangle as shown. Determine the average extensional strains of (*a*) the line AB, (*b*) the line AC, and (*c*) the line AD. Take v/L to be small.

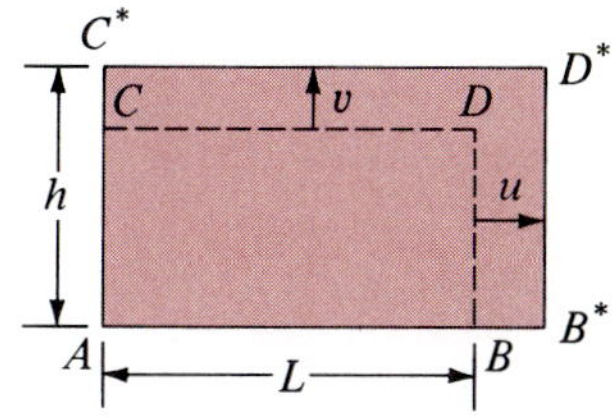

P. 3.23

3.23 A rectangle is deformed into another rectangle as shown. Determine the average extensional strains of (*a*) the line AB, (*b*) the line AC, (*c*) the line BC, and (*d*) the line AD. Take u/L and v/L to be small.

3.24 A rectangle is deformed uniformly into the parallelogram as shown. Determine the average extensional strains of (*a*) the line AB, (*b*) the line AC, and (*c*) the line AD. Take u/L to be small.

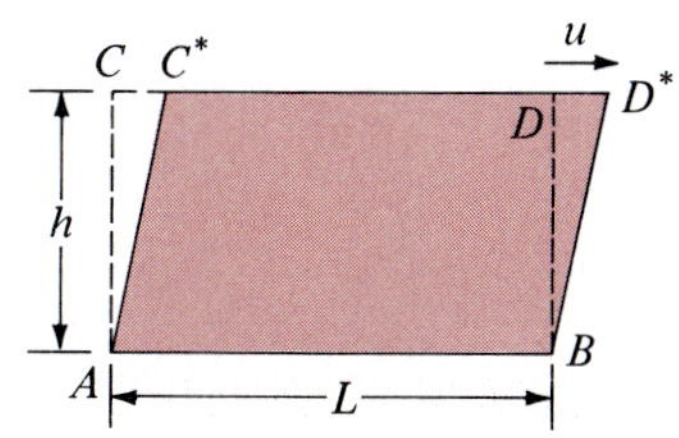

P. 3.24

3.25 A rectangle is deformed uniformly into the rhombus as shown. Determine the average extensional strains of (*a*) the line AB, (*b*) the line AC, (*c*) the line BC, and (*d*) the line AD. Take u/L, U/L, and v/L to be small.

3.26 The definition of strain used in Section 3.2.1, namely

$$\varepsilon = \lim_{\Delta s \to 0} \frac{\Delta s^* - \Delta s}{\Delta s}$$

where Δs and Δs^* are the lengths of a segment before and after deformation, respectively, is the so-called *engineering strain*. Another definition for strain is

$$e = \lim_{\Delta s \to 0} \frac{(\Delta s^*)^2 - (\Delta s)^2}{2(\Delta s)^2}$$

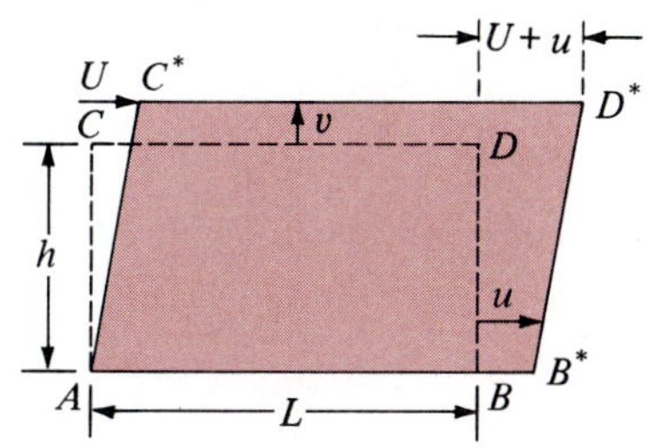

P. 3.25

Show that this definition can be expressed in terms of the engineering strain as $e = \varepsilon(1 + \varepsilon/2)$, so that if ε is small $\varepsilon \approx e$. In order to get some idea as to what "small" means, consider the problem of a bar deformed so that $u(x) = U_0 x/L$ as in Example 3. Plot each of ε and e as a function of U_0/L to see the comparison.

3.27 A straight wire originally of length L is deformed into a straight line in a manner so that there is no displacement in the x direction as shown. Compute the extensional strain at any point along the wire.

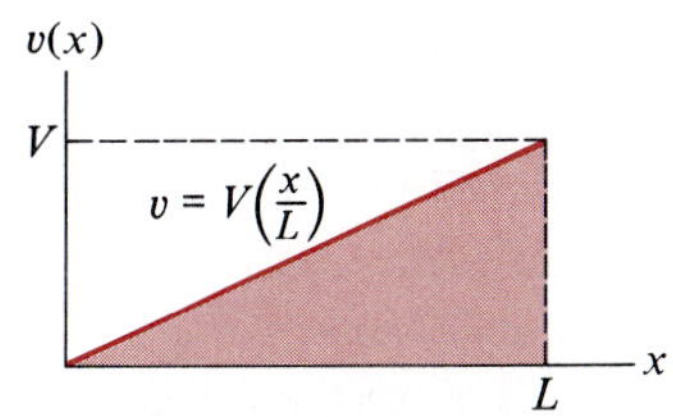

P. 3.27

3.28 Repeat problem 3.27 for the vertical displacement in the form of the parabola indicated.

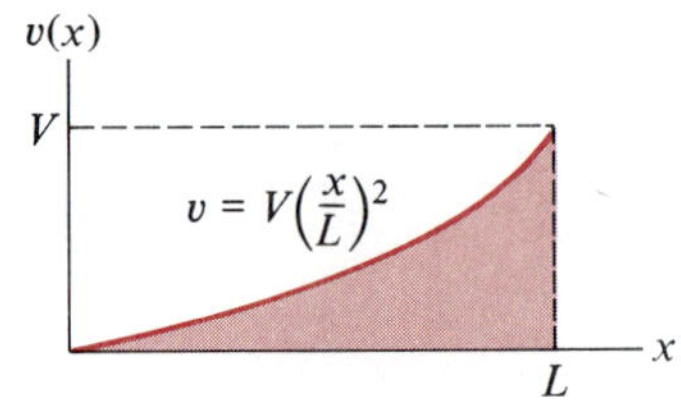

P. 3.28

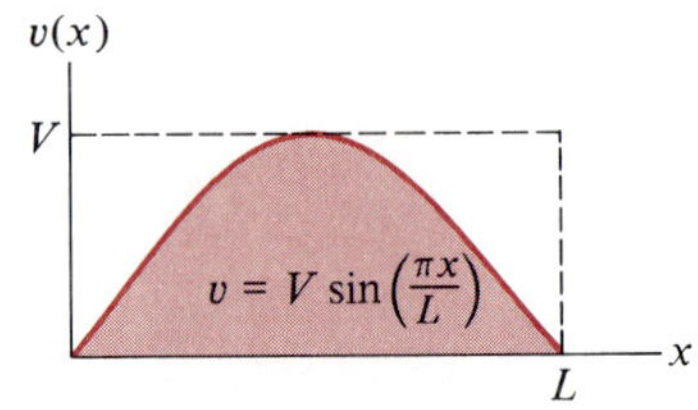

P. 3.29

3.29 Repeat problem 3.27 for the vertical displacement in the form of the sine curve indicated.

3.3.2 Shear Strain

The deformation of a body generally involves both changes in shape and changes in size of the body. Changes in size are naturally associated with the extensions of the body. Changes in the shape of a body result not only from extensional deformations but also from the local changes in angles between lines in the body that are originally perpendicular.

Shown in Fig. 3.12a is a square that is deformed by shearing into a rhombic shape. Figures 3.12b, 3.12c, and 3.12d each picture a different final orientation of the deformed block. The difference in the three orientations of the deformed block is a rigid-body rotation. In each of the deformed blocks, however, the change in the angle between the two sides that were originally perpendicular (the sides originally coinciding with the x- and y-axes), is the angle γ_{xy}. This decrease in angle is an example of a shear strain.

Generally, shear strain at a point is defined as the decrease in angle between two initially perpendicular line segments as a result of a deformation.

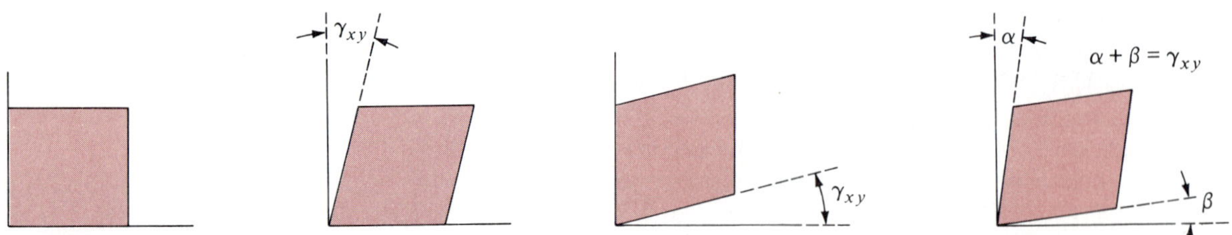

(a) Undeformed (b) Deformed (c) Deformed (d) Deformed

Figure 3.12 Changes in angle associated with a shear deformation

With reference to Fig. 3.13, the shear strain at P for the line segments PQ and PR, parallel to the x- and y-axes, respectively, is denoted by γ_{xy} and is defined as

$$\gamma_{xy}(P) = \lim_{\substack{Q \to P \\ R \to P}} [\text{angle } QPR - \text{angle } Q^*P^*R^*]$$

$$\gamma_{xy}(P) = \lim_{\substack{Q \to P \\ R \to P}} [\pi/2 - \text{angle } Q^*P^*R^*]$$

showing the shear strain to be positive when the angle decreases. Physically one can think of attaching two perpendicular wires to the body at point P, one having the direction of PQ and the other having the direction of PR. The shear strain at P for the directions PQ and PR is the decrease in the angle between the wires as a result of the deformation. Angular changes are measured in radians with the result that shear strain is also dimensionless. Shear strain is a measure of the change in shape of a body or of a portion of a body in the neighborhood of a point P.

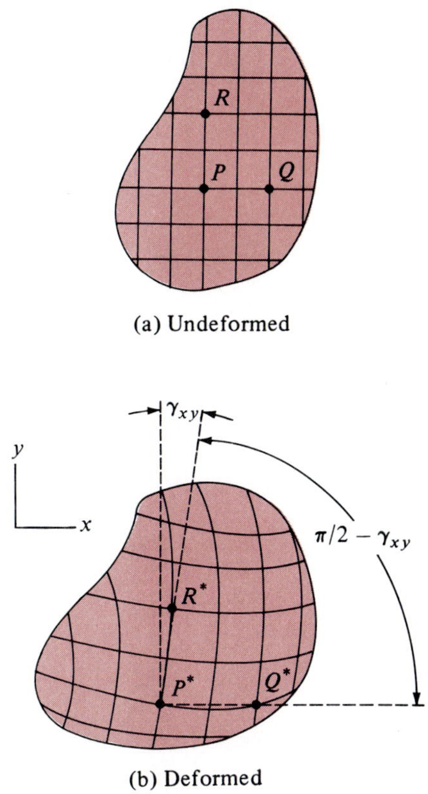

Figure 3.13 Definition of shear strain

Example 3.7 A block undergoes the deformation shown in Fig. 3.14. Determine the shear strain at point P.

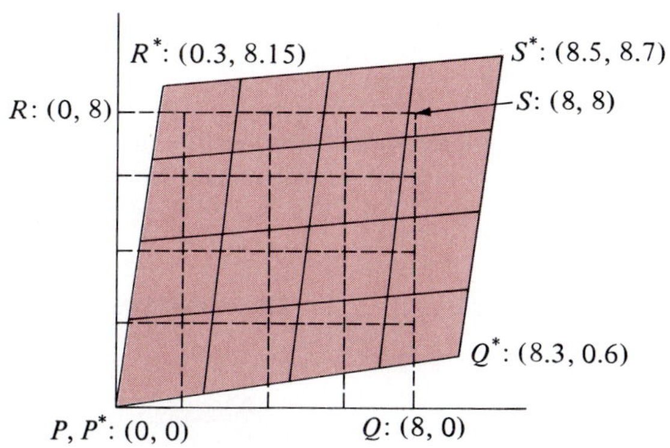

Figure 3.14 Deformation of a block

Solution The shear strain at P is given by

$$\begin{aligned}\gamma &= \text{angle } QPR - \text{angle } Q^*P^*R^* \\ &= \pi/2 - [\pi/2 - \text{angle } R^*PR - \text{angle } Q^*PQ] \\ &= \arctan(0.3/8.15) + \arctan(0.6/8.3) \\ &= 0.0368 + 0.0723 = 0.1091\end{aligned}$$

As we will determine later in our study, this is a very large shear strain for most real situations.

Example 3.8 A rubberlike material is placed between two rigid plates, one of which is displaced horizontally an amount U as shown in Fig. 3.15. Determine the average shear strain in the material.

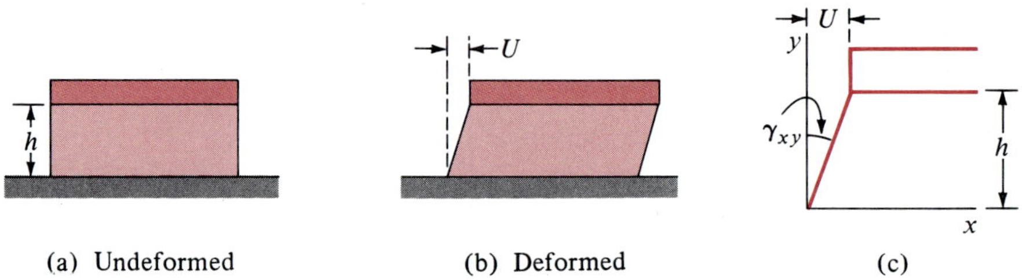

Figure 3.15 Shear deformation of a rubberlike material

Solution Consulting Fig. 3.15c, it is seen that the angle γ_{xy} is given by

$$\tan \gamma_{xy} = \frac{U}{h}$$

from which $\gamma_{xy} = \tan^{-1}(U/h)$. If U/h is small, $\gamma_{xy} = \tan^{-1}(U/h) \approx U/h$.

PROBLEMS

3.30 A plate is fastened to two slabs of rubberlike material and displaced an amount Δ as shown. Determine the average shear strain in the rubber. Take h = 20 mm and Δ = 3 mm.

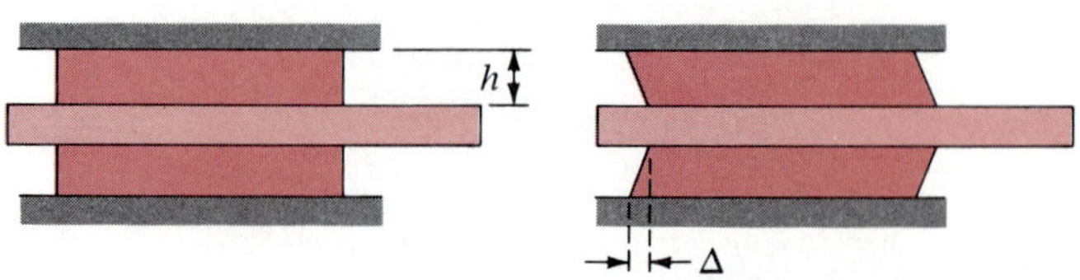

P. 3.30

3.31 Two rigid parallel plates are separated by a rubber block as shown. The block is deformed such that each point moves only in the horizontal direction with $u(y) = U(y/h)$ as shown. Determine the shear strain γ_{xy} at the point $x = L/2$, $y = h/2$.

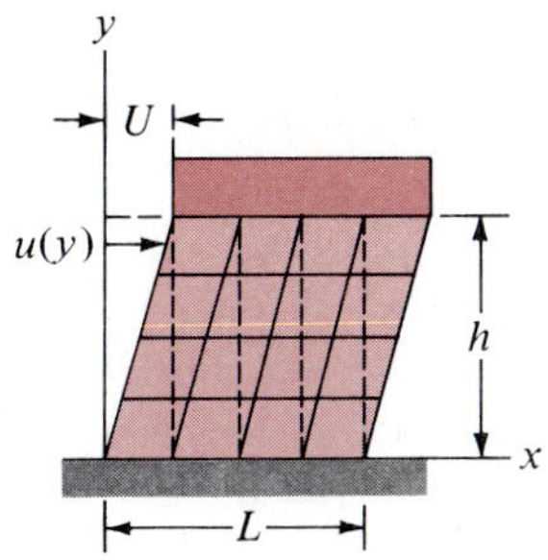

P. 3.31

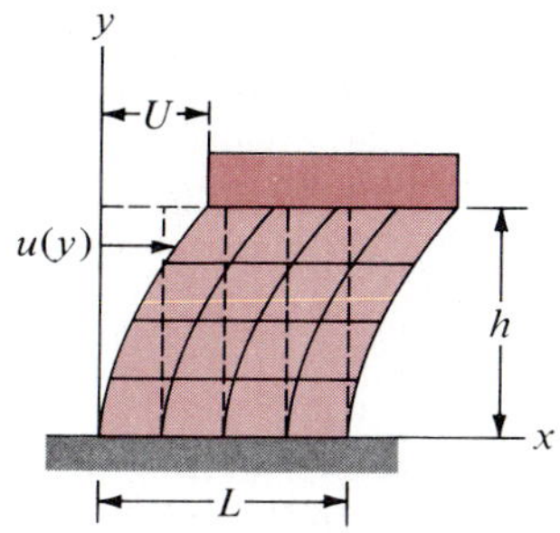

P. 3.32

3.32 Repeat problem 3.31 if the upper plate moves a distance U and if the sides of the block deform into parabolas $u(y) = U(y/h)^2$ rather than straight lines. Assume that there is no displacement in the y-direction.

3.33 A structure that is usually referred to as a shear panel is shown. If points B and E are displaced vertically an amount v as shown, determine the average shear strain in either of the panels. Take $L = 3$ ft and $v = 1.6$ in.

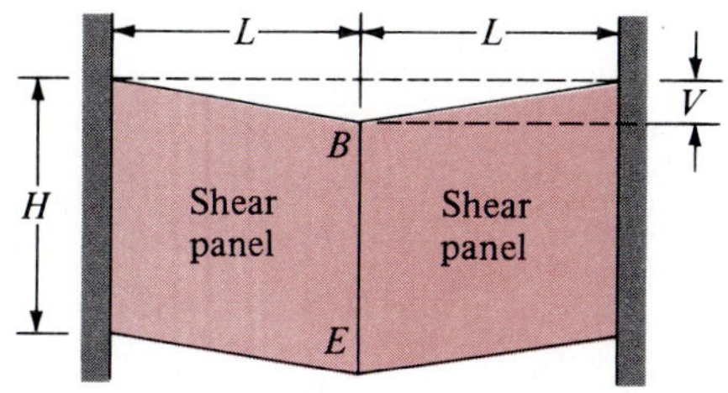

P. 3.33

3.34 A square block is elongated uniformly as shown. Determine the shear strain at C for angle ACB and for angle BCE, assuming no displacement in the vertical direction. Take $u/a = 0.1$.

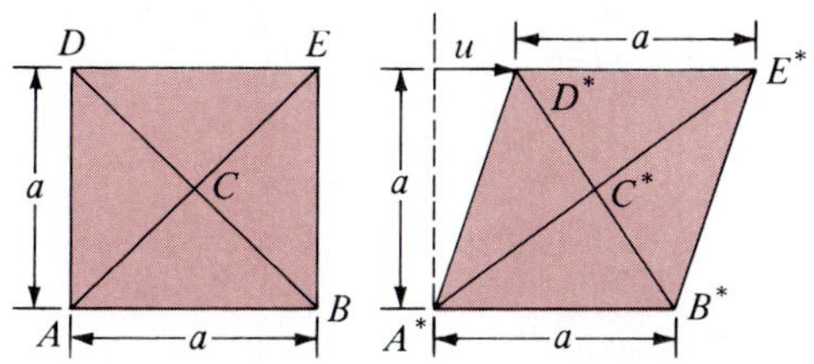

P. 3.34

3.35 A square block is elongated uniformly in the x-direction as indicated. Determine the shear strain at C for angle ACB and for angle BCE. Assume no displacement in the vertical direction and take $u/a = 0.2$.

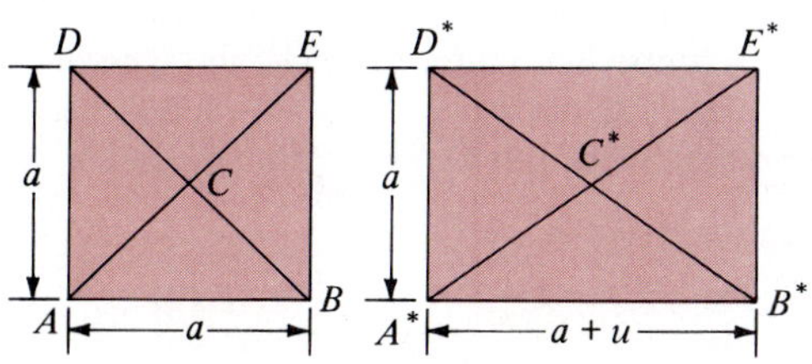

P. 3.35

3.36 With $u/a = v/a = 0.1$, repeat problem 3.35 for the figure shown.

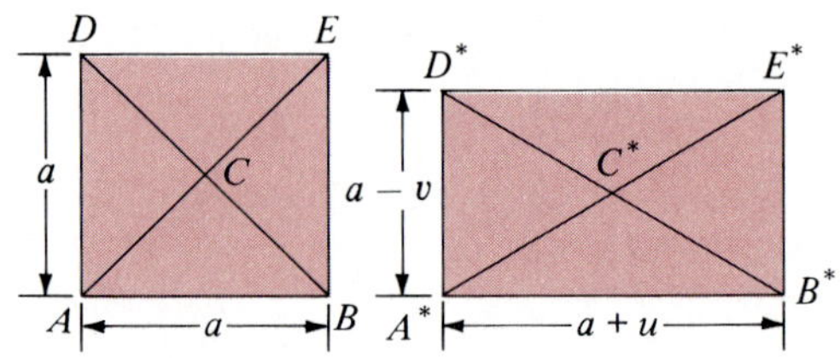

P. 3.36

3.37 Determine the shear strain between the perpendicular s- and t- directions at point C due to the uniform deformation shown.

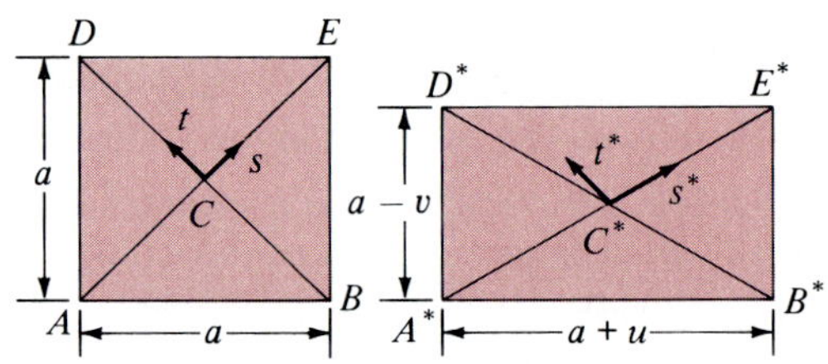

P. 3.37

3.38 Repeat problem 3.37 for the uniform deformation shown.

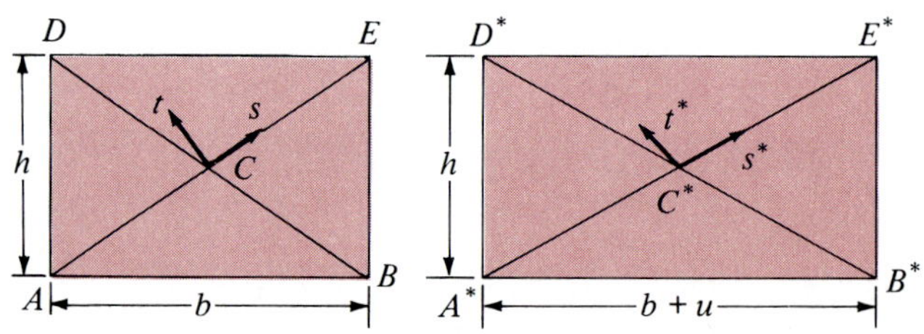

P. 3.38

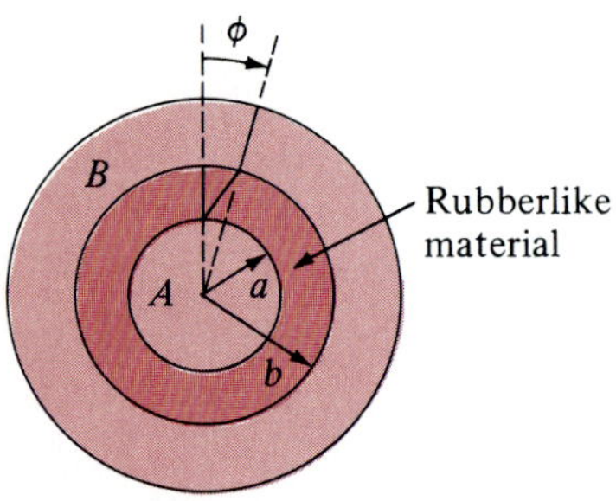

P. 3.39

3.39 A torsional spring consists of a rigid inner shaft A and a rigid outer tube B, and is filled with a rubberlike material as shown. The outer tube is rotated through an angle ϕ relative to the inner tube with each line that was originally a radius remaining straight as indicated. Determine the shear strain at a position r in the rubber.

3.40 A thin-walled cylinder is deformed by rotating one end through an angle ϕ as shown. A plane section perpendicular to the axis remains plane as a result of the deformation. Rectangular segments on the surface deform into rhombical segments as indicated. Develop an expression for the shear strain on the surface of the cylinder.

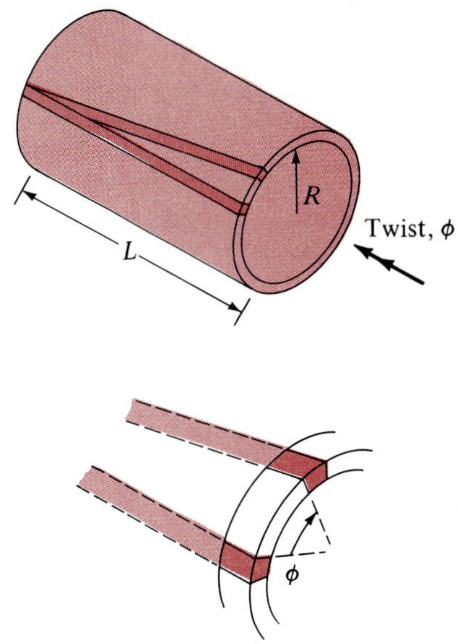

P. 3.40

3.41 An annular region is deformed as shown with the only displacement being in the z-direction. Assuming that each annular plane surface deforms into a conical surface as shown, determine the shear strain γ_{rz} in terms of r, a, b, and Δ.

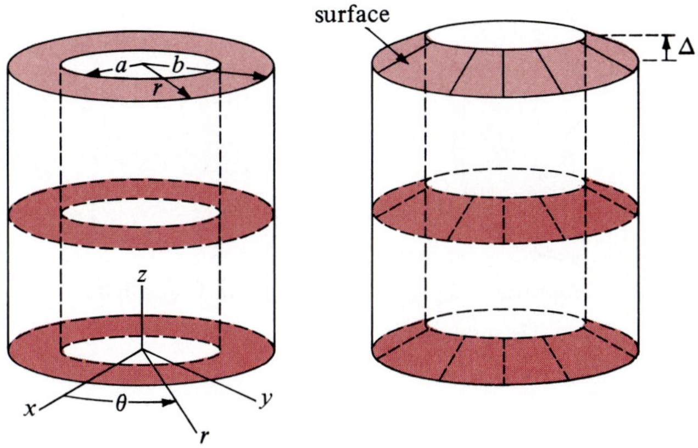

P. 3.41

3.4 GENERAL STRAIN-DISPLACEMENT RELATIONS

The concepts of extensional and shear strains were defined and illustrated in the previous section. The next step in the discussion of kinematics is to relate the strains to the displacements. This step is necessary in order to ultimately be able to relate the stresses to the displacements. We will develop the strain-displacement relations for a two-dimensional situation in order to more easily discern the basic ideas. The deformations will be assumed to be such that the geometry of the deformed configuration is not significantly different than that of the undeformed configuration.

Consider the general two-dimensional body shown in Fig. 3.16. The deformation is assumed to consist only of displacements in the xy-plane; that is, $u = u(x, y)$, $v = v(x, y)$, and $w = 0$. The magnitudes of the displacements are assumed to be small compared to the size of the body. Thus the deformed configuration differs only slightly from the undeformed configuration. We focus on an enlarged portion of the body in the neighborhood of point P as indicated in Fig. 3.17.

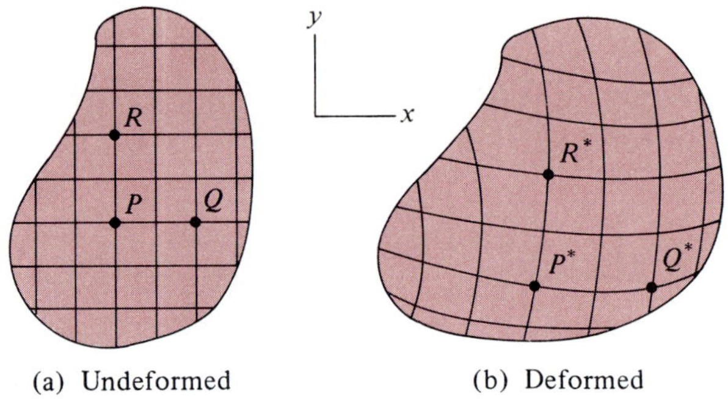

Figure 3.16 Deformation of a two-dimensional body

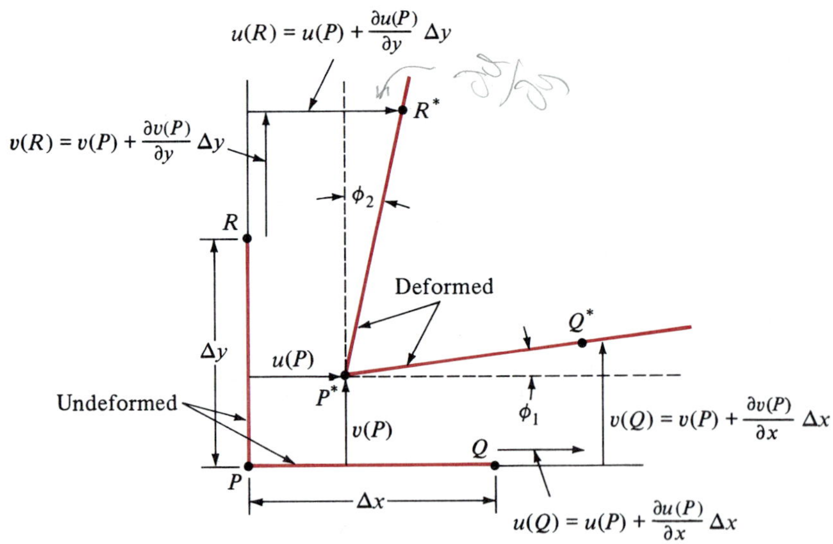

Figure 3.17 Enlarged view of a neighborhood of point P

The displacements at points P, Q, and R are expressed in a fashion that accounts for the possibility of deformation as well as any rigid-body motion that may be present. The extensional strain in the x-direction at P is defined as

$$\varepsilon_x(P) = \lim_{Q \to P} \frac{P^*Q^* - PQ}{PQ} \tag{3.4}$$

Referring to Fig. 3.17, the length P^*Q^* can be represented as

$$P^*Q^* = \sqrt{\left[\left(1 + \frac{\partial u}{\partial x}\right) \Delta x\right]^2 + \left(\frac{\partial v}{\partial x} \Delta x\right)^2}$$

which on the basis of the small-deformation assumption can be simplified, using the binomial expansion, to

$$P^*Q^* = \left(1 + \frac{\partial u}{\partial x}\right) \Delta x$$

Using this in Eqn. (3.4) yields

$$\varepsilon_x(P) = \frac{\partial u}{\partial x}$$

The symbol $\varepsilon_x(P)$ is to indicate the **extensional strain at P in the x-direction,** and is usually written more concisely as ε_x, or

$$\varepsilon_x = \frac{\partial u}{\partial x}$$

The extensional strain in other directions at P is in general different than in the x-direction.

In a completely similar fashion, the extensional strain at P in the y-direction is defined as

$$\varepsilon_y(P) = \lim_{R \to P} \frac{P^*R^* - PR}{PR} \tag{3.5}$$

The length P^*R^* can be represented as

$$P^*R^* = \sqrt{\left[\left(1 + \frac{\partial v}{\partial y}\right) \Delta y\right]^2 + \left(\frac{\partial u}{\partial y} \Delta y\right)^2}$$

which on the basis of the small-deformation assumption can be simplified, using the binomial expansion, to

$$P^*R^* = \left(1 + \frac{\partial v}{\partial y}\right) \Delta y$$

Using this expression in Eqn. (3.5) yields

$$\varepsilon_y(P) = \frac{\partial v}{\partial y}$$

frequently written as

$$\varepsilon_y = \frac{\partial v}{\partial y}$$

The shear strain at P for the directions x and y is the decrease in the right angle QPR as a result of the deformation, and is defined as

$$\gamma_{xy}(P) = \lim_{\substack{Q\to P \\ R\to P}} [\text{angle } QPR - \text{angle } Q^*P^*R^*] \tag{3.6}$$

Angle $Q^*P^*R^*$ is given by

$$\text{angle } Q^*P^*R^* = \frac{\pi}{2} - \phi_1 - \phi_2$$

with the angles ϕ_1 and ϕ_2 given by

$$\tan \phi_1 = \frac{\dfrac{\partial v}{\partial x}\Delta x}{\left(1 + \dfrac{\partial u}{\partial x}\right)\Delta x} \qquad \text{and} \qquad \tan \phi_2 = \frac{\dfrac{\partial u}{\partial y}\Delta y}{\left(1 + \dfrac{\partial v}{\partial y}\right)\Delta y}$$

For small deformations these can be replaced by

$$\phi_1 = \frac{\partial v}{\partial x} \qquad \text{and} \qquad \phi_2 = \frac{\partial u}{\partial y}$$

Eqn. (3.6) for the shear strain can then be expressed in terms of the displacements as

$$\gamma_{xy}(P) = \frac{\partial u}{\partial y} + \frac{\partial v}{\partial x}$$

$\partial v/\partial x$ represents the change in angle associated with the deformation of the x-axis whereas $\partial u/\partial y$ represents the change in angle associated with the deformation of the y-axis.

The collection

$$\varepsilon_x = \frac{\partial u}{\partial x} \qquad \varepsilon_y = \frac{\partial v}{\partial y} \qquad \text{and} \qquad \gamma_{xy} = \gamma_{yx} = \frac{\partial u}{\partial y} + \frac{\partial v}{\partial x}$$

constitute the two-dimensional strain-displacement relations referred to the xy coordinates.

In a general three-dimensional situation where all three displacement components u, v, and w in the x-, y-, and z- directions, respectively, are present, there are six independent strains, three extensional strains and three shear strains given by

$$
\begin{aligned}
\varepsilon_x &= \frac{\partial u}{\partial x} & \gamma_{xy} &= \gamma_{yx} = \frac{\partial u}{\partial y} + \frac{\partial v}{\partial x} \\
\varepsilon_y &= \frac{\partial v}{\partial y} & \gamma_{yz} &= \gamma_{zy} = \frac{\partial v}{\partial z} + \frac{\partial w}{\partial y} \\
\varepsilon_z &= \frac{\partial w}{\partial t} & \gamma_{zx} &= \gamma_{xz} = \frac{\partial w}{\partial x} + \frac{\partial u}{\partial z}
\end{aligned}
$$

where $u(x, y, z)$, $v(x, y, z)$, and $w(x, y, z)$ are three displacements referred to the xyz system. The three-dimensional strain-displacement relations are also based on the assumption that the deformed configuration differs only slightly from the undeformed configuration.

Example 3.9 A block is deformed as indicated in Fig. 3.18. The displacements are given by $u = U_0(x/L)$, $v = w = 0$. A plane that is perpendicular to the x-axis in the undeformed block translates in the x-direction with no point in the plane experiencing any other displacement. This special deformation is referred to as **uniform extension** (in the x-direction). The strains are easily determined to be $= \partial u/\partial x = U_0/L$, with all the rest of the strain components zero.

Note that the displacement of the right end of the bar is U_0 so that the final length of the bar is $L + U_0$. The average extensional strain is computed to be

$$
(\varepsilon_x)_{\text{avg}} = \frac{(U_0 + L) - L}{L} = \frac{U_0}{L}
$$

showing why this deformation is referred to as uniform extension.

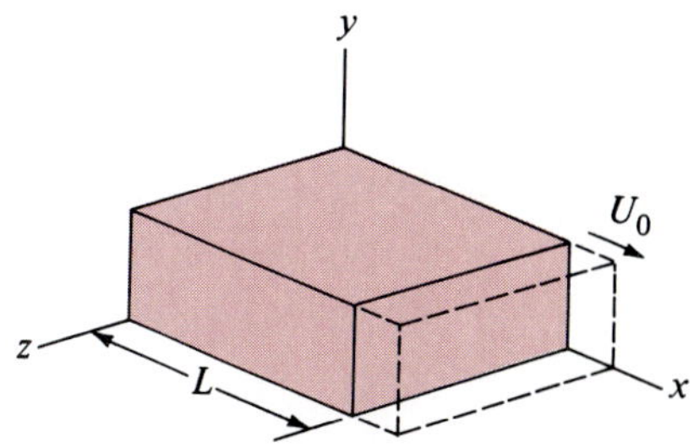

Figure 3.18 Uniform extension in the x-direction

Example 3.10 A block is deformed as indicated in Fig. 3.19. The deformation consists of a displacement in the y-direction given by $v = V_0(z/h)$. This special deformation is known as **pure shear.** The displacements in the x- and z-directions are both zero. For small V_0, the shear strain is computed as $\gamma_{yz} = \partial v/\partial z + \partial w/\partial y = V_0/h$, with all the other strain components zero.

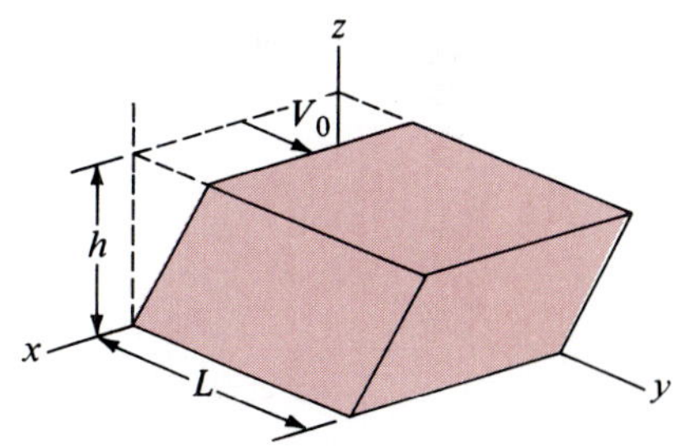

Figure 3.19 Uniform shear of a block

Several of the theories we will develop and use in this text are based on the assumption that the deformations are similar to these simple extension and shear deformations. The student should study these simple deformations and be familiar with their basic characteristics.

PROBLEMS

3.42 A square is elongated uniformly as indicated with no displacements in the vertical direction. Determine (*a*) the average extensional strains ε_x, ε_y, and γ_{xy} at a point in the block, (*b*) the average extensional strain of both the diagonals, and (*c*) the shear strain between the diagonals at Q. Take e/L to be small.

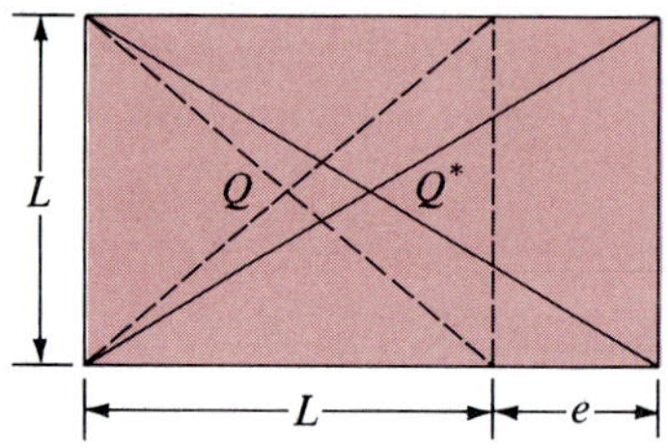

P. 3.42

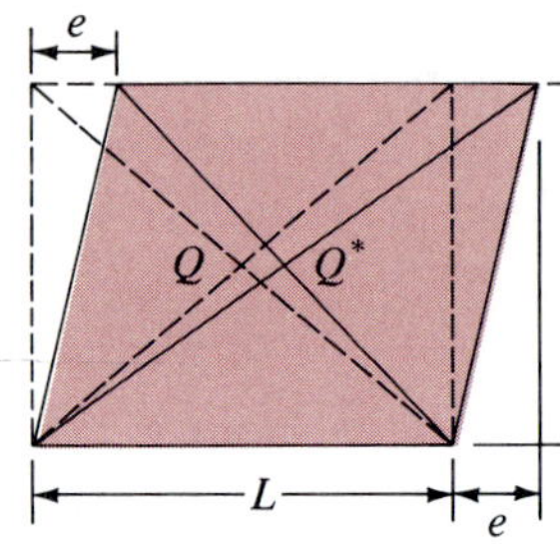

P. 3.43

3.43 The square is deformed into the rhombus as shown. There is no displacement in the y-direction. Determine (*a*) the average extensional strains ε_x, ε_y, and γ_{xy} at a point in the block, (*b*) the average extensional strain of both the diagonals, and (*c*) the shear strain between the diagonals at Q. Take e/L to be small.

3.44 The rectangular plate shown is deformed with the x and y components of displacement given by $u = \alpha xy/bh$ and $v = \alpha x/b$. Sketch the deformed configuration of the plate and determine the strains ε_x, ε_y, and γ_{xy}.

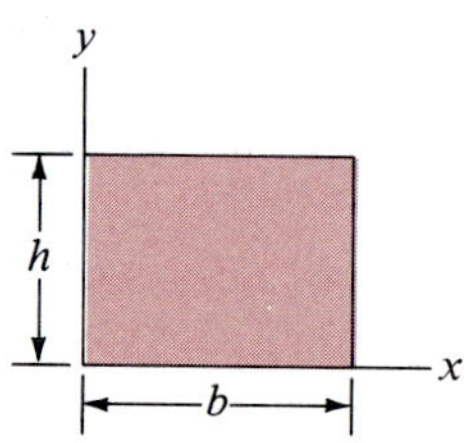

P. 3.44

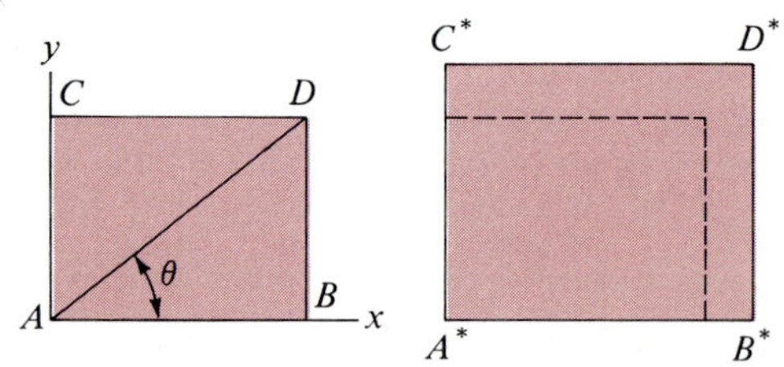

P. 3.45

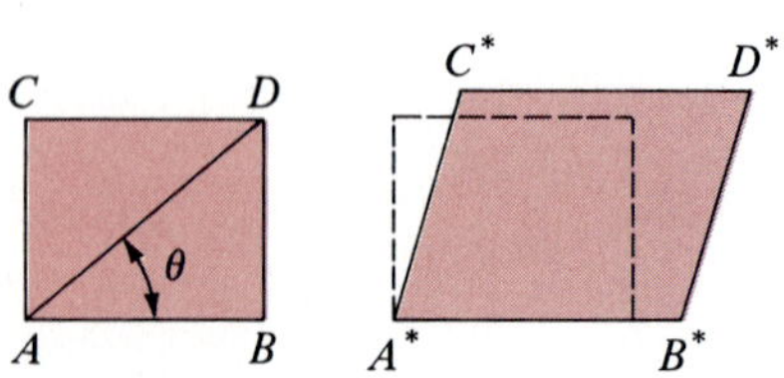

P. 3.46

3.45 A rectangle is deformed into another rectangle as shown. The average extensional strain for lines AB and CD is ε_1 and for lines AC and BD is ε_2. Show that the average extensional strain of the diagonal AD is given by $\varepsilon_1 \cos^2\theta + \varepsilon_2 \sin^2\theta$.

3.46 A rectangle is deformed uniformly into a rhombus as shown. The average extensional strain for lines AB and AC is ε_1 and for lines AC and BD is ε_2. The shear strain between lines AB and AC is γ. Show that the average extensional strain of line AD is given by $\varepsilon_{AD} = \varepsilon_1 \cos^2\theta + \varepsilon_2 \sin^2\theta + \gamma \sin\theta \cos\theta$.

3.47 A square plate is deformed uniformly as shown. There is no displacement of any point in the y-direction and all vertical lines remain straight as shown. Determine ε_x, ε_y, and γ_{xy} as functions of x and y at any point in the plate in terms of the parameters e and L.

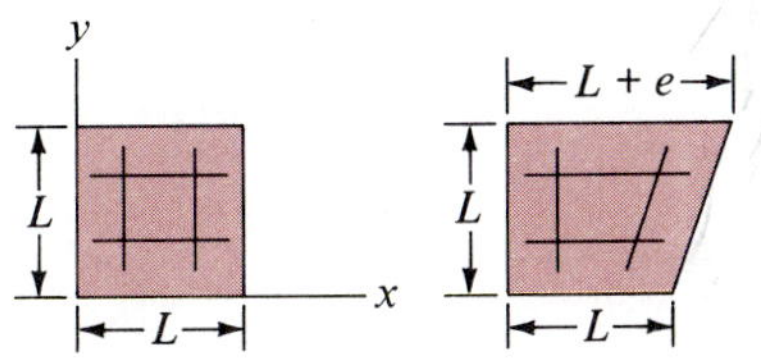

P. 3.47

3.5 TRANSFORMATION OF STRAIN

Strain transformations are a part of the general topic of kinematics and as such are related to the geometry of deformations. There are several specific reasons for looking at the transformation of strain. One is that, as indicated in Fig. 3.20, not everyone formulating a particular problem will choose the same coordinate system for representing the displacements, strains, and stresses. It might be necessary to relate the variables in the xyz coordinate system to the corresponding variables in the $x'y'z'$ coordinate system. Another is that the understanding and use of strain transformations are integral parts of certain strain measurement techniques that will be discussed later in the chapter. With these ideas in mind let us develop the necessary transformations between the components of strain in different coordinate systems.

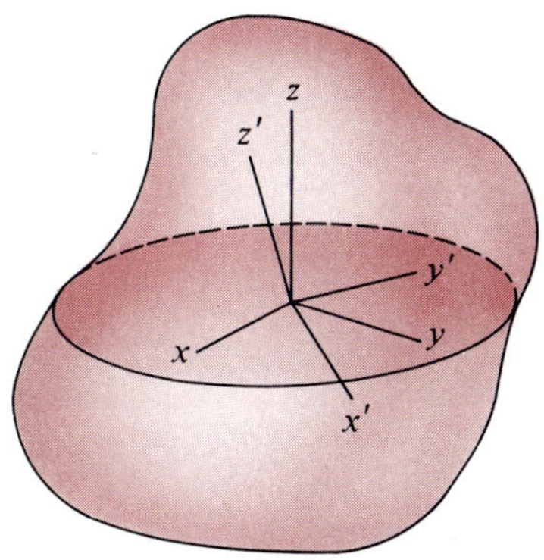

Figure 3.20 Different choices for the coordinate system

Consider a point P in a two-dimensional body as shown in Fig. 3.21. The $x'y'$ system here can be thought of as having rotated the xy system through a positive angle θ around the z-axis. This rotation is an example of an **orthogonal transformation** familiar to the student from linear algebra.

The strains ε_x, ε_y, and γ_{xy} are assumed to be known at P. The idea of strain transformation allows us to determine the corresponding value $\varepsilon_{x'}$, $\varepsilon_{y'}$, and $\gamma_{x'y'}$, measured in the primed coordinate system, that is, in the $x'y'$ system, in terms of the known strains ε_x, ε_y, and γ_{xy}, measured in the xy coordinate system. We accomplish this by using the relationships between strains and the accompanying changes in size and shape associated with the two coordinate systems.

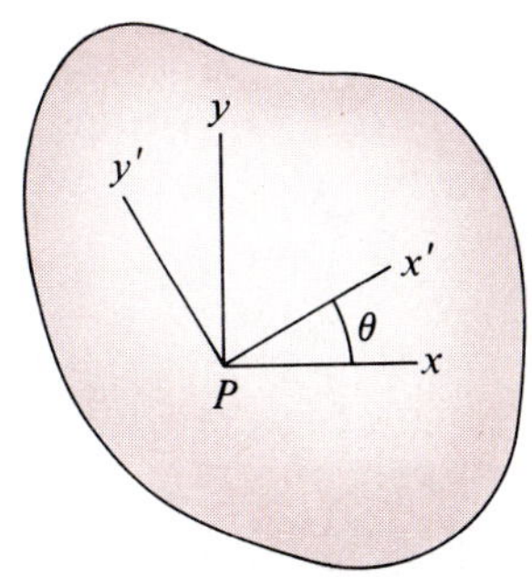

Figure 3.21 Transformation of strain in two dimensions

Shown in Fig. 3.22 is an infinitesimal block showing the effects of the positive strains ε_x, ε_y, and γ_{xy}.

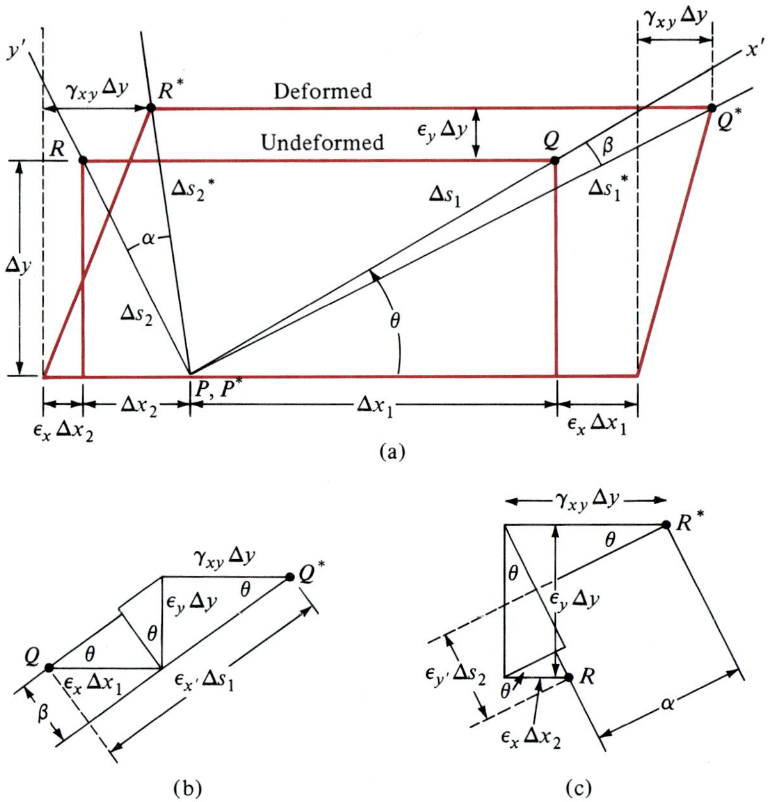

Figure 3.22 Strains, extensions, and shears at a point

As indicated in the figure, the undeformed and deformed configurations are juxtaposed so that P and P^*, as well as the initial and final x-axes, coincide. This involves the rigid-body motions which translate P to P^*, and the rotation of the deformed x-axis into coincidence with the undeformed x-axis. Rigid-body motions neither produce strains in an undeformed body nor do they alter the strains in a body that has already undergone deformations, so that an analysis of the strains using Fig. 3.22 will yield correct conclusions regarding the state of strain at P.

The definition for the strain in the x-direction is

$$\varepsilon_x = \lim_{\Delta x \to 0} \frac{\Delta x^* - \Delta x}{\Delta x}$$

from which $\Delta x^* - \Delta x \approx \varepsilon_x \Delta x$. We specialize this to the portions Δx_1 and Δx_2 of the block as

$$\Delta x_1^* - \Delta x_1 \approx \varepsilon_x \, \Delta x_1 \qquad \text{and} \qquad \Delta x_2^* - \Delta x_2 \approx \varepsilon_x \, \Delta x_2$$

as shown in Fig. 3.22a. Similarly, the change in length of the block in the y-direction is $\varepsilon_y \, \Delta y$. As a result of the rigid-body motion discussed above, the deformation due to the positive shear strain γ_{xy} is as shown. This reduction in the angle γ_{xy} between the x- and y-axes produces the displacement $\gamma_{xy}\Delta y$ as shown. These three deformations each contribute to the changes in the lengths of line segments located along the x'- and y'-axes as well as to a change in the angle between the x'- and y'-axes.

Still referring to Fig. 3.22a, the change in length along the x'-axis is related to the original and final lengths of the line segment Δs_2 and Δs_2^* according to

$$\Delta s_1^* = \Delta s_1 \, (1 + \varepsilon_{x'}) = \Delta s_1 + \Delta s_1 \, \varepsilon_{x'}$$

or

$$\Delta s_1^* - \Delta s_1 = \varepsilon_{x'} \, \Delta s_1$$

The elongation $\Delta s_1 \, \varepsilon_{x'}$ is also equal to the sum of the components of the elongations due to the strains ε_x, ε_y, and γ_{xy} along the diagonal. Referring to Fig. 3.22b, this is written as

$$\varepsilon_{x'} \, \Delta s_1 = \varepsilon_x \, \Delta x_1 \cos\theta + \varepsilon_x \, \Delta x_1 \sin\theta + \gamma_{xy} \, \Delta y \cos\theta$$

Dividing by Δs_1 and recognizing that $\Delta x_1/\Delta s_1 = \cos\theta$ and $\Delta y/\Delta s_1 = \sin\theta$ leads to

$$\varepsilon_{x'} = \varepsilon_x \cos^2\theta + \varepsilon_y \sin^2\theta + \gamma_{xy} \sin\theta \cos\theta$$

Referring to Fig. 3.22c, a similar treatment of the strain in the y' direction produces

$$\varepsilon_{y'} = \varepsilon_x \sin^2\theta + \varepsilon_y \cos^2\theta - \gamma_{xy} \sin\theta \cos\theta$$

Referring to Fig. 3.22a, the change in the angle between the x'- and y'-axes is given by

$$\gamma_{x'y'} = \alpha - \beta$$

From Fig. 3.22c, the angle α and the contributions to the x' motion of point R are related by

$$\alpha \, \Delta s_2^* \approx \alpha \, \Delta s_2 = -\varepsilon_x \, \Delta x_2 \cos\theta + \varepsilon_y \, \Delta y \sin\theta + \gamma_{xy} \, \Delta y \cos\theta$$

Dividing by Δs_2 and using $\Delta x_2/\Delta s_2 = \sin\theta$ and $\Delta y/\Delta s_2 = \cos\theta$, yields

$$\alpha = -\varepsilon_x \sin\theta\cos\theta + \varepsilon_y \sin\theta\cos\theta + \gamma_{xy}\cos^2\theta$$

Similarly, from Fig. 3.22b the angle β and the contributions to the y' motion of point Q are related by

$$\beta\,\Delta s_1^* \approx \beta\Delta s_1 = \varepsilon_x\,\Delta x_1 \sin\theta - \varepsilon_y\,\Delta y\cos\theta + \gamma_{xy}\,\Delta_y \sin\theta$$

Dividing by Δs_1 yields

$$\beta = \varepsilon_x \cos\theta\sin\theta - \varepsilon_y \sin\theta\cos\theta + \gamma_{xy}\sin^2\theta$$

The shear strain $\gamma_{x'y'}$ then becomes

$$\gamma_{x'y'} = \alpha - \beta = -2(\varepsilon_x - \varepsilon_y)\sin\theta\cos\theta + \gamma_{xy}(\cos^2\theta - \sin^2\theta)$$

Collectively the **strain transformation equations** can be written as

$$\varepsilon_{x'} = \varepsilon_x\cos^2\theta + \varepsilon_y\sin^2\theta + \gamma_{xy}\sin\theta\cos\theta \tag{3.7a}$$

$$\varepsilon_{y'} = \varepsilon_x\sin^2\theta + \varepsilon_y\cos^2\theta - \gamma_{xy}\sin\theta\cos\theta \tag{3.7b}$$

$$\gamma_{x'y'} = -2(\varepsilon_x - \varepsilon_y)\sin\theta\cos\theta + \gamma_{xy}(\cos^2\theta - \sin^2\theta) \tag{3.7c}$$

Given the strains ε_x, ε_y, and γ_{xy} referred to the xy-axes, the corresponding strains $\varepsilon_{x'}$, $\varepsilon_{y'}$, and $\gamma_{x'y'}$ referred to the $x'y'$ system can be computed using the strain transformation Eqns. (3.7).

On the basis of the fact that they behave according to the strain transformation Eqns. (3.7), the collection of strains

$$\begin{matrix} \varepsilon_x & \dfrac{\gamma_{xy}}{2} \\ \dfrac{\gamma_{xy}}{2} & \varepsilon y \end{matrix}$$

is referred to as the two-dimensional strain tensor. Note that the **form** of the transformation equations is exactly the same as the corresponding stress transformation Eqns. (2.4–2.6).

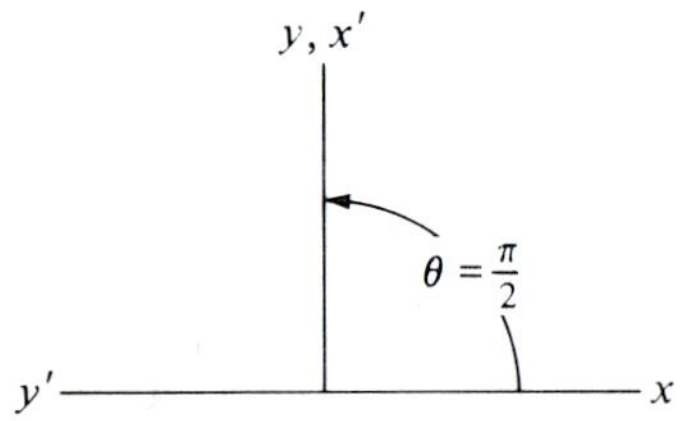

Figure 3.23 Axes rotated 90°

As some simple checks on our development consider the following. Take $\theta = 90°$, corresponding to rotation of the axes 90° as indicated in Fig. 3.23.

The transformation equations should state that with respect to the $x'y'$ system, $\varepsilon_{x'} = \varepsilon_y$ and $\varepsilon_{y'} = \varepsilon_x$ and, since a decrease in the angle between the x'- and y'-axes corresponds to an increase in the angle between the x- and y-axes, $\gamma_{x'y'} = -\gamma_{xy}$. Substituting $\theta = 90°$ into the transformation Eqns. (3.7) yields precisely these results. Notice that if Eqn (3.7a) is evaluated for $\theta = 45°$ there results

$$\varepsilon_{x'} = \frac{(\varepsilon_x + \varepsilon_y + \gamma_{xy})}{2}$$

from which

$$\gamma_{xy} = 2\varepsilon_{x'} - \varepsilon_x - \varepsilon_y$$

which relates the extensional strains in three different directions at a point to the shear strain at the point. This relationship is very important in the experimental measurement of strain and will be discussed in detail in a later section.

Example 3.11 Consider a uniform extensional deformation such as that introduced in Example 3.9. Take the dimensions and deformation of the block as shown. Determine the strains in the $x'y'$ coordinate system in Fig. 3.24b.

Solution First, none of the essentials of this deformation is excluded by considering the view along the z-axis as shown in Fig. 3.24a. From Example 3.9, $\varepsilon_x = U_0/L = 0.01$ in./in. with all the other strains as zero. We will determine the strains in the $x'y'$ system, first using the basic approach of determining changes in geometry, and then using Eqns. (3.7).

Applying basic principles and definitions, let us determine $\varepsilon_{x'}$: Referring to Fig. 3.24b, the lengths of the diagonal AB before and after deformation are given by

$$L_{AB} = \sqrt{3^2 + 2^2}\text{ in.} = 3.6056\text{ in.}$$

$$L_{AB}^* = \sqrt{3.03^2 + 2^2}\text{ in.} = 3.6306\text{ in.}$$

so that

$$\varepsilon_{x'} = (3.6306 - 3.6056)\text{ in.}/3.6056\text{ in.} = 0.0069\text{ in./in.}$$

To determine $\varepsilon_{y'}$: look at the undeformed and deformed lengths of line CD. Since the extension is uniform, the displacement of point D is ratioed to be

$$u_D = \left(\frac{4/3}{3}\right)(0.003)\text{ in.} = 0.0133\text{ in.}$$

Thus

$$L_{CD} = \sqrt{2^2 + \left(\frac{4}{3}\right)^2} = 2.4037\text{ in.}$$

and

$$L_{CD}^* = \sqrt{2^2 + \left(\frac{4}{3} + 0.0133\right)^2} = 2.4111\text{ in.}$$

with

$$\varepsilon_{y'} = (2.4111 - 2.4037)\text{ in.}/2.4037\text{ in.} = 0.0031\text{ in./in.}$$

To determine $\gamma_{x'y'}$: The shear strain for directions $x'y'$ is determined by evaluating the angles through which these directions rotate.

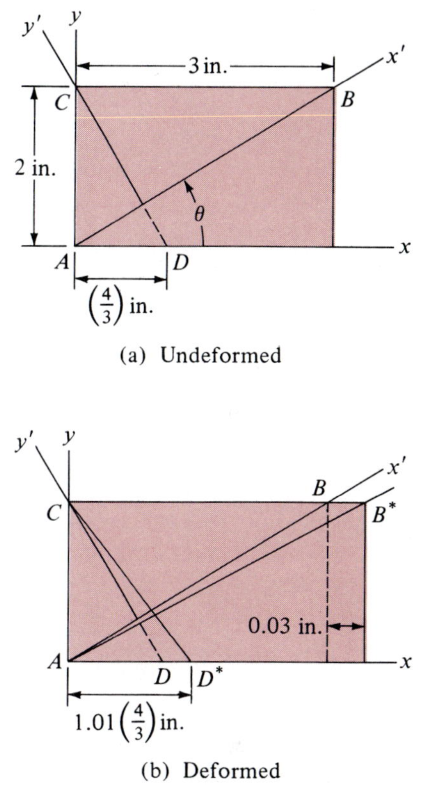

Figure 3.24 Deformations and strains for uniform extension

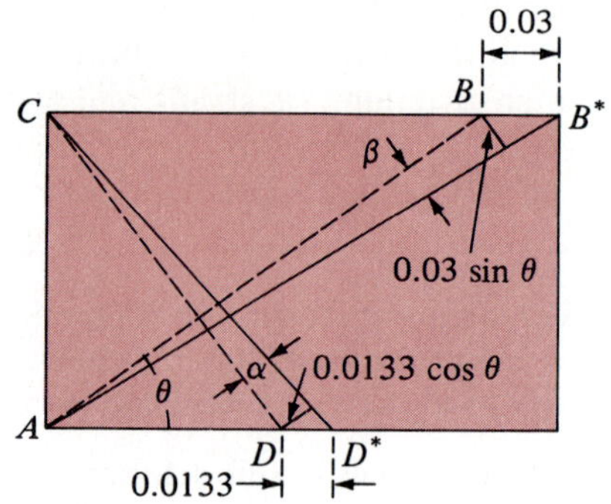

Figure 3.25 Angular changes due to the deformation

Referring to Fig. 3.25, the angle α through which line CD rotates is given by

$$2.4037 \text{ in.} \cdot \alpha = 0.0133 \cos\theta = 0.0111 \text{ in.}$$

or

$$\alpha = 0.0046 \text{ in./in.}$$

Similarly, referring to Fig. 3.25, β is given by the equation

$$3.6056 \text{ in.} \cdot \beta = 0.03 \sin\theta = 0.0166 \text{ in.}$$

or

$$\beta = 0.0046 \text{ in./in.}$$

Both of these angles contribute to an increase in the angle between the x'- and y'-axes so that

$$\gamma_{x'y'} = (-0.0046 - 0.0046) \text{ in./in.} = -0.0092 \text{ in./in.}$$

Now let us solve the problem using Eqns (3.7). The angle θ between the two coordinate systems is given by $\tan\theta = 2/3$ from which $\cos\theta = 0.8321$ and $\sin\theta = 0.5546$. Substitution into Eqns. (3.7) yields

$$\varepsilon_{x'} = 0.01(0.8321)^2 + 0 = 0.0069 \text{ in./in.}$$

$$\varepsilon_{y'} = 0.01(0.5546)^2 + 0 = 0.0031 \text{ in./in.}$$

$$\gamma_{x'y'} = -2(0.01)(0.8321)(0.5546) = -0.0092 \text{ in./in.}$$

This coincides with the results obtained using the first approach.

Note that by adding the first two of Eqns. 3.7 there results

$$\varepsilon_{x'} + \varepsilon_{y'} = \varepsilon_x + \varepsilon_y$$

This is referred to as the first strain invariant. It clearly has the same form as the corresponding relationship between the stresses in rotated coordinate systems. There is another strain invariant that the student is asked to investigate in the problems.

PROBLEMS

The point of the next three problems is twofold; first, to get a better feel for the relationships between deformations and strain, and second, to appreciate the fact that the strains at a point depend upon the coordinate system in which they are expressed.

3.48 For the deformation indicated, namely, $\varepsilon_x = e_1 > 0$, $\varepsilon_y = \gamma_{xy} = 0$, use geometry to determine the extensional strain of line segments oriented along the s and t directions and the shear strain between the s and t directions. Take $e_1 << 1$.

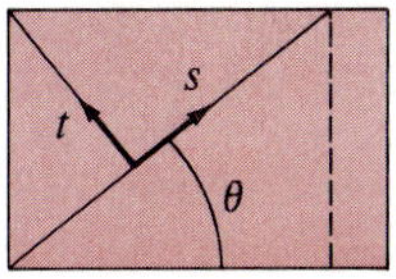

$\epsilon_x = e_1 > 0$

P. 3.48

3.49 Repeat problem 3.48 for the case $\varepsilon_y = e_2 > 0$, $\varepsilon_x = \gamma_{xy} = 0$. Take $e_2 << 1$.

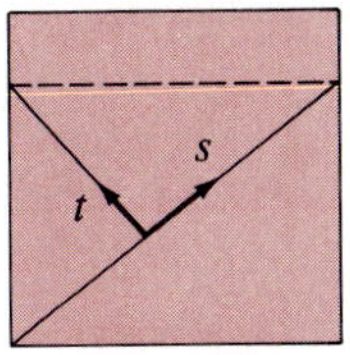

$\epsilon_y = e_2 > 0$

P. 3.49

3.50 Repeat problem 3.48 for the case $\gamma_{xy} = \gamma > 0$, $\varepsilon_x = \varepsilon_y = 0$. Take $\gamma << 1$.

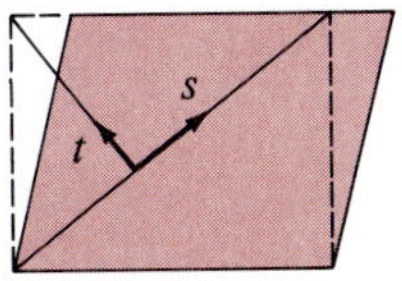

$\gamma_{xy} = \gamma > 0$

P. 3.50

The student should observe that the superposition of the states of strain in the previous three problems, which is valid for small strains, leads to the general transformation equations presented in Section 3.5.

3.51 Using the strain transformation equations (3.7) show that $\varepsilon_{x'} + \varepsilon_{y'} = \varepsilon_x + \varepsilon_y$. This is the first strain invariant for the strains in a two-dimensional situation.

3.52 Using the strain transformation equations (3.7) show that $\varepsilon_{x'} \cdot \varepsilon_{y'} - \gamma_{x'y'}^2/4 = \varepsilon_x \cdot \varepsilon_y - \gamma_{xy}^2/4$. This is the second strain invariant for the strains in a two-dimensional situation.

3.6 PRINCIPAL STRAINS

If the first and last of the strain transformation equations given by Eqns. (3.7) in the previous section are rewritten using the double-angle formulae of trigonometry, there results

$$\varepsilon_{x'} = \frac{\varepsilon_x + \varepsilon_y}{2} + \frac{\varepsilon_x - \varepsilon_y}{2} \cos 2\theta + \frac{\gamma_{xy}}{2} \sin 2\theta$$

$$\frac{\gamma_{x'y'}}{2} = -\frac{\varepsilon_x - \varepsilon_y}{2} \sin 2\theta + \frac{\gamma_{xy}}{2} \cos 2\theta$$

The stress transformation equations given by Eqns. (2.7) and (2.9) are

$$\sigma_{x'} = \frac{\sigma_x + \sigma_y}{2} + \frac{\sigma_x - \sigma_y}{2} \cos 2\theta + \tau_{xy} \sin 2\theta$$

$$\tau_{x'y'} = -\frac{\sigma_x - \sigma_y}{2} \sin 2\theta + \tau_{xy} \cos 2\theta$$

These have essentially the same **form** as the strain transformation equations with σ_x and σ_y replaced by ε_x and ε_y, and with τ_{xy} replaced with $\gamma_{xy}/2$. Thus all the results pertaining to the discussion of principal stresses, principal planes, and the maximum shear stress can be applied to the corresponding problem of principal strains, principal planes, and the maximum shear strain if σ_x is replaced by ε_x, σ_y is replaced by ε_y, and τ_{xy} is replaced by $\gamma_{xy}/2$.

Recalling the principal stress results from Section 2.6, it follows that directions of the axes for which the strains are principal, that is, the **principal planes**, are given by the roots of

$$\tan 2\theta = \frac{\gamma_{xy}}{\varepsilon_x - \varepsilon_y} \tag{3.8}$$

The two roots given by $2\theta_1$ and $2\theta_2$ differ by π so that the values θ_1 and θ_2 differ by $\pi/2$. The corresponding strain values for these directions are given by

$$\varepsilon = \frac{\varepsilon_x + \varepsilon_y}{2} \pm \sqrt{\left(\frac{\varepsilon_x - \varepsilon_y}{2}\right)^2 + \left(\frac{\gamma_{xy}}{2}\right)^2} \tag{3.9}$$

and are referred to as the **principal strains**. The corresponding shear strain between the two principal directions is zero.

The directions of the axes with respect to which the shear strain is maximum are given by

$$\tan 2\theta = -\frac{\varepsilon_x - \varepsilon_y}{\gamma_{xy}} \tag{3.10}$$

The two roots of this equation, denoted by $2\theta_3$ and $2\theta_4$, again differ by π, so that the two perpendicular directions between which the shear strain is maximum are 90° from the two perpendicular directions between which the shear strain is minimum.

Note also that the right-hand sides of Eqns. (3.8) and (3.10) are negative reciprocals, indicating that the directions of the principal planes and the directions for which the shear strain is a maximum are 45° apart.

The values of the corresponding maximum shear strains can be determined by substituting the solutions of Eqn. (3.8) into Eqn. (3.9) and are

$$\frac{\gamma_{\max,\min}}{2} = \pm \sqrt{\left(\frac{\varepsilon_x - \varepsilon_y}{2}\right)^2 + \left(\frac{\gamma_{xy}}{2}\right)^2} \tag{3.11}$$

The positive value is interpreted as the decrease in the angle, whereas the negative corresponds to the increase in the angle. As was the case for the stresses, there are extensional strains in the directions of the axes with respect to which the shear strains are a maximum or a minimum. The extensional strain in the direction of both of these axes is

$$\varepsilon = \frac{\varepsilon_x + \varepsilon_y}{2}$$

Fig. 3.26 shows the relationships between these sets of axes.

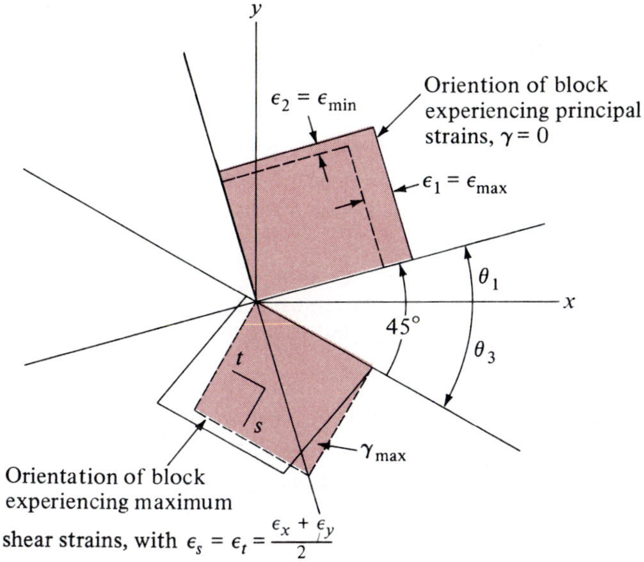

Figure 3.26 Relation between principal planes and planes of maximum shear strain

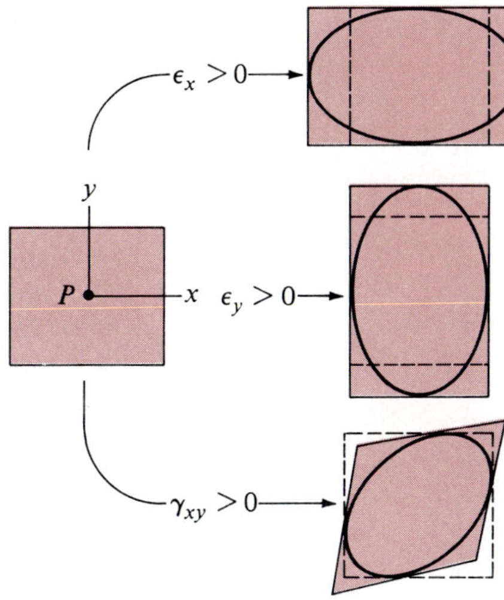

Figure 3.27 Deformations at *P*

Here is another way to picture the results of the relations between deformation and strain that helps in understanding the deformation in the neighborhood of a point. Consider an infinitesimal square in the neighborhood of a point P as indicated in Fig. 3.27. The deformed shapes of the block are shown for each of the three cases $\varepsilon_x > 0$, $\varepsilon_y > 0$, and $\gamma_{xy} > 0$. In each case a circle on the block is deformed into an ellipse. A circle on the block subjected simultaneously to all three deformations would also deform into an ellipse and appear as in Fig. 3.28.

The principal directions for strain correspond to the directions of the semimajor and semiminor axes of the ellipse. These axes correspond to the perpendicular directions in the block that experience no shear strain as a result of the deformation. This method of picturing the deformations at a point can be used effectively to understand the locations of the principal directions and the corresponding directions of maximum shear strain. This mode of representing the deformations at a point will be used frequently in the remainder of this chapter.

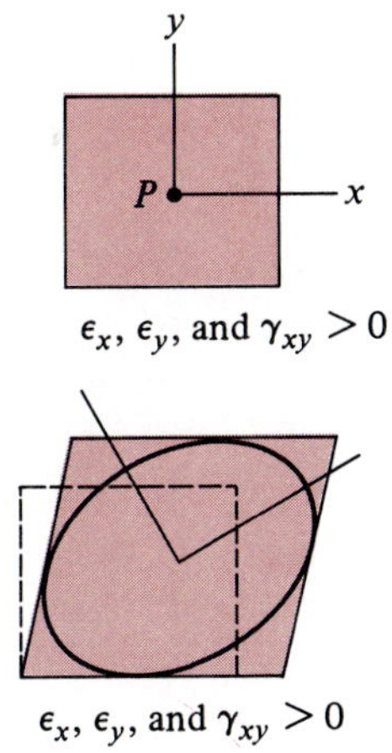

Figure 3.28 Block subjected to extensional and shear strains

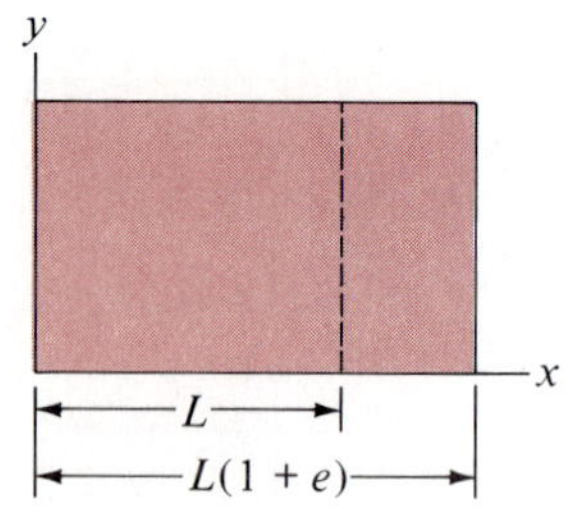

Figure 3.29 Uniform extension

Example 3.12 A block deformed by a uniform extension is shown in Fig. 3.29. Determine the principal strains and the maximum shear strain and sketch deformations of the corresponding blocks.

Solution Substituting into Eqn. (3.8) for the principal directions yields

$$\tan 2\theta = \frac{0}{e - 0} = 0$$

so that $\theta_1 = 0$ and $\theta_2 = \pi/2$, or the original planes. This result should be anticipated since there was no shear strain associated with the x- and y-axes. The principal strains are clearly $\varepsilon_1 = e$ and $\varepsilon_2 = 0$. The maximum shear strains occur where

$$\tan 2\theta = \frac{e - 0}{0}$$

from which $2\theta = \pi/2$, with $\theta = \pi/4$ and γ_{max} given by

$$\frac{\gamma_{max}}{2} = \frac{e}{2}$$

The extensions of these lines is $(e + 0)/2 = e/2$. All this information is shown in Fig. 3.30.

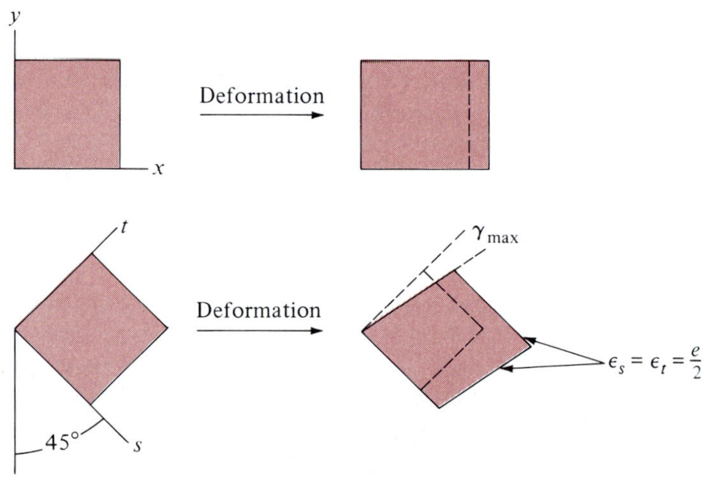

Figure 3.30 Principal strain and maximum shear strain information for uniform extension

The top two figures indicate that a block oriented with sides parallel to the xy-axes elongates in the x-direction with no strain in the y-direction nor any change in the angle between the x- and y-directions, that is, the principal planes. The bottom two figures indicate that a block oriented 45° to the xy-axes undergoes a shear strain of $\gamma_{xy} = e/2$ with the s- and t-axes both experiencing strains of $\varepsilon = +e/2$.

Example 3.13 A two-dimensional block deformed by pure shear given as $\gamma_{xy} = \gamma$, $\varepsilon_x = \varepsilon_y = 0$, is indicated in Fig. 3.31. Determine the principal strains and the maximum shear strain, and sketch deformations of the corresponding blocks.

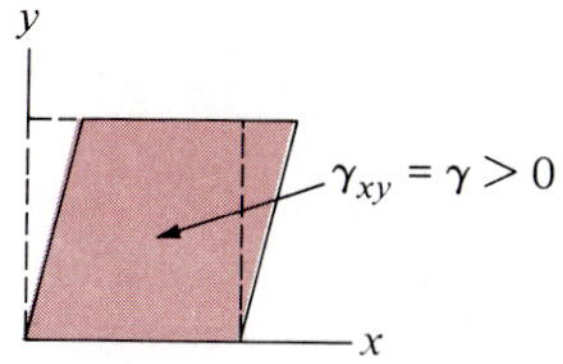

Figure 3.31 Deformation for pure shear

Solution Substituting into Eqn. (3.8) for the principal planes yields

$$\tan 2\theta = \gamma/(0 - 0) = 0$$

so that $2\theta = \pi/2$ with $\theta = \pi/4$. Using Eqn. (3.9) the principal strain for this direction is

$$\varepsilon_1 = \frac{0 + 0}{2} + \sqrt{\left(\frac{0 - 0}{2}\right)^2 + \left(\frac{\gamma}{2}\right)^2} = +\frac{\gamma}{2}$$

For the perpendicular direction

$$\varepsilon_2 = \frac{0 + 0}{2} - \sqrt{\left(\frac{0 - 0}{2}\right)^2 + \left(\frac{\gamma}{2}\right)^2} = -\frac{\gamma}{2}$$

Having computed ε_1, ε_2 could also have been computed using the first strain invariant $\varepsilon_x + \varepsilon_y = \varepsilon_1 + \varepsilon_2$. The maximum shear strains occur where

$$\tan 2\theta = \frac{0}{\gamma}$$

from which $\theta = 0$ and $\pi/2$, that is, the original directions. All this information is shown in Fig. 3.32.

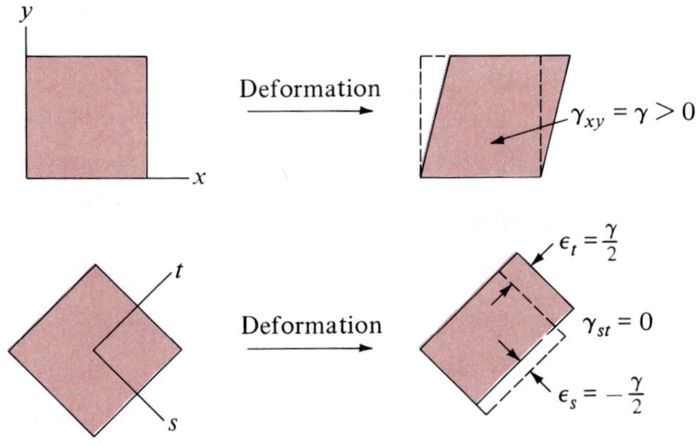

Figure 3.32 Principal strain and maximum shear strain information for pure shear

The upper two figures indicate that there is pure shear for the *xy*-axes, that is, no extensional strain in either of the *x*- or *y*-directions. The bottom two figures indicate the principal planes for which $\varepsilon_s = -\gamma/2$ and $\varepsilon_t = \gamma/2$, with $\gamma_{st} = 0$.

Example 3.14 At a point P in a two-dimensional solid, the strains are indicated as in Fig. 3.33. Determine the principal strains and the maximum shear strain, and sketch deformations of the corresponding blocks.

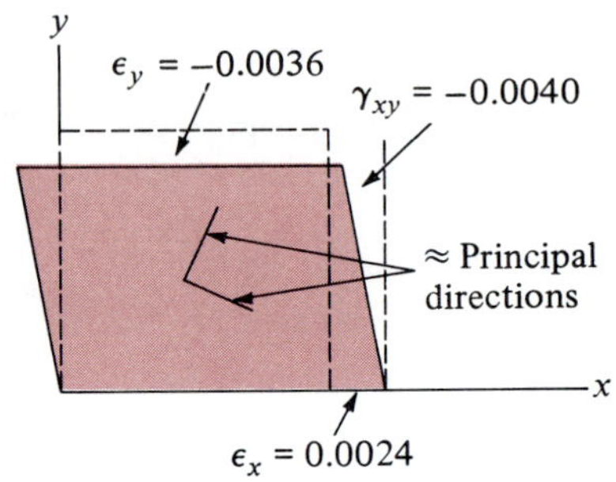

Figure 3.33 A two-dimensional deformation

Solution By sketching the deformed block, a rough idea as to the location of the principal directions emerges. Substituting into Eqn. (3.8) for the locations of the principal strains yields

$$\tan 2\theta = -\ 0.0040/0.0060 = -\ 0.6667$$

from which $2\theta = -33.69°$ or $\theta_1 = -16.85°$. The other principal direction is at right angles to the direction given by θ_1. The values of the principal strains are computed to be

$$\varepsilon_1 = [-0.0006 + \sqrt{0.0030^2 + (-0.0020)^2}]\ \text{in./in.} = 0.0030\ \text{in./in.}$$
$$\varepsilon_2 = [-0.0006 - \sqrt{0.0030^2 + (-0.0020)^2}]\ \text{in./in.} = -\ 0.0042\ \text{in./in.}$$

The location of the maximum shear strain is given by

$$\tan 2\theta = 0.0060/0.0040$$

from which $2\theta = 56.3°$ or $\theta = 28.15°$. The value of the maximum shear strain is given by

$$\gamma_{\max}/2 = \sqrt{0.0030^2 + (-0.0020)^2}\ \text{in./in.} = 0.0036\ \text{in./in.}$$

or

$$\gamma_{max} = 0.0072\ \text{in./in.}$$

and occurs between directions making angles of 45° with the principal directions. The corresponding extensional strain for these two axes is given by $\varepsilon_p = \varepsilon_q = -0.0006$ in./in. All this information is indicated in Fig. 3.34.

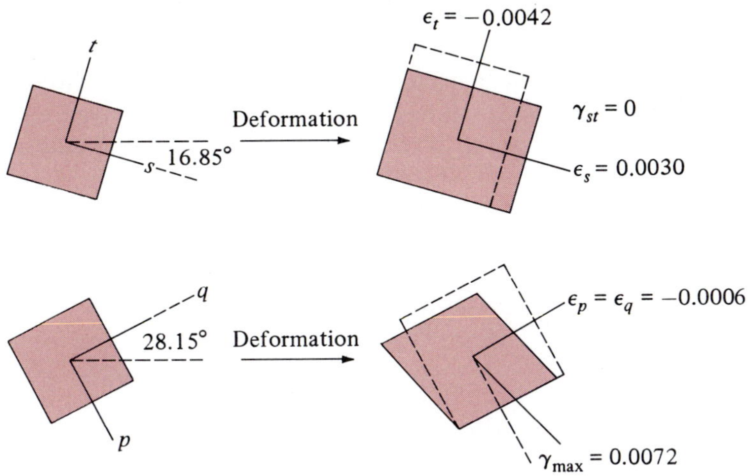

Figure 3.34 Location of principal directions

The upper two figures indicate the directions undergoing only extensional strain whereas the bottom two figures indicate the directions between which the shear strain is a maximum or minimum together with the corresponding mean extensional strain.

PROBLEMS

For each of the following, use the results of Section 3.6 to determine the principal strains and the corresponding maximum shear strain. First sketch the deformed block to get a rough idea as to the location of the principal directions. Use the analytical results to draw, for each problem, the properly oriented blocks that experience the principal strains and the maximum shear strains.

3.53 $\varepsilon_x = 0.000100$, $\varepsilon_y = 0.000060$, $\gamma_{xy} = 0.000080$.

3.54 $\varepsilon_x = 0.000120$, $\varepsilon_y = -0.000060$, $\gamma_{xy} = 0.000040$.

3.55 $\varepsilon_x = -0.000100$, $\varepsilon_y = -0.000060$, $\gamma_{xy} = -0.000080$.

3.56 $\varepsilon_x = -0.000080$, $\varepsilon_y = -0.000060$, $\gamma_{xy} = 0.000060$.

3.57 $\varepsilon_x = 0.000100$, $\varepsilon_y = 0.000100$, $\gamma_{xy} = 0.000100$.

3.58 $\varepsilon_x = -0.000100$, $\varepsilon_y = 0.000100$, $\gamma_{xy} = -0.000080$.

3.7 MOHR'S CIRCLE FOR TWO-DIMENSIONAL STATES OF STRAIN

The construction of the Mohr's circle for two-dimensional states of stress was based upon the double-angle form of the stress transformation equations. As we saw in the previous section, the strain transformation equations have precisely the same form as those for stress, with the important exception that the shear strain is divided by two everywhere in the equations. It follows then that the Mohr's circle for a two-dimensional state of strain is constructed using the following steps. Refer to Fig. 3.35.

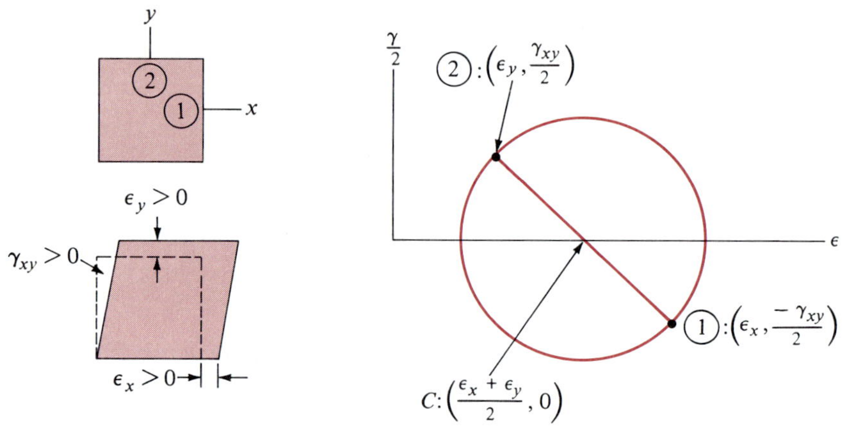

Figure 3.35 Block with directions labeled

1. Sketch the element showing the axes and the extensions ε_x, ε_y, and γ_{xy}. Carefully label the directions, such as directions 1 and 2, as shown.
2. Sketch the coordinate axes for the circle, calling them ε and $\gamma/2$ as shown.
3. Plot *and label* the points $(\varepsilon_x, -\gamma_{xy}/2)$ and $(\varepsilon_y, \gamma_{xy}/2)$ as shown. Signs must be observed: a negative strain must be plotted as such.
4. Draw the diameter of Mohr's circle between points 1 and 2 and sketch in or use a compass to construct the circle. This completes the construction of the circle.

Properties of Mohr's Circle for Two-Dimensional States of Strain (Refer to Fig. 3.36). First note that the points on Mohr's circle that correspond to the principal strains are designated as $P1$ and $P2$. The points that correspond to the maximum shear strain values are designated as $T1$ and $T2$. Referred to the diameter through point 1, the angles corresponding to the principal strains and maximum shear strain are shown as $2\theta_1$ and $2\phi_1$, respectively.

1. Each point $(\varepsilon, \gamma/2)$ on Mohr's circle represents the state of strain for a set of directions in the element. ε is the extensional strain for the

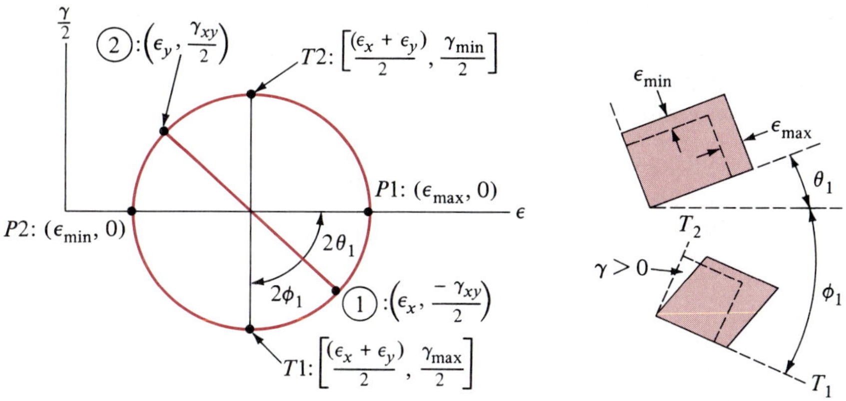

Figure 3.36 Properties of Mohr's circle

direction s in question and γ is the corresponding shear strain between the s- and t-directions, where t is the direction rotated 90° counterclockwise from s. Points 1 and 2 that were used to plot the circle are examples.

2. Rotation of the diameter of Mohr's circle through an angle 2θ corresponds to a rotation through an angle θ, in the same direction, of the corresponding direction in the element.

3. The values of the principal strains are given by the coordinates of the points on the ε-axis labeled $P1$ and $P2$.

4. The locations of the principal planes in the element are determined from Mohr's circle by rotating the directions 1 and 2 in the element through the angles θ_1 and $\theta_1 + \pi/2$, corresponding to the angles $2\theta_1$ and $2\theta_1 + \pi$ on the circle.

5. The values of the maximum shear strains are determined from the coordinates of the points at the top and bottom of the circle labeled $T1$ and $T2$.

6. The locations of the directions between which the shear strains in the element are a maximum, are determined from Mohr's circle by rotating the directions 1 and 2 in the element through angles ϕ_1 and $\phi_1 + \pi/2$ corresponding to the angles $2\phi_1$ and $2\phi_1 + \pi$ on the circle. The direction corresponding to point T_1 is given by the angle ϕ_1 as shown in Fig. 3.36. The positive shear strain associated with this direction and the T_2 direction, which is located 90° counterclockwise from T_1, is indicated by the fact that the angle between these two directions has decreased. The minimum shear strain associated with the T_2 direction and the direction 90° counterclockwise to the T_2 direction is negative, that is, the angle between these two directions decreases. Alternatively, the diagonals of the square element corresponding to the princi-

pal directions can be used to construct the element for which the shear strains are maximum (minimum) as outlined in the previous section; in other words, the diagonals are the directions between which the shear strain is maximum (minimum).

Mohr's circle can be used in place of the transformation equations. Once Mohr's circle has been constructed, the strains for any directions can be determined by following the rules outlined above.

Example 3.15 For the state of strain represented by the block in Fig. 3.37a, use Mohr's circle to determine (*a*) the principal strains and directions, and (*b*) the maximum shear strain and the corresponding directions.

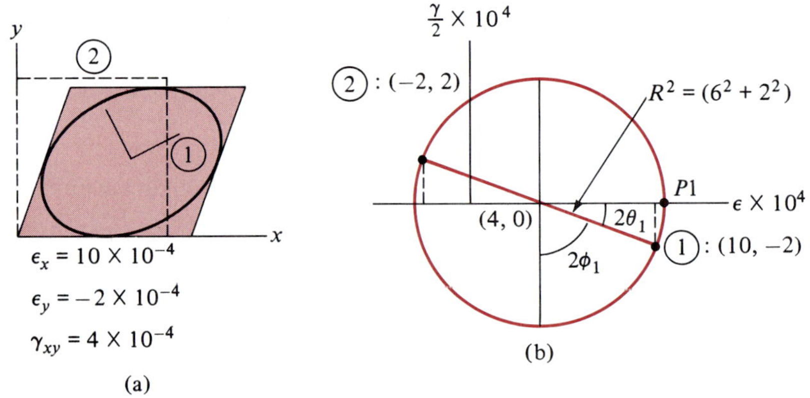

Figure 3.37 Block and Mohr's circle

Solution Point 1 with coordinates $(\varepsilon_x, -\gamma_{xy}/2) = (10, -2) \times 10^{-4}$ in./in. and point 2 with coordinates $(\varepsilon_x, -\gamma_{xy}/2) = (-2, 2) \times 10^{-4}$ in./in. are plotted on Mohr's circle as shown in Fig. 3.37b. The center of the circle has the coordinates $(\varepsilon_x + \varepsilon_y)/2, 0 = [(10 - 2)/2, 0] \times 10^{-4}$ in./in. and the radius is given by $\sqrt{[(10 + 2)/2]^2 + 2^2} \times 10^{-4}$ in./in. $= 6.32 \times 10^{-4}$ in./in. The angle $2\theta_1$ is given by $\tan 2\theta_1 = 2/6$ from which $\theta_1 = 18.44°$. It follows from the figure that $2\phi_1 = 90° - 2\theta_1 = 71.56°$ or $\phi_1 = 35.78°$. The values of the principal strains correspond to points $P1$ and $P2$ on the circle and are

$$\varepsilon_1 = (4.00 + 6.32) \times 10^{-4} \text{ in./in.} = 10.32 \times 10^{-4} \text{ in./in.}$$

and

$$\varepsilon_2 = (4.00 - 6.32) \times 10^{-4} \text{ in./in.} = -2.32 \times 10^{-4} \text{ in./in.}$$

Note that as a check, $\varepsilon_1 + \varepsilon_2 = \varepsilon_x + \varepsilon_y = 8.0 \times 10^{-4}$ in./in. The maximum shear strain is numerically equal to twice the radius of Mohr's circle, that is, $\gamma_{max} = 12.64 \times 10^{-4}$ in./in.

On Mohr's circle the diameter is rotated counterclockwise 18.44° to travel from point 1 to $P1$. Thus in the element, direction 1 is rotated counterclockwise an angle of 9.22° to the direction for ε_1. This is indicated in Fig. 3.38a. The principal strain ε_2 is for the perpendicular direction as shown. These directions are both indicated in Figs. 3.38a and b.

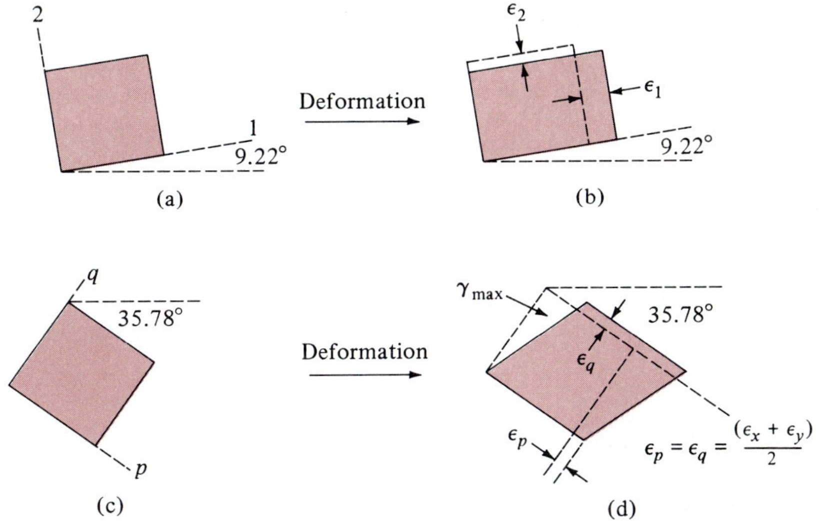

Figure 3.38 Principal planes, principal strains, and maximum shear strain

The two directions between which the shear strain is maximum make angles of 45° with the principal directions. By consulting Fig. 3.37a it is concluded that the orientation and deformation of this element are as indicated in Figs. 3.38c and d. Note again that the diagonals of the square element corresponding to the principal strain directions define the orientation of the element for which the shear strain is maximum.

Example 3.16 For the state of strain defined in Example 3.15 and shown in Fig. 3.39a, use Mohr's circle to determine the extensional strains and the corresponding shear strain for an element corresponding to the $x'y'$ system shown.

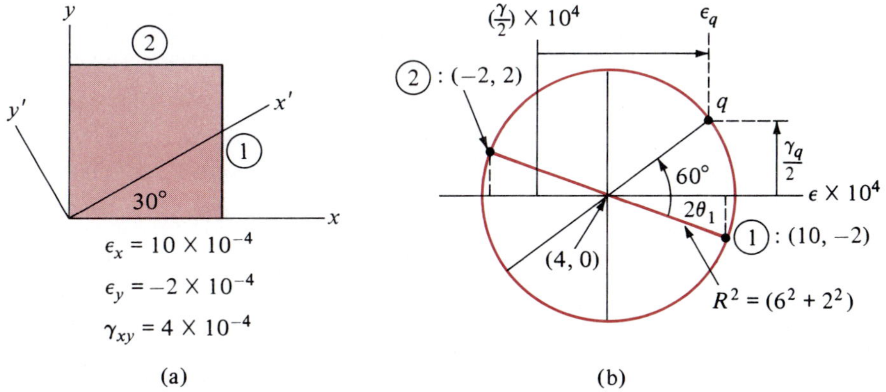

Figure 3.39 Use of Mohr's circle for the transformation of strain

To determine the state of strain for the directions rotated 30° as indicated, direction 1 can be rotated counterclockwise through an angle of 30°. This corresponds to a 60° counterclockwise rotation of the diameter of Mohr's circle to point q. With $2\theta_1 = 18.44°$ from Example 3.15, the coordinates of point q are given by

$$\varepsilon = (4.00 + 6.32 \cos 41.56°) \times 10^{-4} = 8.73 \times 10^{-4}$$

and

$$\frac{\gamma}{2} = 6.32 \sin 41.56° \times 10^{-4} = 4.19 \times 10^{-4}$$

or $\gamma = 8.38 \times 10^{-4}$. The extensional strain in the perpendicular direction is computed using the first strain invariant as $\varepsilon = (8.0 - 8.73) \times 10^{-4} = -0.73 \times 10^{-4}$. The corresponding deformations are indicated in Fig. 3.40.

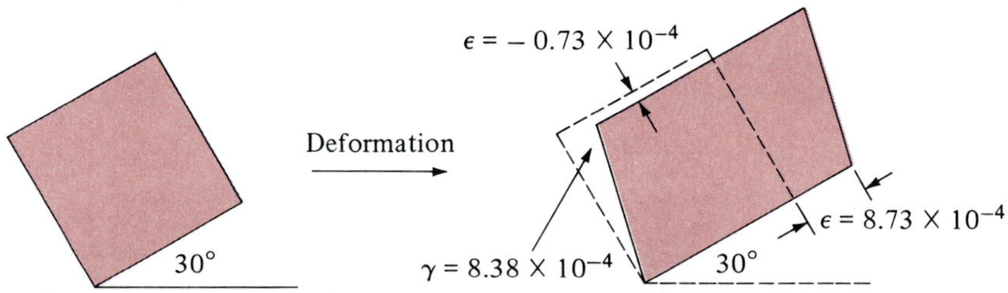

Figure 3.40 Deformations and strains for an element rotated 30°

We can also consider point q as having been obtained by rotating the y direction 60° clockwise. This corresponds to rotating the diameter of Mohr's circle clockwise through an angle of 120° from point 2 which would again put us at point q on Mohr's circle.

PROBLEMS

For each of the following use Mohr's circle to determine the principal strains and the corresponding maximum shear strain. Show these on properly oriented blocks.

3.59 $\varepsilon_x = 0.000100$, $\varepsilon_y = 0.000060$, $\gamma_{xy} = 0.000080$.

3.60 $\varepsilon_x = 0.000120$, $\varepsilon_y = -0.000060$, $\gamma_{xy} = 0.000040$.

3.61 $\varepsilon_x = -0.000100$, $\varepsilon_y = -0.000060$, $\gamma_{xy} = -0.000080$.

3.62 $\varepsilon_x = -0.000080$, $\varepsilon_y = -0.000060$, $\gamma_{xy} = 0.000060$.

3.63 $\varepsilon_x = 0.000100$, $\varepsilon_y = 0.000100$, $\gamma_{xy} = 0.000100$.

3.64 $\varepsilon_x = -0.000100$, $\varepsilon_y = 0.000100$, $\gamma_{xy} = -0.000080$.

For each of the following use Mohr's circle to determine the extensional and shear strains referred to the s and t directions indicated.

3.65 $\varepsilon_x = 0.0010$, $\varepsilon_y = -0.0006$, $\gamma_{xy} = 0.0005$, $\theta = 60°$.

3.66 $\varepsilon_x = -0.0010$, $\varepsilon_y = -0.0006$, $\gamma_{xy} = -0.0005$, $\theta = -30°$.

3.67 $\varepsilon_x = 0.0005$, $\varepsilon_y = 0.0010$, $\gamma_{xy} = 0.0008$, $\theta = 135°$.

3.68 $\varepsilon_x = -0.0010$, $\varepsilon_y = -0.0006$, $\gamma_{xy} = 0.0005$, $\theta = -45°$.

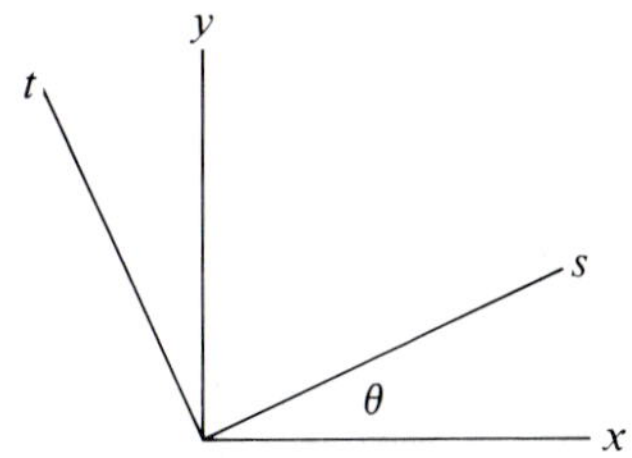

Ps. 3.65–3.68

3.8 MEASUREMENT OF STRAIN: STRAIN GAGES

In situations where the geometry of the structure or the loading on the structure is sufficiently complex it may be necessary or desirable to perform experiments in the process of designing a structure or structural component. One common experimental technique uses what are termed **strain gages** to measure directly the strain at a point on the surface of the structure. The strain information can then be used to determine the corresponding stresses, using ideas that will be discussed in Chapter 4.

Strain gages are constructed of very thin, short wires that are attached to the surface of a structure, usually by a high-strength adhesive. If the gage is properly attached to the structure, it experiences essentially the same extensional strain as the structure to which it is attached. Implicit is the assumption that the presence of the strain gage does not alter the strain at the point in question. The basic property of the wire in the strain gage that makes it useful in this setting is that when the wire is extended or contracted, that is, strained, its resistance changes. By combining the strain gage with properly calibrated electronic equipment that can determine the change in the resistance, it is possible to interpret the change in the resistance of the wire directly as a strain. It is clear that the strain measured in this fashion represents an **average strain** over the length of the strain gage.

Some thought must be given to the placement of the strain gages in order that enough information is obtained regarding the state of strain in the neighborhood of the point of interest. There are two common groupings of strain gages in the neighborhood of a point P known as **rosettes.** There is a **45° strain rosette** and a **60° rosette,** pictured in Fig. 3.41a and b respectively.

For the 45° strain rosette, $\theta_b - \theta_a = \theta_c - \theta_b = 45°$, whereas for the 60° strain rosette, $\theta_b - \theta_a = \theta_c - \theta_b = 60°$. For both rosettes, the gages are shown oriented arbitrarily with respect to an xy coordinate system at the point in question. The strain transformation equation

$$\varepsilon_{x'} = \varepsilon_x \cos^2\theta + \varepsilon_y \sin^2\theta + \gamma_{xy} \sin\theta\cos\theta$$

is used for each of the directions θ_a, θ_b, and θ_c yielding

$$\begin{aligned}
\varepsilon_a &= \varepsilon_x \cos^2\theta_a + \varepsilon_y \sin^2\theta_a + \gamma_{xy} \sin\theta_a\cos\theta_a \\
\varepsilon_b &= \varepsilon_x \cos^2\theta_b + \varepsilon_y \sin^2\theta_b + \gamma_{xy} \sin\theta_b\cos\theta_b \qquad (3.12)\\
\varepsilon_c &= \varepsilon_x \cos^2\theta_c + \varepsilon_y \sin^2\theta_c + \gamma_{xy} \sin\theta_c\cos\theta_c
\end{aligned}$$

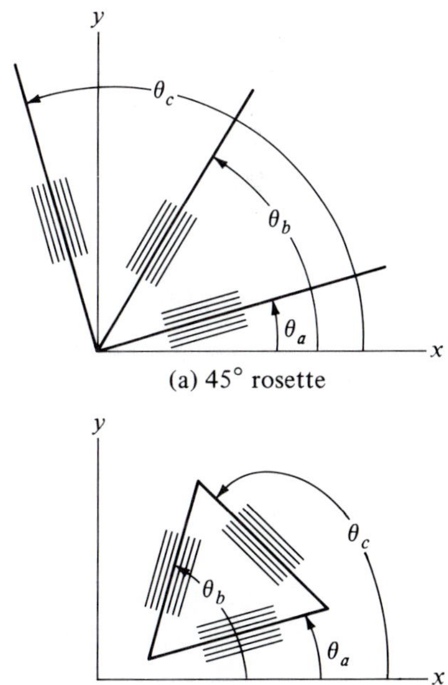

(a) 45° rosette

(b) 60° rosette

Figure 3.41 Strain rosettes

The strains ε_a, ε_b, and ε_c are determined from the experiment so that the unknowns in Eqns. (3.12) are the values ε_x, ε_y, and γ_{xy}. Once Eqns. (3.12) have been solved for ε_x, ε_y, and γ_{xy}, the strains in any other direction in the neighborhood of the point P can then be determined using the strain transformation equations.

Specifically, if $\theta_a = 0°$, $\theta_b = 45°$, and $\theta_c = 90°$, the 45° strain rosette results. The first and third equations yield $\varepsilon_x = \varepsilon_a$ and $\varepsilon_y = \varepsilon_c$, respectively. With $\theta_b = 45°$, the second equation becomes

$$\varepsilon_b = \frac{(\varepsilon_a + \varepsilon_c + \gamma_{xy})}{2}$$

from which

$$\gamma_{xy} = 2\varepsilon_b - \varepsilon_a - \varepsilon_c$$

completely determining the strains at P.

By orienting the gages at 0°, 60°, and 120° (i.e., the 60° strain rosette), the three equations become

$$\varepsilon_a = \varepsilon_x$$

$$\varepsilon_b = \frac{\varepsilon_x}{4} + \frac{3\varepsilon_y}{4} + \gamma_{xy}\frac{\sqrt{3}}{4}$$

$$\varepsilon_c = \frac{\varepsilon_x}{4} + \frac{3\varepsilon_y}{4} - \gamma_{xy}\frac{\sqrt{3}}{4}$$

from which

$$\varepsilon_x = \varepsilon_a$$

$$\varepsilon_y = \frac{2(\varepsilon_b + \varepsilon_c) - \varepsilon_a}{3}$$

and

$$\gamma_{xy} = \frac{2(\varepsilon_c - \varepsilon_b)}{\sqrt{3}}$$

again determining the strains at point P.

Using the information gained from either of the two rosettes, the principal strains can be determined in the neighborhood of the point P.

PROBLEMS

In each of the following, the strains given have been obtained using a 45° rosette. The angle θ_a orients the first gage of the rosette relative to the xy-axes. Determine the strains ε_x, ε_y, and γ_{xy}.

3.69 $\varepsilon_a = 0.0010$, $\varepsilon_b = -0.0020$, $\varepsilon_c = -0.0040$, $\theta_a = 15°$.

3.70 $\varepsilon_a = -0.0020$, $\varepsilon_b = 0.0030$, $\varepsilon_c = -0.0040$, $\theta_a = 40°$.

3.71 $\varepsilon_a = 0.0030$, $\varepsilon_b = -0.0020$, $\varepsilon_c = 0.0020$, $\theta_a = -25°$.

In each of the following, the strains given have been obtained using a 60° rosette. The angle θ_a orients the first gage of the rosette relative to the xy-axes. Determine the strains ε_x, ε_y, and γ_{xy}.

3.72 $\varepsilon_a = 0.0010$, $\varepsilon_b = -0.0020$, $\varepsilon_c = -0.0040$, $\theta_a = 18°$.

3.73 $\varepsilon_a = -0.0020$, $\varepsilon_b = 0.0030$, $\varepsilon_c = -0.0040$, $\theta_a = 35°$.

3.74 $\varepsilon_a = 0.0030$, $\varepsilon_b = -0.0020$, $\varepsilon_c = 0.0020$, $\theta_a = -20°$.

In each of the following, the strains given have been obtained using a 45° rosette. Determine the (*a*) principal strains and directions, and (*b*) the maximum shear strain.

3.75 $\varepsilon_a = 0.0030$, $\varepsilon_b = -0.0042$, $\varepsilon_c = -0.0025$.

3.76 $\varepsilon_a = -0.0015$, $\varepsilon_b = 0.0025$, $\varepsilon_c = -0.0035$.

3.77 $\varepsilon_a = 0.0040$, $\varepsilon_b = -0.0030$, $\varepsilon_c = 0.0055$.

In each of the following, the strains given have been obtained using a 60° rosette. Determine the (*a*) principal strains and directions, and (*b*) the maximum shear strain.

3.78 $\varepsilon_a = 0.0015$, $\varepsilon_b = -0.0045$, $\varepsilon_c = -0.0035$.

3.79 $\varepsilon_a = -0.0022$, $\varepsilon_b = 0.0033$, $\varepsilon_c = -0.0027$.

3.80 $\varepsilon_a = 0.0042$, $\varepsilon_b = -0.0018$, $\varepsilon_c = 0.0035$.

3.9 CLOSURE

In this chapter we have considered kinematics or, more descriptively, the geometry of deformation. Generally, the deformation of a body consists of extensions, shears, and rigid-body motions. Extensions result in a change in the size of a body, shears result in changes in the shape of a body, and rigid-body motions result in a change in the orientation of a body. Shown in Fig. 3.42 is a general deformation of a body originally rectangular in shape.

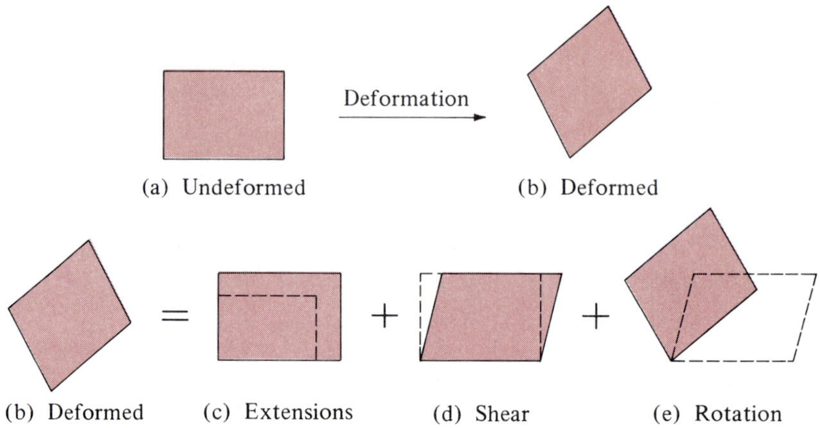

Figure 3.42 Deformation as extensions, shear, and rigid-body rotation

Figs. 3.42a and b indicate the deformation. Figs. 3.42c, d, and e show how the deformation can be considered as a combination of the extensions, the shear, and the rotation, respectively.

Strains measure the relative deformation between different points in a body. The extensional strains, given by

$$\varepsilon = \frac{\Delta s^* - \Delta s}{\Delta s}$$

are a measure of the local changes in size of the body. The shear strains, given by

$$\gamma = \text{angle } PQR - \text{angle } P^*Q^*R^*$$

measure local changes in the shape of a body.

These descriptions of the deformation in terms of strains are indispensible in completely describing the behavior of a deformable solid. By a complete description we mean being able to determine both the stresses in the body and the deformations of the body caused by the external loads acting on the body. With only the information available to us as a result of our discussions of equilibrium and of the geometry of deformation in Chapters 2 and 3, it is not possible to relate the deformations of the body to the external forces acting on the body.

As a very simple illustration, consider the problem of a straight prismatic bar acted upon by tensile forces as indicated in Fig. 3.43.

Using the results of Chapter 2 and referring to Fig. 3.43a, we can compute the average tensile stress at any point in the bar as

$$\sigma = \frac{P}{A} = \frac{P_0}{A}$$

From the present chapter, the deformations of the bar are related to the strains throughout the bar by the strain displacement relation

$$\varepsilon = u'$$

With P_0 as the basic input to the problem we would also like to determine the elongation Δ in terms of the external load P_0. For the axial deformation problem this is not possible at present since σ and P, the force variables, are contained in the equation $\sigma = P/A$, whereas ε and u', the displacement variables, are present only in the strain displacement relation $\varepsilon = u'$. We need an additional equation which, in the mechanics of solids, turns out to be a relation between the stress and the strain.

Strains are a measure of the stretches and shears in the neighborhood of a point P. The character of the deformation associated with a positive ε_x is indicated in Fig. 3.44a. Our intuition should tell us that the corresponding stresses acting on the block that are associated with these strains (deformations) should be somewhat as indicated in Fig. 3.44b: we expect that there are relationships between normal stresses and extensional strains.

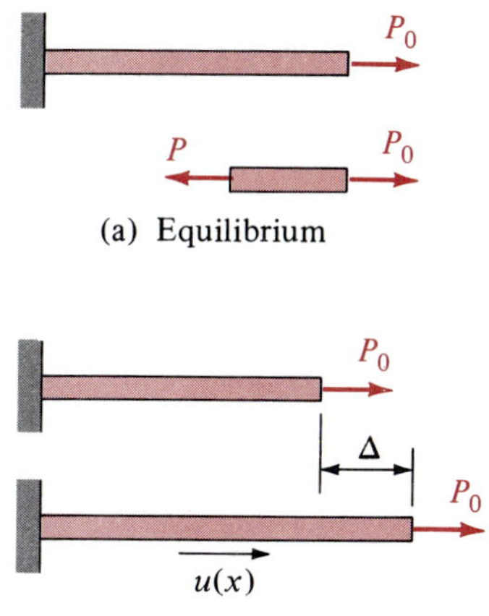

Figure 3.43 Equilibrium and deformation of an axially loaded straight bar

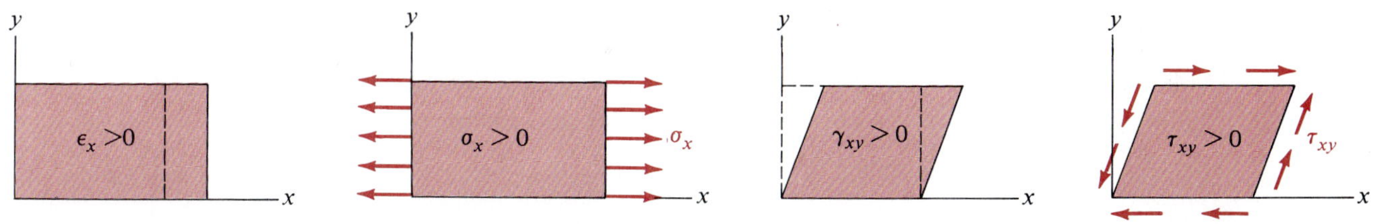

Figure 3.44 Relation between stress and strain

In a completely analogous fashion, there should be some correspondence between the shear deformation $\gamma_{xy} > 0$ shown in Fig. 3.44c and the corresponding shear stresses indicated in Fig. 3.44d. These ideas and the general relations between stress and strain are discussed in detail in Chapter 4.

CHAPTER 4 MATERIAL BEHAVIOR AND PROPERTIES

CHAPTER OUTLINE

4.1 INTRODUCTION

The basic goals of our studies in solid mechanics can be stated roughly as follows. Given the loads applied to the structure, what are *(a)* the stresses and *(b)* the deformations that result? Knowledge of the stresses is used to assess the strength properties, whereas knowledge of the deformations is used to assess the stiffness properties of the structure.

Our discussions in Chapter 2 involved using the equations of equilibrium to determine the internal force resultants. Using equilibrium concepts, the internal force resultants were then used for estimating the stresses in terms of the loads, that is, to answering part *(a)* above. We observed that for statically determinate systems it was possible to solve for the internal force resultants using only the equations of equilibrium, but that for statically indeterminate systems the deformations had to be considered.

Chapter 3 was devoted to the geometry of deformation or, stated differently, to developing the relationships between the deformations of the body and quantities that we defined as extensional and shear strains. As defined, the extensional and shear strains are measures of the changes in the size and the shape of the body. The results were purely geometrical; thus, there was no mention of forces associated with the deformations. As pointed out in the closure for Chapter 3, it seems reasonable that for real materials, there should be some sort of relationship between extensional strains and normal stresses, and also between shear strains and shear stresses.

At this juncture in the study of solid mechanics there are in general no relationships between the kinetic variables (forces and stresses) and the kinematic variables (displacements and strains). If it were possible, using the ideas from Chapter 2, to completely determine the stresses in terms of the external loads and hence answer the strength question, there is currently no way to relate the force variables to the displacement variables and thus determine the displacement variables in terms of the external loads. Thus it is not possible to completely solve the solid mechanics problem from the standpoint of being able to determine both the stresses and the deformations.

From an analytical point of view, it invariably turns out that the total number of unknowns consisting of *(a)* the forces and stresses appearing in the equilibrium equations and *(b)* the displacements and strains appearing in the equations describing the deformations, exceeds the total number of corresponding equations that are available from the same equations of equilibrium and deformation.

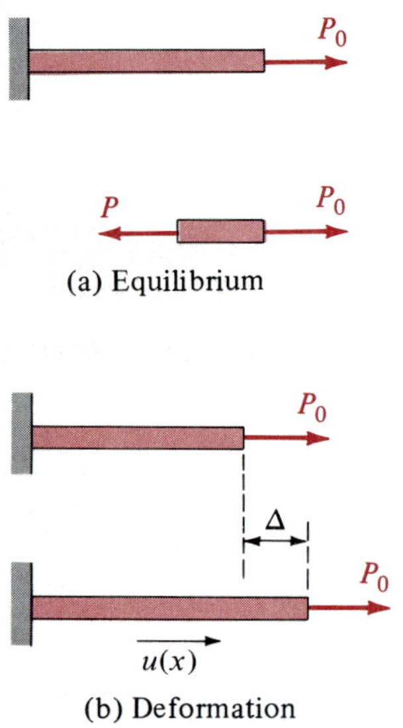

Figure 4.1 Equilibrium and deformation of an axially loaded straight bar

Example 4.1 As a very simple illustration, consider the problem of a straight prismatic bar acted upon by tensile forces as indicated in Fig. 4.1. Referring to the figure, we can use equilibrium to compute the average tensile stress in the bar as

$$\sigma(x) = \frac{P(x)}{A} = \frac{P_0}{A}$$

relating the stress in the bar to the external load P_0. Referring to Fig. 4.1b, the axial deformation of the bar, given by $u(x)$, is related to the strain throughout the bar by the strain displacement relation

$$\varepsilon = u'$$

With P_0 as the known input to the problem there are the three unknowns $\sigma(x)$, $u(x)$, and ε, related by only two equations. Additionally, as mentioned previously, the displacement variables u and ε are interrelated by the strain displacement relation but are completely unrelated to the force variables σ and P. Without further information, there is no way to determine the displacement u in the bar, or specifically, the elongation Δ in terms of the external load P_0.

Material behavior and properties are the means by which the force or stress variables are related to the displacement or strain variables. More precisely, by the study of material behavior we attempt to characterize, qualitatively and quantitatively, the responses of different solids to mechanical and thermal inputs by specifying the form of, as well as the physical constants in, the relations between the force and displacement variables for a particular material. Any and all factors that influence the response of a particular solid in a given environment should be included when deciding on a model for the material behavior of a particular solid.

In studying the qualitative aspects of material behavior, we are concerned with the character of the response of the material. Thus, we try to answer such questions as these: Does the response depend upon the temperature, and if so, how? Does the body return to its original shape and size upon removal of the loads? Is the response influenced by the rate at which the loading occurs? Answers to these questions help to decide the **FORM** of the relationships between the force and displacement variables. These qualitative aspects are sometimes referred to as the **behavior of materials.**

The quantitative aspects of material behavior, on the other hand, pertain more to the values of the physical or material constants used in the equations that describe the response, once the basic form of the equations is known. These quantitative aspects are sometimes referred to as the **properties of materials.**

Note carefully that both the qualitative and quantitative aspects of material behavior must be determined experimentally.

A very important practical concern is with the obvious differences in the basic behavior of different materials. Two different solids respond differently, both quantitatively and qualitatively, to the same level of loading, the same rate of loading, the same temperature, and so forth. A given solid may demonstrate different behavior at varying levels of loading, rates of loading, temperatures, and so on.

Example 4.2 As a simple illustration of different types of material behavior, consider two geometrically identical thin bars loaded transversely as shown in Fig. 4.2.

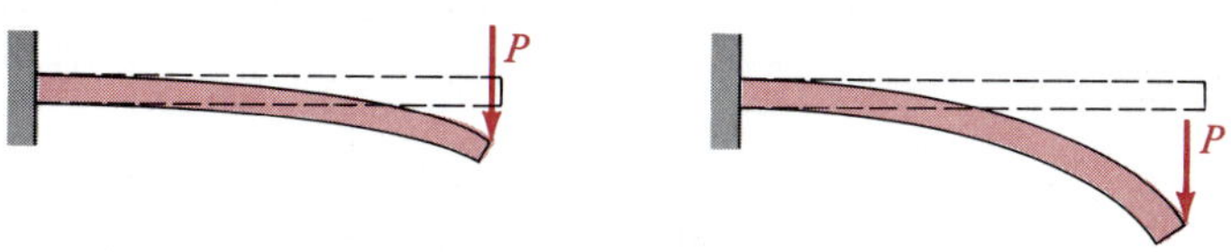

(a) Steel bar (b) Rubber bar

Figure 4.2 Two geometrically identical bars of different materials loaded by a transverse force

Assume one of the bars to be made of steel and the other of rubber. Under the action of the equal external forces P shown, we expect that the response of the two bars will in general be different, both quantitatively and qualitatively.

Quantitatively, the rubber bar will experience significantly larger deformations than the steel bar under the action of the equal loads P.

On the other hand, if each bar is subjected to an unequal force resulting in the same tip deflection as shown in Fig. 4.3, the rubber bar will return to its original straight form upon removal of the load, whereas the steel bar may not—a qualitative difference in the behavior.

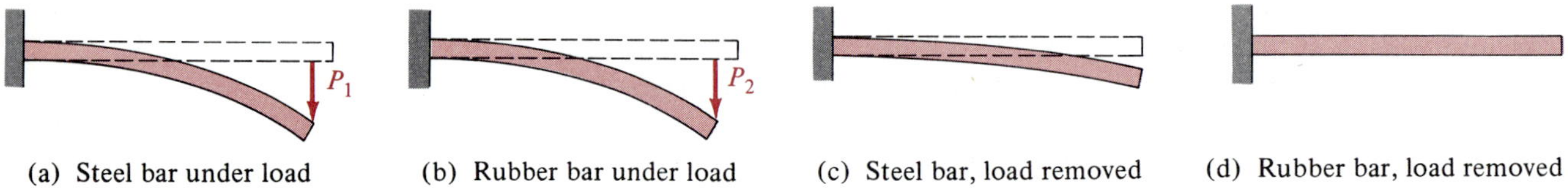

Figure 4.3 Similar deformations of steel and rubber bars with different responses upon removal of the load

---

Example 4.3 As another illustration, consider the material called Silly Putty, a claylike material that clearly exhibits a rate-sensitive behavior. At room temperature, a chunk of Silly Putty will elongate to several times its original length if the tensile force is applied slowly, as indicated in Fig. 4.4a. This type of behavior will be referred to generally as ductile behavior.

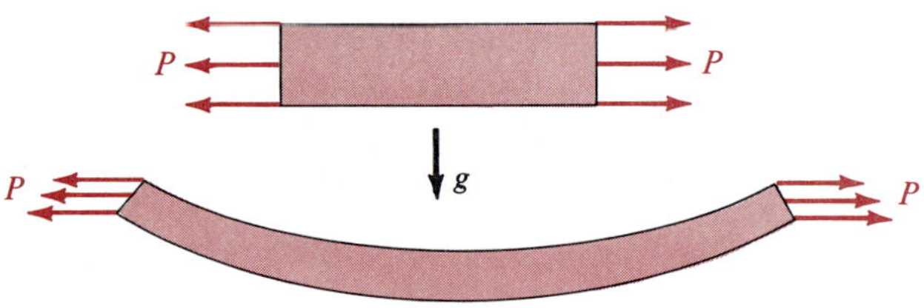

Figure 4.4a Behavior of Silly Putty in response to a slowly applied load

If the same piece of Silly Putty at room temperature is loaded suddenly, as in Fig. 4.4b, the Silly Putty fractures, that is, it behaves in a brittle manner.

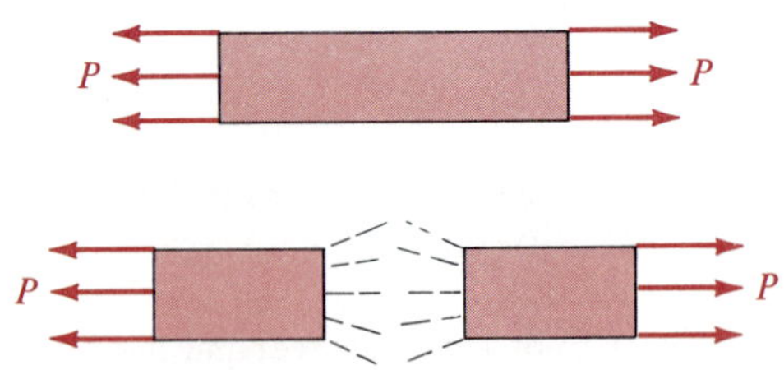

Figure 4.4b Behavior of Silly Putty in response to a rapidly applied load

Similarly, if a piece of Silly Putty is rolled into a ball and bounced easily on the floor, it rebounds in a fashion similar to an ordinary rubber ball; but if the ball is thrown sharply against a hard surface, it tends to shatter into many small pieces. The point here is that the response of this particular material can be strongly dependent upon the rate at which loads are applied.

These examples show qualitative and quantitative differences in the responses of various materials to mechanical and thermal inputs. It is these qualitative and quantitative differences between materials that we will address in this chapter.

The first steps to be taken when investigating the material behavior of a particular solid are to determine the environment in which the solid will operate, and then to determine which aspects of the environment will have a significant influence on the behavior of the solid. There are many potential factors that may influence the behavior of a particular solid, such as temperature, rate of loading, and history of loading. A specific theory in solid mechanics is obtained by systematically combining the results from equilibrium, deformation, and material behavior. The success of the theory depends strongly upon whether or not the material behavior component has been correctly considered. The form of the relation between the force and displacement variables must be correctly modeled, and the physical or material constants appearing in the relations describing the material behavior must be correctly determined, in order for the predictions based on the theory to correspond both qualitatively and quantitatively to what is observed in the real life situation.

In the first part of this chapter we will consider in detail the results of experiments designed to determine the behavior for static situations (i.e., where the loads are applied or vary slowly). The results of experiments performed on many common engineering materials indicate that, to a high degree of approximation, the relationship between the force and displacement variables can be modeled as being linear for a suitably restricted but practically significant range. Materials such as these that behave in a linear fashion will be the focus of our studies in this text.

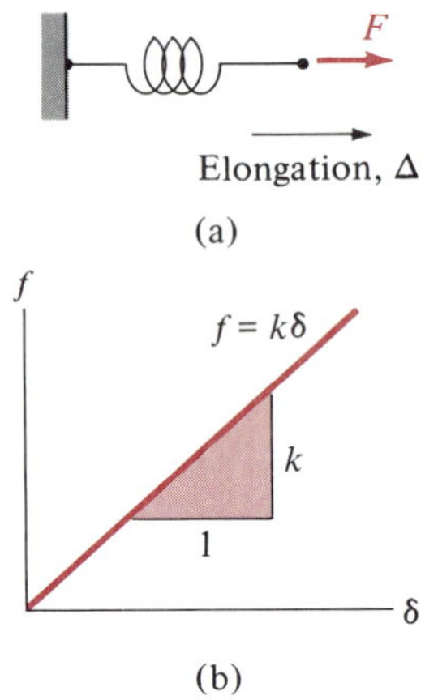

Figure 4.5 Force-displacement relation for a linear spring

Example 4.4 A simple example of a model of linear material behavior is that of a linear spring shown in Fig. 4.5a.

With δ as the elongation of the spring measured from the undeformed configuration, and f as the force transmitted, the material behavior of the spring is represented by the relation

$$f = k\,\delta$$

relating the force variable f and the displacement variable δ. This relationship is shown in Fig. 4.5b. It is assumed that when loaded from an unstretched configuration or when unloaded from a stretched configuration, the relationship between the force f and the elongation δ is given by the linear relationship $f = k\,\delta$.

Under the scrutiny of a high enough power of magnification, any real solid, whether it is composed of steel, aluminium, rubber, or wood, will exhibit a certain amount of structure. There can be many particles or crystals, and even voids apparent in a photograph using even a relatively low power of magnification. It is essentially impossible to take full account of this complex structure when analyzing the stresses and deformations in a solid. In this regard, we make an assumption known as the **continuum hypothesis.** The continuum hypothesis assumes that there are no voids anywhere in the interior of the body—every arbitrarily small portion of the body is assumed to be completely filled with material. The properties of the material may vary from point to point but nowhere is there an absence of material. Any properties of the solid such as mass, stiffness, and strength are assumed to vary gradually throughout the interior. Finite-size holes such as are a regular part of many structural components are not considered as voids in the above sense.

4.2 THE TENSILE TEST

Many applications of solid mechanics involve pseudo-static situations: the external loads are assumed to be applied slowly and then maintained at some constant level. Time is not considered explicitly in the formulation and/or solution of the problem. The experiment or test designed to measure the basic one-dimensional behavior of a material subjected to pseudo-static loadings is called the uniaxial tension test, and is described below.

The test specimen is of course composed of a sample of the material in question and is often constructed in the shape of a slender, uniform, cylindrical cross-section bar as shown in Fig. 4.6a. A rectangular cross-section bar such as shown in Fig. 4.6b is also frequently used. Over the constant cross-section portion of the test specimen a **gage length** L is chosen as indicated.

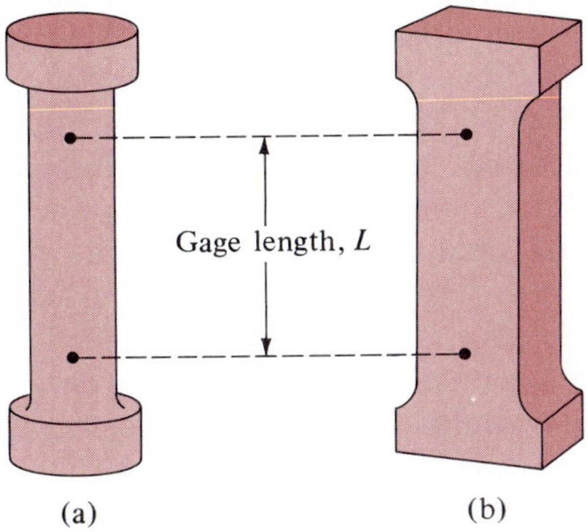

Figure 4.6 Test specimens for uniaxial tension test

As seen in Fig. 4.7, the specimen is placed in a testing machine that is capable of slowly applying a uniaxial tensile load F. Many testing machines are hydraulically activated, with an internal load cell accurately calibrated to indicate the force being transmitted by the machine at any point during the experiment. Gages measuring elongations of the gage length L and the lateral contractions of the diameter d are placed on the specimen as shown in Fig. 4.7.

Figure 4.7 Testing machine and gages for measuring changes in dimension (Photo courtesy MTS Systems Corp., Minneapolis, MN.)

As a result of applying the load F, the gage length L increases an amount ΔL. There is in general an accompanying decrease in the diameter Δd. As F is gradually increased, its value, as well as the values of ΔL and Δd, is measured and recorded. Many modern testing machines do all of this automatically. This part of the experiment is referred to as the loading phase.

At each value of the load F we can compute the average normal stress in the gage-length portion of the bar as

$$(\sigma_x)_{\text{avg}} = \frac{F}{A}$$

where A is the original cross-sectional area of the bar. The average extensional strain in the gage-length portion of the bar is

$$(\varepsilon_x)_{\text{avg}} = \frac{\Delta L}{L}$$

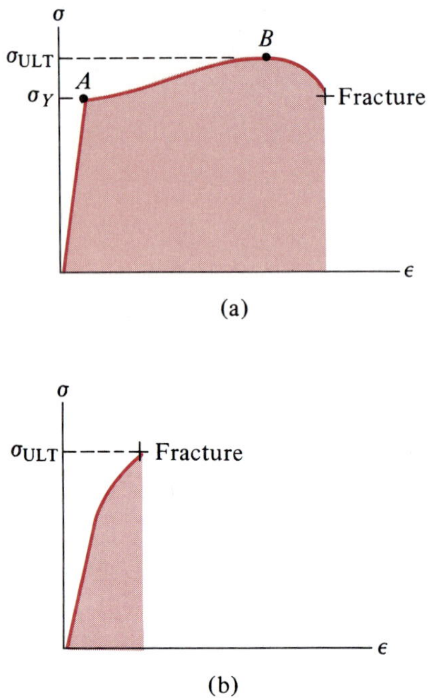

Figure 4.8 Typical experimental stress-strain curves

These values are then plotted to generate a **stress-strain curve.** Two typical results are indicated in Figs. 4.8a and 4.8b.

The curve of Fig. 4.8a is typical of a ductile material whereas the curve of Fig. 4.8b is associated with a brittle material. For a ductile material there is what is called a yield phenomenon at A in Fig. 4.8a, where the stress remains at essentially a constant level while the deformation continues to increase as indicated. The stress at which this phenomenon occurs is referred to as σ_Y, the **yield stress.** The maximum stress at point B on Fig. 4.8a is referred to as σ_{ULT}, the **ultimate stress** or **ultimate strength.** For a typical brittle material as indicated in Fig. 4.8b, there is relatively little deformation before fracture. The stress at which fracture occurs is referred to as the **fracture stress.** We will discuss brittle and ductile behavior in detail in Section 4.4.

If the test is terminated before fracture occurs, and the load is then gradually decreased to zero, a second curve is usually generated. This portion of the test is referred to as the unloading phase and for different materials produces a corresponding **unloading curve** as shown in Fig. 4.9. The shape of this unloading curve depends upon the material, and upon the history of the loading. Several possibilities are indicated. Figs. 4.9a and b depict situations where the strain returns to zero upon removal of the load. Fig. 4.9c shows a situation where there is permanent deformation upon removal of the load. We will discuss both types of behavior in detail in the next section.

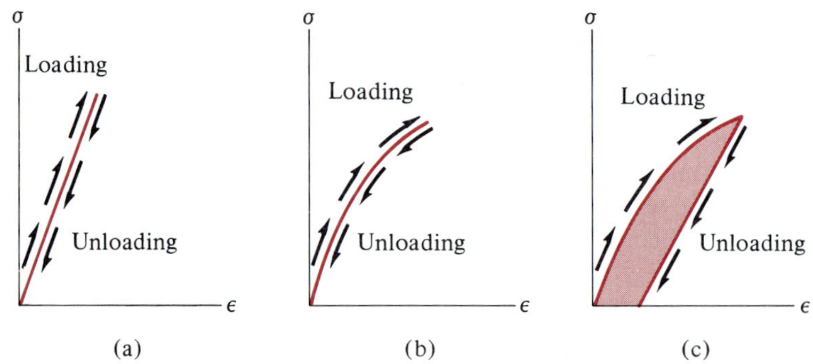

Figure 4.9 Loading and unloading curves

Another essential question to ask when considering the behavior and properties of solids is whether there is any difference in the behavior of a material when loaded in compression as opposed to tension. Either or both the qualitative and quantitative aspects of behavior

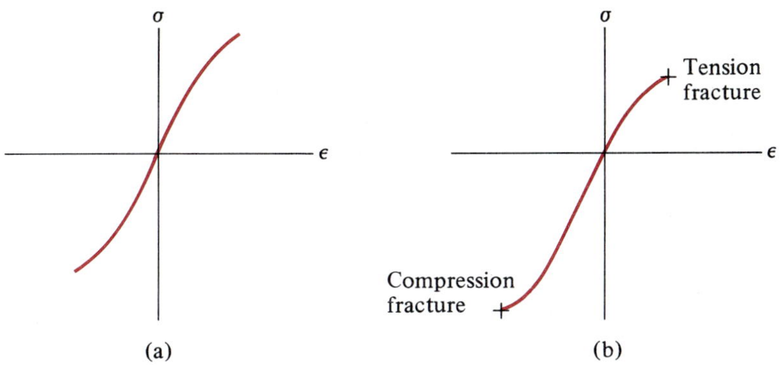

Figure 4.10 Stress-strain curves for tension and compression

might differ. Fortunately, many engineering materials have stress-strain curves such as those in Fig. 4.10a.

For these materials, there is an initial linearly elastic response in compression that is qualitatively and quantitatively the same as in tension. In this text, we will assume that the shape of the stress-strain curve in the linear region is the same for compression and tension. However, quantities such as yield stress and ultimate stress may be different in compression than in tension. For instance, the ultimate stress for a brittle material may depend strongly on whether the stress is tension or compression. The stress-strain curve shown in Fig. 4.10b is typical of certain brittle materials such as glass, grey cast iron, and concrete, materials whose strengths, as measured by the ultimate stress, are much stronger in compression than in tension. As an example, the ultimate compressive stress for concrete is typically in the range of 2 to 5 ksi whereas the ultimate tensile strength is usually less than 10% of this. The ratios of the ultimate stresses in tension and compression for glass and grey cast iron are similar.

4.3 ELASTIC AND INELASTIC BEHAVIOR

Referring to Fig. 4.9 of the previous section, the curves for Figs. 4.9a and b are examples of **elastic behavior:** when the loads are removed, the material returns to its original state. An ordinary rubber band is apparently elastic; it seems to return to its original length upon removal to a tensile load. Fig. 4.9c is typical of **inelastic behavior** in that there is permanent deformation upon removal of the load. A given material may behave elastically over a certain range of stress and strain values, and inelastically over a larger range. Example 4.6 later in this chapter describes this behavior.

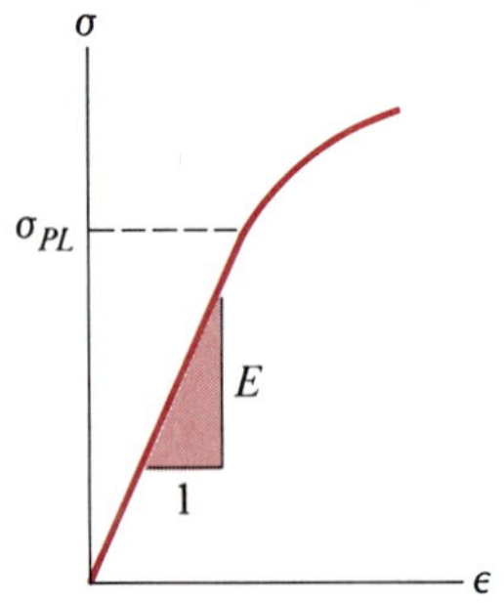

Figure 4.11 Stress-strain curve for a material that is initially linearly elastic

For many common engineering materials there is a portion of the stress-strain curve that is essentially linear, as shown in Fig. 4.11. For these materials the stress σ_{PL} corresponding to the last point on the linear portion of the curve is called the **proportional limit;** it is the last point for which stress is proportional to strain. Within this range the **FORM** (qualitative aspects of material behavior) of the relationship between stress and strain is taken to be

$$\sigma = E\varepsilon \tag{4.1}$$

where E, called **Young's Modulus** or the **modulus of elasticity,** is the physical constant of proportionality (quantitative aspects of material behavior). E has dimensions FL^{-2} and is commonly given in units of lbf/in^2 (psi) or newtons/meter^2 = pascals (Pa). As shown in Fig. 4.11, E is the slope of the stress-strain curve within the linearly elastic range of the material. As long as the material is not subjected to stresses in excess of the proportional limit the material is classified as **linearly elastic.**

To reiterate, notice that the experiment known as the uniaxial tension test has not only provided qualitative information about the behavior of the material (i.e., linearly elastic with $\sigma = E\varepsilon$ below the proportional limit), but has also yielded the necessary quantitative information as to the value of the constant of proportionality E. This was one of the stated goals in our investigation of material behavior.

Recall that in the tension test of a circular cross-section bar, measurements of the change in diameter are also taken. Assuming that the cross section remains circular, the strain ε_T in the transverse direction, that is, the strain measuring the change in diameter, is given by

$$\varepsilon_T = \frac{\Delta d}{d}$$

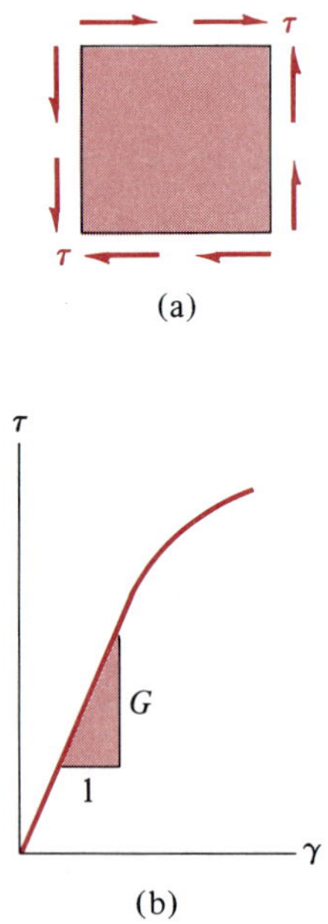

Figure 4.12 Experimental determination of the shear modulus

In the linearly elastic range, the negative of the ratio of this transverse strain to the corresponding axial strain is called **Poisson's ratio** and is given by

$$\nu = -\frac{\text{transverse strain}}{\text{axial strain}} = -\frac{\varepsilon_T}{\varepsilon_x} \tag{4.2}$$

Poisson's ratio is positive for all known materials, indicating that the transverse dimension decreases when the test specimen is subjected to tension.

Another important elastic constant is the shear modulus or modulus of rigidity. It is determined on the basis of the experiment shown in Fig. 4.12. A block is subjected to a stress state of pure shear as indicated in Fig. 4.12a. The value of the shear stress is increased gradually and the resulting shear strain γ is measured. Many materials respond as indicated in Fig. 4.12b, where it is seen that the shear stress τ and the shear strain γ are related in a linear fashion, or

$$\tau = G\gamma \tag{4.3}$$

where G is called the **shear modulus** or the **modulus of rigidity.** The shear modulus G is the elastic constant for shear behavior, and is the counterpart of the modulus of elasticity E for extensional behavior. The shear modulus G also has dimensions FL^{-2} and again is given in units of lbf/in^2 (psi) or newtons/meter^2 = pascals (Pa). We will find later in the chapter that for many materials there is a relationship between E, G, and ν given by $E = 2G(1 + \nu)$.

Another stress value often used is the **elastic limit** σ_{EL}, defined as the greatest stress that the material can withstand without any permanent deformation occurring upon release of the load. In principle, the elastic limit is distinctly different from the proportional limit, with its value potentially higher than the value of the proportional limit. Practically, however, it is very difficult to distinguish between the two values on the basis of data taken from an actual test, hence the values of the elastic and proportional limits are frequently assumed to be the same. As indicated in Fig. 4.13, both values are taken to be equal to a value near the proportional limit.

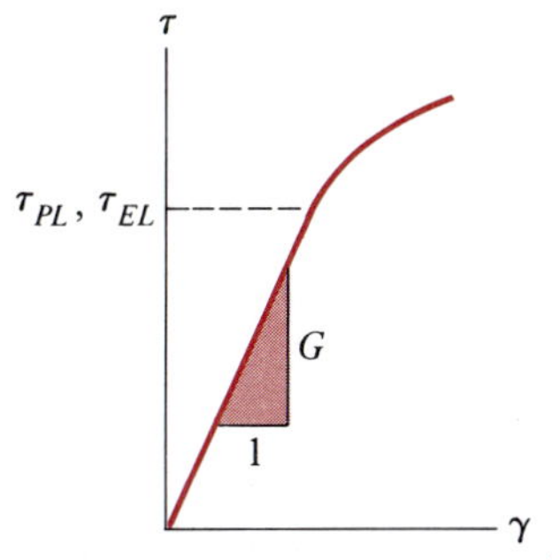

Figure 4.13 Elastic and proportional limits

For nonlinearly elastic materials such as those in Fig. 4.14, there are several measures of the local stiffness. One is the **tangent modulus** E_T: quite simply, the slope of a tangent drawn to the stress-strain curve at the point P in question. The **secant modulus** E_S at the same point P is obtained by drawing a straight line from the origin to the point on the stress-strain curve and measuring the slope. For materials that behave initially in a linear fashion, both of these moduli are meant to be used beyond the proportional limit. Below the proportional limit both coincide with the modulus of elasticity E.

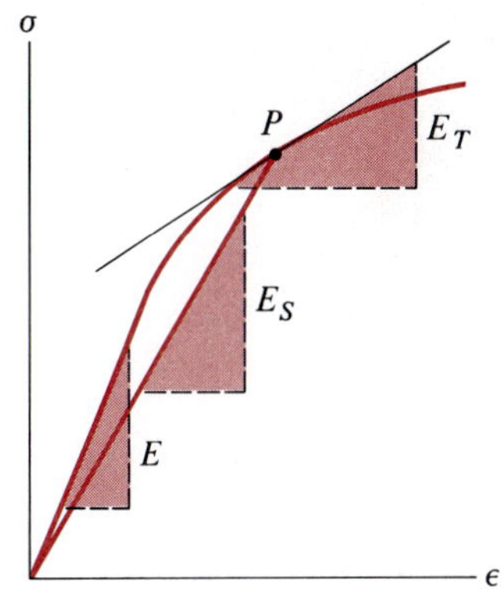

Figure 4.14 Various moduli for nonlinearly elastic materials

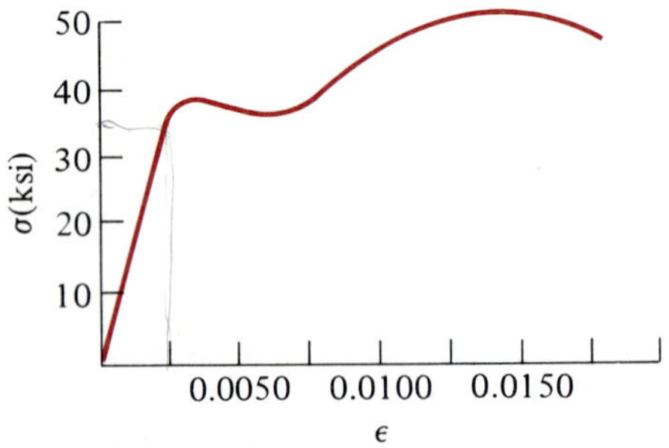

Figure 4.15 Stress-strain data from a uniaxial tension test

Example 4.5 A 0.505-in. diameter test specimen is subjected to a uniaxial test with the results shown in Fig. 4.15. If the diameter of the bar at the proportional limit was 0.5046 in., determine *(a)* Young's modulus, *(b)* the proportional limit, *(c)* the yield point, *(d)* Poisson's ratio, *(e)* the ultimate strength, and *(f)* the tangent modulus at $\sigma = 45$ ksi.

Solution Consulting the stress-strain curve, the proportional limit is approximately $\sigma_{PL} = 35$ ksi. The corresponding strain is approximately $\varepsilon = 0.0025$ so that $E = 35 \text{ ksi}/0.0025 = 14 \times 10^6$ psi. Again consulting the figure, the yield point is about 38 ksi with the ultimate stress as 50 ksi. Drawing a tangent at $\sigma =$ 45 ksi the tangent modulus is estimated to be 2.4×10^6 psi. By definition Poisson's ratio is given by

$$\nu = -\frac{\varepsilon_{tr}}{\varepsilon_{ax}} = -\frac{(0.5046 - 0.505)/0.505}{0.0025} \approx 0.32$$

PROBLEMS

4.1 At the proportional limit the 6-in.-gage length of a ⅜-in.-diameter aluminum bar has elongated 0.012 in. with a decrease in the diameter to 0.37477 in. The corresponding load was 2210 lbf. Determine *(a)* Young's modulus, *(b)* Poisson's ratio, and *(c)* the proportional limit.

4.2 A tensile test bar with a standard diameter of 0.505 in. and a gage length of 2 in. was tested to failure by fracture. Load and deformation data were recorded as follows.

Load (lbf)	ΔL (in.)	Load (lbf)	ΔL (in.)
0	0	9580	0.0420
1395	0.0014	10120	0.0588
2800	0.0028	10760	0.0840
4195	0.0042	10940	0.1176
5600	0.0056	10720	0.1428
7010	0.0070	10140	0.1680
8400	0.0084	9420	fracture
8620	0.0110		
8635	0.0136		
8660	0.0168		
8680	0.0220		
8880	0.0310		

Determine the following:

a. Young's modulus or the modulus of elasticity.

b. The proportional limit σ_{PL}, and the ultimate strength σ_{ULT}.

c. The yield stress.

d. The tangent modulus at a stress of 50.6 ksi.

e. The secant modulus at a stress of 50.6 ksi.

f. The fracture stress.

g. The true fracture stress if the diameter of the bar at fracture was 0.464 in.

h. Would you classify this material as brittle or ductile?

4.3 A ¼-in. by 2-in. by 60-in. bar elongates 0.02 in. under the action of a 5000-lbf tensile load. At this load, the 2-in. dimension is reduced to 1.99981 in. Assuming that these conditions are within the linear range of the stress-strain curve, determine *(a)* Young's modulus, *(b)* Poisson's ratio, *(c)* the proportional limit, and *(d)* the decrease in the ¼-in. dimension.

4.4 A tensile test bar with a standard diameter of 0.505 in. and a gage length of 2 in. was tested to failure by fracture. Load and deformation data were recorded as follows.

Load (lbf)	ΔL (in.)
0	0
1260	0.00043
2520	0.00087
3780	0.00130
5040	0.00174
6300	0.00217
7560	0.00261
8820	0.00341
9540	0.00447
10280	0.00683
10860	fracture

Determine the following:

a. Young's modulus or the modulus of elasticity.

b. The proportional limit σ_{PL} and the ultimate strength σ_{ULT}.

c. The yield strength at 0.2% offset (see Section 4.4).

d. Would you classify this material as brittle or ductile?

4.5 For the bar loaded in tension as shown, the increase in the length b is 0.00185 in. and the decrease in the length a is 0.00028 in. Assuming the corresponding stress is in the elastic range, determine Poisson's ratio. Initially a = 1 in. and b = 2 in.

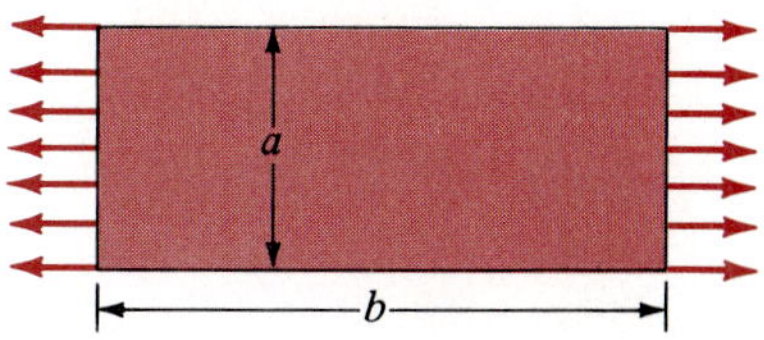

P. 4.5

4.6 At a point in the elastic range a standard 0.505-in.-diameter bar is acted upon by a 6400-lbf tensile load. A 2-in. section is observed to elongate 0.016 in. and the diameter to decrease to 0.5039 in. What are the modulus of elasticity and Poisson's ratio?

4.7 A 25-mm by 100-mm by 1500-mm steel bar elongates 1.05 mm under the action of a 350-kN tensile load. At this load the 100-mm dimension is reduced to 99.98 mm. Assuming that these conditions are within the linear range of the stress-strain curve, determine *(a)* Young's modulus, *(b)* Poisson's ratio, *(c)* the proportional limit, and *(d)* the decrease in the 25-mm dimension.

4.8 A tensile test bar with a standard diameter of 12.83 mm and a gage length of 50 mm was tested to failure by fracture. Load and deformation data were recorded as follows.

Load (kN)	ΔL (mm)	Load (kN)	ΔL (mm)
0	0	42.61	1.0668
6.20	0.0356	45.01	1.4935
12.45	0.0711	47.86	2.1336
18.66	0.1067	48.66	2.9870
24.91	0.1422	47.68	3.6271
31.18	0.1778	45.10	4.2672
37.36	0.2134	41.90	fracture
38.34	0.2794		
38.41	0.3454		
38.52	0.4267		
38.61	0.5588		
39.50	0.7874		

Determine the following:

a. Young's modulus or the modulus of elasticity.

b. The proportional limit, σ_{PL}.

c. The ultimate strength, σ_{ULT}.

d. The yield stress.

e. The tangent modulus at a stress of 349 MPa.

f. The secant modulus at a stress of 349 MPa.

g. The fracture stress.

h. The true fracture stress if the diameter of the bar at fracture was 11.79 mm.

i. Would you classify this material as brittle or ductile?

4.9 A tensile test bar with a standard diameter of 12.83 mm and a gage length of 50 mm was tested to failure by fracture. Load and deformation data were recorded as follows.

Load (kN)	ΔL (mm)	Load (kN)	ΔL (mm)
0	0	33.65	0.0663
5.60	0.0109	39.23	0.0866
11.21	0.0221	42.44	0.1135
16.81	0.0330	45.73	0.1735
22.41	0.0442	48.31	fracture
27.99	0.0551		

Determine the following:

a. Young's modulus or the modulus of elasticity.

b. The proportional limit, σ_{PL}.

c. The ultimate strength, σ_{ULT}.

d. The yield strength at 0.2% offset (see Section 4.4).

e. Would you classify this material as brittle or ductile?

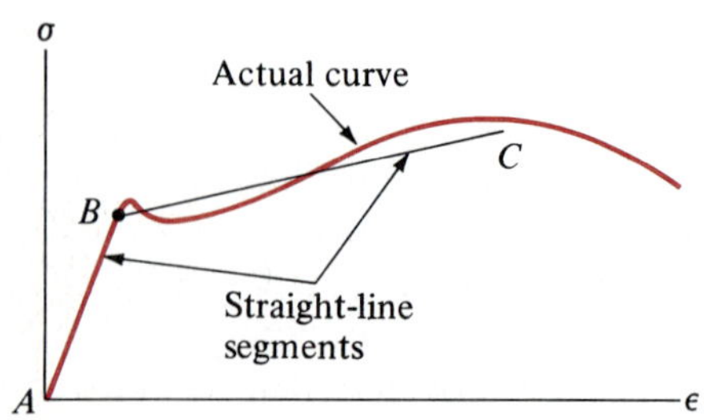

Figure 4.16 Stress-strain and idealized stress-strain curves for steel

4.3.1 Idealizations of Material Behavior

The stress-strain curves for most materials are relatively complicated from an analytical point of view. For a typical stress-strain curve for steel as shown in Fig. 4.16, it would be tedious, at best, to represent the stress σ as a function of the strain ε.

The essential characteristics of the response—the initial linearly elastic behavior *AB* followed by a so-called strain-hardening phase *BC*—can be represented approximately by the two straight-line segments shown in Fig. 4.16. This is an example of an idealized stress-strain curve that is frequently used for analysis when it is known that the stresses do not fall within the elastic range and the plastic behavior in the region *BC* must also be modeled. This particular idealization is referred to as **elastic-linear hardening.** Other idealizations including **elastic–perfectly plastic** and **rigid–perfectly plastic** are indicated in Figs. 4.17a and b, respectively.

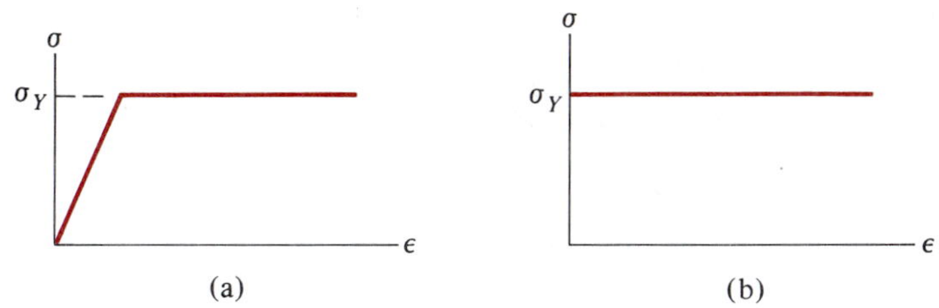

Figure 4.17 Elastic–perfectly plastic and rigid–perfectly plastic idealizations of stress-strain curves

Another idealization that is frequently useful in representing general shapes of some stress-strain curves for elastic materials is shown in Fig. 4.18.

It has been found that curves of this general shape can be represented quite effectively and accurately using the Ramberg-Osgood equation.[1] This is given by

$$\varepsilon = \frac{\sigma}{E} + K\left(\frac{\sigma}{E}\right)^n$$

The parameters K, E, and n can be adjusted to obtain a good fit between the experimental stress-strain curve and the Ramberg-Osgood curve for a particular material. Practically, it turns out that for materials that initially demonstrate linearly elastic behavior (i.e., materials for which there is a $\sigma = E\,\varepsilon$ portion of the stress-strain curve near the origin), the values of K and n are such that the nonlinear term is only significant for points away from the origin.

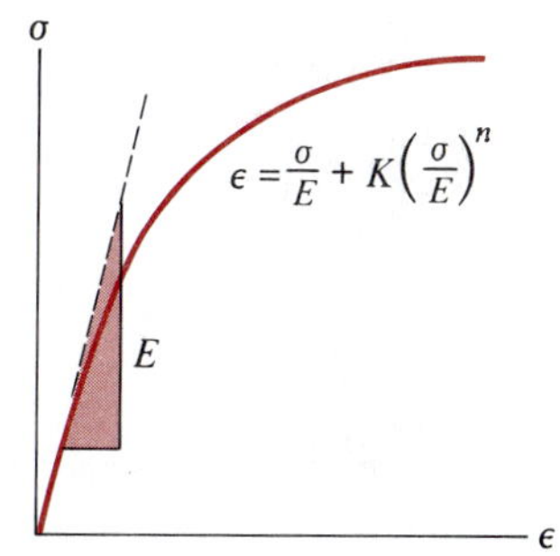

Figure 4.18 Nonlinear elastic stress-strain diagrams

[1] W. A. Ramberg and W. R. Osgood, "Description of stress-strain curves by three parameters," *National Advisory Committee for Aeronautics,* Technical Note #902 (July 1943).

Idealizations such as these can be used very effectively to approximate the stress-strain relations of a variety of real solids, or to model a particular aspect of the behavior of a solid. For applications the student should refer to the sections on nonlinear behavior in Chapters 5, 6, and 7.

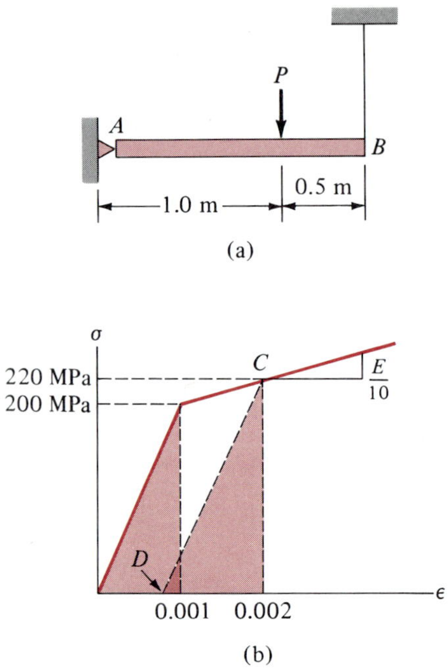

Figure 4.19 Wire supporting a rigid bar and stress-strain curve for the wire

Example 4.6 The rigid bar AB is supported by a wire 6 mm in diameter, which behaves according to the stress-strain curve shown in Fig. 4.19. Determine the load-displacement behavior of the bar and wire.

Solution We are interested in *(a)* the maximum permissible elastic load P—and the corresponding motion of point B—for elastic action of the wire, *(b)* the motion of B if P is increased 10% above the elastic value, and *(c)* the residual displacement at B when the load P is removed. We will consider each of the basic ingredients separately and then combine them for the solution.

Part A *Equilibrium:* Draw a free-body diagram of the bar AB and sum moments about A as shown in Fig. 4.20 to determine that $T = 2P/3$ for the tensile force in the wire.

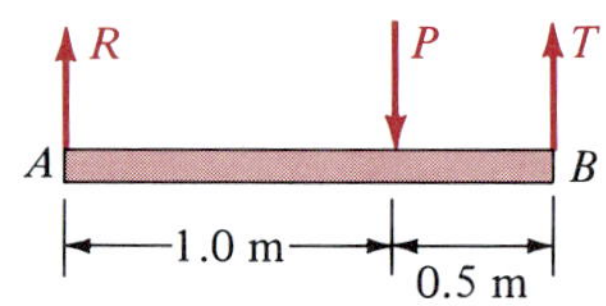

$$\Sigma M_A = 0 = -P(1\text{ m}) + T(1.5\text{ m})$$

Figure 4.20 FBD of bar AB

The corresponding stress is

$$200\text{ MPa} = \frac{2P}{3A}$$

from which

$$P = \frac{200\text{ MPa}(3)(\pi/4)(0.006\text{ m})^2}{2} = 8.482\text{ kN}$$

as the maximum permissible load for elastic behavior of the wire. Note that for this statically determinate problem the strength question was answered without having to involve the deformations. To answer questions about the deformations we must consider the other two basic ingredients.

Strain-displacement: For this simple problem, the strain everywhere in the wire is given by the average strain, namely $\varepsilon = \Delta/L$, where Δ is the elongation of the wire and the displacement at point B.

Stress-strain: When the stress in the wire is at or below the elastic-limit stress of 200 MPa, the relation between stress and strain is

$$\sigma = E\varepsilon$$

where, from Fig. 4.19b, the modulus of elasticity is given by

$$E = \frac{\sigma}{\varepsilon} = \frac{200\ \text{MPa}}{0.001} = 200\ \text{GPa}$$

Combining the three basic ingredients yields

$$\sigma = \frac{E\Delta}{L}$$

from which

$$\Delta_B = \frac{\sigma L}{E} = 0.002\ \text{m}$$

as the elongation of the wire and the displacement at B.

Part B *Equilibrium:* When the load P is increased by 10% to 9.3 kN, the results of Part A show that the stress also increases by 10% to 220 MPa.

Strain-displacement: The strain-displacement component is still given by $\varepsilon = \Delta/L$.

Stress-strain: From Fig. 4.19b the equation for the stress-strain curve beyond the elastic limit is seen from its slope to be

$$\frac{(\sigma - 200)\ \text{MPa}}{\varepsilon - 0.001} = \frac{20\ \text{MPa}}{0.002 - 0.001} = 20\ \text{GPa}$$

Combining the results with $\sigma = 220$ MPa yields

$$\frac{(220 - 200)\ \text{MPa}}{20\ \text{GPa}} = \varepsilon - 0.001$$

or

$$\varepsilon = 0.002$$

The corresponding vertical motion at B is $\Delta_B = \varepsilon L = 0.002\ (2\ \text{m}) = 0.004$ m.

Part C *Equilibrium:* When the load P is removed, $T = 0$ and $\sigma = 0$.

Strain-displacement: Using the strain-displacement relation $\varepsilon = \Delta/L$ yields

$$\Delta_B = \varepsilon L = 0.0009\ (2.0\ \text{m}) = 0.0018\ \text{m}$$

Stress-strain: When the load P is removed, unloading takes place along the line CD parallel to line AB in Fig. 4.19b. The corresponding value of $\varepsilon = \varepsilon_F$ is given by the equation of the line CD as

$$\frac{\sigma_c}{\varepsilon_c} = \frac{220\ \text{MPa}}{0.002 - \varepsilon_F} = 200\ \text{GPa}$$

from which $\varepsilon_F = 0.0009$.

The P versus Δ_B curve for the three cases is shown in Fig. 4.21.

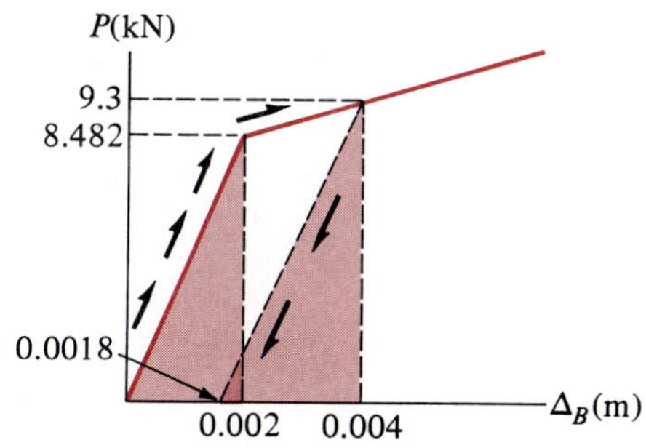

Figure 4.21 Load-deflection curve

PROBLEMS

4.10 Bars AB and BD are rigid with bar CE behaving according to the stress-strain diagram shown, that is, unloading takes place on a straight line parallel to the initial linear portion of the stress-strain curve. Determine *(a)* the vertical displacement of D under the action of the load P, and *(b)* the permanent or residual displacement of point D when the load is removed. Carefully state how each of the basic ingredients of equilibrium, strain-displacement, and stress-strain contribute to the formulation and solution of the problem. Take the area of bar CE to be 600 mm^2, $h = 4$ m, $\sigma_0 = 150$ MPa, $E = 70$ GPa, $k = 10$ GPa, and $P = 55$ kN.

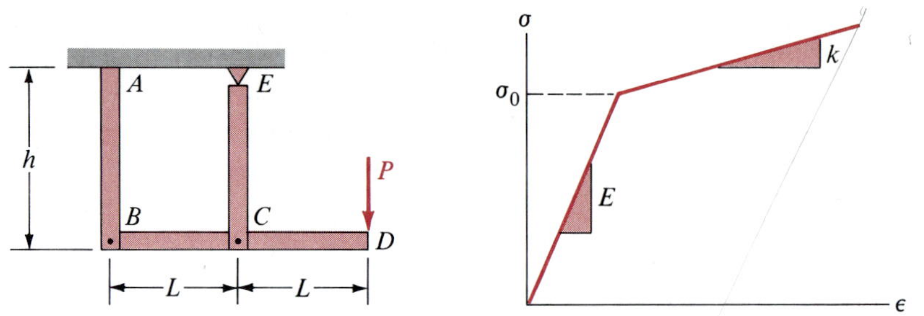

Ps. 4.10 and 4.11

4.11 Repeat problem 4.10 for $\sigma_0 = 20$ ksi, $E = 10^7$ psi, $k = 1.5 \times 10^6$ psi, $P = 15$ kips, $h = 15$ ft, and the area of bar CE as 0.9 in^2.

4.12 A uniform bar hangs under its own weight as shown. The material has weight density γ and behaves according to the relation $(\sigma/\sigma_0)^n = \varepsilon$, where σ_0 and n are parameters determined on the basis of a uniaxial test. Determine *(a)* the stress in the bar as a function of position, *(b)* the displacement $u(x)$, and *(c)* the elongation of the bar. Carefully state how equilibrium, strain-displacement, and material behavior each contribute to the formulation and solution of the problem.

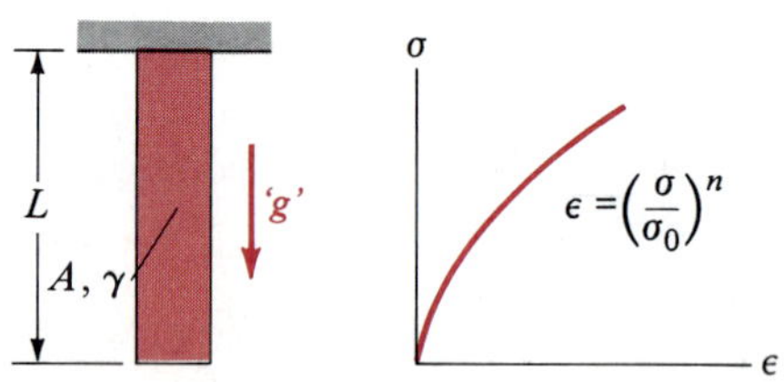

P. 4.12

4.13 The behavior of a material as determined from a uniaxial test is shown in the figure. The curve is to be approximated by the three straight-line segments indicated. Determine the analytical expressions that define the approximate stress strain relation.

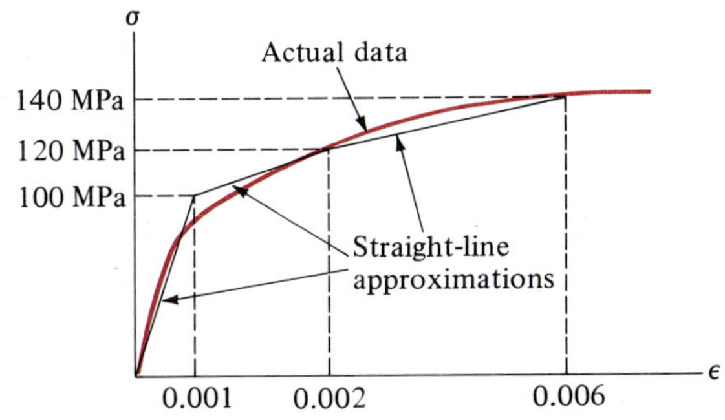

Ps. 4.13–4.16

4.14 A 15-mm-radius bar 2.6 meters in length is made from the material described in problem 4.13. Determine the elongation of the bar if it is subjected to a tensile load of 56 kN. Carefully state how equilibrium, strain-displacement, and material behavior each contribute to the formulation and solution of the problem.

4.15 Repeat problem 4.14 for a load of 78 kN.

4.16 Repeat problem 4.14 for a load of 92 kN.

4.17 The two bars loaded as shown are composed of the material shown. Determine *(a)* the total elongation of the bar, and *(b)* the residual or permanent elongation when the load is removed. Carefully state how equilibrium, strain-displacement, and material behavior each contribute to the formulation and solution of the problem. Take $A_1 = 2\text{ in}^2$, $A_2 = 1\text{ in}^2$, $L_1 = 36$ in., $L_2 = 26$ in., and $P = 12$ k.

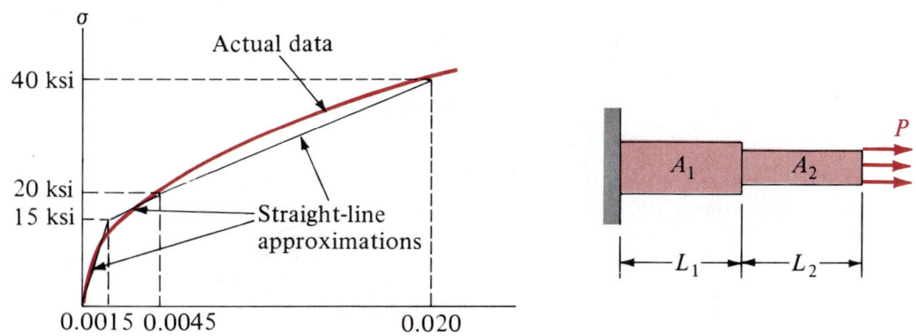

Ps. 4.17–4.19

4.18 Repeat problem 4.17 with $P = 18$ k.

4.19 Repeat problem 4.17 with $P = 36$ k.

4.20 A truss structure is loaded as shown. The material behaves according to the stress-strain curve indicated. Determine the vertical displacement of point B. Carefully state how each of the basic ingredients of equilibrium, strain-displacement, and material behavior enters into the formulation and solution of the problem. Take $L = 3$ ft, $\alpha = 30°$, $P = 20$ kip, $E = 30 \times 10^6$ psi, $k = 5 \times 10^6$ psi, $\sigma_0 = 30$ ksi, and the areas of AB and $BC = 1$ in^2.

4.21 Repeat problem 4.20 if $P = 40$ kip.

4.22 Repeat problem 4.20 if $L = 2$ m, $\alpha = 45°$, $P = 90$ kN, $E = 200$ GPa, $k = 30$ GPa, $\sigma_0 = 200$ MPa, and the areas of AB and $BC = 400$ mm^2.

4.23 Repeat problem 4.20 if $L = 3$ m, $\alpha = 60°$, $P = 150$ kN, $E = 200$ GPa, $k = 30$ GPa, $\sigma_0 = 200$ MPa, and the areas of AB and $BC = 400$ mm^2.

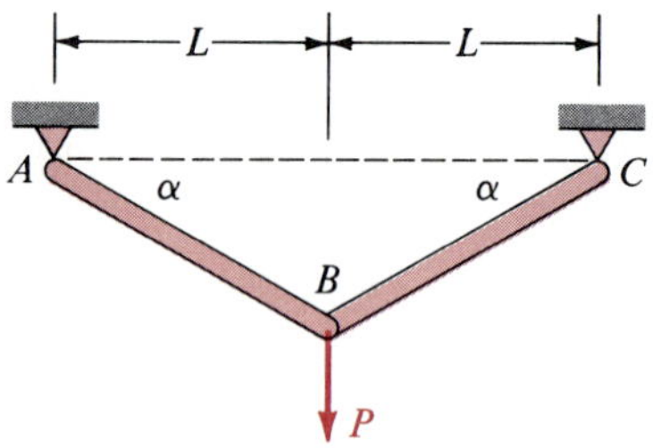

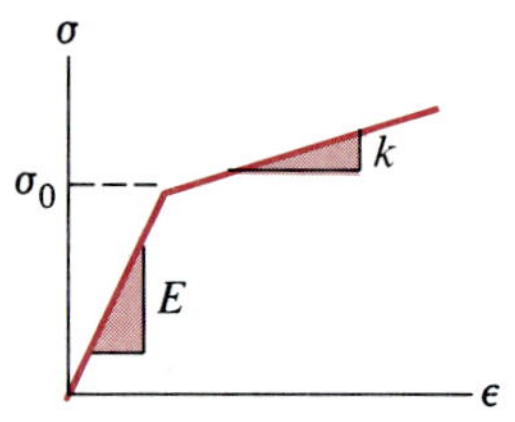

Ps. 4.20–4.23

4.24 A truss structure is loaded as shown. The material behaves according to the stress-strain curve indicated. Determine the vertical displacement of point B. Carefully state how equilibrium, strain-displacement, and material behavior each enters into the formulation and solution of the problem. Take $L = 5$ ft, $\alpha = 36.9°$, $\sigma_0 = 3 \times 10^5$ ksi and the areas of AB and $BC = 0.2$ in^2. Take $P = 1200$ lbf, 2400 lbf, 4800 lbf, and 9600 lbf, and make a plot of the vertical displacement of point B versus P.

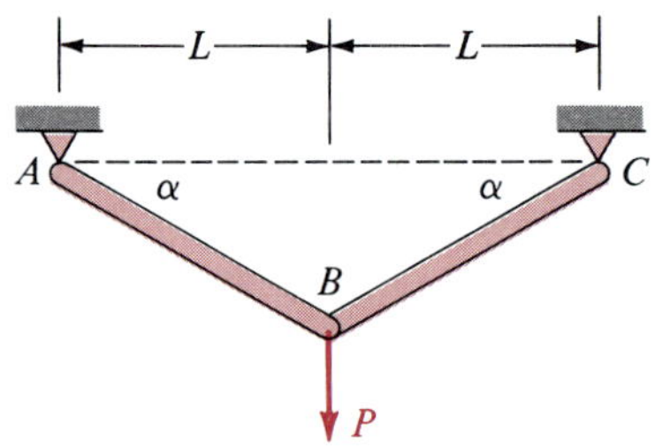

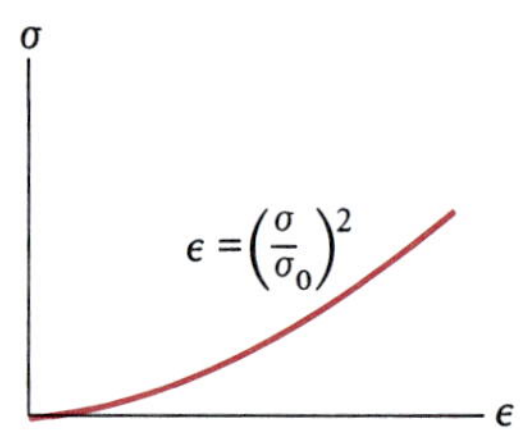

P. 4.24

4.4 DUCTILE AND BRITTLE BEHAVIOR

At this point it is convenient to distinguish between materials that behave in a brittle manner and those that behave in a ductile manner. A material that behaves in a **ductile** manner generally experiences large inelastic strains before fracturing. Stress-strain curves for typical materials behaving in a ductile manner such as steel or aluminum at room temperature are indicated in Fig. 4.22.

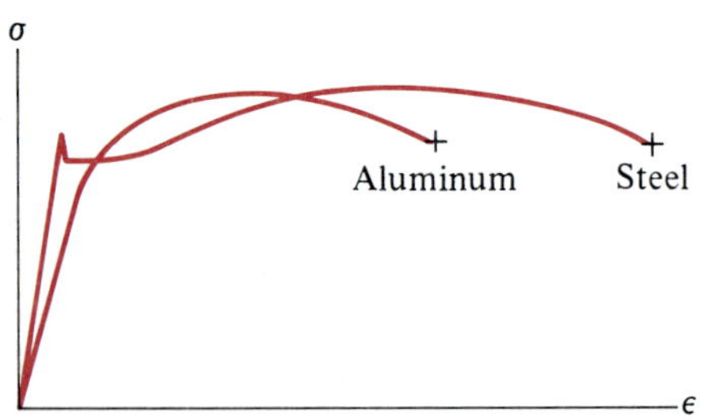

Figure 4.22 Stress-strain curves for steel and aluminium

As an example, an ordinary paper clip at room temperature is composed of an extremely ductile steel in that it can undergo very large permanent bending deformations before fracturing.

A material that behaves in a **brittle** manner, on the other hand, exhibits little permanent deformation before fracture. A stress-strain curve for a typical material behaving in a brittle manner is indicated in

Fig. 4.23. Examples of materials behaving in a brittle manner include chalk, steel at low temperatures, and grey cast iron.

It is inappropriate to refer to a particular material as brittle or ductile. Rather, a material should be described as behaving in a ductile or brittle manner in a particular environment. Temperature and rate of loading are two factors that can influence whether a particular material behaves in a ductile or brittle manner. Steel at room temperature behaves in a ductile manner whereas at low temperatures it can behave in a brittle manner. In particular, a steel bar that has been immersed in liquid nitrogen is so brittle that it will shatter if struck sharply by a hammer. Silly Putty, on the other hand, is very sensitive to the rate at which it is loaded. If a piece of Silly Putty is extended slowly it will elongate to several times its own length, that is, behave in a ductile manner. If it is loaded suddenly in tension or rolled into a ball and thrown sharply against a hard surface, it will fracture with relatively little deformation, that is, behave in a brittle fashion.

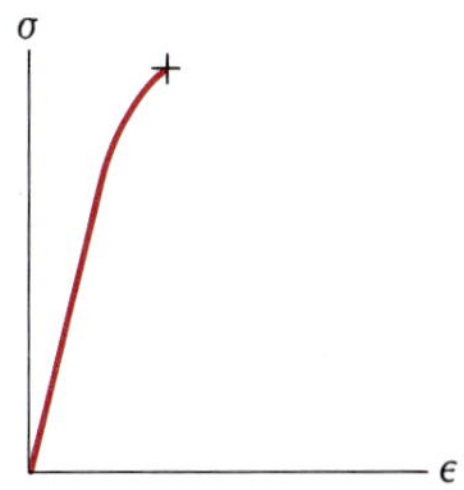

Figure 4.23 Stress-strain diagram for a typical brittle material

Ductile behavior As mentioned, steel and aluminum are materials that behave in a ductile fashion under normal conditions. The stress-strain curve for a typical steel at room temperature is indicated in Fig. 4.24.

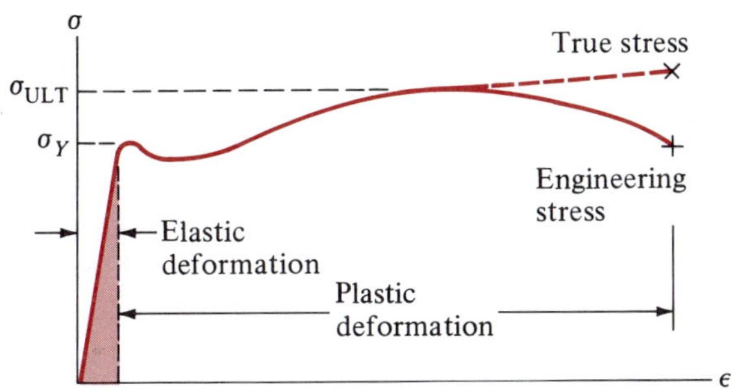

Figure 4.24 Stress-strain curve for steel

There is clearly a linearly elastic range of behavior near the origin. For many steels the phenomenon called yielding occurs at a stress level usually somewhat above the proportional and elastic limits. The stress at which this occurs is called the yield stress, denoted by σ_Y in Fig. 4.24. This yield stress is sometimes referred to as the upper yield stress. At the upper yield stress, the stress actually drops off somewhat as the strain continues to increase. The stress level decreases until the lower yield stress is reached, after which there is a general increase in stress until the ultimate stress σ_{ULT} is reached. There is then a drop in the stress to the point where the bar fractures, that is, breaks into two

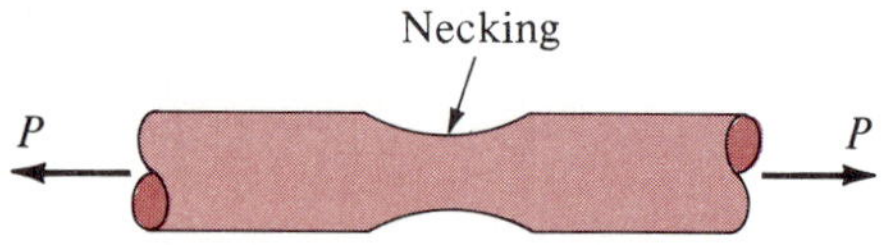

(a)

(b)

Figure 4.25 Reduction of area associated with necking. (*b*) Photo courtesy MTS Systems Corp., Minneapolis, MN.

pieces. The portion of the experiment between the ultimate stress and fracture is usually accompanied by a phenomenon called **necking,** which is shown in Fig. 4.25.

The indicated decrease in the stress level between the ultimate stress and fracture is due to the fact that the original area is used for computing σ. If the actual area of the necked-down portion of the bar is taken into account, the stress-strain curve appears as in the dotted line of Fig. 4.24. The stress computed using the actual area is called the **true stress** whereas the stress computed using the original is called **engineering stress.**

The deformation that takes place over the elastic range of stress and strain is referred to as **elastic deformation.** Any deformation between the elastic range and failure is referred to as **plastic deformation** or inelastic deformation. As mentioned previously, for a ductile material the total deformation before failure consists almost entirely of plastic deformation.

If unloading occurs somewhere within the plastic region (point B) as indicated in Fig. 4.26, the unloading curve is essentially parallel to the straight elastic portion of the stress-strain curve as shown. The permanent plastic deformation corresponds to the plastic strain at point C. If the material is then reloaded from point C, the loading curve retraces the unloading curve in a linear fashion until a new proportional limit is reached. The modulus of elasticity E has not changed but the proportional and elastic limits have changed. The material can now be stressed to a higher level before behaving inelastically. This phenomenon is called strain hardening or work hardening. It is the basis for several processes designed to increase the strength of a material, that is, to produce a material which can be stressed to a higher level while continuing to behave elastically.

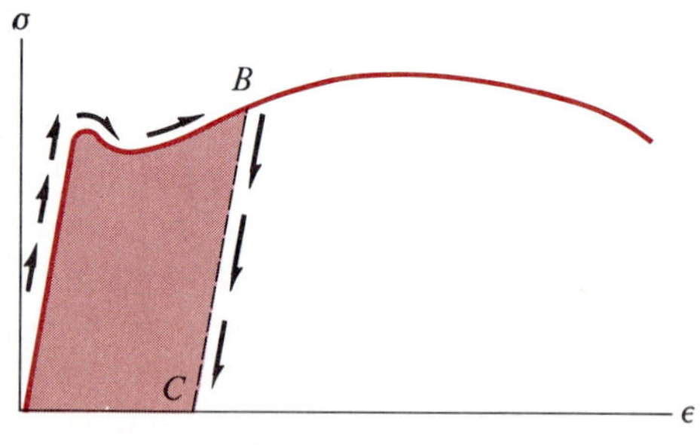

Figure 4.26 Unloading in the plastic region

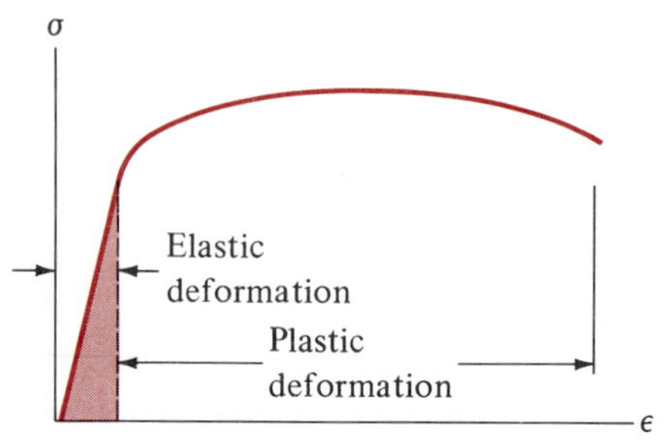

Figure 4.27 Stress-strain curve for ordinary aluminum

Ordinary aluminum at room temperature behaves in a ductile manner as indicated in Fig. 4.27. There is an initial linearly elastic region after which a large amount of plastic deformation occurs before fracture. Aluminum also exhibits the necking phenomenon.

Example 4.7 Take an ordinary paper clip which has been straightened as indicated in Fig. 4.28. Applying a small load produces a small transverse deformation of the paper clip as indicated in Fig. 4.28a. When the load is removed the paper clip returns to its original configuration (i.e., no permanent deformation). This is evidence that the stresses resulting from the applied load were at or below the elastic limit. Applying a large load results in a substantial deflection as indicated in Fig. 4.28b. When this load is removed there is permanent deformation. The material has behaved in an inelastic manner because the stresses from the larger load were in excess of the elastic limit. The deformations past the elastic range take place in a ductile fashion, in a manner that allows large strains past the value of the strain corresponding to the elastic limit. If the wire were replaced by a slender piece of hard plastic, the plastic would not undergo large deformations but would fracture in a brittle manner.

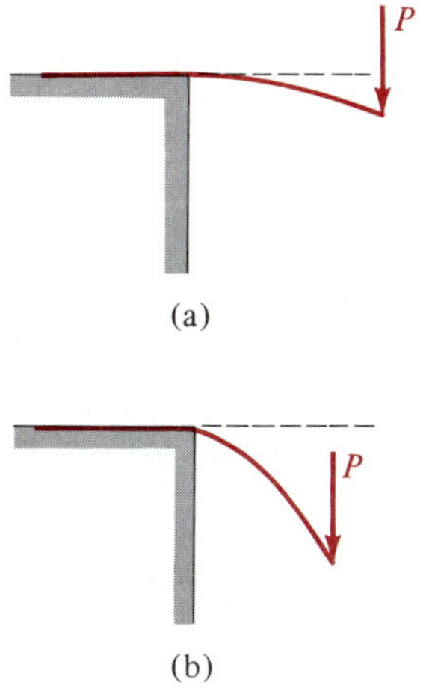

Figure 4.28 Elastic and inelastic deformation of a paper clip

The practical ideas involved in Example 4.7 are basic to the study of solid mechanics. In many situations a structure is intended to withstand loads without any permanent deformation. This means we must be able to determine the stresses caused by the loads and assure that they will not produce permanent deformation.

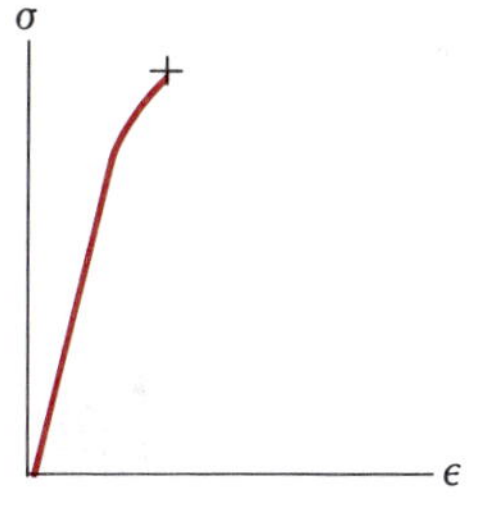

Figure 4.29 Typical stress-strain curve for a material behaving in a brittle manner

Brittle behavior A material that behaves in a brittle manner does not yield, as does a ductile material. As shown in Fig. 4.29, there is a relatively small amount of plastic deformation associated with stresses above the elastic limit before fracture occurs. Grey cast iron and many ceramics are examples of materials that behave in a brittle manner at room temperature.

Primarily in connection with brittle behaviors there is yet another stress value of importance called the **yield strength.** This stress value is defined by selecting an arbitrary strain value referred to as an **offset strain,** usually in the range from 0.05 to 0.30%, and erecting a line parallel to the linear portion of the stress-strain curve as shown in Fig. 4.30.

The value of the stress σ_{YS} at the intersection of this parallel line with the stress-strain curve is the yield strength. The value 0.20% is commonly chosen for determining the yield strength. The method described above for determining the yield strength is often referred to as the **offset yield method.** The yield strength is also used for aluminum where there is not a definite yield stress.

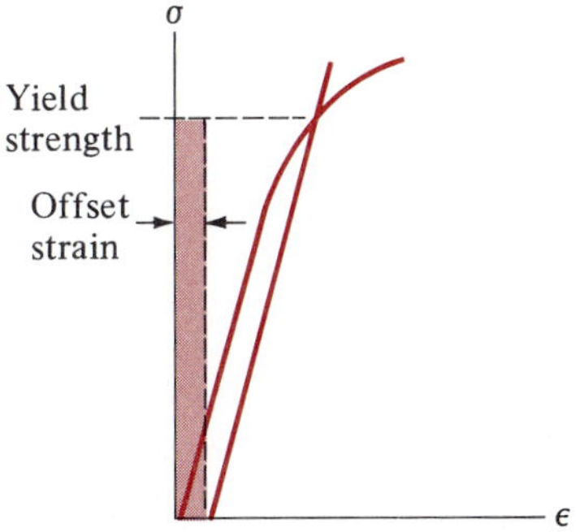

Figure 4.30 Definition of offset yield strength

One of the major differences in ductile versus brittle material behavior is in the ability of a ductile material to undergo relatively large deformation before fracturing. There are practical situations, such as in the crashworthiness of airplanes and automobiles, where this capability can be used to advantage. The basic idea is to convert kinetic energy into potential or strain energy. A car or airplane that crashes possesses a large amount of kinetic energy. A passenger in such a vehicle experiences large accelerations as a result of the crash. Anything that can reduce these accelerations reduces the chances that a passenger will be seriously injured. Rapid accelerations are associated with rapid changes in velocity with respect to time or distance. Increasing the distance over which the velocity change occurs decreases the acceleration. Ultimately this translates into a desire to have large deformations during the deceleration period.

Generally the area under the stress-strain curve, which is an energy per unit volume, is called **strain-energy density.** We designate the strain-energy density by the symbol U_0 defined as

$$U_0 = \int \sigma \, d\varepsilon$$

Strain-energy density has dimensions $FL^{-2} = FL/L^3$ (i.e., energy per unit volume). Common units are lbf/in^2 = in.-lbf/in^3 or newtons/meter^2 = joules/meter^3.

The strain-energy density associated with the linear portion of the stress-strain curve is sometimes referred to as the **modulus of resilience.** The strain-energy density within the linearly elastic range of a solid is given as

$$U_0 = \int_0^{\varepsilon_{PL}} \sigma \, d\varepsilon = \int_0^{\varepsilon_{PL}} E\varepsilon \, d\varepsilon = \frac{E\varepsilon_{PL}^2}{2}$$

The strain-energy density is generally a function of position throughout the body. The total strain energy U in a body is obtained by integrating the strain-energy density over the volume, or

$$U = \int_{\text{Vol}} U_0 \, d\text{Vol}$$

The strain energy that can be recovered upon release of the loads is referred to as elastic strain energy. Strain energy has dimensions of FL, the same as any energy. Common units are in.-lbf or newton-meters = joules. Strain energy and its uses in formulating and solving problems in solid mechanics are discussed in detail in Chapter 10.

The deformation of a solid is related to the strain experienced during the deformation. The more a material can be deformed before fracturing, the more energy it can absorb. The strain-energy density associated with the entire stress-strain curve out to fracture is called the **modulus of toughness** or simply the **toughness** of the material.

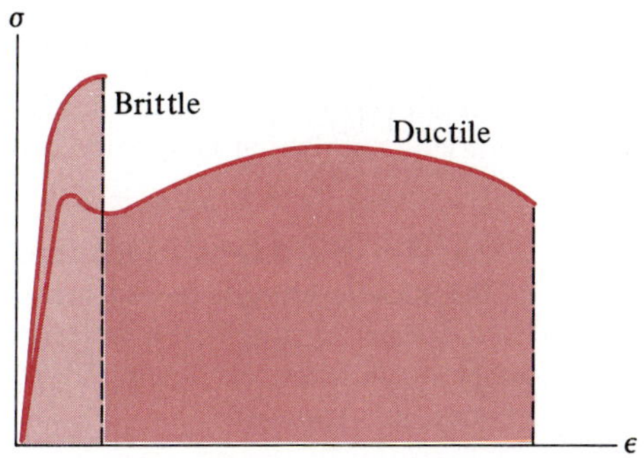

Figure 4.31 Stress-strain diagrams for ductile and brittle materials

Inspection of Fig. 4.31 clearly reveals that the toughness of a ductile material is generally much greater than that of a brittle material. Toughness is an important consideration in any situation where it is necessary to absorb a large amount of energy, particularly in large plastic deformations.

Example 4.8 A uniform bar is loaded by an axial force as indicated in Fig. 4.32. Determine the elastic energy in the bar.

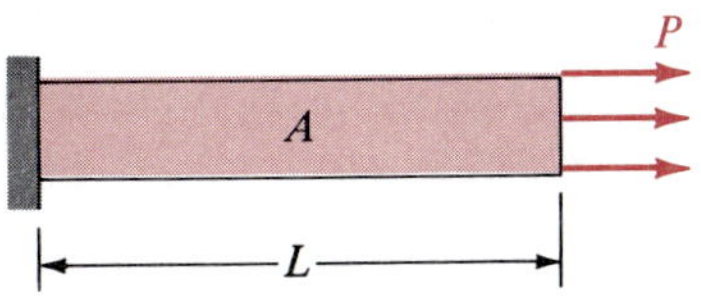

Figure 4.32 Bar loaded by an axial force

Solution Assuming that the stress is below the proportional limit, the strain energy is

$$U = \int_{\mathrm{Vol}} U_0 \, d\mathrm{Vol} = \int_{\mathrm{Vol}} \frac{E\varepsilon^2}{2} \, d\mathrm{Vol}$$

The stress is given by $\sigma = P/A$ and the strain by $\varepsilon = \sigma/E = P/AE$ so that the strain energy density is $P^2/2A^2E$ and the strain energy is

$$U = \int_{\mathrm{Vol}} \left(\frac{P^2}{2A^2E}\right) d\mathrm{Vol}$$

Since the integrand is constant, this reduces to

$$U = \frac{P^2L}{2AE}$$

as the total elastic energy of the bar.

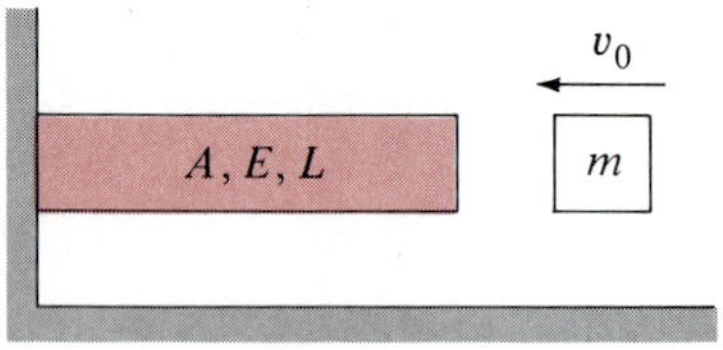

Figure 4.33 Model of an energy-absorbing device

Example 4.9 A structure is to be designed so as to absorb a given amount of kinetic energy associated with a mass m as indicated in Fig. 4.33. Analyze the interrelations between the variables if the bar is to behave in an elastic manner during the process.

Solution We will assume that the structure is a uniform elastic rod as indicated in Fig. 4.33. All of the kinetic energy of the mass is assumed to be transferred into elastic strain energy or potential energy in the structure. The transfer of energy is to take place without permanent deformation of the structure. The equation representing this transfer is

$$\text{K.E.}_{\text{mass}} = \text{P.E.}_{\text{structure}} \tag{a}$$

or

$$\frac{1}{2} m v_0^2 = \int_{\text{Vol}} \frac{E\varepsilon_{PL}^2}{2} \, d\text{Vol}$$

where ε_{PL} is the proportional limit strain. (At this point we make the assumption that the strain-energy density is uniform throughout the body and that the maximum contraction of the structure occurs at the instant the velocity of the mass is reduced to zero. This essentially ignores the fact that stress waves propagate through the structure as a result of the suddenly applied force in the rod.) Eqn. (a) can then be expressed as

$$\frac{1}{2} m v_0^2 = E \frac{\varepsilon_{PL}^2}{2} AL \tag{b}$$

We now take the average extensional strain ε_{PL} to be Δ/L, where Δ is the total shortening of the rod. Solving Eqn. (b) for Δ yields

$$\Delta = v_0 \sqrt{\frac{mL}{AE}} = v_0 \sqrt{\frac{m}{k}}$$

where $k = AE/L$ is termed the stiffness of the rod. Note that k has dimensions FL^{-1}. This result clearly indicates that to make Δ large, and hence reduce the accelerations of the mass, k should be made small. This can be accomplished by making A or E small or by making L large. This must be done keeping in mind that the stress, given by $E\varepsilon$, must remain below σ_{PL}. In terms of the contraction Δ, the stress is given by

$$\sigma = \frac{E\,\Delta}{L} = v_0 \sqrt{\frac{mE}{AL}}$$

At this point the problem becomes a parameter study to produce a design with Δ as large as possible while at the same time keeping σ at or below σ_{PL}.

PROBLEMS

4.25 Each of the bars shown is subjected to the same load P. Determine the energy absorbed by each in the elastic range.

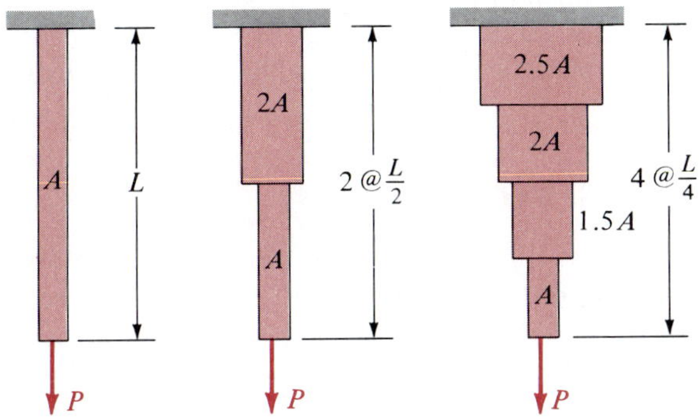

P. 4.25

4.26 The two bars shown are to absorb the same amount of energy in the elastic range. Determine *(a)* the ratio P_1/P_2, and *(b)* the ratio of the maximum stress in the two bars.

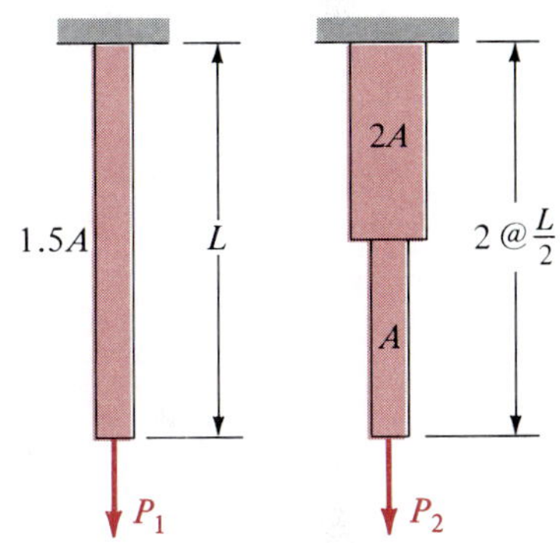

P. 4.26

4.27 With reference to the data of problem 4.4, determine *(a)* the modulus of resilience and *(b)* the modulus of toughness.

4.28 With reference to the data of problem 4.8, determine *(a)* the modulus of resilience and *(b)* the modulus of toughness.

4.29 The two bars shown are composed of a material whose behavior has been approximated by the stress-strain curve indicated. If the energy to be absorbed by the two bars is to be the same, determine P_1.

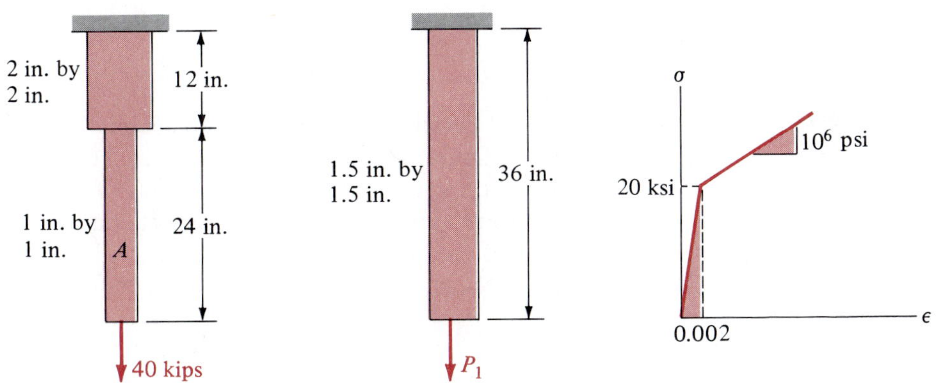

P. 4.29

4.30 An elevator is being lowered at a speed of $v = 3$ m/s. When the extended length of the cable is $L = 75$ m, the elevator is suddenly stopped by the drum to which the cable is attached. Determine the maximum tensile stress σ in the cable. Also determine the percentage of the maximum tensile stress due to having suddenly stopped the elevator. The area A of the steel cable is 700 mm^2 and the mass m of the elevator is 1600 kg. Neglect the mass of the cable and assume that there are no energy losses. (*Suggestion:* First set up the problem in general in terms of σ, m, L, v, A, E, and g before calculating the numerical values.)

4.5 THERMAL EFFECTS

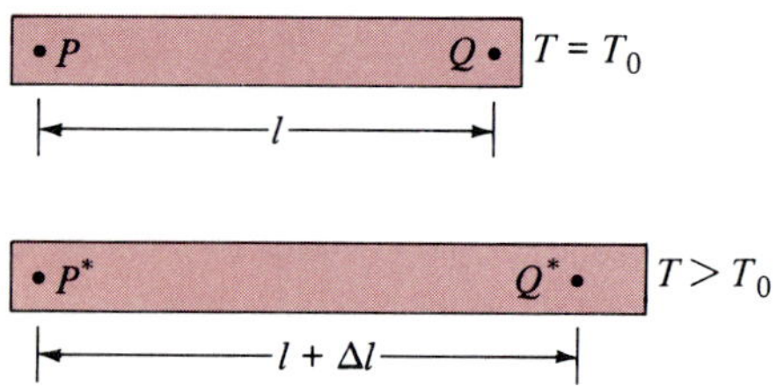

Figure 4.34 Test specimen for thermal effects

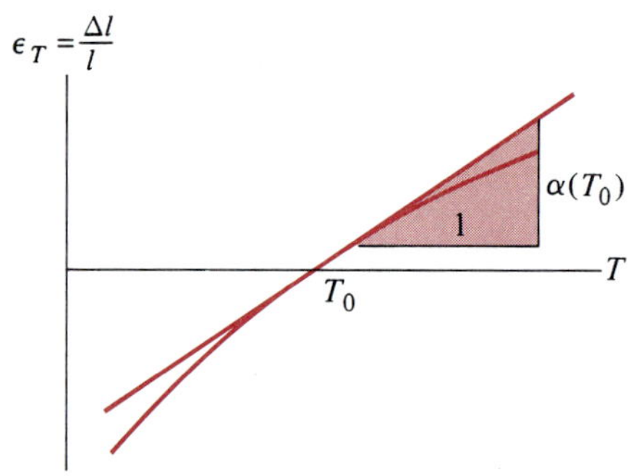

Figure 4.35 Thermal strains as a function of temperature

It is well known that a solid generally changes in size and shape as a result of a change in temperature. Solids generally expand when heated and contract when cooled. The experiment to determine the nature of this effect is as follows. Two marks are made on a specimen initially at a uniform temperature as indicated in Fig. 4.34.

The distance l between the marks is referred to as a gage length. The unrestrained specimen is then subjected to a temperature change ΔT and the corresponding change in the gage length Δl is measured. The results are indicated in Fig. 4.35.

The results of Fig. 4.35 indicate that the thermal strain ε_T is approximately a linear function of temperature in the neighborhood of the initial temperature T_0. This thermal strain versus temperature behavior is represented as

$$\varepsilon_T = \alpha(T_0)\,\Delta T = \alpha(T_0)(T - T_0)$$

where $\alpha(T_0)$ is the slope of the curve at $T = T_0$. The corresponding elongation of the gage length l is given by

$$\varepsilon_T = \frac{\Delta l}{l}$$

or

$$\Delta l = l\,\varepsilon_T = \alpha(T_0)\,\Delta T\,l$$

The symbol $\alpha(T_0)$ is called the **coefficient of thermal expansion.** Typical units associated with the coefficient of thermal expansion are in./in./°F and m/m/°C. For most engineering materials the dependence of α on T_0 is weak and we will in the future assume that α is a constant equal to an average value of the slope of the curve in Fig. 4.35. We will simply refer to α, the coefficient of thermal expansion, and write that

$$\varepsilon_T = \alpha\,\Delta T$$

and

$$\Delta l = l\,\varepsilon_T = \alpha\,\Delta T\,l$$

Any strains resulting from thermal effects are assumed to be independent of any stress or deformation that exists at the time of, or that is imposed during or after, the temperature change. This enables us to use superposition and to write that the total strain resulting from a combination of mechanical and thermal effects can be superposed according to

$$\varepsilon_{\text{TOTAL}} = \varepsilon_{\text{MECH}} + \varepsilon_{\text{THERMAL}}$$

or

$$\varepsilon_{\text{TOTAL}} = \varepsilon_{\text{MECH}} + \alpha\,\Delta T$$

This is the stress-strain temperature relation for a one-dimensional solid. If the solid is linearly elastic the stress-strain temperature relation for a one-dimensional solid can be further specialized to

$$\varepsilon_{\text{TOTAL}} = \varepsilon = \frac{\sigma}{E} + \alpha\,\Delta T \tag{4.4}$$

where E is Young's modulus. Another form of this relation that is frequently useful is

$$\sigma = E(\varepsilon - \alpha\,\Delta T) \tag{4.5}$$

Example 4.10 A bar of length L, fixed at one end and free at the other as indicated in Fig. 4.36a, is subjected to a uniform temperature change ΔT. Determine the stresses and displacements in the bar.

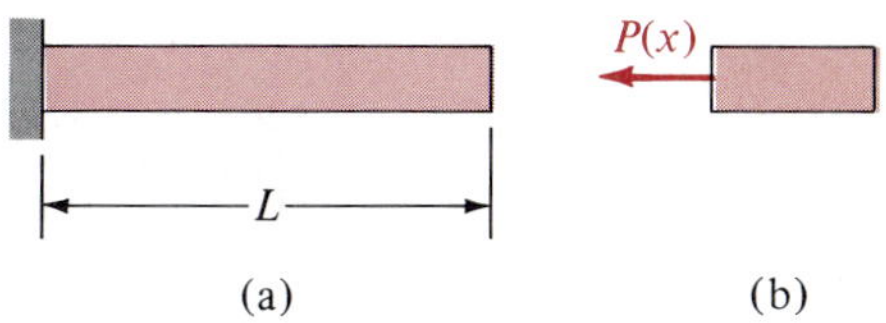

Figure 4.36 Bar subjected to a uniform temperature change

Solution There are no mechanical loads applied so that the free-body diagram of a portion of the bar indicated in Fig. 4.36b yields the result that there is no internal force P transmitted and hence that $\sigma = P/A = 0$. From Eqn. (4.4), the strain is then given by

$$\varepsilon = \alpha\,\Delta T$$

Assuming that α is a constant, the elongation of the bar is obtained from

$$\varepsilon = \frac{\Delta L}{L}$$

or

$$\Delta L = \varepsilon L = \alpha\,\Delta T\,L$$

for the elongation of the bar. The conclusion is that for this statically determinate structure, there is strain ($\varepsilon = \alpha\Delta T$), but no stress!

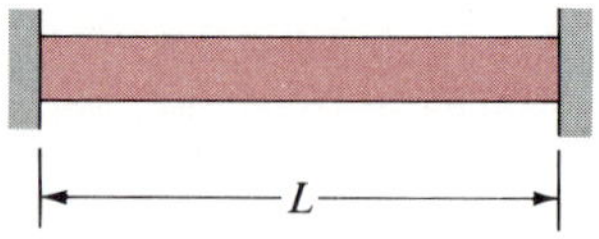

Figure 4.37 Bar completely restrained against motion

Example 4.11 The bar of Example 4.10 is fixed at both ends (i.e., statically indeterminate), so as to prevent motion at either end as shown in Fig. 4.37. It is assumed that the bar is initially free of stress. Determine the stresses and displacements in the bar when the temperature is increased uniformly an amount ΔT.

Solution Using Eqn. (4.4),

$$\varepsilon = u' = \frac{\sigma}{E} + \alpha \, \Delta T$$

An integration yields

$$u(x) = \left(\frac{\sigma}{E} + \alpha \, \Delta T\right) x + C_1$$

Requiring that the displacement vanish at both ends yields

$$u(0) = 0 = 0 + C_1$$

and

$$u(L) = 0 = \left(\frac{\sigma}{E} + \alpha \, \Delta T\right) L + C_1$$

from which $C_1 = 0$ and $\sigma = -E\alpha \, \Delta T$. Thus for this problem $u \equiv 0$, $\varepsilon \equiv 0$; there is stress but no strain!

Examples 4.10 and 4.11, respectively, indicate the differences in the behavior of very simple statically determinate and statically indeterminate structures. It is very important to keep in mind that for a statically determinate structure subjected to a uniform temperature change there are no resulting thermal stresses. In general however, there can be thermal stresses that result when either *(a)* a statically determinate structure is subjected to a nonuniform temperature change, or *(b)* a statically indeterminate structure is subjected to any temperature changes.

PROBLEMS

4.31 A large cylindrical steel shell with a wall thickness of 0.8 in. is used in the processing of cement. The length of the shell is 18 ft and its diameter is 15 ft. Determine the length and diameter if the temperature of the shell is increased by 130°F. Take $\alpha = 6.5 \times 10^{-6}$ in./in.°F.

4.32 A concrete ($E = 3.6 \times 10^6$ psi, $\alpha = 6 \times 10^{-6}$ in./in.°F) bridge for an overpass is 140 ft long and has an area (perpendicular to its length) of 120 ft^2. If the bridge was poured at a temperature of 85°F, determine *(a)* the expansion when the temperature of the concrete reaches 120°F, and *(b)* the corresponding compressive force if the bridge were completely restrained.

4.33 A prismatic bar of length L is placed between two rigid walls separated by a distance of $L + g$ as shown. Determine the stress developed in the bar in response to an increase in temperature ΔT. Take the material to be linearly elastic with a modulus of elasticity E and a coefficient of thermal expansion α. Note that if $g = 0$, your answer should coincide with the answer found in Example 4.11.

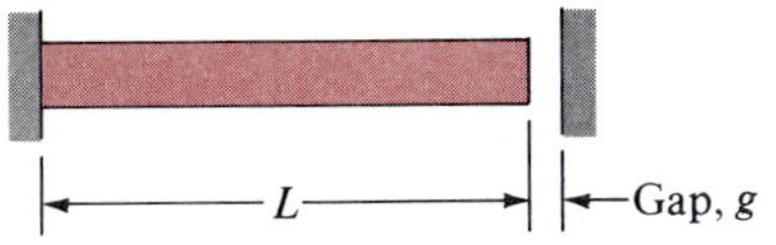

P. 4.33

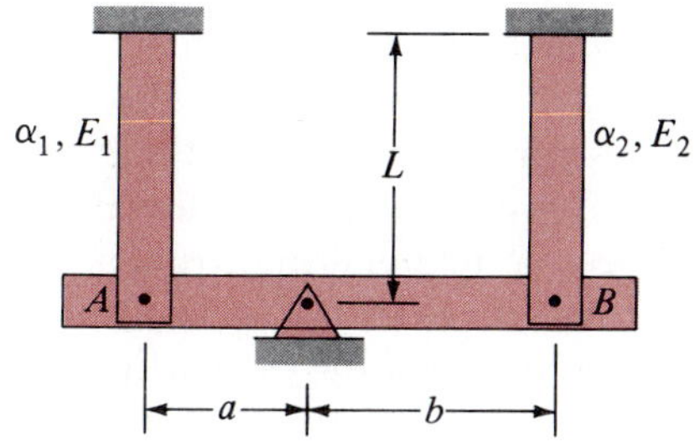

P. 4.34

4.34 The temperatures of the aluminum and steel bars shown are each increased by an amount ΔT. Assuming that bar AB is rigid and that there is no force in either rod initially, determine *(a)* the stress in each bar and *(b)* the rotation of bar AB.

4.35 The assembly shown is to act as a thermometer. At room temperature the arm BE is vertical. Develop the relation between the variables a, b, c, L, α_S, α_A, and ΔT in order that each increment on the scale corresponds to a change of temperature ΔT. Neglect thermal expansion of the vertical arm and pointer and assume that $c/L << 1$. Specifically, if $b = 60$ mm, $a = 5$ mm, $c = 1$ mm, $\Delta T = 1\text{F}°$, what must be the length L? Take $\alpha_S = 6.5 \times 10^{-6}/°\text{F}$ and $\alpha_A = 13 \times 10^{-6}/°\text{F}$.

4.36 A steel ring with an outside diameter of 3.000 in. and an inside diameter of 2.93 in. is to be heated so that it can be slipped with a diametral clearance of 0.002 in. over a rigid 2.935-in.-diameter rod. Determine *(a)* the required temperature increase, and *(b)* the stress in the ring when the temperature is reduced to room temperature. Take $\alpha = 6.5 \times 10^{-6}/°\text{F}$ and $E = 29 \times 10^6$ psi.

4.37 A steel rail 40 ft long has a cross-sectional area of approximately 12 in^2. When exposed to the midday sun the temperature of the rail can reach a temperature of approximately 125°F. What elongation must be allowed for if the rails are initially laid at 75°F? If the rails are actually laid with a 0.125 in. gap between successive rails, what internal force is developed? (Assume that the rail is constrained to remain straight during the temperature increase.) Take $\alpha = 6.5 \times 10^{-6}/°\text{F}$ and $E = 29 \times 10^6$ psi.

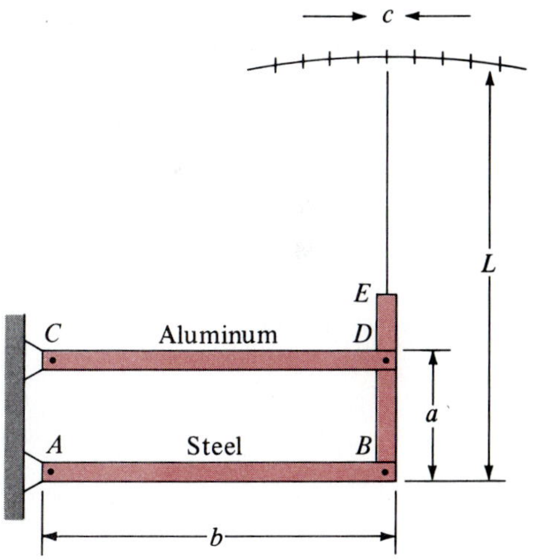

P. 4.35

4.6 LINEARLY ELASTIC BEHAVIOR—HOOKE'S LAW

Because of the simplified nature of the mathematical model of solids for which stress and strain are linearly related, we devote particular attention to formulating and solving problems of this class. Fortunately, many engineering materials behave in a linear manner over a practically important range of stress and deformation. In this section we will discuss the relationship between stress, strain, and temperature for materials subjected to environments such that the material can be considered to be linearly elastic.

The crucial aspect of the assumption that a solid behaves in a linearly elastic fashion is this: If in the uniaxial tension test we apply a stress σ_1, the resulting strain is $\varepsilon_1 = \sigma_1/E$. If another stress σ_2 (resulting in the strain $\varepsilon_2 = \sigma_2/E$) is subsequently applied and is such that $\sigma_1 + \sigma_2$ does not exceed the proportional limit σ_{PL}, the total strain is simply the sum of the individual strains, or

$$\varepsilon_{\text{TOTAL}} = \varepsilon_1 + \varepsilon_2 = \frac{\sigma_1}{E} + \frac{\sigma_2}{E} = \frac{\sigma_1 + \sigma_2}{E}$$

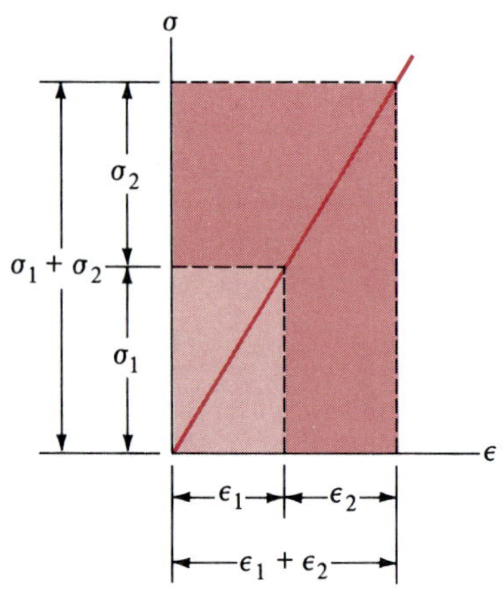

Figure 4.38 Superposition of stresses and strains

This superposition is indicated in Fig. 4.38.

It also follows from linearity that given $\sigma_1 = E\varepsilon_1$ and $\sigma_2 = E\varepsilon_2$, the two relationships can be added or superposed according to

$$\sigma_1 + \sigma_1 = E\varepsilon_1 + E\varepsilon_2 = E(\varepsilon_1 + \varepsilon_2)$$

that is, the stresses arising from the superposition of two strains can be superposed as long as the resulting state is within the proportional range. Note that this is simply a restatement of the superposition as applied to the stresses.

To this point, we have focused entirely on simple one-dimensional states of stress and strain. Due to the nature of the stresses in most real applications, it is necessary to extend these ideas to more complicated states of stress.

> Assume that if the loading on a structure does not result in any stress component exceeding the proportional limit as determined by the uniaxial tension test for the one-dimensional solid, the principle of superposition holds. This means that the effects from the simultaneous application of any and all normal and shear stresses can be superposed.

With this in mind we need to design an experiment to determine the form of the stress-strain relations for a linearly elastic body. Consider a body composed of a linearly elastic material as shown in Fig. 4.39. We describe a *gedanken* (thought) experiment, which is very useful in helping to visualize and understand the idea of stress-strain relations for a two-dimensional plane-stress situation. The corresponding three-dimensional case can be carried out in a similar fashion.

Consider a point P in the body and define a local xy coordinate system at P as shown in Fig. 4.39b. Imagine that an infinitesimal block of material in the neighborhood of P and with sides parallel to the coordinate planes is extracted from the body and subjected to the sequence of tests indicated in Fig. 4.40.

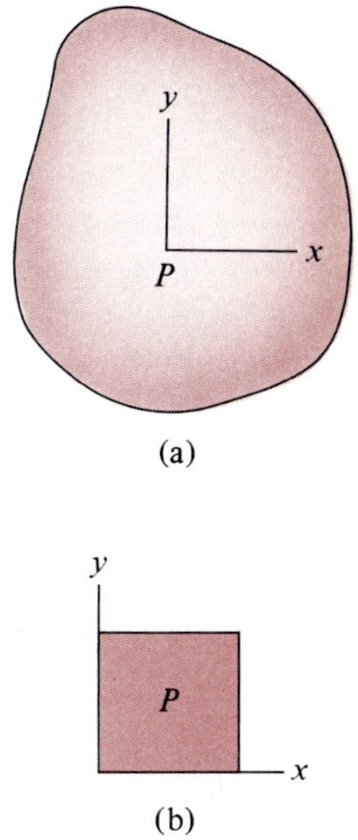

Figure 4.39 Thought experiment for assessing the stress-strain relations at a point

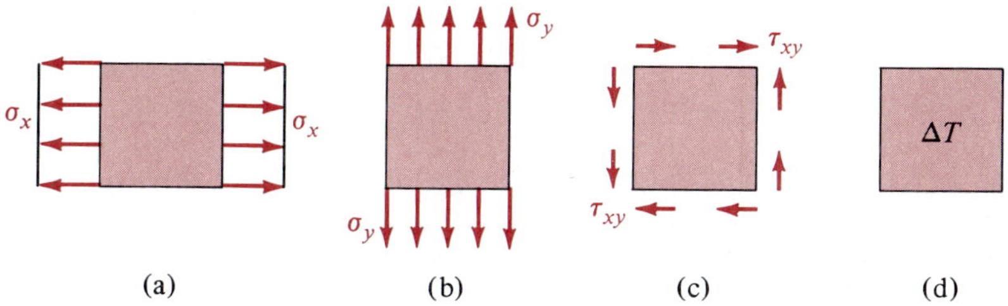

Figure 4.40 Sequence of tests to determine the material behavior of a two-dimensional elastic solid

Test 1 As shown in Fig. 4.40a, apply a known uniform normal stress σ_x to the block, and measure the resulting strains ε_x, ε_y, and γ_{xy}. The results of the experiment enable us to express each of the measured strains in terms of the stress σ_x as

$$\begin{aligned}\varepsilon_x &= S_{11}\sigma_x\\ \varepsilon_y &= S_{21}\sigma_x\\ \gamma_{xy} &= S_{31}\sigma_x\end{aligned}$$

Test 2 As shown in Fig. 4.40b, apply a known uniform normal stress σ_y to the block, and measure the resulting strains, ε_x, ε_y, and γ_{xy}. The three strains can then be expressed in terms of the stress σ_y as

$$\begin{aligned}\varepsilon_x &= S_{12}\sigma_y\\ \varepsilon_y &= S_{22}\sigma_y\\ \gamma_{xy} &= S_{32}\sigma_y\end{aligned}$$

Test 3 As shown in Fig. 4.40c, apply a known uniform shear stress τ_{xy} to the block, and measure the resulting strains ε_x, ε_y, and γ_{xy}. The results can be expressed as

$$\begin{aligned}\varepsilon_x &= S_{13}\tau_{xy}\\ \varepsilon_y &= S_{23}\tau_{xy}\\ \gamma_{xy} &= S_{33}\tau_{xy}\end{aligned}$$

Test 4 As shown in Fig. 4.40d, change the temperature of the block an amount ΔT, and measure the resulting strains ε_x, ε_y, and γ_{xy}. The results of the experiment enable us to express each of the measured strains in terms of the temperature change ΔT as

$$\begin{aligned}\varepsilon_x &= \alpha_1\,\Delta T\\ \varepsilon_y &= \alpha_2\,\Delta T\\ \gamma_{xy} &= \alpha_3\,\Delta T\end{aligned}$$

The fact that the material is linearly elastic means that the principle of superposition is valid; thus, if we were to apply all of the stresses and the temperature increase simultaneously, the strains would be the sum of the strains for each stress and the temperature increase applied separately. This superposition enables us to express the strains resulting from the simultaneous application of the stresses and the temperature increase as

$$\begin{aligned}\varepsilon_x &= S_{11}\sigma_x + S_{12}\sigma_y + S_{13}\tau_{xy} + \alpha_1\,\Delta T\\ \varepsilon_y &= S_{21}\sigma_x + S_{22}\sigma_y + S_{23}\tau_{xy} + \alpha_2\,\Delta T\\ \gamma_{xy} &= S_{31}\sigma_x + S_{32}\sigma_y + S_{33}\tau_{xy} + \alpha_3\,\Delta T\end{aligned} \tag{4.6}$$

The coefficients S_{ij} are known as **flexibility** or **compliance coefficients** and α_i as coefficients of thermal expansion. It is known from the theory of elasticity that the coefficients S_{ij} are such that $S_{12} = S_{21}$, $S_{13} = S_{31}$, . . . ; that is, considered as an array, S is symmetric. Since S is symmetric there are six independent constants S_{ij} for the two-dimensional elastic solid. The three coefficients α_i are called the generalized coefficients of thermal expansion. The form of Hooke's law given in Eqns. (4.6) is general enough to describe the behavior of any linearly elastic two-dimensional solid. Such a completely general material is referred to as **anisotropic** or **orthotropic.**

In the most general situation the coefficients S_{ij} and α_i can depend on the temperature and on the position within the body; thus $S_{ij} = S_{ij}(x, y, T)$ for instance. In many situations the values of the S_{ij} and the α_i are independent of position throughout the body. Such a body is said to be **homogeneous.**

This statement of the relationship between stress, strain, and temperature given by Eqns. (4.6) is generally referred to as the **generalized Hooke's law,** after Robert Hooke, who studied force deflection relations for springs. In connection with his studies he announced in 1676 the anagram ceiiinosssttuv, which when deciphered states *ut tensio sic vis,* which is Latin for "the force varies as the extension."

To a good approximation, many materials behave in a much simpler fashion than represented in Eqns. (4.6). These materials are referred to as isotropic. To understand what this means, return to the gedanken experiment and extract a block of material oriented along the $x'y'$ system as indicated in Fig. 4.41.

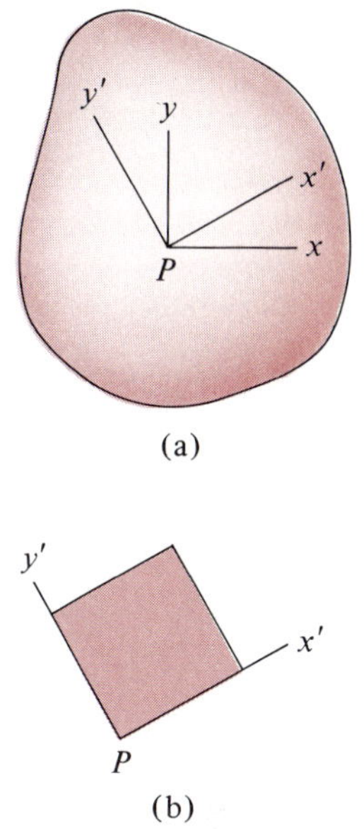

Figure 4.41 Differently oriented block at P

If the sequence of experiments is now performed on the $x'y'$ block, the results for a general anisotropic solid will be

$$\begin{aligned} \varepsilon_{x'} &= S'_{11}\sigma_{x'} + S'_{12}\sigma_{y'} + S'_{13}\tau_{x'y'} + \alpha'_1\,\Delta T \\ \varepsilon_{y'} &= S'_{21}\sigma_{x'} + S'_{22}\sigma_{y'} + S'_{23}\tau_{x'y'} + \alpha'_2\,\Delta T \\ \gamma_{x'y'} &= S'_{31}\sigma_{x'} + S'_{32}\sigma_{y'} + S'_{33}\tau_{x'y'} + \alpha'_3\,\Delta T \end{aligned} \tag{4.7}$$

where in general, all the S'_{ij} will be different than the corresponding S_{ij} obtained from the experiment for the block oriented with respect to the xy coordinate system. Similarly, all the α'_i would, in general, be different than the corresponding α_i.

If, independent of the orientation of the block chosen for the experiment at P, the **FORM** of the stress-strain relations as well as the numerical values of all of the coefficients are the same, the material is said to be **isotropic.** For the isotropic case it can be shown that the strain-stress relations reduce to

$$\varepsilon_x = \frac{\sigma_x - \nu\sigma_y}{E} + \alpha\,\Delta T \tag{4.8a}$$

$$\varepsilon_y = \frac{\sigma_y - \nu\sigma_x}{E} + \alpha\,\Delta T \tag{4.8b}$$

$$\gamma_{xy} = \frac{\tau_{xy}}{G} \tag{4.8c}$$

where E is Young's modulus or the modulus of elasticity, G is the shear modulus, and ν is Poisson's ratio, all of which were defined in Section 4.3. α represents the single coefficient of thermal expansion for an isotropic solid.

Note that for an isotropic material there is no coupling between shear and normal effects, that is, the application of normal stresses produces only extensional strains, but no shear strains. Similarly, the application of shear stresses produces only shear strains, but no extensional strains. Last, a temperature increase results in equal extensional strains and no distortion due to shear. We state without proof that for an isotropic solid, the principal directions for stress coincide with the principal directions for strain.

From the form of the isotropic stress-strain relations it appears as if there are three elastic constants, namely, E, G, and ν, that need to be determined by experiment. That there are in fact only two independent elastic constants is demonstrated as follows. Consider a block of isotropic material acted upon by a state of pure shear as shown in Fig. 4.42.

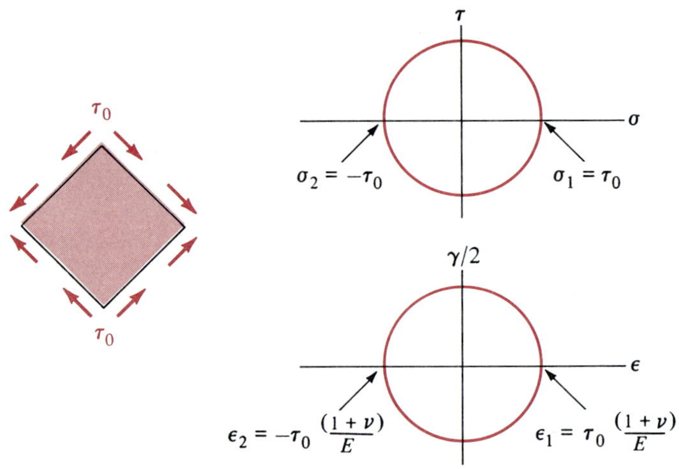

Figure 4.42 Relations between E, G, and ν for an isotropic solid

The Mohr's circle for stress and strain are both indicated. The principal stresses on planes at 45° are $\sigma_1 = \tau_0$ and $\sigma_2 = -\tau_0$, respectively. Using Hooke's law, with all stresses other than τ_{xy} as zero, yields

$$\varepsilon_1 = \frac{\sigma_1 - \nu\sigma_2}{E} = \frac{\tau_0(1 + \nu)}{E}$$

and

$$\varepsilon_2 = \frac{\sigma_2 - \nu\sigma_1}{E} = \frac{-\tau_0(1 + \nu)}{E}$$

for the principal strains. Looking at the Mohr's circle for strain, it follows that the radius R is given by

$$R = \frac{\gamma}{2} = \frac{\varepsilon_1 - \varepsilon_2}{2} = \frac{\tau_0(1 + \nu)}{E}$$

Using the Hooke's law relationship between shear stress and shear strain yields $\gamma = \tau_0/G$. Equating the two expressions for γ yields

$$\frac{\gamma}{2} = \frac{\tau_0}{2G} = \frac{\tau_0(1 + \nu)}{E}$$

from which

$$\frac{1}{2G} = \frac{1 + \nu}{E}$$

or

$$E = 2G(1 + \nu) \tag{4.9}$$

as the relation between the elastic constants E, G, and ν. This clearly indicates that there are two independent elastic constants. An experiment such as the uniaxial tension test on an isotropic solid determines E and ν. Eqn. (4.9) can then be used to compute G according to $G = E/2(1 + \nu)$.

For the corresponding three-dimensional case, the isotropic solid behaves according to the stress-strain relations

$$\begin{aligned}
\varepsilon_x &= \frac{\sigma_x - \nu(\sigma_y + \sigma_z)}{E} + \alpha\,\Delta T \\
\varepsilon_y &= \frac{\sigma_y - \nu(\sigma_z + \sigma_x)}{E} + \alpha\,\Delta T \\
\varepsilon_z &= \frac{\sigma_z - \nu(\sigma_x + \sigma_y)}{E} + \alpha\,\Delta T
\end{aligned} \tag{4.10}$$

and

$$\gamma_{xy} = \frac{\tau_{xy}}{G} \quad \gamma_{yz} = \frac{\tau_{yz}}{G} \quad \text{and} \quad \gamma_{zx} = \frac{\tau_{zx}}{G}$$

For a plane-stress situation, the three stress components σ_z, τ_{yz}, and τ_{xz} are assumed to be zero. The three-dimensional stress-strain equation can be used to conclude that γ_{yz} and γ_{xz} are zero, but that ε_z, the component of strain in the direction perpendicular to the plane, is nonzero, being given by

$$\varepsilon_z = -\frac{\nu(\sigma_x + \sigma_y)}{E}$$

showing that if both σ_x and σ_y are positive at a point in the plate, the strain in the direction perpendicular to the plate is negative, implying that the thickness of the plate decreases. Similarly, if both σ_x and σ_y are negative, the z-dimension of the plate will increase. This change in the z-dimension due to the Poisson's ratio effect should be kept in mind when dealing with isotropic elastic solids.

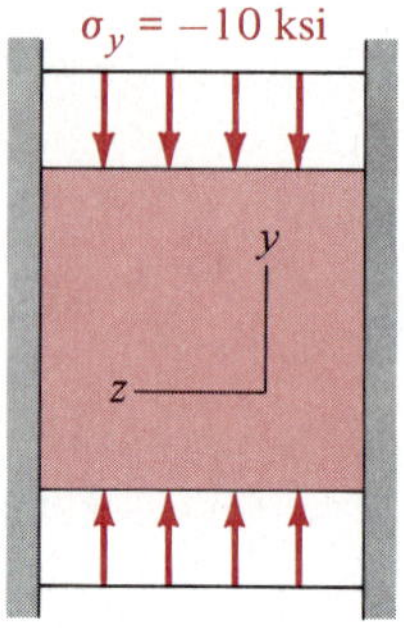

Figure 4.43 Constrained aluminum cube

Example 4.12 A 2-in. aluminum cube is placed between two rigid smooth walls and subjected to the stress $\sigma_y = -10^4$ psi and $\sigma_x = 0$ on the pairs of opposite faces as shown in Fig. 4.43. Determine the stress in the z-direction and the changes in length of the x- and y-dimensions of the cube. Take $\nu = 0.3$ and $\varepsilon = 10^4$ ksi.

Solution The stresses in the block are given by $\sigma_y = -10^4$ psi, $\sigma_x = 0$, and σ_z unknown. The strains ε_x and ε_y are unknown with the strain in the z-direction being zero due to the constraint of the wall. With $\Delta T = 0$, the stress-strain relations of Eqns. (4.10) yield

$$\varepsilon_x = \frac{0 - \nu(\sigma_y + \sigma_z)}{E}$$

$$\varepsilon_y = \frac{\sigma_y - \nu\sigma_z}{E}$$

$$\varepsilon_z = 0 = \frac{\sigma_z - \nu\sigma_y}{E}$$

that is, three equations for the unknowns ε_x, ε_y, and σ_z. From the last of these $\sigma_z = \nu\sigma_y = -3$ ksi. The strains ε_x and ε_y are then given by

$$\varepsilon_x = \frac{-\nu(\sigma_y + \sigma_z)}{E} = \frac{-0.3(-13\text{ ksi})}{10^4\text{ ksi}} = 3.9 \times 10^{-4}$$

and

$$\varepsilon_y = \frac{\sigma_y - \nu\sigma_z}{E} = \frac{-10\text{ ksi} - 0.3(-3\text{ ksi})}{10^4\text{ ksi}} = -9.1 \times 10^{-4}$$

The change in the 2-in. dimensions in the x- and y-directions are $\Delta l_x = 3.9 \times 10^{-4}$ (2 in.) $= 0.78 \times 10^{-3}$ in. and $\Delta l_y = -9.1 \times 10^{-4}$ (2 in.) $= -1.82 \times 10^{-3}$ in., respectively.

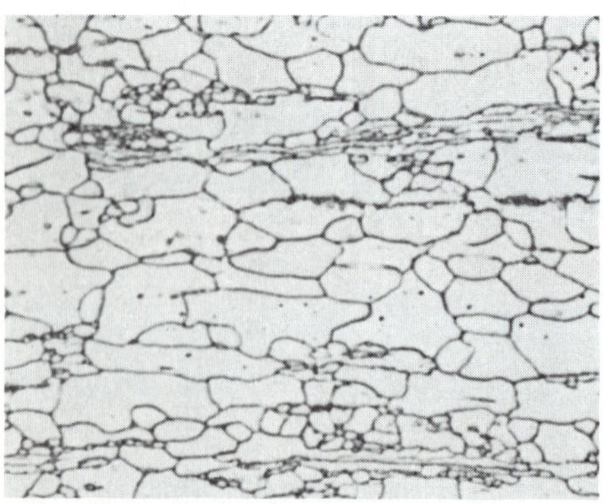

Figure 4.44 Microstructure of steel (Courtesy of J. R. Davis, ASM International, Materials Park, OH.)

If the student has had a course in the area of materials, he or she may be familiar with the details of the microstructure of a typical metal such as steel or aluminum. A greatly magnified photograph of the microstructure of steel, for example, appears as in Fig. 4.44.

The photograph clearly indicates the grain structure and the grain boundaries that have resulted from processing (melting, rolling, cooling) of the steel into its final form. If one considers a small enough chunk of material for the experiment described above, the mechanical properties will be directionally dependent: the material will appear to behave anisotropically. If one instead considers a finite sized chunk, the statistical nature of the orientations of the grains or crystals is often such that the material behaves isotropically. For many steels and aluminum this is in fact the case; in this text we will generally assume metals to behave in an isotropic manner unless there is evidence to the contrary.

As an example of a situation where the assumption of isotropy must be reexamined, consider the process of the forming of steel or aluminum by rolling. By this we mean that the steel is continually forced through a constriction such as depicted in Fig. 4.45 in order to produce a desired shape. This can result in the steel having different properties in directions parallel and perpendicular to the direction of rolling. If these differences are large enough, the isotropic assumption may result in a model that is sufficiently in error so that an appropriate anisotropic model must be employed.

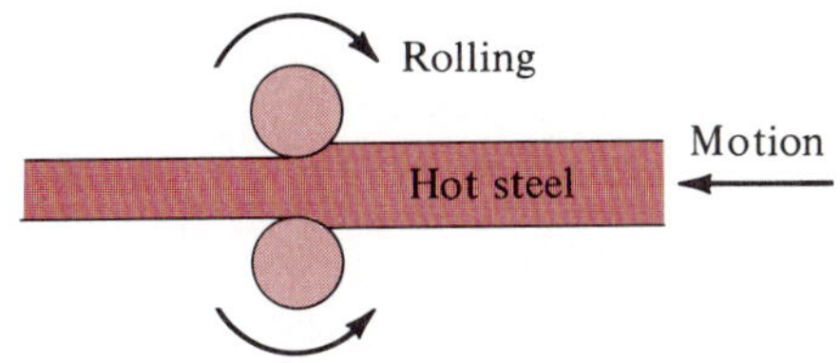

(a)

Figure 4.45 Rolling of steel

Glowing ingot takes on new shape as it is worked by rolls of a blooming mill at a Bethlehem Steel Corporation plant. This ingot eventually will be shaped into a bloom or billet. From these semifinished pieces of steel, finished steel products—such as structural shapes—are rolled. Rolling not only shapes the steel but also improves its mechanical properties. (Photo courtesy of Bethlehem Steel.)

PROBLEMS

4.38 The 1-in.-thick rectangular plate is loaded by stresses on its faces as shown. Determine *(a)* the change in the 100-in. length, *(b)* the change in the 50-in. length, *(c)* the change in the 1-in. length, and *(d)* the change in the volume. Show that $\Delta V/V = \varepsilon_x + \varepsilon_y + \varepsilon_z$. Assume plane stress, $E = 10^7$ psi, and $\nu = 0.3$.

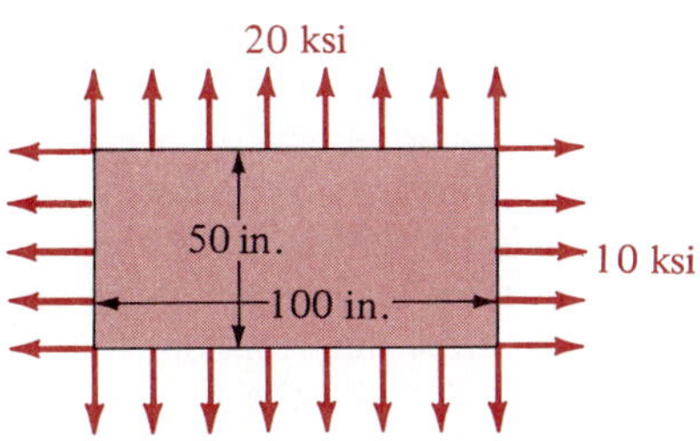

P. 4.38

4.39 A 50-mm by 50-mm by 200-mm steel bar is loaded by a load P along the 200-mm direction. If each of the 50-mm dimensions is observed to decrease to 49.985 mm, determine *(a)* the load P, and *(b)* the increase in the 200-mm dimension. Take $\nu = 0.3$ and $E = 200$ GPa.

4.40 Repeat problem 4.39 if the same decrease in the 50-mm dimensions is observed and if temperature is observed to drop 10°C during the experiment. Take $\alpha = 11.7 \times 10^{-6}$ in./in.°C.

4.41 A uniaxial tension test is run on an aluminum 0.5-in. by 2-in. rectangular specimen with a gage length of 8 in. When the load is 20,000 lbf the gage length has lengthened 0.016 in. What is the value of E? If as a result of an air-conditioning failure the temperature of the test specimen had actually increased 20°F during the test, what is the value of E? Take $\alpha = 13 \times 10^{-6}$ in./in.°F.

4.42 A 2-in. by 2-in. by 2-in. aluminum block is constrained against all motion in the x- and y-directions. It is loaded in the z-direction by a uniform tensile stress of 30,000 psi. Determine *(a)* how much does the block elongate in the z-direction, and *(b)* what are the values of σ_x and σ_y? Take $E = 10^7$ psi and $\nu = 0.3$.

4.43 A block consisting of a rubberlike material is to be tested in shear to determine the shear modulus G by attaching it to ground and to a rigid plate as shown. A load P of 100 kN results in a horizontal displacement of 1 mm. Estimate the shear modulus G. Also develop a general relationship between G, h, w, b, P, and δ.

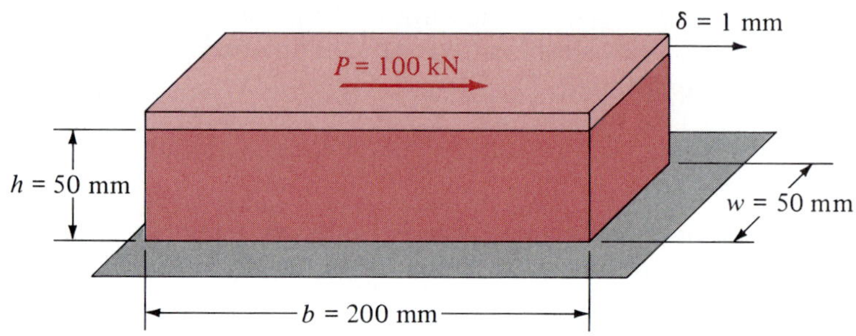

P. 4.43

4.44 Electronic equipment is frequently attached to an airframe by a device that can be idealized as indicated in the figure. The device consists of a rigid shaft connected to a rigid tube by a rubberlike material. Forces are transmitted from the equipment to ground primarily by shear in the rubber. Note that the average shear stress in the rubber is different at $r = a$ than at $r = b$. Using these two shear-stress values, determine an average shear stress through the thickness $b - a$. Then assuming that the deformation is as pictured (i.e., conical), develop an approximate expression for the displacement δ. Take the shear modulus of the material to be G. The result turns out to be quite good as long as $b/a < 2$.

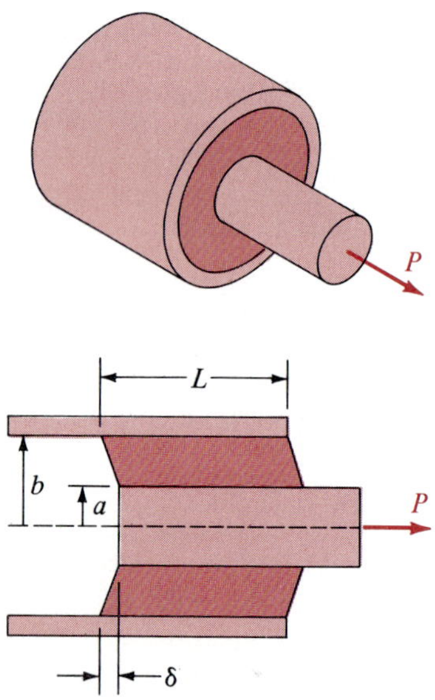

P. 4.44

4.45 Show that if Eqns. (4.8) are solved for the stresses there results

$$\sigma_x = \frac{E}{1-\nu^2}[(\varepsilon_x + \nu\varepsilon_y) - (1+\nu)\alpha T]$$

$$\sigma_y = \frac{E}{1-\nu^2}[(\varepsilon_y + \nu\varepsilon_x) - (1+\nu)\alpha T]$$

$$\tau_{xy} = G\gamma_{xy}$$

4.46 Show that if Eqns. (4.10) are solved for the stresses there results

$$\sigma_x = \frac{E[(1-\nu)\varepsilon_x + \nu(\varepsilon_y + \varepsilon_z) - (1+\nu)\alpha T]}{[(1+\nu)(1-2\nu)]}$$

$$\sigma_y = \frac{E[(1-\nu)\varepsilon_y + \nu(\varepsilon_z + \varepsilon_x) - (1+\nu)\alpha T]}{[(1+\nu)(1-2\nu)]}$$

$$\sigma_z = \frac{E[(1-\nu)\varepsilon_z + \nu(\varepsilon_x + \varepsilon_y) - (1+\nu)\alpha T]}{[(1+\nu)(1-2\nu)]}$$

$$\tau_{xy} = G\gamma_{xy} \qquad \tau_{yz} = G\gamma_{yz} \qquad \text{and} \qquad \tau_{zx} = G\gamma_{zx}$$

4.47 Plane strain refers to a situation where there is no deformation in one direction, usually chosen as the z-direction. It is then assumed that ε_z, γ_{yz}, and γ_{zx} are zero. Show that the corresponding isotropic stress-strain relations for plane strain can be written as

$$\varepsilon_x = \frac{1 - \nu^2}{E}\left(\sigma_x - \frac{\nu\sigma_y}{1 - \nu}\right) + \alpha(1 + \nu)\,\Delta T$$

$$\varepsilon_y = \frac{1 - \nu^2}{E}\left(\sigma_y - \frac{\nu\sigma_x}{1 - \nu}\right) + \alpha(1 + \nu)\,\Delta T$$

$$\gamma_{xy} = \frac{\tau_{xy}}{G}$$

or equivalently

$$\sigma_x = \frac{E[(1 - \nu)\varepsilon_x + \nu\varepsilon_y - (1 + \nu)\alpha T]}{[(1 + \nu)(1 - 2\nu)]}$$

$$\sigma_y = \frac{E[(1 - \nu)\varepsilon_y + \nu\varepsilon_x - (1 + \nu)\alpha T]}{[(1 + \nu)(1 - 2\nu)]}$$

$$\sigma_z = \nu(\sigma_x + \sigma_y) - E\alpha\,\Delta T \qquad \text{and} \qquad \tau_{xy} = G\gamma_{xy}$$

4.6.1 Composite Materials

One of the most common and important applications for linearly elastic materials that are not isotropic is in the field of composites. A composite typically consists of at least two different materials that are combined to produce a useful result. The usefulness stems from the fact that composites are generally quite strong in comparison to their weight (a fact that makes composites particularly useful in the aircraft industry). A simple example of a composite that predates recorded history is the use of straw to reinforce and thus strengthen an adobe brick.

A fibrous composite is a common type that is frequently idealized as in Fig. 4.46. The fibers usually consist of a very high-strength carbon or graphite material that are embedded in what is called the matrix. The matrix is usually a relatively low-modulus, low-strength epoxy that

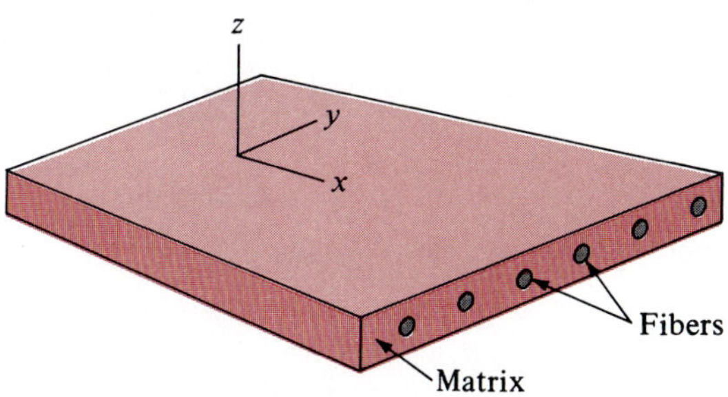

Figure 4.46 Idealization of a fibrous composite

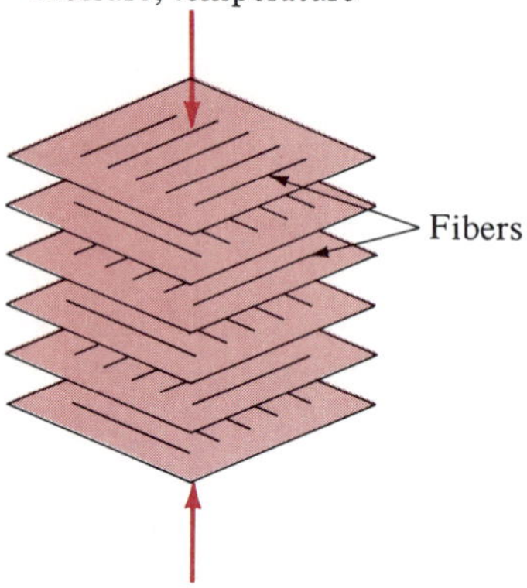

Figure 4.47 Construction of a laminated composite from fibrous layers

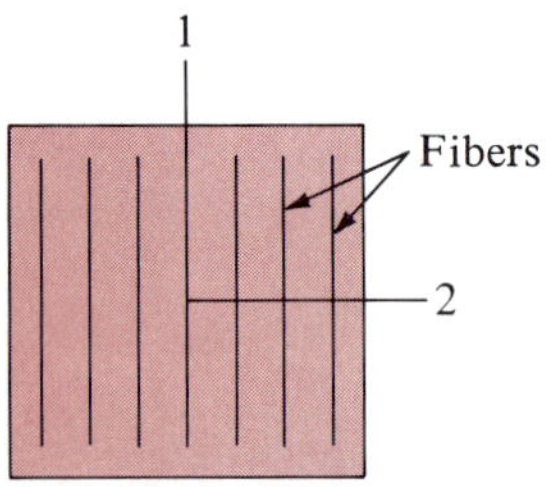

Figure 4.48 Definitions of directions for experimental determination of elastic constants

serves primarily to preserve the position and orientation of the fibers. A fibrous composite will exhibit directional or anisotropic material properties.

In particular, one would suspect that the successive application of equal-valued stresses σ_x, σ_y, and σ_z would result in different responses ε_x, ε_y, and ε_z for the fibrous composite of Fig. 4.46.

The primary use of such a material is in the construction of laminated composites. As indicated in Fig. 4.47, laminated composites consist of several layers of a fibrous composite usually bonded in a high-pressure, high-temperature environment to form the final laminate. The fiber directions of the individual layers can be arranged in several different "stacking sequences" to produce a variety of different desired results.

In order to determine analytical expressions for the properties of the laminates, the properties of each of the individual fibrous composite layers must be known. The fibers, which are assumed to behave in a linearly elastic manner, provide almost all of the stiffness. Thus the composite behaves essentially as a linearly elastic material.

Referring to Fig. 4.48, the appropriate elastic constants for a fibrous layer are determined on the basis of the following set of experiments.

Experiment 1 With $\sigma_1 \neq 0$, $\sigma_2 = \tau_{12} = 0$, measure ε_1 and ε_2 as indicated in Fig. 4.49. Then compute $E_1 = \sigma_1/\varepsilon_1$ and $\nu_{12} = -\varepsilon_2/\varepsilon_1$. E_1 is called Young's modulus in the fiber direction and ν_{12} is called the major Poisson's ratio.

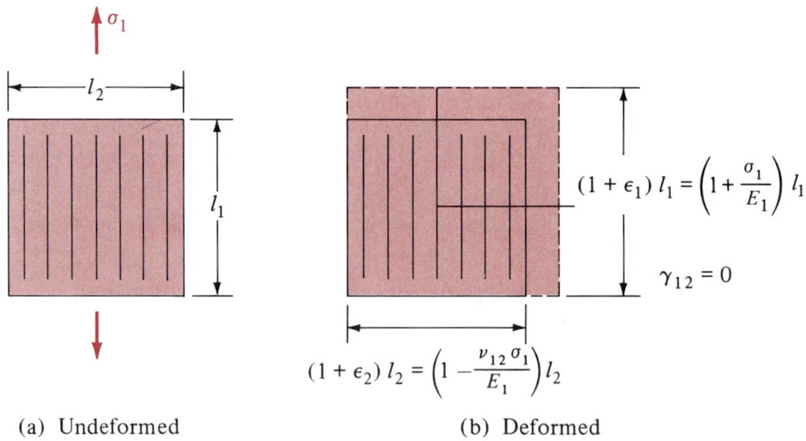

Figure 4.49 Uniform stress in the fiber direction

Experiment 2 With $\sigma_2 \neq 0$, $\sigma_1 = \tau_{12} = 0$, measure ε_1 and ε_2 as indicated in Fig. 4.50. Then compute $E_2 = \sigma_2/\varepsilon_2$ and $\nu_{21} = -\varepsilon_1/\varepsilon_2$. E_2 is called Young's modulus in the transverse direction and ν_{21} is called the minor Poisson's ratio.

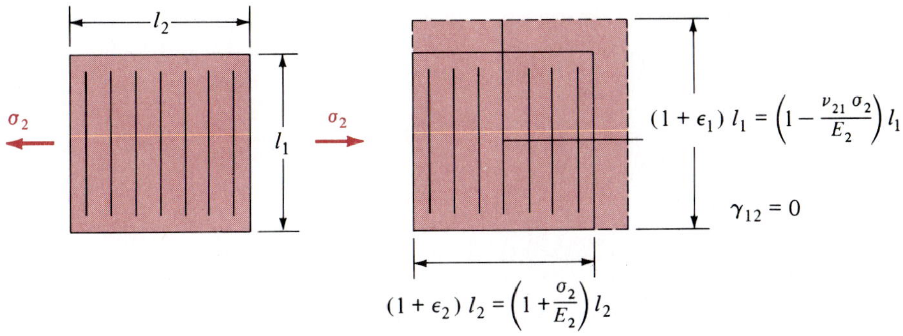

(a) Undeformed (b) Deformed

Figure 4.50 Uniform stress in the transverse direction

Experiment 3 With $\tau_{12} \neq 0$, $\sigma_1 = \sigma_2 = 0$, measure γ_{12} as indicated in Fig. 4.51. Then compute $G_{12} = \tau_{12}/\gamma_{12}$.

Superposing the results of the three experiments, the stress-strain relations for the fibrous layer are seen to be

$$\begin{aligned} \varepsilon_1 &= \frac{\sigma_1}{E_1} - \nu_{21}\left(\frac{\sigma_2}{E_2}\right) \\ \varepsilon_2 &= -\nu_{12}\left(\frac{\sigma_1}{E_1}\right) + \frac{\sigma_2}{E_2} \\ \gamma_{12} &= \frac{\tau_{12}}{G_{12}} \end{aligned} \tag{4.11}$$

From the theory of elasticity, the equality $\nu_{21}/E_2 = \nu_{12}/E_1$ must be satisfied by the elastic coefficients. A good check on the validity of the experiments is the accuracy with which the equality is satisfied. Thus there are four independent elastic constants for the fibrous layer: E_1, ν_{12}, E_2, and G_{12}.

Note that the stress-strain equations given by Eqns. (4.11) are valid only for a specimen oriented with respect to the fiber directions as

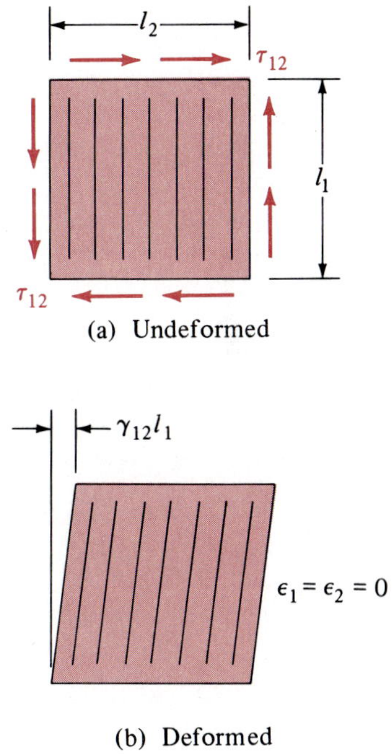

Figure 4.51 Uniform shear stress

indicated in Fig. 4.52. A test piece formed by cutting parallel to the fiber directions of the layer will respond to normal and shear stresses in exactly the same fashion as in the three experiments described above.

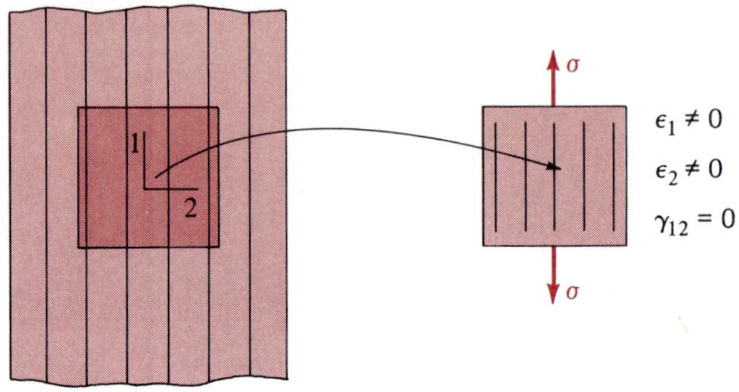

Figure 4.52 Response of a test specimen oriented along the fiber directions

By contrast, a piece cut in an arbitrary direction such as indicated in Fig. 4.53 will elongate and distort in response to a normal stress on one of the faces.

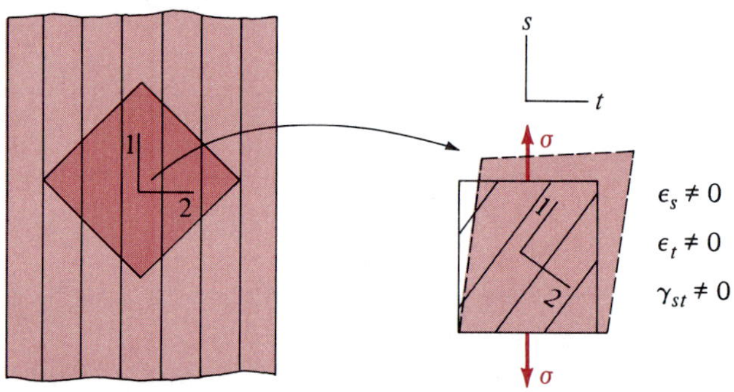

Figure 4.53 Response of a test specimen rotated with respect to the fiber directions

Recall that an isotropic solid is such that experiments performed on differently oriented test specimens at a point result in identical stress-strain relations. The response of test specimens oriented differently as outlined above clearly shows the anisotropic nature of a typical fibrous layer in that normal stresses generally result in both extensional and shear strains, and shear stresses also result in both extensional and shear strains.

PROBLEMS

4.48 For the lamina shown, determine the change in the length of *(a)* the 100-mm dimension, *(b)* the 120-mm dimension, and *(c)* the diagonal AC. Take $E_L = 14.0$ GPa, $E_T = 3.5$ GPa, $\nu_{LT} = 0.36$, $\nu_{TL} = 0.09$, and $G_{LT} = 4.2$ GPa.

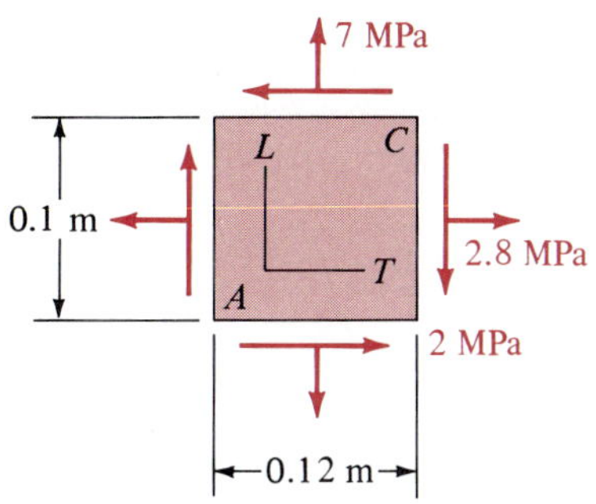

P. 4.48

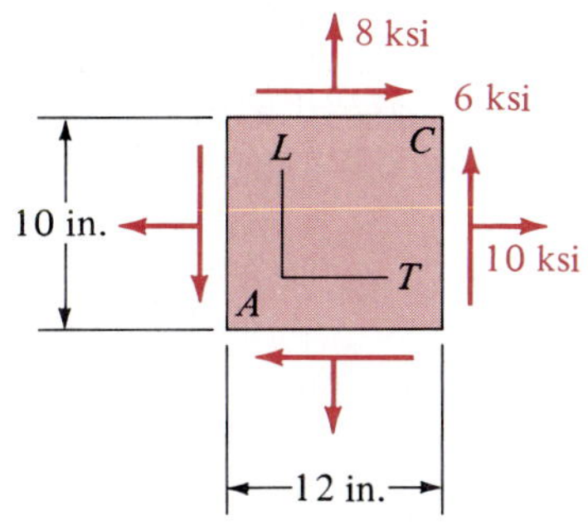

P. 4.49

4.49 For the lamina shown, determine the change in the length of *(a)* the 10-in. dimension, *(b)* the 12-in. dimension, and *(c)* the diagonal AC. Take $E_L = 44 \times 10^6$ psi, $E_T = 15 \times 10^6$ psi, $\nu_{LT} = 0.28$, $\nu_{TL} = 0.095$, and $G_{LT} = 5 \times 10^6$ psi.

4.50 An experiment is conducted on the single lamina shown.

Test 1: $\sigma_L = 5.0$ MPa, $\sigma_T = \tau_{LT} = 0$
$\Delta l_L = 0.0667$ mm, $\Delta l_T = -0.036$ mm, Δ angle$_{LT} = 0$

Test 2: $\sigma_T = 2.0$ MPa, $\sigma_L = \tau_{LT} = 0$
$\Delta l_T = 0.06$ mm, $\Delta l_L = -0.024$ mm, Δ angle$_{LT} = 0$

Test 3: $\sigma_L = \sigma_T = 0$, $\tau_{LT} = 2.5$ GPa
$\Delta l_T = \Delta l_L = 0$ and Δ angle$_{LT} = 0.0286°$

Determine the constants E_L, E_T, ν_{LT}, ν_{TL}, and G_{LT}, and comment on the results.

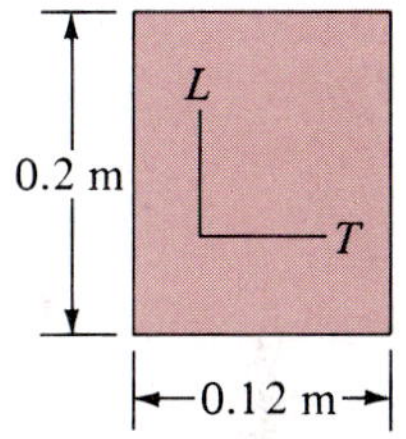

P. 4.50

4.51 An experiment is conducted on the single lamina shown.

Test 1: $\sigma_L = 10$ ksi, $\sigma_T = \tau_{LT} = 0$
$\Delta l_L = 0.00227$ in., $\Delta l_T = -0.00040$ in., Δ angle$_{LT} = 0$

Test 2: $\sigma_T = 4$ ksi, $\sigma_L = \tau_{LT} = 0$
$\Delta l_T = 0.00225$ in., $\Delta l_L = -0.00028$ in., Δ angle$_{LT} = 0$

Test 3: $\sigma_L = \sigma_T = 0$, $\tau_{LT} = 5$ ksi
$\Delta l_T = \Delta l_L = 0$, Δ angle$_{LT} = 0.0573°$

Determine the constants E_L, E_T, ν_{LT}, ν_{TL}, and G_{LT}, and comment on the results.

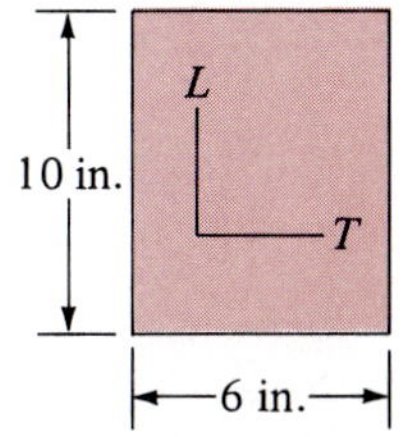

P. 4.51

4.7 RATE-DEPENDENT BEHAVIOR

The example of the Silly Putty in Section 4.1 makes us aware of materials for which rate effects must be included when considering material behavior, in order for the theoretical and experimental results to be approximately the same, both qualitatively and quantitatively. In this section we will discuss several different situations where the response of a static or quasi-static problem is time dependent.

In the case of a problem that is strictly static, it is assumed that the time rates of change of both stress and strain are zero, or

$$\frac{d\sigma}{dt} = 0 \qquad \text{and} \qquad \frac{d\varepsilon}{dt} = 0$$

To see that this can never actually be the case for a structure that is initially unloaded, consider a bar as shown in Fig. 4.54a.

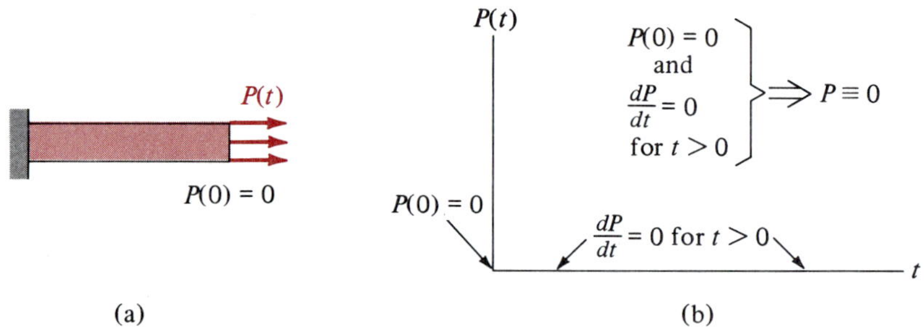

Figure 4.54 Stress and strain rates for an axially loaded bar

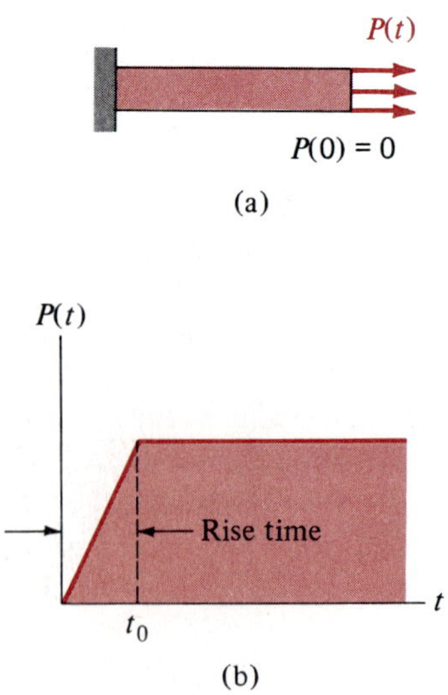

Figure 4.55 Slowly applied axial load P

The bar is initially unloaded so that $\sigma = P/A = 0$. If dP/dt is also zero for all time then, as indicated in Fig. 4.54b, P must always be zero, that is, we can never apply any load and satisfy $dP/dt = 0$.

For a linearly elastic material for instance, the theory assumes that the load P is increased very slowly and that the stress (given by P/A) and the strain both increase very slowly in accordance with the relation $\sigma = E\varepsilon$. Thus the rates of changes of stress and strain must be small for the static theory to be strictly valid. We refer to this as quasi-static. For a bar loaded in tension as in Fig. 4.55a, a force applied as in Fig. 4.55b would qualify as quasi-static.

For a load to be considered as quasi-static, the load must be applied slowly enough to avoid generating significant shock or stress waves. Thus for a typical engineering material the rise time t_0 shown in Fig. 4.55b could be as small as a few tenths of a second and still qualify as quasi-static. Frequently we will simply state that a load is applied without bothering to specify precisely the time interval involved. Unless otherwise indicated, it will always be assumed that the loading is gradual enough so that no shocks or stress waves are produced.

One of the most frequent situations where displacements and stresses are time dependent is associated with what is commonly called **creep.** Creep is a general term referring to continued deformation under load. Some metals at elevated temperatures and most plastics will continue to elongate under the action of constant sustained stresses. This situation is indicated in Fig. 4.56, where a bar is loaded by an axial force P.

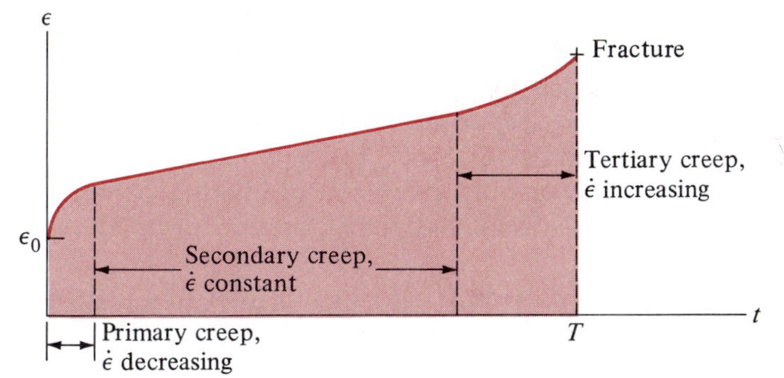

Figure 4.56 Time-dependent behavior known as creep

The curve of Fig. 4.55b indicates that the load P has been applied over a short time interval t_0 and maintained at that value for all $t > t_0$. It is assumed that the time T required for the specimen to fracture is many orders of magnitude larger than the time t_0 so that t_0 is essentially ignored in Fig. 4.56. Thus the initial value of the strain in Fig. 4.56 is taken to be the strain associated with the final constant value of the load P, namely, $\varepsilon_0 = \sigma/E = P_0/AE$.

The curve of Fig. 4.56 is divided into three distinct regions. In the first region the strain rate is decreasing from the value associated with the application of the load; this is called **primary creep.** The second stage is associated with a region for which there is a constant strain rate; this is called **secondary creep.** The third stage, called **tertiary creep,** is associated with the beginning of the failure process and generally ends in fracture of the bar. The strain rates for materials exhibiting creep are largely a function of the stress level and the temperature. The larger the stress or the temperature, the larger the strain rate. Creep is one of the many phenomena that should be considered in any rational analysis and design.

A phenomenon that is closely related to creep is called **relaxation.** Consider a test specimen subjected to an elongation Δ as indicated in Fig. 4.57a. For the relaxation test the elongation Δ is maintained at a constant value and the corresponding stress is monitored. Typical results are shown in Fig. 4.57b.

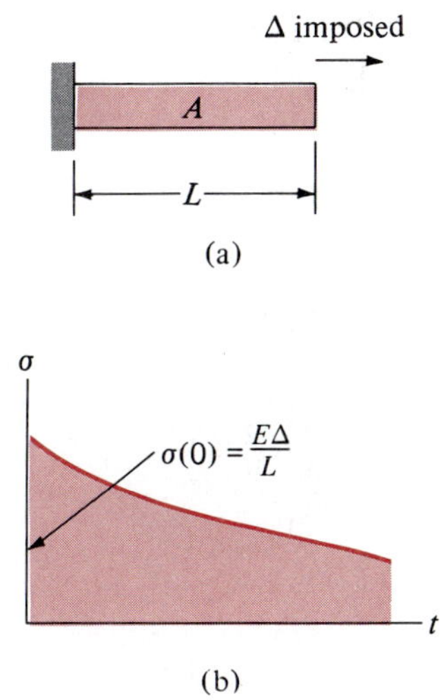

Figure 4.57 Typical relaxation curve

The initial value of the stress for the relaxation test is taken in the same vein as the initial strain for the creep test, namely, as the value associated with the stress at the end of the short loading process, or $\sigma_0 = E\varepsilon_0 = E\Delta/L$. As indicated, the stress and hence the load being transmitted drop off gradually with time, and may eventually approach zero.

4.8 FATIGUE

Many of the failures that occur in structures are associated with cyclic or repeated loadings. Examples include the blade in a turbine engine, the region neighboring a rivet hole in a structural component of an air or spacecraft, and the suspension system of an automobile that must pass over rough stretches of road. These failures usually occur after several millions, tens or even hundreds of millions of cycles and are referred to as **fatigue failures.** Fatigue failures occur in both ductile and brittle materials.

As a result of cyclic loadings, the time history of a particular stress component at a point in a structure is often idealized as indicated in Fig. 4.58.

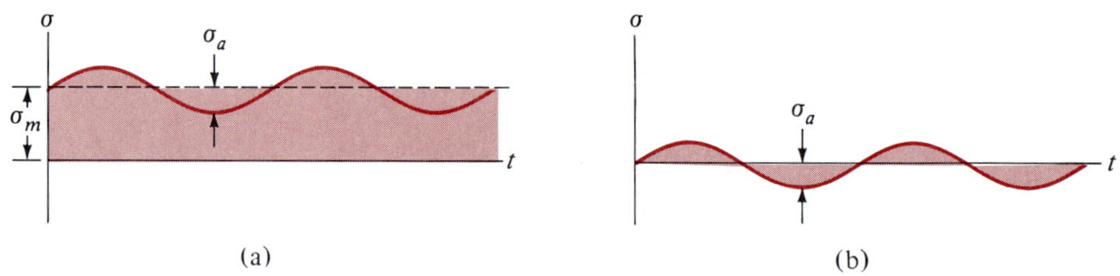

Figure 4.58 Cyclic nature of stress

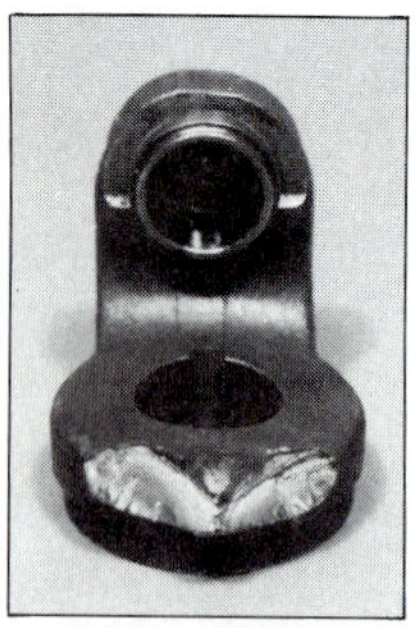

Figure 4.59 Photograph of the fracture surface associated with a typical fatigue failure. (Photo by Michael Hruby.)

Fig. 4.58a shows a situation where there is a **mean stress** σ_m upon which an **alternating stress** σ_a is superimposed. Fig. 4.58b shows a situation where the mean stress is zero.

Fig. 4.59 is a photograph of a fatigue failure. The rough surfaces at the front are the fracture surfaces. From the photo it is frequently possible with magnification to determine a point in the fracture surface at which the crack initiates.

Although the precise mechanism of fatigue is not known, we assume on the basis of inspection of many failed components in cyclic loading environments that failure occurs in a progressive manner according to roughly the following scenario. As a result of the various processes to which the material constituting a structure or structural component is subjected in order to ready it for its intended use, there are many microscopic flaws present. These flaws are usually small enough to be undetectable using flaw-detection techniques such as an X-ray machine. (There may also be larger, detectable flaws. When evidence indicates a detectable flaw, the structure or structural component may be discarded, or if possible reprocessed to remove the flaw.) These microscopic flaws act as stress concentrations or stress risers; that is, the stresses in the neighborhood of such a flaw can be several times the value that would exist if the flaw were not present. The ratio of the maximum stress to the average stress is called a **stress concentration factor.** Upon repeated or cyclic loading at a sufficiently high level, a small crack develops. If the cyclic loading continues, the small crack can develop into a larger crack, eventually growing to the size of the member, resulting in a fatigue failure.

Fatigue failures are largely dependent upon the specific material involved. Experimental data are required as a means for assessing the effect of cyclic loading on the design of a structure or structural component. The experimental results are used to construct a so-called *S–N* curve as shown in Fig. 4.60, where the alternating stress level *S* is plotted against the number of cycles *N* to failure.

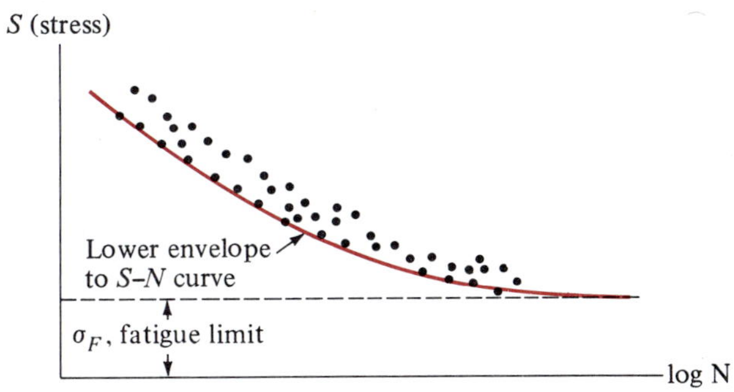

Figure 4.60 Typical *S–N* curve

A statistically significant number of these tests are run at different alternating stress levels to produce the curve shown. As indicated, there is usually a great deal of scatter in the data. For this reason it is necessary to conduct a reasonably large number of tests in order to correctly identify the *S–N* curve for a particular material. This scatter in the data is a fact of life regardless of the care taken in preparing the test specimens to insure that they are all initially the same, that is, that all of the experiments are identical.

Fatigue strength refers to the stress level required to produce failure at a given number of cycles. Given the stress level, the *S–N* curve can be used to estimate the number of cycles to failure, or vice versa. As indicated in Fig. 4.60, there is a range of stresses at a given *N* due to the nature of the experiment. Selecting the lower envelope as the *S–N* curve is a conservative way of dealing with the data.

For most materials, the S–N curve levels out for lower stress levels so that a horizontal asymptote can be drawn as indicated in Fig. 4.60. This lower limit is called the **endurance limit** and is denoted by σ_F. If the alternating stress is below this value the specimen is taken to have infinite life: the number of cycles the specimen can endure at that stress level is infinite. The endurance limit is sometimes referred to as the **fatigue limit.** In the event that there is no available fatigue data that can be used, it is customary to take the fatigue limit as half the value of the yield stress or as half the value of the ultimate stress.

Shown in Fig. 4.61 are results of fatigue tests for several common metals and alloys.

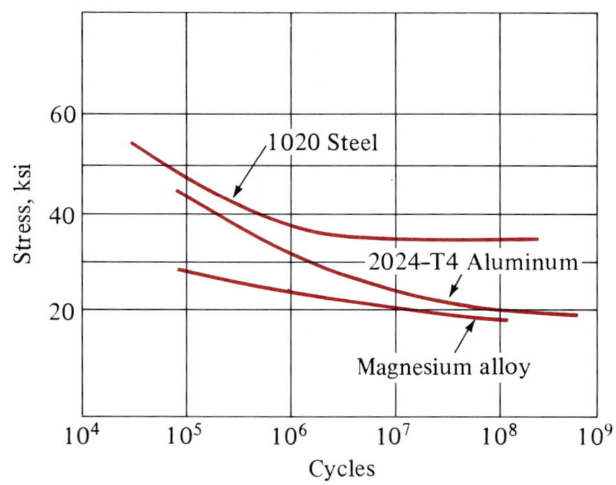

Figure 4.61 Fatigue data for common metals

There are many different aluminum and steel alloys. In addition there are numerous treatments to which a structure or structural component can be subjected. Some of these processes in fact are designed to improve fatigue behavior. The most important point to be made in connection with fatigue is that any time there are cyclic loadings to which a structure or structural component is subjected, the possibility of a fatigue failure must be considered. Care must be taken to use fatigue data that corresponds exactly to the specific material and to the specific type of material processing in question.

4.9 THEORIES OF FAILURE—FAILURE CRITERIA

In Chapter 1, we introduced the two basic problems in solid mechanics; the first is assessing whether or not the strength of a structure is adequate. In the present chapter we have seen from the uniaxial tension test that real materials can fail when they are subjected to loads resulting in stresses that are too large. In this section we will be concerned with the modes of failure associated *(a)* with fracture, and *(b)* with excessive deformations in the form of yielding. In the case of the uniaxial test of a brittle specimen, failure is due to fracture, at which point the average axial stress in the specimen is the ultimate stress σ_{ULT}. For a ductile specimen, failure is usually defined to occur with the onset of yielding, that is, when the average axial stress is the yield stress σ_Y.

Any structure or structural component will fail if the loads are sufficiently high. Tests could be performed on each structure to determine the failure loads but would be prohibitively time-consuming and expensive. In lieu of testing it is necessary to have some sort of rational analytical process to assist in making decisions as to the adequacy of the strength of a structure. Before presenting the theories that have been developed in this regard, let us review the results of several standard tests.

Example 4.13 Consider a tension member containing a central hole as indicated in Fig. 4.62a. The specimen is made from a ductile material and is polished before testing. When the load is increased to a sufficient level, visible regions called Luder bands will appear on the front and back surfaces as indicated in Fig. 4.62b. These bands, marking the regions over which yielding occurs, make an angle of approximately 45° with the axis of loading as shown. It is on these 45° planes that the shear stress is maximum. Luder bands can also appear when there is no hole. This is evidence to support the proposition that for ductile materials, failure occurs when the maximum shear stress reaches a critical value.

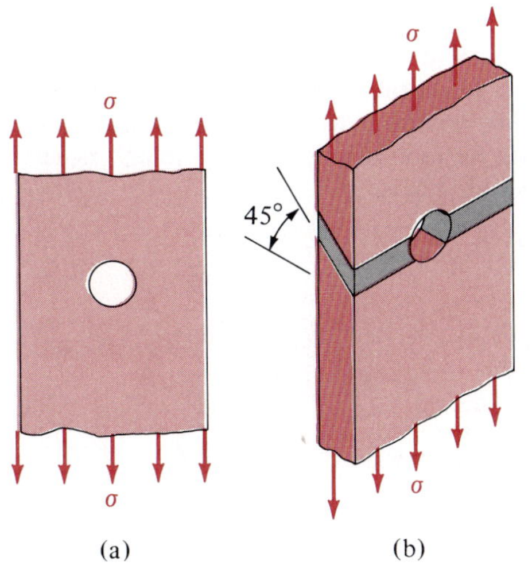

Figure 4.62 Luder bands appearing in the tensile test of a ductile material

Example 4.14 If a specimen made from a brittle material is subjected to the uniaxial tension test the failure or fracture surface appears as shown in Fig. 4.63.

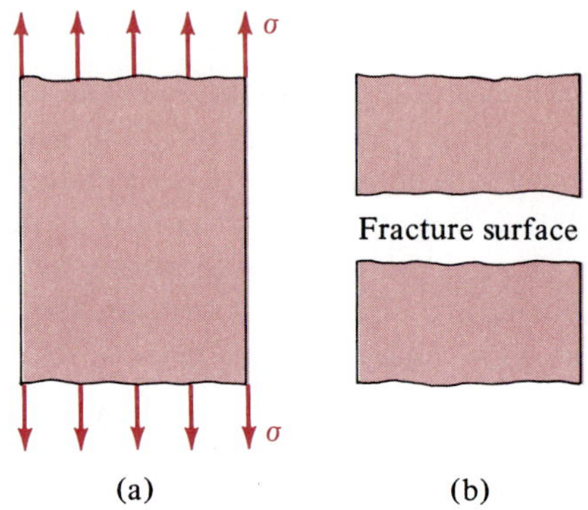

Figure 4.63 Failure surface for the uniaxial tension test of a brittle material

In view of the relationship between the fracture surface and the normal stress $\sigma_{avg} = P/A$ acting on the fracture surface, one would suspect that failure is related to the fact that the normal stress has exceeded some critical value. This observation is basic to the theories of failure for brittle materials.

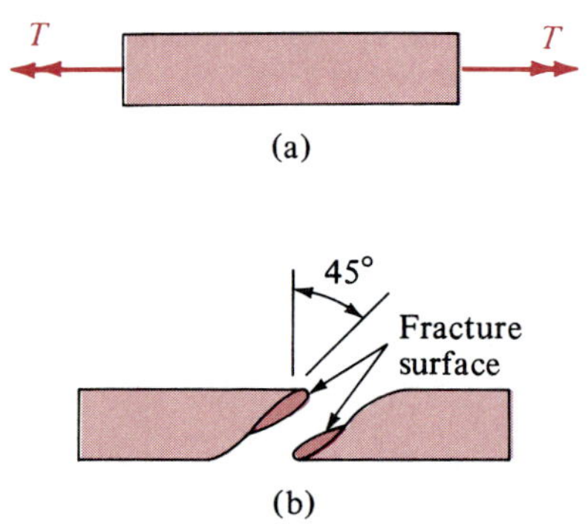

Figure 4.64 Failure surface for the torsion test of a piece of chalk

Example 4.15 Next consider a simple experiment consisting of the twisting of a piece of chalk as shown in Fig. 4.64a. Failure occurs as a result of a brittle-type fracture indicated in Fig. 4.64b.

As we will learn in Chapter 6, the stress resulting for the torque is a shear stress on planes perpendicular to the axis as indicated in Fig. 4.65a. An element of the lateral surface of the piece of chalk is subjected to the shear stresses shown in Fig. 4.65b.

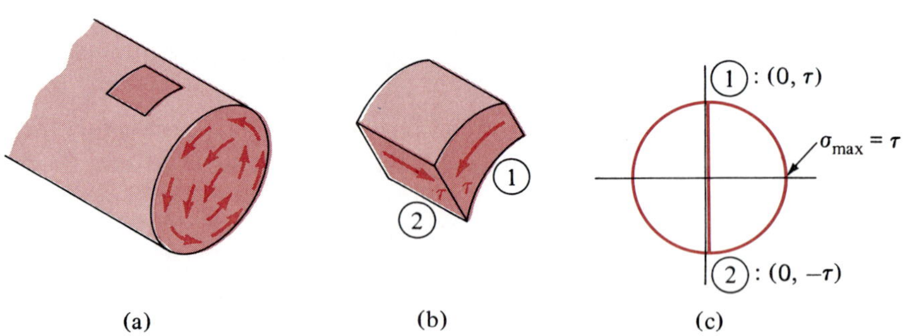

Figure 4.65 Shear and normal stress for torsion

Using Mohr's circle for stress as in Fig. 4.65c, it is seen that maximum tensile stress occurs on planes at 45°, that is, precisely the fracture planes. The results of this experiment further support the idea that brittle materials fail when the maximum normal stress reaches a critical value.

The results of these tests seem to indicate that there is a connection between failure and stress level. Whether the specimen is loaded in tension, compression, or shear (torsion), there is a value of the load and hence a stress or stresses at which failure occurs. The precise nature of the mechanism involved in either a ductile or brittle fracture is actually a very complicated function of the microstructure of the material (i.e., grain size, grain orientation, etc). Based on the assumption that we are dealing with a continuum as discussed in Section 4.1, we will restrict ourselves to making decisions about the mechanism of failure on the basis of observable or measurable quantities that are a part of the theory of solid mechanics. Stress is the primary quantity that is used for correlating the experimental data with the predictions of the theories of failure in solid mechanics.

The careful consideration of all the experimental data has resulted in several commonly used theories of failure. These theories are based on the assumption that failure occurs when the maximum shear or normal stress in a structure or structural component reaches a critical value as determined from the results of a uniaxial tension test.

Just as we distinguished between ductile and brittle behavior, it is appropriate to distinguish between failure criteria for materials that fail in a brittle fashion and those that fail in a ductile fashion. For ductile materials we will introduce *(a)* the maximum shear stress, and *(b)* the Hencky-von Mises failure criterion. For brittle materials we will introduce *(a)* the maximum normal stress, and *(b)* the Coulomb-Mohr failure criterion. Each of these failure criteria attempts to predict failure in an actual structure based on knowledge of normal stress values on failure surfaces for tests such as the uniaxial tension test and the torsion test described above.

It must be recognized that a structure has a certain "intelligence" in that, if the structure is going to fail, for instance because the maximum normal stress has reached a critical value σ_{cr}, it will do so as soon as the σ_{cr} value is reached *at any point on any plane in the structure.* In other words, the structure, in a sense, "knows" when the critical value is reached irrespective of any analysis that is performed. Thus it is essential that the analysis be able to indicate the location and magnitude of the maximum stresses and that these values are the ones compared to the chosen critical values in assessing failure.

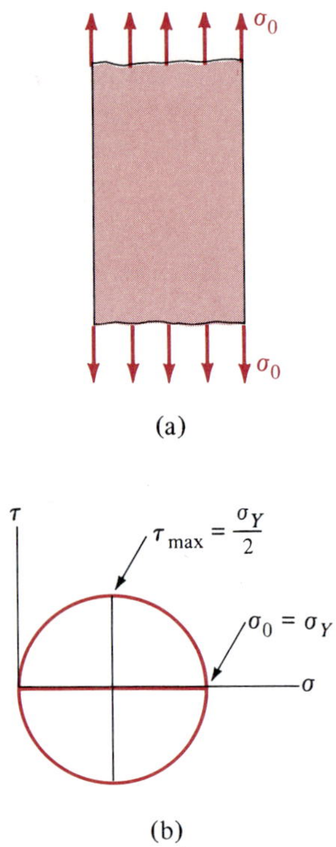

Figure 4.66 Uniaxial test specimen

Maximum Shear Stress Failure Criterion Charles Coulomb (1773) and Henri Tresca (1868) suggested the maximum shear stress theory, which states that for all members composed of the same material, failure, as indicated by yielding, occurs when the maximum shear stress in the material reaches the value of the shear stress at yielding as determined by a uniaxial tension (or compression) test. Consider a uniaxial specimen as indicated in Fig. 4.66.

For the coordinate system shown, the stresses are $\sigma_Y = \sigma_0$, with all the other components of stress vanishing. Mohr's circle for stress, Fig. 4.66b, clearly shows that for the uniaxial test, the maximum shear stress at yielding is

$$\tau_{max} = \tau_Y = \frac{\sigma_Y}{2} \tag{4.11}$$

as the critical value to be used in the maximum shear stress criterion of failure. In Eqn. (4.11), τ_Y indicates the value of the maximum shear stress at yield.

The maximum shear stress theory extends this result to the general case involving an arbitrary state of stress by stating that failure occurs when the maximum shear stress on any plane at any point in the structure reaches the value τ_Y. When the three principal stresses $\sigma_1 > \sigma_2 > \sigma_3$ are known at a point, the maximum shear stress is given by

$$\tau_Y = \frac{\sigma_1 - \sigma_3}{2}$$

and failure occurs when

$$\frac{\sigma_1 - \sigma_3}{2} = \tau_Y = \frac{\sigma_Y}{2}$$

For a plane stress state with $\sigma_z = \tau_{yz} = \tau_{xz} = 0$, one of the principal stresses, say σ_3, is always zero. We will refer to the other two principal stresses σ_1 and σ_2 as in-plane principal stresses. For clarity, we will look in detail at three different combinations of in-plane principal stresses, namely, *(a)* $\sigma_1 > \sigma_2 > 0$, $\sigma_3 = 0$, *(b)* $\sigma_2 < \sigma_1 < 0$, $\sigma_3 = 0$, and *(c)* $\sigma_1 > 0$, $\sigma_2 < 0$, $\sigma_3 = 0$.

Case a: $\sigma_1 > \sigma_2 > 0$, $\sigma_3 = 0$. For this case the Mohr's circle and the block on which the principal stresses act are shown in Fig. 4.67a and b, respectively. The maximum in-plane shear stress, given by $(\sigma_1 - \sigma_2)/2$ and shown in Fig. 4.67a, acts on the plane that bisects the principal planes. For this case however, $(\sigma_1 - 0)/2 > (\sigma_1 - \sigma_2)/2$, so that the absolute maximum shear stress is given by

$$\tau_{abs} = \frac{\sigma_1}{2}$$

and occurs on the plane bisecting the 1 and 3 principal directions and also on the shaded plane indicated in Fig. 4.67b.

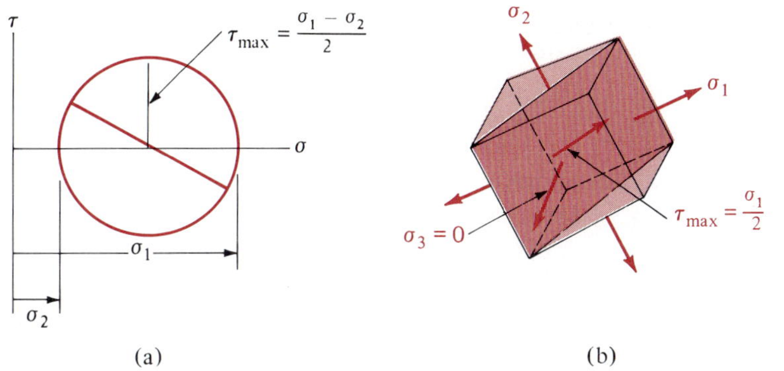

Figure 4.67 Maximum shear stresses for $\sigma_1 > \sigma_2 > 0$, $\sigma_3 = 0$

Case b: $\sigma_2 < \sigma_1 < 0$, $\sigma_3 = 0$. For this case the Mohr's circle and the block on which the principal stresses act are shown in Fig. 4.68a and b, respectively. The maximum in-plane shear stress, given by $(\sigma_1 - \sigma_2)/2$ and shown in Fig. 4.68a, acts on the plane that bisects the principal planes. For this case however, $|\sigma_2 - 0|/2 > (\sigma_1 - \sigma_2)/2$, so that the absolute maximum shear stress is given by

$$\tau_{abs} = \frac{|\sigma_2|}{2}$$

and occurs on the plane bisecting the 2 and 3 principal directions and also on the shaded plane indicated in Fig. 4.68b.

Case c: $\sigma_1 > 0$, $\sigma_2 < 0$, $\sigma_3 = 0$. For this case the Mohr's circle and the block on which the principal stresses act are shown in Fig. 4.69a and b, respectively.

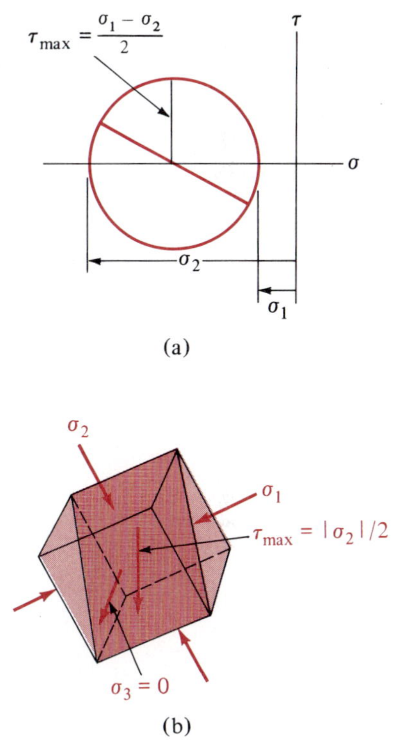

Figure 4.68 Maximum shear stresses for $\sigma_2 < \sigma_1 < 0$, $\sigma_3 = 0$

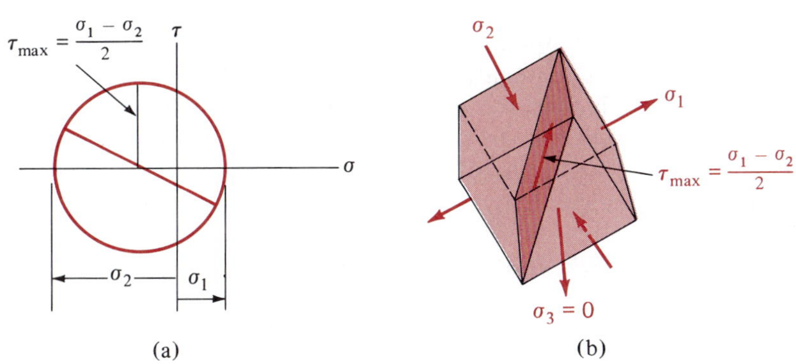

Figure 4.69 Maximum shear stresses for $\sigma_1 > 0$, $\sigma_2 < 0$, $\sigma_3 = 0$

For this case, $(\sigma_1 - \sigma_2)/2 > \sigma_1/2$ and $(\sigma_1 - \sigma_2)/2 > |\sigma_2|/2$ so that the absolute maximum shear stress is given by

$$\tau_{\text{abs}} = \frac{\sigma_1 - \sigma_2}{2}$$

and occurs on the plane bisecting the 1 and 2 principal directions as indicated in Fig. 4.69b.

Thus for a state of plane stress where $\sigma_3 = 0$, the results of cases *a*, *b*, and *c* can be summarized by the three corresponding sets of Mohr's circles shown in Fig. 4.70a, b, and c, respectively.

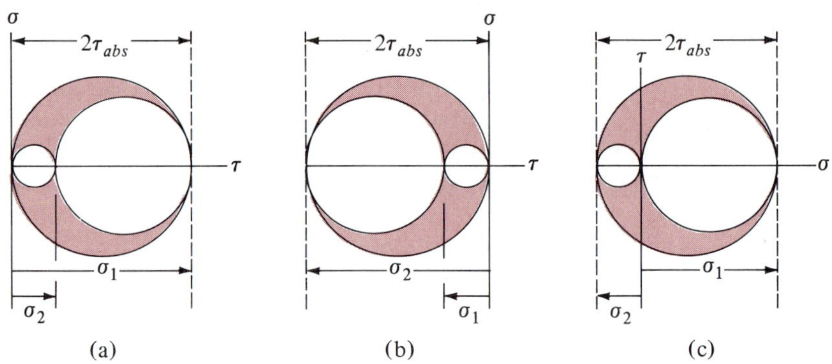

Figure 4.70 Mohr's circles for cases *a*, *b*, and *c*

In each case three circles are drawn. One is constructed for the plane associated with the $(\sigma_1, 0)$ principal stresses, one for the $(\sigma_2, 0)$ principal stresses and one for the (σ_1, σ_2) principal stresses. For each case, the absolute maximum shear stress at the point is equal in value to half the diameter of the largest of the three circles. For case *a* this is seen to be the circle associated with the $(\sigma_1, 0)$ principal stresses. For case *b* the circle having the largest diameter is the circle associated with the $(\sigma_2, 0)$ principal stresses, and for case *c* it is the circle associated with the (σ_1, σ_2) principal stresses.

Finally then for the case where $\sigma_3 = 0$, the process for determining the largest shear stress is to use Mohr's circle to determine the in-plane principal stresses σ_1 and σ_2. The maximum shear stress at the point is then given by

$$\tau_{\text{abs}} = \max\left(\frac{|\sigma_1|}{2}, \frac{|\sigma_2|}{2}, \frac{|\sigma_1 - \sigma_2|}{2}\right)$$

According to the maximum shear stress failure criterion, failure will occur if any of the above values exceeds the allowable shear stress given by $\sigma_Y/2$. These collective requirements can be presented very nicely in terms of the maximum shear stress hexagon in Fig. 4.71. Thus as long as the in-plane principal stresses σ_1 and σ_2 for a biaxial state of stress lie within the shaded region of Fig. 4.71, the maximum shear stress criterion of $\tau_{abs} < \sigma_Y/2$ is satisfied.

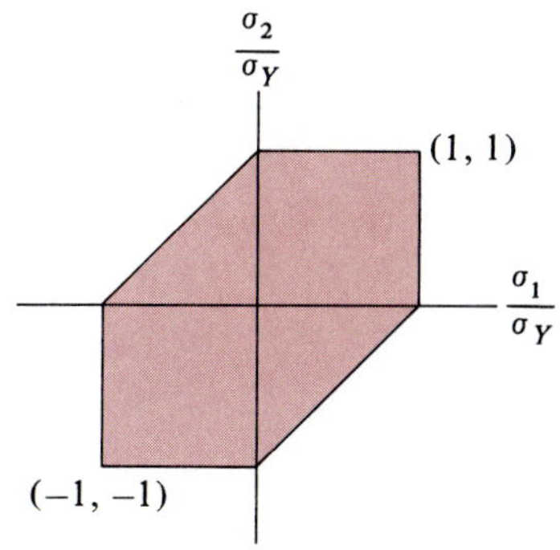

Figure 4.71 Maximum shear stress hexagon

Example 4.16 As indicated in Fig. 4.72a, the stresses as a point P in a body are given by $\sigma_x = \sigma_x$ ksi, $\sigma_Y = 2$ ksi, $\tau_{xy} = 5$ ksi, with $\sigma_z = \tau_{xz} = \tau_{yz} = 0$. A uniaxial test of the same material indicates yielding at the normal stress value of 16 ksi. What is the allowable value of σ_x?

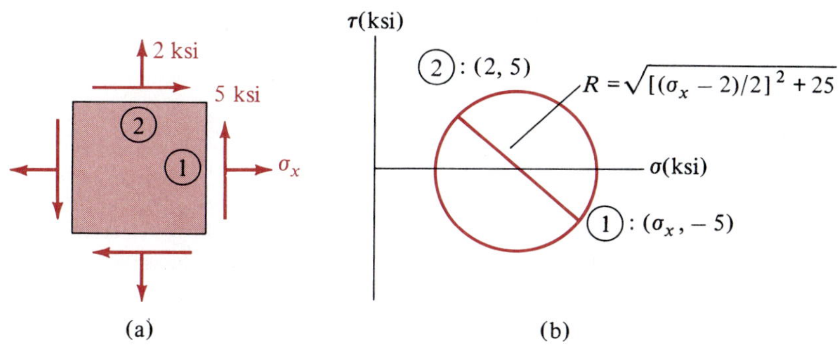

Figure 4.72 State of stress at a point in a body

Solution Mohr's circle for this state of stress in shown in Fig. 4.72b. The maximum shear stress is given by

$$\tau_{max} = \sqrt{\left(\frac{\sigma_x - 2}{2}\right)^2 + 25} \text{ ksi}$$

so that failure occurs when

$$\tau_{max} = \sqrt{\left(\frac{\sigma_x - 2}{2}\right)^2 + 25} \text{ ksi} = 16 \text{ ksi}/2 = 8 \text{ ksi}$$

Solving for σ_x yields $\sigma_x = +14.49$ ksi and $\sigma_x = -10.49$ ksi; thus, either a tensile stress of 14.49 ksi or a compressive stress of 10.49 ksi will produce a state of stress for which failure is indicated using the maximum shear stress criterion.

Hencky-von Mises Failure Criterion The Hencky-von Mises failure theory is often referred to as the maximum distortion strain-energy theory. The idea for the theory is best motivated as follows. Experiments have shown that homogeneous materials can withstand very high hydrostatic stresses ($\sigma_x = \sigma_y = \sigma_z = \sigma_0$, $\tau_{xy} = \tau_{yz} = \tau_{xz} = 0$) without yielding. This leads to the thought that failure in a material subjected to a general state of stress may be associated with the **difference** between the actual state of stress and a hydrostatic state having the value σ equal to the average of the principal stresses at the point, namely, $\sigma = (\sigma_1 + \sigma_2 + \sigma_3)/3$. The Hencky-von Mises criterion states that failure will occur when the energy associated with the difference in the actual and hydrostatic states is equal to the corresponding energy in a uniaxial tensile test at yielding. The resulting criterion is identical to what is called the octahedral shear stress theory, which is much easier to develop and understand. For this reason we discuss the results of the octahedral shear stress theory.

Fig. 4.73a indicates the block at a point on which the principal stresses act. Fig. 4.73b indicates four of the eight directions that make equal angles with adjacent triads of principal directions. Each of these eight directions is perpendicular to what is termed an octahedral plane.

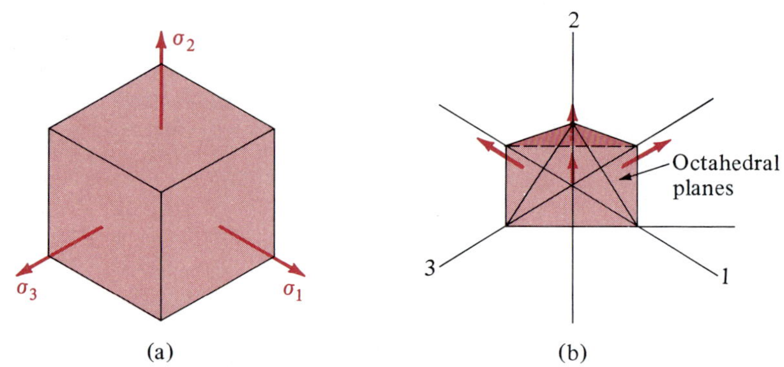

Figure 4.73 Principal planes and octahedral planes

Using the three-dimensional stress-transformation equations, the normal stress on each of the octahedral planes can be computed as

$$\sigma_{\text{octahedral}} = \frac{\sigma_1 + \sigma_2 + \sigma_3}{3}$$

or the average of the principal stresses at the point. Also, the shear stress on **each** of the octahedral planes can be computed as

$$\tau_{\text{oct}} = \frac{1}{2}\sqrt{(\sigma_1 - \sigma_2)^2 + (\sigma_2 - \sigma_3)^2 + (\sigma_3 - \sigma_1)^2}$$

The octahedral shear-stress theory states that failure occurs when this octahedral shear-stress value is equal to the value of the octahedral shear stress at yielding in a uniaxial tensile test. For the uniaxial tensile test $\sigma_1 = \sigma_Y$ (the yield stress), $\sigma_2 = \sigma_3 = 0$, so that the octahedral shear stress in the uniaxial tensile test is given by

$$\tau_{oct} = \frac{1}{2}\sqrt{(\sigma_Y - 0)^2 + (0 - 0)^2 + (0 - \sigma_Y)^2} = \frac{\sigma_Y}{\sqrt{2}}$$

Equating the two values for τ_{oct} yields

$$\sigma_Y = \frac{1}{\sqrt{2}}\sqrt{(\sigma_1 - \sigma_2)^2 + (\sigma_2 - \sigma_3)^2 + (\sigma_3 - \sigma_1)^2} \qquad (4.12)$$

as the octahedral shear-stress criterion for failure in terms of the principal stresses. For the case where one of the principal stresses, say σ_3, is zero, Eqn. (4.12) reduces to

$$\sigma_Y = \sqrt{\sigma_1^2 - \sigma_1\sigma_2 + \sigma_2^2}$$

which is frequently written as

$$(\sigma_1/\sigma_Y)^2 - (\sigma_1/\sigma_Y)(\sigma_2/\sigma_Y) + (\sigma_2/\sigma_Y)^2 = 1$$

This is the equation of the ellipse shown in Fig. 4.74.

Failure due to the initiation of yielding is assumed to take place if the state of stress given by the principal stresses σ_1 and σ_2 lies on or outside the ellipse. The Hencky-von Mises criterion is generally accepted as the best criterion to use for judging failure, as indicated by initial yielding, for ductile materials. The dotted lines in Fig. 4.74 indicate the boundaries for the maximum shear-stress failure criterion.

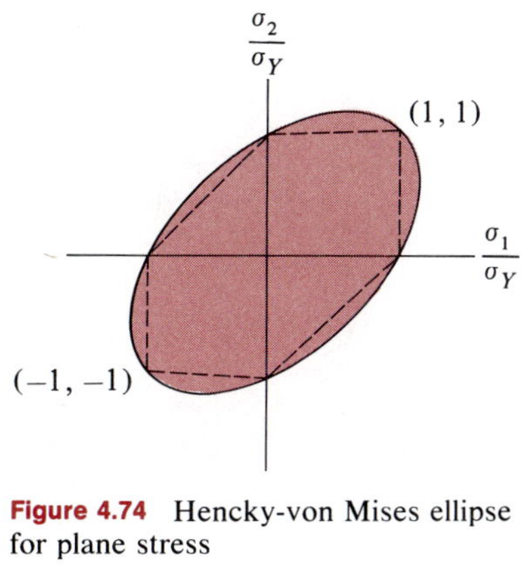

Figure 4.74 Hencky-von Mises ellipse for plane stress

Example 4.17 For a material with a tensile yield of 16 ksi subjected to the stresses $\sigma_x = 10$ ksi, $\sigma_y = -2$ ksi, and $\tau_{xy} = 8$ ksi, assess failure using the Hencky-von Mises criterion.

Solution The principal stresses are determined to be

$$\sigma_{1,2} = \left[\frac{10-2}{2} \pm \sqrt{\left(\frac{10+2}{2}\right)^2 + 8^2}\right] \text{ksi} = (4 \pm 10) \text{ ksi}$$

or $\sigma_1 = 14$ ksi and $\sigma_2 = -6$ ksi. Substituting these into the equation for the ellipse yields

$$(14/16)^2 - (14/16)(-6/16) + (-6/16)^2 = 316/256 > 1$$

so that failure is indicated using the Hencky-von Mises criterion.

Maximum Normal-Stress Failure Criterion The maximum normal-stress theory or, as it is sometimes called, the maximum principal-stress theory, is generally credited to William Rankine. This theory predicts that failure in a brittle material will occur when the maximum principal stress in the material reaches the value of the normal stress σ_{ULT}, as determined by a uniaxial tension test. For a plane-stress situation, with $\sigma_3 = 0$ and the in-plane principal stresses given by σ_1 and σ_2, failure is predicted when the principal stresses σ_1 and σ_2 lie on or outside the boundaries of the unit square with axes σ_1/σ_{ULT} and σ_2/σ_{ULT} as indicated in Fig. 4.75. This theory is presented only because of its historical interest; predictions of failure using this theory are not in good agreement with experimental results.

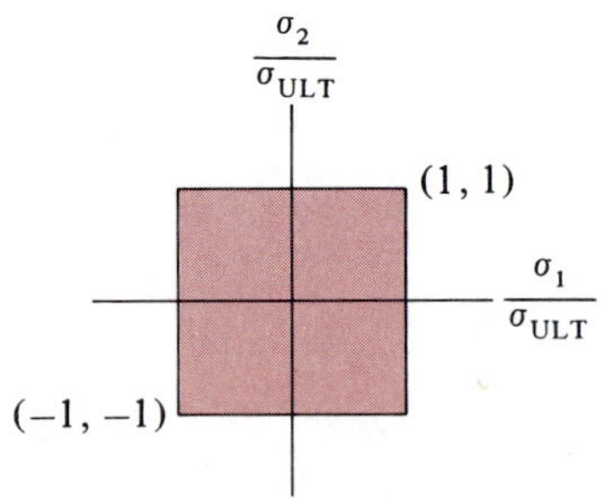

Figure 4.75 Failure for the maximum normal-stress theory

Coulomb-Mohr Failure Criterion Many materials behaving in a brittle manner exhibit markedly different ultimate strengths in tension and compression. As determined by uniaxial tension and compression tests, respectively, denote these ultimate strengths as σ_{UT} and σ_{UC}. When the in-plane principal stresses σ_1 and σ_2 are both positive or both negative, the safe regions in terms of the principal stresses σ_1 and σ_2 are assumed to be indicated by the shaded squares in quadrants I and III in Fig. 4.76.

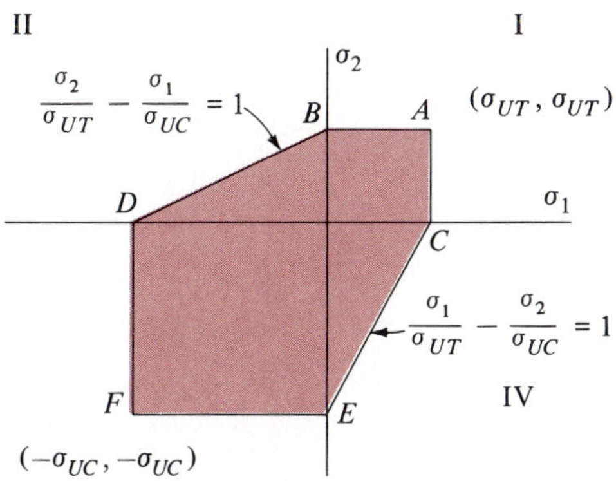

Figure 4.76 Safe regions for the Coulomb-Mohr criterion

When σ_1 and σ_2 are of opposite sign, the theory and experimental results are in good agreement by taking failure to be predicted when the in-plane principal stresses σ_1 and σ_2 lie on or outside of the lines *BD* and *CE* in quadrants II and IV, respectively. In quadrant IV, the equation of the line *BD* is

$$\frac{\sigma_1}{\sigma_{UT}} - \frac{\sigma_2}{\sigma_{UC}} = 1$$

In the second quadrant the equation of the line connecting points *C* and *E* is given by

$$\frac{\sigma_2}{\sigma_{UT}} - \frac{\sigma_1}{\sigma_{UC}} = 1$$

Failure using the Coulomb-Mohr criterion is predicted when the state of stress given in terms of the in-plane principal stresses σ_1 and σ_2 lies on or outside of the boundaries of the shaded region in Fig. 4.76. We note in passing that the Coulomb-Mohr failure criterion is also sometimes referred to as the internal-friction theory.

Example 4.18 The stresses at a point P on the surface of a structure made of a brittle material are $\sigma_x = 12.5$ ksi, $\sigma_y = -8.7$ ksi, and $\tau_{xy} = 6.0$ ksi. If a uniaxial tension test of the same material indicates failure at 20 ksi, and a uniaxial compression test indicates failure at 40 ksi, will failure occur according to the Hencky-von Mises criterion?

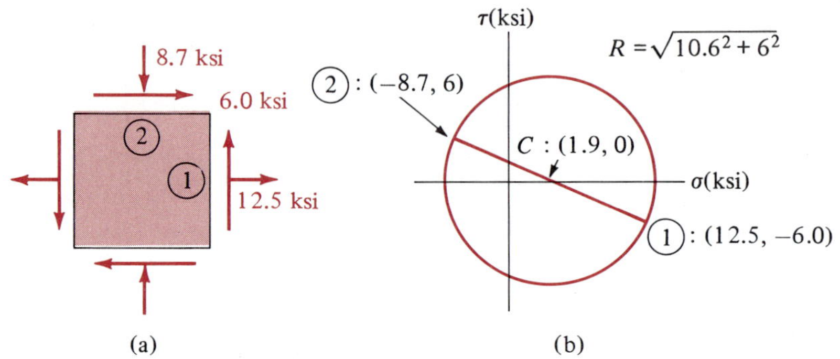

Figure 4.77 Mohr's circle for a state of stress

Solution As indicated in Fig. 4.77, the principal stresses using Mohr's circle are

$$\sigma_{\max,\min} = \left[\frac{12.5 - 8.7}{2} \pm \sqrt{\left(\frac{12.5 + 8.7}{2}\right)^2 + 6^2}\right] \text{ ksi}$$

or $\sigma_1 = 14.08$ ksi and $\sigma_2 = -10.28$ ksi. Using the Coulomb-Mohr theory equation in the fourth quadrant results in

$$\frac{\sigma_1}{\sigma_{UT}} - \frac{\sigma_2}{\sigma_{UC}} = \frac{14.08}{20} - \frac{-10.28}{40} = 0.961 < 1$$

so that failure will not occur.

Last it should be emphasized that these failure criteria are only applicable to isotropic materials—each criterion predicts failure when the state of stress at a point reaches a critical level independent of the actual orientation of the stresses at the point.

In summary, the maximum shear stress and Hencky-von Mises failure criterion are used for ductile materials, the Hencky-von Mises criterion generally considered to more accurately predict results consistent with experiment. For brittle materials, the Coulomb-Mohr criterion generally shows the best agreement with experiment.

PROBLEMS

For each of the following plane stress states ($\sigma_3 = \sigma_z = 0$) recall that the absolute maximum shear stress is given by the largest of $|\sigma_1 - \sigma_2|/2$, $|\sigma_2|/2$, and $|\sigma_1|/2$, where σ_1 and σ_2 are the in-plane principal stresses.

4.52 A steel specimen is tested in pure shear and observed to yield when the shear stress reaches 12 ksi. At what extensional stress would you expect yielding if the specimen of the same material were subjected to a uniaxial tension test? Use *(a)* the maximum shear stress criterion and *(b)* the Hencky-von Mises criterion.

4.53 Uniaxial tension and compression tests on a brittle isotropic material show failure strengths of $\sigma_{UT} = 3$ ksi and $\sigma_{UC} = 18$ ksi. At a point in a member subjected to plane stress, the stress components are $\sigma_x = -3$ ksi, $\sigma_y = 0$, and $\tau_{xy} = 16$ ksi. Determine whether the state of stress satisfies the Coulomb-Mohr criterion.

4.54 Uniaxial tension and compression tests on a brittle isotropic material show failure strengths of $\sigma_{UT} = 5$ ksi and $\sigma_{UC} = 25$ ksi. At a point in a member subjected to plane stress the stress components are $\sigma_x = \sigma_0$ ksi, $\sigma_y = 0$, and $\tau_{xy} = 4$ ksi. Determine the range of permissible values of σ_0 in order that the Coulomb-Mohr criterion is satisfied.

4.55 A plane stress state at a point is $\sigma_x = 64$ MPa, $\sigma_y = -32$ MPa, and $\tau_{xy} = 20$ MPa. Assuming the yield stress in tension is 200 MPa, determine whether yielding occurs according to *(a)* the maximum shear stress criterion, and *(b)* the Hencky-von Mises criterion.

4.56 A steel that has a tensile yield stress of 400 MPa is subjected to a state of plane stress with $\sigma_y = -160$ MPa, and $\tau_{xy} = 90$ MPa. Using the maximum shear-stress criterion, determine whether yielding occurs when *(a)* $\sigma_x = 150$ MPa, *(b)* $\sigma_x = 180$ MPa, and *(c)* $\sigma_x = 210$ MPa.

4.57 Repeat problem 4.56 using the Hencky-von Mises criterion.

4.58 A structural component with a tensile yield stress of 40 ksi is subjected to a state of plane stress with $\sigma_x = 32$ ksi and $\sigma_y = 19$ ksi. Using the maximum shear-stress criterion, determine whether yielding occurs when *(a)* $\tau_{xy} = 8$ ksi, *(b)* $\tau_{xy} = 16$ ksi, and *(c)* $\tau_{xy} = 24$ ksi.

4.59 Repeat problem 4.58 using the Hencky-von Mises criterion.

4.60 A ductile material with $\sigma_Y = 36$ ksi is subjected to the state of stress shown. Determine whether yielding occurs on the basis of *(a)* the maximum shear-stress criterion, and *(b)* the Hencky-von Mises criterion.

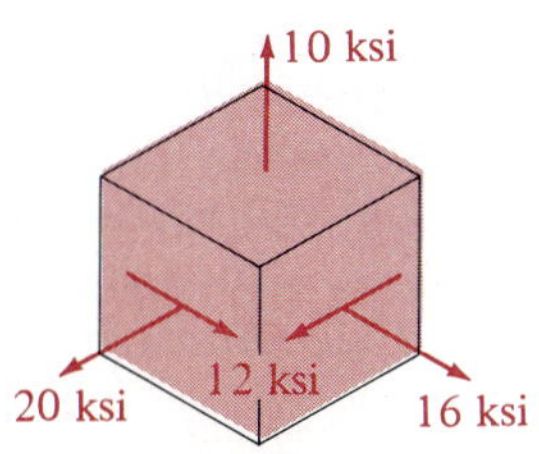

P. 4.60

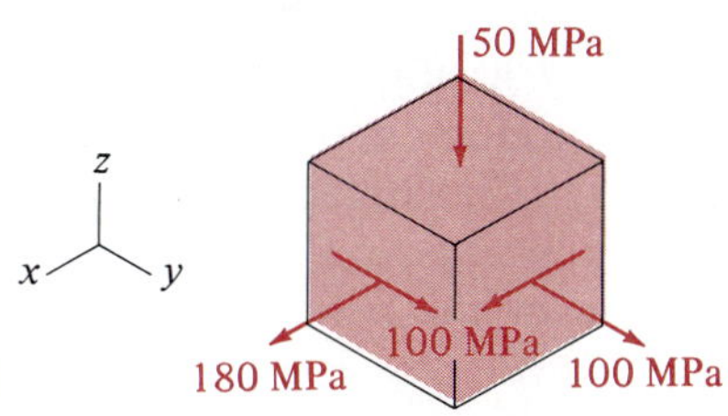

P. 4.61

4.61 Repeat problem 4.60 for the state of stress shown if $\sigma_Y = 240$ MPa.

4.62 A ductile material with $\sigma_Y = 32$ ksi is subjected to the state of stress shown. Determine the permissible range of σ_z on the basis of *(a)* the maximum shear-stress criterion, and *(b)* the Hencky-von Mises criterion.

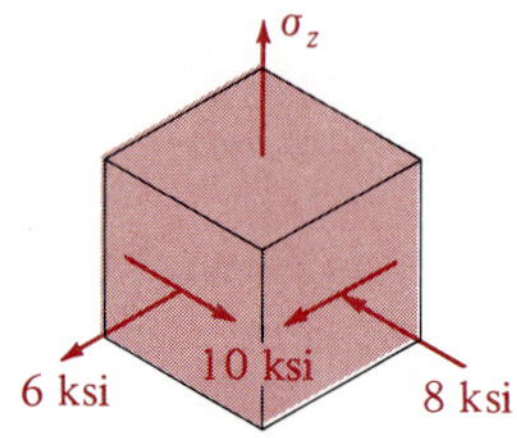

P. 4.62

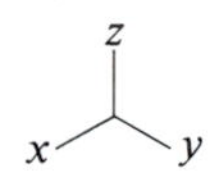

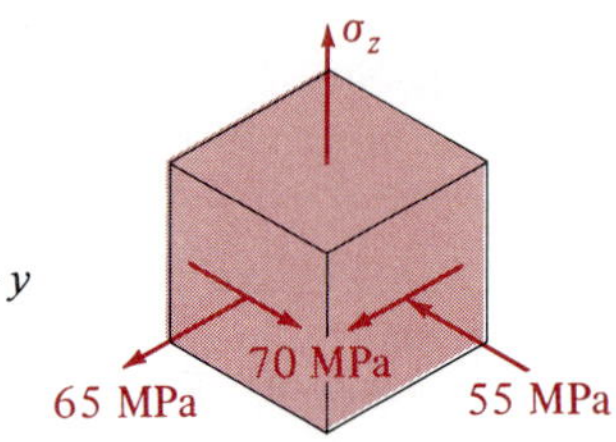

P. 4.63

4.63 Repeat problem 4.62 for the state of stress shown if $\sigma_Y = 210$ MPa.

4.64 A ductile material with $\sigma_Y = 36$ ksi is subjected to the state of stress shown. Determine the largest permissible value of τ_{xy} if there is to be no yielding according to the maximum shear-stress criterion.

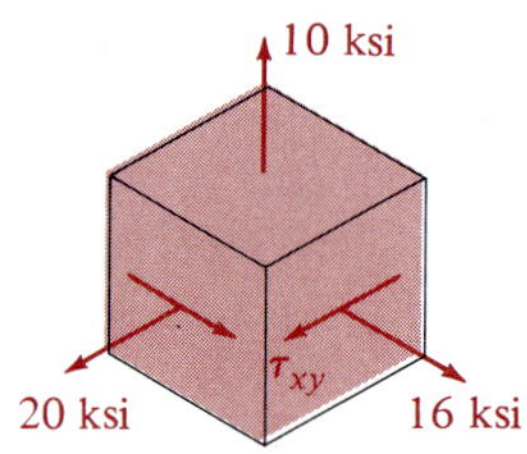

P. 4.64

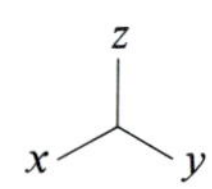

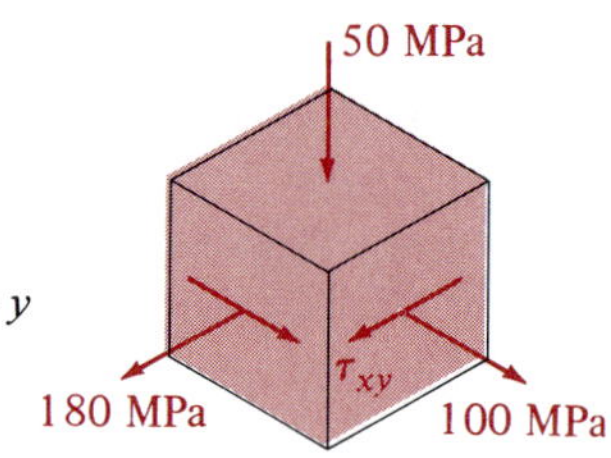

P. 4.65

4.65 Repeat problem 4.64 for $\sigma_Y = 240$ MPa.

4.66 Repeat problem 4.64 using the Hencky-von Mises criterion.

4.67 Repeat problem 4.65 using the Hencky-von Mises criterion.

4.10 WORKING STRESSES AND FACTOR OF SAFETY

Chapters 2, 3, and 4 have been devoted to the three basic ideas of solid mechanics. In Chapters 5 through 9 these three basic ideas will be combined to produce mathematical models for analyzing a variety of different types of structures. The mathematical model, which is intended to represent both the environment and the structure, is only an approximation of the actual environment and structure. If we could create and solve an exact mathematical model, the predictions of the model would exactly match the behavior of the corresponding real-world structure. In such a hypothetical situation, the structure would be such that when the loads were applied, the stresses and deformations in the actual structure would be exactly as predicted.

In some situations it may be possible to develop a mathematical model that represents the original physical problem with a great deal of accuracy: the environment and structure may be simple enough so that there are very few (if any) doubts about the mathematical model. Much more frequently, however, it is not possible to anticipate the loads precisely nor to know precisely the material behavior and properties. In short, there are simply too many factors about which we have imprecise knowledge, to be able to design a structure strictly on the basis of the mathematical model; that is, there is too great a chance that failure of some sort will occur.

If the probability of failure is to be reduced to an acceptable level, the loads a structure is actually capable of sustaining must be greater than the loads the structure will be required to sustain in the actual application. As mentioned in Section 1.3, the ability of a structure to withstand loads is called strength. Thus in terms of strength, the design of a structure must be such that the ultimate strength exceeds, to some extent, the allowable strength required of the structure in order to accomplish its mission. The ultimate strength is usually based upon a yield stress or upon an ultimate stress. The ratio of the ultimate strength to the allowable strength is called the **factor of safety** and is denoted by FS.

$$FS = \frac{\text{Ultimate strength}}{\text{Allowable strength}}$$

The factor of safety must clearly be greater than unity. Depending upon the many details of the particular situation in question, the factor of safety is generally chosen as a value from slightly above 1.0 up to perhaps as high as 5.0–10.0. Note carefully that for a structure consisting of a number of components, the factor of safety will in general be different for each of its components.

The strength of a structure or component of a structure is often evaluated by determining the stress levels in the structure. It is customary to represent the factor of safety in terms of the stresses rather than the less specific concept of strength. Let the maximum stress a component can withstand be designated as the ultimate stress. We define a corresponding **working stress** σ_W according to

$$\sigma_W = \frac{\text{Ultimate stress}}{FS}$$

The working stress is also referred to as the allowable stress or the design stress.

For ductile materials the ultimate stress would be the yield stress σ_Y as determined by the uniaxial tensile test. The working stress would then be

$$\sigma_W = \frac{\sigma_Y}{FS}$$

with either of the maximum shear stress or the Hencky-von Mises criteria used to assess failure using σ_W rather than σ_Y. For a brittle material the ultimate stress would be the stress at fracture σ_{ULT} as determined by the uniaxial tensile test with the working stress as

$$\sigma_W = \frac{\sigma_{ULT}}{FS}$$

The Coulomb-Mohr criterion is used to assess failure using σ_W rather than σ_{ULT}.

Note that it is also possible to express ultimate and working loads on a component in terms of the factor of safety as

$$P_W = \frac{P_{ULT}}{FS}$$

where P_W and P_{ULT} are the working or allowable and ultimate loads, respectively.

Example 4.19 For the maximum shear-stress failure criterion, the failure envelope with a factor of safety FS would be associated with the hexagon indicated in Fig. 4.78.

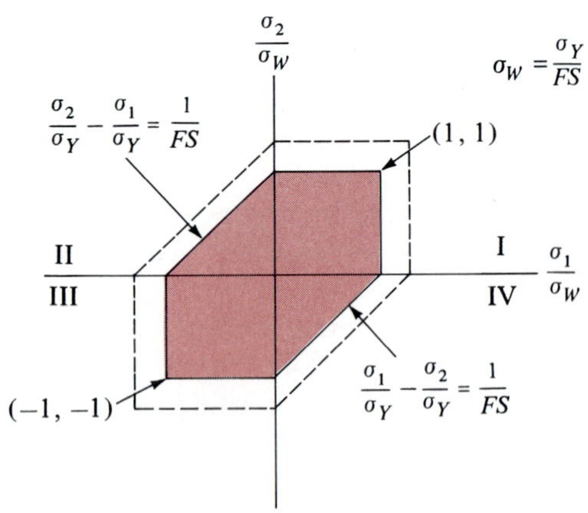

Figure 4.78 Maximum shear-stress hexagon with a factor of safety

The equation of the straight line in quadrant II could be written as

$$\frac{\sigma_2}{(\sigma_Y/FS)} - \frac{\sigma_1}{(\sigma_Y/FS)} = 1 \qquad \text{or} \qquad \frac{\sigma_2}{\sigma_Y} - \frac{\sigma_1}{\sigma_Y} = \frac{1}{FS}$$

and the equation in quadrant IV as

$$\frac{\sigma_1}{\sigma_Y} - \frac{\sigma_2}{\sigma_Y} = \frac{1}{FS}$$

The corresponding hexagon with $FS = 1$ is indicated by the dotted lines in Fig. 4.78, showing that the use of the factor of safety simply shrinks the size of the maximum shear stress hexagon.

As a specific example, consider a plane-stress state with $\sigma_x = 17$ ksi, $\sigma_y = -4$ ksi, and $\tau_{xy} = 6$ ksi. Assume a ductile material with $\sigma_y = 40$ ksi. The principal stresses are calculated as

$$\sigma_{1,2} = \left[\frac{17-4}{2} \pm \sqrt{\left(\frac{17+4}{2}\right)^2 + 36}\right] \text{ksi} = 18.6 \text{ ksi}, \ -5.6 \text{ ksi}$$

The factor of safety is determined from the equation

$$\frac{18.6}{40} - \frac{-5.6}{40} = \frac{1}{F}$$

from which

$$FS = 1.65$$

Example 4.20 For the Hencky-von Mises failure criterion, the failure envelope with a factor of safety FS would be associated with the ellipse indicated in Fig. 4.79.

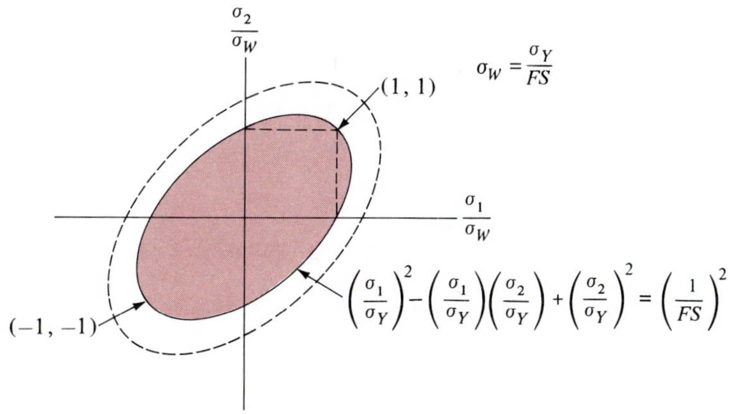

Figure 4.79 Hencky-von Mises ellipse with a factor of safety

The equation of the boundary of the elliptical area could be written as

$$\left(\frac{\sigma_1}{\sigma_Y/FS}\right)^2 - \left(\frac{\sigma_1}{\sigma_Y/FS}\right)\left(\frac{\sigma_2}{\sigma_Y/FS}\right) + \left(\frac{\sigma_2}{\sigma_Y/FS}\right)^2 = 1$$

or

$$\left(\frac{\sigma_1}{\sigma_Y}\right)^2 - \left(\frac{\sigma_1}{\sigma_Y}\right)\left(\frac{\sigma_2}{\sigma_Y}\right) + \left(\frac{\sigma_2}{\sigma_Y}\right)^2 = \frac{1}{(FS)^2}$$

The corresponding ellipse with $FS = 1$ is indicated by the dotted lines in Fig. 4.79, showing that the use of the factor of safety reduces the size of the ellipse within which the Hencky-von Mises criterion is satisfied.

For a specific example use the state of stress given in Example 4.19, namely $\sigma_x = 17$ ksi, $\sigma_y = -4$ ksi, $\tau_{xy} = 6$ ksi, $\sigma_1 = 18.6$ ksi, $\sigma_2 = -5.6$ ksi, and $\sigma_Y = 40$ ksi. The factor of safety FS is determined from the equation

$$\left(\frac{18.6}{40}\right)^2 - \left(\frac{18.6}{40}\right)\left(\frac{-5.6}{40}\right) + \left(\frac{-5.6}{40}\right)^2 = \frac{1}{(FS)^2}$$

from which $FS = 1.82$.

Example 4.21 For the Coulomb-Mohr failure criterion with the same factor of safety FS applied to both σ_{UT}, the ultimate tensile strength in tension, and σ_{UC}, the ultimate tensile strength in compression, the failure envelope for a brittle material appears as in Fig. 4.80.

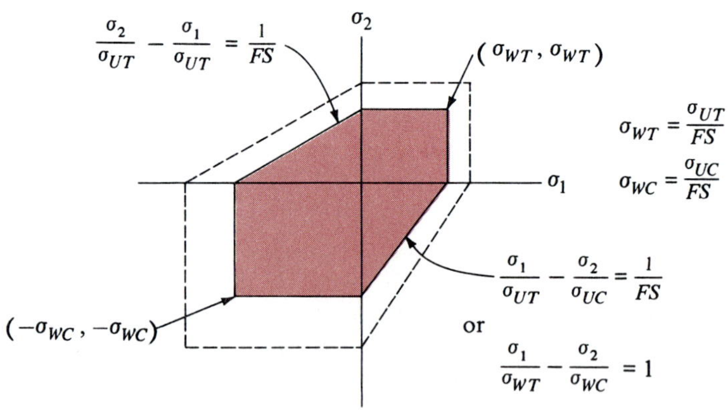

Figure 4.80 Coulomb-Mohr failure criterion with a factor of safety

The equations of the lines bounding the shaded region in the second and fourth quadrants are also indicated. In addition the dotted lines in Fig. 4.80 indicate the boundaries of the corresponding diagram for $FS = 1$, showing the reduction in the size of the region corresponding to safe combinations of the in-plane principal stresses.

As a specific example, assume a brittle material with $\sigma_{UT} = 20$ ksi and $\sigma_{UC} = 40$ ksi subjected to the plane stress state $\sigma_x = 14$ ksi, $\sigma_y = -8$ ksi, and $\tau_{xy} = 10$ ksi. The principal stresses are given by

$$\sigma_{1,2} = \left[\frac{14 - 8}{2} \pm \sqrt{\left(\frac{14 + 8}{2}\right)^2 + 10^2}\right] \text{ksi} = 17.9 \text{ ksi}, \ -11.9 \text{ ksi}$$

The factor of safety is determined from

$$\frac{17.9}{20} - \frac{-11.9}{40} = \frac{1}{FS}$$

from which $FS = 0.84$, indicating failure since $FS < 1.0$!

There are many situations to be considered that influence the decision as to the value to be used for the factor of safety. These include anticipated mode(s) of failure, the degree of confidence in the mathematical model used in the design and the many aspects involved in the manufacturing process that produces the structure or its components.

The mode of failure for a particular component of the structure may be sudden (no advance warning) or gradual. If a component of a structure fails it may be possible for a redistribution of the loads to take place internally and for the structure to continue to function. On the other hand the failure of a single component may result in failure of the entire structure. In the first scenario, the factor of safety could be lower than in the second.

There are many unknowns involved in the mathematical model. The inputs or loads may be static or dynamic in nature. Dynamic loads are frequently less well defined than static loads. The actual nature of the materials used is such that the flexibility and strength properties must be considered to be somewhat variable. In simple models there is generally a larger degree of approximation involved in the process of converting the physical model into the mathematical model than for more complex models. On the other hand, a complex mathematical model may have to be analyzed approximately.

To actually create the structure or device in a form suitable for testing or for actual use, the results of the analysis and design must be converted from the drawing board to the actual configuration by some sequence of manufacturing/assembly processes. The manufacturing/assembly process generally involves many steps, each of which can result in a slight deviation of the structure from what is intended in the mathematical model. These deviations can result from a variety of sources, including misinterpretation of the drawings, poor workmanship, failure to produce components with the specified geometric tolerances, and incorrect processing of materials. All of these deviations essentially introduce further unknowns into the analysis and design process.

In the final analysis it is often the case that the factor of safety is chosen by an experienced engineer or group of engineers after considerable evaluation of all the available information about the physical and mathematical models, as well as previous experience in the area.

PROBLEMS

4.68 A 0.5-in.-diameter structural steel bar transmits a tensile load of 3200 lbf. If $\sigma_Y = 24$ ksi, determine the factor of safety with respect to failure by yielding using *(a)* the maximum shear stress criterion, and *(b)* the Hencky-von Mises criterion.

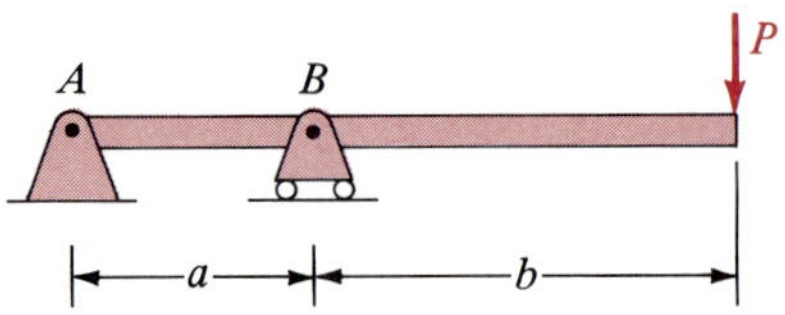

P. 4.69

4.69 The rigid bar shown is supported by a pin and a roller at A and B. If the 0.75-in.-diameter pins are made of a steel with a tensile yield stress of 30 ksi, what is the allowable load P if the factor of safety with respect to yielding for the pins is to be 2.0? Take $a = 1$ ft, $b = 2$ ft, and assume that the pins act in double shear.

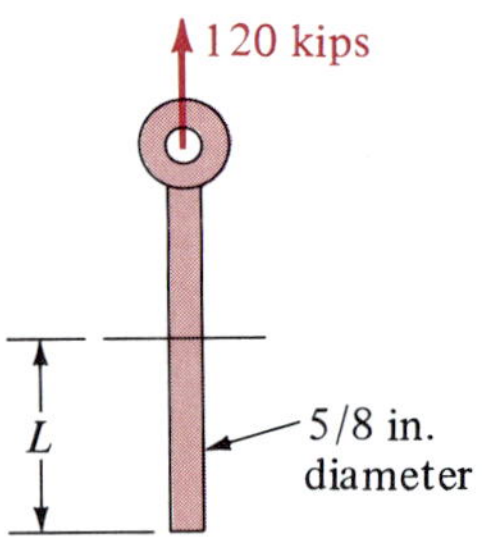

P. 4.70

4.70 A tie rod that secures a cable to ground is shown. The load to be carried by the cable is 120 kips. Determine the minimum length the tie rod must be imbedded in the concrete is order to provide a factor of safety of 2.5 with respect to the shear stress ultimate of 600 psi between the tie rod and the concrete.

4.71 The truss connected by pins in double shear at A, B, and C is acted upon by a 1200 lbf load as shown. If the ⅝-in.-diameter pins are capable of withstanding an average shear stress in double shear of 15 ksi, what is the factor of safety with respect to failure by shear, for the pin at B. Use the maximum shear-stress criterion. Take $a = b = c = 2$ ft.

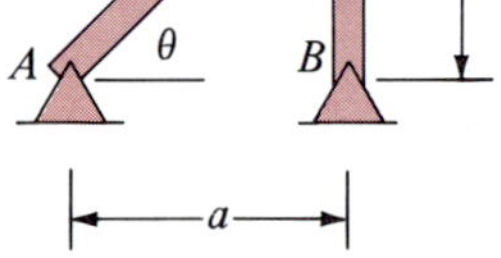

P. 4.71

4.72 Two bars are connected by splice plates as shown in the figure. The 0.5-in.-diameter rivets yield when the average shear stress reaches a value of 18 ksi. The critical bearing stress between the rivets and the plate is 16 ksi and the ultimate stress in tension for the plates is 24 ksi. Using the maximum shear stress criterion, determine the maximum load P that the joint is capable of transmitting if the factor of safety with respect to all types of failure is 2.5. Take the stress concentration factor for the tensile stress at the holes to be 2.5. Which of the stresses is critical? Take $t = 0.5$ in. and the width of each of the plates to be 2 in. Neglect friction between the plates.

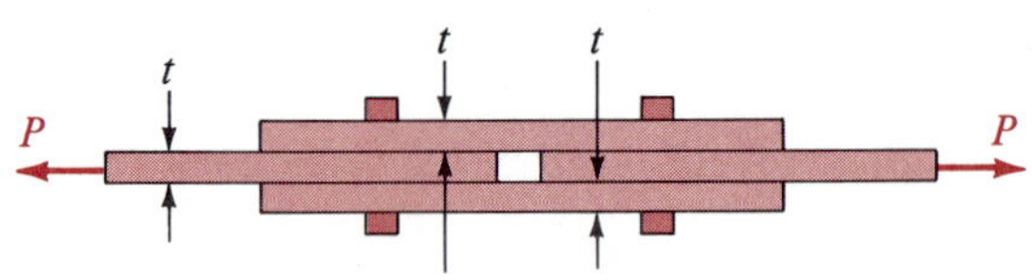

P. 4.72

4.73 A short piece of pipe made of a material for which the tensile yield stress is 280 MPa is to carry a 1.5 MN load with a factor of safety against yielding of 2.2. If the pipe is to have a wall thickness of 15 mm, what must be the outside diameter? Use the Hencky-von Mises criterion.

4.74 A T joint in a high-pressure pipeline is capped by a circular plate secured by 6 high-strength 25-mm-diameter bolts (σ_{ULT} = 1400 MPa) as shown. If the pressure in the line is 5.0 MPa, determine the factor of safety with respect to fracture for the bolts using the Coulomb-Mohr criterion.

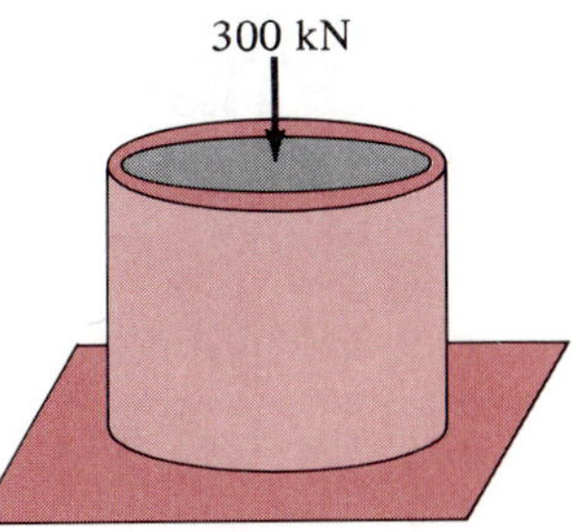

P. 4.73

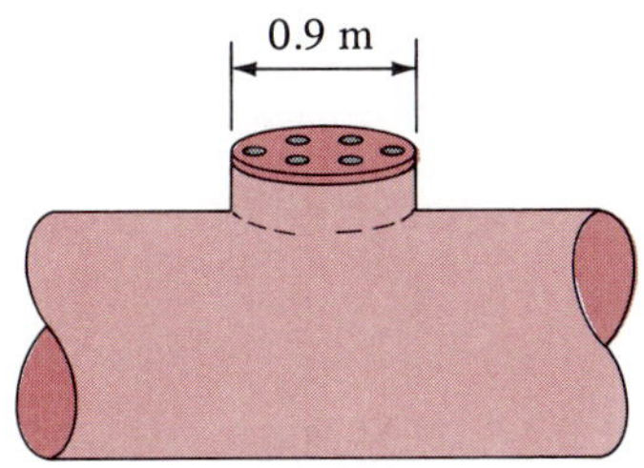

P. 4.74

4.11 CLOSURE

The three basic ingredients of solid mechanics are equilibrium, deformation, and material behavior. A valid theory is based upon the proper combination of these three basic ingredients for a particular geometry and type of loading. It is necessary to consider material behavior in order to provide the link between the force and displacement variables. The consideration of material behavior is necessary in order to determine the necessary information about both the *form* of the relation between the force and displacement variables (elastic, linear, rate independent, and so forth), and the *values* of the material constants (elastic constants, yield and ultimate stresses, and so on). In a practical vein, the investigation of material behavior indicates that the behavior is local from the standpoint that for most materials, there is usually a restricted range of values for both stress and strain in which the relationship is linear.

The investigation of material behavior provides the connection between the force variables and the displacement variables, from considerations of equilibrium and deformation, respectively. Experiments or tests that are properly conceived and executed can provide both the qualitative information necessary to decide upon the form of the mathematical model, and the quantitative information as to the values of the constants that appear in the constitutive relations. Other constants determined from experiments performed on engineering materials serve as limits on force and displacement variables used in the theories of failure.

The theories that result from the combination of the three basic ingredients are only a part of the overall design process. The theories serve to generate a mathematical model that serves as a tool in the analysis phase of the process. The overall design process also involves the assessment of, and the guarding against, failure. Failure is generally judged on the basis of the so-called failure criteria. These failure criteria are based upon a combination of experience and experimental results and are an indispensable part of the design process. In addition to these more or less rational components of the design process, it is frequently necessary to employ factors of safety to account for the many unknowns associated with the analysis, design, and manufacture of a structure or structural component.

CHAPTER 5 AXIAL DEFORMATIONS

CHAPTER OUTLINE

5.1 Introduction
5.2 Equilibrium, Deformation, and Material Behavior
5.3 Classical Solutions
5.4 Statically Indeterminate Problems
  5.4.1 Flexibility Formulations
  5.4.2 Stiffness Formulations
5.5 Exceptions to the Theory
  5.5.1 Stress Concentrations
  5.5.2 Loading
  5.5.3 Inelastic Behavior
5.6 Dynamic Loading
5.7 Closure

5.1 INTRODUCTION

In many practical situations, the primary purpose of a structure or portion of a structure is to transmit or assist in transmitting a single force along the structure. Examples include a strut of a small single-engine aircraft, a simple truss, and a cable in an elevator, all pictured in Fig. 5.1.

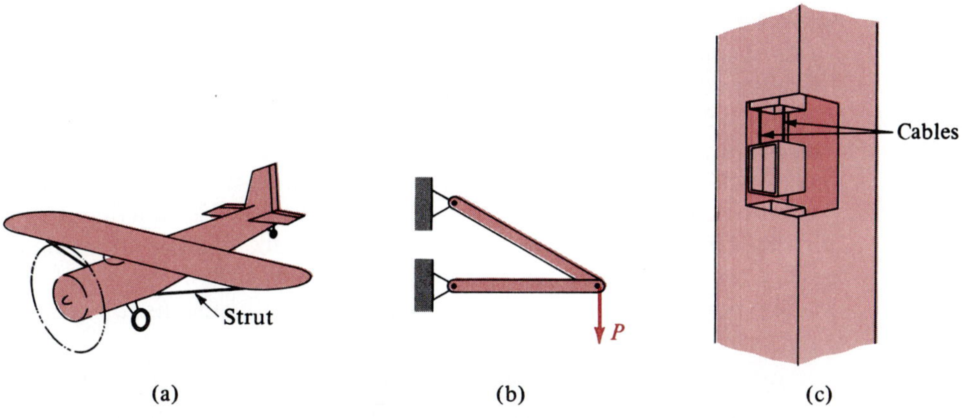

Figure 5.1 Examples of members transmitting an axial force

The members involved are often called **axial members:** the forces they transmit are along the axis of the member. By an axial member we mean the following:

1. The member is straight and untwisted: it has the same cross section, such as circular or rectangular, at all points along its length.
2. The cross-sectional area of the bar is either constant or slowly varying with respect to distance along its length.

A typical member satisfying these conditions is shown in Fig. 5.2.

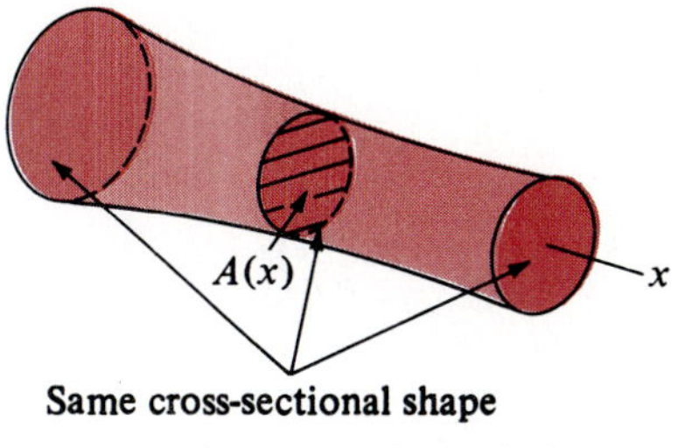

Figure 5.2 Geometry of a typical axial member

In this chapter we will consider in detail the two basic problems of solid mechanics, strength and stiffness, as well as the development and combination of the three basic ideas, equilibrium, deformation, and material behavior, as they pertain to the axial deformation problem. Additionally, practical considerations such as the effects of several important exceptions to the theory of axial deformation will be covered.

It must be kept in mind that the conditions (restrictions) on the geometry of the bar are a necessary part of the theory—if the conditions of the geometry are not satisfied, the theory cannot be considered valid.

5.2 EQUILIBRIUM, DEFORMATION, AND MATERIAL BEHAVIOR

The basic ideas to use for analyzing any solid mechanics problem are equilibrium, deformation, and material behavior. The equations resulting from each of these three considerations are then combined to form the governing equations for the theory.

Equilibrium Consider a typical axial member shown in Fig. 5.3. We will take the x-axis to be the axis of the bar.

As discussed in Section 2.2, any external forces are either body forces or surface forces. Surface forces will be assumed to act as shear stresses parallel to the lateral surface, or as normal stresses perpendicular to the faces at the ends of the bar, as indicated in Fig. 5.3a. Body forces will be assumed to act only in the direction of the axis of the bar. The effect of all the external forces is assumed to be equivalent to a distributed loading $q(x)$, as indicated in Fig. 5.3b. The line of action of $q(x)$ is, of course, parallel to the axis of the bar. Note that $q(x)$ is a loading per unit of length along the bar and has dimensions $[q] = FL^{-1}$.

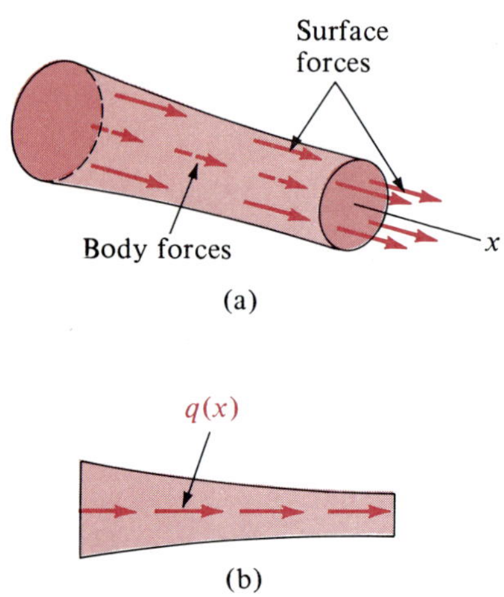

Figure 5.3 External forces on a typical axial member

> Any external loading must be such that the line of action of the internal force P is through the centroid of the cross section at all locations along the length of the bar.

Fig. 5.4a shows a typical axial member in equilibrium under the action of end forces P_1, P_2, and the distributed loading $q(x)$. The typical section AA, as marked in Fig. 5.4a, will expose the internal force P transmitted as shown in Fig. 5.4b.

Assuming P_1 and q to be known, the free-body diagram shown in Fig. 5.4b is used to satisfy equilibrium and determine P. As indicated, the distribution of the normal stress σ_x must be statically equivalent to the internal force P according to

$$P = \int_{\text{Area}} \sigma_x \, dA$$

If σ_x is assumed to be uniform (e.g., $\sigma_x = \sigma$), this equation can be expressed as

$$P = \sigma A \qquad \text{or} \qquad \sigma = \frac{P}{A} \tag{5.1}$$

which is also the average value of the normal stress on the exposed face. If σ_x is not uniform the maximum value will exceed the average value. Several technically important situations where the normal-stress distributions in axial members are not uniform, that is, where the average stress $\sigma = P/A$ does not give an accurate picture of the distribution of the normal stresses acting on a cross section, are discussed in Section 5.5.

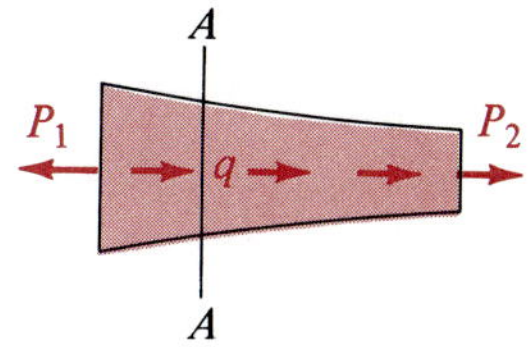

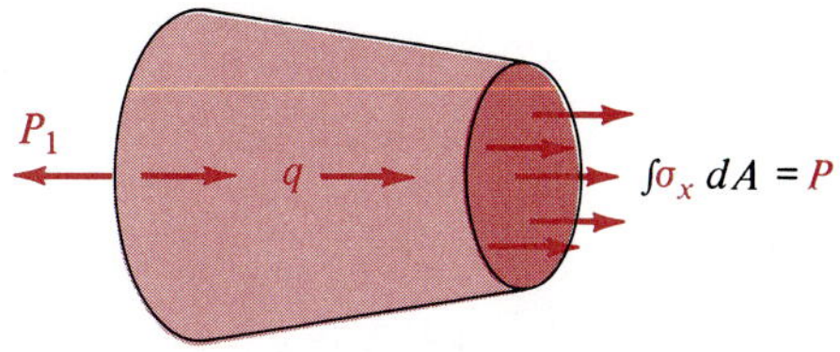

Figure 5.4 Internal force and stresses on an internal section

One of the primary goals of the study of equilibrium is to determine the internal force P in terms of the external loads. In this regard, consider a free-body diagram of a segment Δx of the bar as indicated in Fig. 5.5. The resultant force P and the possibly different resultant force $P + \Delta P$ are indicated on the left and right faces, respectively. Note carefully that the change in the positive internal force P is assumed to take place in the positive coordinate direction. The single equilibrium equation is

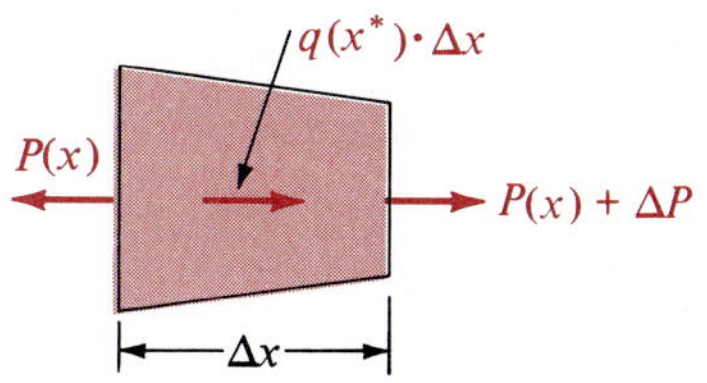

Figure 5.5 Free-body diagram of a differential element of the bar

$$\Sigma F_x = -P + q(x^*)\,\Delta x + P + \Delta P = 0$$

where $q(x^*)$ indicates the value of q at a point x^* in the differential element such that $q(x^*)\,\Delta x$ is numerically equal to the total load being input from the external loading $q(x)$. Canceling like terms, dividing by Δx and allowing Δx to approach zero in the equilibrium equation yields

$$\frac{dP}{dx} + q(x) = 0$$

or

$$P' + q = 0 \tag{5.2}$$

as the equilibrium equation for the axial problem. Eqn. (5.2) relates the internal force P to the external loading $q(x)$, and together with

Eqn. (5.1), that is, $P = \sigma A$, constitutes the contribution to the axial deformation theory from the basic idea of equilibrium.

Deformation We will assume the axial deformation to take place as a simple translation of any particular cross section as indicated in Fig. 5.6.

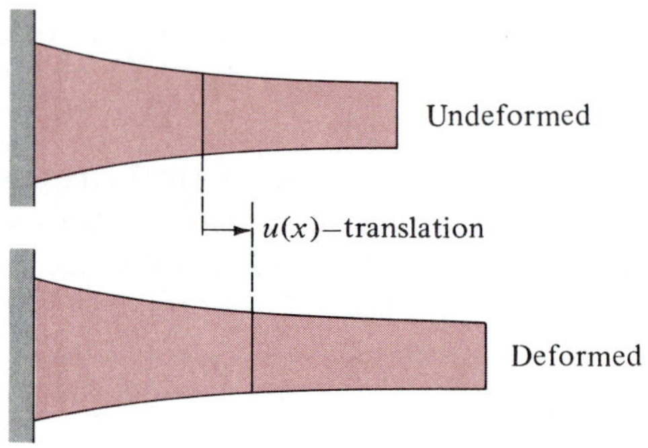

Figure 5.6 Axial deformation as a translation of a particular cross section

We developed, in Section 3.2, the strain-displacement relation for this simple deformation as

$$\varepsilon_x = \frac{du}{dx} = u' \tag{5.3}$$

where ε_x is the extensional strain and $u(x)$ is the axial displacement.

Notice the fact that Eqns. (5.1) and (5.2) relate the unknowns P and σ_x, and that Eqn. (5.3) relates the unknowns ε_x and u. Not only are there only three equations for the four unknowns P, σ_x, ε_x, and u, but the force variables P and σ_x are contained in Eqns. (5.1) and (5.2), with the displacement variables ε_x and u present only in Eqn. (5.3), making it impossible to relate the force variables and the displacement variables. We need to consider the basic idea of material behavior, which relates stress to strain, in order to bridge the gap between the force and displacement variables.

Material Behavior We will assume the bar to be composed of an elastic material operating in the linear range, so that the relation between the stress σ_x and the strain ε_x is given by

$$\varepsilon_x = \frac{\sigma_x}{E} + \alpha \, \Delta T \qquad (5.4)$$

where the temperature change ΔT is assumed to be known. There are now four equations for the four variables. These equations can be combined to yield a rational theory for the axial deformation problem.

Combination The first step is to eliminate the strain between Eqns. (5.3) and (5.4) to obtain

$$\sigma_x = E(u' - \alpha T)$$

Eqn. (5.1) is then used to write

$$P = A\sigma_x = AE(u' - \alpha T) \qquad (5.5)$$

Eqn. (5.5), which relates P, the force transmitted, to the term $u' - \alpha T$, is referred to as the **force-displacement relation.** Given $P(x)$, Eqn. (5.5) is integrated to determine the displacement $u(x)$. The last step in combining the variables is to substitute the force-displacement relation into the equilibrium equation, Eqn. (5.2), to obtain

$$[AE(u' - \alpha T)]' + q = 0 \qquad (5.6)$$

as the equilibrium equation expressed in terms of the displacement $u(x)$.

The two constants arising upon integration of Eqn. (5.6) are to be determined from two boundary conditions, *one at each end of the bar.* Appropriate conditions at a boundary include specifying either

(a) the displacement u at a boundary.

or

(b) the force $P = AE(u' - \alpha T)$ at a boundary.

Several examples of the application of boundary conditions are contained in the sections that follow.

The use of the equilibrium equation $[AE(u' - \alpha T)]' + q = 0$ assumes that all of the conditions on the geometry of the bar, the loading, and the material behavior have been satisfied. If any of these assumptions are not satisfied, the results for the stresses and displacements obtained by solving the equilibrium equation and satisfying the boundary conditions necessarily cannot be expected to compare well with the actual values in the corresponding physical problem.

PROBLEMS

5.1 A bar with rectangular cross section is composed of two materials of moduli E_1 and E_2 as shown. Assuming $u = u(x)$, use basic principles of statics to show that if there is to be no bending about a z-axis, the line of action of the load P is given by $\bar{y} = (A_1E_1\bar{y}_1 + A_2E_2\bar{y}_2)/(A_1E_1 + A_2E_2)$, and the stresses in the two bars as $\sigma_1 = PE_1/(A_1E_1 + A_2E_2)$ and $\sigma_2 = PE_2/(A_1E_1 + A_2E_2)$. Show also that the elongation of the bar is given by $\Delta = PL/(A_1E_1 + A_2E_2)$.

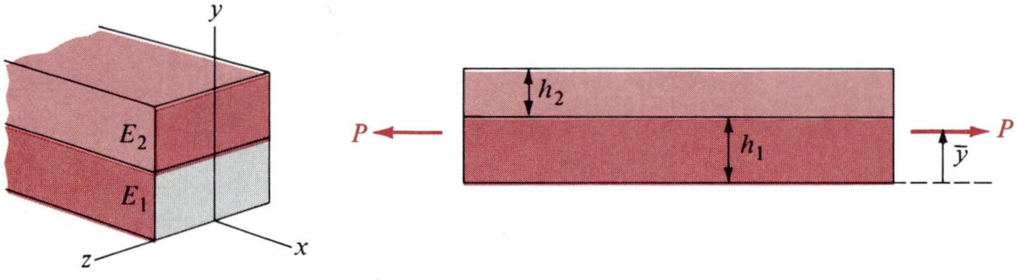

P. 5.1

5.2 Two bars, one steel and one aluminum, are bonded as shown. Determine the line of action of the axial load P if there is to be no bending. If the allowable stresses in the steel and aluminum are 20 ksi and 14 ksi, respectively, determine *(a)* the largest permissible axial load P and *(b)* the elongation of the bar. (*Hint:* See problem 5.1.)

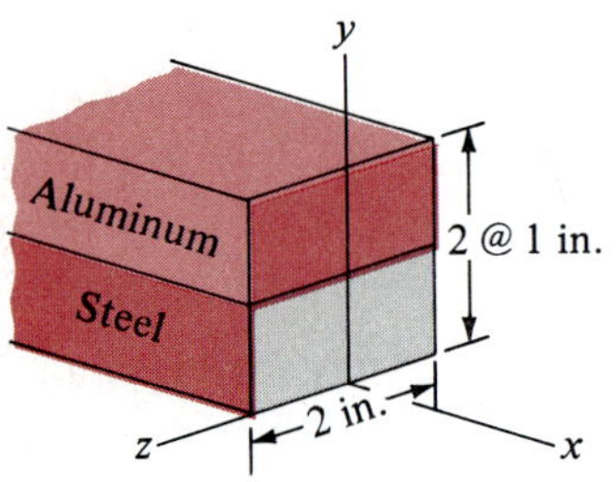

P. 5.2

5.3 A bar with rectangular cross section is composed of a material with a Young's modulus that can be approximated as $E(y) = E_1 + E_2y$ as shown in the figure. Assuming $u = u(x)$, use basic principles of statics to show that if there is to be no bending about a z-axis, the line of action of the load P is given by $\bar{y} = h(E_1/2 + hE_1/3)/(E_1 + hE_2/2)$ and that the stress in the bar is $\sigma = P(E_1 + yE_2)/[bh(E_1 + hE_2/2)]$. Show also that the elongation of the bar is given by $\Delta = PL/[bh(E_1 + hE_2/2)]$, where L is the length of the bar.

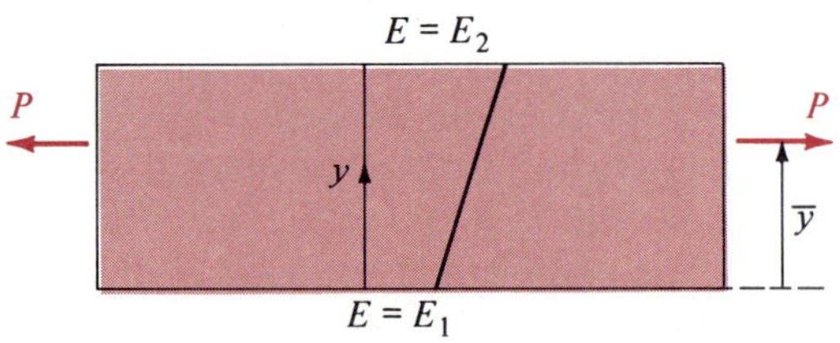

P. 5.3

5.4 A bar with a variable elastic constant E as shown in the figure is to be loaded axially with no bending. Determine *(a)* the location of the line of action of the load P, *(b)* the maximum load P if the stress is not to exceed 120 MPa, and *(c)* the elongation of the bar. Take $b = 20$ mm. (*Hint:* See problem 5.3.)

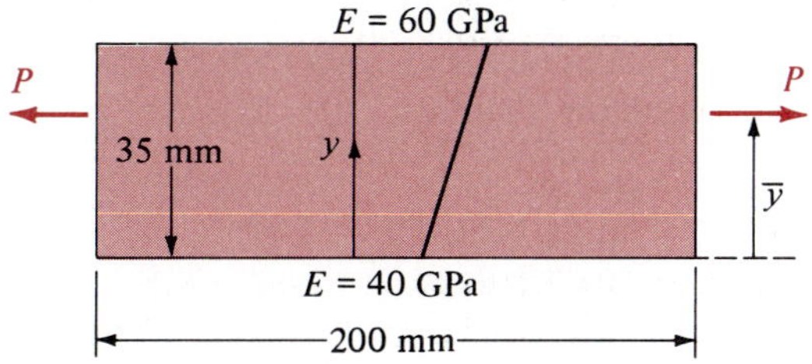

P. 5.4

5.5 Present an argument to show that the result σ_x = constant, with all the other stresses zero, cannot possibly hold when there is a sudden change in area such as shown in the figure. (*Hint:* Draw a free-body diagram including a portion of the bar containing the sudden change in cross-sectional area.)

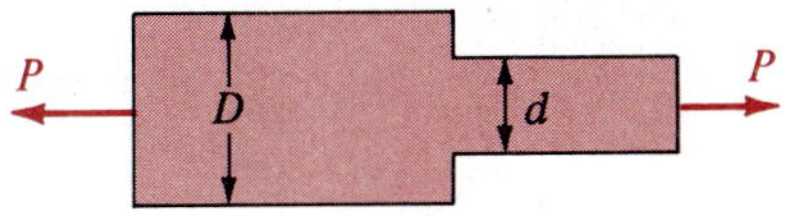

P. 5.5

5.6 Repeat problem 5.5 for the figure shown.

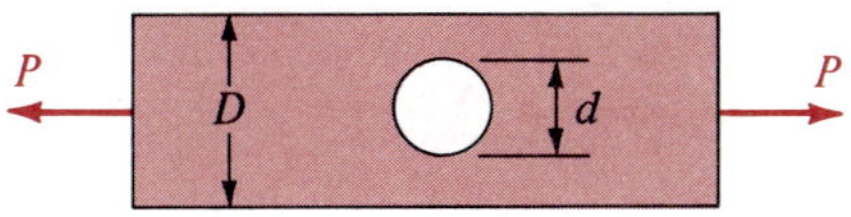

P. 5.6

5.7 Integrate the equilibrium equation $P' + q = 0$ between $x = x_1$ and $x = x_2$, and interpret the result in terms of a free-body diagram of the bar between $x = x_1$ and $x = x_2$.

5.8 Integrate the equilibrium equation $P' + q = 0$ between $x = 0$ and $x = L$, and interpret the result in terms of a free-body diagram of the bar between $x = 0$ and $x = L$.

5.9 Show that if u, the axial component of displacement, is assumed to depend only upon x, and if the moments about the y- and z-axes are both assumed to be zero, it follows that the line of action of P, the axial load transmitted, must be through the centroid. (The point of this problem is that the result does not have to be assumed: it can be derived.)

5.3 CLASSICAL SOLUTIONS

In this section, we will apply the theory of axial deformation developed in Section 5.2 to a variety of important technical problems. As stated in the introduction, these include the strut in a small aircraft, a simple truss, and the cable in an elevator. There are several classical solutions concerning axial deformations that are significant from the standpoint that many problems can be modeled in a fashion that uses these classical solutions as part of the solution.

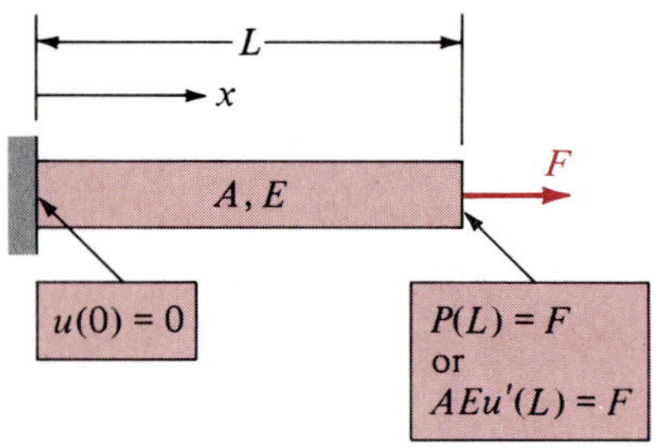

Figure 5.7 Bar subjected to a force F

Example 5.1 Consider the problem of a uniform bar with $\Delta T = 0$, acted upon by a single load F as shown again in Fig. 5.7. Determine the force $P(x)$ transmitted along the bar and the displacement $u(x)$.

Solution The support at the left side of the figure indicates that there is no displacement, that is, the boundary condition at $x = 0$ is $u(0) = 0$. At the right end, $x = L$, the displacement is not prescribed; thus, the appropriate boundary condition is on the force, namely

$$AEu'(L) = P(L) = F$$

In order to determine the internal force transmitted as a function of position along the bar, draw a free-body diagram of a portion of the bar as shown in Fig. 5.8.

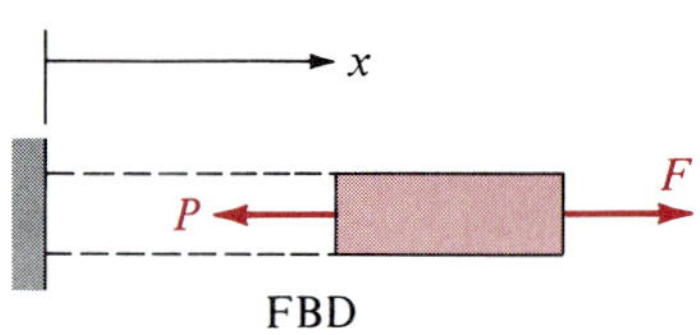

Figure 5.8 Free-body diagram of a portion of the bar

Requiring equilibrium yields

$$\Sigma F_x = -P(x) + F = 0$$

or

$$P(x) = F = \text{constant} \qquad (a)$$

The average stress can be determined as $\sigma = F/A$, to be used in evaluating the strength of the bar.

In order to determine the displacements, the force-displacement relation $AEu' = P$ is used. With $P(x) = F$, there results $AEu' = F$, which can be integrated as

$$AEu = Fx + C_1$$

The constant C_1 is evaluated by setting $x = 0$ on both sides of the equation and using the boundary condition $u(0) = 0$, thus

$$AEu(0) = AE(0) = F(0) + C_1$$

from which $C_1 = 0$ and

$$u(x) = \frac{Fx}{AE} \qquad (b)$$

The results for $P(x)$, the force transmitted, and $u(x)$, the displacement, are given by Eqns. (a) and (b), respectively, and shown in Fig. 5.9. Note carefully that the slope of the $u(x)$ versus x curve is proportional to the ordinate of the $P(x)$ versus x curve, that is, $u'(x) = P(x)/AE$. This fact can be very helpful in plotting the $u(x)$ versus x curve using the $P(x)$ versus x curve.

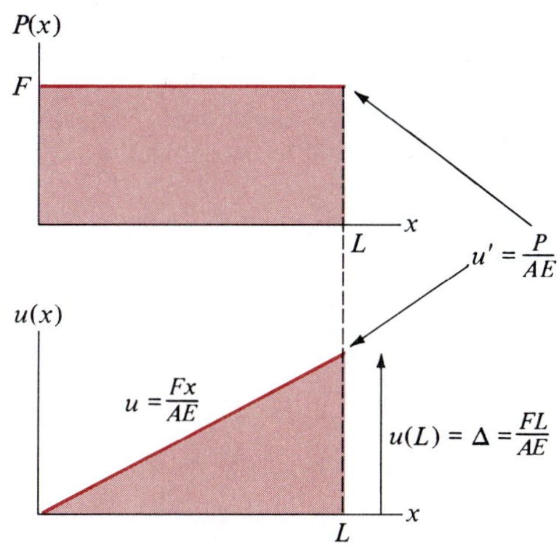

Figure 5.9 Internal force P and displacement u in a uniform axially loaded bar

As indicated in the figure, the displacement at $x = L$ is the total elongation Δ given by

$$\Delta = \frac{FL}{AE} = \frac{F}{(AE/L)} = \frac{F}{k} \qquad (c)$$

where $k = AE/L$ is termed the **axial stiffness** of the bar. As a matter of fact, a check of the dimensions of k reveals that $[k] = FL^{-1}$. This means that $k = AE/L$ has the same dimensions as an ordinary spring constant in the equation for a linear spring, $F = kx$. In this regard, Eqn. (c) can be rewritten as

$$F = \left(\frac{AE}{L}\right) \Delta = k\,\Delta \qquad (d)$$

This is one of the most fundamental and useful results from the mechanics of solids, namely that as shown in Fig. 5.10, a constant-cross section, linearly elastic bar undergoing axial deformation under the action of a single axial force F, behaves as a linear spring, or $F = k\,\Delta$, with spring constant given by $k = AE/L$.

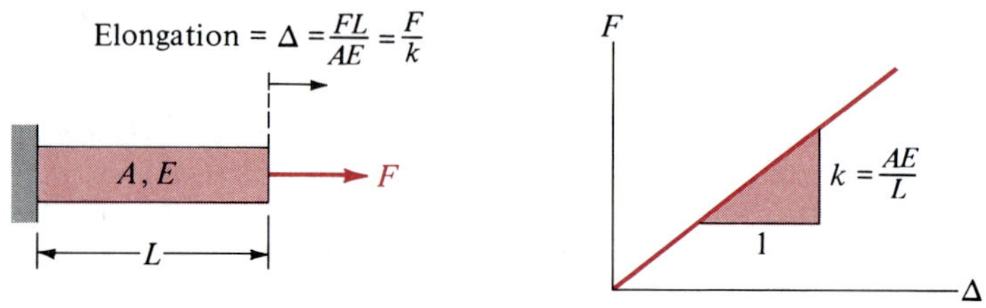

Figure 5.10 Uniform elastic bar considered as a linear spring

It can also be shown that for a suitably defined spring constant, a variable-area, linearly elastic bar loaded by a single force F can also be considered as a linear spring. The student is referred to the problems at the end of the section in this regard.

A second approach to this problem is as follows. With $q = \Delta T = 0$, the equilibrium equation and boundary conditions can be written as

$$(AEu')' = P' = 0$$

$$u(0) = 0$$

$$AEu'(L) = F$$

Integrating the equilibrium equation yields

$$AEu' = P = C_1$$

Use the boundary condition at $x = L$ to evaluate C_1 according to

$$AEu'(L) = C_1 = F$$

from which $C_1 = F$. Eqn. (a) can then be written as

$$AEu'(x) = P = F \qquad (e)$$

stating that P, the internal force transmitted, is everywhere equal to F, the force applied at the end of the bar. Note that this is precisely the same conclusion as was reached by drawing a free-body diagram of a portion of the bar in the first approach to this problem. Note also for this statically determinate problem that after one integration of the equilibrium equation, or equivalently, after drawing the appropriate free-body diagram, $P(x)$ is known and it is possible to evaluate the strength by computing $\sigma = P/A$. The determination of the displacements using equation (e) proceeds exactly as in the first approach.

Example 5.2 The elongation of a segmented bar such as in Fig. 5.11a, acted upon by a single axial load P_0, can easily be determined by using the results of Example 5.1.

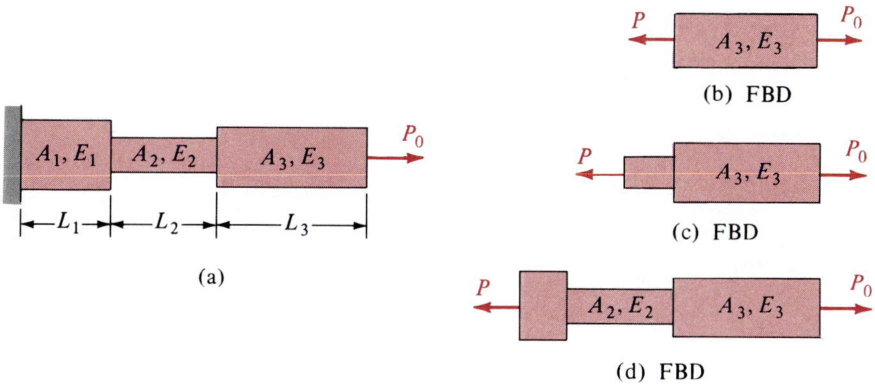

Figure 5.11 Elongation of a segmented bar

Referring to Figs. 5.11b, c, and d, it is seen that the internal force transmitted by each segment of the bar is the load P_0. Using the results of Example 5.1, the corresponding elongation e_i of each segment is given by

$$e_i = \frac{P_0 L_i}{A_i E_i}$$

with the total elongation of the bar as

$$\Delta = P_0 \sum_{i=1}^{3} \frac{L_i}{A_i E_i}$$

The $P(x)$ and $u(x)$ curves are shown in Fig. 5.12.

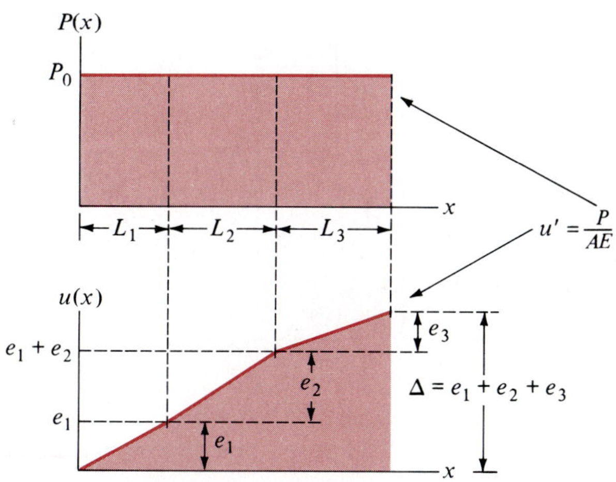

Figure 5.12 Force transmitted and displacement in a segmented bar

Example 5.3 Consider the problem of three uniform aluminum bars loaded by three axial loads at the points where the bars join as shown in Fig. 5.13.

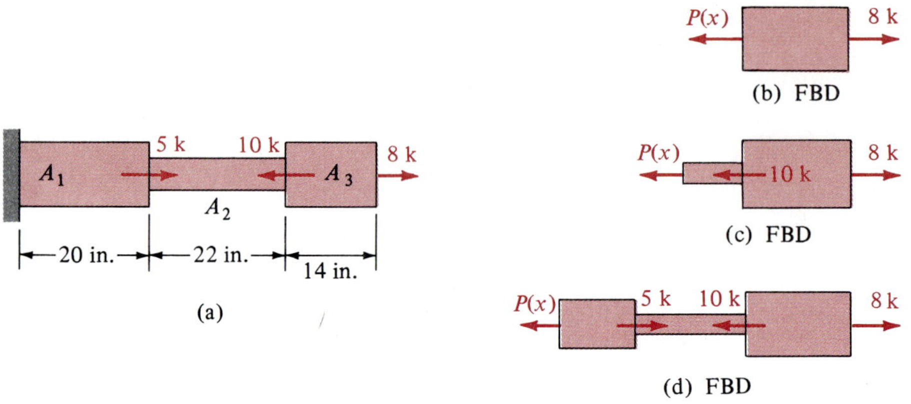

Figure 5.13 Loading of a segmented bar

Referring to the free-body diagrams shown in Figs. 5.13b, c, and d, equilibrium is used to determine that the internal force in each of the segments is 8 k, −2 k, and 3 k, respectively, as shown in Fig. 5.14. With $A_1 = 3$ in^2, $A_2 = 1$ in^2, and $A_3 = 2$ in^2, the respective values of $u' = P/AE$ are 10^{-4}, -2×10^{-4}, and 4×10^{-4}.

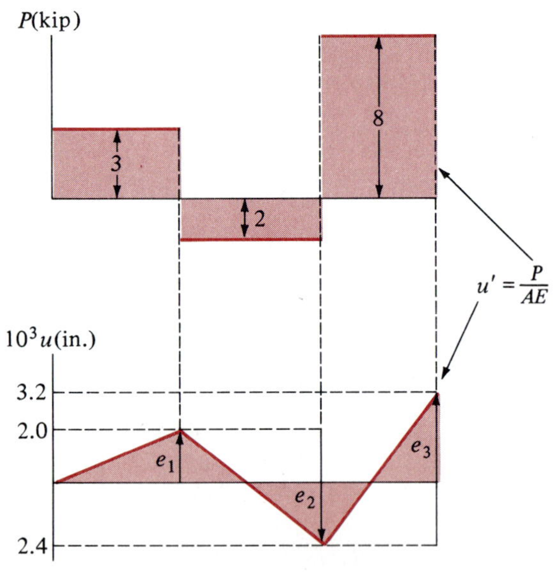

Figure 5.14 Internal forces and displacements in a segmented bar

The values e_1, e_2, and e_3 shown in Fig. 5.14 are the individual elongations of the three segments calculated according to $e_i = P_i/k_i$, that is,

$$e_1 = \frac{(3\text{ k})(20\text{ in.})}{(3\text{ in}^2)(10^7\text{ psi})} = 0.0020\text{ in.}$$

$$e_2 = \frac{(-2\text{ k})(22\text{ in.})}{(1\text{ in}^2)(10^7\text{ psi})} = -0.0044\text{ in.}$$

$$e_3 = \frac{(8\text{ k})(14\text{ in.})}{(2\text{ in}^2)(10^7\text{ psi})} = 0.0056\text{ in.}$$

The total elongation of the bar is the sum of these elongations; thus, $\Delta = u(L) = e_1 + e_2 + e_3 = 0.0032$ in.

Example 5.4 Consider the problem of a uniform bar hanging under its own weight as shown in Fig. 5.15a. Determine the transmitted force and the displacement. Take the mass density of the material of the bar to be ρ with the corresponding weight density as ρg.

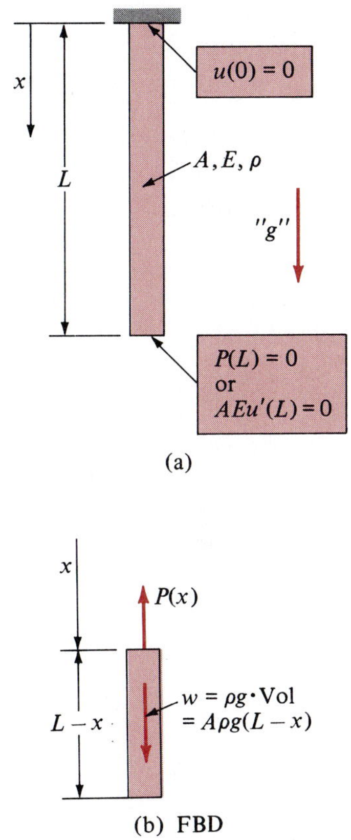

Figure 5.15 Bar hanging under its own weight

Solution 1 Draw a free-body diagram of a portion of the bar as shown in Fig. 5.15b, and require equilibrium to obtain

$$P(x) = A\rho g(L - x) = AEu'(x) \qquad (a)$$

Note that this satisfies the boundary condition $P(L) = 0$ shown in Fig. 5.15a. Note also that it is possible to evaluate the strength at this juncture by computing

$$\sigma = \frac{P}{A} = \rho g(L - x)$$

The maximum stress occurs at $x = 0$ and is $\sigma_{max} = \rho g L$.

Canceling the common A in Eqn. (a) and integrating the result yields

$$u(x) = -\frac{(\rho g/E)(L - x)^2}{2} + C_1$$

Satisfying the boundary condition $u(0) = 0$ leads to

$$0 = -\frac{(\rho g/E)(L)^2}{2} + C_1$$

from which $C_1 = \rho g L^2/2E$ and

$$u(x) = \frac{\rho g}{E}\left(Lx - \frac{x^2}{2}\right) \qquad (b)$$

From Eqn. (b) the elongation of the bar is given by $u(L) = \rho g L^2/2E$. The force transmitted ($P = AEu'$) and the displacement are plotted in Fig. 5.16. Note that the slope of the u versus x curve is zero at $x = L$, verifying that there is no load ($P = AEu'$) transmitted at that point.

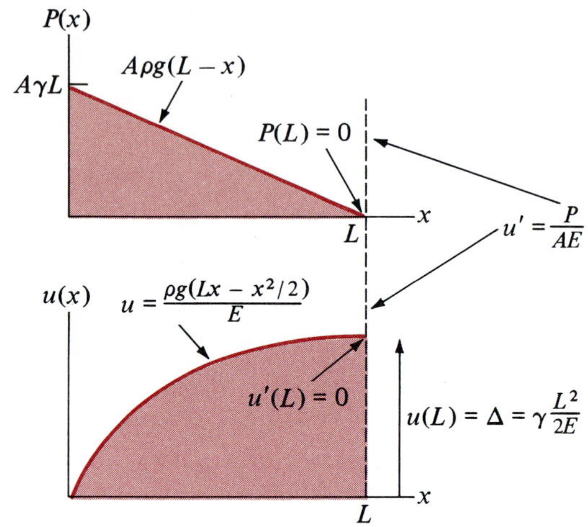

Figure 5.16 Force transmitted and displacement

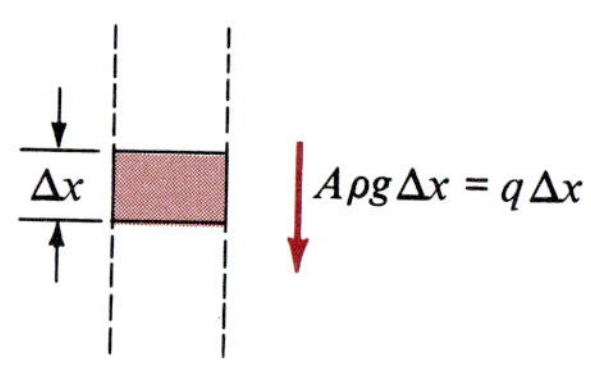

Figure 5.17 Free-body diagram of a portion of the bar

Solution 2 For this approach we will use the equilibrium equation (5.6) as the starting point. To determine the external loading $q(x)$ it is advisable to draw a free-body diagram of a portion of the bar as shown in Fig. 5.17.

The total external force acting on the element is $A\rho g\ \Delta x$. Comparing this to $q\Delta x$ yields $q = A\rho g$, so that with $\Delta T = 0$, the equation of equilibrium and boundary conditions become

$$(AEu')' + A\rho g = 0$$

$$u(0) = 0$$

$$AEu'(L) = 0$$

Integrating the equilibrium equation leads to

$$AEu' + A\rho g x = C_1 \tag{c}$$

Enforcing the boundary condition on $P(L) = AEu'(L) = 0$ yields

$$0 + A\rho g L = C_1$$

so that from Eqn. (c)

$$AEu' = P = A\rho g(L - x)$$

as per Eqn. (a) in the first approach to the problem. The remainder of solution 2 is carried out in precisely the same manner as solution 1.

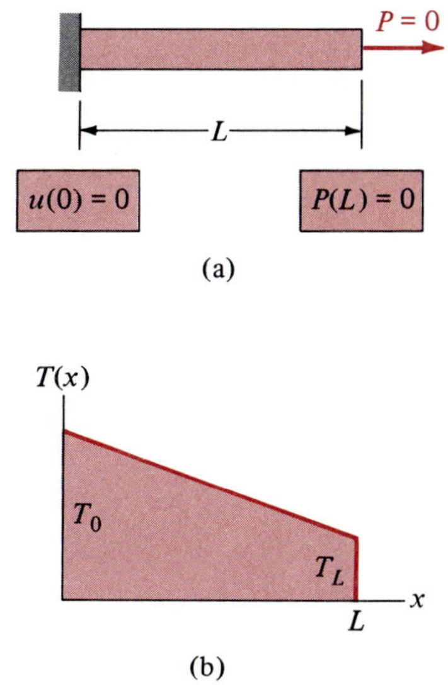

Figure 5.18 Thermal expansion of a heated bar

Example 5.5 A uniform elastic bar is placed in an environment resulting in the temperature increase shown in Fig. 5.18b.

Determine *(a)* the elongation of the bar, and *(b)* the force required to prevent it from elongating.

Solution Since there is no external mechanical loading, $q = 0$, and the equilibrium equation becomes

$$[AE(u' - \alpha T)]' = 0$$

An integration yields

$$AE(u' - \alpha T) = C_1$$

This equation can also be written as $P(x) = C_1$. Enforcing the boundary condition at $x = L$ yields

$$AE(u' - \alpha T) = 0 = C_1$$

or

$$AE(u' - \alpha T) = P(x) = 0$$

Dividing by AE and integrating again yields

$$u = \int \alpha T \, dx + C_2 = \alpha \left[T_0 x - \frac{(T_0 - T_L)x^2}{2L} \right] + C_2$$

Applying the boundary condition at $x = 0$ yields

$$0 = 0 + C_2$$

so that

$$u(x) = \alpha \left[T_0 x - \frac{(T_0 - T_L)x^2}{2L} \right]$$

The elongation of the bar is given by the displacement at $x = L$, namely

$$u(L) = \alpha \frac{T_0 + T_L}{2} L = \alpha T_{\text{avg}} L$$

that is, the *average temperature* times α times the length. Note that if $T_0 = T_L$ this reduces to the result developed previously in example 4.11.

The second part of the problem is easily solved using superposition as follows. We ask, what force applied at $x = L$ is required to shorten the bar to its original length? From Example 5.1, the elongation of a uniform bar acted upon by a single axial load F is given by

$$\delta = \frac{FL}{AE} \qquad \text{or} \qquad F = \frac{AE\,\delta}{L} = k\delta$$

Requiring that the elongation due to the temperature change plus the elongation due to the force F sums to zero, gives

$$\alpha T_{avg} L + \frac{F}{k} = 0$$

from which

$$F = -k\alpha T_{avg} L = -AE\alpha T_{avg} = -AE\alpha \frac{T_0 + T_L}{2}$$

which is compressive as expected.

Note in passing that for a general temperature distribution $T(x)$ it follows when $P(x) = 0$, that $u' = \alpha T(x)$, and that

$$u = \int_0^x \alpha T(x)\, dx + C_1$$

which, with $u(0) = 0$, leads to $C_1 = 0$ and

$$u = \int_0^x \alpha T(x)\, dx$$

with $u(L)$ then given by

$$u(L) = \alpha L \left[\frac{\int_0^L T(x)\, dx}{L} \right] = \alpha L T_{avg}$$

thus, the elongation of the unrestrained bar is the product of α, L, and the average temperature, irrespective of the particular temperature distribution.

PROBLEMS

5.10 A 1-in.-diameter steel bar 30 in. long transmits a load of 18 kips. Determine the average normal stress and the elongation of the bar. What is the maximum average shear stress in the bar? Take $E = 3 \times 10^7$ psi.

5.11 A load of 80 kN is applied to a 30-mm-diameter steel bar 30 m in length. Determine the average normal stress and the elongation of the bar. What is the maximum average shear stress in the bar? Take $E = 200$ GPa.

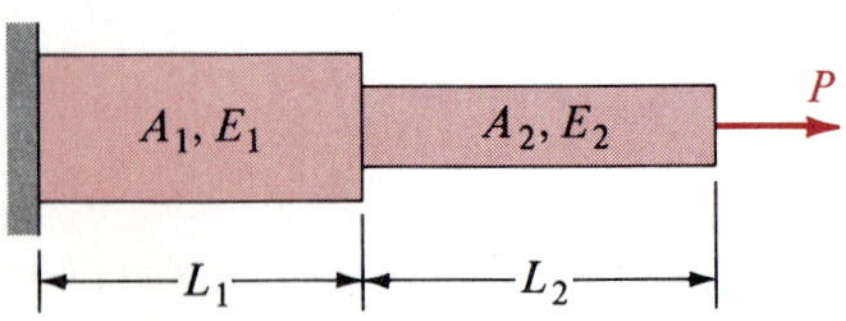

Ps. 5.12 and 5.13

5.12 Determine the allowable load P if the maximum permissible elongation of the bar is Δ and the allowable normal stress is σ_{all}. Take $A_1 = 2$ in^2, $A_2 = 1.5$ in^2, $E_1 = E_2 = 10^6$ psi, $L_1 = 10$ in., $L_2 = 12$ in., $\Delta = 0.013$ in., and $\sigma_{all} = 14$ ksi.

5.13 Repeat problem 5.12 for $A_1 = 600$ mm^2, $A_2 = 400$ mm^2, $E_1 = E_2 = 70$ GPa, $L_1 = 0.25$ m, $L_2 = 0.4$ m, $\Delta = 0.35$ mm, and $\sigma_{all} = 100$ MPa.

5.14 Determine and plot the internal force $P(x)$. Also determine and plot the displacement $u(x)$. What is the total change in length in each case? What is the maximum normal stress? Take $A_1 = 4$ in^2, $A_2 = 1$ in^2, $L_1 = 15$ in., $L_2 = 12$ in., $E_1 = 30 \times 10^6$ psi, $E_2 = 10 \times 10^6$ psi, $P_1 = 4$ kips, and $P_2 = 2$ kips.

5.15 Repeat problem 5.14 with $A_1 = 630$ mm^2, $A_2 = 900$ mm^2, $L_1 = 200$ mm, $L_2 = 180$ mm, $E_1 = 200$ GPa, $E_2 = 70$ GPa, $P_1 = 40$ kN, and $P_2 = -90$ kN.

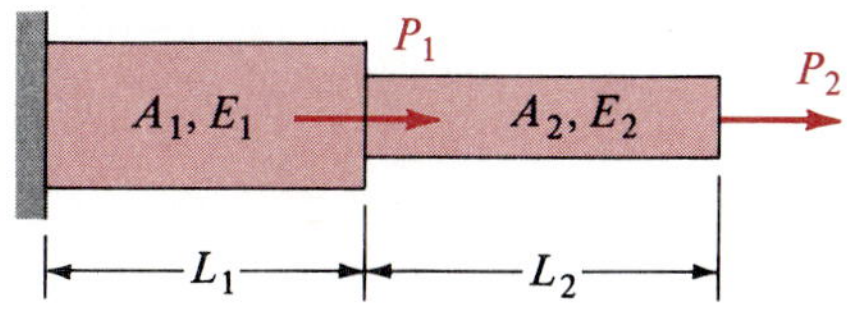

Ps. 5.14 and 5.15

5.16 Determine and plot the internal force $P(x)$. Also determine and plot the displacement $u(x)$. What is the total change in length in each case? What is the maximum normal stress? Take $A_1 = 0.04$ in^2, $A_2 = 0.08$ in^2, $A_3 = 0.135$ in^2, $L_1 =$ 1.5 in., $L_2 = 0.8$ in., $L_3 = 1.1$ in., $E_1 = 30 \times 10^6$ psi, $E_2 = 10 \times 10^6$ psi, $E_3 =$ 15×10^6 psi, $P_1 = 240$ lbf, $P_2 = -360$ lbf, and $P_3 = 1080$ lbf.

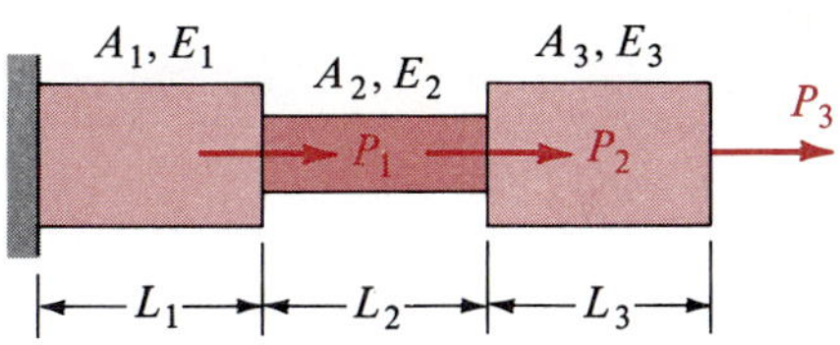

Ps. 5.16 and 5.17

5.17 Repeat problem 5.16 with $A_1 = 60$ mm^2, $A_2 = 40$ mm^2, $A_3 = 90$ mm^2, $L_1 = 50$ mm, $L_2 = 60$ mm, $L_3 = 110$ mm, $E_1 = 200$ GPa, $E_2 = 70$ GPa, $E_3 =$ 120 GPa, $P_1 = 2.4$ kN, $P_2 = -3$ kN, and $P_3 = 6$ kN.

5.18 Determine the elongation of the bar subjected to the forces and temperature distribution shown. Take $A_1 = A_2 = 200$ mm^2, $E_1 = 210$ GPa, $E_2 =$ 100 GPa, $\alpha_1 = 12 \times 10^{-6}/°C$, $\alpha_2 = 23 \times 10^{-6}/°C$, $P = 100$ kN, $L_1 = 180$ mm, $L_2 = 160$ mm, and ΔT for both bars is 80°C. What are the stresses in the two bars?

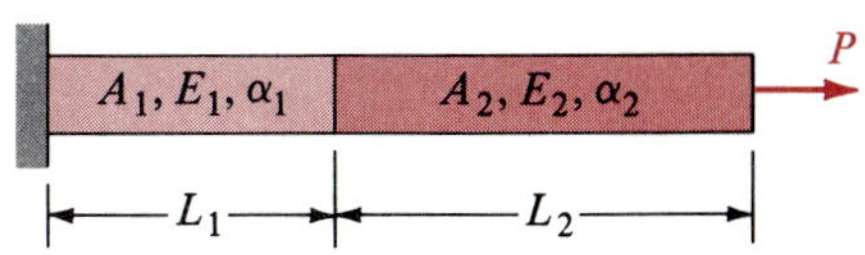

Ps. 5.18 and 5.19

5.19 Repeat problem 5.18 if $A_1 = A_2 = 1.5$ in^2, $E_1 = 30 \times 10^6$ psi, $E_2 = 10 \times$ 10^6 psi, $\alpha_1 = 6.5 \times 10^{-6}/°F$, $\alpha_2 = 12 \times 10^{-6}/°F$, $P = 20$ kips, $L_1 = 6$ in., $L_2 =$ 4 in., and ΔT for both bars is 140°F.

5.20 For the bar shown, use superposition to determine the displacement of point B and the stresses in portions AB and BC. Take $A_1 = 4$ in^2, $A_2 = 3$ in^2, $E_1 = 30 \times 10^6$ psi, $E_2 = 10 \times 10^6$ psi, $\alpha_1 = 6.5 \times 10^{-6}/°F$, $\alpha_2 = 12 \times 10^{-6}/°F$, $L_1 = 4$ in., $L_2 = 6$ in., and ΔT for both bars is 80°F.

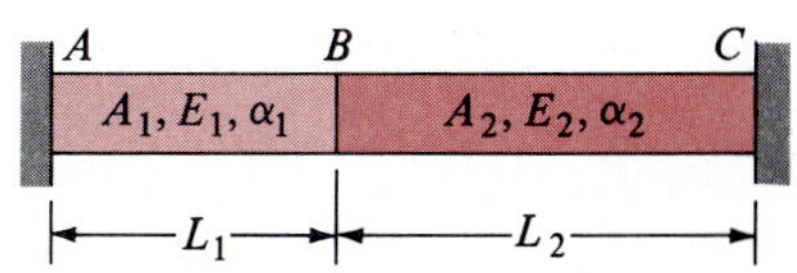

Ps. 5.20 and 5.21

5.21 Repeat problem 5.20 if $A_1 = 200$ mm^2, $A_2 = 140$ mm^2, $E_1 = 200$ GPa, $E_2 = 70$ GPa, $\alpha_1 = 13 \times 10^{-6}/°C$, $\alpha_2 = 23 \times 10^{-6}/°C$, $L_1 = 120$ mm, $L_2 = 90$ mm, and ΔT for both bars is 50°C.

5.22 Repeat problem 5.20 for the nonuniform temperature distribution shown. Take $T_1 = 100°F$ and $T_2 = 30°F$.

5.23 Repeat problem 5.21 for the nonuniform temperature distribution shown. Take $T_1 = 60°C$ and $T_2 = 25°C$.

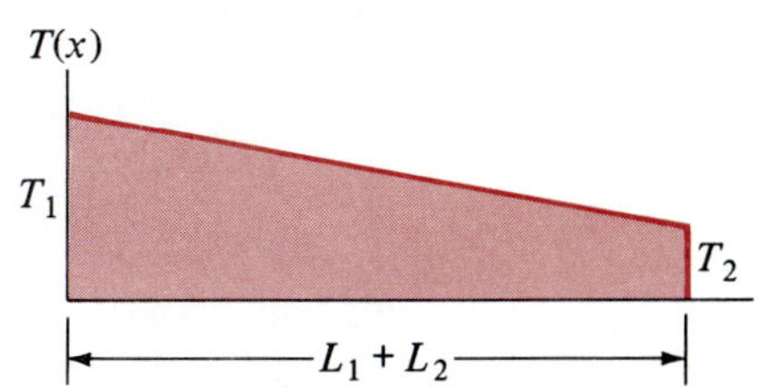

Ps. 5.22 and 5.23

5.24–5.29 For each of the following, draw the necessary free-body diagrams to plot $P(x)$, and then use the relation $P = AEu'$ to assist in plotting $u(x)$. Determine the total elongation of the bar in each case.

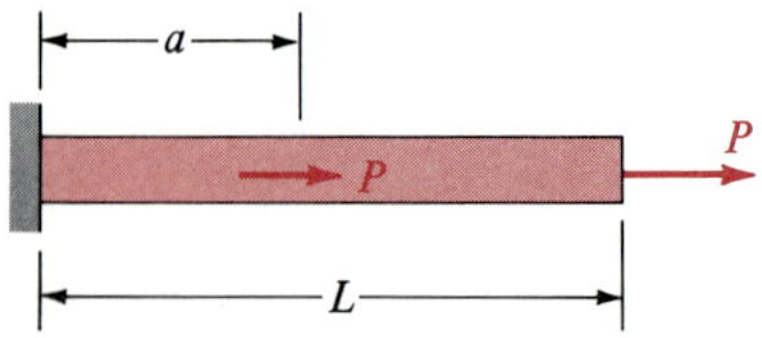

P. 5.24

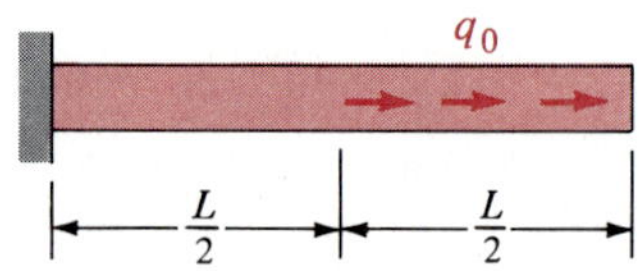

P. 5.25

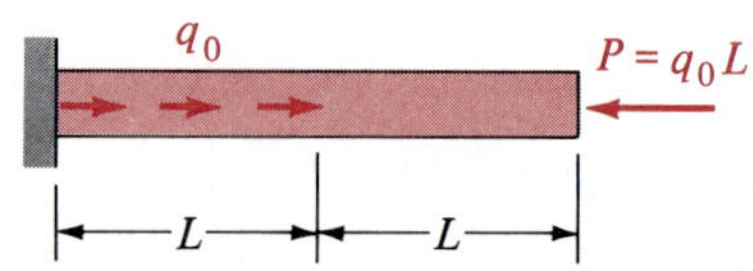

P. 5.26

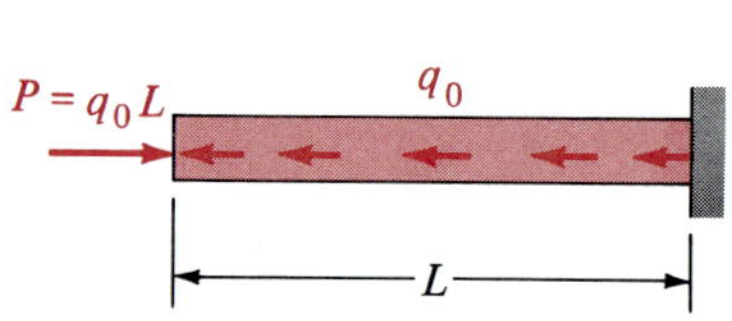

P. 5.27

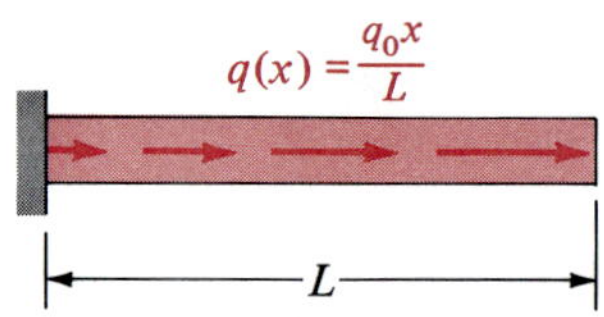

P. 5.28

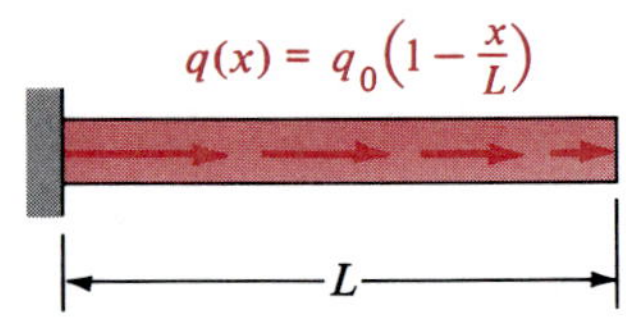

P. 5.29

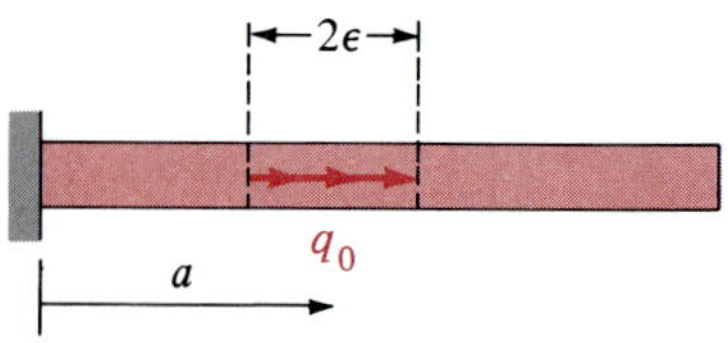

P. 5.30

5.30 The uniform bar shown is loaded by a uniform intensity loading q_0 over a short distance 2ε as shown. Sketch the $P(x)$ and $u(x)$ versus x curves for several increasingly smaller values of the distance ε while requiring that the total external load $2q_0\varepsilon = P$ remain constant. Interpret the results as $\varepsilon = 0$.

5.31 The transverse dimension of a rectangular-cross-section bar varies linearly from $2b$ to b as shown. Develop an expression for the total elongation of the bar under the action of the tensile load P. Take the width to be w.

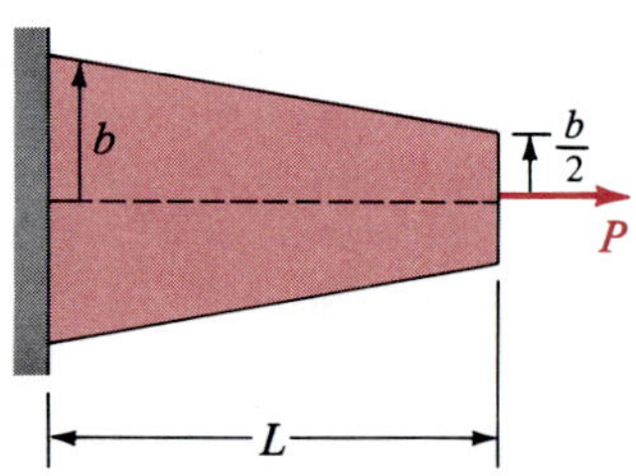

P. 5.31

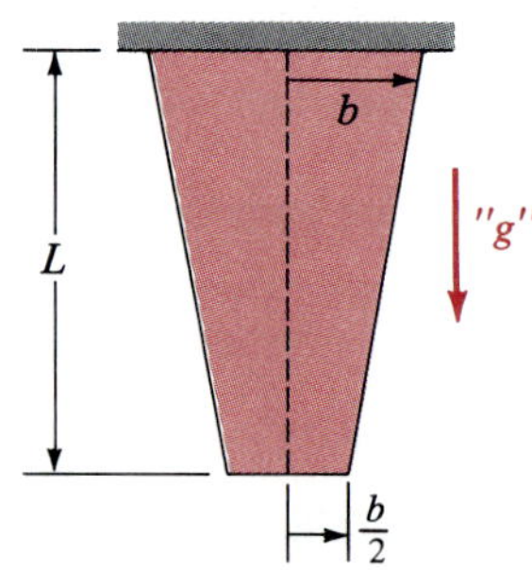

P. 5.32

5.32 Repeat the previous problem when the external loading is due to the weight of the bar as shown.

5.33 A bar with circular cross section that is a conical frustum is loaded by a single axial force as shown. Determine the elongation by *(a)* integrating the equilibrium equation, and *(b)* replacing the bar by the two constant-cross-section bars indicated. Compare the results.

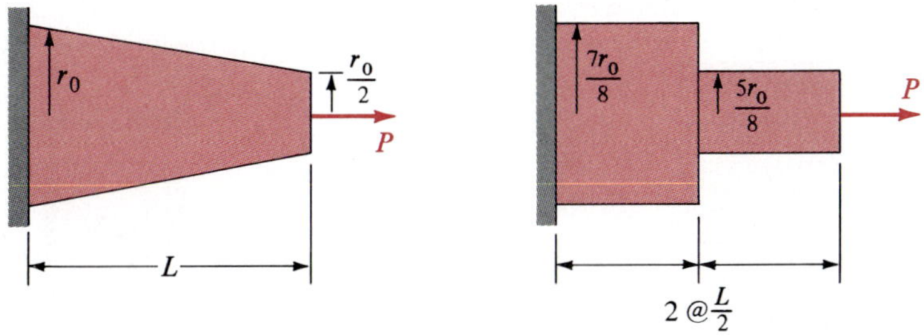

Ps. 5.33–5.35

5.34 Repeat problem 5.33 for four equal-length segments whose radii have average values of $15r_0/16$, $13r_0/16$, $11r_0/16$, and $9r_0/16$, respectively.

5.35 Repeat problem 5.33 for eight equal-length segments whose radii have average values of $(31r_0, 29r_0, 27r_0, 25r_0, 23r_0, 21r_0, 19r_0$, and $17r_0)/32$, respectively.

5.36 Determine the elongation of the conical-frustum bar of problem 5.33 when the bar hangs under its own weight. Take the specific weight to be ρg. Replace the actual weight of the bar by the loads shown in the figure, that is, by placing half of the weight of each segment at the top and bottom of the corresponding segment.

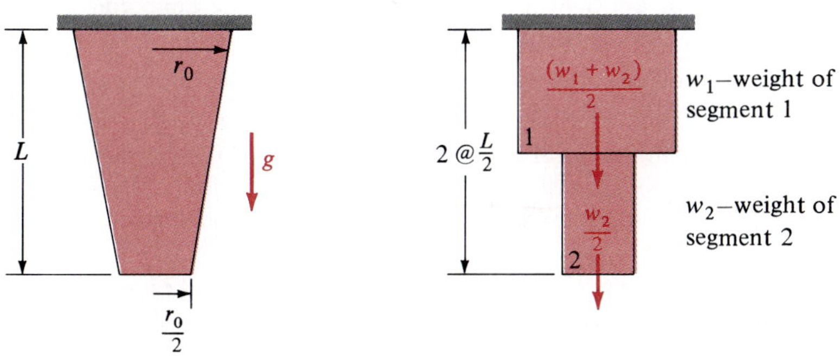

Ps. 5.36–5.38

5.37 Repeat problem 5.36 for four equal-length segments.

5.38 Repeat problem 5.36 for eight equal-length segments.

5.39 A concrete ($E = 6 \times 10^8$ lbf/ft^2) column is shaped as shown. Estimate the total shortening by replacing the variable-cross-section portion by a constant-cross-section portion whose radius is the average of the radii at the ends of the section as indicated in the figure. Neglect the weight of the concrete.

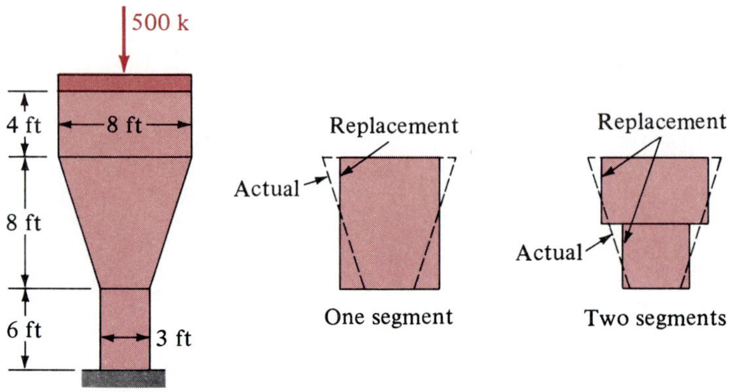

Ps. 5.39–5.44

5.40 Repeat problem 5.39 replacing the conical segment by two constant cross segments whose radii are the average of the radii at the ends of the sections as indicated in the figure.

5.41 Repeat problem 5.39 replacing the conical segment by four constant cross segments whose radii are the average of the radii at the ends of the sections.

5.42 Repeat problem 5.39 including the weight of the concrete. Take the specific weight of the concrete to be 150 lbf/ft^2.

5.43 Repeat problem 5.40 including the weight of the concrete. Take the specific weight of the concrete to be 150 lbf/ft^2.

5.44 Repeat problem 5.41 including the weight of the concrete. Take the specific weight of the concrete to be 150 lbf/ft^2.

5.45 A bar in the form of a conical frustum is loaded by the two equal concentrated forces $P/2$ as shown. Determine the elongation of the bar. (*Hint:* Integrate $AEu' = P$ in each segment and require that the displacements be the same at $x = L/2$.)

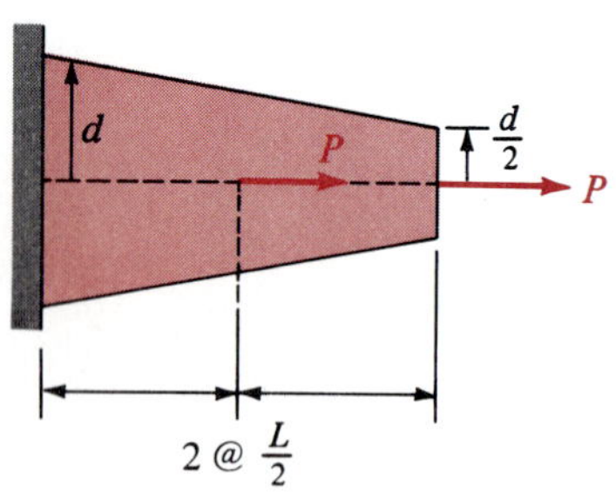

P. 5.45

5.46 A uniform rigid bar of weight W is suspended by two identical wires as shown in the figure. Determine the vertical displacement of point C.

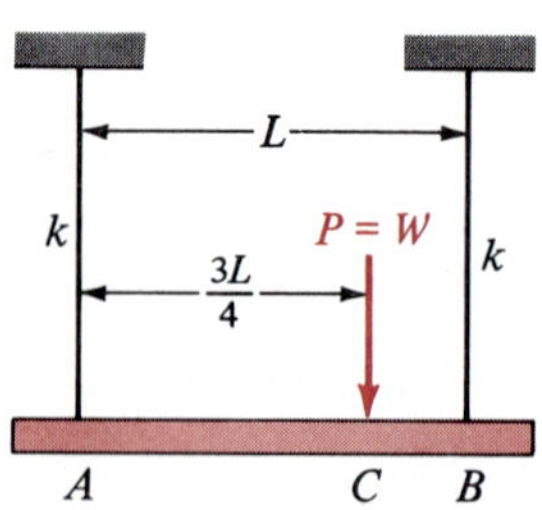

P. 5.46

5.47 A rigid bar of weight W is suspended by two identical wires as shown in the figure. For $x = 3L/4$, determine the vertical displacement of point D and the angle through which the bar rotates. Are there any limits on the value of x in order that your solution be valid? Explain.

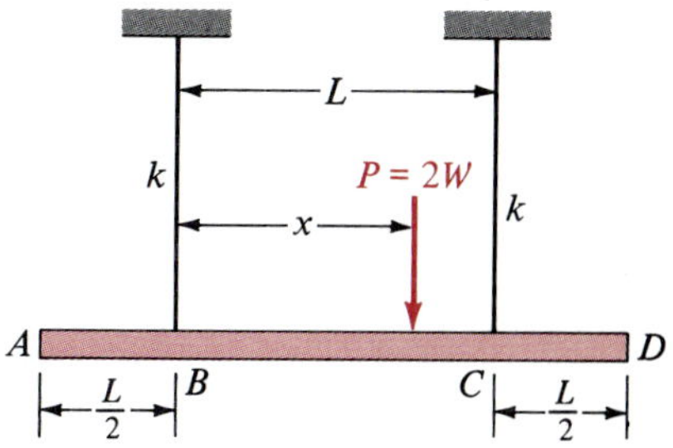

P. 5.47

5.48 Two rigid bars are connected by axial members as shown in the figure. Determine *(a)* the forces in the bars, *(b)* the displacement of the load P, and *(c)* the angle through which the lower bar rotates.

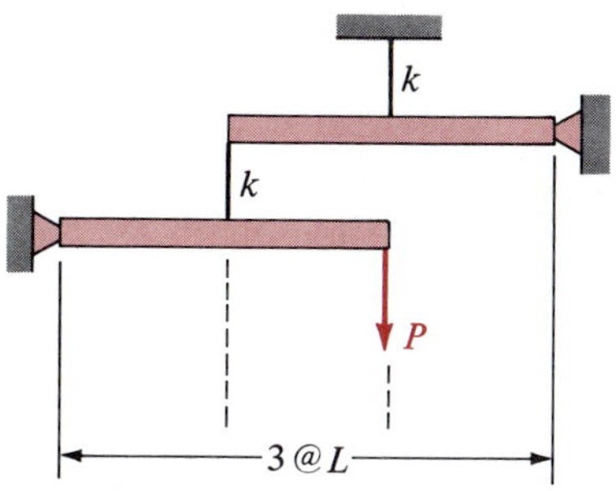

P. 5.48

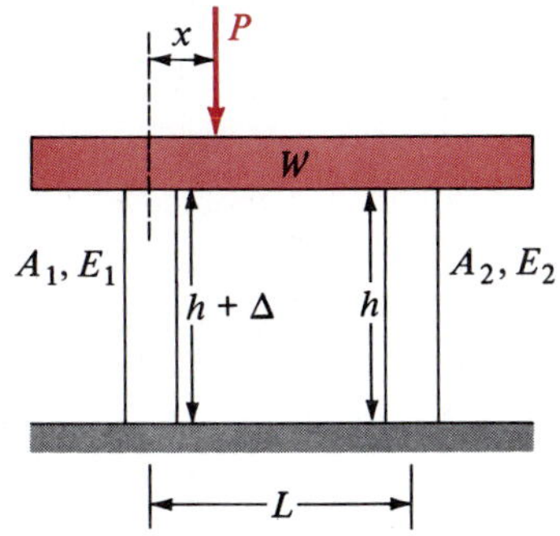

P. 5.49

5.49 Two bars whose initial lengths differ by an amount Δ support a heavy rigid plate of weight W as shown in the figure. Where should the load $P = W$ be applied so as to level the rigid plate? What is the range of values of x such that the angle the plate makes with the horizontal is less than 1 second of arc? Take $h = 4$ ft, $\Delta = 0.02$ in., $A_1 = A_2 = 40$ in^2, $E_1 = E_2 = 30 \times 10^6$ psi, $L = 10$ ft, and $W = 500$ tons.

5.50 A uniform rigid platform of weight W_2, supported by two compression members as shown, supports some satellite tracking equipment of weight W_1 located nominally at the center of the platform. Develop a general expression for the angle θ through which the platform will rotate if the weight W_1 is incorrectly positioned a distance ε from the center. If $W_2 = 800$ kN, $W_1 = 300$ kN, $L = 5$ m, $h = 3$ m, and ε, the error in positioning, can be as much as 0.03 m, determine the required area of the steel piles if θ is not to exceed 3 seconds of arc. If the allowable compressive stress in the steel is 120 MPa, what is the required area?

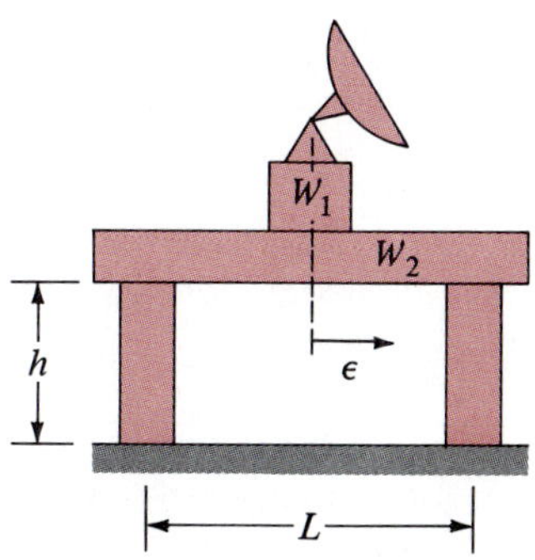

P. 5.50

5.51 A rigid bar is pinned at A and supported by an elastic wire at B as shown in the figure. The allowable stress in the wire is σ_{all}. Determine the angle α such that the volume of the wire is a minimum. What is the corresponding displacement at point C?

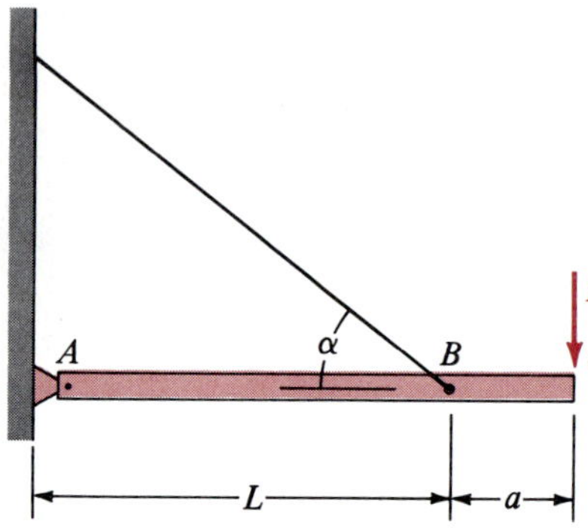

P. 5.51

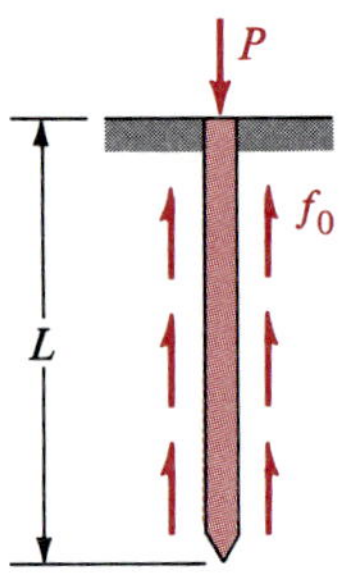

P. 5.52

5.52 A pile is driven into the ground and supports a load P as shown in the figure. Assuming that the interaction of the pile with the ground is equivalent to a uniform shear over the length of the pile, determine the internal force transmitted in the pile and the change in length in the pile. Assume that there is no force applied to the pile from the earth over the bottom surface of the pile.

5.53 Repeat problem 5.52 if the interaction is assumed to be linearly distributed as shown in the figure.

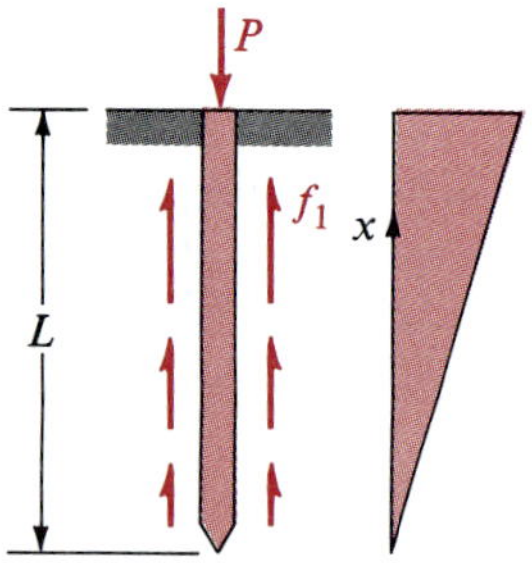

P. 5.53

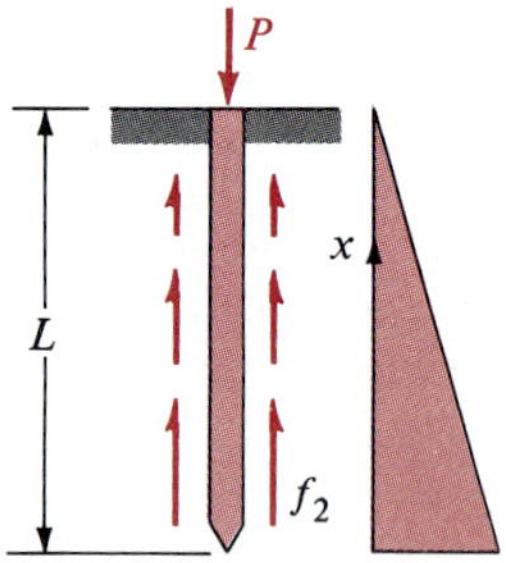

P. 5.54

5.54 Repeat problem 5.52 if the interaction is assumed to be linearly distributed as shown in the figure.

5.55 A uniform bar of mass density ρ and area A rotates at constant angular velocity ω about an axis through its center. Determine and plot $P(x)$, the internal force transmitted. Also determine and plot the displacement $u(x)$.

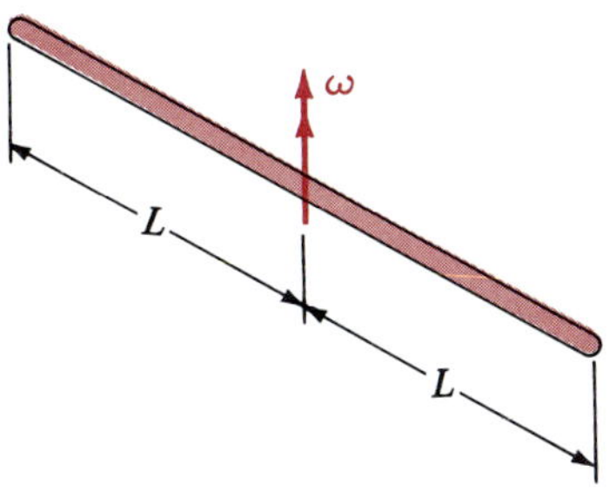

P. 5.55

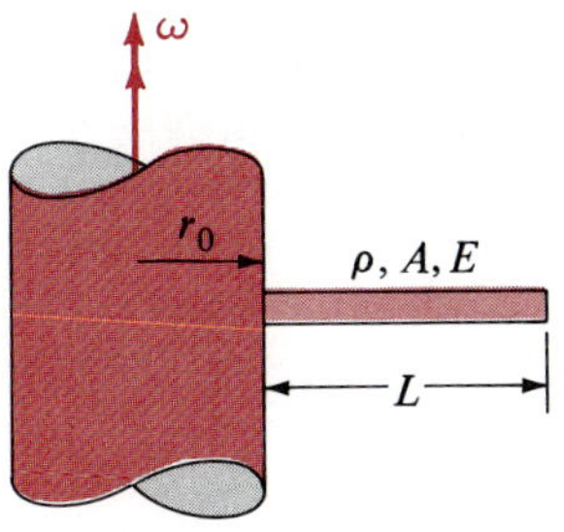

P. 5.56

5.56 Repeat problem 5.55 for a bar attached to a shaft of radius r_0 as shown in the figure.

5.57 Two initially unstressed uniform linearly elastic wires are loaded by a vertical force P as shown in the figure. With E as the modulus of elasticity and A as the area show that the nonlinear relation between the load P and the vertical displacement of the point P is given by $\delta = l(P/AE)^{1/3}$.

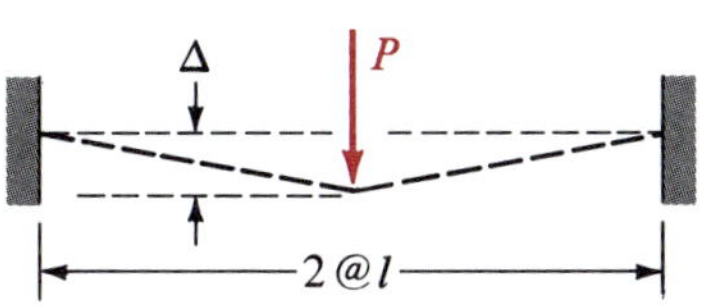

P. 5.57

5.58 A steel bar is placed near a magnet whose center of attraction is located a distance x_0 from the end of the bar as shown in the figure. The force experienced by a differential mass of the bar is proportional to the differential mass and inversely proportional to the square of the distance from the "pole" to the differential mass. With β as the constant of proportionality, determine and plot $P(x)$, the internal force, and $u(x)$, the displacement.

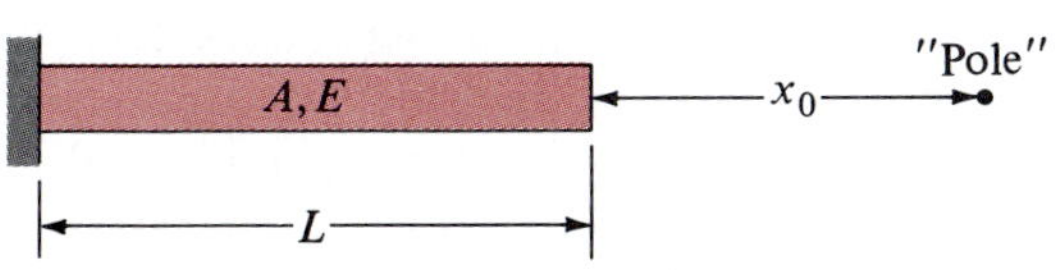

P. 5.58

5.59 For a bar with a slowly varying cross-sectional area $A(x)$, show that the elongation can be expressed as $\Delta = PL/EA_{avg}$, where A_{avg} is given by the equation $L/A_{avg} = \int_0^L dx/A(x)$. This result shows that a variable-cross-section bar can also be considered to be a linear spring with spring constant $k = A_{avg}E/L$ and that the force deflection relation is $P = (A_{avg}E/L)\,\Delta = k\Delta$.

5.60 A common assumption used in the analysis of certain suspension bridges is that the loading (due to the rest of the structure) on the suspension cable is vertical with a uniform horizontal distribution as shown. Show that this assumption leads to the conclusion that the shape of the cable is a parabola. Determine *(a)* the maximum tension in the cable, and *(b)* the elongation of the cable. (*Hint:* Consider the horizontal and vertical equilibrium of a segment Δx of the cable.)

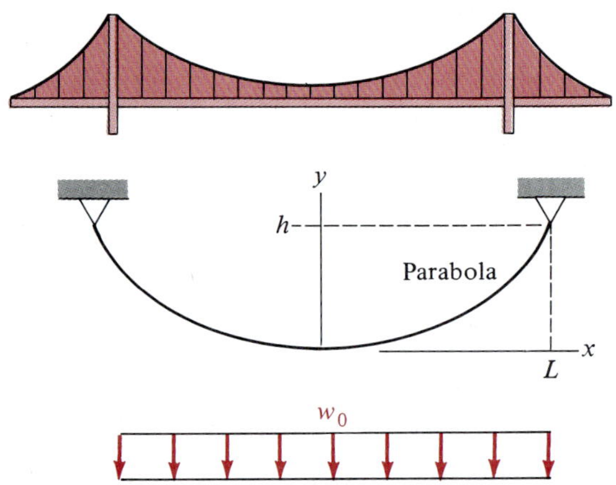

P. 5.60

5.4 STATICALLY INDETERMINATE PROBLEMS

By definition, a statically determinate problem is one for which the number of unknown reactions and/or the number of internal forces is equal to the number of independent equations of equilibrium that can be written. For a statically determinate problem, free-body diagrams are drawn, after which the internal forces can be determined and used to answer the strength question, based on computing $\sigma = P/A$. The force-displacement relations $AEu' = P$ can then be integrated to determine the displacements, after which the stiffness question can be answered. These ideas are illustrated in Fig. 5.19.

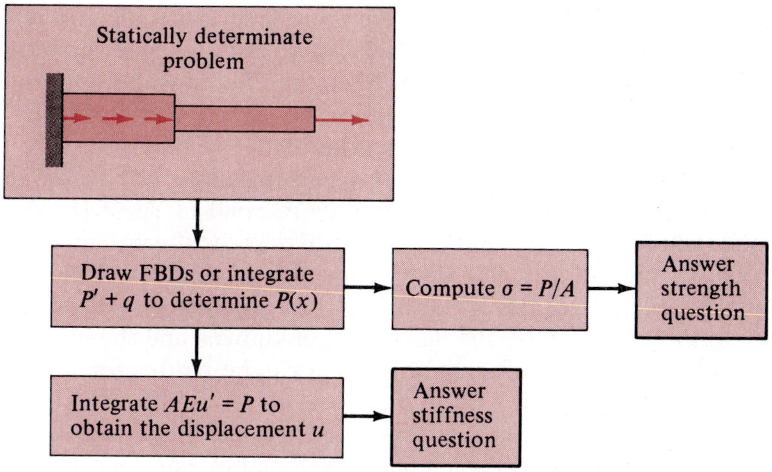

Figure 5.19 Flowchart indicating the steps for a statically determinate problem

For a statically indeterminate problem, the number of unknown reactions and/or internal forces exceeds the number of equations of equilibrium, so that the internal forces cannot be determined from the equations of equilibrium alone. Two typical examples involving axial members are shown in Fig. 5.20.

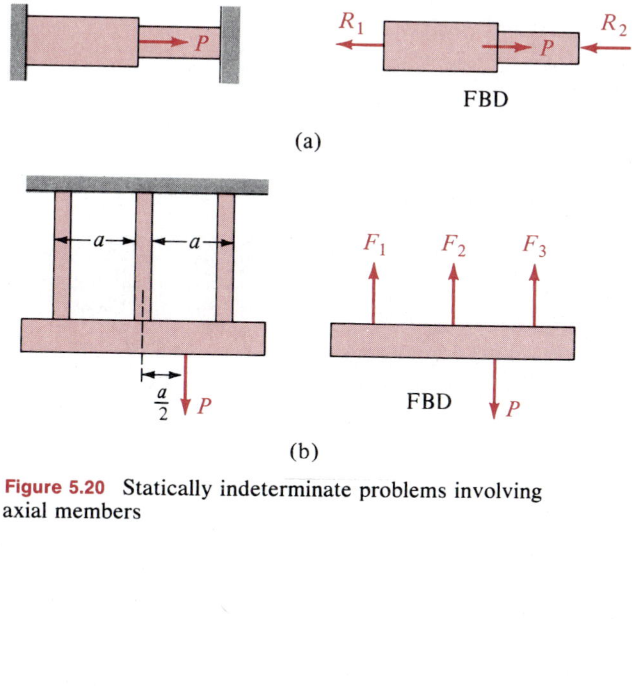

Figure 5.20 Statically indeterminate problems involving axial members

Fig. 5.20a shows a situation where there is an extra support. There are two unknown reactions R_1 and R_2, with only a single equation of equilibrium. The structure in Fig. 5.20b can be considered from the standpoint that any of the three members could be removed leaving the structure still capable of supporting the load P. There are three unknown forces F_1, F_2, and F_3 and only two equations of equilibrium.

Generally, the additional reactions or internal member forces are termed **redundants.** Note that the choice of the reaction or member that is to be considered the redundant is not unique in either of the above examples. Note also that by definition, the number of redundants is equal to the difference in the number of unknowns and the number of independent equations of equilibrium that can be written for the structure in each example.

The additional equations necessary to complete the formulation of a statically indeterminate problem are developed by considering the deformations. These additional equations are then combined with the equilibrium equations, yielding a set of equations that can be used to solve for all of the force and displacement variables. The strength and stiffness questions can be answered. These ideas are illustrated in Fig. 5.21.

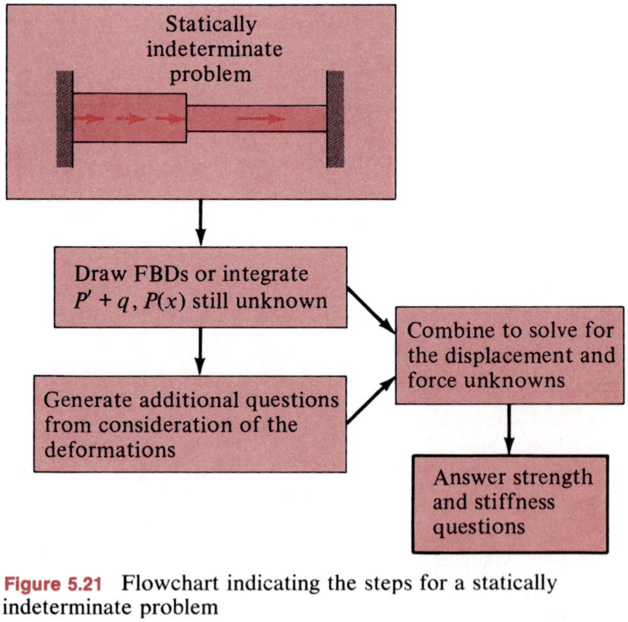

Figure 5.21 Flowchart indicating the steps for a statically indeterminate problem

There are two basic approaches for obtaining the additional equations: the **flexibility** or **force method,** and the **stiffness** or **displacement method.** The flexibility method is based on considering the forces as the unknowns, whereas the stiffness method is based on considering the displacements as the unknowns. In this section both methods will be discussed as applied to typical statically indeterminate axial problems.

5.4.1 Flexibility Formulations

As mentioned, the flexibility method is based on considering the forces as the unknowns. The equations of equilibrium for the structure are written in terms of the applied forces and the reaction forces. For a statically indeterminate structure, the number of unknown reactions or internal forces will exceed the number of independent equations of equilibrium by the number of redundants. A set of redundants is selected and removed from the structure to produce a statically determinate structure. One equation requiring that the displacements be such that the structure "fit together" is then written for each of the points in the structure where a redundant was removed. These equations are expressed in terms of the redundants, which can then be determined. We will demonstrate the application of these ideas in the examples that follow.

Example 5.6 The structure of Fig. 5.20a is shown again in Fig. 5.22a. The corresponding free-body diagram is shown in Fig. 5.22b.

The single equation of equilibrium is

$$R_1 + R_2 = P$$

Note that the structure will be in equilibrium under the action of any R_1 and R_2 that satisfy this equation. Selecting the reaction R_2 as the redundant, R_1 is expressed in terms of the redundant using the equation of equilibrium as

$$R_1 = P - R_2$$

For this problem the two statically determinate problems associated with removing the redundant R_2 are shown in Fig. 5.23.

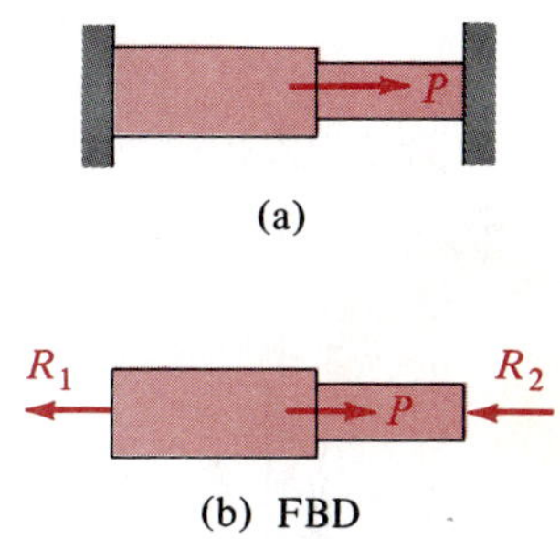

Figure 5.22 Indeterminate structure

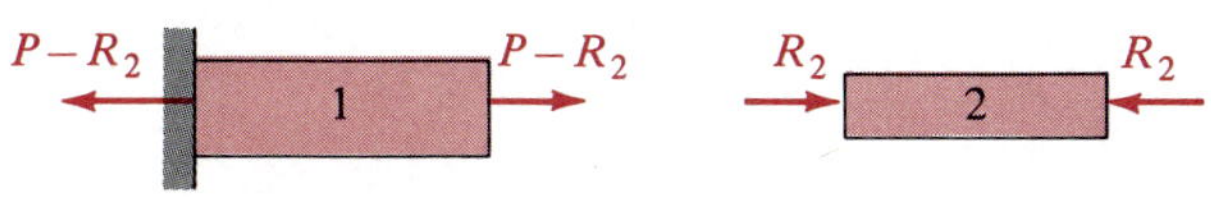

Figure 5.23 Corresponding statically determinate problems

Continuity of displacement is enforced in this example by requiring the elongation of member 1 be the same as the shortening of member 2. Using the force-deflection relation for the axial-deformation problem, the elongation of member 1 and the shortening of BC can be written as

$$e_1 = \frac{R_1}{k_1} = \frac{P - R_2}{k_1}$$

and

$$e_2 = \frac{R_2}{k_2}$$

respectively. We use k_1 and k_2 to represent the axial stiffnesses given by $k_1 = A_1E_1/L_1$ and $k_2 = A_2E_2/L_2$. Unless e_1 and e_2 are equal, the structure will not "fit together," that is, the displacements will not be continuous. Equating the two expressions for the displacements yields

$$\frac{P - R_2}{k_1} = \frac{R_2}{k_2} \quad \text{or} \quad R_2 = \frac{k_2P}{k_1 + k_2}$$

The corresponding value of R_1 is given by

$$R_1 = \frac{k_1P}{k_1 + k_2}$$

Note that the total load P is split between the reactions according to the relative values of the stiffness: if $k_1 > k_2$, then $R_1 > R_2$. The displacement at the point of application of the load P is given by

$$\Delta = \frac{R_1}{k_1} = \frac{R_2}{k_2} = \frac{P}{k_1 + k_1}$$

The flexibility formulation for this problem can be interpreted to state that of all the possible values of R_1 and R_2 that satisfy equilibrium, the correct values are those that also satisfy continuity, or $e_1 = e_2$.

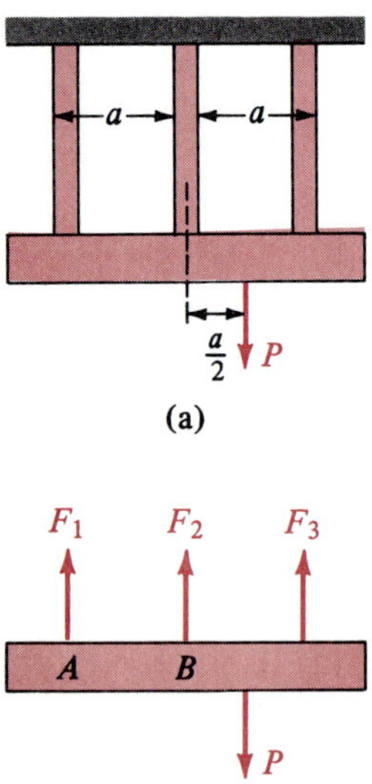

(a)

(b) FBD

Figure 5.24 Indeterminate structure

Example 5.7 Consider again the example shown in Fig. 5.24a, with the corresponding free-body diagram in Fig. 5.24b.

The two equations of equilibrium are

$$\Sigma F_y = F_1 + F_2 + F_3 - P = 0$$

and

$$\Sigma M_A = aF_2 + 2aF_3 - \frac{3a}{2}P = 0$$

which can be rewritten as

$$F_1 + F_3 = P - F_2$$

$$2F_3 = \frac{3P}{2} - F_2$$

For this problem there are three unknowns with only two equations, and therefore one redundant, which as indicated will be chosen as F_2. The additional equation is obtained by considering the deformations of the problem. We require that the vertical displacement of point B on the rigid bar be the same as the elongation of bar 2. The two statically determinate problems that are to be solved are shown in Fig. 5.25.

For the first problem indicated in Fig. 5.25a, solve the equilibrium equations for F_1 and F_3 in terms of the redundant F_2 to obtain

$$F_1 = \frac{P}{4} - \frac{F_2}{2} \quad \text{and} \quad F_3 = \frac{3P}{4} - \frac{F_2}{2} \tag{a}$$

Using the force-displacement relations for members 1 and 2, we can compute the elongations of bars 1 and 2 and hence the vertical displacements of points A and C as

$$\Delta_A = \frac{F_1}{k} = \frac{P}{4k} - \frac{F_2}{2k}$$

and

$$\Delta_C = \frac{F_3}{k} = \frac{3P}{4k} - \frac{F_2}{2k}$$

Based on the fact that the bar is rigid, we can express the displacement of point B in terms of those at A and C as shown in Fig. 5.26. The displacements have been exaggerated for clarity.

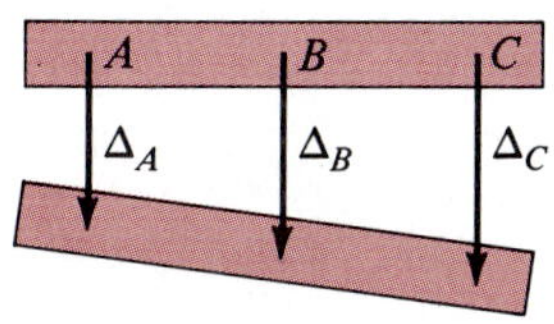

Figure 5.26 Rigid body motion of a bar

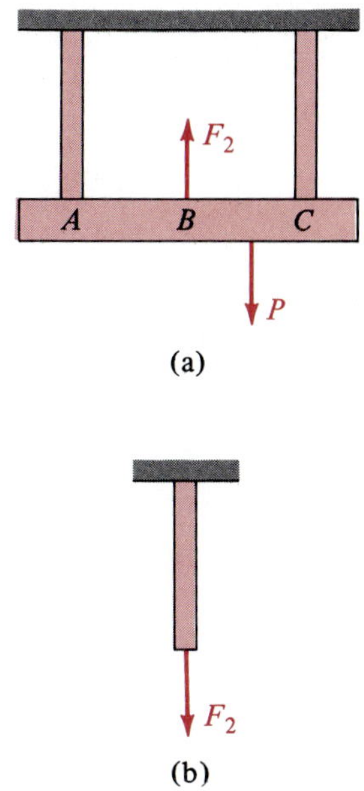

Figure 5.25 Corresponding statically determinate problems

The displacement at B is clearly the average of the displacements at A and C, thus

$$\Delta_B = \frac{\Delta_A + \Delta_C}{2} = \frac{P}{2k} - \frac{F_2}{2k}$$

The second statically indeterminate problem is shown in Fig. 5.25b. If the structure is to fit together when the two statically indeterminate problems are recombined, the displacement at B for the first problem must be the same as the elongation of the middle member for the second problem, or F_2/k. Equating the two values yields

$$\frac{P}{2k} - \frac{F_2}{2k} = \frac{F_2}{k}$$

or $F_2 = P/3$. F_1 and F_2 can then be determined from Eqns. (a) as $F_1 = P/12$ and $F_2 = 7P/12$. The student should check to see that these forces actually satisfy the original equations of equilibrium. The elongation of each of the members can be determined very simply from $e_i = F_i/k$.

In summary, the flexibility approach as applied to statically indeterminate axial deformation problems can be outlined as follows:

1. Write the appropriate equilibrium equations for the structure and determine the number of redundants. Designate this number as NR.
2. Select a suitable set of NR redundants and remove these to produce a statically determinate structure, called the new structure.
3. From the equations of equilibrium, solve for the remaining unknowns in terms of the NR redundants.
4. Require the displacements at the points in the new structure, corresponding to the points where the redundant forces were removed, to be the same as the corresponding displacements associated with the redundants. This set of NR equations has the redundant forces as the unknowns.
5. Solve the NR equations for the redundants and determine also the remaining forces. Then determine the displacements if desired.

The ideas associated with the flexibility approach are fundamental to solid mechanics and can be applied to a wide variety of structural analysis problems, as we will see in subsequent chapters.

PROBLEMS

In each of the following, use the flexibility method to set up and solve the problem.

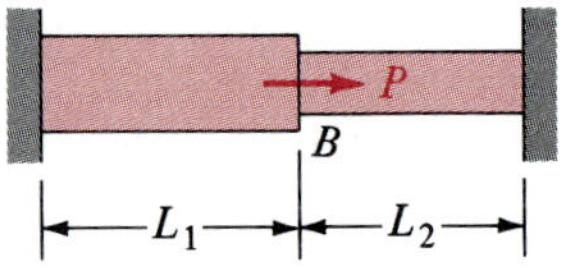

Ps. 5.61, 5.62, and 5.87

5.61 Determine *(a)* the reactions, and *(b)* the displacement at B. Take $A_1 = 2\ \text{in}^2$, $A_2 = 1\ \text{in}^2$, $E_1 = E_2 = 30 \times 10^6$ psi, $L_1 = 15$ in., $L_2 = 9$ in., and $P = 30$ kips.

5.62 Repeat problem 5.61 if $A_1 = 900\ \text{mm}^2$, $A_2 = 600\ \text{mm}^2$, $E_1 = E_2 = 70$ GPa, $L_1 = 0.4$ m, $L_2 = 0.3$ m, and $P = 80$ kN.

5.63 Determine *(a)* the reactions, and *(b)* the displacements at B and C. Take $A_1 = 3\ \text{in}^2$, $A_2 = 1.5\ \text{in}^2$, $A_3 = 4.0\ \text{in}^2$, $E_1 = E_2 = 30 \times 10^6$ psi, $E_3 = 20 \times 10^6$ psi, $L_1 = 12$ in., $L_2 = 8$ in., $L_3 = 18$ in., $P_1 = 40$ kips, and $P_2 = 36$ kips.

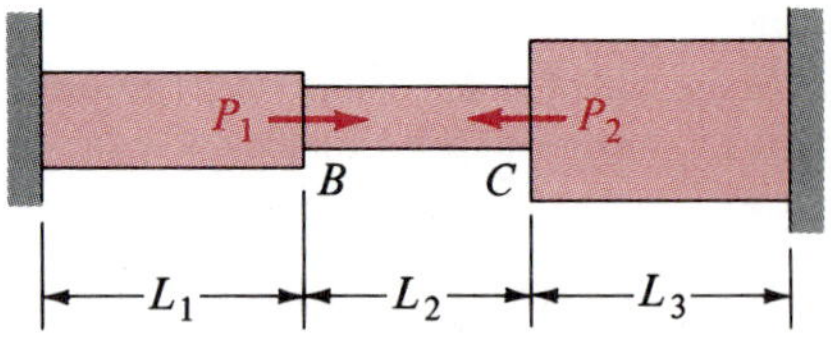

Ps. 5.63, 5.64, and 5.88

5.64 Repeat problem 5.63 for $A_1 = 0.18\ \text{m}^2$, $A_2 = 0.1\ \text{m}^2$, $A_3 = 0.24\ \text{m}^2$, $E_1 = E_2 = 70$ GPa, $E_3 = 120$ GPa, $L_1 = 0.28$ m, $L_2 = 0.20$ m, $L_3 = 0.45$ m, $P_1 = 150$ kN, and $P_2 = 110$ kN.

5.65 What will be the displacement of the rigid bar when the load P is twice the value required to close the gap of 0.125 mm at the steel bar. What are the corresponding stresses in the brass and steel bars? Take E_{st} = 210 GPa, E_{br} = 105 GPa, $A_{br} = A_{st}$ = 0.003 m^2, and H = 2 m.

5.66 Repeat problem 5.65 if the gap is 0.040 in. and the areas of the brass and steel are 5 in^2. Take h = 5 ft, $E_{br} = 15 \times 10^6$ psi, and $E_{st} = 30 \times 10^6$ psi.

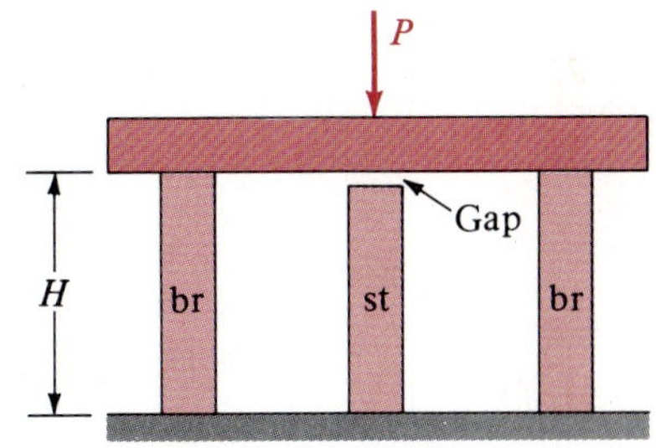

Ps. 5.65, 5.66, and 5.89

5.67 Determine the force in each of the bars when the temperature is raised 20°C. Take E_{br} = 105 GPa, E_{st} = 210 GPa, $A_{br} = A_{st}$ = 0.0018 m^2, l_1 = 0.4 m, l_2 = 0.6 m, $\alpha_{br} = 19 \times 10^{-6}$ m/m/°C, and $\alpha_{st} = 12 \times 10^{-6}$ m/m/°C.

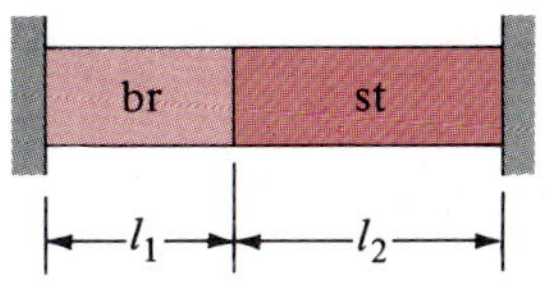

Ps. 5.67 and 5.68

5.68 Repeat problem 5.67 if the temperature is raised 40°F. Take l_1 = 15 in., l_2 = 25 in., $A_{br} = A_{st}$ = 12 in^2, $E_{br} = 15 \times 10^6$ psi, $E_{st} = 30 \times 10^6$ psi, α_{br} = 10.5 $\times 10^{-6}$ in/in/°F, and $\alpha_{st} = 6.5 \times 10^{-6}$ in/in/°F.

5.69 The steel rod 20 ft in length is attached to the 300-lbf weight as shown and carries no load at room temperature. Neglect the weight of the rod. If the temperature of the rod is dropped by 36°F, what will be the force between the floor and the mass? What temperature decrease is required to raise the weight 0.05 in.? Take $E_{st} = 30 \times 10^6$ psi and $\alpha_{st} = 6.5 \times 10^{-6}$ in/in/°F, and the area of the rod to be 0.02 in^2.

5.70 Repeat problem 5.69 if the temperature is dropped 20°C. What temperature decrease is required to raise the weight 2.5 mm? Take A = 12.5 mm^2, W = 1200 N, L = 5 m, E = 210 GPa, and $\alpha = 12 \times 10^{-6}$ m/m/°C.

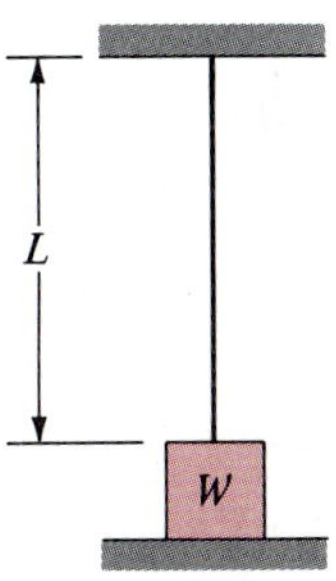

P. 5.69 and 5.70

5.71 A steel pipe with outside diameter 4 in. and inside diameter 3.5 in. is filled with concrete. If a load of 50 kips is applied as shown, determine the stress in the steel and in the concrete. $E_{st} = 30 \times 10^6$ psi and $E_c = 4 \times 10^6$ psi. Does L need to be known?

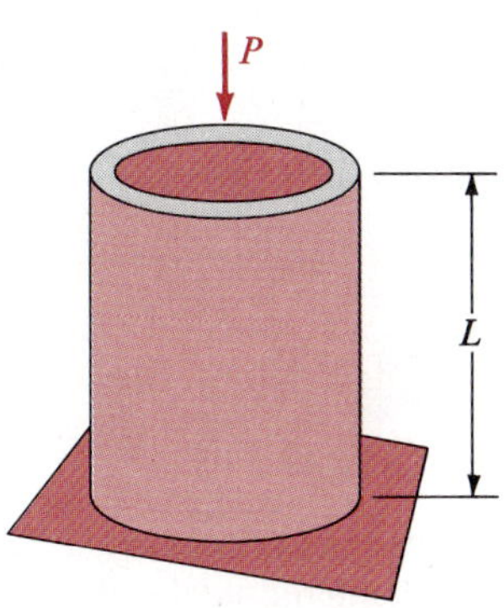

Ps. 5.71 and 5.90

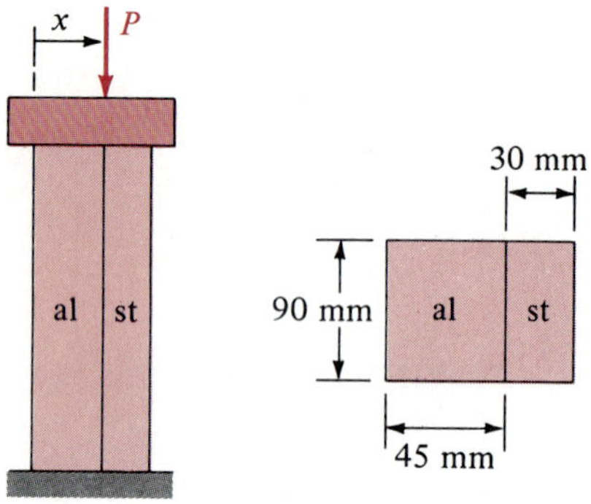

Ps. 5.72 and 5.91

5.72 Where should the load P be applied in order that the rigid plate on top of the bar remain horizontal? Take $E_{al} = 70$ GPa, $E_{st} = 205$ GPa. If $P = 500$ kN, what are the stresses in the aluminum and in the steel?

5.73 The rigid bar AB is supported by the three rods of equal stiffness and loaded as shown. If there are initially no forces in the rods, determine *(a)* the force in each rod, and *(b)* the displacement at B.

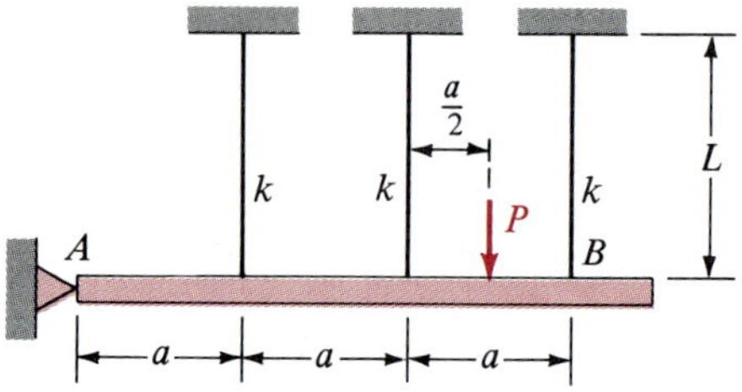

Ps. 5.73, 5.74, 5.92, and 5.93

5.74 Due to a manufacturing error, the length of the middle bar is $L - \delta$, $\delta/L \ll 1$. What are the forces in the bars after assembly but before the load P is applied? What are the forces in the bars after the load P is applied?

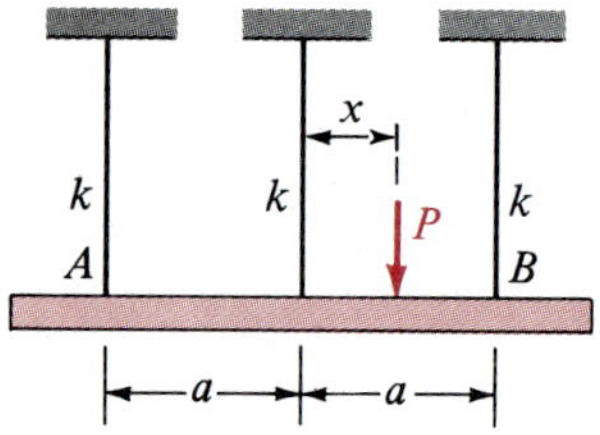

P. 5.75

5.75 The rigid bar AB is supported by the identical wires and loaded as shown. The wires can sustain only tension. If there are initially no forces in the rods, determine *(a)* the force in each rod, and *(b)* the displacement at B when x is taken to be $a/2$. For what range of x is your analysis valid?

5.76 A rigid bar ABC is pinned at C, supported by a 30-mm-diameter steel rod BE and by a 20-mm-diameter aluminum rod AD as shown. The nut at A is tightened one turn on a bolt with a 1-mm pitch. If there are initially no loads in any of the members, determine the stresses in the aluminum and steel rods. Take $a = b = 0.3$ m, $H_1 = 0.4$ m, $H_2 = 0.25$ m, $E_s = 205$ GPa, and $E_a = 70$ GPa.

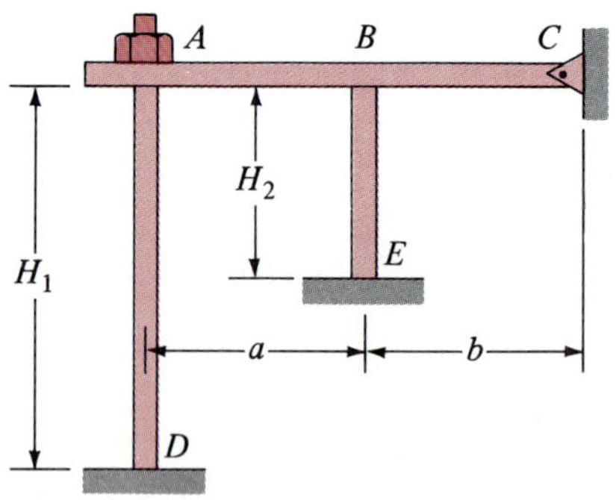

Ps. 5.76, 5.77, and 5.94

5.77 A rigid bar ABC is pinned at C, supported by a 1.5-in.-diameter aluminum rod BE and by a 1.0-in.-diameter steel rod AD as shown. The nut at A is tightened one turn on a bolt with a 0.125-in. pitch. If there are initially no loads in any of the members, determine the stresses in the aluminum and steel rods. Take $a = b = 15$ in., $H_1 = 20$ in., $H_2 = 14$ in., $E_s = 29 \times 10^6$ psi, and $E_a = 10 \times 10^6$ psi.

5.78 A concrete column is reinforced by nine 0.5-in.-diameter steel rods and loaded by a centrally applied force P as shown. Determine *(a)* the stresses in the steel and in the concrete, and *(b)* the amount the column shortens. Take $E_{st} = 30 \times 10^6$ psi, $E_c = 4 \times 10^6$ psi, $P = 180$ kips, $b = 8$ in., and $L = 10$ ft.

5.79 Repeat problem 5.78 for nine 15-mm-diameter steel rods, $b = 0.2$ m, $L = 4$ m, $P = 800$ kN, $E_c = 28$ GPa and $E_{st} = 210$ GPa.

5.80 Determine *(a)* the stresses in the aluminum and steel rods as a result of a 20°C temperature rise, and *(b)* the displacement of point B. Take $E_{al} = 70$ GPa, $E_{st} = 210$ GPa, $\alpha_{al} = 23 \times 10^{-6}$ m/m/°C, $\alpha_{st} = 12 \times 10^{-6}$ m/m/°C, $A_{al} = 600$ mm², $A_{st} = 1300$ mm², $l_1 = 0.3$ m, $l_2 = 0.35$ m, and the gap to be 0.05 mm.

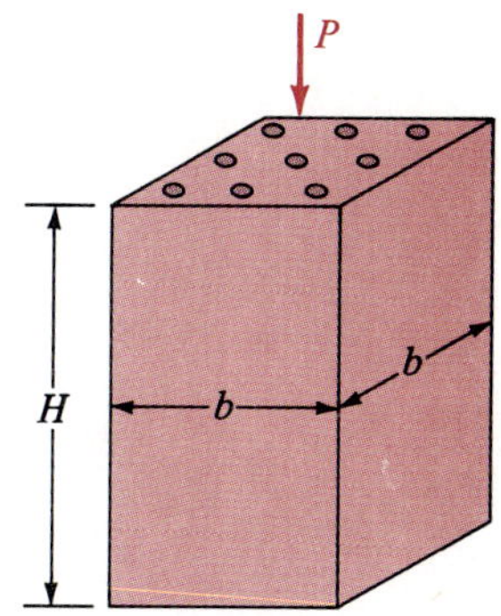

Ps. 5.78, 5.79, and 5.95

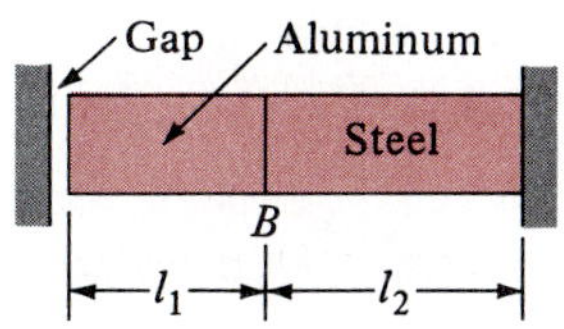

Ps. 5.80, 5.81, and 5.96

5.81 Repeat problem 5.80 for a temperature rise of 40°F and a gap of 0.002 in. Take $A_{al} = 2$ in², $A_{st} = 4$ in², $l_1 = 10$ in., $l_2 = 14$ in., $E_{st} = 30 \times 10^6$ psi, $E_{al} = 10^7$ psi, $\alpha_{al} = 13 \times 10^{-6}$ in/in/°F, and $\alpha_{st} = 6.5 \times 10^{-6}$ in/in/°F.

5.82 A concrete column is to be reinforced by a doubly symmetric pattern of 0.5-in.-diameter steel bars as shown. For a load of 250 kips, how many bars are required in order that the stress in the concrete not exceed 4 ksi? What is the corresponding stress in the steel? Take $E_{st} = 30 \times 10^6$ psi and $E_c = 4 \times 10^6$ psi.

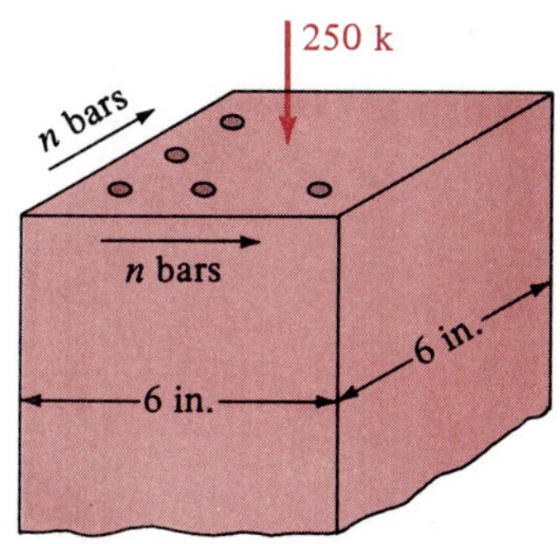

P. 5.82

5.83 The uniform rigid bar AB of weight W is pinned at A and supported by equal stiffness wires as shown. What are the forces in the wires? Assume the wires are originally just taut before they are required to support the weight W. Take $h = 1.5a$.

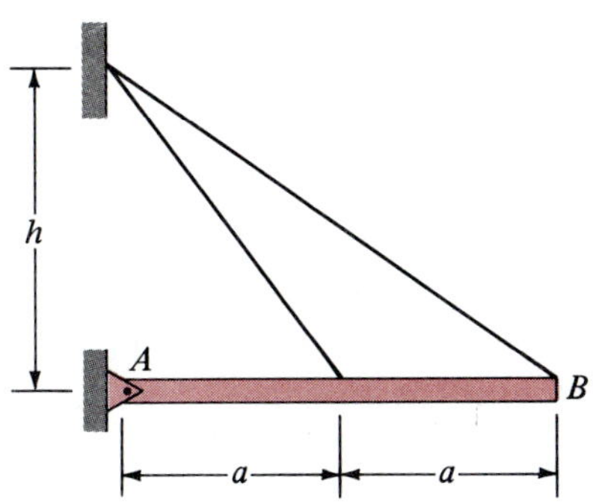

Ps. 5.83 and 5.97

5.84 The solid steel rod is manufactured so that it is too long as indicated. When the load of 50 kips is applied, determine *(a)* the stresses in the brass pipe and steel rod, and *(b)* the shortening of the brass. Take $d_1 = 2$ in., $d_2 = 2.25$ in., $d_3 = 2.5$ in., $H = 120$ in., $g = 0.01$ in., $G_b = 15 \times 10^6$ psi, and $E_s = 30 \times 10^6$ psi.

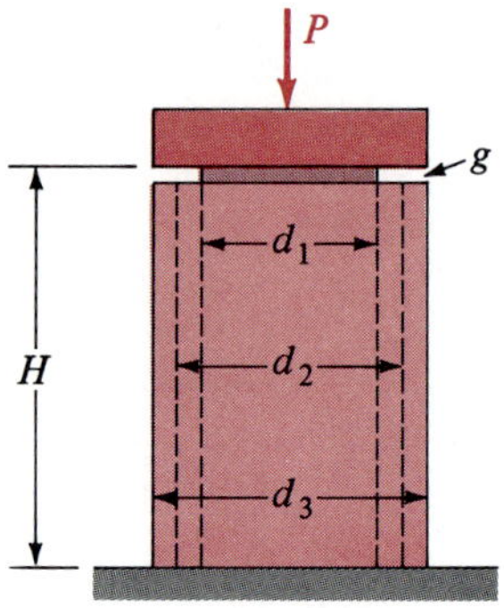

Ps. 5.84, 5.85, and 5.98

5.85 Repeat problem 5.84 for $P = 250$ kN, $d_1 = 45$ mm, $d_2 = 55$ mm, $d_3 = 75$ mm, $H = 4$ m, $g = 0.325$ mm, $E_b = 105$ GPa, and $E_s = 210$ GPa.

5.86 Assuming the bars to be made of linearly elastic materials, express the force in each of the bars in terms of the load P. Take all the bars to have the same A and E. Take $k = AE/h$.

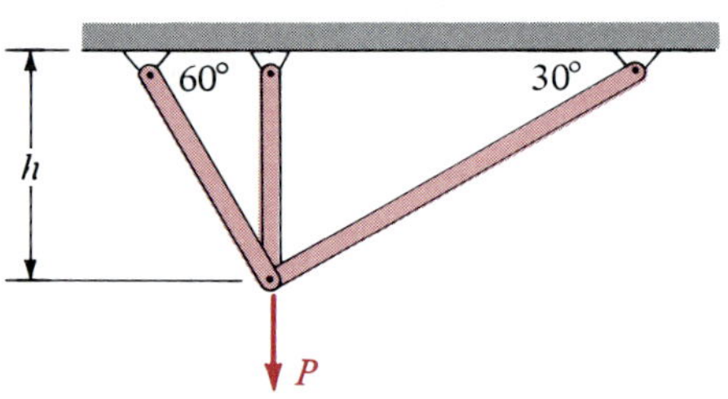

Ps. 5.86 and 5.99

5.4.2 Stiffness Formulations

In stiffness formulations, the equilibrium equations are expressed in terms of the displacements by eliminating the forces using the force-displacement relations for the individual members. For stiffness formulations, the displacements are the basic unknowns or dependent variables for the problem. Any conditions that are necessary to enforce continuity of displacement between the individual members are observed when writing the force-displacement relations. After solving the equilibrium equations for the displacements, the member forces can be determined by reusing the force-displacement relations. The approach is demonstrated in the following examples.

Example 5.8 The structure from Fig. 5.20a, along with the corresponding free-body diagram, is shown in Fig. 5.27.

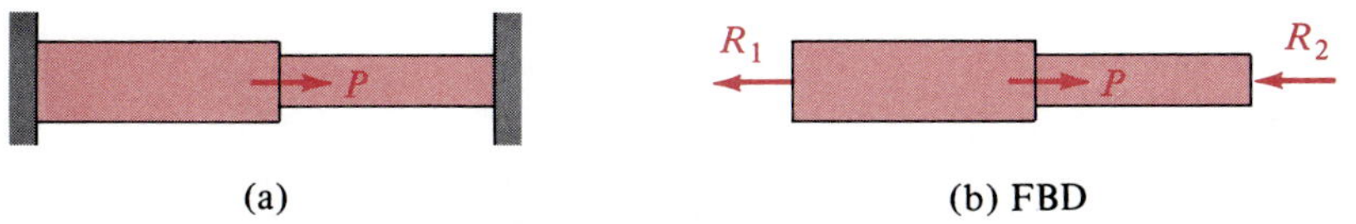

Figure 5.27 Axial statically indeterminate problem

The equilibrium equation is

$$R_1 + R_2 = P$$

The first step for the stiffness formulation is to express the unknown forces in the equilibrium equation in terms of the displacement of the problem. This is accomplished by expressing R_1 and R_2 as

$$R_1 = k_1\delta_1 \qquad \text{and} \qquad R_2 = k_2\delta_2$$

where δ_1 and δ_2 represent the elongation and contraction of bars AB and BC, respectively, with k_1 and k_2 the stiffnesses of the two bars. Denoting Δ_B as the displacement at point B, the continuity statement for this problem is then

$$\delta_1 = \delta_2 = \Delta_B$$

Eliminating δ_1 and δ_2, the equilibrium equation can be written as

$$k_1 \Delta_B + k_2 \Delta_B = P$$

or

$$\Delta_B = \frac{P}{k_1 + k_2}$$

The forces can now be determined as

$$F_1 = k_1\delta_1 = k_1 \Delta_B = \frac{k_1 P}{k_1 + k_2}$$

and

$$F_2 = k_2\delta_2 = k_2 \Delta_B = \frac{k_2 P}{k_1 + k_2}$$

and are seen to be the same as we obtained using the flexibility formulation.

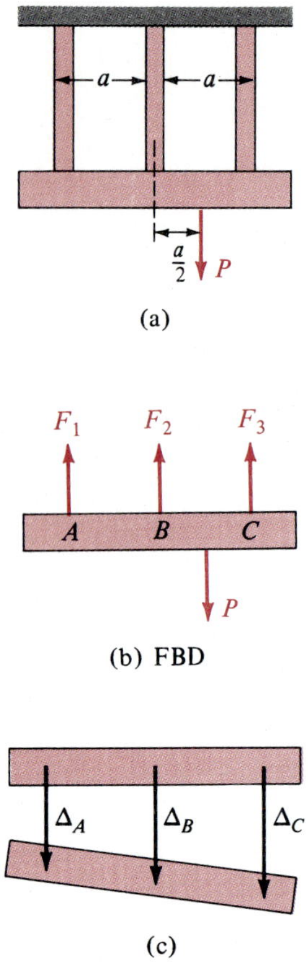

Figure 5.28 Rigid bar supported by three axial members

Example 5.9 The second example is shown in Fig. 5.28a along with the free-body diagram in Fig. 5.28b.

The equations of equilibrium can be written as

$$F_1 + F_2 + F_3 = P$$

$$F_2 + 2F_3 = \frac{3P}{2}$$

To carry out the stiffness formulation, the forces in all the members must be expressed in terms of the displacements of the structure. Consider again the deformations of the rigid bar as indicated in Fig. 5.28c. Although the figure indicates three displacements Δ_A, Δ_B, and Δ_C, only two of these are independent due to the fact that the bar displaces as a rigid body. Consulting Fig. 5.28c, the equation that relates the rigid-body displacements is

$$\Delta_B = \frac{\Delta_A + \Delta_C}{2}$$

This is the equation that enforces continuity of displacements for the structure. The force-displacement relations for the members are

$$F_1 = k\,\Delta_A \qquad F_2 = k\,\Delta_B \qquad F_3 = k\,\Delta_C \tag{a}$$

Eliminating Δ_B and then substituting into the equilibrium equations yields

$$k\,\Delta_A + k\left(\frac{\Delta_A + \Delta_C}{2}\right) + k\,\Delta_C = P$$

and

$$k\left(\frac{\Delta_A + \Delta_C}{2}\right) + 2k\,\Delta_C = \frac{3P}{2}$$

which can be written as

$$\Delta_A + \Delta_C = \frac{2P}{3k}$$

$$\Delta_A + 5\Delta_C = \frac{3P}{k}$$

Solving yields $\Delta_C = 7P/12k$ and $\Delta_A = P/12k$, with $\Delta_B = 4P/12k$. Using Eqn. (a), the member forces can then be computed as $F_1 = k\,\Delta_A = P/12$, $F_2 = k\,\Delta_B = 4P/12$, and $F_3 = k\,\Delta_C = 7P/12$ reproducing the results obtained using the flexibility method.

In summary, the stiffness formulation consists of the following steps:

1. Using any necessary intermember continuity statements along with the appropriate force-displacement relations, write the equations of equilibrium in terms of an independent set of displacements.
2. Solve the resulting set of algebraic equations for the displacements.
3. Use the force-displacement relations to determine the forces.

As they have been applied to the axial-deformation problems of this chapter, the flexibility and stiffness approaches appear to be simply restatements of the same basic ideas. This is due to the fact that for all of the examples considered, all the forces and the corresponding displacements were unidirectional. If, in contrast, the geometry or loading of the second example were altered so that symmetry was no longer present, the problem would be somewhat more complex and the similarities of the two approaches would be less evident. Nevertheless, the same ideas are used in different fashions for the two formulations.

The flexibility and stiffness approaches are intimately related to much of the basic material that will be covered in Chapter 10 on energy methods. At that time the student should be able to see the relationships between the so-called complementary energy principles and the flexibility method on the one hand, and between the strain-energy principles and the stiffness method on the other hand.

PROBLEMS

For each of the following, use a stiffness formulation to set up the problem.

5.87 Solve problem 5.61 using the stiffness method.

5.88 Solve problem 5.64 using the stiffness method.

5.89 Solve problem 5.65 using the stiffness method.

5.90 Solve problem 5.71 using the stiffness method.

5.91 Solve problem 5.72 using the stiffness method.

5.92 Solve problem 5.73 using the stiffness method.

5.93 Solve problem 5.74 using the stiffness method.

5.94 Solve problem 5.76 using the stiffness method.

5.95 Solve problem 5.78 using the stiffness method.

5.96 Solve problem 5.80 using the stiffness method.

5.97 Solve problem 5.83 using the stiffness method.

5.98 Solve problem 5.84 using the stiffness method.

5.99 Solve problem 5.86 using the stiffness method.

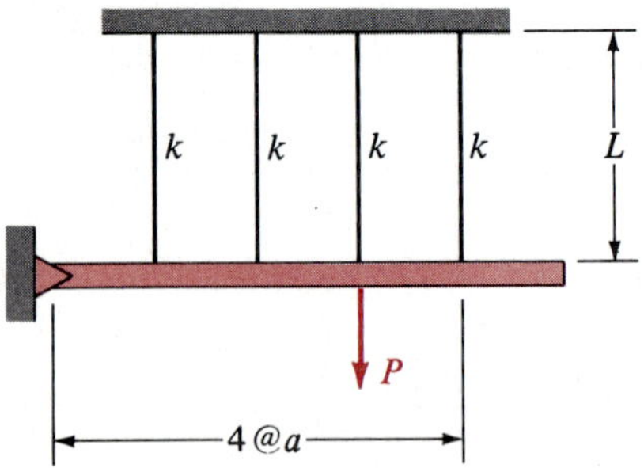

Ps. 5.100–5.102

5.100 The rigid bar AB is pinned at A and supported by the four identical wires as shown. Determine *(a)* the deflection at the load, and *(b)* the forces in the wires.

5.101 With the lengths all the same, is it possible to choose the areas of the wires in problem 5.100 so that the stress in each wire is the same?

5.102 With the areas all the same, is it possible to choose the lengths of the wires in problem 5.100 so that the stress in each wire is the same?

5.103 A structure is formed from three axial members as shown. There is a mismatch due to manufacturing, resulting in the middle bar being δ units shorter. What will be the residual stresses in the members after the rigid plate is attached? What will be the stresses in the members if a load of 600 kips is then applied as shown? Take the areas of the members to be 3 in^2, $L = 20$ in., $\delta = 0.1$ in., and $E = 20 \times 10^6$ psi.

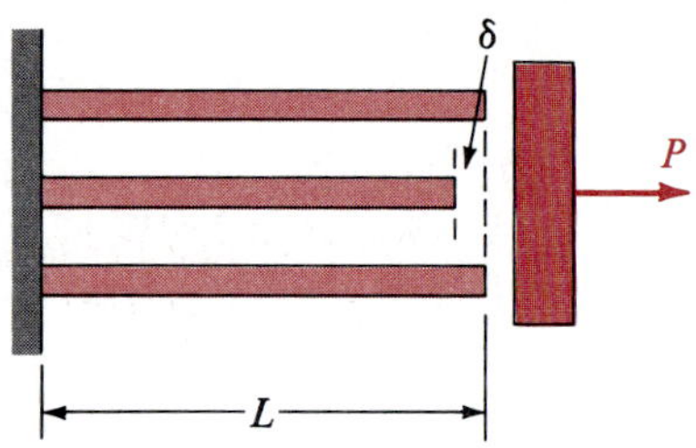

P. 5.103

5.104 The device shown, referred to as a turnbuckle, consists of a 1.0-in. outside diameter, 0.9-in. inside diameter brass sleeve, and a 0.5-in.-diameter steel bolt. The pitch of the threads on the bolt is 0.08 in. per turn of the nut. Determine the stresses in the bolt and sleeve for each turn of the nut. Take $E_{br} = 15 \times 10^6$ psi, $E_{st} = 30 \times 10^6$ psi, and $L = 12$ in.

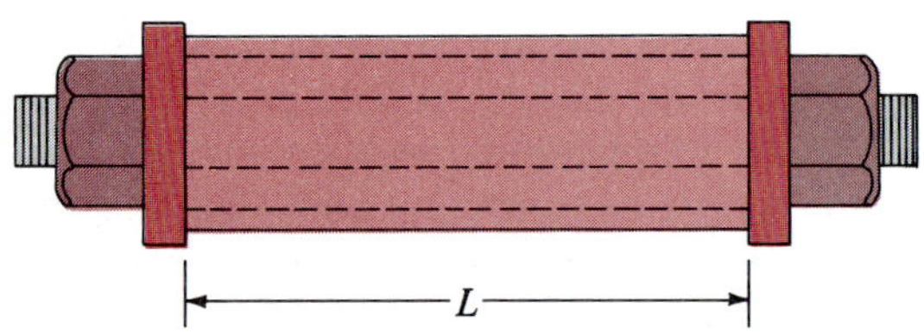

Ps. 5.104–5.106

5.105 The turnbuckle shown consists of a 30-mm outside diameter, 28-mm inside diameter brass sleeve, and a 20-mm-diameter steel bolt. The pitch of the threads on the bolt is 2 mm per turn of the nut. Determine the stresses in the bolt and sleeve for each turn of the nut. Take $E_{br} = 105$ GPa, $E_{st} = 210$ GPa, and $L = 400$ mm. What is the final length of the sleeve?

5.106 The turnbuckle shown consists of a 40-mm outside diameter, 37-mm inside diameter aluminum sleeve, and a 15-mm-diameter steel bolt. The pitch of the threads on the bolt is 3 mm per turn of the nut. If the nut is turned twice and the temperature is increased 5°C, what are the stresses in the sleeve and bolt? Take $E_{al} = 70$ GPa, $E_{st} = 210$ GPa, $\alpha_{al} = 23 \times 10^{-6}$/°C, $\alpha_{st} = 12 \times 10^{-6}$/°C, and $L = 1.0$ m.

5.107 A prestressed concrete column is constructed by applying tensile forces to steel reinforcing rods by means of rigid plates attached at the ends, pouring the concrete around the rods, allowing the concrete to ''set up,'' and then releasing the tension in the rods as indicated in the figure. Assuming that the N wires initially stressed to a value of σ_0 constitute 4% of the total area, what are the final values of the stresses in the wires and the concrete? Take $E_{st}/E_c = 8$.

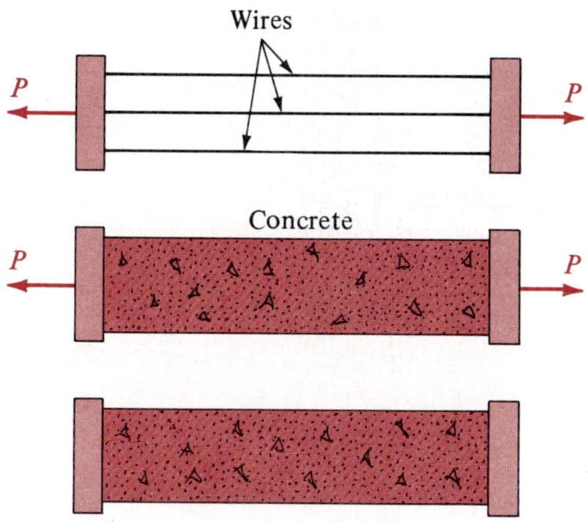

Ps. 5.107 and 5.108

5.108 A prestressed concrete column (see previous problem) with a diameter of 400 mm contains 16 steel reinforcing bars of 25-mm diameter, that were initially stressed to 500 MPa. What is the maximum tensile load that the column can transmit before the stress in the concrete becomes tensile? What is the corresponding stress in the steel?

5.109 The middle member of a symmetric three-member truss experiences a temperature change ΔT as shown. Determine *(a)* the forces in each of the members, and *(b)* the vertical motion of point B. Take each of the areas to be A and the modulii of elasticity to be E.

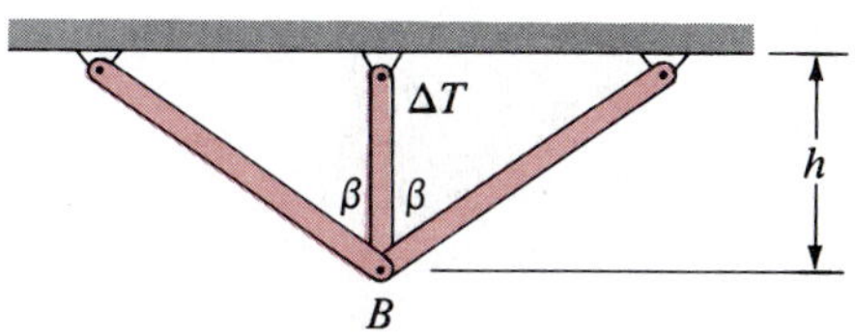

P. 5.109

5.110 A rigid bar is supported by $2n + 1$ identical rods of stiffness k as shown. Assuming that the load P is applied at the position of the jth rod from the middle, determine *(a)* the force in each rod, and *(b)* the angle through which the rod rotates.

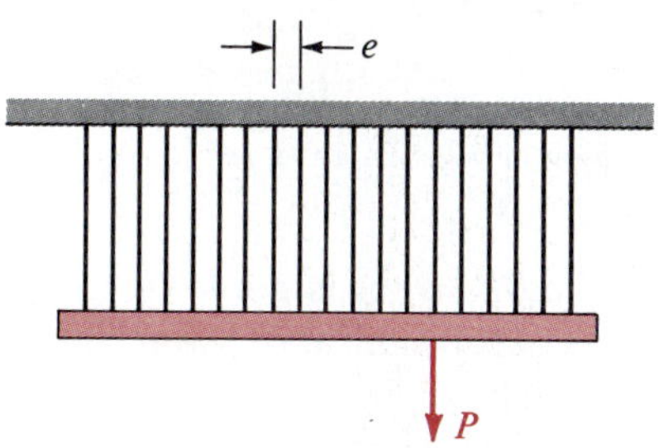

P. 5.110

5.5 EXCEPTIONS TO THE THEORY

The considerations that went into forming the theory for the axial deformation theory were *(a)* geometry, *(b)* equilibrium, *(c)* deformation, and *(d)* material behavior. If any of the assumptions associated with these considerations are not satisfied, there is reason to suspect that the predictions of the theory will be inadequate, either quantitatively, qualitatively, or both. In this section, we will present some technically important exceptions, and indicate the changes that must be made to the results of the elementary theory.

5.5.1 Stress Concentrations

If the cross-sectional area changes abruptly with respect to distance along the bar, the assumption that the bar has a slowly varying area is no longer valid. Several typical examples of violations of this sort are indicated in Fig. 5.29.

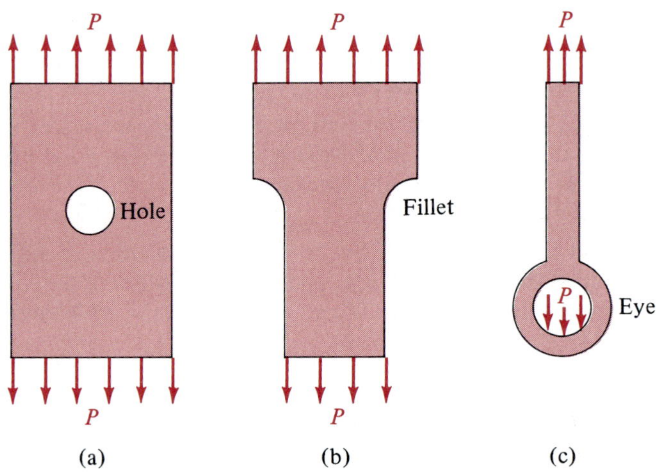

Figure 5.29 Sudden changes in cross-sectional area

In order to help visualize what is going on in the neighborhood of a sudden change in cross-sectional area, consider the deformation of a bar with a central hole as shown in Fig. 5.30. We imagine that a net of mutually perpendicular lines is inscribed on the bar as indicated in Fig. 5.30a. The deformed bar, with the extent of the deformation exaggerated for emphasis, is indicated in Fig. 5.30b. In the neighborhood of section *AA*, the deformation is "uniform" as indicated by the fact that

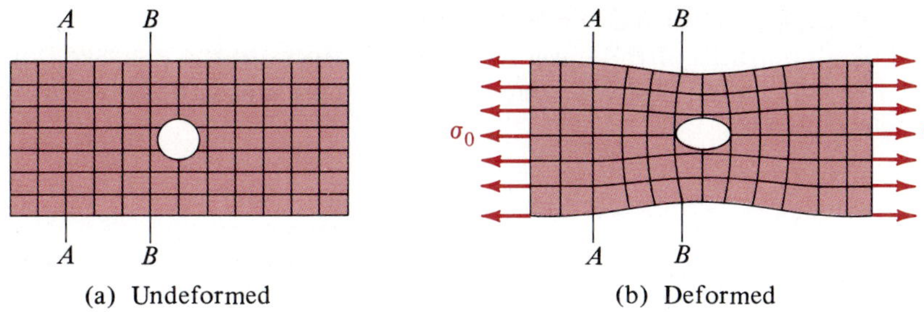

Figure 5.30 Deformations of an axially loaded bar with a central hole

the lines remain parallel and perpendicular, whereas in the neighborhood of the section through the hole there is substantial distortion from the original sets of lines. This distortion is an indication that the stresses are not uniform in the neighborhood of the hole.

Another way to view the effect of the hole is as follows. Consider a bar loaded by the constant stress σ_0 and containing a central hole as shown in Fig. 5.31a. A free-body diagram of the shaded portion of the figure is indicated in Fig. 5.31b. Last, a free-body diagram of the portion shaded in Fig. 5.31b is indicated in Fig. 5.31c.

Assuming that the normal stress on segment AB of Fig. 5.31c is the uniform value σ_0, it is clear that there must be shear stresses on segment BC in order to have equilibrium in the x-direction; thus, the stresses in the neighborhood of the hole are not given by the single simple relationship $\sigma_x = P/A = \sigma_0$, with σ_y and τ_{xy} both zero, but rather are much more complicated. Results from a theory that accounts for the two-dimensional character of the problem (i.e., for the presence of the hole) indicate that the actual normal stress distribution on the section through the hole is as indicated in Fig. 5.32.

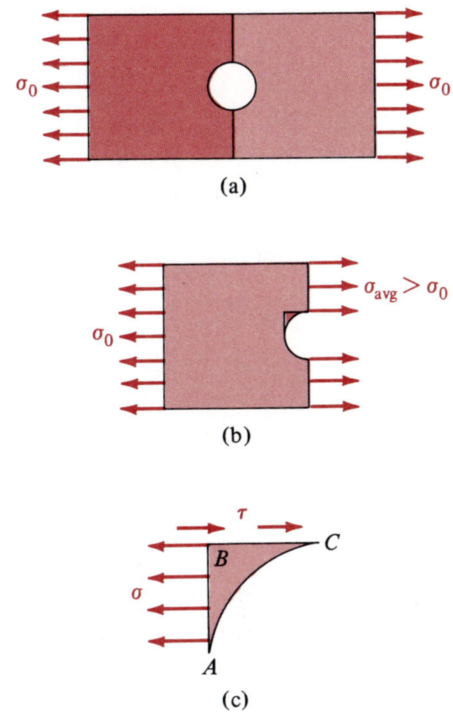

Figure 5.31 Equilibrium in the neighborhood of the hole

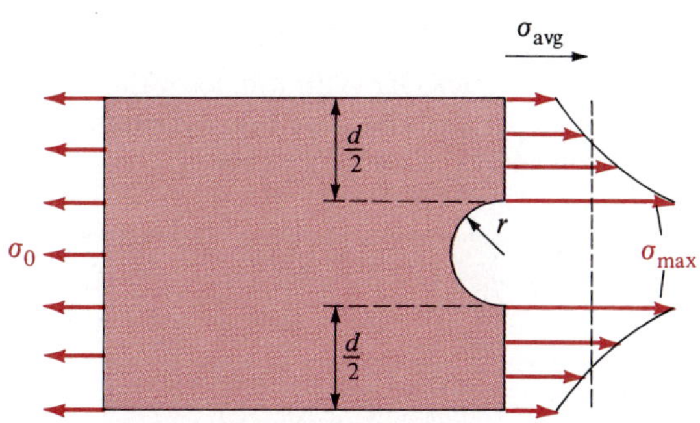

Figure 5.32 Normal stress distribution on a section through a central hole in a plate

The maximum normal stress denoted by σ_{max} occurs at the edge of the hole. The ratio of the maximum normal stress to the average normal stress is called a **stress concentration factor** and is designated by K_T, thus

$$K_T = \frac{\sigma_{max}}{\sigma_{avg}}$$

where σ_{avg} refers to the average normal stress on the section through the hole, that is, to the value computed on the basis of the reduced area $t \cdot d$. The value of the stress concentration factor depends upon the size of the hole relative to the depth of the bar and can be read from the graph shown in Fig. 5.33.

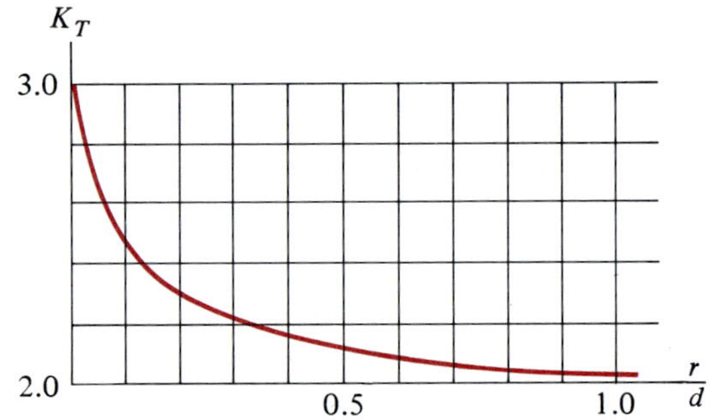

Figure 5.33 Stress concentration factor for a central hole in a plate subjected to tension

The parameter governing the value of K_T is seen to be the ratio r/d. The maximum occurs for a small hole ($r/d \to 0$) and is approximately 3.0. For a hole that is large (i.e., $r \approx d$), K_T is approximately 2.0.

Example 5.10 A 10-mm-thick bar with a hole of the dimensions shown in Fig. 5.34 is subjected to a tensile load of 5 kN. Determine the average and maximum stresses on a section through the hole.

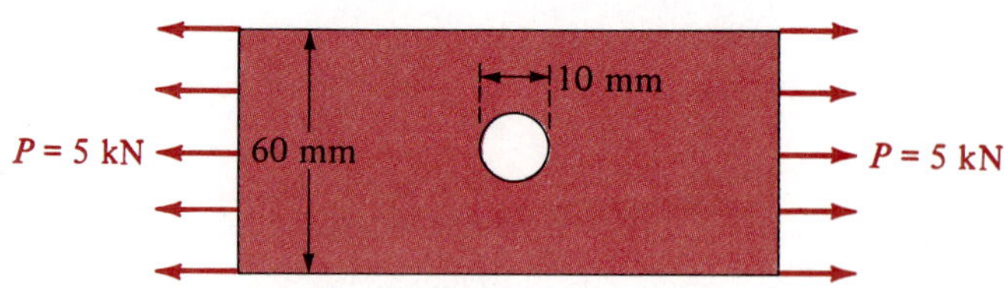

Figure 5.34 Bar with a central hole

Solution The average stress is easily computed to be

$$\sigma_{avg} = \frac{5 \text{ kN}}{(0.060 - 0.010)(0.01) \text{ m}^2} = 10 \text{ MPa}$$

The value of r/d is

$$\frac{r}{d} = \frac{0.005}{0.060 - 0.010} = 0.1$$

Consulting Fig. 5.31, $K_T \approx 2.5$, from which the maximum normal stress is

$$\sigma_{max} = K_T \sigma_{avg} = (2.5)(10 \text{ MPa}) = 25 \text{ MPa}$$

a significant increase over the value of 10 MPa given by the average stress $\sigma = P/A$.

Another type of sudden change in cross section is known as a fillet, indicated in Fig. 5.35.

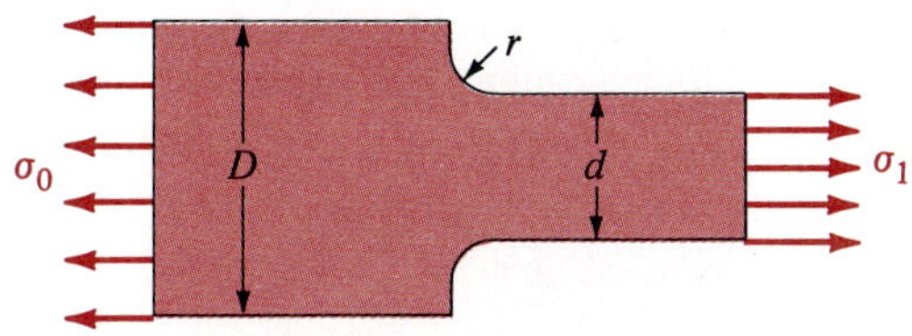

Figure 5.35 Sudden change in cross section due to the presence of a fillet

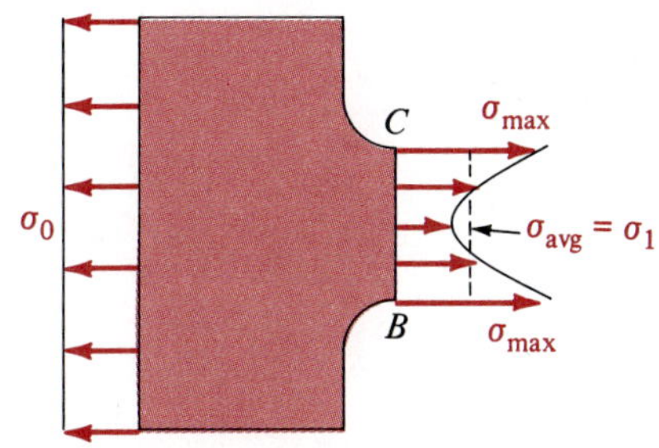

Figure 5.36 Normal-stress distribution on a section in the neighborhood of a fillet

The sudden change in cross-sectional area results in a maximum normal stress that can be substantially larger than the average normal stress on the section. As determined on the basis of a two-dimensional stress analysis, the maximum normal stress occurs at point B or C through the section as shown in Fig. 5.36.

The stress concentration factor is again defined to be

$$K_T = \frac{\sigma_{max}}{\sigma_{avg}}$$

where σ_{avg} refers to the average on the section through BC, that is, to the value σ_1 shown in Fig. 5.36. The value of the stress concentration

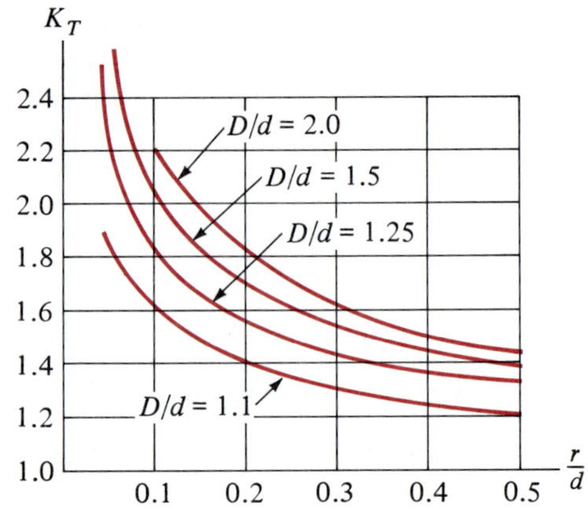

Figure 5.37 Stress concentration factor for fillets

factor, shown in Fig. 5.37, depends upon the two parameters r/d and D/d indicated in Fig. 5.35. The value of D/d is used to select the appropriate curve after which the K_T corresponding to the value of r/d is read from the K_T-axis of Fig. 5.37.

Example 5.11 A 0.25-in.-thick bar with a fillet transmits a tensile load of 10 kips as shown in Fig. 5.38. Determine the average and maximum stresses on the section AA.

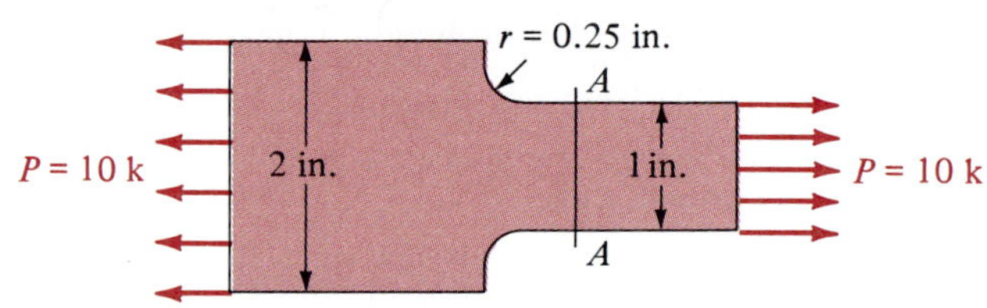

Figure 5.38 Bar with fillet

Solution From Fig. 5.38, the necessary geometrical parameters are $r/d = 0.25$ and $D/d = 2.0$. Selecting the $D/d = 2.0$ curve and choosing $r/d = 0.25$ from Fig. 5.37, yields $K_T \approx 1.70$. The average normal stress on the reduced section is computed to be

$$\sigma_{\text{avg}} = \frac{10000 \text{ lbf}}{0.25 \text{ in}^2} = 40 \text{ ksi}$$

with the maximum given by

$$\sigma_{\max} = K_T \sigma_{\text{avg}} = (1.7)(40 \text{ ksi}) = 68 \text{ ksi}$$

a substantially larger value than the average.

Another less obvious situation is the change in geometry indicated in Fig. 5.39, where there is an apparent gradual change in area but where the ratio of the areas at the end, namely $A_0/A_L = N$, is relatively large. For simplicity, assume a rectangular cross section so that $A_0 = h_0 \cdot t$, $A_L = h_L \cdot t$, and $N = h_0/h_L$.

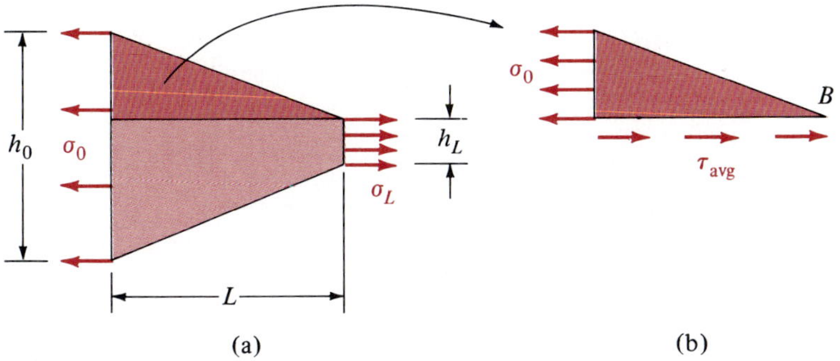

Figure 5.39 Gradual change in area

The average stress at the end $x = L$ is given by $\sigma_{avg} = \sigma_L = P/A_L$, whereas the average stress at the end $x = 0$ is given by $\sigma_{avg} = \sigma_0 = P/A_0, = (P/A_L) \div (A_0/A_L) = \sigma_1/N$. A free-body diagram of the shaded region of Fig. 5.39a is indicated in Fig. 5.39b. It is clear that in order for the shaded region to be in equilibrium, there must be shear stresses on the face AB. The average value of these shear stresses can be computed from

$$\tau_{avg} \cdot Lt = \frac{A_0 - A_L}{2} \cdot \frac{\sigma_1}{N}$$

or

$$\tau_{avg} = \frac{h_0}{2L}\left(1 - \frac{1}{N}\right)\sigma_0$$

If $h_0/2L$ is approximately unity, and $A_L \ll A_0$ (i.e., N is large), the value of the average shear stress on the face AB approaches the average normal stress at the left end, namely σ_0. The elementary theory, which computes a single average normal stress as P/A with the other stress components as zero, is not adequate to describe the actual state of stress in general.[1]

[1] For stress concentrations associated with a variety of geometries and loadings, the reader is referred to the following: R. E. Peterson, *Stress Concentration Factors* (NY: John Wiley & Sons, 1974); and R. Roark and W. C. Young, *Formulas for Stress and Strain,* 5th ed. (NY: McGraw-Hill, 1975).

PROBLEMS

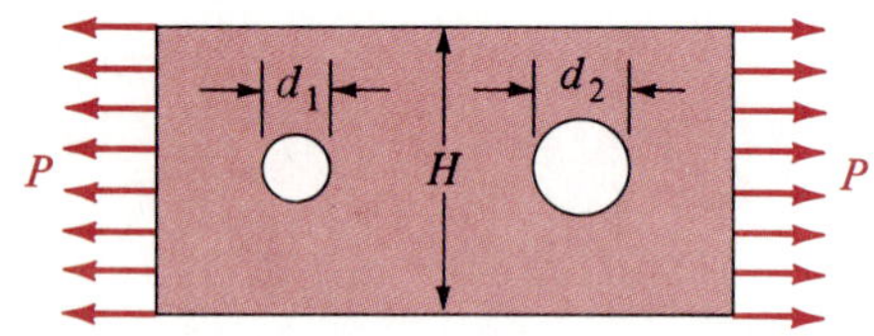

P. 5.111

5.111 The axial member contains the two central holes as shown. If P = 40 kN, determine the maximum normal stress at each of the holes. Take H = 60 mm, d_1 = 15 mm, d_2 = 25 mm, and the thickness to be 12 mm.

5.112 If the ultimate normal stress in the member shown is 24 ksi, determine the allowable value of P for a factor of safety of 2.0. Take the thickness to be 0.5 in.

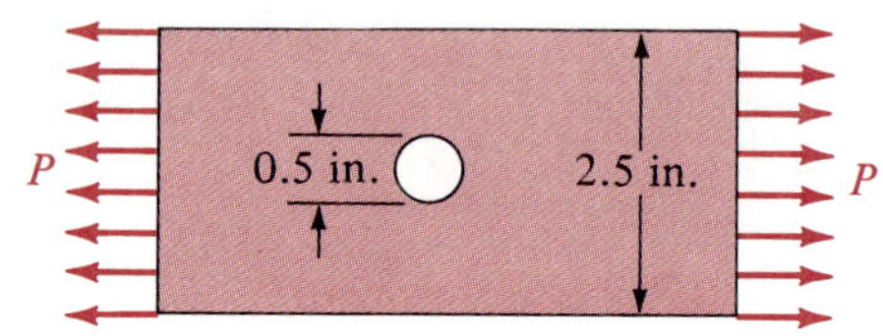

P. 5.112

5.113 The axial member contains the symmetric fillet shown. If P = 20 kips, determine the maximum stress. Take the thickness to be 0.375 in.

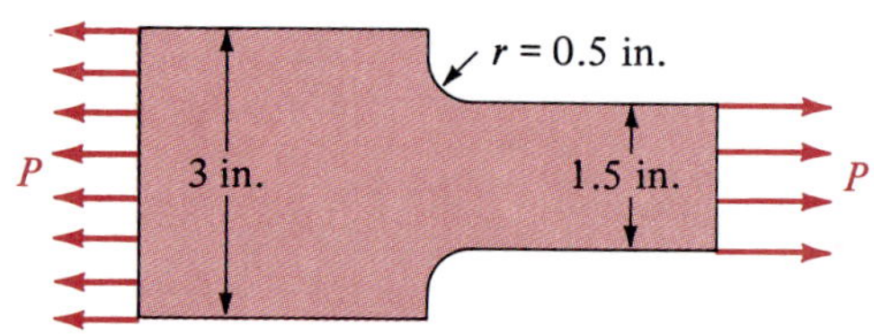

P. 5.113

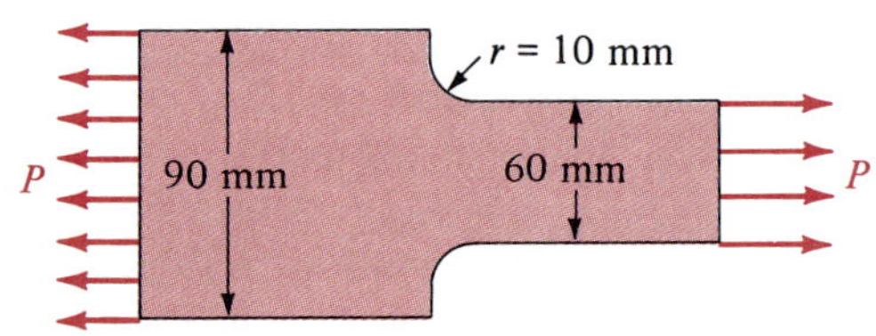

P. 5.114

5.114 If the allowable tensile stress in the member shown is 150 MPa, determine the allowable value of P. Take the thickness to be 10 mm.

5.115 The axial member contains the symmetric fillets and central hole shown. If P = 10 kips, determine the maximum stress at *(a)* the fillet, and *(b)* the hole. Take H = 3 in., h = 1.5 in., r = 0.5 in., d = 1 in., and the thickness to be 0.5 in.

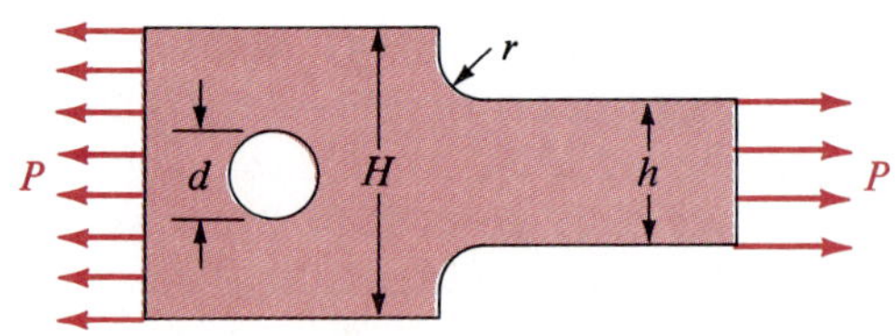

Ps. 5.115 and 5.116

5.116 If the allowable stress in the member shown is 18 ksi, determine the allowable value of P. Take H = 3.5 in., h = 2.0 in., r = 0.75 in., d = 0.875 in., and the thickness to be 0.375 in.

5.5.2 Loading

One of the requirements of the theory for the axial deformation problem is that the resultant of the loading at any section should be a single force parallel to the axis of the straight bar and passing through the centroid of the area. One of the most obvious exceptions to this requirement would be a noncentral force applied as indicated in Fig. 5.40a.

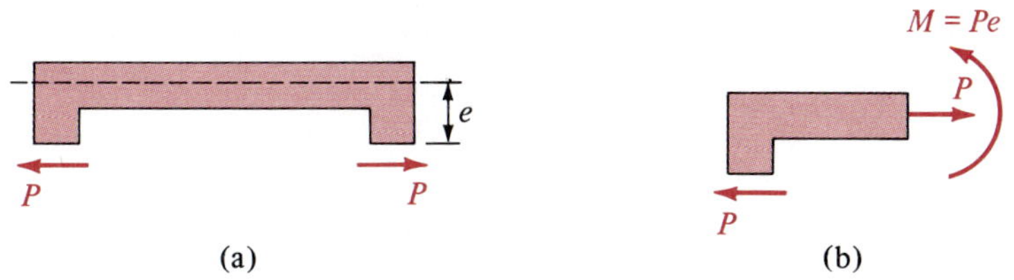

Figure 5.40 Noncentroidal loading of a straight prismatic bar

In addition to the axial load P there would clearly be a moment $M = Pe$ transmitted at any section along the bar as shown in Fig. 5.40b. Computation of the stresses on the basis of the simplified relation $\sigma = P/A$ would in general grossly underestimate the actual stresses due to presence of the moment. The moment produces what are called bending stresses, which we will discuss in detail in Chapters 7 and 8.

Another situation where the use of the equation $\sigma = P/A$ would in general result in significantly underestimating the actual stresses is portrayed in Fig. 5.41. Fig. 5.41a indicates that the force P transmitted by the bar is applied over a segment of length $b/10$, where b is the width of the rod. This corresponds roughly to what is called a concentrated load. Greatly exaggerated, the corresponding deformation is indicated in Fig. 5.41b. The deformation is decidedly nonuniform in the neighborhood of the end of the bar. However, at a distance away from the end of the bar roughly equal to the width of the bar, the deformation appears to be essentially uniform.

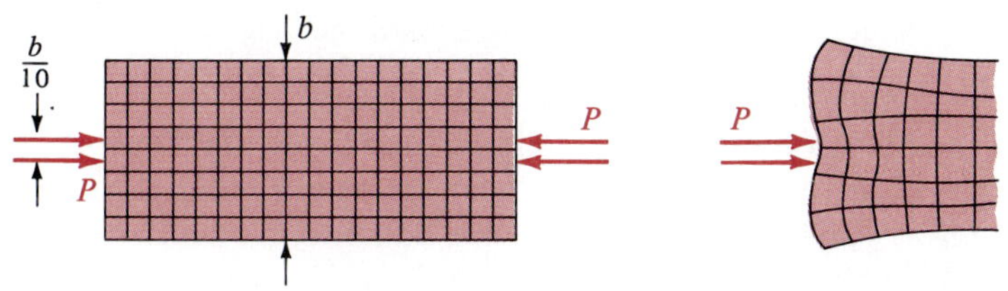

(a) Undeformed (b) Deformed

Figure 5.41 Centrally applied concentrated load

For comparison, consider also the deformation pattern when the load P is applied as a uniform normal stress at the ends of the bar as indicated in Fig. 5.42. In this case, the axial component of stress at any point along the bar is equal to the same uniform normal stress applied at the ends.

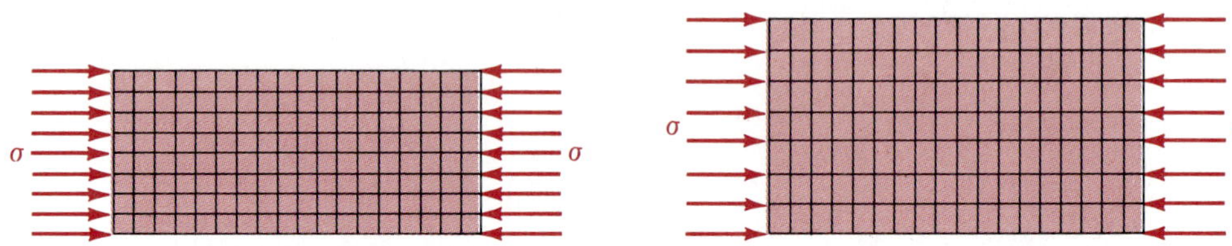

Figure 5.42 Uniform stress application of the load P

As determined by a two-dimensional analysis, the normal-stress distributions at different distances from the end are indicated in Fig. 5.43. As we see, by the time a section a distance b from the end of the bar is reached, the error involved in using the average stress is less than 3%. At positions closer to the point of application of the load P, results from a two-dimensional analysis should be used to determine the stresses and deformations.

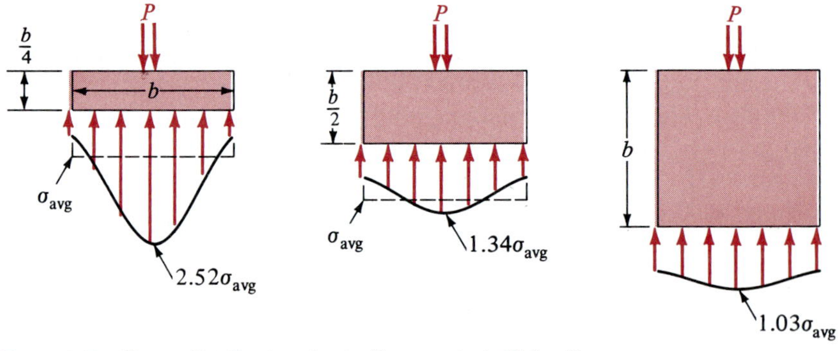

Figure 5.43 Stress distribution due to ''concentrated'' loading

For the case of the concentrated load, the fairly rapid transition from the nonuniform stress distribution near the end of the bar, to the essentially uniform distribution at a distance equal to the width of the bar, is associated with the name of Barre de Saint-Venant who first enunciated the idea. The essence of the idea, known as **St. Venant's principle,** is that if two statically equivalent force systems are applied to a body over a portion of the surface of the body, the stresses that result at points sufficiently far removed from the surface will be independent of the actual details of the two force systems.

In the present situation, the two force systems are indicated in Fig. 5.44. Each force system is statically equivalent to a force P applied through the centroid of the area and, as discussed above, results in the same stresses on a section removed roughly a distance b from the surface. This transition distance b corresponds to the width b of the surface over which the two statically equivalent force systems are applied.

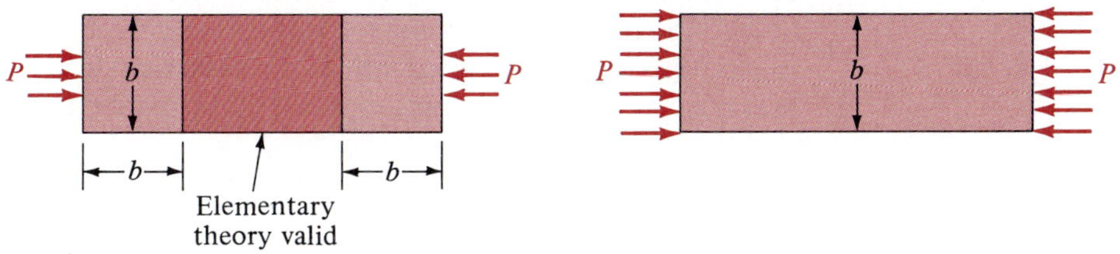

Figure 5.44 Statically equivalent force systems

5.5.3 Inelastic Behavior

In developing the elementary theory in the first sections of this chapter, we assumed the material behavior to be linearly elastic. There are practical situations where it is desirable to relax this assumption and to consider that a bar (or bars) serving as a component of a structure is stressed beyond the proportional limit into the inelastic range of the material. Consider for example the schematic of a passenger aircraft seat shown in Fig. 5.45.

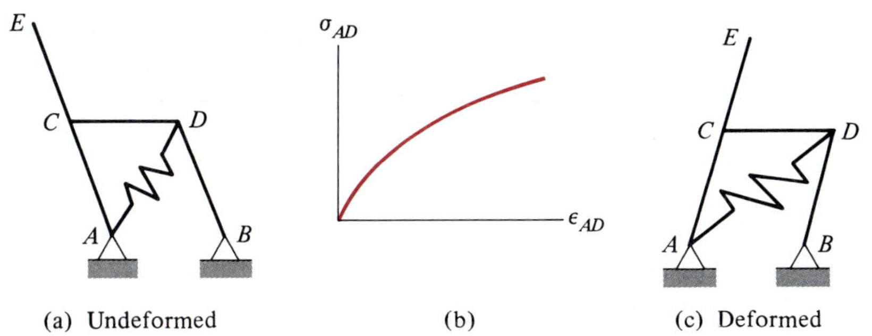

Figure 5.45 Schematic of a passenger aircraft seat

In order to reduce the forces to which an occupant of the seat is subjected if the aircraft is suddenly decelerated, it is desirable to design the seat so as to undergo large deformations as indicated in Fig. 5.45c.

Members ACE, CD, and BD are assumed to be rigid, whereas member AD is assumed to be deformable and to behave according to the stress-strain curve shown in Fig. 5.45b. The idea is that member AD can absorb a great deal more energy if the deformations are allowed to extend into the inelastic range as discussed in Section 4.4.

Regardless of assumptions made regarding the material behavior, it continues to be necessary to consider the other two basic ingredients, equilibrium and deformation. For the applications discussed in this section, we will generally continue to assume small deformations, so that the strain for a bar is given by $\varepsilon = \Delta/L$, and the stress by P/A, with Δ as the elongation of the bar and P as the load transmitted.

For the material behavior we will use either *(a)* the elastic–perfectly plastic stress-strain curve, or *(b)* the elastic–strain-hardening stress-strain curve, shown in Figs. 5.46a and b, respectively.

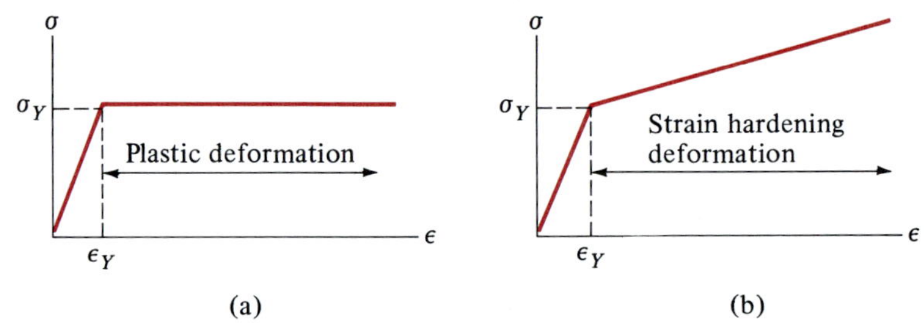

Figure 5.46 Elastic–perfectly plastic and elastic–strain-hardening stress-strain behavior

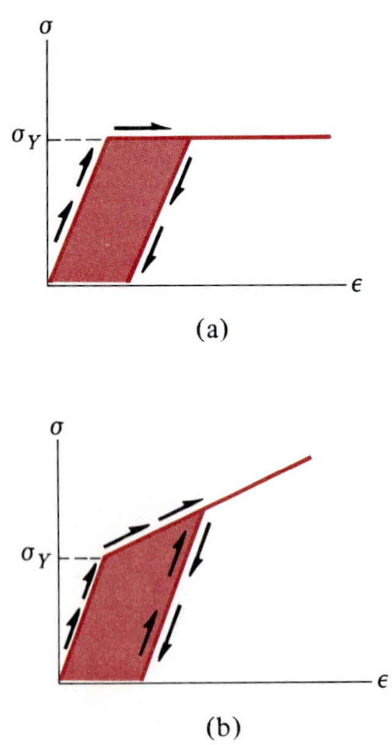

Figure 5.47 Loading and unloading behavior for elastic–perfectly plastic and elastic–strain-hardening materials

Recall from the discussions in Section 4.4 that if unloading occurs, the relation between stress and strain is as indicated in Fig. 5.47. For either of these two cases there is permanent deformation when the material is stressed beyond the proportional limit and then unloaded. Further, if reloading takes place, the relation between change in stress and strain is given by the reloading portion of the stress-strain curve as indicated.

Axial members are more likely to be designed to operate in the inelastic range when they constitute a component of a statically indeterminate structure. This idea is best illustrated by an example such as the two trusses in Fig. 5.48. The truss of Fig. 5.48a is statically determinate whereas the truss of Fig. 5.48b is statically indeterminate. In the first figure, if member AB is composed of an elastic–perfectly plastic material and the load P is sufficient to stress member AB to the yield point, the truss will fail due to the fact that there is nothing to stop the continued straining of member AB until failure occurs. For the statically indeterminate truss of Fig. 5.48b, if loading increases until the stress in member BD reaches the yield point, members AB and BC can

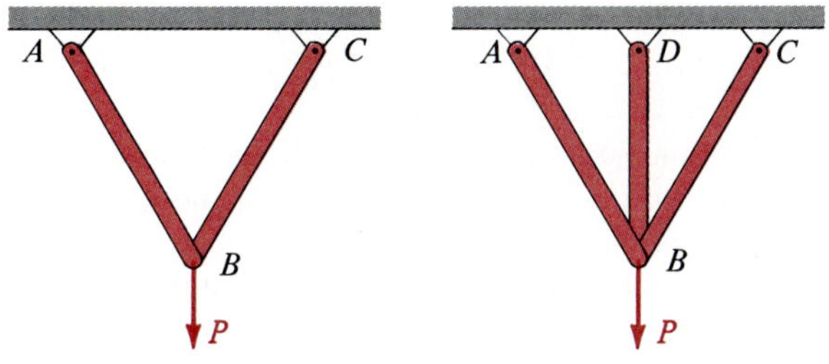

(a) (b)

Figure 5.48 Statically determinate and indeterminate structures

"take up the slack," allowing the load to be increased beyond the yield point for member BD; thus the structure is able to carry a larger load.

In this section we will consider the analysis of structures that behave inelastically, carefully indicating just where the three basic ideas of equilibrium, deformation, and material behavior enter into the formulation and solution of the problem.

Example 5.12 The rigid bar AB is supported by two identical rods as shown in Fig. 5.49a. The rods are composed of an elastic–perfectly plastic material as indicated in Fig. 5.49b. Determine, first, the value of the force P that produces yielding in rod 2, and second, the value that also produces yielding in rod 1. Also plot P versus the vertical displacement at B.

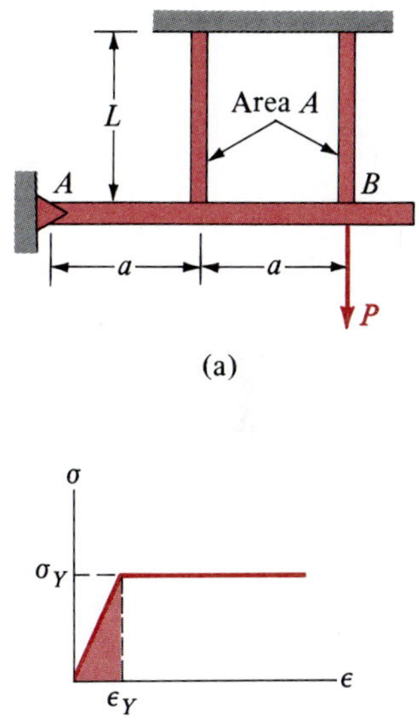

Figure 5.49 Structure and stress-strain curve

Solution Phase 1 of the problem is to investigate the portion of the loading for which the response is linear, that is, before the stress in either rod reaches σ_Y.

Equilibrium: The free-body diagram member AB is shown in Fig. 5.50a.

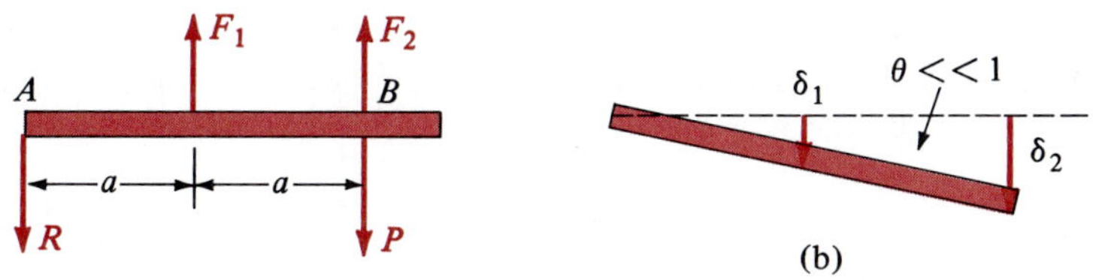

(a) (b)

Figure 5.50 Equilibrium and deformation of the rigid bar

Summing forces and equating the sum of the moments about A to zero yields

$$\Sigma F = 0 = F_1 + F_2 - P - R \qquad (a)$$

and

$$\Sigma M_A = 0 = aF_1 + 2aF_2 - 2aP$$

or

$$F_1 + 2F_2 = 2P \qquad (b)$$

Note that there are two equations for the three unknowns R, F_1, and F_2; thus, the problem is statically indeterminate. The ΣF in the vertical direction only serves to introduce the other unknown R into the problem, so that we will only use the moment equation (one equation and two unknowns) in solving the problem.

Deformation: Since the bar is rigid and rotates through an angle θ about point A as shown in Fig. 5.50b, it follows that

$$\delta_1 = a\theta \qquad \text{and} \qquad \delta_2 = 2a\theta \qquad (c\ \&\ d)$$

At this point we have six unknowns, F_1, F_2, R, δ_1, δ_2, and θ, with the four Eqns. (a) through (d). There are clearly two more equations that must be developed in order to be able to solve for all six unknowns.

Stress-strain relations: At or below the value of P that produces yielding in rod 2, $\sigma = E\varepsilon$ in each rod, that is, each rod behaves as a linear spring. The corresponding force-deflection equations are

$$F_1 = k_1\delta_1 \qquad \text{and} \qquad F_2 = k_2\delta_2 \qquad (e\ \&\ f)$$

where $k_1 = k_2 = AE/L$. These are the two additional equations needed to solve for the six unknowns.

Combining Eqns. (c), (d), (e), and (f) yields $F_2 = 2F_1$. Substitution into (b) yields

$$F_1 + 4F_1 = 2P$$

from which

$$F_1 = \frac{2P}{5} \qquad \text{and} \qquad F_2 = \frac{4P}{5}$$

Note that the corresponding elongations of the rods are $\delta_2 = 4P/5k$ and $\delta_1 = 2P/5k = \delta_2/2$, where k is the spring constant of either rod.

In order to determine the maximum elastic load the structure is capable of carrying we require the stress in rod 2 to be σ_Y. If the stress in rod 2 is to be σ_Y, the load in the rod is $F_2 = A\sigma_Y$ leading to

$$P = P_Y = \frac{5A\sigma_Y}{4} \qquad (a)$$

as the required value of the external load P. This completes the elastic analysis and is valid as long as the stress in neither rod exceeds σ_Y. For $P < P_Y$, the relationship between the force P and the displacement δ_B is as indicated in Fig. 5.52 over the first straight-line portion of the curve.

Phase 2 considers the load required to raise the stress in member 1 to the yield point. In order to investigate the inelastic deformations that result if P is increased beyond P_Y, we use the same basic ideas.

Equilibrium: Continuing to designate the forces in the members F_1 and F_2 there again results by summing moments about A,

$$F_1 + 2F_2 = 2P$$

Deformation: The bar is still rigid so that $\delta_2 = 2\,\delta_1$.

Stress-strain relations: If P is increased beyond the value that produces yielding in rod 2, the stress and strain are related as in Fig. 5.51b; that is, there is no increase in stress in rod 2 due to the additional straining.

At the load $P = P_Y$, the stress and strain in rod 1 correspond to point C on Fig. 5.51a. Any increase in the load P beyond P_Y will result in an increase in the load in rod 1 until point Y in Fig. 5.51a is reached. At this point, the stress in both rods is σ_Y with the force in each rod given by $\sigma_Y A$. Substituting into the equilibrium equation yields

$$A\sigma_Y + 2A\sigma_Y = 2P \qquad \text{or} \qquad P = \frac{3A\sigma_Y}{2}$$

which, with $A\sigma_Y = 4P_Y/5$ from eqn (a), can be expressed as

$$P = \frac{6P_Y}{5}$$

thus, it only takes a 20% increase in P from $P = P_Y$ to $P = 6P_Y/5$ to increase the stress in rod 1 from $\sigma_Y/2$ to σ_Y.

For the displacements, we see from Fig. 5.51a (point Y) that the strain in rod 1 is now ε_Y, and that the elongation is $\varepsilon_Y L$. Using the fact that the bar AB is rigid leads to the result that $\delta_2 = 2\,\delta_1 = 2\varepsilon_Y L$. This is twice the value of the displacement at the end of phase 1. All of these results are indicated in Fig. 5.52.

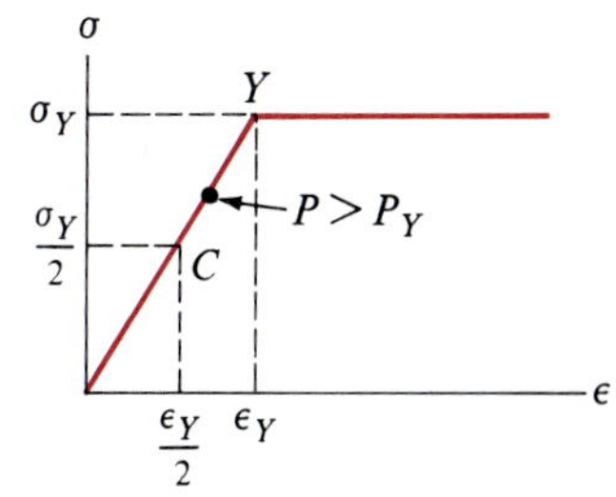

(a) Rod 1

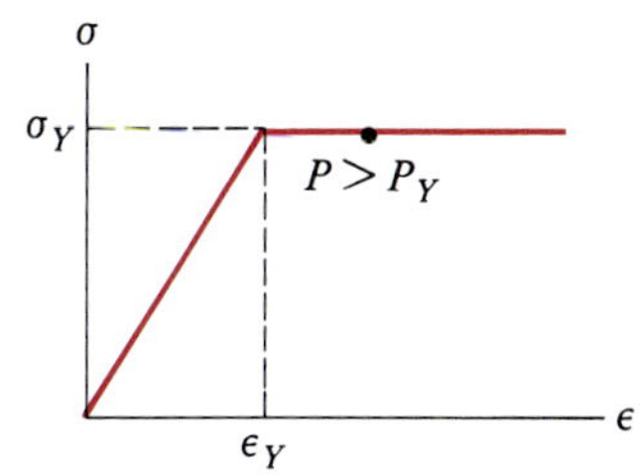

(b) Rod 2

Figure 5.51 Stress-strain behavior for rods 1 and 2

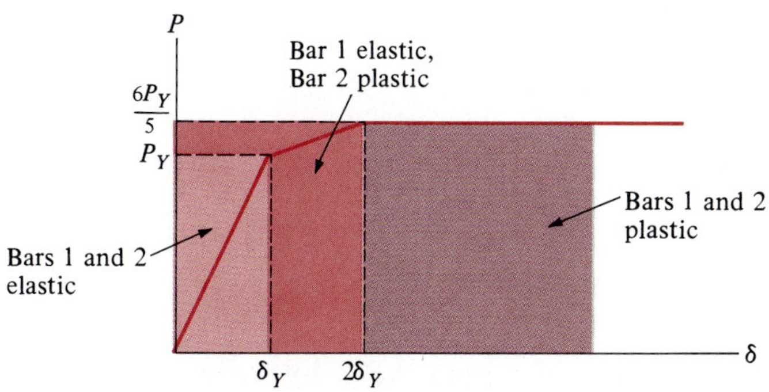

Figure 5.52 Load-deflection curve

Up until $P = P_Y$, both bars are in the elastic range and the force-deflection is linear as shown. For P between P_Y and $6P_Y/5$, bar 1 continues to act elastically while the force in bar 2 remains at the value corresponding to $P = P_Y$, resulting in the different straight-line behavior indicated. Finally, as indicated, the load P cannot be increased beyond $6P_Y/5$. According to the theory, any attempts to increase the load would only result in increased strains and deformations, but no additional load could be supported.

Example 5.13 Analyze Example 5.12 if the load P is removed after reaching the value of $6P_Y/5$.

Solution The application of the three basic ideas will continue to be emphasized in setting up and solving the problem. Rather than the problem posed, we will consider a slightly more general problem: that of gradually reducing the load P from the maximum value of $6P_Y/5$ to zero.

Equilibrium: Equilibrium continues to be exactly as in the two previous situations. Thus we have

$$F_1 + 2F_2 = 2P$$

Deformation: The bar is still rigid so that $\delta_2 = 2\delta_1$.

Stress-strain relations: The stress-strain curves for the two rods are shown in Fig. 5.53a and b, respectively. Unloading is assumed to take place along straight lines parallel to the elastic portion of the curve as indicated. For rod 1 the unloading stress-strain relation is $\sigma_1 = E\varepsilon_1$. For rod 2 the unloading stress-strain curve is linear but does not pass through the origin; rather

$$\sigma_2 = E\varepsilon_2 - \sigma_Y$$

is the equation of the line along which the unloading occurs for rod 2. Multiplying the two stress-strain equations by the area and using $k = AE/L$ yields

$$F_1 = k\,\delta_1 \qquad \text{and} \qquad F_2 = k\,\delta_2 - A\sigma_Y$$

Thus we have the four equations

$$F_1 + 2F_2 = 2P \tag{a}$$

$$\delta_2 = 2\,\delta_1 \tag{b}$$

$$F_1 = k\,\delta_1 \tag{c}$$

$$F_2 = k\,\delta_2 - A\sigma_Y \tag{d}$$

to solve for the four unknowns F_1, F_2, δ_1, and δ_2. Substituting Eqns. (b), (c), and (d) into (a) yields

$$5k\,\delta_1 - 2A\sigma_Y = 2P \tag{e}$$

Note as a check that for $P = 6P_Y/5 = A\sigma_Y$ (the value at the beginning of the process), eqn. (e) becomes

$$5k\,\delta_1 = 2A\sigma_Y + 2A\sigma_Y$$

or

$$\delta_1 = L\varepsilon_Y$$

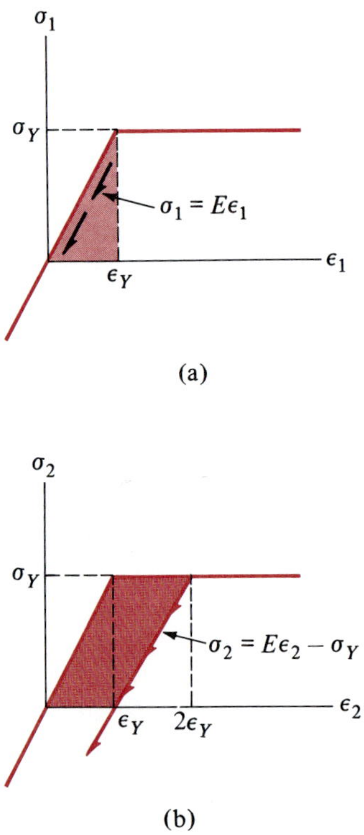

Figure 5.53 Unloading curves for rods 1 and 2

which is correct. Also, when $P = 0$,

$$\delta_1 = \frac{2A\sigma_Y}{5k} = \frac{\delta_2}{2}$$

Substituting into Eqns. (*c*) and (*d*) yields $F_1 = 2A\sigma_Y/5$ and $F_2 = -A\sigma_Y/5$ which, together with the reaction, are shown on the free-body diagram in Fig. 5.54. These results are based on assuming that the rods are capable of sustaining the compression loads without buckling.

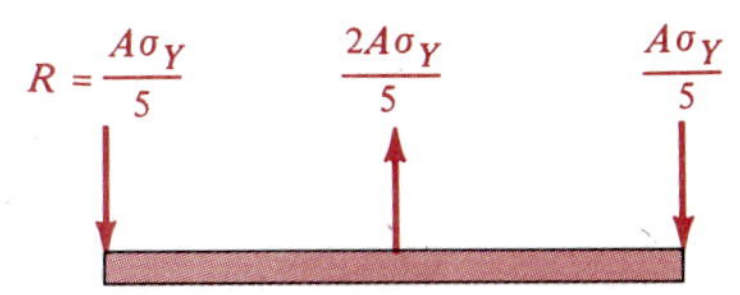

Figure 5.54 Member forces and reaction when the load is removed

PROBLEMS

5.117 The rigid bar is suspended by two wires and loaded by the force P as shown. The wires are made of a material whose stress-strain curve is as indicated. Determine *(a)* the angle through which the bar rotates, and *(b)* the angle that the bar makes with the horizontal when the force P is removed. Take the areas to be 400 mm^2, $L = 3.5$ m, $a = 2.5$ m, $H = 4.0$ m, $P = 70$ kN, $\sigma_1 = 100$ MPa, $\sigma_2 = 120$ MPa, $\varepsilon_1 = 0.001$, and $\varepsilon_2 = 0.002$.

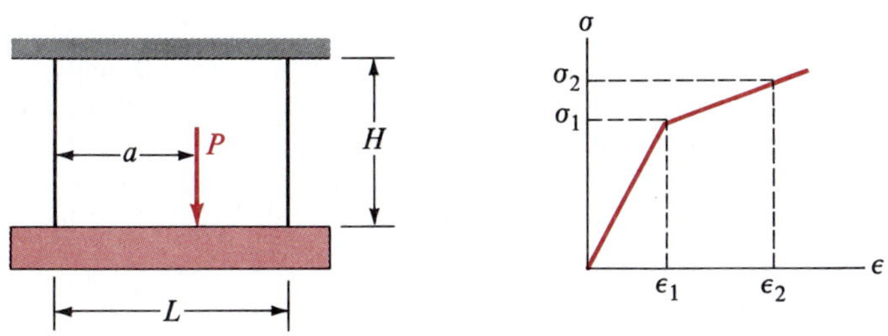

Ps. 5.117 and 5.118

5.118 Repeat problem 5.117 if the areas of the wires are 0.7 in^2, $L = 10$ ft, $a = 7$ ft, $H = 15$ ft, $P = 18$ kips, $\sigma_1 = 15$ ksi, $\sigma_2 = 18$ ksi, $\varepsilon_1 = 0.001$, and $\varepsilon_2 = 0.002$.

5.119 The composite axial member is loaded by the force P. The bars are made of a material whose stress-strain curve is as indicated. The displacement at C is observed to be 0.0274 in. Determine the load P and the residual deformation when the load P is removed. Take $A_1 = 2$ in^2, $A_2 = 1$ in^2, $L_1 = 15$ in., $L_2 = 12$ in., $\sigma_1 = 40$ ksi, $\sigma_2 = 60$ ksi, $\varepsilon_1 = 0.001$, and $\varepsilon_2 = 0.002$.

5.120 Repeat problem 5.119 if the displacement at C is 0.7 mm. Take $A_1 = 1200$ mm^2, $A_2 = 600$ mm^2, $L_1 = 0.4$ m, $L_2 = 0.3$ m, $\sigma_1 = 260$ MPa, $\sigma_2 = 400$ MPa, $\varepsilon_1 = 0.001$, and $\varepsilon_2 = 0.002$.

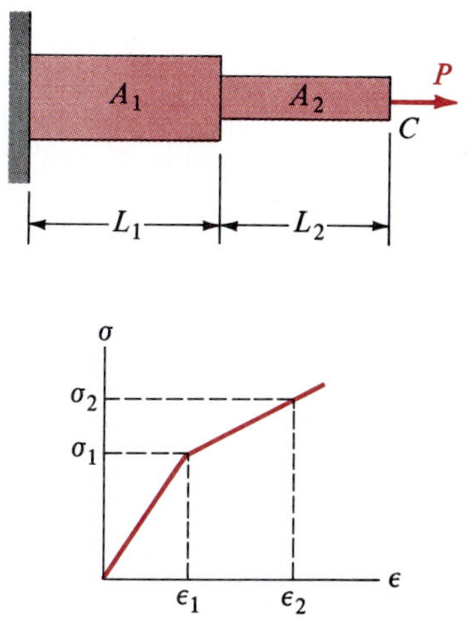

Ps. 5.119 and 5.120

5.121 The composite axial member is loaded by the forces shown. The bars are made of a material whose stress-strain curve is as indicated. Determine the displacement at D and the residual deformation at D when the load P is removed. Take $A_1 = 3\ \text{in}^2$, $A_2 = 1\ \text{in}^2$, $A_3 = 2\ \text{in}^2$, $L_1 = 10$ in., $L_2 = 8$ in., $L_3 = 8$ in., $P_1 = 10$ kips, $P_2 = 20$ kips, $P_3 = 30$ kips, $\sigma_1 = 40$ ksi, $\sigma_2 = 60$ ksi, $\varepsilon_1 = 0.001$, and $\varepsilon_2 = 0.002$.

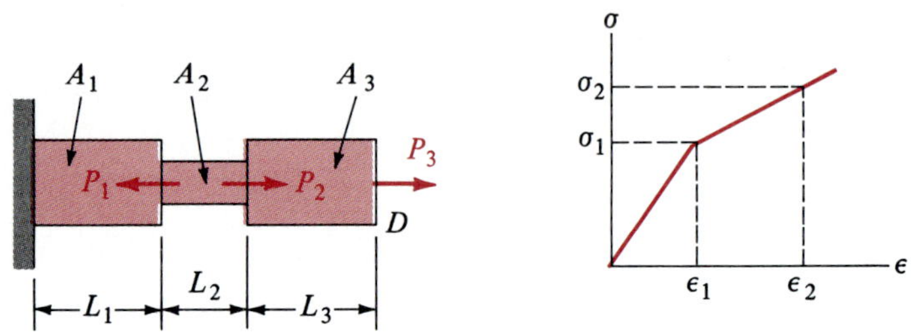

Ps. 5.121 and 5.122

5.122 Repeat problem 5.121 if $A_1 = 1800\ \text{mm}^2$, $A_2 = 700\ \text{mm}^2$, $A_3 = 1300\ \text{mm}^2$, $L_1 = 0.25$ m, $L_2 = 0.2$ m, $L_3 = 0.3$ m, $P_1 = 40$ kN, $P_2 = 80$ kN, $P_3 = 100$ kN, $\sigma_1 = 250$ MPa, $\sigma_2 = 370$ MPa, $\varepsilon_1 = 0.001$, and $\varepsilon_2 = 0.002$.

5.123 The members of the truss shown are made from a material whose behavior is indicated. Determine *(a)* the displacement of point B, and *(b)* the residual displacement if the load is removed. Take $A_1 = A_2 = 100\ \text{mm}^2$, $L = 2$ m, $P = 9$ kN, $\sigma_1 = 100$ MPa, $\sigma_2 = 140$ MPa, $\varepsilon_1 = 0.001$, and $\varepsilon_2 = 0.002$.

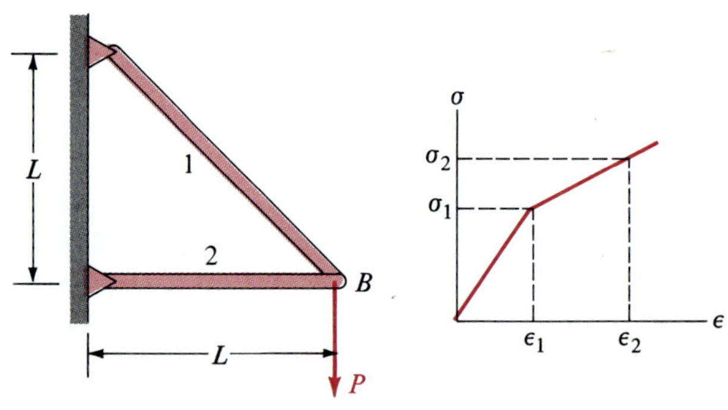

Ps. 5.123 and 5.124

5.124 Repeat problem 5.123 if $A_1 = A_2 = 0.2\ \text{in}^2$, $L = 6$ ft, $P = 2.5$ kips, $\sigma_1 = 15$ ksi, $\sigma_2 = 20$ ksi, $\varepsilon_1 = 0.001$, and $\varepsilon_2 = 0.002$.

5.125 Determine *(a)* the stresses in the solid circular bar and in the surrounding pipe, *(b)* the shortening of the pipe, and *(c)* the residual shortening of the pipe after the load is removed. Take $d_1 = 20$ mm, $d_2 = 27$ mm, $L = 1.0$ m, and $d_3 = 32$ mm. The solid bar behaves elastically with $E = 120$ GPa and the pipe behaves according to the stress-strain curve shown.

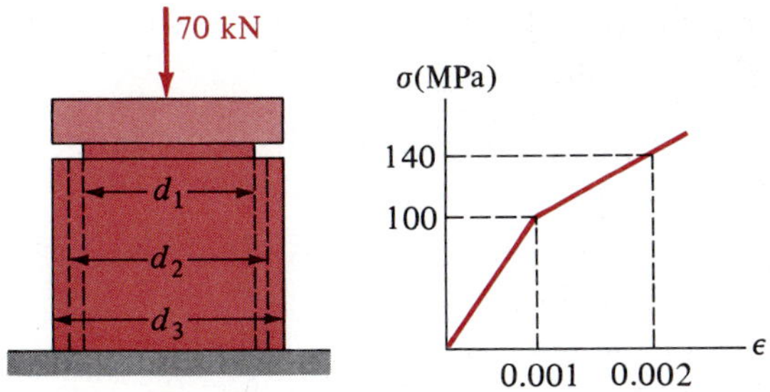

Ps. 5.125 and 5.126

5.126 Repeat problem 5.125 if the material of the pipe behaves in an elastic–perfectly plastic manner for strains above $\varepsilon = 0.001$.

5.6 DYNAMIC LOADING

Structures are frequently subjected to time-dependent loads. Aerospace structures, automobiles, machines of all kinds are just a few of the numerous examples that come to mind. One of the major concerns in such situations is the sudden application of a load, or **dynamic loading.** The corresponding stresses and displacements can be many times higher than if the load were applied slowly, that is, in a static manner.

As a simple example, consider an elastic bar onto which a rigid mass M is dropped as shown in Fig. 5.55. The loading produced in the bar is frequently referred to as an impact load.

In falling through the height h the mass acquires a velocity v, and hence kinetic energy. At impact (Fig. 5.55b), we will assume that the mass impacts and "sticks" to the bar with no loss in energy. The kinetic energy of the mass is then converted into potential energy in the bar with the assumption that during the resulting deformation of the bar (Fig. 5.55c), the bar behaves as an elastic spring with spring constant $k = AE/L$. With these assumptions it is possible to use conservation of energy to determine approximate values of deformations and stresses in the bar. The actual behavior of the system is very complicated and requires a much more detailed model to adequately describe the response.

Considering the mass and the bar to be the system, the initial potential energy V_0 can be taken as

$$V_0 = Mg \cdot h$$

where h is the height of the mass above the top of the bar. The initial kinetic energy is assumed to be zero. The final configuration of the system (Fig. 5.55c) is taken to be at the state when the velocity of the mass M has been reduced to zero, that is, when the kinetic energy

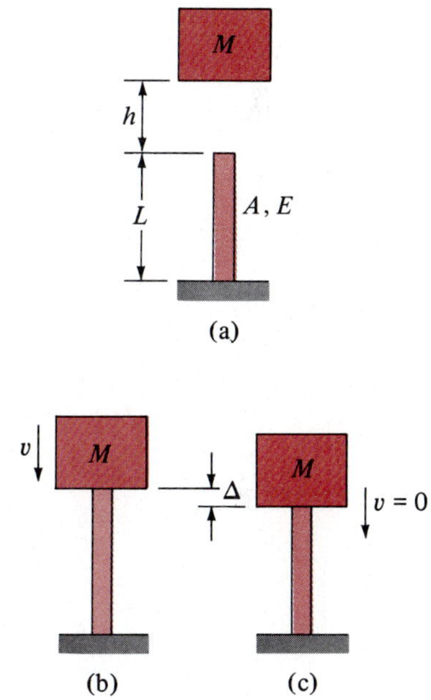

Figure 5.55 Rigid mass impacting an elastic bar

of the system again has a zero value. At this state the potential energy of the system is

$$V_1 = -Mg\,\Delta + \frac{k\,\Delta^2}{2}$$

where $k = AE/L$ is the stiffness of the elastic bar and Δ is its elongation. Equating these two values yields

$$\frac{k\,\Delta^2}{2} - Mg\,\Delta - Mg\cdot h = 0$$

which can be solved to yield

$$\Delta = \frac{Mg}{k} + \frac{Mg}{k}\sqrt{1 + \frac{2kh}{Mg}}$$

With $\Delta_{st} = Mg/k$ as the corresponding static deflection of the bar, the maximum elongation of the bar can be represented as

$$\frac{\Delta}{\Delta_{st}} = 1 + \sqrt{1 + \frac{2h}{\Delta_{st}}} \tag{5.7}$$

which can be quite large depending upon the ratio h/Δ_{st}. The quantity on the right side of Eqn. (5.7) is referred to as the **impact factor** or the **dynamic load factor.** Note that when $h = 0$, $\Delta = 2\Delta_{st}$, that is, when the weight is placed in contact with the bar and suddenly released, the resulting displacement of the bar is twice the corresponding static value.

The corresponding stress in the bar is taken to be

$$\begin{aligned}\sigma = E\varepsilon = \frac{E\Delta}{L} &= \frac{Mg}{A} + \frac{Mg}{A}\sqrt{1 + \frac{2kh}{Mg}} \\ &= \frac{Mg}{A}\left[1 + \sqrt{1 + \frac{2h}{\Delta_{st}}}\right] \\ &= \sigma_{st}\left[1 + \sqrt{1 + \frac{2h}{\Delta_{st}}}\right]\end{aligned}$$

where σ_{st} is the corresponding static stress caused by the weight. Thus, on the basis of this simple model of the mass and bar, both the displacement and the stress are magnified by the same dynamic load factor.

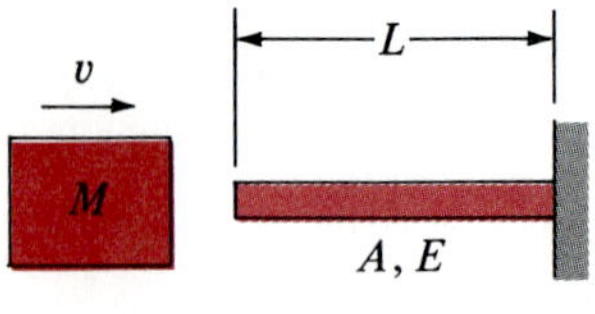

Figure 5.56 Mass impacting a bar

If the mass is moving in a horizontal direction as shown in Fig. 5.56, the kinetic energy is converted into potential energy in the bar according to

$$\frac{Mv^2}{2} = \frac{k\,\Delta^2}{2}$$

so that

$$\Delta = v\sqrt{\frac{M}{k}} = v\sqrt{\frac{ML}{AE}}$$

and

$$\sigma = E\frac{\Delta}{L} = v\sqrt{\frac{ME}{AL}}$$

The essence of these analyses is that the elastic part of the system—the bar—is assumed to behave as an elastic spring during the application of a dynamic load. This sort of replacement of the actual bar by a linear spring gives reasonable estimates of the maximum displacements and stresses, as long as the mass of the bar is small compared to the impacting mass, and as long as the impacting velocities are not too large. It is, of course, assumed that the resulting stresses remain in the linearly elastic range.

Example 5.14 A mass of 10 kg is dropped from a height of 100 mm onto a uniform circular bar of diameter 10 mm and length 300 mm. Determine the maximum compression and the maximum compressive stress in the bar. Take $E = 70$ GPa.

Solution The static compression is given by

$$\Delta_{st} = \frac{(10)(9.81)(0.3)}{\pi(0.005)^2(70 \times 10^9)} \text{ m} = 5.35 \times 10^{-6} \text{ m}$$

with the static stress as

$$\sigma_{st} = \frac{(10)(9.81)}{\pi(0.005)^2} \text{ MPa} = 1.25 \text{ MPa}$$

From Eqn. (5.7), the dynamic load factor is

$$DLF = 1 + \sqrt{1 + \frac{2h}{\Delta_{st}}} = 1 + \sqrt{1 + \frac{2(0.1)}{5.35 \times 10^{-6}}} = 194.3$$

from which

$$\Delta_{max} = 194.3(5.35 \times 10^{-6} \text{ m}) = 1.04 \times 10^{-3} \text{ m}$$

and

$$\sigma_{max} = 194.3(1.25 \text{ MPa}) = 242.9 \text{ MPa}$$

In summary, if it is necessary to analyze the response of a bar to dynamic loads, a reasonable "ball park" estimate can be obtained by replacing the bar with a linear spring. More accurate models and their corresponding solutions are covered in a course in vibrations.

PROBLEMS

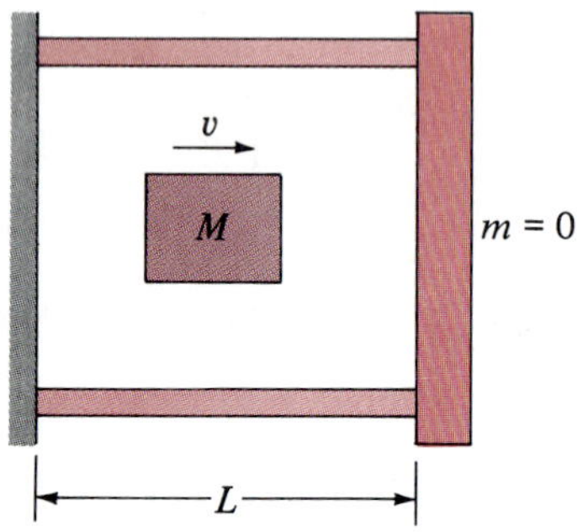

Ps. 5.127 and 5.128

5.127 A 5-kg block strikes the rigid bar with a velocity $v = 10$ m/s as shown. Assuming that there is no energy loss due to the impact, and that the block A "sticks" to the rigid bar, determine *(a)* the maximum stress in one of the bars, *(b)* the equivalent static load for one of the bars, and *(c)* the maximum elongation of one of the bars. For each of the bars, $A = 300$ mm^2, $L = 1.5$ m, and $E =$ 70 GPa.

5.128 An $M = 10$ lbm block strikes the rigid bar with a velocity of 30 ft/s as shown. Assuming that there is no loss in energy due to the impact and that the block A sticks to the rigid bar, determine *(a)* the maximum stress in one of the bars, *(b)* the equivalent static load for one of the bars, and *(c)* the maximum elongation of one of the bars. For each of the bars, $A = 0.5$ in^2, $E = 10^7$ ksi, and $L = 3$ ft.

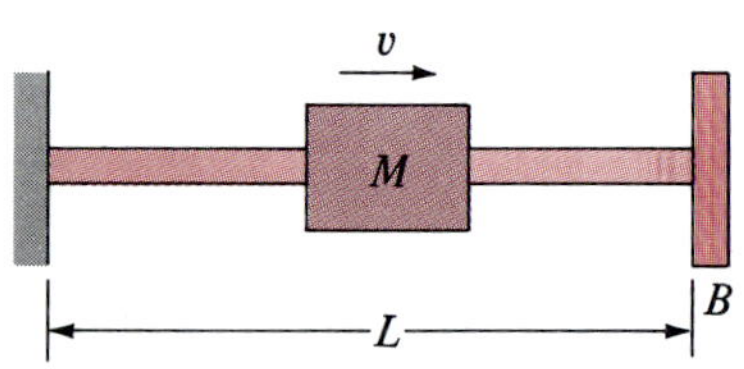

Ps. 5.129 and 5.130

5.129 The 1.0-in.-diameter uniform bar 3 ft in length is made of aluminum, for which the offset yield strength is 30 ksi and $E = 10^7$ psi. The mass M moves along the bar at a speed of 15 ft/s and impacts the collar at B. Assuming a 25% energy loss due to the impact, but that the mass M sticks to the collar B, determine the largest allowable mass M if there is to be no permanent deformation of the bar. Repeat the calculations if the length L is *(a)* doubled, and *(b)* halved.

5.130 The 20-mm-diameter uniform bar of length 2.0 m is made of aluminum for which the offset yield strength is 210 MPa and $E = 100$ GPa. The mass M moves along the bar at a speed of 5 m/s and impacts the collar at B. Assuming a 25% energy loss due to the impact, but that the mass M sticks to the collar B, determine the largest allowable mass M if there is to be no permanent deformation of the bar. Repeat the calculations if the length L is *(a)* doubled, and *(b)* halved.

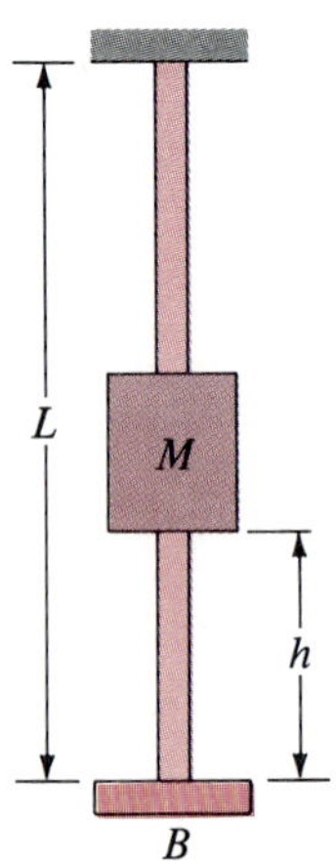

Ps. 5.131–5.134

5.131 The 20-kg-mass M is released from rest when $h = 1$ m, and is stopped by the plate attached at B. Knowing that $E = 70$ GPa, determine *(a)* the maximum stress in the bar, *(b)* the equivalent static load, and *(c)* the maximum elongation of the bar. Assume a 20% energy loss during impact but that the mass sticks to the plate. Take $L = 2$ m and the diameter of the bar to be 40 mm.

5.132 The 40-lbm-mass M is released from the rest when $h = 5$ ft, and is stopped by the plate attached at B. Knowing that $E = 10^7$ ksi, determine *(a)* the maximum stress in the bar, *(b)* the equivalent static load, and *(c)* the maximum elongation of the bar. Assume no energy loss during impact but that the mass sticks to the plate. Take $L = 10$ ft and the diameter of the bar to be 1.0 in.

5.133 The mass M is released from rest when $h = 6$ ft, and is stopped by the plate attached at B. Knowing that $E = 10^7$ psi, determine the maximum allowable mass in order that the stress in the bar does not exceed 20 ksi. Take $L =$ 8 ft and $d = 1$ in.

5.134 The mass M is released from rest when $h = 2$ m, and is stopped by the plate attached at B. Knowing that $E = 70$ GPa, determine the maximum allowable mass in order that the stress in the bar does not exceed 150 MPa. Take $L = 3.2$ m and $d = 30$ mm.

5.7 CLOSURE

In this chapter we developed a theory for the axial deformation of straight prismatic bars, using the basic ideas of equilibrium, deformation, and material behavior. We then combined the results from these three ideas to yield an equilibrium equation expressed in terms of the displacement u and appropriate boundary conditions. Several practical problems were solved using the theory.

Generally the difficulty is not in applying the theory to solve problems but rather in making sure that all the assumptions used to develop the theory are in fact satisfied. Many failures occur because one or more of the assumptions have not been satisfied. For this reason these assumptions are repeated:

1. **Geometry.** The bar is prismatic, straight, and untwisted, with an area that is constant or a slowly varying function of position along the axis of the bar.
2. **Equilibrium.** Any external loads are such that the internal force resultant at any section along the length of the bar is a single force parallel to the axis of the bar and passing through the centroid of the section.
3. **Deformation.** Plane sections originally perpendicular to the axis of the bar remain plane and translate, without rotating, in the direction of the axis of the bar.
4. **Material behavior.** The bar is composed of a linearly elastic material that is stressed at or below the proportional limit of the material.

By far the most frequent situation where the assumptions are not satisfied occurs where the area varies too rapidly with position along the bar. Holes, fillets, and many modes of connection between a tension member and the rest of a structure make it necessary to use a two- or three-dimensional theory to assess the stresses and deformations in such cases. The responsible engineer keeps in mind all of the assumptions so as to be on guard against using the theory in situations where it does not apply.

When the assumptions for geometry and loading are satisfied for a bar made of a material behaving in a linearly elastic fashion, the simple formulae for stress and elongation of the bar are given by

$$\sigma = \frac{P}{A} \quad \text{and} \quad \Delta = \frac{PL}{AE} = \frac{P}{k}$$

respectively; these can be used with confidence to provide information about the stress levels and deformations in many situations.

Finally, as outlined in Section 5.4, statically indeterminate problems can be solved using either a flexibility or a stiffness approach wherein the three basic ideas are brought together simultaneously, to generate the equations necessary to solve for the force and displacement variables.

COMPUTER EXERCISES

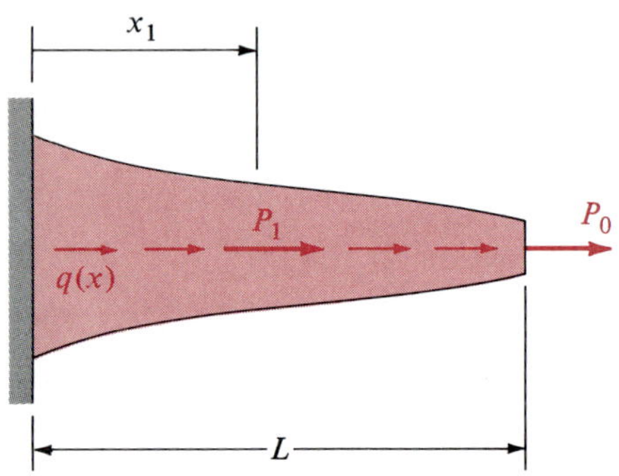

CE 5.1

CE 5.1 Write a general computer program to determine approximate values for the displacement $u(x)$ and the normal stress $\sigma(x)$ for a bar supported at $x = 0$ as shown. Include an option for several concentrated loads P_i at locations x_i, and an option for a distributed loading $q(x)$. Allow the code to determine the reaction at the support, and output the displacement and normal stress at positions along the axis of the bar. Check your code on a number of test problems for which you already know the solution.

CE 5.2 Use the code of CE 5.1 to solve the following problem. Take the cross section to be circular and the weight density to be 100 lbf/ft^3.

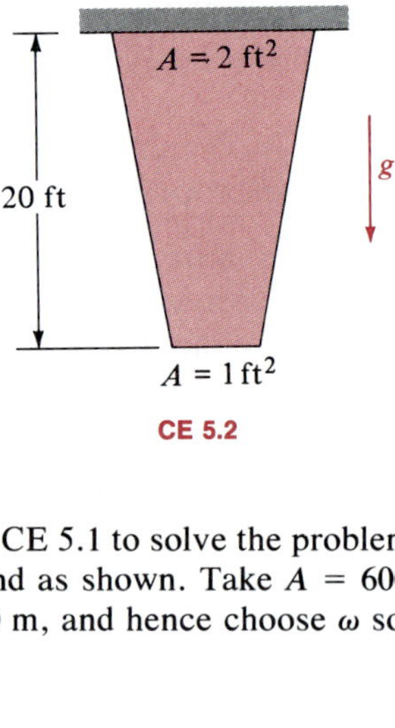

CE 5.2

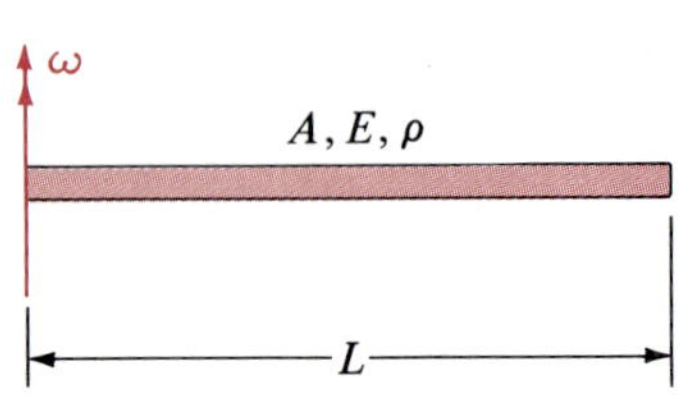

CE 5.3

CE 5.3 Use the code of CE 5.1 to solve the problem of a bar rotating about a fixed axis through one end as shown. Take $A = 600$ mm^2, $E = 70$ GPa, $\rho = 2.8 \times 10^3$ kg/m^3, $L = 1.0$ m, and hence choose ω so that $\sigma \leq 70$ MPa.

CE 5.4 Repeat CE 5.3 when a mass is attached as shown. How should the area depend upon x if the stress is to be a constant? Take $M = 10$ slugs, $L = 5$ ft, $E = 3 \times 10^7$ psi, $\rho = 8.8 \times 10^{-3}$ slugs/in^3, $\omega = 10^4$ rpm, and $\sigma \leq 9$ ksi.

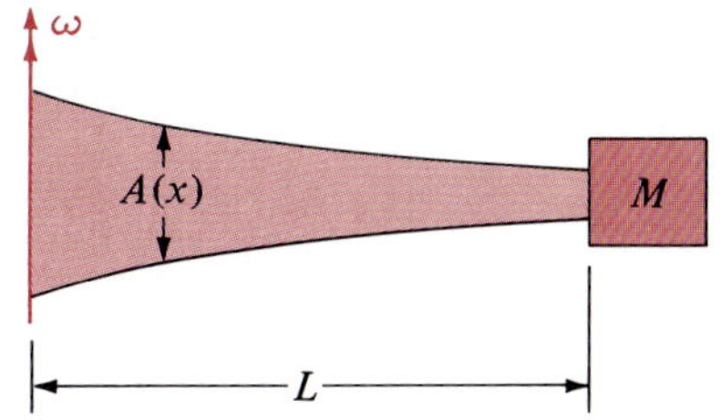

CE 5.4

CE 5.5 Write a general computer program to determine approximate values for the displacement $u(x)$ and the normal stress $\sigma(x)$ for a bar supported at $x = 0$ and $x = L$ as shown. Include an option for several concentrated loads P_i at locations x_i, and an option for a distributed loading $q(x)$. Allow the code to determine the reactions at the supports, and output the displacement and normal stress at positions along the axis of the bar. Check your code on a number of test problems for which you already know the solution.

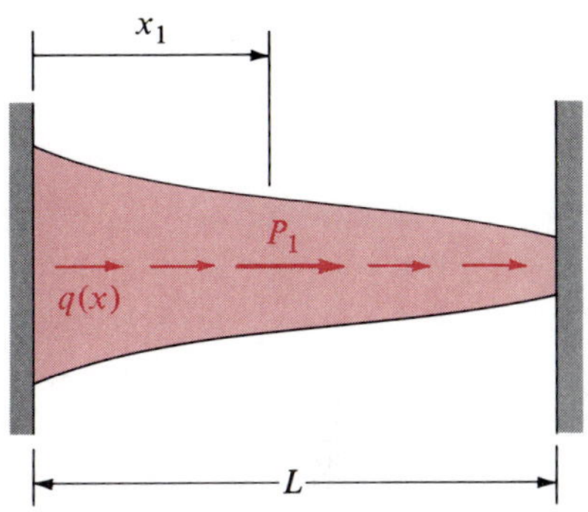

CE 5.5

CE 5.6 Use the code of CE 5.5 to solve the following problem. Take the cross section to be rectangular and the weight density to be 25 kN/m^3.

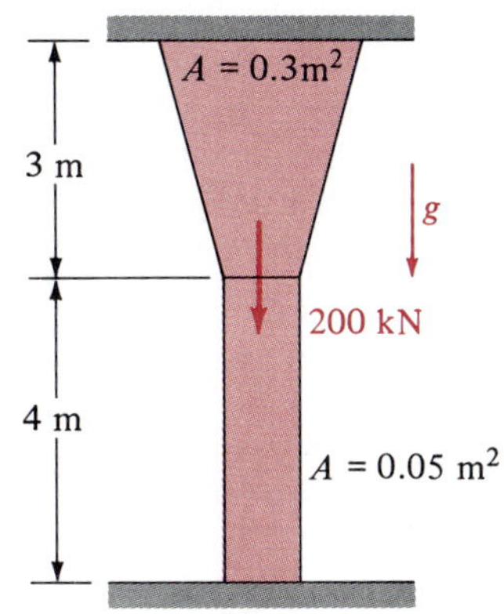

CE 5.6

CE 5.7 Repeat CE 5.1 for a material behaving according to a Ramberg-Osgood–type stress-strain relation, that is, $\varepsilon = \alpha\sigma + \beta\sigma^n$.

CE 5.8 Use the code developed in CE 5.7 to solve the problem of a bar loaded as shown. Take $P = 20{,}000$ lbf, and $\varepsilon = \sigma/10^7 + (1/650)(\sigma/40000)^9$, where σ is measured in psi.

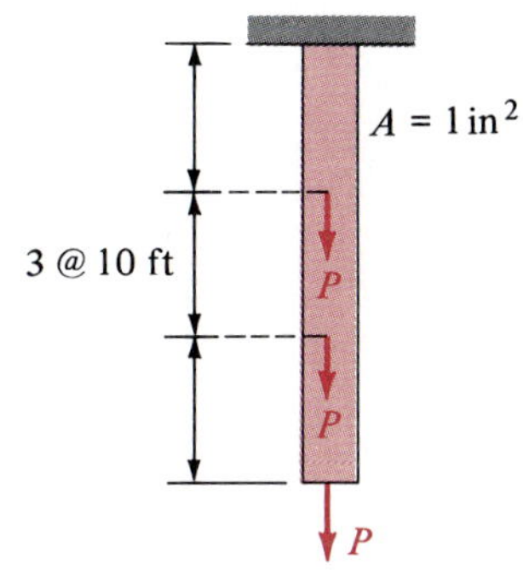

CE 5.8

ANALYSIS/DESIGN PROBLEMS

AD 5.1 A rectangular member must have a central hole of diameter 2.0 in. as shown. Design the minimum-area cross section that will transmit an axial load of 200 kips if the allowable normal stress is 18 ksi.

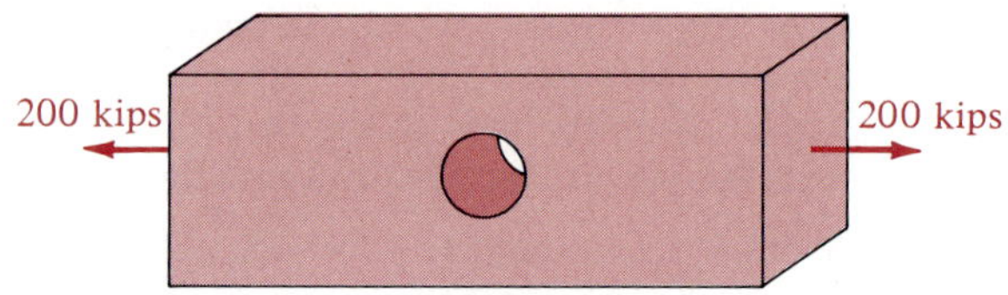

AD 5.1

AD 5.2 Design a minimum weight structure that will support the three loads that are to be applied at the points indicated. Take the yield stress for the material to be 20 ksi. Use the maximum shear-stress theory of failure with a safety factor of 2.0, and assume that the point of application of the 60 k load must not displace more than 0.1 in. Neglect stress concentrations.

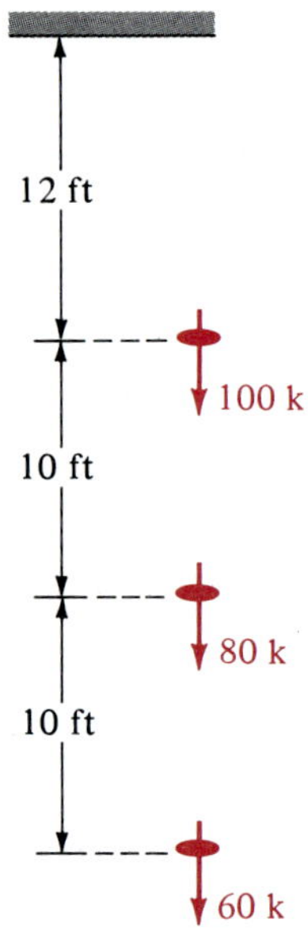

AD 5.2

AD 5.3 A pin joint has the configuration shown. If the allowable shear stress in the pin is 60 MPa and the allowable normal stress (including bearing) in the other components is 75 MPa, design the connection. Take the stress concentration factor for the tensile stress in the neighborhood of the holes in the plate to be 5.0 and the stress concentration for bearing to be 3.0.

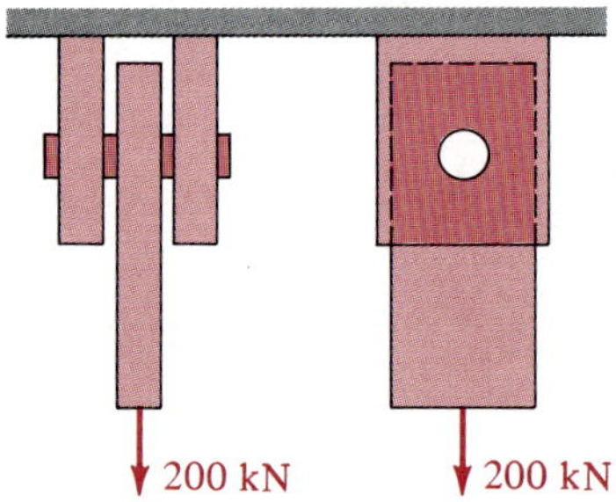

AD 5.3

AD 5.4 Design the key in the tension member so that failure will occur at a load of 8 kN. Account for the shear in the key, the bearing in the tubes, and the tension in the hollow circular tubes. Account for the slots in the tubes by assuming a stress concentration factor of 7.5 for the maximum tensile stress.

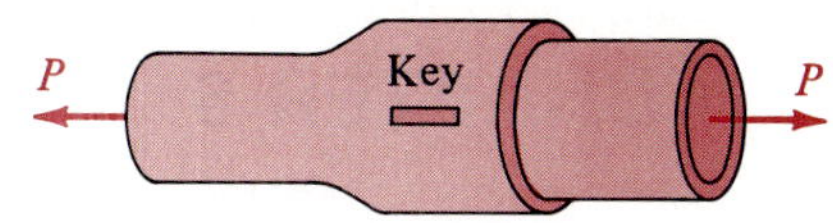

AD 5.4

AD 5.5 A reinforced concrete member 8 m in length is to transmit a tensile load of 1.4 mN. Design the member using 20-mm-diameter reinforcing rods so that the concrete does not experience tension. Take the allowable compression in the concrete to be 18 MPa and the allowable tension in the steel to be 100 MPa.

AD 5.6 A 6-kg mass is to be rotated in a circle of radius 1.2 m at a constant rate of 10^4 rpm. Design a tension member such that the stress is approximately constant at 70 MPa along the length. Include the mass of the member.

AD 5.7 A segmented shaft is subjected to 5000 lbf as shown. Design efficient members that will transmit the load P with stresses in the elastic range, and so that the load P does not displace more than 0.03 in. Assume each bar is solid and carries half the load.

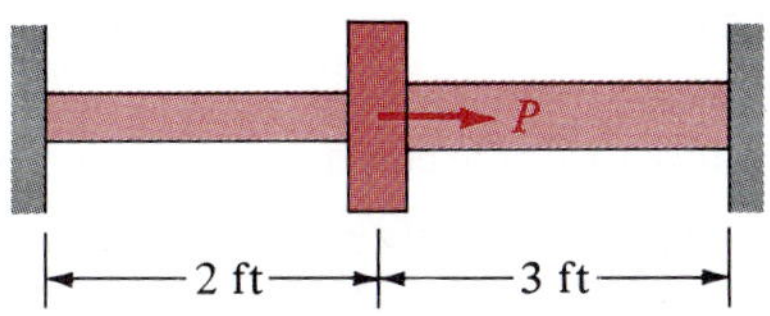

AD 5.7

AD 5.8 Repeat AD 5.7 if both shafts must have the same cross section.

AD 5.9 Repeat AD 5.7 if the condition that each bar carry half the load is replaced by the condition that the least amount of material be used.

AD 5.10 Repeat AD 5.7 if both shafts can have hollow circular cross sections with $t/R_{outer} = 0.2$.

AD 5.11 Repeat AD 5.8 if both shafts can have hollow circular cross sections with $t/R_{outer} = 0.2$.

AD 5.12 Repeat AD 5.9 if both shafts can have hollow circular cross sections with $t/R_{outer} = 0.2$.

CHAPTER 6 TORSIONAL DEFORMATIONS

CHAPTER OUTLINE

6.1 INTRODUCTION

In many machines there are rotating components between which torques must be transmitted. These torques are frequently transmitted by straight solid or straight hollow shafts with circular cross sections. Any such shaft must be "strong enough" to transmit the required torque. Additionally, because the transmission of power by shafts involves motion, care must be taken to ensure that the shaft is "stiff enough" to avoid unwanted vibrations that result from the fluctuations in the torque being transmitted. The primary purpose of this chapter is to develop and apply the theory of the deformation of straight, solid and hollow circular shafts. Several examples of transmission of torque by circular and noncircular members are presented here.

Example 6.1 One of the most common situations involving the transmission of a torque by a circular shaft is in the drive train of an automobile (Fig. 6.1).

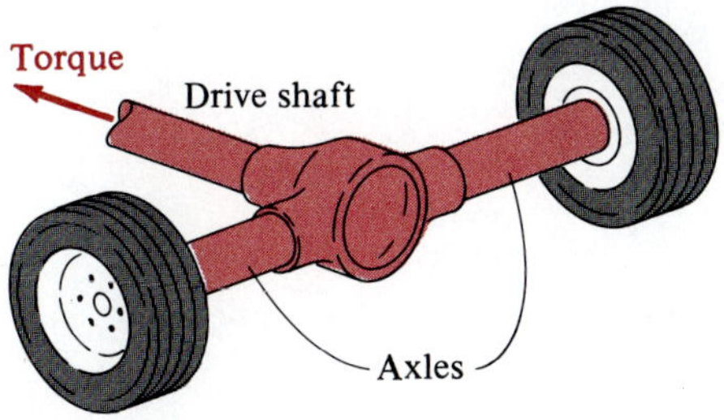

Figure 6.1 Transmission of power by the drive shaft and axles

The drive shaft as well as the axles in a rear-drive configuration are examples of circular shafts used to transmit torques. In a front-wheel-drive configuration, there are two axles to the front wheels, as shown in Fig. 6.2; each axle transmits a torque. In both of these applications, the power is transmitted from the engine to the wheels by a circular shaft or shafts.

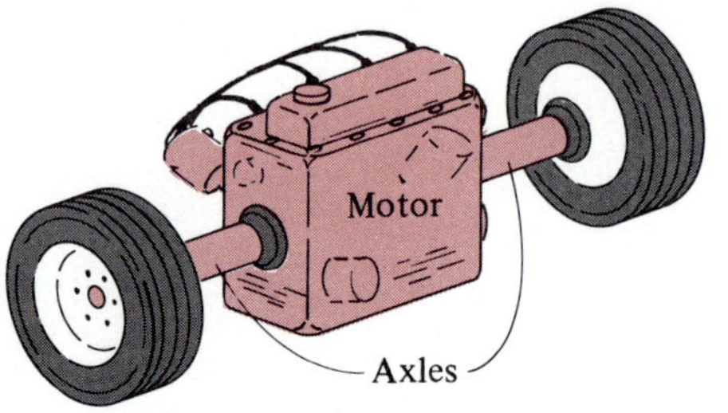

Figure 6.2 Front-wheel-drive axles

Example 6.2 The screws of any large ocean-going vessel are connected to the power plant by large-diameter circular shafts, sketched in Fig. 6.3. Although there are other force resultants being transmitted due to the weight of the shaft and the thrust, the primary purpose of the shaft is to transmit the torque.

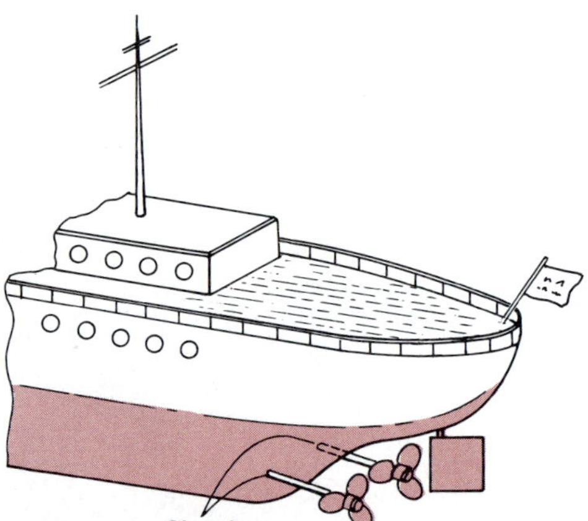

Figure 6.3 Circular shafts transmitting power to screws

In many other applications, noncircular structural components are required to transmit torques. Commonly used cross sections such as rectangular shafts are also considered in this chapter.

Example 6.3 As an example of a noncircular member transmitting torques, consider the typical traffic sign in Fig. 6.4. In the presence of a wind acting in the direction shown, there is clearly a torque that must be transmitted at a typical section pictured in Fig. 6.4b. The other internal force resultants that must be transmitted are also indicated.

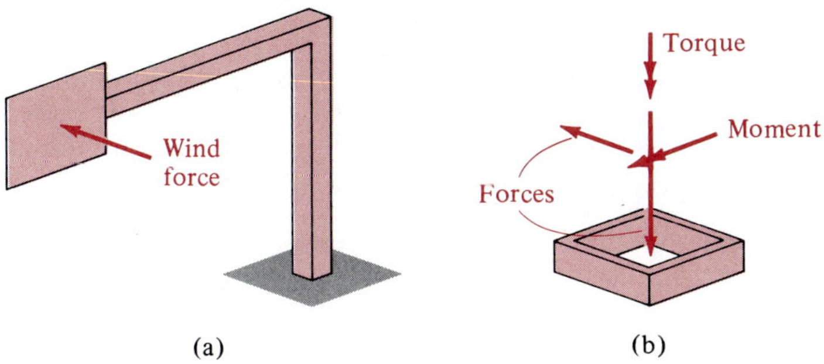

Figure 6.4 Noncircular member transmitting torques

Example 6.4 In flight, a large commercial jetliner experiences a variety of input loads. The lift, drag, weight, and thrust forces are indicated in the free-body diagram of Fig. 6.5. The swept-wing configuration and the offset of the engine combine to produce an internal torque that must be transmitted by the structure of the wing. This is a relatively complicated example of the transmission of torque and is only presented to indicate the wide-ranging areas of application for the general problem of the transmission of torques by structures.

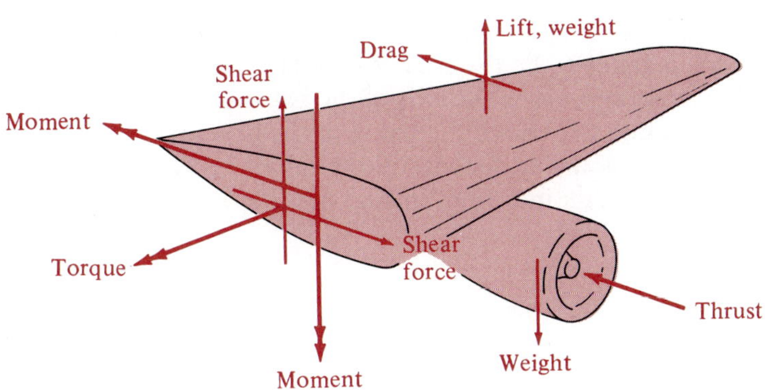

Figure 6.5 Torque transmitted by a wing

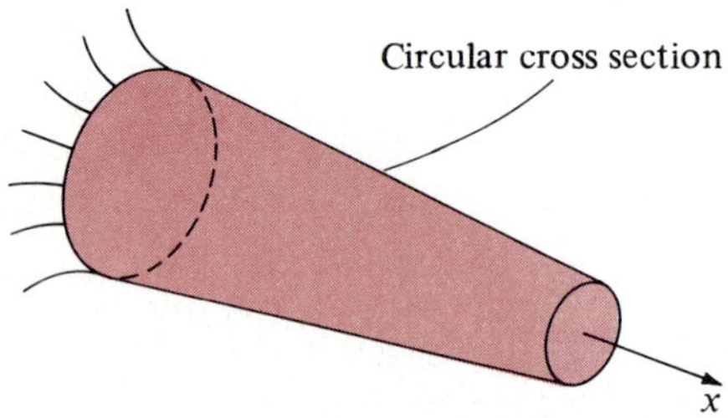

Figure 6.6 Geometry for the torsional member

In this chapter, we will consider in detail the two basic problems of strength and stiffness as they pertain to the transmission of torque(s) by straight circular and noncircular members. The three basic ideas will be discussed and combined to yield an elementary theory for circular torsional members; the theory will then be applied to a variety of practical applications. Additionally, several important exceptions to the theory will be outlined—for example, situations where there are relatively sudden changes in geometry, such as stepped shafts, are treated in detail in Section 6.5, on exceptions to the theory. The analysis of rectangular members will be presented in Section 6.6.

The geometry for which we will develop our theory is indicated in Fig. 6.6. We will refer to such a member as a circular torsional member; our assumptions as to the geometry are these:

a. A circular torsional member is straight and circular in cross section.

b. The diameter of the bar is constant or varies slowly with respect to position along the axis.

These assumptions must be regarded as an integral part of the theory: if they are not satisfied, the theory cannot be expected to be valid.

6.2 EQUILIBRIUM, DEFORMATION, AND MATERIAL BEHAVIOR

In this section, we will outline the three basic ideas of solid mechanics as they apply to the torsion of members with circular cross section. The equations associated with equilibrium, deformation, and material behavior will be developed and combined to form the theory of torsional deformation.

Equilibrium Consider the circular torsional member shown in Fig. 6.7. The x-axis is taken to be the axis of the bar. External forces are shown as surface tractions along the axis of the bar, and as a torque applied as shear stresses on the end of the bar. For the general case, it is assumed that the effects of the surface tractions can be represented as a distributed loading $t(x)$, as indicated in Fig. 6.8a.

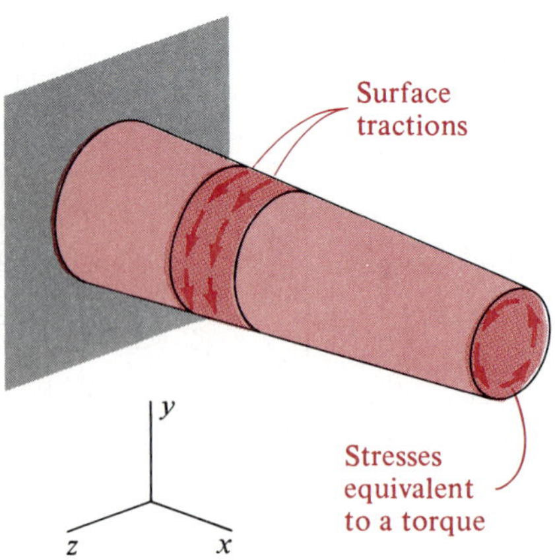

Figure 6.7 External forces on a typical torsional member

The internal force resultant, a torque or moment in this situation, is exposed by passing a section through the bar as shown in Fig. 6.8b.

It is assumed that the external forces are such that the only internal force resultant transmitted is a single torque $T(x)$ parallel to the x-axis, as indicated in Fig. 6.8b.

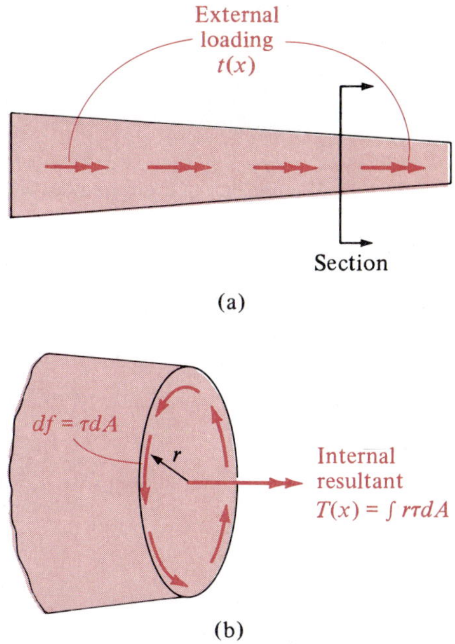

Figure 6.8 External torque loading $t(x)$ and internal resultant $T(x)$

The shear stresses, also shown in Fig. 6.8b, must be statically equivalent to the internal torque $T(x)$; that is, there is no resultant force arising from the shear stresses on the exposed face.

Assuming that the shear stresses on the exposed face are everywhere tangent to the circle, the internal torque and the shear stresses are related according to

$$T = \int_{\text{Area}} r\, df = \int_{\text{Area}} r\tau\, dA \tag{6.1}$$

In addition to the shear stresses τ acting on an exposed face, there are also shear stresses τ acting on longitudinal planes as indicated in Fig. 6.9.

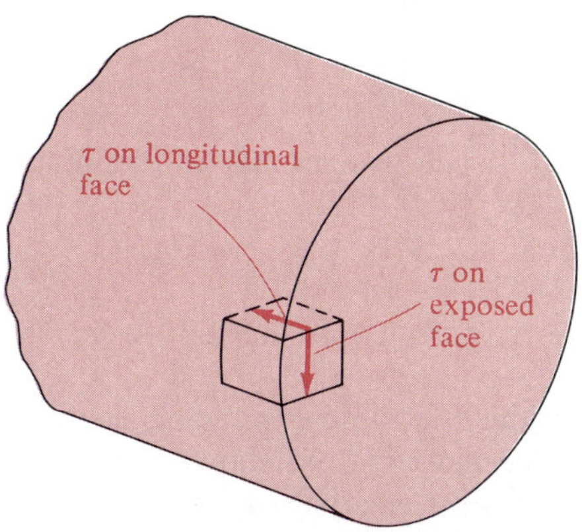

Figure 6.9 Shear stresses on longitudinal planes

This follows directly from the symmetry of the stresses as we discussed in Section 2.3. The existence of such stresses can also be appreciated on the basis of the following argument. Take a cardboard tube such as used in a package of paper towels (Fig. 6.10a). Cut a slit along the length of the tube as indicated in Fig. 6.10b. Apply a torque and notice the deformation along the slit. When there is no slit, it is the longitudinal shear stresses that act along the line corresponding to the slit to resist the torque.

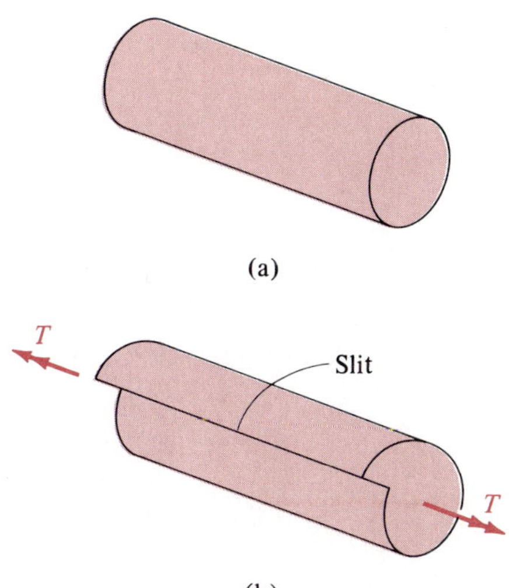

Figure 6.10 Deformations in a slit cardboard tube

The equilibrium equation that relates the internal torque $T(x)$ to the externally applied torque $t(x)$ is developed by considering a free-body

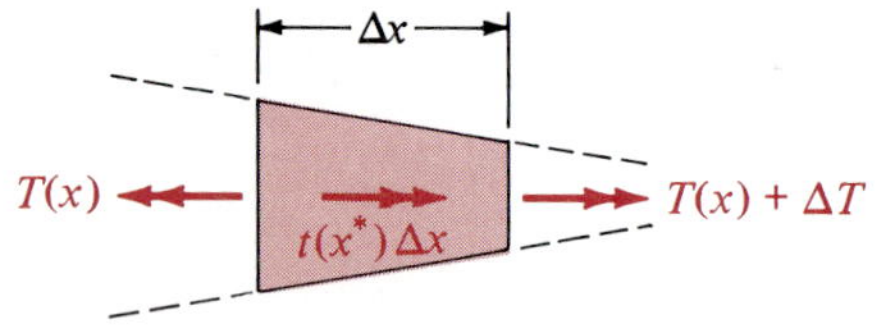

Figure 6.11 Equilibrium of a differential element

diagram of a typical differential element of the bar as in Fig. 6.11. Shown are the internal torque $T(x)$ at the location x, the resultant of the external loading $t(x^*)\Delta x$, and the internal torque $T(x) + \Delta T(x)$ at the location $x + \Delta x$. The single equation of equilibrium for the differential element is obtained by summing moments about the x-axis to obtain

$$\Sigma Tx = -T(x) + t(x^*)\,\Delta x + T(x) + \Delta T(x) = 0$$

Passing to the limit as Δx approaches zero yields

$$\frac{dT}{dx} + t(x) = 0 \tag{6.2}$$

as the equation of equilibrium relating the internal force resultant $T(x)$ to the external loading $t(x)$. Note that for dimensional homogeneity, $[t] = [T'] = F$.

Fig. 6.11 defines the positive direction for the torque T; thus, the double-headed arrows point in the positive x-direction when the normal to the face is in the positive x-direction. By Newton's third law the positive direction for the torque T on the opposite face is in the negative x-direction, also indicated in Fig. 6.11.

In summary, the two equations that arise from considerations of equilibrium are

$$\frac{dT}{dx} + t(x) = 0$$

relating the internal torque to the external loading, and

$$T(x) = \int_{\text{Area}} r\tau\, dA$$

relating T, the internal torque transmitted, to the shear stresses τ.

Deformation Consider the typical torsional member acted upon by a single torque T_0 at the end $x = L$, as shown in Fig. 6.12. Due to the fact that it is the variable most easily observed, it is customary to take the

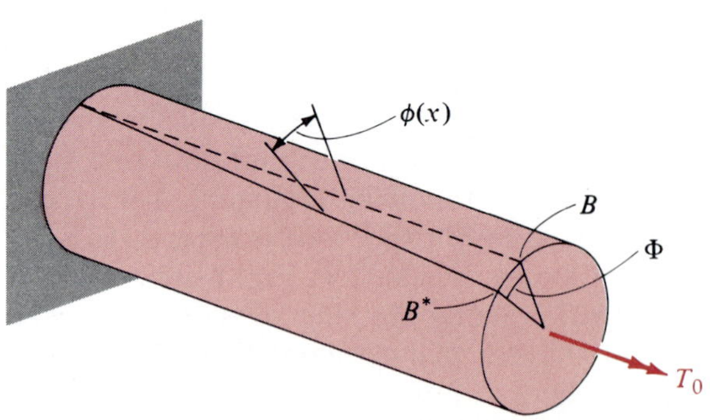

Figure 6.12 Rotation of a bar subjected to a torque T_0

angle of rotation or twist $\phi(x)$, pictured in Fig. 6.12, to be the primary deformation variable. At the end of the bar, the application of the torque T_0 results in point B deforming to point B^* through the circular arc BB^*. Φ is the corresponding angle of rotation at the point of application of the torque T_0.

To see more clearly the character of the deformation, consider a solid circular bar subjected to a torque T_0 as shown in Fig. 6.13a. The circles on the surface of the undeformed bar represent the visible intersections of the cylindrical surface with planes perpendicular to the x-axis of the bar. The longitudinal lines are all parallel to the x-axis. Fig. 6.13b shows the circles and lines after the application of the torque T_0 (i.e., after the deformation). Actual experiments show that the circles remain circles and that the lines become helices. On the basis of these observations we make the following fundamental assumption.

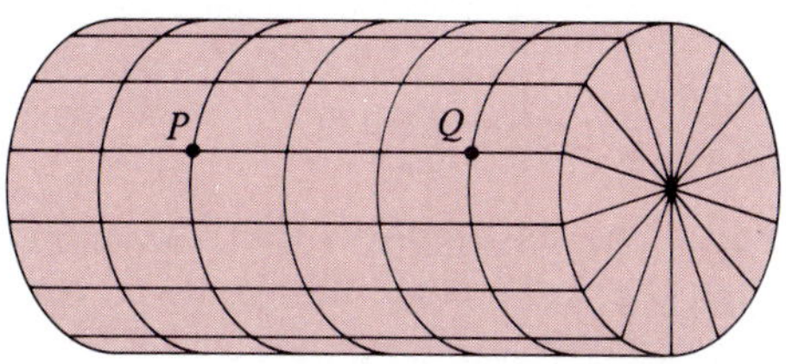

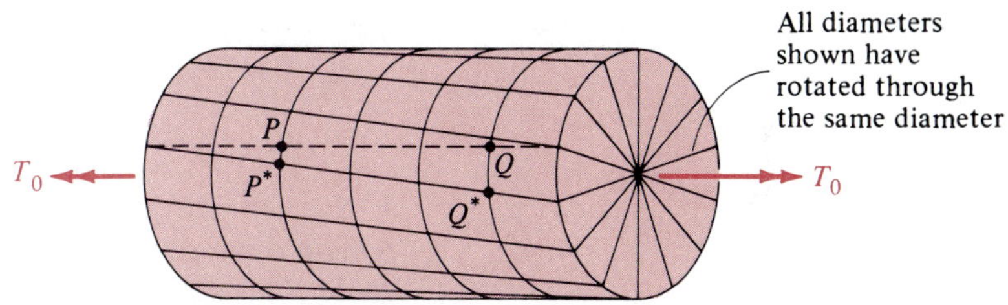

Figure 6.13 Torsional deformation of a circular bar

> Points contained in a plane section perpendicular to the bar behave as a rigid body during deformation. All diameters at a given position x remain diameters of the same length and rotate through the same local angle $\phi(x)$. As a result of the local rotation $\phi(x)$, a given circular cross section does not change in size or shape, nor does it deform in the direction of the axis of the bar.

This assumption is traditionally known as the "Plane Sections Remain Plane" assumption. On the basis of this assumption, the only deformation that occurs is associated with the *difference* in the values of the local rotation between neighboring planes. This difference is indicated in Fig. 6.13b, by the two arc lengths PP^* and QQ^*. As also seen from the figure, the deformation results in a change in the original right angles between the longitudinal and circumferential lines. This change is recognized as a shear strain that will be designated as γ.

In order to relate the rotation $\phi(x)$ to the corresponding shear strain γ, consider Fig. 6.14, where a segment of the bar between typical locations x and $x + \Delta x$ is shown. The extent of the deformations is

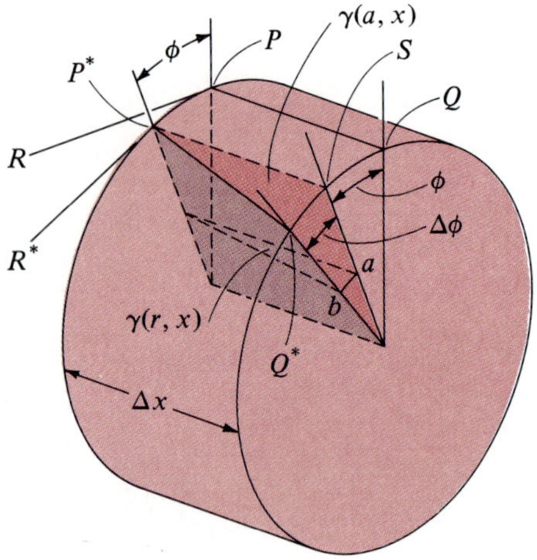

Figure 6.14 Relation between deformation and strain for torsion of a circular bar

greatly exaggerated for clarity. The strain-displacement results that we will develop will be valid only as long as the shear strain is small.

To help visualize what is taking place, imagine that there is a wire attached in a tangential manner to the bar at P and indicated by line PR in Fig. 6.14. As a result of the assumption that plane sections remain plane, PR deforms into P^*R^* lying in the same circumferential plane as PR. The shear strain is associated with the decrease in the right angle QPR. By definition $\gamma(a, x)$, the shear strain at the surface of the bar is given by

$$\gamma(a, x) = \lim_{\substack{Q \to P \\ R \to P}} (\text{angle } QPR - \text{angle } Q^*P^*R^*)$$

as indicated in Fig. 6.14. The arc length SQ^*, which subtends the angle $\Delta\phi$, can be represented in two different ways according to

$$SQ^* = \gamma(a, x)\, \Delta x = a\, \Delta\phi$$

from which

$$\gamma(a, x) = \lim_{\Delta x \to 0} \left(a \frac{\Delta\phi}{\Delta x} \right) = a \frac{d\phi}{dx} = a\phi'(x) \tag{6.3}$$

at the outside surface of the shaft. The assumption that diameters remain diameters and rotate through the same angle allows us to conclude that at an intermediate radius r, given by $0 \le r \le a$,

$$\text{arc } ab = \gamma(r, x)\, \Delta x = r\, \Delta\phi$$

from which

$$\gamma(r, x) = \lim_{\Delta x \to 0} \left(r \frac{\Delta\phi}{\Delta x} \right) = r \frac{d\phi}{dx} = r\phi'(x) \tag{6.4}$$

Eqn. (6.4) is the strain-displacement relation for the torsional deformation problem.

Eliminating $\phi'(x)$ from Eqns. (6.3) and (6.4) yields

$$\gamma(r, x) = \frac{r}{a}\, \gamma(a, x)$$

thus, the shear strain is a linear function of the radial position at any location along the bar. This is a direct result of the assumption that "plane sections remain plane." We state without proof that as a result of the assumption that plane sections remain plane, the only nonzero strain is the shear strain γ and that all the extensional strains and the other shear strains are zero.

Note that as a result of considering equilibrium and deformation, there are the three equations $T' + t = 0$, $T = \int r\tau\, dA$, and $\gamma = r\phi'$, connecting the four variables T, τ, γ, and ϕ, that is, an insufficient number of equations. Equally important, there is no way to relate the unknown force variables T and τ to the unknown displacement variables γ and ϕ. The bridge between the force variables and the displacement variables, which will depend upon the material used for the torsional member, is provided by the description of the material behavior.

Material Behavior The third basic idea—material behavior—is necessary to supply the additional equation that will enable us to complete the formulation of the theory for the torsional deformation problem. We will assume that the material from which the bar is composed is linearly elastic and isotropic, with the stress-strain relation given by

$$\tau = G\gamma \tag{6.5}$$

where G is the shear modulus discussed in Chapter 4.

Finally, then, the four equations relating the four unknowns T, τ, γ, and ϕ are

Equilibrium: $$T(x) = \int r\tau \, dA \qquad (a)$$

$$T' + t = 0 \qquad (b)$$

Deformation: $$\gamma = r\phi' \qquad (c)$$

Material behavior: $$\tau = G\gamma \qquad (d)$$

Eliminating γ from Eqns. (c) and (d) yields

$$\tau = Gr\phi' \tag{6.6}$$

showing that at a particular location along the bar, the stresses are a linear function of the radial position r, vanishing at the origin, as shown in Fig. 6.15.

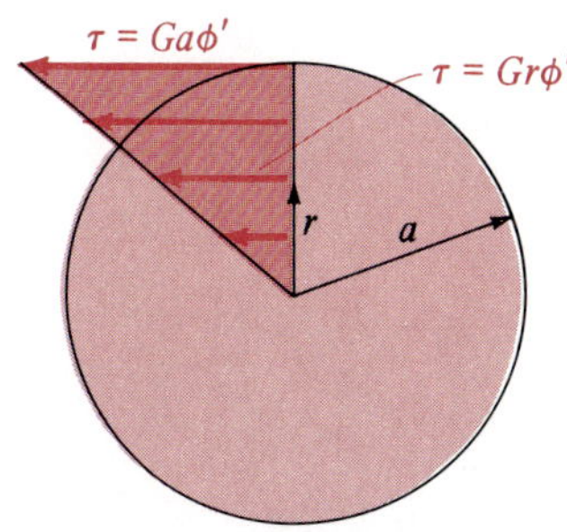

Figure 6.15 Distribution of shear stress for the torsion problem

Then substitute τ from Eqn. (6.6) into Eqn. (a) to obtain

$$T(x) = \int_{\text{Area}} r\tau \, dA = \int_{\text{Area}} r^2 G\phi' \, dA = G\phi' \int_{\text{Area}} r^2 \, dA = JG\phi' \tag{6.7}$$

$T(x) = JG\phi'$ will be referred to as the **torque-twist relation.** The integral

$$J = \int_{\text{Area}} r^2 \, dA$$

is called the **polar moment of the area** of the circular section and is evaluated as

$$J = \int_{\text{Area}} r^2 \, dA = \int_0^{2\pi} \int_0^a r^2 r \, dr \, d\theta = \frac{\pi a^4}{2}$$

For a hollow circular bar of inside radius b and outside radius a, the integral for the polar moment of the area becomes

$$J = \int_0^{2\pi} \int_b^a r^2 r \, dr \, d\theta = \frac{\pi(a^4 - b^4)}{2}$$

Note that Eqn. (6.7) now provides a connection between the force variable T and the deformation variable ϕ', namely

$$T(x) = JG\phi' \qquad \text{or} \qquad \phi' = \frac{T}{JG} \tag{6.8}$$

Once the equilibrium equation $T' + t = 0$ has been solved for T, Eqn. (6.8) can be integrated to obtain the angle of rotation ϕ.

Eliminating ϕ' between Eqns. (6.6) and (6.8) yields

$$\tau = rG\phi' = r\left(\frac{T}{J}\right) = \frac{Tr}{J} \tag{6.9}$$

which is the celebrated **elastic torsion formula.** It enables us to evaluate the stresses in terms of T, the torque transmitted, and thus to answer the strength question once the torque T is known.

The equations

$$T(x) = JG\phi' \qquad \text{and} \qquad \tau = \frac{Tr}{J}$$

are valid for either the solid circular bar, with $J = \pi a^4/2$, or the hollow circular cylinder, with $J = \pi(a^4 - b^4)/2$. For the solid bar the shear stress τ is distributed as shown in Fig. 6.16a. For the hollow bar, the stress τ is distributed as shown in Fig. 6.16b.

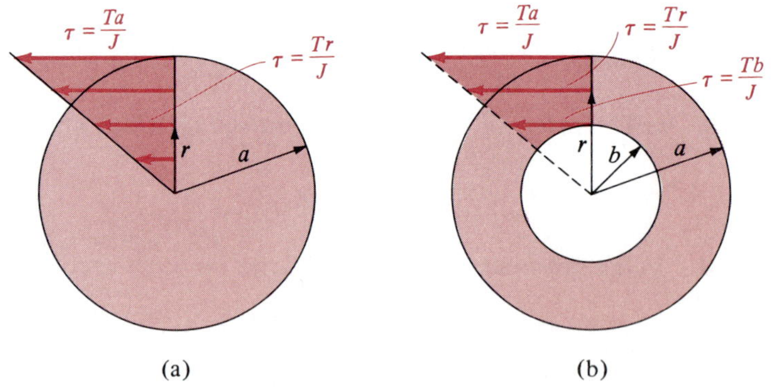

Figure 6.16 Shear-stress distribution for solid and hollow circular bars

For the solid shaft, the maximum value of the shear stress at the outer radius a can be expressed as

$$\tau_{max} = \frac{Ta}{J} = \frac{2T}{\pi a^3}$$

showing the strong dependence on the radius, in that if the radius a is doubled, τ_{max} is reduced by a factor of 8! Similar reductions occur for a hollow shaft. The details are left for the student to investigate in the problems.

If the torque-twist relation given by Eqn. (6.8) as $T(x) = JG\phi'$ is substituted into the equilibrium equation $T' + t = 0$, there results

$$(JG\phi')' + t(x) = 0 \tag{6.10}$$

as the equation of equilibrium for determining the rotation ϕ of the circular or hollow bar. Given the loading $t(x)$, Eqn. (6.10) is to be integrated subject to two boundary conditions, *one at each end.* Appropriate boundary conditions include specifying

either

a. the rotation ϕ,

or

b. the torque $T = JG\phi'$.

A type *(a)* boundary condition is imposed on the deformation variable ϕ, whereas a type *(b)* boundary condition is imposed on the force variable T. One type of boundary condition must be imposed at each end of the bar. We observe that since the equilibrium equation and boundary conditions are linear, *superposition is applicable.*

For a statically determinate problem, the equation

$$T' + t = 0$$

can be integrated once, or the appropriate free-body diagram drawn, to determine T, after which the strength question can be answered using

$$\tau = \frac{Tr}{J}$$

Material properties need not enter into a statically determinate problem unless the rotations are also desired. If the rotations are required, $\phi' = T/JG$ must also be integrated. As will be discussed in detail in Section 6.4, the material behavior and properties also come into play for statically indeterminate torsional problems.

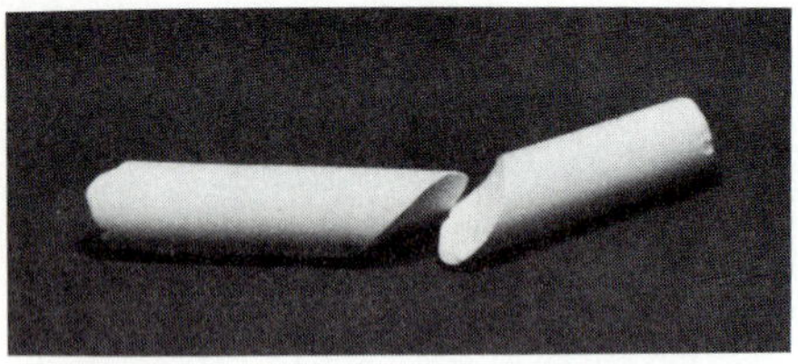

Figure 6.17 Brittle fracture of a piece of chalk. (Photo by Michael Hruby.)

The last item that needs mention before we look at applications of the theory is the state of stress associated with the torsion problem. As pointed out in Example 4.16, a brittle material, such as an ordinary piece of chalk, will fracture as shown in Fig. 6.17 when subjected to a large enough torque. On the basis of the elastic torsion formula, the shear stress τ is the only stress that acts on an element at the surface as shown in Fig. 6.18a. Fig. 6.18b shows the corresponding block for setting up Mohr's circle. The Mohr's circle for this state of stress is shown in Fig. 6.18c; it indicates that the principal stresses, equal in magnitude to the shear stress $\tau = Ta/J$, are tensile and compressive on the 45° planes shown in Fig. 6.18d.

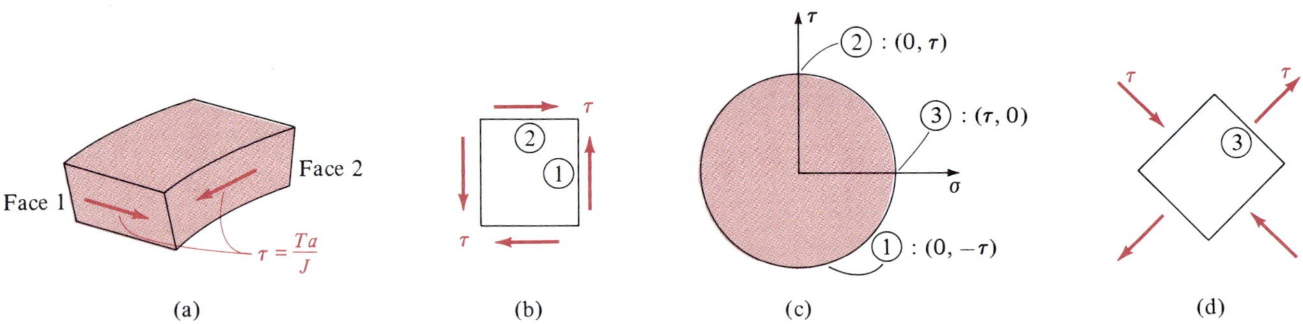

Figure 6.18 State of stress for a circular shaft in torsion

The location of the plane on which the positive principal stress occurs coincides with the failure plane shown in Fig. 6.17. This observation provides excellent correlation between experiment and theory. It also serves as a reminder that it is always important to investigate all aspects of the state of stress at a point by evaluating the principal stresses, the maximum shear stress, and the planes upon which they act.

Before applying this theory to practical problems of interest, it is important to review all the assumptions that have been used. These are:

1. *Geometry:* The bar is straight, with a solid or hollow circular cross section. For a solid bar the radius is a slowly varying function of position along the axis of the bar; also, for a hollow bar both the inner and outer radii vary slowly with position along the axis of the bar.

2. *Equilibrium:* Any external loading is such that the internal force resultant at any section along the axis of the bar is a torque or couple parallel to the axis of the bar.

3. *Deformation:* The relative deformation given by γ or ϕ' must be small.

4. *Material behavior:* The bar is composed of a linearly elastic material that is stressed at or below the proportional limit of the material.

The user of the theory is responsible for seeing that all the assumptions are satisfied. Violaticn of any of the assumptions may result in substantially inaccurate predictions by the theory.

PROBLEMS

6.1 Determine the maximum torque that a solid circular bar can transmit if the allowable shear stress is 20 ksi and the diameter is *(a)* 0.25 in., *(b)* 0.5 in., *(c)* 1.0 in., and *(d)* 2.0 in.

6.2 For an allowable shear stress of 180 MPa, determine the maximum torque that can be transmitted for a bar with a diameter of *(a)* 6 mm, *(b)* 12 mm, *(c)* 25 mm, and *(d)* 50 mm.

6.3 Determine the maximum torque that a hollow circular bar can transmit if the allowable shear stress is 22 ksi and the outer diameter is *(a)* 0.5 in., *(b)* 1.0 in., *(c)* 2.0 in., and *(d)* 4.0 in. Take the inner diameter to be 75% of the outer diameter.

6.4 For an allowable shear stress of 160 MPa, determine the maximum torque that can be transmitted for a hollow circular bar with an outer diameter of *(a)* 10 mm, *(b)* 20 mm, *(c)* 40 mm, and *(d)* 80 mm. Take the inner diameter to be 80% of the outer diameter.

6.5 A frustum of a cone is loaded by shear stresses $\tau = Tr/J$ applied as shown to the flat face at $x = L$. Assuming that the stresses at $x = 0$ are also distributed according to the theory, what is the maximum shear stress at $x = 0$?

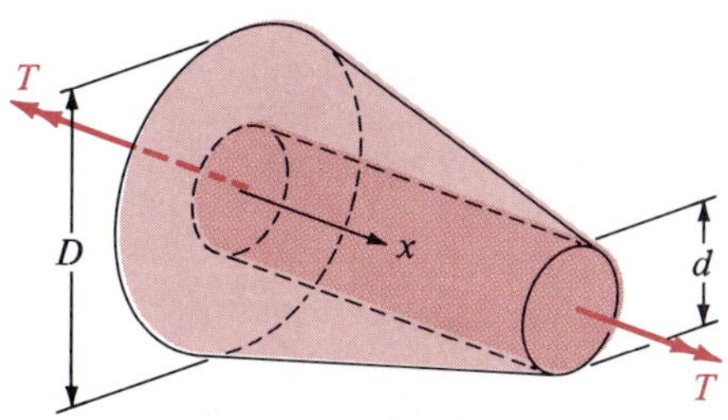

Ps. 6.5 and 6.6

6.6 For the conical frustum shown, draw a free-body diagram of the inner cylindrical cone indicated and discuss what stresses and stress resultants must be present for equilibrium. Under what conditions could it be argued that these stresses are small?

6.7 When tractions are applied as shown in the figure, which stress component is involved at the boundary? Why is this not in accordance with the theory?

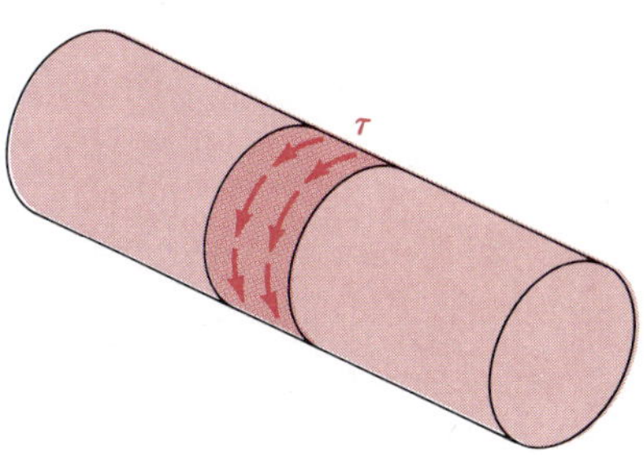

P. 6.7

6.8 Show that if the only stress acting on the face of a circular bar is $\tau_{x\theta} = \tau = Tr/J$ as shown, the force components X, Y, and Z as well as the moments M_y and M_z are zero, that is, the only force resultant transmitted is the torque T.

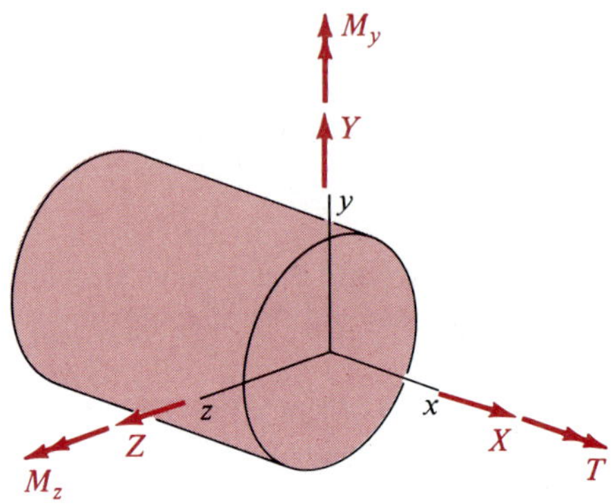

P. 6.8

6.9 Integrate $T' + t = 0$ from the location $x = x_1$ to location $x = x_2$ along the bar and interpret the results using a free-body diagram of the same portion of the bar. Comment specifically on the case $t = 0$.

6.10 Integrate $T' + t = 0$ from the location $x = 0$ to location $x = L$ (i.e., the length of the bar), and interpret the results using a free-body diagram of the entire bar. Comment specifically on the case $t = 0$.

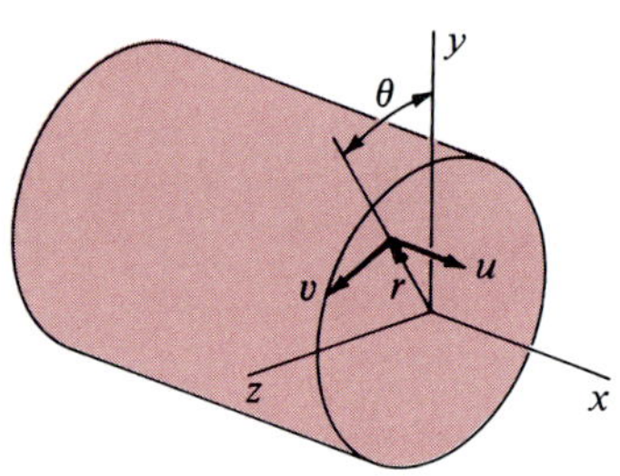

P. 6.11

6.11 The shear strain γ corresponding to the shear stress τ is given by

$$\gamma = \frac{1}{r}\frac{\partial u}{\partial \theta} + \frac{\partial v}{\partial x}$$

where u and v are the displacements in the x and θ directions, respectively, and r is the radius, as shown in the figure. Assuming $u \equiv 0$ (i.e., no axial displacement) and using the strain-displacement relation given by Eqn. (6.4), determine $v(r, x)$.

6.3 CLASSICAL SOLUTIONS

As discussed in Section 1.3, the two basic problems of solid mechanics involve the strength and stiffness of the structure in question. For circular shafts, the equation for evaluating strength was developed in Section 6.2 as

$$\tau = \frac{Tr}{J}$$

and the equation for evaluating stiffness as

$$\phi' = \frac{T}{JG}$$

In this section these basic equations will be applied to a variety of classical problems. In addition, the technically important problem of the transmission of power by shafts will be treated. In all cases the student should always bear in mind the assumptions that must be satisfied in order that the predictions of the elementary theory be valid.

Example 6.5 A uniform circular bar fixed against rotation at one end transmits a torque T_0 as shown in Fig. 6.19a. Determine *(a)* the torque $T(x)$, *(b)* the rotation $\phi(x)$, *(c)* the maximum torsional stress, and *(d)* plot the T versus x and ϕ versus x diagrams. Note the practice of placing the appropriate boundary condition in a box at the boundaries $x = 0$ and $x = L$.

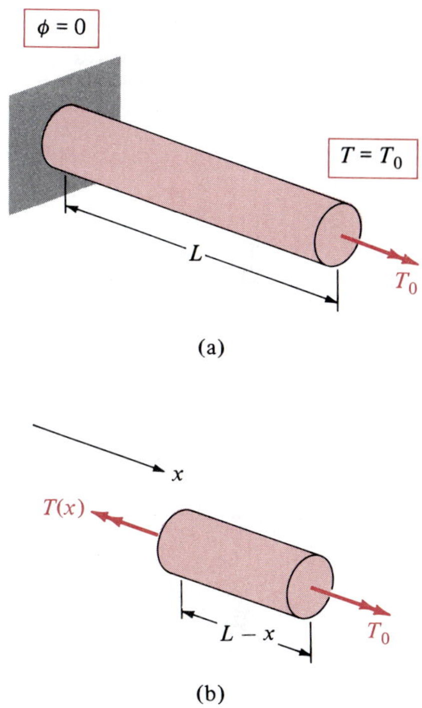

Figure 6.19 Uniform bar subjected to a torque

Solution By drawing a free-body diagram as shown in Fig. 6.19b, and requiring equilibrium, it is seen that the torque transmitted is T_0 everywhere along the bar. For this statically determinate problem, the maximum shear stress at the outside radius can be computed as

$$\tau = \frac{T_0 a}{J} = \frac{2T_0}{\pi a^3}$$

and can be used to answer questions of strength. Note, as was indicated in Section 6.2, that by doubling the radius a, the maximum shear stress can be reduced by a factor of 8!

The rotation $\phi(x)$ can then be determined by integrating the torque-twist equation

$$\phi(x)' = \frac{T_0}{JG}$$

to obtain

$$\phi(x) = \frac{T_0 x}{JG} + C_1$$

Since the bar is fixed against rotation at $x = 0$, $\phi(0) = 0$ leading to

$$\phi(0) = \frac{T_0(0)}{JG} + C_1$$

from which $C_1 = 0$. Eqn. *(a)* then states that

$$\phi(x) = T_0 x / JG$$

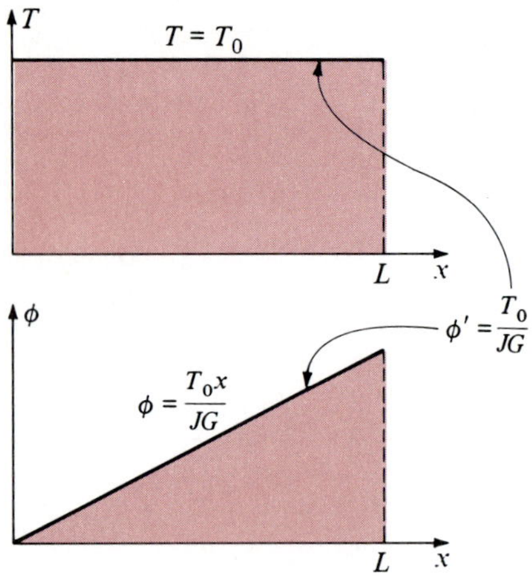

Figure 6.20 T versus x and ϕ versus x diagrams

which is plotted as a function of x in Fig. 6.20. The diagrams indicating the torque and angle of twist as functions of position along the bar are shown in Fig. 6.20. Note that for all points along the bar, the slope of the twist diagram is proportional to the value of the corresponding torque diagram. This idea can be useful in relating the shapes of the T and ϕ diagrams.

The value of the rotation at the right end of the bar is given by

$$\phi(L) = \frac{T_0 L}{JG} = \Phi \tag{b}$$

frequently rewritten as

$$T_0 = \frac{JG\Phi}{L} = C\Phi \tag{c}$$

where $C = JG/L$ is a torsional spring constant. With $J = \pi a^4/2$, it is seen from Eqn. *(c)* that if the radius of a shaft is doubled, everything else being equal, the rotation is reduced by a factor of 16!

The relationship stated in Eqn. *(c)* is a fundamental result for the torsional deformation problem, stating that there is a linear relationship between the torque and the angle of twist for a uniform bar, thus that the bar acts as a linear torsional spring with spring constant $C = JG/L$. This relationship is shown in Fig. 6.21.

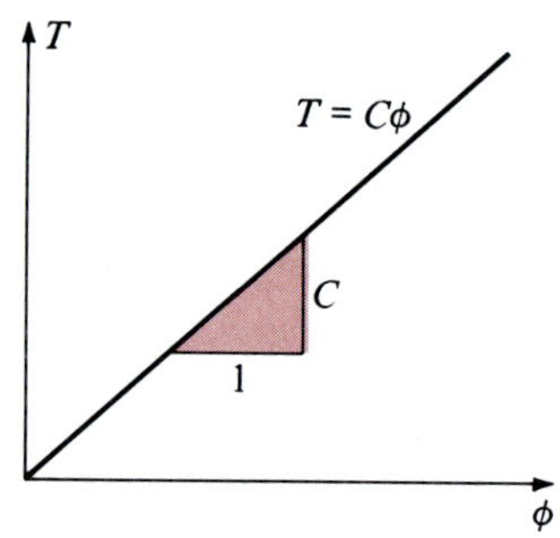

Figure 6.21 Relationship between the torque T and the twist ϕ for a linear torsional spring

The other approach to determining the torques and rotations for this example problem is to begin with the equilibrium equation and two boundary conditions (*one at each end*) stated as

$$(JG\phi')' + t = 0 \tag{d}$$

$$\phi(0) = 0 \tag{e}$$

$$T(L) = JG\phi'(L) = T_0 \tag{f}$$

There are no distributed torques along the bar so that the equilibrium equation (d) integrates to

$$JG\phi' = C_1$$

Satisfying the second boundary condition given by Eqn. (f) yields

$$JG\phi'(L) = C_1 = T_0$$

Eliminating C_1 yields

$$JG\phi' = T_0 = T$$

exactly as obtained by drawing the free-body diagram to satisfy equilibrium. The determination of the rotation ϕ is carried out in precisely the same fashion as in the previous approach given in this example.

For a better understanding of the magnitudes of the stresses and twists in relation to the size of the bar for a typical torsion problem, consider the specific case shown in Fig. 6.22.

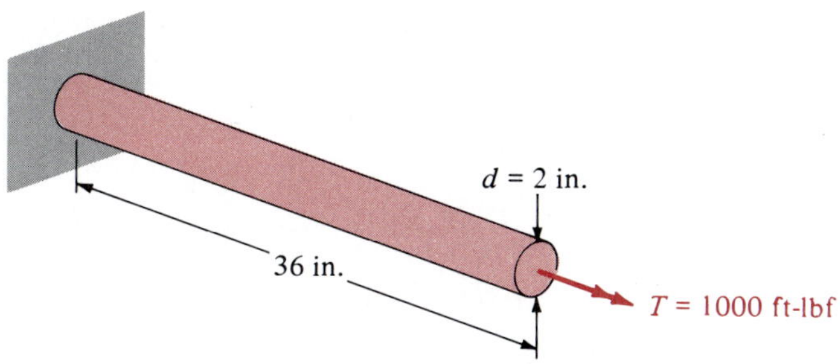

Figure 6.22 Aluminum bar subjected to a torque

The polar moment is evaluated to be $J = \pi a^4/2 = 1.57\ \text{in}^4$. The stress at the outside radius is

$$\tau = \frac{Tr}{J} = \frac{(12{,}000\ \text{in.-lbf})(1\ \text{in.})}{1.57\ \text{in}^4} = 7.64\ \text{ksi}$$

The rotation at $x = L$ is given by

$$\Phi = \frac{TL}{JG} = \frac{(12{,}000\ \text{in.-lbf})(36\ \text{in.})}{(1.57\ \text{in}^4)(10^7\ \text{psi})} = 0.0275\ \text{rad} = 1.58°$$

Note that if the radius of the bar is changed to 0.5 in.,

$$\tau = \frac{(12{,}000\ \text{in.-lbf})(0.5\ \text{in.})}{(\pi\,\text{in}^4/32)} = 61.1\ \text{ksi} = 8(7.64\ \text{ksi})$$

and

$$\Phi = \frac{(12{,}000 \text{ in.-lbf})(36 \text{ in.})}{(\pi \text{in}^4/32)(10^7 \text{ psi})} = 25.2° = 16(1.58°)$$

indicating the a^{-3} dependence of the stresses and the a^{-4} dependence of the rotation, on the radius, respectively.

Example 6.6 If the bar of the previous example is to be replaced with a hollow bar with a 0.25-in. wall thickness and with the same amount of material being used, determine the stresses and maximum rotation.

Solution The area of the hollow section can be expressed as

$$A = \pi(r^2 - (r - 0.25)^2) \text{ in}^2$$

which must be equal to $\pi 1^2$ in^2. Solving for r yields $r = 2.125$ in. and $r - t = 1.875$ in. The corresponding J is

$$J = (\pi(2.125^4 - 1.875^4)/2) \text{ in}^4 = 12.62 \text{ in}^4$$

The maximum shear stress is

$$\tau = \frac{Tr}{J} = \frac{(12{,}000 \text{ in.-lbf})(2.125 \text{ in.})}{12.62 \text{ in}^4} = 2.02 \text{ ksi}$$

whereas the end rotation is given by

$$\phi(1) = \frac{(12{,}000 \text{ in.-lbf})(36 \text{ in.})}{(12.62 \text{ in}^4)(10^7 \text{ psi})} = 0.00342 \text{ rad} = 0.196°$$

The numbers associated with this example and the previous example illustrate a very important practical consideration: both the stress and the rotation are strong functions of the outer radius of the shaft. If it is possible to increase the radius of a solid shaft or to use an equivalent hollow shaft, the benefits, in terms of the increased strength and stiffness, can be substantial.

Example 6.7 A solid circular bar is acted upon by a uniformly distributed torque t_0 as shown in Fig. 6.23. Determine $T(x)$ and $\phi(x)$.

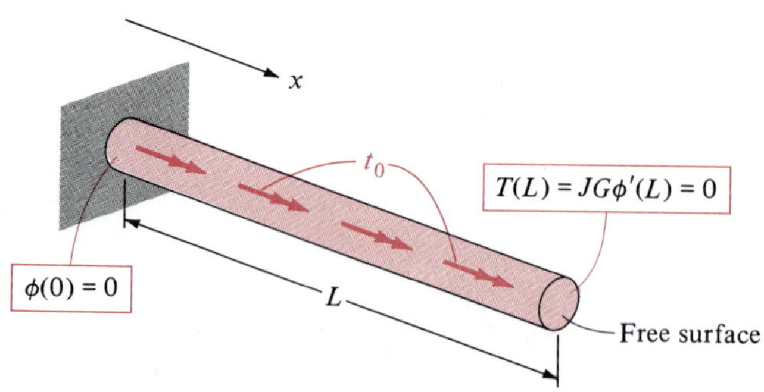

Figure 6.23 Distributed torque loading t_0

Solution Method 1 Draw a free-body diagram of a portion of the bar as shown in Fig. 6.24.

Writing the equilibrium equations yields

$$\Sigma Tx = -T(x) + t_0(L - x) = 0$$

or

$$T(x) = t_0(L - x)$$

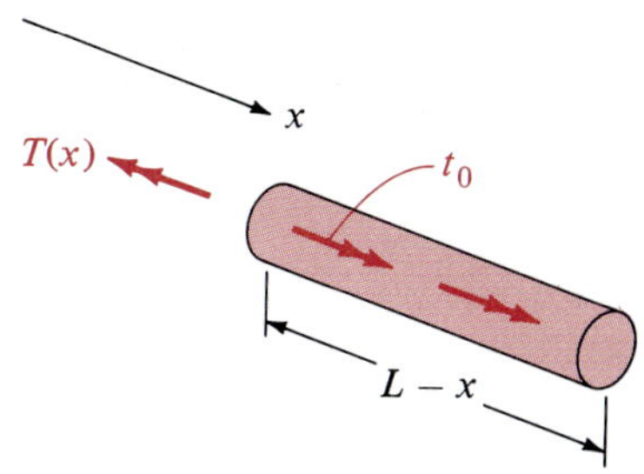

Figure 6.24 Free-body diagram of a portion of the uniformly load bar

The maximum internal torque occurs at $x = 0$ and has the value $T(0) = t_0 L$, which is the total applied external torque. The corresponding maximum stress at $x = 0$ would be $\tau = Tr/J = 2t_0 L/\pi a^3$, where a is the radius of the bar.

Integrating the torque-twist relation $T = JG\phi'$ yields

$$JG\phi = -\frac{t_0(L - x)^2}{2} + C_2 \qquad (f)$$

Satisfying the boundary condition $\phi(0) = 0$ on the rotation at $x = 0$ yields

$$JG\phi(0) = 0 = -\frac{t_0 L^2}{2} + C_2$$

from which $C_2 = t_0 L^2/2$. The rotation ϕ is then given from Eqn. (f) by

$$\phi(x) = \frac{t_0(2Lx - x^2)}{2JG}$$

with the maximum value of

$$\phi(L) = \frac{t_0 L^2}{2JG}$$

The results for $T(x)$ and $\phi(x)$ are plotted in Fig. 6.25. Note that the shapes of the two curves are related by $T = JG\phi'$, that is, the slope of the ϕ diagram is proportional to the value of the corresponding T diagram.

Solution Method 2 With $t(x) = t_0$, the equilibrium equation and boundary conditions can be stated as

$$(JG\phi')' + t_0 = 0 \qquad (a)$$

$$\phi(0) = 0 \qquad (b)$$

$$T(L) = JG\phi'(L) = 0 \qquad (c)$$

Integrating Eqn. (a) yields

$$JG\phi' + t_0 x = C_1$$

Evaluating this equation at $x = L$ and using Eqn. (c) yields

$$JG\phi'(L) + t_0 L = C_1 \qquad \text{or} \qquad 0 + t_0 L = C_1$$

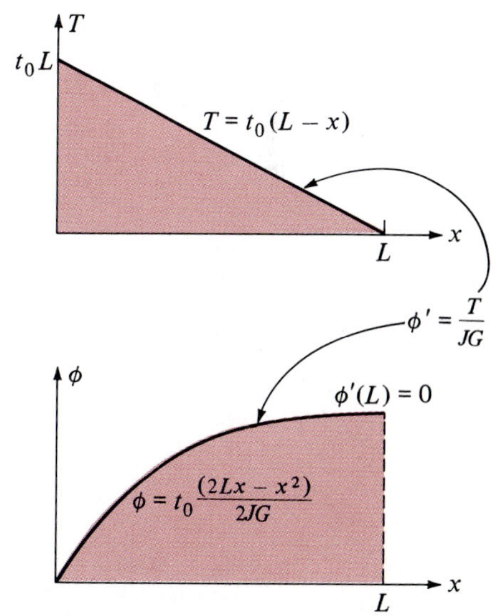

Figure 6.25 T versus x and ϕ versus x for the uniformly distributed torque

leading to

$$JG\phi' = t_0(L - x) \tag{d}$$

Note that since $JG\phi' = T$, this equation can be restated as

$$T(x) = t_0(L - x) \tag{e}$$

the same as in Solution Method 1. The rest of the solution is carried out exactly as in Solution Method 1, showing the two procedures to be equivalent. The student should endeavor, wherever possible, to carry through and be comfortable with both approaches.

Example 6.8 The gears of a shaft are acted upon by the torques shown in Fig. 6.26. Draw the T versus x and ϕ versus x diagrams and thus determine the magnitude and location of the maximum shear stress and of the maximum angle of twist.

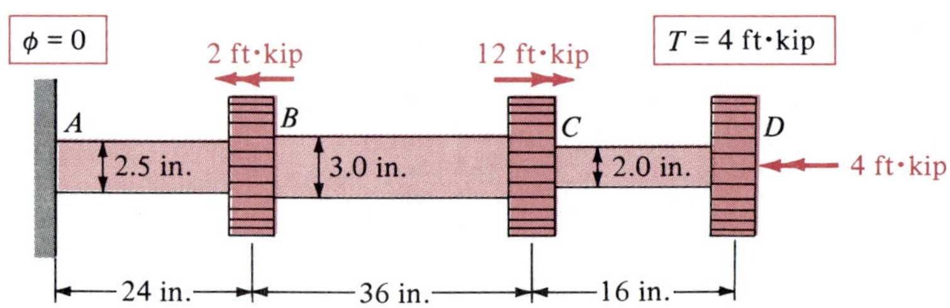

Figure 6.26 System of shafts and gears

Solution The first step is to draw the appropriate free-body diagrams and determine the torques in each of the segments. The free-body diagrams taken through segments CD, BC, and AB are shown in Figs. 6.27a, b, and c.

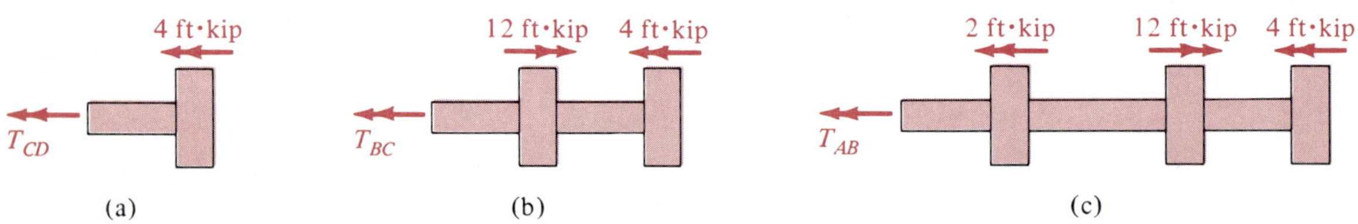

Figure 6.27 Free-body diagrams for segments CD, BC, and AB

Requiring equilibrium for each of the free-body diagrams yields:

Shaft CD: $$\Sigma T = -T_{CD} - 4000 \text{ ft-lbf} = 0$$

or $$T_{CD} = -4000 \text{ ft-lbf}$$

Shaft BC: $$\Sigma T = -T_{BC} + 12{,}000 \text{ ft-lbf} - 4000 \text{ ft-lbf} = 0$$

or $$T_{BC} = 8000 \text{ ft-lbf}$$

Shaft AB: $$\Sigma T = -T_{AB} - 2000 \text{ ft-lbf} + 12{,}000 \text{ ft-lbf} - 4000 \text{ ft-lbf} = 0$$

or $$T_{AB} = 6000 \text{ ft-lbf}$$

These are shown in the torque diagram of Fig. 6.28 where T is plotted as a function of the position x along the shaft. Note that the difference in the torque in two segments is exactly equal to the torque applied to the disk joining the two segments.

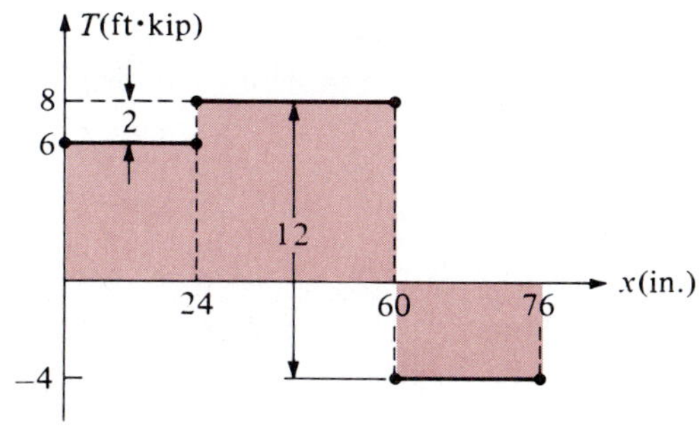

Figure 6.28 T versus x diagram

With the internal torques all determined from equilibrium, we can compute the stresses in each of the shafts and hence answer the strength question.

Shaft CD: $$\tau_{max} = \frac{2(-4000 \text{ ft-lbf})(12 \text{ in./ft})}{\pi(1 \text{ in.})^3} = -30.6 \text{ ksi}$$

Shaft BC: $$\tau_{max} = \frac{2(8000 \text{ ft-lbf})(12 \text{ in./ft})}{\pi(1.5 \text{ in.})^3} = 18.1 \text{ ksi}$$

Shaft AB: $$\tau_{max} = \frac{2(6000 \text{ ft-lbf})(12 \text{ in./ft})}{\pi(1.25 \text{ in.})^3} = 23.5 \text{ ksi}$$

The negative value for the maximum shear stress in shaft CD simply means that the stresses act in the direction shown in Fig. 6.29a, rather than in the opposite direction for shaft AB as shown in Fig. 6.29b. The positive shear stress corresponds to the positive torque defined in Fig. 6.11.

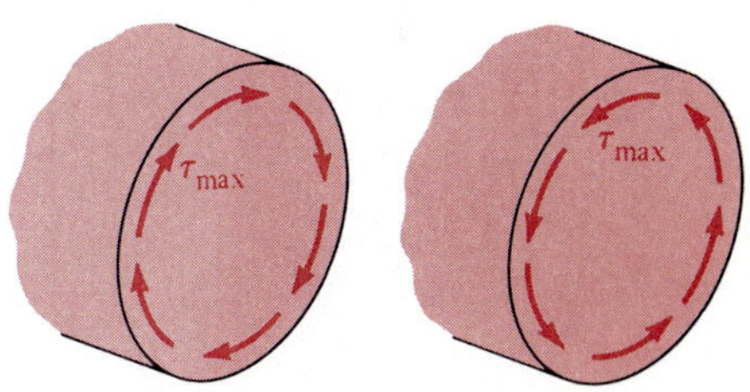

Figure 6.29 Direction of shear stresses in shafts AB and CD

To determine the rotations we proceed as follows. The basic result discussed in Example 6.5 is that for uniform shaft of length L transmitting a single torque T, the relative angle of rotation between the ends is given by

$$\phi = \frac{T}{C}$$

where $C = JG/L$ is the torsional stiffness. As seen from Fig. 6.28 the torque in each of the segments is constant so that we can write

Shaft *AB:*

$$\phi(B) - \phi(A) = \phi(B) = \frac{T_{AB}}{C_{AB}}$$

$$= \frac{(72{,}000 \text{ in.-lbf})(24 \text{ in.})}{(\pi/2)(1.25 \text{ in.})^4(30 \times 10^6 \text{ psi})} = 0.015 \text{ rad} = 0.86°$$

Shaft *BC:*

$$\phi(C) - \phi(B) = \frac{T_{BC}}{C_{BC}}$$

$$= \frac{(96{,}000 \text{ in.-lbf})(36 \text{ in.})}{(\pi/2)(1.5 \text{ in.})^4(30 \times 10^6 \text{ psi})} = 0.014 \text{ rad} = 0.83°$$

so that

$$\phi(C) = \phi(B) + 0.83° = 1.69°$$

Shaft *CD:*

$$\phi(D) - \phi(C) = \frac{T_{CD}}{C_{CD}}$$

$$= \frac{(-48{,}000 \text{ in.-lbf})(16 \text{ in.})}{(\pi/2)(1.0 \text{ in.})^4(30 \times 10^6 \text{ psi})} = -0.016 \text{ rad} = -0.93°$$

from which

$$\phi(D) = \phi(C) - 0.93° = 0.76°$$

Since T is a constant in each segment, $\phi' = T/JG$ is a constant with ϕ a linear function of x. The ϕ versus x diagram, with the T versus x diagram repeated, is shown in Fig. 6.30.

Note carefully that the T versus x and ϕ versus x curves are related by $\phi' = T/JG$; thus, the slope of the ϕ versus x curve is proportional to the value of the corresponding T versus x curve. Once the T versus x curve has been drawn, this idea can be very helpful in deciding on the shape of the corresponding ϕ versus x curve.

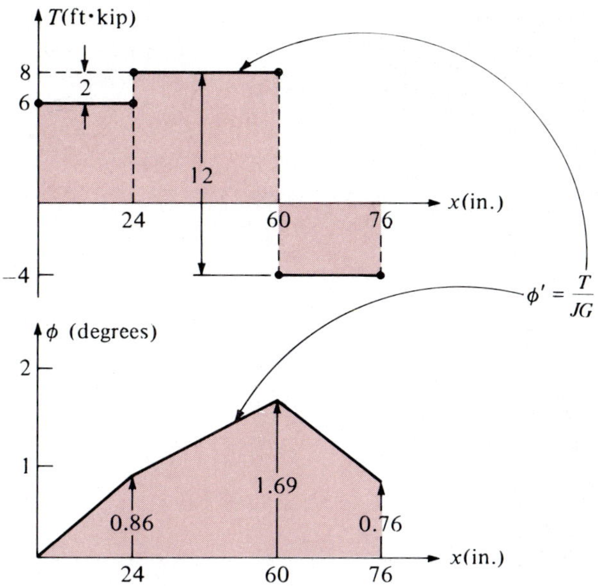

Figure 6.30 T versus x and ϕ versus x diagrams

PROBLEMS

6.12–6.18 For each of the following, determine the internal torque transmitted and the angle of rotation at all points along the bar. Plot the results on T versus x and ϕ versus x diagrams. For problems 6.15–6.17, determine the maximum shear stress in each of the segments.

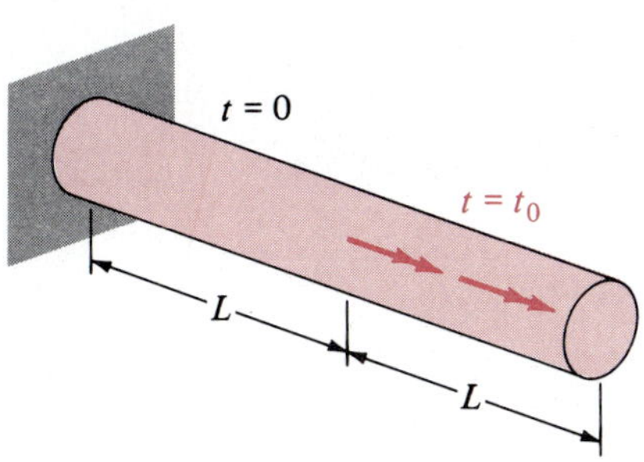

P. 6.12

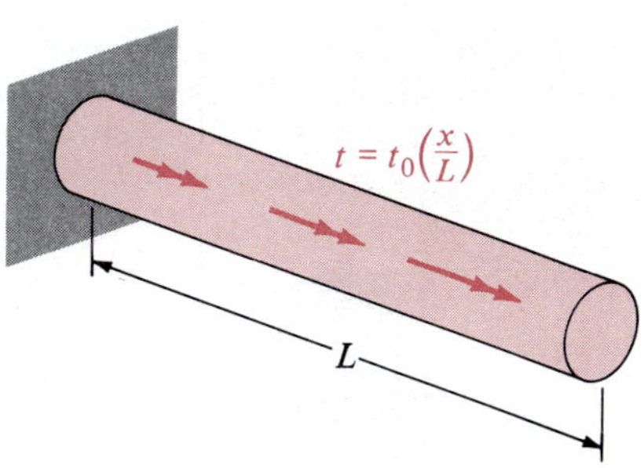

P. 6.13

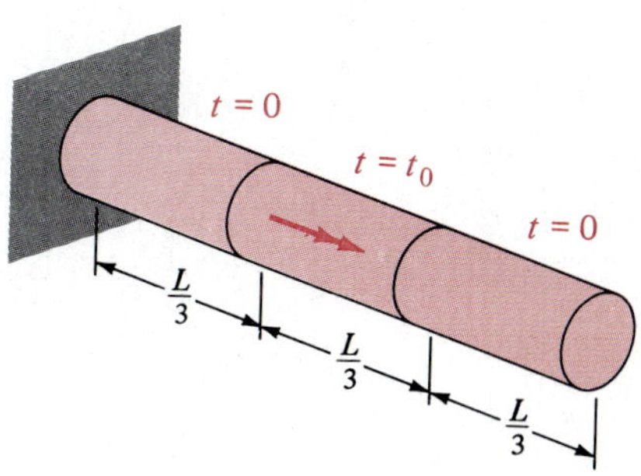

P. 6.14

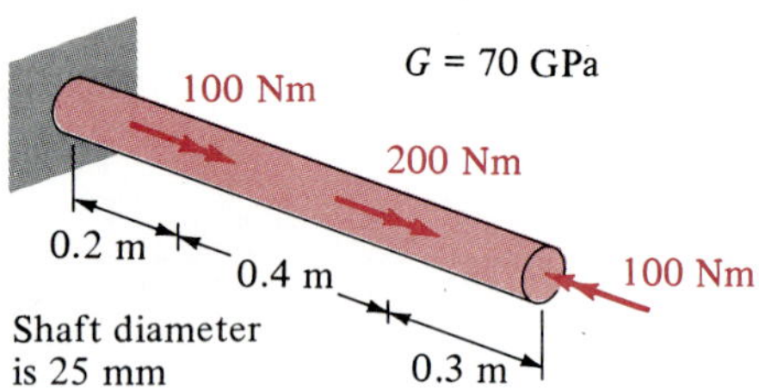

P. 6.15

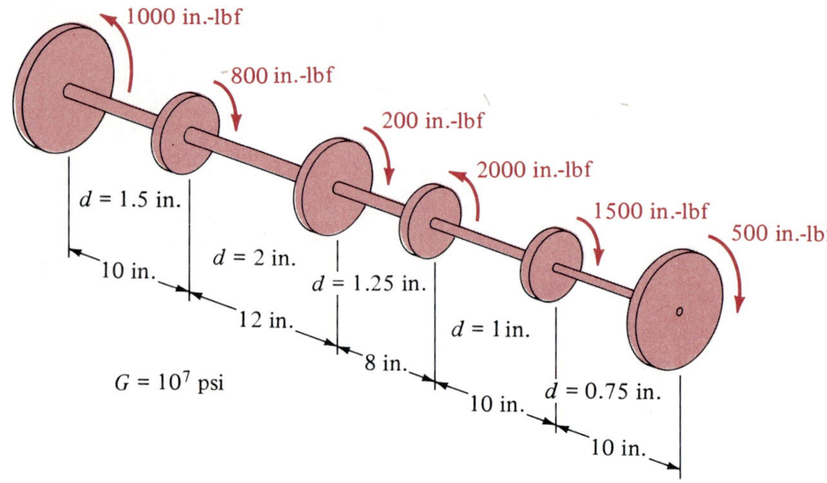

P. 6.16

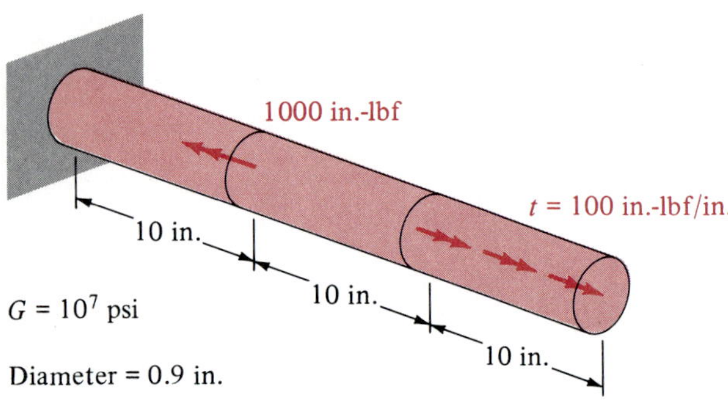

P. 6.17

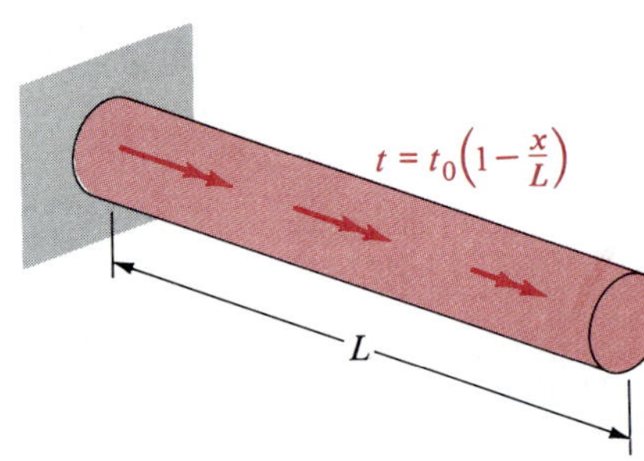

P. 6.18

6.19 The motor shown supplies a torque of 60 kN·m. The torques that are taken off through the gears are shown in the figure. Determine the torques transmitted in each of the sections and plot the internal torque diagram. If the allowable shear stress is 90 MPa, what uniform shaft diameter is required?

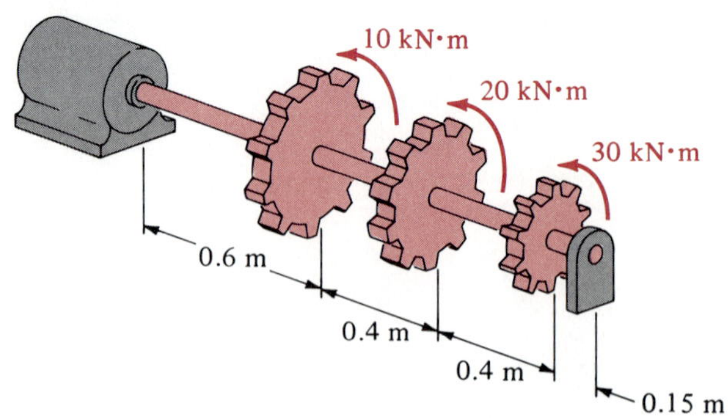

P. 6.19

6.20 Bars AD and EG are connected through the gears F and C. Plot the internal torque in bars AD and EG. Assume frictionless bearings at B, D, and E. What diameters are required for shafts EG and AD if the allowable shear stress is 120 MPa?

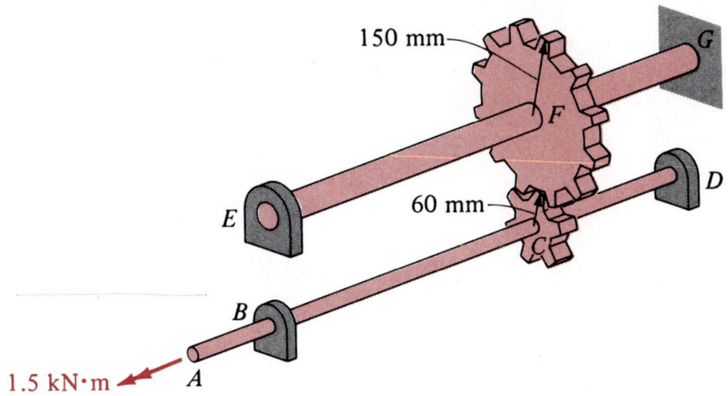

P. 6.20

6.21 The solid shaft shown is composed of a brittle material with $\sigma_{ult} = 40$ ksi. If the diameter is 1 in., what is the maximum torque T the shaft can transmit if the factor of safety against fracture is to be 2.0?

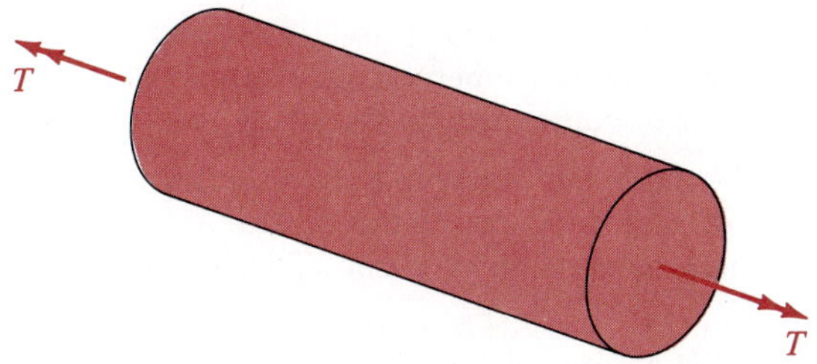

Ps. 6.21 and 6.22

6.22 The solid shaft shown is composed of a brittle material with $\sigma_{ult} =$ 300 MPa. What is the maximum torque T the shaft can transmit if the factor of safety against fracture is to be 1.8? Take the diameter to be 30 mm.

6.23 The hollow shaft shown is composed of a brittle material with σ_{ult} = 36 ksi. What is the maximum torque T the shaft can transmit if the factor of safety against fracture is to be 2.4? Take d_0 = 1.5 in. and d_i = 1.25 in.

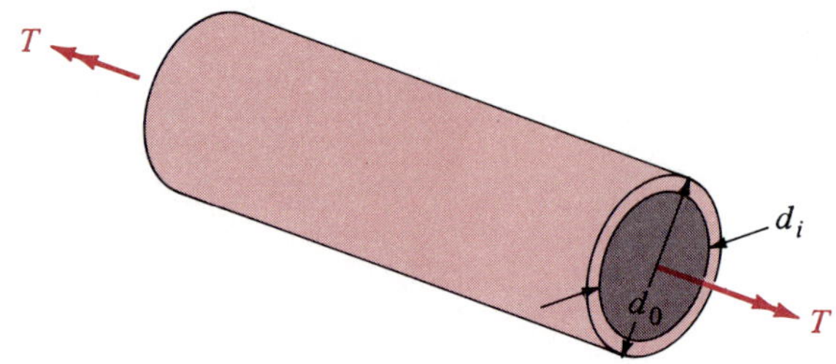

Ps. 6.23 and 6.24

6.24 The hollow shaft shown is composed of a brittle material with σ_{ult} = 320 MPa. If the inside and outside diameters are 21 mm and 28 mm, respectively, what is the maximum torque T the shaft can transmit if the factor of safety against fracture is to be 1.9?

6.25 A solid shaft of radius a and length L is subjected to a torque T_0. If a central hole of radius ka is drilled from the shaft, determine *(a)* the percent increase in the maximum stress, and *(b)* the percent increase in the relative angle of twist between the ends of the shaft.

6.26 Show that the polar moment of the area of a uniform hollow shaft can be approximated by $J = 2\pi R^3 t$, where $R = (a + b)/2$ and $t = b - a$. As a function of the outer radius a, what is the percent error involved in using the approximate expression for J in computing *(a)* the stress, and *(b)* the rotation?

6.27 Show that the maximum stress in a solid shaft is increased by only 6.6% by drilling a hole of radius $r/2$ out of the center. What is the corresponding increase in the angle of rotation of a bar of length L subjected to a single constant torque T?

6.28 The segmented bar is subjected to the torques shown. Construct the torque diagram and the rotation diagram. Determine the location and value of the maximum shear stress and the angle of rotation at C. Take $T_1 = 10^4\pi$ in·lbf, $T_2 = 2 \times 10^4\pi$ in·lbf, $G_{al} = 4 \times 10^6$ psi, and $G_{st} = 12 \times 10^6$ psi.

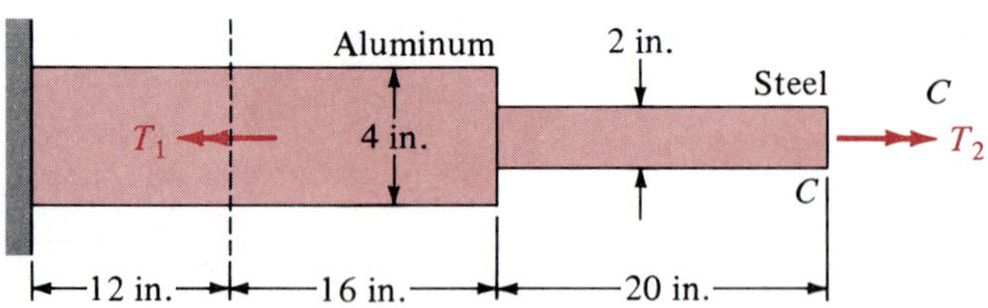

P. 6.28

6.29 The small lathes are driven by a motor as shown in the figure. The two pulleys exert the moments shown. Plot the T versus x and ϕ versus x diagrams and hence determine *(a)* the maximum shear stress, and *(b)* the maximum angle of twist relative to the motor end of the assembly. Take $G = 12 \times 10^6$ psi.

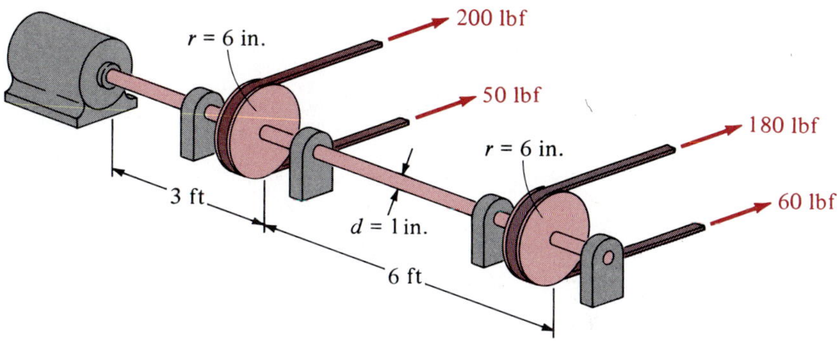

P. 6.29

6.30 The steel ($G = 80$ GPa) shaft is subjected to a torque of 1400 N·m. If the allowable shear stress is 90 MPa and the allowable angle of twist is 3° per meter of shaft, determine the minimum permissible diameter d. Which criterion governs?

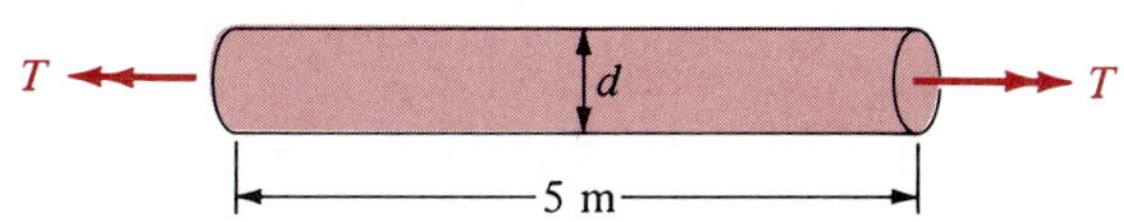

P. 6.30

6.31 Each of two different shafts carries a torque of 150 ft-lbf. One shaft is solid with a 0.875-in. diameter and the other shaft has an outside diameter of 1.75 in. and an inside diameter of 1.625 in. Compare *(a)* the maximum shear stresses in the two shafts, and *(b)* the twist per unit length in the two shafts.

6.32 Each of two different shafts carries a torque T. One shaft is hollow with a 1.50-in. outside diameter and a 1.25-in. inside diameter. What would be the diameter of a solid shaft with the same maximum shear stress? What is the ratio of the corresponding areas?

6.33 Each of two equal-length shafts carries a torque of 200 ft-lbf. One of the shafts is solid with a diameter of 1.25 in. The other is hollow with an outside diameter of 1.625 in. and an inside diameter such that the areas of the two shafts are the same. What is the ratio of the maximum torsional stresses for the two shafts? What is the ratio of the corresponding angles of twist?

6.34 Two solid shafts are to be welded as shown. Estimate the shear stress in the weld if a torque of 2500 ft-lbf is to be transmitted. State which of the assumptions of the theory is violated in using the elastic torsion formula.

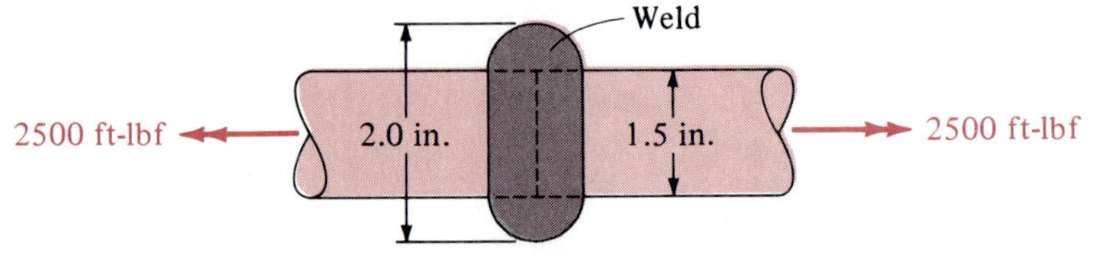

P. 6.34

6.35–6.38 In each of the following, *(a)* plot the T versus x and ϕ versus x diagrams, *(b)* determine the maximum shear stress and its location, and *(c)* determine the maximum angle of twist and its location.

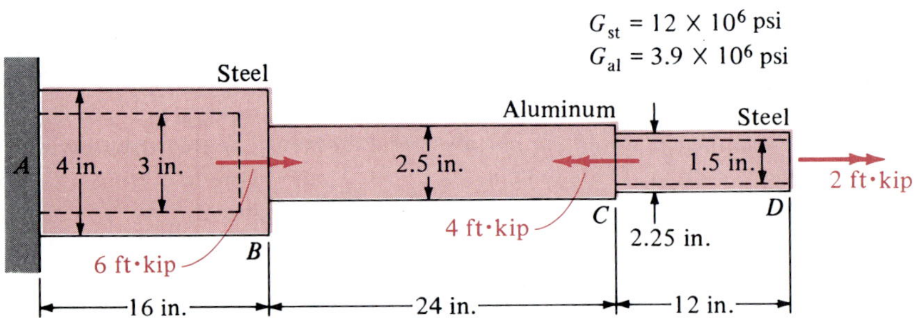

P. 6.35

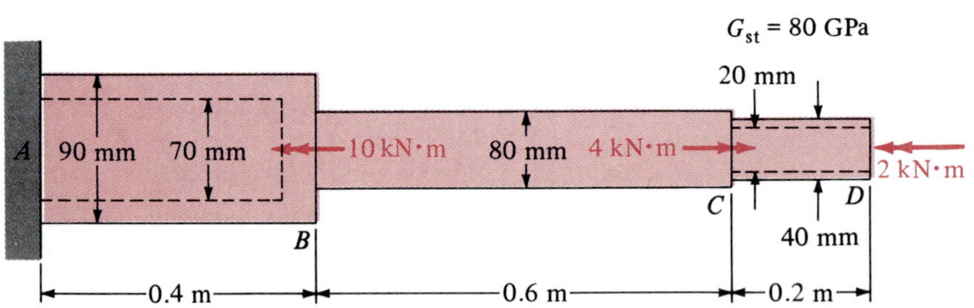

P. 6.36

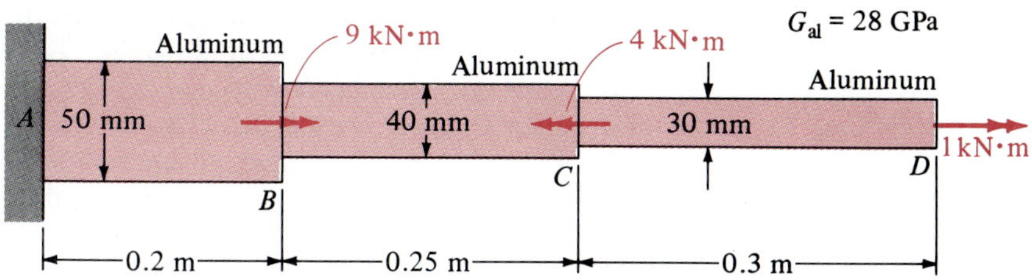

P. 6.37

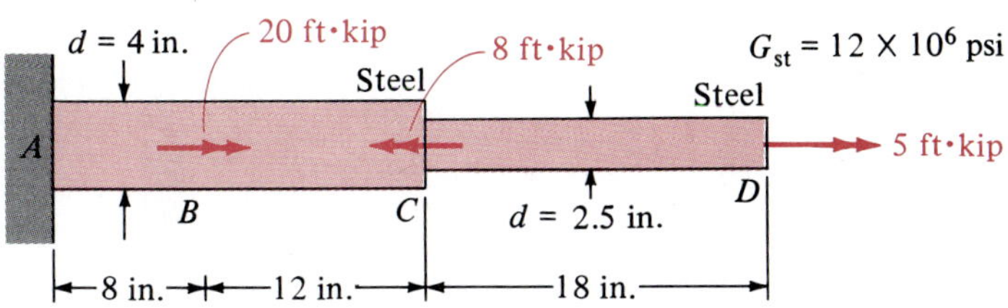

P. 6.38

6.39 A segmented solid steel shaft (G = 11,400 ksi) is subjected to the torques shown. What must be the diameters of segments AB and BC in order that the maximum shear stress nowhere exceeds 20 ksi? What is the corresponding angle of twist at the bearing D? If there is an additional condition requiring the angle of twist at D to be less than or equal to 2°, to what must the diameter of segment BC be increased? Construct the T versus x and ϕ versus x diagrams to assist in answering the questions.

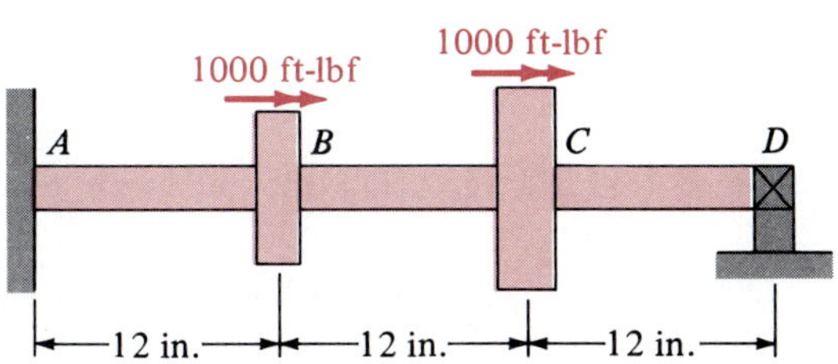

P. 6.39

6.40 A segmented hollow aluminum shaft (G = 40 GPa) is subjected to the torques shown. If the inside diameters of segments AB and BC are to be 75% of the outside diameters, what must be the outside diameters of segments AB and BC in order that the maximum shear stress nowhere exceeds 130 MPa? What is the corresponding angle of twist at the bearing D? If there is an additional condition requiring the angle of twist at D to be less than or equal to 2.5 degrees, to what must the outside diameter of segment BC be increased? Construct the T versus x and ϕ versus x diagrams to assist in answering the questions.

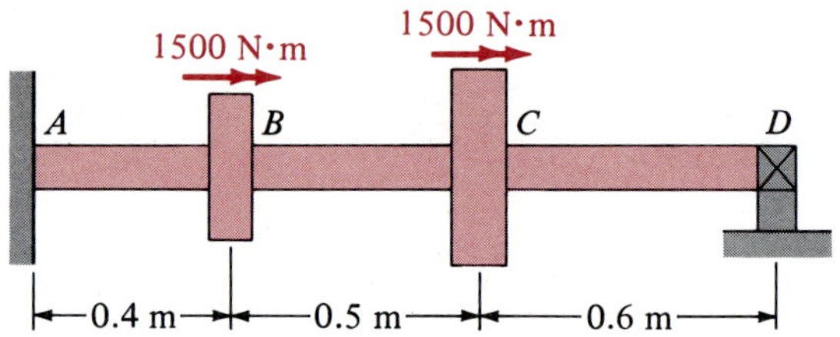

P. 6.40

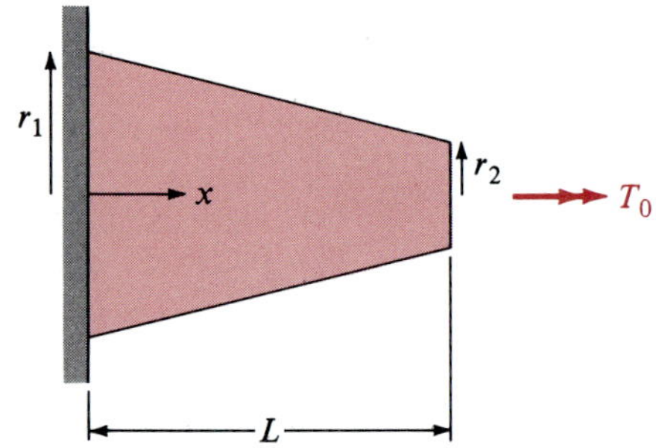

Ps. 6.41–6.46

6.41 A solid shaft in the shape of a conical frustum is subjected to a torque T_0 as shown in the figure. Integrate $T_0 = JG\phi'$ to determine $\phi(L)$.

6.42 Estimate $\phi(L)$ for the previous problem by replacing the bar with a constant-cross-section member whose radius is the average radius given by $(r_1 + r_2)/2$. Compare your solution with the corresponding exact solution for $r_1/r_2 = 2$.

6.43 Repeat problem 6.42 by replacing the bar with two equal-length segments.

6.44 Repeat problem 6.42 by replacing the bar with four equal-length segments.

6.45 Repeat problem 6.42 by replacing the bar with eight equal-length segments.

6.46 Repeat problem 6.42 by replacing the bar with eight equal-length segments whose radii are equal to *(a)* the actual value at the right end of each segment, and *(b)* the actual value at the left end of each segment.

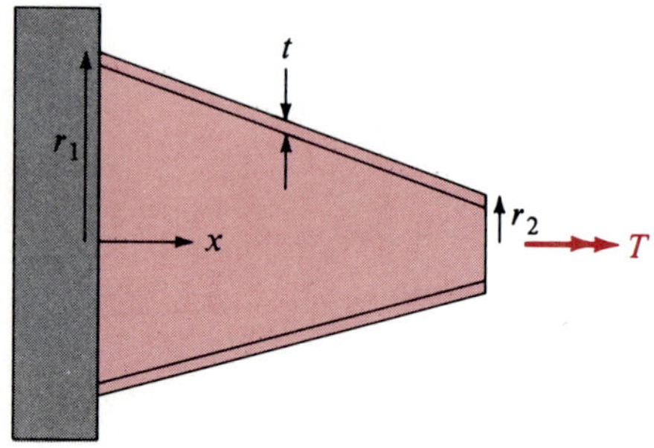

Ps. 6.47–6.52

6.47 A hollow shaft in the shape of a conical frustum is subjected to a torque T_0 as shown in the figure. Integrate $T_0 = JG\phi'$ to determine $\phi(L)$. Take t = cnst and $t/r_2 \ll 1$.

6.48 Estimate $\phi(L)$ for the previous problem by replacing the bar with a constant-cross-section member whose radius is the average of the radii given by $(r_1 + r_2)/2$. Compare your solution with the corresponding exact solution for $r_1/r_2 = 2$.

6.49 Repeat problem 6.48 by replacing the bar with two equal-length segments.

6.50 Repeat problem 6.48 by replacing the bar with four equal-length segments.

6.51 Repeat problem 6.48 by replacing the bar with eight equal-length segments.

6.52 Repeat problem 6.48 by replacing the bar with eight equal-length segments whose radii are equal to *(a)* the actual value at the right end of each segment, and *(b)* the actual value at the left end of each segment.

6.53 The segmented steel (G = 12,000 ksi) and brass (G = 5600 ksi) shaft is subjected to a torque of 1000 ft-lbf. If the ultimate shear stresses in the steel and brass are both 30 ksi and the allowable relative angle of twist between the ends is 6.5 degrees, what should the radius be?

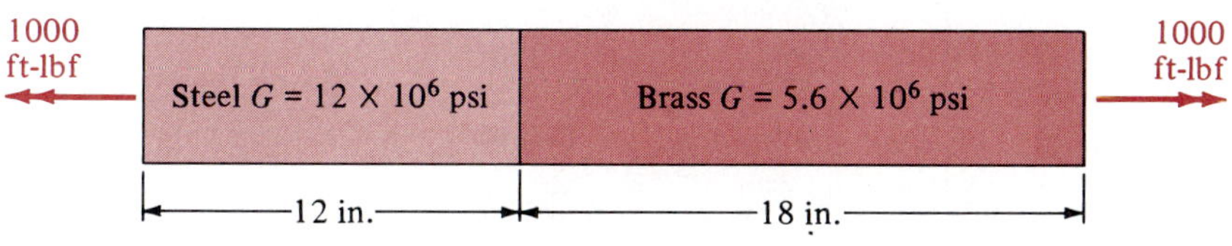

P. 6.53

6.54 A uniform solid steel (G = 80 GPa) shaft is subjected to the distributed torque that varies from 200 N·m/m to 800 N·m/m as shown. Determine the maximum shear stress in the shaft and the angle of twist at the free end.

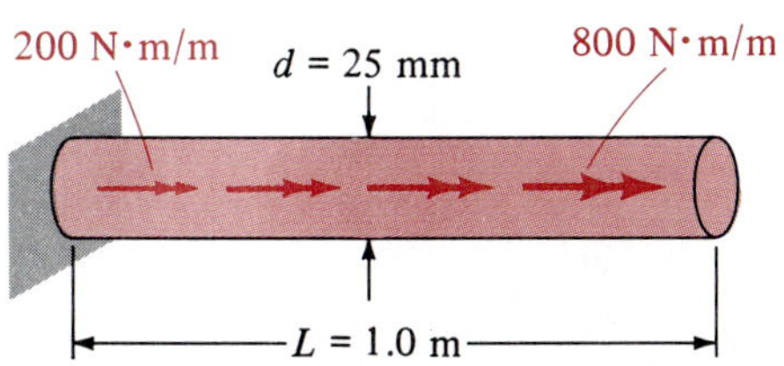

P. 6.54

6.55 For the system of gears and shafts shown, determine T_{AB}, T_{CD}, $\phi(B)$, and $\phi(D)$. Take G = 78 GPa.

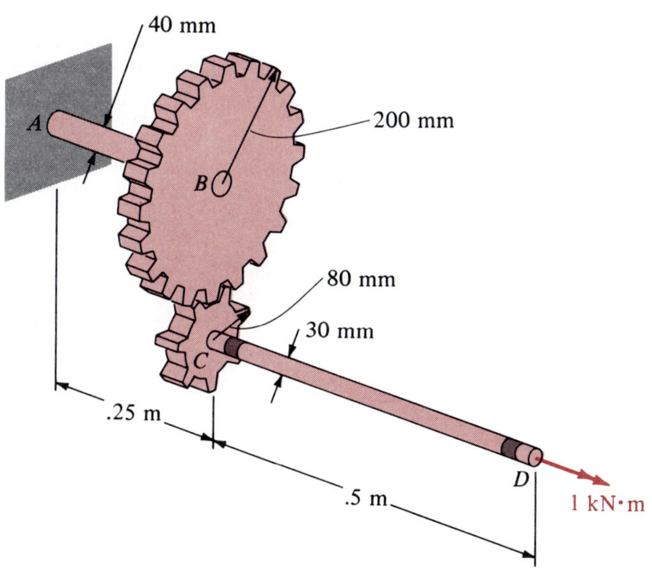

P. 6.55

6.56 For the torsional system shown, determine T_{AB}, T_{CD}, $\phi(B)$, and $\phi(D)$. Take $G = 12 \times 10^6$ psi.

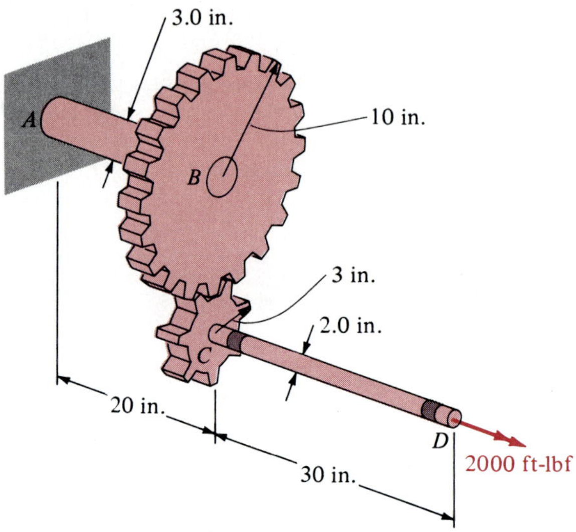

P. 6.56

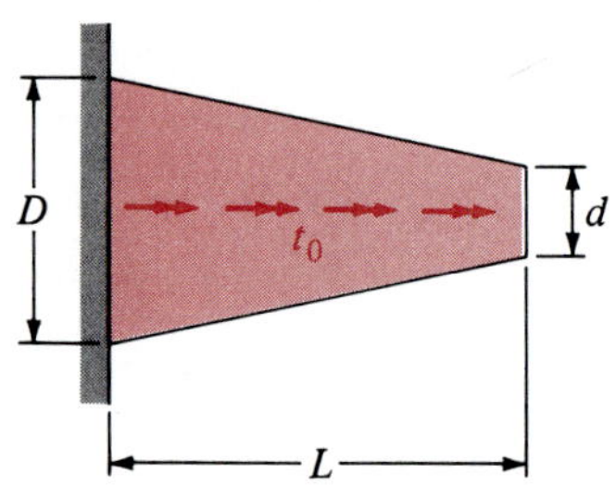

P. 6.57

6.57 Determine the angle of twist between the ends for the variable-cross-section solid shaft loaded by the constant distributed torque loading t_0 as shown.

6.58 Determine the angle of twist between the ends for the variable-cross-section, constant-thickness t hollow shaft loaded by the constant distributed torque loading t_0 as shown. Take D and d to be the average diameters at the ends.

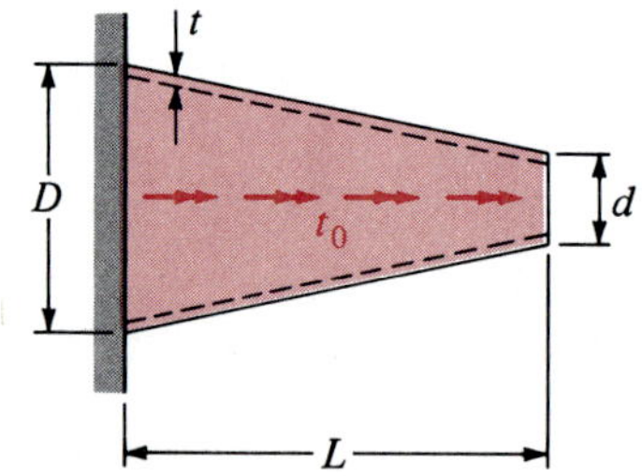

P. 6.58

6.59 A solid circular bar of radius a and length L, subjected to a single torque T_0, is composed of a material that behaves as shown in the figure. Develop the theory needed to be able to determine the stresses and the rotations in terms of the torque T. State which of the assumptions that were made for the linearly elastic material continue to be valid for the present situation. Specifically, *(a)* what is the stress distribution as a function of the radius, and *(b)* what is the relative angle of twist between the ends of a uniform bar of radius a transmitting a torque T_0?

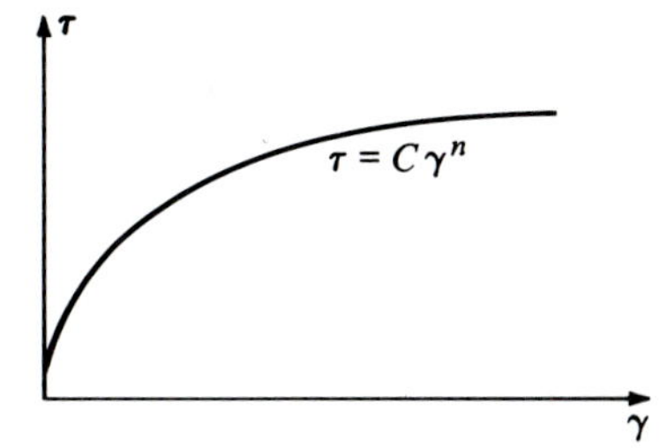

P. 6.59

6.3.1 Transmission of Power by Shafts

One of the most important areas of application for the elementary torsion theory is in systems where power is transmitted by rotating circular shafts. In a rotating system, the power to be transmitted by the shaft as well as the angular speed of the shaft are generally specified. When the power and speed are specified, the corresponding torque T that will be transmitted by the shaft is determined uniquely on the basis of the definition of power. When the torque is known, the stresses and twists can then be determined.

Power is defined as the rate at which work is done. The power P generated by the two forces in the belts shown in Fig. 6.31 is

$$P = F_1 v_1 - F_2 v_2 = F_1(r\omega) - F_2(r\omega) = \omega(F_1 r - F_2 r) = \omega T \qquad (6.11)$$

where $T = F_1 r - F_2 r$ is the torque transmitted by the shaft. The angular velocity ω in Figs. 6.31a and b must be constant in order that $\Sigma T = 0$ for the free-body diagram of Fig. 6.31b. Thus the torque in a rotating shaft is related to the power being transmitted by the shaft according to

$$T = \frac{P}{\omega}$$

It should be remembered that the constant angular speed ω must be expressed as radians (rad) per unit time, not as revolutions per unit time.

Power has dimensions of work per unit time, that is, FL/T. Common units are N·m/sec = joules/sec = watts (W) in the SI system, and ft-lbf/sec or horsepower (1 hp = 550 ft-lbf/sec) in the U.S. customary system of units. The conversion between the two systems is

$$1 \text{ hp} = \frac{550 \text{ ft-lbf}}{\text{sec}} \frac{0.3048 \text{ m}}{\text{ft}} \frac{4.448 \text{ N}}{\text{lbf}} = 745.7 \text{ N·m/sec} = 745.7 \text{ W}$$

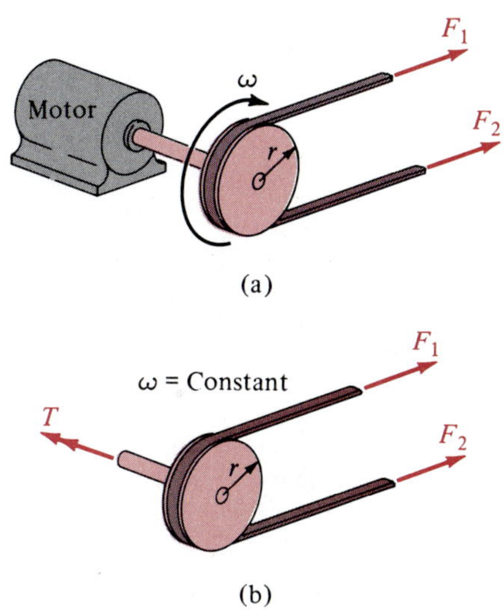

Figure 6.31 Power due to a rotating torque

Example 6.9 An 850-rpm motor supplies 6 hp that is distributed through gears as indicated in Fig. 6.32. The diameter of each of the shafts is 1.0 in. with $G = 12 \times 10^6$ psi throughout. Determine the torque transmitted by each shaft and the relative angles of twist between the gears.

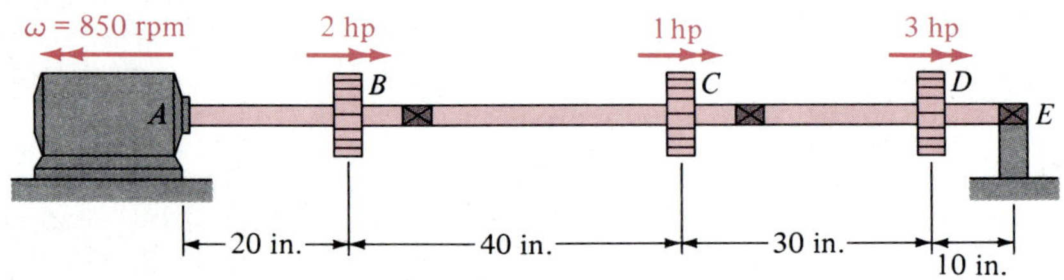

Figure 6.32 Motor-shaft-gear system

Solution At 850 rpm = $(170\pi/6)$ rad/sec the torque corresponding to 1 hp is

$$T = 1\ \text{hp}/\omega = \frac{550\ \text{ft-lbf/sec}}{(170\pi/6)\text{rad/sec}} = 6.18\ \text{ft-lbf}$$

Fig. 6.32 can be redrawn to show the input torques associated with the loads as indicated in Fig. 6.33a.

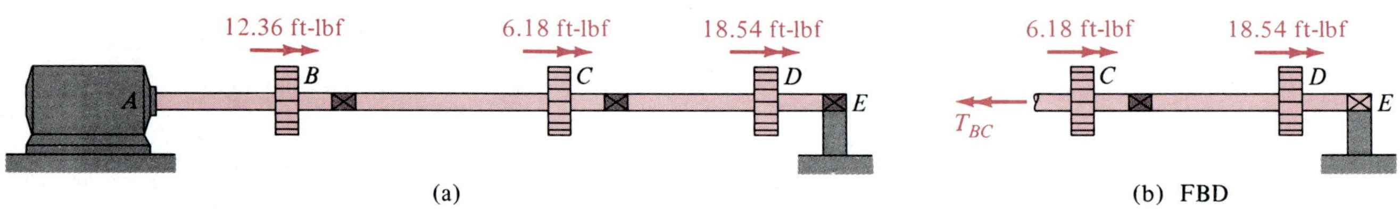

Figure 6.33 External torques arising from designated horsepowers

The T versus x diagram is easily constructed by drawing free-body diagrams exposing the internal torques between adjacent gears. For example, the free-body diagram required to determine the torque between pulleys B and C is shown in Fig. 6.33b. The equilibrium equation is

$$\Sigma T = -T_{BC} + 6.18\ \text{ft-lbf} + 18.54\ \text{ft-lbf} = 0$$

or

$$T_{BC} = 24.72\ \text{ft-lbf}$$

Repeating this step for each of the other segments, the corresponding T versus x diagram can be constructed as shown in Fig. 6.34. Using the general expression $\tau_{\max} = 2T/\pi a^3$ for a solid shaft, the maximum shear stress in each of the segments is given by

Shaft *AB*: $$\tau_{AB} = \frac{2(37.08\ \text{ft-lbf})(12\ \text{in./ft})}{\pi(0.5\ \text{in.})^3} = 2.27\ \text{ksi}$$

Shaft *BC*: $$\tau_{BC} = \frac{2(24.72\ \text{ft-lbf})(12\ \text{in./ft})}{\pi(0.5\ \text{in.})^3} = 1.51\ \text{ksi}$$

Shaft *CD*: $$\tau_{AB} = \frac{2(18.54\ \text{ft-lbf})(12\ \text{in./ft})}{\pi(0.5\ \text{in.})^3} = 1.13\ \text{ksi}$$

With $\phi(A) = 0$, the angle of twist diagram can be constructed from the torque diagram using the basic force-displacement relation $T = JG\phi'$ or $\phi' = T/JG$. Since T is a constant within any segment, ϕ' is a constant and ϕ is linear within the segment. Using the general relation $\Delta\phi = TL/JG$ for each segment, the values are

Shaft *AB*: $$\phi' = \frac{(37.08\ \text{ft-lbf})(12\ \text{in./ft})}{(\pi/32)\text{in}^4\cdot 12 \times 10^6\ \text{psi}} = 3.78 \times 10^{-4}\ \text{rad/in.}$$

from which $\Delta\phi_{AB} = \phi(B) - \phi(A) = (3.78 \times 10^{-4}\ \text{rad/in.})(20\ \text{in.})$

or $$\phi(B) = 7.55 \times 10^{-3}\ \text{rad} = 0.43°$$

Shaft *BC*: $$\phi' = \frac{(24.72\ \text{ft-lbf})(12\ \text{in./ft})}{(\pi/32)\text{in}^4\cdot 12 \times 10^6\ \text{psi}} = 2.52 \times 10^{-4}\ \text{rad/in.}$$

with $$\Delta\phi_{BC} = \phi(C) - \phi(B) = (2.52 \times 10^{-4}\ \text{rad/in.})(40\ \text{in.})$$

or $$\phi(C) - \phi(B) = 10.07 \times 10^{-3}\ \text{rad} = 0.58°$$

and $$\phi(C) = 0.43° + 0.58° = 1.01°$$

Shaft *CD*: $$\phi' = \frac{(18.54\ \text{ft-lbf})(12\ \text{in./ft})}{(\pi/32)\text{in}^4\cdot 12 \times 10^6\ \text{psi}} = 1.89 \times 10^{-4}\ \text{rad/in.}$$

so that $$\Delta\phi_{CD} = \phi(D) - \phi(C) = (1.89 \times 10^{-4}\ \text{rad/in.})(30\ \text{in.})$$

or $$\phi(D) - \phi(C) = 5.67 \times 10^{-3}\ \text{rad} = 0.32°$$

with $$\phi(D) = 1.01° + 0.32° = 1.33°$$

Shaft *DE*: $$T = 0 \rightarrow \phi' = 0 \rightarrow \phi(E) = \phi(D) = 1.33°$$

The ϕ versus x diagram is shown in Fig. 6.34.

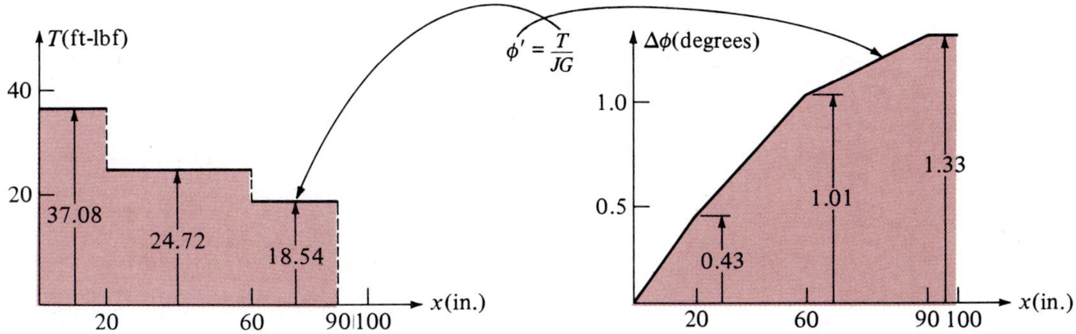

Figure 6.34 T versus x and ϕ versus x diagrams

PROBLEMS

6.60 Design a solid shaft for a 12-hp motor operating at 2000 rpm. Take the allowable shear stress to be 16 ksi.

6.61 For an allowable shear stress of 70 MPa, design a solid steel shaft to transmit 12 kW at 20 Hz.

6.62 A steel shaft rotating at 1600 rpm transmitting 0.8 hp can withstand a shear stress of 10 ksi. Determine the required diameter of the shaft.

6.63 A shaft 3 m in length delivers 2000 kW at 6 Hz. Design the shaft so that the maximum shear stress does not exceed 70 MPa and the relative angle of twist between the ends does not exceed 3°. Take $G = 80$ GPa.

6.64 A solid shaft is composed of a steel with a yield point in tension of $\sigma_Y =$ 40 ksi. If the shaft rotates at 2000 rpm and transmits 1 hp, determine the required diameter for a factor of safety of 1.5.

6.65 A hollow ($r_0 = 1.25r_i$) steel shaft is made from a brittle steel with an ultimate stress of 400 MPa. For a factor of safety of 1.8, design the shaft to transmit 500 kW at 10 Hz.

6.66 The simple motor-belt-pulley system (bearings not shown) is to deliver power at A. If the motor is rated at 5 kW and rotates at 3000 rpm, what are the required diameters of shafts AB and CD if the allowable shear stress in each is 110 MPa? Take $r_1 = 36$ mm and $r_2 = 20$ mm.

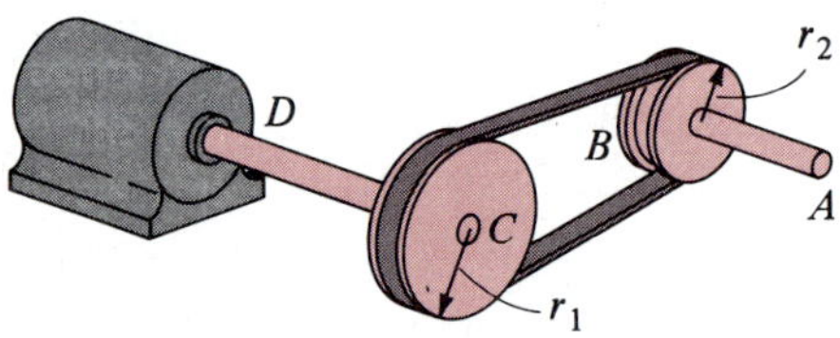

Ps. 6.66 and 6.67

6.67 The simple motor-belt-pulley system (bearings not shown) is to deliver power at A. If the motor is rated at 7 hp and rotates at 3000 rpm, what are the required diameters of shafts AB and CD if the allowable shear stress in each is 18 ksi? Take $r_1 = 15$ in. and $r_2 = 7.5$ in.

6.68 Design each of the solid shafts for the gear system shown. The motor supplies 3 hp at 3600 rpm and the allowable shear stress in each of the shafts is 12 ksi. Take $r_1 = 3$ in., $r_2 = 2$ in., $r_3 = 4$ in., $r_4 = 5$ in., and $r_5 = 4$ in.

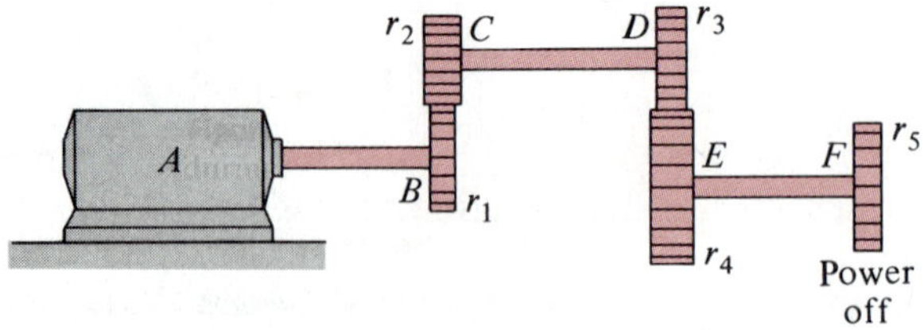

Ps. 6.68 and 6.69

6.69 Design each of the solid shafts for the gear system shown. The motor supplies 4 kW at 2400 rpm and the allowable shear stress in each of the shafts is 90 MPa. Take $r_1 = 80$ mm, $r_2 = 60$ mm, $r_3 = 150$ mm, $r_4 = 160$ mm, and $r_5 = 120$ mm.

6.70 A 100-kW motor delivers power to the shaft with the power takeoffs also indicated. Assume the hollow shafts ($r_0 = 1.1r_i$) to be made of a brittle material with $\sigma_{ULT} = 60$ MPa. If the motor shaft rotates at 4 Hz, design the diameters for factors of safety of 3.0. What are the relative angles of twist between C and A, and between C and E? Take $P_1 = 30$ kW, $P_2 = 20$ kW, $P_3 = 10$ kW, $P_4 = 40$ kW, $L_1 = 2.5$ m, $L_2 = 3.0$ m, $L_3 = 2.5$ m, $L_4 = 3.6$ m, and the radii of the gears to be $r_A = 0.15$ m, $r_B = 0.10$ m, $r_C = 0.20$ m, $r_D = 0.175$ m, $r_E = 0.225$ m, $r_M = 0.1$ m, and $G = 80$ GPa.

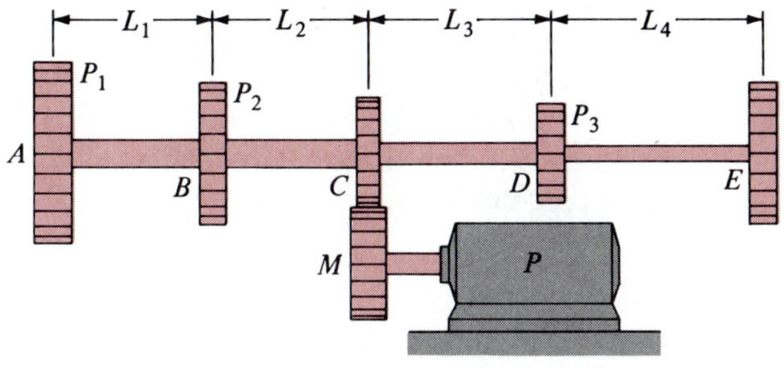

Ps. 6.70 and 6.71

6.71 A 120-hp motor delivers power to the shaft with the power takeoffs also indicated. Assume the hollow shafts ($r_0 = 1.1r_i$) to be made of a brittle material with $\sigma_{ULT} = 20$ ksi. If the motor shaft rotates at 5 Hz, design the diameters for factors of safety of 2.5. What are the relative angles of twist between C and A, and between C and E? Take $P_1 = 20$ hp, $P_2 = 30$ hp, $P_3 = 15$ hp, $P_4 = 55$ hp, $L_1 = 8$ ft, $L_2 = 10$ ft, $L_3 = 7.5$ ft, $L_4 = 14$ ft, and the radii of the gears to be $r_A = 6$ in., $r_B = 4.5$ in., $r_C = 8$ in., $r_D = 7.5$ in., $r_E = 9$ in., $r_M = 4$ in., and $G = 12 \times 10^6$ psi.

6.72 For a solid shaft, develop a general relation for determining the required radius in terms of the power P, the angular speed ω, and the allowable shear stress τ.

6.73 For a hollow shaft with outside radius r_0 and inside radius $r_1 = kr_0$, develop a general relation for determining the required radius in terms of the power P, the angular speed ω, the constant k, and the allowable shear stress τ.

6.4 STATICALLY INDETERMINATE PROBLEMS

Situations arise frequently in which torsional systems are designed to be statically indeterminate. Such systems are stiffer, and hence generally stronger, than their statically determinate counterparts. In this section, we will illustrate the use of the flexibility method and then the stiffness method for setting up and solving typical statically indeterminate torsional problems. We will accomplish this using examples representative of those encountered in practice.

6.4.1 Flexibility Formulations

Recall that for flexibility formulations, the force resultants are the unknowns. The number of unknown force resultants exceeds the number of equations by the number of redundants. The continuity equations, requiring that the structure "fit together," along with the force-displacement equations, are used to generate the additional equations needed to determine the unknown forces. The displacement and force variables can then be calculated and the strength and stiffness questions answered.

Example 6.10 Consider two shafts, connected to a disk at B and constrained against rotation at the ends. We assume that the shafts are initially unloaded and that a torque is then applied to the disk as shown in Fig. 6.35a. Determine the reaction torques at the supports and the rotation of the disk.

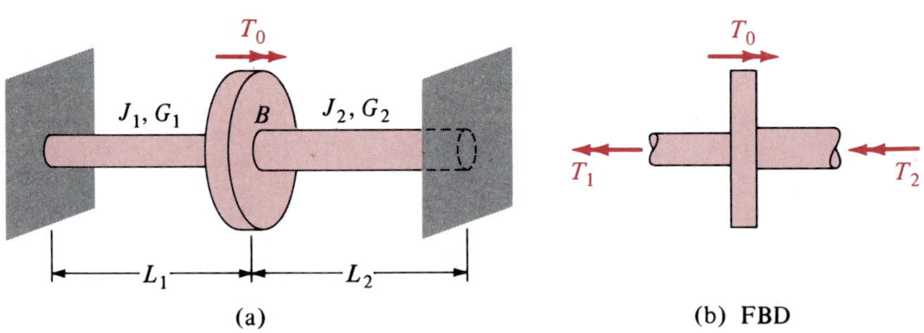

Figure 6.35 Indeterminate torque problem

Solution Referring to the free-body diagram shown in Fig. 6.35b, the single equation of equilibrium is

$$\Sigma T = -T_1 + T_0 - T_2 = 0 \qquad \text{or} \qquad T_1 + T_2 = T_0$$

that is, only one equation for the two unknowns T_1 and T_2. For this example there is one redundant, so there will be one unknown force in setting up the equations using the flexibility method. Recall that the strategy for the flexibility method is to consider a set of redundant internal forces or reactions as the unknowns. These redundants are then chosen by requiring that the displacements within the structure are continuous, that is, that the structure "fit together" under the actions of the loads and internal forces. For the present example the torques T_1 and T_2, the unknowns, can have any values that satisfy the equilibrium equation $T_0 = T_1 + T_2$. However, there is only one set of torques T_1 and T_2 satisfying equilibrium that *also* produce twists of the shafts that have the same values at the disk. It is this one set T_1 and T_2 that we must determine.

Select the torque T_2 as the redundant. Using the equation of equilibrium, T_1 is expressed in terms of the redundant as $T_1 = T_0 - T_2$. The two statically determinate problems associated with removing the redundant T_2 are shown in Fig. 6.36.

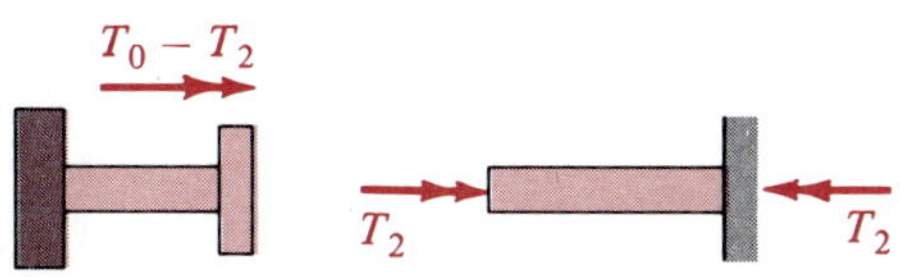

Figure 6.36 Corresponding statically determinate problems

Continuity of displacement is enforced by requiring the angle of twist ϕ_1 at the right end of member 1 be the same as the angle of twist ϕ_2 at the left end of member 2. Using the torque-twist relations for the two shafts, $\phi_1 = (T_0 - T_2)/C_1$ and $\phi_2 = T_2/C_2$, so that $\phi_1 = \phi_2$ leads to

$$\frac{T_0 - T_2}{C_1} = \frac{T_2}{C_2} \quad \text{or} \quad T_2 = \frac{C_2 T_0}{C_1 + C_2}$$

The corresponding T_1 is

$$T_1 = \frac{C_1 T_0}{C_1 + C_2}$$

and the twist is

$$\phi_1 = \phi_2 = \frac{T_0}{C_1 + C_2}$$

Note that $T_1/C_1 = T_2/C_2$, showing that the amount of torque carried by each shaft is proportional to its torsional stiffness.

Example 6.11 For the system of shafts and gears shown in Fig. 6.37a, determine *(a)* the torques transmitted by shafts 2 and 3, and *(b)* the rotations of gears 1, 2, and 3.

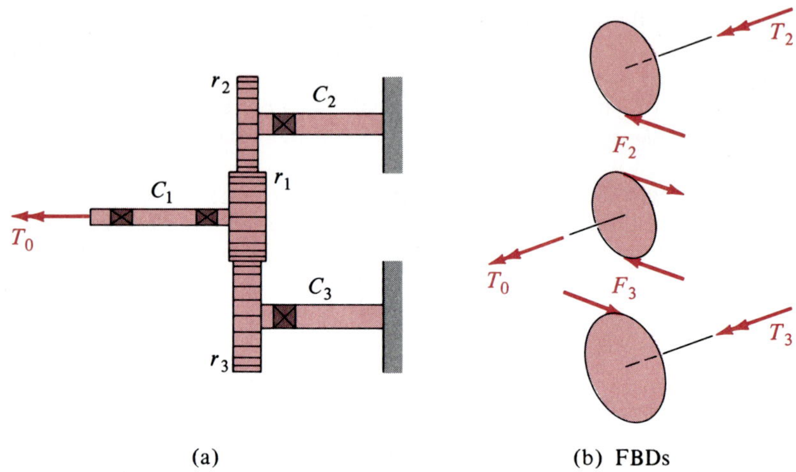

Figure 6.37 Statically indeterminate torsional system

Solution Referring to the free-body diagrams of Fig. 6.37b, the equilibrium equations are

$$T_0 = r_1(F_2 + F_3) \qquad T_2 = r_2 F_2 \qquad T_3 = r_3 F_3$$

that is, three equations for the four unknowns T_2, T_3, F_2, and F_3. Eliminating F_2 and F_3 yields the one equation

$$\frac{T_0}{r_1} = \frac{T_2}{r_2} + \frac{T_3}{r_3}$$

for the two unknowns T_2 and T_3. Either T_2 or T_3 can be considered as the redundant. (Note that the bearing forces have not been shown in the free-body diagrams of Fig. 6.37b, as none of the bearing forces produce any torques about the axis of the corresponding shaft.)

The additional equation must be generated on the basis of the displacements. Select T_2 as the redundant, with the two statically determinate problems to be solved shown in Fig. 6.38.

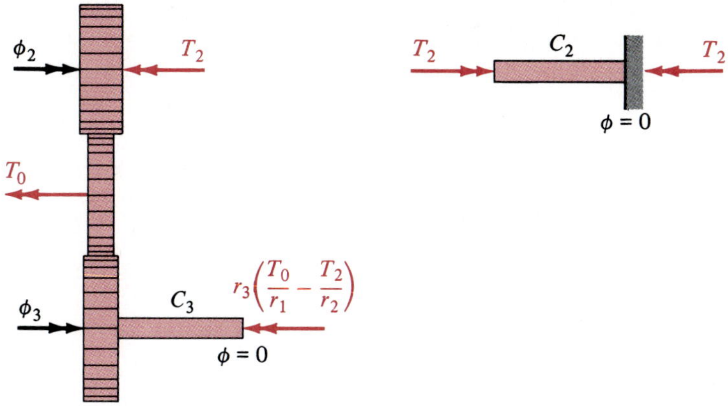

(a) Problem 1 (b) Problem 2

Figure 6.38 Corresponding statically determinate problems

The necessary extra equation is obtained by requiring the twist at gear 2 to be the same as the twist of shaft 2. Denote the rotations of the three gears by ϕ_1, ϕ_2, and ϕ_3 as shown. Continuity of the tangential gear displacements at the interface between gears 1 and 2 is expressed by $r_1\phi_1 = r_2\phi_2$. The corresponding equation relating the tangential interface displacement at gears 1 and 3 is $r_1\phi_1 = r_3\phi_3$. Eliminating ϕ_1 yields $r_2\phi_2 = r_3\phi_3$. The rotation ϕ_3 at gear 3 is due to the deformation of shaft 3, namely

$$\phi_3 = \frac{r_3}{C_3}\left(\frac{T_0}{r_1} - \frac{T_2}{r_2}\right)$$

with the corresponding rotation at gear two as

$$\phi_2 = \frac{r_3\phi_3}{r_2} = \frac{r_3^2}{r_2 C_3}\left(\frac{T_0}{r_1} - \frac{T_2}{r_2}\right)$$

This value must coincide with the angle of twist in shaft 2, namely $\phi_2 = T_2/C_2$ or

$$\frac{T_2}{C_2} = \frac{r_3^2}{r_2 C_3}\left(\frac{T_0}{r_1} - \frac{T_2}{r_2}\right)$$

from which

$$T_2 = \frac{r_2}{r_1}\frac{C_2 r_3^2}{C_2 r_3^2 + C_3 r_2^2}T_0$$

Using the equilibrium equation, T_3 can be determined as

$$T_3 = \frac{r_3}{r_1} \frac{C_3 r_2^2}{C_2 r_3^2 + C_3 r_2^2} T_0$$

The rotations are then given by

$$\phi_2 = \frac{T_2}{C_2} = \frac{r_2}{r_1} \frac{r_3^2}{C_2 r_3^2 + C_3 r_2^2} T_0$$

$$\phi_3 = \frac{T_3}{C_3} = \frac{r_3}{r_1} \frac{r_2^2}{C_2 r_3^2 + C_3 r_2^2} T_0$$

$$\phi_1 = \left(\frac{r_2}{r_1}\right) \phi_2 = \left(\frac{r_2}{r_1}\right)^2 \frac{r_3^2}{C_2 r_3^2 + C_3 r_2^2} T_0$$

As a check, let $C_2 = C_3$ and $r_2 = r_3$, leading to

$$T_2 = T_3 = \frac{r_2}{r_1} \frac{T_0}{2}$$

If further, $r_2 = r_1$, then $T_2 = T_0/2$ as would be expected from symmetry.

Example 6.12 Using a flexibility approach, extend the elementary torsion theory to the case of a shaft consisting of two concentric materials, and transmitting a torque T as shown in Fig. 6.39.

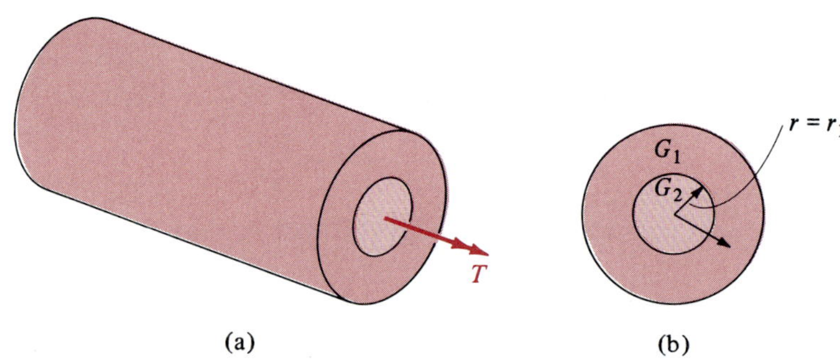

Figure 6.39 Geometry of a composite shaft

Solution The formulation will be carried out by combining equilibrium and deformation as per a flexibility formulation.

Equilibrium: Let the torque carried by portions 1 and 2 be T_1 and T_2, respectively, so that equilibrium requires that

$$T_1 + T_2 = T \qquad (a)$$

Deformation: Assume that the assumption regarding the deformation made in Section 6.2 (plane sections remain plane) continues to hold so that

$$\gamma = r\phi' \qquad (b)$$

Also assume that each material behaves in a linearly elastic fashion so that the stresses in the two portions are

$$\tau_1 = G_1\gamma \qquad \text{and} \qquad \tau_2 = G_2\gamma \qquad (c)$$

The total torque T carried by the two portions is then given by

$$\begin{aligned} T &= \int_{\text{material 1}} rG_1r\phi'\,dA + \int_{\text{material 2}} rG_2r\phi'\,dA \\ &= \phi'(J_1G_1 + J_2G_2) \qquad (d) \\ &= T_1 + T_2 \qquad (e) \end{aligned}$$

where J_1 and J_2 are, respectively, the polar moments of the area of the regions occupied by materials 1 and 2. Solving from ϕ' from Eqn. *(d)* and substituting into Eqns. *(b)* and *(c)* leads to

$$\tau_1 = Tr\frac{G_1}{J_1G_1 + J_2G_2} \qquad \text{and} \qquad \tau_2 = Tr\frac{G_2}{J_1G_1 + J_2G_2}$$

From Eqns. *(d)* and *(e)* it follows that the amount of the total torque carried by each portion is

$$T_1 = \frac{J_1G_1}{J_1G_1 + J_2G_2}T \qquad \text{and} \qquad T_2 = \frac{J_2G_2}{J_1G_1 + J_2G_2}T$$

As a specific example consider a composite bar consisting of a solid brass ($G = 5.6 \times 10^6$ psi) core and a hollow steel ($G = 12.0 \times 10^6$ psi) cylinder as shown in Fig. 6.40. If a torque of 5000 ft-lbf is applied to the bar, determine the stresses in the brass and steel portions and the relative angle of twist between the ends.

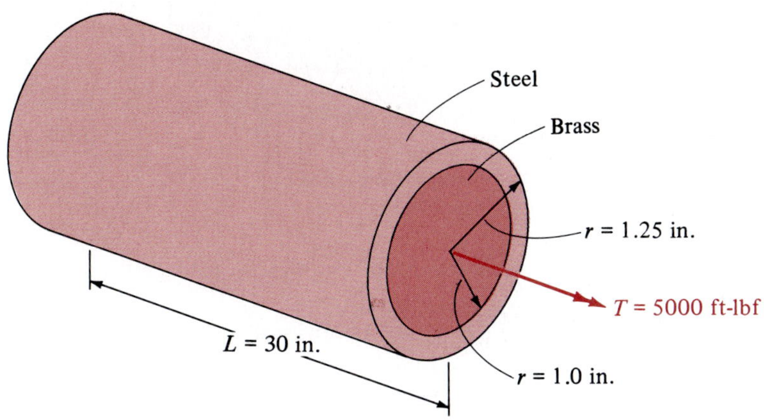

Figure 6.40 Composite bar of brass and steel

The constant JG is given by

$$JG = J_1G_1 + J_2G_2$$

$$= \frac{\pi}{2}[(1^4\ \text{in}^4)(5.6 \times 10^6\ \text{psi}) + (1.25^4\ \text{in}^4 - 1^4\ \text{in}^4)(12.0 \times 10^6\ \text{psi})]$$

$$= (8.80 \times 10^6 + 27.17 \times 10^6)\ \text{lbf·in}^2 = 35.97 \times 10^6\ \text{lbf·in}^2$$

The torques carried by the brass and steel, respectively, are

$$T_{\text{brass}} = \frac{J_1G_1}{J_1G_1 + J_2G_2}T = \frac{8.80}{35.97}T = 0.24T$$

and

$$T_{\text{steel}} = \frac{J_2G_2}{J_1G_1 + J_2G_2}T = \frac{27.17}{35.97}T = 0.76T$$

The maximum stress in the brass occurs at the interface and has the value

$$\tau_{\text{brass}} = \frac{(60{,}000\ \text{in.-lbf})(1\ \text{in.})(5.6 \times 10^6\ \text{lbf/in}^2)}{35.97 \times 10^6\ \text{lbf·in}^2} = 9.34\ \text{ksi}$$

and in the steel at the outer radius with the value

$$\tau_{\text{steel}} = \frac{(60{,}000\ \text{in.-lbf})(1.25\ \text{in.})(12.0 \times 10^6\ \text{lbf/in}^2)}{35.97 \times 10^6\ \text{lbf·in}^2} = 25.02\ \text{ksi}$$

This stress distribution is shown in Fig. 6 .41.

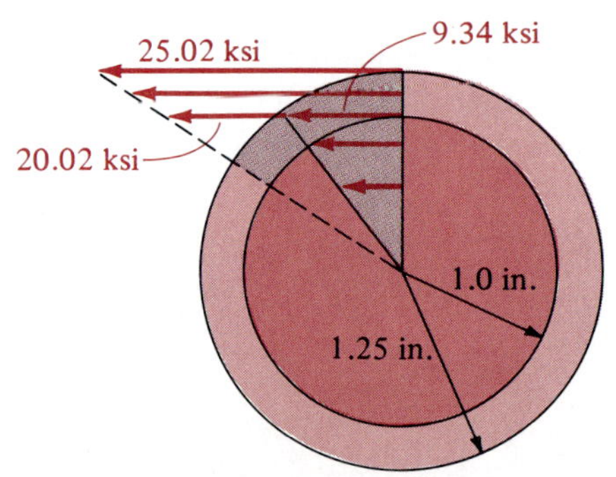

Figure 6.41 Stress distribution in a brass/ steel composite bar

The constant angle of twist per in. of the shaft is given by

$$\phi' = \frac{60{,}000\ \text{in.-lbf}}{35.97 \times 10^6\ \text{lbf-in}^2} = 0.000167\ \text{rad/in.} = 0.096°/\text{in.}$$

so that the total angle of twist between the ends of the shaft is

$$\Phi = (30\ \text{in.})(0.096°/\text{in.}) = 2.88°$$

In summary, the basic strategy of the flexibility method is to use the redundant forces as the unknowns; the extra equations are then obtained by requiring that the displacements in the system be continuous.

PROBLEMS

In each of the problems, carefully draw free-body diagrams and write the equations of equilibrium to identify the number of equations and the number of unknowns, thus the degree of statical indeterminacy. Use the flexibility method to carry out the formulation and solution of the problems.

6.74 Determine the torques in shafts AB and BC. Also determine the rotation of disk B. $G_{al} = 3.9 \times 10^6$ psi and $G_{st} = 12 \times 10^6$ psi.

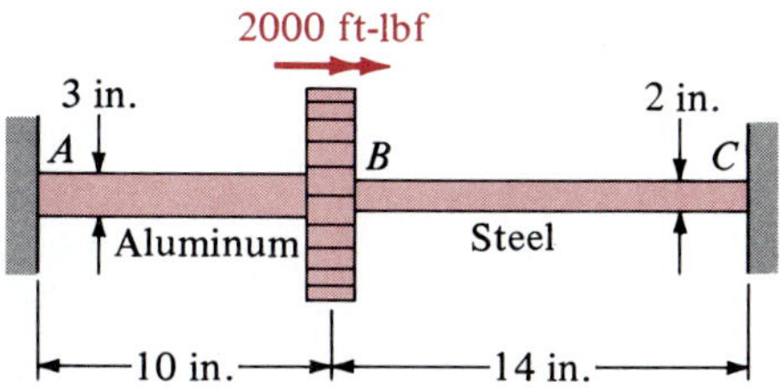

Ps. 6.74 and 6.98

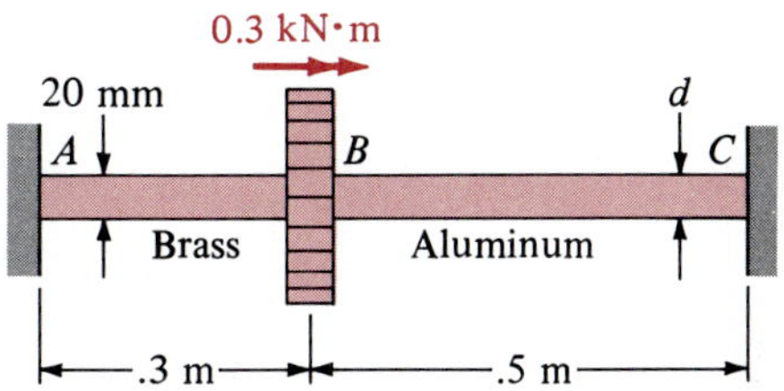

Ps. 6.75 and 6.99

6.75 Determine the required diameter of the aluminum shaft so that it carries the same torque as the brass shaft. Take $G_{al} = 28$ GPa and $G_{br} = 40$ GPa. What are the stresses in the shafts?

6.76 What is the allowable torque that can be applied to the gear shown in order that the maximum shear stress not exceed 200 MPa in either shaft? What is the corresponding angle of twist of the gear? Take G for both shafts to be 28 GPa.

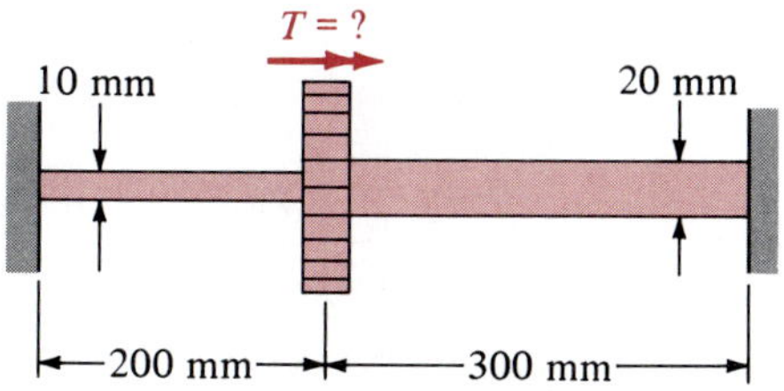

Ps. 6.76 and 6.100

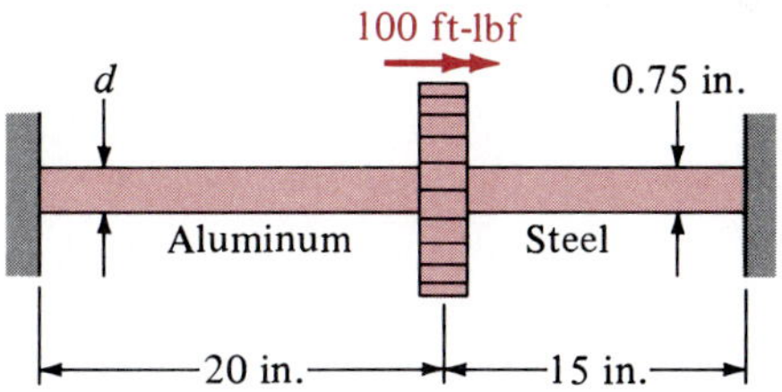

Ps. 6.77 and 6.101

6.77 If the rotation of the gear is to be less than one degree, determine the minimum required diameter d. What are the corresponding stresses in the two shafts? Take $G_{st} = 12 \times 10^6$ psi and $G_{al} = 3.9 \times 10^6$ psi.

6.78 For the system of gears and shafts indicated, determine *(a)* T_{AB}, *(b)* T_{BC}, *(c)* T_{CD}, *(d)* ϕ_B, and *(e)* ϕ_C. Take $G = 12 \times 10^6$ psi for all shafts.

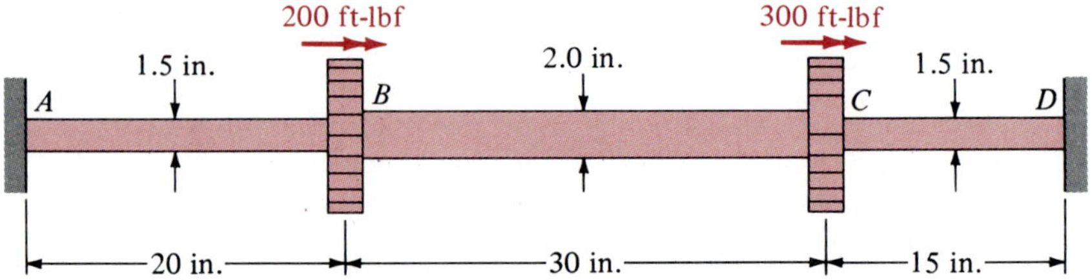

Ps. 6.78 and 6.102

6.79 For the system of gears and shafts shown, determine *(a)* the rotation at *A*, *(b)* the rotation of gear at *C*, *(c)* the torque in shaft *CD*, and *(d)* the torque in shaft *EF*. Take G = 28 GPa for all shafts.

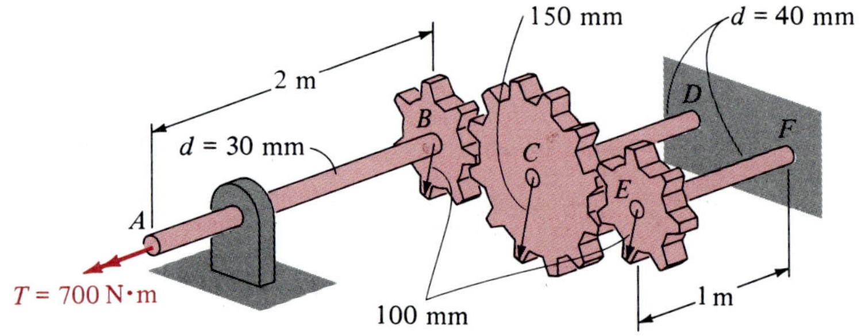

Ps. 6.79 and 6.103

6.80 For the system of gears and shafts shown, determine *d*, the diameter of shafts *CD* and *EF* in order that the maximum rotation of gear *A* not exceed 5°. What is the corresponding maximum shear stress in *(a)* shaft *CD* and *(b)* shaft *EF*? Take G = 80 GPa for all shafts.

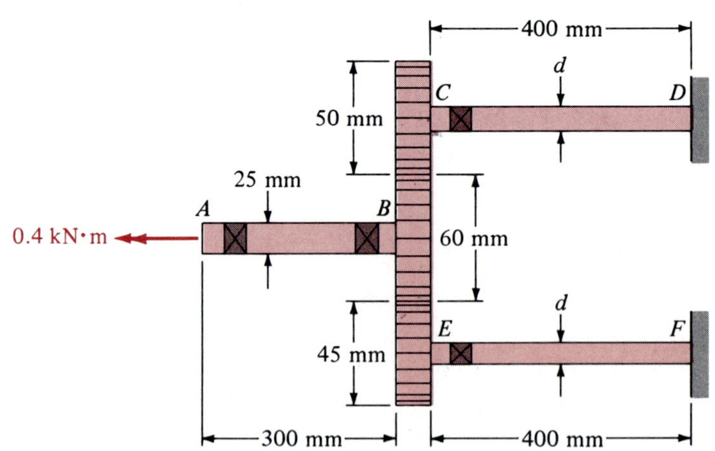

Ps. 6.80 and 6.104

6.81 For the system shown, gear A is rotated through an angle of 3° as indicated. The two gears are then brought into mesh and released. Determine the torque in each shaft and the rotation of gear C. Take $G = 80$ GPa for both shafts. The radii of gears A and C are 100 mm and 120 mm, respectively.

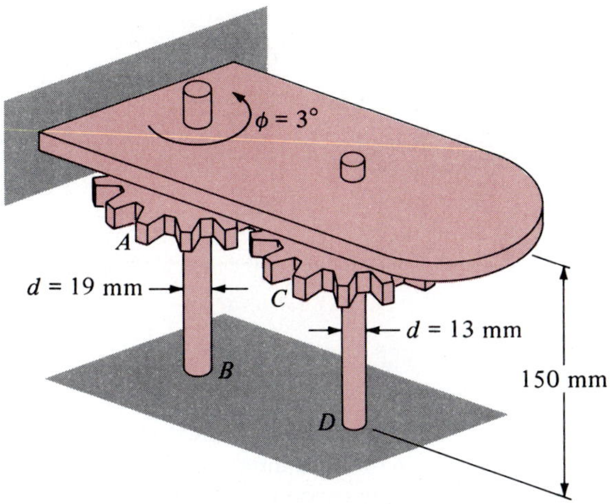

Ps. 6.81 and 6.105

6.82 Determine the torques in shafts 2 and 3 in terms of d_1, d_2, d_3, C_1, C_2, C_3, and T.

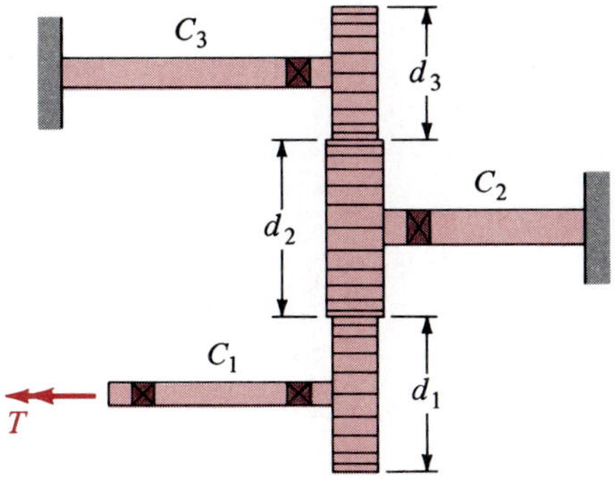

Ps. 6.82 and 6.106

6.83 For the system shown, determine *(a)* the torque in shaft *AB*, *(b)* the torque in shaft *BC*, *(c)* the torque in shaft *CD*, *(d)* the torque in shaft *EF*, *(e)* the rotation of disk *B*, and *(f)* the rotation of disk *C*. Take $G = 12 \times 10^6$ psi for all the shafts.

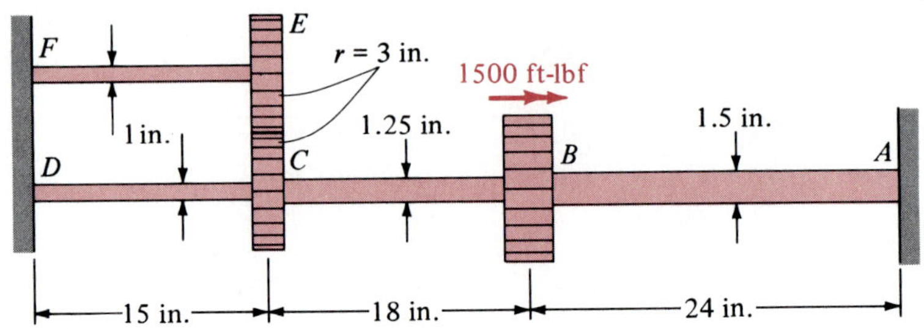

Ps. 6.83 and 6.107

Composite Shaft Problems

6.84 For the composite torsional member indicated in the figure, determine *(a)* the maximum shear stress in the aluminum, *(b)* the maximum shear stress in the steel, *(c)* the part of the torque carried by the aluminum, *(d)* the part of the torque carried by the steel, and *(e)* the relative angle of twist between the ends of the bar.

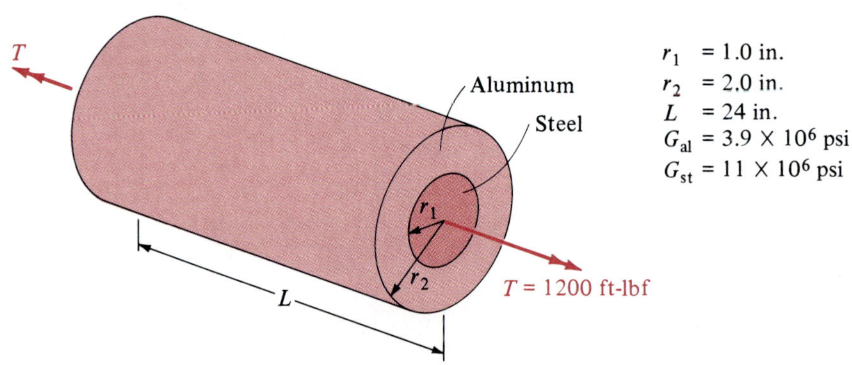

Ps. 6.84 and 6.85

6.85 Repeat problem 6.84 with the positions of aluminum and steel reversed.

6.86 The steel core and the aluminum shell are attached to a rigid plate at the ends as indicated in the figure. What should be the diameter of the steel core in order that the torque carried by the aluminum be twice that carried by the steel? What are the corresponding maximum shear stresses in the steel and the aluminum portions? What is the relative angle of twist between the ends of the bar?

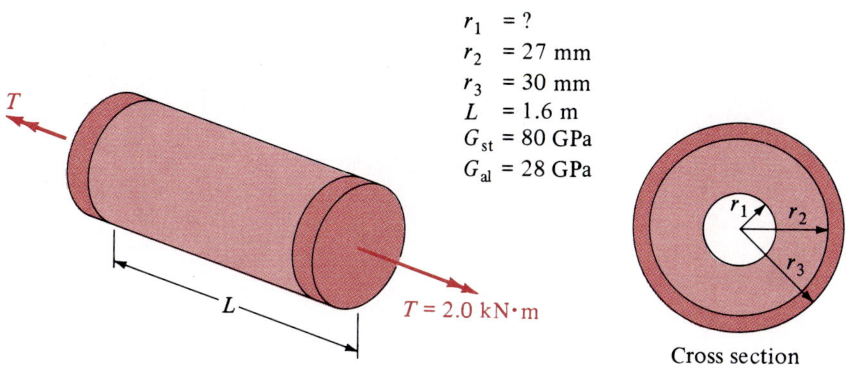

P. 6.86

6.87 A composite bar consists of a steel core and a brass shell as indicated. What should the interface radius be in order that each component carry half the torque? What are the corresponding maximum shear stresses in the steel and the brass portions?

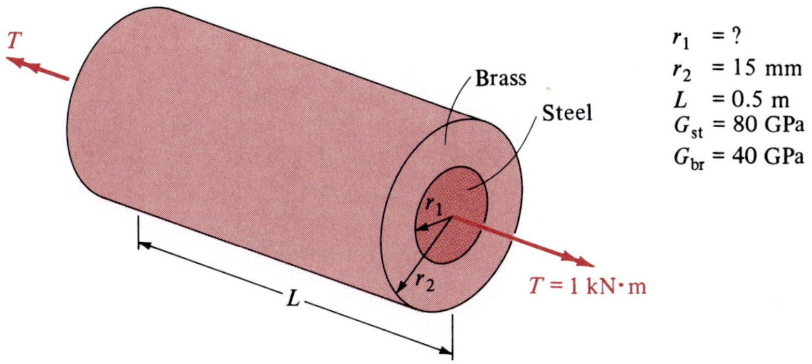

Ps. 6.87 and 6.88

6.88 Repeat problem 6.87 with the positions of brass and steel reversed.

6.89 The composite torsional member is composed of an inner aluminum core and an outer steel shell. If the allowable shear stresses in the aluminum and steel are 16 ksi and 20 ksi, respectively, determine the allowable torque that can be transmitted. What is the stress in the noncritical material? What is the relative angle of twist between the ends of the bar?

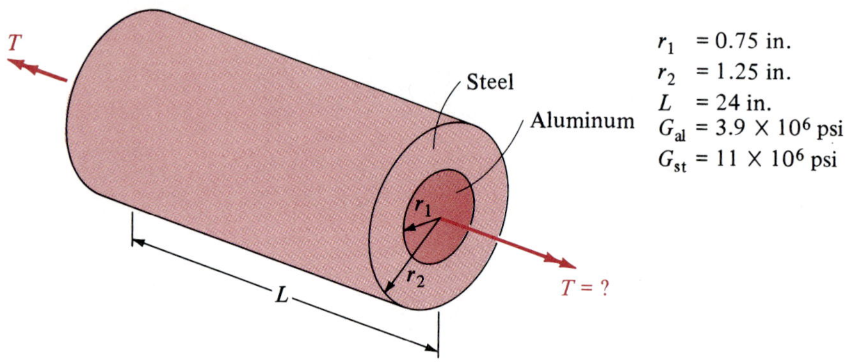

Ps. 6.89 and 6.90

6.90 Repeat problem 6.89 with the positions of aluminum and steel reversed.

6.91 The steel core and the aluminum shell are attached to a rigid plate at the ends as indicated in the figure. Determine *(a)* the maximum shear stress in the steel, *(b)* the maximum shear stress in the aluminum, *(c)* the part of the torque carried by the steel, *(d)* the part of the torque carried by the aluminum, and *(e)* the relative angle of twist between the ends of the bar.

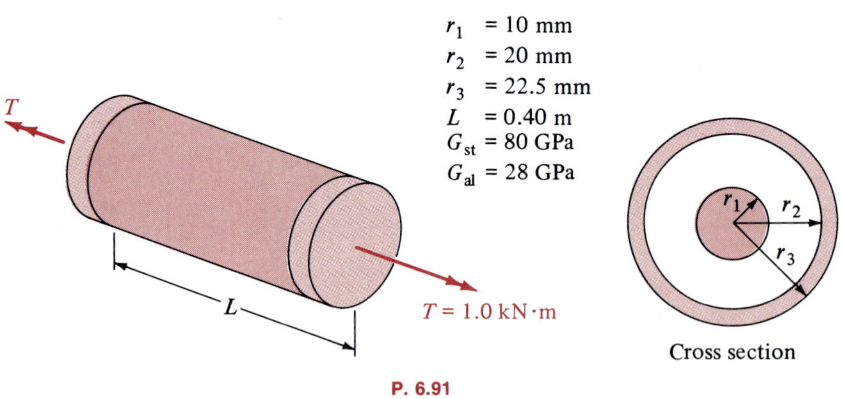

P. 6.91

6.92 A composite bar consists of a brass core and an aluminum shell as indicated. What should the interface radius be in order that each component carry half the torque? What are the corresponding maximum shear stresses in the aluminum and the brass portions? What is the relative angle of twist between the ends.

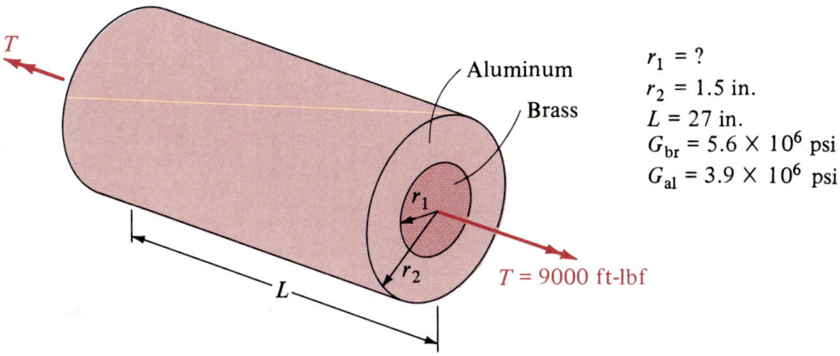

Ps. 6.92 and 6.93

6.93 Repeat problem 6.92 with the positions of aluminum and brass reversed.

6.94 The composite torsional member is composed of an inner aluminum core and an outer steel shell. If the allowable shear stresses in the aluminum and steel are 110 MPa and 160 MPa, respectively, determine the allowable torque that can be transmitted. What is the stress in the noncritical material? What is the relative angle of twist between the ends of the bar?

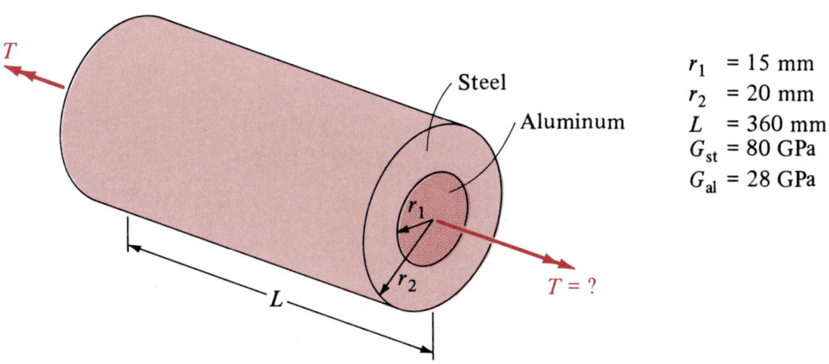

Ps. 6.94 and 6.95

6.95 Repeat problem 6.94 with the positions of aluminum and steel reversed.

6.96 The steel core and the aluminum shell are attached to a rigid plate at the ends as indicated in the figure. Determine *(a)* the maximum shear stress in the steel, *(b)* the maximum shear stress in the aluminum, *(c)* the part of the torque carried by the steel, *(d)* the part of the torque carried by the aluminum, and *(e)* the relative angle of twist between the ends of the bar.

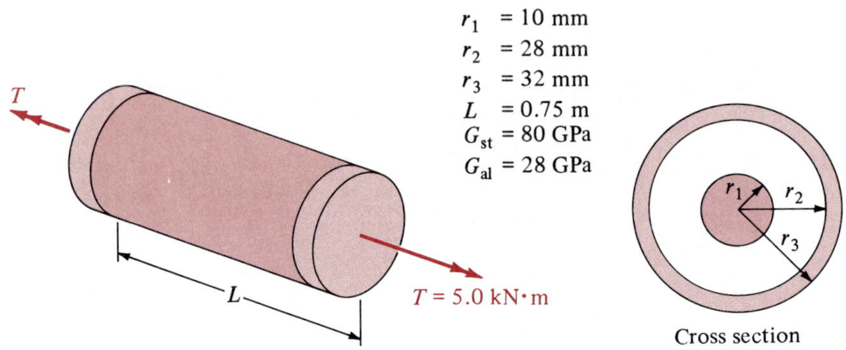

P. 6.96

6.97 The steel core and the brass shell are attached to a rigid plate at the ends as indicated in the figure. What should be the diameter of the steel core in order that the torque carried by the brass be twice that carried by the steel? What are the corresponding maximum shear stresses in the steel and the brass portions? What is the relative angle of twist between the ends of the bar?

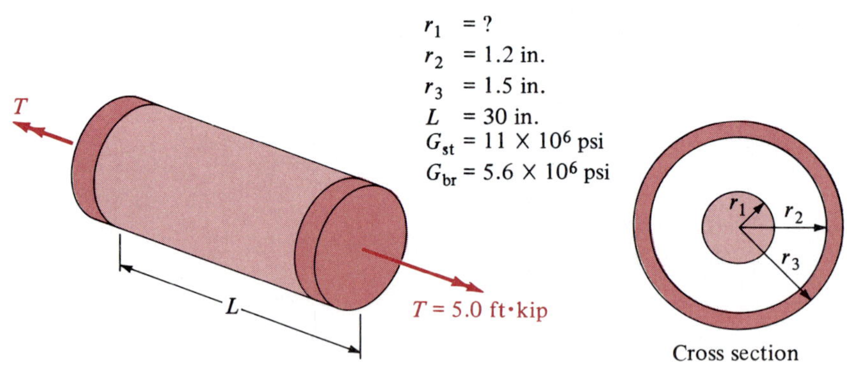

P. 6.97

6.4.2 Stiffness Formulations

As we stated in Section 5.4.2, the stiffness method is based on considering the displacements as the unknowns. The strategy of the stiffness method is to express the internal forces in terms of the unknown displacements. The internal forces are then eliminated from the equilibrium equations, resulting in a set of equilibrium equations expressed in terms of the unknown displacements. These equations are then solved for the displacements, after which the internal forces can be determined.

Example 6.13 Consider again the problem of Example 6.10, shown in Fig. 6.42a. Determine the torques and rotations using a stiffness formulation.

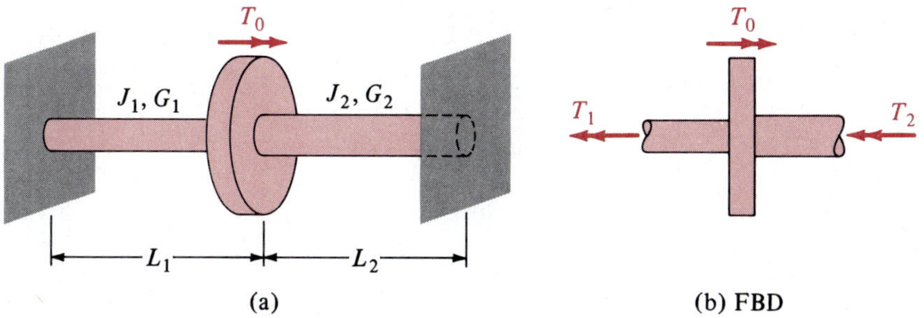

Figure 6.42 Two shafts with an intermediate torque

Solution Referring to the free-body diagram of Fig. 6.42b, the equilibrium equation is

$$T_0 = T_1 + T_2$$

The appropriate force-displacement relation for the torsion problem is the torque-twist equation

$$T = C\phi$$

where $C = JG/L$.

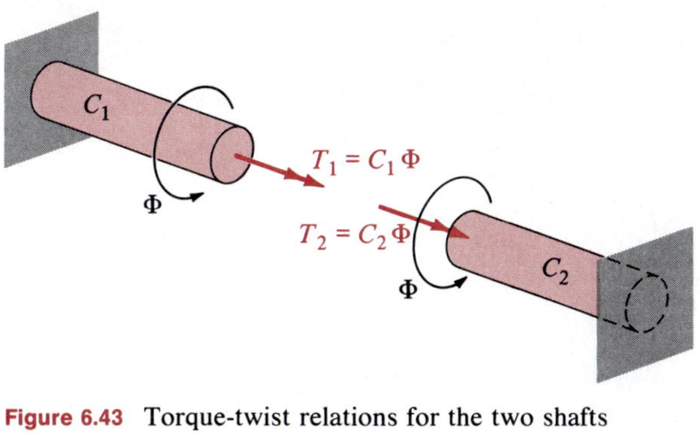

Figure 6.43 Torque-twist relations for the two shafts

Denoting Φ as the angle of rotation at the disk as indicated in Fig. 6.43, the force-displacement relations for the two shafts are

$$T_1 = C_1\Phi \qquad \text{and} \qquad T_2 = C_2\Phi$$

Note that continuity of rotation is ensured by virtue of having used Φ for both T_1 and T_2. Substituting these two relations into the equation of equilibrium, $T_0 = T_1 + T_2$, yields

$$C_1\Phi + C_2\Phi = T_0$$

from which

$$\Phi = \frac{T_0}{C_1 + C_2}$$

The individual torques T_1 and T_2 can then be determined as

$$T_1 = C_1\Phi = \frac{C_1 T_0}{C_1 + C_2}$$

and

$$T_2 = C_2\Phi = \frac{C_2 T_0}{C_1 + C_2}$$

All results are seen to be precisely the same as obtained using the flexibility method.

Example 6.14 Use a stiffness formulation to solve the statically indeterminate problem of Example 6.11.

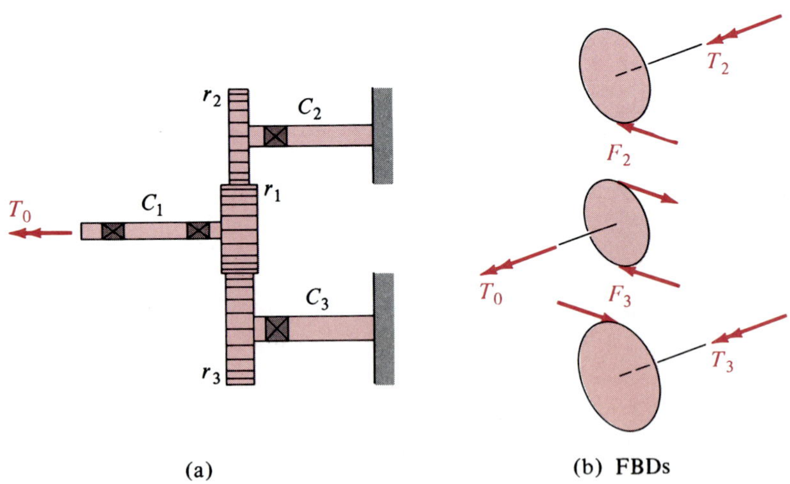

Figure 6.44 Statically indeterminate torsional system

Solution Referring to Example 6.11 and its figure, repeated in Fig. 6.44, the equilibrium equation is

$$\frac{T_0}{r_1} = \frac{T_2}{r_2} + \frac{T_3}{r_3} \tag{a}$$

and the equations relating the displacements are

$$r_1\phi_1 = r_2\phi_2 \qquad \text{and} \qquad r_1\phi_1 = r_3\phi_3 \tag{b}$$

The torque-twist relations are

$$T_2 = C_2\phi_2 \qquad \text{and} \qquad T_3 = C_2\phi_3 \tag{c}$$

Denoting ϕ_1 as Φ, and using Eqns. *(b)*, the rotations ϕ_2 and ϕ_3 can be expressed in terms of Φ as

$$\phi_2 = \frac{r_1}{r_2}\Phi \qquad \text{and} \qquad \phi_3 = \frac{r_1}{r_3}\Phi \tag{d}$$

Substituting the expressions for ϕ_2 and ϕ_3 from Eqns. *(d)* into Eqns. *(c)*, and then substituting T_2 and T_3 into the equilibrium Eqn. *(a)*, yields

$$\frac{T_0}{r_1} = C_2\frac{r_1}{r_2^2}\Phi + C_3\frac{r_1}{r_3^2}\Phi$$

from which

$$\Phi = \frac{r_2^2}{r_1^2}\frac{r_3^2}{C_2r_3^2 + C_3r_2^2}T_0$$

coinciding with the value of ϕ_1 obtained using the flexibility formulation in Example 6.11. Using Eqns. *(c)* and *(d)*, the values for T_2, T_3, ϕ_2, and ϕ_3 are, respectively,

$$T_2 = \frac{r_2}{r_1}\frac{C_2r_3^2}{C_2r_3^2 + C_3r_2^2}T_0 \qquad T_3 = \frac{r_3}{r_1}\frac{C_3r_2^2}{C_2r_3^2 + C_3r_2^2}T_0$$

$$\phi_2 = \frac{r_2}{r_1}\frac{r_3^2}{C_2r_3^2 + C_3r_2^2}T_0 \qquad \phi_3 = \frac{r_3}{r_1}\frac{r_2^2}{C_2r_3^2 + C_3r_2^2}T_0$$

the same as in Example 6.11.

PROBLEMS

6.98–6.107 Resolve problems 6.74–6.83 using the stiffness method.

6.4.3 Overview of the Torsional Deformation Problem

At this point it is helpful to consider an overview of the torsional deformation problem. Fig. 6.45 is a flowchart of the steps necessary to solve a statically determinate torsional deformation problem.

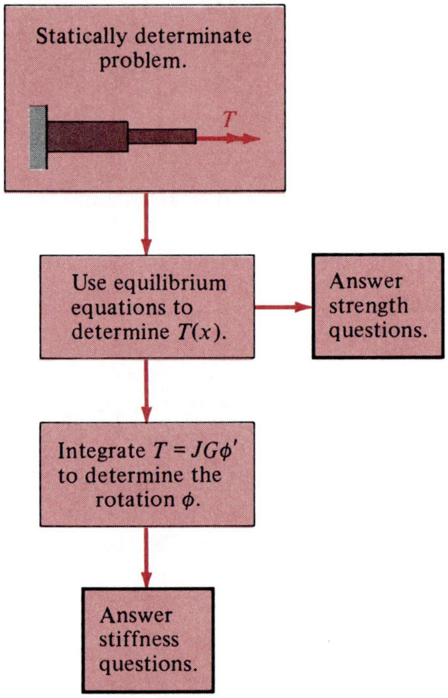

Figure 6.45 Flowchart of the steps for a statically determinate torsion problem

Equilibrium is satisfied using either free-body diagrams or the integration of $T' + t$ to determine the torque $T(x)$. Questions of strength can then be answered using the elastic torsion formula $\tau = Tr/J$. The torque-twist equation $T = JG\phi'$ can then be integrated to determine the rotations and thus answer any questions regarding stiffness.

Fig. 6.46 is a flowchart of the steps necessary to solve a statically indeterminate torsional deformation problem.

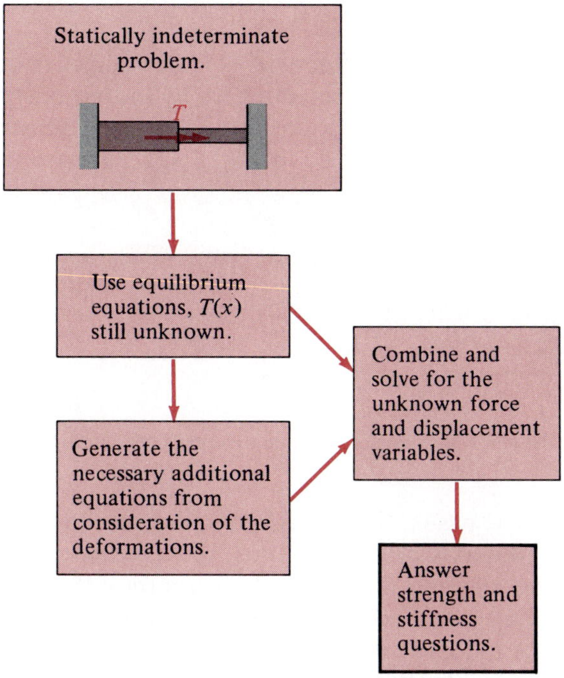

Figure 6.46 Flowchart of the steps for a statically indeterminate problem

Equilibrium is satisfied using either free-body diagrams or the integration of $T' + t$, at which point there are still unknown torques appearing in the equilibrium equations. The necessary additional equations are obtained by considering the deformations. The total number of equilibrium and deformation equations must be solved for the unknowns, after which both the strength and stiffness questions can be answered.

6.5 EXCEPTIONS TO THE THEORY

The development of the theory for the torsional problem involved requirements regarding *(a)* the geometry of the region, *(b)* equilibrium, *(c)* geometry of deformation, and *(d)* material behavior or stress-strain relations. If one or more of these assumptions is not satisfied, the predictions based on the theory can be substantially in error, either quantitatively or qualitatively. In this section we will consider several technically important situations in which one or more of the requirements of the theory are not satisfied, and will present the corresponding changes in the results of the theory.

6.5.1 Stress Concentrations

Recall that in Chapter 5, the effects of sudden changes in cross section such as holes and fillets were accounted for by the use of stress concentration factors. Analogous situations that are technically important for the torsion problem are indicated in Fig. 6.47.

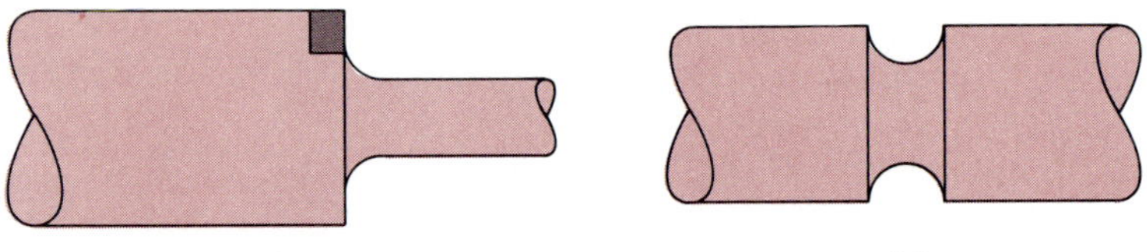

Figure 6.47 Typical geometries resulting in stress concentrations

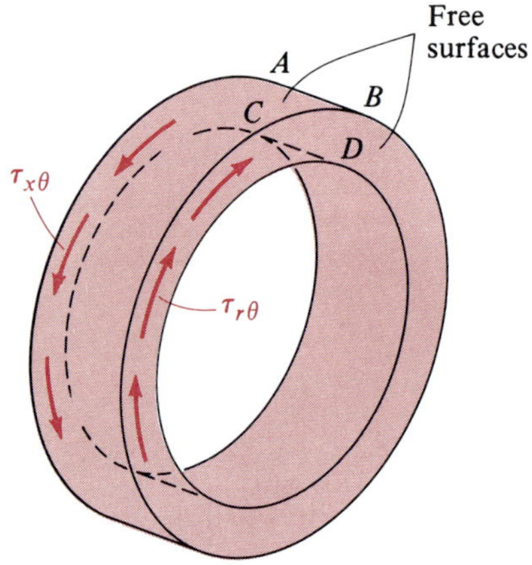

Figure 6.48 Equilibrium of a segment for a stepped geometry

In each of these cases, the radius is not a slowly varying function of position along the axis of the bar; thus, the assumption regarding the geometry of the region is not satisfied. For both of the geometries it is easily shown that it is necessary to have stresses other than the $\tau_{x\theta}$ predicted by the theory. For example, consider a free-body diagram of the shaded area for the fillet of Fig. 6.47a, as shown in Fig. 6.48.

The resultant of any shear stresses $\tau_{x\theta}$ acting on the surface corresponding to segment AC is a torque that can only be equilibrated by a torque composed of stresses $\tau_{r\theta}$ on the surface corresponding to segment CD; thus, there must be stresses other than those predicted by the theory. A similar analysis of a properly chosen free-body diagram for the geometry of Fig. 6.47b produces a similar conclusion.

For a stepped shaft, the geometry is characterized by the parameters D, d, and r shown in Fig. 6.49.

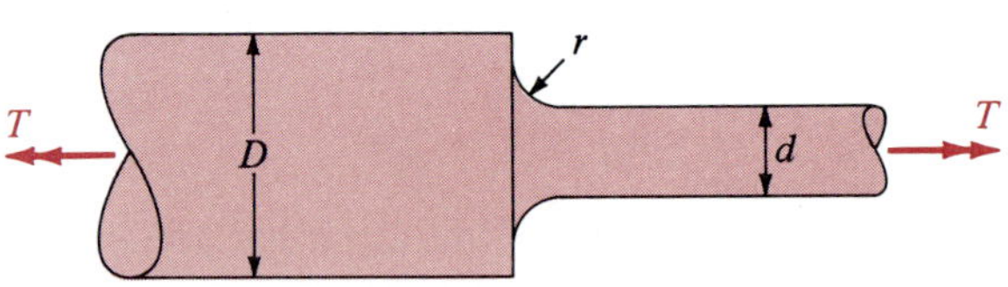

Figure 6.49 Standard geometry for a stepped shaft

The maximum shear stress is computed in terms of a stress concentration factor K given by

$$K = \frac{\tau_{max}}{\tau_{avg}}$$

or

$$\tau_{max} = K\tau_{avg}$$

where τ_{avg} is the shear stress $\tau_{x\theta}$ computed on the basis of the reduced section, or

$$\tau_{avg} = \frac{Tr}{J} = \frac{T\left(\frac{d}{2}\right)}{\frac{\pi}{2}\left(\frac{d}{2}\right)^4} = \frac{2T}{\pi\left(\frac{d}{2}\right)^3}$$

The maximum value occurs in the smaller shaft at the fillet radius. The curves that are used to determine K are shown in Fig. 6.50. The values of r/d and D/d are determined from the geometry, after which the value of K is read from the appropriate curve of the graph.

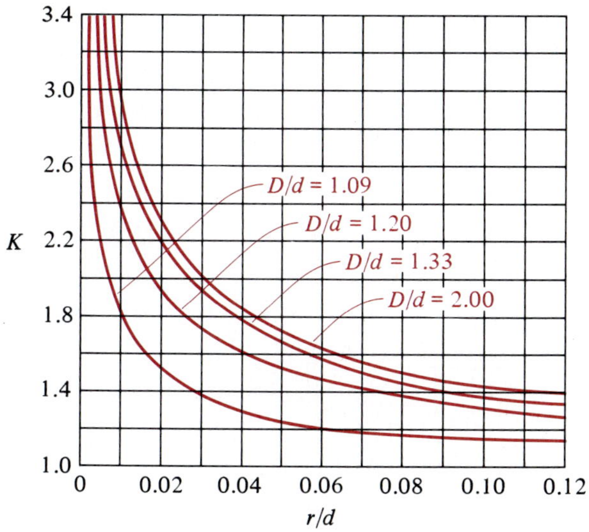

Figure 6.50 Stress concentration factors for a stepped shaft in torsion

Example 6.15 The stepped shaft shown in Fig. 6.51 transmits a torque of 4000 in.-lbf. Determine the maximum shear stress based on using the corresponding stress concentration factor.

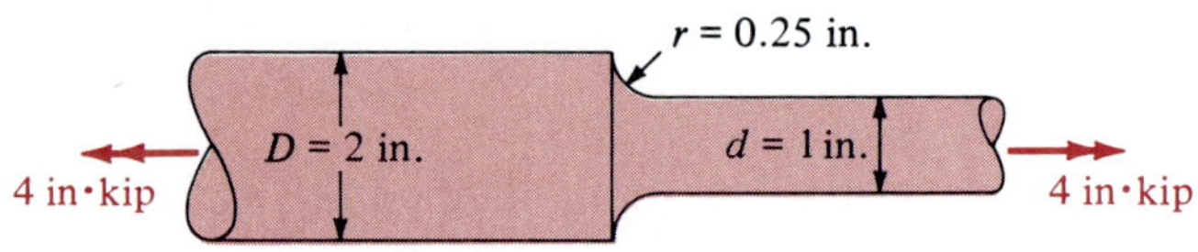

Figure 6.51 Stepped shaft transmitting a torque

Solution The average value of the shear stress based on the reduced section is

$$\tau_{avg} = \frac{2 \cdot 4000 \text{ in.-lbf}}{\pi (0.5)^3 \text{ in}^3} = 20.4 \text{ ksi}$$

The r/d ratio is $0.25/1.00 = 0.25$ and the D/d ratio is $2.0/1.0 = 2.0$. The corresponding stress concentration factor is determined from Fig. 6.50 as $K \approx 1.4$. The corresponding maximum shear stress is then

$$\tau_{max} = 1.4\tau_{avg} = 28.5 \text{ ksi}$$

For the case of a shaft with a groove, the geometry is defined by the parameters r, d, and D as shown in Fig. 6.52.

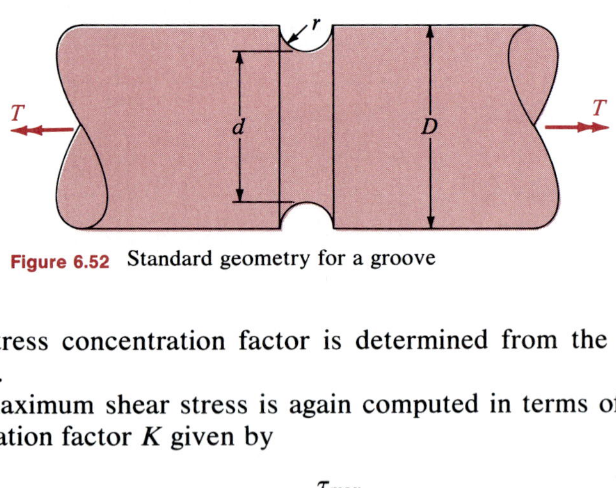

Figure 6.52 Standard geometry for a groove

The stress concentration factor is determined from the graph of Fig. 6.53.

The maximum shear stress is again computed in terms of a stress concentration factor K given by

$$K = \frac{\tau_{max}}{\tau_{avg}}$$

or

$$\tau_{max} = K\tau_{avg}$$

where τ_{avg} is the shear stress $\tau_{x\theta}$ computed on the basis of the reduced section. The maximum value occurs at the point of minimum radius. K is determined from Fig. 6.53 on the basis of the r/d and D/d ratios.

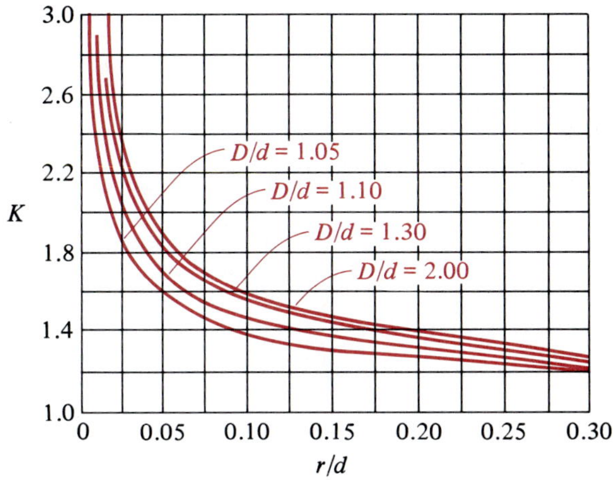

Figure 6.53 Stress concentration factor for a grooved shaft in torsion

Example 6.16 The shaft shown in Fig. 6.54 transmits a torque of 6000 in.-lbf. Determine the maximum shear stress based on using the corresponding stress concentration factor.

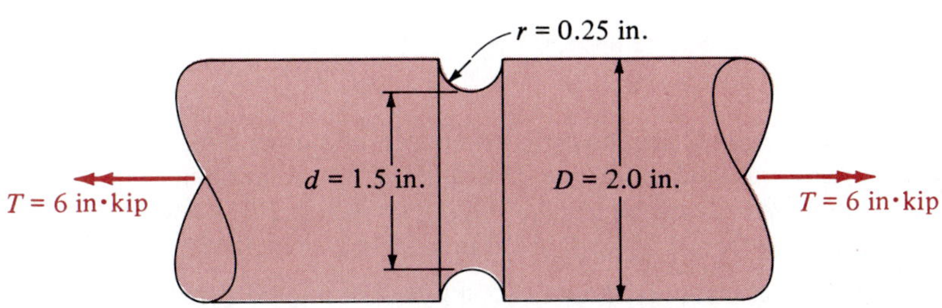

Figure 6.54 Grooved shaft transmitting a torque

Solution The average value of the shear stress based on the reduced section is

$$\tau_{avg} = \frac{2(6000) \text{ in.-lbf}}{\pi(0.75)^3 \text{ in}^3} = 9.05 \text{ ksi}$$

The r/d ratio is $0.25/1.50 = 0.167$ and the D/d ratio is $2.0/1.5 = 1.33$. From the curve of Fig. 6.53, the corresponding stress concentration factor is estimated as $K \approx 1.32$. The corresponding maximum shear stress is then

$$\tau_{max} = 1.32\tau_{avg} = 11.9 \text{ ksi}$$

PROBLEMS

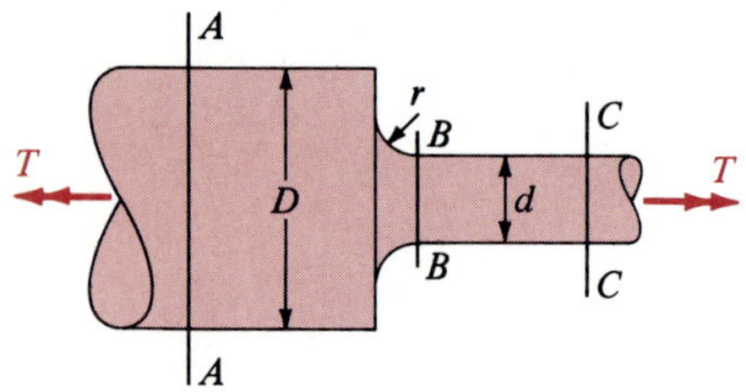

Ps. 6.108 and 6.109

6.108 The stepped shaft shown transmits a torque of 250 ft-lbf. Determine the maximum shear stress at sections AA, BB, and CC. Take $D = 2$ in., $d = 1.0$ in., and $r = 0.1$ in.

6.109 The stepped shaft shown transmits a torque of 25 N·m. Determine the maximum shear stress at sections AA, BB, and CC. Take $D = 16$ mm, $d = 10$ mm, and $r = 1$ mm.

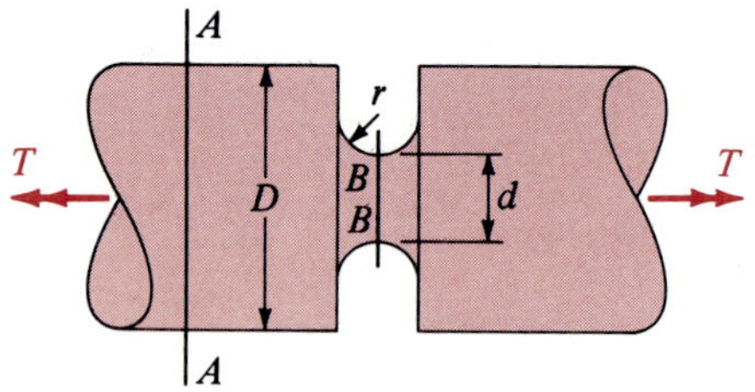

Ps. 6.110 and 6.111

6.110 The grooved shaft shown transmits a torque of 900 ft.-lbf. Determine the maximum shear stress at sections AA and BB. Take $D = 1.75$ in., $d = 1.25$ in., and $r = 0.5$ in.

6.111 The grooved shaft shown transmits a torque of 2.0 kN·m. Determine the maximum shear stress at sections AA and BB. Take $D = 60$ mm, $d = 35$ mm, and $r = 5$ mm.

6.112 The stepped shaft shown is composed of a ductile material with $\sigma_Y =$ 40 ksi, and transmits a torque of 1100 ft-lbf. Based on computing the maximum shear stress according to $\tau_{\max} = K\tau_{\text{avg}}$ at section BB, what is the factor of safety against yielding? Take $D = 1.8$ in., $d = 0.9$ in., and $r = 0.1$ in.

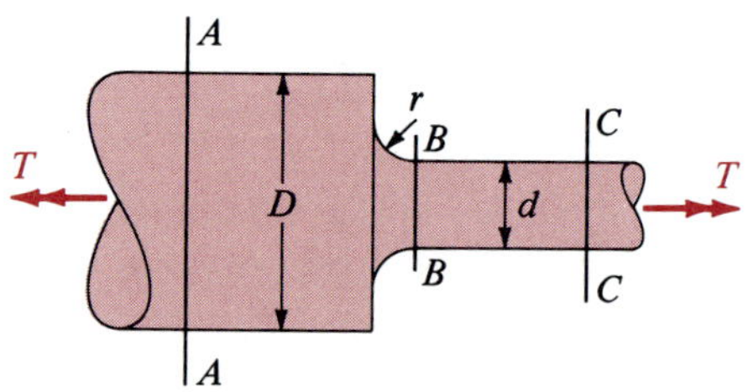

Ps. 6.112 and 6.113

6.113 The stepped shaft shown is composed of a ductile material with $\sigma_Y =$ 320 MPa, and transmits a torque of 470 N·m. Based on computing the maximum shear stress according to $\tau_{\max} = K\tau_{\text{avg}}$ at section BB, what is the factor of safety against yielding? Take $D = 50$ mm, $d = 28$ mm, and $r = 3$ mm.

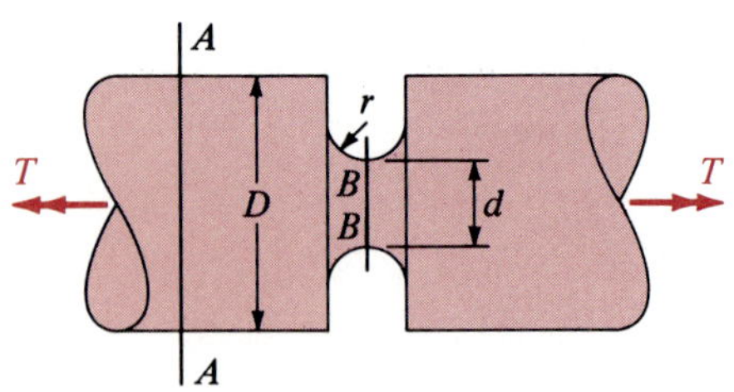

Ps. 6.114 and 6.115

6.114 The grooved shaft shown is composed of a ductile material with $\sigma_Y =$ 36 ksi, and transmits a torque of 100 ft-lbf. Based on computing the maximum shear stress according to $\tau_{\max} = K\tau_{\text{avg}}$ at section BB, what is the factor of safety against yielding? Take $D = 1.75$ in., $d = 0.9$ in., and $r = 0.375$ in.

6.115 The grooved shaft shown is composed of a ductile material with $\sigma_Y =$ 280 MPa, and transmits a torque of 1.65 kN·m. Based on computing the maximum shear stress according to $\tau_{\max} = K\tau_{\text{avg}}$ at section BB, what is the factor of safety against yielding? Take $D = 90$ mm, $d = 50$ mm, and $r = 10$ mm.

6.116 The stepped shaft shown is composed of a ductile material with $\sigma_Y =$ 40 ksi, and transmits a torque of 200 ft-lbf. Based on computing the maximum shear stress according to $\tau_{max} = K\tau_{avg}$ at section BB, what is the minimum allowable r if the factor of safety against yielding is 2.4? Take $D = 2.5$ in. and $d = 1.5$ in.

6.117 The stepped shaft shown is composed of a ductile material with $\sigma_Y =$ 280 MPa, and transmits a torque of 0.75 kN·m. Based on computing the maximum shear stress according to $\tau_{max} = K\tau_{avg}$ at section BB, what is the minimum allowable r if the factor of safety against yielding is to be 1.6? Take $D = 60$ mm and $d = 40$ mm.

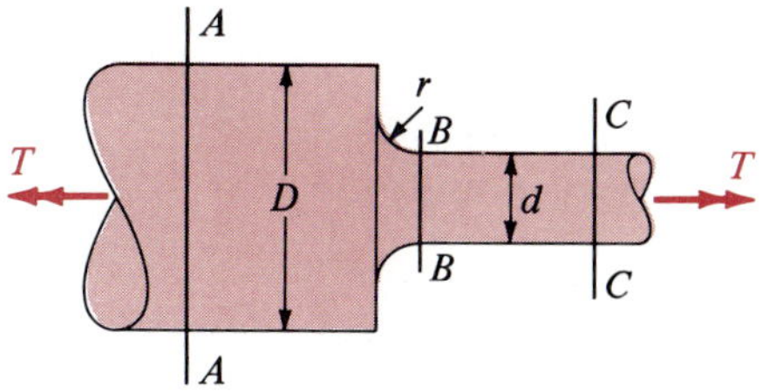

Ps. 6.116 and 6.117

6.118 The grooved shaft shown is composed of a ductile material with $\sigma_Y =$ 40 ksi, and transmits a torque of 500 ft-lbf. Based on computing the maximum shear stress according to $\tau_{max} = K\tau_{avg}$ at section BB, what is the minimum allowable r if the factor of safety against yielding is to be 1.8? Take $D = 3.0$ in. and $d = 1.5$ in.

6.119 The grooved shaft shown is composed of a ductile material with $\sigma_Y =$ 280 MPa and transmits a torque of 0.25 kN·m. Based on computing the maximum shear stress according to $\tau_{max} = K\tau_{avg}$ at section BB, what is the minimum allowable r if the factor of safety against yielding is to be 2.1? Take $D = 45$ mm and $d = 30$ mm.

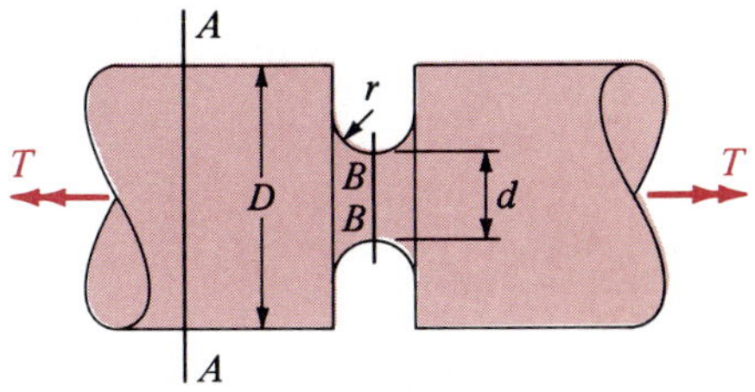

Ps. 6.118 and 6.119

6.5.2 Inelastic Behavior

If the torque transmitted by a circular bar produces stresses above the elastic range, the governing equations must be altered accordingly. In this section we will investigate situations where Hooke's Law is no longer valid, and will consider in detail the case of the elastic–perfectly plastic material. Each of the three basic ideas will be considered in turn while we develop the resulting theory.

Elastic–Perfectly Plastic Behavior As outlined in Section 4.3.1, the behavior of ductile materials is frequently idealized as shown in Fig. 6.55.

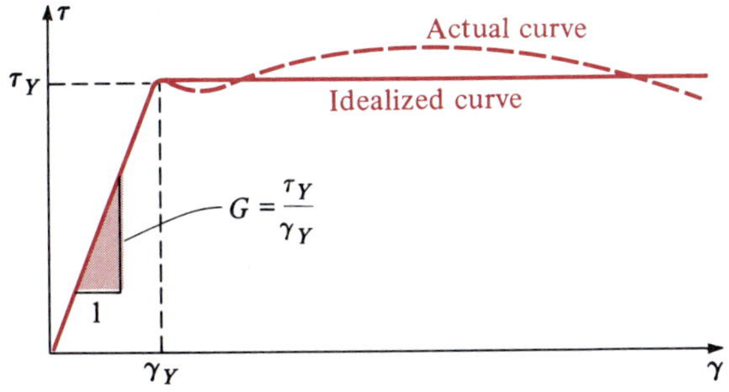

Figure 6.55 Ductile material idealized as elastic–perfectly plastic

If the applied torque is such that the shear strain γ_{max} does not exceed γ_Y, the maximum shear stress is given by $\tau = G\gamma_{max}$ as shown in Fig. 6.56a.

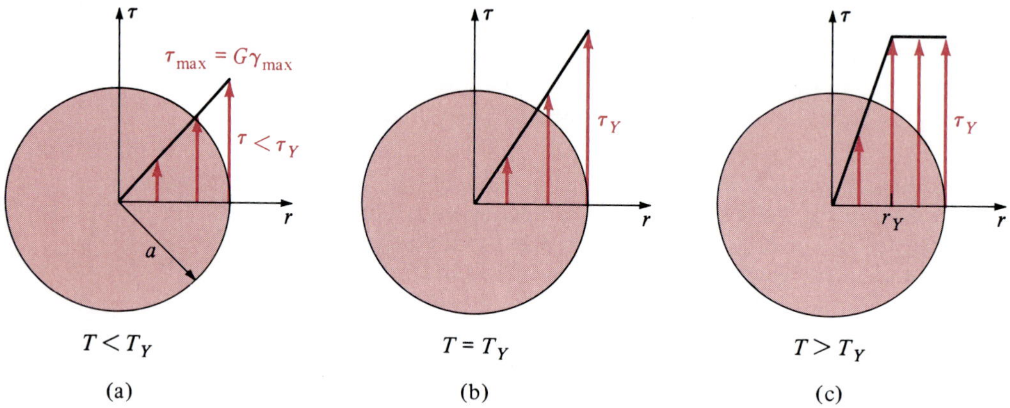

Figure 6.56 Shear-stress distributions for various torques T

If the applied torque is such that the shear strain γ_{max} is equal to γ_Y, the maximum shear stress is given by $\tau_Y = G\gamma_Y$ as shown in Fig. 6.56b. The corresponding torque is designated by T_Y and is referred to as the **maximum elastic torque.** Using the equation $\tau = Tr/J$, which is still valid since the stresses are still in the elastic range, it follows that for the maximum elastic torque

$$\tau_Y = \frac{T_Y a}{J}$$

from which

$$T_Y = \frac{\pi a^3 \tau_Y}{2} \tag{6.14}$$

When $T > T_Y$, the shear-stress distribution is as indicated in Fig. 6.56c. The inner elastic core corresponds to the shear strain not having exceeded γ_Y, whereas outside of the inner core the shear strain has exceeded γ_Y so that on the basis of the assumed elastic–plastic behavior, the shear stress is everywhere equal to τ_Y. The actual location of the boundary between the elastic and plastic regions, indicated by the radius r_Y, depends upon the amount by which T has exceeded T_Y.

The relation between T_Y and r_Y is an equilibrium relation, obtained by requiring that the moment of the stresses be balanced by the torque

T. Using the general expression for the torque in terms of the integral of the shear stresses and using

$$\tau = \left(\frac{r}{r_Y}\right)\tau_Y \qquad 0 \le r \le r_Y$$

and

$$\tau = \tau_Y \qquad r_Y \le r \le a$$

there results

$$\begin{aligned} T &= \int_{\text{area}} r\tau \, dA = 2\pi \left[\int_0^{r_Y} r\left(\frac{r}{r_Y}\right)\tau_Y r\, dr + \int_{r_Y}^{a} r\tau_Y r\, dr\right] \\ &= \frac{2\pi}{4} r_Y^3 \tau_Y + \frac{2\pi}{3}(a^3 - r_Y^3)\tau_Y = \frac{2\pi a^3}{3}\tau_Y\left[1 - \frac{1}{4}\left(\frac{r_Y}{a}\right)^3\right] \end{aligned} \tag{6.15}$$

With $T_Y = \pi a^3 \tau_Y / 2$, it follows that

$$T = \frac{4}{3} T_Y \left[1 - \frac{1}{4}\left(\frac{r_Y}{a}\right)^3\right] \tag{6.16}$$

Eqn. (6.16) can be used to determine the region over which the stresses are plastic when $T > T_Y$. Note as a check that as $r_Y \rightarrow a$, $T \rightarrow T_Y$ as required. Also note that as $r_Y \rightarrow 0$, $T \rightarrow (4/3)T_Y = T_P$, where T_P is defined as the **fully plastic torque.**

In order to determine the deformations in a circular shaft composed of an elastic–perfectly plastic material and acted upon by a single torque T, consider Fig. 6.57.

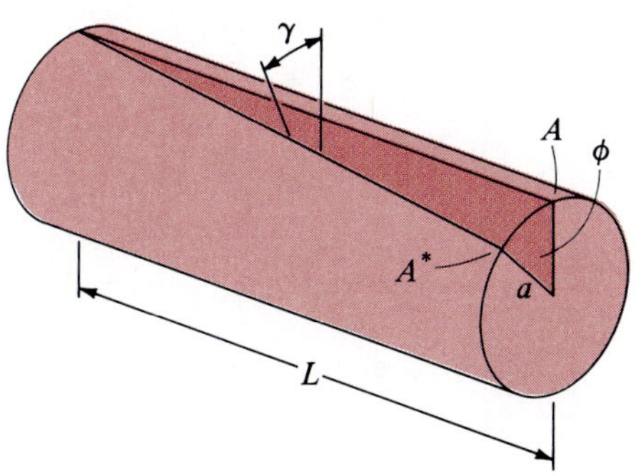

Figure 6.57 Geometry of deformation in a circular shaft

Continuing to assume that plane sections remain plane, the arc length AA^* can be represented in two ways in terms of the shear strain γ and the rotation ϕ, as

$$\text{arc}(AA^*) = L\gamma = a\phi \tag{6.17}$$

In particular, when $T = T_Y$ then $\gamma = \gamma_Y$, and Eqn. (6.17) becomes

$$L\gamma_Y = a\phi_Y \tag{6.18}$$

where ϕ_Y is the twist of the bar when yielding first occurs. On the basis of the basic relationship

$$\gamma = r\phi' = r\frac{\phi}{L}$$

it follows that for $T > T_Y$, γ_Y occurs at a location $r_Y < a$ given by

$$L\gamma_Y = r_Y\phi \tag{6.19}$$

where ϕ is the rotation corresponding to the torque $T > T_Y$. Dividing Eqns. (6.18) and (6.19) yields

$$\phi = \left(\frac{a}{r_Y}\right)\phi_Y \tag{6.20}$$

Eqn. (6.16) for the torque can then be written as

$$T = \frac{4}{3}T_Y\left[1 - \frac{1}{4}\left(\frac{\phi_Y}{\phi}\right)^3\right] \tag{6.21}$$

which is valid for $\phi \geq \phi_Y = T_Y L/JG$. Eqn. (6.21) can be used to determine the rotation ϕ when $T > T_Y$. A plot of the relationship between T and ϕ is shown in Fig. 6.58.

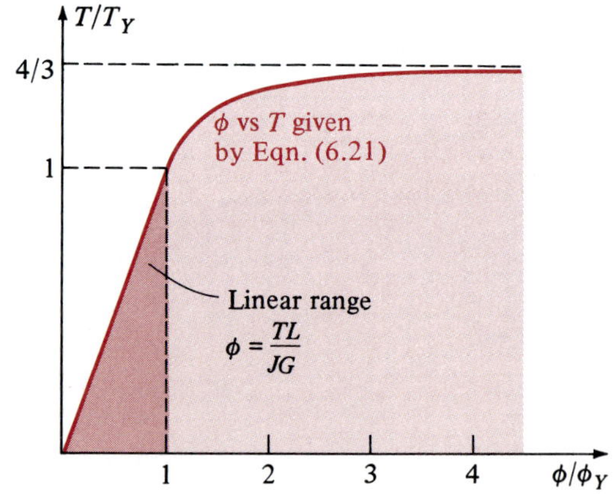

Figure 6.58 Torque-twist relation for a bar made of an elastic–perfectly plastic material

Table 6.1 gives some of the corresponding numerical values and shows clearly that the fully plastic torque is realized quickly for rotations beyond the elastic range, that is, when $\phi = 2\phi_Y$, $T \approx 0.97T_P$.

Table 6.1 Numerical values of the T versus ϕ relation for a bar composed of an elastic–perfectly plastic material

ϕ/ϕ_Y	T/T_Y	T/T_P
1.0	1.00	0.750
1.25	1.163	0.872
1.5	1.235	0.926
2.0	1.292	0.969
3.0	1.321	0.991
4.0	1.328	0.996

Example 6.17 The steel shaft in Fig. 6.59a has been idealized to have the stress-strain relation shown in Fig. 6.59b. Determine the maximum elastic torque and the corresponding angle of twist. If the torque is then increased by 10%, determine the stress distribution and the twist.

Solution The maximum elastic torque is given by Eqn. (6.14) as

$$T_Y = \frac{1}{2}\pi a^3 \tau_Y = \frac{1}{2}\pi(0.020\text{ m})^3(130 \times 10^6\text{ N/m}^2) = 1634\text{ N·m}$$

Using Eqn. (6.18) it follows that

$$\phi_Y = \frac{L}{a}\gamma_Y = (1.0\text{ m})\left(\frac{0.001625}{0.020\text{ m}}\right) = 0.0813\text{ rads} = 4.66°$$

If the torque is then increased 10% to (1.1)(1634 N·m) = 1797 N·m, the corresponding radius inside of which the stresses are elastic is given by Eqn. (6.16), namely

$$T = \frac{4}{3}T_Y\left[1 - \frac{1}{4}\left(\frac{r_Y}{a}\right)^3\right]$$

or

$$1797\text{ N·m} = \frac{4}{3}(1634\text{ N·m})\left[1 - \frac{1}{4}\left(\frac{r_Y}{a}\right)^3\right]$$

from which

$$\frac{r_Y}{a} = 0.888 \qquad \text{or} \qquad r_Y = 17.8\text{ mm}$$

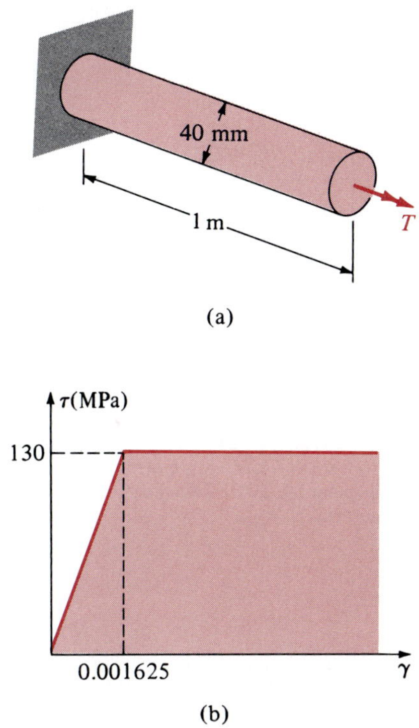

Figure 6.59 Torsion of an elastic–perfectly plastic shaft

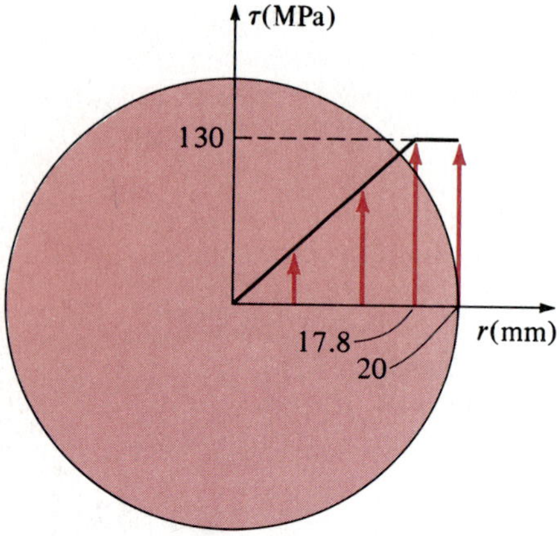

Figure 6.60 Stresses for $T = 1.1T_Y$

The corresponding stress diagram is shown in Fig. 6.60. The corresponding value of ϕ for $T = 1.1T_Y$ is obtained from Eqn. (6.19) as

$$\phi = (a/r_Y)\phi_Y = \frac{20}{17.8} 4.66 \text{ degrees} = 5.24 \text{ degrees}$$

The student should show that ϕ can also be obtained from Eqn. (6.21), namely

$$T = \frac{4}{3} T_Y \left[1 - \frac{1}{4} \left(\frac{\phi_Y}{\phi} \right)^3 \right]$$

the equation relating T and ϕ.

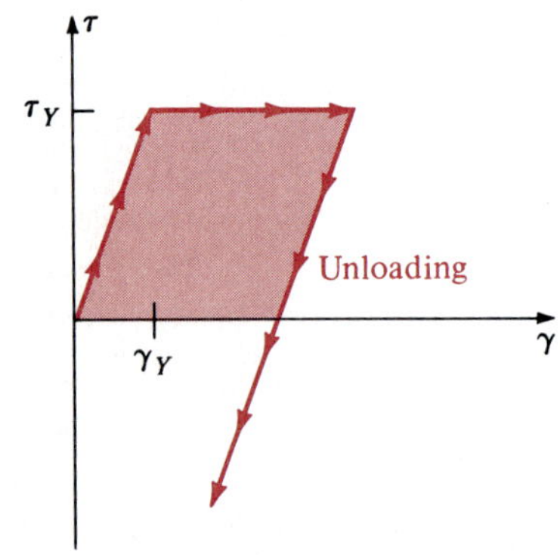

Figure 6.61 Unloading stress-strain diagram for an elastic–perfectly plastic material

Residual Stresses If the torque is removed from an elastic–perfectly plastic shaft that has been stressed above the yield point, there will be permanent deformation in the form of a residual rotation ϕ_R. There will also be corresponding residual shear stresses. We can address these stresses by recalling from Chapter 4 that unloading from stresses above the elastic limit occurs in a fashion indicated in Fig. 6.61.

We assume that the unloading curve remains linear over a range of $2\tau_Y$, that is, until the stress has dropped to $-\tau_Y$. On the corresponding T versus ϕ diagram the unloading takes place as indicated in Fig. 6.62.

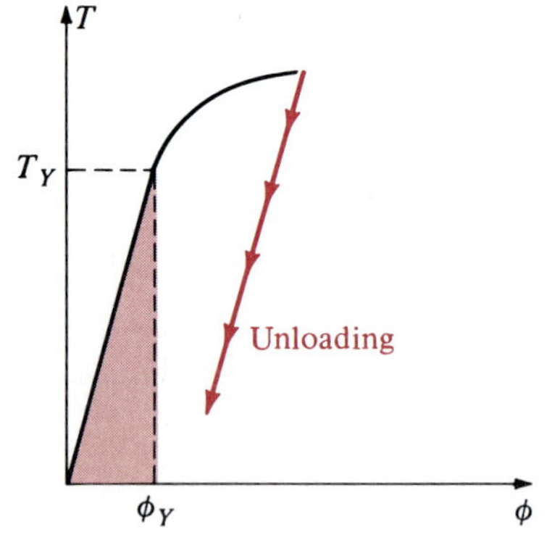

Figure 6.62 Relation between T and ϕ during unloading

The basic equilibrium idea is that when the torque is removed, the moment of the stresses acting on an exposed face must also vanish. At the beginning of the unloading process, the stresses appear as shown in Fig. 6.63a.

The additional stresses that result during the removal of the external torque are indicated in Fig. 6.63b; they are linearly dependent upon r as given by the linear theory. The stresses at the end of the unloading process appear in Fig. 6.63c and must be such that

$$T = \int r\tau \, dA = 0$$

As indicated in the figure, the residual stress distribution is obtained by superposing the elastic unloading stresses of Fig. 6.63b upon the elastic–perfectly plastic stresses of Fig. 6.63a to yield the final bilinear stress distribution shown in Fig. 6.63c.

The value of the shear stress τ_E is determined by noting that the value of the superposed torque T_E associated with the stress τ_E must be equal and opposite to the torque T_{EP} that was present at the beginning of the unloading process, or

$$T_{EP} = T_E$$

or with $\tau_E = T_E 2/(\pi a^3)$,

$$\tau_E = \frac{2T_{EP}}{\pi a^3} \tag{6.22}$$

The corresponding residual rotation is obtained by superposing the rotation due to the torque T_E on the initial rotation corresponding to the torque T_{EP}. With ϕ_{EP} the rotation corresponding to the torque T_{EP}, the residual rotation can be expressed as

$$\phi_R = \phi_{EP} - \frac{T_E L}{JG}$$

or with $T_E = J\tau_E/a$

$$= \phi_{EP} - \frac{\tau_E}{G}\frac{L}{a} \tag{6.23}$$

where ϕ_R is the residual angle of twist.

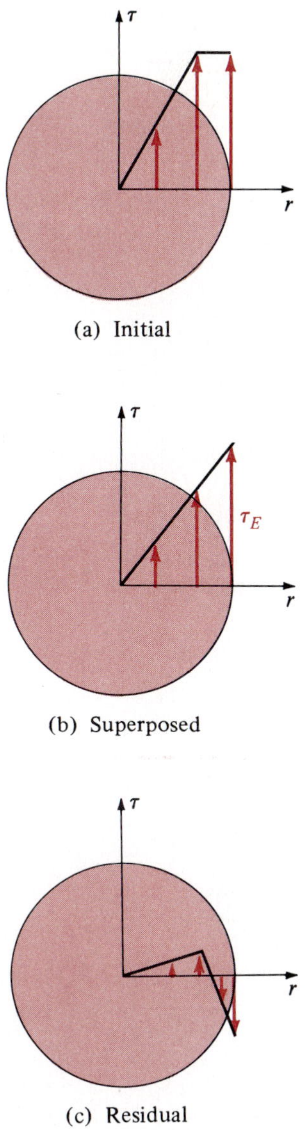

Figure 6.63 Superposition of elastic–perfectly plastic and elastic stresses

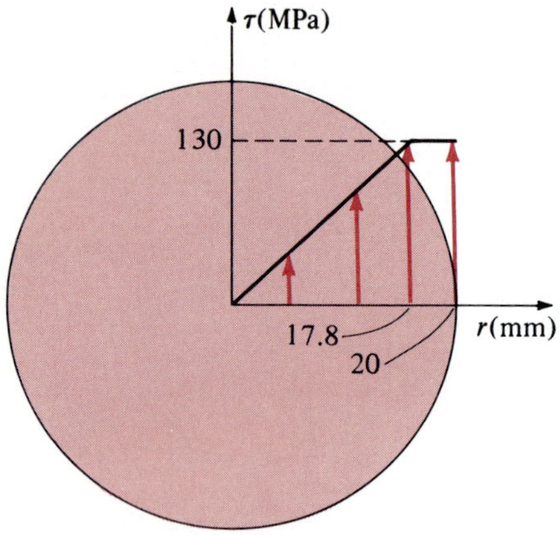

Figure 6.64 Elastic–perfectly plastic initial stresses

Example 6.18 If the torque $T = 1.1\,T_Y$ is removed from the shaft of Example 6.17, determine the distribution of residual stresses and the final angle of twist.

Solution Consulting Example 6.17, the initial stress distribution is as indicated in Fig. 6.64.

Using Eqn. (6.22) with $T_{EP} = 1.1\,T_Y = 1797$ N·m there results

$$\tau_E = \frac{2(1797 \text{ N}\cdot\text{m})}{\pi(0.020 \text{ m})^3} = 143 \text{ MPa}$$

Although this stress is above the yield point of 130 MPa, it is less than $2\tau_Y$ and thus satisfies the assumption regarding the allowable magnitude of the unloading stresses. Superposing the two stress distributions results in the residual distribution indicated in Fig. 6.65.

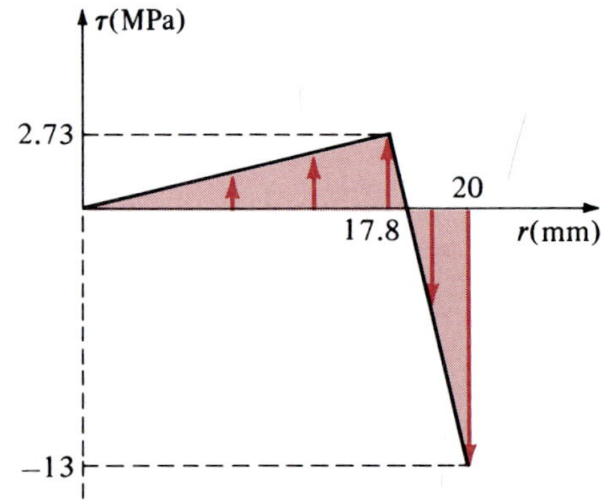

Figure 6.65 Residual stress distribution

Also, from Example 6.17, the initial rotation was $\phi = 5.24° = 0.0915$ radians, so that using Eqn. (6.32),

$$\phi_R = 0.0915 \text{ rad} - \frac{143 \text{ MPa}}{80 \text{ GPa}}\,\frac{1.0 \text{ m}}{0.02 \text{ m}} = 0.0021 \text{ rad} = 0.12°$$

PROBLEMS

6.120 For a 1.0 in. diameter steel shaft with $\tau_Y = 28$ ksi, what is T_Y? For what torque will the stresses in the outer half of the rod be plastic?

6.121 For a 30 mm diameter steel shaft with $\tau_Y = 150$ MPa, what is T_P? For $T = 0.9T_P$, what is the radius of the elastic portion?

6.122 A bar along with the corresponding shear-stress–shear-strain curve is shown in the figure. Determine *(a)* the maximum elastic torque T_Y, and *(b)* the fully plastic torque T_P.

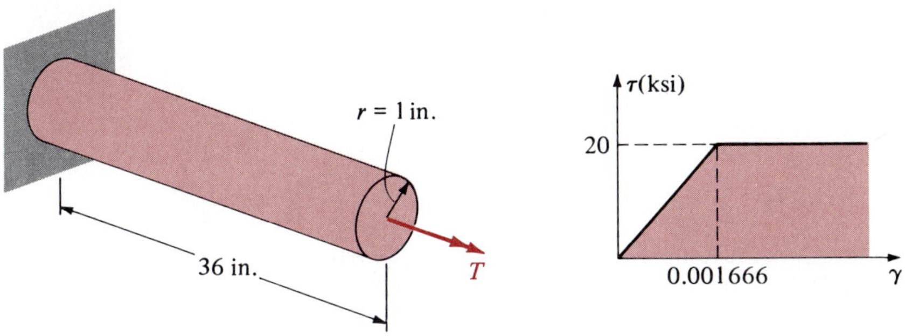

Ps. 6.122 and 6.123

6.123 The bar of the previous exercise is subjected to a torque of 37.5 in.-kip. Determine *(a)* the elastic radius r_Y, *(b)* the angle of twist, *(c)* the residual stress distribution if the torque is removed, and *(d)* the corresponding residual angle of twist.

6.124 For the bar shown in the figure, what is the torque necessary to produce yielding in the outer half of the bar? What is the corresponding angle of twist?

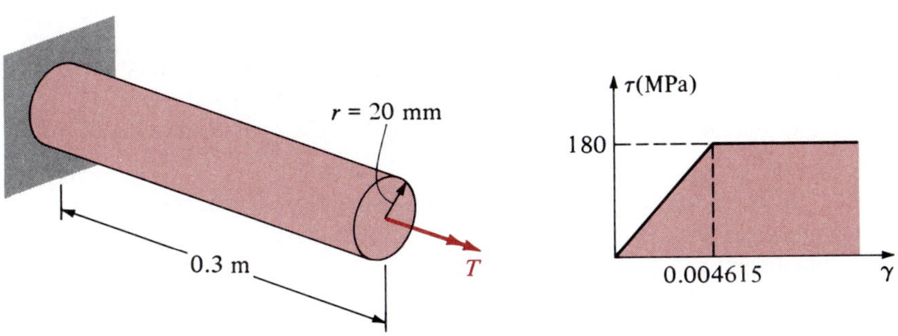

P. 6.124

6.125 The solid elastic–perfectly plastic bar shown is subjected to a torque that produces an angle of twist $\phi = 1.25\phi_Y$. What is the corresponding torque? To what value should the torque be reduced in order to reduce ϕ to ϕ_Y? What is the corresponding stress distribution?

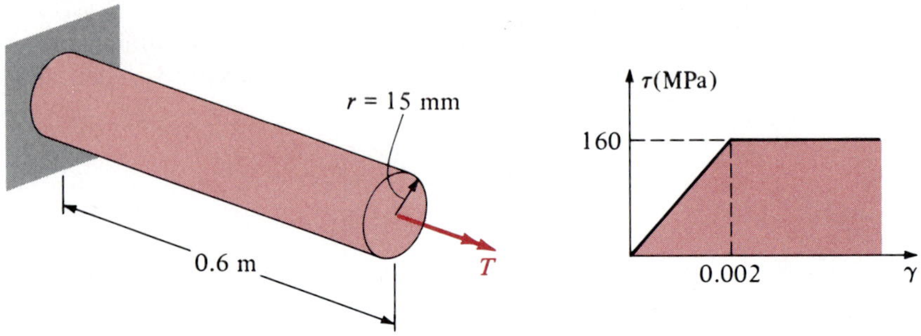

P. 6.125

6.126 For the solid elastic–perfectly plastic bar shown, the angle of twist is $2\phi_Y$. Determine *(a)* the torque, and *(b)* the corresponding stress distribution.

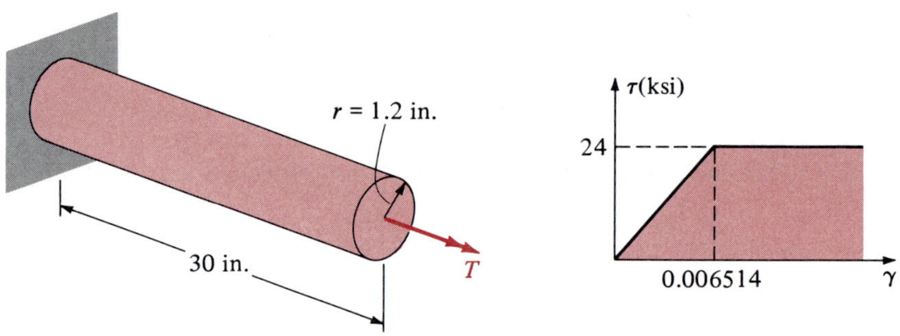

P. 6.126

6.127 For a hollow elastic–perfectly plastic shaft of inside radius b and outside radius a, develop the relationship between *(a)* the yield stress τ_Y and the maximum elastic torque T_Y, and *(b)* the yield stress τ_Y and the plastic torque T_P.

6.128 A bar along with the corresponding shear-stress–shear-strain curve is shown in the figure. Determine *(a)* the maximum elastic torque T_Y, and *(b)* the fully plastic torque T_P.

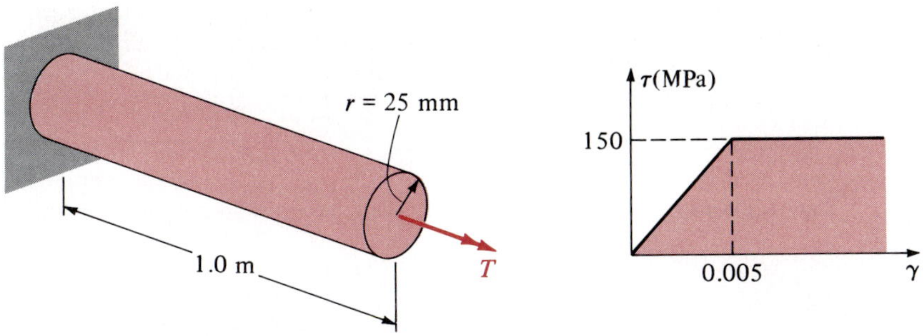

Ps. 6.128 and 6.129

6.129 The bar of the problem 6.128 is subjected to a torque of 4.5 kN·m. Determine *(a)* the elastic radius r_Y, *(b)* the angle of twist, *(c)* the residual stress distribution if the torque is removed, and *(d)* the corresponding residual angle of twist.

6.130 A solid circular shaft of radius 13 mm is composed of a material that behaves according to $\tau = C\gamma^{0.65}$, with $C = 5.35$ GPa. If the allowable shear stress is 60 MPa, determine the allowable torque. What is the angle of twist in a shaft of length 1.0 m?

6.131 A solid circular shaft of radius 6 in. is composed of a material for which $\tau = C\gamma^{1.2}$, with $C = 39.8 \times 10^6$ psi. If the allowable shear stress is 8 ksi, determine the allowable torque. What is the angle of twist in a shaft of length 6 ft?

6.132 A hollow circular shaft of outside radius 1.5 in. and inside radius 1.25 in. is composed of a material for which $\tau = C\gamma^{1.15}$, with $C = 42.3 \times 10^6$ psi. If the torque transmitted is 1200 in.-lbf, determine the maximum shear stress. What is the angle of twist in a shaft of length 40 in.?

6.133 A hollow circular shaft of outside radius 30 mm and inside radius 27 mm is composed of a material that behaves according to $\tau = C\gamma^{0.89}$, with $C = 46.8$ GPa. If the torque transmitted is 200 N·m, determine the maximum shear stress. What is the angle of twist in a shaft of length 0.8 m?

6.6 TORSION OF RECTANGULAR AND THIN-WALLED OPEN SECTIONS

Recall that for members with circular cross sections, the assumption was made in Section 6.2 that the deformations take place in a manner such that plane sections remain plane as indicated in Fig. 6.66a.

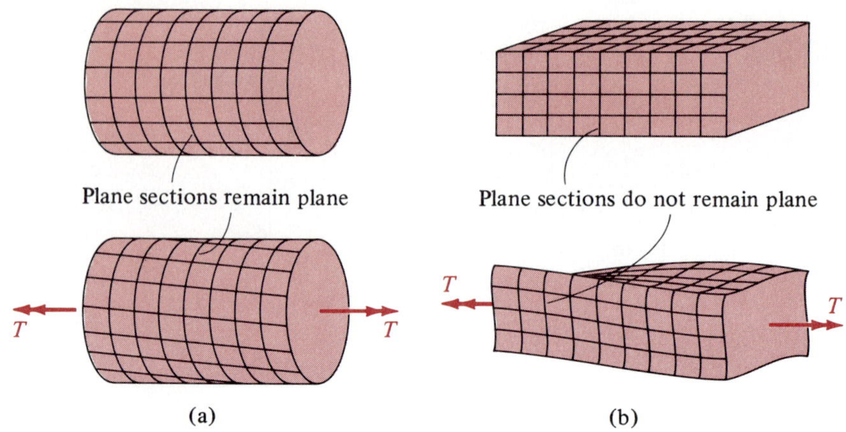

Figure 6.66 Character of the deformations for circular and rectangular cross sections

The deformations in a member that is rectangular in cross section, on the other hand, are such that plane sections *do not* remain plane. This distortion of the sections that are originally plane is referred to as warping, and appears as indicated in Fig. 6.66b. The mathematics necessary to formulate and solve the torsion problem for the rectangular cross section, or any other solid noncircular section, is beyond the scope of this text. Instead of developing the theory, we will present and discuss some of the results from the theory. In particular, results for shear stresses and angles of twist in a member with a rectangular cross section will be presented.

Consider a member with a rectangular cross section that has sides a, $b \leq a$, and length L as shown in Fig. 6.67a.

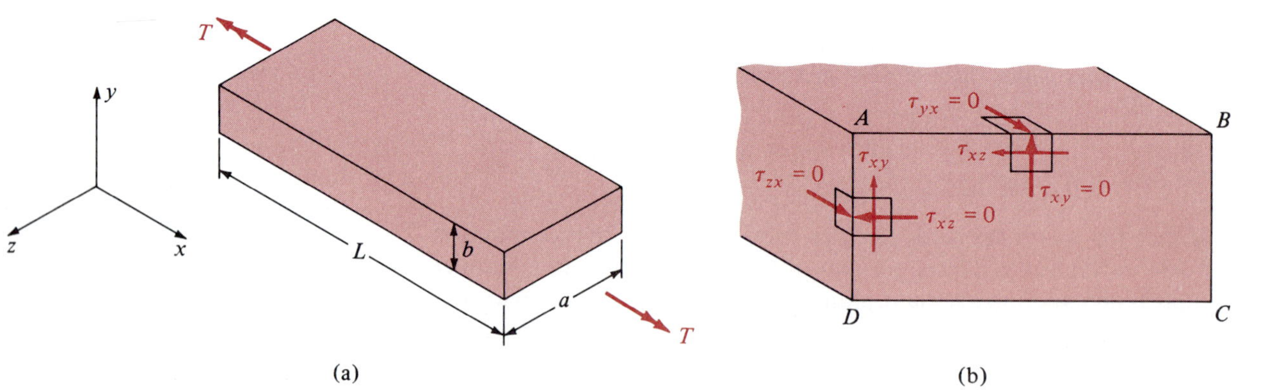

Figure 6.67 Geometry of and stresses in a rectangular-cross-section member

The shear stresses τ_{xy} and τ_{xz} acting on a typical exposed segment, such as shown in Fig. 6.67b, must be statically equivalent to the resultant torque T shown. In view of the fact that the lateral surface of the bar is assumed to be stress free, it follows that $\tau_{xy} = 0$ along the portions of the lateral surface corresponding to segments AB and CD and that $\tau_{xz} = 0$ along the portions of the lateral surface corresponding to segments BC and DA.

The major results from the theory are:

1. The maximum shear stress occurs at the midpoint of the longest side and can be represented as

$$\tau_{\max} = \frac{T}{(C_1 ab^2)} \tag{6.24}$$

where T is the torque transmitted and C_1 is a constant depending on the ratio of the lengths of the sides a and b. Refer to Table 6.2 for values of C_1. The distribution of stresses on a typical cross section is shown in Fig. 6.68.

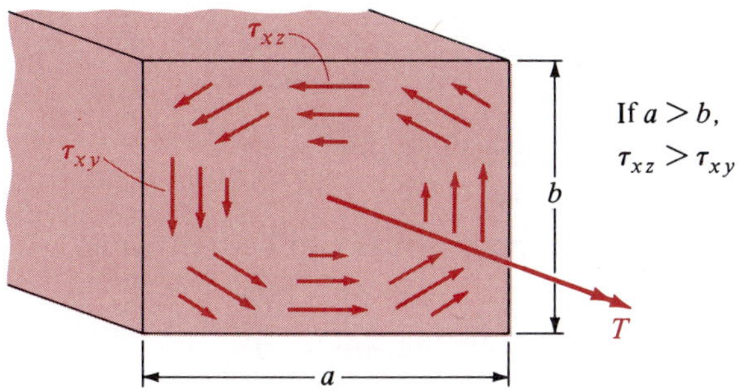

Figure 6.68 Shear stresses for a rectangular member in torsion

2. The angle of twist between the ends of the bar is given by

$$\Phi = \frac{TL}{JG} \tag{6.25}$$

where $J = C_2 ab^3$, with C_2 also a constant depending on the ratio of the sides of the cross section. Refer to Table 6.2 for values of C_2.

Knowing the aspect ratio a/b, the data of Table 6.2 can be used to determine the maximum shear stress for evaluating the strength and the torsional constant for evaluating the stiffness.

Table 6.2 Coefficients for torsion of bars with rectangular cross section

a/b	C_1	C_2
1.0	0.208	0.141
1.2	0.219	0.166
1.5	0.231	0.196
2.0	0.246	0.229
3.0	0.267	0.263
4.0	0.281	0.281
5.0	0.291	0.291
10.0	0.312	0.312
∞	0.333	0.333

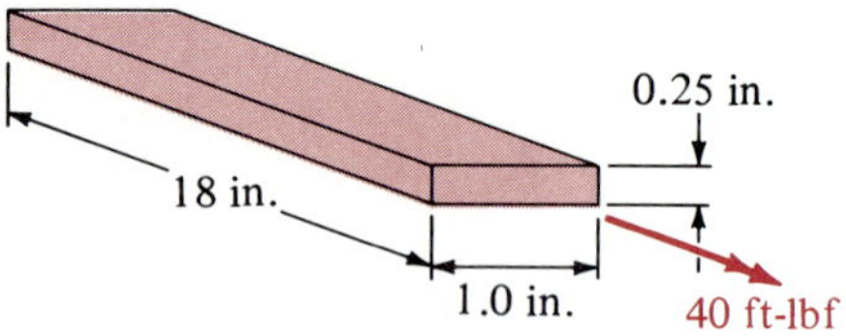

Figure 6.69 Torsion of a thin steel strip

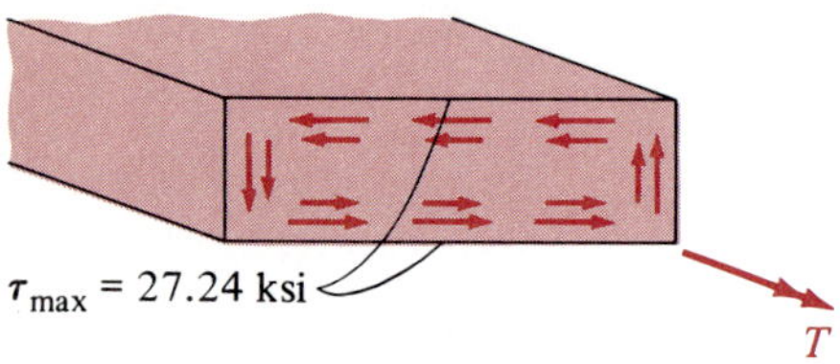

Figure 6.70 Shear stresses in a steel strip

Example 6.19 A thin steel ($G = 12 \times 10^6$ psi) strip with dimensions shown in Fig. 6.69 transmits a torque of 40 ft-lbf.

The ratio of a to b is 4.0 so that from Table 6.2, $C_1 = 0.281$ and $C_2 = 0.281$. The maximum shear stress τ_{xz}, shown in Fig. 6.70, occurs at the midpoint of the 1.0 in. side, and has the value

$$\tau_{max} = \frac{480 \text{ in.-lbf}}{(0.281)(1.0 \text{ in.})(0.25 \text{ in.})^2} = 27.24 \text{ ksi}$$

The angle of twist between the ends is given by

$$\Phi = \frac{(480 \text{ in.-lbf})(18 \text{ in.})}{(0.281)(1.0 \text{ in.})(0.25 \text{ in.})^3(12 \times 10^6 \text{ psi})} = 0.164 \text{ rad} = 9.40°$$

Note from Table 6.2 that in general as $a/b \to \infty$ (i.e., for a thin rectangle), $C_2 \to 0.333$, so that J becomes

$$J = \frac{ab^3}{3}$$

Eqn. (6.24) for the maximum stress can be written as

$$\tau_{max} = \frac{3T}{ab^2} = \frac{Tb}{J} \tag{6.26}$$

with Eqn. (6.25) for the angle of twist still given by

$$\Phi = \frac{3TL}{ab^3G} = \frac{TL}{JG} \tag{6.27}$$

The results expressed by Eqns. (6.25) and (6.26) continue to be valid for any of the cross sections shown in Fig. 6.71. These sections are commonly referred to as **thin-walled open sections.** For these sections the dimension a refers to the "length" of the section whereas the dimension b still refers to the thickness. For this treatment of thin-walled open sections we will assume b to be a constant.

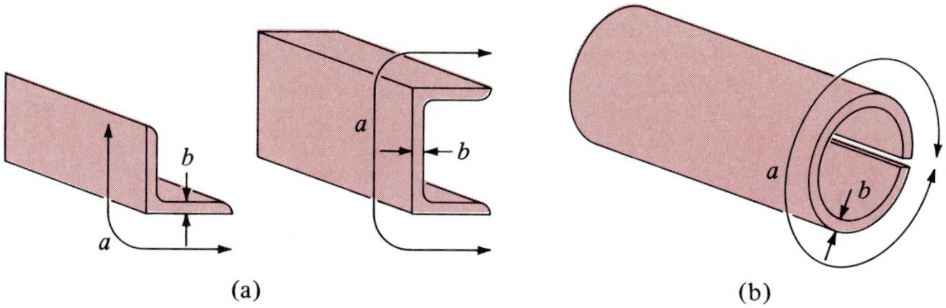

Figure 6.71 Thin-walled open sections

For the angle and channel sections shown in Fig. 6.71a, the stresses in the neighborhood of a reentrant corner at A, shown in Fig. 6.72, can be significantly higher than those given by the nominal value of Tb/J. The increase above the nominal value depends upon the ratio of r to the thickness b. Generally such a section is manufactured so that $r \approx b$ with the maximum shear stress at A approximately 1.4 times the nominal value.

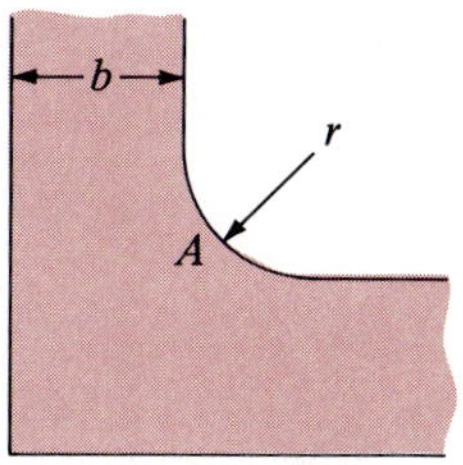

Figure 6.72 Geometry of a reentrant corner

In the absence of any reentrant corners, the stresses on a typical thin-walled section such as indicated in Fig. 6.73a are shown in Fig. 6.73b. Except near the ends, the maximum shear stress of $3T/ab^2$ is present at all points along the longest "side" as shown in Fig. 6.73b.

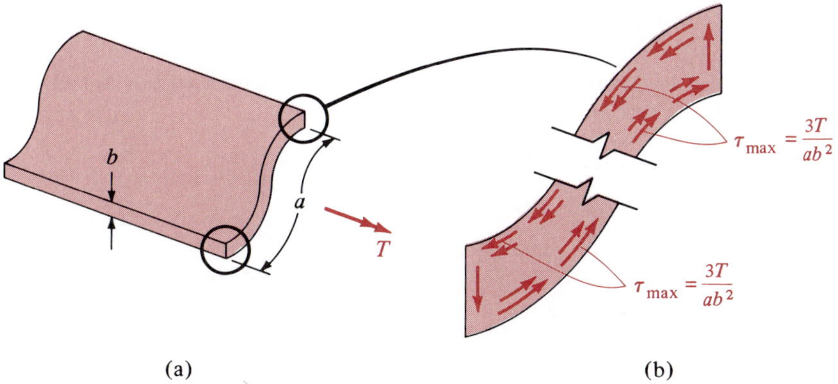

Figure 6.73 Stresses on a typical thin-walled section

Example 6.20 The steel (G = 80 GPa) angle section is subjected to a torque of 200 N·m as shown in Fig. 6.74.

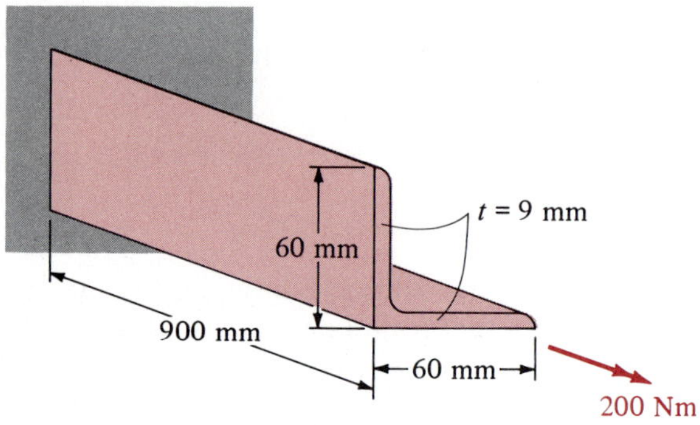

Figure 6.74 Torsion of an angle section

The torsional constant J is computed to be

$$J = \frac{ab^3}{3} = \frac{1}{3}(0.0555 \text{ m} + 0.0555 \text{ m})(0.009 \text{ m})^3 = 2.70 \times 10^{-8} \text{ m}^4$$

The nominal shear stress in the section is computed to be

$$\tau_{\max} = \frac{Tb}{J} = \frac{(200 \text{ Nm})(0.009 \text{ m})}{(2.70 \times 10^{-8} \text{ m}^4)} = 66.67 \text{ MPa}$$

Assuming the radius of the reentrant corner to be approximately 9 mm, the maximum shear stress would be

$$\tau_{\max} \approx 1.4(66.67 \text{ MPa}) = 93.33 \text{ MPa}$$

The corresponding angle of twist is given by

$$\Phi = \frac{TL}{JG} = \frac{(200 \text{ Nm})(0.9 \text{ m})}{(2.70 \times 10^{-8} \text{ m}^4)(80 \text{ GPa})} = 0.0833 \text{ rad} = 4.77°$$

and is not appreciably affected by the presence of the reentrant corner.

PROBLEMS

6.134 Each of the bars transmits a 1 kN·m torque as shown. Determine the minimum dimension d in order that the maximum shear stress not exceed 150 MPa in each bar. Show the relative sizes on a sketch.

6.135 Repeat problem 6.134 when the torque is 9.0 in.-kip and the maximum shear stress is 20 ksi.

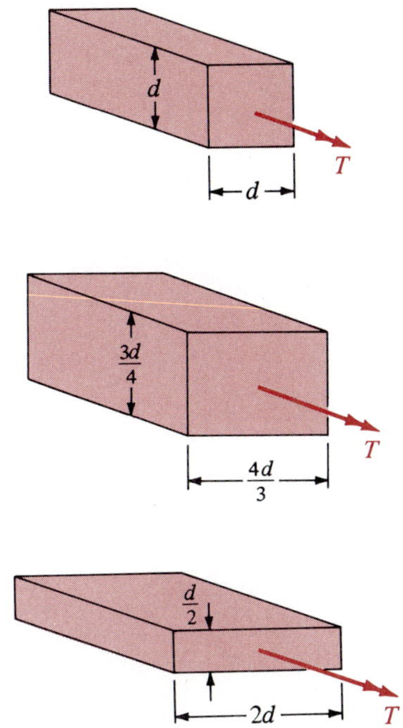

Ps. 6.134 and 6.135

6.136 If each of the bars of equal area shown in the figure is composed of a ductile material for which the yield stress in shear is 20 ksi, determine the allowable torque for a factor of safety of 2.4.

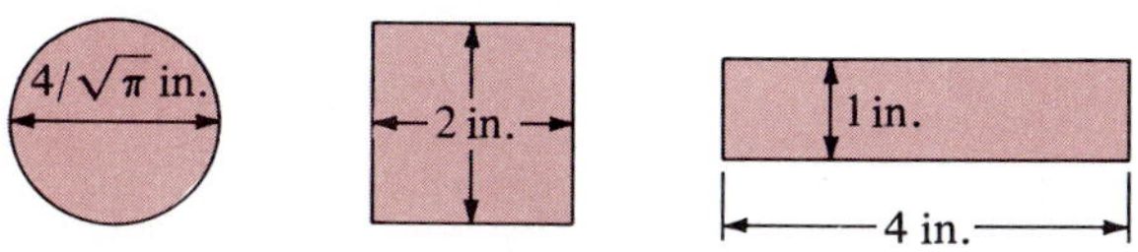

P. 6.136

6.137 For the rectangular shaft shown, the torque results in a rotation of 3.6 degrees at end A. If $G = 80$ GPa, determine the maximum shear stress in the bar.

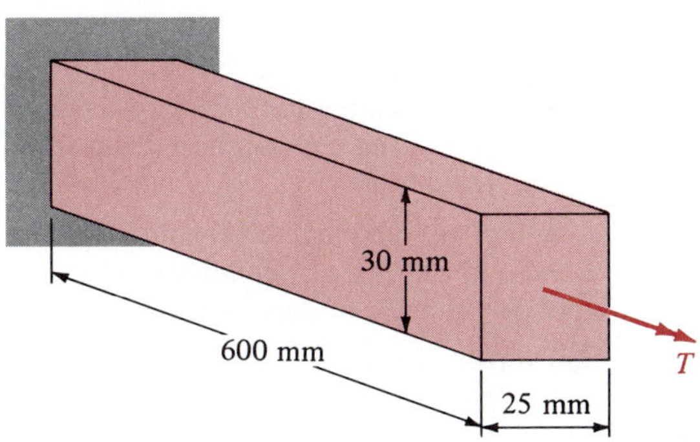

P. 6.137

6.138 Two shafts of the same length, one circular in cross section and the other square, have the same area and transmit the same torque. Determine the ratio of the maximum shear stresses in the two bars.

6.139 For the two shafts of problem 6.138, determine the ratio of the relative angles of twist for torques that produce the same maximum shear stress in each bar.

6.140 A 25-mm square steel ($G = 80$ GPa) shaft transmits a torque resulting in a maximum shear stress of 150 MPa. What must be the length of the shaft if the relative rotation between the ends is to be one revolution?

6.141 Two tubes, one with a slit and one without, are each subjected to a torque T. Determine *(a)* the ratio of the maximum shear stresses, and *(b)* the ratio of the angles of twist between the ends. Take $t << r$.

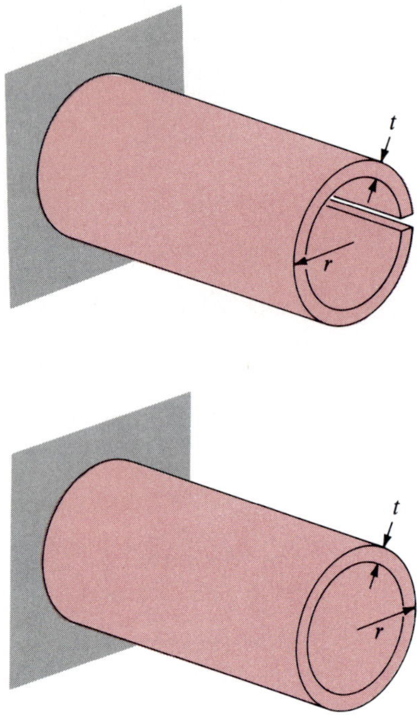

P. 6.141

6.142 The two thin-walled open sections shown each transmit a torque T. According to the theory, in which member is the shear stress the largest? In which is the angle of twist the largest? Neglect the stress concentration from the reentrant corner in the channel section.

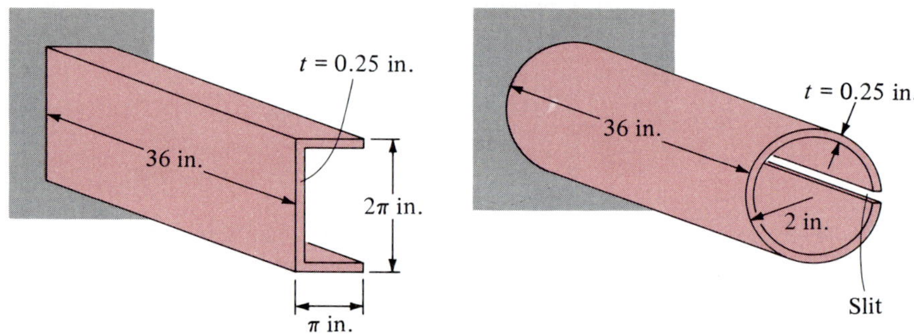

Ps. 6.142 and 6.143

6.143 Repeat problem 6.142 accounting for the stress concentration associated with the reentrant corner in the channel section, assumed to have a radius of 0.25 in.

6.144 Two thin strips with cross-sectional dimensions shown each transmit a torque T. What is the ratio of *(a)* the maximum shear stresses, and *(b)* the relative angles of twist if the bars are of equal length?

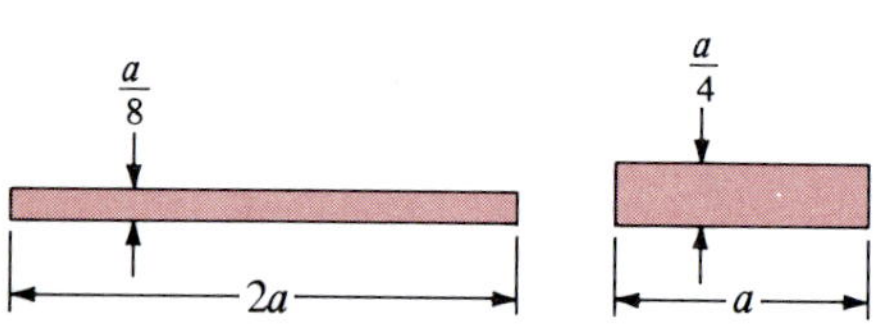

P. 6.144

6.145 The angle section shown is made of a steel ($G = 12 \times 10^6$ psi) for which the yield stress in tension is $\sigma_Y = 36$ ksi. What is the maximum torque that can be transmitted if the factor of safety is 2.0? What is the corresponding angle of twist? Neglect the effects of the reentrant corner.

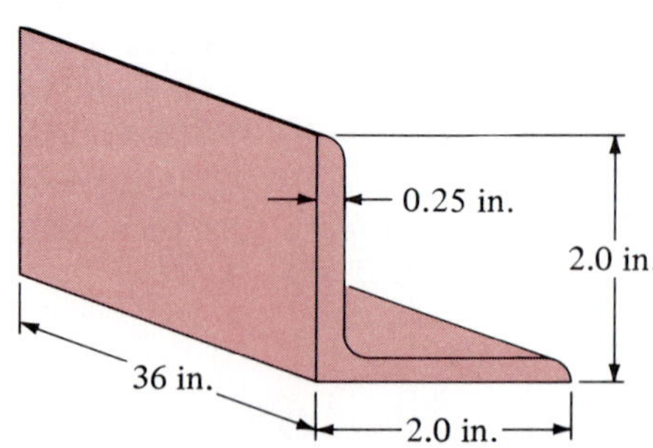

P. 6.145

6.146 The channel section shown is made of an aluminum ($G = 27$ GPa) for which the yield stress in tension is $\sigma_Y = 270$ MPa. What is the maximum torque that can be transmitted for a factor of safety of 1.8? What is the corresponding angle of twist? Neglect the effects of the reentrant corner.

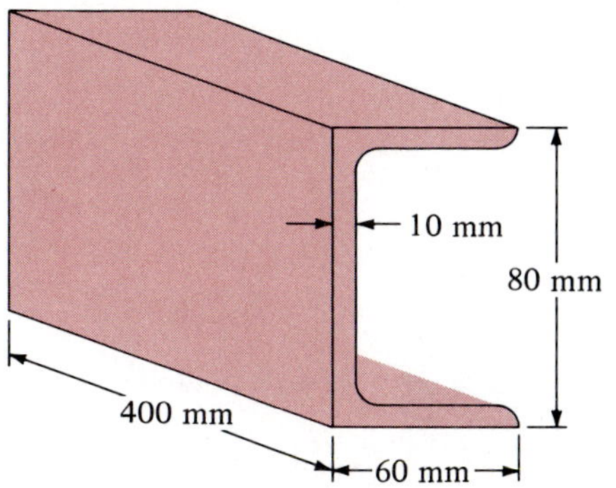

P. 6.146

6.147 The angle section shown is made of an aluminum ($G = 28$ GPa) for which the yield stress in tension is $\sigma_Y = 240$ MPa. If the torque transmitted is 400 N·m, what is the factor of safety with respect to yield? What is the corresponding angle of twist? Neglect the effects of the reentrant corner.

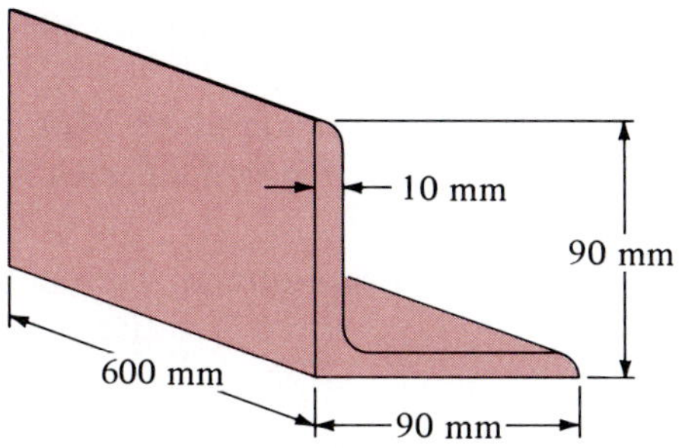

P. 6.147

6.148 The channel section shown is made of an aluminum ($G = 3.8 \times 10^6$ psi) for which the yield stress in tension is $\sigma_Y = 40$ ksi. What is the maximum torque that can be transmitted for a factor of safety of 2.2? What is the corresponding angle of twist? Neglect the effects of the reentrant corners.

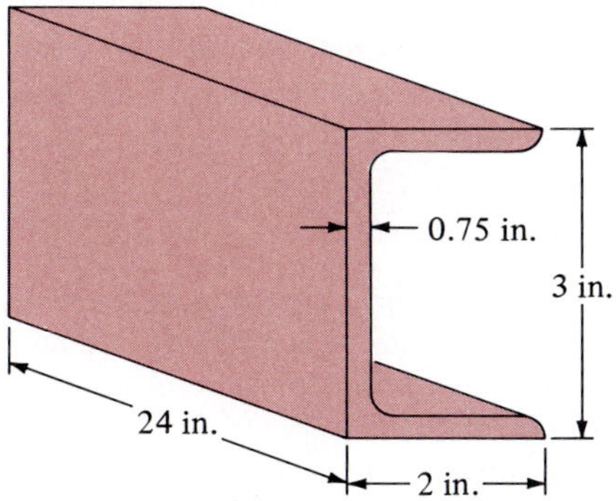

P. 6.148

6.149 The hollow tube is made of an aluminum ($G = 80$ GPa) for which the yield stress in tension is $\sigma_Y = 220$ MPa. What is the maximum torque that can be transmitted for a factor of safety of 1.6? What is the corresponding angle of twist?

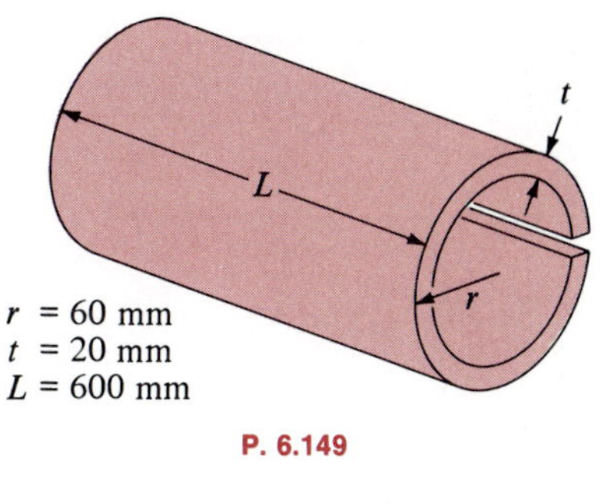

P. 6.149

6.7 CLOSURE

Several aspects of the problem of transmitting torques by structural members have been investigated in this chapter. For the elementary theory that we developed, the basic assumptions are:

1. *Geometry:* The bar is straight with either a solid or hollow circular cross section, and such that the radius of the bar varies at most slowly as a function of position along the bar.
2. *Equilibrium:* The external loading is such that the resultant internal force at any location along the axis of the bar is a single torque parallel to the axis of the bar.
3. *Deformation:* Plane sections remain plane with a plane section at any location x along the axis of the bar rotating as a rigid body through an angle $\phi(x)$ such that $\phi' \ll 1$.
4. *Material behavior or stress and strain:* The material is isotropic and linearly elastic with $\tau = G\gamma$, where G is the shear modulus of the material.

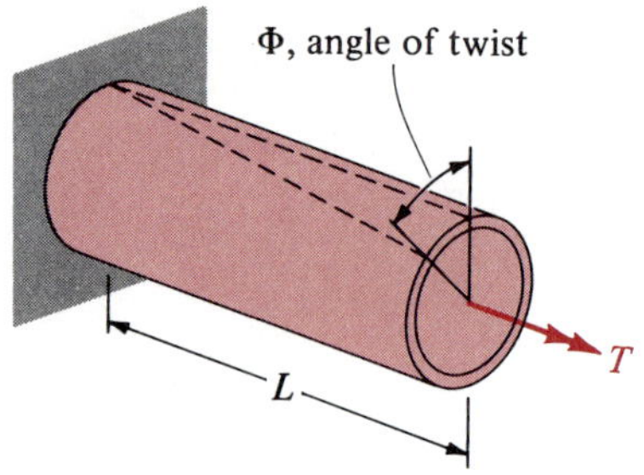

(a)

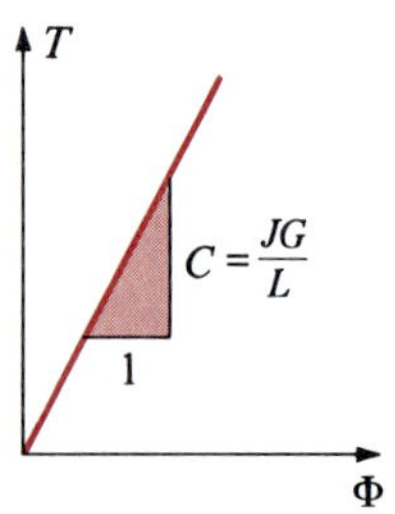

(b)

Figure 6.75 Special results for the uniform circular bar

Combining these basic ideas produces a theory that yields the very important special results

$$\tau = \frac{Ta}{J} \qquad \text{and} \qquad \Phi = \frac{TL}{JG} = \frac{T}{C}$$

for a solid or annular bar of constant radius and length L subjected to a torque T at the ends, as shown in Fig. 6.75a.

Under these circumstances, the bar behaves as a linear torsional spring as indicated in Fig. 6.75b, a fact that can be very useful in a number of applications. It is the responsibility of the engineer to make sure that all of the assumptions of the theory are satisfied before using any of these special results.

COMPUTER EXERCISES

CE 6.1 Write a general computer program to determine approximate values for the rotation $\phi(x)$ and the shear stress $\tau(x)$ for a circular bar supported at $x = 0$ as shown. Include an option for several concentrated torques T_i at locations x_i and an option for a distributed loading $t(x)$. Allow the code to determine the reaction at the support, and output the rotation and shear stress at positions along the axis of the bar. Check your code on a number of test problems for which you already know the solution.

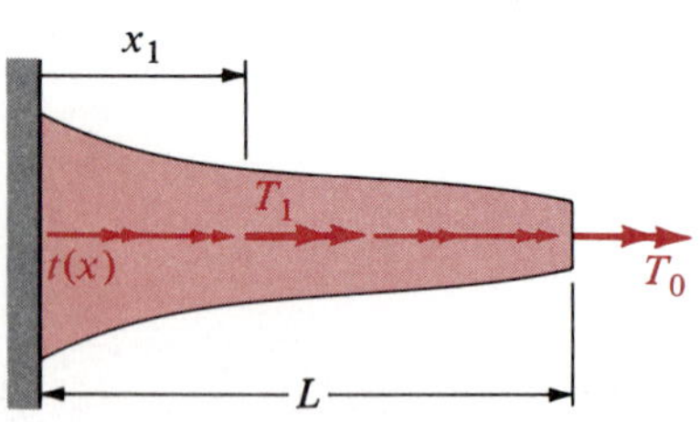

CE 6.1

CE 6.2 Write a general computer program to determine approximate values for the rotation $\phi(x)$ and the shear stress $\tau(x)$ for a circular bar supported at the ends as shown. Include an option for several concentrated torques T_i at locations x_i and an option for a distributed loading $t(x)$. Allow the code to determine the reaction at the support and output the rotation and shear stress at positions along the axis of the bar. Check your code on a number of test problems for which you already know the solution.

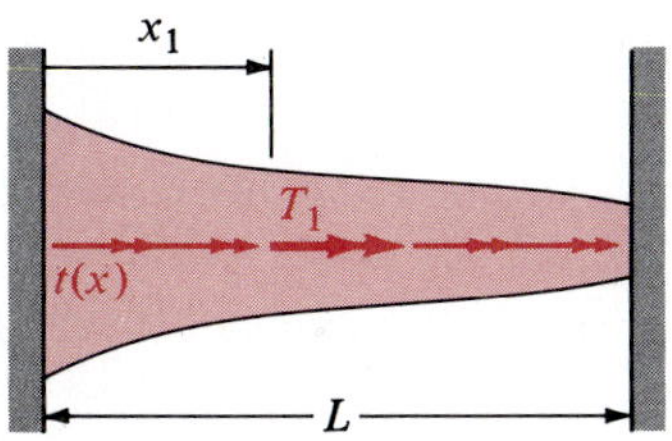

CE 6.2

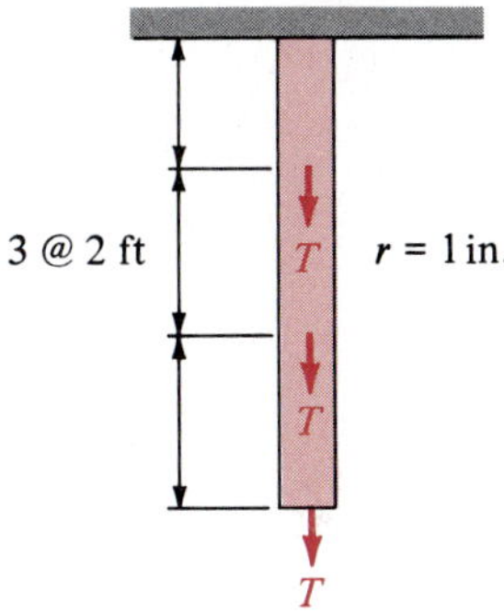

CE 6.4

CE 6.3 Repeat problem CE 6.1 for a material behaving according to a Ramberg-Osgood type stress-strain relation, that is, $\gamma = \alpha\tau + \beta\tau^n$.

CE 6.4 Use the code developed in problem CE 6.3 to solve the problem of a bar loaded as shown. Take T = 20,000 in.-lbf and $\gamma = \tau/10^7$ psi + (1/650) (τ/40,000 psi)^3, where τ is measured in psi.

ANALYSIS/DESIGN PROBLEMS

AD 6.1 A straight uniform shaft of circular or annular cross section and composed of a ductile material whose tensile yield stress is σ_Y is subjected to a torque T as shown. The allowable rotation between the ends of the shaft is Φ radians. Design the shaft.

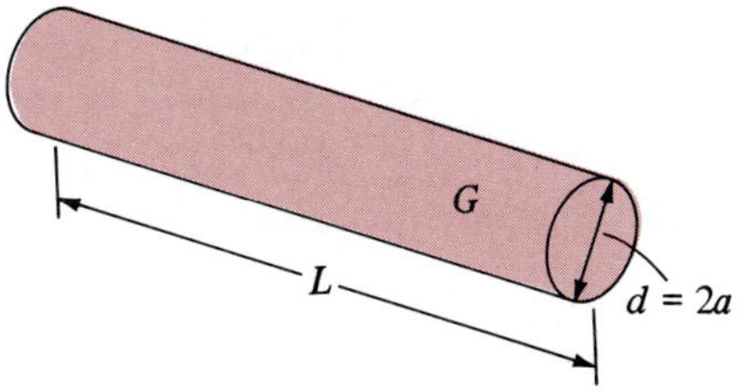

AD 6.1

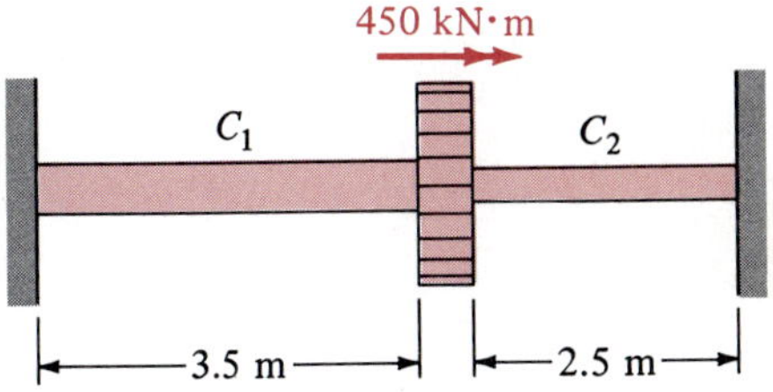

AD 6.2

AD 6.2 A gear is subjected to a torque of 450 kN·m as shown. Design efficient shafts C_1 and C_2 that will transmit the torque with stresses in the elastic range and restrict the rotation of the gear to 4°. Assume that each shaft is solid and carries half the torque. Take τ_{PL} = 100 MPa.

AD 6.3 Repeat problem AD 6.2 if the condition that each shaft must carry half the torque is replaced by the condition that the two solid shafts have the same diameters.

AD 6.4 Repeat problem AD 6.2 if the condition that each shaft must carry half the torque is replaced by the condition that the least amount of material be used. Take both shafts to be solid.

AD 6.5 Repeat problem AD 6.2 if the shafts can be hollow with $t/R_{\text{outer}} = 0.25$.

AD 6.6 Repeat problem AD 6.3 if the shafts can be hollow with $t/R_{\text{outer}} = 0.25$.

AD 6.7 Repeat problem AD 6.4 if the shafts can be hollow with $t/R_{\text{outer}} = 0.25$.

AD 6.8 For the system shown, the motor delivers 12 hp at 3000 rpm. The outputs to the three gears are 6 hp, 4 hp, and 2 hp as shown. Design each of the solid shafts to operate in the elastic range, and so that the relative angle of twist at D does not exceed 6°. Take the material to be steel with $\tau_Y = 25$ ksi and $G = 12 \times 10^6$ psi. Take an additional condition to be that the diameters of all the shafts are the same.

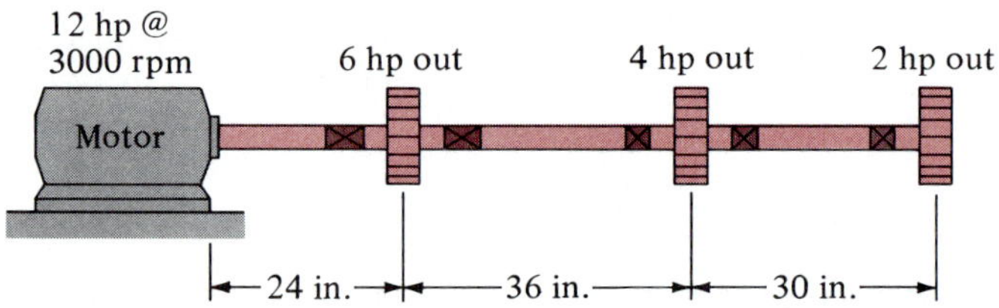

AD 6.8 and AD 6.13

AD 6.9 Repeat problem AD 6.8 if the condition on the diameters is replaced by one requiring that the minimum amount of material be used.

AD 6.10 Repeat problem AD 6.8 if the condition on the diameters is replaced by one requiring that the shear stress in each shaft be the same.

AD 6.11 Repeat problem AD 6.8 if the shafts can be hollow rather than solid. Take the ratio of the wall thickness to outside radius to be 0.2 (an additional condition).

AD 6.12 Repeat problem AD 6.11 if the condition on the diameters is replaced by one requiring that the minimum amount of material be used.

AD 6.13 Repeat problem AD 6.11 if the condition on the diameters is replaced by one requiring that the shear stress in each shaft be the same.

AD 6.14 Design the system shown so that the stress in each of the shafts is in the elastic range and so that the rotation at A does not exceed 10°. Take the material to be aluminum with $\tau_Y = 55$ MPa and $G = 26$ GPa. Take as an additional condition that all the solid shafts have the same diameter.

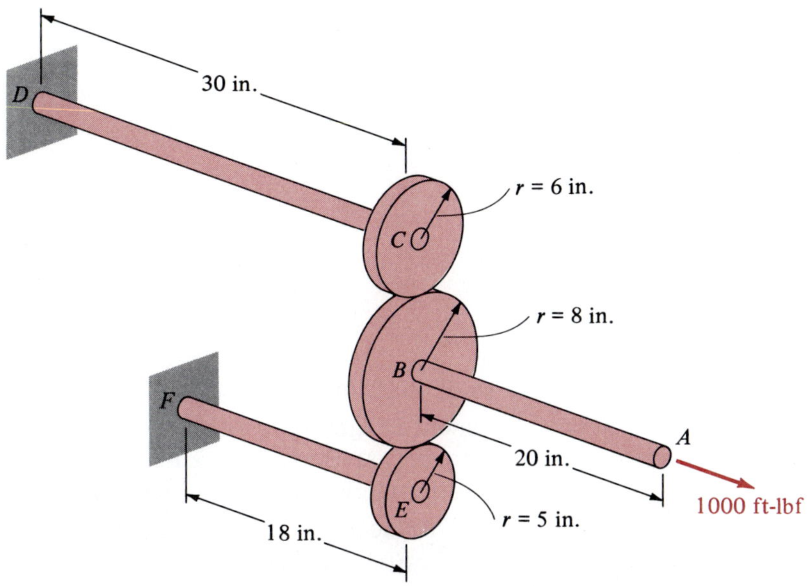

AD 6.14

AD 6.15 Repeat problem AD 6.14 replacing the condition regarding the diameters with one that uses the minimum amount of material.

AD 6.16 Repeat problem AD 6.14 if hollow shafts are allowed. Take the ratio of the wall thickness to outside radius to be 0.25.

CHAPTER 7 BENDING DEFORMATIONS: QUESTIONS OF STRENGTH

CHAPTER OUTLINE

7.1 INTRODUCTION

There are many situations in the world around us where structures function primarily by what is commonly referred to as bending. Roughly speaking, bending refers to a change in shape from a straight configuration to a nonstraight configuration, as indicated in Fig. 7.1a. Bending can also refer to a change in a nonstraight configuration as indicated in Fig. 7.1b. The members shown in Fig. 7.1 are frequently referred to as "beams." Generally, a beam refers to a member that resists transverse loading by transmitting shear forces and bending moments.

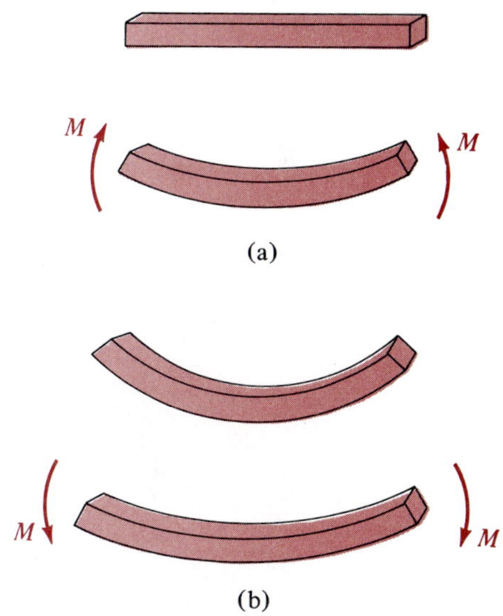

Figure 7.1 Bending of straight and nonstraight members

As typical of a bending deformation, consider a straight uniform beam with a rectangular cross section that is placed upon a flat surface and acted upon by a transverse force F as shown in Fig. 7.2a.

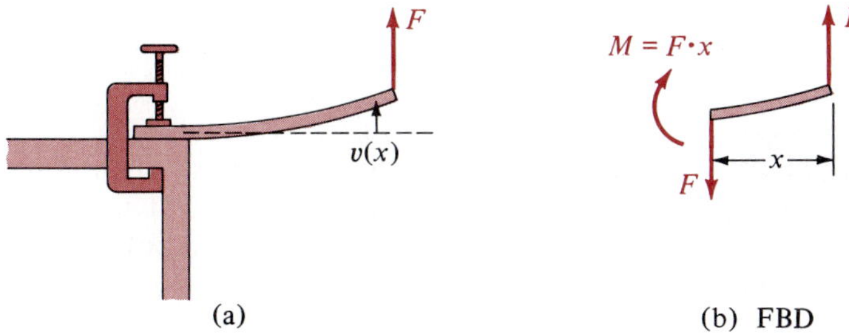

Figure 7.2 Bending of a straight uniform beam by a transverse force

In such a situation we are interested in answering the two basic questions: *(a)* is the beam strong enough to transmit the load F? and *(b)* is the beam stiff enough so that the transverse displacements ($v(x)$ as shown in Fig. 7.2a) are not too large? Based on our experience in the previous two chapters, we expect that the answer to the strength question will be related to the stresses in the bar. These stresses will in turn depend upon the cross-sectional dimensions of the beam and the magnitude of the transverse force F. We expect that the answer to the stiffness question will depend upon the magnitude of the load F, the dimensions of the beam, and the mechanical properties of the material used for the beam.

The theory we will develop in this chapter begins with a set of assumptions, just as in the cases of the axial and torsional deformations of the previous two chapters. For the bending deformation we make the following assumptions about the geometry of the beam:

1. The axis of the beam is straight.
2. Cross-sectional dimensions (such as b and h for the rectangular section shown in Fig. 7.3) are small compared to the length of the bar (i.e., $h/L \ll 1$ and $b/L \ll 1$). In addition, cross-sectional dimensions are either constant or vary slowly with position along the axis of the beam.

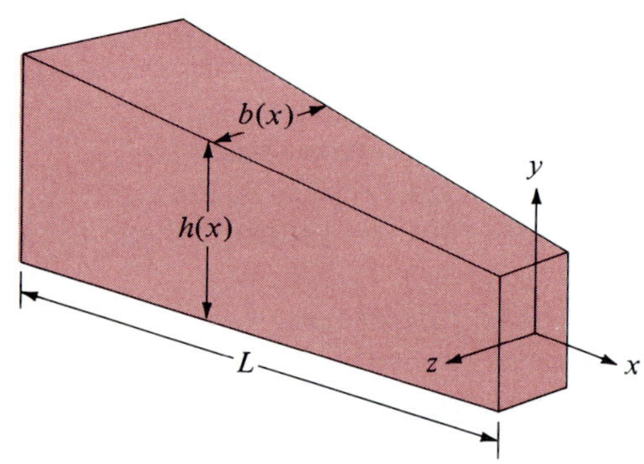

Figure 7.3 Geometry of the region

The axis along the length L of the beam is taken to be the x-axis. The y-axis is taken to be "up," and the z-axis chosen in accordance with right-hand rule as in Fig. 7.3. For reasons that will be outlined in Section 7.2, we assume for the elementary theory that the xy-plane is a plane of symmetry. Examples of permissible cross sections with these axes are indicated in Fig. 7.4.

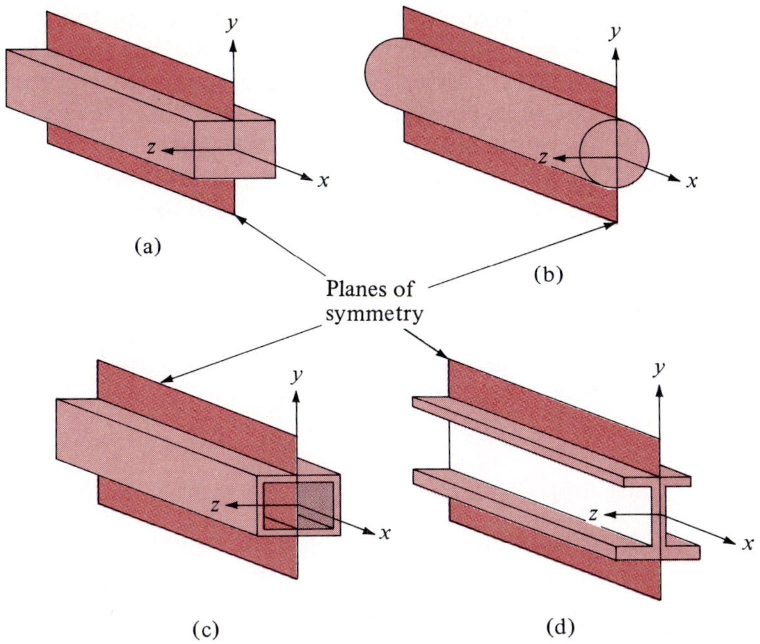

Figure 7.4 Examples of permissible cross sections

Typical cross sections that are not permissible for the choice of axes shown are indicated in Fig. 7.5.

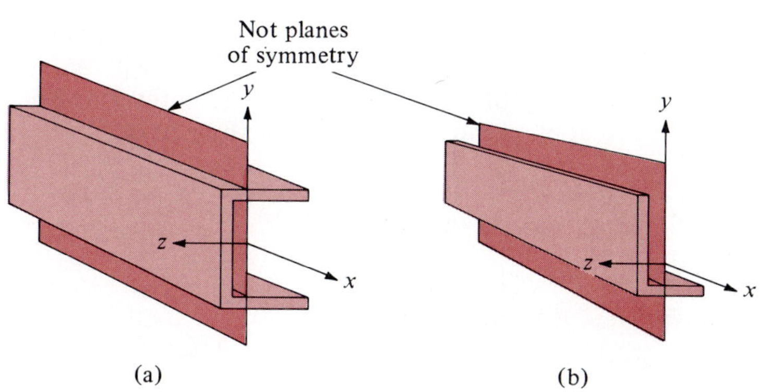

Figure 7.5 Examples of cross sections that are not permissible

A beam is generally required to transmit both a shear force and a bending moment. This situation is basically different than in the simple axial problem, where the only resultant transmitted is a force parallel to the axis of the bar and passing through the centroid, or in the torsional problem, where the only resultant transmitted is a torque parallel to the axis of the bar. Although the bending deformation problem is somewhat more complicated than either the axial or torsional deformation problems, we will nevertheless develop the theory by the same method used in the previous two chapters: we will systematically apply the basic ideas of equilibrium, deformation, and material behavior. When complete, the theory can be used very successfully to answer many of the questions about both the strength and the stiffness of beams.

7.2 LOADING AND EQUATIONS OF EQUILIBRIUM

A primary goal of this section is to establish the equilibrium relation(s) between the external loads and the internal force resultants. Logically, one of the first questions to ask is: What are the permissible types of external loads for the bending problem? In this section we will outline the scope of the bending problem from the standpoint of the permissible external loads and the types of internal force resultants that must be transmitted for equilibrium. Additionally, we will develop the equations of equilibrium that establish the relationships between the external loads and the internal force resultants.

Loading Generally we consider that external loads are applied either as forces or as couples. Referring to Fig. 7.6, we will require any external forces to be parallel to the y-axis and lie in the xy-plane. Any moments that are applied must be parallel to the z-axis.

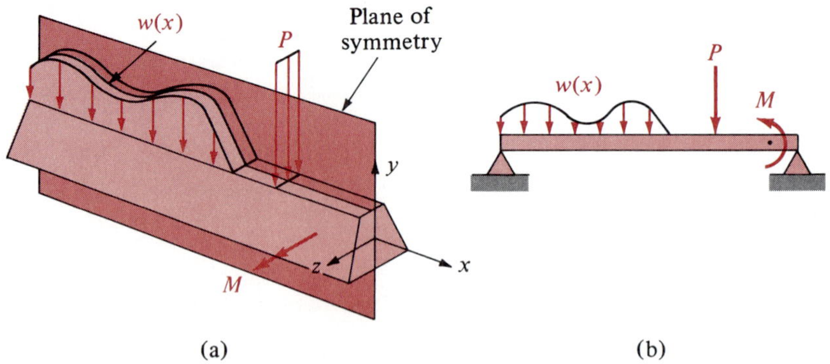

Figure 7.6 Permissible loadings for the bending problem

Fig. 7.6a is intended to show that the loadings are actually applied over the width of the beam in a manner that satisfies the assumption that the line of action be in the xy-plane or in the plane of symmetry. The usual way such loadings are represented, however, is indicated in Fig. 7.6b, where only the front view is shown. The student should always keep in mind that there is also a z-dimension of the beam, and that any loadings should be symmetric with respect to the xy-plane. Coupled with the assumption from Section 7.1 that the cross section is symmetric with respect to the xy-plane, these assumptions on the loading are necessary to ensure that bending deformations will occur only in the xy-plane.

Several common types of loading are shown in Fig. 7.7. Fig. 7.7a shows a beam under the action of loading that has been idealized as a uniform loading w_0. The dimension of w_0 is $[w_0] = F/L$. Typical units are N/m, N/mm, or kN/m in the SI system and lbf/ft, lbf/in., or kip/ft in the U.S. customary system. Fig. 7.7b shows a beam under the action of loading that has been idealized as concentrated. As we shall show later, this is a convenient idealization in situations where a load w_0 has a high intensity distributed over a relatively short distance. Figs. 7.7c and d show situations where the effect of the loading is equivalent to a couple being applied at the location of the attach points between the vertical members and the beams. We will introduce other, less common types of loading when necessary.

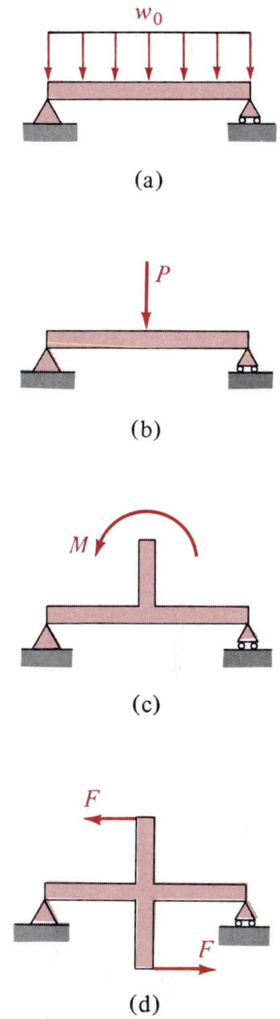

Figure 7.7 Common types of loading

Equilibrium Equations Consider a beam loaded as prescribed, that is, with forces in the plane of symmetry parallel to the y-axis, and with couples parallel to the z-axis, as shown in Fig. 7.8a.

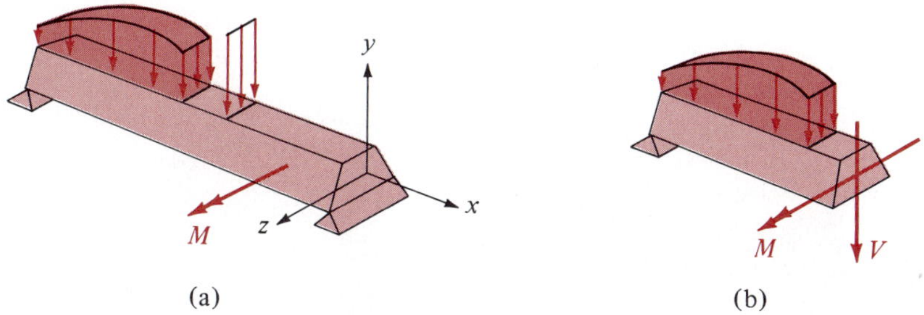

Figure 7.8 Equilibrium of a segment

Draw a free-body diagram of a typical segment as indicated in Fig. 7.8b. It is left for the student to investigate in the problems that, for loading to satisfy the assumptions listed above, the four equilibrium equations

$$\Sigma F_z \equiv 0 \qquad \Sigma F_x \equiv 0 \qquad \Sigma M_x \equiv 0 \qquad \Sigma M_y \equiv 0$$

can be satisfied identically as long as the internal force V is parallel to the y-axis and lying in the xy-plane and as long as the internal moment M is parallel to the z-axis. The other two equilibrium equations

$$\Sigma F_y = 0 \qquad \text{and} \qquad \Sigma M_z = 0$$

will be used to determine the magnitudes of the internal force V and the internal moment M shown in Fig. 7.8b.

For this text, the defined positive directions for the internal shear force V and the internal moment M are taken to be as shown in Fig. 7.9. The definitions for the positive directions of V and M are referred to as the **sign convention** for shear and moment. The sign convention for V and M used in this text is chosen to be consistent with the convention most frequently used throughout the engineering profession. Other choices are possible for positive shear and moment, but a different choice changes only the appearance of some of the equilibrium equations involving V and M. Equilibrium must be satisfied irrespective of the definitions of positive shear V and positive moment M.

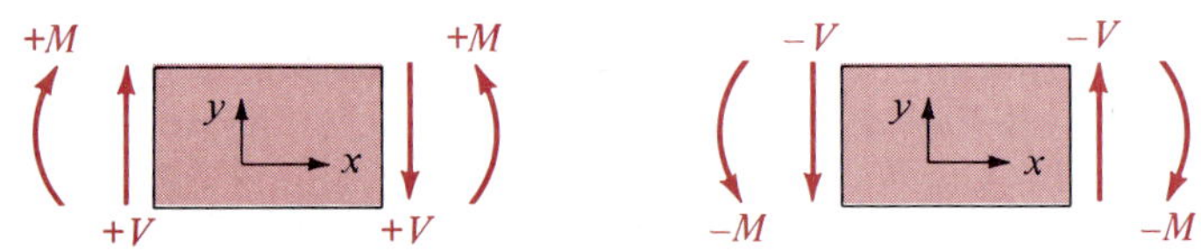

Figure 7.9 Positive and negative directions for V and M

Our next step is to represent V and M in terms of the stresses present on a face that is exposed by a section such as indicated in Fig. 7.10.

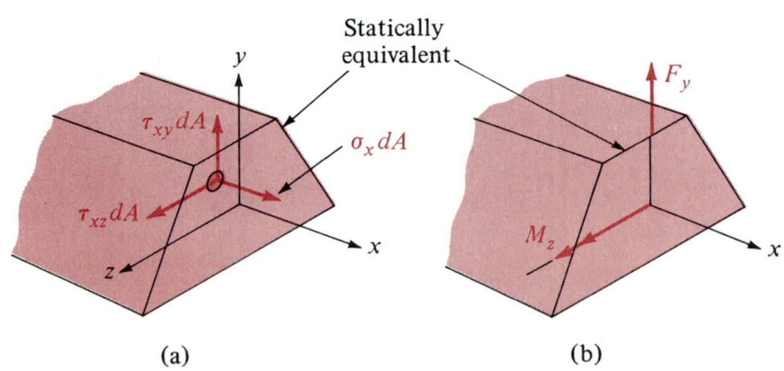

Figure 7.10 Statical equivalence of internal and external forces

As pictured in Figs. 7.10a and b, the idea is that the stresses σ_x, τ_{xy}, and τ_{xz} acting on the exposed face must be distributed in a manner such that they are statically equivalent to the force $\mathbf{F} = \mathbf{i}F_x + \mathbf{j}F_y + \mathbf{k}F_z$ and to the moment $\mathbf{M} = \mathbf{i}M_x + \mathbf{j}M_y + \mathbf{k}M_z$ transmitted at the location in question. For the most general case, the six equations relating the resultants of the stresses to the external forces are

$$F_x = \int \sigma_x \, dA \qquad (7.1a) \qquad M_x = \int (y\tau_{xz} - z\tau_{xy}) \, dA \qquad (7.1d)$$

$$F_y = \int \tau_{xy} \, dA = -V \qquad (7.1b) \qquad M_y = \int z\sigma_x \, dA \qquad (7.1e)$$

$$F_z = \int \tau_{xz} \, dA \qquad (7.1c) \qquad M_z = -\int y(\sigma_x \, dA) = M \qquad (7.1f)$$

For the present class of problems where the only two nonzero resultants are a shear force V in the negative y-direction and a moment M about the z-axis, we need to require that the stresses be distributed so that $F_x = 0$, $F_z = 0$, $M_x = 0$, and $M_y = 0$. A complete treatment of Eqns. (7.1) is beyond the scope of the elementary theory. At this point we will simply observe that Eqns. (7.1c–e) are satisfied by taking $\tau_{xz} \equiv 0$ and by further assuming that both σ_x and τ_{xy} are distributed symmetrically about the xy-plane.

The other three stress components, σ_y, σ_z, and τ_{yz}, that act on the lateral faces are shown in Fig. 7.11 for the case of a rectangular cross section. These faces are often unloaded; thus, they are stress-free faces, in which case σ_y, σ_z, and τ_{yz} are zero on the surface. Any nonzero values of σ_y, σ_z, and τ_{yz} in the interior are then assumed to be negligibly small. Alternatively, the values of any nonzero stresses applied to these surfaces, such as the normal stress $\sigma_y = -w(x)$, as shown in Fig. 7.12, will be small compared to σ_x, the bending stresses, with the corresponding stresses σ_y, σ_z, and τ_{yz} in the interior of the bar assumed to be small.

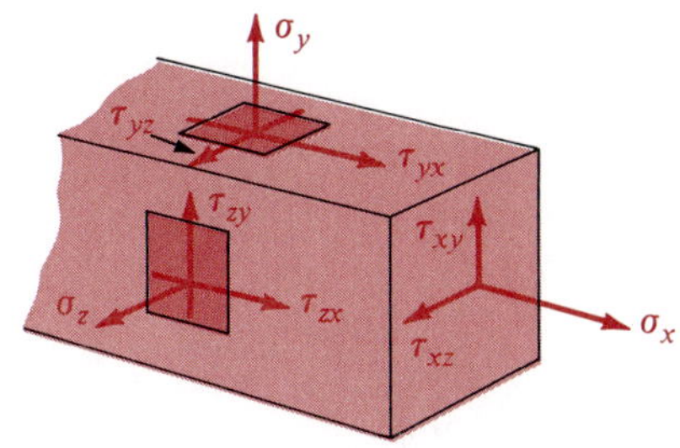

Figure 7.11 Stress components σ_y, σ_z, and τ_{yz}

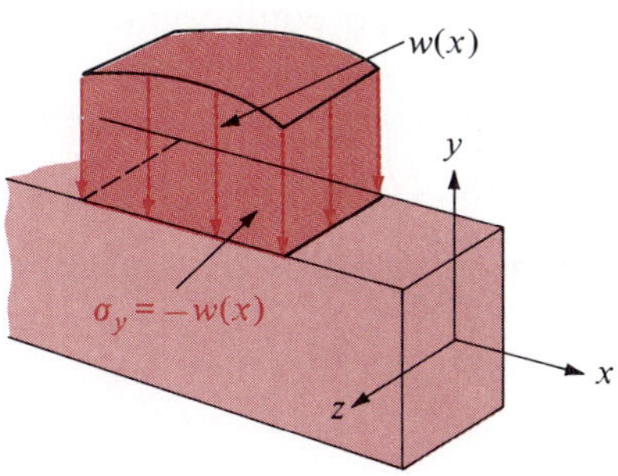

Figure 7.12 Normal-stress loading on the top surface

The student should realize that arguments such as these regarding the stresses σ_y, σ_z, τ_{xz}, and τ_{yz} are really only plausibility arguments: they do not constitute any sort of proof that the assumptions are true. The ultimate justification for such assumptions is found if the theory that results from the assumptions can be used to successfully predict stresses and deformations for the bending deformation problem.

Finally, accounting for all these assumptions, namely, that $\sigma_y = \sigma_z = \tau_{xz} = \tau_{yz} = 0$, it follows that $F_z = M_x = M_y = 0$. The three equilibrium equations that are not satisfied identically and that must be considered further are

$$\int \sigma_x \, dA = 0$$
$$V = -\int \tau_{xy} \, dA$$
$$M = -\int y(\sigma_x \, dA)$$

Note that the negative sign on the shear V is necessary in order to adhere to the sign convention adopted previously. Also, in what follows the subscripts on the shear force $F_y = -V$ and the moment M_z will be dropped. The shear and moment will simply be denoted by V and M, respectively.

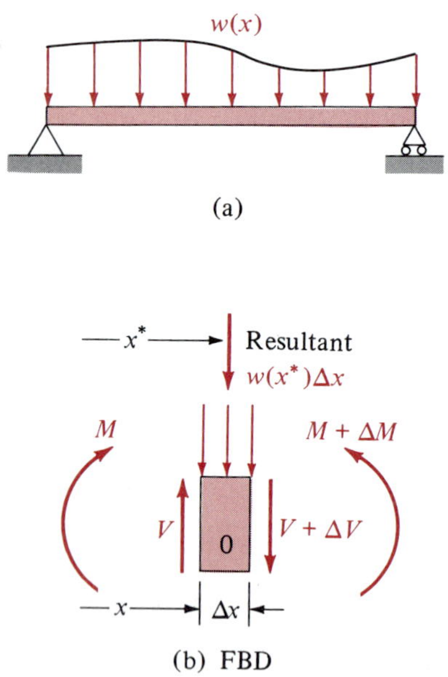

Figure 7.13 Loading, shear, and moment in a general situation

The last step in satisfying equilibrium is to relate the shear V and the moment M to the external loading $w(x)$ indicated in Fig. 7.13a. Fig. 7.13a defines the positive direction for the external loading $w(x)$. This is not completely consistent with the idea of having external loads acting in the positive coordinate direction, as in the previous two chapters. However, since a majority of loads carried by beams are due to a weight and hence act "downward," that is, in the negative y-direction, it will be practical to take the positive direction for external loadings to be in the negative y-direction.

Any segment of the beam between two positions x and $x + \Delta x$, as shown in Fig. 7.13b, must be in equilibrium under the action of the external loading $w(x)$ and the internal shears and moments acting on the two exposed faces. Equilibrium of forces requires

$$\Sigma F_y = +V - w(x^*)\,\Delta x - (V + \Delta V) = 0$$

After canceling like terms, dividing by Δx, and allowing Δx to approach zero, there results

$$\frac{dV}{dx} + w(x) = 0 \qquad (7.2)$$

The corresponding moment equilibrium equation is

$$\Sigma M_0 = -M - V\,\Delta x + [w(x^*)\,\Delta x](\alpha\,\Delta x) + M + \Delta M = 0$$

where $[w(x^*)\,\Delta x]$ is the resultant force due to the external loading and $\alpha\,\Delta x$ is the distance from the right end of the segment to the line of action of the resultant. Canceling like terms, dividing by Δx, and allowing Δx to approach zero lead to

$$\frac{dM}{dx} - V(x) = 0 \tag{7.3}$$

These two equations, (7.2) and (7.3), are the equations of equilibrium that relate the internal force resultants V and M to the external loading w for the bending problem. They are the counterparts of $P' + q = 0$ and $T' + t = 0$ for the axial and torsional problems, respectively. Note again that there are two equilibrium equations to determine the two internal resultants V and M, as opposed to the single equation of equilibrium for both the axial and torsional deformation problems, where there is only one internal resultant. Several useful results can be developed from Eqns. (7.2) and (7.3); these will be discussed and applied in detail in Section 7.3, which is devoted to shear and moment diagrams.

PROBLEMS

7.1–7.3 For each of the following, state which of the assumptions regarding geometry is not satisfied.

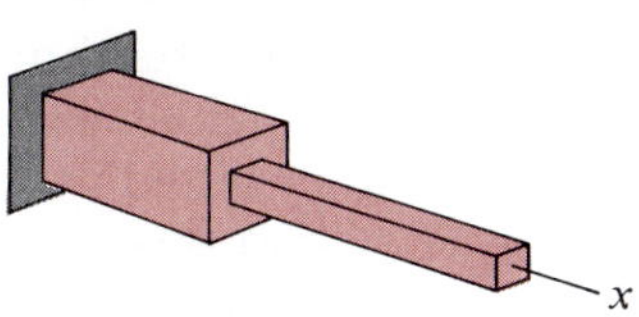

P. 7.1

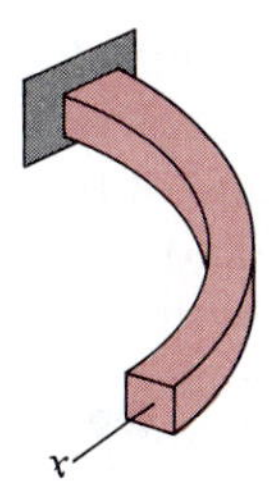

P. 7.2

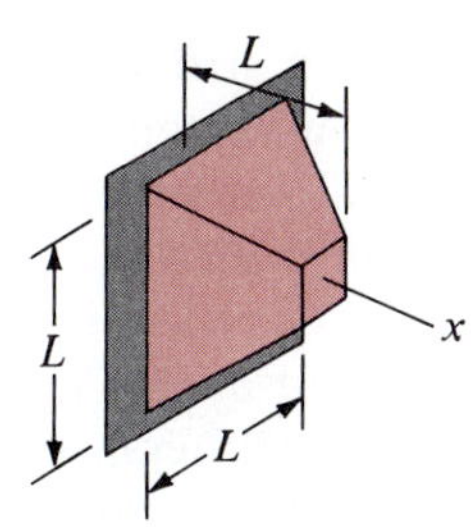

P. 7.3

7.4–7.6 For each of the following, state which of the assumptions regarding geometry or loading is not satisfied.

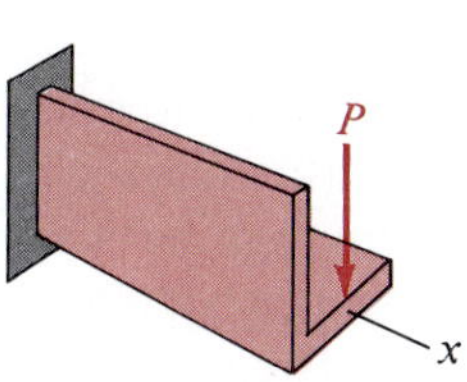

P. 7.4

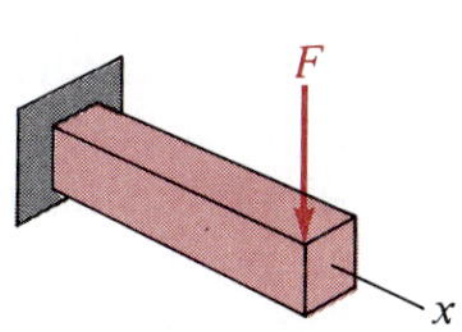

P. 7.5

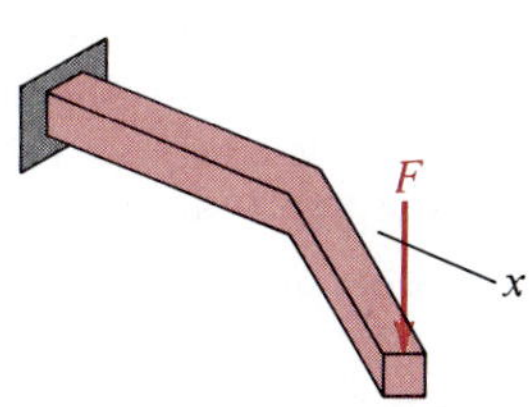

P. 7.6

7.7 Show that if the normal stress σ_x is assumed to be a linear function of the y coordinate, i.e., $\sigma_x = \alpha + \beta y$, the two equations $\int \sigma_x \, dA = 0$ and $-\int y\sigma_x \, dA = M$ can be satisfied by taking $\alpha = 0$, $\beta = -M/I$, where the z-axis $(y = 0)$ is to be located at the centroid and I is the second moment of the area of the section about the z-axis. Compare this with the results of Section 7.6.

7.8 Show that if the normal stress σ_x is assumed to be a linear function of y and z, i.e., $\sigma_x = \alpha + \beta y + \gamma z$, the equations $\int \sigma_x \, dA = 0$, $-\int y\sigma_x \, dA = M$, and $\int z\sigma_x \, dA = 0$ can be satisfied by taking $\alpha = \gamma = 0$, $\beta = -M/I$, where the z-axis $(y = 0)$ is to be located at the centroid and I is the second moment of the area of the section about the z-axis.

7.9 Show that by taking $\tau_{xz} = 0$ with σ_x and τ_{xy} both symmetric functions of z, M_x in Eqn. (7.1d) is identically zero.

7.10 Show that by taking $\tau_{xz} = 0$ with σ_x and τ_{xy} both symmetric functions of z, M_y in Eqn. (7.1e) is identically zero.

7.3 SHEAR AND BENDING MOMENT DIAGRAMS

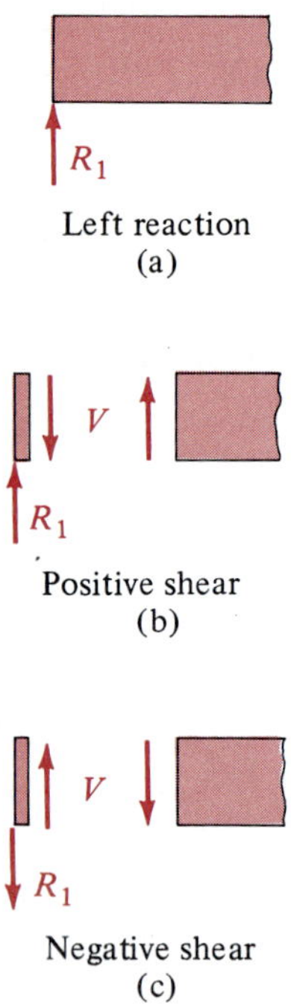

Figure 7.14 Shear for a reaction at the left end of a beam

In the previous section on equilibrium it was shown that the shear V and moment M were related to the stresses τ_{xy} and σ_x according to

$$V = -\int \tau_{xy} \, dA \qquad \text{and} \qquad M = -\int y(\sigma_x \, dA)$$

showing that τ_{xy} depends directly on V and that σ_x depends directly on M. In view of the fact that the evaluation of the strength of a beam will depend strongly on the magnitudes of the stresses, it is important to have a systematic way to determine the locations and magnitudes of the maximum moment M and of the maximum shear V transmitted by a beam under a given set of loads. In this section we will outline and discuss procedures for determining the shear V and the bending moment M at any location along the length of the beam.

Let us begin with two basic procedures for constructing shear and bending moment diagrams. The first consists of sectioning the beam at a number of points along the beam, drawing free-body diagrams, using the equations of equilibrium to determine V and M, and then plotting the results. The second procedure consists of a systematic use of the equilibrium equations $V' + w = 0$ and $M' - V = 0$, along with results based upon integrating these equations to construct the shear and bending moment diagrams.

Before outlining either of these procedures, consider the situation regarding the shear transmitted at an end support such as shown in Fig. 7.14. A free-body diagram of a thin slice of the beam together with the corresponding forces exerted on the rest of the beam is shown in Fig. 7.14b. Recalling the definition for positive shear, it is clear that when the reaction is in the direction shown, the shear is equal to the value of the reaction and is positive in sign. Similarly, the free-body diagram of Fig. 7.14c shows that the shear is negative the value of the reaction when the reaction is oppositely directed. For a support at the right end of the beam as shown in Fig. 7.15, the free-body diagrams of Fig. 7.15b and c picture the cases where the shear is negative and positive, respectively.

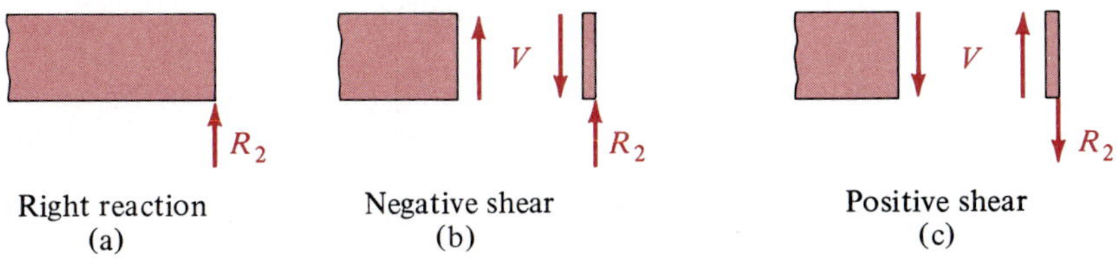

Figure 7.15 Shear for a reaction at the right end of a beam

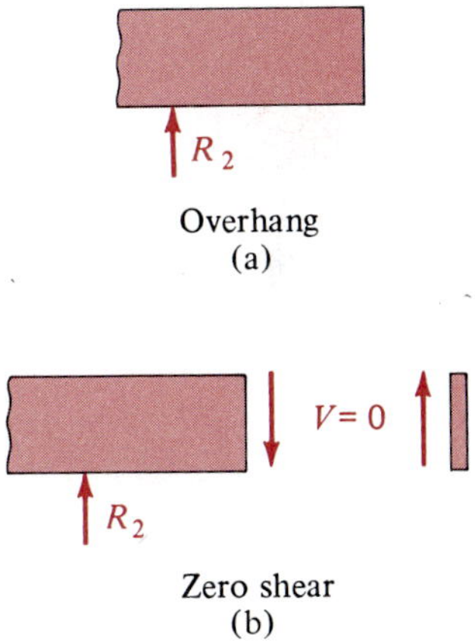

Figure 7.16 Determination of the shear for an overhang

Thus the shear at the ends of a beam can be determined by inspection on the basis of the values and directions of the reactions. Note also that at an unsupported end such as the unloaded overhang shown in Fig. 7.16, the shear must be zero on the basis of equilibrium for the free-body diagram of a thin slice of the beam, as shown in Fig. 7.16b. The shear for an unloaded overhang at the left end of the beam would also be zero. These observations regarding the value of the shear at the ends of a beam are valid regardless of the specific method used to construct the shear and bending moment diagrams.

7.3.1 Use of Sections and Free-Body Diagrams

The process is most easily demonstrated by setting up and working through several examples. In Examples 7.1 and 7.2, we will demonstrate the process of sectioning at a number of locations along the length of the beam, and using the corresponding free-body diagrams to determine the internal shear V and the internal moment M.

Example 7.1 Consider a beam acted upon by a uniformly distributed load w_0 as shown in Fig. 7.17a.

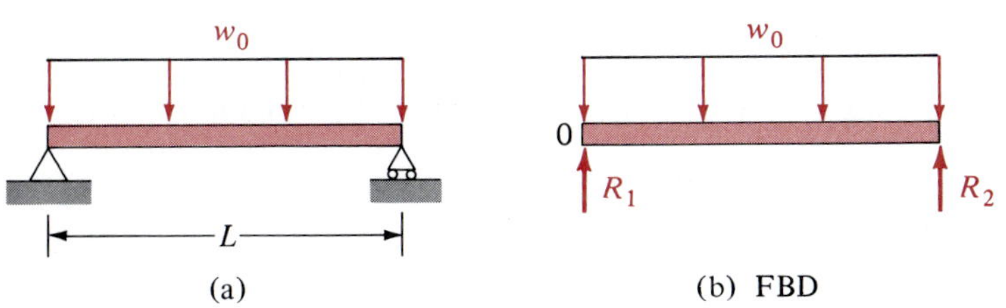

Figure 7.17 Uniformly loaded beam

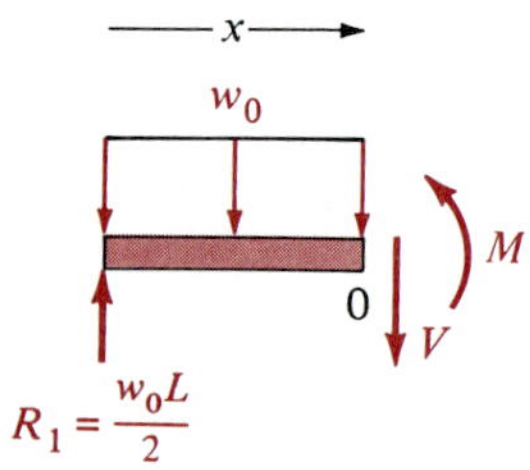

Figure 7.18 Free-body diagram of a segment of a uniformly loaded beam

Equilibrium A free-body diagram of the beam is drawn in Fig. 7.17b. The two equilibrium equations are

$$\Sigma F_y = R_1 + R_2 - w_0 L = 0$$

and

$$\Sigma M_0 = R_2 L - w_0 L \left(\frac{L}{2}\right) = 0$$

from which $R_1 = R_2 = w_0 L/2$. A free-body diagram of a segment of the beam is then drawn as in Fig. 7.18, showing the internal shear force V and the internal moment M.

Note carefully that the directions of the unknown internal shear V and the unknown internal moment M are both assumed to be positive when drawing the free-body diagram. The equilibrium equations for this segment are

$$\Sigma F_y = \frac{w_0 L}{2} - w_0 x - V = 0$$

$$\Sigma M_0 = -\frac{w_0 L}{2} x + w_0 x \frac{x}{2} + M = 0$$

leading to

$$V = w_0 \left(\frac{L}{2} - x\right)$$

$$M = w_0 x \frac{L - x}{2}$$

These are plotted in Fig. 7.19.

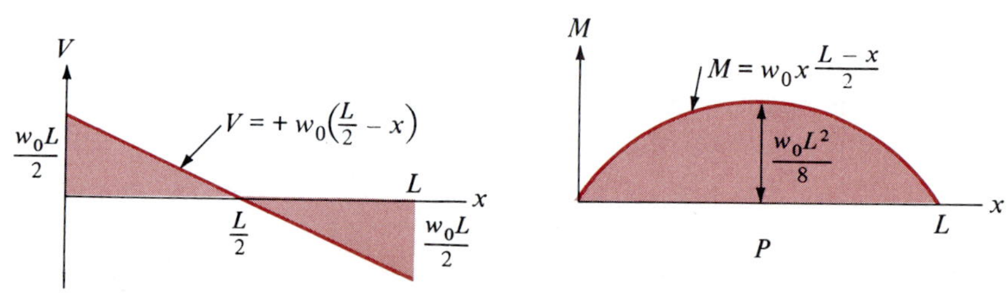

Figure 7.19 Shear and moment diagrams for a uniformly loaded beam

As a check, $M(L) = 0$ is verified by the expression $M = w_0 x(L - x)/2$. The plots or graphs of V and M are referred to as **shear and moment diagrams.**

Example 7.2 Consider a beam loaded by a single force P as indicated in Fig. 7.20a.

Equilibrium Referring to Fig. 7.20b, the equilibrium equations for determining the reactions R_1 and R_2 are

$$\Sigma F_y = R_1 + R_2 - P = 0 \qquad \text{and} \qquad \Sigma M_0 = R_2 L - Pa = 0$$

from which $R_1 = Pb/L$ and $R_2 = Pa/L$. A free-body diagram taken for a section $0 < x < a$, a point to the left of the load P, is indicated in Fig 7.21a.

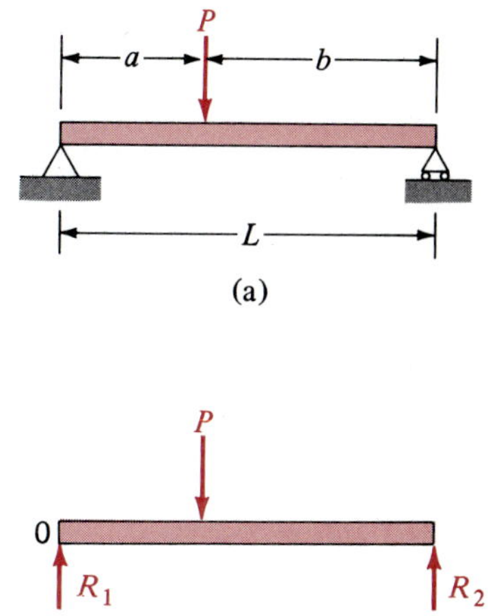

Figure 7.20

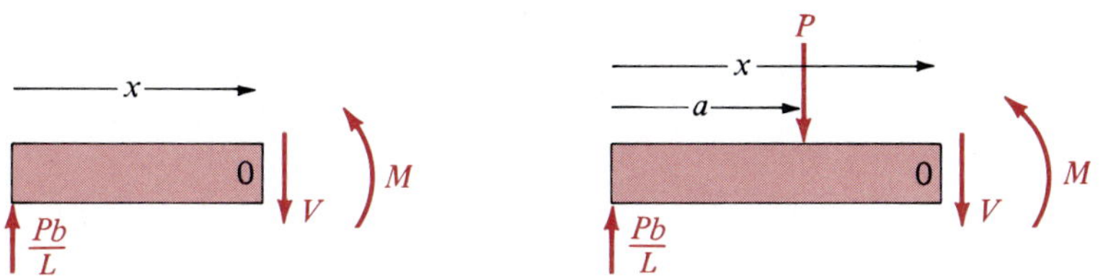

FBD for $0 < x < a$

(a)

FBD for $a < x < L$

(b)

Figure 7.21 Free-body diagrams for various segments of the beam

Equilibrium for this segment takes the form

$$\Sigma F_y = \frac{Pb}{L} - V = 0 \qquad \text{or} \qquad V = \frac{Pb}{L}$$

and

$$\Sigma M_0 = M - \frac{Pb}{L} x = 0 \qquad \text{or} \qquad M = \frac{Pb}{L} x$$

For the segment shown in Fig. 7.21b, that is, for $a < x < L$, a point to the right of the load P, the equilibrium equations are

$$\Sigma F_y = \frac{Pb}{L} - P - V = 0 \qquad \text{or} \qquad V = -\frac{Pa}{L}$$

and

$$\Sigma M_0 = -\frac{Pb}{L} x + P(x - a) + M = 0 \qquad \text{or} \qquad M = Pa\left(1 - \frac{x}{L}\right)$$

Note again that the directions of the unknown internal shear V and the unknown internal moment M are both assumed to be positive when drawing the free-body diagram. The corresponding graphs for V and M are shown in

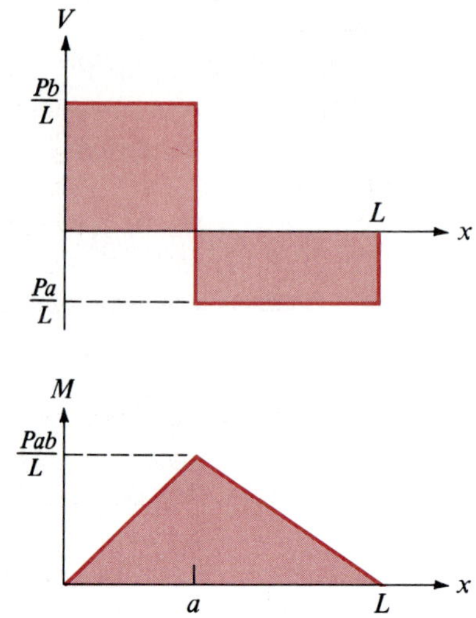

Figure 7.22 Shear and moment diagrams for a load P

Fig. 7.22. The maximum internal shear force will depend upon the values of a and b, with the maximum occurring in the shorter segment. The maximum moment, with the value Pab/L, occurs at the location of the load P.

In summary, for this method of constructing the shear and moment diagrams, free-body diagrams are used in conjunction with the two equations of equilibrium, to determine the shear V and the moment M at a number of points along the length of the beam. A critical step in drawing the free-body diagrams is to assume the *positive* directions for V and M. Note that this method results in analytical expressions for the shear $V(x)$ and moment $M(x)$.

PROBLEMS

7.11–7.44 For each of the following, draw the necessary free-body diagrams and construct the shear and moment diagrams and label all critical values. In problem 7.29, assume that the reactive loading from the base is uniform.

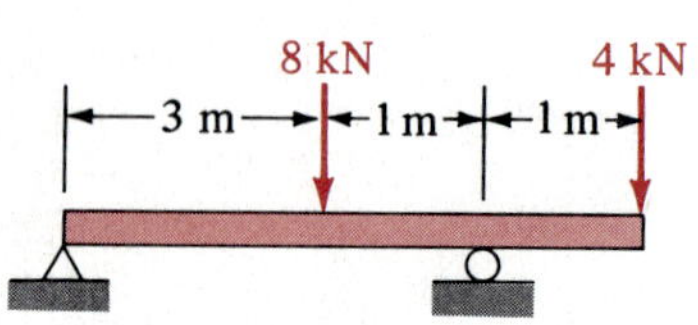

P. 7.11

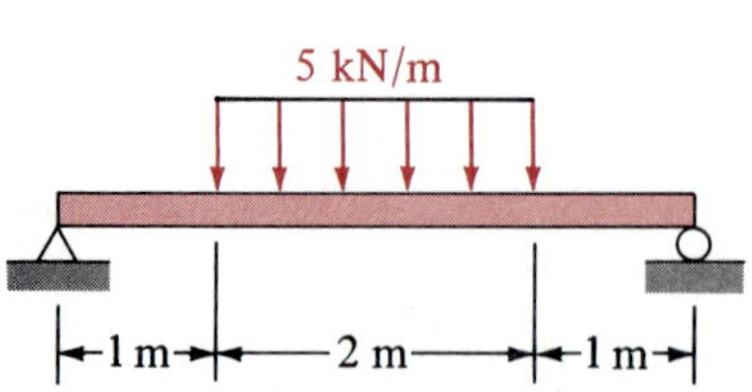

P. 7.12

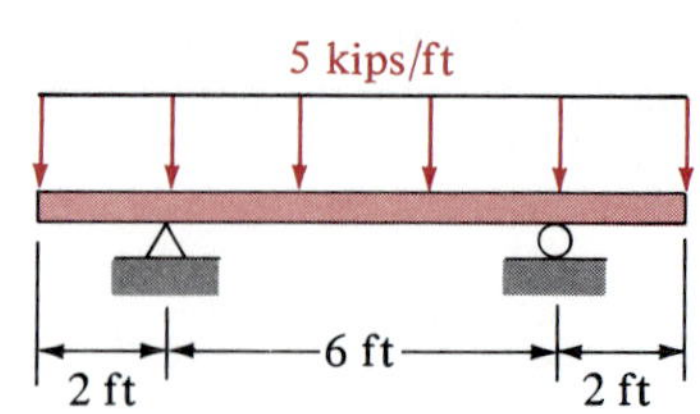

P. 7.13

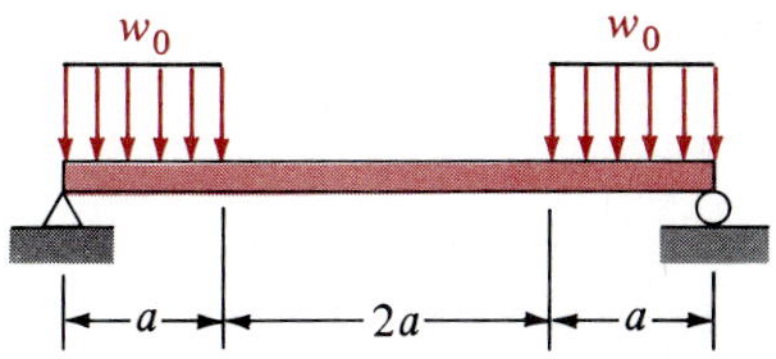

P. 7.14

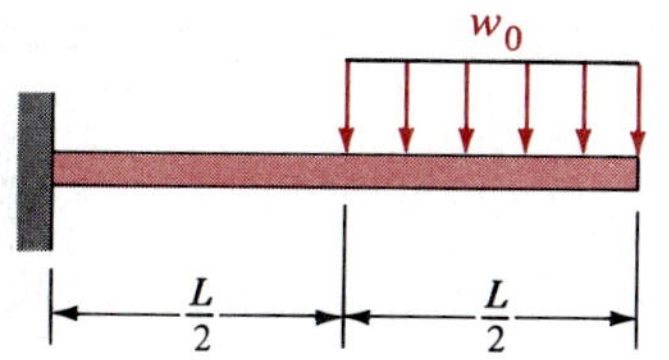

Ps. 7.15 and 7.67

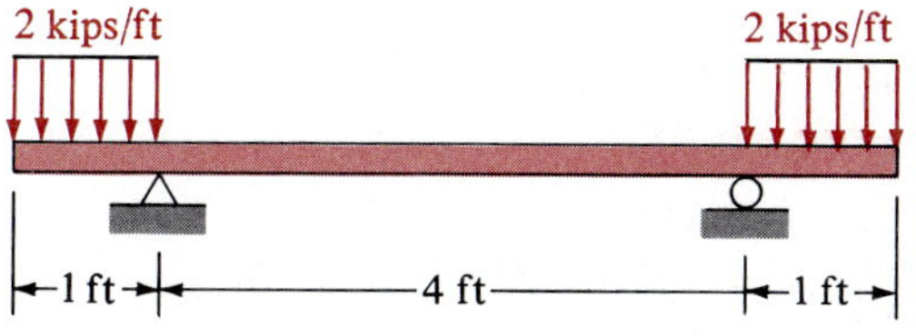

P. 7.16

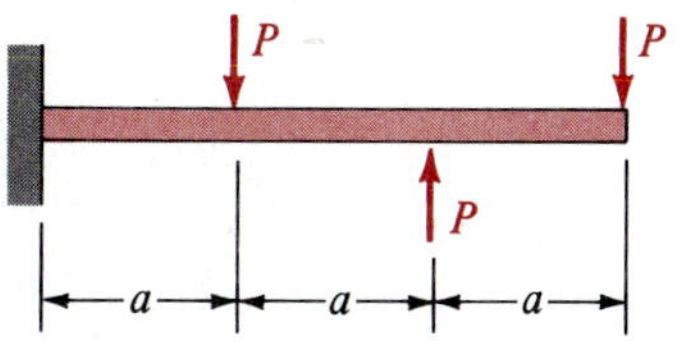

P. 7.17

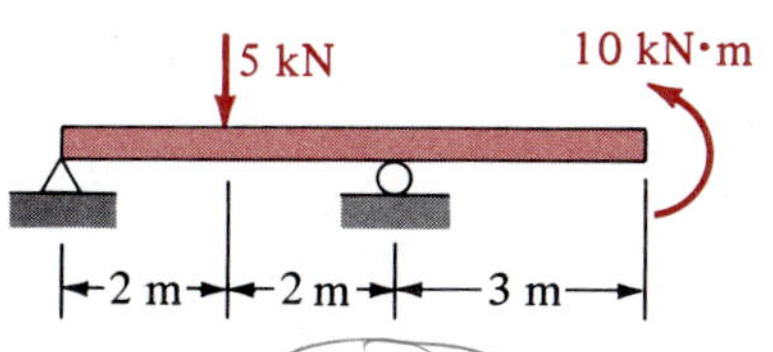

P. 7.18

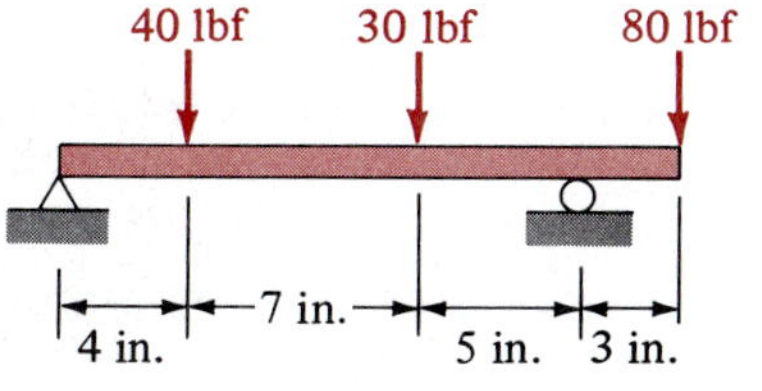

P. 7.19

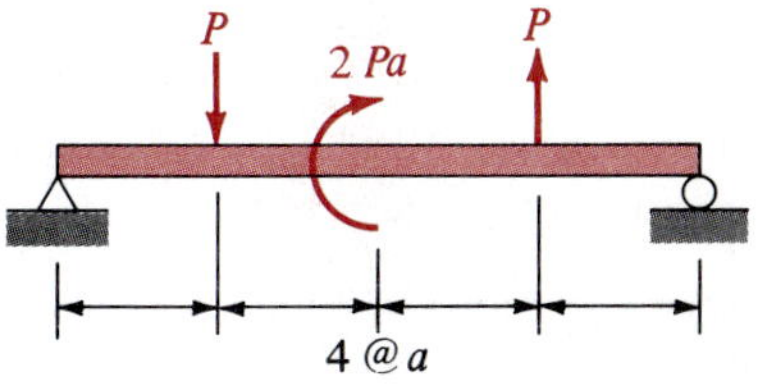

P. 7.20

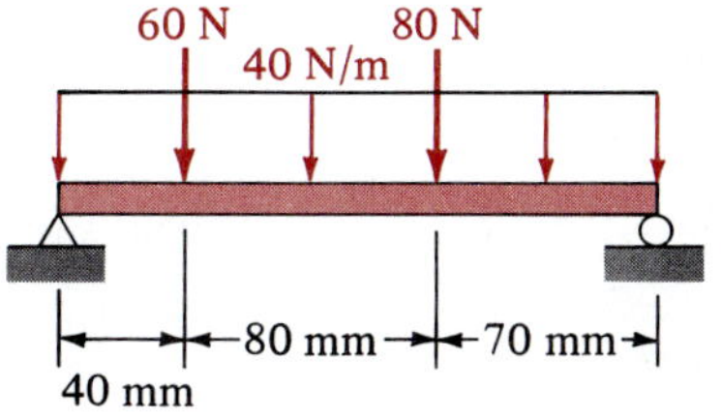

P. 7.21

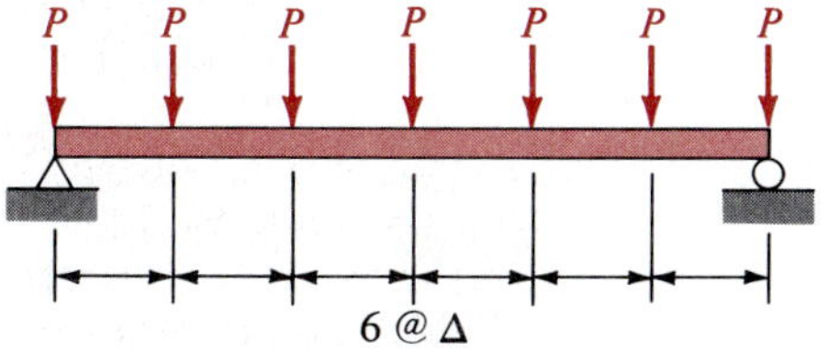

P. 7.22

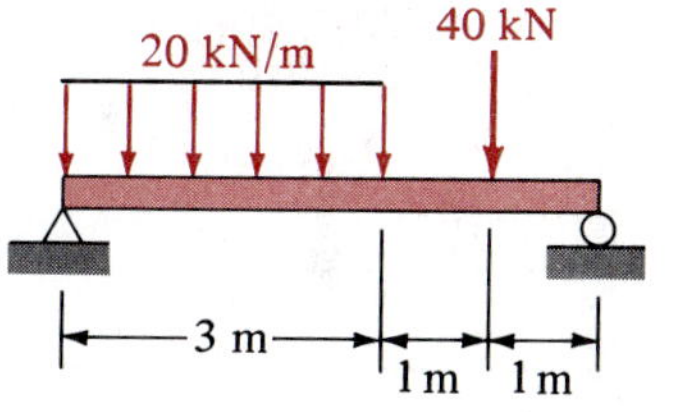

Ps. 7.23 and 7.68

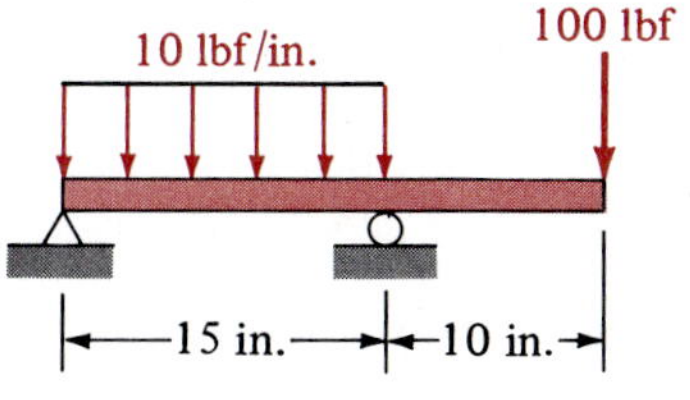

Ps. 7.24 and 7.69

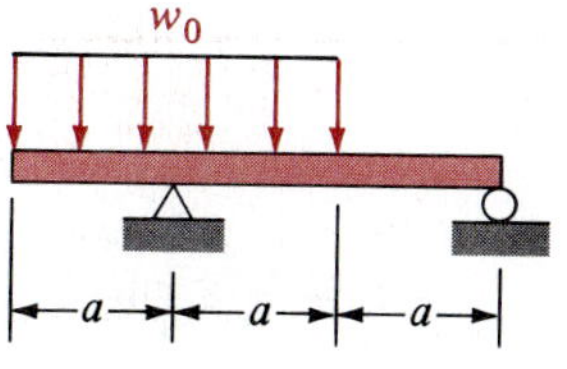

Ps. 7.25 and 7.70

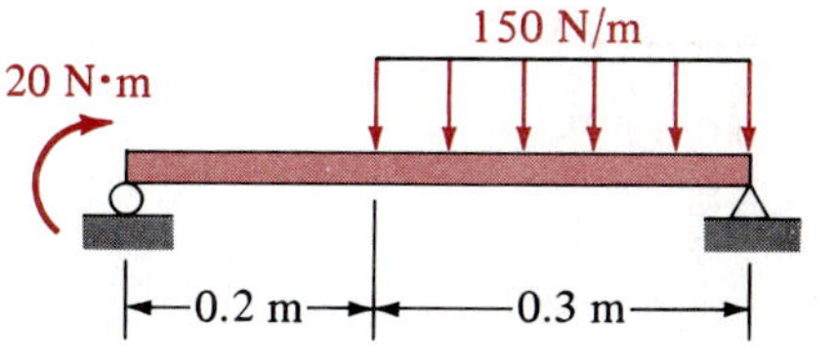

P. 7.26

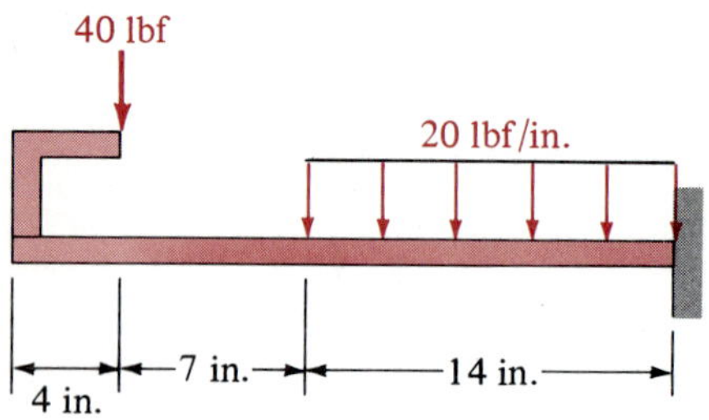

P. 7.27

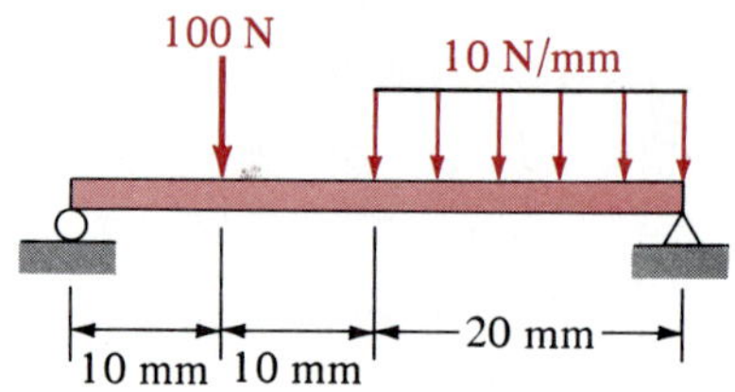

Ps. 7.28 and 7.71

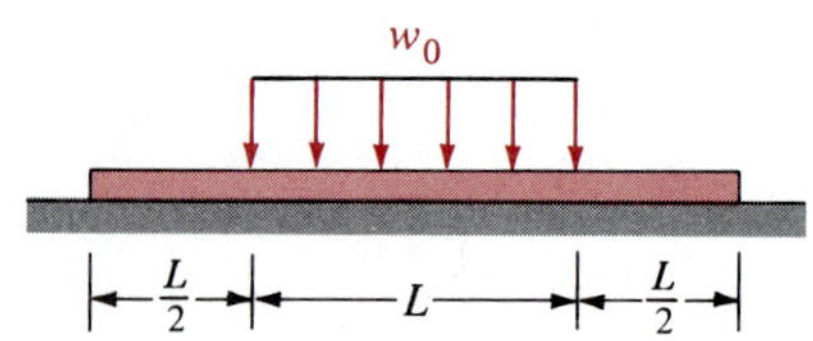

P. 7.29

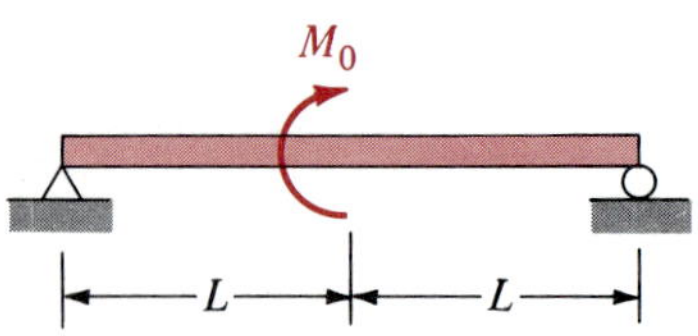

P. 7.30

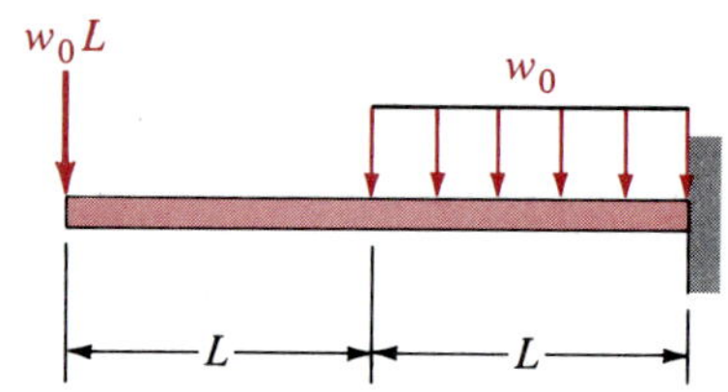

Ps. 7.31 and 7.72

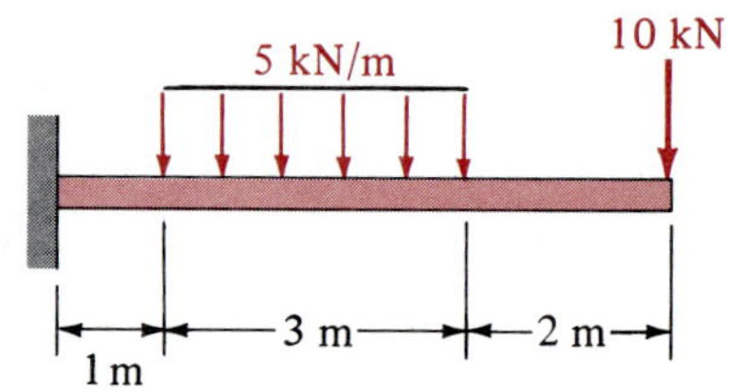

P. 7.32

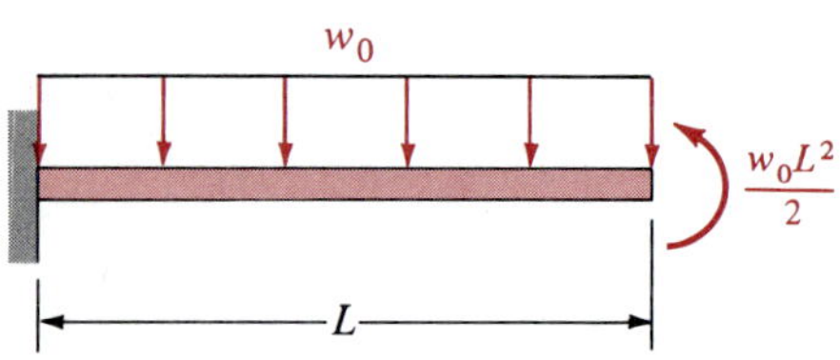

P. 7.33

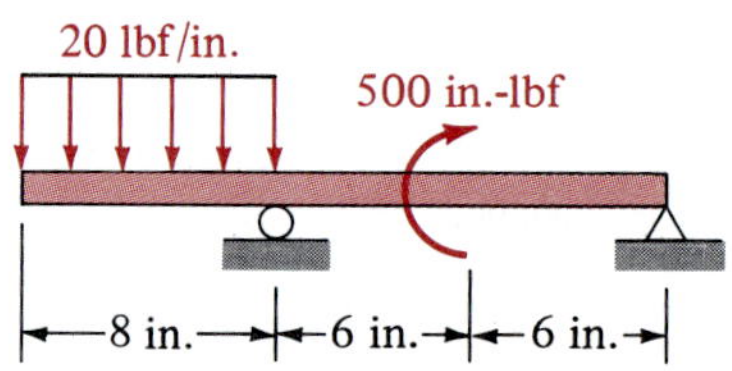

P. 7.34

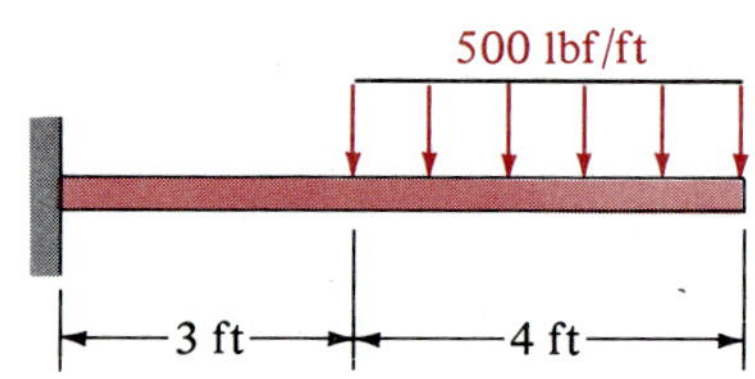

P. 7.35

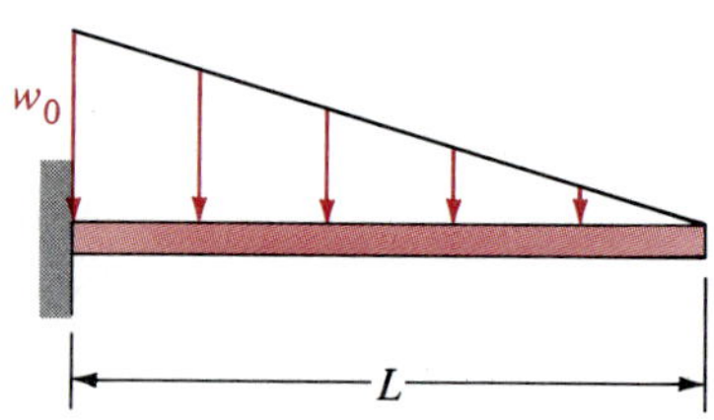

P. 7.36

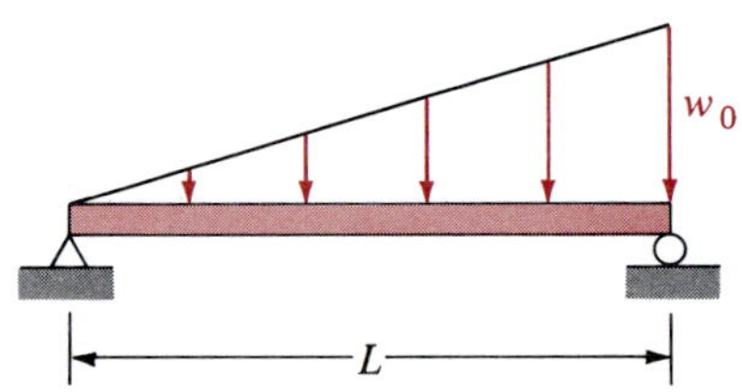

P. 7.37

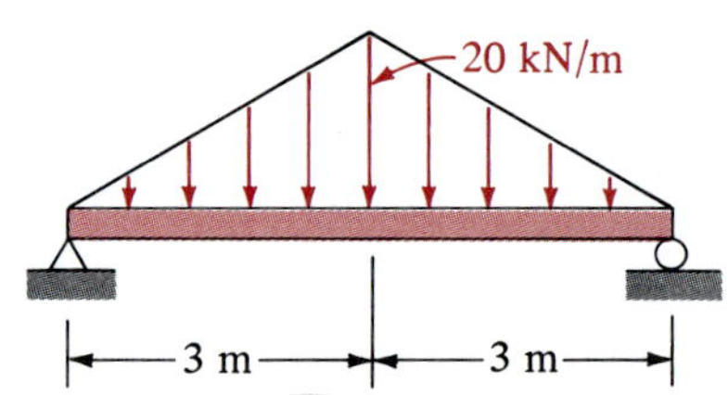

Ps. 7.38 and 7.73

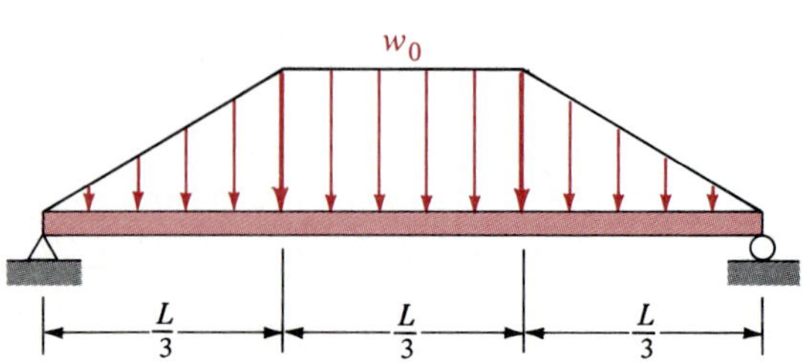

Ps. 7.39 and 7.74

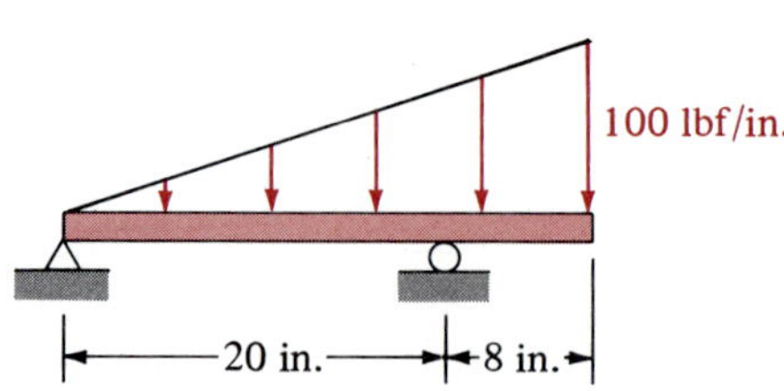

Ps. 7.40 and 7.75

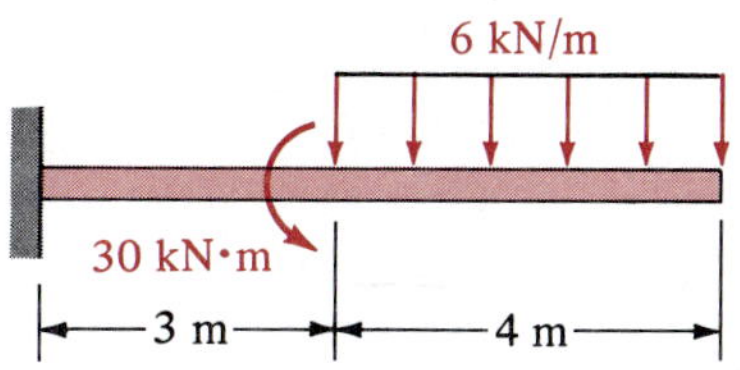

P. 7.41

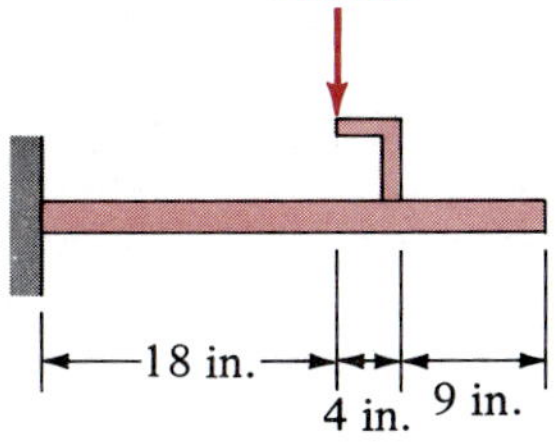

P. 7.42

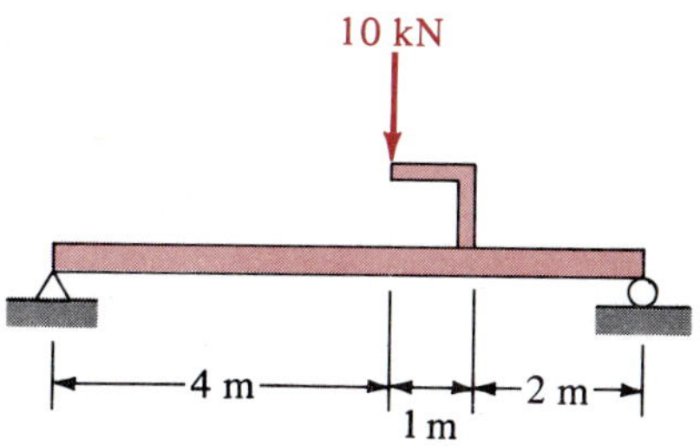

P. 7.43

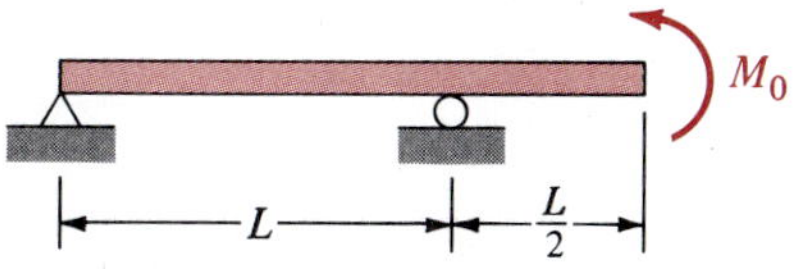

P. 7.44

7.3.2 Integration of the Equations of Equilibrium

For this approach, we make use of the equations of equilibrium, $V' + w = 0$ and $M' - V = 0$, and certain integrated forms of these equations, to assist in constructing the shear and moment diagrams. Before we look at examples demonstrating the use of these equations, consider the force equation of equilibrium,

$$\frac{dV}{dx} = -w(x) \tag{7.4}$$

which can be interpreted to state that *the slope of the shear diagram at any point along the beam is the negative of the value of the loading diagram at that point*. Thus there is a geometric relation between the value of the loading diagram and the corresponding slope of the shear diagram at any point along the beam.

Eqn. (7.4) can be integrated between any two values x_1 and x_2 to obtain

$$\Delta V = V(x_2) - V(x_1) = -\int_{x_1}^{x_2} w(x)\, dx \tag{7.5}$$

which can be interpreted to state that *the change in the shear between any two locations x_1 and x_2 is the negative of the area under the loading diagram between the same two locations.*

As indicated in Fig. 7.23, this can be very useful when plotting the shear diagram using information from the loading diagram. The two graphs for the loading w and the shear V in Fig. 7.23 show this idea—that the change in V is the negative of the area under the loading diagram.

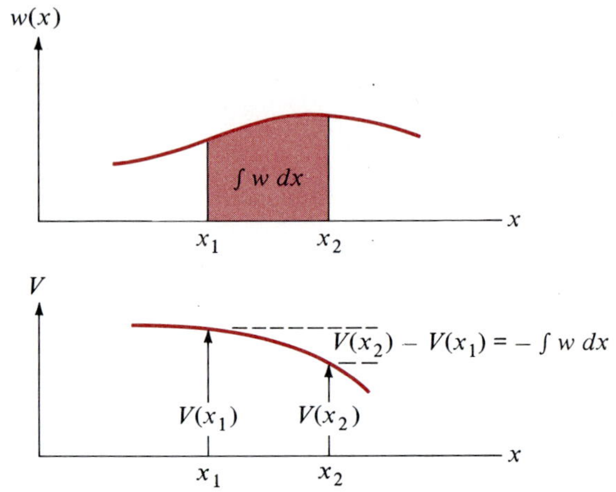

Figure 7.23 Graphical interpretation of $\Delta V = -\int w\,dx$

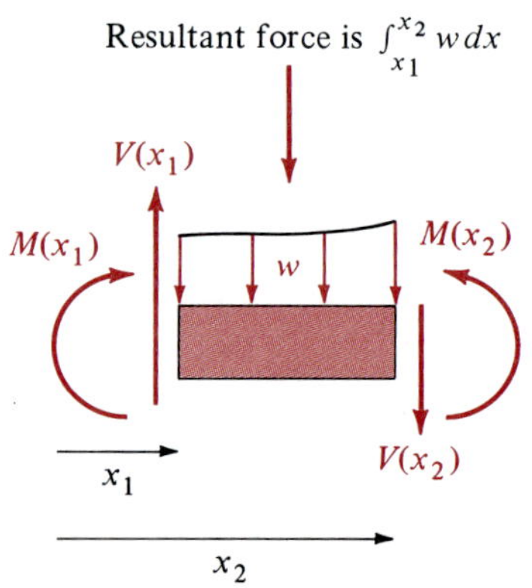

Figure 7.24 Equilibrium of a portion of a beam between x_1 and x_2

Eqn. (7.5) is easily interpreted physically by drawing a free-body diagram of the portion of the beam between the locations x_1 and x_2 as shown in Fig. 7.24. The equilibrium equation for forces in the y-direction is

$$\Sigma F_y = +V(x_1) - \int_{x_1}^{x_2} w(x)\,dx - V(x_2) = 0$$

or

$$V(x_2) - V(x_1) = -\int_{x_1}^{x_2} w(x)\,dx$$

that is, precisely Eqn. (7.5). This shows that integrating $V' = -w$ is equivalent to drawing a free-body diagram and summing forces in the y-direction!

Several specific loadings are indicated in Fig. 7.25; in each case, the loading diagram together with the corresponding shear diagram is pictured. We observe that if the degree of the polynomial representing the loading w on some interval is p, the degree of the polynomial representing the shear V on that same interval is $p + 1$.

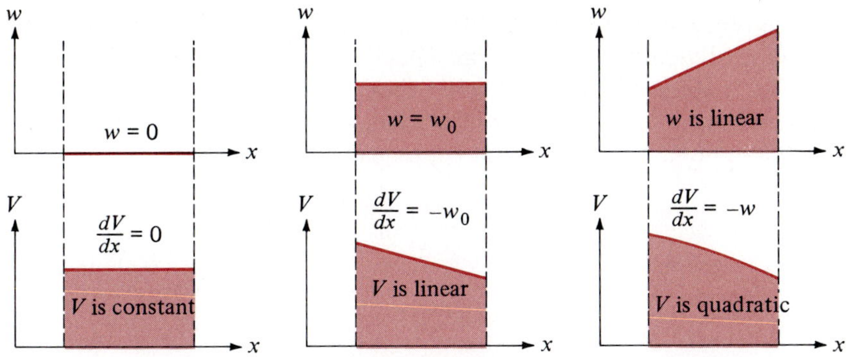

Figure 7.25 Geometric relations between w and V for several specific loadings

Next consider the moment equilibrium equation

$$\frac{dM}{dx} = V \tag{7.6}$$

which can be interpreted to state that *the slope of the moment diagram is equal to the corresponding value of the shear at every point along the beam*. Integrating Eqn. (7.6) between x_1 and x_2 produces

$$\Delta M = M(x_2) - M(x_1) = \int_{x_1}^{x_2} V(x)\, dx \tag{7.7}$$

Eqn. (7.7) can be interpreted to state that *the change in the moment between two values x_1 and x_2 is equal to the area under the shear diagram between those two points*. As indicated in Fig. 7.26, this result can be very useful in the course of plotting the moment diagram. It is left in the problems for the student to show that this is equivalent to summing moments about $x = x_2$ for the free-body diagram shown in Fig. 7.24.

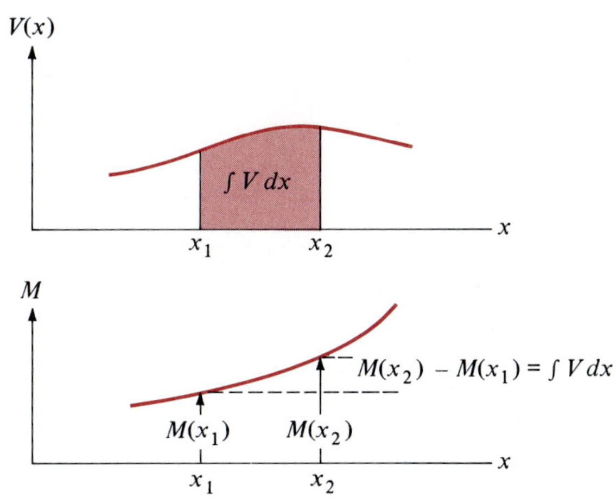

Figure 7.26 Graphical interpretation of $\Delta M = \int V\, dx$

The relations between the loading diagram, the shear diagram, and moment diagrams for several specific loadings are indicated in Fig. 7.27. Here again, we observe that if the order of the polynomial representing V on a particular interval is p, the order of the corresponding polynomial representing M on that interval is $p + 1$.

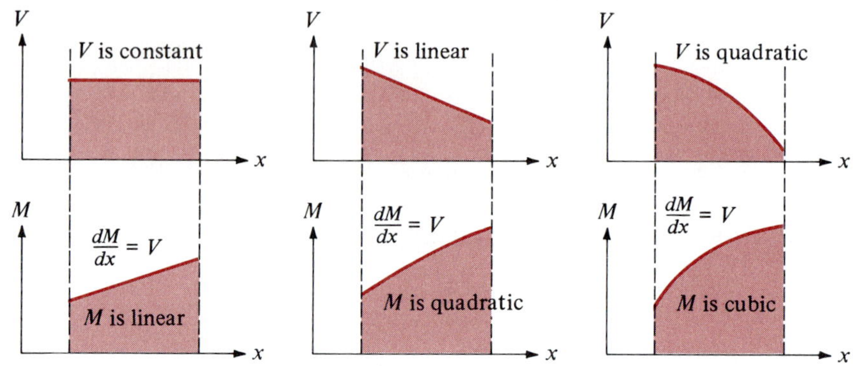

Figure 7.27 Relations between V and M for several specific loadings

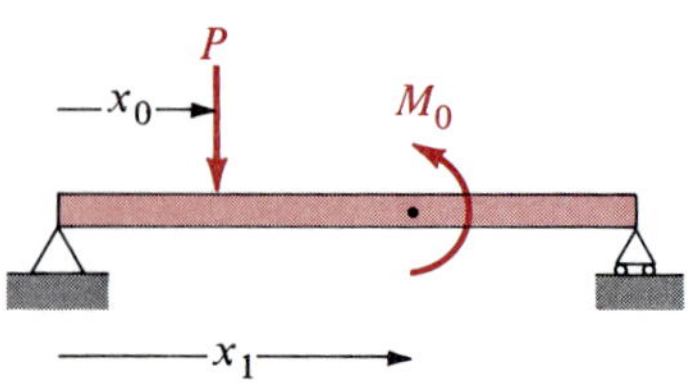

Figure 7.28 Beam loaded by a concentrated force P and a concentrated moment M_0

Before looking at examples illustrating the use of these ideas, it is useful to consider the effect of concentrated loads P and concentrated moments M_0 on the shear force and bending moment. A positive load P and positive moment M_0 are pictured in Fig. 7.28.

For the concentrated load P, let us draw a free-body diagram of a short segment of the beam containing the load P as indicated in Fig. 7.29.

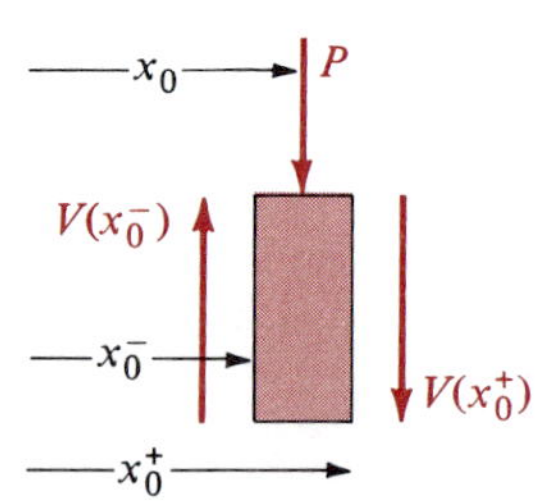

Figure 7.29 Free-body diagram of a segment containing the load P

As shown in the figure, x_0 is the location of the load P with x_0^- and x_0^+ indicating locations just to the left and right, respectively, of the load. The result of requiring equilibrium in the y-direction is

$$V(x_0^+) - V(x_0^-) = -P \tag{7.8}$$

This equation indicates that there is an abrupt drop in the shear by the amount of the positive load P. The corresponding shear and moment diagrams are shown in Fig. 7.30.

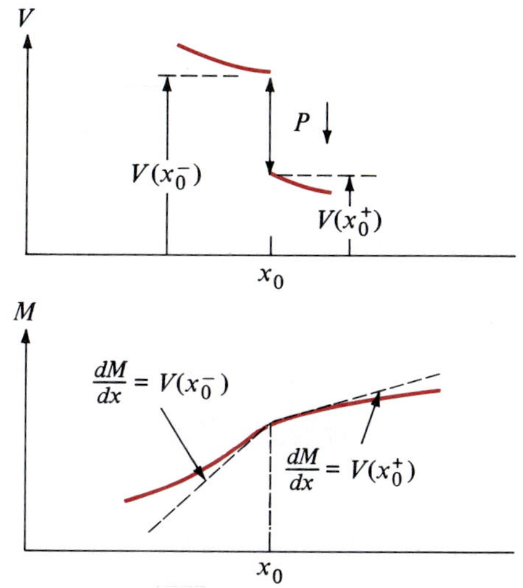

Figure 7.30 Effect on the shear force and moment from a concentrated load P

Note in Fig. 7.30 that since $dM/dx = V$, there must be a difference in the slopes of the moment diagram just to the left and right of the point x_0. This is indicated by the dotted lines, showing the decrease in the slopes of the moment curve to the left and right of the point x_0.

For the concentrated moment located at a point x_1, we construct the free-body diagram shown in Fig. 7.31. The width of the segment considered is taken to be $2a$. Summing moments yields

$$-M(x_1^-) + M_0 + M(x_1^+) - aV(x_1^-) - aV(x_1^+) = 0$$

or, in the limit as a approaches zero,

$$M(x_1^+) - M(x_1^-) = -M_0 \tag{7.9}$$

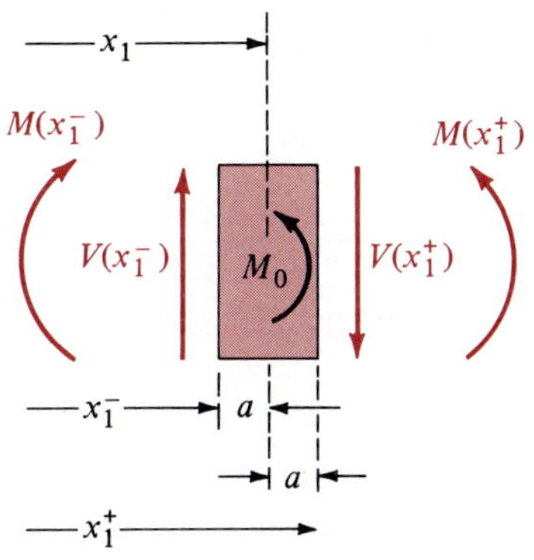

Figure 7.31 Free-body diagram of a segment containing the moment M_0

The moment diagram in the neighborhood of x_1 is shown in Fig. 7.32, where it is seen that there is an abrupt decrease M_0 in the moment in passing from x_1^- to x_1^+.

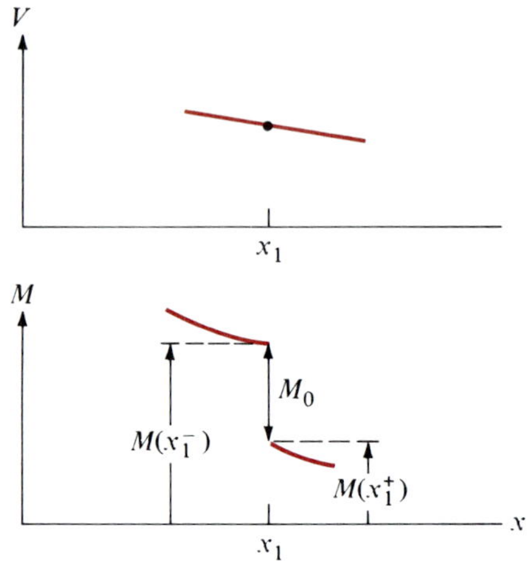

Figure 7.32 Effect on the bending moment from a concentrated moment M_0

The results indicated in Figs. 7.30 and 7.32 can be very helpful in constructing the details of the shear and moment diagrams near a concentrated load or moment.

Example 7.3 A beam is loaded as shown in Fig. 7.33a. Use the equations of equilibrium, $V' + w = 0$ and $M' = V$, along with the corresponding integrated forms, to construct the shear and moment diagrams.

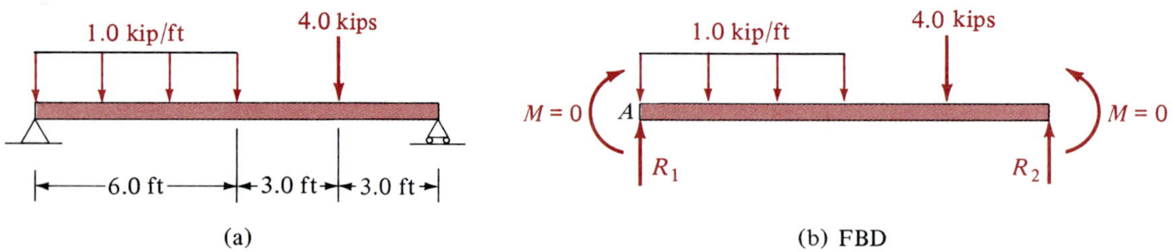

Figure 7.33 Loaded beam and free-body diagram

Solution Referring to the free-body diagram shown in Fig. 7.33b, the equilibrium equations for determining the reactions R_1 and R_2 are

$$\Sigma F_y = R_1 + R_2 - (1.0 \text{ kip/ft})(6 \text{ ft}) - 4.0 \text{ kips} = 0$$

$$\Sigma M_A = -(1.0 \text{ kip/ft})(6 \text{ ft})(3 \text{ ft}) - (4.0 \text{ kips})(9 \text{ ft}) + R_2(12 \text{ ft}) = 0$$

Solving yields $R_1 = 5.5$ kips and $R_2 = 4.5$ kips. The loading diagram is shown in Fig. 7.34, and consists of the constant 1 kip/ft loading for $0 < x < 6$, the concentrated load of 4.0 kips at $x = 9$ ft, and the reactions $R_1 = 5.5$ kips and $R_2 = 4.5$ kips at $x = 0$ and $x = 12$ ft, respectively. Note that dotted lines are extended downward from the loading diagram at the ends of the beam and at any other point where there is a change in the type of loading, or at the location of any concentrated load or concentrated moment. The shear and moment diagrams are constructed as follows.

0 < x < 6 ft With $w = 1.0$ kip/ft, the constant slope of the shear diagram is a negative 1.0 kip/ft. The change in the shear is given by Eqn. (7.5) as

$$V(x) - V(0) = -\int_0^x (1.0 \text{ kip/ft})\, dx$$

or, with x measured in ft, $V(x) = (5.5 - x)$ kips, as indicated in Fig. 7.34. Note that the value of x at which the shear passes through zero is $x = 5.5$ ft, and that at $x = 6$ ft, $V = -0.5$ kip. For the bending moment, Eqn. (7.7) is used to express

$$M(x) - M(0) = \int_0^x (5.5 - x) \text{ kips}\, dx$$

or, with x measured in ft,

$$M(x) = 0 + \left(5.5x - \frac{x^2}{2}\right) \text{ ft·kip}$$

As shown in Fig. 7.34, the maximum value of the moment occurs at $x = 5.5$ ft ($V = M' = 0$) and is $M(5.5 \text{ ft}) = 15.125$ ft·kip, with the value at 6.0 ft as $M(6.0 \text{ ft}) = 15.0$ ft·kip.

6 ft < x < 9 ft With $w = 0$ in this interval, $V' = 0$, so that in this interval, V has the constant value $V = -0.5$ kip, as shown in the V diagram in Fig. 7.34. Using the equation $M' = V = -0.5$ kip and Eqn. (7.7), the moment is computed to be

$$M(x) = M(6 \text{ ft}) + \int_{6.0 \text{ ft}}^x (-0.5 \text{ kip})\, dx$$

or

$$M(x) = 15.0 \text{ ft·kip} - (0.5 \text{ kip})(x - 6.0) \text{ ft} = (18.0 - 0.5x) \text{ ft·kip}$$

The values of the shear and moment at $x = 9$ ft are $V = -0.5$ kip and $M = 13.5$ ft·kip.

At x = 9 ft Using Eqn. (7.8), we conclude that the concentrated load at $x = 9$ ft produces an abrupt drop of 4.0 kips to a negative 4.5 kips in the shear diagram, and also a corresponding abrupt drop in the slope of the moment diagram at $x = 9.0$ ft. These are both indicated in the V and M diagrams of Fig. 7.34.

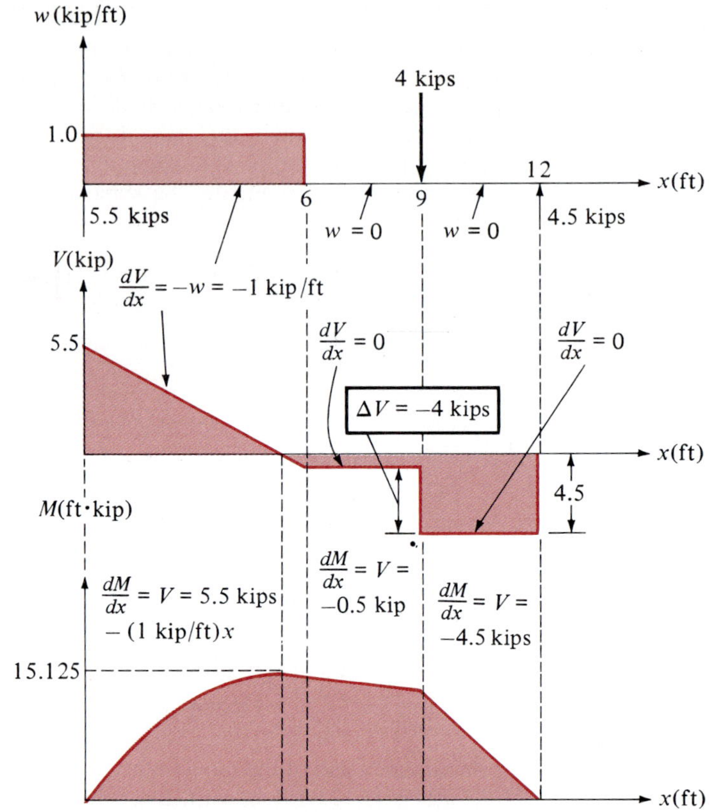

Figure 7.34 Loading, shear, and moment diagrams

9 ft < x < 12 ft In this interval w is zero, with the corresponding V having the constant value of -4.5 kips. Note that this magnitude is exactly the value of the reaction at $x = 12$ ft as must be the case according to previous discussions. Using $M' = V = -4.5$ kips for the slope of the moment diagram and Eqn. (7.7), we can write

$$M(x) - M(9 \text{ ft}) = \int_{9.0 \text{ ft}}^{x} (-4.5 \text{ kips})\, dx$$

or

$$M(x) = (54.0 - 4.5x) \text{ ft·kip}$$

Note that $M(12 \text{ ft}) = 0$, checking the fact that there is no moment at $x = 12$ ft.

The values arrived at for the shear and moment at $x = 12$ ft are very important from the standpoint that they serve as a check on the previous work. If the shear at the end of the beam does not match the known value of the reaction, or if the moment at the end does not match the known value, there is a mistake somewhere!

Example 7.4 A cantilever beam is loaded as in Fig. 7.35a. Use the equations of equilibrium, $V' + w = 0$ and $M' = V$, along with the corresponding integrated forms, to construct the shear and moment diagrams.

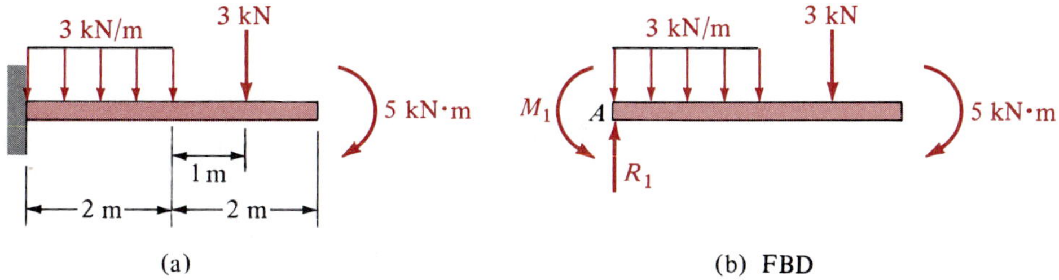

Figure 7.35 Cantilever beam and corresponding free-body diagram

Solution Referring to the free-body diagram shown in Fig. 7.35b, the equilibrium equations for determining the reactions R_1 and M_1 are

$$\Sigma F_y = R_1 - (3.0 \text{ kN/m})(2 \text{ m}) - 3.0 \text{ kN} = 0$$

$$\Sigma M_A = M_1 - (6 \text{ kN})(1 \text{ m}) - (3 \text{ kN})(3 \text{ m}) - 5 \text{ kN·m} = 0$$

Solving yields $R_1 = 9$ kN and $M_1 = 20$ kN·m. The loading diagram is shown in Fig. 7.36, and consists of the reactions $R_1 = 9$ kN and $M_1 = 20$ kN·m at $x = 0$, the constant 3 kN/m loading for $0 < x < 2$ m, and the concentrated load of 3 kN at $x = 3$ m. The shear and moment diagrams are constructed as follows.

0 < x < 2 m With $w = 3$kN/m, the constant slope of the shear diagram is a negative 3 kN/m. The change in the shear is given by Eqn. (7.5) as

$$V(x) - V(0) = -\int_0^x (3 \text{ kN/m})\, dx$$

or

$$V(x) = (9 - 3.0x) \text{ kN}$$

as indicated in Fig. 7.36; x is to be measured in meters. For the bending moment, Eqn. (7.7) is used to express

$$M(x) - M(0) = \int_0^x (9 - 3.0x) \text{ kN}\, dx$$

or

$$M(x) = (-20 + 9x - 1.5x^2) \text{ kN·m}$$

as shown in Fig. 7.36. The values of the shear and moment at $x = 2$ m are $V(2 \text{ m}) = (9 - 3.0 \cdot 2.0)$ kN $= 3$ kN and $M(2) = [-20 + 9(2) - 1.5(4)]$ kN·m $= -8.0$ kN·m, respectively.

2 m < x < 3 m With $w = 0$ in this interval, $V' = 0$, so that V has the constant value 3 kN. Using the equation $M' = V = 3$ kN and Eqn. (7.7), the moment is computed to be

$$M(x) = M(2 \text{ m}) + \int_{2 \text{ m}}^{x} (3 \text{ kN}) \, dx$$

or

$$M(x) = (-14 + 3x) \text{ kN·m}$$

The values of the shear and moment at a point just to the left of $x = 3$ m are $V(3 \text{ m}) = 3$ kN and $M(3 \text{ m}) = -5$ kN·m.

At x = 3 m Using Eqn. (7.8), we conclude that the concentrated load at $x = 3$ m produces an abrupt drop of 3 kN to a value of zero for the shear, and also a corresponding abrupt drop in the slope of the moment diagram at $x = 3$ m. These are both indicated in the V and M diagrams of Fig. 7.36.

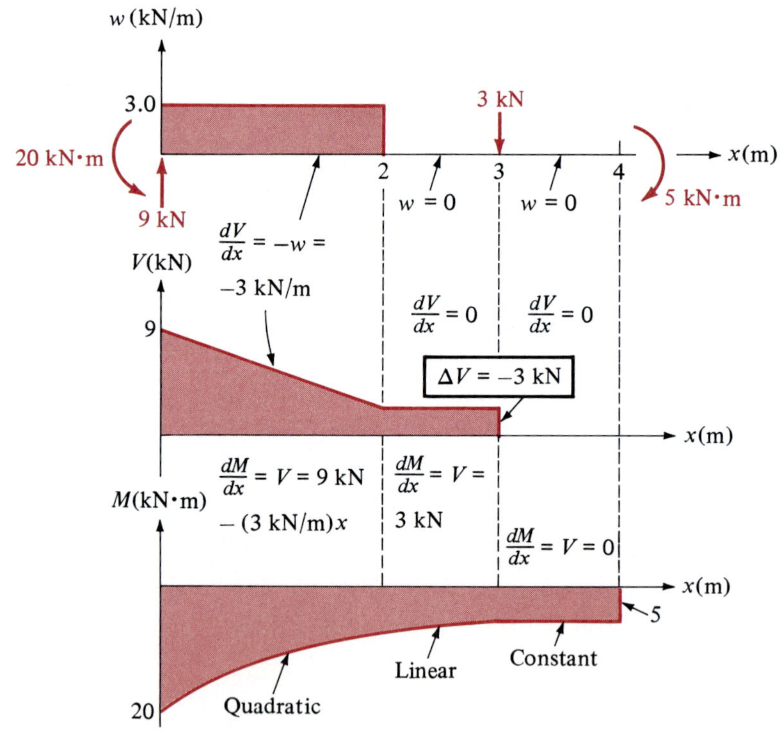

Figure 7.36 Loading, shear, and moment diagrams

3 m < x < 4 m In this interval, w is zero, with the corresponding V having the constant value of 0 kip. Note that this magnitude is exactly the value of the reaction at $x = 4$ m, providing the check mentioned previously. Using $M' = V = 0$ for the slope of the moment diagram, and Eqn. (7.7), we can write

$$M(x) - M(3\text{ m}) = \int_{3\text{ m}}^{x} (0)\, dx$$

or

$$M(x) = -5\text{ kN·m} - 0\text{ kN·m} = -5\text{ kN·m}$$

providing the check on the moment at $x = 4$ m!

It is worth mentioning again that the values obtained for V and M at the right end of the beam serve as a valuable check on the calculations.

Example 7.5 A beam is loaded as in Fig. 7.37a. Use the equations of equilibrium, $V' + w = 0$ and $M' = V$, along with the corresponding integrated forms, to construct the shear and moment diagrams.

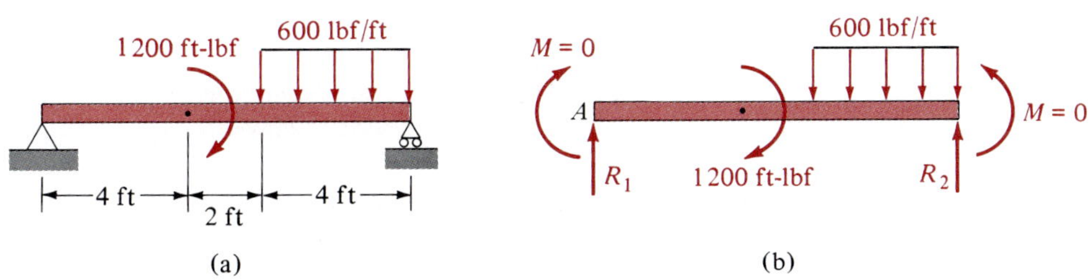

Figure 7.37 Simply supported beam and corresponding free-body diagram

Solution Referring to the free-body diagram shown in Fig. 7.37b, the equilibrium equations for determining the reactions R_1 and R_2 are

$$\Sigma F_y = R_1 + R_2 - (600\text{ lbf/ft})(4\text{ ft}) = 0$$

$$\Sigma M_A = -1200\text{ ft-lbf} - (2400\text{ lbf})(8\text{ ft}) + R_2(10\text{ ft}) = 0$$

Solving yields $R_1 = 360$ lbf and $R_2 = 2040$ lbf.

The loading diagram is shown in Fig. 7.38; it consists of the reactions $R_1 = 360$ lbf and $R_2 = 2040$ lbf at $x = 0$ and $x = 10$ ft respectively, together with the concentrated moment of -1200 ft-lbf at $x = 4$ ft, and the constant 600 lbf/ft loading for 6 ft $< x <$ 10 ft. The shear and moment diagrams shown in Fig. 7.38 are constructed as follows.

0 < x < 4 ft With $w = 0$, the slope of the shear diagram is zero, with the shear having the constant value of 360 lbf. For the bending moment, $M' = V =$ 360 lbf, indicating that the slope of the moment diagram is constant. Eqn. (7.7) is used to express

$$M(x) - M(0) = \int_0^x 360 \text{ lbf } dx$$

or

$$M(x) = 0 + 360x \text{ ft-lbf}$$

with x measured in ft. The value of M just to the left of $x = 4$ ft is $M(4 \text{ ft}) = 1440$ ft-lbf.

x = 4 ft Using Eqn. (7.9), we can see that the negative 1200 ft-lbf moment results in an abrupt positive change in the moment of 1200 ft-lbf as shown in Fig. 7.38. The value of the moment just to the right of $x = 4$ ft is then 2640 ft-lbf.

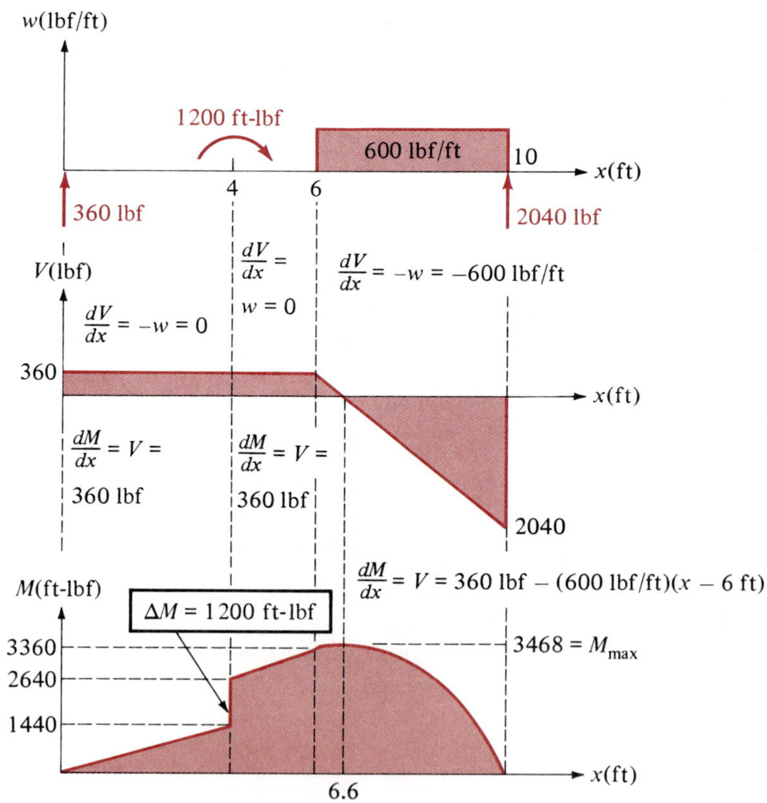

Figure 7.38 Loading, shear, and moment diagrams

4 ft < x < 6 ft With $w = 0$ in this interval, $V' = 0$, so that in this interval, V has the constant value of 360 lbf. Using Eqn. (7.7), the moment in this interval is computed to be

$$M(x) = M(4\text{ ft}) + \int_{4\text{ ft}}^{x} (360\text{ lbf})\,dx$$

or

$$M(x) = 2640\text{ ft-lbf} + 360\text{ lbf}(x - 4\text{ ft}) = (1200 + 360x)\text{ft-lbf}$$

with $M(6\text{ ft}) = 3360$ ft-lbf.

6 ft < x < 10 ft In this interval, w is 600 lbf/ft, with the corresponding slope of the V diagram the negative of this value. Using Eqn. (7.5), the shear in this interval is given by

$$V(x) - V(6\text{ ft}) = -\int_{6\text{ ft}}^{x} (600\text{ lbf/ft})\,dx$$

or

$$V(x) = (3960 - 600x)\text{lbf}$$

with $V(10\text{ ft}) = -2040$ lbf, checking the value of the reaction at the right end of the beam. For 6 ft $\leq x \leq$ 10 ft, M is given by

$$M(x) - M(6\text{ ft}) = \int_{6\text{ ft}}^{x} (3960 - 600x)\text{lbf}\,dx$$

or

$$M(x) = [3360 + 3960(x - 6) - 300(x^2 - 6^2)]\text{ft-lbf}$$

The reader should verify that $M(10\text{ ft}) = 0$ as required.

Note that the shear is zero at $x = 6.6$ ft; thus, this is the point ($M' = V = 0$) at which the moment will be a maximum. The corresponding change in the moment over the interval between $x = 6.0$ ft and $x = 6.6$ ft is given by

$$M(6.6\text{ ft}) - M(6\text{ ft}) = \int_{6\text{ ft}}^{6.6\text{ ft}} 360\text{ lbf}\,dx = \frac{360\text{ lbf}(0.6\text{ ft})}{2} = 108\text{ ft-lbf}$$

or

$$M(6.6\text{ ft}) = 3468\text{ ft-lbf} = M_{\text{max}}$$

PROBLEMS

7.45–7.66 Use the ideas of this section to draw the loading, shear, and moment diagrams for each of the following. Label all critical values. For problem 7.57, assume that the reactive pressure from the base is uniformly distributed.

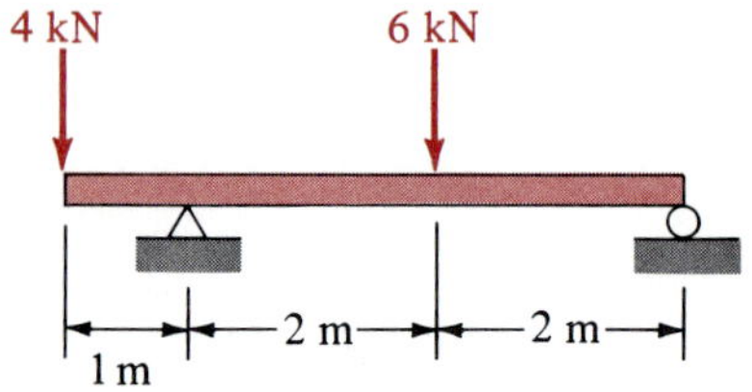

P. 7.45

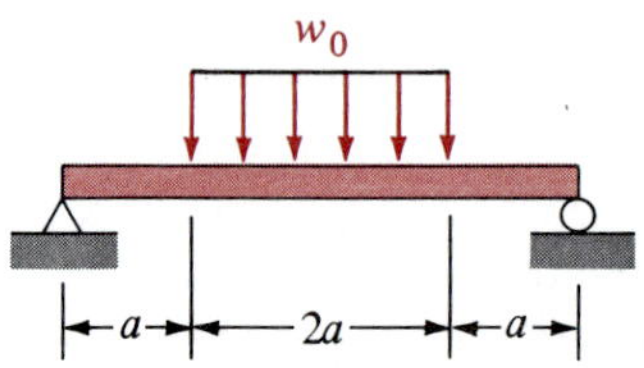

P. 7.46

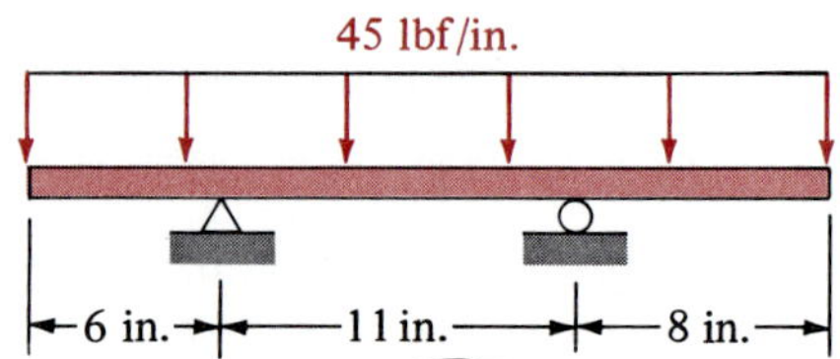

P. 7.47

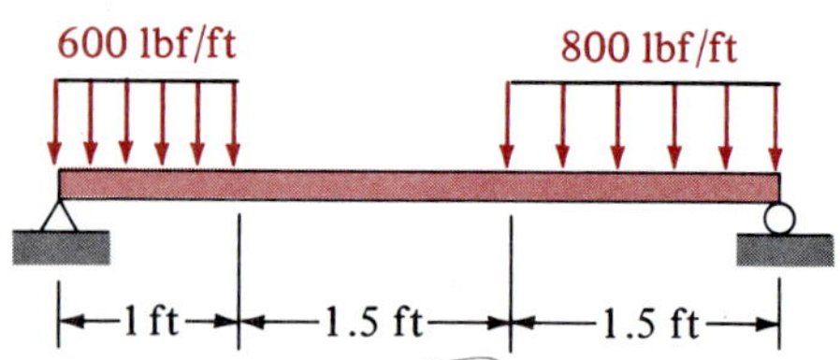

P. 7.48

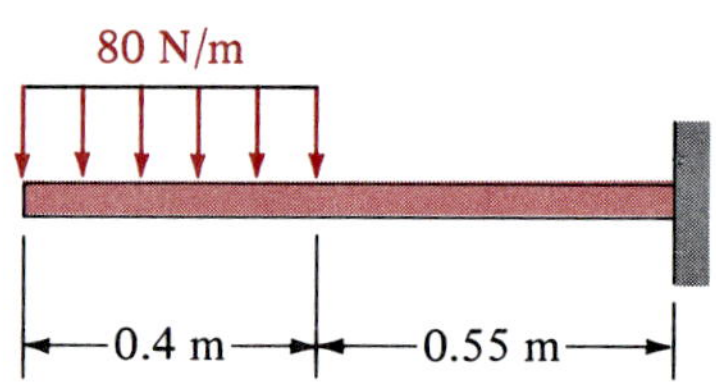

P. 7.49

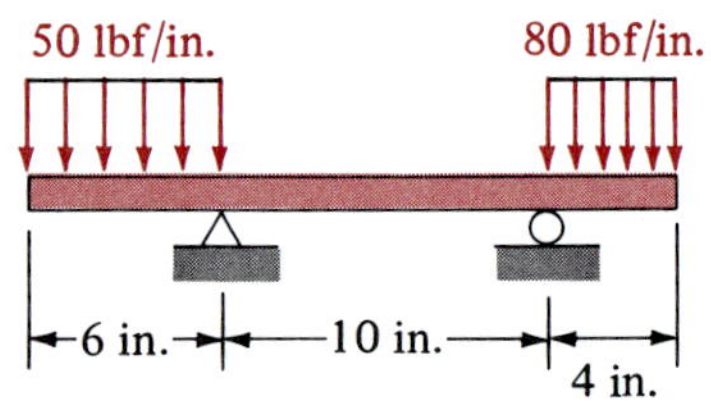

P. 7.50

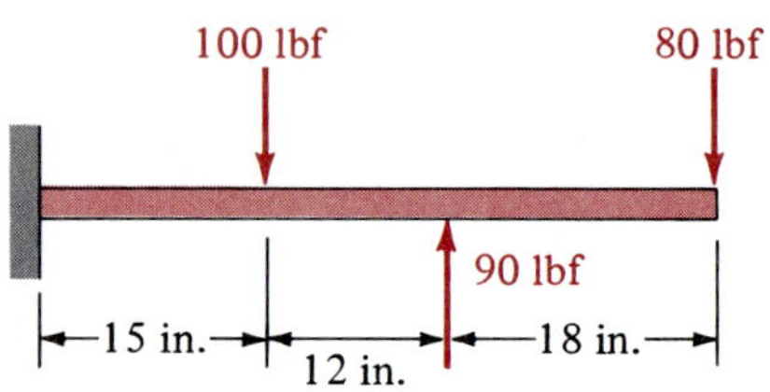

P. 7.51

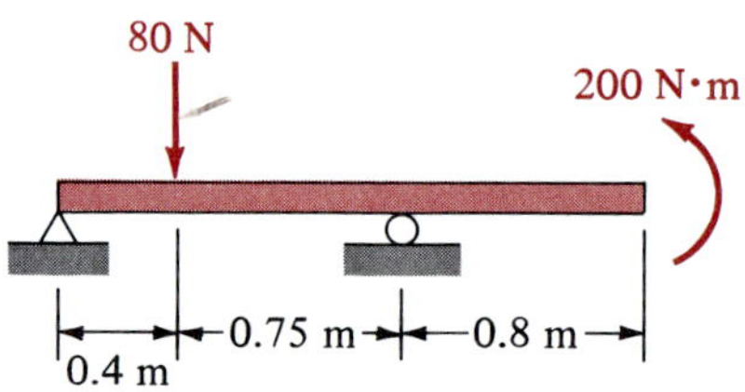

P. 7.52

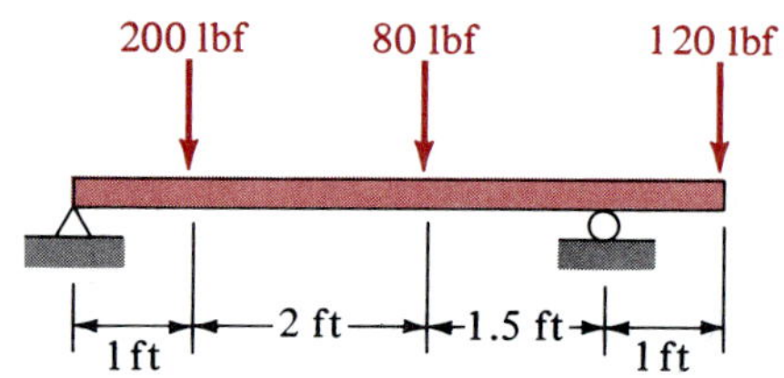

P. 7.53

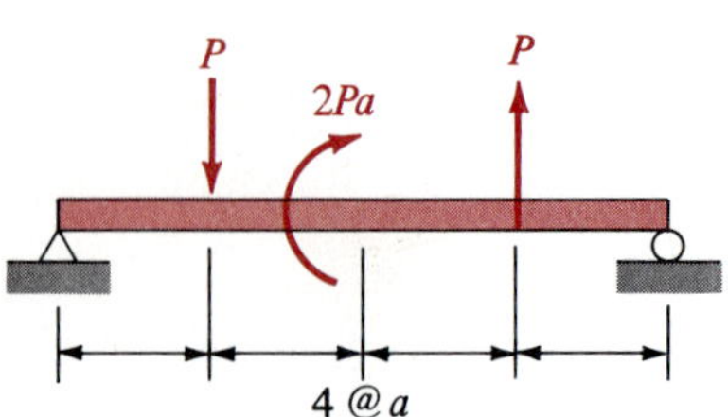

P. 7.54

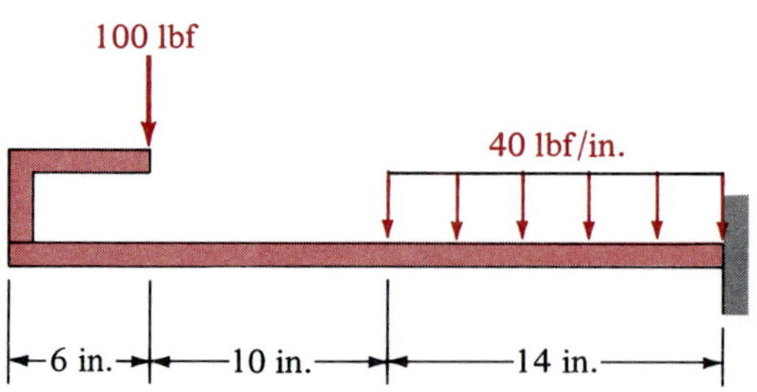

P. 7.55

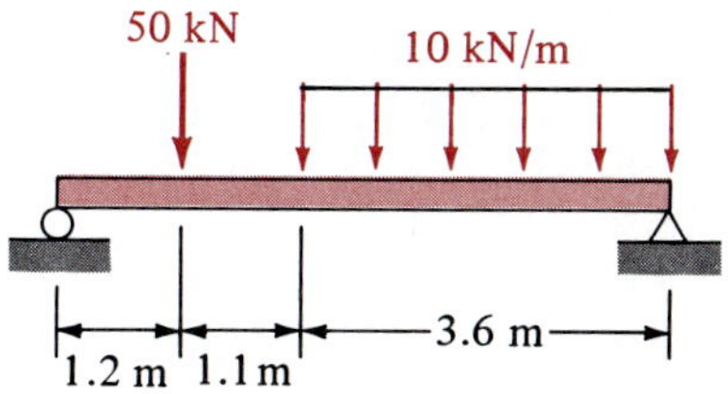

P. 7.56

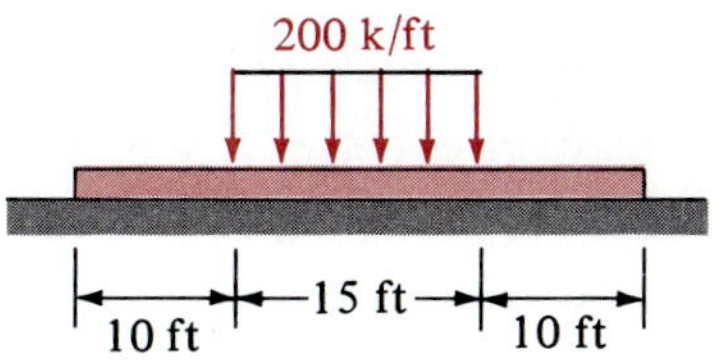

P. 7.57

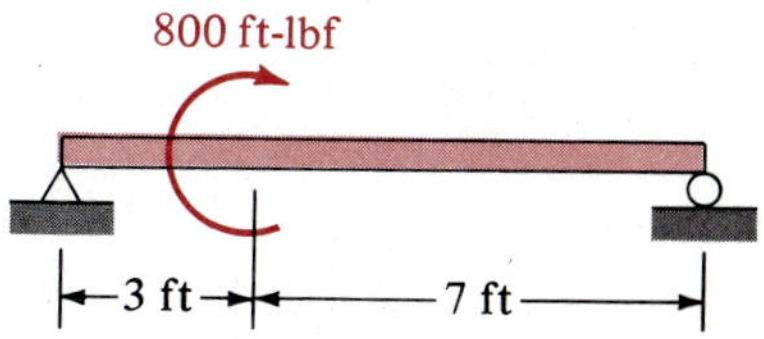

P. 7.58

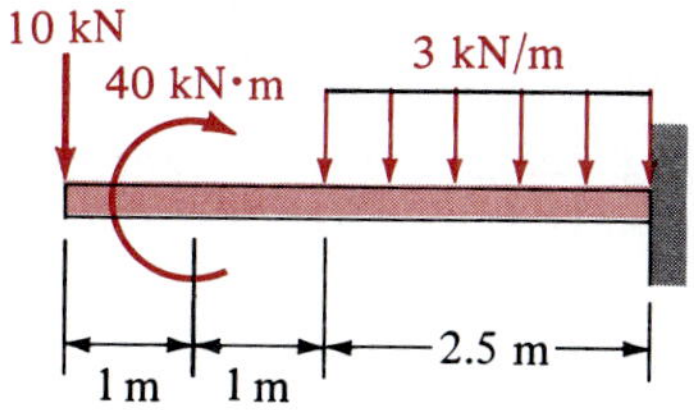

P. 7.59

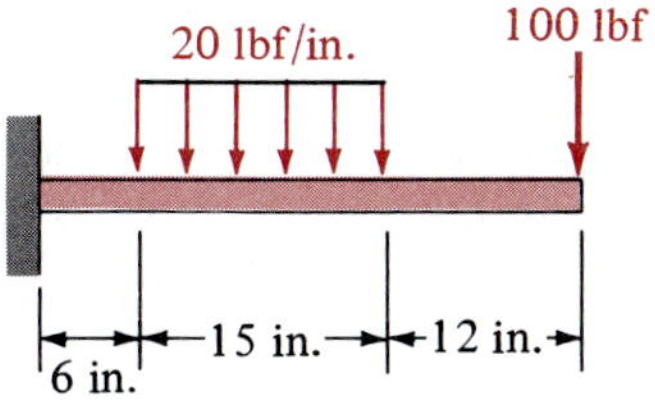

P. 7.60

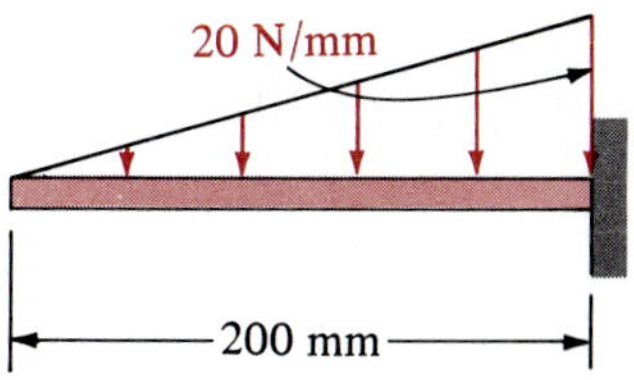

P. 7.61

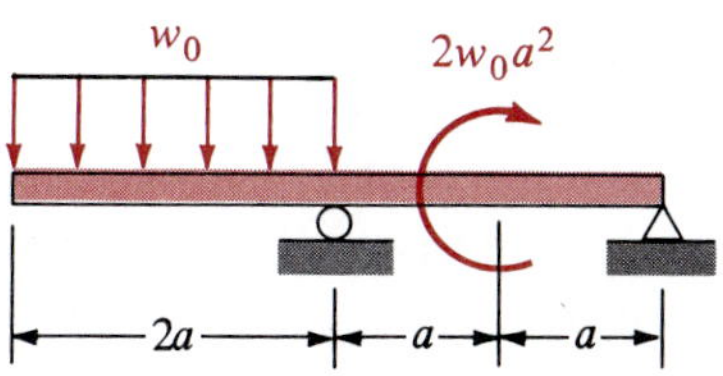

P. 7.62

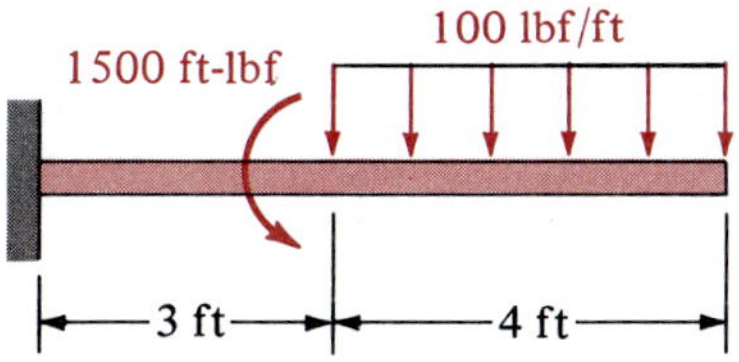

P. 7.63

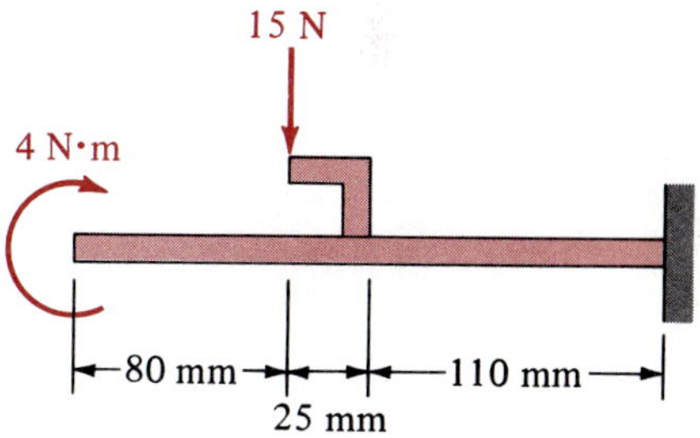

P. 7.64

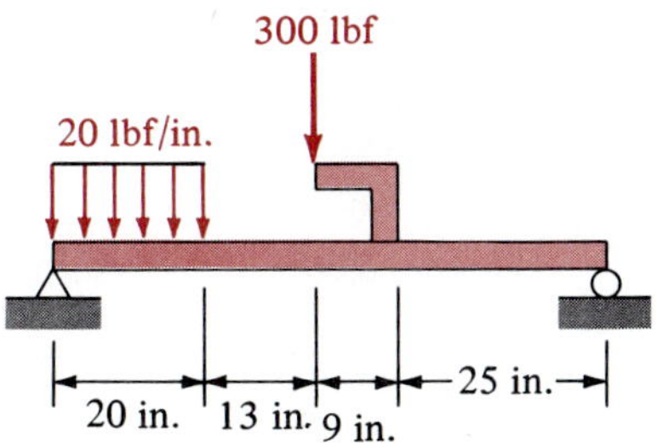

P. 7.65

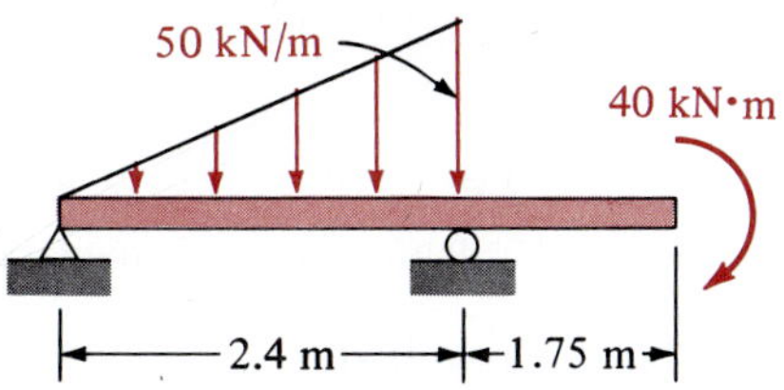

P. 7.66

7.67 Work problem 7.15 by integrating the equations of equilibrium.

7.68 Work problem 7.23 by integrating the equations of equilibrium.

7.69 Work problem 7.24 by integrating the equations of equilibrium.

7.70 Work problem 7.25 by integrating the equations of equilibrium.

7.71 Work problem 7.28 by integrating the equations of equilibrium.

7.72 Work problem 7.31 by integrating the equations of equilibrium.

7.73 Work problem 7.38 by integrating the equations of equilibrium.

7.74 Work problem 7.39 by integrating the equations of equilibrium.

7.75 Work problem 7.40 by integrating the equations of equilibrium.

7.76 Show by summing moments about the right edge of the free-body diagram of Fig. 7.24, and using Eqn. (7.4), that Eqn. (7.7) results.

7.4 DEFORMATION AND STRAIN-DISPLACEMENT RELATIONS

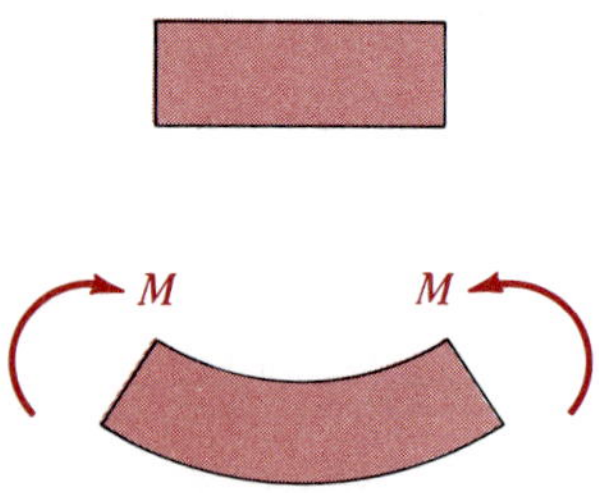

Figure 7.39 Bending deformations

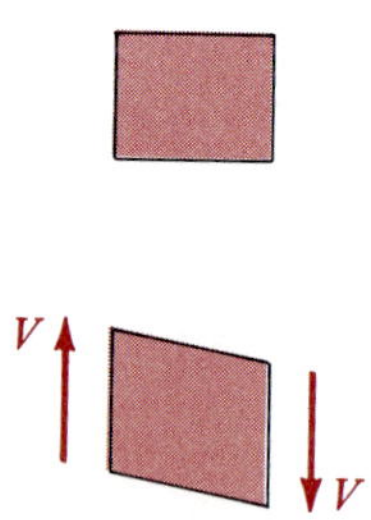

Figure 7.40 Shear deformations

When subjected to a positive moment, a beam will respond by deforming as shown in Fig. 7.39, whereas positive shear forces will produce a different type of deformation as shown in Fig. 7.40. The two types (both greatly exaggerated here) are referred to as bending deformations and shear deformations, respectively. Except in the case of short beams, it will be shown in Section 8.8 that essentially all the transverse displacement is due to the bending deformations.

Consider a uniform bar with a cross section that is symmetric with respect to the xy-plane, as shown in Fig. 7.41b.

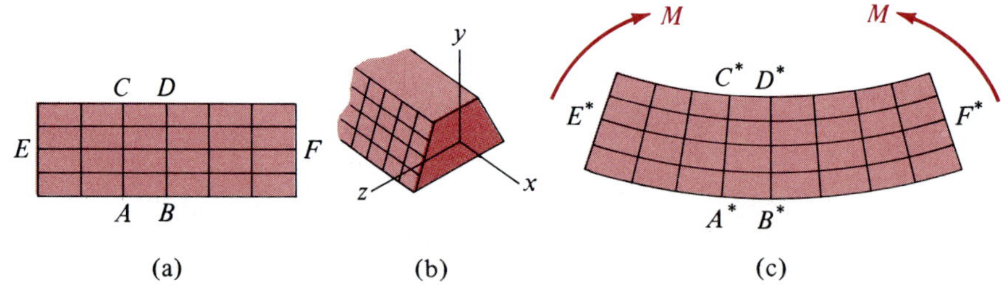

Figure 7.41 Pure bending of a uniform symmetric bar

As a result of being acted upon by the couples M, we will assume that the bar deforms as shown in Fig. 7.41c. Straight lines EF are assumed to deform into circular arcs such as E^*F^*. Typical lines A^*C^* and B^*D^* are assumed to remain straight and perpendicular to the circular arcs E^*F^*. These assumptions are the basis for the classical statement regarding the deformation associated with **pure bending:**

> Planes originally perpendicular to the x-axis remain plane and perpendicular to the deformed x-axis upon bending.

As a result of the bending it is clear that the upper portions of the beam shorten while those at the bottom lengthen; thus, there is a negative extensional strain at the top and a positive extensional strain at the bottom of the beam. It then follows that there is a location somewhere in between that undergoes neither elongation nor shortening. This location, indicated by $y = 0$ in Fig. 7.42a, is referred to as the **neutral axis.** The dotted line in Fig. 7.42b indicates the location of the corresponding **neutral surface.**

We can determine the extensional strains by considering the deformed geometry shown enlarged in Fig. 7.43b.

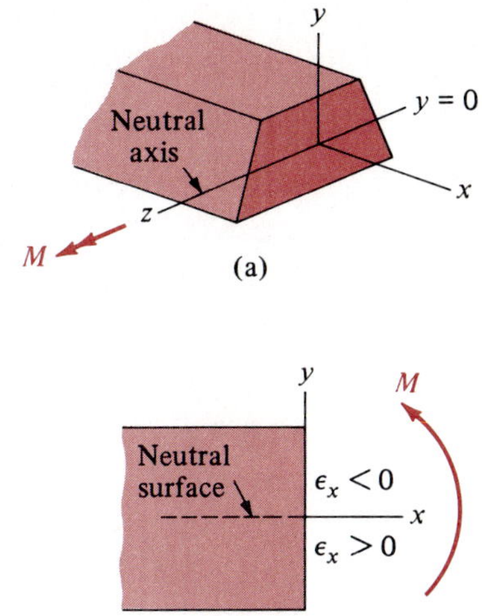

(b)

Figure 7.42 Neutral axis in pure bending

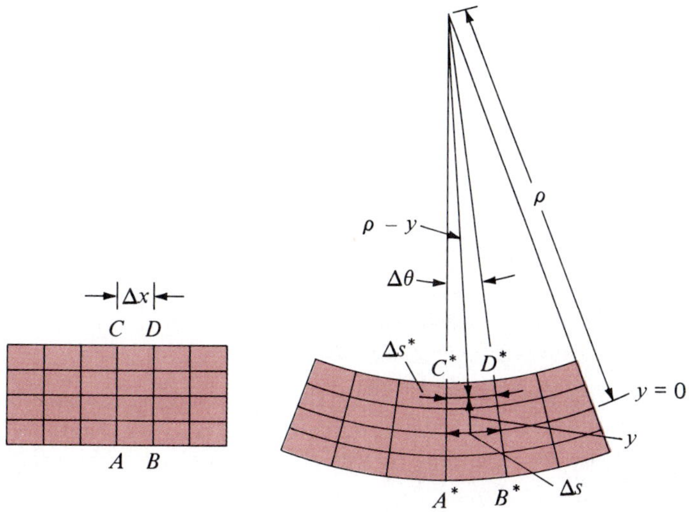

(b)

Figure 7.43 Figures for developing the strain-displacement relations

By definition, the neutral axis, $y = 0$, is unchanged in length during the bending. Denoting the radius of curvature of the circular arc corresponding to $y = 0$ as ρ, it follows that

$$\Delta s = \Delta x = \rho \Delta\theta$$

where $\Delta\theta$ is the angle subtended by the segment Δs. Referring to Fig. 7.43b, the deformed length Δs^* of a segment originally of length Δx located a distance y above the neutral axis is given by

$$\Delta s^* = (\rho - y)\,\Delta\theta = \Delta x - y\,\Delta\theta = \Delta x - y\frac{\Delta x}{\rho}$$

$$= \Delta x\left(1 - \frac{y}{\rho}\right) = \Delta s\left(1 - \frac{y}{\rho}\right)$$

The extensional strain is defined as

$$\varepsilon_x = \lim_{\Delta s \to 0} \left(\frac{\Delta s^* - \Delta s}{\Delta s} \right)$$

resulting in

$$\varepsilon_x = \lim \left(\frac{\Delta s(1 - y/\rho) - \Delta x}{\Delta s} \right) = -\frac{y}{\rho} \tag{7.10}$$

which will be referred to as the **strain-curvature relation.** This linear dependence of the strain on the coordinate y is a direct consequence of the "plane sections remain plane" assumption and is indicated in Fig. 7.44.

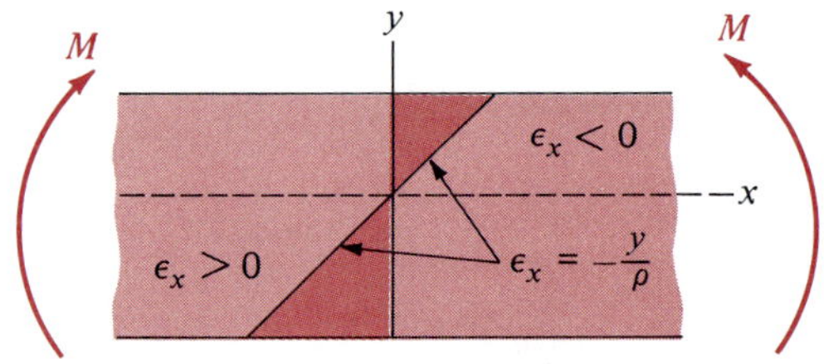

Figure 7.44 Linear strain distribution

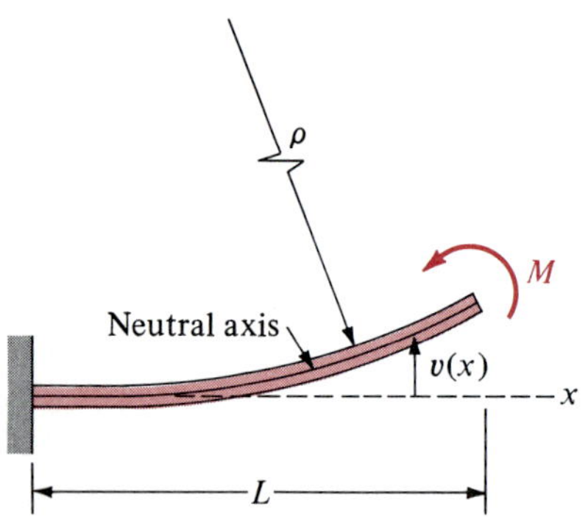

Figure 7.45 Transverse displacement $v(x)$

Note that the assumption about the deformation leading to $\varepsilon_x = -y/\rho$ also leads directly to the conclusion that $\gamma_{xy} = 0$, since lines that were originally perpendicular to the x-axis remain so when the deformation is into a circular arc. The topic of shear deformation when a shear force V is being transmitted is covered in Section 8.8.

The form of the relationship between the strains and the deformation given by Eqn. (7.10) does not relate the strains directly to a displacement component. To put the strain-curvature relation $\varepsilon_x = -y/\rho$ into a form involving the displacements, consider Fig. 7.45.

Shown is a bar deformed into a curve defined by the transverse displacement $v(x)$ of the neutral axis of the beam. As commonly shown in an elementary calculus course, the curvature κ, the radius of curvature ρ, and the transverse displacement v are related by

$$\kappa = \frac{1}{\rho} = \frac{v''}{(1 + v'^2)^{3/2}}$$

where $' = d/dx$. As long as the slope v' is small, this can be approximated very accurately by

$$\kappa = \frac{1}{\rho} \approx v'' \tag{7.11}$$

that is, the curvature κ is given approximately by the second derivative v''. Eliminating ρ between Eqns. (7.10) and (7.11) yields

$$\varepsilon_x = -yv'' \tag{7.12}$$

relating the strains to the displacement derivative v''. The relationship between the strain ε_x and the displacement v given in Eqn. (7.12) is called the **strain-displacement relation.** In some instances of pure bending it is appropriate to use the exact expression relating strain to the deformations, namely $\varepsilon_x = -y/\rho$, whereas the approximate expression $\varepsilon_x = -yv''$ is normally used when a beam is subjected to the simultaneous action of bending and shear.

PROBLEMS

7.77 A length of 0.02-in.-diameter wire is coiled around a 6-in.-diameter drum. Determine the extensional strain at the outer and inner surfaces of the wire.

7.78 Determine the extensional strain at the outer and inner surfaces of a 3-mm-diameter wire that has been coiled around a 1.50-m-diameter drum.

7.79 A cantilever beam of length 0.3 m and thickness 2 mm is subjected to a couple as shown. The tip displacement is measured to be 1.5 mm. Determine *(a)* the radius of curvature, and *(b)* the extensional strains at the top and bottom surfaces.

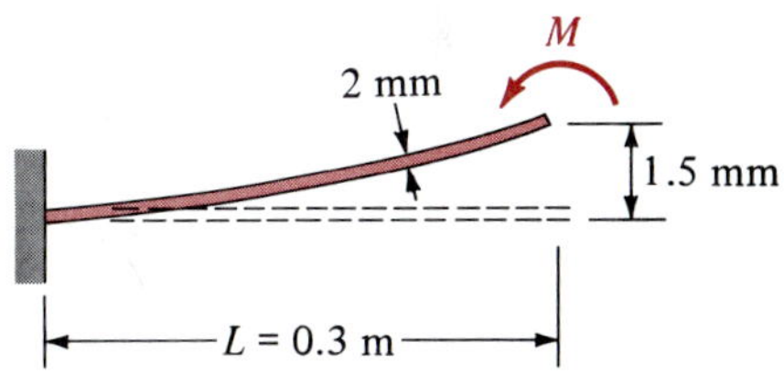

P. 7.79

7.80 A 1-in.-diameter steel cable is to be wrapped around a drum of diameter D. If the allowable extensional strain in the cable is known to be 0.0005, determine the minimum permissible diameter of the drum. For the same allowable extensional strain in the cable, does the answer change if the cable is made of aluminum?

7.81 A straight ⅛-in.-diameter aluminum wire of original length L is to be bent into a circle by couples as shown. If the allowable extensional strain is 0.0001, determine the minimum value of L.

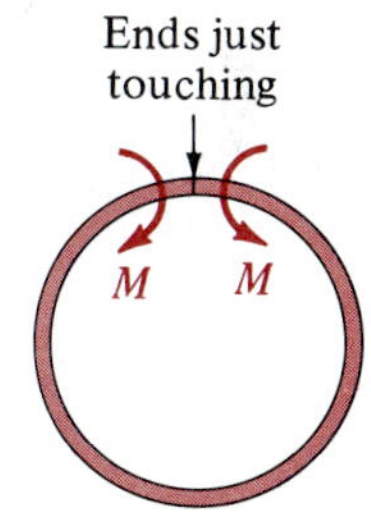

P. 7.81

7.82 Determine the maximum value of h in order that the straight strip can be bent into a circle if the allowable extensional strain is ε.

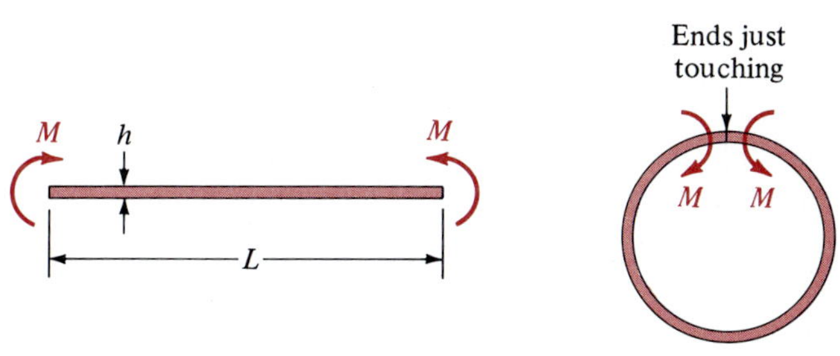

P. 7.82

7.5 MATERIAL BEHAVIOR—STRESS-STRAIN RELATIONS

In the course of our discussions of equilibrium, we developed the equilibrium equations relating the internal resultants V and M to the stresses τ_{xy} and σ_x as

$$-\int \tau_{xy}\, dA = V \qquad \text{and} \qquad -\int y\sigma_x\, dA = M$$

and the equations of equilibrium between the external loading w and the internal force resultants V and M as

$$\frac{dV}{dx} + w = 0 \qquad \text{and} \qquad \frac{dM}{dx} - V = 0$$

The strain-displacement relation developed in Section 7.4 is

$$\varepsilon_x = -yv''$$

This brings the total number of equations generated by considering equilibrium and deformation to five, whereas the number of unknowns is six: M, V, σ_x, τ_{xy}, ε_x, and v. From this simple count of knowns and unknowns, it is clear that additional information is needed. Even if it were possible by some means to solve the first four equations for the four force variables M, V, σ_x, and τ_{xy}, there would then be no way to connect the results to the strains or deformations in the fifth equation, $\varepsilon_x = -yv''$. Thus, without some additional information relating the force variables to the deformation variables, it would not be possible in general to answer any questions regarding stiffness.

The sixth equation, specifying the material behavior, is based on assuming that the beam is composed of a one-dimensional elastic isotropic solid, a solid for which the relation between the normal stress σ_x and the corresponding extensional strain ε_x is given by

$$\varepsilon_x = \frac{\sigma_x}{E} - \frac{\nu\sigma_y}{E} - \frac{\nu\sigma_z}{E}$$

which, on the basis of the assumptions made in Section 7.2 regarding σ_y and σ_z, can be written

$$\varepsilon_x = \frac{\sigma_x}{E} \tag{7.13}$$

where E is Young's modulus determined on the basis of a uniaxial tension test. The combination of the six equations

$$-\int \tau_{xy}\, dA = V \qquad -\int y\sigma_x\, dA = M$$

$$\frac{dV}{dx} + w = 0 \qquad \frac{dM}{dx} - V = 0$$

$$\varepsilon_x = -yv'' \qquad \varepsilon_x = \frac{\sigma_x}{E}$$

for the unknowns σ_x, ε_x, τ_{xy}, V, M, and v will be discussed in the next section.

Because the shear stress τ_{xy} also appears in the equilibrium equations, one might be tempted to include the equation $\tau_{xy} = G\gamma_{xy}$. However, γ_{xy} does not appear elsewhere in the equations. More importantly, in the development of the strain-displacement relation $\varepsilon_x = -yv''$ we assumed that the deformations were entirely due to bending, with no shear deformations being considered, that is, $\gamma_{xy} = 0$. Thus, a basic contradiction is present in that it is clear on the basis of equilibrium requirements that V, the shear force transmitted, is in general not zero. This definitely contradicts developing the shear force V by using the equations $\tau_{xy} = G\gamma_{xy} = G(0)$, and $V = -\int \tau_{xy}\, dA = -\int (0) dA = 0$, or $V = 0$! This contradiction can be resolved in a rational manner (see Section 8.8) by including deformations due to shear.

Anticlastic Bending When a bar is bent by couples as shown in Fig. 7.46, there is curvature $\kappa = 1/\rho$ in the xy-plane as we discussed in the previous section, with the extensional strain in the x-direction given by $\varepsilon_x = -y/\rho$.

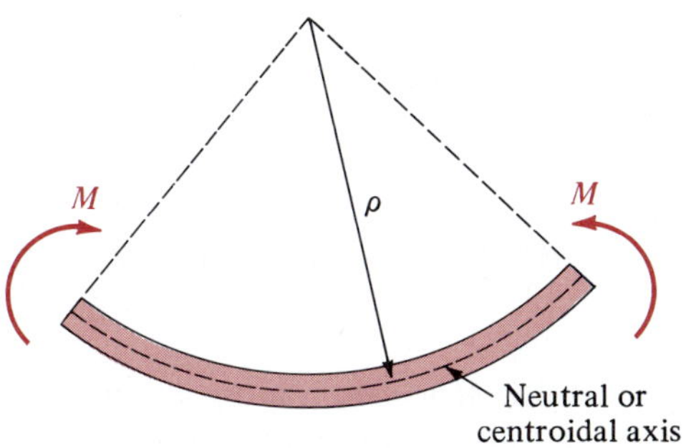

Figure 7.46 Deformations associated with pure bending

As discussed in Section 7.2, it is reasonable to assume that the two normal stresses σ_y and σ_z are zero for the bending problem. This does not, however, lead to the conclusion that the corresponding strains ε_y and ε_z are zero. In fact, the assumption that the material is linear, elastic, and isotropic enables us to conclude that neither

$$\varepsilon_y = \frac{\sigma_y - \nu(\sigma_z + \sigma_x)}{E} = -\frac{\nu\sigma_x}{E} = -\nu\varepsilon_x = \nu\frac{y}{\rho}$$

nor

$$\varepsilon_z = \frac{\sigma_z - \nu(\sigma_x + \sigma_y)}{E} = -\frac{\nu\sigma_x}{E} = -\nu\varepsilon_x = \nu\frac{y}{\rho}$$

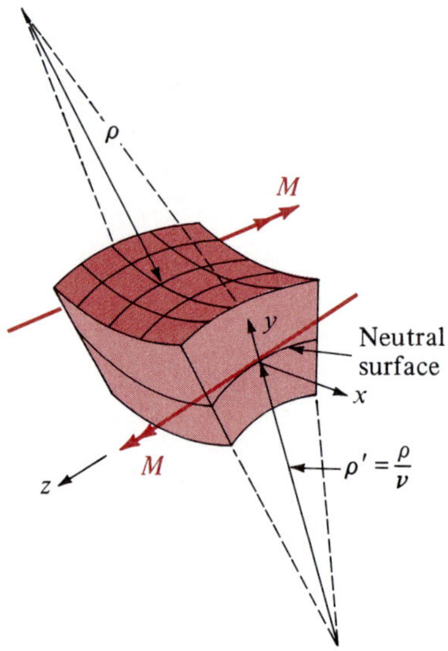

Figure 7.47 Deformations associated with anticlastic bending

is zero. The deformation corresponding to these strains is indicated in Fig. 7.47, where ε_z is positive with a corresponding increase in the z-dimension when y is positive, and negative with a corresponding decrease in the z-dimension when y is negative. The y-dimension also increases above the neutral axis and decreases below the neutral axis.

Rewriting $\varepsilon_y = \varepsilon_z = y/(\rho/\nu) = y/\rho'$, we see that the neutral surface is deformed so that it has the curvature $\rho' = \rho/\nu$. This curvature, shown in Fig. 7.47, is referred to as **anticlastic curvature.**

Thus although the only stress in a beam subjected to pure bending is assumed to be $\sigma_x = -Ey/\rho$, with $\sigma_y = \sigma_z = \tau_{xy} = \tau_{xz} = \tau_{yz} = 0$, there are, because of the behavior of an isotropic elastic material, additional strains and deformations other than those associated with ε_x.

7.6 COMBINATION—GOVERNING EQUATIONS FOR THE BENDING DEFORMATION PROBLEM

The six equations we developed from the basic ideas of equilibrium, deformation, and material behavior are repeated below and for convenience denoted by *(a)* through *(f)*.

$$-\int \tau_{xy}\, dA = V \qquad (a) \qquad \frac{dM}{dx} - V = 0 \qquad (d)$$

$$-\int y\sigma_x\, dA = M \qquad (b) \qquad \varepsilon_x = -yv'' \qquad (e)$$

$$\frac{dV}{dx} + w = 0 \qquad (c) \qquad \varepsilon_x = \frac{\sigma_x}{E} \qquad (f)$$

Combining *(e)* and *(f)* by eliminating the strain ε_x results in

$$\sigma_x = -Eyv'' \tag{7.14}$$

indicating that the bending stress σ_x is a linear function of the distance from the neutral axis at any location along the beam.

Recall from Section 7.2 that one of the equilibrium requirements was that any stress distribution σ_x must be such that there be no resultant axial force, or

$$\int \sigma_x\, dA = 0 \tag{7.1a}$$

Substituting Eqn. (7.14) into (7.1*a*) results in

$$0 = \int (-Eyv'')dA$$

Assuming E to be a constant, we can express this equation as

$$0 = Ev'' \int y\, dA$$

or

$$0 = \int y\, dA$$

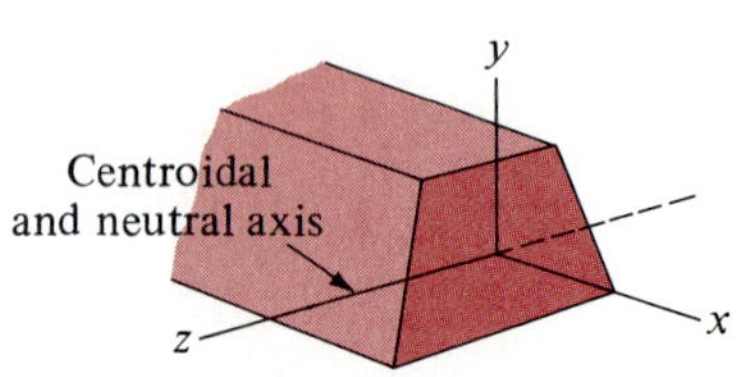

Figure 7.48 Coincidence of the neutral and centroidal axes for pure bending

indicating that, as pictured in Fig. 7.48, y is to be measured from the centroidal axis; thus, *the neutral and centroidal axes coincide.*

With the location of the axis from which y is measured now known, substitute Eqn. (7.14) into Eqn. (b) to obtain

$$M = \int Ey^2 v'' \, dA \tag{7.15}$$

Continuing to assume that E is constant, Eqn. (7.9) can be written as

$$M = Ev'' \int y^2 \, dA = EIv'' \tag{7.16}$$

where $I = \int y^2 \, dA$ is the second moment of the area computed about the centroidal (or neutral) axis. Recall that v'' is an approximation for the curvature $\kappa = 1/\rho$, so that the corresponding exact version of Eqn. (7.16) is

$$M = \frac{EI}{\rho} \tag{7.17}$$

and is the celebrated **moment-curvature relation.** Note that Eqn. (7.17) can also be expressed as

$$\frac{1}{\rho} = \frac{M}{EI}$$

indicating that a uniform straight symmetric beam subjected to a pure moment deforms into a circle as shown in Fig. 7.49.

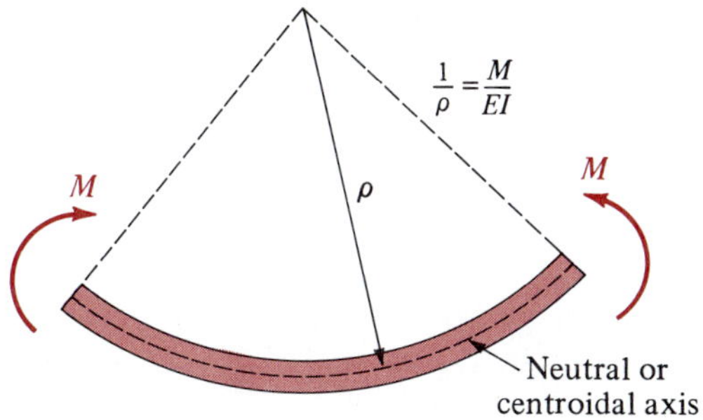

Figure 7.49 Bending into a circle by a moment

This result can actually be verified by experiment, meaning that the assumptions we made in Section 7.4 regarding the character of the deformations for pure bending are in fact correct. In situations where the moment is constant, the exact moment-curvature relation $1/\rho = M/EI$ can be used to determine the deformed configuration of the beam. In a general case where $M = M(x)$ the approximate expression $M(x) = EIv''$ given by Eqn. (7.16) can be integrated to determine $v(x)$, the transverse displacement. The determination of the transverse displacement $v(x)$ by integration of $EIv'' = M$ and by other techniques is discussed in detail in Chapter 8.

It is now possible to eliminate the deformation variable v'' from Eqns. (7.14) and (7.16) to obtain

$$\sigma_x = -\frac{My}{I} \tag{7.18}$$

which is referred to as the **flexure formula,** stating that the bending stresses vary linearly over the cross section as indicated in Fig. 7.50. The uses of the flexure formula will be discussed in detail in the next section.

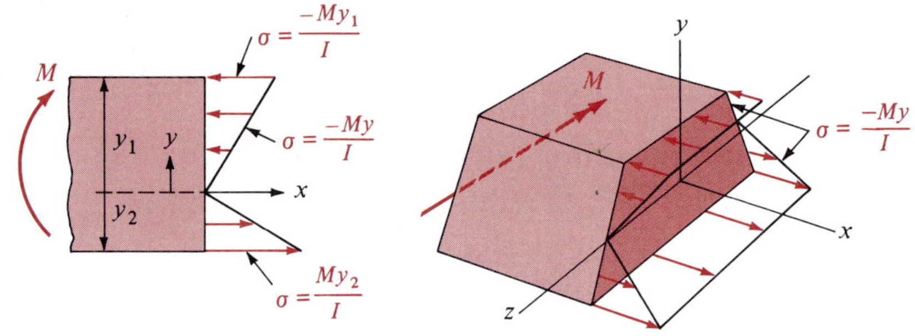

Figure 7.50 Linear distribution of bending stress

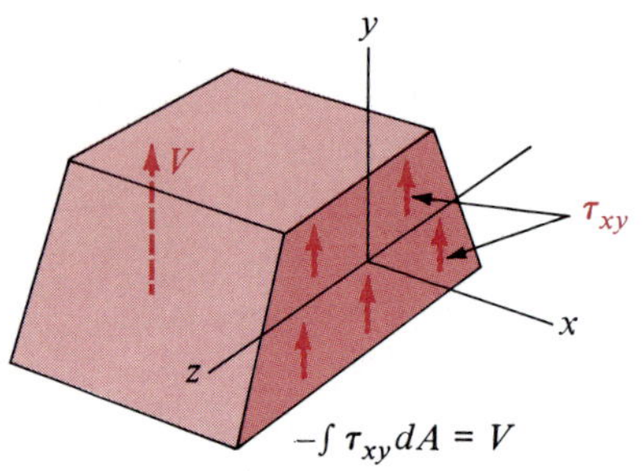

Figure 7.51 Relation between V and τ_{xy}

Observe that at this point in the development of the theory, the shear stress τ_{xy} is only related to the shear force V by the equilibrium relation given by Eqn. (7.16), $V = -\int \tau_{xy}\, dA$. This relation does not provide any direct means for determining the distribution of τ_{xy} on an exposed face by a knowledge of V, the corresponding shear force. Thus the final step in completing the theory is to develop a relation from which, given V, the shear stress τ_{xy} on an exposed face such as shown in Fig. 7.51 can be determined. We postpone this task until Section 7.8, where shear stresses in beams subjected to combined bending and shear loading are discussed in detail.

The theory we have been developing, relating the stresses and displacements to the external loads, is based on the following set of assumptions:

1. The axis of the bar, usually chosen to be the x-axis, is straight. Cross-sectional dimensions are small compared to the length of the bar, and vary slowly along the axis of the bar. The shape of the cross section is such that the y-axis can be chosen as a plane of symmetry.
2. Loading consists of forces applied in the y-direction and lying in the plane of symmetry, and of moments about the z-axis.
3. Slopes associated with the transverse displacements are small.
4. The beam is composed of material behaving in a linearly elastic fashion.

We will refer to the theory associated with these assumptions as the **elementary theory of bending.** If any of these assumptions are not satisfied, the elementary theory cannot be expected to predict stresses or displacements with accuracy. The remainder of this chapter is devoted to the many useful applications of this theory.

PROBLEMS

The following problems are intended to help the student become familiar with the relation between the applied moment M, the bending stress σ, and the size of the beam.

7.83 The beam shown is composed of a steel for which the allowable bending stress is 10 ksi. Determine the minimum permissible depth h if *(a)* $M = 1$ in.-kip, *(b)* $M = 10^2$ in.-kips, and *(c)* $M = 10^4$ in.-kips. Do your answers depend upon L?

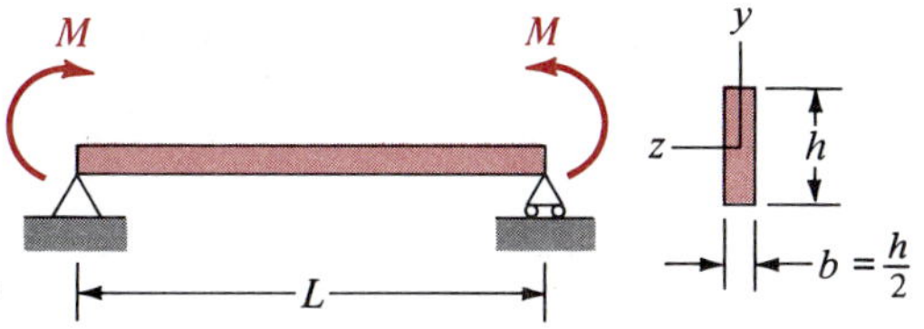

Ps. 7.83 and 7.84

7.84 The allowable bending stress in the beam shown is 15 ksi. What is the allowable bending moment if *(a)* $h = 0.5$ in., *(b)* $h = 1.0$ in., *(c)* $h = 2.0$ in., and *(d)* $h = 4.0$ in.

7.85 The beam shown is composed of a steel for which the allowable bending stress is 70 MPa. Determine the minimum permissible depth h if *(a)* $M = 10$ N·m, *(b)* $M = 1000$ N·m, and *(c)* $M = 10^5$ N·m. Do your answers depend upon L?

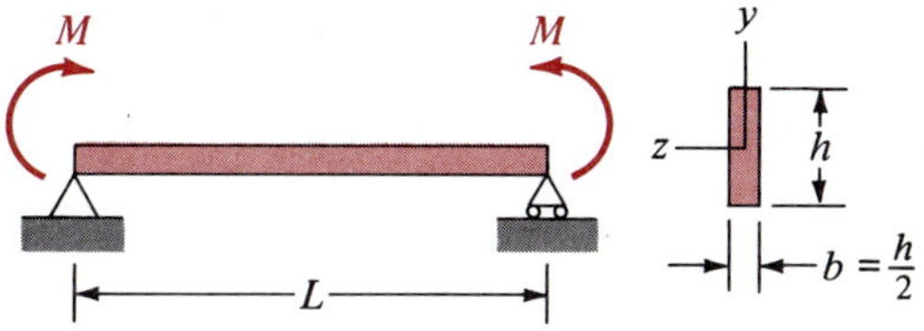

Ps. 7.85 and 7.86

7.86 The allowable bending stress in the beam shown is 100 MPa. What is the allowable bending moment if *(a)* $h = 12$ mm, *(b)* $h = 25$ mm, *(c)* $h = 50$ mm, and *(d)* $h = 200$ mm.

7.87 The beam shown is composed of a steel for which Young's modulus is 200 GPa. Use the moment-curvature relation in the form $1/\rho = M/EI$ to determine the radius of curvature and the vertical displacement at the center of the beam. Take $h = 50$ mm, $L = 1.0$ m, and $M = 1$ kN·m.

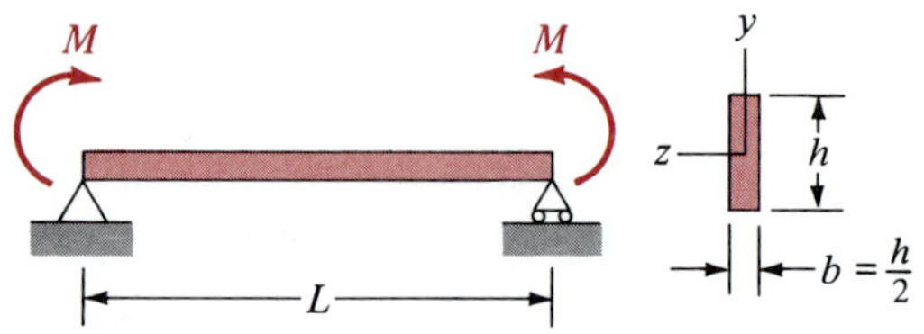

Ps. 7.87 and 7.88

7.88 For the beam of problem 7.87, integrate the moment-curvature relation $EIv'' = M$ to determine the displacement $v(x)$. Use the conditions $v(0) = 0$ and $v(1.0 \text{ m}) = 0$ to evaluate the two constants of integration.

7.89 A 25-mm square-cross-section beam is subjected to a moment of 200 N·m as shown. Determine *(a)* the change in length of segment *AB*, and *(b)* the change in length of segment *CD*. Take $E = 70$ GPa.

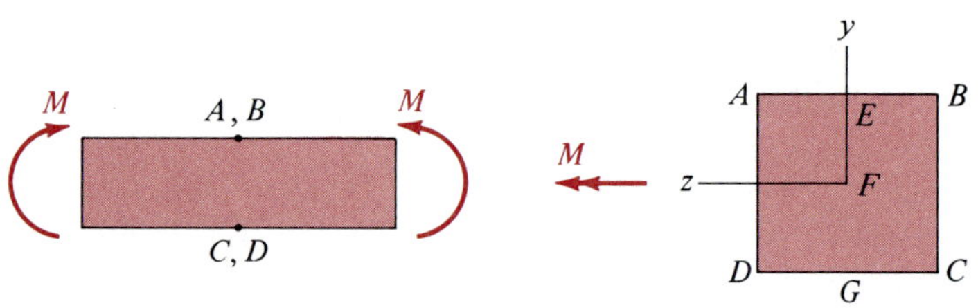

Ps. 7.89 and 7.90

7.90 For the beam of problem 7.89, determine *(a)* the change in length of segment *EF*, and *(b)* the change in length of segment *FG*.

7.91 A beam for which the *xy*-plane is a plane of symmetry is composed of a material for which Young's modulus depends on the coordinate y according to $E = E_0 + E_1 y$ as shown. Assuming that plane sections remain plane and that the loading is pure bending about the z-axis, develop *(a)* the moment-curvature relation, and *(b)* the corresponding flexure formula.

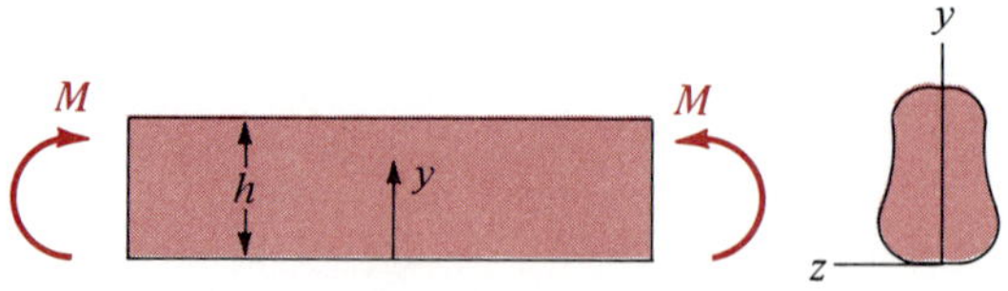

Ps. 7.91 and 7.92

7.92 Reconsider problem 7.91 if $E = E_2 + E_3 z$.

7.7 BENDING STRESSES IN BEAMS

In Section 7.6, the expression for the bending stress associated with a bending moment M was developed as

$$\sigma_x = -\frac{My}{I}$$

commonly referred to as the flexure formula. The stresses themselves are referred to as flexural stresses.

As indicated in Fig. 7.52, M is the moment about the centroidal or neutral axis, which is often referred to as the **axis of bending.** The variable y is the distance to the point at which the bending stress is desired, and is measured from the neutral or centroidal axis of the symmetric cross section. The quantity $I = \int y^2\, dA$ represents the corresponding second moment of the area about the centroidal axis.

The bending stress is clearly a linear function of y and appears as shown in Fig. 7.53 for several typical cross sections.

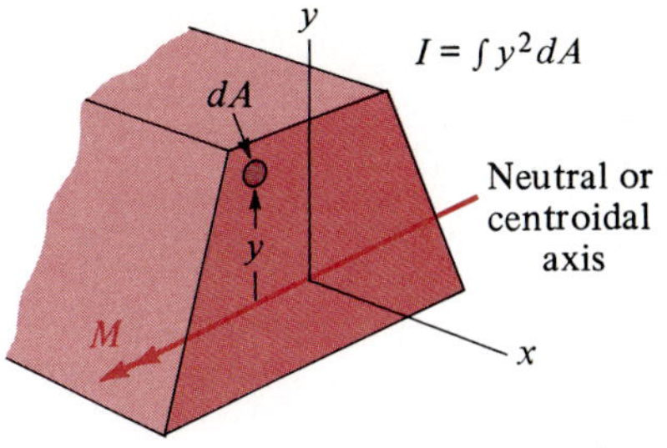

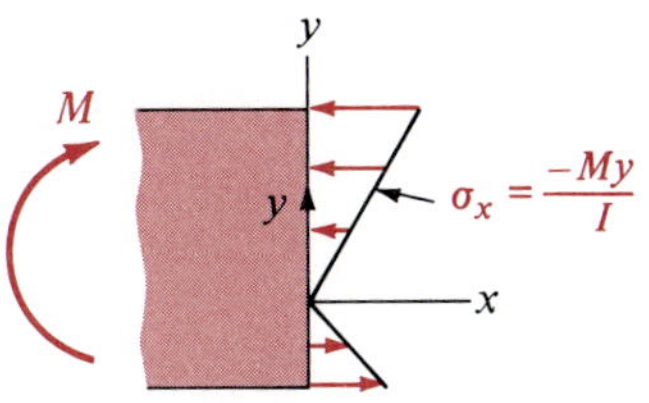

Figure 7.52 Variables in the flexure formula

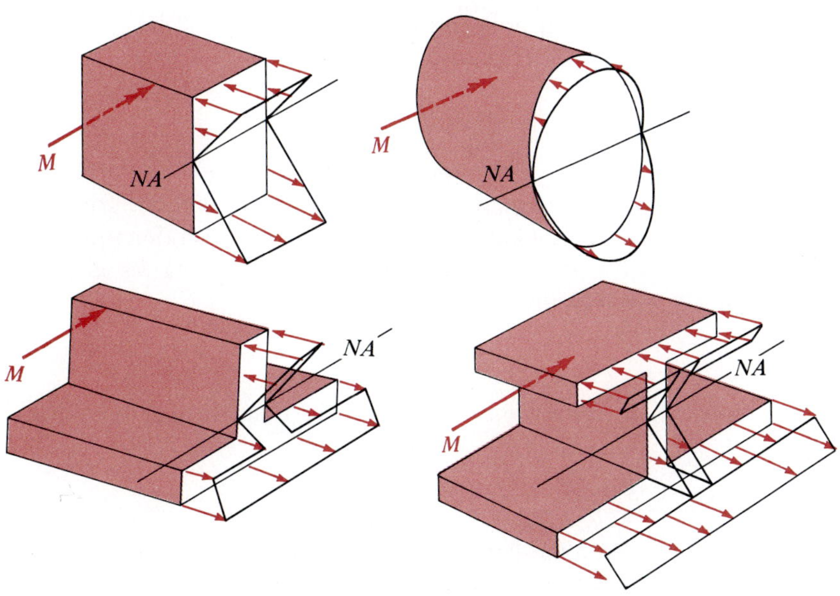

Figure 7.53 Bending stresses for typical symmetric cross sections

At a given location along the x-axis, the maximum bending stress obviously occurs at the point furthest from the neutral axis.

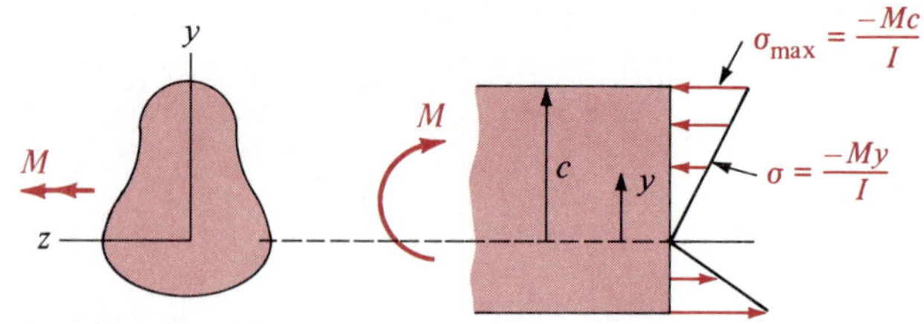

Figure 7.54 Maximum bending stress

As shown in Fig. 7.54, the distance to this furthest point is represented by the symbol c, so that the maximum absolute value of the bending stress is given by

$$\sigma_{max} = \frac{Mc}{I} \tag{7.19}$$

The maximum value may be either tensile or compressive depending upon the orientation of the cross section and the direction of the moment. In case the z-axis is an axis of symmetry, the maximum values of the bending stress occur at both the top and bottom of the section with opposite signs. Eqn. (7.19) can be written as

$$\sigma_{max} = \frac{M}{I/c} = \frac{M}{S}$$

where $S = I/c$ depends only upon the cross section. The quantity S, called the **section modulus,** is frequently used in practice. Tables that give section properties such as I, often also give the section modulus S. If, for instance, the moment M and allowable stress σ_{all} are given, the beam can be selected on the basis of computing

$$S = \frac{M}{\sigma_{all}} \tag{7.20}$$

This approach is somewhat more efficient than dealing with both I and c.

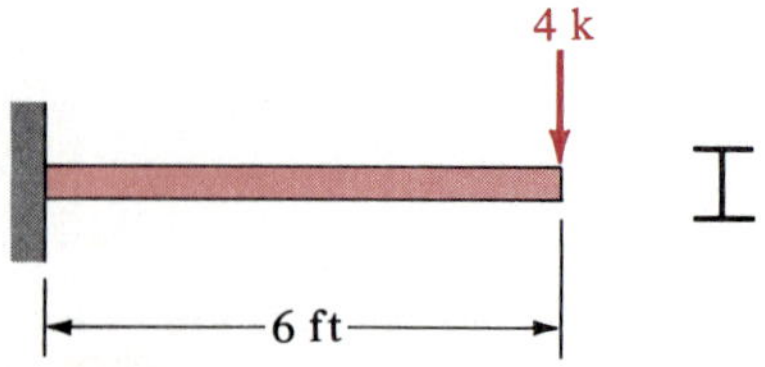

Figure 7.55 Stresses in a cantilever beam

Example 7.6 If the allowable bending stress for the cantilever beam shown in Figure 7.55 is 14 ksi, select an appropriate S section.

Solution The largest bending moment occurs at the support and has the value M = 24 kip-ft = 288,000 in.-lbf. Using Eqn. (7.20) yields

$$S = \frac{288{,}000 \text{ in.-lbf}}{14 \text{ ksi}} = 20.57 \text{ in}^3$$

Using App. C, an S 10 × 25.4 section with a section modulus of $S = 24.7$ in^3 could be used. The corresponding actual maximum bending stress would be σ = 14 ksi (20.57/24.7) = 11.66 ksi, providing an additional margin of safety.

Determination of Centroidal Axis and *I* For doubly symmetric cross-sectional shapes such as squares, rectangles, and circles, the location of the centroid coincides with the geometric center; all that is required is to evaluate I. Fig. 7.56 shows several common shapes, each with the corresponding I. The verification of the Is is left as review to be carried out in the problems. See Appendix A for a review of area properties.

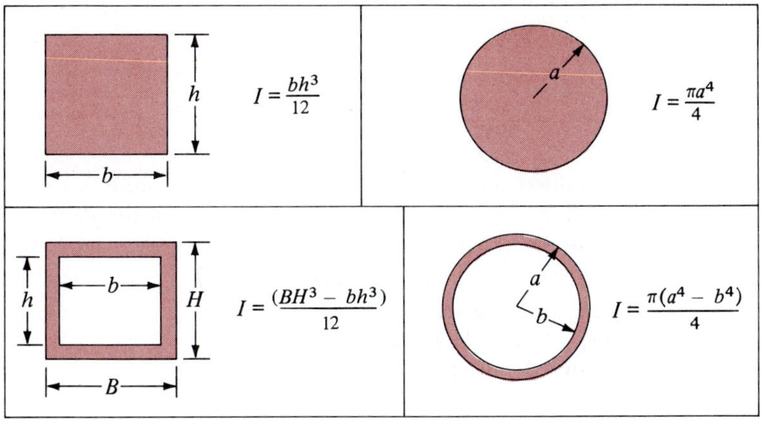

Figure 7.56 I for selected doubly symmetric sections

Example 7.7 A rectangular member is subjected to a bending moment M as shown in Fig. 7.57a. Determine the maximum bending stress.

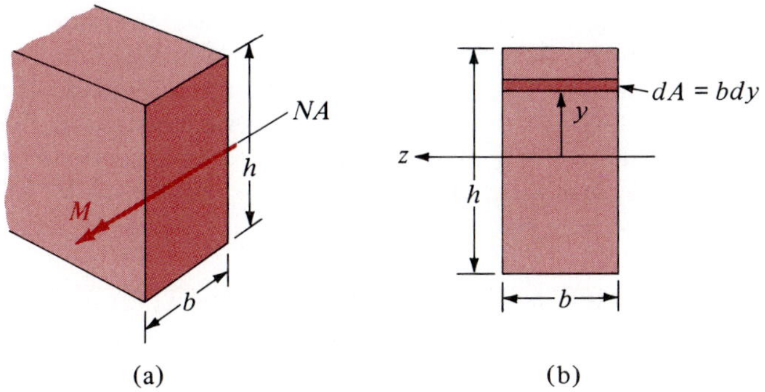

Figure 7.57 Beam subjected to a moment M

Solution In connection with Fig. 7.57, the second moment of the area for the rectangular section is evaluated as

$$I = \int_{-h/2}^{h/2} y^2 b\, dy = \frac{bh^3}{12}$$

so that

$$\sigma_x\left(y = -\frac{h}{2}\right) = -\frac{M(-h/2)}{bh^3/12} = \frac{6M}{bh^2}$$

and

$$\sigma_x\left(y = \frac{h}{2}\right) = -\frac{M(h/2)}{bh^3/12} = -\frac{6M}{bh^2}$$

Note that if h is doubled, the maximum bending stress is reduced by 75%, showing the strong dependence of the bending stresses on the depth of the cross section. For the rectangular section the section modulus is given by

$$S = \frac{I}{c} = \frac{bh^3/12}{h/2} = \frac{bh^2}{6}$$

For a specific case let $b = 1.5$ in., $h = 3.5$ in. (the approximate dimensions of a two-by-four), and $M = 1000$ ft-lbf. The maximum bending stress is then

$$\sigma_x = \pm \frac{6(1000 \text{ ft-lbf})(12 \text{ in./ft})}{(1.5 \text{ in.})(3.5 \text{ in.})^2} = \pm 3.92 \text{ ksi}$$

at the bottom and top of the section, respectively. If bending takes place about the long side of the cross section (i.e., with $b = 3.5$ in. and $h = 1.5$ in.), the stresses are

$$\sigma_x = \pm \frac{6(1000 \text{ ft-lbf})(12 \text{ in./ft})}{(3.5 \text{ in.})(1.5 \text{ in.})^2} = \pm 9.14 \text{ ksi}$$

The plus and minus again refer to the sign of the stress at the bottom and top of the beam, respectively. Both these cases are shown in Fig. 7.58.

This example clearly shows the importance of the orientation of the cross section with respect to the axis of bending. By this we mean that in the absence of other considerations, it is a good idea to choose a cross section that has an I as large as possible. This idea cannot be carried to extreme, as we will discuss in Chapter 11 concerning stability.

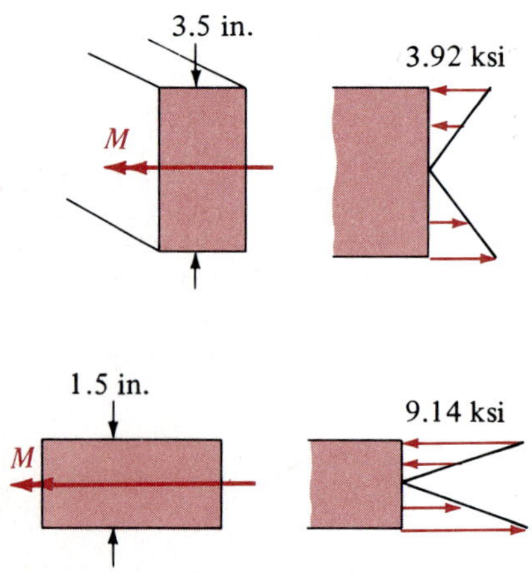

Figure 7.58 Bending about both axes in a two-by-four

For a general case where there is symmetry only about the y-axis, it is generally necessary to locate the centroidal axis and to then compute the corresponding I. Recall from statics that in connection with Fig. 7.59, the location of the centroidal axis is given by

$$\bar{y} = \frac{1}{A} \int y \, dA$$

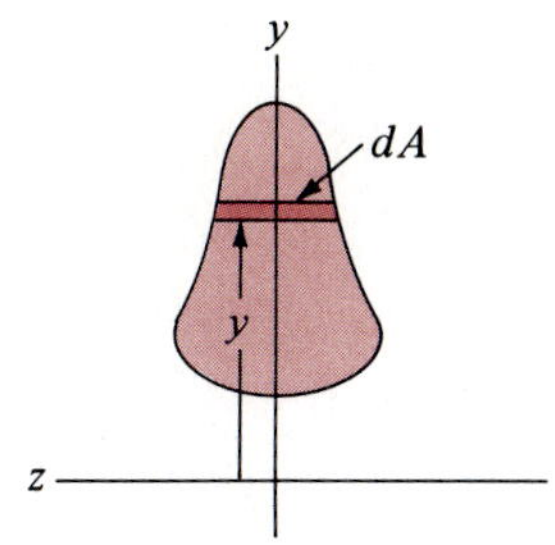

Figure 7.59 Locating the centroidal axis

and that it is often convenient to think of the cross section as composed of a number of subareas whose areas and centroidal locations are known. It is then possible to consider the area as a composite of the individual areas and write

$$\bar{y} = \frac{\Sigma A_i y_i}{\Sigma A_i}$$

and to construct a table to carry out the calculations. Once the location of the centroid is known, the centroidal second moments of the area are determined using the transfer equation

$$I_{\text{cent}} = \Sigma(I_{0i} + A_i d_i^2)$$

where I_{0i} is the second moment of the ith area about its own centroidal axes, A_i is the ith area, and d_i is the distance between the centroidal axis of the ith area and the centroid of the section. These ideas are reviewed and illustrated in the examples that follow.

Example 7.8 Determine the location of the centroid and the corresponding centroidal I for the T section shown in Fig. 7.60a.

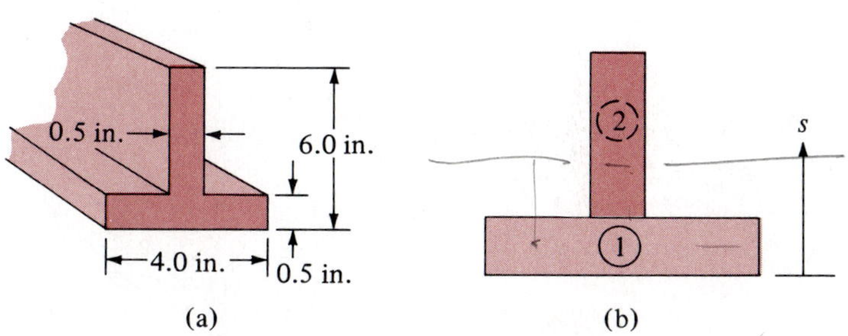

Figure 7.60 T section

Solution In connection with Fig. 7.60b, the most systematic approach for determining the location of the centroidal axis and the corresponding I is to construct the table shown below.

Table for calculation of section properties

i	A_i (in^2)	s_i (in.)	$A_i s_i$ (in^3)	d_i (in.)	I_0 (in^4)	$A_i d_i^2$ (in^4)
1	2.0000	0.2500	0.5000	−1.7368	0.0417	6.0329
2	2.7500	3.2500	8.9375	1.2362	6.9323	4.3881
	Σ = 4.7500		Σ = 9.4375		Σ = 6.9740	Σ = 10.4210

Using the second and fourth columns

$$\bar{s} = 9.4375\ \text{in}^3/4.7500\ \text{in}^2 = 1.9868\ \text{in.}$$

The fifth, sixth, and seventh columns can then be constructed and I calculated as the sum of the sixth and seventh columns as

$$I = (6.9740 + 10.4210)\ \text{in}^4 = 17.3950\ \text{in}^4$$

The student should review carefully the ideas on determining the location of the centroidal axis and the second moment of the area I, particularly the use of a table to carry out the calculations as demonstrated in the previous example. The student is referred to Appendix A for additional information on centroids and second moment of areas.

Example 7.9 The section of Example 7.8 is subjected to a bending moment of 3000 ft-lbf as indicated in Fig. 7.61a. Determine the distribution of the bending stresses.

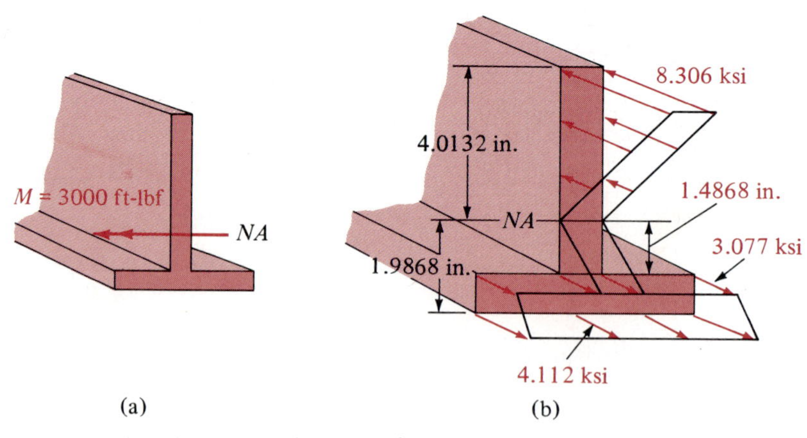

Figure 7.61 Bending stresses in a T section

Solution At the top, where $y = 4.0132$ in.,

$$\sigma_x = -\frac{(3000 \text{ ft-lbf})(12 \text{ in/ft})(4.0132 \text{ in.})}{17.3950 \text{ in}^4} = -8.306 \text{ ksi}$$

and at the bottom, where $y = -1.9868$ in.,

$$\sigma_x = -\frac{(3000 \text{ ft-lbf})(12 \text{ in/ft})(-1.9868 \text{ in.})}{17.3950 \text{ in}^4} = 4.112 \text{ ksi}$$

The bending stress distribution appears in Fig. 7.61b. Note that the resultant force for the stresses above the neutral axis is given by

$$F_a = \frac{(-8306 + 0) \text{ psi}}{2} [(4.0132)(0.5) \text{ in}^2] = -8333 \text{ lbf}$$

acting at a distance of 2(4.0132 in.)/3 = 2.6755 in. above the neutral axis, with the corresponding resultant force for the stresses below the neutral axis given by

$$F_b = \frac{(3077 + 0) \text{ psi}}{2} [(1.4868)(0.5) \text{ in}^2] + \frac{(3077 + 4112) \text{ psi}}{2} [(4.0)(0.5) \text{ in}^2]$$
$$= 8333 \text{ lbf}$$

acting 1.6447 in. below the neutral axis. The resultant of the two forces is a couple of magnitude

$$M = 8333 \text{ lbf}(2.6755 + 1.6447) \text{ in.} = 36{,}000 \text{ in.-lbf}$$

checking that the resultant is a couple as shown in Fig. 7.62.

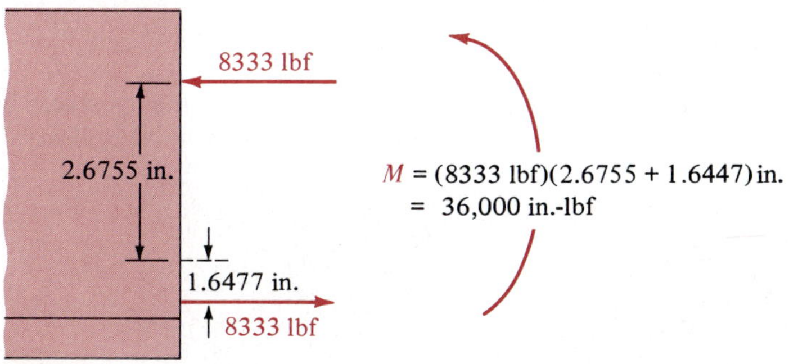

Figure 7.62 Resultant of the bending stresses as a moment

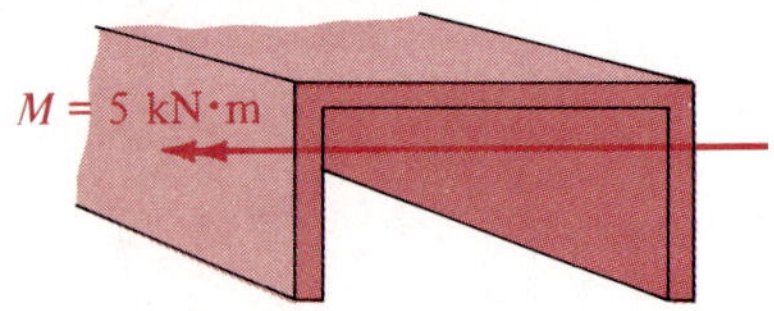

(a)

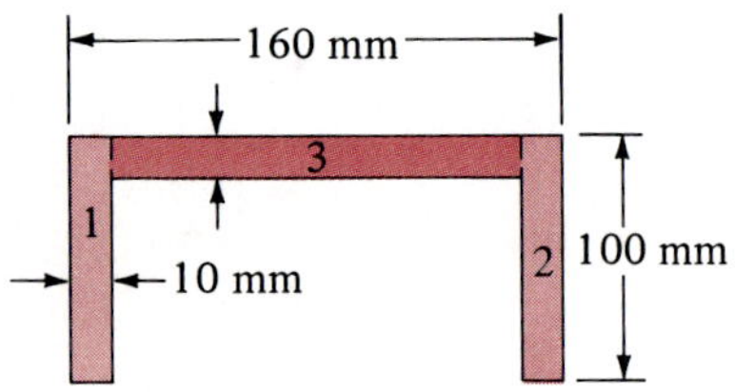

(b)

Figure 7.63 Bending of a channel section

Example 7.10 The channel section is subjected to a bending moment of 5 kN·m as shown in Fig. 7.63a. Determine the bending stresses.

Solution The first step is to locate the centroidal axis and to determine I. Referring to Fig. 7.63b, the table below is constructed for this purpose.

Table for calculation of section properties

i	A_i (mm²)	s_i (mm)	$A_i s_i$ (mm³)	d_i (mm)	I_0 (mm⁴)	$A_i d_i^2$ (mm⁴)
1	1000	50	5×10^4	−18.5	8.333×10^5	0.342×10^6
2	1000	50	5×10^4	−18.5	8.333×10^5	0.342×10^6
3	1400	95	1.33×10^5	26.5	1.167×10^4	0.983×10^6
	$\Sigma = 3400$		$\Sigma = 2.33 \times 10^5$		$\Sigma = 1.678 \times 10^6$	$\Sigma = 1.667 \times 10^6$

Using the second and fourth columns

$$\bar{s} = \frac{2.33 \times 10^5 \text{ mm}^3}{3400 \text{ mm}^2} = 68.5 \text{ mm}$$

The fifth, sixth, and seventh columns can then be constructed and I calculated as the sum of the sixth and seventh columns as

$$I = (1.678 + 1.667) \times 10^6 \text{ mm}^4 = 3.345 \times 10^{-6} \text{ m}^4$$

The bending stress at the top of the section is given by

$$\sigma_x = -\frac{(5000 \text{ N·m})(0.0315 \text{ m})}{3.345 \times 10^{-6} \text{ m}^4} = -47.09 \text{ MPa}$$

and at the bottom of the section by

$$\sigma_x = -\frac{(5000 \text{ N·m})(-0.0685 \text{ m})}{3.345 \times 10^{-6} \text{ m}^4} = 102.39 \text{ MPa}$$

As mentioned previously and as indicated by Fig. 7.64, the maximum bending stress occurs at the point in the section furthest from the neutral axis.

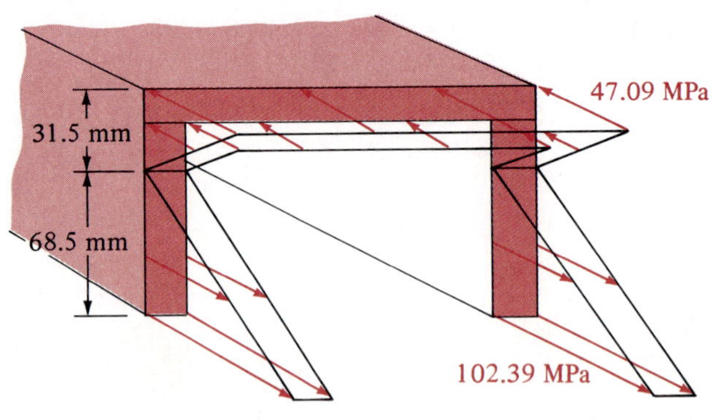

Figure 7.64 Bending stresses in a channel section

PROBLEMS

7.93 Verify each of the Is given in Fig. 7.56.

7.94 Verify that the Is given for the W 14 × 82 section in Appendix C are approximately correct.

7.95 Verify that the Is given for the W 360 × 122 section in Appendix C are approximately correct.

7.96 Verify that the Is given for the C 10 × 25 section in Appendix C are approximately correct.

7.97 Verify that the Is given for the C 250 × 37 section in Appendix C are approximately correct.

7.98 Verify that the Is given for the L 4 × 4 × $\frac{1}{2}$ section in Appendix C are approximately correct.

7.99 Verify that the Is given for the L 102 × 102 × 12.7 section in Appendix C are approximately correct.

7.100 Verify that the Is given for the L 5 × 3 × $\frac{1}{2}$ section in Appendix C are approximately correct.

7.101 Verify that the Is given for the L 127 × 76 × 12.7 section in Appendix C are approximately correct.

7.102 Determine I about the z-axis for the section shown.

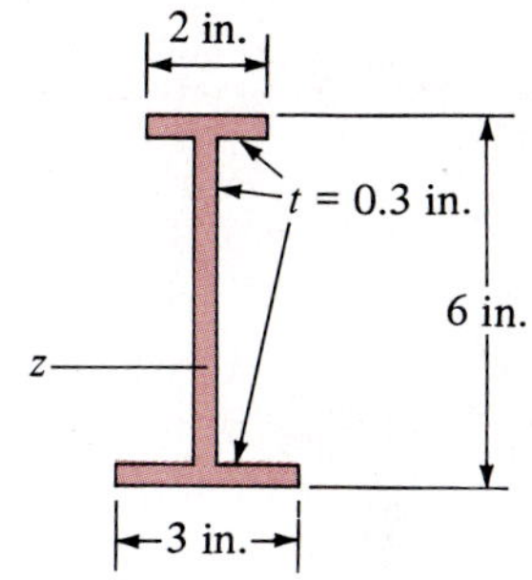

P. 7.102

7.103–7.108 For each of the following shapes, determine *(a)* the maximum compressive bending stress, and *(b)* the maximum tensile bending stress in response to a positive bending moment M.

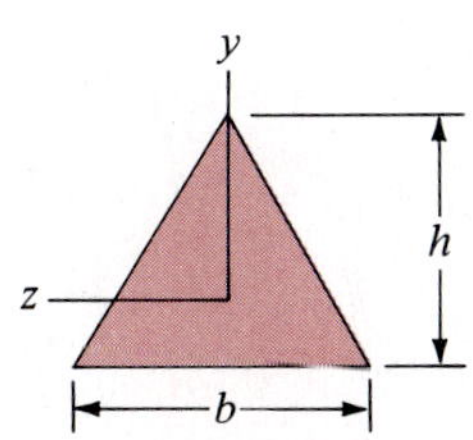

P. 7.103

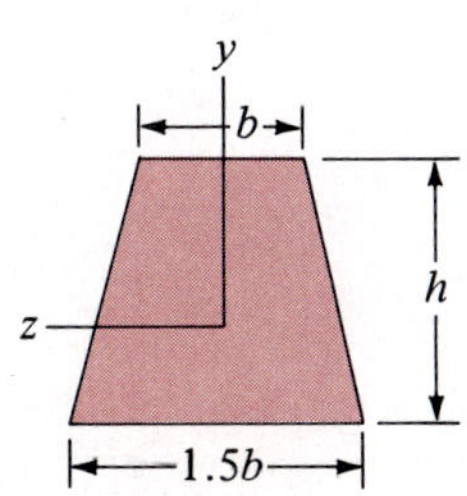

P. 7.104

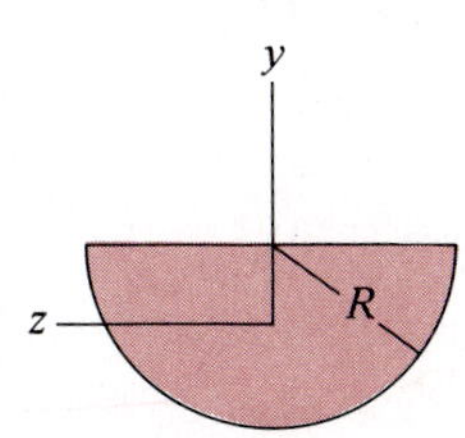

P. 7.105

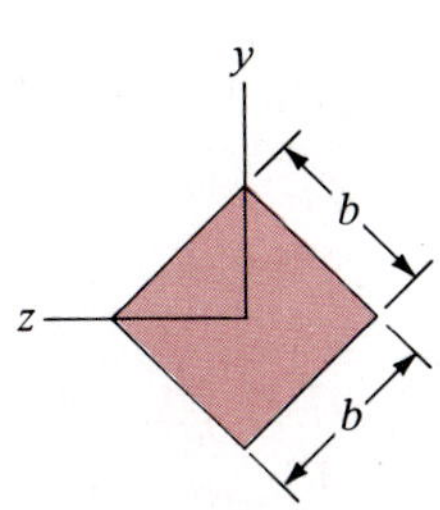

P. 7.106

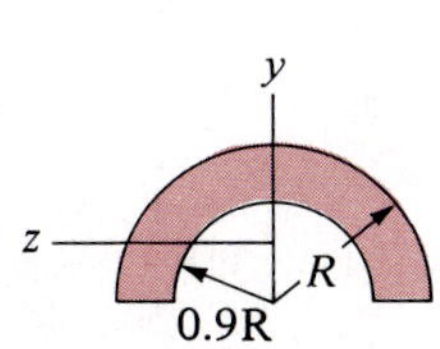

P. 7.107

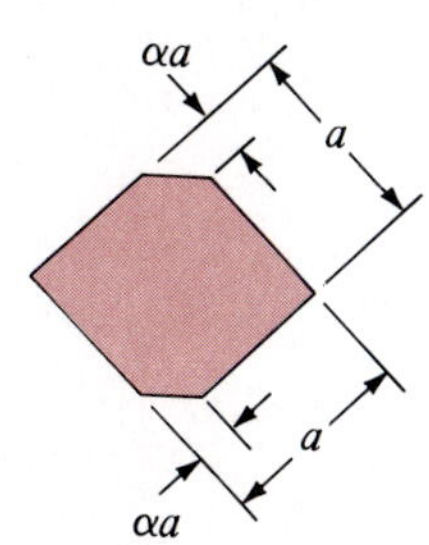

Ps. 7.108 and 7.109

7.109 For the shape of problem 7.108, determine the value of α that gives the smallest bending stress.

7.110 Determine the bending stress *(a)* at point A, and *(b)* at point B.

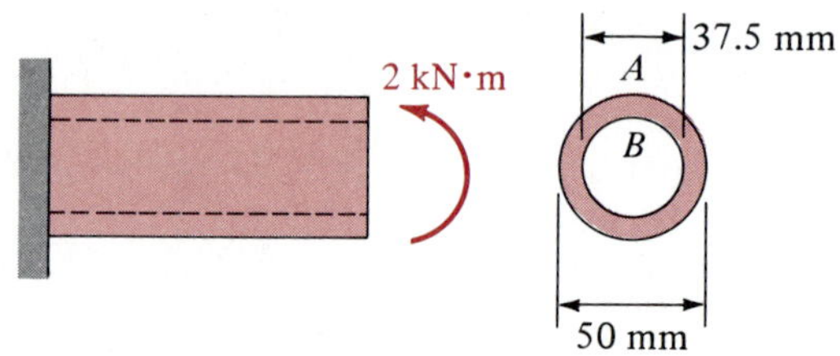

P. 7.110

7.111 Determine the bending stress at *(a)* point A, and *(b)* point B for the square hollow section shown.

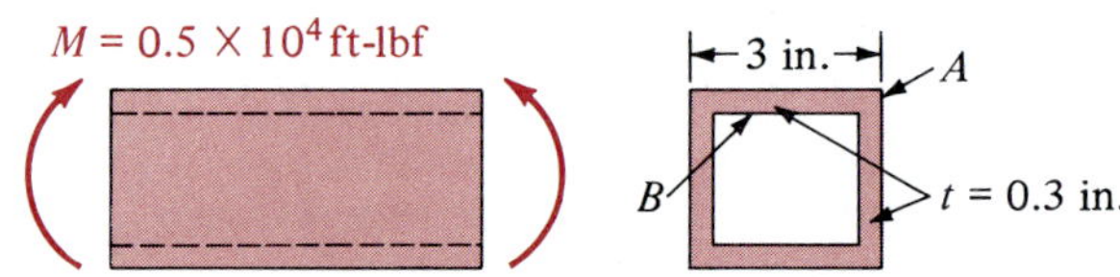

P. 7.111

7.112 Determine the maximum bending stress for bending about the z-axis.

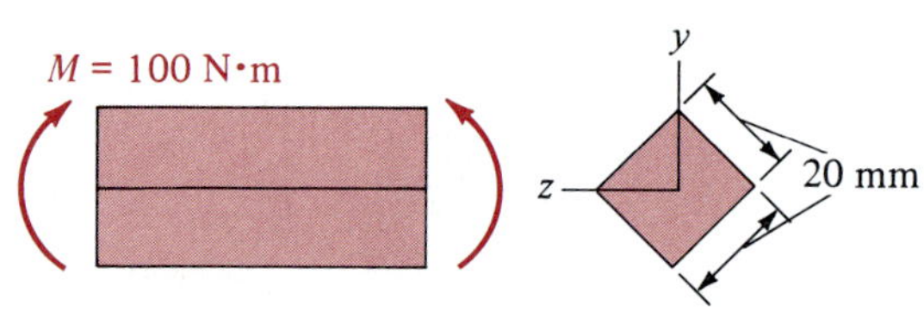

P. 7.112

7.113 Repeat problem 7.112 if the sides of the section are parallel to the axes as shown.

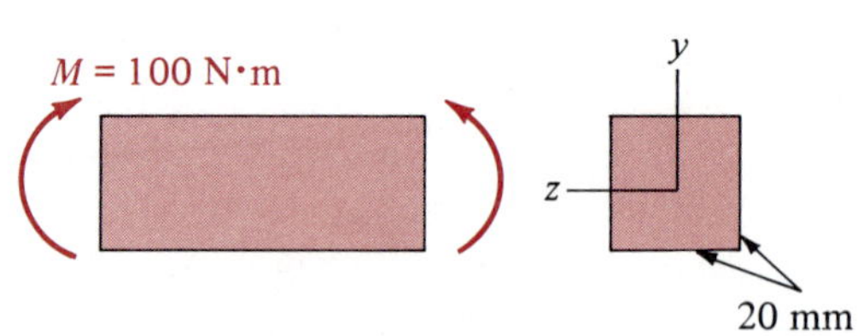

P. 7.113

7.114 For the section shown, determine the maximum bending moment for which the bending stress does not exceed 120 MPa. Bending is about the z-axis.

7.115 Repeat problem 7.114 for bending about the y-axis.

7.116 The cantilever beam is subjected to the load shown. Determine the largest value of P for which the maximum bending stress does not exceed 70 MPa.

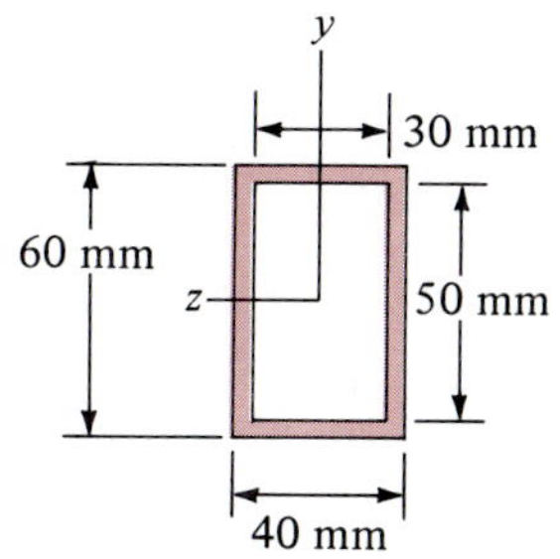

Ps. 7.114 and 7.115

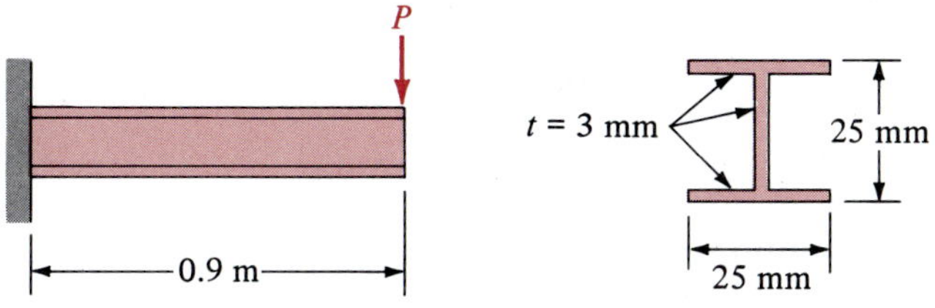

P. 7.116

7.117 For the W 10 × 15 cantilever beam shown, determine the allowable load P if the bending stress is not to exceed 20 ksi. Neglect the weight of the beam.

7.118 Repeat problem 7.117 if the weight is included.

7.119 The simply supported beam is subjected to a uniform load of 120 lbf/ft as shown. Determine the location and magnitude of *(a)* the largest positive bending stress, and *(b)* the largest negative bending stress.

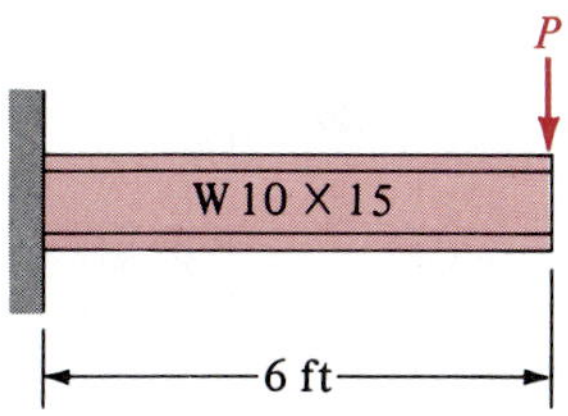

Ps. 7.117 and 7.118

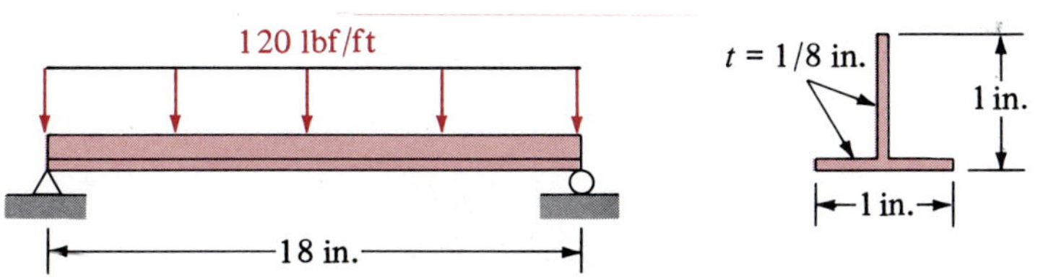

P. 7.119

7.120 Determine the magnitude of the force arising from the compressive stresses in a rectangular cross section of the dimensions shown and acted upon by a moment of 8 in.-kip. Also determine the distance h between the compressive and tensile forces.

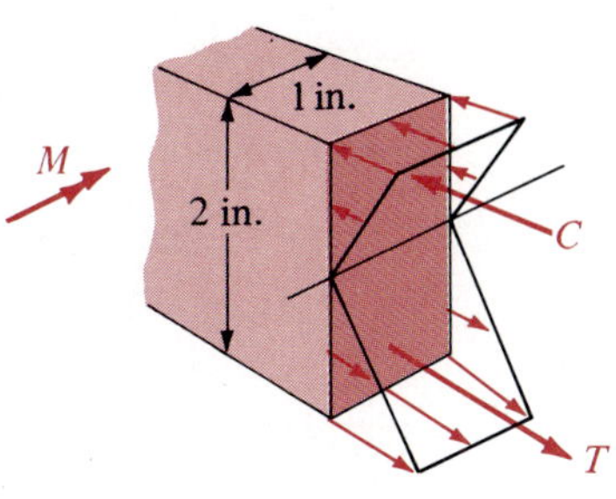

P. 7.120

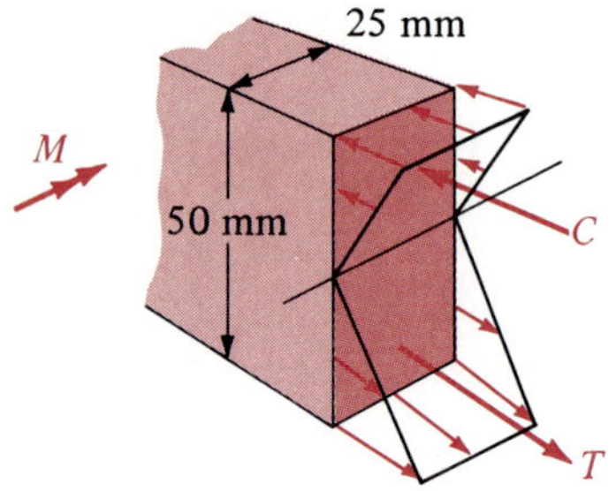

P. 7.121

7.121 A rectangular-cross-section beam is subjected to a moment of 1 kN·m as shown. Determine the magnitude and location of the associated compressive force C. Also determine the distance between C and T.

7.122 The 10-mm-thick T section shown is subjected to a positive moment of 2.0 kN·m about the z-axis. Determine and show on a sketch the magnitude and location of the forces that constitute the couple.

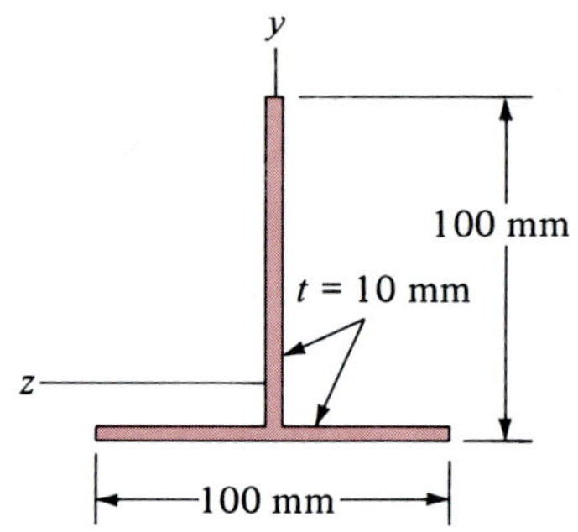

P. 7.122

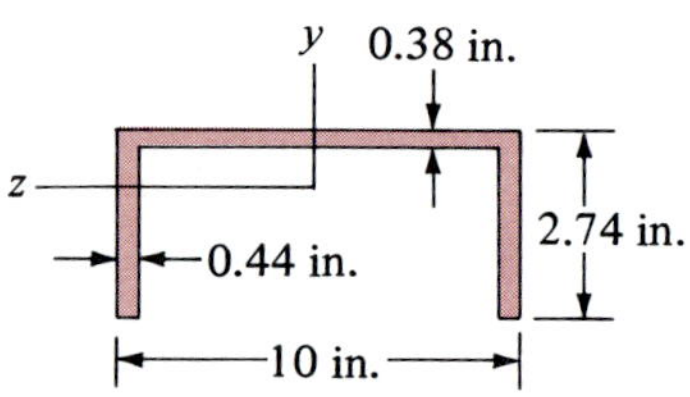

P. 7.123

7.123 The channel section shown is subjected to a bending moment that produces a tensile stress of 17.5 ksi at the bottom. Determine and show on a sketch the magnitude and location of the forces C and T associated with the bending.

7.124 A W 10 × 22 section is subjected to a bending moment of 400 in.-kip as shown. Determine the location and magnitude of the compressive force in the upper flange. Thus determine what percentage of the moment is supplied by the stresses in the flanges.

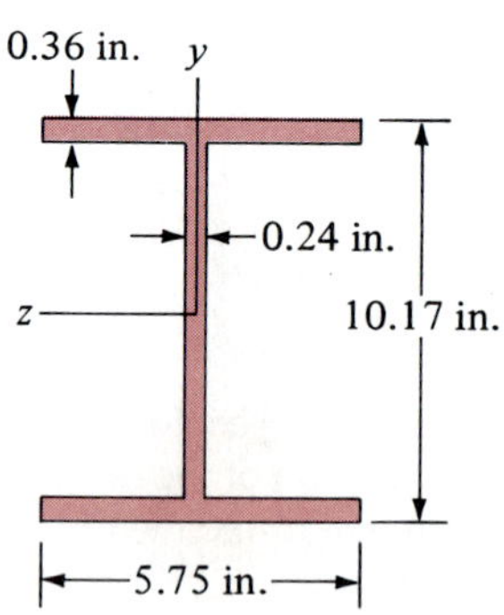

P. 7.124

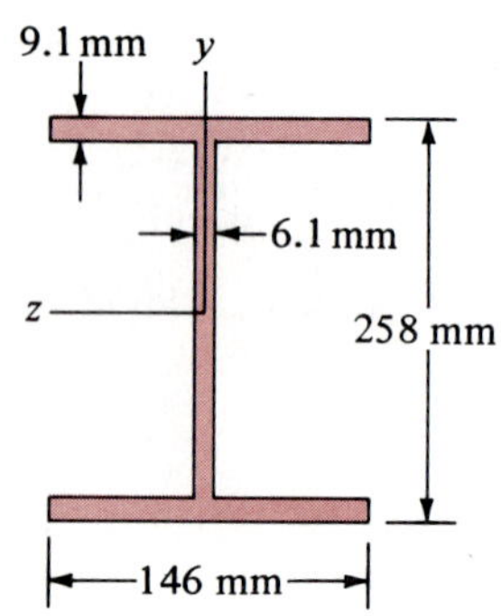

P. 7.125

7.125 A W 250 × 32.7 section is subjected to a bending moment of 45 kN·m about the z-axis. Determine *(a)* the location and magnitude of the tensile force in the lower flange, and *(b)* what percentage of the moment is supplied by the stresses in the flanges.

7.126–7.139 For each of the following, draw the shear and moment diagrams and thus determine the location and magnitude of *(a)* the largest tensile bending stress, and *(b)* the largest compressive bending stress. Neglect the weight of the beam in all cases.

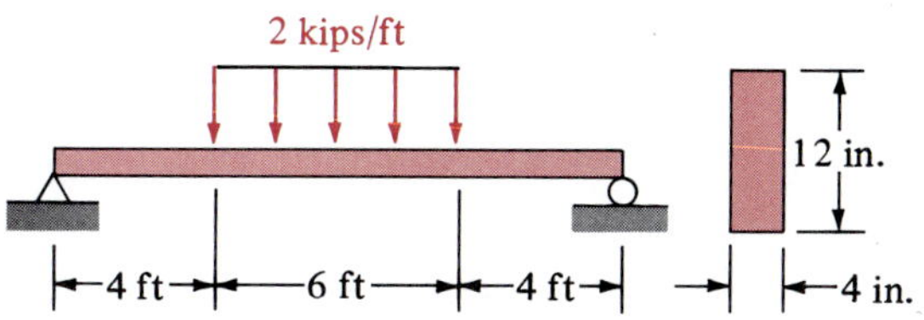

P. 7.126

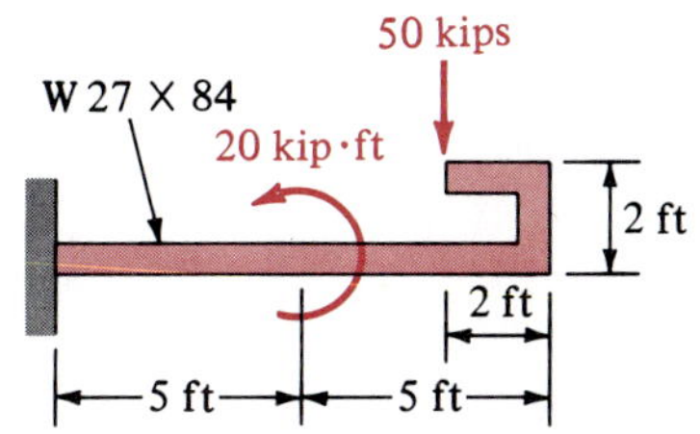

Ps. 7.127 and 7.140

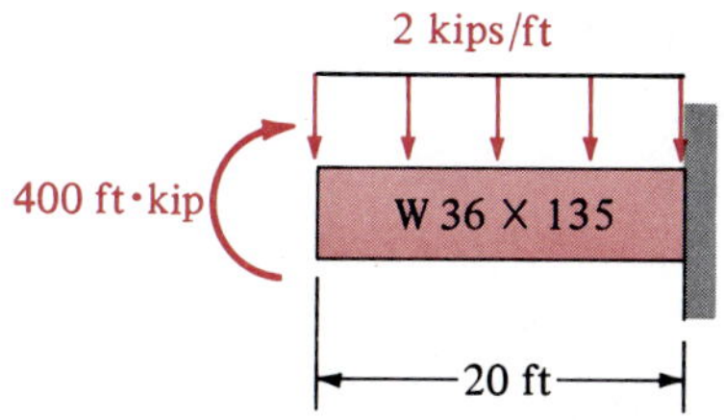

Ps. 7.128 and 7.141

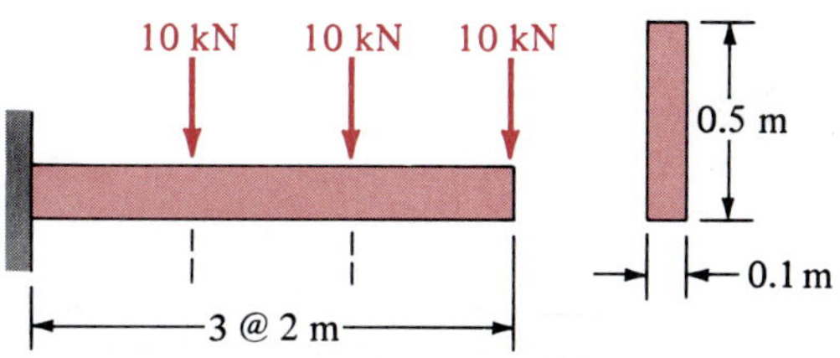

P. 7.129

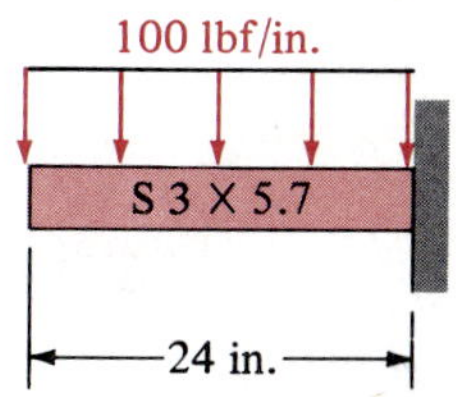

Ps 7.130 and 7.142

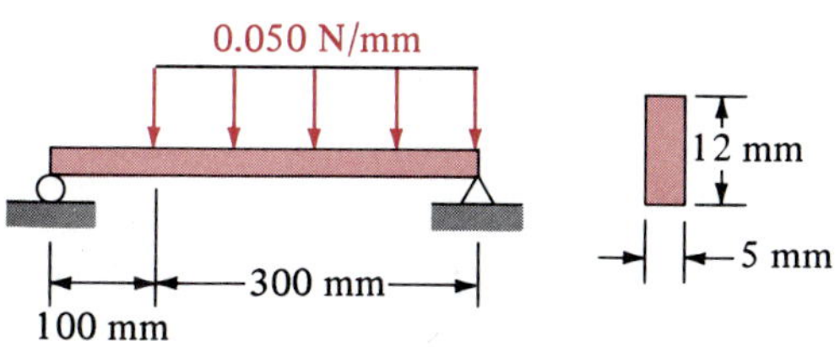

P. 7.131

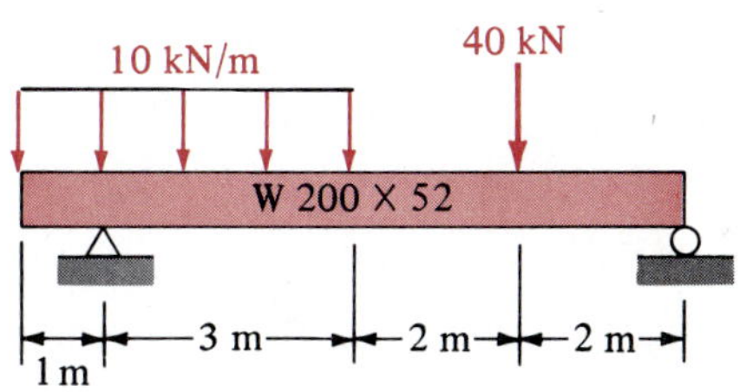

Ps. 7.132 and 7.143

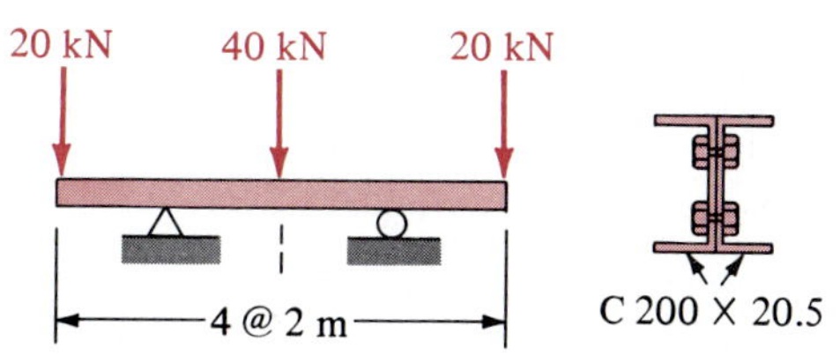

Ps. 7.133 and 7.144

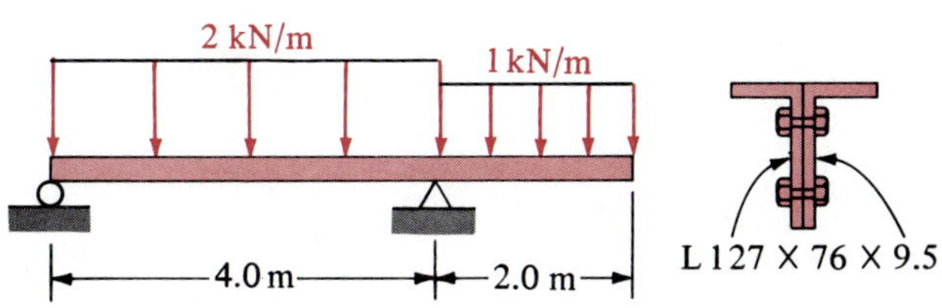

Ps. 7.134 and 7.145

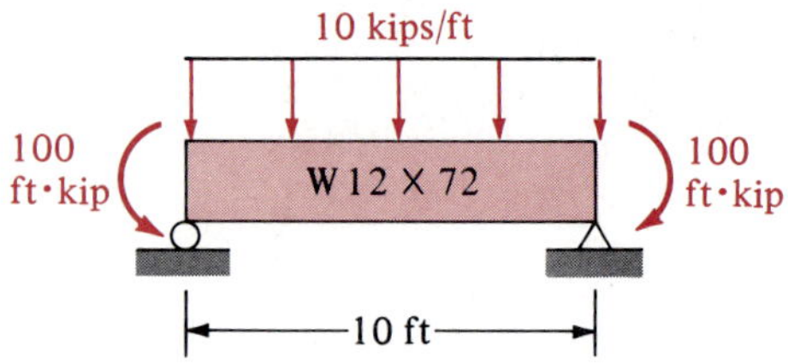

Ps. 7.135 and 7.146

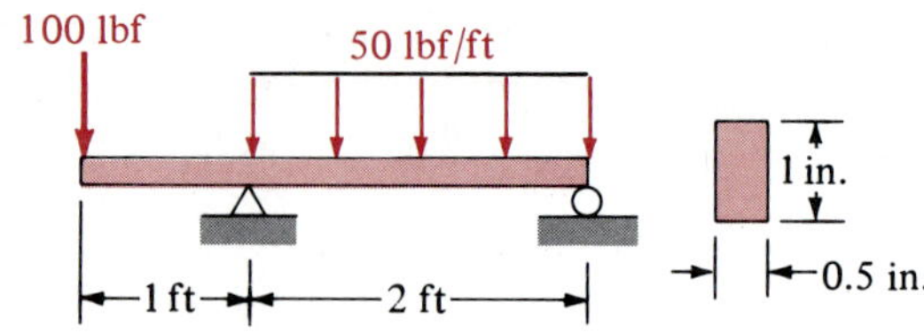

P. 7.136

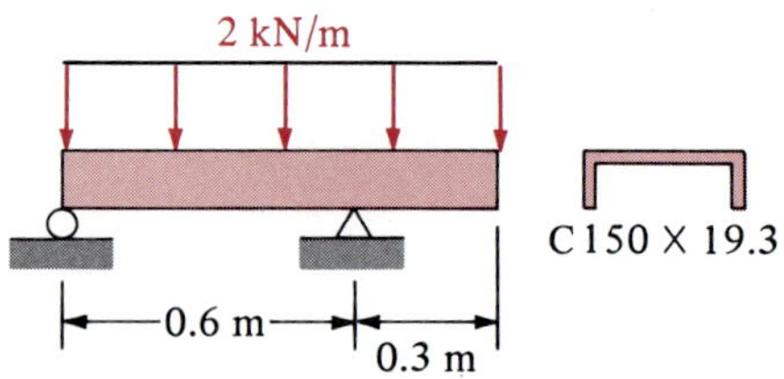

Ps. 7.137 and 7.147

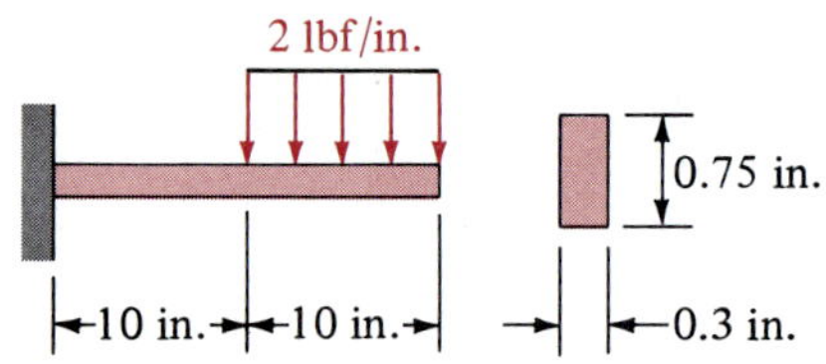

P. 7.138

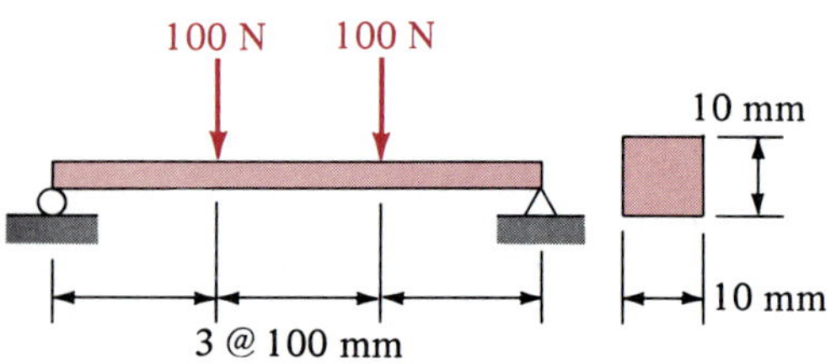

P. 7.139

7.140 Resolve problem 7.127 including the weight of the beam.

7.141 Resolve problem 7.128 including the weight of the beam.

7.142 Resolve problem 7.130 including the weight of the beam.

7.143 Resolve problem 7.132 including the weight of the beam.

7.144 Resolve problem 7.133 including the weight of the beam.

7.145 Resolve problem 7.134 including the weight of the beam.

7.146 Resolve problem 7.135 including the weight of the beam.

7.147 Resolve problem 7.137 including the weight of the beam.

7.148 The steel T-section beam carries a concentrated load P as shown. If the allowable bending stresses in tension and compression are $\sigma_t = 20$ ksi and $\sigma_c = 15$ ksi, respectively, determine the maximum allowable value for P. Take $b = 2$ in., $h = 3$ in., $t = 0.25$ in., and $L = 2$ ft.

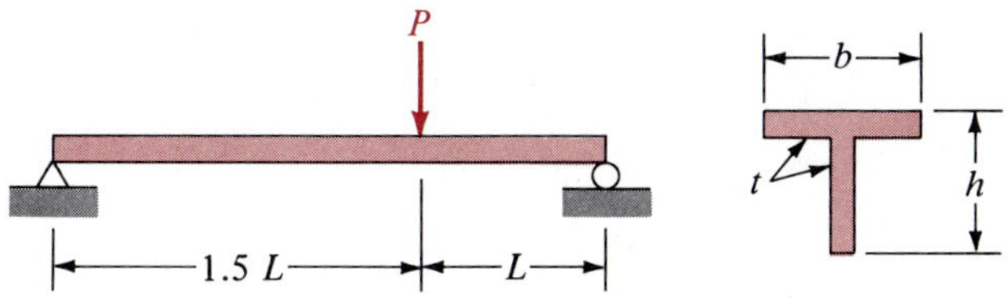

Ps. 7.148–7.151

7.149 Repeat problem 7.148 if the direction of the load P is reversed.

7.150 Repeat problem 7.148 with σ_t = 130 MPa, σ_c = 88 MPa, b = 45 mm, h = 70 mm, t = 6 mm, and L = 0.6 m.

7.151 Repeat problem 7.150 if the direction of the load P is reversed.

7.152 A beam with a rectangular cross section is subjected to the loading shown. Determine the maximum bending stress. Take w = 100 lbf/ft, L = 10 ft, b = 1.5 in., and h = 3.5 in.

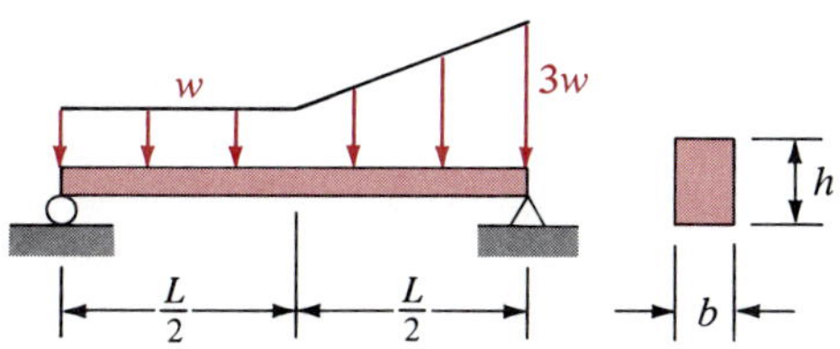

Ps. 7.152 and 7.153

7.153 Repeat problem 7.152 if L = 3 m, b = 45 mm, h = 90 mm, and W = 1.5 kN/m.

7.154 The T-section beam is loaded by the uniformly distributed loadings as shown. Determine *(a)* the location and magnitude of the maximum tensile bending stress, and *(b)* the location and magnitude of the maximum compressive bending stress. Take L_1 = 3.0 m, L_2 = 6.0 m, w_1 = 4 kN/m, w_2 = 12 kN/m, b = 0.15 m, h = 0.2 m, and t = 0.05 m.

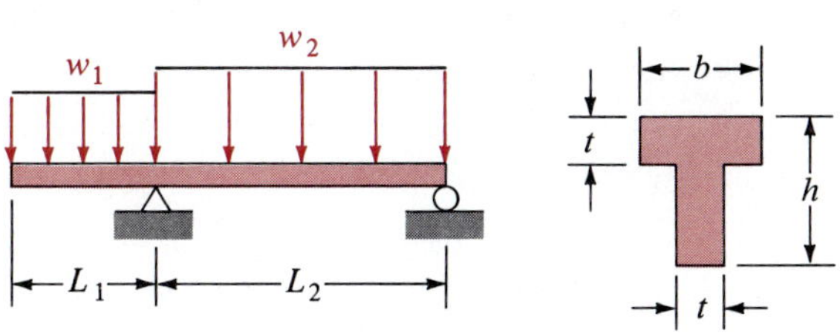

Ps. 7.154 and 7.155

7.155 Repeat problem 7.154 with L_1 = 10 ft, L_2 = 20 ft, w_1 = 300 lbf/ft, w_2 = 900 lbf/ft, b = 6 in., h = 8 in., and t = 2 in.

7.156 A W 360 × 216 section is used as a simply supported beam spanning a distance of 8 m. If the maximum bending stress is not to exceed 105 MPa, determine the maximum allowable uniformly distributed load w that the beam can carry. Neglect the weight of the beam.

7.157 Repeat problem 7.156 including the weight of the beam.

7.158 Repeat problem 7.156 for a W 14 × 145 section spanning a distance of 26 ft. Take the maximum allowable bending stress to be 15 ksi. Neglect the weight of the beam.

7.159 Repeat problem 7.158 including the weight of the beam.

7.160 The C 10 × 20 channel section is used to support the loads shown. Determine *(a)* the maximum tensile bending stress, and *(b)* the maximum compressive bending stress if $P = 500$ lbf. Take $L = 6$ ft.

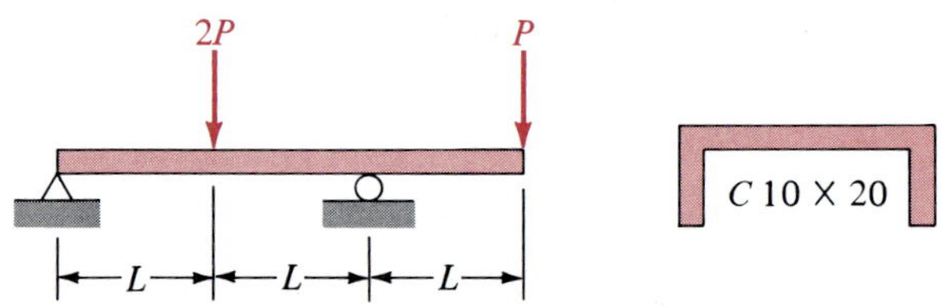

Ps. 7.160 and 7.161

7.161 Repeat problem 7.160 for a C 250 × 30 channel section with $L = 2$ m and $P = 2$ kN.

7.162 A group of wooden beams is used as a dam as shown. To what depth h can the water be raised if the maximum bending stress σ_{max} is not to exceed 2 ksi? Take $t = 6$ in. and the water to weigh 62.4 lbf/ft³.

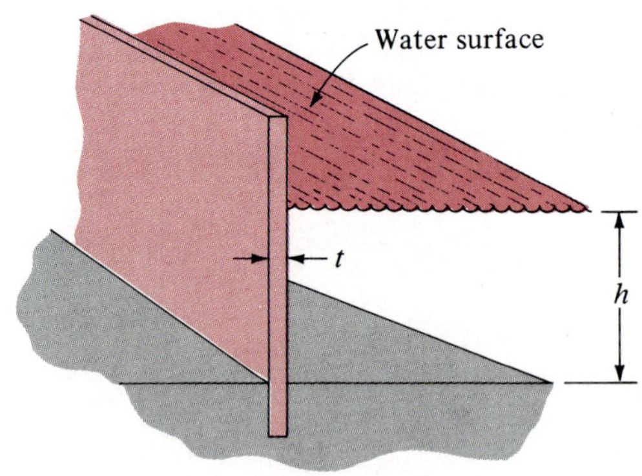

Ps. 7.162 and 7.163

7.163 Repeat problem 7.162 if $\sigma_{max} = 15$ MPa, $t = 150$ mm, and the weight of the water is 9.81 kN/m³.

7.164 A hollow circular shaft of outside diameter D and inside diameter d is acted upon by the belt tensions shown. Determine the location and magnitude of the maximum bending stress. Take $D = 0.2$ m, $d = 0.175$ m, $L = 1$ m, and $P = 8$ kN. Assume each bearing provides a simple support.

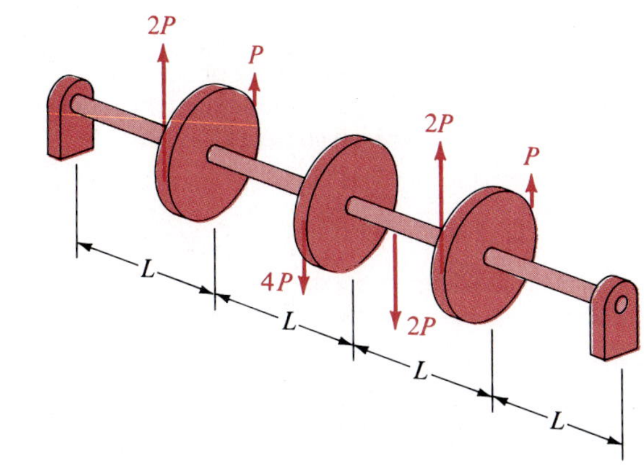

Ps. 7.164 and 7.165

7.165 Repeat problem 7.164 if $D = 8$ in., $d = 7$ in., $L = 3$ ft, and $P = 2$ kips.

7.166 A roofing system consists of simply supported W 360 × 39 sections of length L on a spacing s as shown. The weight of the roofing material per unit area is w. Neglecting the weight of the beams, determine the maximum bending stress in one of the beams. Take $L = 6$ m, $s = 3$ m, and $w = 8.0$ kN/m^2.

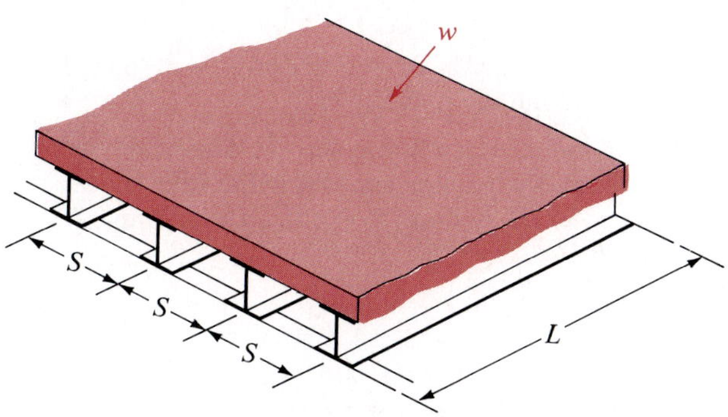

Ps. 7.166–7.169

7.167 Repeat problem 7.166 including the weights of the beams.

7.168 Repeat problem 7.166 for a W 14 × 26 section with $L = 20$ ft, $s = 10$ ft, and $w = 160$ lbf/ft^2.

7.169 Repeat problem 7.168 including the weights of the beams.

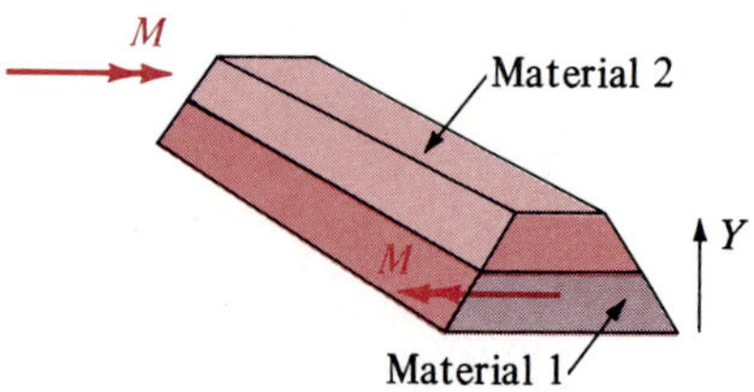

P. 7.170

7.170 For the two-material composite beam subjected to a bending moment as shown, assume that plane sections remain plane and that each material is linearly elastic; show, using the appropriate equations of equilibrium, that the location of the neutral axis is given by

$$\bar{y} = \frac{A_1 E_1 \bar{Y}_1 + A_2 E_2 \bar{Y}_2}{A_1 E_1 + A_2 E_2}$$

where A_1 and A_2 are the cross-sectional areas of materials 1 and 2, respectively, and where $\bar{Y}_1$ and $\bar{Y}_2$ are the corresponding centroidal positions. Also show that the stresses in the two materials are given by

$$\sigma_1 = -My \frac{E_1}{E_1 I_1 + E_2 I_2} \qquad \text{and} \qquad \sigma_2 = -My \frac{E_2}{E_1 I_1 + E_2 I_2}$$

where I_1 and I_2 are the second moments of areas 1 and 2 about the neutral axis, and where y is measured from the neutral axis.

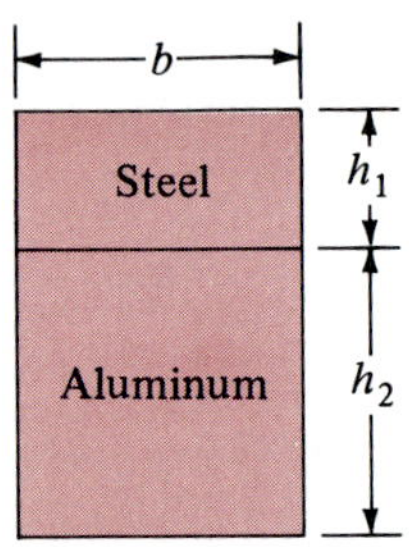

Ps. 7.171 and 7.172

7.171 A composite beam is formed by securely bonding the steel and aluminum bars shown. If the allowable stresses in the aluminum and steel are 15 ksi and 24 ksi, respectively, determine the maximum allowable moment about a horizontal axis. Take $E_{st} = 30 \times 10^6$ psi, $E_{al} = 10 \times 10^6$ psi, $b = 1.0$ in., $h_1 = 0.5$ in., and $h_2 = 1.0$ in.

7.172 The steel and aluminum bars shown are securely bonded to form a composite beam as shown. If the allowable stresses in the aluminum and steel are 100 MPa and 170 MPa, respectively, determine the maximum allowable moment about a horizontal axis. Take $E_{st} = 200$ GPa, $E_{al} = 70$ GPa, $b = 25$ mm, $h_1 = 15$ mm, and $h_2 = 30$ mm.

7.173 The two pipes shown are securely attached by shrinking the brass pipe into the aluminum pipe. Determine *(a)* the stress in the aluminum, and *(b)* the stress in the brass when $M = 1$ ft·kip. Take $a = 1.25$ in., $b = 1.0$ in., $c = 0.75$ in., $E_{al} = 10 \times 10^6$ psi, and $E_{br} = 15 \times 10^6$ psi.

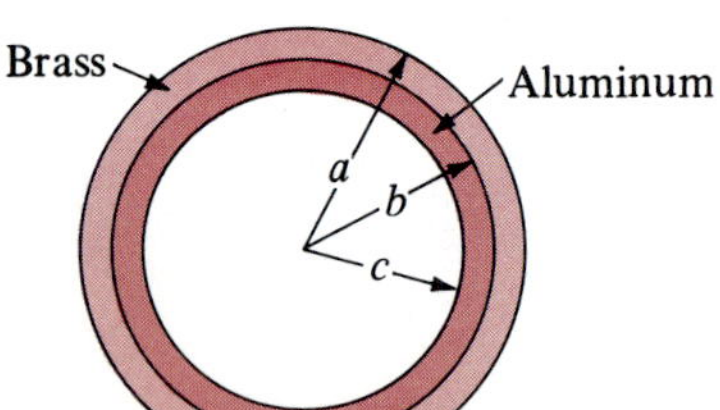

Ps. 7.173 and 7.174

7.174 Repeat problem 7.173 if $M = 5$ kN·m, $a = 36$ mm, $b = 29$ mm, $c = 22$ mm, $E_{al} = 70$ GPa, and $E_{br} = 105$ GPa.

7.175 For the wood and steel section shown, determine the allowable bending moment if the allowable bending stresses in the steel and wood are 2 ksi and 20 ksi, respectively. Take $b = 6$ in., $h = 8$ in., $t = 0.75$ in., $E_w = 2 \times 10^6$ psi, and $E_{st} = 29 \times 10^6$ psi.

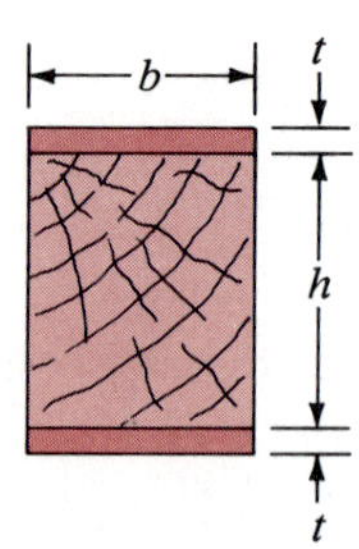

Ps. 7.175 and 7.176

7.176 Repeat problem 7.175 if the allowable stresses in the wood and steel are 15 MPa and 150 MPa, respectively. Take b = 150 mm, h = 250 mm, t = 18 mm, E_w = 14 GPa, and E_{st} = 200 GPa.

7.177 The square wooden member is securely bonded to the two steel plates as shown. Determine *(a)* the maximum bending stress in the steel, and *(b)* the maximum bending stress in the wood. Compare these to the same values if bending is about the vertical rather than the horizontal axis. Take $b = h = 6$ in., t = 0.625 in., $E_w = 2 \times 10^6$ psi, $E_{st} = 29 \times 10^6$ psi, and M = 500 in.-kip.

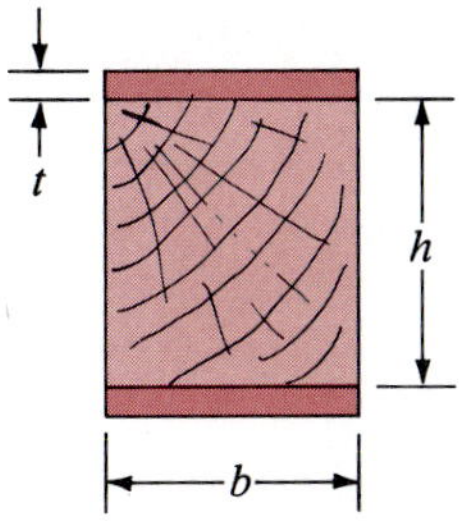

Ps. 7.177 and 7.178

7.178 Repeat problem 7.177 with h = 12 in., b = 6 in., and t = 0.625 in.

7.179 A wooden member is bonded securely to the C 4 × 5.4 section as shown. The allowable bending stresses in the wood and steel are 1.2 ksi and 18 ksi, respectively. If the composite section is subjected to a moment of 50 in.-kips, what are the factors of safety *(a)* for the wood, and *(b)* for the steel? Take $E_w = 2 \times 10^6$ psi, and $E_{st} = 29 \times 10^6$ psi.

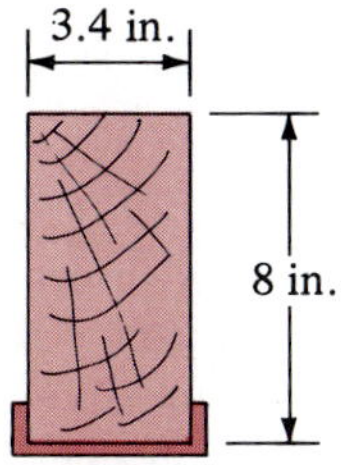

P. 7.179

7.180 A cantilever beam is constructed from a wooden member and a steel plate as shown. Determine the allowable value of P if the ultimate bending stresses in the wood and steel are 24 MPa and 245 MPa, respectively, and the factors of safety for the wood and steel are 1.5 and 1.2, respectively. Take L = 2 m, b = 150 mm, h = 200 mm, t = 25 mm, E_w = 15 GPa, and E_{st} = 200 GPa.

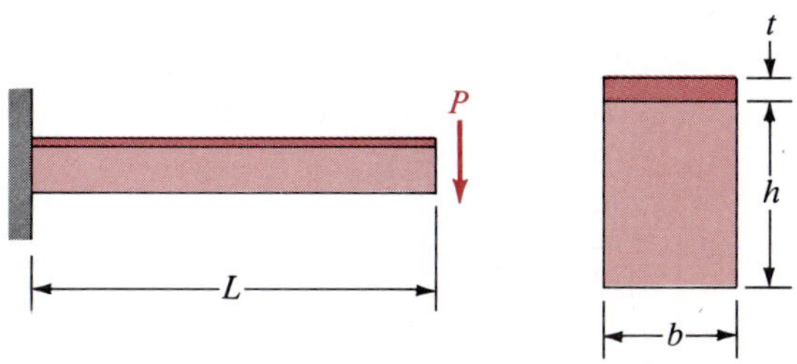

Ps. 7.180 and 7.181

7.181 Repeat problem 7.180 if the ultimate bending stresses in the wood and steel are 3.5 ksi and 35 ksi, respectively, and the factors of safety for the wood and steel are 1.5 and 1.2, respectively. Take L = 7 ft, b = 6 in., h = 8 in., t = 1 in., $E_W = 2 \times 10^6$ psi, and $E_{st} = 29 \times 10^6$ psi.

7.182 A simply supported beam is constructed from a wooden member and two steel plates as shown. What are the maximum bending stresses *(a)* in the wood, and *(b)* in the steel? Take $L = 4$ m, $b = 100$ mm, $h = 150$ mm, $t = 20$ mm, $w_0 = 10$ kN/m, $P = 12.5$ kN, $E_w = 16$ GPa, and $E_{st} = 200$ GPa.

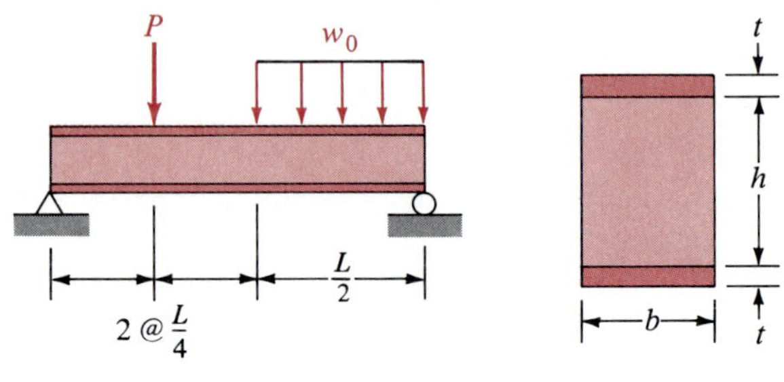

Ps. 7.182 and 7.183

7.183 Repeat problem 7.182 for $L = 14$ ft, $b = 4$ in., $h = 6$ in., $t = 0.75$ in., $w_0 = 750$ lbf/ft, $P = 3$ kips, $E_w = 2 \times 10^6$ psi, and $E_{st} = 29 \times 10^6$ psi.

7.184 The reinforced concrete beam shown is subjected to a positive bending moment of 0.3×10^6 in.-lbf. Determine *(a)* the tensile stress in the steel bars, and *(b)* the maximum compressive stress in the concrete. Take $E_c = 3 \times 10^6$ psi and $E_{st} = 29 \times 10^6$ psi. (*Hint:* Assume that the concrete can transmit only compression.)

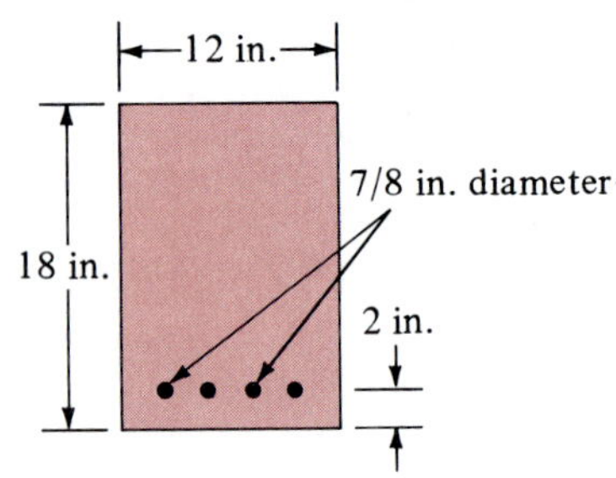

P. 7.184

7.185 A 6-in.-thick concrete slab is reinforced by ⅝-in.-diameter steel rods placed on 6-in. centers as shown. If the allowable stresses for concrete and steel are 1.2 ksi and 18 ksi, respectively, determine the allowable bending moment per foot of the slab. Take $E_c = 3 \times 10^6$ psi and $E_{st} = 29 \times 10^6$ psi. (*Hint:* Assume that the concrete can transmit only compression.)

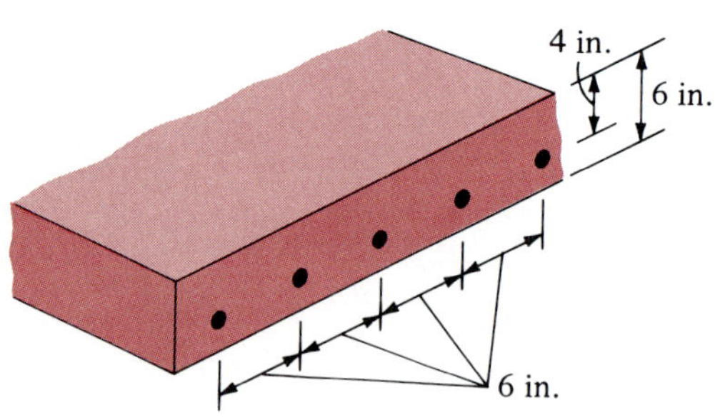

Ps. 7.185 and 7.186

7.186 Repeat problem 7.185 for a 150-mm-thick slab with 15-mm-diameter rods on 150-mm centers, 50 mm above the bottom of the slab. Take the allowable stresses for concrete and steel to be 10 MPa and 120 MPa, respectively. Also take $E_c = 20$ GPa and $E_{st} = 200$ GPa.

7.187 If the allowable bending stresses in the reinforced T section shown are 1.35 ksi for the concrete and 18 ksi for the steel, determine the maximum uniformly distributed load a 30-ft-long simply supported beam can support. Include the weight of the concrete. Take the specific weight of the concrete to be 150 lbf/ft^3, $E_c = 3 \times 10^6$ psi, and $E_{st} = 29 \times 10^6$ psi. (*Hint:* Assume that the concrete can transmit only compression.)

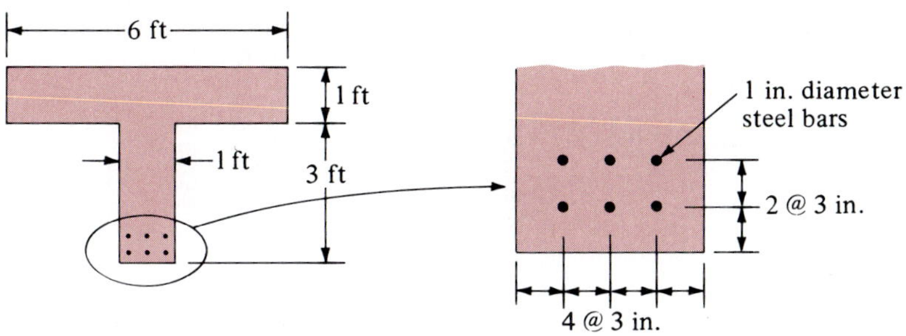

P. 7.187

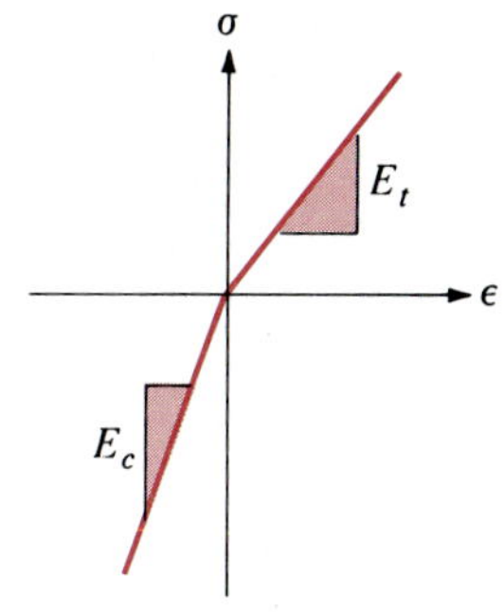

P. 7.188

7.188 A beam composed of a material that behaves differently in compression and tension is subjected to a bending moment as shown. Develop expressions for *(a)* the tensile and compressive bending stresses, and *(b)* the moment-curvature relationship.

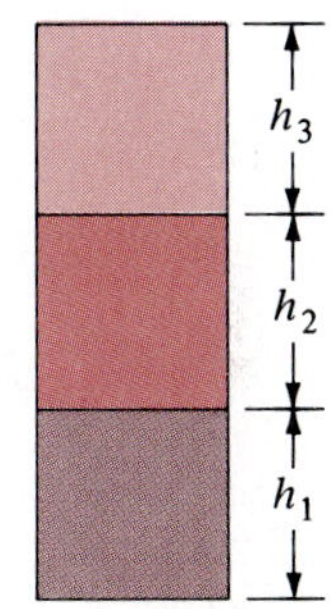

P. 7.189

7.189 Three different materials with moduli E_1, E_2, and E_3, respectively, are to be bonded to form a beam as shown. Develop expressions for *(a)* the bending stresses in each of the materials, and *(b)* the moment-curvature relationship.

7.8 TRANSVERSE SHEAR STRESSES IN BEAMS

The equation of equilibrium relating the shear stress τ_{xy} to the shear force V was developed in Section 7.2 as

$$V = -\int \tau_{xy}\, dA$$

In this section we will use equilibrium ideas to develop an approximate expression for the shear stresses arising from the requirement of transmitting a shear force V.

To motivate the discussion of shear stress in beams, consider the situation shown in Fig. 7.65.

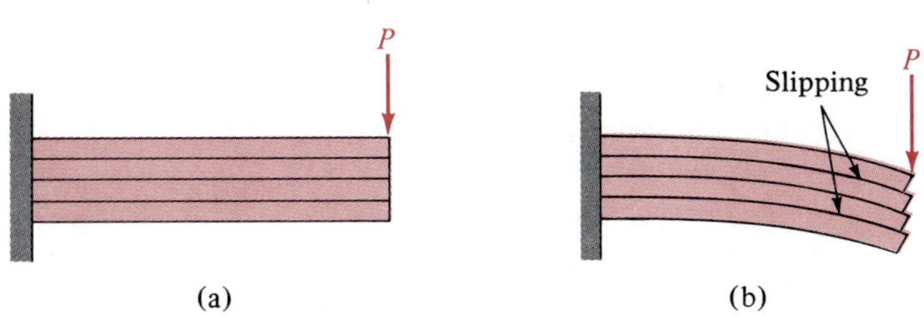

Figure 7.65 Deformations in unconnected strips

Shown are several unconnected thin strips, each fixed to the support, and laid upon one another. When loaded by the shear load P, there is clearly a relative slipping between the beams as indicated in Fig. 7.65b. If on the other hand the strips are connected by gluing before the load P is applied as shown in Fig. 7.66a, the slipping is for the most part prevented by the shear stresses developed in the glue.

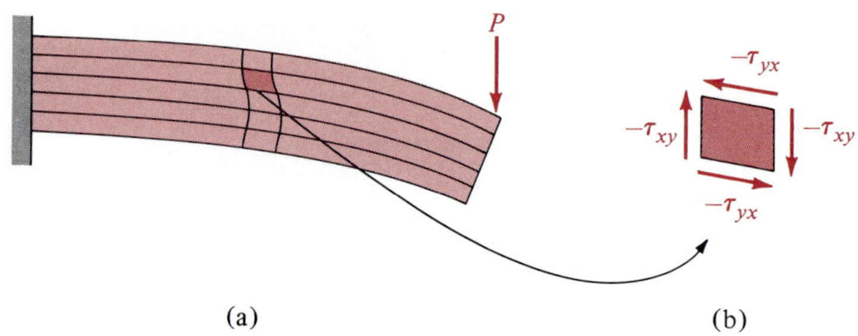

Figure 7.66 Strips connected by gluing

Drawing a free-body diagram of a typical element of one of the strips as shown in Fig. 7.66b, we see that there is horizontal shear $-\tau_{yx}$ in the glue and vertical shear $-\tau_{xy}$ in the strip.

In a solid beam of the same dimensions, both the horizontal shear τ_{yx} and the vertical shear τ_{xy} are present as shown in Fig. 7.67.

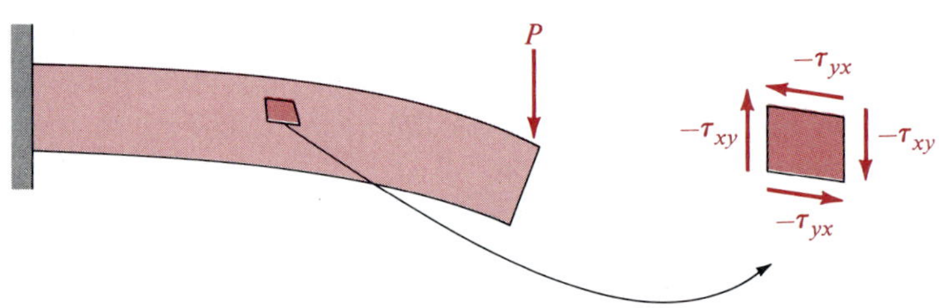

Figure 7.67 Vertical and horizontal shear in a beam

Our objective is to determine just how the vertical shear stress τ_{xy} is distributed over the depth of the cross section in a solid-cross-section beam in order to satisfy equilibrium within the beam, and in order to be statically equivalent to the resultant shear force V.

As a typical application, consider a cantilever beam of rectangular cross section, subjected to an end load P as shown in Fig. 7.68.

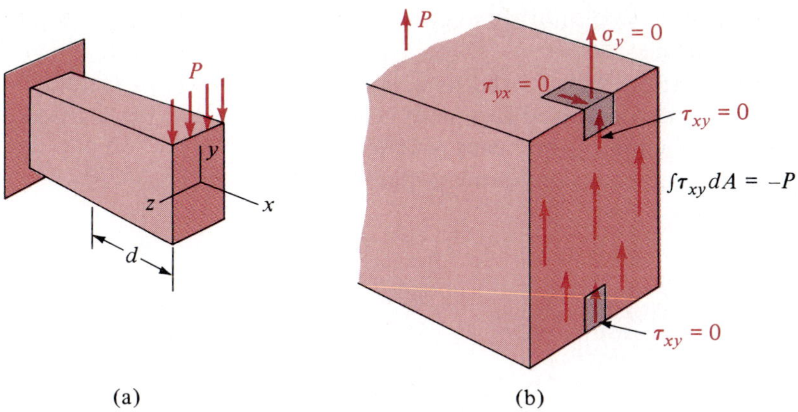

Figure 7.68 Rectangular cross section transmitting a shear $V = P$

The shear stresses acting on an exposed face a distance d from the end of the beam are shown in Fig. 7.68b. Several important observations are possible: First, as indicated, the distribution of shear stresses must be statically equivalent to the shear force transmitted, according to the equilibrium relation $V = P = -\int \tau_{xy}\, dA$. Second, the top and bottom surfaces of the beam are stress free; thus, σ_y and τ_{yx} $(=\tau_{xy})$ are both zero on the top and bottom faces of the bar. This means that whatever the distribution the shear stress τ_{xy} has on the exposed vertical face, it must approach a zero value at the top and at the bottom surfaces. Two such *possible* distributions are indicated in Fig. 7.69.

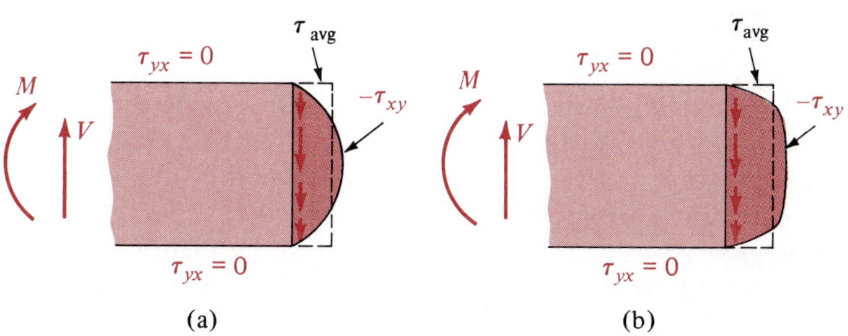

Figure 7.69 Possible distributions for the transverse shear stress

An estimate of the magnitude of the shear stress due to the shear force $V = P$ can be obtained by computing the average shear stress

$$\tau_{avg} = \frac{V}{A} = \frac{P}{A}$$

as shown in Fig. 7.69. In view of the fact that τ_{xy} must vanish at the top and bottom of the rectangular section, this average value will clearly be less than the maximum that occurs somewhere between the top and bottom surfaces.

The equation that is used to compute the distribution of the shear stress τ_{xy} on an exposed face is developed by considering the equilibrium of a portion of the beam of length Δx above a section parallel to and at a distance y above the xz-plane as shown in Fig. 7.70a.

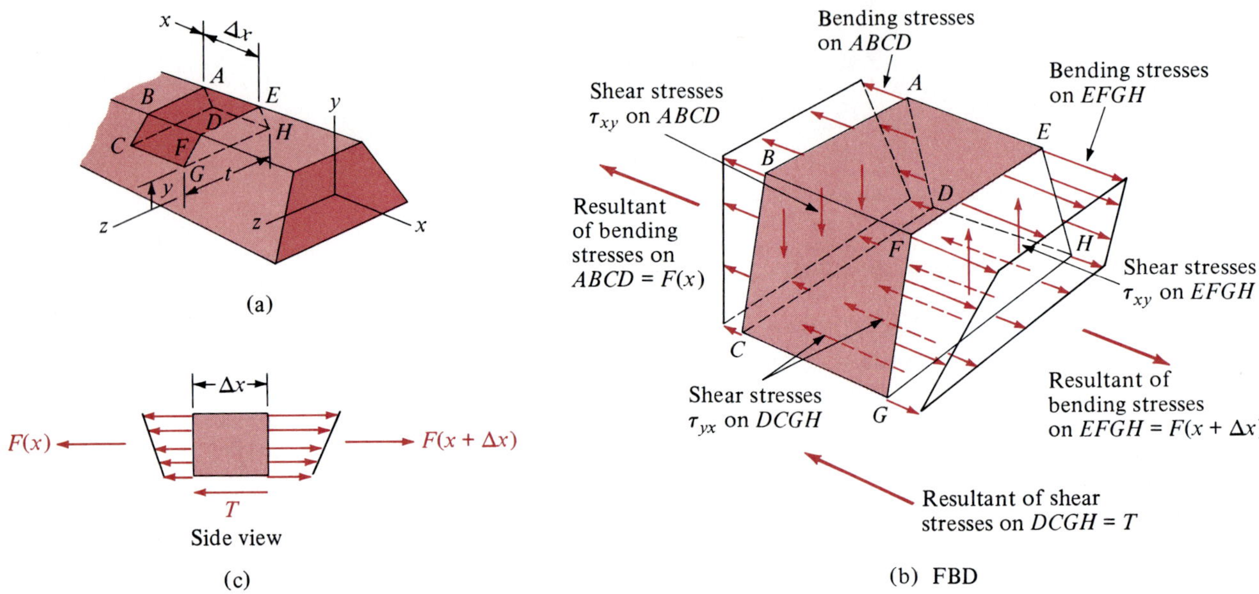

Figure 7.70 Free-body diagram for determining the transverse shear stress

The forces acting in the x-direction on the free-body diagram shown in Fig. 7.70b are $F(x)$, due to the bending stresses acting on the area $ABCD$ at location x; $F(x + \Delta x)$, due to the bending stresses acting on the corresponding area $EFGH$ at location $x + \Delta x$; and T, the force due to the shear stresses acting on the bottom face $DCGH$ of the free-body diagram. Summing forces in the x-direction as indicated in Fig. 7.70c yields

$$-F(x) - T + F(x + \Delta x) = 0$$

Recalling that the bending stresses are given generally as $\sigma_x = -My/I$, the force $F(x)$ can be represented in terms of the bending stresses on the face at location x as

$$F(x) = \int_{A_s} \sigma_x(x)\, dA = -\int_{A_s} \frac{M(x)y}{I}\, dA$$

Similarly,

$$F(x + \Delta x) = \int_{A_s} \sigma_x(x + \Delta x)\, dA = -\int_{A_s} \frac{M(x + \Delta x)y}{I}\, dA$$

where A_s is the shaded area above the section in question as shown in Fig. 7.71. The force T is represented as

$$T = (\tau_{yx})_{\text{avg}} t\, \Delta x$$

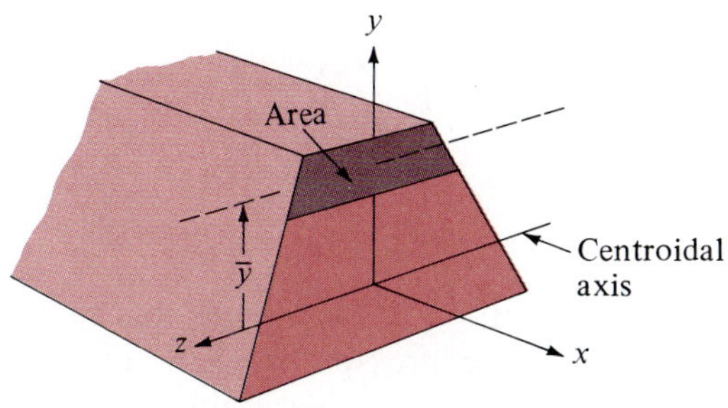

Figure 7.71 Interpretation of $Q = A\bar{y}$ or $\Sigma A_i \bar{y}_i$

where $(\tau_{yx})_{\text{avg}}$ is the *average* shear stress acting on the bottom face of the free-body diagram. Substituting these into the equilibrium equation yields

$$\int_{A_s} \frac{M(x)y}{I}\, dA - (\tau_{yx})_{\text{avg}} t\, \Delta x - \int_{A_s} \frac{M(x + \Delta x)y}{I}\, dA$$

Combining the integrals and writing $M(x + \Delta x) - M(x)$ as ΔM yields

$$(\tau_{yx})_{\text{avg}} = -\frac{1}{t\, \Delta x} \int_{A_s} \frac{\Delta M y}{I}\, dA = -\frac{1}{It} \frac{\Delta M}{\Delta x} \int_{A_s} y\, dA$$

In the limit as $\Delta x \rightarrow$ zero, $\Delta M / \Delta x \rightarrow dM/dx = V$, so that there results

$$(\tau_{yx})_{\text{avg}} = -\frac{V}{It} \int_{A_s} y\, dA = -\frac{VQ}{It} \tag{7.21}$$

where $Q = \int_{A_s} y\, dA$ is the first moment of the shaded area with respect to the centroidal axis, $y = 0$, as indicated in Fig. 7.71. It is frequently convenient to think of the first moment of the area Q as $A\bar{y}$ or as $\Sigma A_i \bar{y}_i$.

Then, since $\tau_{yx} = \tau_{xy}$, the average shear stress τ_{xy} acting on the exposed face at the location of the bottom of the free-body diagram is also given by this same expression, that is, $(\tau_{xy})_{\text{avg}} = -VQ/It$. This is shown clearly in Fig. 7.72.

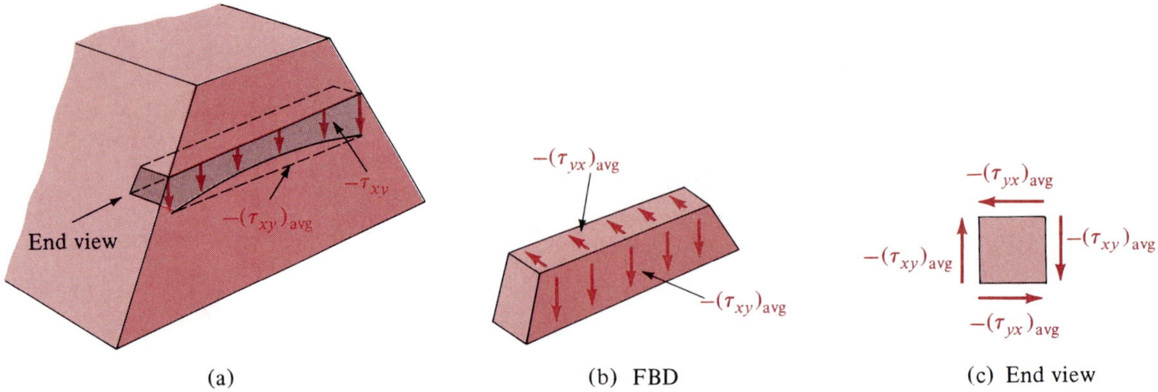

Figure 7.72 Equality of τ_{yx} on the horizontal face and τ_{xy} on the vertical face

The negative sign in Eqn. (7.21) is clearly due to the fact that the positive resultant of the shear stresses, $\int \tau_{xy}\, dA$, and the positive shear force V are oppositely directed. In what follows, we will simply write

$$\tau = \frac{VQ}{It} \tag{7.22}$$

and take the direction of the shear stress τ to be the same as the direction of V on the face in question. As indicated in Fig. 7.72a, the actual distribution of τ_{xy} across the width of the cross section is not uniform. The deviation from the average is small for solid cross sections with depth greater than width.

When applying the general expression $\tau = VQ/It$, it helps to remember its derivation, that is, to recall that once the location for computing the shear stress is selected, a free-body diagram of a portion of the beam above that section is drawn. Q represents the first moment of the area above that section with respect to the centroidal ($y = 0$) axis, with t the total thickness over which the average shear stress τ acts on that section.

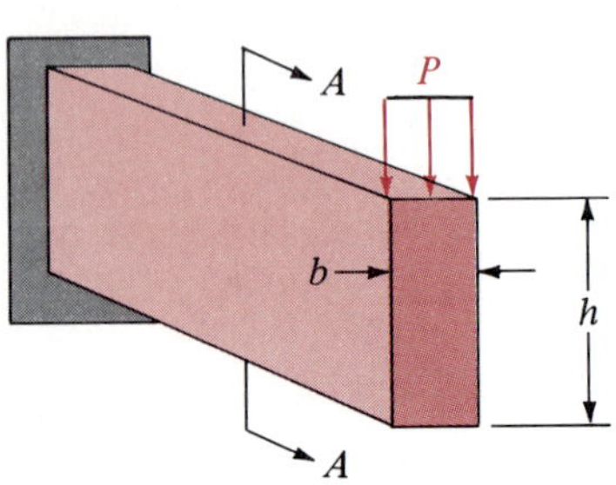

Figure 7.73 A rectangular beam transmitting a shear force P

Example 7.11 A rectangular-cross-section beam transmits a shear force $V = P$ as shown in Fig. 7.73. Determine the distribution of shear stress in the beam.

Solution At a typical section AA, the shear force V is clearly equal to the applied load P. It follows that the shear stresses will be the same at any location along the bar. The shear stresses are computed using $\tau = VQ/It$, where $t = b$, $V = P$, and $I = bh^3/12$. It is only necessary to evaluate Q. As indicated in Fig. 7.74a, Q is the first moment of the area above the location at which the shear stress is being computed.

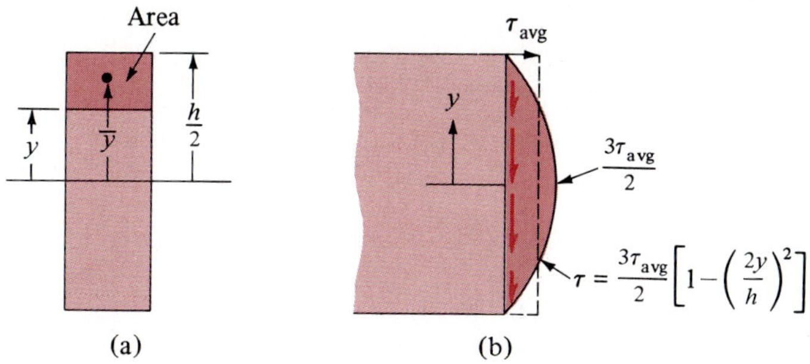

Figure 7.74 First moment of the area for a rectangular cross section

The expression for Q is easily evaluated as

$$Q = A\bar{y} = \left[b\left(\frac{h}{2} - y\right)\right]\left[\frac{(h/2 + y)}{2}\right]$$

so that the shear stress becomes

$$\tau = \frac{Pb(h^2/4 - y^2)12}{bh^3b} = \frac{3P}{2A}\left[1 - \left(\frac{2y}{h}\right)^2\right]$$

Note that the shear-stress distribution, shown in Fig. 7.74b, is parabolic with τ vanishing at the bottom ($y = -h/2$) and at the top ($y = h/2$) of the cross section as required. The maximum value occurs at $y = 0$ and has the value

$$\frac{3P}{2A} = \frac{3}{2}\frac{P}{A} = \frac{3}{2}\tau_{avg}$$

Note also that strictly speaking the load P must be applied as shear stresses distributed parabolically at the end of the beam.

Example 7.12 A solid circular pipe transmits a shear load as shown in Fig. 7.75. Determine the average shear stress across the width of the cross section at the neutral axis.

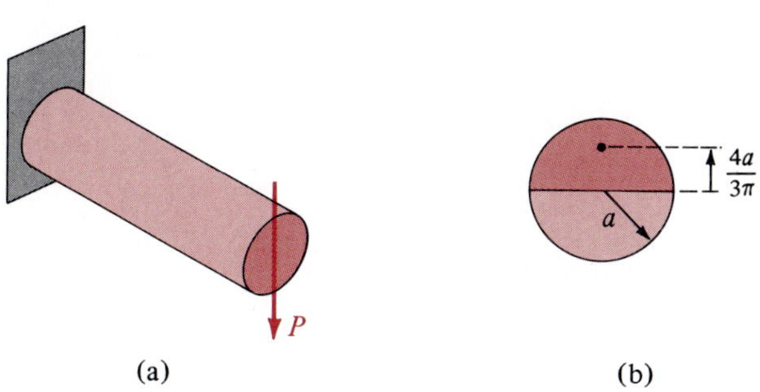

Figure 7.75 A solid circular pipe transmitting a shear load

Solution The average shear stress across the width of the cross section is given by the basic expression $\tau = VQ/It$, where Q is the first moment of the area above the neutral axis shown in Fig. 7.75b, calculated as

$$Q = \int y\, dA = A\bar{y} = \frac{\pi a^2}{2}\frac{4a}{3\pi} = \frac{2a^3}{3}$$

The average shear stress τ along the neutral axis is then given by

$$\tau = \frac{P\dfrac{2a^3}{3}}{\dfrac{\pi a^4}{4}\,2a} = \frac{4}{3}\frac{P}{\pi a^2} = \frac{4}{3}\tau_{avg}$$

where τ_{avg} is the average shear stress over the entire cross section. Both τ and τ_{avg} are shown in Fig. 7.76.

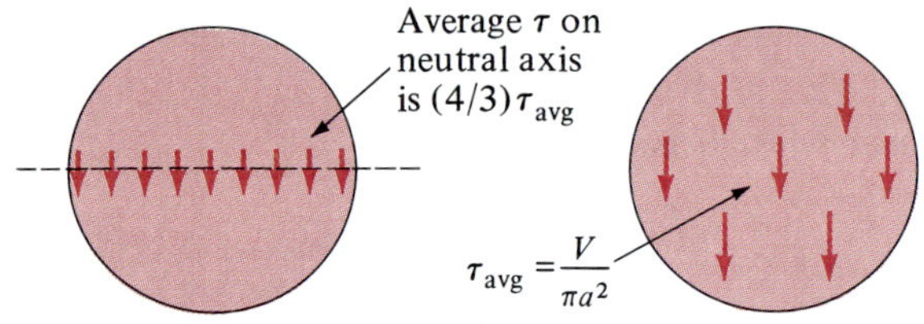

Figure 7.76 Average shear stresses

Finally, in Fig. 7.77, the sketch of the deformed configuration indicates that there is warping of a typical cross section from its original plane configuration; thus, in the presence of shear, plane sections do not remain plane, as we assumed for the development of the flexure formula, $\sigma = -My/I$.

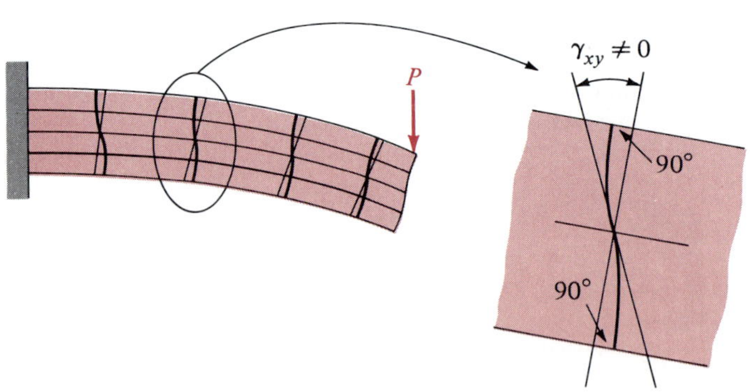

Figure 7.77 Warping due to the presence of shear

The extent of the warping in this figure is greatly exaggerated for clarity. In most situations the influence of shear on the bending stresses is negligible. Thus we will continue to compute bending stresses on the basis of the flexure formula $\sigma = -My/I$, even when the beam is simultaneously transmitting a shear V. This issue will be discussed further in Section 8.8.

Glued Members It is common to build up beams by bonding several members as indicated in Fig. 7.78. The bonding is often accomplished using some type of glue. For the theory we will use, all of the individual members are assumed to be of the same material.

The glue in the joints must be strong enough to transmit the shear stresses at those locations. An estimate of the shear stresses in the glue at each of the joints can be obtained by using the transverse shear stress equation $\tau = VQ/It$. This essentially assumes that the thickness of the glue is negligible and that the shear stress in the glue is the same as would exist at that location in a solid section with no glue-joint present. For this application, Q represents the first moment of the area above the glue joint in question. The calculations are best illustrated by an example.

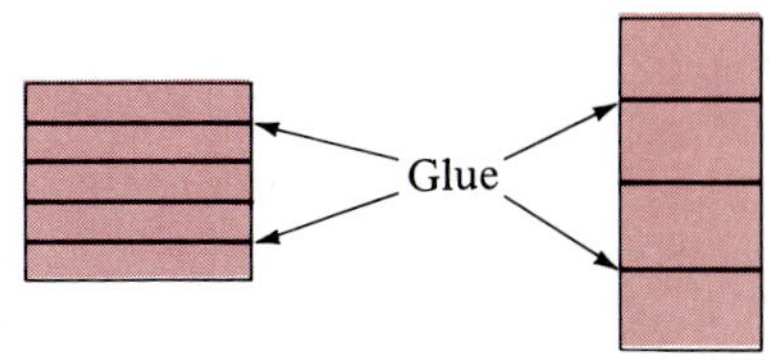

Figure 7.78 Typical glued members

Example 7.13 The simply supported beam of Fig. 7.79a is to be formed by gluing three pieces of plastic as shown in Fig. 7.79b. If the allowable shear stress that the glue can transmit is 135 psi, what is the allowable value of P?

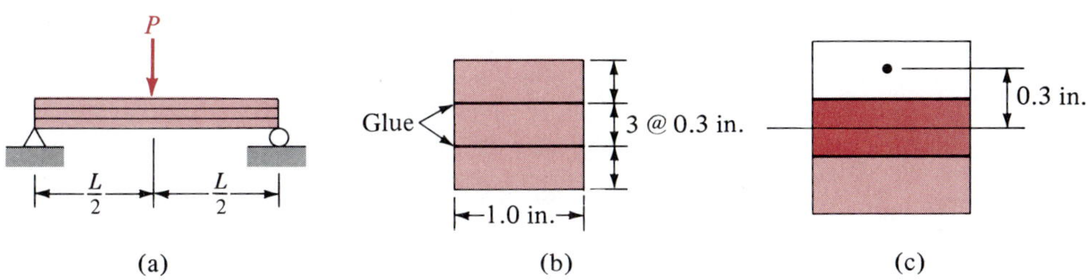

Figure 7.79 Shear in a glued plastic beam

Solution The average shear stress in the upper glue-joint is computed according to $\tau = VQ/It$, where Q is the first moment of the shaded area shown in Fig. 7.79c, or

$$Q = A\bar{y} = [(0.3 \text{ in.})(1.0 \text{ in.})](0.3 \text{ in.}) = 0.09 \text{ in}^3$$

With $I = (1.0 \text{ in.})(0.9 \text{ in.})^3/12 = 0.06075 \text{ in}^4$, $t = 1.0$ in., and $V = P/2$, the average shear stress is

$$\tau = \frac{\frac{P}{2}(0.09 \text{ in}^3)}{(0.06075 \text{ in}^4)(1.0 \text{ in.})} = 135 \text{ psi}$$

from which $P = 182.25$ lbf. For the lower glue-joint $Q = A\bar{y} = [(0.6 \text{ in.})(1.0 \text{ in.})](0.15 \text{ in.}) = 0.09 \text{ in}^3$, yielding the same result.

PROBLEMS

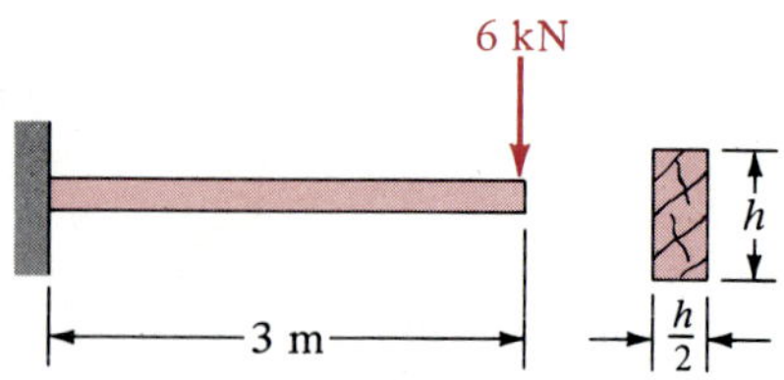

Ps. 7.190 and 7.191

7.190 A rectangular-cross-section wooden member is loaded as shown. If the allowable horizontal shear is 1 MPa, determine the minimum dimension for h.

7.191 Repeat problem 7.190 for a circular-cross-section member with radius a.

7.192 A rectangular wooden member is loaded as shown. If the allowable horizontal shear is 200 psi, determine the minimum dimension h.

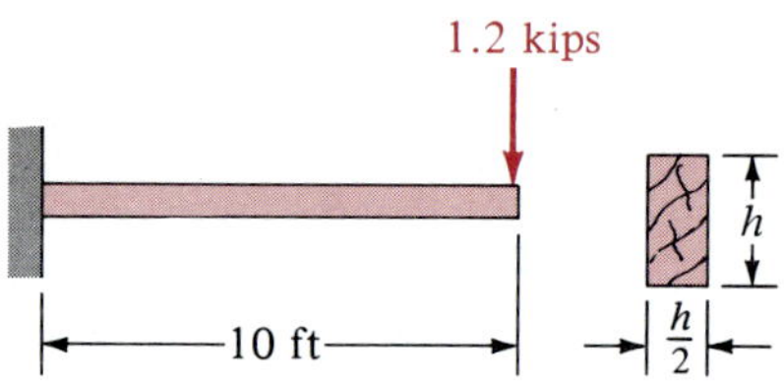

Ps. 7.192 and 7.193

7.193 Repeat problem 7.192 for a circular-cross-section member with radius a.

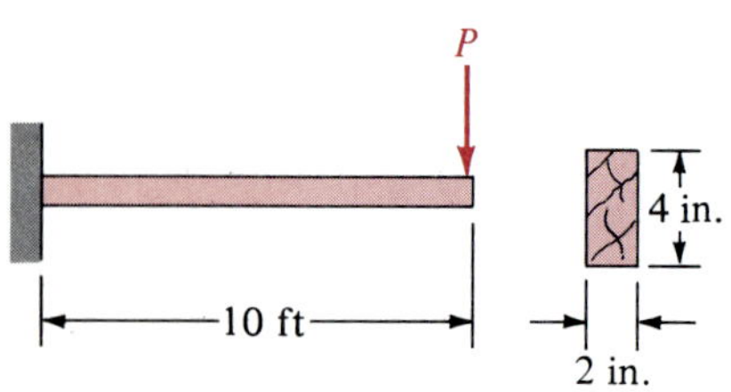

Ps. 7.194 and 7.195

7.194 Determine the largest value of P for which the maximum transverse shear stress in the beam will not exceed 300 psi.

7.195 Repeat problem 7.94 for a beam with a circular cross section of radius 1.5 in.

7.196 The expression $\tau = (3P/2bh)[1 - (2y/h)^2]$ for the transverse shear stress in a beam of rectangular cross section was developed in Example 7.11. Show by integrating the shear stress over the cross section that there results $\int \tau \, dA = P$, verifying that the shear-stress distribution is statically equivalent to the shear force P.

7.197 A rectangular-cross-section cantilever beam is loaded as shown. Develop an expression for the maximum transverse shear stress in the beam. Also compute the ratio of the maximum bending stress in the beam to the maximum transverse shear stress in the beam. What conclusions are possible regarding the relative magnitudes of bending and transverse shear stresses?

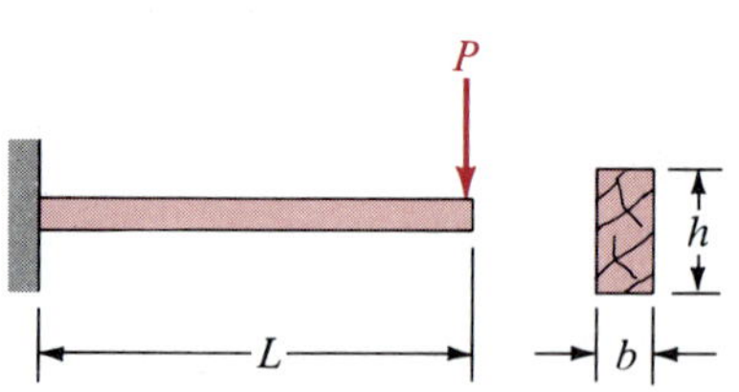

Ps. 7.197 and 7.198

7.198 Repeat problem 7.197 for a circular-cross-section beam of radius a.

7.199 A simply supported rectangular-cross-section beam is loaded uniformly as shown. Develop an expression for the maximum transverse shear stress in the beam. Also compute the ratio of the maximum bending stress in the beam to the maximum transverse shear stress in the beam. What conclusions are possible regarding the relative magnitudes of bending and transverse shear stresses?

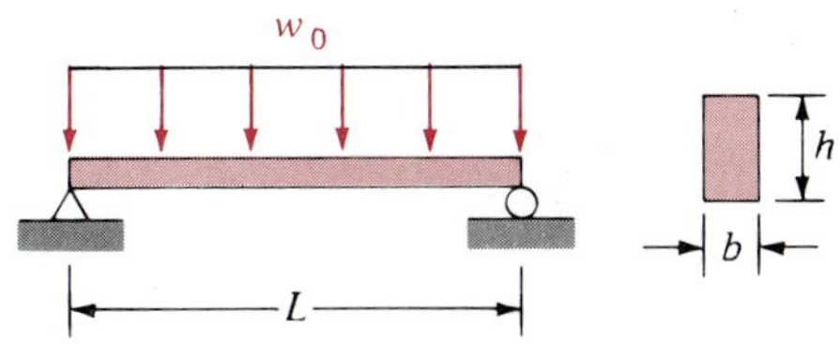

Ps. 7.199 and 7.200

7.200 Repeat problem 7.199 for a circular-cross-section beam of radius a.

7.201 A simply supported rectangular-cross-section beam is loaded uniformly as shown. If the maximum allowable transverse shear stress is 4 ksi and the maximum allowable bending stress is 16 ksi, determine the maximum length of the beam. Take $w_0 = 5$ kips/ft, $b = 2$ in., and $h = 4$ in.

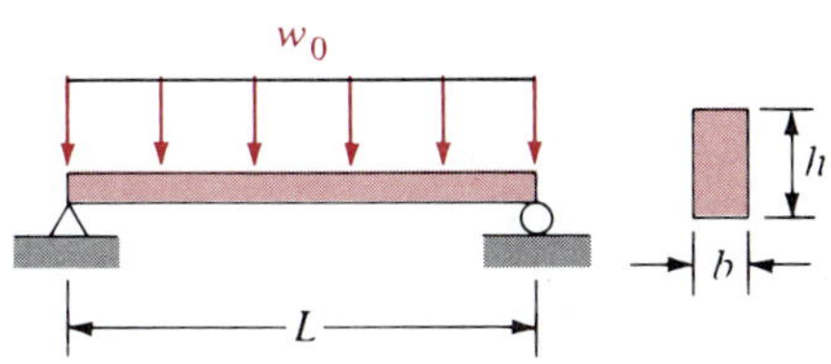

Ps. 7.201–7.204

7.202 Repeat problem 7.201 if the maximum allowable transverse shear stress is 30 MPa and the maximum allowable bending stress is 120 MPa. Take $w_0 =$ 70 kN/m, $b = 50$ mm, and $h = 100$ mm.

7.203 Repeat problem 7.201 for a circular-cross-section beam with $w_0 =$ 7 kips/ft and a radius of 3 in.

7.204 Repeat problem 7.202 for a circular-cross-section beam with $w_0 =$ 80 kN/m and a radius of 70 mm.

7.205–7.208 For each of the cross sections shown, determine the location and magnitude of the maximum average transverse shear stress.

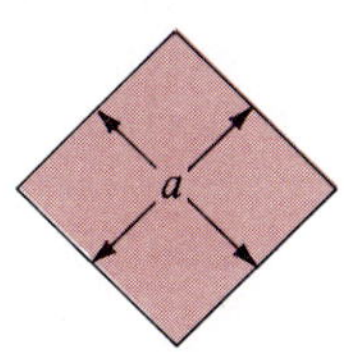

P. 7.205

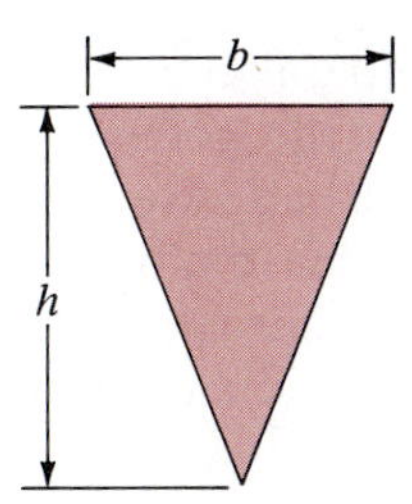

P. 7.206

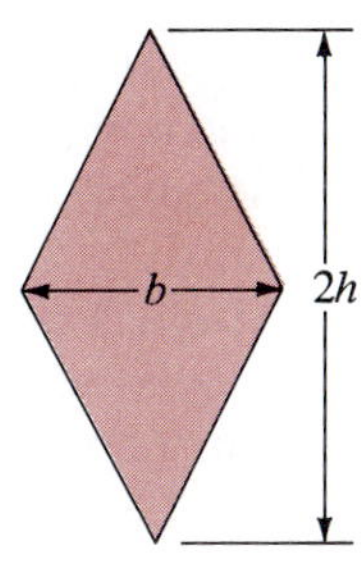

P. 7.207

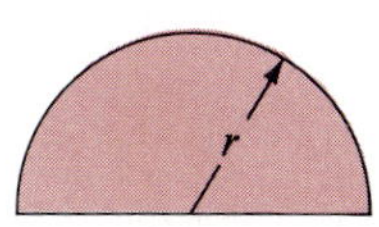

P. 7.208

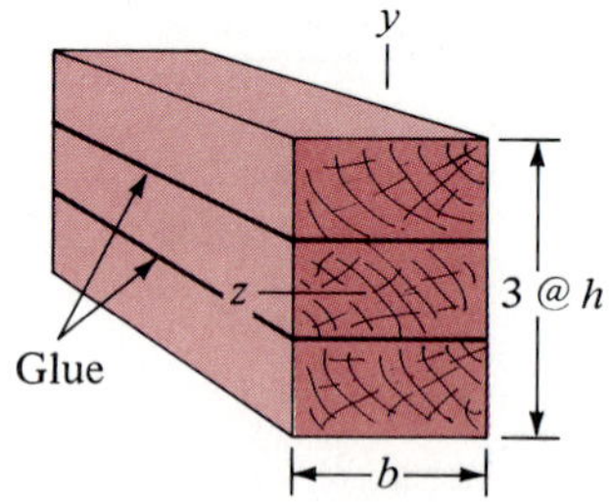

Ps. 7.209 and 7.210

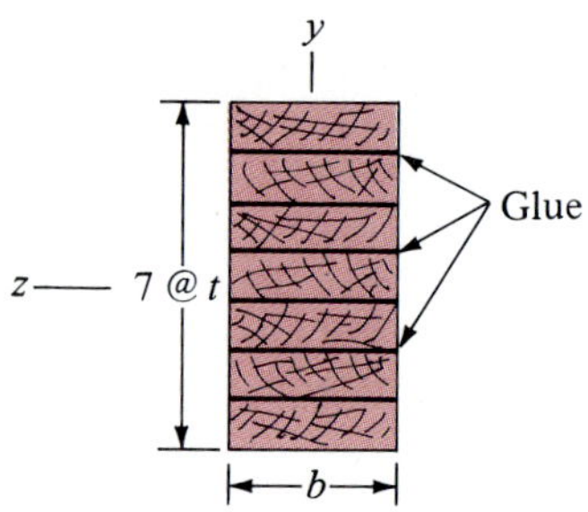

Ps. 7.211 and 7.212

7.209 A beam is to be constructed from wooden members by gluing as shown. If the glue is capable of withstanding a shear stress of 150 psi, what is the maximum shear force the beam can transmit? What is the corresponding maximum transverse shear stress in the wood? Take $b = 3.5$ in. and $h = 1.5$ in.

7.210 Repeat problem 7.209 if the allowable shear stress in the glue is 1.0 MPa. Take $b = 85$ mm and $h = 35$ mm.

7.211 A laminated beam is built up from wooden members as indicated. If the allowable shear stress in the glue is 170 psi, what is the maximum shear force V that can be transmitted? What is the corresponding maximum transverse shear stress in the wood? Take $b = 10$ in. and $t = 3.0$ in.

7.212 Repeat problem 7.211 if the allowable shear stress in the glue is 1.1 MPa. Take $b = 240$ mm and $t = 75$ mm.

7.8.1 Shear Flow in Thin-Walled Symmetrical Sections

The shapes of many commonly used beams are designed to be more efficient in bending and in shear than solid sections such as rectangles and circles. Several typical sections are shown in Fig. 7.80.

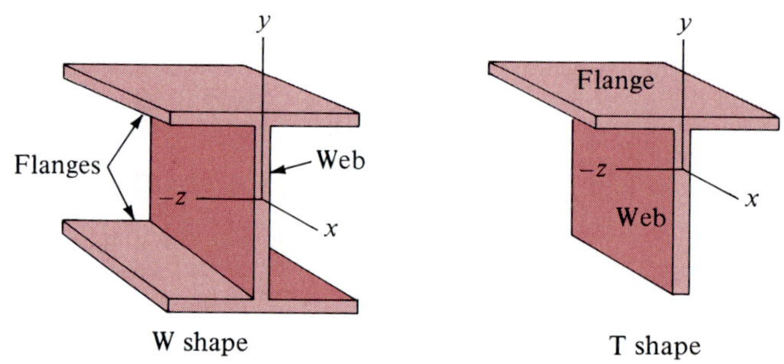

Figure 7.80 Typical thin-walled symmetrical sections

As indicated in the figure, the portions of the cross sections away from and generally parallel to the neutral axis are frequently referred to as flanges, whereas the portions parallel to the y-axis are called the webs. As we will see in this section, the webs carry essentially all of the vertical shear V, while the flanges transmit horizontal shear stresses associated with V. Other commonly used sections include built-up sections consisting of wood members that are nailed or glued as shown in Fig. 7.81.

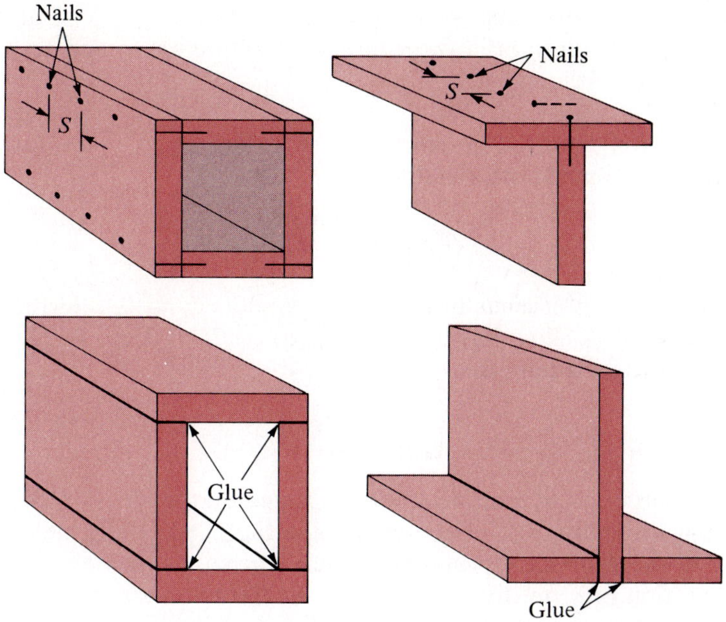

Figure 7.81 Typical glued or nailed symmetrical sections

For all of the sections shown in Figs. 7.80 and 7.81, it is convenient to develop expressions for what is called shear flow, for discussing the manner in which the shear stresses are distributed over the face of the cross section while transmitting a shear force V. The basic ideas can be illustrated by considering the T section of Fig. 7.82.

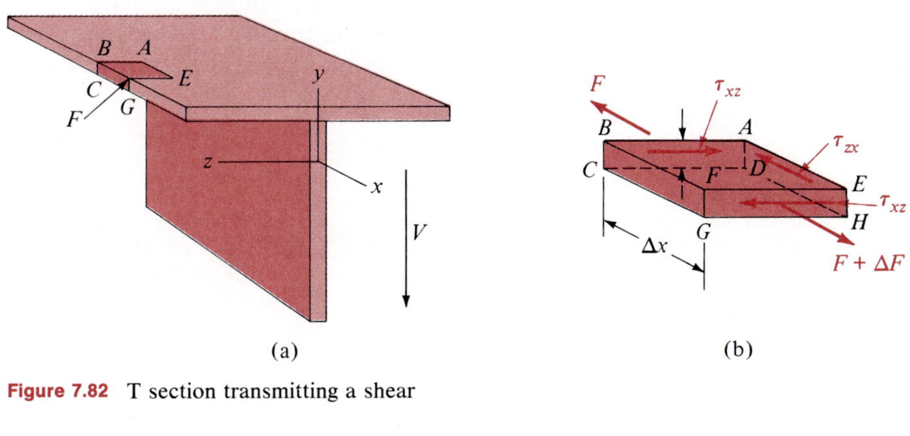

Figure 7.82 T section transmitting a shear

A free-body diagram of a piece of the flange is constructed in Fig. 7.82b. Equilibrium in the x-direction can be expressed as

$$\Sigma F_x = -F + F + \Delta F - \tau_{zx} t \, \Delta x = 0$$

With $F = \int \sigma_x(x)\, dA$, $F + \Delta F = \int \sigma_x(x + \Delta x)\, dA$, $\sigma_x = -My/I$, and $dM/dx = V$, this can be simplified to

$$\tau_{zx} t = -\frac{VQ}{I} \tag{7.23}$$

The quantity $\tau_{zx} t$ is usually given the symbol q and is referred to as **shear flow.** Shear flow has dimensions $[q] = FL^{-1}$. Eqn. (7.23) is frequently written simply as

$$q = \frac{VQ}{I} \tag{7.24}$$

with the direction of q ultimately determined on the basis of the corresponding direction of V. Also indicated in Fig. 7.82b is the idea that by virtue of the symmetry property of the shear stress, namely $\tau_{zx} = \tau_{xz}$, the direction of q on the exposed face $EFGH$ is horizontal.

At a particular location x along the beam, V and I are constant, so that from Eqn. (7.24), $q = \tau t$ depends upon Q. As shown in Fig. 7.83a, $Q = A\bar{y}$, with $A = ts$. As $s \to 0$, $Q \to 0$ and $q \to 0$. The centroidal distance $\bar{y}$ is the same for any s along the flange with the result that, for any value of s up to the web.

$$q = \frac{Vts\bar{y}}{I}$$

thus, q is a linear function of s as s ranges from the outside edge of the flange to the midpoint of the flanges. Exactly the same conclusion is reached if the flange on the right side of the web is considered. The resulting shear flow in the flanges is shown in Fig. 7.83b.

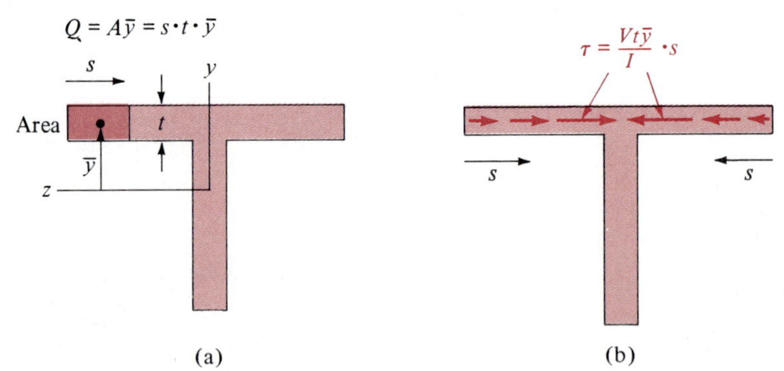

Figure 7.83 Shear flow in a T section

For a typical location in the web, the free-body diagram shown in Fig. 7.84 is drawn, exposing the horizontal shear stresses τ_{yx} in the web. Equilibrium in the x-direction again yields

$$\tau_{yx} t = -\frac{VQ}{I}$$

where Q is the first moment of the shaded area, that is, the area above the position y at which the shear stress is being computed. The shear flow q is again taken to be $\tau = VQ/I$, with the direction determined from the direction of V. The linear dependence of the shear flow on the position in the flanges and the quadratic dependence of the shear flow on the position in the web are indicated in Figs. 7.85a and b.

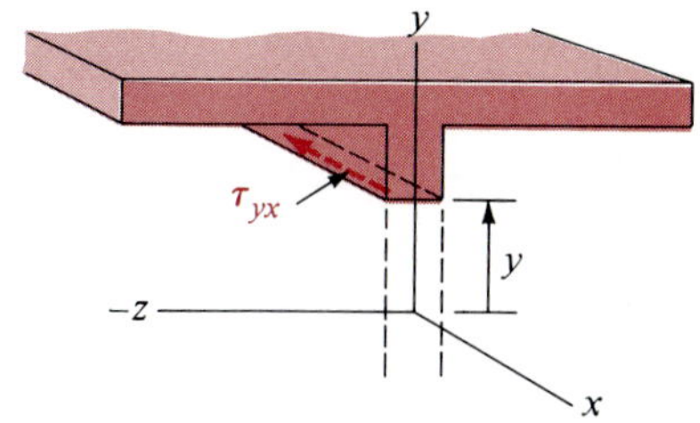

Figure 7.84 Horizontal shear stresses in the web of a T section

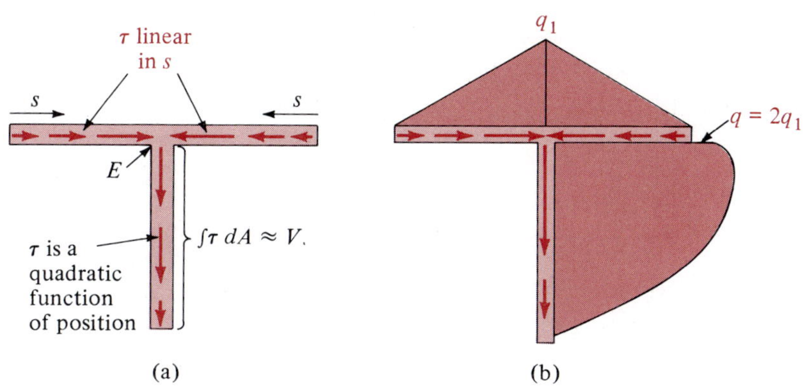

Figure 7.85 Shear flow in a T section

Note that at the bottom of the web, $Q = \int y \, dA$ is evaluated over the entire cross section of the T section, with the result that $Q = 0$, so that $\tau_{yx} = \tau_{xy} = 0$, as we know to be the case on the bottom, stress-free surface. Note also that the sum of the shear flows in the flanges at the point directly above the centerline of the web is equal to the shear flow in the web at point E.

As indicated in Fig. 7.85a, the resultant of the shear stresses acting on the web is approximately equivalent to the shear force V. Computing the shear flows in the flanges up to the midpoint misses the fact that there is actually a small contribution to V from the stresses in the flanges. The actual shear stress distribution at the junction of the flanges and web would appear as in Fig. 7.86, showing the nature of the small contribution to the shear force V from the shear stresses in the flanges.

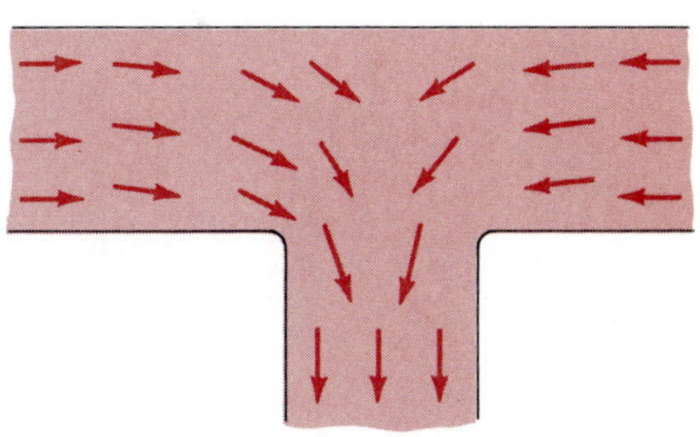

Figure 7.86 Distribution of shear stresses near the junction of the flanges and web in a T section

The approximate shear flows and corresponding shear stresses computed using $\tau t = q = VQ/I$ are quite accurate when the thickness t is small compared to the overall depth and width of the T section. The distribution of the shear flow in two other common sections is indicated in Fig. 7.87.

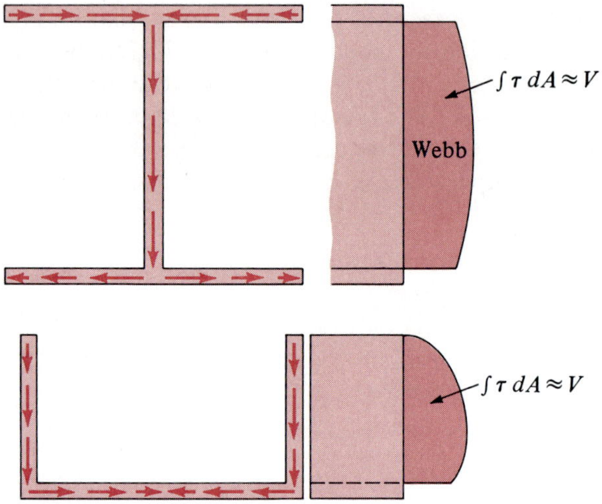

Figure 7.87 Shear flow for W and C sections

Example 7.14 The W 8 × 31 section shown in Fig. 7.88 transmits a shear force of 30 kips. Determine the shear-stress distribution and the maximum transverse shear stress.

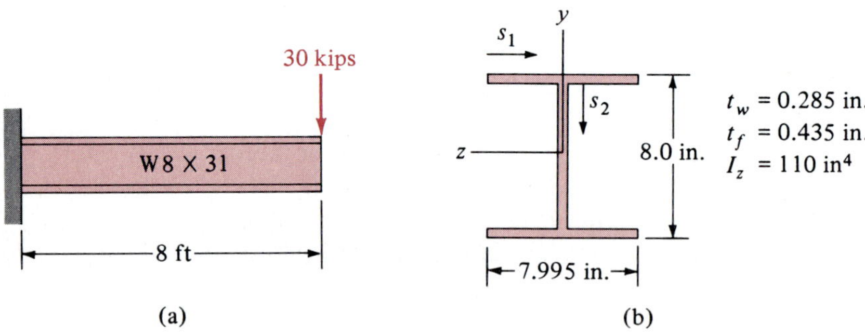

Figure 7.88 W section in shear

Solution From Appendix C, the dimensions of the W 8 × 31 section are shown in Fig. 7.88b. As a measure of the shear stress in the web, we compute $\tau_{avg} = 30{,}000$ lbf/[(8.0 − 0.435) in.](0.285 in.) = 13.91 ksi. We expect that the maximum, occurring at the neutral axis, will be within a few percentage points of this average value.

Along the top flange, as shown in Fig. 7.89a, the first moment Q is given in terms of the coordinate s by

$$Q = t_f s\bar{y} = (0.435\text{ in.})s\left(4.0 - \frac{0.435}{2}\right)\text{ in.}$$

and is maximum as s approaches 3.997 in., at the midpoint of the flanges. The shear flow at that location is

$$q = \frac{(30{,}000\text{ lbf})(0.435\text{ in.})(3.997\text{ in.})(3.783\text{ in.})}{110\text{ in}^4} = 1794\text{ lbf/in.}$$

The shear flow just below the flange in the web is given by

$$q = \frac{(30{,}000\text{ lbf})(0.435\text{ in.})(7.995\text{ in.})(3.783\text{ in.})}{110\text{ in}^4} = 3588\text{ lbf/in.}$$

Note that the total of the two shear flows in the flange at the point where they intersect the web is 2(1794 lbf/in.) = 3588 lbf/in. which, as mentioned, is the same as the 3588 lbf/in. value in the web just below the flange.

At a location in the web between the bottom of the flange and the neutral axis, as indicated by the coordinate s shown in Fig. 7.89b, the first moment Q is given by

$$Q = Q_{\text{flange}} + Q_{\text{web}}(s)$$

$$Q = \left[(0.435)(7.995)(3.783) + (0.285s)\left(3.565 - \frac{s}{2}\right)\right]\text{ in}^3$$

and is seen to vary parabolically with s, measured in inches. The maximum occurs at $s = 3.565$ with the value $Q = 14.97\text{ in}^3$, and the corresponding shear flow as

$$q = \frac{(30{,}000\text{ lbf})(14.97\text{ in}^3)}{110\text{ in}^4} = 4082\text{ lbf/in.}$$

The corresponding shear stress at the neutral axis is $\tau = q/t = (4082\text{ lbf/in.})/(0.285\text{ in.}) = 14.33$ ksi, approximately 3% higher than the average value computed above. The shear-stress distribution is pictured in Fig. 7.90.

The total force arising from the shear stresses in the web is calculated to be

$$F_{\text{web}} = \left[\left(12.59 + \left(\frac{2}{3}\right)\left(14.33 - 12.59\right)\right)\text{ ksi}\right](7.13\text{ in.})(0.285\text{ in.}) = 28.6\text{ kips}$$

an indication of the amount carried by the flanges, as compared to the total shear of 30.0 kips.

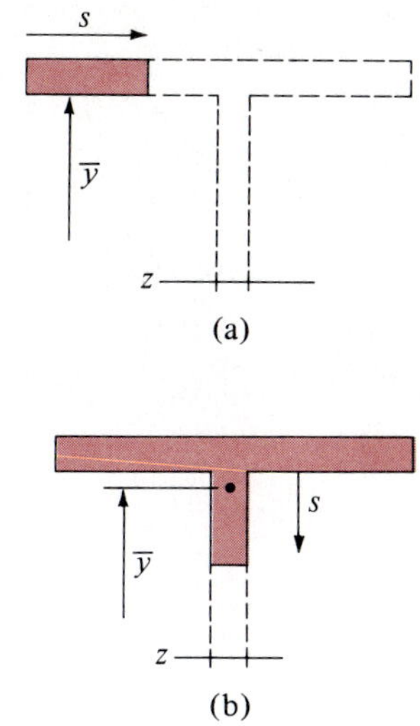

Figure 7.89 First moments for flange and web

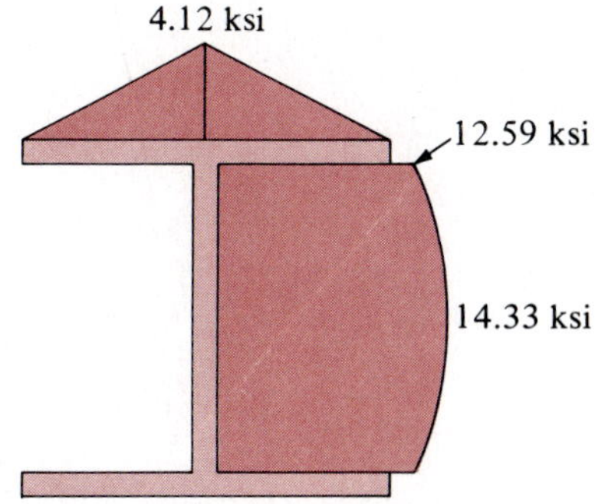

Figure 7.90 Shear-stress distribution in a W section

Example 7.15 A hollow circular pipe transmits a shear load as shown in Fig. 7.91. Determine the shear flow and the average shear stress at the neutral axis.

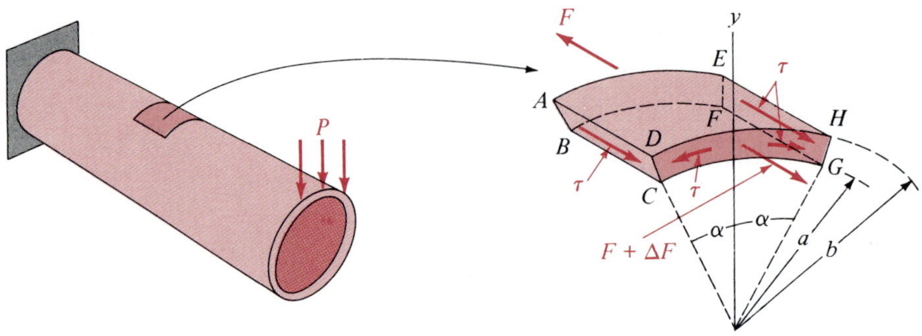

(a) (b)

Figure 7.91 A hollow circular pipe transmitting a shear load

Solution To determine the shear flow, we consider the free-body diagram drawn in Fig. 7.91b. The shear flow is given by the basic expression $q = VQ/I$, where Q is the first moment of the area above the neutral axis shown in the figure, and is calculated to be

$$Q = \int y\, dA = \int_{-\alpha}^{\alpha} \int_{a}^{b} r \cos\theta\, r\, dr\, d\theta = \frac{2 \sin\alpha (b^3 - a^3)}{3}$$

Recall from the derivation that t represents the total thickness over which the average shear stress acts, so that in this case $t = 2(b - a)$. The shear stress is then given by

$$\tau = \frac{P[2 \sin\alpha (b^3 - a^3)/3]}{[\pi/4(b^4 - a^4)][2(b - a)]} = \frac{4P \sin\alpha}{3A} \frac{a^2 + ab + b^2}{a^2 + b}$$

and is distributed over the cross section as shown in Fig. 7.92.

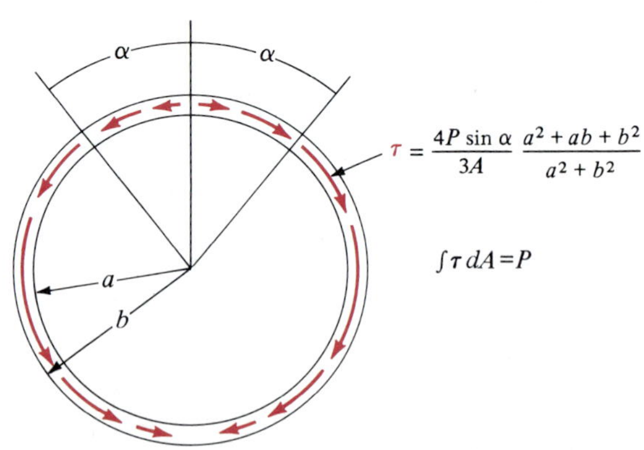

Figure 7.92 Distribution of transverse shear stresses in a circular pipe

The maximum value occurs at $\alpha = \pi/2$, that is, at the neutral axis, and has the value

$$\tau_{\max} = \frac{4P}{3A}\frac{a^2 + ab + b^2}{a^2 + b^2}$$

For a solid section, $a = 0$, and the maximum average shear stress is given by

$$\tau_{\max} = \frac{4P}{3A} = \frac{4}{3}\tau_{\text{avg}}$$

coinciding with the result of Example 7.12. For a thin-walled section such as a pipe, it is left in the problems to show that $\tau_{\max} \approx 2\tau_{\text{avg}}$.

Glued Members Consider the typical glued members shown in Fig. 7.93, showing two different construction modes.

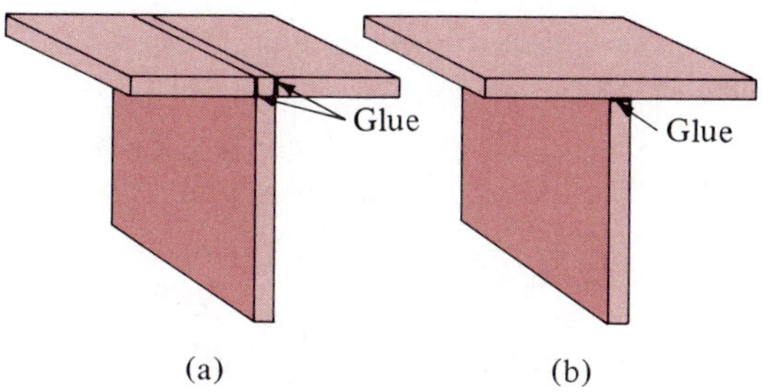

Figure 7.93 Glued T sections

For the member shown in Fig. 7.93a, the shear stress in the glue is determined by evaluating $\tau = VQ/It$, where Q is the first moment of the area of the shaded portion shown in Fig. 7.94a. Similarly, for the member of Fig. 7.93b, the shear stress is determined on the basis of the Q of the shaded area shown in Fig. 7.94b.

Assuming the thicknesses of the flange and the web to be equal, the shear stress in the glue-joints of the member in Fig. 7.94a would be roughly half that of the glue-joint of the member in Fig. 7.94b. In either case, the strength of the glue would have to be adequate to resist the shear stresses transmitted across the joint, at the location in question. The assumption we make is that the thickness of the glue is negligible, with the stress in the glue the same as would exist at that location in a solid section of the same shape and dimensions.

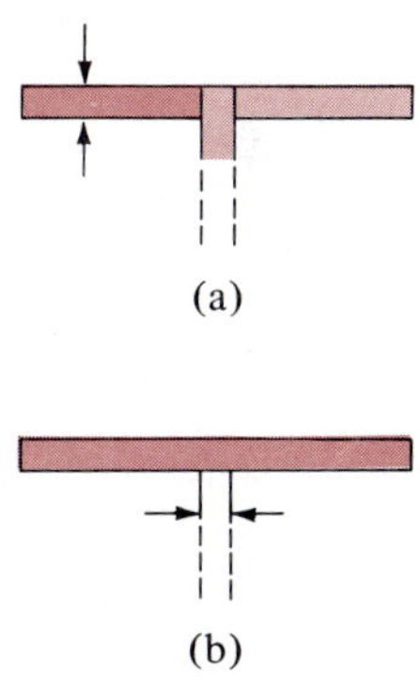

Figure 7.94 Q's for shear stresses in glued members

Example 7.16 A box section is constructed by gluing as shown in Fig. 7.95a. If the ultimate shear stress the glue is capable of transmitting is 1 MPa, determine the allowable shear force that can be transmitted for a factor of safety of 2. Also determine the corresponding maximum shear stress in the wood.

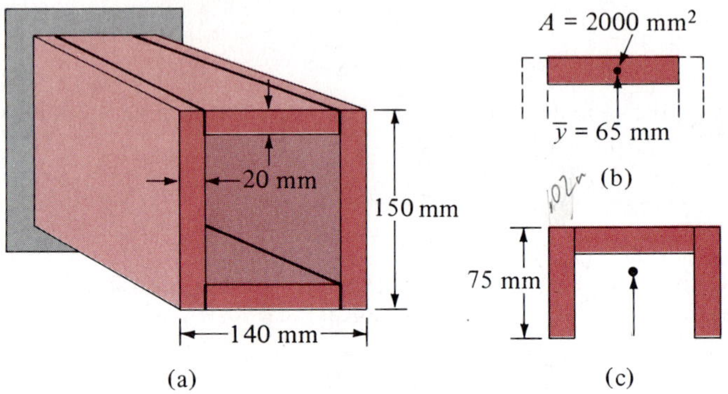

Figure 7.95 Glued box section transmitting a shear load

Solution The allowable shear stress in the glue is given by $\tau_{\text{all}} = 1\ \text{MPa/FS} = 1\ \text{MPa}/2 = 0.5\ \text{MPa}$. The second moment of the area is calculated as

$$I = \frac{[(0.14)(0.15)^3 - (0.1)(0.11)^3]}{12}\ \text{m}^4 = 2.83 \times 10^{-5}\ \text{m}^4$$

and the first moment Q for the shaded area indicated in Fig. 7.95b is given by $Q = (0.020)(0.10)(0.065)\ \text{m}^3 = 1.3 \times 10^{-4}\ \text{m}^3$. The total thickness over which the shear stress acts in the glue is 2(0.02 m) so that the basic equation $\tau = VQ/It$ becomes

$$0.5\ \text{MPa} = \frac{V(1.3 \times 10^{-4}\ \text{m}^3)}{(2.83 \times 10^{-5}\ \text{m}^4)(0.04\ \text{m})}$$

Solving for V yields $V = 4.35$ kN. Referring to Fig. 7.95c, the first moment for the shear stress at the neutral axis is

$$Q = 1.3 \times 10^{-4}\ \text{m}^3 + 2\,\frac{(0.075\ \text{m})(0.02\ \text{m})(0.075\ \text{m})}{2}\ \text{m}^3 = 2.425 \times 10^{-4}\ \text{m}^3$$

so that the maximum shear stress at the neutral axis is calculated to be

$$\tau = \frac{(4.35\ \text{kN})(2.425 \times 10^{-4}\ \text{m}^3)}{(2.83 \times 10^{-5}\ \text{m}^4)[2(0.02\ \text{m})]} = 0.93\ \text{MPa}$$

Bolted and Nailed Members Consider the typical nailed members shown in Fig. 7.96.

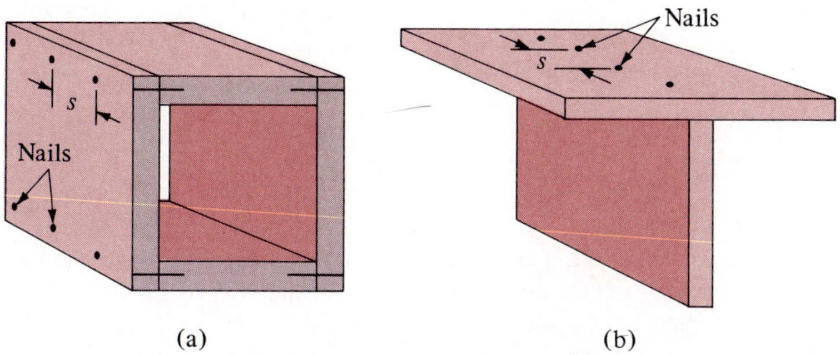

Figure 7.96 Box and T sections constructed by nailing

Such members are formed by connecting wooden pieces with nails that are usually spaced at regular intervals as shown. For this type of member, the shear connection between the individual pieces of wood is supplied by the nails. Each nail is assumed to be capable of resisting a given shear force F. For the box section of Fig. 7.96a, the shear flow, assumed to act equally between the two side members and the top member, is determined from $q = VQ/I$, where Q is the first moment of the area of the top piece shown in Fig. 7.97.

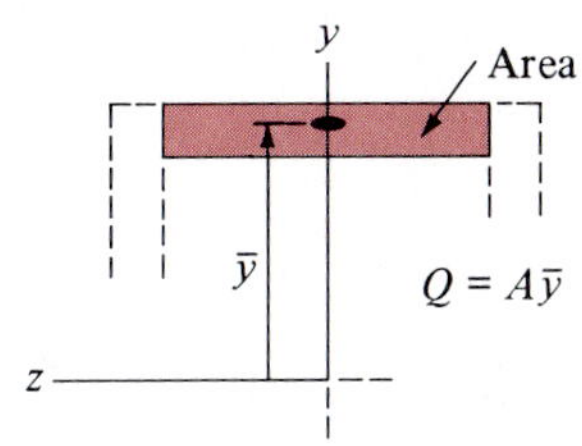

Figure 7.97 First moment for a nailed section

Half of this shear flow must be supplied by the nails on each side according to

$$\frac{q}{2} = \frac{F}{s} \qquad \text{or} \qquad q = \frac{2F}{s}$$

where F is the force in the nail, and s is the spacing of the nails. With both q (= VQ/I) and F known, the required spacing s can be determined as $s = 2F/q$.

Example 7.17 A box section is built up from two-by-six and two-by-ten members as shown in Fig. 7.98a. If the shear force to be transmitted by the section is 2000 lbf, determine the required nail spacing if the nails can withstand 150 lbf in shear. Also determine the maximum shear stress in the wood.

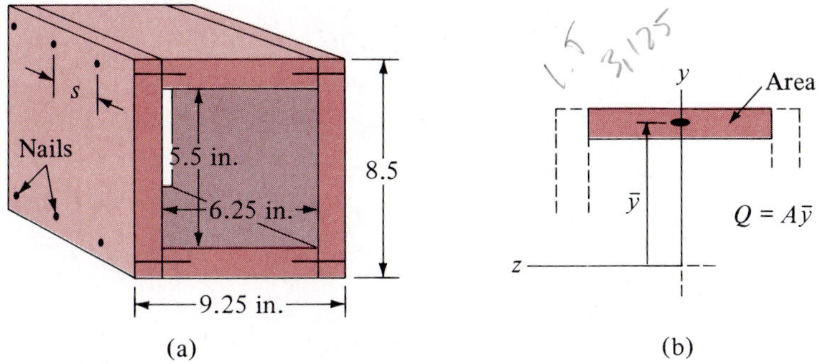

Figure 7.98 Box section of two-by-sixes and two-by-tens

Solution The second moment I for the box section is computed as $I = [(9.25(8.5)^3 - 6.25(5.5)^3)/12]\ \text{in}^4 = 387.6\ \text{in}^4$. The shear flow at the locations of the nails is determined on the basis of the first moment Q shown in Fig. 7.98b. Q has the value

$$Q = A\bar{y} = (6.25\ \text{in.})(1.5\ \text{in.})(4.25 - 0.75)\ \text{in.} = 32.81\ \text{in}^3$$

The shear flow is then computed to be

$$q = \frac{(2000\ \text{lbf})(32.81\ \text{in}^3)}{386.7\ \text{in}^4} = 169.7\ \text{lbf/in.}$$

indicating that a shear flow of $q = (169.7\ \text{lbf/in.})/2 = 84.8\ \text{lbf/in.}$ must be transmitted between the two-by-ten and each of the two-by-sixes by the nails. The spacing is then determined by

$$84.8\ \text{lbf/in.} = \frac{150\ \text{lbf}}{s}$$

from which $s = 1.77$ in. A spacing of 1.75 in. would likely be used. The maximum transverse shear stress in the wood will occur at the neutral axis. At that location, the first moment of the area is given by

$$Q = \left[(6.25)(1.5)(3.5) + 2(4.25)(1.5)\left(\frac{4.25}{2}\right)\right]\ \text{in}^3 = 59.91\ \text{in}^3$$

The total thickness over which the average shear stress acts is $t = 2(1.5\ \text{in.}) = 3.0$ in., so that

$$\tau = \frac{(2000\ \text{lbf})(59.91\ \text{in}^3)}{(386.7\ \text{in}^4)(3.0\ \text{in.})} = 103.3\ \text{psi}$$

Consider also typical bolted members as shown in Fig. 7.99.

In these situations the bolts supply the shear connection that permits the individual sections to act as a unit in bending. To determine the required size and spacing of the bolts, one must determine the shear flow at the interface between the two portions being connected. For the section of Fig. 7.99b, for instance, the shear flow is computed between the shaded portion of Fig. 7.100 and the rest of the section.

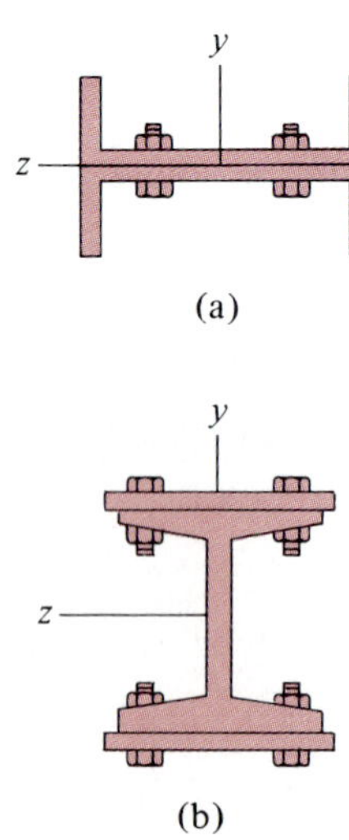

Figure 7.99 Typical bolted members

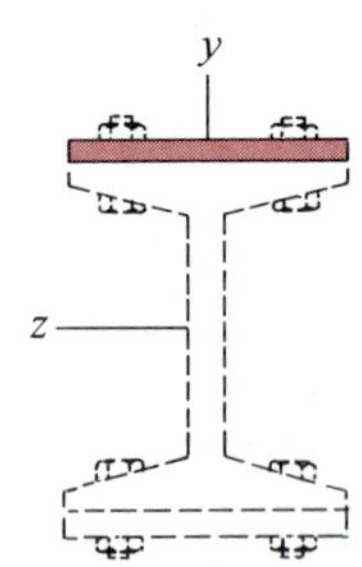

Figure 7.100 Shear flow calculations for shear in bolts

The spacing for the *two* rows of bolts is again determined by requiring that $q/2 = F/s$ from which, with F and q known,

$$s = \frac{2F}{q}$$

Other situations are handled in a similar fashion.

Example 7.18 Plates are bolted to the box section as shown in Fig. 7.101. If the two rows of $\frac{5}{8}$-in.-diameter bolts are spaced on 6-in. intervals, what is the largest shear force that can be transmitted? Assume the bolts are made of a steel with an allowable shear stress of 18 ksi. What is the corresponding maximum shear stress in the webs?

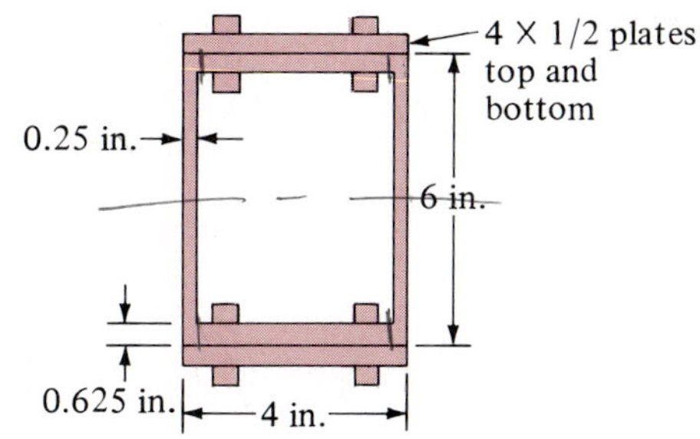

Figure 7.101 Bolted box section

Solution The section properties necessary for the calculations are

$$I = \left[\frac{4(6^3) - 3.5(4.75^3)}{12} + 2(4)(0.5)(3.25)^2\right]\text{in}^4 = (40.74 + 42.25)\ \text{in}^4$$

$$= 82.97\ \text{in}^4$$

for the entire section and

$$Q = A\bar{y} = [(4)(0.5)](3.25)\ \text{in}^3 = 6.5\ \text{in}^3$$

for the plate that is to be connected by the bolts to the box. The corresponding shear flow per row of bolts is

$$\frac{q}{2} = V\ 6.5\ \text{in}^3/(2 \cdot 82.97\ \text{in}^4) = 0.0392V/\text{in.}$$

The allowable force in a single bolt is $F = \tau A = (18\ \text{ksi})(\pi(0.625)^2/4) = 5520$ lbf, so that on 6-in. spacing, each row of bolts can supply a shear flow of 5520 lbf/6 in. = 920 lbf/in. Equating the two values yields

$$0.0392V/\text{in.} = 920\ \text{lbf/in.}$$

from which $V = 23.47$ kips. With Q given by

$$Q = \{6.5 + (3.5)(0.625)(2.6875) + 2[3(0.25)(1.5)]\}\ \text{in}^3 = 14.63\ \text{in}^3$$

for the first moment of the portion of the section above the neutral axis, the corresponding maximum shear stress is

$$\tau = \frac{(23.47\ \text{kips})(14.73\ \text{in}^3)}{(82.97\ \text{in}^4)(2 \cdot 0.25\ \text{in.})} = 8.28\ \text{ksi}$$

PROBLEMS

Thin-Walled Sections

7.213–7.217 For each of the following, state the reason(s) that the elementary equation for the shear flow $\tau = VQ/I$ cannot be used. Assume that the resultant shear V transmitted has the direction and line of action shown.

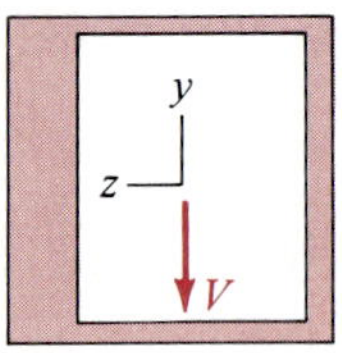

P. 7.213

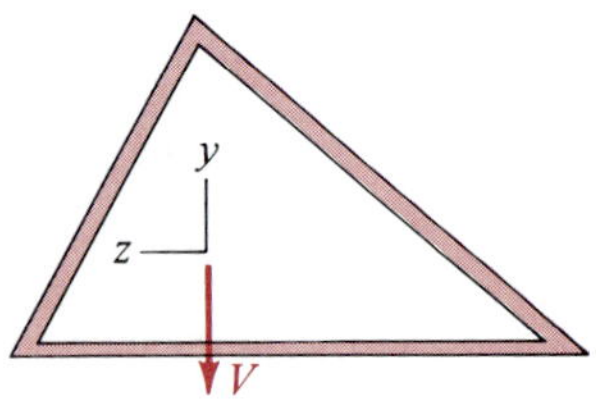

P. 7.214

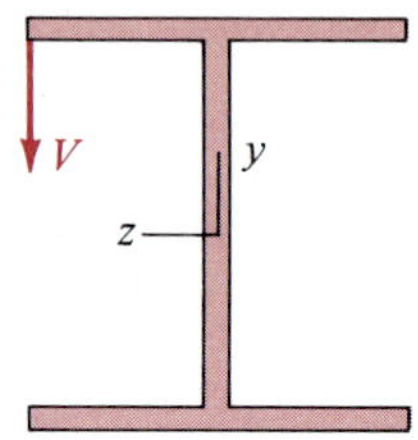

P. 7.215

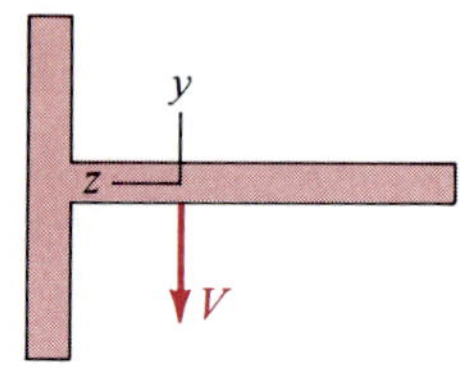

P. 7.216

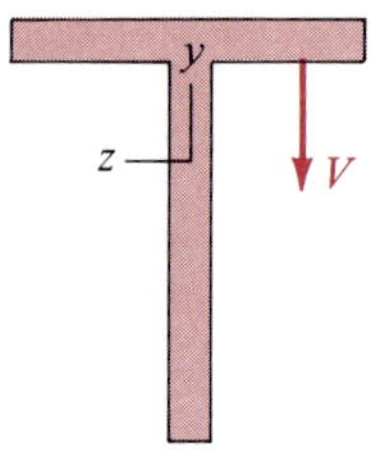

P. 7.217

7.218–7.222 For each of the following constant-thickness sections, determine the maximum transverse shear stress τ_{max} and hence the constant k in representing $\tau_{max} = kV/A$. Take $t/b \ll 1$ in all cases.

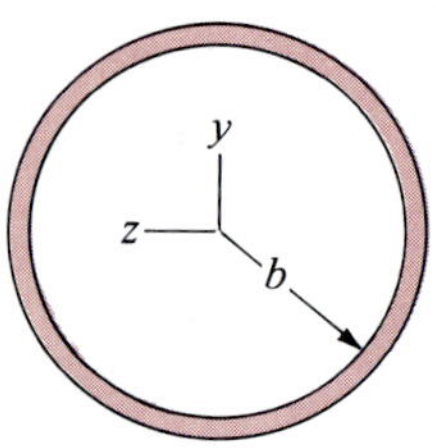

P. 7.218

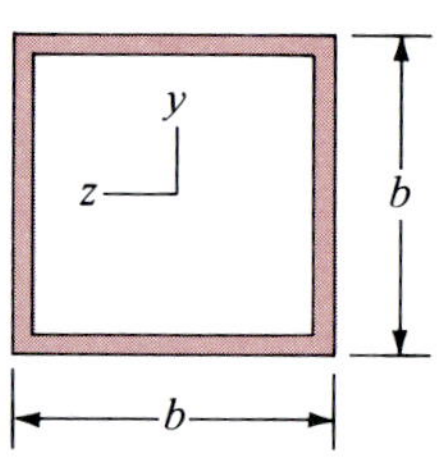

P. 7.219

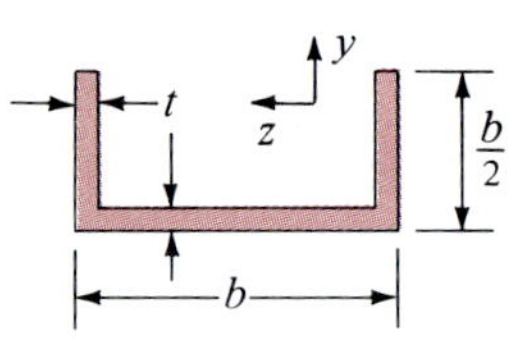

P. 7.220

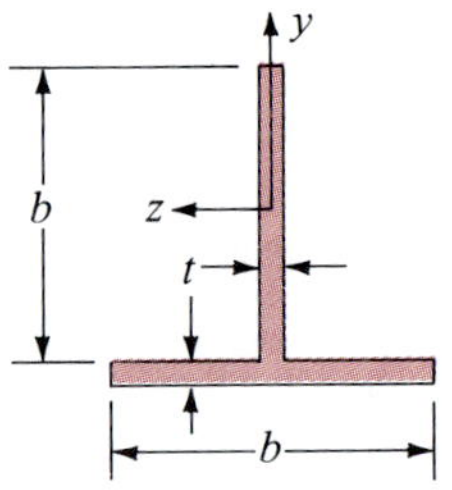

P. 7.221

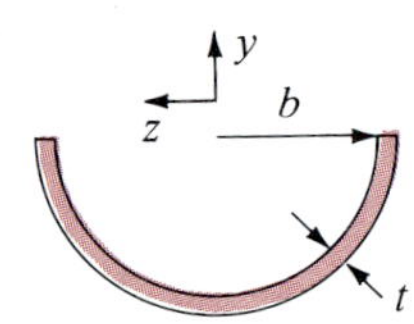

P. 7.222

7.223 The section shown transmits a shear load of 1700 lbf. Determine the shear stress at the four points indicated. Also sketch the shear flow over the cross section. Take $b = 3$ in. and $t = 0.25$ in.

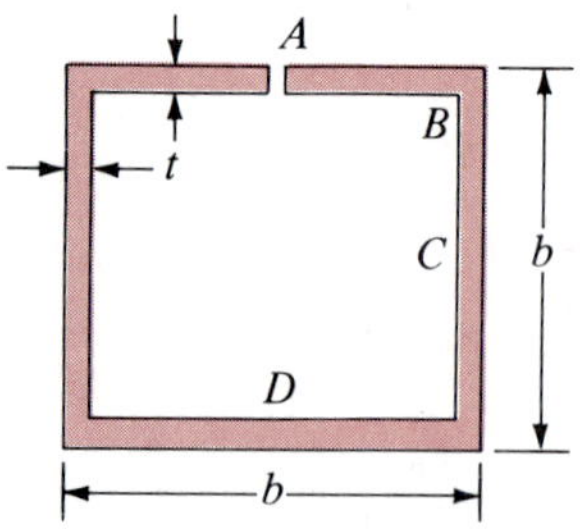

P. 7.223

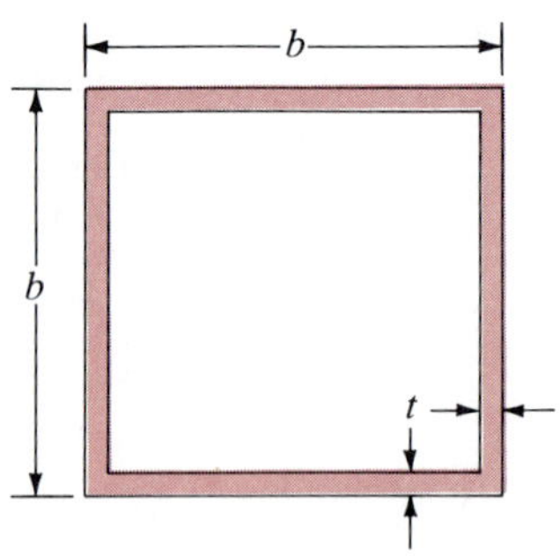

P. 7.224

7.224 The box section transmits a shear load of 11 kN. Determine the maximum shear stress. Take $b = 100$ mm and $t = 10$ mm.

7.225 The extruded box section shown is made from an aluminum with an allowable shear stress of 90 MPa. What is the maximum shear force that the section can transmit?

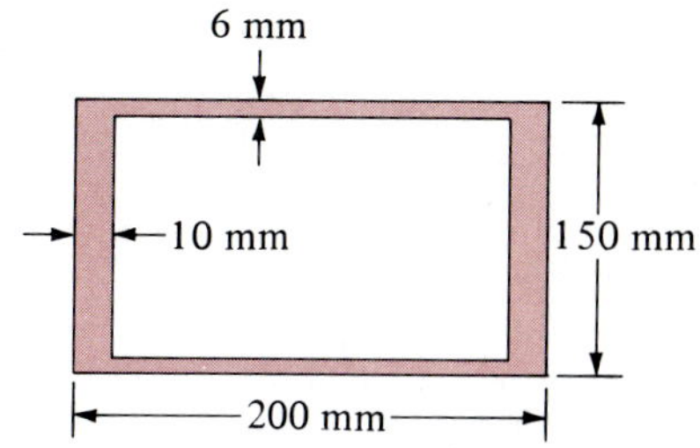

P. 7.225

7.226 A simply supported W 10 × 54 section beam 20 ft in length carries an 80-kip load 6 ft from one end. Neglecting the weight of the beam, determine the magnitude and location of the maximum transverse shear stress in the beam.

7.227 Repeat problem 7.226 including the weight of the beam.

7.228 A simply supported W 250 × 80 section beam 6 m in length carries a 360-kN load 4 m from one end. Neglecting the weight of the beam, determine the magnitude and location of the maximum transverse shear stress in the beam.

7.229 Repeat problem 7.228 including the weight of the beam.

7.230 A beam composed of a C 200 × 20.5 section is loaded as shown. Determine the magnitude and location of the maximum transverse shear stress in the beam. Neglect the weight of the beam.

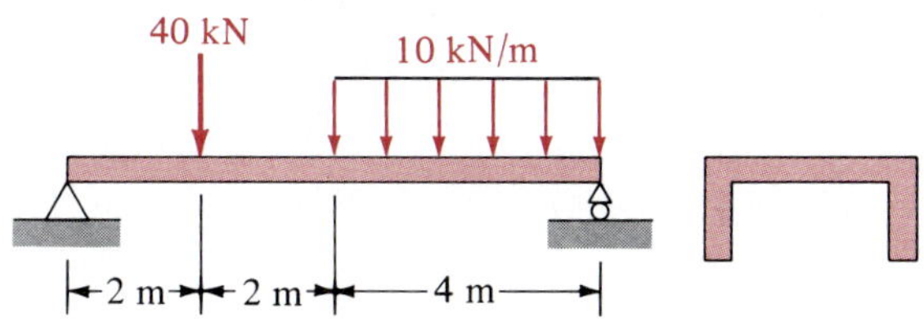

Ps. 7.230 and 7.231

7.231 Repeat problem 7.230 including the weight of the beam.

7.232 A beam composed of a C 8 × 13.75 section is loaded as shown. Determine the magnitude and location of the maximum transverse shear stress in the beam. Neglect the weight of the beam.

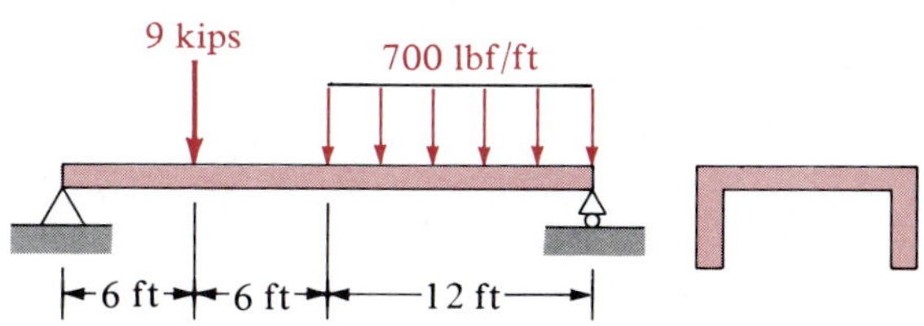

Ps. 7.232 and 7.233

7.233 Repeat problem 7.232 including the weight of the beam.

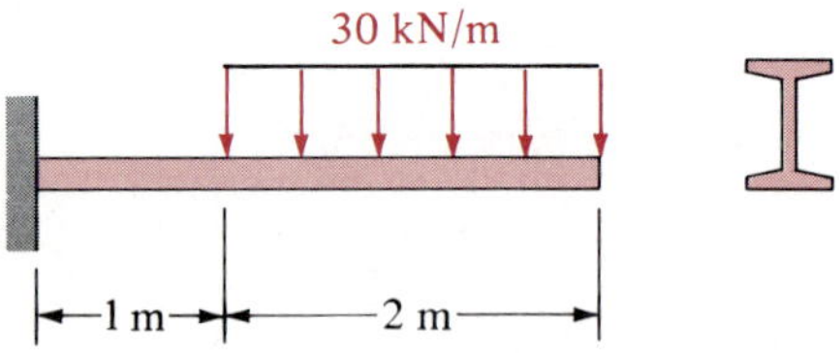

Ps. 7.234 and 7.235

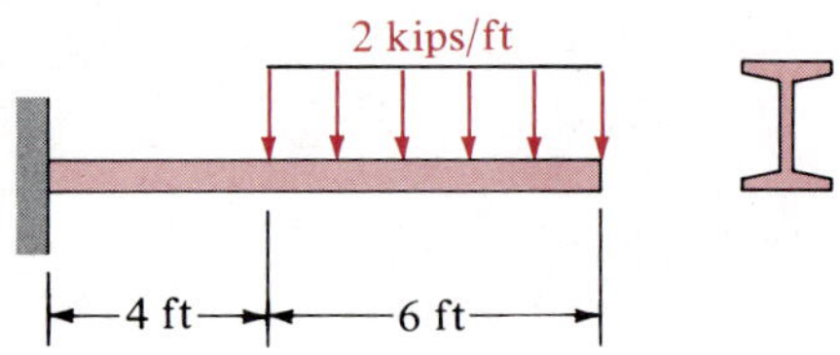

Ps. 7.236 and 7.237

7.234 An S 250 × 52 cantilever beam is loaded as shown. Determine the magnitude and location of the maximum transverse shear stress in the beam. Neglect the weight of the beam.

7.235 Repeat problem 7.234 including the weight of the beam.

7.236 An S 12 × 35 cantilever beam is loaded as shown. Determine the magnitude and location of the maximum transverse shear stress in the beam. Neglect the weight of the beam.

7.237 Repeat problem 7.236 including the weight of the beam.

7.238 A beam made from a T section is loaded as shown. Determine the magnitude and location of the maximum transverse shear stress in the beam. Take $w_0 = 50$ lbf/ft, $L_1 = 5$ ft, $L_2 = 3$ ft, $L_3 = 4$ ft, $b = h = 2$ in., and $t = 0.25$ in. Neglect the weight of the beam.

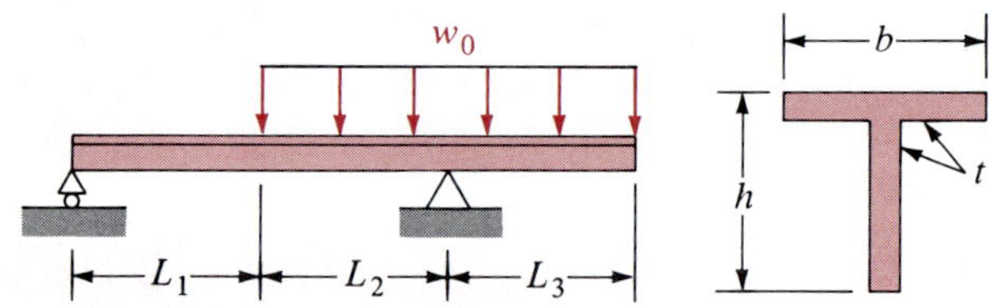

Ps. 7.238 and 7.239

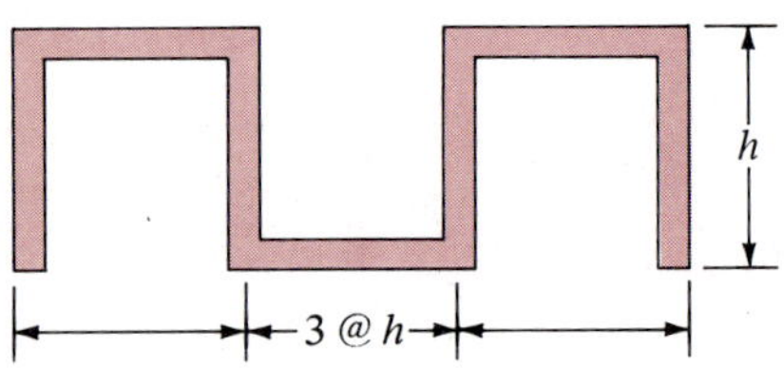

P. 7.240

7.239 Repeat problem 7.238 with $w_0 = 700$ N/m, $L_1 = 2$ m, $L_2 = 1$ m, $L_3 = 1.3$ m, $b = h = 60$ mm, and $t = 7$ mm.

7.240 The extruded section shown is made from an aluminum with an allowable shear stress of 12 ksi. What is the maximum shear force that the section can transmit? Take the uniform thickness to be 0.06 in. and $h = 3$ in. Plot the shear stress distribution.

7.241 For the thin-walled section shown, determine *(a)* the maximum shear stress, and *(b)* the shear-stress distribution. Take $t/h \ll 1$.

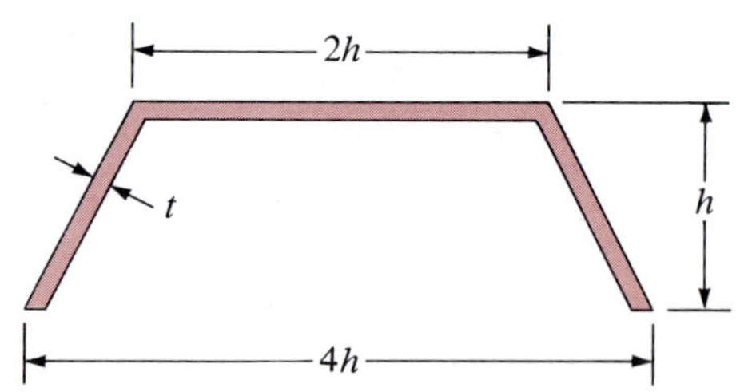

P. 7.241

7.242 The W section with dimensions as shown is to transmit a shear V. Develop a general expression for the percentage of the shear carried by the web. For $t_f/d = \frac{1}{20}$ and $d = b$, plot the percentage as a function of $\alpha = t_f/t_w$ between $1 \leq \alpha \leq 2$. Neglect the I of the flanges about their own centroidal axes.

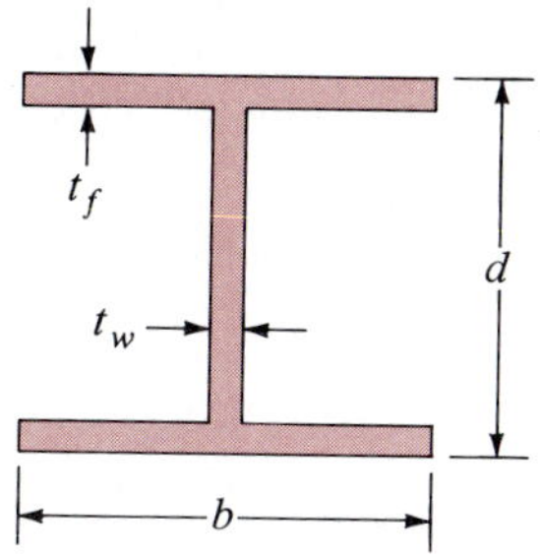

Ps. 7.242–7.244

7.243 Repeat problem 7.242 if $d = 1.5b$.

7.244 Repeat problem 7.242 if $d = 2.0b$.

Bolted Sections

7.245 The two C 180 × 22 sections are connected by two rows of 30-mm-diameter bolts as shown and required to transmit a shear load of 40 kN. Determine the required spacing of the bolts if the allowable shear stress in a bolt is 70 MPa.

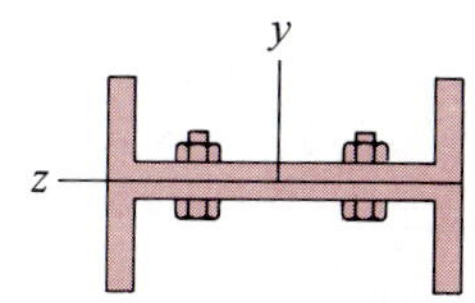

Ps. 7.245 and 7.246

7.246 The two C 7 × 14.75 sections are connected by two rows of $1\frac{1}{4}$-in.-diameter bolts as shown and required to transmit a shear load of 9 kips. Determine the required spacing of the bolts if the allowable shear stress in a bolt is 10 ksi.

7.247 A beam made of an S 200 × 34 section with 10-mm by 120-mm plates connected by bolts as shown is required to transmit a transverse shear of 80 kN. Determine the spacing of the bolts if the ultimate shear stress in each 25-mm-diameter bolt is 100 MPa and a factor of safety of 2.5 is specified.

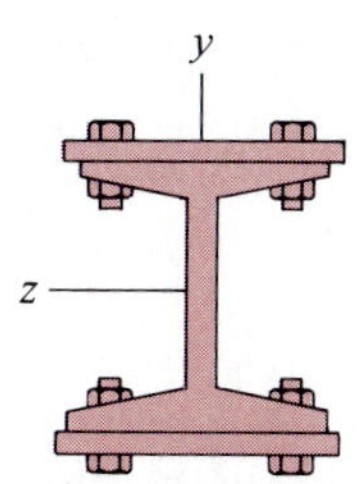

Ps. 7.247 and 7.248

7.248 A beam made of an S 8 × 23 section with $\frac{3}{8}$-in. by 5-in. plates connected by bolts as shown is required to transmit a transverse shear of 20 kips. Determine the spacing of the bolts if the ultimate shear stress in each 1-in.-diameter bolt is 15 ksi and a factor of safety of 2.25 is specified.

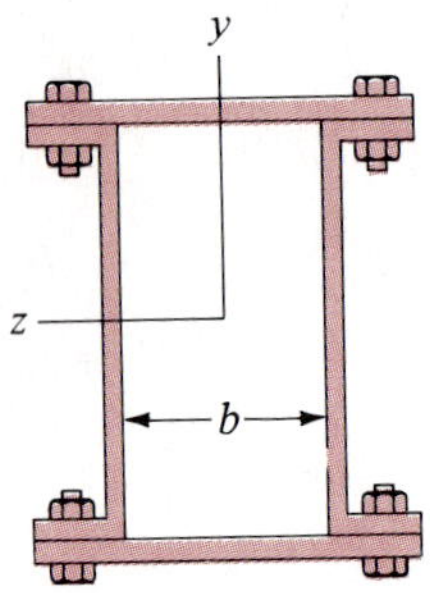

Ps. 7.249 and 7.250

7.249 A box beam required to transmit a shear of 10 kips is constructed by connecting two C 9 × 20 sections and two 0.5-in.-thick plates by bolts as shown. If the ultimate shear stress in each $\frac{5}{8}$-in.-diameter bolt is 30 ksi, determine the required bolt spacing for a factor of safety of 2.25. Take $b = 4.0$ in.

7.250 A box beam required to transmit a shear of 50 kN is constructed by connecting two C 230 × 30 sections and two 13-mm-thick plates by bolts as shown. If the ultimate shear stress in each 15-mm-diameter bolt is 200 MPa, determine the required bolt spacing for a factor of safety of 2.5. Take b = 90 mm.

7.251 Two 0.5-in. by 5.0-in. plates and two L 2 × 2 × $\frac{3}{8}$ angles are connected by 0.5-in.-diameter bolts as shown. If the allowable shear stress in each bolt is 8 ksi and the section is required to transmit a shear load of 3 kips, determine *(a)* the required spacing of the two rows of bolts A, and *(b)* the required spacing of the single row of bolts B.

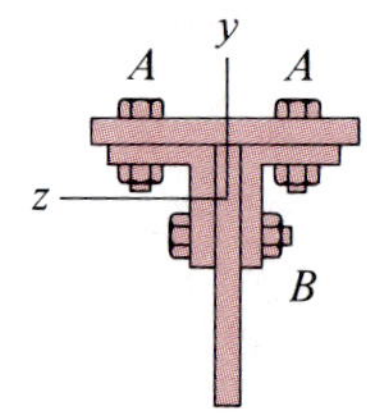

Ps. 7.251 and 7.252

7.252 Two 13-mm by 125-mm plates and two L 51 × 51 × 9.5 angles are connected by 13-mm-diameter bolts as shown. If the allowable shear stress in each bolt is 60 MPa and the section is required to transmit a shear load of 15 kN, determine *(a)* the required spacing of the two rows of bolts A, and *(b)* the required spacing of the single row of bolts B.

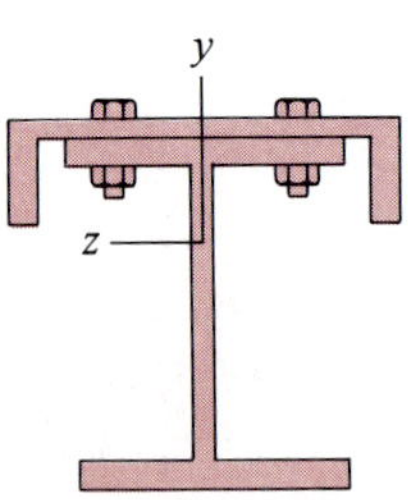

Ps. 7.253 and 7.254

7.253 The W 6 × 25 section is attached to a C 8 × 11.5 section by the two rows of bolts shown. If the beam is to transmit a vertical shear of 15 kips, determine the spacing of the 0.4375-in.-diameter bolts for which the allowable shear stress is 11 ksi.

7.254 The W 150 × 37.1 section is attached to a C 200 × 17.1 section by the two rows of bolts shown. If the beam is to transmit a vertical shear of 70 kN, determine the spacing of the 12-mm-diameter bolts for which the allowable shear stress is 75 MPa.

Glued Sections

7.255 A W section is constructed from wooden members by gluing as indicated. If the allowable shear stress in the glue is 170 psi, what is the maximum shear force V that can be transmitted by the section? What is the corresponding maximum transverse shear stress in the wood? Take h = 7.5 in. and t = 1.6 in.

7.256 Wooden members are to be glued to form a W section as shown. What is the maximum shear force V that can be transmitted by the section if the allowable shear stress in the glue is 1.0 MPa? What is the corresponding maximum transverse shear stress in the wood? Take h = 190 mm and t = 40 mm.

7.257 The T section is constructed by gluing the members as shown. If the allowable shear stress in the glue is 140 psi, what shear force can the section transmit? What is the corresponding maximum transverse shear stress in the wood? Take h = 6 in. and t = 0.875 in.

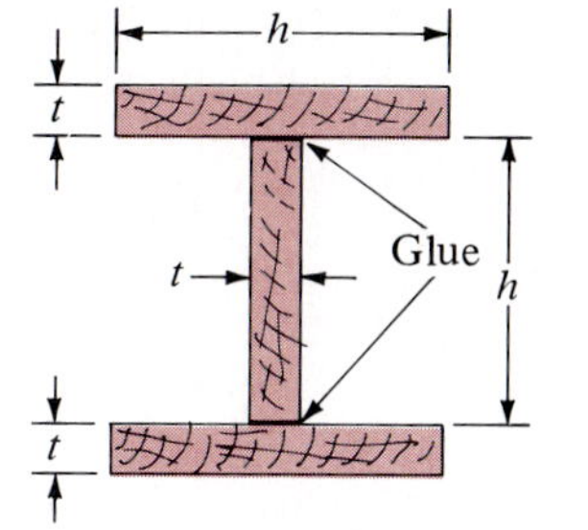

Ps. 7.255 and 7.256

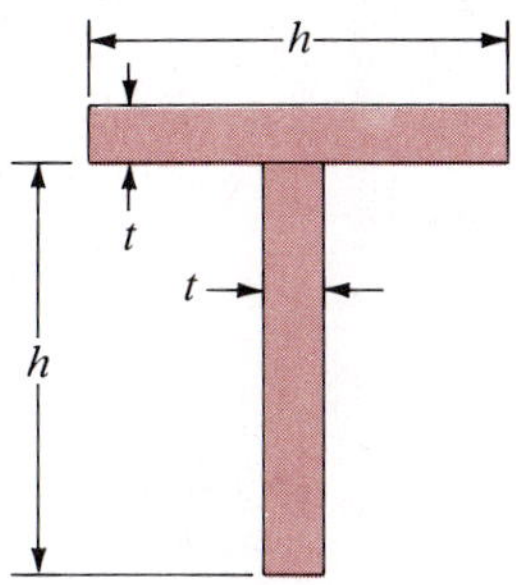

Ps. 7.257 and 7.258

7.258 Repeat problem 7.257 if the allowable shear stress in the glue is 0.95 MPa, h = 160 mm, and t = 20 mm.

7.259 The box section is constructed by gluing the members as shown. If the allowable shear stress in the glue is 1.1 MPa, what shear force can the section transmit? What is the corresponding maximum transverse shear stress in the wood? Take h = 250 mm and t = 40 mm.

7.260 Repeat problem 7.259 if the allowable shear stress in the glue is 150 psi, h = 10 in., and t = 1.625 in.

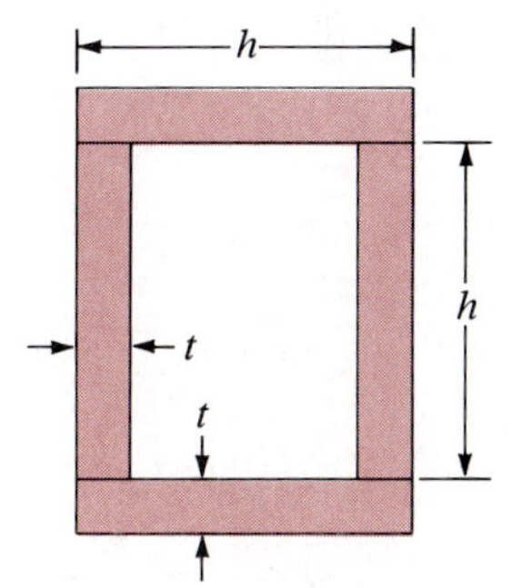

Ps. 7.259 and 7.260

Nailed Sections

7.261 A beam is to be constructed by nailing wooden members as shown. If each nail can withstand a shear force of F and the spacing of the double rows of nails is s, what is the maximum shear force V that the beam can transmit? What is the corresponding maximum transverse shear stress in the wood? Take F = 75 lbf, b = 3.625 in., h = 1.625 in., and s = 2 in.

7.262 Wooden members are used to construct a beam by nailing as shown. If the spacing of the double rows of nails is s and each nail can withstand a shear force of F, what is the maximum shear force V that the beam can transmit? What is the corresponding maximum transverse shear stress in the wood? Take F = 300 N, b = 95 mm, h = 42 mm, and s = 50 mm.

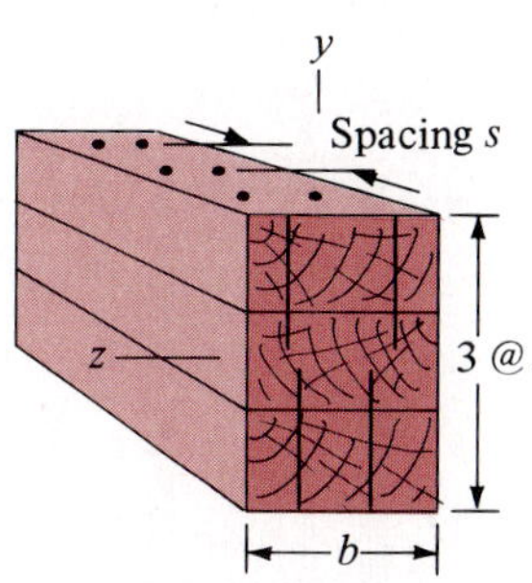

Ps. 7.261 and 7.262

7.263 A box beam is to be formed from wooden members by nailing in two different ways as shown. If the beam is to transmit a shear force of 3 kN, what must be the spacing of the nails for each mode of construction? Each nail can transmit a shear force of 300 N. What is the corresponding maximum transverse shear stress in the wood? Take b = 150 mm, h = 180 mm, and t = 15 mm.

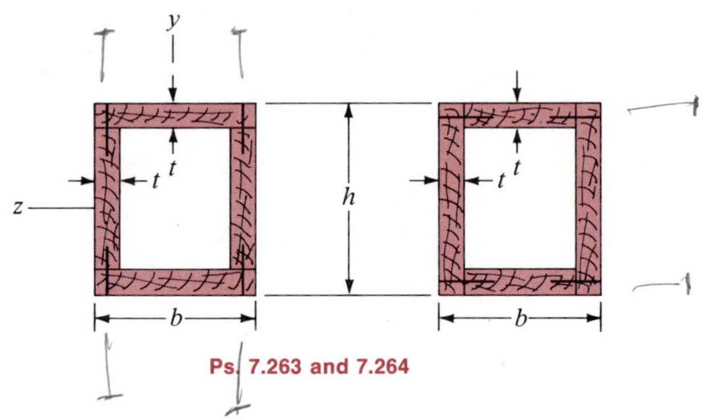

Ps. 7.263 and 7.264

7.264 Wooden members are to be used to form a box beam by two different modes of nailing as shown. If the beam is to transmit a shear force of 700 lbf, what must be the spacing of the nails for each mode of construction? Each nail can transmit a shear force of 70 lbf. What is the corresponding maximum transverse shear stress in the wood? Take b = 6 in., h = 7 in., and t = 0.6 in.

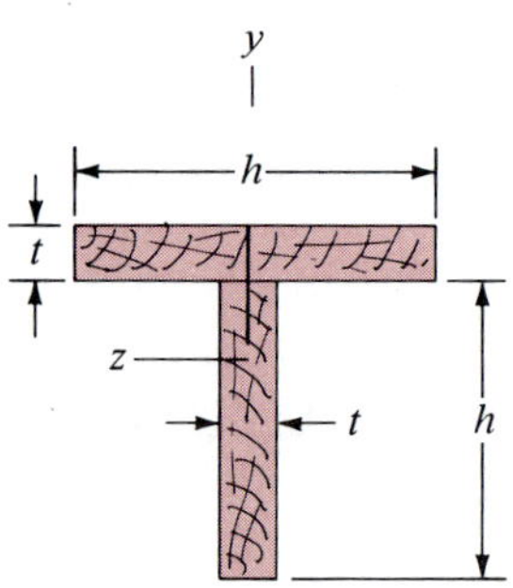

Ps. 7.265 and 7.266

7.265 A T section is to be formed by nailing two wooden members as shown. Determine the required spacing s if the section is to transmit a shear force of 500 lbf and each nail can withstand a shear force of 110 lbf. What is the corresponding maximum transverse shear stress in the wood? Take h = 9 in. and t = 1.75 in.

7.266 Two wooden members are to be joined by nailing to form a T section as shown. If each nail can transmit a shear force of 500 N and the section is required to transmit a shear force of 1.3 kN, determine the required spacing s of the nails. What is the corresponding maximum transverse shear stress in the wood? Take h = 220 mm and t = 45 mm.

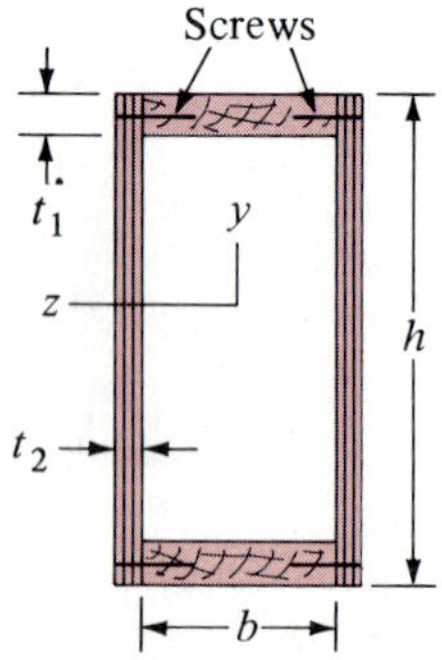

Ps. 7.267 and 7.268

7.267 A box section is to be formed by using screws to attach pieces of plywood to wooden members as shown. If the section is to transmit a shear force of 2000 lbf and the allowable shear force in each screw is 150 lbf, what is the required spacing? What is the corresponding maximum transverse shear stress in the plywood? Take t_1 = 1.625 in., t_2 = 0.75 in., b = 8 in., and h = 20 in.

7.268 Wooden members are to be attached to pieces of plywood by screws to form a box section as shown. If the allowable shear force in each screw is 700 N and the section is to transmit a shear force of 8.6 kN, what is the required spacing? What is the corresponding maximum transverse shear stress in the plywood? Take t_1 = 40 mm, t_2 = 20 mm, b = 200 mm, and h = 500 mm.

Welded Sections

7.269 A beam is formed by welding three 15-mm by 150-mm plates as shown. If the allowable shear stress in the 10-mm-thick weld is 120 MPa, determine the allowable shear load that can be transmitted. Neglect the contribution from the welds in computing I.

7.270 A beam is formed by welding the three $\frac{1}{2}$-in. by 7-in. plates as shown. If the allowable shear stress in the $\frac{1}{2}$-in. weld is 20 ksi, determine the allowable shear load that can be transmitted. Neglect the contribution from the welds in computing I.

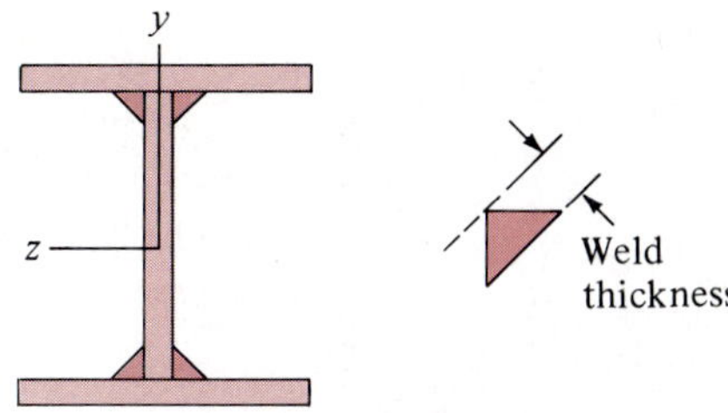

Ps. 7.269 and 7.270

7.9 SKEW LOADING AND BENDING OF UNSYMMETRIC SECTIONS

Earlier in the chapter, when we developed the equations for determining bending and shear stresses and the transverse deflection, two main assumptions were that the section should possess a plane of symmetry, and that the loads should be applied so that bending deformations take place in the plane of symmetry. If either of these assumptions is not satisfied the theory must be altered accordingly.

Fig. 7.102a shows a typical situation where both of these assumptions are satisfied. Fig. 7.102b shows a situation where a beam that has two planes of symmetry is not loaded in either plane of symmetry. As we will show in this section, such a case can be handled by a superposition of results obtained using the usual elementary theory. Fig. 7.102c shows a case where there is no symmetry of the cross section. For this case, as well as the case of a beam with one plane of symmetry that is loaded out of the plane of symmetry, we will discuss the equations necessary for determining the bending stresses.

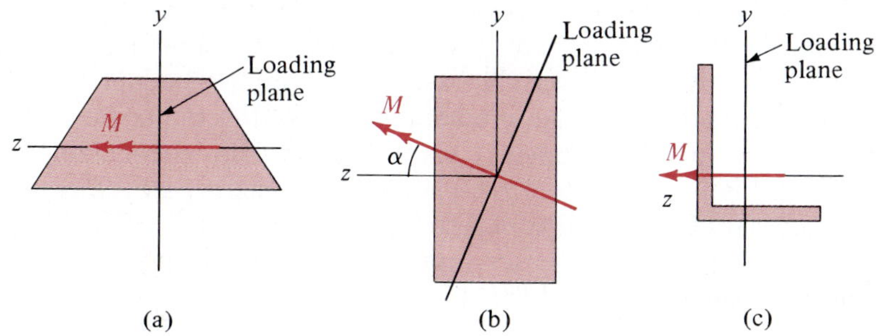

Figure 7.102 Various cases of cross-section symmetry and loading

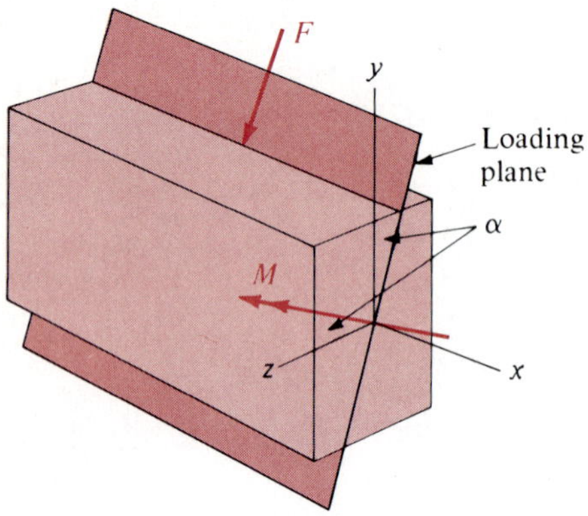

Figure 7.103 Skewed plane of loading

Skew Loading of a Doubly Symmetric Section Consider bending of a beam whose cross section is doubly symmetric, that is, both the xy- and xz-planes are planes of symmetry for the cross section of the beam. By skew loading, we mean that the plane of the loads lies in neither of the planes of symmetry, but rather makes an angle α with the vertical as shown in Fig. 7.103.

Any forces F or distributed loadings are applied in the loading plane, with any moments M applied perpendicular to the plane of loading as shown. Thus at any point along the bar, the moment transmitted makes an angle of α with the z-axis as shown in Fig. 7.104a.

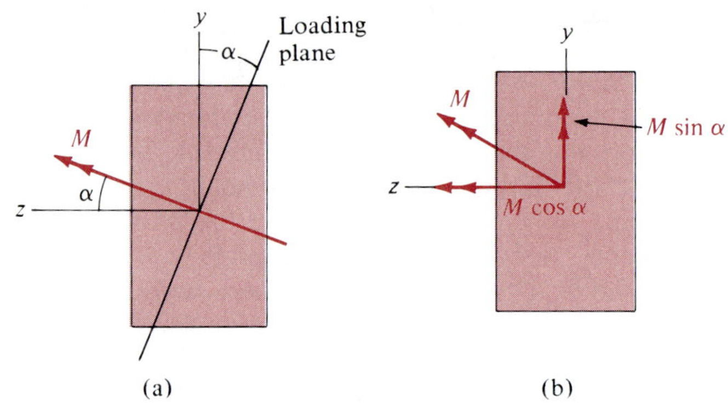

Figure 7.104 Skew loading of a doubly symmetric section

The moment M, inclined at an angle α to the z-axis as shown in Fig. 7.104a, can be resolved into the components $M_y = M \sin \alpha$ and $M_z = M \cos \alpha$ about the y- and z-axes, respectively, as shown in Fig. 7.104b. We then use the **principle of superposition** to state that the stresses arising from the application of the moment $M_z = M \cos \alpha$ about the z-axis, and the stresses arising from the application of the moment $M_y = M \sin \alpha$ about the y-axis, can simply be added to determine the stresses due to the moment M inclined at the angle α. That superposition in this manner is valid follows from recognizing that since both the y- and z-axes are planes of symmetry, all the assumptions necessary for pure bending about either the y- or z-axis are in fact satisfied as long as the stresses remain within the elastic range. With the proper signs accounted for, the stresses for bending about the z- and y-axes are, respectively,

$$\sigma = -\frac{(M \cos \alpha)y}{I_{zz}} \tag{a}$$

and

$$\sigma = +\frac{(M \sin \alpha)z}{I_{yy}} \tag{b}$$

and are pictured in Fig. 7.105. I_{zz} and I_{yy} are the second moments of the area about the z- and y-axes, respectively.

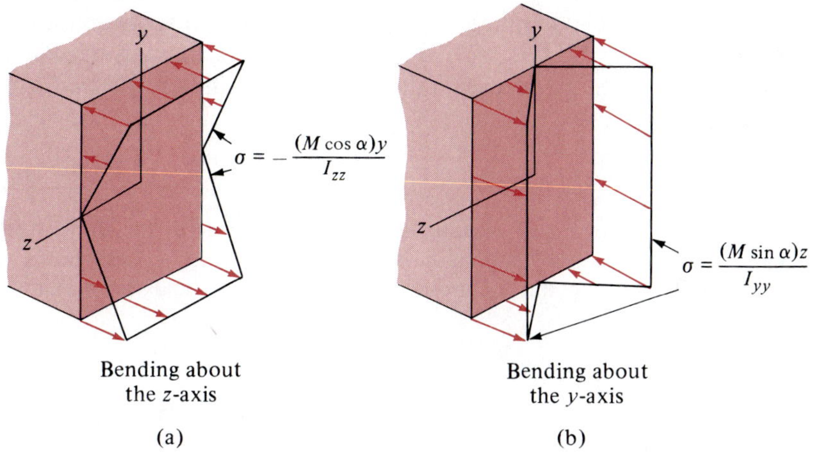

Figure 7.105 Superposition for a doubly symmetric section

Note that for bending about the z-axis, the z-axis is the neutral axis and that for bending about the y-axis, the y-axis is the neutral axis. Superposing the two bending stresses given by Eqns. (a) and (b) yields

$$\sigma = -\frac{(M \cos \alpha)y}{I_{zz}} + \frac{(M \sin \alpha)z}{I_{yy}} = -M\left(\frac{y \cos \alpha}{I_{zz}} - \frac{z \sin \alpha}{I_{yy}}\right) \tag{7.25}$$

The stresses clearly depend upon both the y and z locations within the cross section. The line on which the bending stress is zero for skew loading is still referred to as the **neutral axis of bending,** and is determined by setting $\sigma = 0$, thus

$$0 = \frac{y \cos \alpha}{I_{zz}} - \frac{z \sin \alpha}{I_{yy}}$$

so that the equation of the line that is the neutral axis can be expressed as

$$y = \left[\frac{I_{zz}}{I_{yy}} \tan \alpha\right] z$$

or with $y/z = \tan \beta$,

$$\tan \beta = \left[\frac{I_{zz}}{I_{yy}} \tan \alpha\right] \tag{7.26}$$

This is pictured in Fig. 7.106.

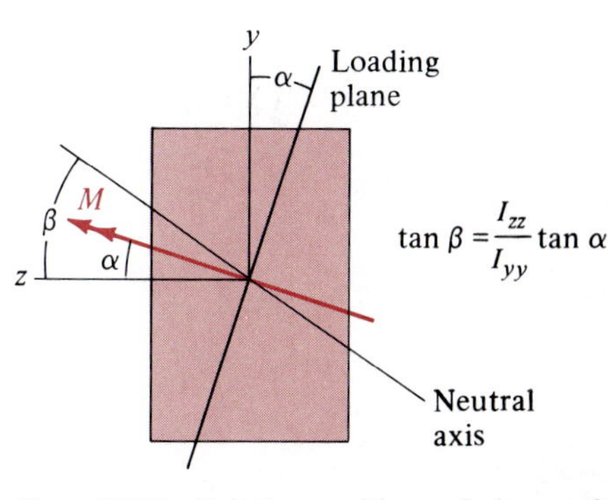

Figure 7.106 Relative positions of planes of loading and the neutral axis

The location of the neutral axis depends upon α and on the ratio of the second moments and is in general no longer perpendicular to the plane of loading, as was the case for the elementary theory of bending. Only if $I_{zz} = I_{yy}$ does the neutral axis coincide with the axis of the bending moment. The bending stresses are of different signs on the two sides of the neutral axis, with the bending stress having maximum absolute value at the point within the cross section that is furthest from the neutral axis.

Example 7.19 A beam of rectangular cross section with the dimensions shown is subjected to a bending moment M inclined at 45° as shown in Fig. 7.107a. Determine the location of the neutral axis and the maximum bending stress.

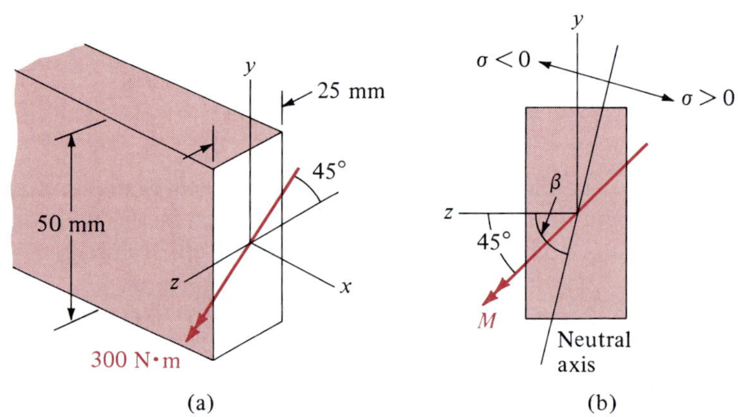

Figure 7.107 Beam with rectangular cross section subjected to a skew moment

Solution The second moments of the area are given by $I_{zz} = (0.025\ \text{m})(0.050\ \text{m})^3/12$ and $I_{yy} = (0.050\ \text{m})(0.025\ \text{m})^3/12$. The angle α is $-45°$, so that from Eqn. (7.26), the equation for the neutral axis is given by

$$\tan\beta = \left(\frac{0.050}{0.025}\right)^2(-1) = -4$$

from which $\beta = -75.96°$, shown in Fig. 7.107b. Referring to Fig. 7.108, points A and B are furthest from the neutral axis so that the maximum bending stresses are

$$\sigma(A) = -\frac{0.707(300\ \text{N}\cdot\text{m})(0.025\ \text{m})}{2.6\times10^{-7}\ \text{m}^4} + \frac{(-0.707)(300\ \text{N}\cdot\text{m})(0.0125\ \text{m})}{6.51\times10^{-8}\ \text{m}^4}$$

$$= -20.4\ \text{MPa} - 40.7\ \text{MPa} = -61.1\ \text{MPa}$$

It follows that the stress at B is given by $\sigma(B) = -\sigma(A) = 61.1$ MPa. The bending stresses at points C and D are $\sigma(C) = -20.3$ MPa and $\sigma(D) = 20.3$ MPa, respectively. The distribution of the bending stress over the cross section is indicated in Fig. 7.108.

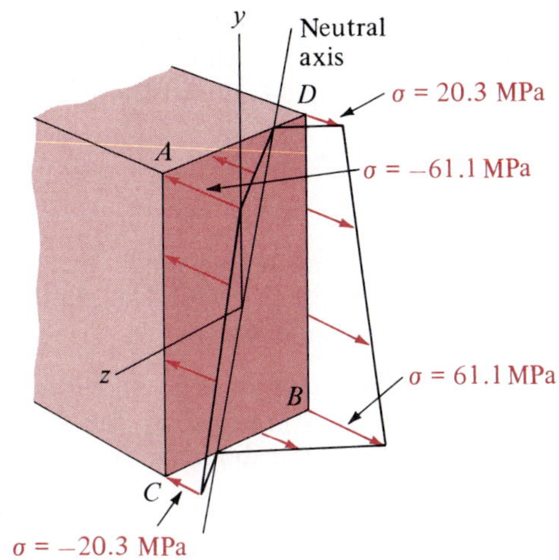

Figure 7.108 Representation of the bending stress

Bending of Unsymmetrical Sections In situations where the cross section of the beam possesses no planes of symmetry as in Fig. 7.109a, or when a beam with only one plane of symmetry is loaded in a skewed fashion, such as in Fig. 7.109b, the equation

$$\sigma = -\frac{M_z y}{I_{zz}} + \frac{M_y z}{I_{yy}}$$

can be used for the bending stress *as long as the y- and z-axes are principal axes.*

Note that in either case, if a pure moment M is applied about the y-axis, the corresponding bending stress is given by

$$\sigma = -\frac{M_y z}{I_{yy}}$$

Computing the corresponding bending moment about the z-axis for these stresses yields

$$M_z = \int_{\text{area}} y\sigma \, dA = \int_{\text{area}} y\left(-\frac{M_y z}{I_{yy}}\right) dA = -\frac{M_y}{I_{yy}}\left[\int_{\text{area}} yz \, dA\right]$$

The integral in the square brackets is the product of the area of the cross section and is zero since y and z are assumed to be principal axes.

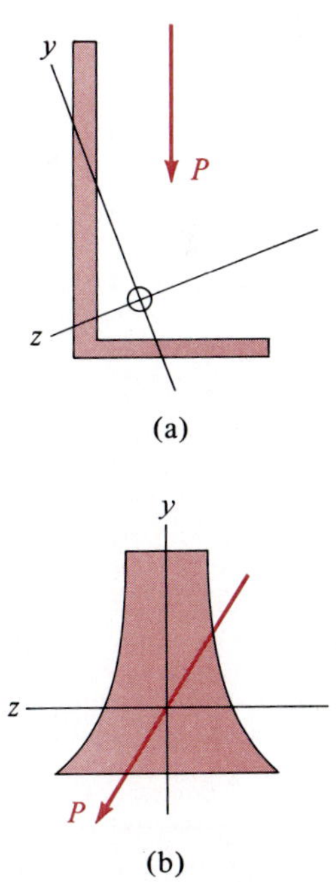

Figure 7.109 Bending with respect to principal axes

Thus $M_z = 0$ and there is no bending about the z-axis. The same argument can be applied to pure bending about the z-axis, that is, if a bending moment is applied about the z-axis, assumed to be principal, the bending stresses will be given by $\sigma = M_y z/I_{yy}$.

In the general case where the moment transmitted has both y and z components, the bending stress continues to be given by Eqn. (7.25) as long as y and z are principal axes, *and,* as long as the moment is resolved into components M_y and M_z about the y- and z-axes, respectively.

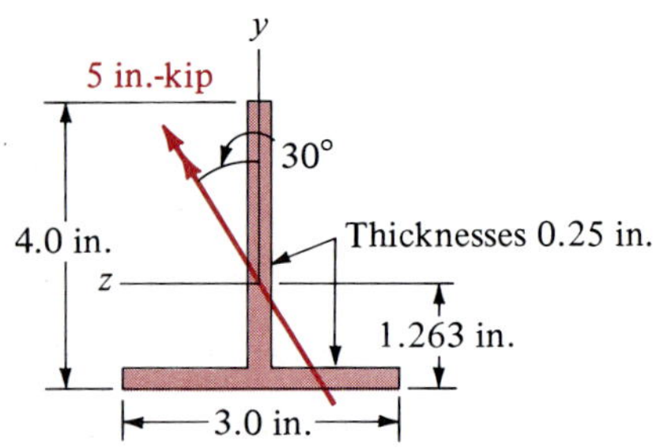

Figure 7.110 Angle section subjected to a moment about the long leg

Example 7.20 A T section with dimensions shown is subjected to a moment $M = 5$ in.-kips inclined 30° to the leg (Fig. 7.110). Determine the location of the neutral axis and the maximum bending stress.

Solution The location of the centroid is easily determined to be 1.236 in. above the bottom of the section as indicated in the figure. Using the axis transfer formulae, the second moments about the z- and y-axes are computed as

$$I_{yy} = \left[\frac{0.25(3)^3}{12} + \frac{3.75(0.25)^3}{12}\right] \text{in}^4 = 0.567 \text{ in}^4$$

and

$$I_{zz} = \left[\frac{3(0.25)^3}{12} + 3(0.25)(-1.236 + 0.125)^2\right] \text{in}^4 + \left[\frac{0.25(3.75)^3}{12} + 0.25(3.75)(2.125 - 1.236)^2\right] \text{in}^4 = 2.770 \text{ in}^4$$

The moments M_y and M_z are given, respectively, by $M_y = 0.866(5000$ in.-kips) and $M_z = 0.5(5000$ in.-kips), so that using Eqn. (7.25) the bending stress can be written as

$$\sigma = \left[-\frac{0.5(5000)y}{2.770} + \frac{0.866(5000)z}{0.567}\right] \text{lbf/in}^2 = [-903y + 7732z] \text{ lbf/in}^2$$

where y and z are to be measured in inches. The location of the neutral axis, shown in Fig. 7.111, is determined by setting $\sigma = 0$ resulting in $\tan\beta = y/z = 7732/903 = 8.56$. The maximum stress of 12.71 ksi occurs at the point furthest from the neutral axis, which in the present problem is point A, where $y = -1.236$ in. and $z = 1.5$ in.

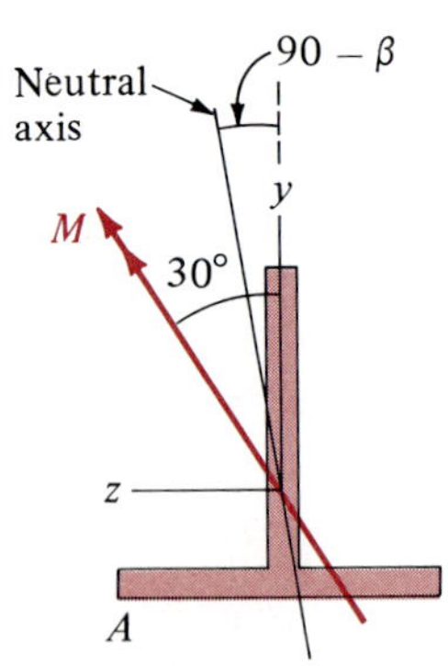

Figure 7.111 Bending stresses in a T section

In the most general case, where there is no symmetry and with the y- and z-axes not chosen to coincide with the principal axes, the bending stresses in the elastic range can be expressed as

$$\sigma = -\frac{M_z I_{yy} + M_y I_{yz}}{I_{yy} I_{zz} - I_{yz}^2} y + \frac{M_y I_{zz} + M_z I_{yz}}{I_{yy} I_{zz} - I_{yz}^2} z \tag{7.27}$$

The location of the neutral axis is obtained by setting $\sigma = 0$ yielding

$$\tan \beta = y/z = \frac{M_y I_{zz} + M_z I_{yz}}{M_z I_{yy} + M_y I_{yz}} \tag{7.28}$$

Note that if y and z are principal axes, $I_{yz} = 0$, with both Eqns. (7.27) and (7.28) reducing to Eqns. (7.25) and (7.26) discussed previously.

Example 7.21 An angle section with dimensions shown is acted upon by a moment of 1 kN·m (Fig. 7.112a). Use Eqns. (7.27) and (7.28) to determine the location of the neutral axis, and the location and magnitude of the maximum bending stress.

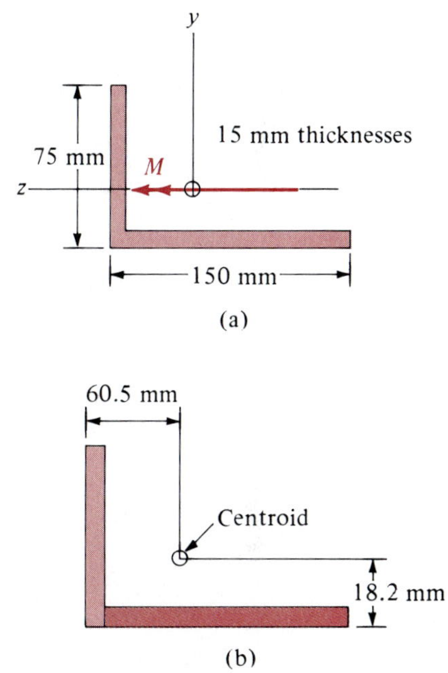

Figure 7.112 Bending of an angle section

Solution In order to use Eqns. (7.27) and (7.28), the three second moments of the area I_{yy}, I_{zz}, and I_{yz} must be evaluated, or the principal axes of the angle section must be located. The location of the centroid and the choice for the composite areas are shown in Fig. 7.112b. Using the transfer formulae for rectangles we compute

$$I_{yy} = \left[\frac{15(135)^3}{12} + 15(135)(60.5 - 90)^2 + \frac{75(15)^3}{12} + 75(15)(60.5 - 7.5)^2\right] \text{mm}^4 = 8.12 \times 10^{-6} \text{ m}^4$$

$$I_{zz} = \left[\frac{135(15)^3}{12} + 135(15)(-18.2 + 7.5)^2 + \frac{15(75)^3}{12} + 15(75)(-18.2 + 37.5)^2\right] \text{mm}^4 = 1.23 \times 10^{-6} \text{ m}^4$$

$$I_{yz} = [0 + 135(15)(-18.2 + 7.5)(60.5 - 90) + 0 + 75(15)(-18.2 + 37.5)(60.5 - 7.5)] \text{ mm}^4 = 1.79 \times 10^{-6} \text{ m}^4$$

Prior to evaluating the bending stresses, we compute

$$I_{yy} I_{zz} - I_{yz}^2 = (8.12(1.23) - 1.79^2) \times 10^{-12} \text{ m}^8 = 6.78 \times 10^{-12} \text{ m}^8$$

so that with $M_z = 1$ kN·m and $M_y = 0$, Eqn. (7.27) becomes

$$\sigma = -\frac{1000 \text{ N·m}(8.12 \times 10^{-6} \text{ m}^4)}{6.78 \times 10^{-12} \text{ m}^8} y + \frac{1000 \text{ N·m}(1.79 \times 10^{-6} \text{ m}^4)}{6.78 \times 10^{-12} \text{ m}^8} z$$

$$= (-1198y + 264z) \text{ MPa}$$

where y and z are in meters. The location of the neutral axis can be computed directly from

$$0 = -1198y + 264z$$

or

$$\frac{y}{z} = \tan\beta = \frac{264}{1198} = 0.220$$

from which $\beta = 12.43°$, as shown in Fig. 7.113.

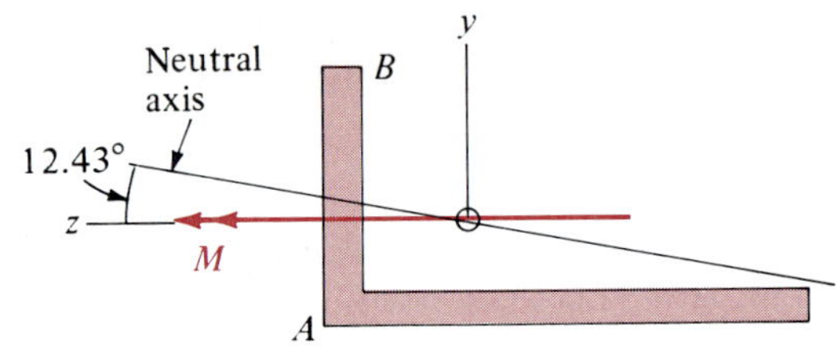

Figure 7.113 Location of the neutral axis

The stresses at points A and B, shown in Fig. 7.113, are, respectively,

$$\sigma(A) = [-1198(-0.0182 \text{ m}) + 264(0.0605 \text{ m})] \text{ MPa} = 37.8 \text{ MPa}$$

and

$$\sigma(B) = [-1198(0.0568 \text{ m}) + 264(0.0455 \text{ m})] \text{ MPa} = -56.0 \text{ MPa}$$

If the problem had been solved naively, ignoring the fact that there is no plane of symmetry (i.e., by simply evaluating the elementary result $\sigma = -M_z y/I_{zz}$), the stresses at A and B would have been computed as

$$\sigma(A) = -\frac{1000 \text{ N·m}(-0.0182 \text{ m})}{1.23 \times 10^{-6} \text{ m}^4} = 14.8 \text{ MPa}$$

and

$$\sigma(B) = -\frac{1000 \text{ N·m}(0.0568 \text{ m})}{1.23 \times 10^{-6} \text{ m}^4} = -46.2 \text{ MPa}$$

which are 61% and 18% low, respectively, compared to the correct values! Other loadings can result in an even more striking difference between computing the stresses naively and according to the correct theory.

PROBLEMS

7.271 For the rectangular-cross-section beam shown, determine the bending stresses at points A, B, C, and D. Take $b = 100$ mm, $h = 150$ mm, $M = 4$ kN·m, and $\alpha = 20°$.

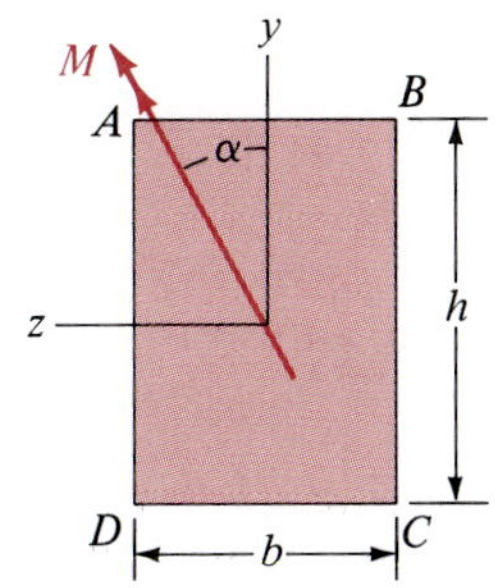

Ps. 7.271 and 7.272

7.272 Repeat problem 7.271 with $b = 4$ in., $h = 6$ in., $M = 3000$ ft-lbf, and $\alpha = 30°$.

7.273 The W 6 × 16 section is simply supported at both ends about the y- and z-axes, and loaded by the uniformly distributed load as shown. Assuming that the loading w_0 is vertical, determine the location and magnitude of the maximum bending stress if the beam is tipped at $\theta = 2°$ to the vertical as shown. What is the % increase compared to $\theta = 0°$? Take $w_0 = 60$ lbf/in., $L = 3$ ft, and include the weight of the beam.

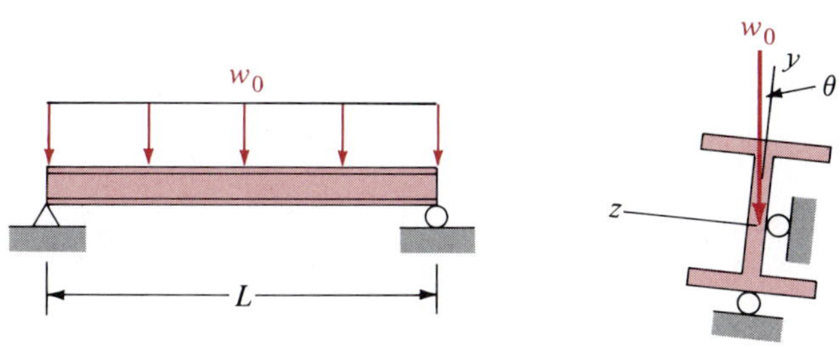

Ps. 7.273 and 7.274

7.274 Repeat problem 7.273 for a W 150 × 24 section, with $L = 1$ m and $w_0 = 10^4$ kN/m.

7.275 The S 130 × 15 section is used for a cantilever beam. Assuming that the loading P is vertical, determine the location and magnitude of the maximum bending stress if the beam is tipped at $\theta = 3°$ to the vertical as shown. What is the % increase compared $\theta = 0°$? Take $P = 9$ kN, $L = 1.2$ m, and neglect the weight of the beam.

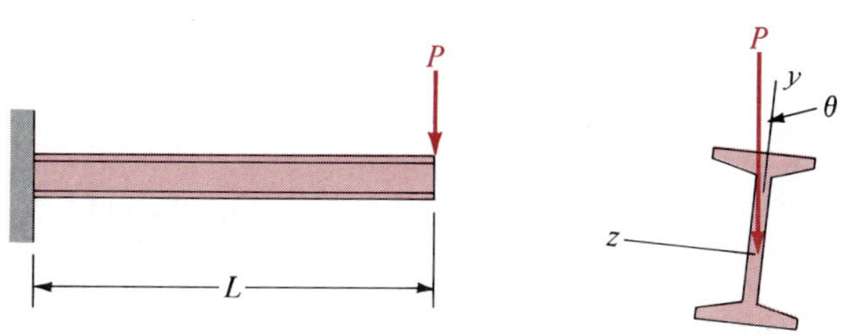

Ps. 7.275 and 7.276

7.276 Repeat problem 7.275 for an S 5 × 10 section, with $P = 600$ lbf, $L = 40$ in., and $\theta = 2.5°$.

7.277 The two-by-four cantilever beam is subjected to the skewed load shown. Determine the location and magnitude of the maximum bending stress at the clamped end.

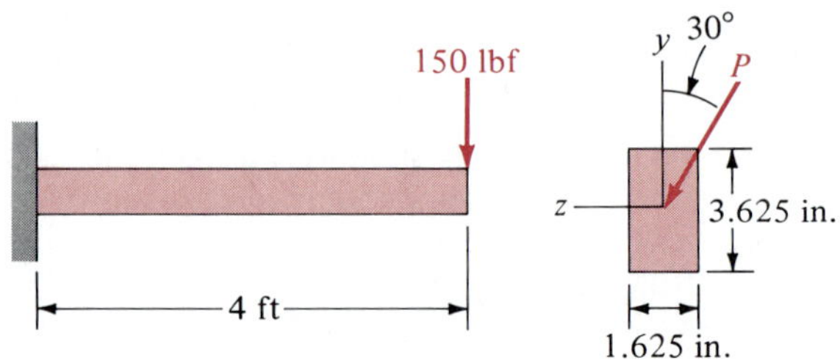

P. 7.277

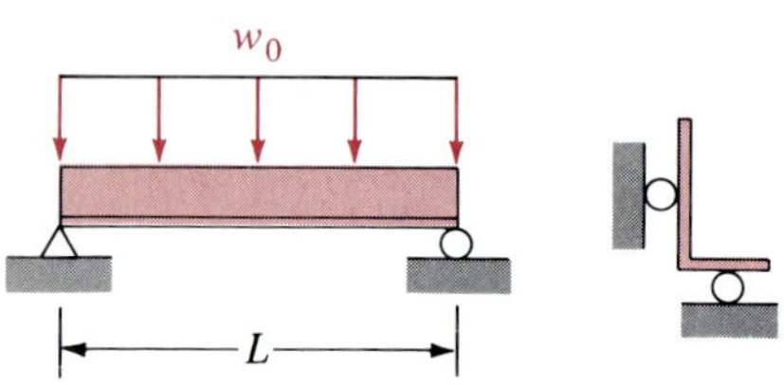

Ps. 7.278 and 7.279

7.278 The L 3 × 2 × $\frac{1}{4}$ section is simply supported at both ends about the y- and z-axes, and loaded by the uniformly distributed load as shown. Determine the location and magnitude of the maximum bending stress. Take L = 36 in. and w_0 = 50 lbf/in.

7.279 Repeat problem 7.278 for an L 89 × 64 × 9.5 section, with L = 900 mm and w_0 = 10 N/mm.

7.280 The Z section shown is to be used as a cantilever beam. What is the allowable load P if the ultimate bending stress for the material is 200 MPa and the factor of safety is to be 2? Take b = 40 mm, t = 6 mm, and L = 600 mm.

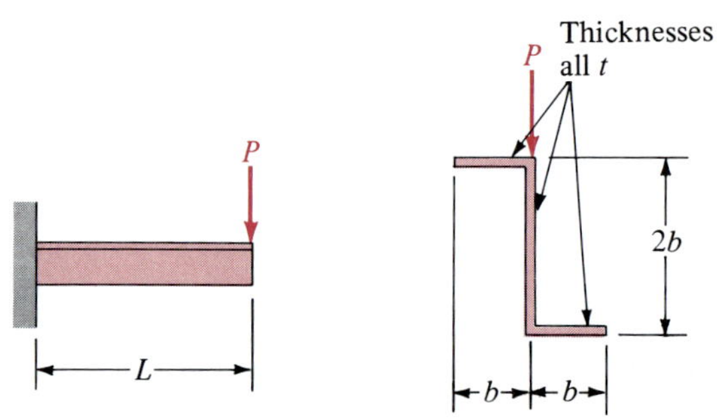

Ps. 7.280 and 7.281

7.281 Repeat problem 7.280 if b = 1.75 in., t = 0.25 in., and L = 24 in. Take the ultimate bending stress to be 28 ksi and the factor of safety to be 2.0.

7.10 SHEAR CENTER

When a doubly symmetric section is loaded in one of its planes of symmetry, such as in Fig. 7.114a, or when a section with a single axis of symmetry is loaded in its plane of symmetry as in Fig. 7.114b, the deformations take place in the xy-plane with no twisting of the beam about the x-axis.

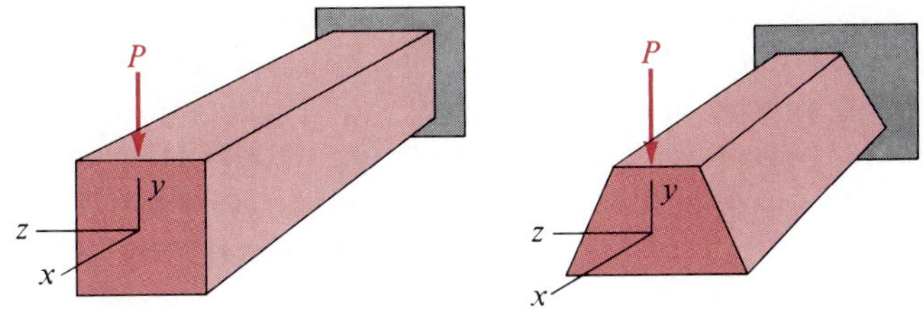

Figure 7.114 Bending without twisting

If, for either of these cases, the loads are *offset* from the xy-plane, there will be stresses other than the bending stresses $\sigma_x = -My/I$ about the z-axis and the shear stresses $\tau_{xy} = -VQ/It$ in the y-direction, due to the fact that there are torques about the x-axis. The deformations will consist of a combination of bending about the z-axis and twisting about the x-axis.

For a section with a single plane of symmetry that is not loaded in the plane of symmetry, as in Fig. 7.115a, or for the loading of a section with no planes of symmetry, as in Fig. 7.115b, there will in general be bending about both the y- and z-axes and twisting about the x-axis. In order to avoid the twisting, the loads must pass through a particular point that we can determine by requiring that the combination of the loading and the transverse shear stresses produce no resultant moment about the x-axis at any cross section along the beam. Thus the problem reduces to one of statics, and we will solve it on the basis of considering equilibrium. We will restrict our attention to thin-walled sections here.

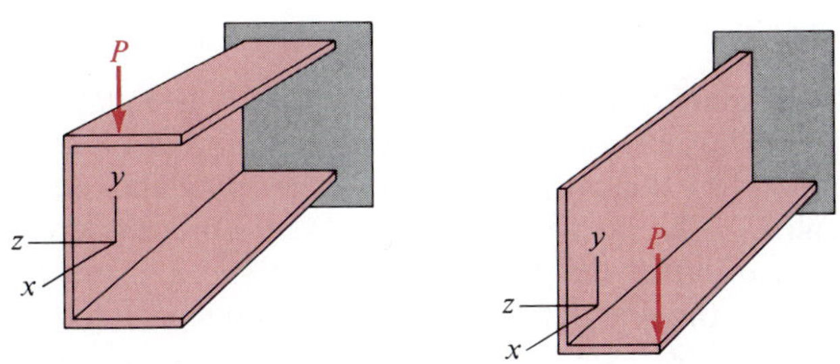

Figure 7.115 Loading of thin-walled sections

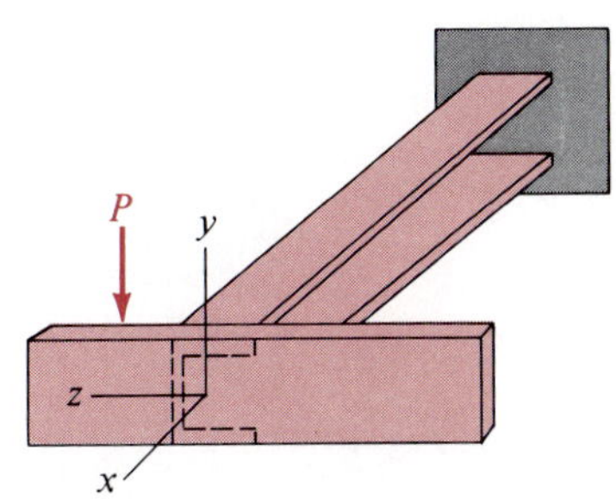

Figure 7.116 Basic idea of the shear center

The basic idea of the shear center can be best demonstrated by referring to Fig. 7.116, where a rigid bar connected to a channel section is acted upon by a force parallel to the y-axis.

If the position of the load P is moved far enough to the left, there will clearly be twisting about the x-axis with the positive z-axis moving downward. Similarly, if the position of the load P is moved far enough to the right, there will clearly be twisting about the x-axis with the positive z-axis moving upward. It then follows that there is some intermediate line of action of the vertical load P that results in no twisting about the x-axis.

Consider specifically a typical thin-walled section with a single plane of symmetry such as the channel section shown in Fig. 7.117a. A free-body diagram of the portion of the beam associated with the section AA is shown in Fig. 7.117b, where F_1, F_2, and F_3 are the resultants of the shear stresses, and $M = Pd$.

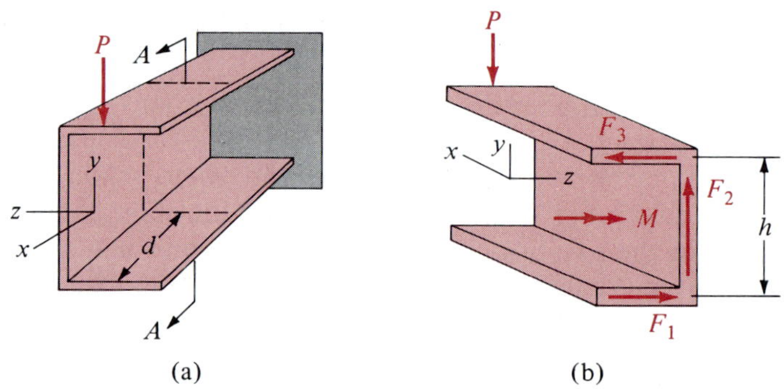

Figure 7.117 Loading of a channel section

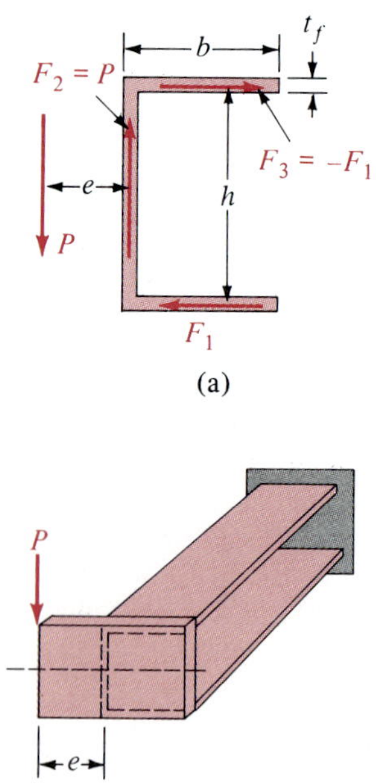

Figure 7.118 Location of plane of loading

Requiring equilibrium in the y- and z-directions, respectively, yields

$$\Sigma F_y \equiv F_2 - P = 0 \qquad \text{and} \qquad \Sigma F_z \equiv F_1 - F_3 = 0$$

showing that F_1 and F_3 must be equal, and that F_2 is equal to the applied shear load P. Thus F_1 and F_3 form a couple about the x-axis that must be counteracted by the couple formed by the equal and opposite pair of forces P and F_2. The only way that the direction of the couple formed by P and F_2 can offset the couple due to the shear forces F_1 and F_3 in the flanges is for the line of action of the applied load V to be located as shown in Fig. 7.118.

Requiring the moment of the two couples about the x-axis to vanish yields

$$F_1 h = Pe$$

defining the distance e from the centerline of the web to the line of action of the shear load P. When the load P is applied at this location, there is moment about the x-axis and consequently no twisting, so that the transverse shear stresses can be computed using $\tau = VQ/It =$

PQ/It, based upon bending about the z-axis. It is left in the problems for the student to show that in general for a channel section

$$e = \frac{b^2h^2t_f}{4I}$$

where b, h, and t_f are shown in Fig. 7.118.

Example 7.22 The channel is loaded in shear as shown in Fig. 7.119. Determine the location of the shear center, and the shear-stress distribution. Assume a constant 3-mm thickness.

Solution Since the y- and z-axes shown in Fig. 7.119b are principal axes, we can still use the basic expression $\tau = VQ/It$ for computing the horizontal and transverse shear stresses in the channel.

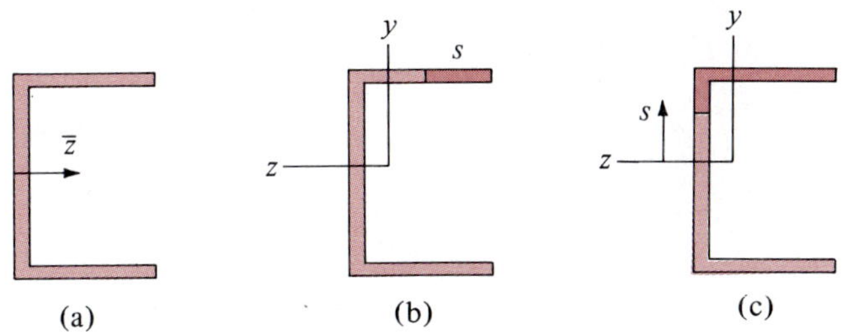

Figure 7.120 Calculation of shear stresses

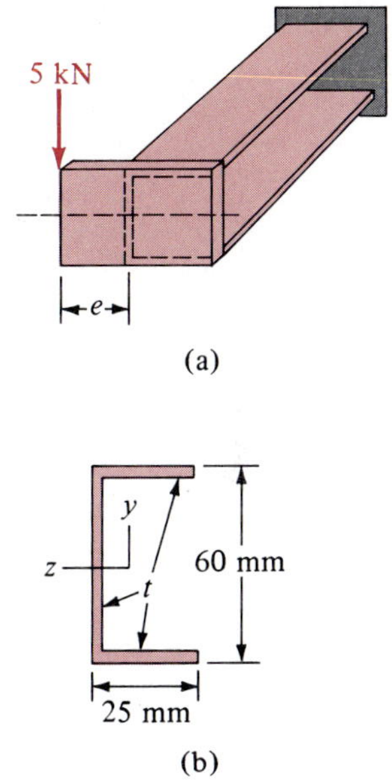

Figure 7.119 Channel loaded in shear

Referring to Fig. 7.120a, the centroid of the cross section is determined from

$$\bar{z} = \frac{2(25\text{ mm})(3\text{ mm})(12.5\text{ mm}) + (54\text{ mm})(3\text{ mm})(1.5\text{ mm})}{2(25\text{ mm})(3\text{ mm}) + (54\text{ mm})(3\text{ mm})} = 6.79\text{ mm}$$

The second moment about the z-axis is given by

$$I = \frac{25\text{ mm}(60\text{ mm})^3 - 22\text{ mm}(54\text{ mm})^3}{12}\text{ mm}^4 = 1.613 \times 10^5\text{ mm}^4$$

Referring to Fig. 7.120b, the first moment of the area in the upper flange about the z-axis is,

$$Q = 3\text{ mm}(s)28.5\text{ mm} = 85.5s\text{ mm}^2$$

for $0 \le s \le 23.5$ mm, so that

$$\tau_{\text{flange}} = \frac{5\text{ kN}(85.5s\text{ mm}^2)}{1.613 \times 10^5\text{ mm}^4\text{ (3 mm)}} = 0.883s\text{ N/mm}^3$$

shown in Fig. 7.121a. For computing the shear stresses in the web, the first moment about the z-axis of the shaded area shown in Fig. 7.120c is

$$Q = 3\text{ mm}(25\text{ mm})(28.5\text{ mm}) + (28.5\text{ mm} - s)(3\text{ mm})\,\frac{28.5\text{ mm} + s}{2}$$

for $0 \le s \le 28.5$ mm. The corresponding shear stress calculated from $\tau = VQ/It$ is shown in Fig. 7.121a.

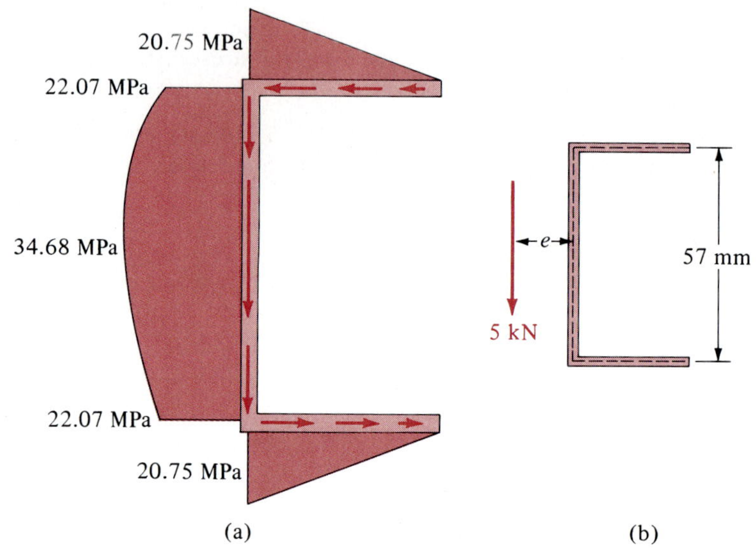

Figure 7.121 Shear-stress distribution in a channel

The force in each flange is

$$F = \frac{(20.75\text{ MPa})(0.003\text{ m})(0.0235\text{ m})}{2} = 0.731\text{ kN}$$

so that the distance from the web to the location of the shear load should be

$$e(5\text{ kN}) = (0.731\text{ kN})(0.057\text{ m})$$

or

$$e = 8.33\text{ mm}$$

shown in Fig. 7.121b. We note in passing that if we use the general expression $e = b^2h^2t_f/4I$ given earlier, there results

$$e = \frac{(23.5\text{ mm})^2(57\text{ mm})^2(3\text{ mm})}{4(1.613 \times 10^5\text{ mm}^4)} = 8.34\text{ mm}$$

essentially verifying the location of the shear load.

In general, the point S through which the shear load P must pass, shown in Fig. 7.122, is called the **shear center.** For a doubly symmetric cross section, the shear center is located at the intersection of the two planes of symmetry, the location of the centroid. Several examples are shown. For any of these sections, the plane of the loading must pass through the centroid or shear center in order to produce bending and shear without twisting.

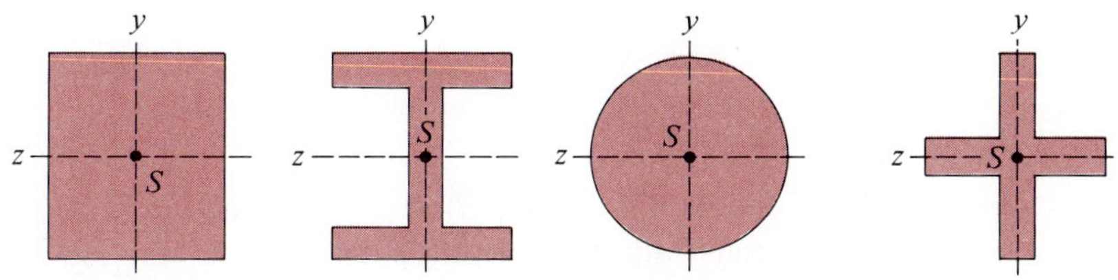

Figure 7.122 Shear center and centroid locations for doubly symmetric sections

For a section with a single plane of symmetry, the centroid and the shear center are both located on the axis of symmetry but do not necessarily coincide. As shown in Fig. 7.123 the load can be resolved into components lying in and perpendicular to the plane of symmetry. F_1, the component lying in the plane of symmetry, produces bending without twisting about the z-axis. In order to prevent twisting, F_2, the component of the load parallel to the z-axis, must be applied at the shear center.

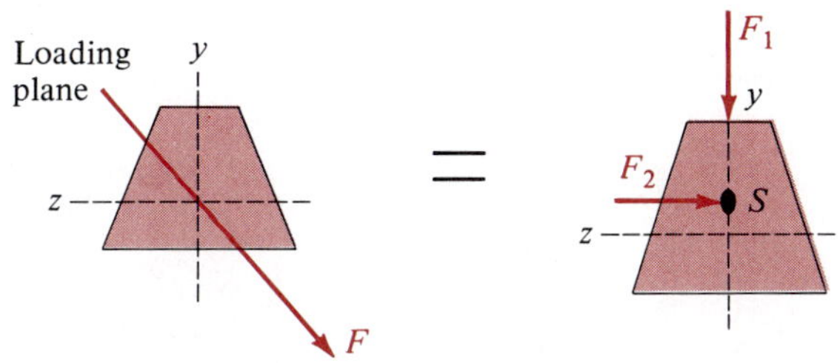

Figure 7.123 Loading and shear center for a section with a single plane of symmetry

For a section with no planes of symmetry, the shear stresses are evaluated on the basis of either Eqn. (7.25) or Eqn. (7.27) for the bending stresses, after which the location of the plane of loading is determined by requiring that there be no resultant moment about the

x-axis. The locations of the shear center for several different thin-walled sections are shown in Fig. 7.124.

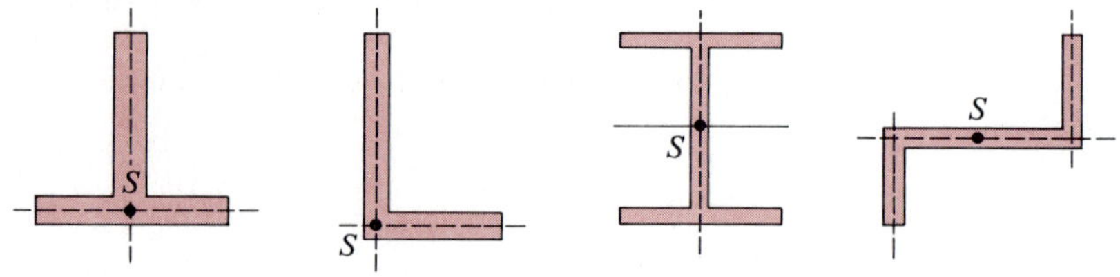

Figure 7.124 Locations of shear centers for various thin-walled sections

As discussed in Section 6.6, thin-walled sections are notoriously weak in torsion. It is thus important to be aware of the concept and meaning of the shear center, so as to be able to apply the loads through the shear center and avoid any torques leading to twisting and torsional stresses.

PROBLEMS

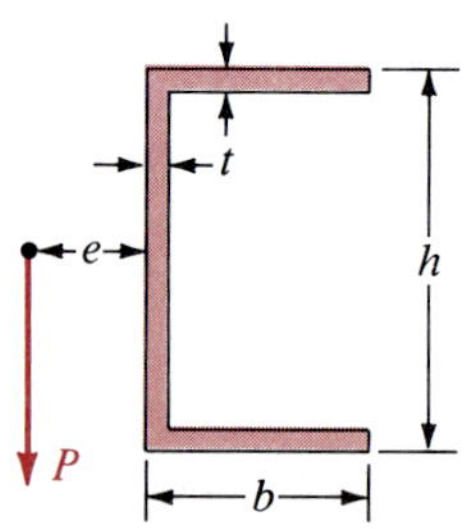

Ps. 7.282 and 7.283

7.282 For the C section loaded as shown, determine *(a)* the distance e to the shear center, and *(b)* the location and magnitude of the maximum transverse shear stress. Take b = 1.5 in., h = 3 in., t = 3⁄16 in., and P = 2000 lbf.

7.283 Repeat problem 7.282 for b = 40 mm, h = 75 mm, t = 4 mm, and P = 9 kN.

7.284–7.287 For each of the following, determine the location of the shear center. Assume the thickness to be small.

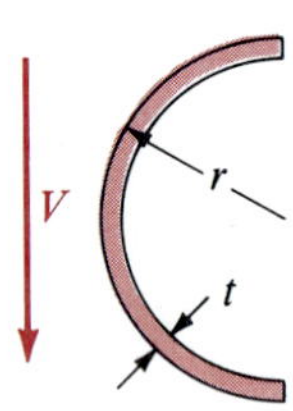

P. 7.284

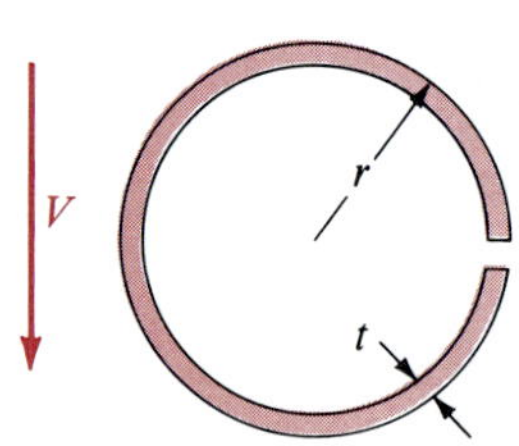

P. 7.285

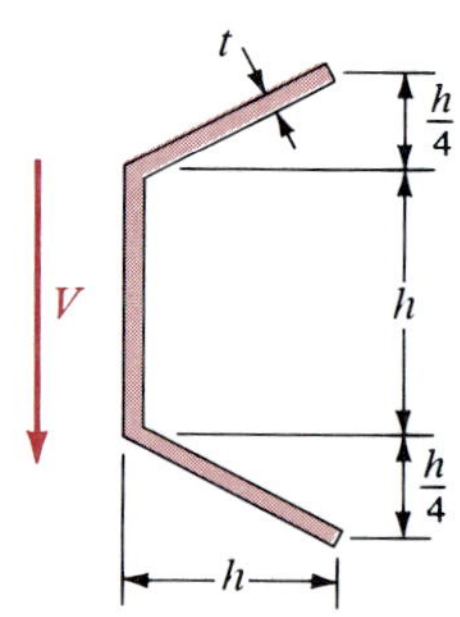

P. 7.286

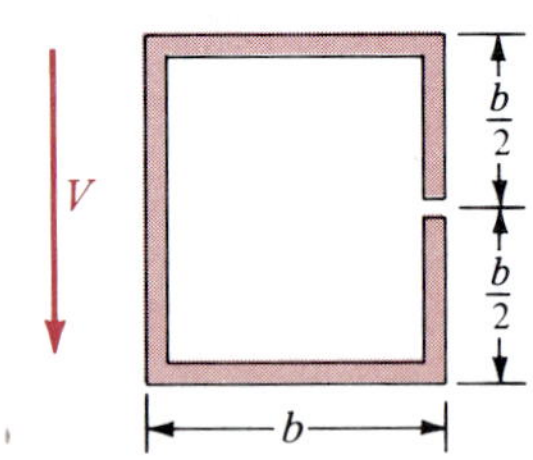

P. 7.287

7.288 Show that for the C section of Fig. 7.118, $e = b^2h^2t_f/4I$.

7.289 Determine the distance e to the shear center for the unsymmetrical I beam shown. (*Hint:* Assume that all the bending and shear stresses are carried by the flanges.)

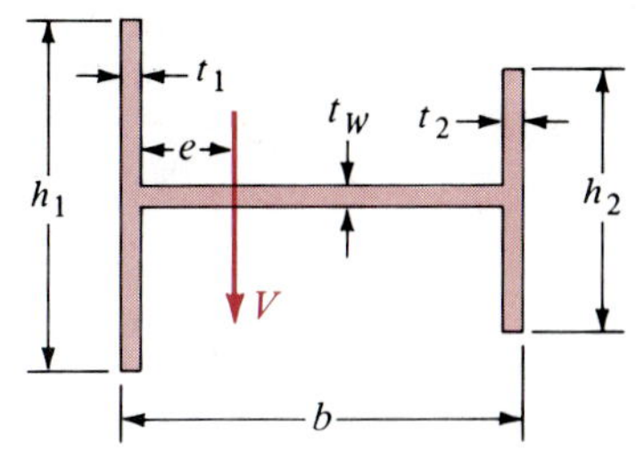

P. 7.289

7.290 For the section shown, determine (*a*) the location of the shear center, and (*b*) the principal stresses and the absolute maximum shear stress at locations A and B. Take $L = 2$ m, $h = 90$ mm, $b = 60$ mm, $t = 5$ mm, $P = 500$ N.

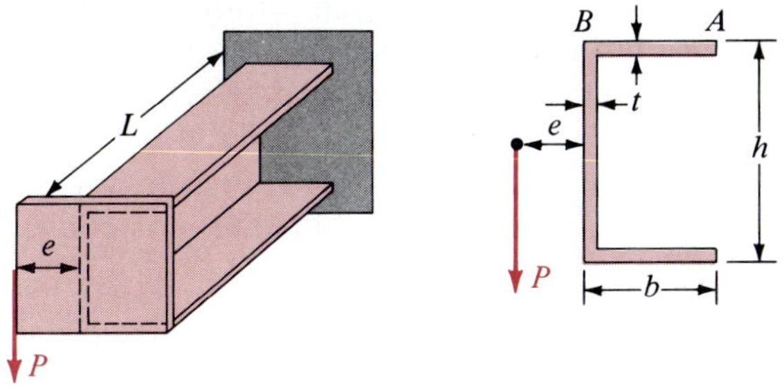

Ps. 7.290 and 7.291

7.291 Repeat problem 7.290 for $L = 7$ ft, $h = 4.25$ in., $b = 2.3$ in., $P = 600$ lbf, and $t = 0.25$ in.

7.11 EXCEPTIONS TO THE THEORY

In this part of the chapter we will consider several commonly encountered technical applications of beam theory, where one or more of the assumptions that are a part of the elementary theory of bending are not satisfied. These exceptions include stress concentrations, nonlinear, and inelastic behavior of beams. In most of these situations, it will be possible to augment or extend the theory in a simple fashion to account for the exception to, or the extension of, the theory. These extensions or augmentations make it possible to use the theory of bending deformations in a much wider range of applications.

7.11.1 Stress Concentrations

One of the most important assumptions of the theory is that the dimensions of the cross section of the beam should vary slowly with position along the axis of the bar as indicated in Figs. 7.125a and b.

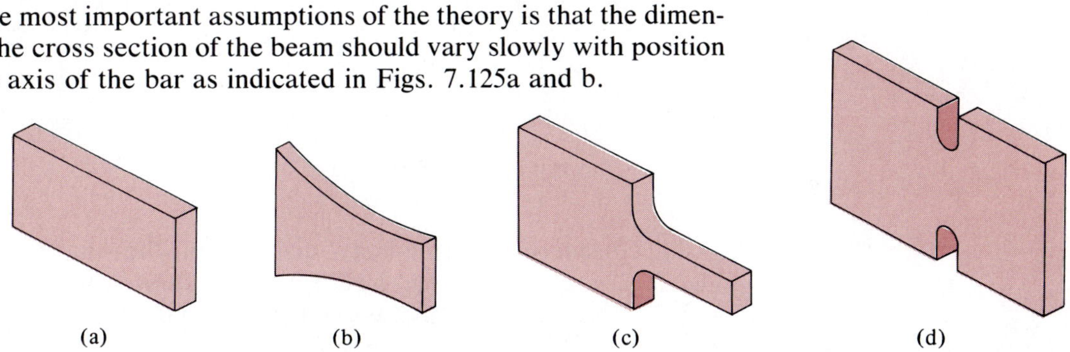

Figure 7.125 Typical beam geometries

However, it is frequently necessary to analyze a situation where there are relatively sudden changes in the cross-sectional dimensions, shown typically in Figs. 7.125c and d. The result is that the bending stresses $\sigma = -My/I$, as given by the elementary theory, are no longer valid in the neighborhood of these sudden changes in cross section. Generally, the stress levels are substantially increased by the presence of these "stress risers" or **stress concentrations.** Particularly with brittle materials subjected to long-term cyclic loadings, knowledge of the states of stress associated with these sudden changes of geometry is critical for investigating fatigue, as discussed in Section 4.8.

Consider the typical case of a fillet, shown in Fig. 7.126. As the thickness of the bar is assumed to be constant, it is commonly called a flat bar.

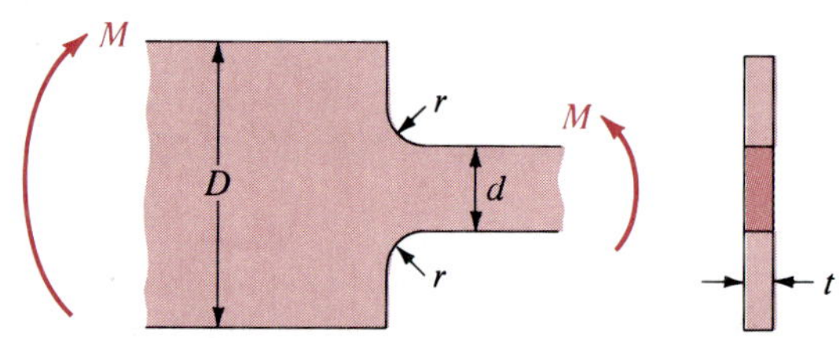

Figure 7.126 Typical fillet geometry

According to St. Venant's principle, the stresses away from the region of sudden change in cross section, such as sections AA and BB in Fig. 7.127a, will be given with good accuracy by $\sigma_x = -My/I$.

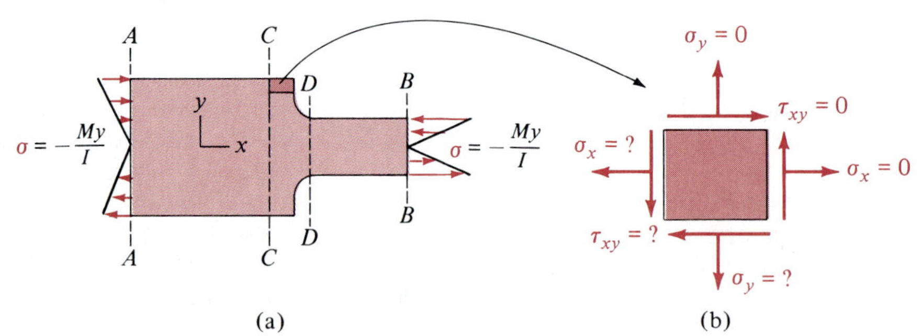

Figure 7.127 Stresses in the neighborhood of a fillet

By considering a free-body diagram of the shaded region of Fig. 7.127a, shown in Fig. 7.127b, it is easily seen that through section CC, the stresses cannot possibly be given by the elementary bending theory. The quantity that is usually of the most interest is the maximum bending stress in the neighborhood of the step change in depth. As shown in Fig. 7.128, the maximum stress occurs at the top and bottom surfaces through the section *DD*.

The value of this maximum stress depends upon the geometry of the step change and is given by

$$\sigma_{\text{max}} = K\sigma_{\text{bending}}$$

where σ_{bending} is the bending stress at section *DD* as given by the elementary theory, that is, $\sigma_{\text{bending}} = \pm 6M/td^2$, and *K* is a stress concentration factor. A graph from which the stress concentration factor for the fillet can be obtained is shown in Fig. 7.129.

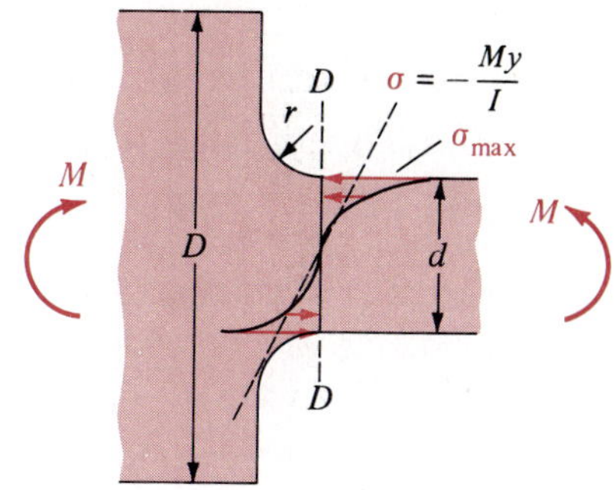

Figure 7.128 Stress distribution through the critical section

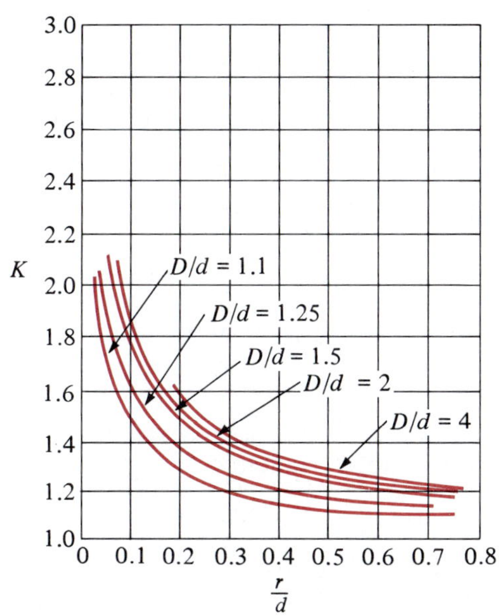

Figure 7.129 Stress concentration factor for a fillet

To determine *K*, the value of *r*/*d*, which has been determined from the geometry, is projected up from the horizontal axis to the proper *D*/*d* curve, after which the stress concentration factor is determined by projecting horizontally to the *K* axis.

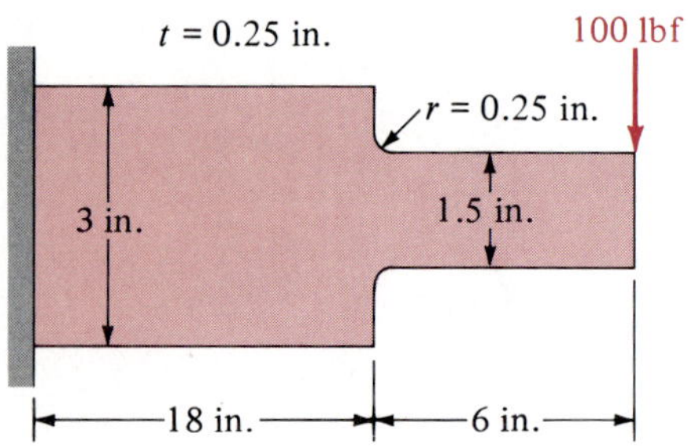

Figure 7.130 Nonprismatic beam subjected to bending

Example 7.23 The beam is subjected to a load of 100 lbf as shown in Figure 7.130. Determine the maximum bending stress.

Solution Two values should be checked: the value at the fillet, due to the stress concentration; and the value at the root, due to the occurrence of the maximum bending moment.

Root The bending stress near the root is given by the elementary theory as

$$\sigma = -\frac{(-2400\text{ in.-lbf})(1.5\text{ in.})}{[(0.25\text{ in.})(3\text{ in.})^3/12]} = 6.4\text{ ksi}$$

Fillet The nominal bending stress through the reduced section is given by

$$\sigma = -\frac{(-575\text{ in.-lbf})(0.75\text{ in.})}{[(0.25\text{ in.})(1.5\text{ in.})^3/12]} = 6.13\text{ ksi}$$

For the fillet, the r/d value is $0.25/1.5 = 0.167$ and the D/d value is 2. Entering the graph of Fig. 7.129 at $r/d = 0.167$, and projecting up to the $D/d = 2$ curve, the K value is determined as $K \approx 1.6$, so that at the fillet

$$\sigma_{\text{max}} \approx (1.6)(6.13\text{ ksi}) = 9.81\text{ ksi}$$

clearly the larger of the two values.

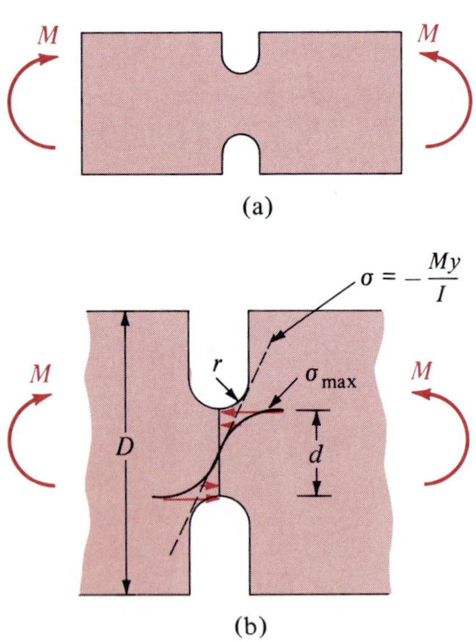

Figure 7.131 Geometry and stress distribution through a section with grooves

Another type of stress concentration is associated with the pure bending of a flat bar containing two symmetrically shaped grooves as shown in Fig. 7.131a.

The maximum bending stress, shown in Fig. 7.131b, occurs at the bottom of the groove and is evaluated according to

$$\sigma_{\text{max}} = K\sigma_{\text{bending}}$$

where σ_{bending} is the bending stress based on the section properties through the groove, or

$$\sigma_{\text{bending}} = \pm\frac{6M}{td^2}$$

with t the thickness of the bar. K is again a stress concentration factor and is determined by referring to Fig. 7.132. Using the graph, the value of K is determined by projecting up from the value of r/d to the appropriate D/d curve and then over to the value of K, the stress concentration factor.

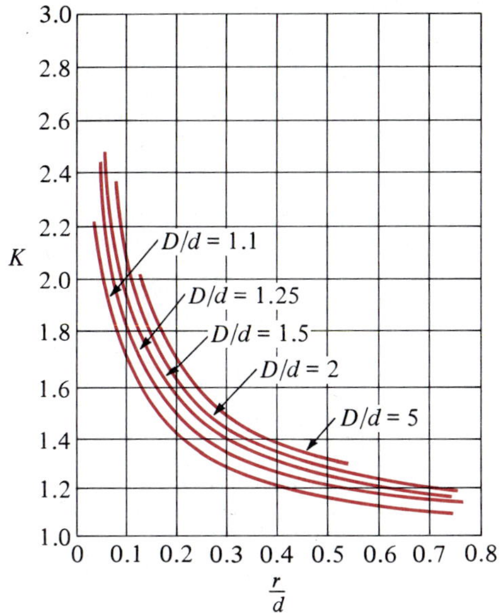

Figure 7.132 Geometry and stress concentration factors for a flat bar with grooves

---

Example 7.24 A beam with grooves is loaded as shown in Fig. 7.133a. Determine the maximum bending stress.

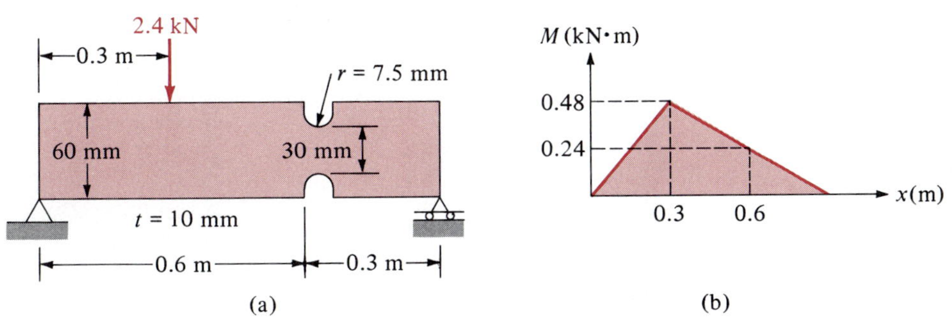

Figure 7.133 Bending of a beam with a groove

Solution The moment diagram is shown in Fig. 7.133b. There are two locations to check: at the load, where the moment is maximum; and at the location of the groove.

Maximum moment location: Away from the grooves the bending stress can be determined using the elementary result

$$\sigma = \frac{(0.48 \text{ kN}\cdot\text{m})(0.03 \text{ m})}{[(0.01 \text{ m})(0.06 \text{ m})^3/12]} = 80 \text{ MPa}$$

Groove location: The nominal bending stress at the groove is given by

$$\sigma = \frac{(0.24 \text{ kN}\cdot\text{m})(0.015 \text{ m})}{[(0.01 \text{ m})(0.03 \text{ m})^3/12]} = 160 \text{ MPa}$$

The value of r/d is 0.25, with $D/d = 2.0$, so that from Fig. 7.132, $K \approx 1.48$, with

$$\sigma_{\max} = 1.48(160 \text{ MPa}) = 236.8 \text{ MPa}$$

The results for the stress concentration factor values given in Figs. 7.129 and 7.132 are, strictly speaking, valid only for pure bending. However, as shown in Examples 7.23 and 7.24, these results are frequently used to estimate the effect that sudden changes in cross section have on bending stresses when there is also a shear force V present. These estimations are usually quite accurate because the maximum bending stress occurs at the top or bottom of a beam, where the shear stress vanishes: the top and bottom surfaces are usually stress-free surfaces.

PROBLEMS

7.292 The allowable bending stress for the 0.5-in.-thick bar shown is 16 ksi. Determine the allowable value of P.

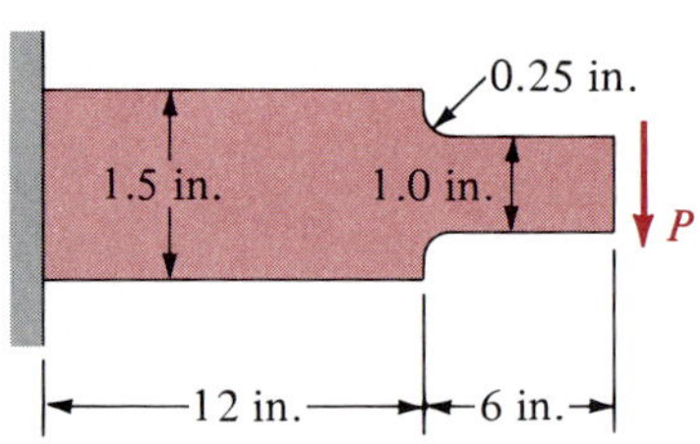

P. 7.292

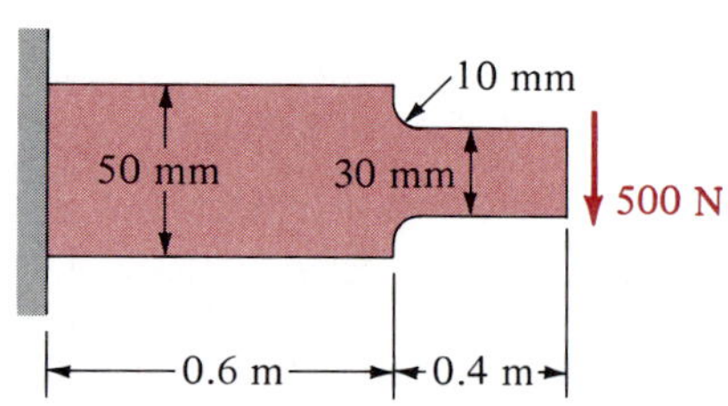

P. 7.293

7.293 Determine the maximum bending stress produced in the 10-mm-thick bar by the loading shown.

7.294 Determine the maximum bending stress produced in the 0.50-in.-thick bar by the loading shown.

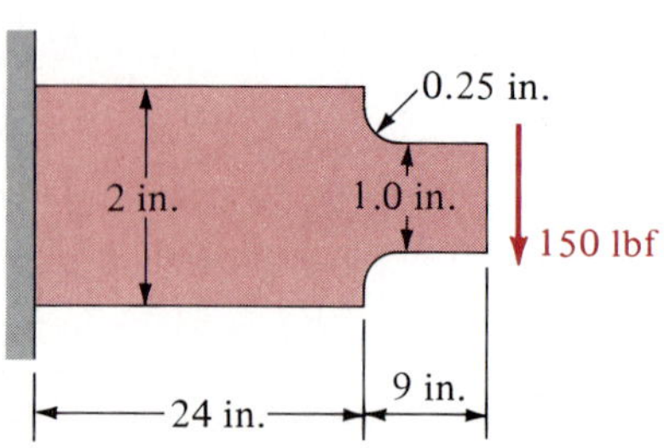

P. 7.294

7.295 What is the allowable moment so that the maximum bending stress does not exceed 48 MPa in the 10-mm thick bar shown?

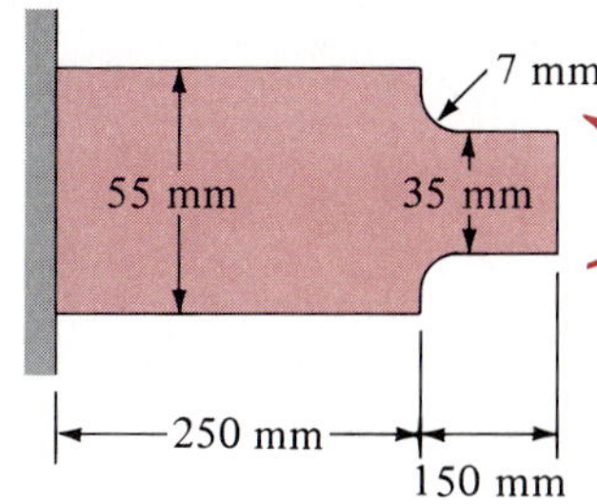

P. 7.295

7.296 The 0.5-in.-thick beam with two symmetrically placed notches is subjected to the bending moment shown. Determine the maximum bending stress if $r =$ *(a)* 0.25 in., *(b)* 0.5 in., and *(c)* 0.75 in.

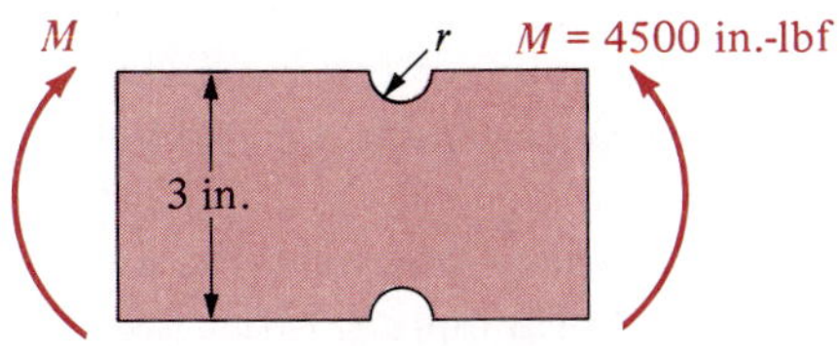

Ps. 7.296 and 7.317

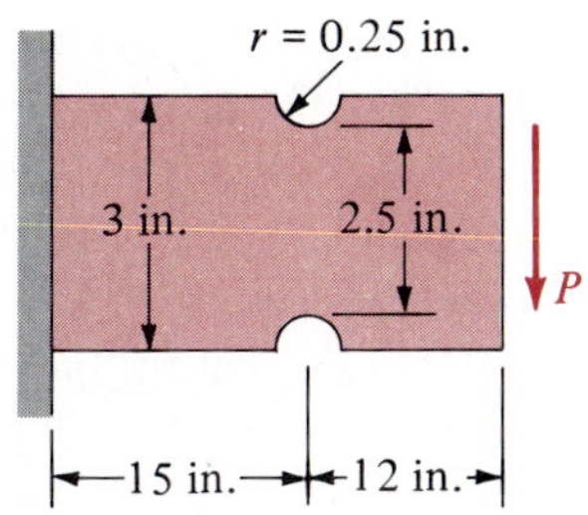

P. 7.297

7.297 What is the allowable load P for the 1-in.-thick bar shown if the allowable bending stress the material can withstand is 20 ksi? Take the factor of safety to be 2.

7.298 The flat bar is subjected to a moment M as shown. For an allowable bending stress of 70 MPa, determine the allowable moment M for *(a)* $d =$ 30 mm, *(b)* $d = 40$ mm, and *(c)* $d = 50$ mm. Take $D = 70$ mm, $r = 5$ mm, and $t =$ 10 mm.

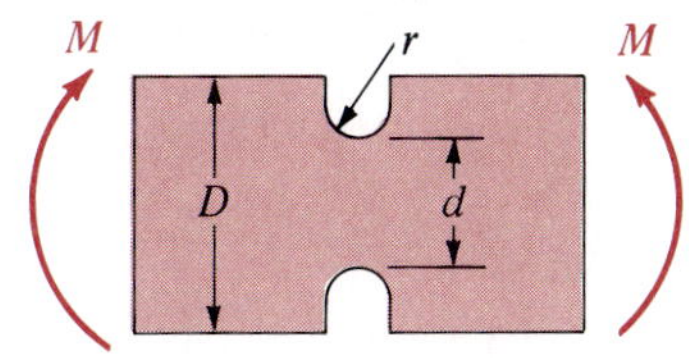

Ps. 7.298 and 7.299

7.299 The flat bar is subjected to a moment M as shown. For an allowable bending stress of 12 ksi, determine the allowable moment M for *(a)* $d = 1.2$ in., *(b)* $d = 1.6$ in., and *(c)* $d = 2.0$ in. Take $D = 2.75$ in., $r = 0.25$ in., and $t = 0.5$ in.

7.300 The simply supported beam has semicircular notches of 1.0-in. radii located at the center of the beam as shown. Take $t = 1$ in., $w_0 = 2$ kips/ft, and determine the minimum h in order that the maximum bending stress not exceed 15 ksi.

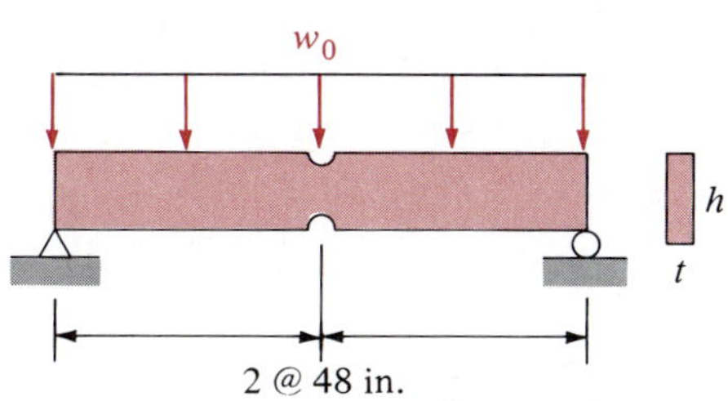

P. 7.300

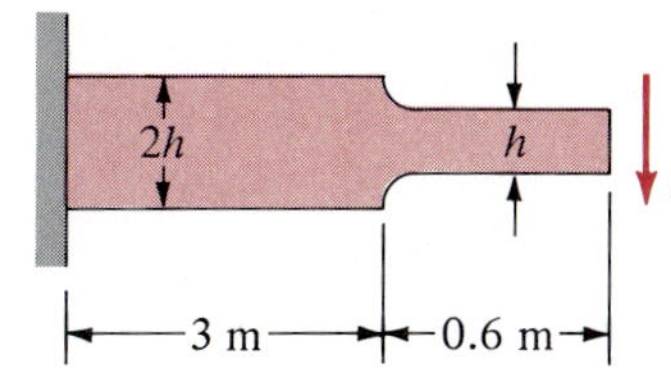

P. 7.301

7.301 The cantilever beam has quarter-circle fillet notches as shown. Take $t = 25$ mm, $P = 40$ kN, and determine the minimum h in order that the maximum bending stress not exceed 60 MPa.

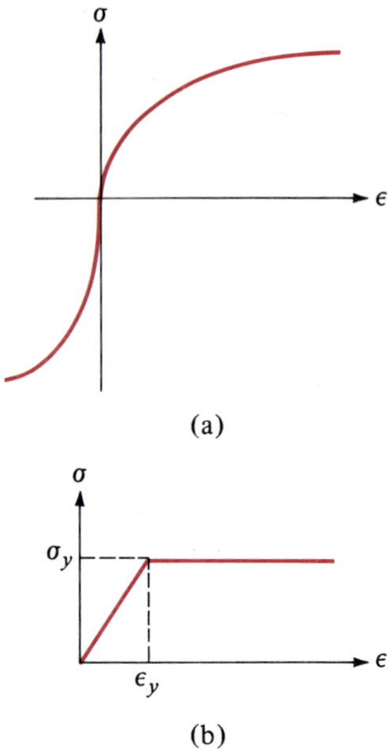

Figure 7.134 Inelastic and nonlinear material behavior

7.11.2 Nonlinear and Inelastic Behavior of Beams

The derivation of the equations for the elementary theory of beams was based, in part, on the assumption that the material behaves in a linearly elastic fashion, that is, the normal stress σ and the normal strain ε are proportional according to $\sigma = E\varepsilon$, where E, called Young's modulus, is the constant of proportionality. As long as the stresses are below the proportional limit, this linear relationship is assumed to be valid. When there is no linear portion of the stress-strain curve as indicated in Fig. 7.134a, or when, for a material that behaves linearly for small stresses and strains, the strains exceed the value of the yield strain as indicated in Fig. 7.134b, the elementary theory must be modified to reflect the change in material behavior.

In this section, we will develop the equations necessary to analyze the bending response of a beam composed of a material that has been idealized as elastic–perfectly plastic, when the beam is loaded beyond the elastic range.

Elastic–Perfectly Plastic Behavior A technically important problem occurs when a beam made from a ductile material is stressed beyond the elastic range, as indicated in Fig. 7.135a. For a typical ductile material, the essential load-carrying capabilities for stresses beyond the yield point can be approximated nicely by assuming the material to behave in an elastic–perfectly plastic manner as shown by the dotted line in Fig. 7.135a. The horizontal value at which the stress remains constant with further straining is taken to be approximately the yield stress σ_Y, obtained from the uniaxial tension test. The resulting idealized stress-strain curve is pictured in Fig. 7.135b.

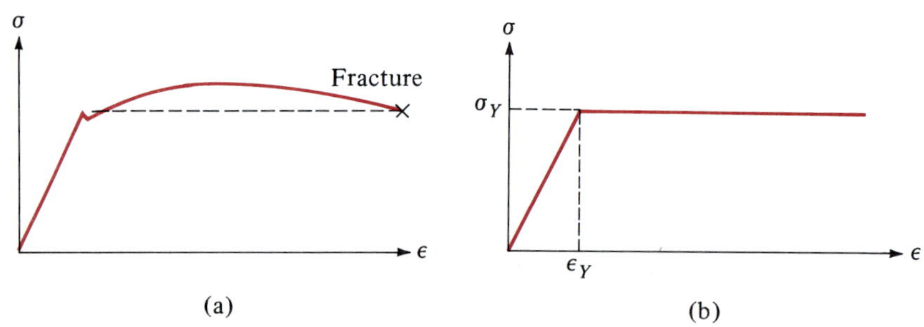

Figure 7.135 Idealization of a ductile material as elastic–perfectly plastic

Consider then the stresses in a rectangular-cross-section beam subjected to a moment M as indicated in Fig. 7.136a.

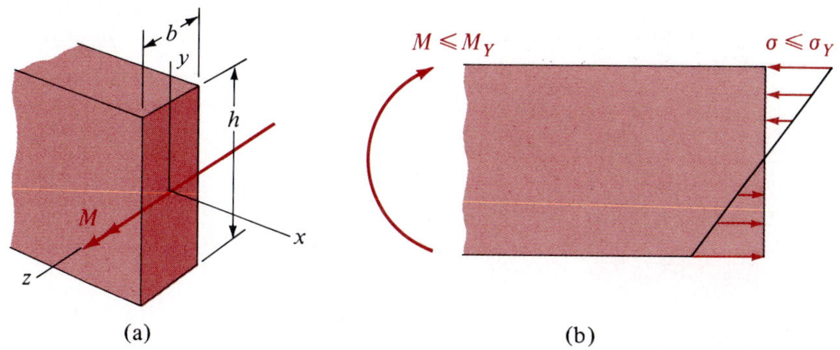

Figure 7.136 Rectangular-cross-section beam subjected to a bending moment M

As long as the maximum stresses at the top and bottom do not exceed the yield stress σ_Y, the stresses are given by $\sigma = -My/I$, per the elementary theory for a linearly elastic material. The moment M_Y, corresponding to a maximum stress of σ_Y at $y = \pm h/2$, is then given by

$$\sigma_Y = M_Y \frac{h}{2I} \qquad \text{or} \qquad M_Y = \frac{bh^2\sigma_Y}{6}$$

where M_Y is called the **maximum elastic moment.** In order to analyze the behavior of the beam when $M > M_Y$, we need to review the basic assumptions that lead to the elementary bending theory.

Equilibrium For pure bending, the equilibrium relations

$$\int \sigma_x \, dA = 0 \qquad \text{and} \qquad \int y\sigma_x \, dA = -M$$

must be satisfied, independent of the material behavior.

Deformation For pure bending, the arguments leading to the assumption that plane sections remain plane and perpendicular to the axis of the beam are also independent of the type of material. As pictured in Fig. 7.137, we will continue to assume that the strain is directly proportional to the distance from the neutral axis, that is, $\varepsilon_x = ky$.

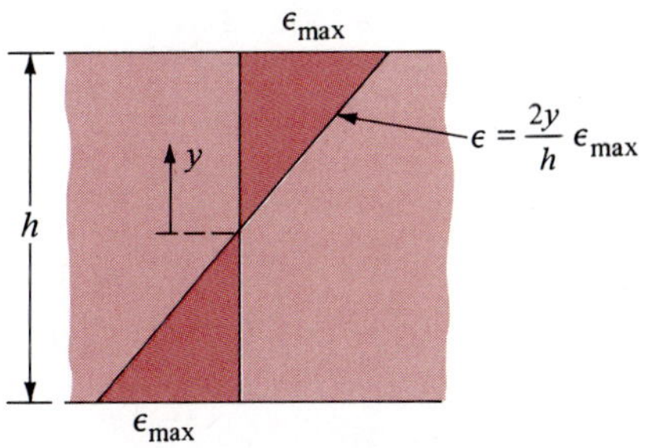

Figure 7.137 Linear variation of strain with position relative to the neutral axis

Material Behavior The idealized material behavior of Fig. 7.135b indicates that even when the strain ε at a point in the beam exceeds the value ε_Y, the stress cannot exceed σ_Y, the yield-stress value. Shown in Fig. 7.138 are the strain distributions, the stress-strain diagrams, and the corresponding stress distributions for several values of the applied moment M.

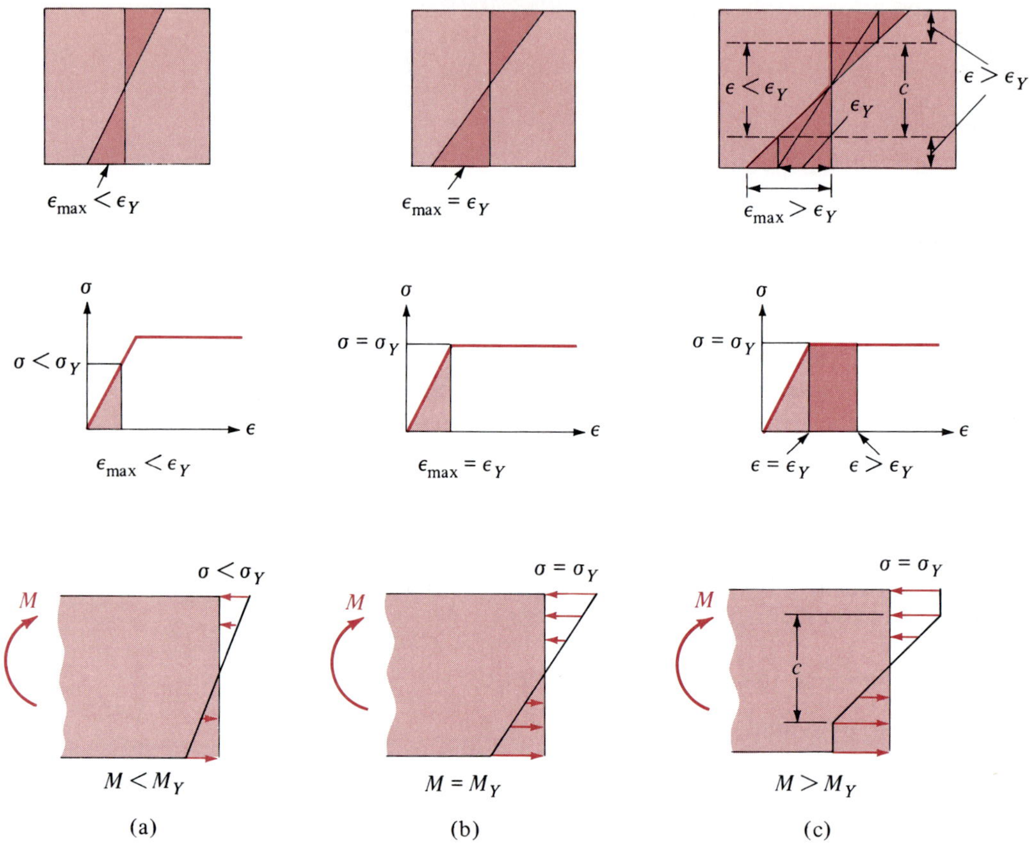

Figure 7.138 Strain distributions, stress-strain diagrams, and stress distributions corresponding to various applied moments M

For the case shown in Fig. 7.138a, where $M < M_Y$, the strains are everywhere less than the yield strain ε_Y, with the result that the stresses, which are proportional to the strains, are entirely in the elastic range and vary linearly with position y. For the second case shown in Fig. 7.137b, with $M = M_Y$, the maximum elastic moment, the maximum strain is equal to ε_Y with the stress again varying linearly up to the maximum value of σ_Y. For the case shown in Fig. 7.137c, the moment has been increased beyond M_Y, with the result that the strains at $y = \pm h/2$ are greater than the yield strain ε_Y. The strains are greater than ε_Y outside of an **elastic core** of depth c given by

$$\frac{(c/2)}{(h/2)} = \frac{\varepsilon_Y}{\varepsilon_{\max}}$$

For those points outside of the elastic core, where $\varepsilon > \varepsilon_Y$, the stress remains at the yield level σ_Y. Inside the elastic core the stresses are proportional to the strains and can be represented as

$$\sigma = -\frac{2y}{h}\sigma_Y$$

Thus, for the case of Fig. 7.137c, the moment is evaluated as

$$\frac{M}{2} = -\left[\int_0^{c/2} -\frac{2\sigma_Y y}{c} yb\,dy + \int_{c/2}^{h/2} (-\sigma_Y) yb\,dy\right]$$

$$= \frac{\sigma_Y b}{3}\left[3\left(\frac{h}{2}\right)^2 - \left(\frac{c}{2}\right)^2\right]$$

Using the fact that $M_Y = bh^2\sigma_Y/6$, the moment M can be expressed as

$$M = M_Y\left[\frac{3}{2} - \frac{1}{2}\left(\frac{c}{h}\right)^2\right] \tag{7.29}$$

In the limit as c approaches zero, that is, when there is no elastic core remaining, M approaches $1.5M_Y = M_P$, where M_P is referred to as the **fully plastic moment.**

To put Eqn. (7.29) in a more useful form, recall that for pure bending the strain ε is related to the radius of curvature ρ by the relation

$$\varepsilon = \frac{-y}{\rho}$$

Taking the radius of curvature corresponding to the maximum elastic moment M_Y to be ρ_Y, and the radius of curvature corresponding to the moment $M > M_Y$ to be ρ, it follows that

$$\varepsilon_Y = -\frac{h}{2\rho_Y}$$

Similarly, since the strain at $y = c/2$, the boundary of the elastic core, is equal to ε_Y, the corresponding strain is given by

$$\varepsilon_Y = -\frac{c}{2\rho}$$

Equating these two expressions yields

$$\frac{c}{h} = \frac{\rho}{\rho_Y}$$

Thus for $\rho \leq \rho_Y$, the moment expression given by Eqn. (7.29) can be expressed in terms of the curvatures as

$$\frac{M}{M_Y} = \frac{3}{2}\left[1 - \frac{1}{3}\left(\frac{\rho}{\rho_Y}\right)^2\right] \tag{7.30}$$

and is plotted in Fig. 7.139. When $M \leq M_Y$, the curvature is given by the linear relationship $1/\rho = M/EI$.

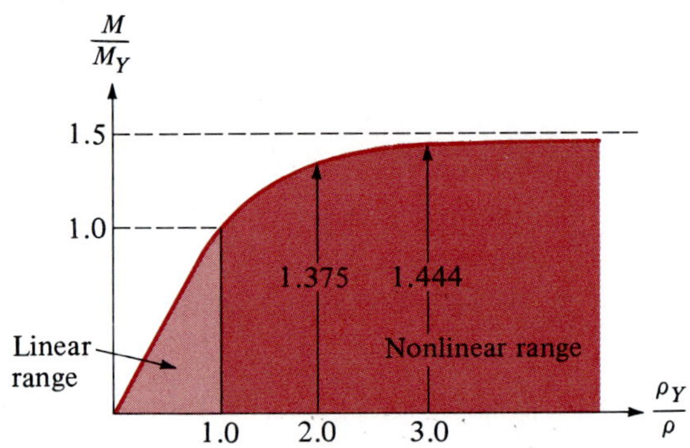

Figure 7.139 Moment-curvature relation for an elastic–perfectly plastic beam

Given the value of the applied moment $M_Y < M < 1.5M_Y$, the radius of curvature can be determined from the curve or by solving Eqn. (7.30). Note, however, that in the limit as M approaches the fully plastic moment M_P, Eqn. (7.30) predicts that the radius of curvature ρ must approach zero, a physical impossibility indicating that, in reality, the value of M_P cannot be reached.

Example 7.25 The rectangular beam composed of a ductile material whose stress-strain curve has been idealized according to Fig. 7.140b is subjected to pure bending as shown in Fig. 7.140a. Determine *(a)* the maximum elastic moment, *(b)* the fully plastic moment, *(c)* the radius of curvature when $M = M_Y$, and *(d)* the radius of curvature when $M = 1.25M_Y$.

Solution The maximum elastic moment M_Y is given by

$$M_Y = \frac{bh^2\sigma_Y}{6} = \frac{(0.015 \text{ m})(0.060 \text{ m})^2(200 \text{ MPa})}{6} = 1.8 \text{ kN·m}$$

The fully plastic moment is then $M_P = 1.5(1.8 \text{ kN·m}) = 2.7 \text{ kN·m}$. The radius of curvature for the maximum elastic moment M_Y is given by

$$\rho_Y = \frac{EI}{M_Y} = \frac{(200 \text{ GPa})(0.015 \text{ m})(0.060 \text{ m})^3}{12(1.8 \text{ kN·m})} = 30 \text{ m}$$

The radius of curvature for $M = 1.25M_Y$ is obtained by solving

$$\frac{M}{M_Y} = 1.5 - 0.5\left(\frac{\rho}{\rho_Y}\right)^2$$

from which $(\rho/\rho_Y)^2 = 0.5$, so that $\rho = 0.707\rho_Y = 21.21$ m. Note also that $c/h = 0.707$, so that the stress distribution corresponding to $M = 1.25M_Y$ is as indicated in Fig. 7.141.

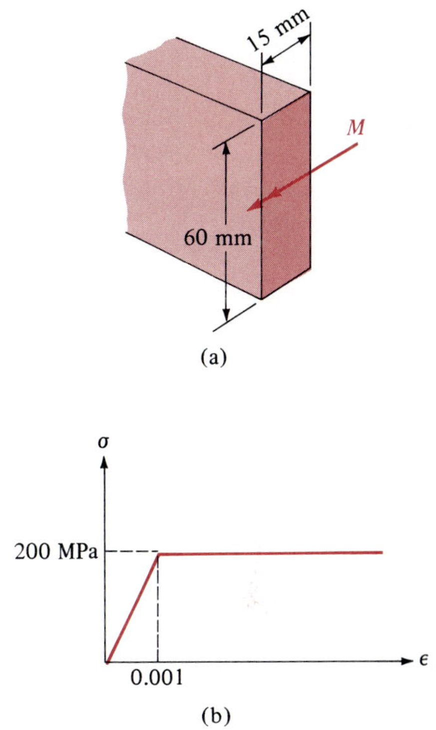

Figure 7.140 Bending of a rectangular beam

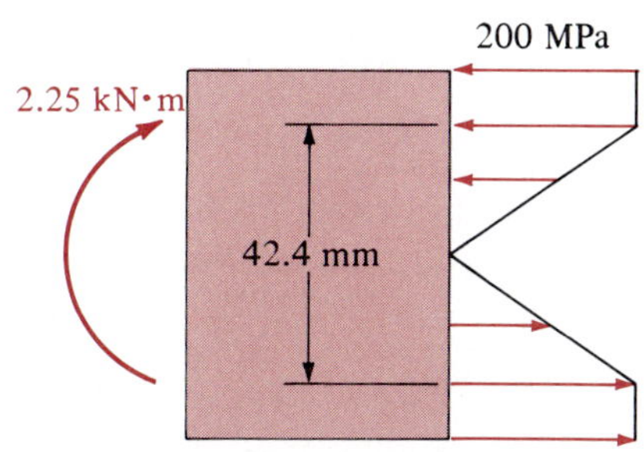

Figure 7.141 Elastic–perfectly plastic stress distribution

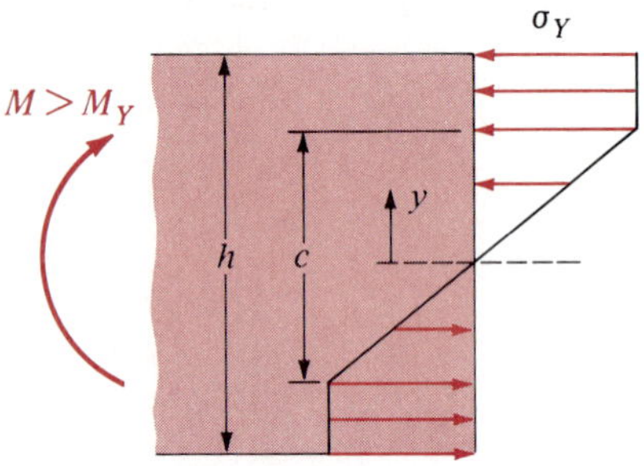

Figure 7.142 Bending-stress distribution in a beam subjected to a moment $M > M_Y$

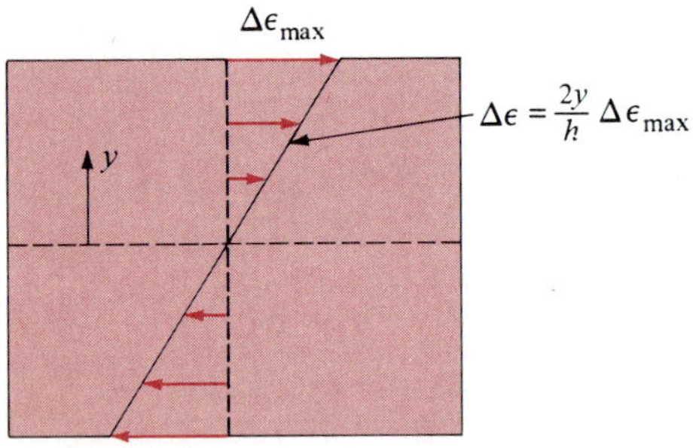

Figure 7.143 Plane sections remain plane with additional strains upon unloading

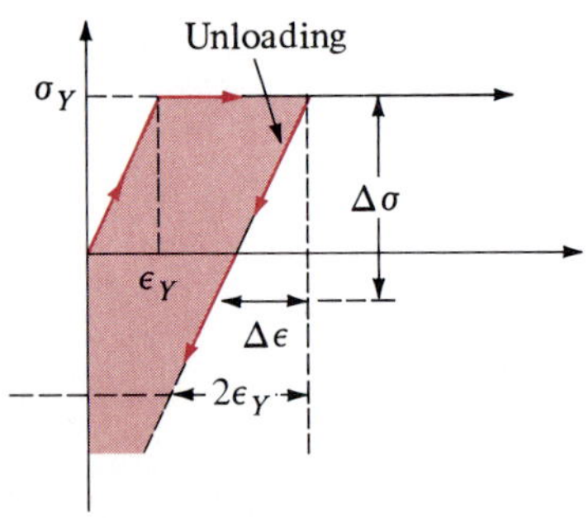

Figure 7.144 Stress-strain relations during unloading

Residual Stresses As indicated by the results of Example 7.25, the bending-stress distribution in an elastic–perfectly plastic beam subjected to a moment $M > M_Y$ appears generally as indicated in Fig. 7.142. If the moment is subsequently removed, there will be **residual stresses** that must satisfy the equilibrium equations

$$\int \sigma \, dA = 0 \qquad \text{and} \qquad -\int y\sigma \, dA = M = 0$$

that is, even though the bending stresses themselves are not identically zero, the resultant force and resultant moment arising from the bending stresses must both be zero.

We assume that the additional deformations that occur upon release of the moment $M > M_Y$ take place according to the assumption that plane sections remain plane; thus, as indicated in Fig. 7.143, the additional strains associated with the unloading can be expressed as

$$\Delta\varepsilon = ky$$

where k is a constant of proportionality.

Referring to Fig. 7.144, the basic idea associated with the material behavior is that upon unloading, the relationship between the additional strains and stresses at points outside of the elastic core is determined from curves parallel to the loading curve.

As long as the magnitudes of the additional strains $\Delta\varepsilon$ are less than $2\varepsilon_Y$, the relation between the additional strains and stresses is linear. Hence the corresponding additional stresses are also a linear function of y, that is,

$$\Delta\sigma = E\Delta\varepsilon = Eky = \beta y$$

where β is another constant of proportionality. Thus the release of the moment can be considered as a superposition of the stresses shown in Fig. 7.145.

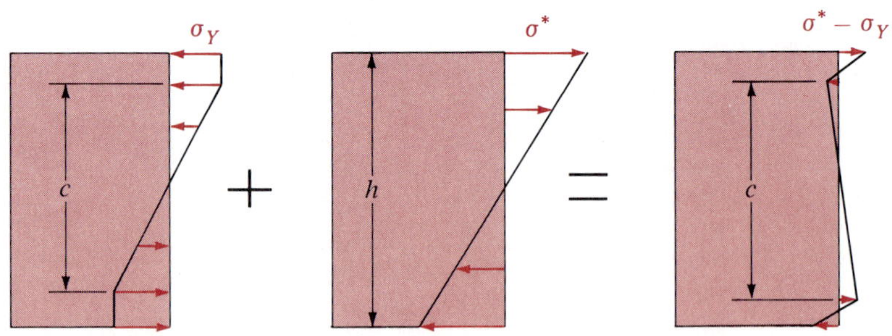

Figure 7.145 Superposition of initial and unloading stresses

The resultant force arising from the superposition of the two stress distributions is obviously zero, with the value of σ^* chosen so that $M + M^* = 0$; thus, so that

$$\frac{bh^2\sigma_Y}{6}\left(\frac{3}{2}\right)\left[1 - \frac{1}{3}\left(\frac{c}{h}\right)^2\right] - \frac{bh^2\sigma^*}{6} = 0$$

or

$$\frac{\sigma^*}{\sigma_Y} = \left(\frac{3}{2}\right)\left[1 - \frac{1}{3}\left(\frac{c}{h}\right)^2\right]$$

The final distribution shown in Fig. 7.145c is a function of the yield stress σ_Y and of the ratio c/h.

To determine the deformations associated with the residual stresses, we note that within the elastic core, the relation $\sigma = E\varepsilon$ is valid. Thus the residual stress σ at the outer boundary of the elastic core is given by

$$\sigma = \sigma_Y - \sigma^*\left(\frac{c}{h}\right) = \sigma_Y\left[1 - \frac{3}{2}\frac{c}{h} + \frac{1}{2}\left(\frac{c}{h}\right)^3\right]$$

The corresponding strain at the boundary of the elastic core is $\varepsilon = \sigma/E$, with the corresponding residual radius of curvature given by

$$\rho = \frac{c}{2\varepsilon} = \frac{c}{2\varepsilon_Y\left[1 - \frac{3}{2}\frac{c}{h} + \frac{1}{2}\left(\frac{c}{h}\right)^3\right]}$$

Note that as c approaches h (i.e., as M approaches M_Y), the radius of curvature approaches infinity, corresponding to the straight unloaded configuration of the beam as expected.

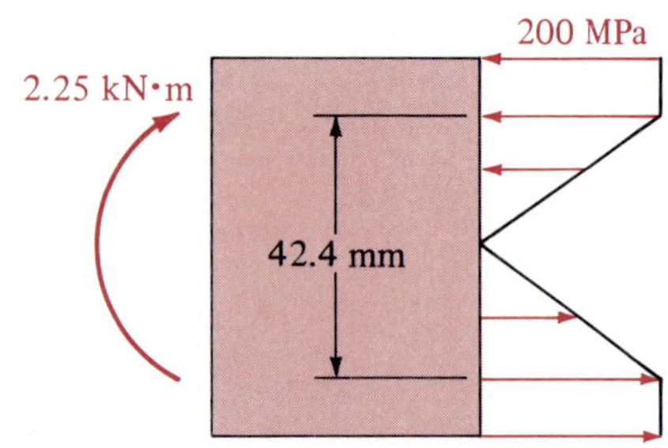

Figure 7.146 Stress distribution corresponding to $M = 1.25M_Y$

Example 7.26 The stress distribution corresponding to $M = 1.25M_Y$ for Example 7.25 is shown in Fig. 7.146. Determine the residual-stress distribution and the final radius of curvature if the moment is released.

Solution The maximum stress associated with the moment to be superposed is given by

$$\sigma^* = \sigma_Y \left(\frac{3}{2}\right)\left(1 - \frac{1}{3}\left(\frac{c}{h}\right)^2\right) - (200 \text{ MPa}) \frac{3}{2}\left[1 - \frac{1}{3}(0.707)^2\right] = 250 \text{ MPa}$$

The stress distributions corresponding to $M = 1.25M_Y$ and σ^* are shown in Figs. 7.147a and b, respectively. The sum, shown in Fig. 7.147c, is the residual-stress distribution.

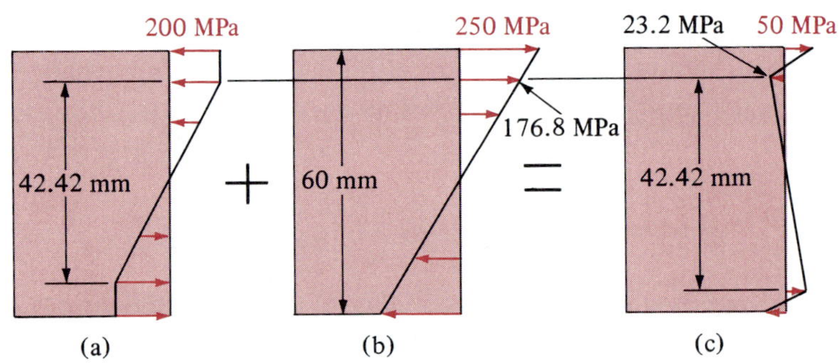

Figure 7.147 Superposition of elastic–perfectly plastic and elastic stresses to yield the residual stress

The value of the final radius of curvature is obtained by evaluating the expression

$$\rho = \frac{c}{2\varepsilon_Y[1 - 3/2(c/h) + 1/2(c/h)^3]}$$

$$= \frac{(0.707)(0.060 \text{ m})}{2(0.001)[1 - 3/2(0.707) + 1/2(0.707)^3]} = 182.6 \text{ m}.$$

To get an idea of the magnitude of these residual deformations, consider Fig. 7.148, where the bar is taken to have a length of 1.0 m.

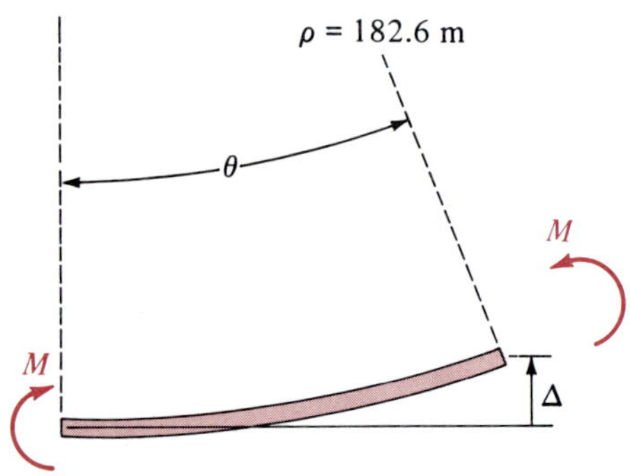

Figure 7.148 Residual deformations

The subtended angle θ can be determined from $s = R\theta$ as

$$\theta = \frac{1.0 \text{ m}}{182.6 \text{ m}} = 0.005476 \text{ rad}$$

The corresponding deflection Δ shown is calculated from

$$\Delta = \rho(1 - \cos\theta) = 0.0027 \text{ m}$$

It is left for the student to show that before the release of the moment, the corresponding deflection was $\Delta = 0.0236$ m, indicating the relative magnitudes of the deformations before and after release of the moment.

PROBLEMS

7.302 Verify that the stress distribution shown in Fig. 7.141 of Example 7.25 is equivalent to the moment of 2.25 kN·m.

7.303 Verify in Example 7.26, that the tip deflection of the 1-m-long bar under the action of the 2.25 kN·m moment is 0.0236 m.

7.304–7.309 For each of the cross sections shown, determine *(a)* the maximum elastic moment M_Y, *(b)* the fully plastic moment M_P, and *(c)* the ratio M_P/M_Y. For the hollow-box section take $b = h$ and $t = h/8$. For the hollow circle take $b = 0.75 \times a$.

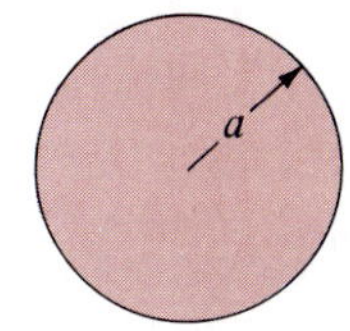

P. 7.304

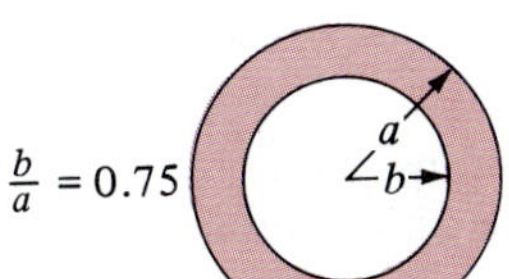

Ps. 7.305 and 7.310

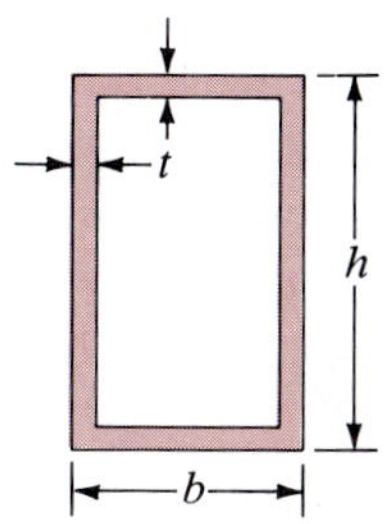

Ps. 7.306 and 7.311

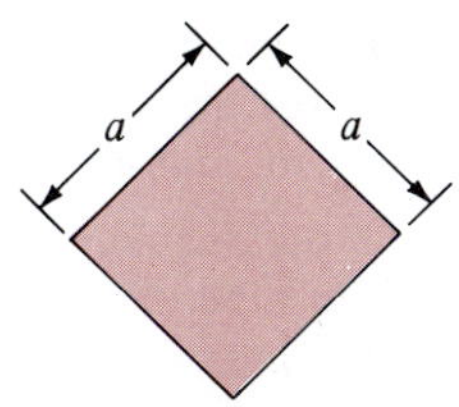

P. 7.307

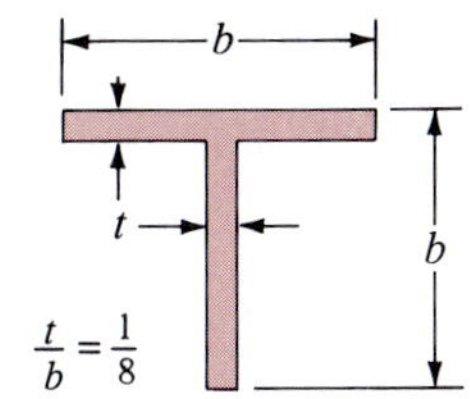

P. 7.308

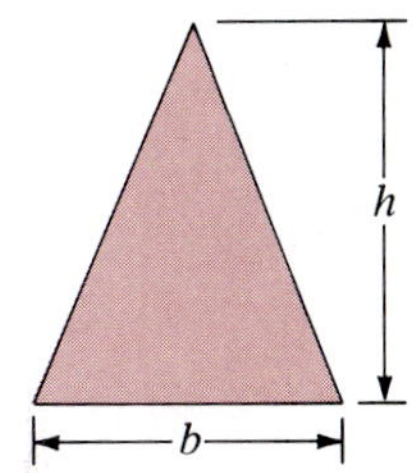

P. 7.309

7.310 For the hollow circular cross section of problem 7.305, specialize the result when the thickness $t = a - b$ is small.

7.311 Repeat problem 7.306 if $h = 2b$ and $t = h/16$.

7.312 For a W 10 × 30 section made from a structural steel with $\sigma_Y = 32$ ksi, determine *(a)* M_Y, *(b)* M_P, and *(c)* M_P/M_Y.

7.313 For a W 250 × 44.8 section made from a structural steel with $\sigma_Y =$ 220 MPa, determine *(a)* M_Y, *(b)* M_P, and *(c)* M_P/M_Y.

7.314 A hollow square bar is subjected to a moment of $1.2M_Y$ as shown. Determine *(a)* the depth of the elastic core, *(b)* the radius of curvature, *(c)* the residual stresses upon release of the moment, and *(d)* the tip deflection upon release of the moment. Take $a = 40$ mm, $t = 4$ mm, $L = 0.8$ mm, $E = 70$ GPa, and $\varepsilon_Y = 0.001$.

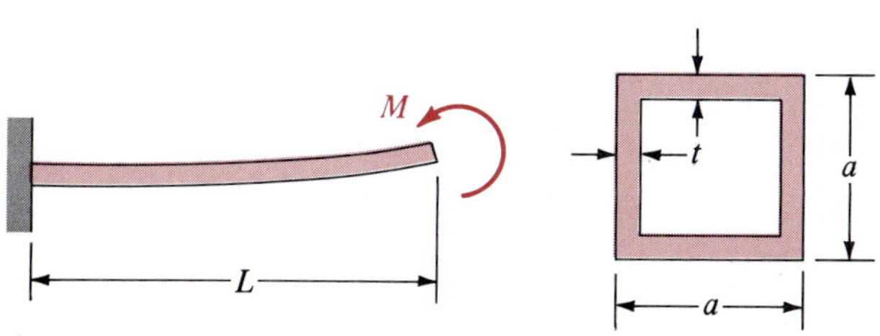

P. 7.314

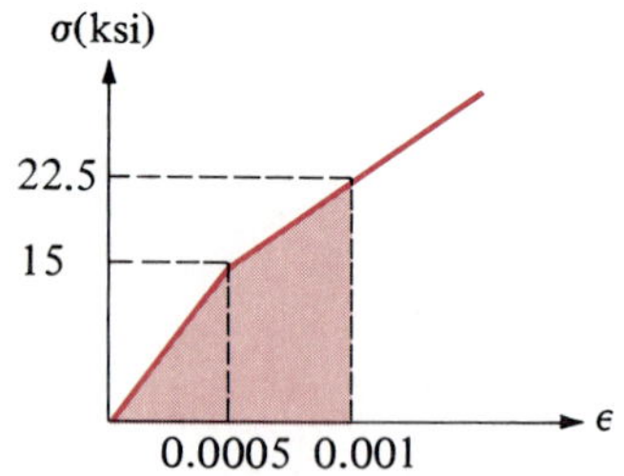

P. 7.318

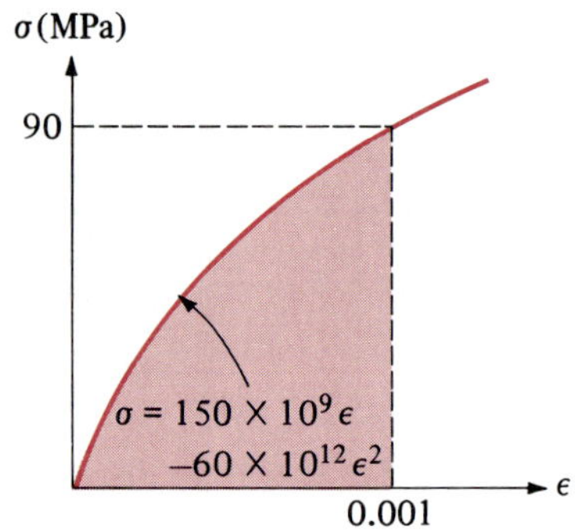

P. 7.319

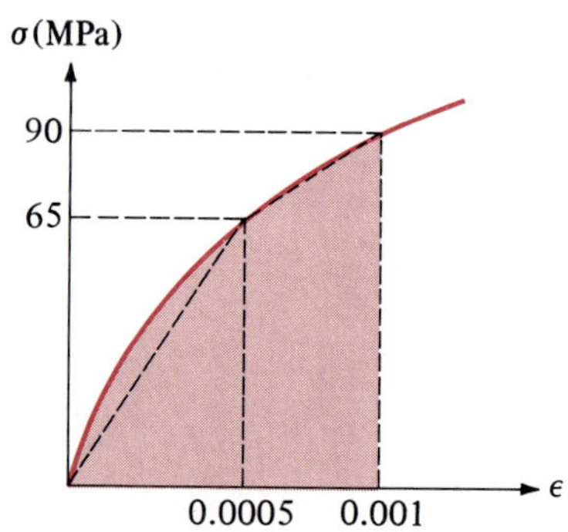

P. 7.320

7.315 A W 10 × 30 section is made from an elastic–perfectly plastic steel with $\sigma_Y = 32$ ksi and subjected to a pure moment M resulting in yielding in the flanges. Determine *(a)* the moment M, and *(b)* the residual stresses upon release of the moment.

7.316 A rectangular-cross-section beam of width b and height h is composed of an elastic material that behaves in tension according to $\sigma = \beta\varepsilon^n$. Assuming that the behavior is the same in compression and that plane sections remain plane, develop *(a)* the moment curvature relationship, and *(b)* an expression for the maximum bending stress as a function of M, n, b, and h.

7.317 Repeat problem 7.296 for a circular cross section with $n = 0.5$. Express the maximum bending stress as a function of M and the radius of the circular cross section.

7.318 A rectangular-cross-section beam composed of the material with the stress-strain curve shown is subjected to a bending moment of 1.5×10^6 in.-lbf. Determine *(a)* the stress distribution, and *(b)* the maximum bending stress. Take $b = 5$ in. and $h = 8$ in.

7.319 A rectangular-cross-section beam with $b = 40$ mm and $h = 90$ mm is subjected to a bending moment of 5 kN·m. For the elastic material with the stress-strain curve shown, determine *(a)* the maximum bending stress, and *(b)* the radius of curvature.

7.320 Replace the stress-strain curve of problem 7.319 with the bilinear stress-strain curve shown and resolve. Compare your answer with problem 7.319.

7.12 CLOSURE

In this chapter we have shown that the elementary theory of beams can be used to begin to answer the strength problem. Ascertaining the strength of beams consists in part of drawing the shear and moment diagrams to determine the moment $M(x)$ and the shear $V(x)$. The bending stresses can then be determined using

$$\sigma = -\frac{My}{I}$$

and the transverse shear stresses using

$$\tau = \frac{VQ}{It}$$

These two results are based on a set of assumptions:

1. The axis of the bar, usually chosen to be the x-axis, is straight. Cross-sectional dimensions are small compared to the length of the bar, and vary slowly along the axis of the bar. The shape of the cross section is such that the y-axis can be chosen as a plane of symmetry.
2. Loading consists of forces applied in the y-direction and lying in the plane of symmetry, and of moments about the z-axis.
3. Slopes associated with the transverse displacements are small.
4. The beam is composed of material behaving in a linearly elastic fashion.

If any of these assumptions is not satisfied, the elementary theory cannot necessarily be expected to predict the bending and shear stresses with accuracy. It is the responsibility of the user to ensure that, when all necessary assumptions are not satisfied, any exceptions are taken into account in a rational manner.

COMPUTER EXERCISES

CE 7.1 Using the two equilibrium equations $V_2 - V_1 = -\int q\,dx$ and $M_2 - M_1 = \int V\,dx$ discussed in Section 7.6, write a computer code that determines the shear and moment as a function of position along a beam. Define nodes at regular intervals as indicated and input the following:

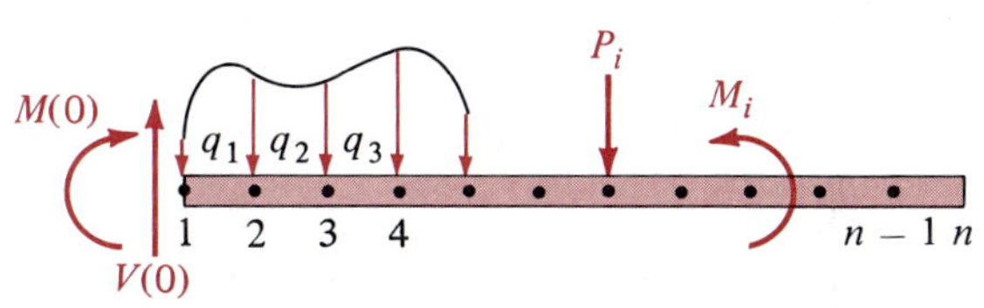

CE 7.1

$V(0) = V_1$, the shear at $x = 0$.

$M(0) = M_1$, the moment at $x = 0$.

q_i: the values of the loading at the nodes, for $i = 1, 2, \ldots, n$.

NC: the number of concentrated loads and/or reactions and the locations and values of each of the concentrated loads.

NM: the number of concentrated moments and the locations and values of each of the concentrated moments.

Output the shear and moment at the nodes.

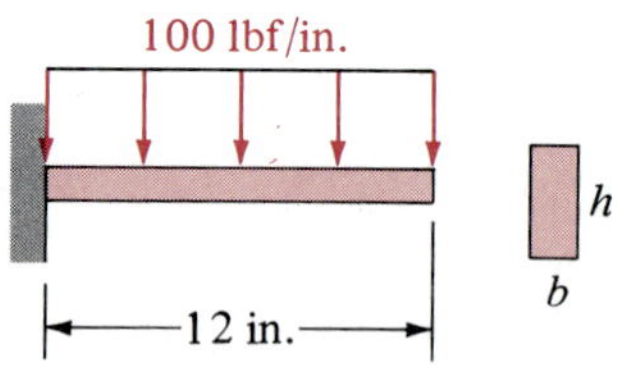

AD 7.1

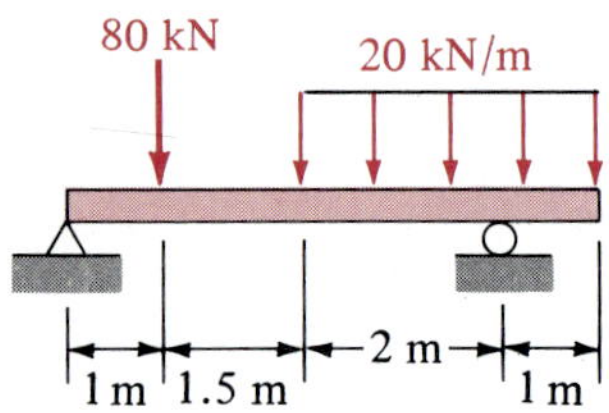

AD 7.2

ANALYSIS/DESIGN PROBLEMS

AD 7.1 The cantilever beam is to carry a distributed load of 100 lbf/in. as shown. With $\sigma_Y = 13.5$ ksi, determine a suitable rectangular cross section using the maximum shear stress criterion.

AD 7.2 The overhang beam is to support the loads shown. Given that $\sigma_{ult} = 150$ MPa, select a suitable W section for the beam using the maximum normal stress criterion with a factor of safety of 1.75.

AD 7.3 A circular pipe is to be used as a cantilever beam loaded as shown. If $\sigma_Y = 18$ ksi and $E = 29 \times 10^6$ psi, select a steel pipe that will suffice using the maximum shear stress criterion.

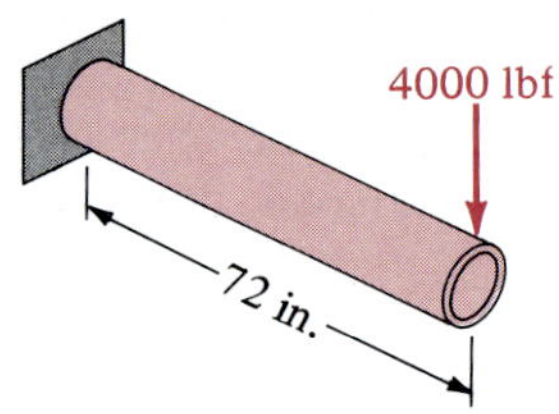

AD 7.3

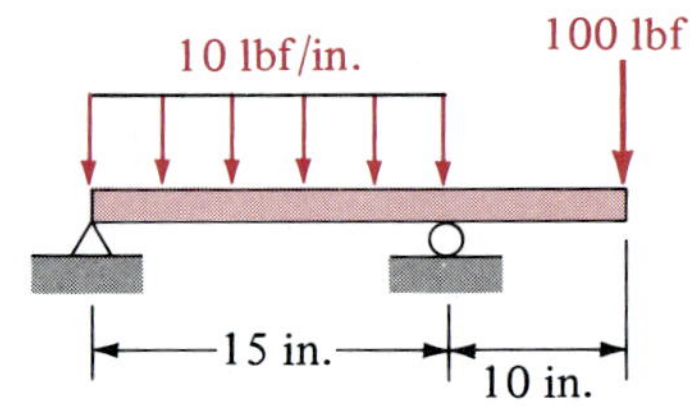

AD 7.4

AD 7.4 The beam shown is to be constructed from a hollow square section whose wall thickness is one-tenth the outside dimension and is made from an aluminum with a yield stress in tension of 18 ksi and a Young's modulus of 10×10^6 psi. Using the maximum shear stress criterion, select the outside diameter of the square.

AD 7.5 A beam is to support the 70-lbf/ft loading over a 40-ft span as shown. If the beam is to be formed by gluing available 4-in. by 6-in. sections, evolve a suitable design that includes the weight of the beam. Take the allowable bending stress in the wood to be 1.5 ksi, the allowable horizontal shear stress in the wood to be 150 psi, the allowable shear stress in the glue to be 100 psi, and the weight density of the wood to be 40 lbf/ft^3.

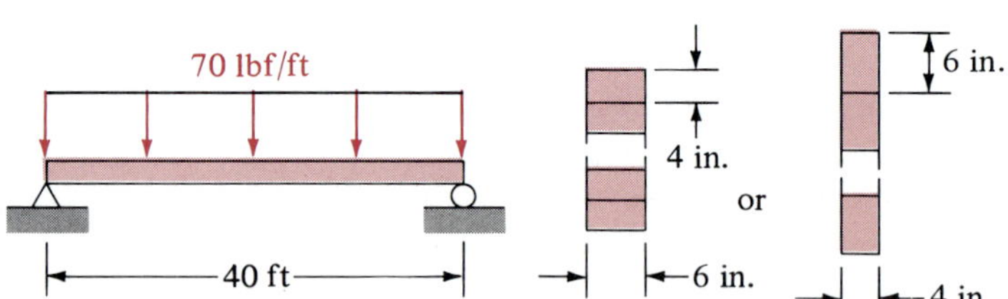

AD 7.5

AD 7.6 Using the maximum normal stress criterion with a factor of safety of 2.2, select a hollow circular section that will carry the loading shown. The material has a tensile ultimate strength of 70 MPa.

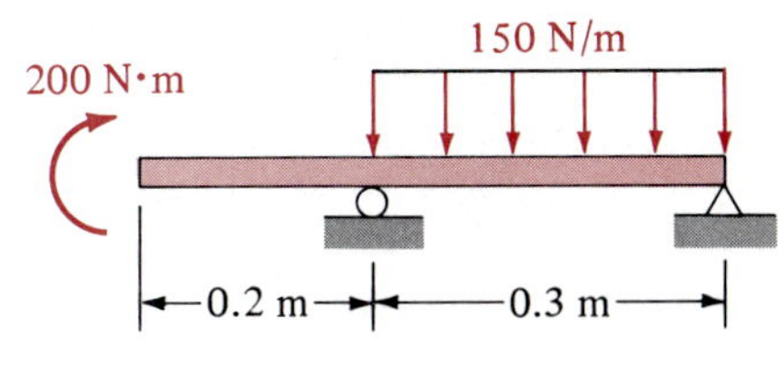

AD 7.6

AD 7.7 The system of beams shown is intended to support a deck that results in a 1750-lbf/ft loading on each of the 15-ft-long beams. Each of the beams rests on the surrounding footings and on the main beam of length 24 ft. If the yield stress for all members is $\sigma_Y = 15$ ksi, and $E = 29 \times 10^6$ psi, select W sections for *(a)* the main beam, and *(b)* each of the identical supporting beams. Use the maximum shear stress theory with a factor of safety of 1.9.

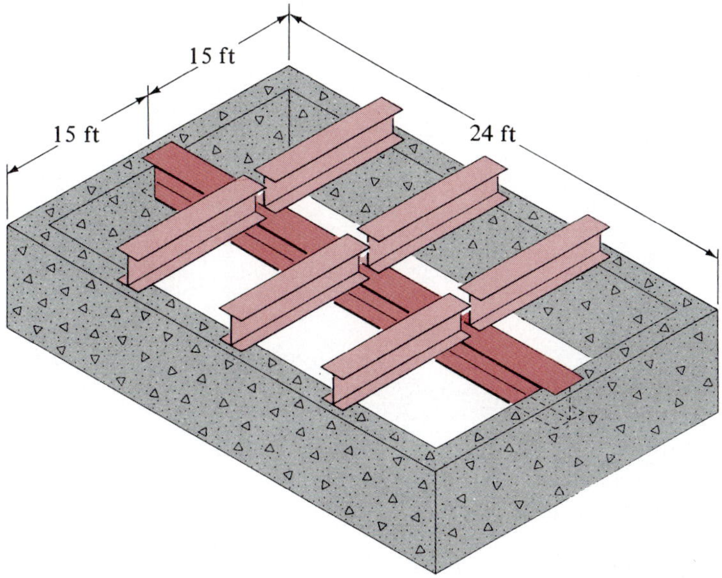

AD 7.7

AD 7.8 The simply supported beam shown is to support a variable position load consisting of two 40-kip loads separated by the 2-ft distance as shown. Given that $\sigma_Y = 9$ ksi, select a suitable W section for the beam using the Hencky-von Mises criterion with a factor of safety of 3.0.

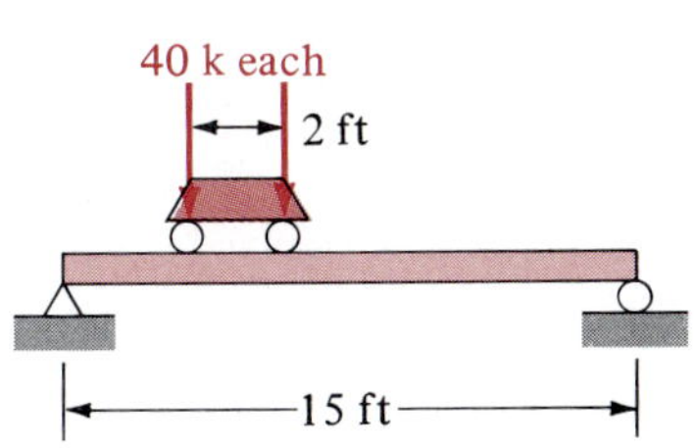

AD 7.8

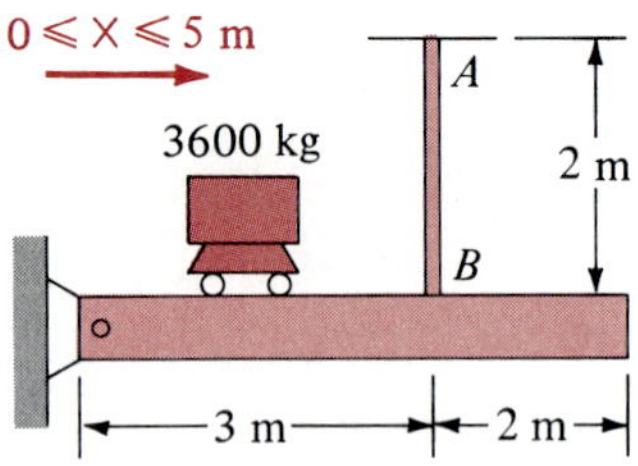

AD 7.9

AD 7.9 The 3600-kg load moves along the 5-m length of the beam as shown. If $\sigma_Y = 60$ MPa for the beam and $\sigma_Y = 70$ MPa for the tension member AB, select an S section and a required diameter for the rod of the tension member. Take E for both the beam and rod as 200 MPa and use the maximum shear stress criterion with a factor of safety of 2.5.

AD 7.10 For the two beams shown there is contact but no contact force at zero load. Use the maximum shear stress criterion to select S sections for each beam. For both beams, take $\sigma_Y = 110$ MPa and $E = 200$ GPa. Also attempt to satisfy the criterion that the maximum bending stresses in the two beams are the same.

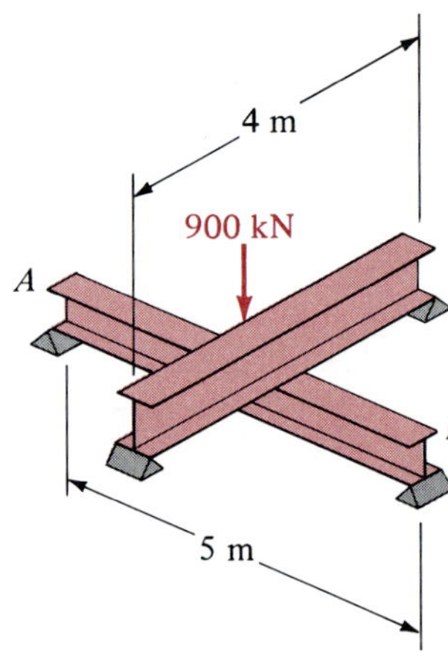

AD 7.10

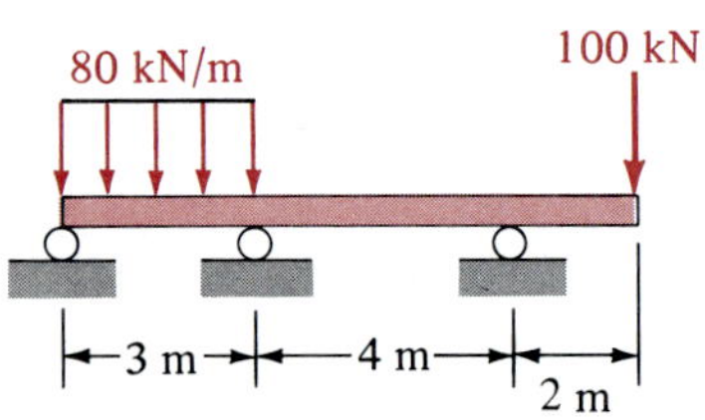

AD 7.11

AD 7.11 A beam is to be used to support the loads shown in the figure. Assuming that the material has a tensile yield strength of 22 ksi and that the Hencky-von Mises criterion is to be used, select an appropriate W section. Take $E = 30 \times 10^6$ psi and the factor of safety to be 2.25.

AD 7.12 A hollow circular section with a wall thickness that is 20% of the outside diameter is to carry the loads shown. Assuming $\sigma_Y = 19$ ksi and $E = 30 \times 10^6$ psi, determine the required outside diameter on the basis of the Hencky-von Mises criterion with a factor of safety of 2.75.

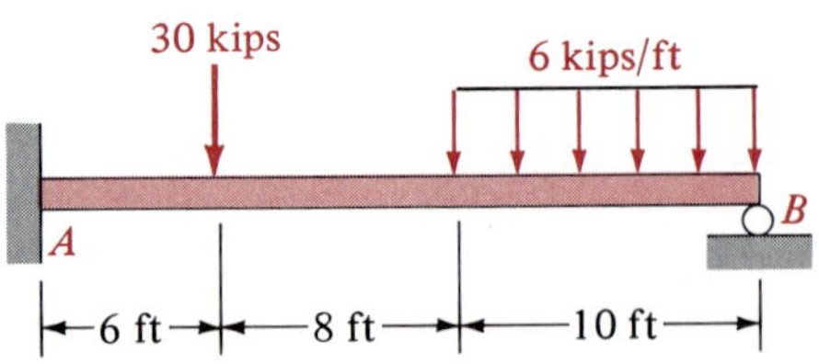

AD 7.12

AD 7.13 A beam is to be designed to support the loading shown. Given that $\sigma_Y = 110$ MPa, use the maximum shear stress criterion with a factor of safety of 2.1 to select an appropriate S section.

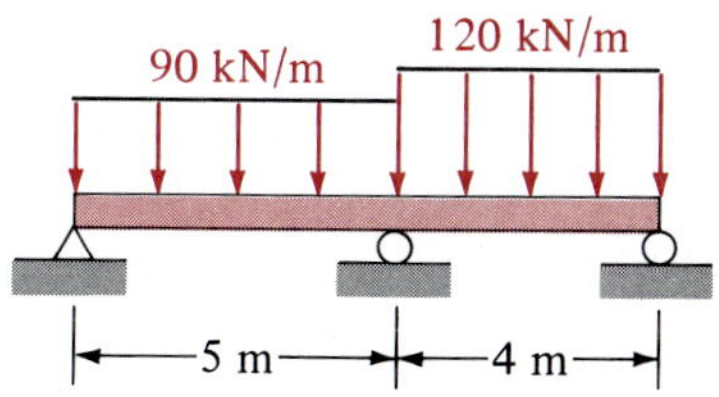

AD 7.13

AD 7.14 Select a W section to carry the loading shown if the beam is composed of a material for which $\sigma_{ult} = 190$ MPa and the maximum normal stress criterion is to be used. Take the factor of safety to be 3.0.

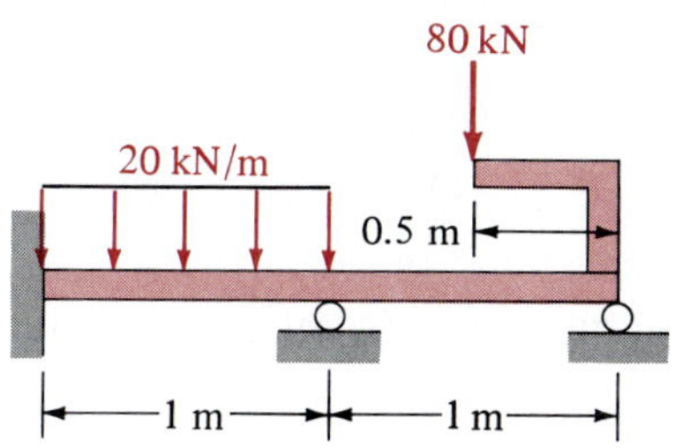

AD 7.14

CHAPTER 8 BENDING DEFORMATIONS: QUESTIONS OF STIFFNESS

CHAPTER OUTLINE

8.1 TRANSVERSE DEFLECTIONS OF BEAMS

When we speak of stiffness, we are generally concerned with the ratio of an applied force to the corresponding displacement. The simplest example is the constant k in the force-displacement relation for a linear spring, $F = kx$. In order to answer the stiffness question for beams, it is necessary to determine the transverse displacements $v(x)$. Generally, if the magnitudes of the displacements are too large, the beam may be judged "not stiff enough." Alternatively, the stiffness of the beam has a major influence on dynamic characteristics and may need to be altered to produce a specific dynamic behavior. To determine the transverse displacement $v(x)$, the approximate moment-curvature relation $M = EIv''$ developed in Section 7.6 can be integrated subject to appropriate boundary conditions. The moment $M(x)$ can be determined by constructing the shear and moment diagrams. Integration of $EIv'' = M$ then yields successively

$$EIv' = \int M\,dx + C_1$$

and

$$EIv = \int \left(\int M\,dx\right) dx + C_1 x + C_2$$

where C_1 and C_2 are constants of integrations that must be determined from conditions on the slope (rotation) v' and/or the displacement v at the ends or intermediate points of the beam.

Alternatively, it is possible to determine v in the following manner. From Section 7.2 we recall the equilibrium equations

$$\frac{dV}{dx} + w(x) = 0 \qquad (8.1)$$

and

$$\frac{dM}{dx} - V = 0 \qquad (8.2)$$

Eliminating V by taking the derivative of Eqn. (8.2) and adding the result of Eqn. (8.1) yields

$$\frac{d^2M}{dx^2} + w(x) = 0 \tag{8.3}$$

Then using $M = EIv''$, Eqn. (8.3) can be written as

$$(EIv'')'' + w(x) = 0$$

which is the equilibrium equation expressed entirely in terms of the transverse displacement $v(x)$. When EI is constant

$$EIv'''' = -w(x) \tag{8.4}$$

By combining $EIv'' = M$ and $M' = V$, the shear force V can be expressed in terms of the displacements as

$$V = EIv''' \tag{8.5}$$

Four successive integrations of Eqn. (8.4) yield

$$EIv''' = V = -\int w(x)\,dx + C_1$$

$$EIv'' = M = -\int[\int w(x)\,dx]dx + C_1x + C_2$$

$$EIv' = -\int\{\int[\int w(x)\,dx]dx\}dx + \frac{C_1x^2}{2} + C_2x + C_3$$

$$EIv = -\int(\int\{\int[\int w(x)\,dx]dx\}dx)dx + \frac{C_1x^3}{6} + \frac{C_2x^2}{2} + C_3x + C_4$$

The constants of integration C_1, C_2, C_3, and C_4 must be determined from conditions on the shear V, the moment M, the slope or rotation v', and the displacement v at the ends or intermediate points of the beam. These conditions on the force resultants V and M and on the slope or rotation v' and displacement v at the ends of the beam, are referred to as boundary conditions. In this chapter we will concentrate on the integration process necessary to determine the displacements so that the stiffness question can be answered.

8.1.1 Boundary Conditions

When integrating either $EIv'' = M$ or $EIv'''' = -w$ to determine the transverse displacement $v(x)$, we need to be able to evaluate the constants of integration from the boundary conditions. In this section we outline the different types of supports along with the corresponding boundary conditions that are commonly encountered in practice.

Simply Supported End A simply supported end, also called a **pinned end,** can occur at either or both ends of the beam and is indicated by either of the two symbols shown in Fig. 8.1.

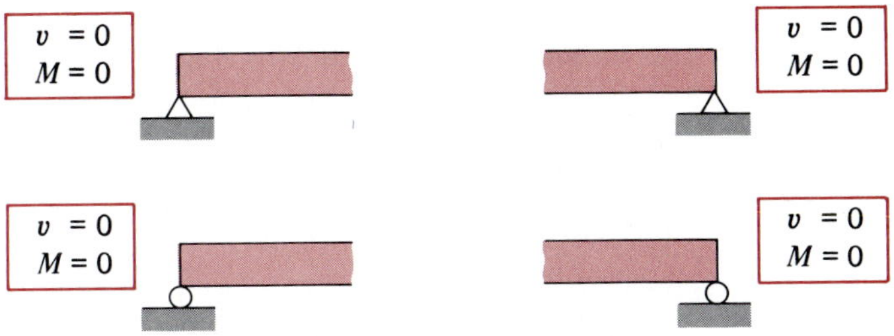

Figure 8.1 Symbols and boundary conditions for a simply supported end

The boundary conditions appropriate for a simply supported end are

$$M = EIv'' = 0 \qquad \text{and} \qquad v = 0$$

thus, there can be no moment transmitted and no transverse displacement. Throughout this text, we will follow the practice of including the boundary conditions as a part of the sketch of the beam by placing them in boxes at the end of the beam, as in Fig. 8.1. This step should be taken before the process of integration begins.

If a moment M_0 is applied at a pinned end, the boundary condition is changed to reflect the fact that the moment is not zero (i.e., $M = EIv'' = \pm M_0$) with the sign chosen on the basis of the direction of the applied moment. As an example consider the beam on pinned supports shown in Fig. 8.2.

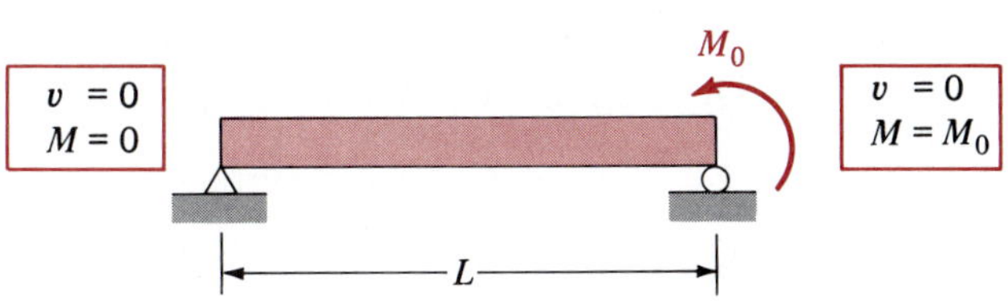

Figure 8.2 Boundary condition for an end moment at a pinned support

For the moment applied as shown, the boundary conditions are:

$x = 0$: $\quad v(0) = 0 \quad M = EIv''(0) = 0$

$x = L$: $\quad v(0) = 0 \quad M = EIv''(L) = M_0$

As mentioned, these boundary conditions are placed in boxes at the ends of the beam, and included in the sketch in Fig. 8.2.

Free End A free end, also called an unsupported end, is shown in Fig. 8.3.

Figure 8.3 Symbol and boundary conditions for a free or unsupported end

The basic idea is that the end surface of the beam is assumed to be stress free, that is, $\sigma_x = \tau_{xy} = 0$, so that the shear and moment given, respectively, by $V = -\int \tau_{xy}\, dA$ and $M = -\int y\sigma_x\, dA$ must both vanish. The corresponding boundary conditions at a free end are

$$M = EIv'' = 0 \qquad \text{and} \qquad V = EIv''' = 0$$

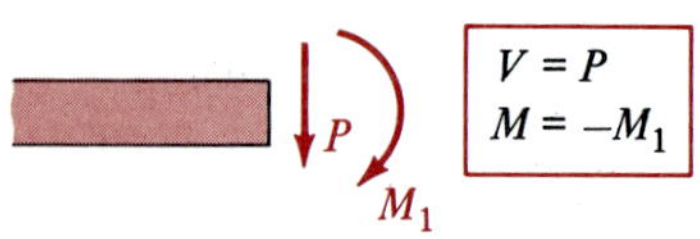

Figure 8.4 Boundary conditions for a beam loaded by an end moment M_1 and an end shear P

If a shear force or moment is applied at a free end, the boundary conditions must be altered accordingly. As an example consider the beam loaded as shown in Fig. 8.4. At the loaded end, the boundary conditions are

$x = L$: $\quad M = EIv''(L) = -M_1$

$$V = EIv'''(L) = P$$

Clamped or Fixed End The symbol for a **clamped** or **fixed end** is shown in Fig. 8.5. The boundary conditions for the clamped end are

$$v = 0 \qquad \text{and} \qquad v' = 0$$

where v' is the slope or the rotation of the centerline of the beam.

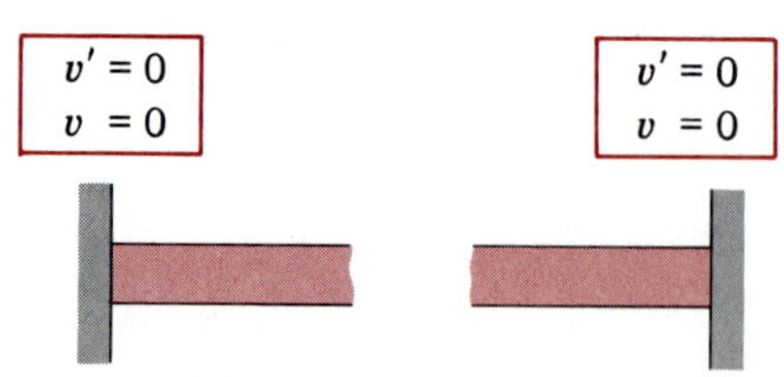

Figure 8.5 Symbol and boundary conditions for a clamped or fixed end

When we determine the displacements by integrating $M = EIv''$, M is assumed to be known. Thus any boundary conditions on M, and also on V, will already have been enforced prior to the last two integrations necessary to determine $v(x)$. The two conditions on v and v' at the ends of the beam are then used to determine the two constants of integration. In some instances, it may be possible to determine one of the constants of integration after the first integration. In other situations, both integrations are required before being able to solve for either C_1 or C_2.

When determining $v(x)$ by integrating $EIv^{iv} = -w(x)$, four boundary conditions, *two at each end,* are used to determine the four constants of integration. Here again, the number of integrations required to evaluate a particular constant of integration depends upon the types of support at the boundaries.

PROBLEMS

8.1–8.9 In each of the following drawings, indicate the boundary conditions using boxes at the ends.

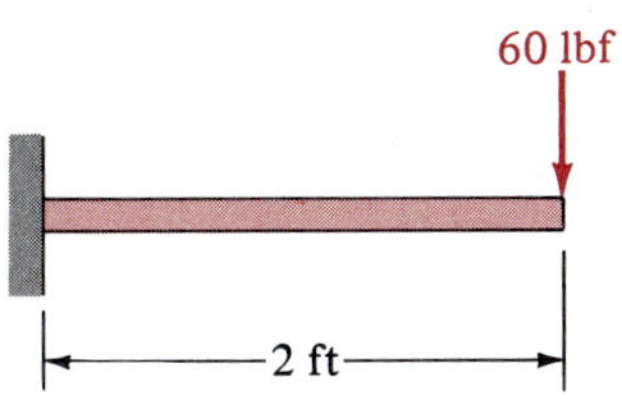

P. 8.1

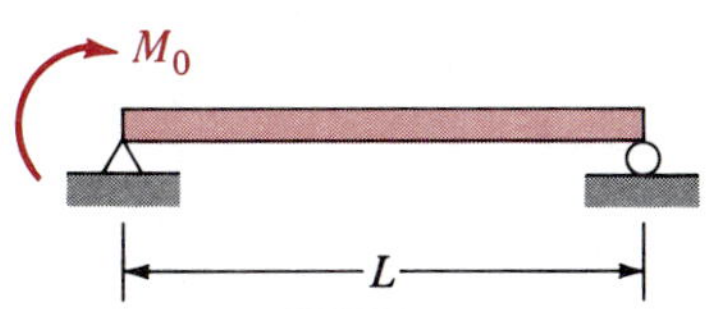

P. 8.2

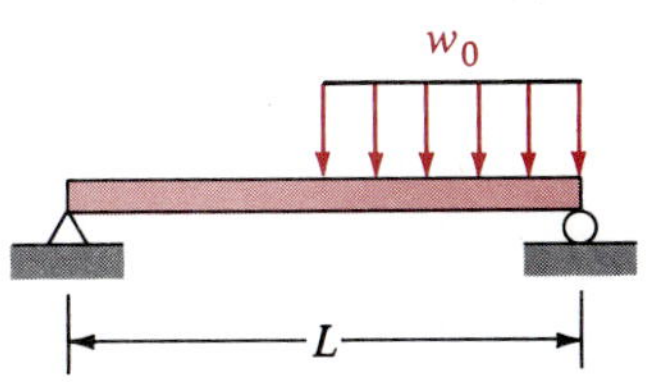

P. 8.3

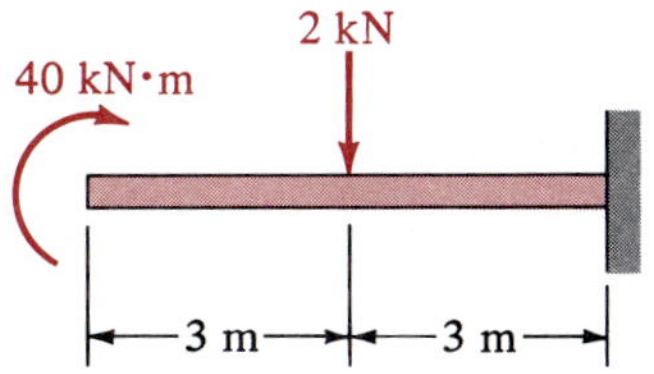

P. 8.4

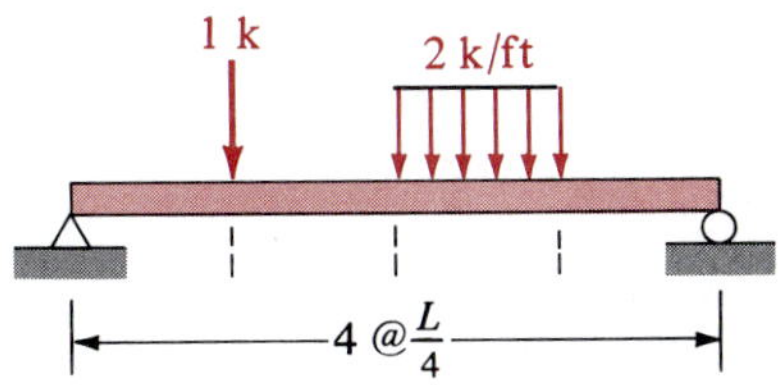

P. 8.5

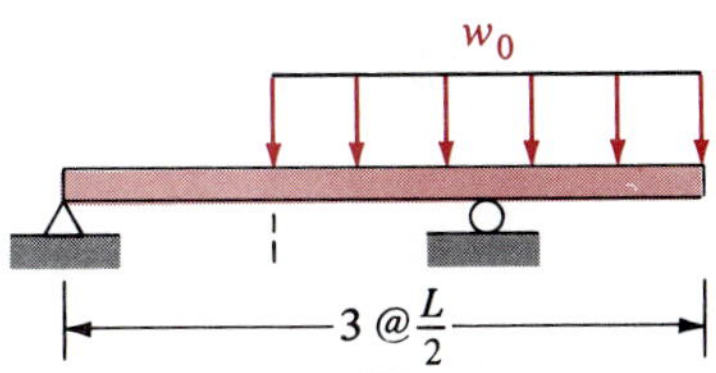

P. 8.6

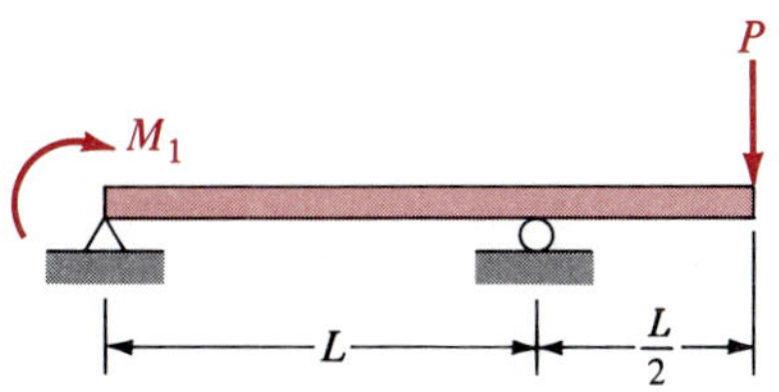

P. 8.7

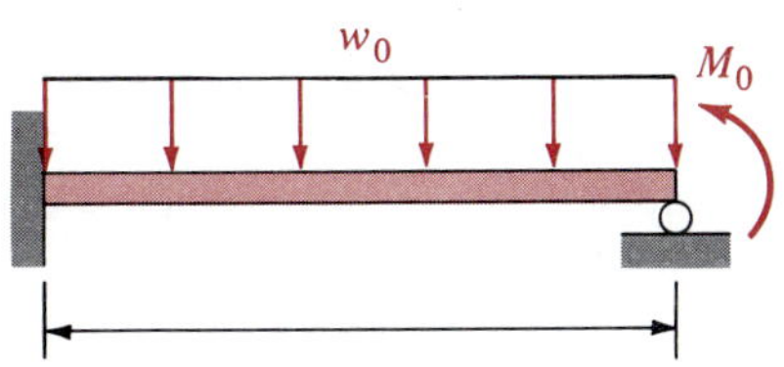

P. 8.8

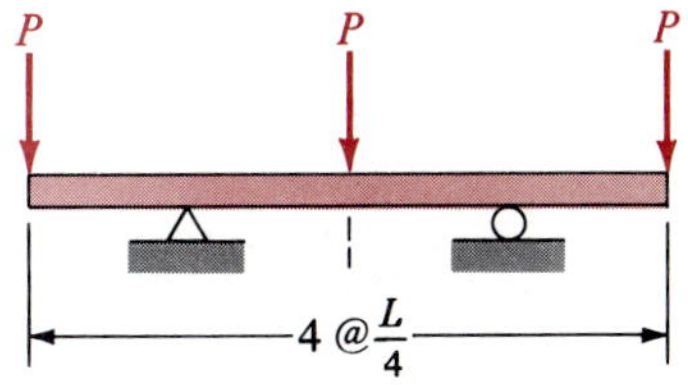

P. 8.9

8.2 DEFLECTIONS USING INTEGRATION OF $EIv'' = M$

In this section, we will demonstrate by means of several examples the process of determining the transverse displacement $v(x)$ by integration of the moment-curvature relation $EIv'' = M$. The first step in the process is to draw the free-body diagrams and determine the moment $M(x)$ using either of the two approaches discussed in Section 7.3. We can then integrate the equation $EIv'' = M(x)$ with the appropriate boundary conditions on v and v' used to determine the two constants.

Example 8.1 A simply supported beam is loaded by a uniform load w_0 as shown in Fig. 8.6. Determine the transverse displacement $v(x)$ and the location and value of the largest displacement.

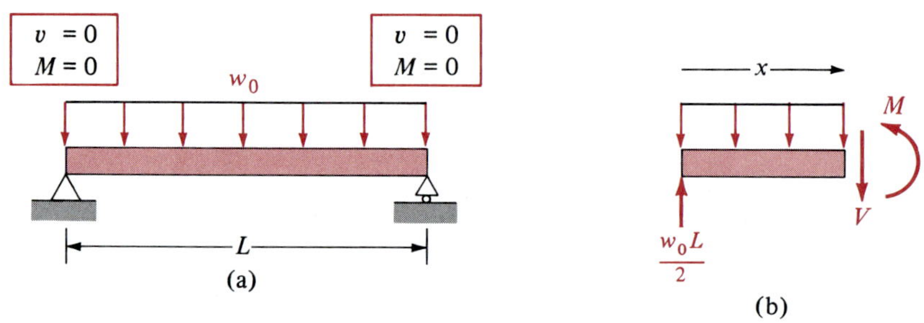

Figure 8.6 Simply supported beam loaded uniformly

Solution Referring to the free-body diagram of Fig. 8.6b, we can determine the shear force V and the internal moment M at a general location x, respectively, as

$$V(x) = w_0\left(\frac{L}{2} - x\right) \qquad \text{and} \qquad M(x) = \frac{w_0x(L - x)}{2}$$

The corresponding shear and bending moment diagrams are shown in Fig. 8.7. As mentioned, the two boundary conditions on the moment at the ends of the beam, namely $M(0) = M(L) = 0$, as indicated in Fig. 8.7, have already been satisfied in the course of correctly determining M.

To determine the displacements we must integrate the equation

$$EIv'' = M = \frac{w_0x(L - x)}{2}$$

and satisfy the two boundary conditions $v(0) = v(L) = 0$. Integrating a first time yields

$$EIv' = \frac{w_0Lx^2}{4} - \frac{w_0x^3}{6} + C_1 \tag{a}$$

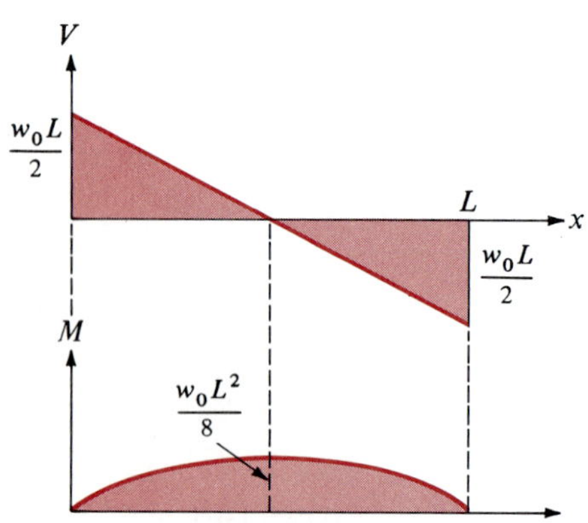

Figure 8.7 Shear and moment diagrams for the uniformly loaded, simply supported beam

There is no boundary condition on v'; thus, we need to integrate again before applying the two boundary conditions on v. A second integration yields

$$EIv = \frac{w_0 L x^3}{12} - \frac{w_0 x^4}{24} + C_1 x + C_2 \qquad (b)$$

Satisfying the boundary condition $v(0) = 0$ gives

$$EIv(0) = 0 = 0 - 0 + C_1(0) + C_2$$

or $C_2 = 0$. Enforcing the boundary condition $V(L) = 0$ yields

$$EIv(L) = 0 = \frac{w_0 L^4}{12} - \frac{w_0 L^4}{24} + C_1 L$$

from which $C_1 = -\frac{w_0 L^3}{24}$. Substituting for C_1 and C_2 into Eqns. (a) and (b) yields

$$v' = -\frac{w_0}{24EI}(L^3 - 6Lx^2 + 4x^3)$$

and

$$v = -\frac{w_0 x}{24EI}(L^3 - 2Lx^2 + x^3)$$

as the equations for the slope v' and the transverse displacement $v(x)$. The minus sign indicates displacement downward, as we would expect. The curves for the slope $v'(x)$ and the displacement $v(x)$ are shown in Fig. 8.8. The equation for v is frequently referred to as **the equation of the elastic curve.**

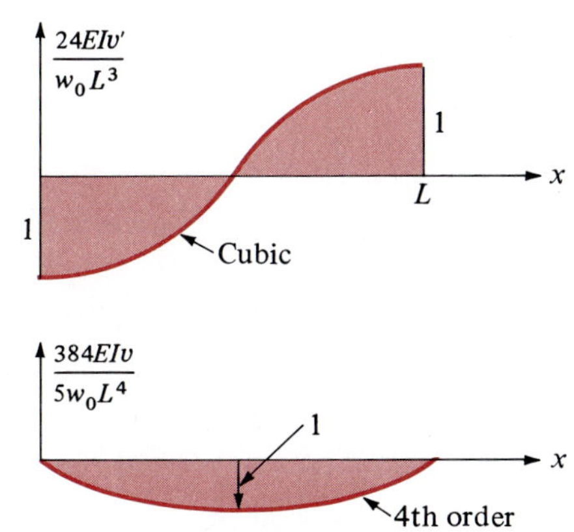

Figure 8.8 Slope and rotation for the uniformly loaded, simply supported beam

Example 8.2 Consider the problem of a cantilever beam loaded by the uniform load w_0 as pictured in Fig. 8.9. Determine the displacement $v(x)$ and the deflection and slope v' at the end $x = L$.

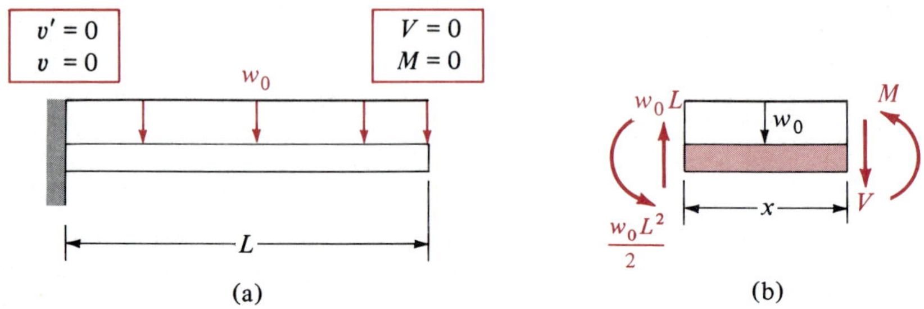

Figure 8.9 Cantilever beam uniformly loaded

Solution Referring to the free-body diagram of Fig. 8.9b, the equilibrium equations $\Sigma F_y = 0$ and $\Sigma M = 0$ yield

$$V = w_0(L - x) \qquad \text{and} \qquad M = -\frac{w_0}{2}(L - x)^2$$

Note that $V(L) = M(L) = 0$, as required by the boundary conditions. The shear and moment diagrams are shown in Fig. 8.10.

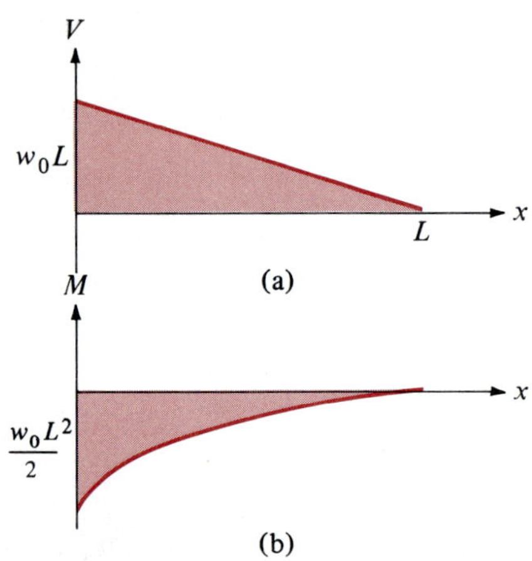

Figure 8.10 Shear and moment diagrams

The moment-curvature relation $EIv'' = M$ becomes

$$EIv'' = -\frac{w_0}{2}(L - x)^2$$

An integration yields

$$EIv' = \frac{w_0}{6}(L - x)^3 + C_1 \qquad (a)$$

The boundary condition on v' at $x = 0$ is used according to

$$EIv'(0) = 0 = \frac{w_0}{6}L^3 + C_1 \qquad \text{or} \qquad C_1 = -\frac{w_0}{6}L^3$$

and

$$EIv' = \frac{w_0}{6}(L - x)^3 - \frac{w_0}{6}L^3$$

Integrating again yields

$$EIv = -\frac{w_0}{24}(L - x)^4 - \frac{w_0}{6}L^3x + C_2 \qquad (b)$$

which with $v(0) = 0$ gives $C_2 = w_0L^4/24$. Substituting the expressions for C_1 and C_2 into Eqns. (a) and (b) gives the equations for the slope and displacement as

$$v' = \frac{w_0}{6EI}[(L - x)^3 - L^3]$$

and

$$v = -\frac{w_0}{24EI}(6L^2x^2 - 4Lx^3 - x^4)$$

The largest displacement occurs at $x = L$ and has the value $v(L) = -w_0L^4/8EI$. The value of the slope at $x = L$ is $v'(L) = -w_0L^3/6EI$. The slope and displacement are shown in Fig. 8.11.

The maximum values for both the slope and displacement occur at the free end $x = L$.

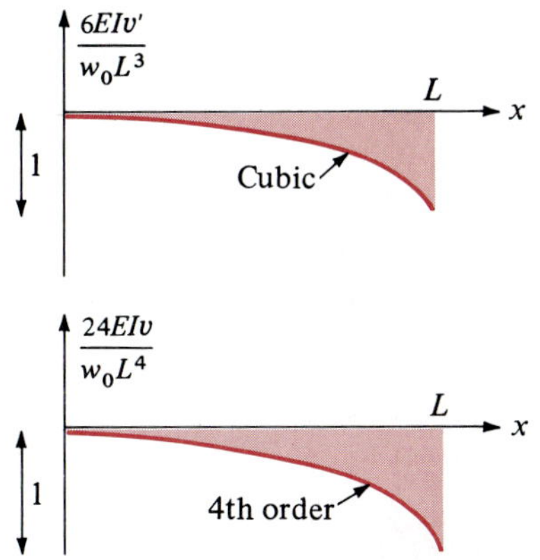

Figure 8.11 Slope and displacement for a uniformly loaded cantilever

In each of these examples illustrating the use of integrating $EIv'' = M$ for determining the displacement $v(x)$, the loading was such that the moment $M(x)$ could be expressed as a single continuous function over the length of the beam. When the loading $w(x)$ consists of concentrated loads, concentrated moments, or distributed loadings that begin and end at locations between the ends of the beam, the moment $M(x)$ is expressed by different functions in several intervals of the beam. The use of $EIv'' = M$ in solving this type of problem is illustrated in the examples that follow.

Example 8.3 Use $EIv'' = M$ to determine the displacements in the beam shown in Fig. 8.12a.

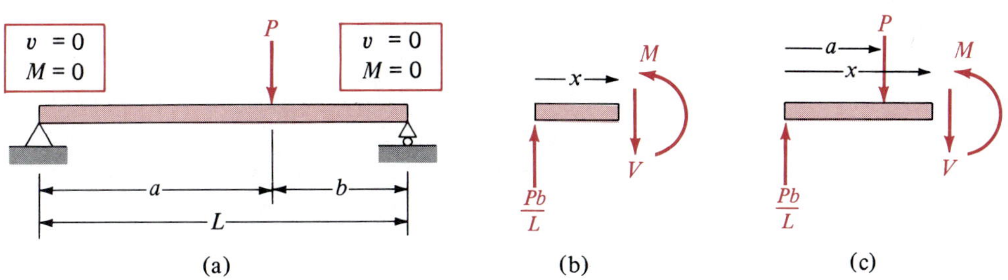

Figure 8.12 Beam loaded by a concentrated load

Solution For this example, we split the problem of integrating $EIv'' = M$ into two problems, one in each of the intervals $0 \le x \le a$ and $a \le x \le L$. Let the displacement in the interval $0 \le x \le a$ be denoted by v_1, and the displacement in the interval $a \le x \le L$ be denoted by v_2. As determined from Figs. 8.12b and c, the expressions for the moments in these two segments are, respectively,

$$M = \frac{Pb}{L}x = EIv_1'' \qquad \text{and} \qquad M = \frac{Pb}{L}x - P(x-a) = EIv_2''$$

Integrating these two equations yields

$$EIv_1 = \frac{Pbx^3}{6L} + C_1x + C_2 \qquad EIv_2 = \frac{Pbx^3}{6L} - P\frac{(x-a)^3}{6} + C_3x + C_4$$

Two of the conditions for evaluating the four constants C_1, C_2, C_3, and C_4 are the boundary conditions $v(0) = 0$ and $v(L) = 0$, where $v(0)$ refers to $v_1(0)$ and $v(L)$ refers to $v_2(L)$. The other conditions come from requiring that the slope $v'(x)$ and the displacement $v(x)$ given by $v_1(x)$ and $v_2(x)$ have the same values at $x = a$, or

$$v_1'(a) = v_2'(a) \qquad \text{and} \qquad v_1(a) = v_2(a)$$

These two conditions, called continuity conditions, ensure that the slope and displacement, respectively, are continuous, and thus that there are no gaps or corners in the displacement curve. Satisfying all four conditions leads to

$$EIv_1(0) = C_2 = 0$$

$$EIv_2(L) = \frac{PbL^2}{6} - \frac{Pb^3}{6} + C_3L + C_4 = 0$$

$$\frac{Pba^2}{2L} + C_1 = \frac{Pba^2}{2L} + C_3$$

$$\frac{Pba^3}{6L} + C_1a = \frac{Pba^3}{6L} + C_3a + C_4$$

from which $C_1 = C_3 = \frac{Pb}{6L}(b^2 - L^2)$, and $C_2 = C_4 = 0$. Eliminating the expressions for C_1 and C_3, the displacements v_1 and v_2 can then be written as

$$0 \le x \le a: \qquad v_1 = -\frac{Pbx}{6EIL}(L^2 - b^2 - x^2)$$

and

$$a \le x \le L: \qquad v_2 = -\frac{Pb}{6EIL}[(L^2 - b^2)bx + L(x-a)^3 - bx^3]$$

The corresponding elastic curve is shown in Fig. 8.13 for the case where the position of the load is to the right of center, that is, when $a > L/2$.

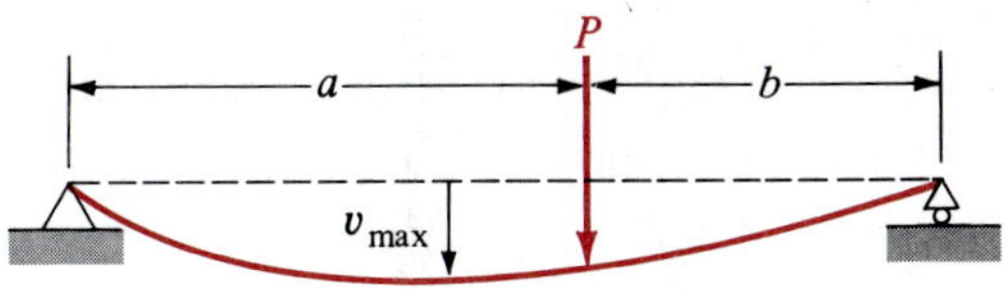

Figure 8.13 Elastic curve for concentrated load P

For $a > L/2$, the maximum value occurs to the left of the load at the value x determined from

$$v_1'(x) = -\frac{Pb}{6EIL}(L^2 - 3x^2 - b^2) = 0$$

or

$$x^2 = \frac{(L^2 - b^2)}{3}$$

The maximum displacement can be expressed as

$$v_{max} = -\left(\frac{Pb[(L^2 - b^2)/3]^{1/2}}{6EIL}\right) 2\left(\frac{L^2 - b^2)}{3}\right) = -\frac{Pb(L^2 - b^2)^{3/2}}{3^{5/2}EIL}$$

For the special case of $b = L/2$, $v_{max} = -PL^3/48EI$.

Example 8.4 Use $EIv'' = M$ to determine the displacements in the overhang beam shown in Fig. 8.14a.

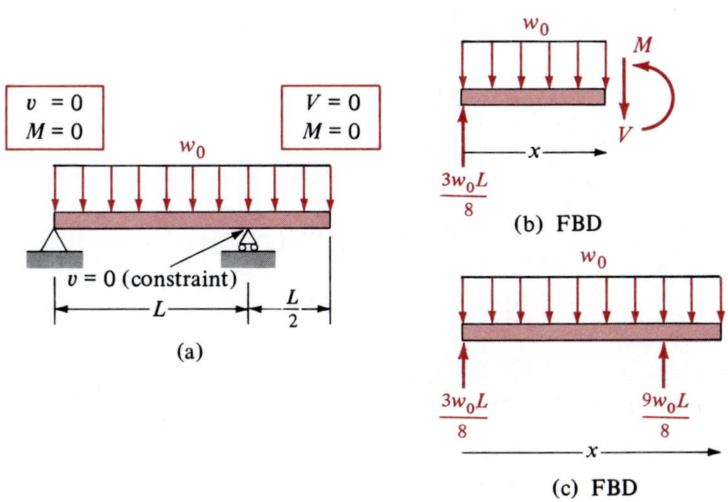

Figure 8.14 Deflections for a uniformly loaded overhang beam

Solution The interval $0 \le x \le 3L/2$ is split into two intervals, one for $0 \le x \le L$ and one for $L \le x \le 3L/2$. Consulting the free-body diagrams of Figs. 8.14b and c, respectively, the two equations to be integrated are

$$0 \le x \le L: \qquad EIv_1'' = \frac{3w_0L}{8}x - \frac{w_0x^2}{2}$$

$$L \le x \le 3L/2: \qquad EIv_2'' = \frac{3w_0L}{8}x - \frac{w_0x^2}{2} + \frac{9w_0L}{8}(x - L)$$

subject to the boundary condition $v(0) = v_1(0) = 0$, the continuity condition $v_1'(L) = v_2'(L)$, and the constraint condition $v(L) = 0$. The constraint condition must be imposed on both v_1 and v_2, leading to the two conditions $v_1(L) = v_2(L) = 0$. Thus the four conditions necessary to determine the four constants of integration are

$$v_1(0) = 0 \qquad v_1'(L) = v_2'(L) \qquad v_1(L) = 0 \qquad v_2(L) = 0$$

Additionally, note from the equation for the interval $0 \le x \le L$ that $EIv_1''(0) = M(0) = 0$, and from the equation for the interval $L \le x \le 3L/2$, that both $EIv_2''(3L/2) = M(3L/2) = 0$ and $EIv_2'''(3L/2) = V(3L/2) = 0$, as required by the boundary conditions shown in Fig. 8.14a. Proceeding with the integrations gives

$$EIv_1' = \frac{3w_0L}{16}x^2 - \frac{w_0x^3}{6} + C_1$$

$$EIv_2' = \frac{3w_0L}{16}x^2 - \frac{w_0x^3}{6} + \frac{9w_0L}{16}(x - L)^2 + C_3$$

Satisfying the continuity condition on the slope at $x = L$, i.e., $v_1'(L) = v_2'(L)$, gives

$$\frac{w_0L^3}{48} + C_1 = \frac{w_0L^3}{48} + C_3$$

from which $C_1 = C_3$. Eliminating the C_3 and integrating again yields

$$EIv_1 = \frac{w_0L}{16}x^3 - \frac{w_0x^4}{24} + C_1x + C_2$$

$$EIv_2 = \frac{w_0L}{16}x^3 - \frac{w_0x^4}{24} + \frac{3w_0L}{16}(x - L)^3 + C_1x + C_4$$

The boundary condition $v_1(0) = 0$ yields $C_2 = 0$. The constraint conditions $v_1(L) = 0$ and $v_2(L) = 0$ then yield

$$\frac{w_0L}{16}L^3 - \frac{w_0L^4}{24} + C_1L = 0$$

and

$$\frac{w_0 L}{16} L^3 - \frac{w_0 L^4}{24} + \frac{3w_0 L}{16}(L - L)^3 + C_1 L + C_4 = 0$$

from which $C_4 = 0$ and $C_1 = -w_0 L^3/8$. The equations for the elastic curve are

$$0 \le x \le L: \qquad v_1 = \frac{w_0}{48EI}(3Lx^3 - 2x^4 - L^3 x)$$

$$L \le x \le 3L/2: \quad v_2 = \frac{w_0}{48EI}(3Lx^3 - 2x^4 - L^3 x + 9(x - L)^2)$$

and are shown in Fig. 8.15.

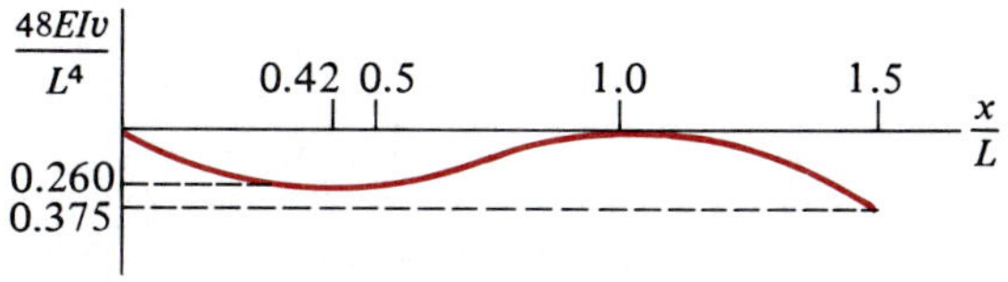

Figure 8.15 Elastic curve for a uniformly loaded overhang beam

The maximum value of the transverse displacement between the supports occurs at $x = 0.42L$ and has the value $0.260w_0L^4/48EI$ as shown.

PROBLEMS

8.10 Use the results of Example 8.1 to determine the maximum deflection of a simply supported W 10 × 45 beam with $w_0 = 2$ kips/ft, $E = 30 \times 10^6$ psi, and $L = 12$ ft. Also determine the rotations at the ends of the beam.

8.11 A rectangular-cross-section cantilever beam is uniformly loaded with $w_0 = 2.1$ kips/ft. Use the results of Example 8.2 to determine deflection and slope at $x = L$ if $L = 13$ ft, $b = 3.625$ in., $h = 10.625$ in., and $E = 4 \times 10^6$ psi.

8.12 Use the results of Example 8.1 to determine the maximum deflection of a simply supported W 250 × 67 beam with $w_0 = 28$ kN/m, $E = 200$ GPa, and $L = 3.6$ m. Also determine the rotations at the ends of the beam.

8.13 A rectangular-cross-section cantilever beam is uniformly loaded with $w_0 = 32$ kN/m. Use the results of Example 8.2 to determine deflection and slope at $x = L$ if $L = 4$ m, $b = 90$ mm, $h = 260$ mm, and $E = 30$ GPa.

8.14–8.21 In each of the following, state and indicate the boundary conditions with boxes, draw the necessary free-body diagrams, set up the expression for the moment $M(x)$, and integrate $EIv'' = M(x)$ to determine the transverse displacement $v(x)$.

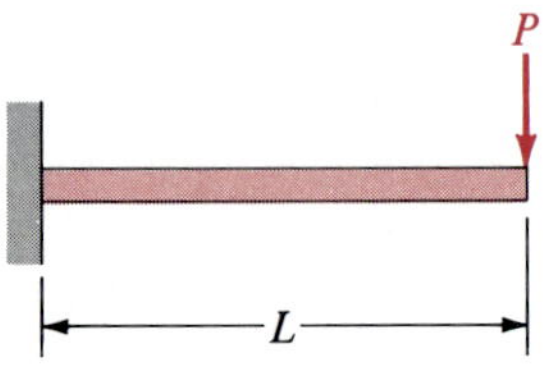

Ps. 8.14 and 8.38

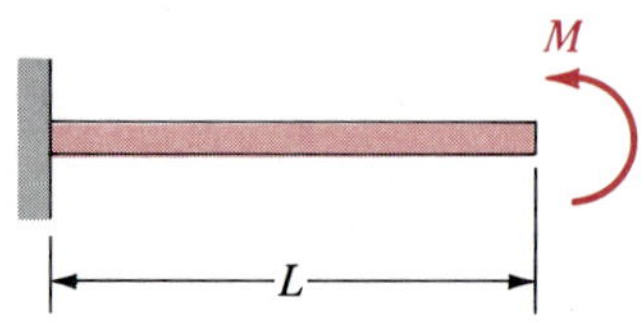

Ps. 8.15 and 8.39

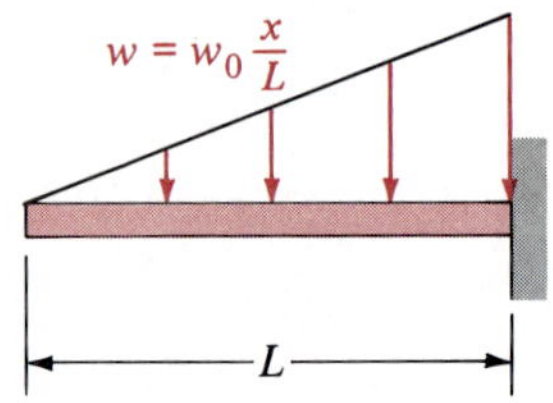

Ps. 8.16 and 8.40

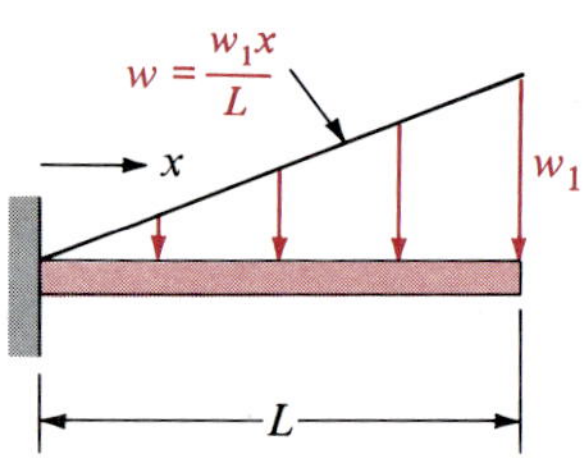

Ps. 8.17 and 8.41

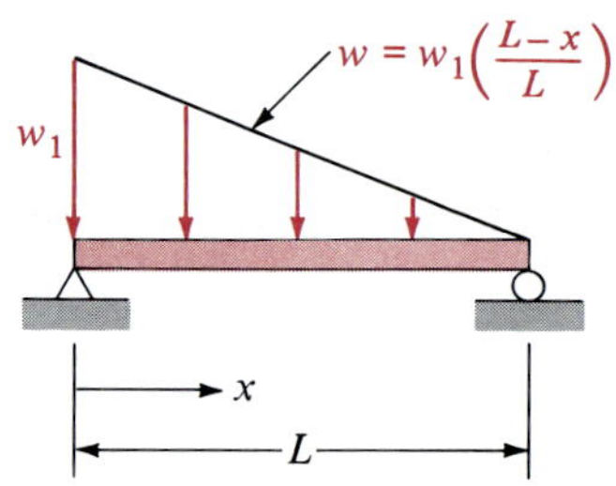

Ps. 8.18 and 8.42

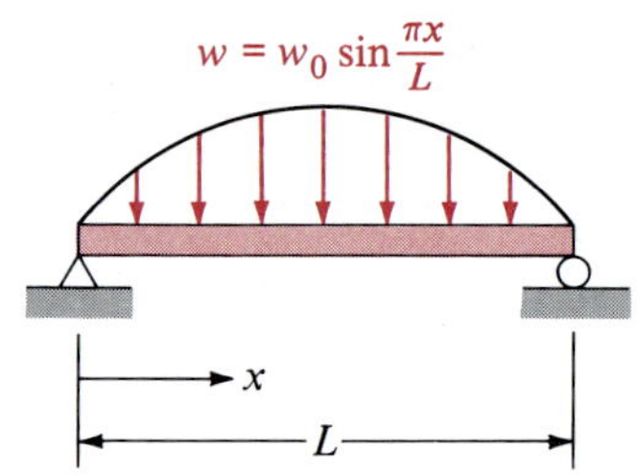

Ps. 8.19 and 8.43

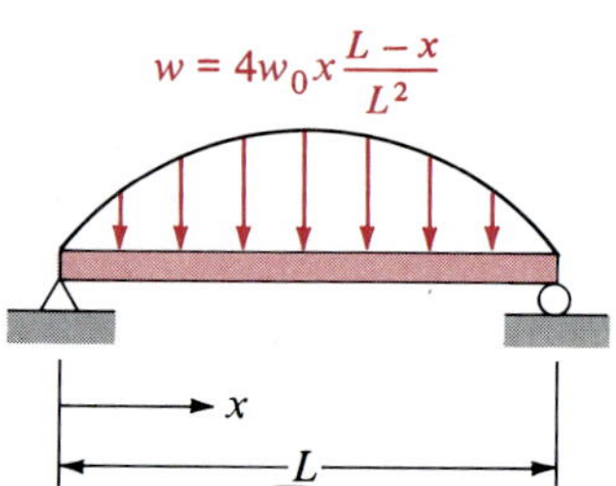

Ps. 8.20 and 8.44

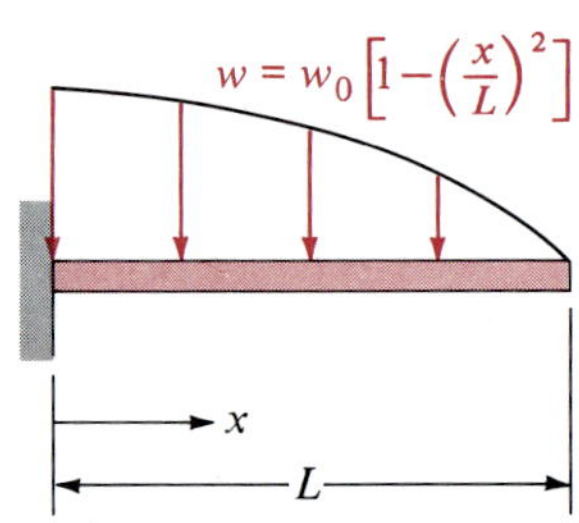

Ps. 8.21 and 8.45

8.22–8.29 In each of the following, draw free-body diagrams, indicate the boundary conditions in boxes, set up the expressions for the moment in the two intervals in question, state the continuity conditions, and integrate $EIV'' = M(x)$ to determine the transverse displacement $v(x)$. Take EI to be a constant.

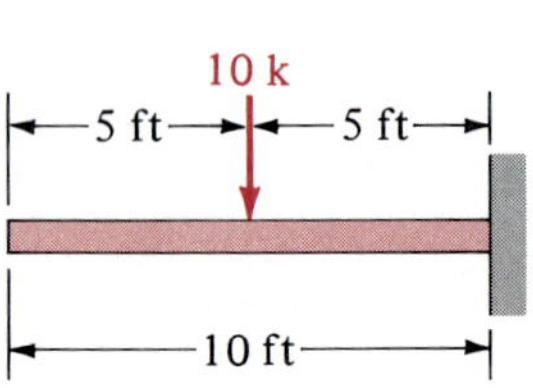

Ps. 8.22 and 8.46

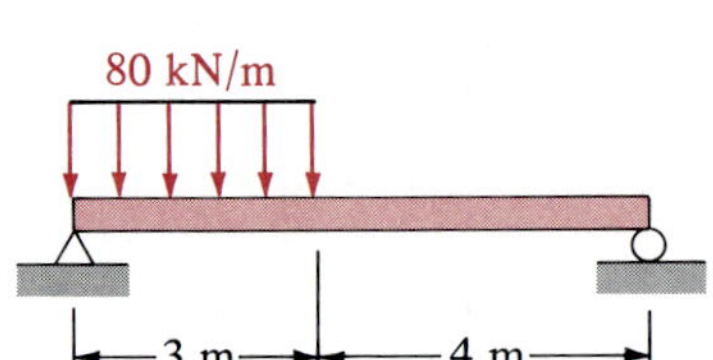

Ps. 8.23 and 8.47

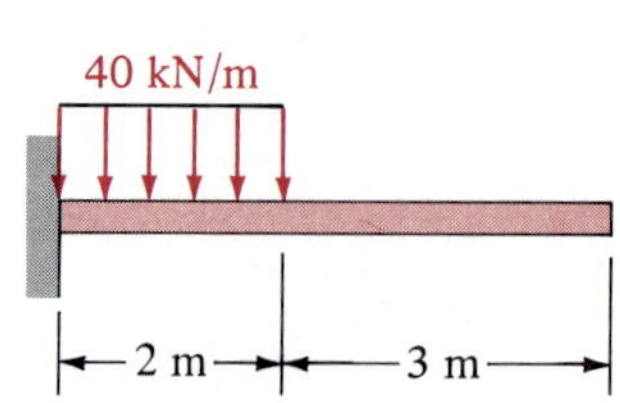

Ps. 8.24 and 8.48

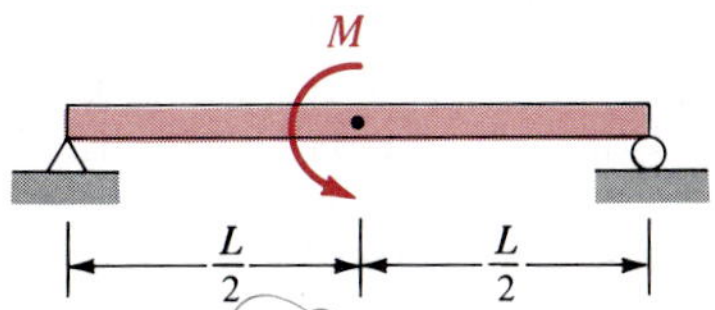

Ps. 8.25 and 8.49

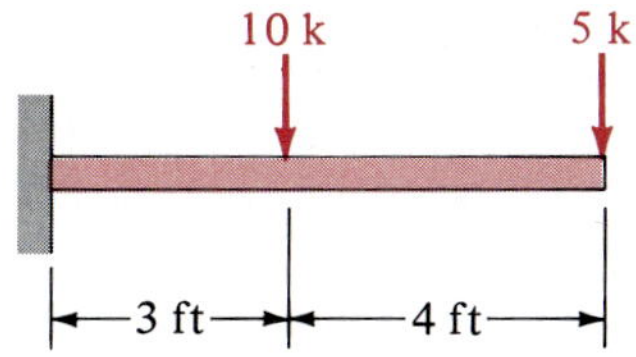

Ps. 8.26 and 8.50

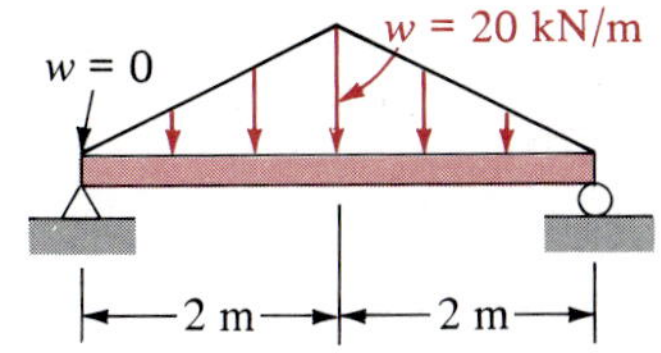

Ps. 8.27 and 8.51

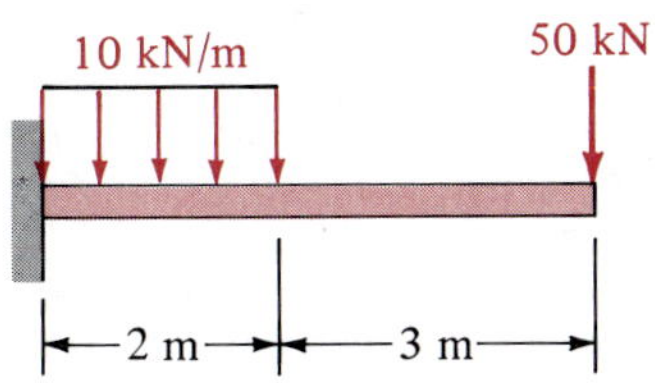

Ps. 8.28 and 8.52

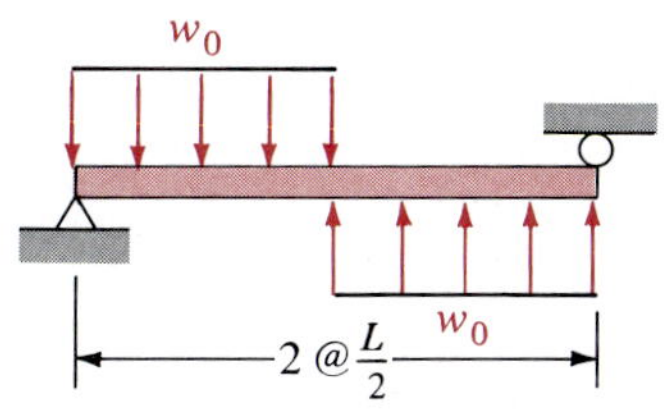

Ps. 8.29 and 8.53

8.30–8.37 For each of the following overhang situations, split the problem into one for the simply supported portion and one for the overhang portion. Draw the necessary free-body diagrams to determine the boundary conditions for each, $M(x)$ for each, and integrate $EIv'' = M$ to obtain the transverse displacement in each portion of the beam. Determine *(a)* the maximum deflection in the simply supported portion, *(b)* the rotation at the point B, and *(c)* the deflection at the end of the overhang.

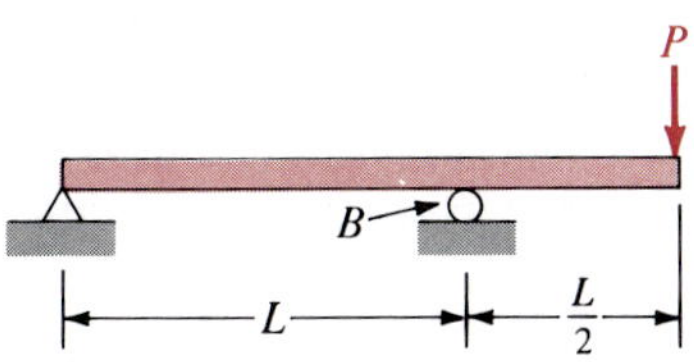

Ps. 8.30 and 8.54

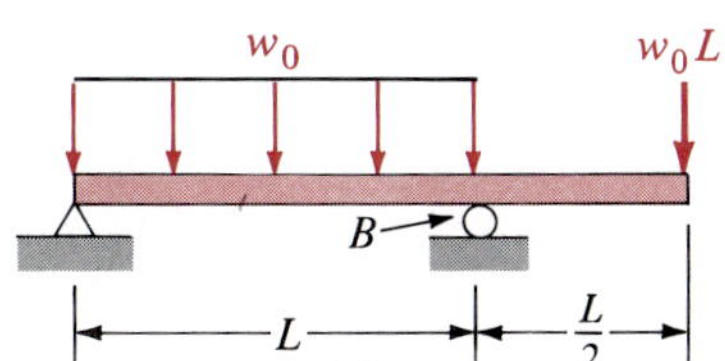

Ps. 8.31 and 8.55

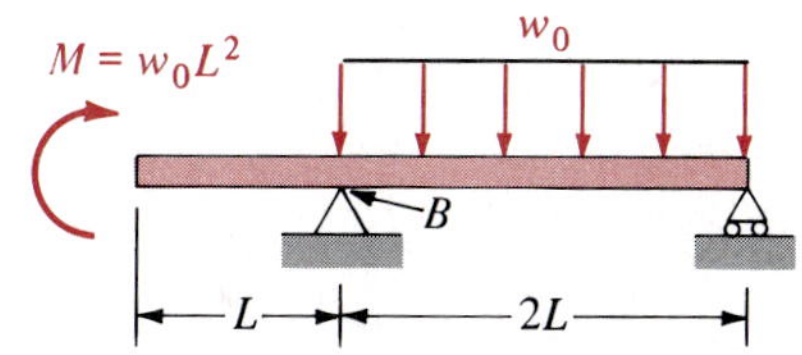

Ps. 8.32 and 8.56

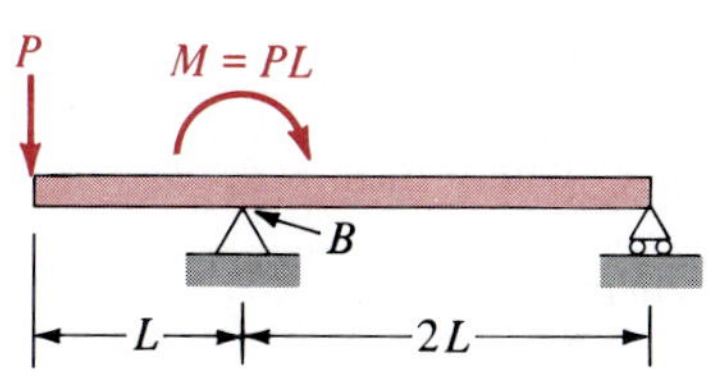

Ps. 8.33 and 8.57

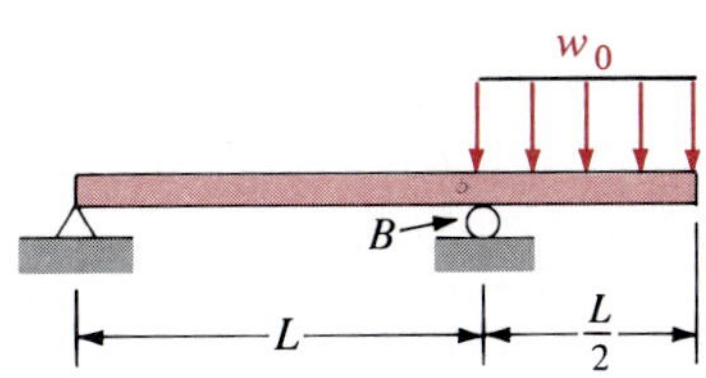

Ps. 8.34 and 8.58

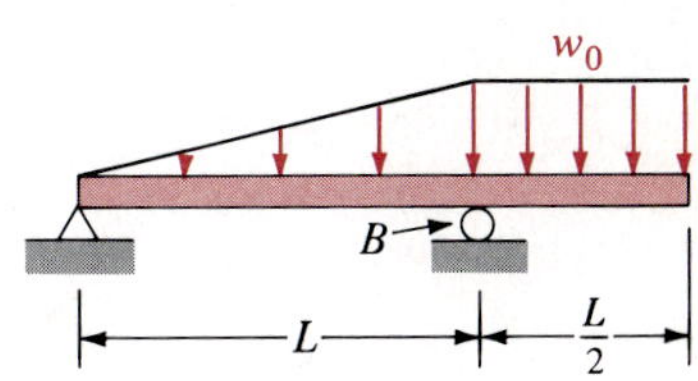

Ps. 8.35 and 8.59

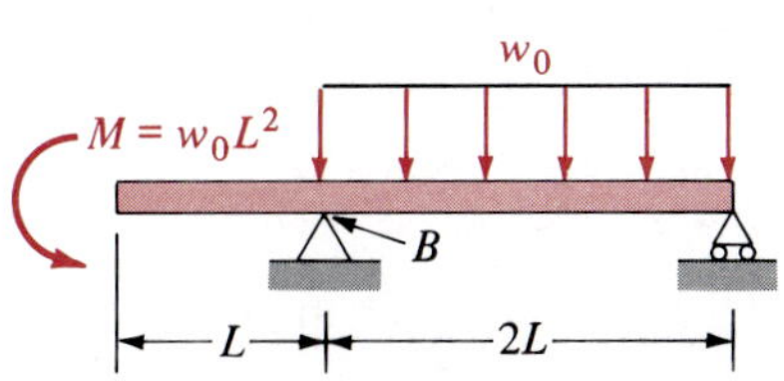

Ps. 8.36 and 8.60

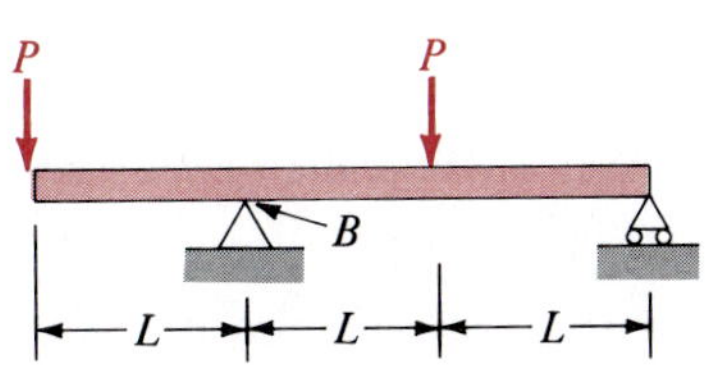

Ps. 8.37 and 8.61

8.3 DEFLECTIONS USING INTEGRATION OF $EI\, d^4v/dx^4 = -w(x)$

As outlined in Section 8.1, it is possible to represent the equilibrium equation as a fourth-order equation in terms of the transverse displacement $v(x)$ according to

$$EI \frac{d^4v}{dx^4} = -w(x)$$

The first two integrations yield

$$EI \frac{d^3v}{dx^3} = V(x) = -\int w(x)\, dx + C_1$$

and

$$EI \frac{d^2v}{dx^2} = M(x) = -\int\int w(x)\, dx + C_1 x + C_2$$

For a statically determinate problem, two of the boundary conditions will be on the shear V and the moment M, so that the two constants C_1 and C_2 can be determined. The resulting shear V and moment M will coincide with those used for the starting point of the process of integrating $EIv'' = M$ outlined in the previous section. Several examples of this process are presented below.

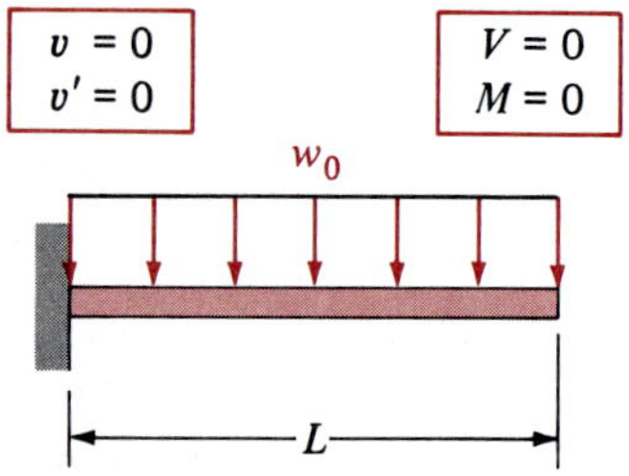

Figure 8.16 Uniformly loaded cantilever beams

Example 8.5 Determine the shear V, the moment M, the rotation v' and the deflection v of the beam shown in Fig. 8.16, using the integration of $EIv'''' = -w$.

Solution With $w(x) = w_0$ as the loading on the beam, the equation to be integrated is $EIv'''' = -w_0$ subject to the four boundary conditions shown in the boxes in Fig. 8.16. A first integration gives

$$EIv''' = V = -w_0 x + C_1$$

Using the boundary condition on shear at $x = L$, namely $V(L) = 0$, yields $C_1 = w_0 L$ and

$$EIv''' = V = w_0(L - x) \qquad (a)$$

A second integration yields

$$EIv'' = M = -\frac{w_0(L - x)^2}{2} + C_2$$

Using the boundary condition on moment at $x = L$, namely $M(L) = 0$, yields $C_2 = 0$, so that

$$EIv'' = M = -\frac{w_0(L - x)^2}{2} \qquad (b)$$

The student should verify that Eqns. (*a*) and (*b*) for the shear and moment are correct. Continuing the integration of Eqn. (*b*) yields

$$EIv' = \frac{w_0(L-x)^3}{6} + C_3$$

Using the boundary condition on v' at $x = 0$ yields $C_3 = -w_0L^3/6$. A last integration gives

$$EIv = -\frac{w_0(L-x)^4}{24} - \frac{w_0L^3x}{6} + C_4$$

which, with the boundary condition $v(0) = 0$, yields $C_4 = w_0L^4/24$ and

$$v' = \frac{w_0x}{6EI}(3Lx - 3L^2 - x^2) \qquad \text{and} \qquad v = \frac{w_0x^2}{24EI}(4Lx - 6L^2 - x^2)$$

The displacement curve is shown in Fig. 8.17.

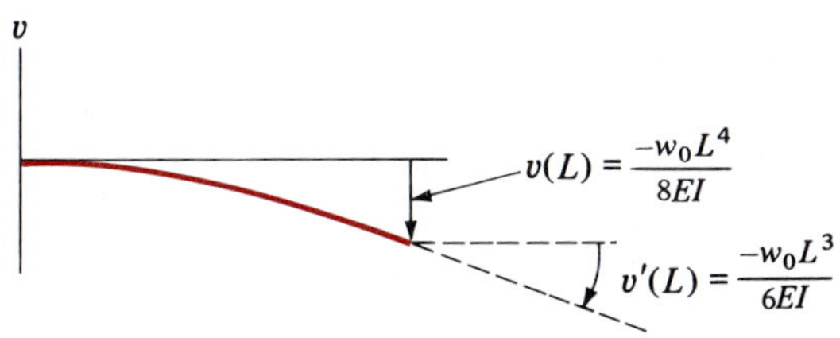

Figure 8.17 Elastic curve for a uniformly loaded cantilever

Example 8.6 Determine the shear V, the moment M, the rotation v', and the deflection v of the beam shown in Fig. 8.18, using integration of $EIv'''' = -w$.

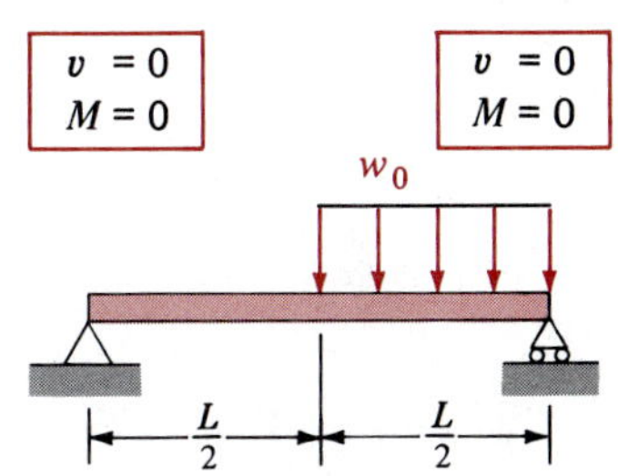

Figure 8.18 Simply supported beam, partially loaded

Solution It is necessary to integrate the equilibrium equation on each half of the beam due to the character of the loading $w(x)$. With v_1 and v_2 being the displacements in the left and right halves of the beam, respectively, the two equations to be integrated are

$$0 \leq x \leq L/2: \qquad EIv_1'''' = 0$$

$$L/2 \leq x \leq L: \qquad EIv_2'''' = -w_0$$

Recall from the discussions of shear and moment in Section 7.3 that in the absence of concentrated loads and moments, the shear and moment are continuous functions. In the present situation this means that at $x = L/2$ the two values of the shear given by integrating the two equations must be the same. Integrating yields

$$EIv_1''' = V_1 = C_1 \qquad EIv_2''' = V_2 = -w_0x + C_2$$

Requiring $V_1(L/2) = V_2(L/2)$ yields

$$C_1 = -\frac{w_0L}{2} + C_2$$

Eliminating C_2 then yields

$$EIv_1''' = C_1 \qquad EIv_2''' = w_0(L/2 - x) + C_1$$

Integrating again gives

$$EIv_1'' = M_1 = C_1 x + C_3 \qquad EIv_2'' = M_2 = -\frac{w_0(L/2 - x)^2}{2} + C_1 x + C_4$$

Requiring continuity of the moment at $x = L/2$ gives

$$C_1 \frac{L}{2} + C_3 = C_1 \frac{L}{2} + C_4$$

from which $C_3 = C_4$. The two boundary conditions on the moment M_1 at $x = 0$ and M_2 at $x = L$ yield $C_3 = C_4 = 0$ and $C_1 = w_0 L/8$, so that

$$EIv_1'' = \frac{w_0 L x}{8} \qquad EIv_2'' = \frac{w_0 L x}{8} - \frac{w_0(L/2 - x)^2}{2}$$

Another integration yields

$$EIv_1' = \frac{w_0 L x^2}{16} + C_5 \qquad EIv_2' = \frac{w_0 L x^2}{16} + \frac{w_0(L/2 - x)^3}{6} + C_6$$

Requiring the slope to be continuous at $x = L/2$ gives

$$\frac{w_0 L^4}{64} + C_5 = \frac{w_0 L^4}{64} + C_6$$

from which $C_5 = C_6$. Eliminating C_6 and integrating again gives

$$EIv_1 = \frac{w_0 L x^3}{48} + C_5 x + C_7$$

$$EIv_2 = \frac{w_0 L x^3}{48} - \frac{w_0(L/2 - x)^4}{24} + C_5 x + C_8$$

Requiring continuity of v at $x = L/2$ yields

$$\frac{w_0 L^4}{384} + C_5 \frac{L}{2} + C_7 = \frac{w_0 L^4}{384} + C_5 \frac{L}{2} + C_8$$

Combining this result with the two equations resulting from the other two boundary conditions, $v_1(0) = 0$ and $v_2(L) = 0$, gives $C_7 = C_8 = 0$ and $C_5 = -7w_0 L^3/384$, and thus

$$v_1 = \frac{w_0}{384EI}(8Lx^3 - 7L^3 x)$$

$$v_2 = \frac{w_0}{384EI}\left(8Lx^3 - 7L^3 x - 16\left(\frac{L}{2} - x\right)^4\right)$$

for the transverse displacement.

PROBLEMS

8.38–8.45 Resolve problems 8.14 through 8.21 using $EIv'''' = -w$. In each, indicate the four boundary conditions and then determine V, M, v', and v.

8.46–8.53 Resolve problems 8.22 through 8.29 using $EIv'''' = -w$. In each, indicate the four boundary conditions and any continuity conditions, and then determine V, M, v', and v.

8.54–8.61 Resolve the overhang problems 8.30 through 8.37 using $EIv'''' = -w$. For each, indicate the four boundary conditions and any continuity conditions, and then determine V, M, v', and v.

8.4 DEVELOPMENT AND USE OF SINGULARITY FUNCTIONS

In Examples 8.3 and 8.4, the process of determining the transverse displacements using integration of $EIv'' = M$, and subsequent satisfaction of the two boundary conditions and the two continuity conditions, resulted in a set of four simultaneous equations for the four constants of integration. Similarly, in Example 8.6, the integration of $EIv'''' = -w$ produced eight constants of integration that were subsequently determined on the basis of four boundary conditions and four continuity conditions. In a situation where there are several intervals for which separate moment or load expressions are necessary, such as the situation pictured in Fig. 8.19, the number of constants and equations can become practically unmanageable.

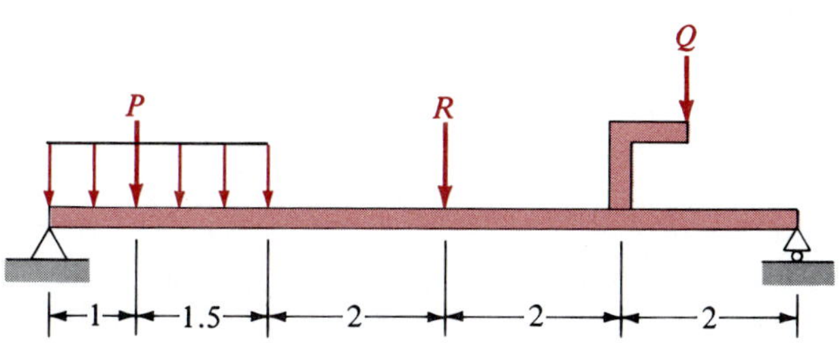

Figure 8.19 Example of several discontinuous load inputs

For the problem of Fig. 8.19, the usual approach of integrating $EIv'' = M$ would necessitate that moment expressions be constructed for each of the five intervals $0 \le x \le 1$, $1 \le x \le 2.5$, $2.5 \le x \le 4.5$, $4.5 \le x \le 6.5$, and $6.5 \le x \le 8.5$ with the result that there would be 10 constants of integration and a corresponding 10 by 10 set of linear equations to be solved to determine the 10 constants. In this section, we will discuss an approach that permits us to write a single expression for the moment that is valid over the entire length of the beam, so that

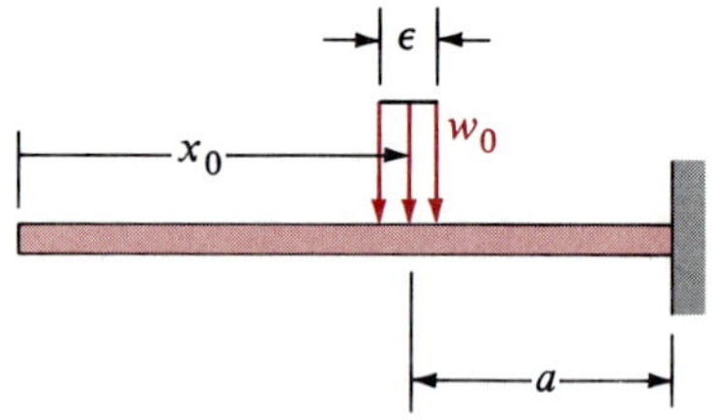

Figure 8.20 Distributed load w_0 on a small segment ε of a beam

the moment-curvature relation $EIv'' = M$ can be integrated with the two constants of integration determined from the boundary conditions on the slope v' and/or the displacement v, thus averting the necessity of setting up and solving a large set of linear algebraic equations.

Consider the problem of a distributed load w_0, acting over a relatively small segment ε of a beam as indicated in Fig. 8.20. With the total load $P = w_0\varepsilon$ held constant, the w, V, and M diagrams for several values of the length ε are as pictured in Fig. 8.21.

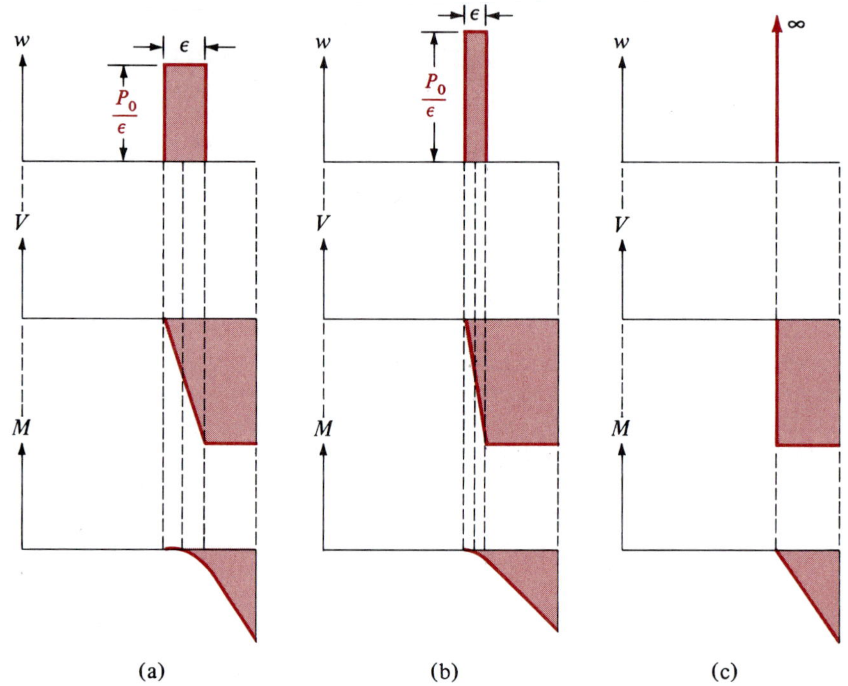

Figure 8.21 The w, V, and M diagrams for several values of ε

When $P = w_0\varepsilon$ is required to be a constant, independent of the value of ε, the value of w_0 must become very large. In the limit, the intensity of w_0 actually approaches infinity, with the w, V, and M curves of Fig. 8.21c representing the idealization referred to as a **concentrated load.** The symbol $\delta(x, x_0)$, also denoted by $\delta(x - x_0)$ or by $\langle x - x_0\rangle^{-1}$, is often referred to as the **Dirac delta function** or the **unit impulse function.** The dirac delta function is not actually a "function," rather only an indication that something rather unusual is being modeled, namely, a finite load P applied over an extremely small distance ε.

The shear diagram of Fig. 8.21c indicates that as the point $x = x_0$ is crossed, V drops abruptly an amount equal to the load P, or

$$V = \begin{cases} 0 & \text{if } x < x_0 \\ -P & \text{if } x > x_0 \end{cases}$$

This abrupt change can be represented by

$$V = -P\langle x - x_0 \rangle^0$$

where $\langle x - x_0 \rangle^0$ is the so-called **step function** given by

$$\langle x - x_0 \rangle^0 = \begin{cases} 0 & \text{if } x < x_0 \\ 1 & \text{if } x \geq x_0 \end{cases}$$

The corresponding moment diagram of Fig. 8.21c indicates that the moment is given by

$$M = \begin{cases} 0 & \text{if } x < x_0 \\ -P(x - x_0) & \text{if } x > x_0 \end{cases}$$

which can be represented as $M = -P\langle x - x_0 \rangle^1$, where $\langle x - x_0 \rangle^1$ is a symbol representing the function

$$\langle x - x_0 \rangle^1 = \begin{cases} 0 & \text{if } x < x_0 \\ x - x_0 & \text{if } x \geq x_0 \end{cases}$$

In view of the fact that the loading, shear, and moment functions are related by the equilibrium equations according to $dV/dx = -w$ and $dM/dx = V$, the corresponding shear and moment diagrams indicate that when the delta function $\langle x - x_0 \rangle^{-1}$ is integrated, the result is the step function $\langle x - x_0 \rangle^0$, or

$$\int^x \langle x - x_0 \rangle^{-1}\, dx = \langle x - x_0 \rangle^0$$

Similarly, when the step function is integrated there results

$$\int^x \langle x - x_0 \rangle^0\, dx = \langle x - x_0 \rangle^1$$

sometimes referred to as the **ramp function.** Successive integrations of these singularity functions are carried out according to the integration formulae

$$\int^x \langle x - x_0 \rangle^n\, dx = \frac{\langle x - x_0 \rangle^{n+1}}{(n + 1)} \qquad n = 0, 1, 2, \ldots$$

We note in passing that the dimensions of the delta function are $[\langle x - x_0\rangle^{-1}] = 1/L$, whereas the step function is dimensionless. Each of the singularity functions $\langle x - x_0\rangle^n$ has dimension L^n. Several of the singularity functions are shown in Fig. 8.22.

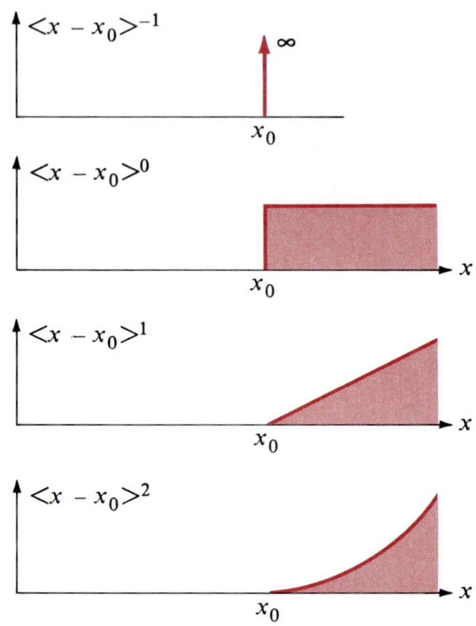

Figure 8.22 Successive singularity functions

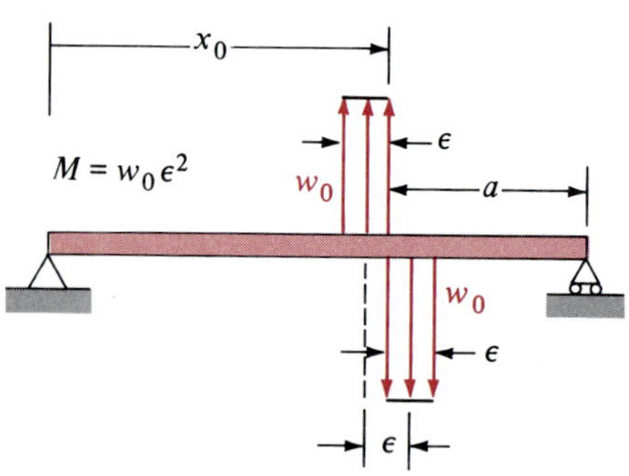

Figure 8.23 Moment loading on a beam

Consider also the loading shown in Fig. 8.23, where two adjacent segments of a beam are subjected to loads $w_0\varepsilon$ that are oppositely directed. The force associated with each of the loads is $w_0\varepsilon$. As indicated in Fig. 8.23, the corresponding moment is given by

$$M = Fd = (w_0\varepsilon)\varepsilon = w_0\varepsilon^2$$

Allowing ε to approach zero while requiring M to be a constant results in what is referred to as a **concentrated moment** applied at the point x_0. The w, V, and M diagrams for several values of ε are shown in Fig. 8.24.

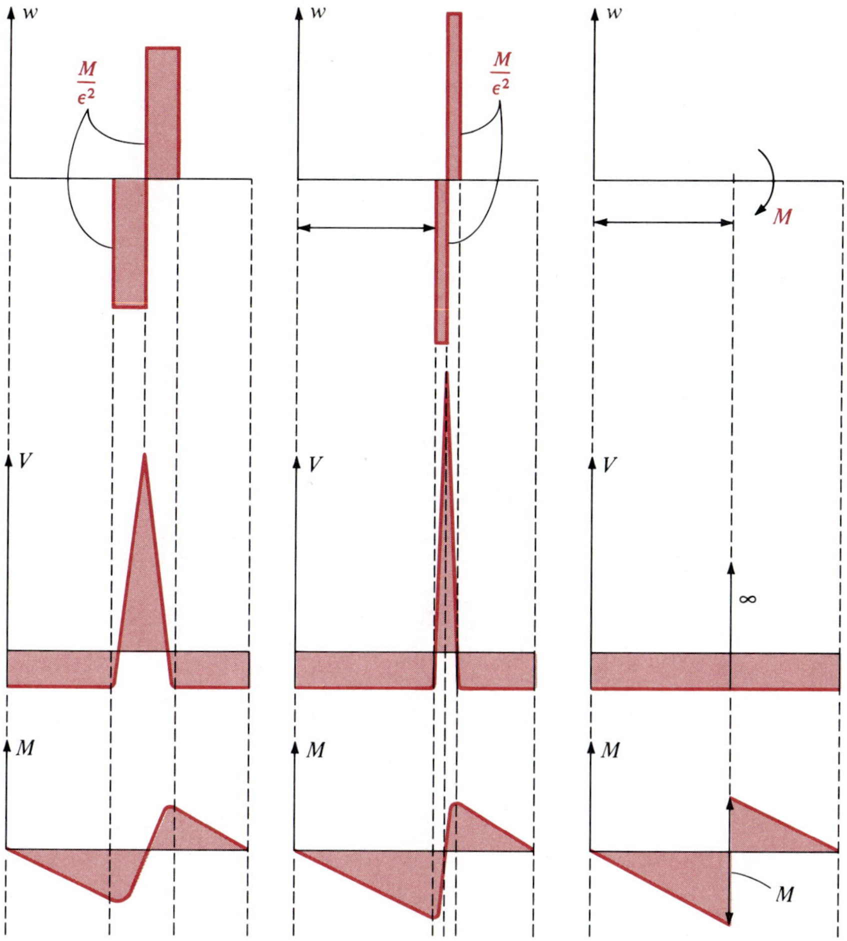

Figure 8.24 The w, V, and M diagrams for several values of ε

The symbol that is used to represent the concentrated moment is $m(x, x_0) = \langle x - x_0 \rangle^{-2}$ and will be referred to as the **moment function** or **unit dipole function.**

Again consulting the loading, shear, and moment curves, and recalling that $dV/dx = -w$ and $dM/dx = V$, it follows that an integral relation between the delta function $\langle x - x_0 \rangle^{-1}$ and the moment function $\langle x - x_0 \rangle^{-2}$ is given by

$$\int^x \langle x - x_0 \rangle^{-2}\, dx = \langle x - x_0 \rangle^{-1}$$

or

$$\int^x m(x, x_0)\, dx = \delta(x, x_0)$$

that is, integrating the moment function produces the delta function.

Collecting all these results we have

$$\int^x \langle x - x_0 \rangle^{-2}\, dx = \langle x - x_0 \rangle^{-1}$$

$$\int^x \langle x - x_0 \rangle^{-1}\, dx = \langle x - x_0 \rangle^{0}$$

$$\int^x \langle x - x_0 \rangle^{n}\, dx = \frac{\langle x - x_0 \rangle^{n+1}}{(n+1)} \qquad n \geq 0$$

The student should keep firmly in focus the fact that all the singularity functions

$$\langle x - x_0 \rangle^{n} \qquad n \geq 0$$

have zero values for $x < x_0$.

The singularity functions turn out to be very useful in constructing shear and moment diagrams and in integrating $EIv'' = M$ to determine $v(x)$. Shown in Fig. 8.25 are several common loadings and their moment representations in terms of the singularity functions. In the examples that follow we will illustrate the use of the singularity functions in integrating $EIv'' = M$.

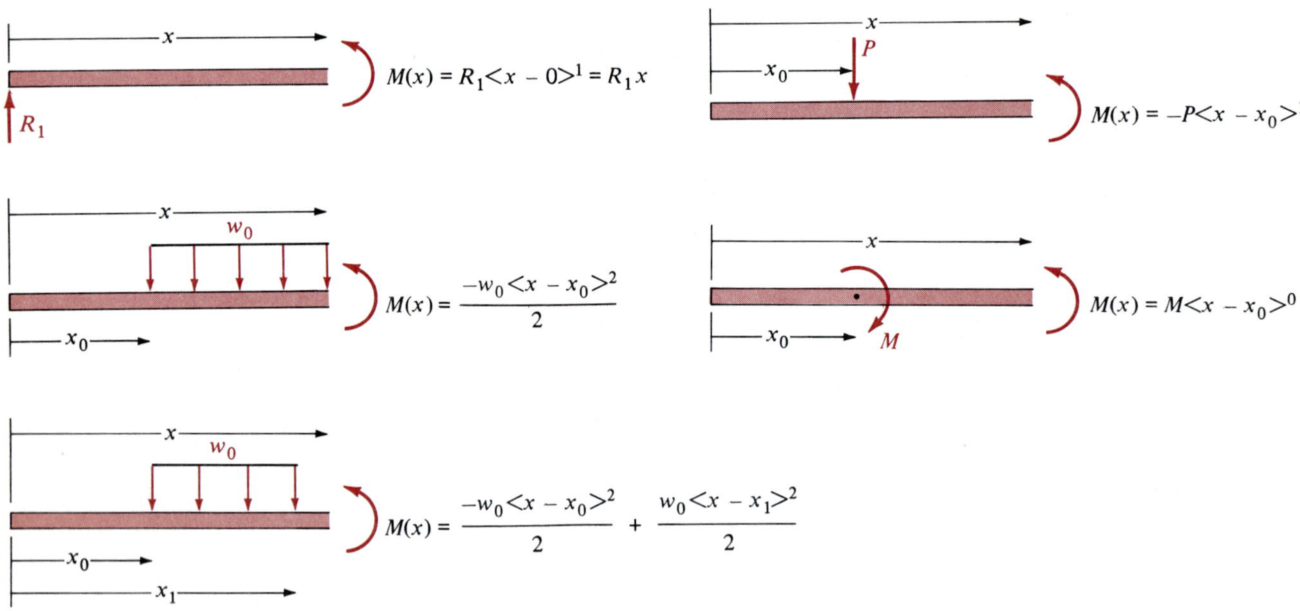

Figure 8.25 Moment function for several different common loadings

Example 8.7 Consider again the problem of Example 8.3, shown in Fig. 8.26. We wish to illustrate the use of the singularity functions to carry out the process of integrating $EIv'' = M$, to determine the displacement $v(x)$.

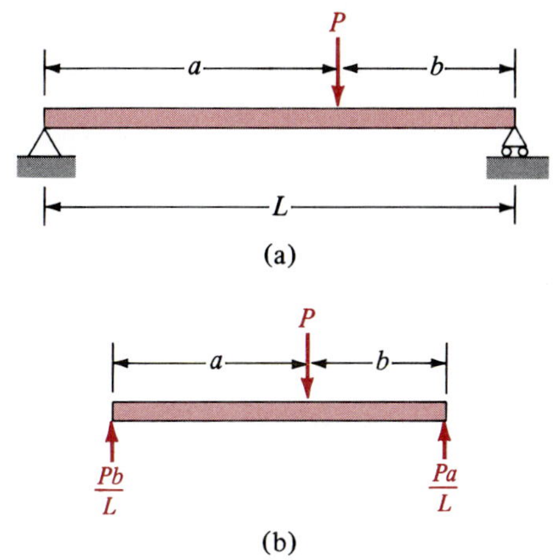

Figure 8.26 Beam acted upon by a concentrated load

Solution Consulting the free-body diagram of Fig. 8.26b and the information shown in Fig. 8.25, we see that the moment can be expressed as

$$M(x) = \frac{Pbx}{L} - P\langle x - a\rangle^1$$

The student should check to see that the two boundary conditions $M(0) = 0$ and $M(L) = 0$ are satisfied by the moment expression. The moment-curvature relation can then be written as

$$EIv'' = \frac{Pbx}{L} - P\langle x - a\rangle^1$$

which is valid for the entire length of the beam. Integrating twice, using the results for the integration of the singularity functions developed previously, yields

$$EIv = \frac{Pbx^3}{6L} - \frac{P\langle x - a\rangle^3}{6} + C_1x + C_2$$

Satisfying the boundary condition at $x = 0$ yields

$$0 = 0 - 0C_1(0) + C_2$$

so that $C_2 = 0$. The boundary condition $v(L) = 0$ gives

$$0 = \frac{PbL^2}{6} - \frac{Pb^3}{6} + C_1L$$

or $C_1 = Pb(b^2 - L^2)/6L$. The equation for the elastic curve is then

$$v = -\frac{P}{6EIL}[L\langle x - a\rangle^3 + bx(L^2 - b^2) - bx^3]$$

So we see that for the segments to the left and right of the load, respectively,

$$0 \le x \le a: \qquad v = -\frac{P}{6EIL}[0 + bx(L^2 - b^2) - bx^3]$$

$$a \le x \le L: \qquad v = -\frac{P}{6EIL}[L(x - a)^3 + bx(L^2 - b^2) - bx^3]$$

precisely as in Example 8.3. The use of the singularity functions substantially simplifies the process of integrating $EIv'' = M$ to determine v.

Example 8.8 The steel W 8 × 28 beam is loaded as shown in Fig. 8.27. Determine *(a)* the equation for the elastic curve, *(b)* the maximum transverse displacement, and *(c)* the slope at the left end of the beam. Neglect the weight of the beam.

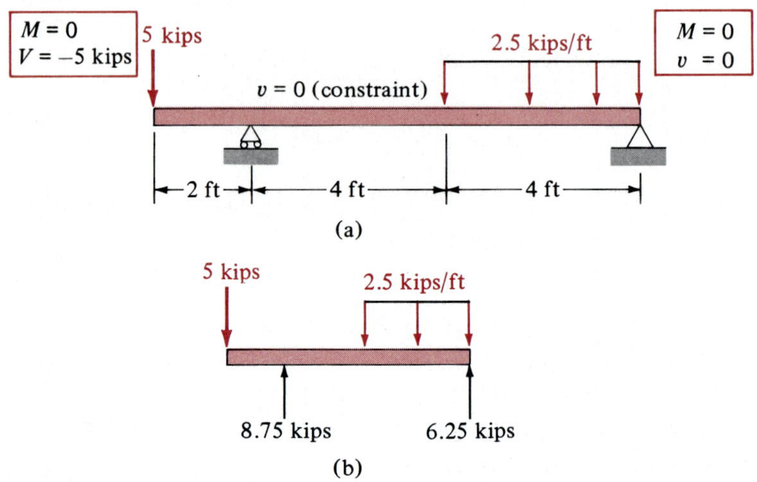

Figure 8.27 Beam loaded by concentrated and distributed loadings

Solution The free-body diagram with the reactions determined is shown in Fig. 8.27b. Referring to Fig. 8.25, we can express the moment as

$$M(x) = (-5000\langle x - 0\rangle^1 + 8750\langle x - 2\rangle^1 - 1250\langle x - 6\rangle^2 + 6250\langle x - 10\rangle^1) \text{ ft-lbf}$$

where x is measured in feet. Note that the term $\langle x - 10\rangle^1$ and any of its integrals will always be zero for any $0 \leq x \leq 10$ ft, and can be discarded if desired. Also note that all of the boundary conditions on shear and moment, namely $V(0) = -5000$ lbf, $M(0)$, and $M(10 \text{ ft}) = 0$, are satisfied.

Dropping the $\langle x - 10\rangle^1$ term, the moment-curvature equation becomes

$$EIv'' = (-5000\langle x - 0\rangle^1 + 8750\langle x - 2\rangle^1 - 1250\langle x - 6\rangle^2) \text{ ft-lbf}$$

Integrating twice gives

$$EIv = \left(-\frac{2500\langle x - 0\rangle^3}{3} + \frac{4375\langle x - 2\rangle^3}{3} - \frac{1250\langle x - 6\rangle^4}{12}\right) \text{ft}^3\text{-lbf} + C_1 x + C_2$$

The two constants of integration are evaluated on the basis of the constraint given as $v(2 \text{ ft}) = 0$, and from the boundary condition $v(10 \text{ ft}) = 0$, leading to

$$0 = -\left(\frac{2500 \times 8}{3}\right) \text{ft}^3\text{-lbf} + C_1 2 \text{ ft} + C_2$$

and

$$0 = \left(-\frac{2500 \times 10^3}{3} + \frac{4375 \times 8^3}{3} - \frac{1250 \times 4^4}{12}\right) \text{ft}^3\text{-lbf} + C_1 10 \text{ ft} + C_2$$

or

$$C_1 2 \text{ ft} + C_2 = 6667 \text{ ft}^3\text{-lbf}$$

and

$$C_1 10 \text{ ft} + C_2 = 113{,}333 \text{ ft}^3\text{-lbf}$$

from which $C_1 = 13{,}333$ ft^2-lbf and $C_2 = -20{,}000$ ft^3-lbf. The equations for the slope v' and the displacement v can then be expressed as

$$EIv' = \left(-2500\langle x - 0\rangle^2 + 4375\langle x - 2\rangle^2 - \frac{1250\langle x - 6\rangle^3}{3} + 13{,}333\right) \text{ft}^2\text{-lbf}$$

and

$$EIv = \left(\frac{-2500\langle x - 0\rangle^3}{3} + \frac{4375\langle x - 2\rangle^3}{3} - \frac{1250\langle x - 6\rangle^4}{12} + 13{,}333x - 20{,}000\right) \text{ft}^3\text{-lbf}$$

Since $\langle x - 0\rangle^2$ and $\langle x - 0\rangle^3$ are simply x^2 and x^3, respectively, for $x \geq 0$, the expressions for v' and v can be rewritten as

$$EIv' = \left(-2500x^2 + 4375\langle x - 2\rangle^2 - \frac{1250\langle x - 6\rangle^3}{3} + 13{,}333\right) \text{ft}^2\text{-lbf}$$

and

$$EIv = \left(-\frac{2500x^3}{3} + \frac{4375\langle x - 2\rangle^3}{3} - \frac{1250\langle x - 6\rangle^4}{12} + 13{,}333x - 20{,}000\right) \text{ft}^3\text{-lbf}$$

It is left in the problems for the student to show that the zeros of v' occur at $x = 2.36$ ft and at $x = 7.03$ ft, so that the maximum displacement occurs either at $x = 0$, $x = 2.36$ ft, or $x = 7.03$ ft.

From the tables in Appendix C, the second moment of the area for a W 8 × 28 section beam is 98 in^4. Then with $E = 30 \times 10^6$ psi

$$v(0) = \frac{(-20{,}000 \text{ ft}^3\text{-lbf})(12 \text{ in./ft})^3}{(30 \times 10^6 \text{ psi})(98 \text{ in}^4)} = -0.01176 \text{ in.}$$

$$v(2.36 \text{ ft}) = \frac{(3780 \text{ ft}^3\text{-lbf})(12 \text{ in./ft})^3}{(30 \times 10^6 \text{ psi})(98 \text{ in}^4)} = 0.00222 \text{ in.}$$

$$v(7.03 \text{ ft}) = \frac{(-30{,}318 \text{ ft}^3\text{-lbf})(12 \text{ in./ft})^3}{(30 \times 10^6 \text{ psi})(98 \text{ in}^4)} = -0.01782 \text{ in.}$$

The shear V, the moment M, and the shape of the elastic curve are shown in Fig. 8.28.

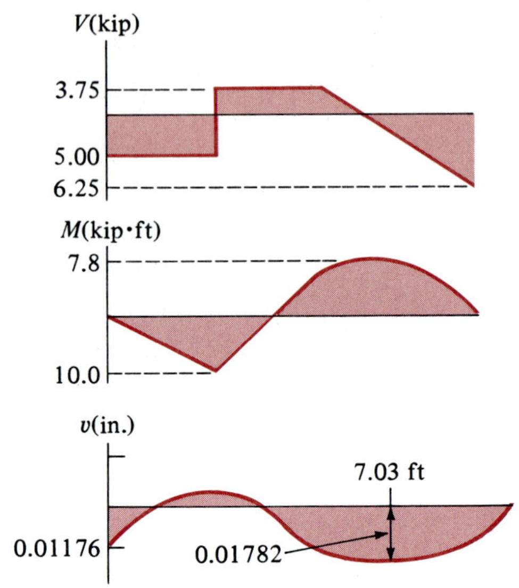

Figure 8.28 Shear V, moment M, and displacement v

PROBLEMS

8.62–8.79 For each of the following, determine the reactions and express the moment $M(x)$, including the reactions, in terms of the singularity functions.

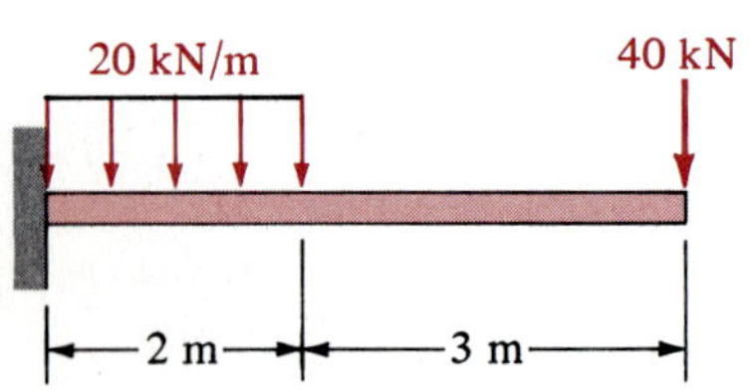

P. 8.62

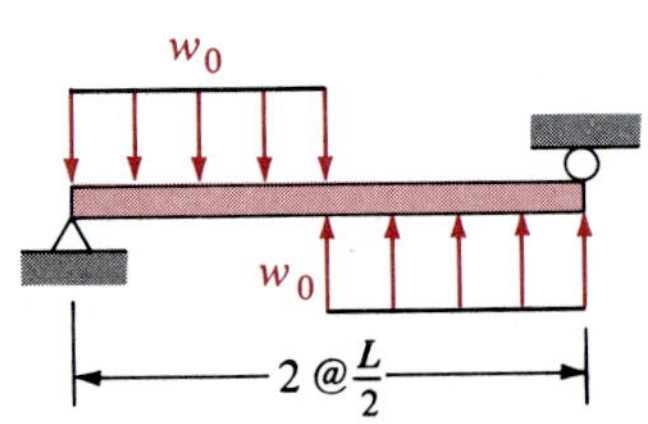

P. 8.63

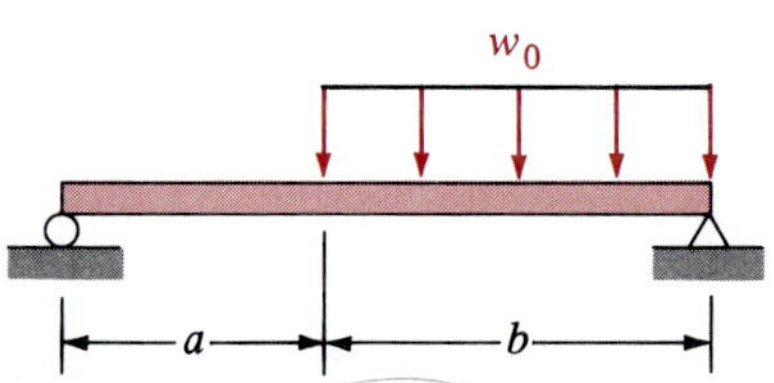

P. 8.64

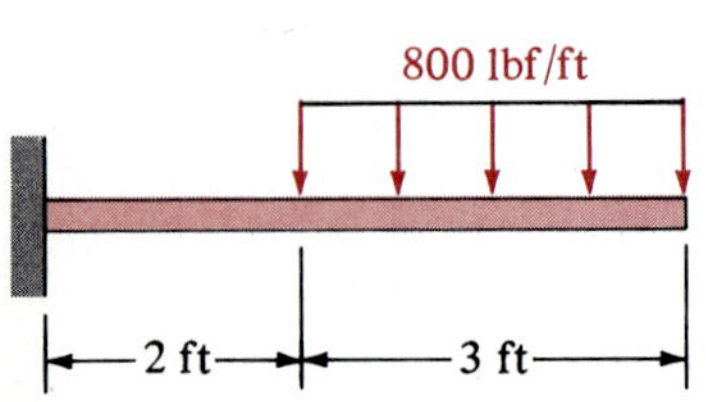

P. 8.65

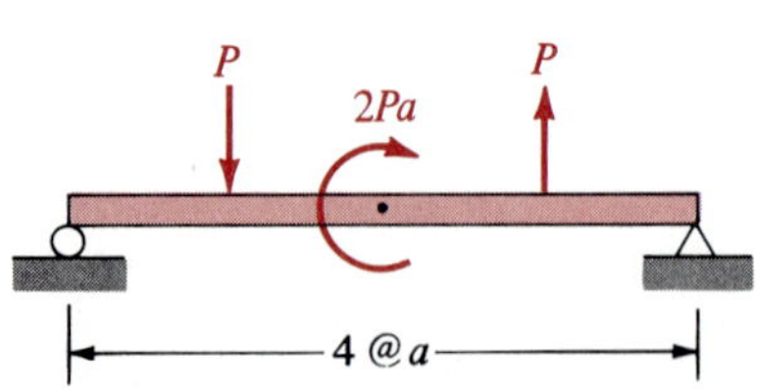

P. 8.66

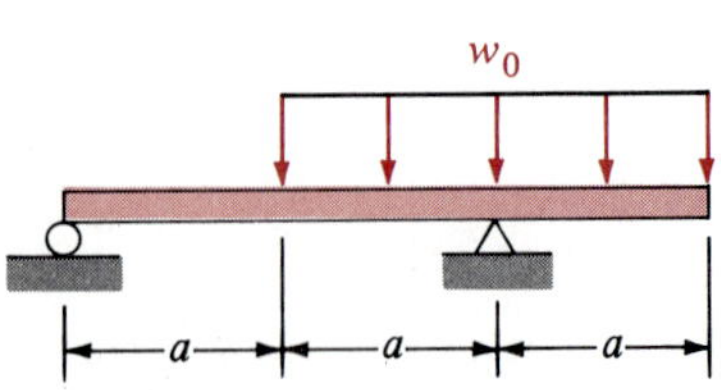

P. 8.67

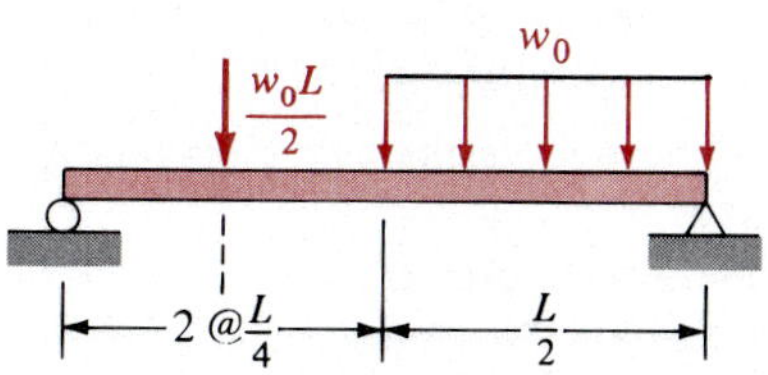

P. 8.68

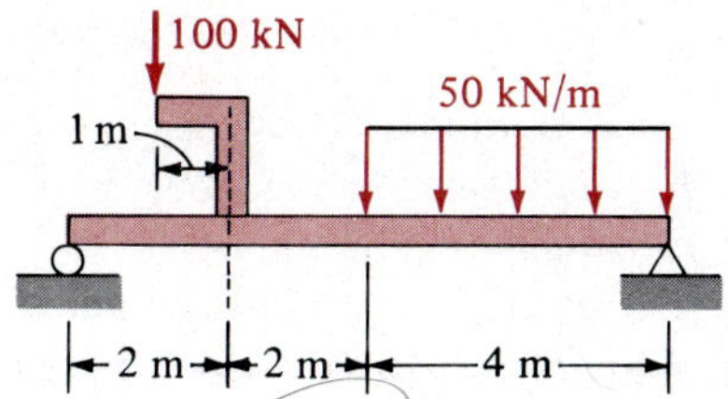

P. 8.69

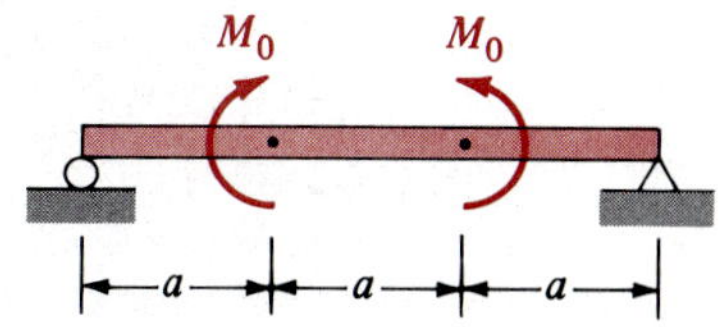

P. 8.70

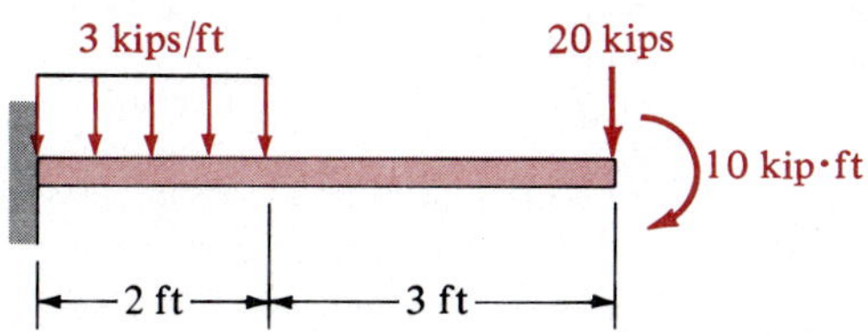

P. 8.71

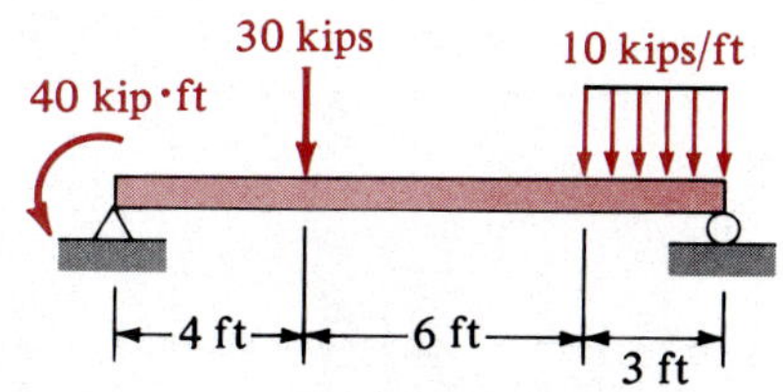

P. 8.72

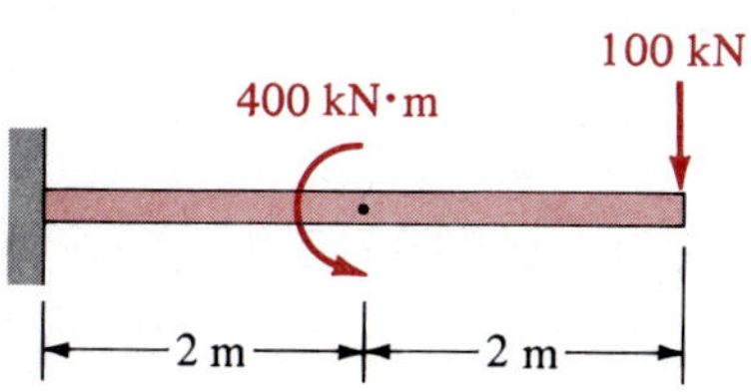

P. 8.73

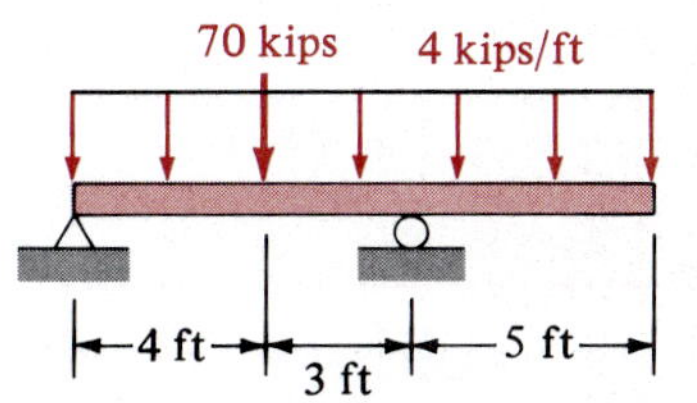

P. 8.74

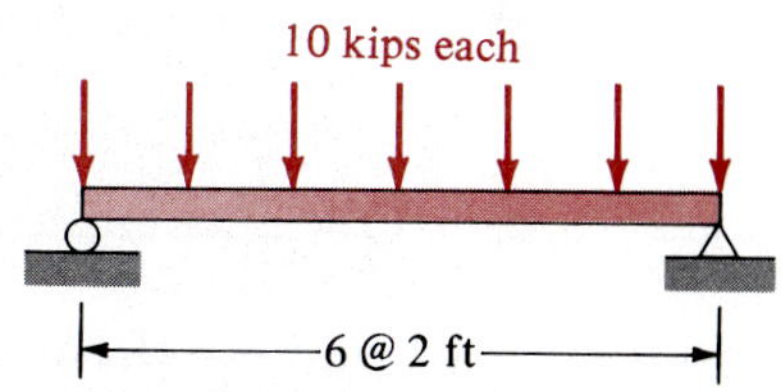

P. 8.75

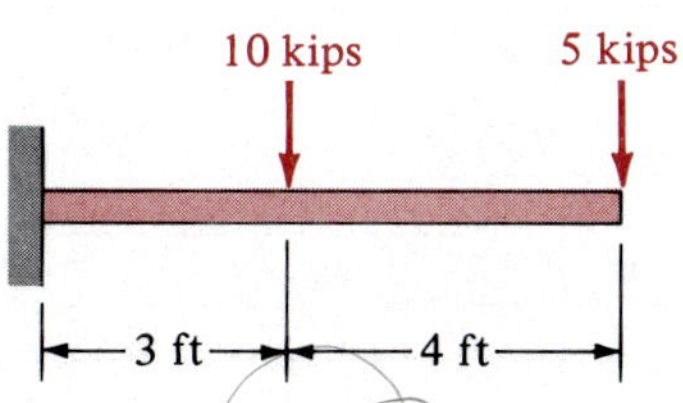

P. 8.76

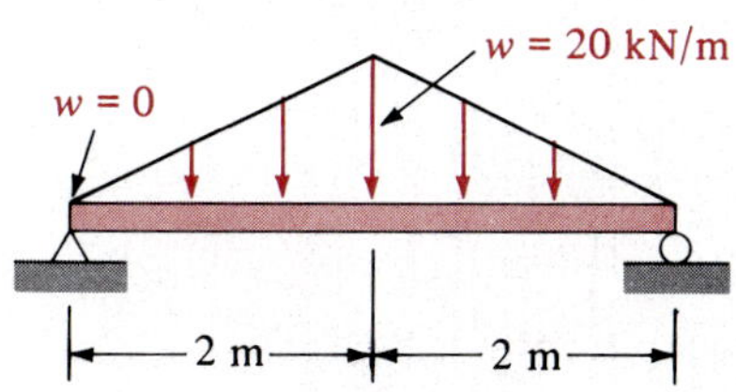

P. 8.77

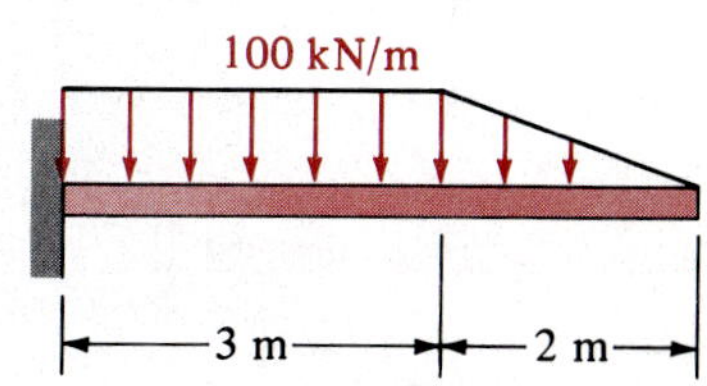

P. 8.78

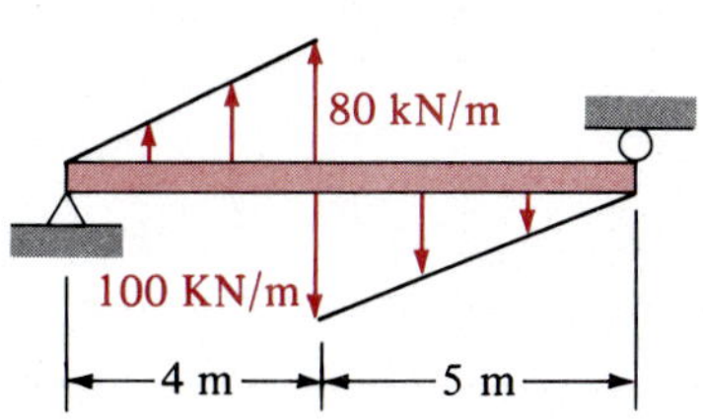

P. 8.79

8.80–8.92 For each of the following figures,

a. Determine the reactions.

b. Including the reactions, express the loading w in terms of the singularity functions.

c. Writing the equilibrium equation as $M'' + w = 0$, integrate once to obtain $V + \int w\,dx = C_1$, and show by evaluating the shear just past the right end of the beam that $C_1 = 0$.

d. Integrate again to obtain $M + \int(\int w\,dx)\,dx = C_2$, and show by evaluating the moment just past the right end of the beam that $C_2 = 0$.

e. Integrate the results of part *(d)* and apply the remaining boundary and constraint conditions to determine v' and v.

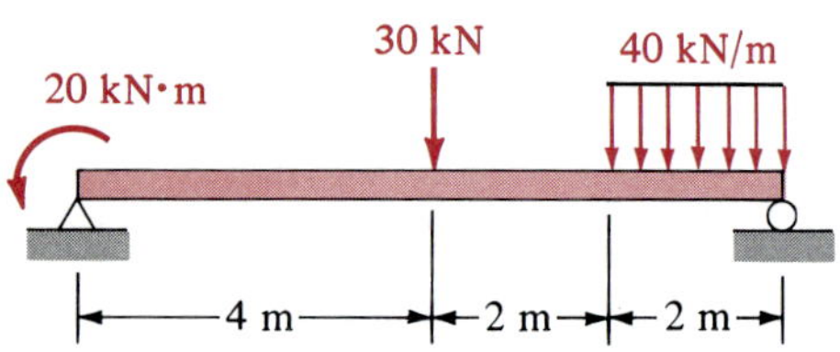

P. 8.80

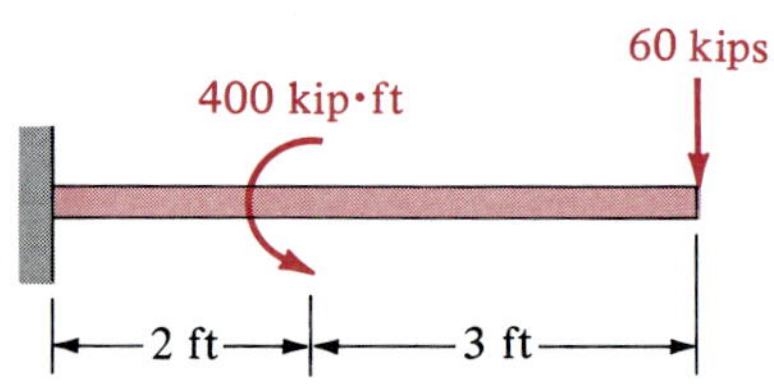

P. 8.81

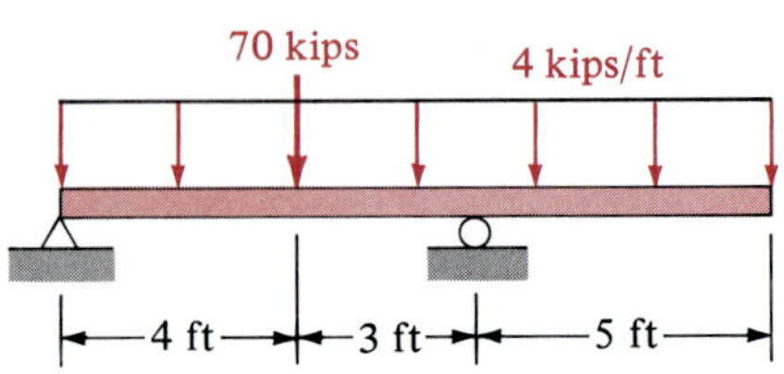

P. 8.82

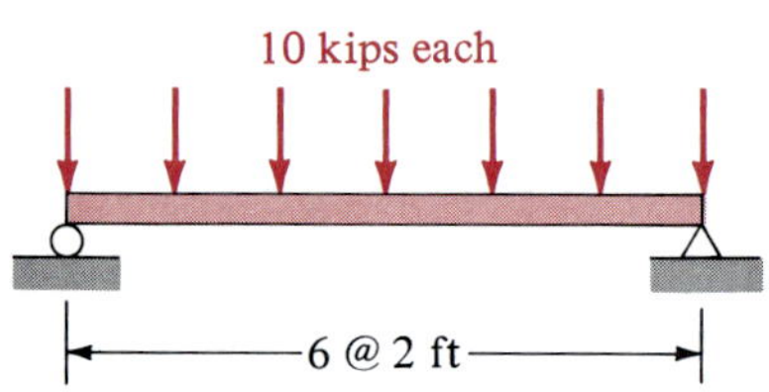

P. 8.83

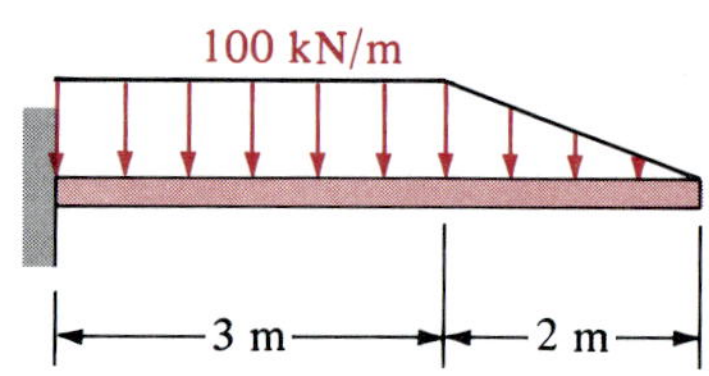

P. 8.84

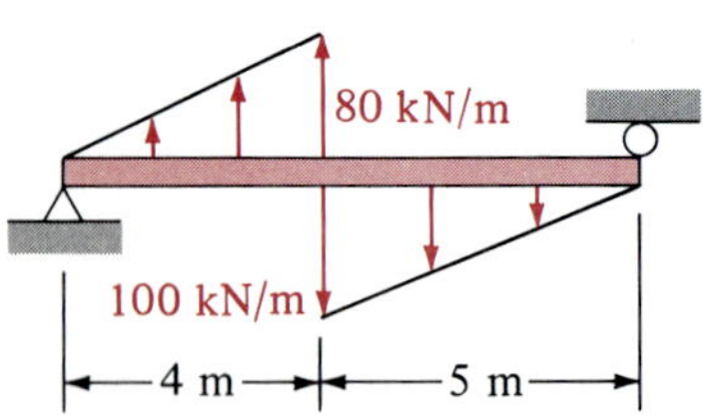

P. 8.85

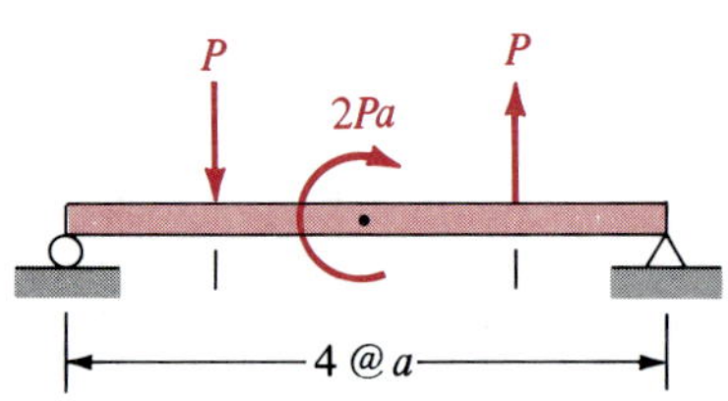

P. 8.86

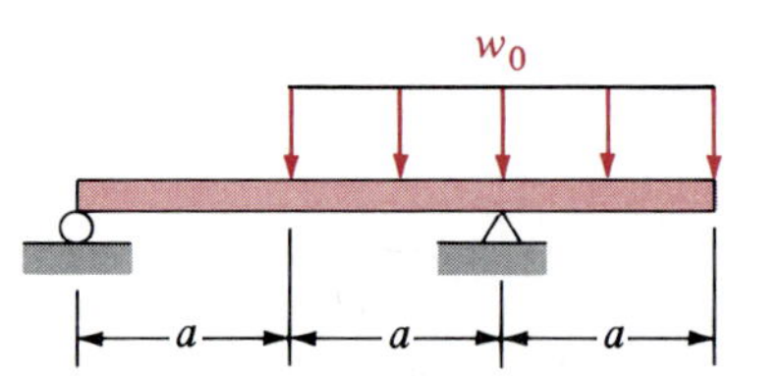

P. 8.87

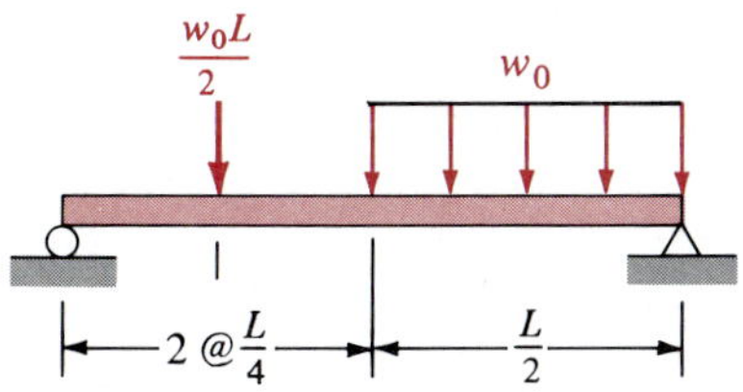

P. 8.88

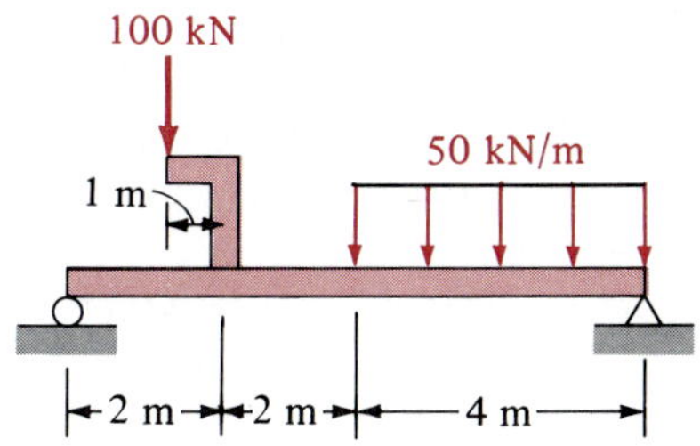

P. 8.89

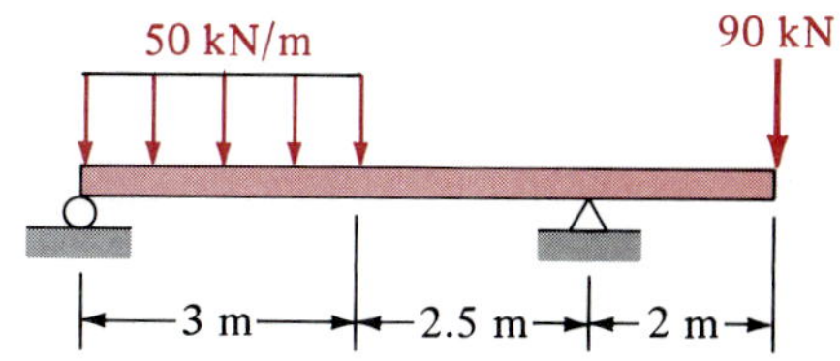

P. 8.90

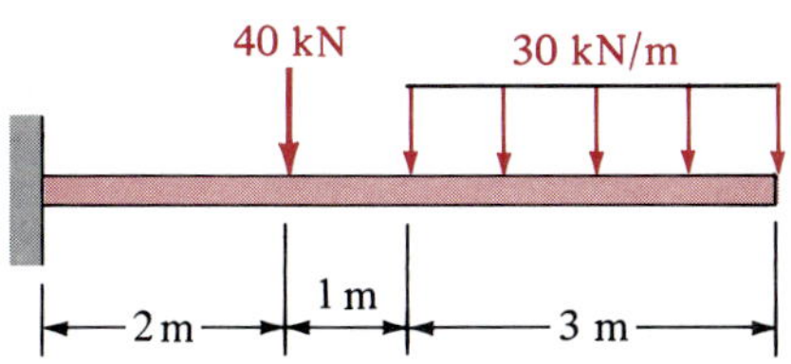

P. 8.91

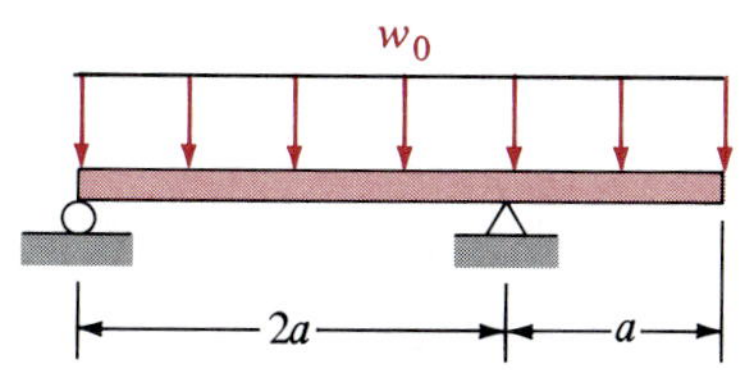

P. 8.92

8.5 DEFLECTIONS USING SUPERPOSITION

In Sections 8.2 and 8.3, we used integration of $EIv'' = M$ and $EIv'''' = -w$, respectively, together with the boundary conditions, to determine the transverse displacement $v(x)$. When using $EIv'' = M$, our purpose was to develop an expression for the moment $M(x)$, including all the loads and reactions, that we could then integrate to determine $v(x)$. When using $EIv'''' = -w$, the starting point was the loading $w(x)$, formulated to include the reactions and the loads. In either case, we considered one equation and one set of boundary conditions. In this section we will demonstrate that, based on the idea that both $EIv'' = M$ and $EIv'''' + w = 0$ are linear differential equations, we can use *superposition* of known cases of loading to accomplish the integration.

Again, the basic idea is that both $EIv'' = M$ and $EIv'''' + w = 0$ are linear differential equations. When using $EIv'' = M$, for instance, this fact enables us to write $v = v_1 + v_2 + \dots + v_n$ and $M = M_1 + M_2 + \dots + M_n$, where each of $M_1, M_2, \dots$, represents the moment due to a particular part of the loading, and each of $v_1, v_2, \dots, v_n$, represents the corresponding displacement. The resulting equation,

$$EIv_1'' + EIv_2'' + \dots + EIv_n'' = M_1 + M_2 + \dots + M_n$$

can be considered to be the collection of equations

$$EIv_i'' = M_i \qquad i = 1, 2, \dots, n$$

each of which is to be integrated subject to the boundary conditions on the original problem.

In practice, one considers that the particular loading in question consists of the superposition of common loadings for which each of the solutions, subject to the same set of boundary conditions, is known. Several such loadings, together with the corresponding solutions for the transverse displacement $v(x)$, are indicated in Appendix D. The actual process of superposition leading to the results of Appendix D is best demonstrated by way of the several examples that follow.

Example 8.9 A simply supported beam is subjected to the loading shown in Fig. 8.29. Select a set of loads that can be superposed to determine the transverse displacement, and hence determine the deflection at the center of the beam.

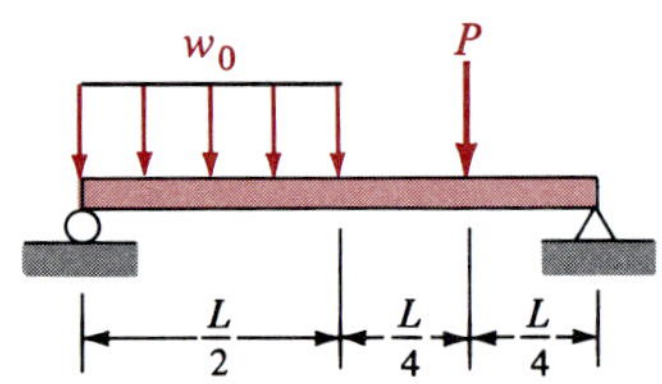

Figure 8.29 Simply supported beam loaded by a force and a distributed loading

Solution Consider the problem to be composed of the two problems shown in Fig. 8.30.

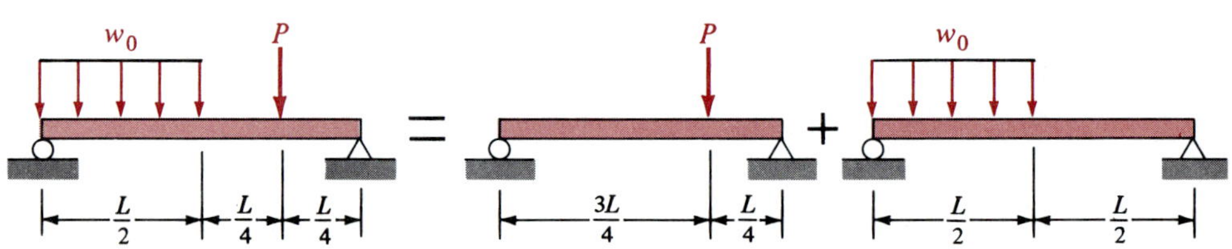

Figure 8.30 Superposition of two loadings

In addition to the two loads adding up to the original load, the boundary conditions on all three problems are the same for this example. From Appendix D, the problem of the concentrated load corresponds to case *(e)*, with $a = 3L/4$, for which the central deflection is

$$v_1\left(\frac{L}{2}\right) = -\frac{11PL^3}{768EI}$$

The distributed loading corresponds to case *(h)*, with $a = L/2$ in Appendix D, for which the central deflection is

$$v_2\left(\frac{L}{2}\right) = -\frac{5w_0L^4}{768EI}$$

so that the total deflection at the center is

$$v_1\left(\frac{L}{2}\right) + v_2\left(\frac{L}{2}\right) = -\frac{5w_0L^4 + 11PL^3}{768EI}$$

The individual shear and moment diagrams can also be added pointwise along the beam to determine the total shear and moment throughout the beam.

Example 8.10 Determine the rotation and displacement at point D for the loading shown in Fig. 8.31. Take $I = 2 \times 10^{-4}$ m^4 and $E = 70$ GPa.

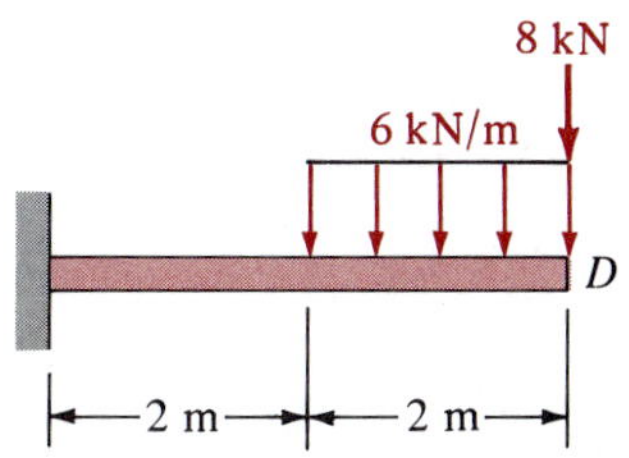

Figure 8.31 Loading of a cantilever

Solution Consider the original problem to be composed of the two problems shown in Fig. 8.32.

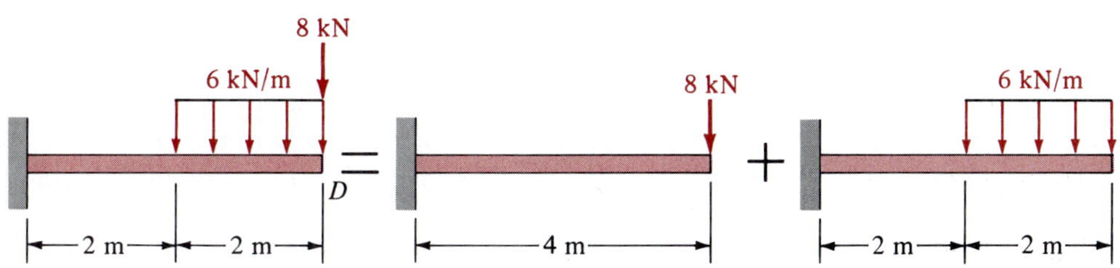

Figure 8.32 Superposition of two loadings

Using case *(a)* with $a = L$ from Appendix D, the rotation and displacement at D due to the load P are given by

$$\theta(D) = -\frac{PL^2}{2EI} = -\frac{(8000\text{ N})(4\text{ m})^2}{2(70\text{ GPa})(2 \times 10^{-4}\text{ m}^4)} = -0.00457\text{ rad}$$

$$v(D) = -\frac{PL^3}{3EI} = -\frac{(8000\text{ N})(4\text{ m})^3}{3(70\text{ GPa})(2 \times 10^{-4}\text{ m}^4)} = -0.0122\text{ m}$$

From case *(d)* with $a = L/2$ the rotation and displacement at D due to the distributed loading are

$$\theta(D) = -\frac{7w_0L^3}{48EI} = -\frac{7(6000\ \text{N/m})(4\ \text{m})^3}{48(70\ \text{GPa})(2 \times 10^{-4}\ \text{m}^4)} = -0.004\ \text{rad}$$

$$v(D) = -\frac{41w_0L^4}{384EI} = -\frac{41(6000\ \text{N/m})(4\ \text{m})^4}{384(70\ \text{GPa})(2 \times 10^{-4}\ \text{m}^4)} = -0.0117\ \text{m}$$

Thus the total rotation and displacement at D are, respectively,

$$\theta(D) = -0.0046\ \text{rad} - 0.0040\ \text{rad} = -0.0086\ \text{rad}$$

and

$$v(D) = -0.0122\ \text{m} - 0.0117\ \text{m} = -0.0239\ \text{m}$$

One of the most frequent and important uses of superposition is in the treatment of what are termed two-dimensional **frames,** two examples of which are shown in Fig. 8.33.

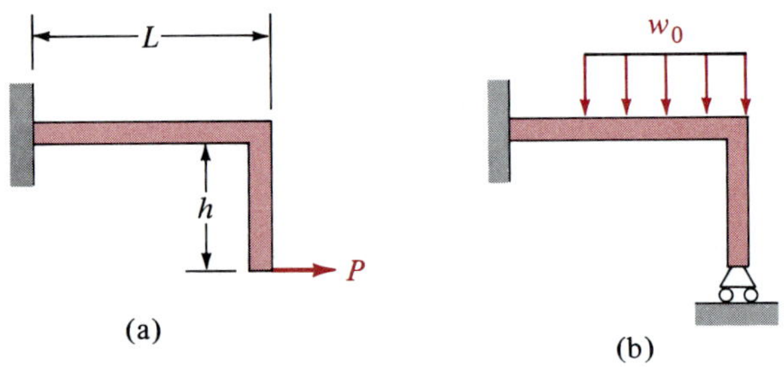

Figure 8.33 Typical two-dimensional frames

Fig. 8.33a indicates a situation where the problem is statically determinate, and Fig. 8.33b indicates a statically indeterminate situation. In this section we will consider statically determinate two-dimensional frames. (Statically indeterminate frames will be considered in Chapter 10.) Let us consider several examples that demonstrate the use of previous results to determine displacements of statically determinate frames.

Example 8.11 For the two-dimensional frame shown in Fig. 8.34, determine *(a)* the vertical deflection at B, *(b)* the rotation at C, and *(c)* the horizontal displacement at C. Neglect the presence of axial forces in the members.

Solution To see what internal forces are being transmitted by the members, consider the free-body diagrams shown in Fig. 8.35.

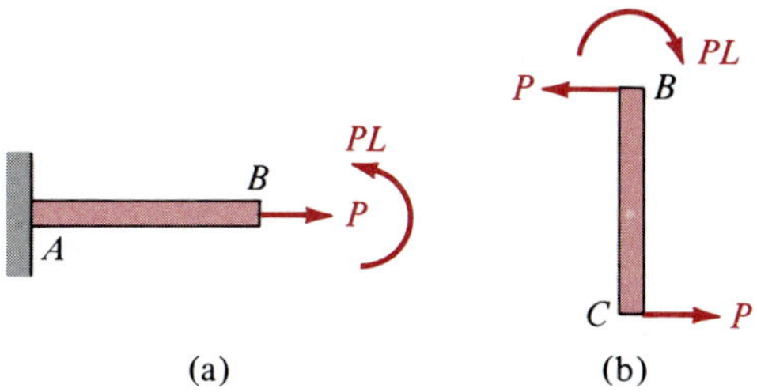

Figure 8.35 Free-body diagrams of the frame members

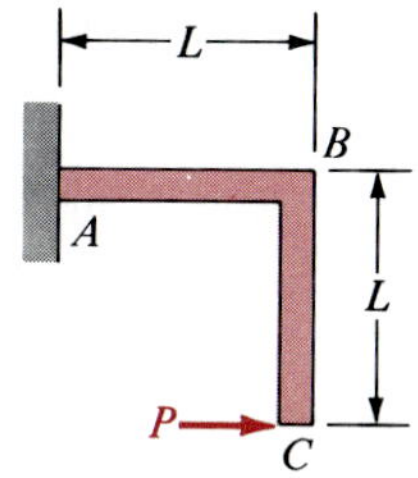

Figure 8.34 Deformations of a two-dimensional frame

As indicated in Fig. 8.35a, the portion AB of the frame can be considered to be a cantilever beam loaded by an end moment PL. Using case *(b)* with $a = L$ and $M = PL$ from Appendix D, the values of θ_B and δ_B shown in Fig. 8.36a are, respectively,

$$\theta_B = \frac{PL^2}{EI} \quad \text{and} \quad \delta_B = \frac{PL^3}{2EI}$$

As shown in Fig. 8.36b, these values must be considered as nonzero boundary conditions when analyzing the deformations of the portion BC of the frame.

Neglecting the effect of the axial load P in the portion AB of the frame, the horizontal displacement of point B is zero. The portion BC of the frame can be considered as a cantilever beam loaded by an end force P as shown in Fig. 8.36b. Using case *(a)* with $a = L$ from Appendix D, the value of the additional rotation due to the deformation of BC is given by

$$\theta = \frac{PL^2}{2EI}$$

The horizontal deflection relative to the tangent inclined at the rotation θ_B at B is given by

$$\delta = \frac{PL^3}{3EI}$$

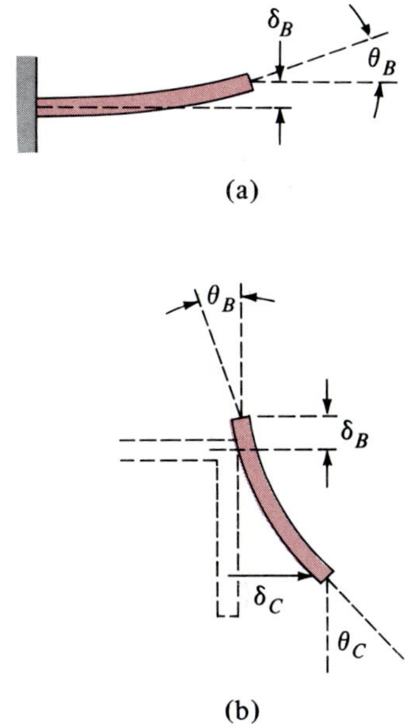

Figure 8.36 Exaggerated deformations of a frame

Finally, then, the total rotation at C is the sum of the rotations at B due to the deformation of AB and the additional rotation due to the deformation of BC, namely

$$\theta_C = \theta_B + \theta = \frac{PL^2}{EI} + \frac{PL^2}{2EI} = \frac{3PL^2}{2EI}$$

The horizontal displacement at C consists of the portion due to the deformation of AB plus the portion due to the deformation of BC, namely

$$\delta_C = (\delta)_{AB} + (\delta)_{BC} = L\theta_B + \delta = L\,\frac{PL^2}{EI} + \frac{PL^3}{3EI} = \frac{4PL^3}{3EI}$$

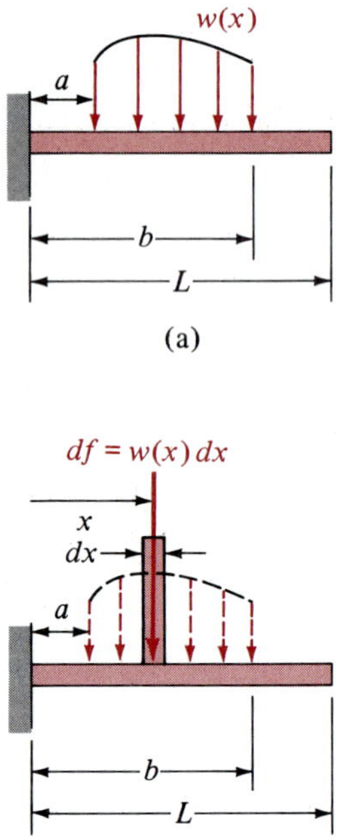

Figure 8.37 Cantilever beam loaded by a distributed loading $w(x)$

Superposition by Integration It is also possible to generate superposition results using an integration process. To illustrate, consider the cantilever beam loaded by the distributed loading $w(x)$ as shown in Fig. 8.37a. Using case *(a)* from Appendix D, the deflection at $x = L$ due to a concentrated load P located a distance x from the support is

$$v(L) = -\frac{Px^2(3L - x)}{6EI}$$

We can determine the contribution to the displacement at $x = L$ from the infinitesimal portion of the distributed loading $df = w(x)\,dx$ located at x in Fig. 8.37b by replacing P by df. This gives

$$dv(L) = -\frac{df\,x^2(3L - x)}{6EI} = -\frac{w(x)\,dx\,x^2(3L - x)}{6EI}$$

The total displacement at $x = L$ due to the distributed loading $w(x)$ over the interval $a \leq x \leq a + b$ of the beam is then obtained by superposing all the infinitesimal contributions using integration, thus

$$v(L) = -\int_a^{a+b} \frac{w(x)\,dx\,x^2(3L - x)}{6EI} \qquad (a)$$

In a completely similar manner, the slope at $x = L$ due to a force P located at x is given by Appendix D as

$$v' = -\frac{Px^2}{2EI}$$

The contribution to the slope at $x = L$ due to the infinitesimal portion of the distributed loading $df = w(x)\,dx$ located at x is given by

$$dv'(L) = -\frac{df\,x^2}{2EI} = -\frac{w(x)\,dx\,x^2}{2EI}$$

so that superposing by integration yields

$$v'(L) = -\int_a^{a+b} \frac{w(x)\,dx\,x^2}{2EI} \qquad (b)$$

These expressions for $v(L)$ and $v'(L)$ can be evaluated once $w(x)$ is specified. These same ideas can be applied to beams with other types of supports.

Example 8.12 Consider the uniformly loaded cantilever beam shown in Fig. 8.38. Determine the slope and deflection at $x = L$ by integration.

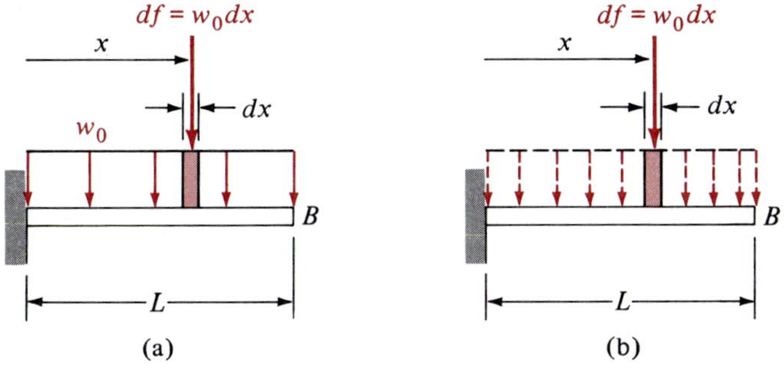

Figure 8.38 Uniformly loaded cantilever beam

Solution Using Eqn. (a) with $a = 0$, $b = L$, and $w(x) = w_0$ yields

$$v' = -\int_0^L \frac{w_0\, dx\, x^2(3L - x)}{6EI} = -\frac{w_0 L^4}{8EI}$$

for the displacement at $x = L$ due to the uniformly distributed load. Using Eqn. (b) with $a = 0$, $b = L$, and $w = w_0$ yields

$$v' = -\int_0^L \frac{w_0\, dx\, x^2}{2EI} = -\frac{w_0 L^3}{6EI}$$

for the slope at $x = L$. These results coincide with those from case *(c)* of Appendix D as expected, showing clearly that superposition by integration can be used as an effective alternate approach for determining rotations and slopes in beams.

PROBLEMS

8.93 For the simply supported beam supported by a spring at point C as shown, determine *(a)* $v'(A)$, *(b)* $v'(C)$, and *(c)* $v(B)$.

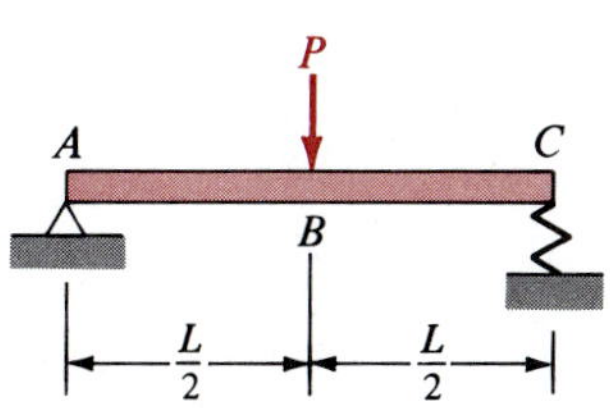

P. 8.93

8.94 For the two identical beams shown, determine *(a)* the vertical displacement at A, *(b)* the angle of rotation at B, and *(c)* the vertical displacement at C.

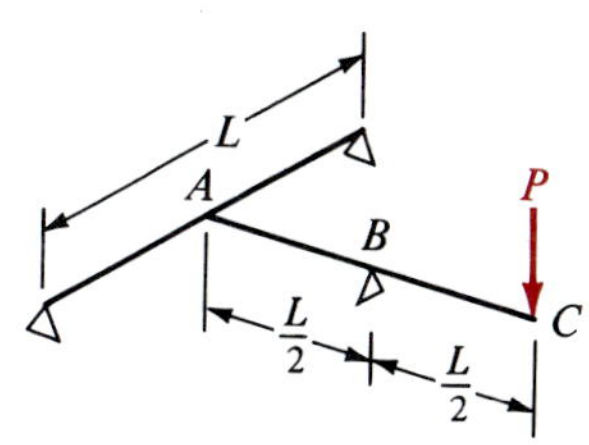

P. 8.94

8.95 Use cases *(a)* and *(c)* of Appendix D to determine *(a)* $v'(B)$, *(b)* $v(B)$, *(c)* $v'(C)$, and *(d)* $v(C)$.

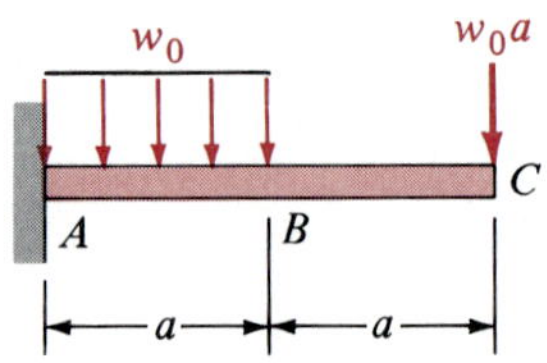

P. 8.95

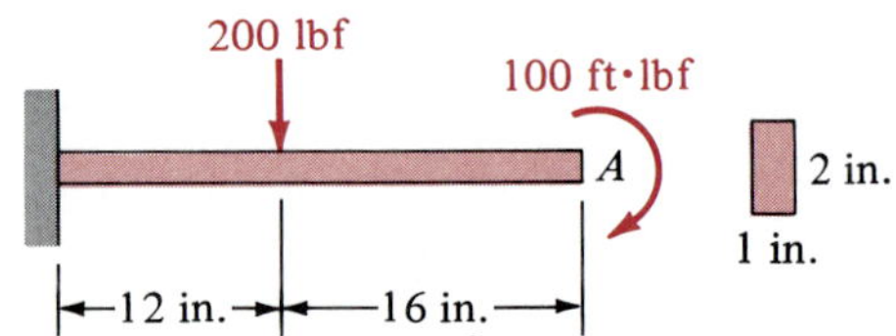

P. 8.96

8.96 Use cases *(a)* and *(b)* of Appendix D to determine *(a)* $v'(A)$, and *(b)* $v(A)$. Take $E = 10^7$ psi.

8.97 For the beam loaded by the oppositely directed loads shown, determine *(a)* $v'(A)$, *(b)* $v(B)$, and *(c)* $v(C)$.

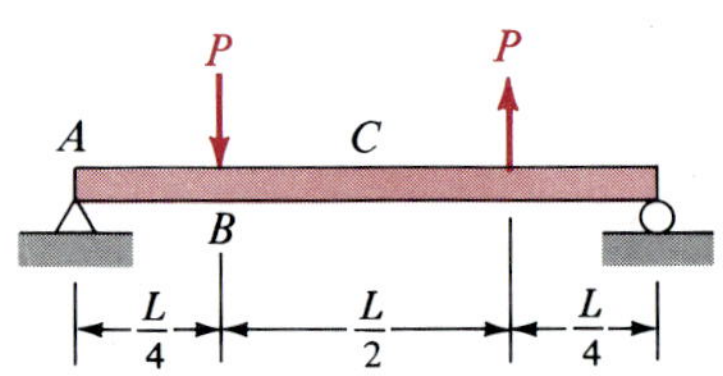

P. 8.97

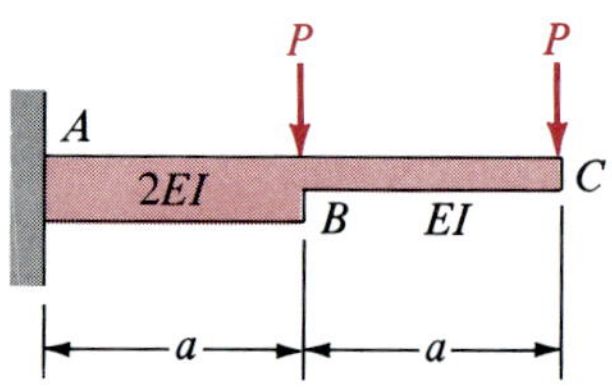

P. 8.98

8.98 Use cases *(a)* and *(b)* of Appendix D to determine *(a)* $v'(C)$, and *(b)* $v(C)$.

8.99 Use cases *(a)*, *(b)*, and *(c)* of Appendix D to determine *(a)* $v'(B)$, and *(b)* $v(C)$.

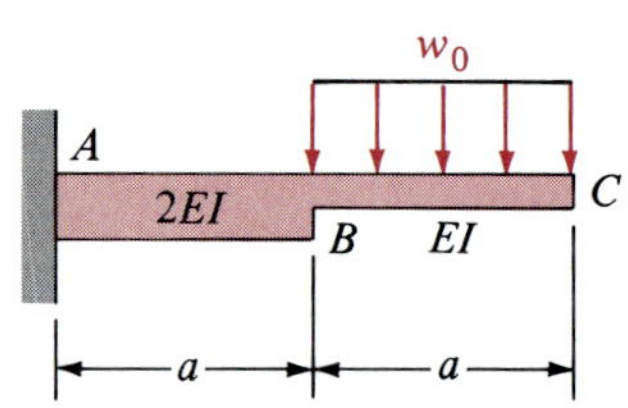

P. 8.99

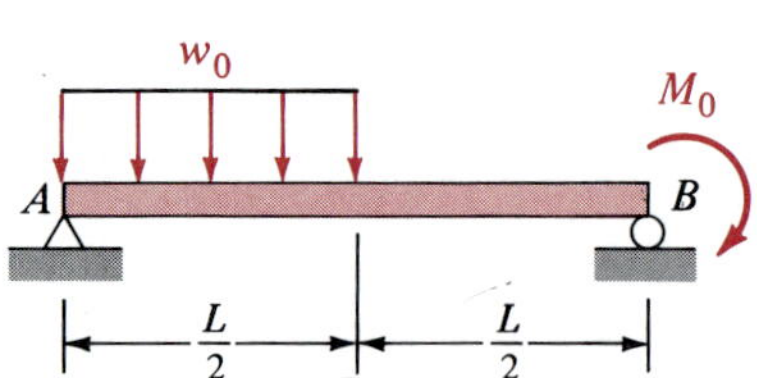

Ps. 8.100 and 8.101

8.100 Use cases *(g)* and *(h)* of Appendix D to determine the moment M_0 that will result in no rotation at B. Take $w_0 = 5$ kN/m, $L = 4$ m, $E = 200$ GPa, and $I = 2.4 \times 10^{-5}$ m^4.

8.101 Repeat problem 8.100 with $w_0 = 300$ lbf/ft, $L = 11$ ft, $E = 29 \times 10^6$ psi, and $I = 60$ in^4.

8.102 For the beam loaded as shown, determine $v'(A)$ and $v(B)$.

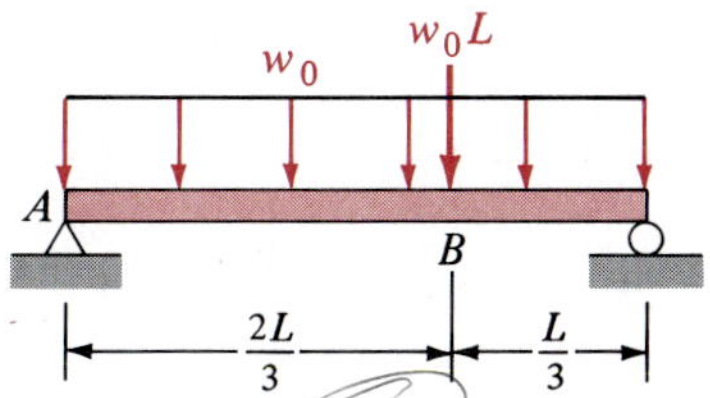

P. 8.102

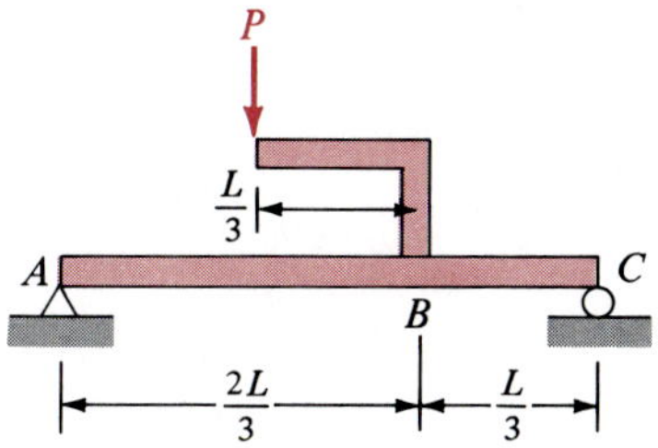

P. 8.103

8.103 The simply supported beam is loaded through a rigid member as shown. Determine $v'(A)$ and $v(B)$.

8.104 Determine the relationship between the moment M_0 and the force R at C that will result in $v(C) = 0$. Also determine $v(B)$ and $v'(C)$.

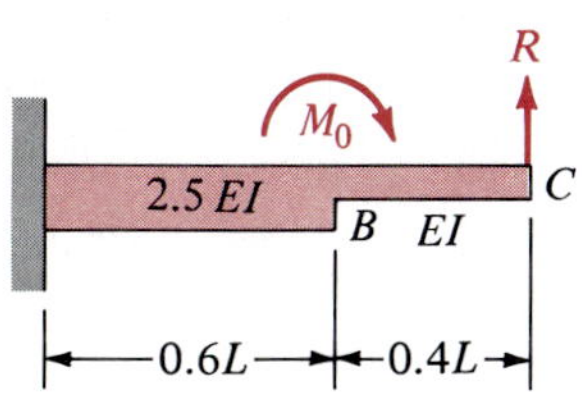

P. 8.104

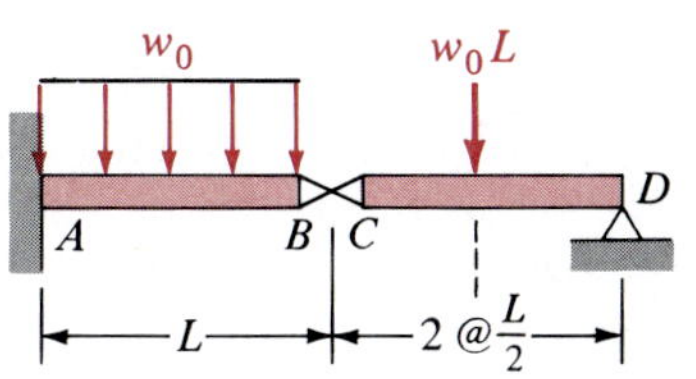

P. 8.105

8.105 The figure indicates the presence of what is called a hinge connecting the cantilever and simply supported beams. A hinge can transmit no moment. Determine *(a)* $v(B)$, *(b)* $v'(D)$, $v'(C)$, and $v'(B)$.

8.106 Use superposition to determine *(a)* the force in the linear spring, and *(b)* $v(C)$.

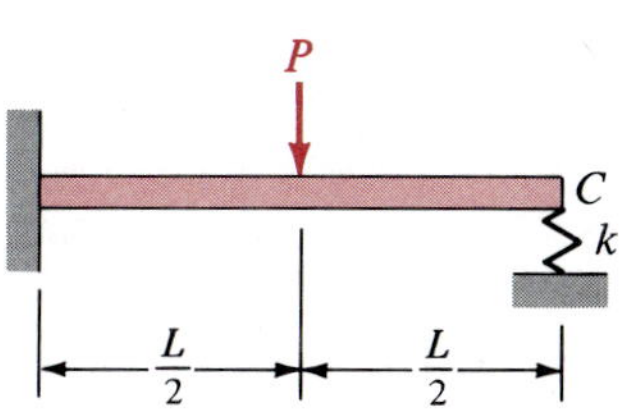

P. 8.106

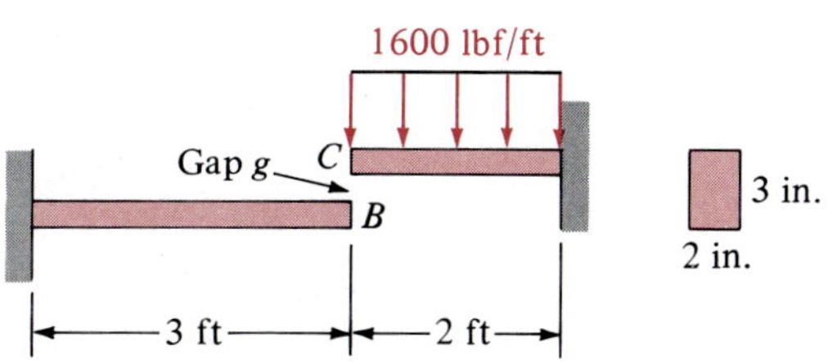

P. 8.107

8.107 The two cantilever beams with cross section shown and separated by the gap g, are loaded as shown. Determine *(a)* the reaction between the beams, *(b)* $v(B)$, and *(c)* $v(C)$. Take $g = 0.1$ in. and $E = 2 \times 10^6$ psi.

8.108 For the rigidly connected beams shown, determine *(a)* the vertical displacement at D, *(b)* the vertical displacement at B, *(c)* the angle of rotation θ at D, and *(d)* the horizontal displacement at C. Neglect axial effects.

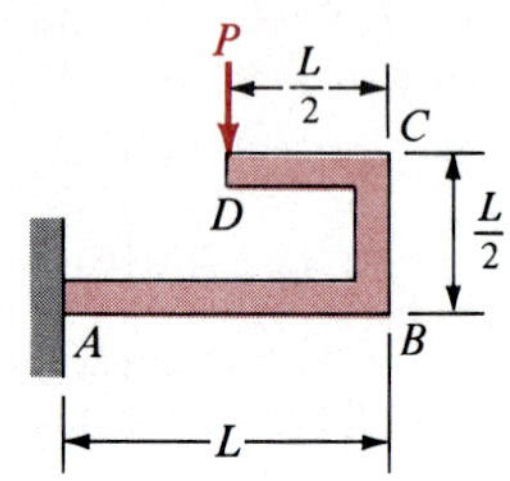

P. 8.108

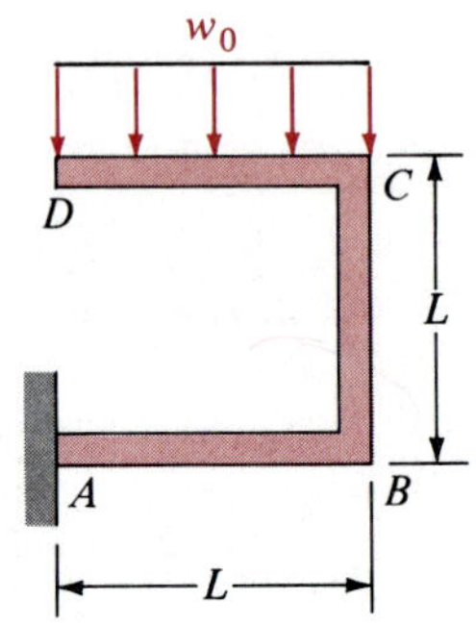

P. 8.109

8.109 For the rigidly connected beams shown, determine *(a)* the vertical displacement at D, *(b)* the horizontal displacement at D, *(c)* the vertical displacement at B, and *(d)* the angle of rotation θ at D. Neglect axial effects.

8.110 For the rigidly connected beams shown, determine *(a)* the horizontal displacement at B, *(b)* the vertical displacement at A, and *(c)* the angle of rotation θ at D.

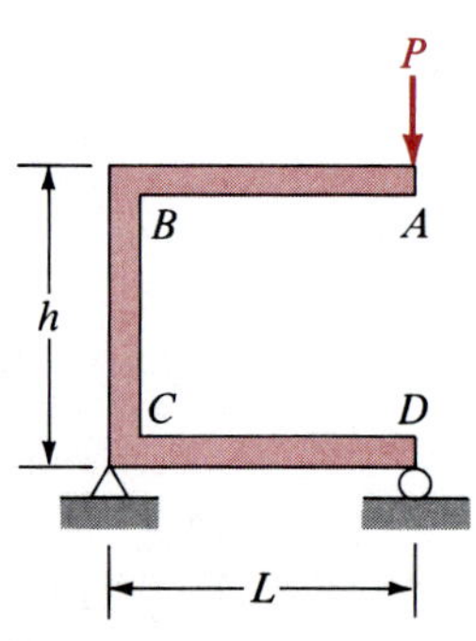

P. 8.110

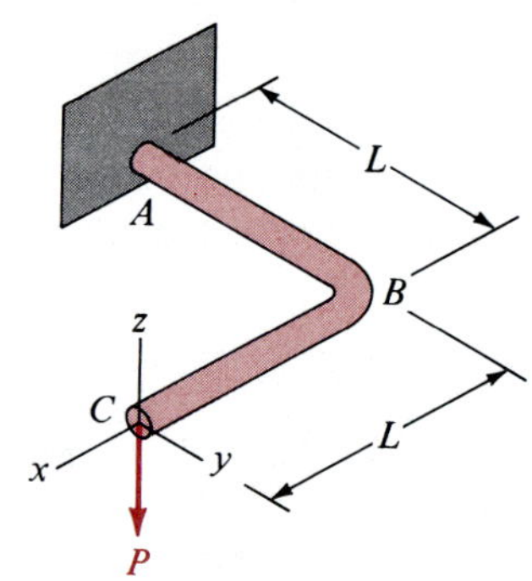

P. 8.111

8.111 An L-shaped beam with a circular cross section of radius r is loaded perpendicular to its plane as shown. Taking into account the torsional deformations, determine *(a)* the displacement in the z-direction at B, *(b)* the displacement in the z-direction at C, and *(c)* the rotation about the y-axis at C. Take $E = 2G(1 + \nu)$.

8.112 A T-shaped beam with a circular cross section of radius r is loaded perpendicular to its plane as shown. Taking into account the torsional deformations, determine *(a)* the displacement in the z-direction at B, *(b)* the displacement in the z-direction at A, and *(c)* the displacement in the z-direction at C. Take $E = 2G(1 + \nu)$.

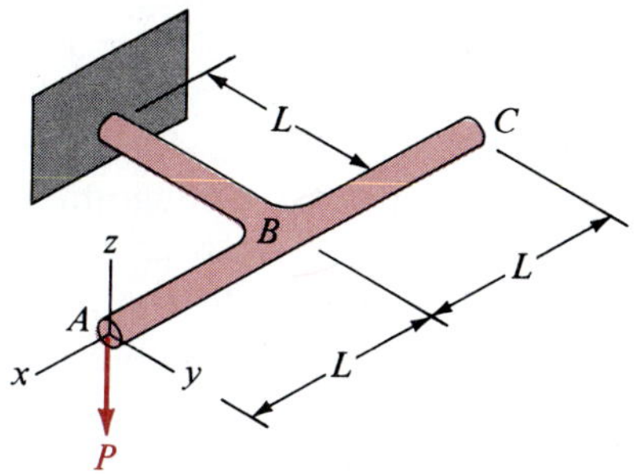

P. 8.112

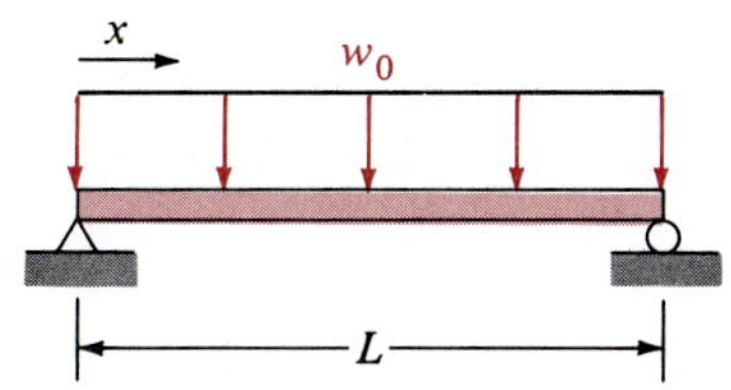

P. 8.113

8.113 Use integration of a concentrated load to verify that $v'(0) = -w_0L^3/24EI$ and $v(L/2) = -5w_0L^4/384EI$ for a simply supported beam subjected to a uniform load.

8.114 Use integration of a concentrated load to determine $v'(A)$ and $v(B)$ for the simply supported beam loaded as shown.

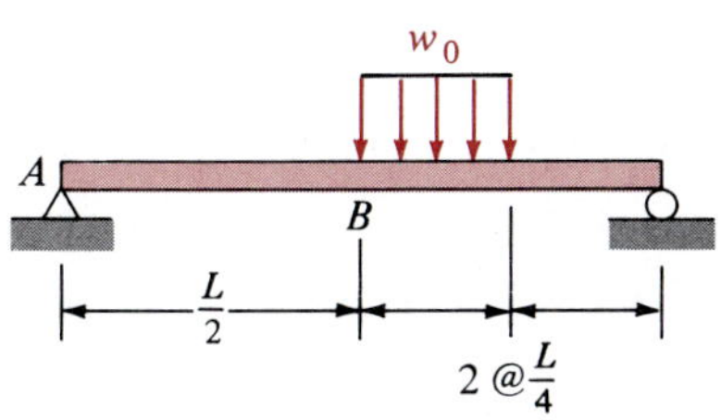

P. 8.114

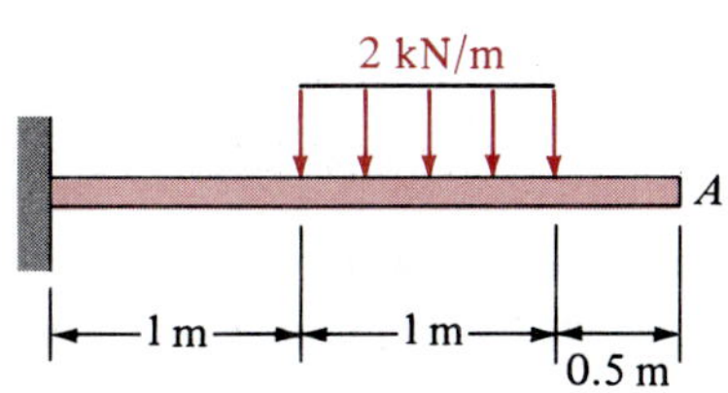

P. 8.115

8.115 For the cantilever beam partially loaded as shown, use integration of a concentrated load to determine *(a)* $v'(A)$, and *(b)* $v(A)$. Take $E = 200$ GPa and $I = 1.2 \times 10^{-6}$ m^4.

8.116 Use integration to determine *(a)* $v'(0)$ and *(b)* $v(L/2)$ for the simply supported beam under the action of the partial uniform load shown.

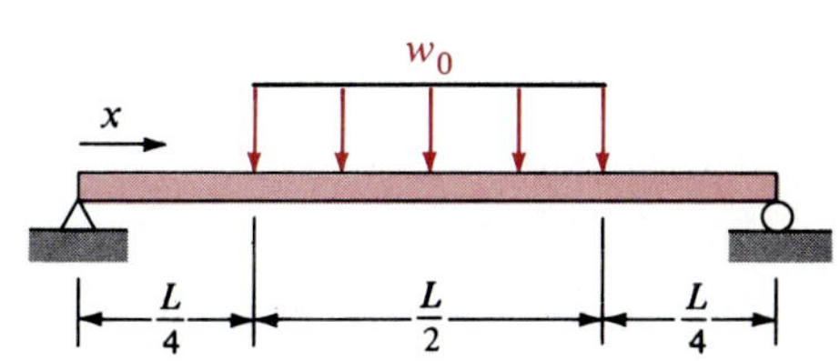

P. 8.116

8.6 INTEGRATION–MOMENT-AREA METHOD FOR DETERMINING DEFLECTIONS

In Sections 8.2 and 8.3, we obtained the equation of the elastic curve $v(x)$ by integrating either $EIv'' = M$ or $EIv'''' = -q(x)$. The results could then be used to determine v and the slope v' at any point in the beam. In many situations, only the displacement or slope at a certain point in the beam may be desired, in which case other approaches may offer advantages. In this section, we will develop and apply what will be called the **integration–moment-area method** for determining slopes and deflections. This method is essentially based on successive integrations of the moment-curvature relation $EIv'' = M$, together with graphical interpretations of the results. It is very closely related to the more familiar moment-area method, but is somewhat more general than that method.

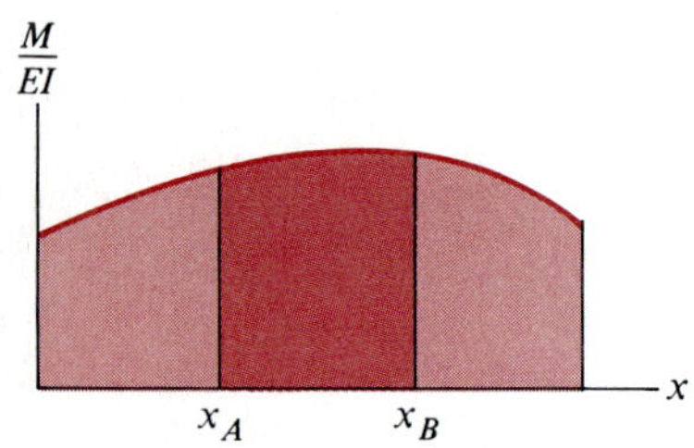

Figure 8.39 M/EI diagram

In a statically determinate problem, the first step in using the moment-area method is to plot a diagram representing the moment $M(x)$ divided by the product EI, as indicated in Fig. 8.39. The basic equation we will use to develop the technique is the moment-curvature relation $EIv'' = M$ or

$$v'' = M/EI$$

Integrating between the values x_A and $x_B > x_A$ yields

$$v'(x_B) = v'(x_A) + \int_{x_A}^{x_B} \frac{M\,ds}{EI} \tag{8.6}$$

We can state this as

> The change in slope between two points $x = x_A$ and $x = x_B$ is equal to the area under the curve in the M/EI diagram between those two points.

We will refer to this as Theorem I. More generally,

$$v'(x) = v'(x_A) + \int_{x_A}^{x} \frac{M\,ds}{EI} \tag{8.6a}$$

can be used to determine the slope at any value of x once $v'(x_A)$ is known.

Example 8.13 A cantilever beam is loaded by a concentrated load at $x = L$ as shown. Determine *(a)* the slope at the end $x = L$ of the beam, and *(b)* $v'(x)$.

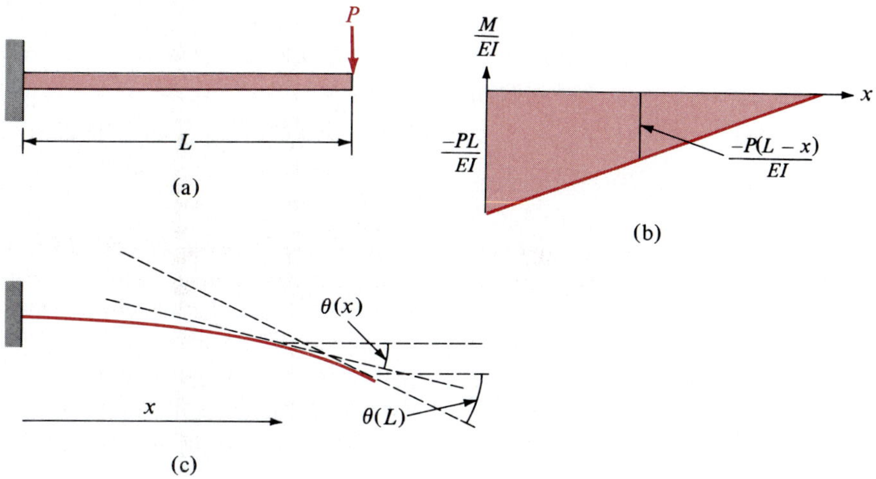

Figure 8.40 Cantilever beam with a concentrated load

Solution The M/EI diagram is simply the moment diagram divided by EI and is shown in Fig. 8.40b. Choosing A as $x = 0$ and B as $x = L$, Theorem I yields

$$v'(L) - v'(0) = \frac{(-PL/EI)(L)}{2} = -\frac{PL^2}{2EI}$$

or since $v'(0) = 0$, $v'(L) = \theta(L) = -PL^2/2EI$, shown in Fig. 8.40c. Note that by taking B to correspond to a value of $0 \leq x \leq L$, Theorem I yields

$$v'(x) - v'(0) = \frac{(-PL/EI - P(L - x)/EI)}{2}x = -\frac{Px(2L - x)}{2EI}$$

or

$$\theta(x) = v'(x) = -\frac{Px(2L - x)}{2EI}$$

which is the equation for the slope of the elastic curve.

Integrating Eqn. (8.6*a*) yields

$$v(x) = v(x_A) + (x - x_A)v'(x_A) + \int_{x_A}^{x}\int_{x_A}^{y}\frac{M\,ds}{EI}\,dy$$

Using a well-known result from integral calculus, we can represent this as

$$v(x) = v(x_A) + (x - x_A)v'(x_A) + \int_{x_A}^{x}(x - s)\frac{M\,ds}{EI} \qquad (8.7)$$

Referring to Fig. 8.41, the integral can be interpreted as follows.

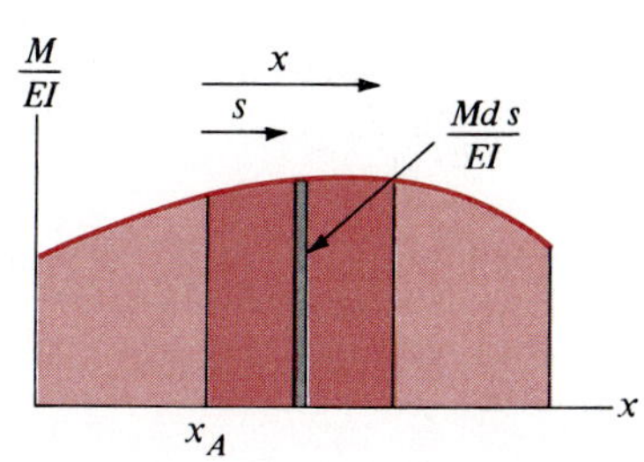

Figure 8.41 Interpretation of the integral

The quantity $(x - s)M\,ds/EI$ is the moment of $M\,ds/EI$ with respect to the location x, with $x - s$ the moment arm, so that

$$v(x) = v(x_A) + (x - x_A)v'(x_A) + A\bar{x}$$

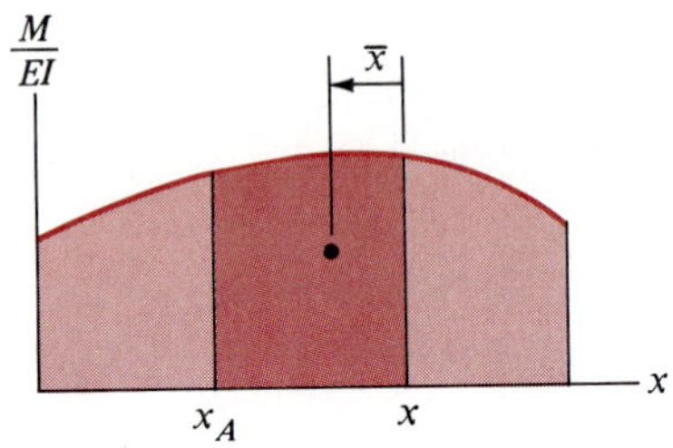

Figure 8.42 Centroidal location

where A is the area between x and x_A, and $\bar{x}$, shown in Fig. 8.42, is the distance to the centroid of the area measured from position x.

Specifically, for $x = x_B$

$$v(x_B) = v(x_A) + (x_B - x_A)v'(x_A) + A\bar{x}$$

We can state this as

> The change in displacement between locations x_A and x_B is the slope at A times the distance $x_B - x_A$ plus the moment of the area under the curve of the M/EI diagram with respect to x_B.

We will refer to this as Theorem II. Note that the process of determining the moment $A\bar{x}$ is entirely analogous to the process of determining resultants in statics, with the loading $q(x)$ replaced by M/EI.

Example 8.14 Consider again the problem of Example 8.13, shown in Fig. 8.43a. Determine $v(L)$ and $v(x)$.

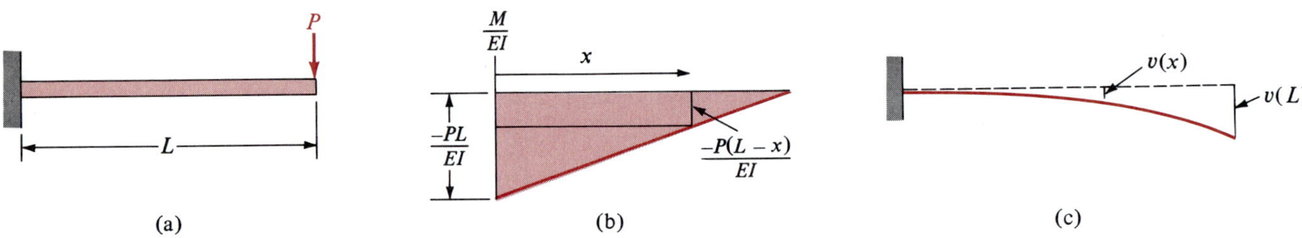

Figure 8.43 Cantilever beam with end load

Solution The M/EI diagram is shown in Fig. 8.43b. Application of Theorem II between $x = 0$ and $x = L$ yields

$$v(L) = v(0) + Lv'(0) + \left[\left(-\frac{PL}{EI}\right)\left(\frac{L}{2}\right)\right]\left(\frac{2L}{3}\right) = -\frac{PL^3}{3EI}$$

For the elastic curve in general (shown in Fig. 8.43c), Theorem II yields

$$v(x) = v(0) + xv'(0) + \left(-\frac{P(L - x)}{EI}\right) x \left(\frac{x}{2}\right) + \left(-\frac{Px}{EI}\right)\left(\frac{x}{2}\right)\left(\frac{2x}{3}\right)$$

$$= -\frac{Px^2(3L - x)}{6EI}$$

To use Theorems I and II efficiently, one must have a knowledge of the centroids of several types of loadings and their corresponding moment diagrams. Typical areas and centroidal locations are shown in Table 8.1.

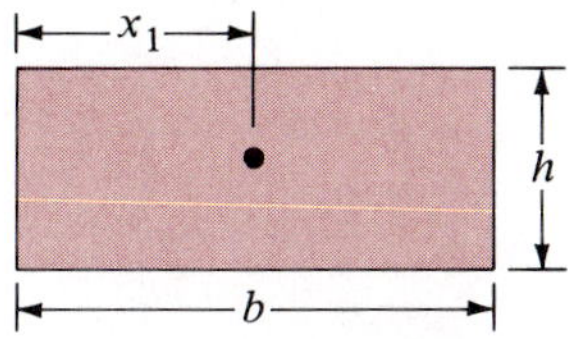	$A = bh$ $x_1 = \frac{b}{2}$	
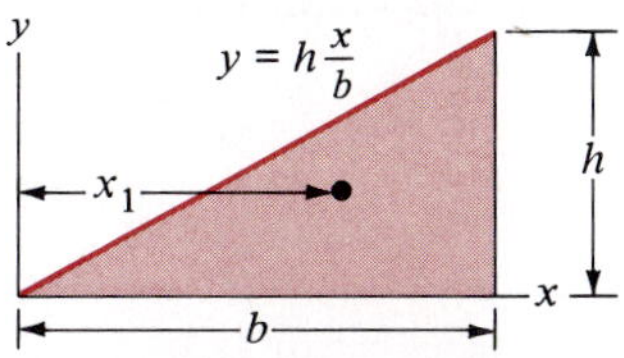	$A = \frac{bh}{2}$ $x_1 = \frac{2b}{3}$	
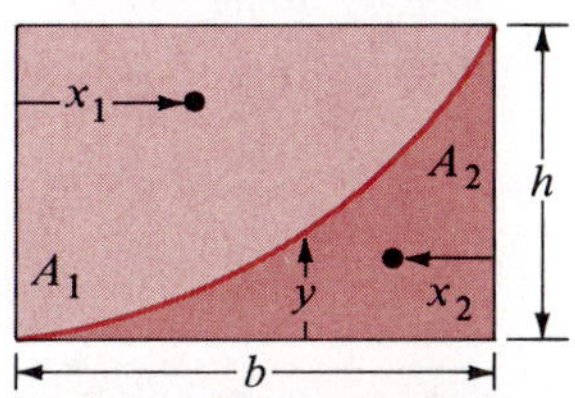	$y = h\left(\frac{x}{b}\right)^2$ $A_1 = \frac{2bh}{3}$ $A_2 = \frac{bh}{3}$	$x_1 = \frac{3b}{8}$ $x_2 = \frac{b}{4}$
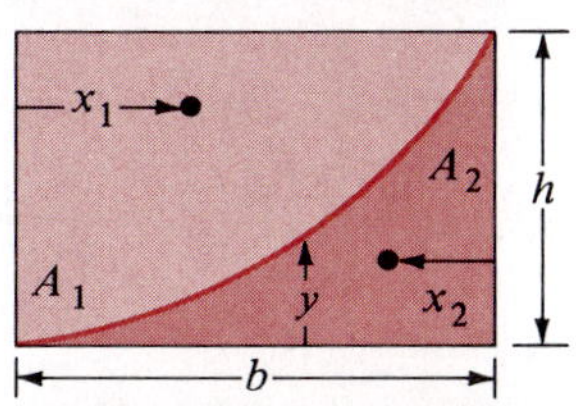	$y = h\left(\frac{x}{b}\right)^3$ $A_1 = \frac{3bh}{4}$ $A_2 = \frac{bh}{4}$	$x_1 = \frac{2b}{5}$ $x_2 = \frac{b}{5}$
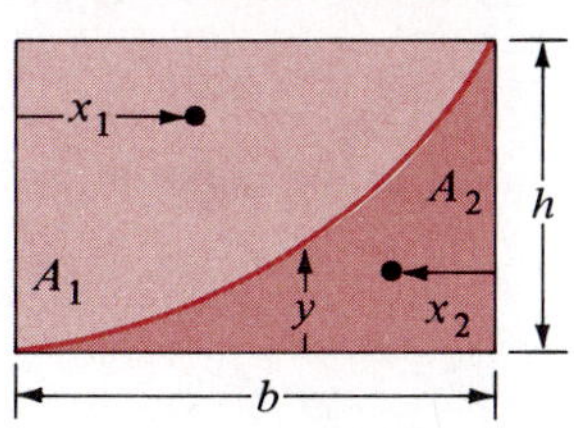	$y = h\left(\frac{x}{b}\right)^n$ $A_1 = \frac{nbh}{(n+1)}$ $A_2 = \frac{bh}{(n+1)}$	$x_1 = (n+1)\frac{b}{2(n+2)}$ $x_2 = \frac{b}{n+2}$

Table 8.1 Typical areas and centroids

We will demonstrate the use of these results in the examples that follow.

Example 8.15 Use Theorems I and II to determine the slope at the ends and the displacement at the midpoint for the beam loaded over a central portion as shown in Fig. 8.44a.

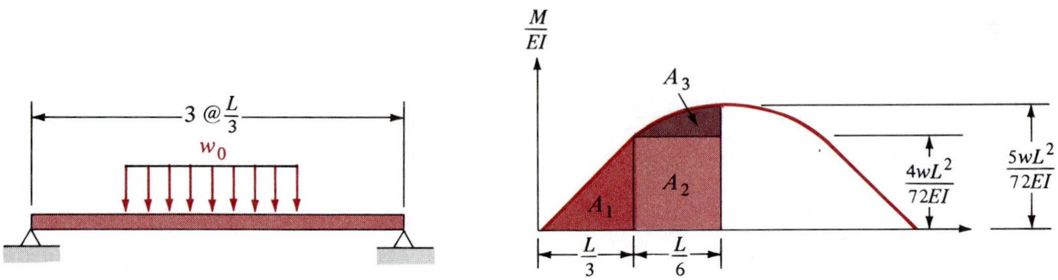

Figure 8.44 Simply supported beam loaded over a central portion

Solution We can determine the slope at $x = 0$ by using symmetry and Theorem I between $x = 0$ and $x = L/2$ according to

$$v'\left(\frac{L}{2}\right) = v'(0) + \int_0^{L/2} \frac{M\,ds}{EI}$$

or

$$0 = v'(0) + A_1 + A_2 + A_3$$

where A_1, A_2, and A_3 are shown in Fig. 8.44b. Using the results of Table 8.1,

$$-v'(0) = \left(\frac{wL^2}{18EI}\right)\left(\frac{L}{3}\right)\left(\frac{1}{2}\right) + \left(\frac{wL^2}{18EI}\right)\left(\frac{L}{6}\right) + \left(\frac{wL^2}{72EI}\right)\left(\frac{2}{3}\right)\left(\frac{L}{6}\right) = \frac{13wL^3}{648EI}$$

or

$$v'(0) = -\frac{13wL^3}{648EI}$$

Using Theorem II and Table 8.1, the displacement at the midpoint is

$$\begin{aligned} v\left(\frac{L}{2}\right) &= v(0) + \frac{L}{2}v'(0) + A_1\bar{x}_1 + A_2\bar{x}_2 + A_3\bar{x}_3 \\ &= \frac{L}{2}\left(-\frac{13wL^3}{648EI}\right) + \frac{wL^3}{108EI}\left(\frac{L}{6} + \frac{L}{8}\right) \\ &\quad + \frac{wL^3}{108EI}\left(\frac{L}{12}\right) + \frac{wL^3}{648EI}\left(\frac{3}{8}\right)\left(\frac{L}{6}\right) \\ &= -0.00659\, wL^4/EI \end{aligned}$$

The previous examples were straightforward: there was either special symmetry, as in the case of the simply supported beam loaded centrally, or, for the case of the cantilever beam, the displacement and slope were both known at one end. For either of these cases, the use of Theorems I and II to determine rotations and deflections is somewhat simplified. As can be seen from the example that follows, the technique is as easily applied to the case of arbitrary loading.

Example 8.16 A simply supported beam is loaded by a noncentral load P as shown in Fig. 8.45a. Determine the end slopes, the displacement at the load, and the maximum displacement.

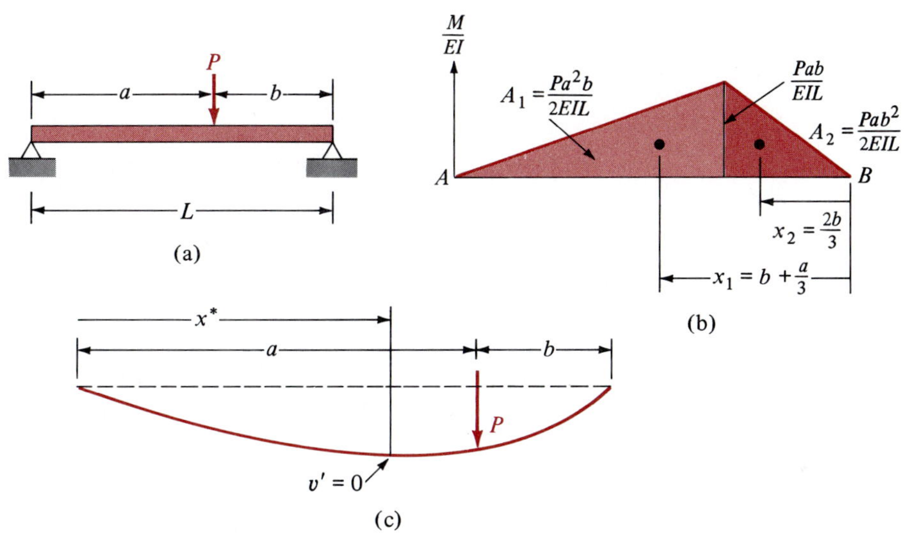

Figure 8.45 Noncentrally loaded beam

Solution The M/EI diagram and a sketch of the elastic curve are shown in Figs. 8.45b and c. To determine the end slopes, apply Theorem II between $x = 0$ and $x = L$ to obtain

$$v(L) = v(0) + Lv'(0) + \Sigma A_i \bar{x}_i$$

or

$$0 = 0 + Lv'(0) + \frac{Pa^2b}{2EIL}\left(b + \frac{a}{3}\right) + \frac{Pab^2}{2EIL}\frac{2b}{3}$$

thus

$$v'(0) = -\frac{Pb(L^2 - b^2)}{6EIL}$$

Again using Theorem II between $x = 0$ and $x = a$, we write

$$v(a) = v(0) + av'(0) + A_1\bar{x}_1$$

or

$$v(a) = 0 + a\left(-\frac{Pb(L^2 - b^2)}{6EIL}\right) + \frac{Pa^2b}{2EIL}\frac{a}{3} = -\frac{Pa^2b^2}{3EIL}$$

For the maximum deflection, we use Theorem I to yield

$$v'(x^*) = v'(0) + A(x^*) = -\frac{Pb(L^2 - b^2)}{6EIL} + \frac{Pab}{EIL}\frac{x^*}{a}\frac{x^*}{2}$$

Our requirement that $v'(x^*) = 0$ leads to $x^* = \sqrt{(L^2 - b^2)/3}$. Theorem II then gives

$$v(x^*) = v(0) + x^*v'(0) + A(x^*)\bar{x}_1 = x^*\left(-\frac{Pb(L^2 - b^2)}{6EIL}\right) + \frac{Pab}{EIL}\frac{x^*}{a}\frac{x^*}{2}\frac{x^*}{3}$$

$$= -\frac{Pb(L^2 - b^2)^{3/2}}{9\sqrt{3}EIL}$$

as we obtained earlier using double integration in Example 8.3.

PROBLEMS

8.117 A cantilever beam is loaded by an end shear and end moment as shown. Develop expressions for the rotation and displacement at the loaded end.

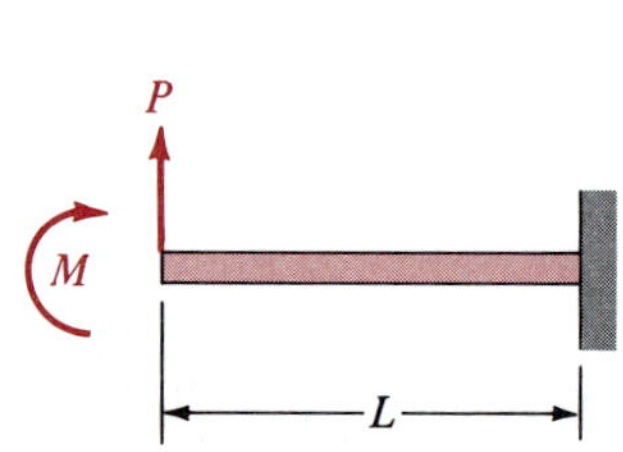

P. 8.117

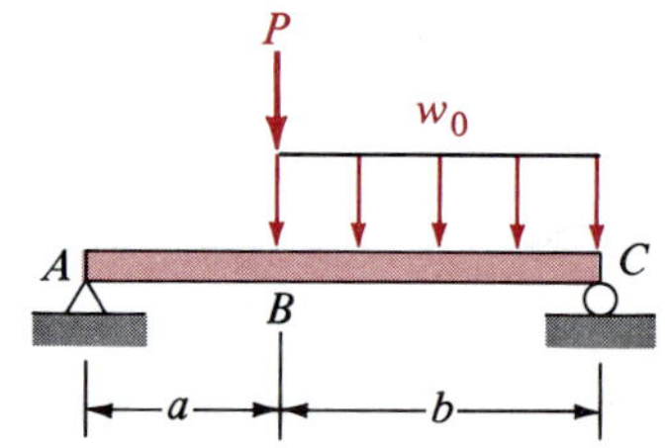

Ps. 8.118 and 8.119

8.118 For the W 12 × 50 section loaded as shown, determine *(a)* the slope at location C, and *(b)* the displacement at point B. Take $w_0 = 6$ kips/ft, $P = 24$ kips, $a = 5$ ft, $b = 5$ ft, and $E = 30 \times 10^6$ psi.

8.119 For the W 310 × 74 section loaded as shown, determine *(a)* the slope at location C, and *(b)* the displacement at point B. Take $w_0 = 80$ kN/m, $P = 100$ kN, $a = 2$ m, $b = 2$ m, and $E = 200$ GPa.

8.120 For the uniformly loaded cantilever beam shown, determine *(a)* the slope at B, and *(b)* the deflection at B.

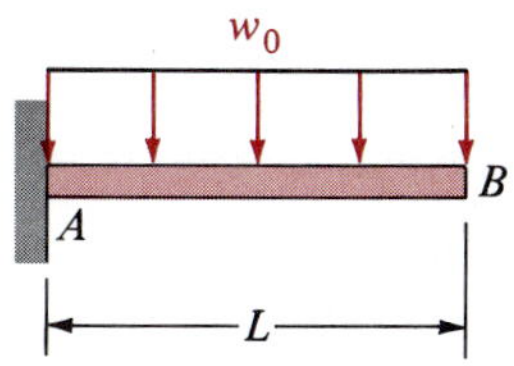

P. 8.120

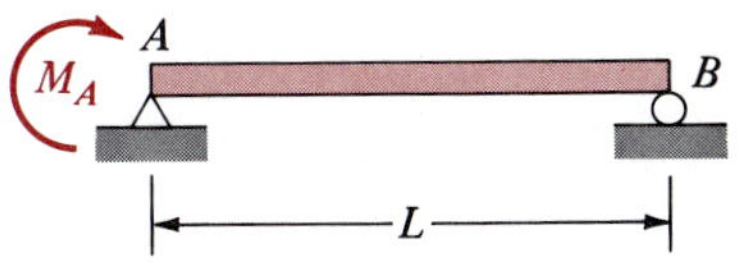

P. 8.121

8.121 For the simply supported beam shown, determine *(a)* the slope at A, *(b)* the slope at B, and *(c)* the deflection at the midpoint of the beam.

8.122 For the overhang beam shown, determine *(a)* the slope at B, and *(b)* the displacement at B.

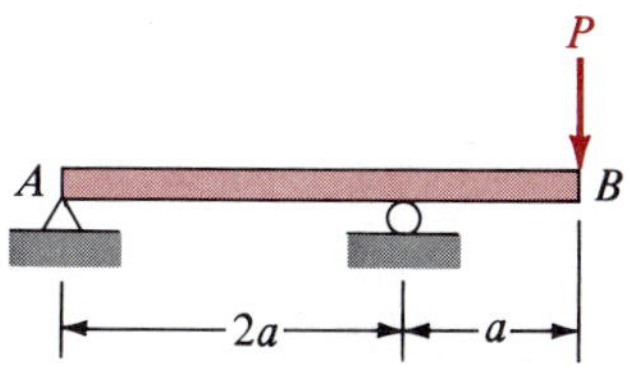

P. 8.122

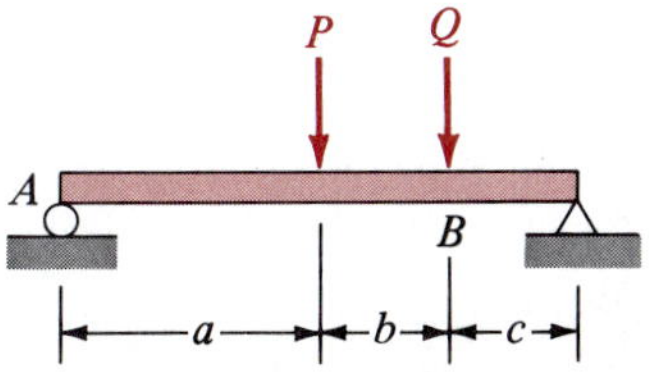

Ps. 8.123 and 8.124

8.123 For the W 250 × 44.8 section loaded as shown, determine *(a)* the slope at A, and *(b)* the displacement at B. Take $a = 0.7$ m, $b = 1$ m, $C = 1.6$ m, $P = 100$ kN, $Q = 60$ kN, and $E = 70$ GPa.

8.124 For the W 10 × 30 section loaded as shown, determine *(a)* the slope at A, and *(b)* the displacement at B. Take $a = 2$ ft, $b = 3$ ft, $c = 5$ ft, $P = 15$ kips, $Q = 10$ kips, and $E = 29 \times 10^6$ psi.

8.125 For the simply supported beam loaded by the two moments shown, determine *(a)* the slope at B, and *(b)* the displacement at the center of the beam.

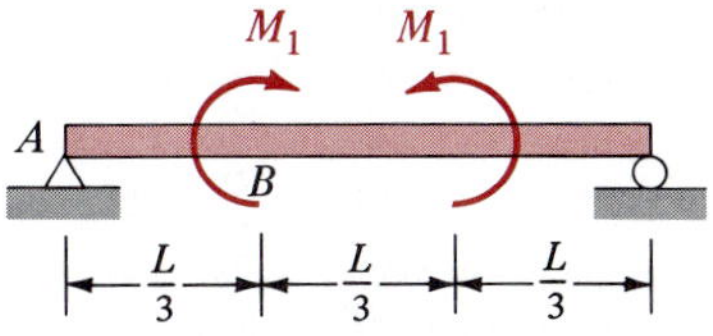

P. 8.125

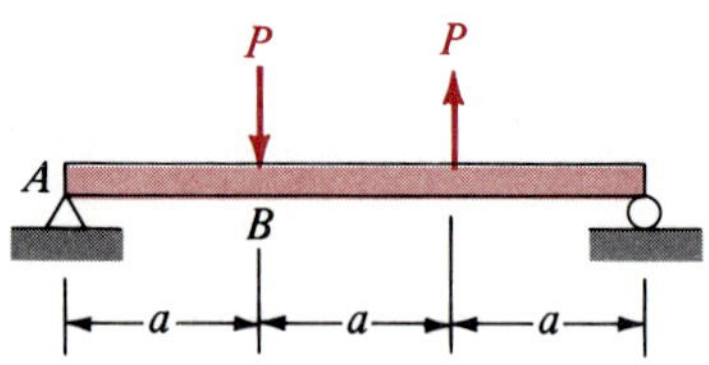

P. 8.126

8.126 For the simply supported beam loaded by the oppositely directed forces shown, determine *(a)* the slope at A, *(b)* the slope at B, and *(c)* the displacement at B.

8.127 For the overhang beam loaded uniformly as shown, determine *(a)* the slope at A, *(b)* the slope at C, and *(c)* the displacement at C.

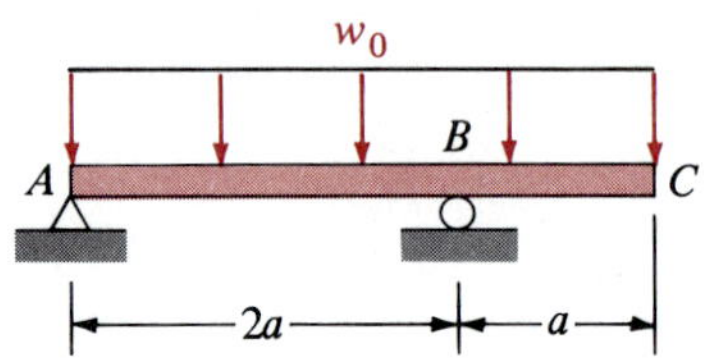

P. 8.127

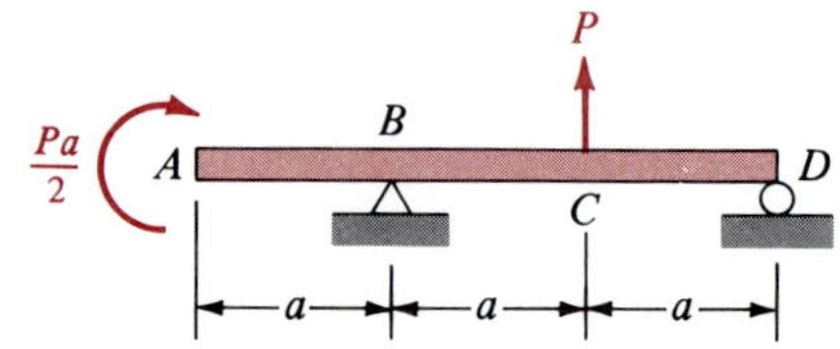

P. 8.128

8.128 For the beam loaded as shown, determine *(a)* the slope at A, *(b)* the displacement at A, and *(c)* the displacement at C.

8.129 For the overhang beam shown, determine *(a)* the slope at A, *(b)* the displacement at A, and *(c)* the slope at B.

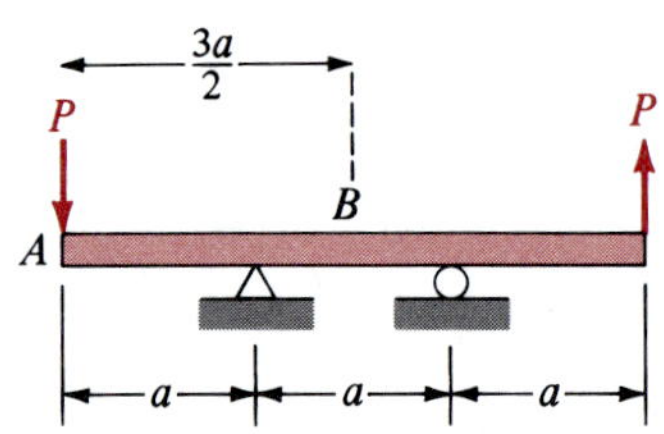

P. 8.129

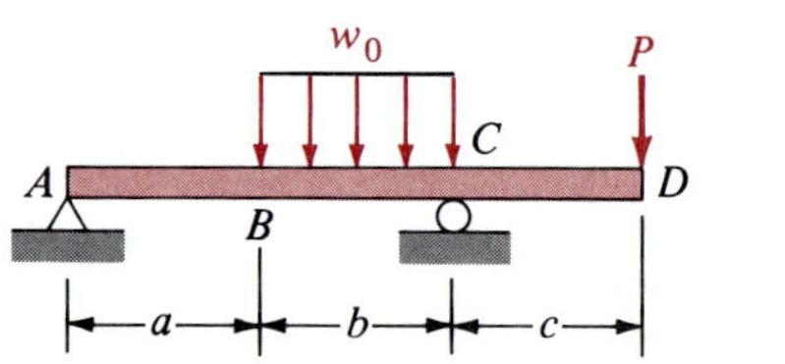

Ps. 8.130 and 8.131

8.130 The W 12 × 26 beam is loaded as shown. Determine *(a)* the slope at C, and *(b)* the displacement at D. Take $a = 6$ ft, $b = 5$ ft, $c = 4$ ft, $P = 14$ kips, $w_0 = 6$ kips/ft, and $E = 29 \times 10^6$ psi.

8.131 The W 310 × 38.7 beam is loaded as shown. Determine *(a)* the slope at C, and *(b)* the displacement at D. Take $a = 1.6$ m, $b = 2$ m, $c = 1.3$ m, $P = 90$ kN, $w_0 = 80$ kN/m, and $E = 200$ GPa.

8.132 For the variable-cross-section beam loaded by the central force as shown, determine *(a)* the slope at A, *(b)* the slope at C, and *(c)* the displacement at B.

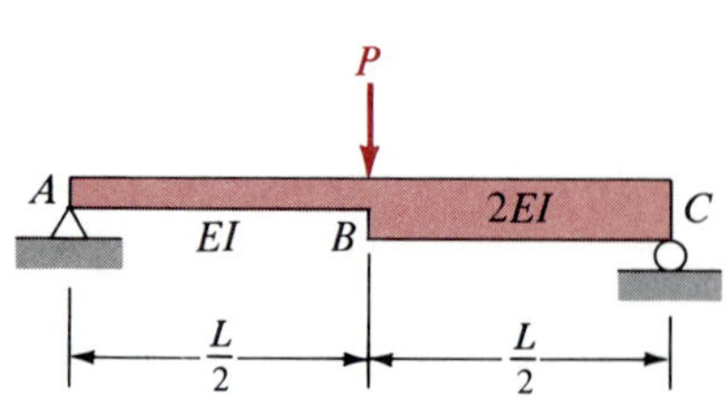

P. 8.132

8.133 For the variable-cross-section beam loaded uniformly as shown, determine *(a)* the slope at A, *(b)* the slope at C, and *(c)* the displacement at B.

8.134 For the variable-cross-section beam loaded by the central force as shown, determine *(a)* the slope at the left end, and *(b)* the displacement at the center.

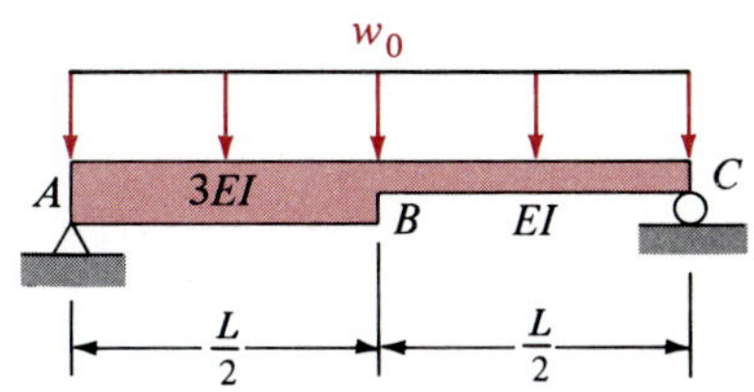

P. 8.133

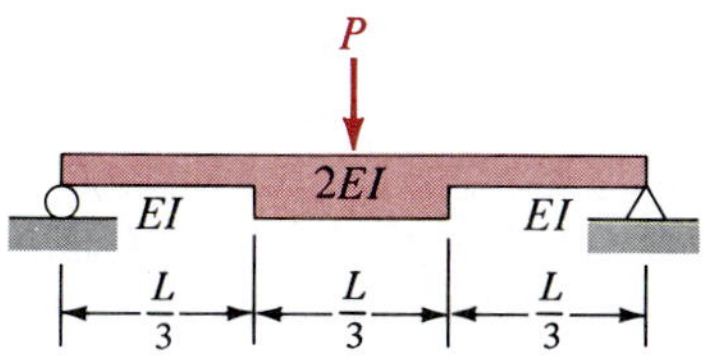

P. 8.134

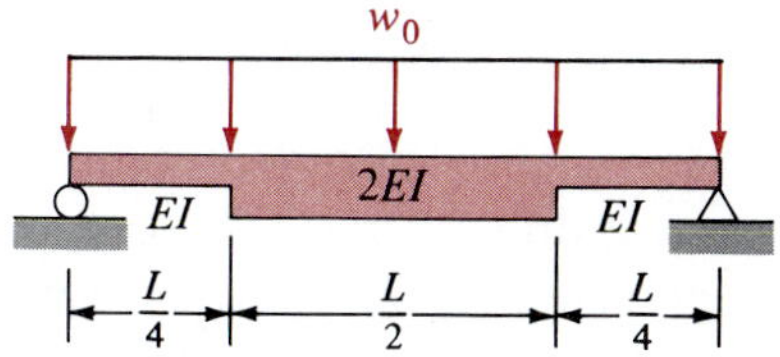

P. 8.135

8.135 For the variable-cross-section beam loaded uniformly as shown, determine *(a)* the slope at the left end, and *(b)* the displacement at the center.

8.136–8.141 For each of the following, use the integration–moment-area method to determine the location and magnitude of the largest transverse displacement.

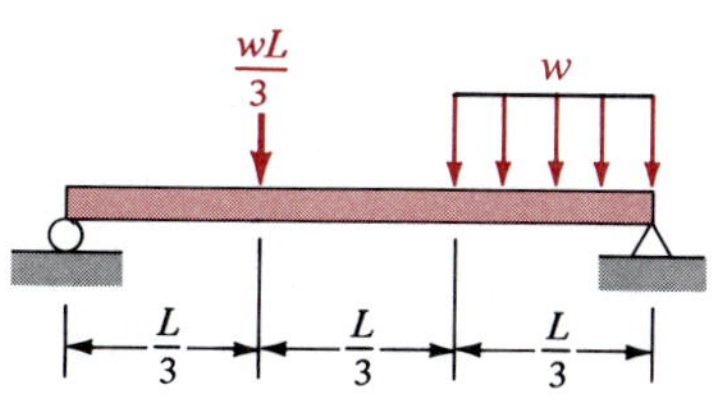

P. 8.136

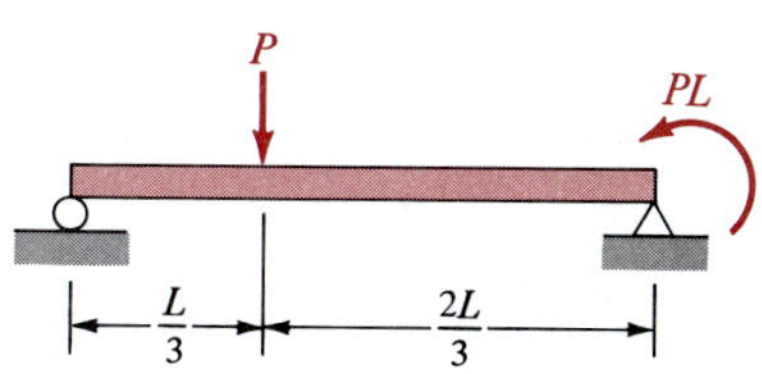

P. 8.137

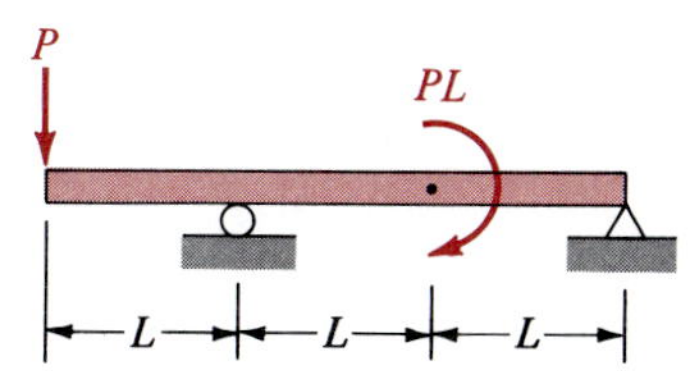

P. 8.138

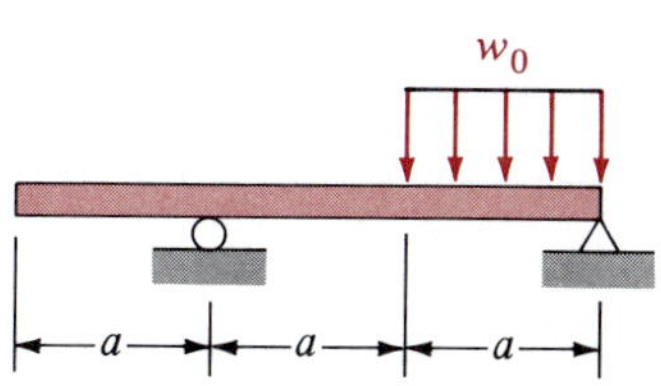

P. 8.139

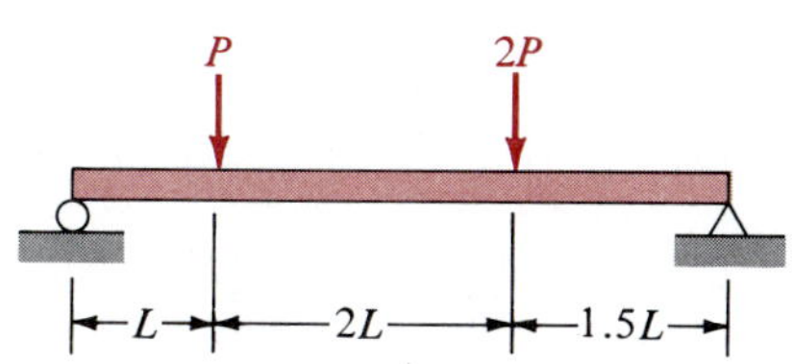

P. 8.140

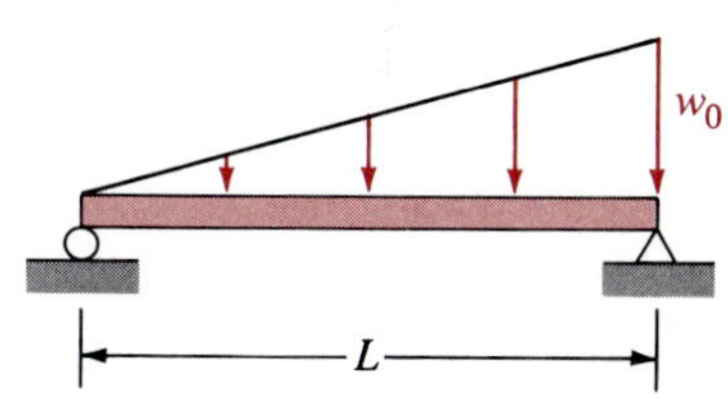

P. 8.141

8.7 STATICALLY INDETERMINATE PROBLEMS

In this section, we will study beams for which the number and type of supports exceed the number and type required for equilibrium, making them statically indeterminate problems. For statically determinate problems, where the number of equations of equilibrium is equal to the number of reactions, it is possible to determine the reactions and hence complete the construction of the shear and moment diagrams, in order to answer the strength question by evaluating the stresses using $\sigma = -My/I$ and $\tau = VQ/It$. It is then possible, by using such techniques as integration of $EIv'' = M$ or $EIv'''' + w = 0$, to determine the required displacements. In contrast, for statically indeterminate problems, the information necessary to construct the shear and moment diagrams cannot be determined from the statics alone: it is necessary to consider the displacements to determine the extra reactions, and to ultimately be able to determine the stresses and displacements.

By definition, statically indeterminate problems have more supports than are necessary to ensure equilibrium of the structure. These additional supports, called redundants, can be located at the boundaries as in Figs. 8.46a and b, or at interior points as in Figs. 8.46c and d.

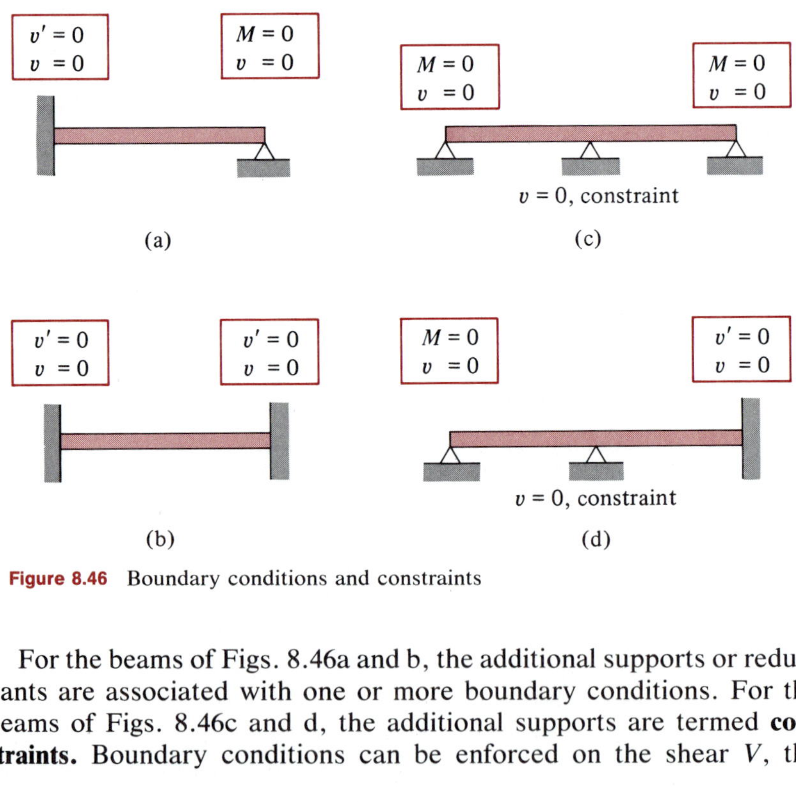

Figure 8.46 Boundary conditions and constraints

For the beams of Figs. 8.46a and b, the additional supports or redundants are associated with one or more boundary conditions. For the beams of Figs. 8.46c and d, the additional supports are termed **constraints.** Boundary conditions can be enforced on the shear V, the moment M, the slope v', and the displacement v. Constraints, on the other hand, are only enforced on the slope or the displacement. It is important to distinguish between boundary conditions and constraints.

The number of extra reactions beyond that required for equilibrium is called the **degree of redundancy** or the **degree of statical indeterminacy.** Several examples are shown in Fig. 8.47.

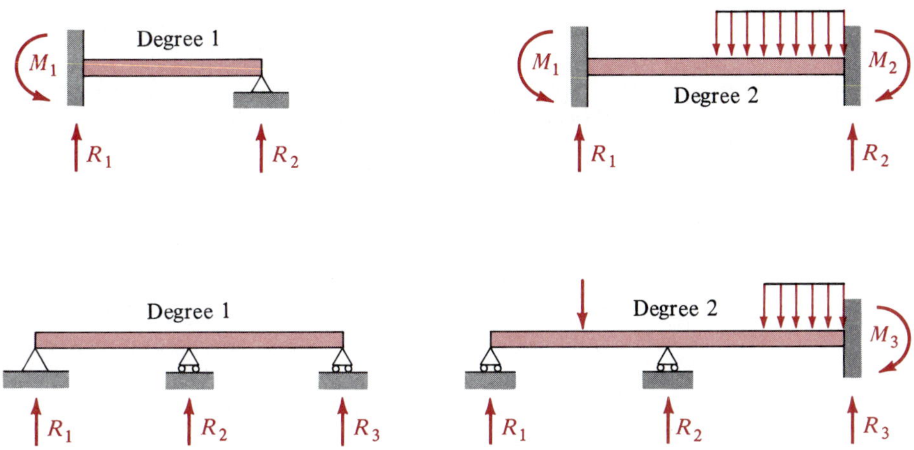

Figure 8.47 Examples of statically indeterminate problems and the corresponding degree of redundancy

For loading that is transverse, there are in each case two equations of equilibrium, $\Sigma F_y = 0$ and $\Sigma M_z = 0$, so that the degree of redundancy is equal to $N - 2$, where N is the number of reactions. Any of the techniques that generate the extra equations do so by constraining the displacements or slopes that are associated with the redundant forces. These ideas will be discussed in detail in the sections that follow.

8.7.1 Use of Superposition

The basic idea of superposition is to decompose a statically indeterminate problem into two or more statically determinate problems whose solutions are known or can be determined. This is generally done by removing the forces or moments associated with the redundants and then determining the values of the redundants that are consistent with actual deformations of the beam, that is, a flexibility method. As an example, consider a beam that is fixed at one end and simply supported at the other, as shown in Fig. 8.48.

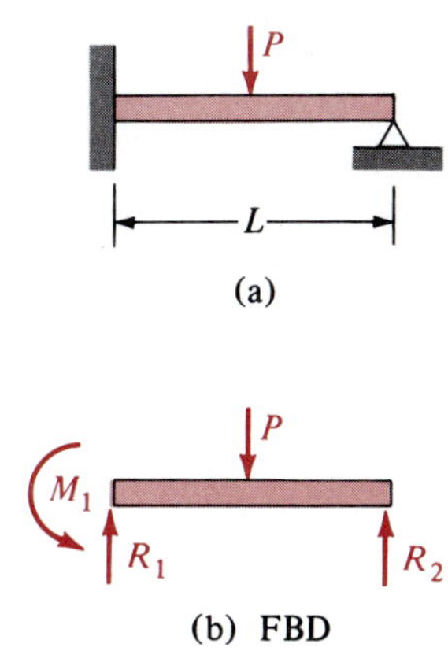

Figure 8.48 Beam with one redundant

Referring to the free-body diagram in Fig. 8.48b, we can see that there are three unknown reactions, R_1, M_1, and R_2; thus there is $3 - 2 = 1$ redundant. If M_1 is selected as the redundant, the two problems to be superposed are shown in Fig. 8.49, where the deflection curves for the two problems must be such that the slope at $x = 0$ is zero; that is, M_1 must be selected so that the sum of the slopes at $x = 0$, due to the load P and due to the moment M_1, adds to zero.

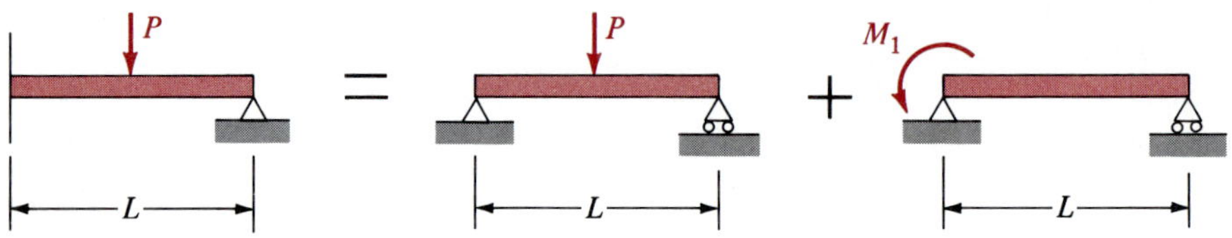

Figure 8.49 Superposition with M_1 as the redundant

If R_2 is selected as the redundant, the two problems to be superposed are shown in Fig. 8.50, where the superposition of the two problems must be such that the displacement at $x = L$ is zero; thus, R_2 must be selected so that the sum of the displacements at $x = L$ due to the concentrated load, and due to the reaction R_2, sum to zero.

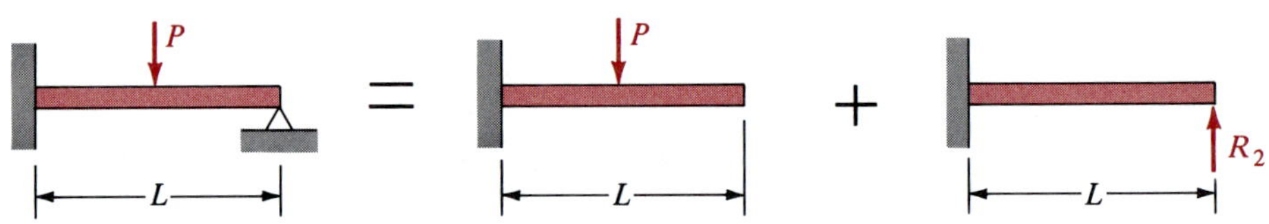

Figure 8.50 Superposition with R_2 as the redundant

As mentioned, superposition is essentially a flexibility method in that the forces (reaction forces and/or reaction moments) are considered as redundants or unknowns that are determined so as to satisfy a particular condition on the deformations. Let us demonstrate the use of the method of superposition with several typical examples. A collection of results that is frequently helpful when applying the superposition method is presented in Appendix D.

Example 8.17 A uniformly loaded cantilever beam is supported at the right end as shown in Fig. 8.51. Determine the reactions, and hence construct the shear and bending moment diagrams.

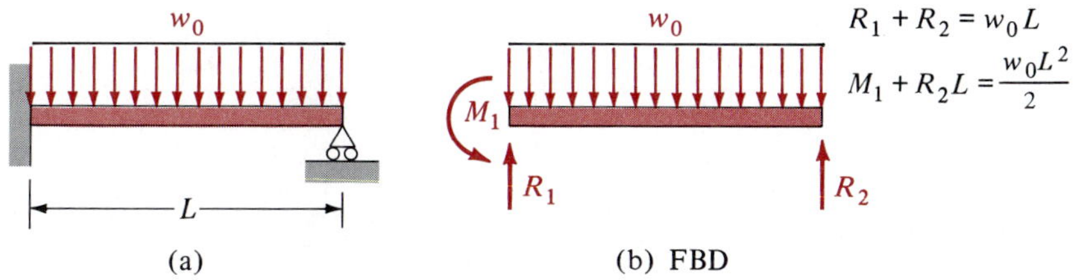

Figure 8.51 Cantilever beam supported at the right end

Solution 1 The first step is to determine the degree of redundancy. Referring to the free-body diagram of Fig. 8.51b, it is seen that there are the three reactions, R_1, M_1, and R_2; thus there is a single redundant. Choosing R_2 as the redundant, the two problems to be superposed are shown in Fig. 8.52.

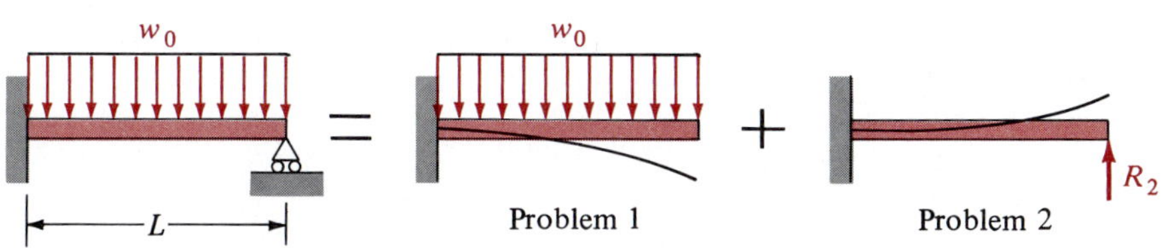

Figure 8.52 Superposition with R_2 as the redundant

Using cases *(c)* and *(a)* from Appendix D, the two deflections at $x = L$ are

Problem 1: $$v_1(L) = \frac{-w_0L^4}{8EI}$$

Problem 2: $$v_2(L) = \frac{R_2L^3}{3EI}$$

The actual displacement at the support is zero, so that

$$v_1(L) + v_2((L) = 0$$

or

$$\frac{-w_0L^4}{8EI} + \frac{R_2L^3}{3EI} = 0$$

from which

$$R_2 = \frac{3w_0L}{8}$$

The equilibrium equations can then be used to determine that $R_1 = 5w_0L/8$ and $M_1 = w_0L^2/8$.

Solution 2 For the second solution, choose M_1 as the redundant. The two problems that must be superposed are shown in Fig. 8.53.

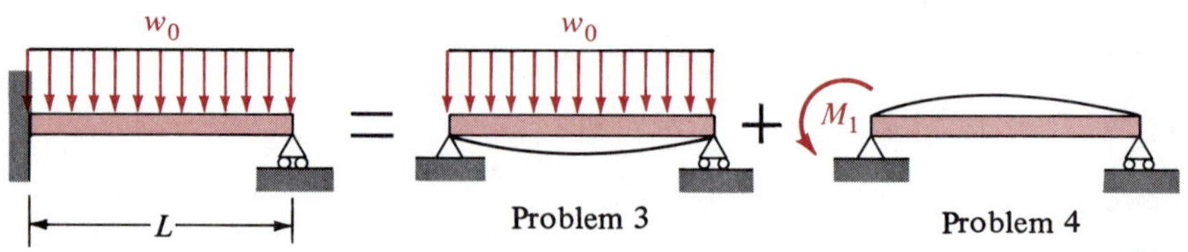

Figure 8.53 Superposition with M_1 as the redundant

The redundant moment M_1 must be selected so that the slope at $x = 0$ from the uniformly distributed loading is reduced to zero by the superposition of the moment M_1. Using cases (f) and (g) from Appendix D, the two slopes at $x = 0$ are

Problem 3: $$v_3'(0) = \frac{-w_0L^3}{24EI}$$

Problem 4: $$v_4'(0) = \frac{M_1L}{3EI}$$

The actual slope at $x = 0$ is given by

$$v_3'(0) + v_4'(0) = 0$$

or

$$\frac{-w_0L^3}{24EI} + \frac{M_1L}{3EI} = 0$$

from which

$$M_1 = \frac{w_0L^2}{8}$$

precisely the same as for our first choice for the redundant.

Example 8.18 A beam is supported at three equidistant points and loaded at the middle of one of the spans as shown in Fig. 8.54a. Neglecting the weight of the beam, determine the reactions.

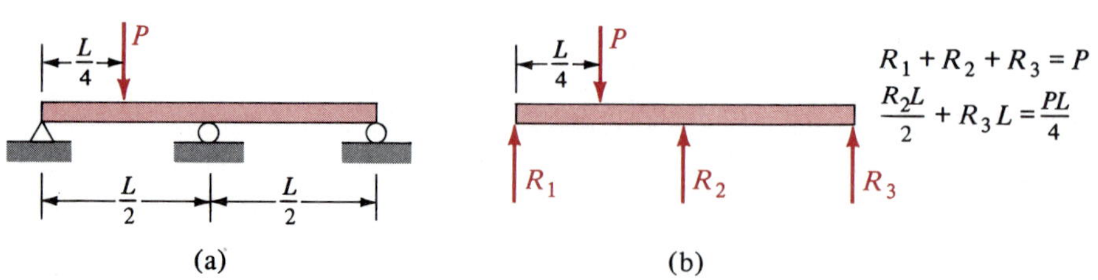

Figure 8.54 Three-span beam

Referring to the equilibrium equations indicated in Fig. 8.54b, there are three unknowns with only two equations, hence one redundant. For this example, any one of the three unknowns can be selected as the redundant. We will develop solutions for two of the possibilities.

Solution 1 Choose the middle reaction R_2 shown in the free-body diagram of Fig. 8.54b as the redundant. The two problems that must then be superposed are shown in Fig. 8.55.

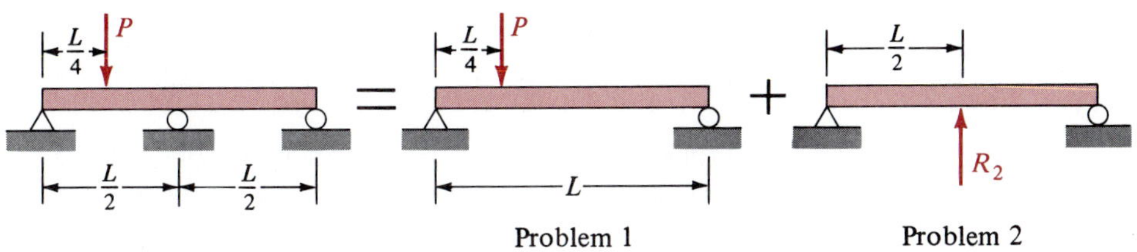

Figure 8.55 Problems to be superposed

The idea is that the magnitude of R_2 must be chosen so that the central displacement for problem 2 is equal and opposite to the central displacement for problem 1. From Appendix D using case *(e)* with $a = L/4$ for problem 1, and case *(e)* with $a = L/2$ for problem 2, the two displacements at $x = L/2$ are

Problem 1: $$v_1\left(\frac{L}{2}\right) = -\frac{11PL^3}{(32)(24)EI}$$

Problem 2: $$v_2\left(\frac{L}{2}\right) = +\frac{R_2L^3}{48EI}$$

Requiring that the actual displacement at the center be zero leads to

$$v_1\left(\frac{L}{2}\right) + v_2\left(\frac{L}{2}\right) = 0$$

or

$$\frac{-11PL^3}{(32)(24)EI} + \frac{R_2L^3}{48EI} = 0$$

from which $R_2 = 11P/16$. The equilibrium equations can then be used to determine that $R_1 = 13P/32$ and $R_3 = -3P/32$.

Solution 2 Choose the right reaction R_3 shown in the free-body diagram of Fig. 8.54b as the redundant. The two problems that must then be superposed are shown in Fig. 8.56.

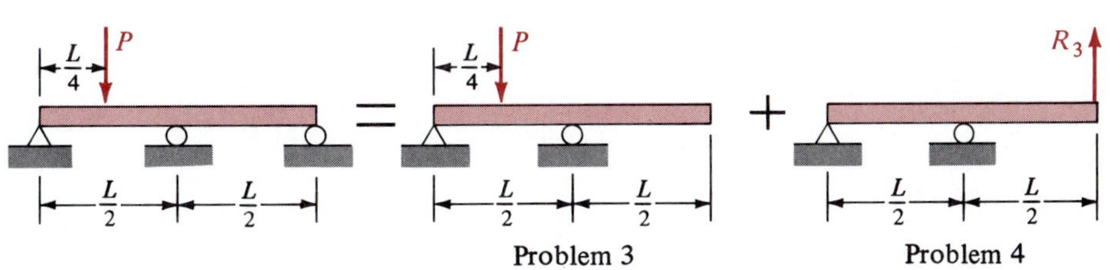

Figure 8.56 Superposition with R_3 as the redundant

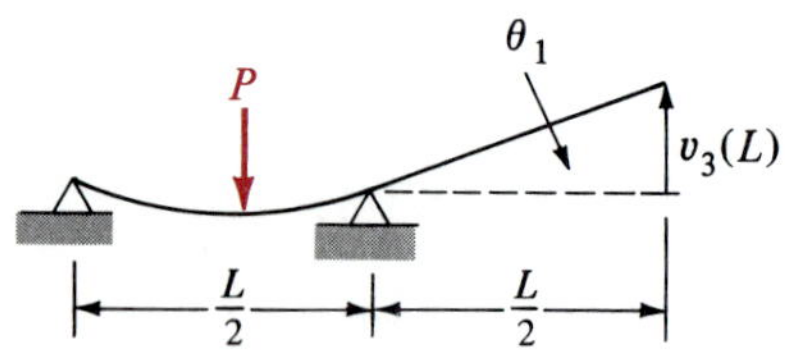

Figure 8.57 Analysis of problem 3

The idea here is that the redundant R_3 must have the value that reduces the displacement $v(L)$ to zero. Although it is clear that R_3 will actually be directed downwards, we will consider that R_3 has the direction shown in Fig. 8.56, and let the method determine the direction of R_3.

For problem 3, we consider a beam of length $L/2$ loaded by a central load P, as in Fig. 8.57. From case *(e)* in Appendix D, the slope at the right end of the beam is given by $v_3'(L/2) = \theta_1 = P(L/2)^2/16EI$. The portion of the beam beyond the support is unloaded, thus straight, so that $v_3(L)$ as shown in Fig. 8.57 is given by

$$v_3(L) = \left(\frac{L}{2}\right)\theta = \frac{PL^3}{128EI}$$

For problem 4, the total displacement at the right end (greatly exaggerated) can be considered as the superposition of two problems as shown in Fig. 8.58. The first problem consists of the simply supported beam loaded by the end moment $M = R_3L/2$, and the second problem consists of a cantilever beam loaded by the force R_3.

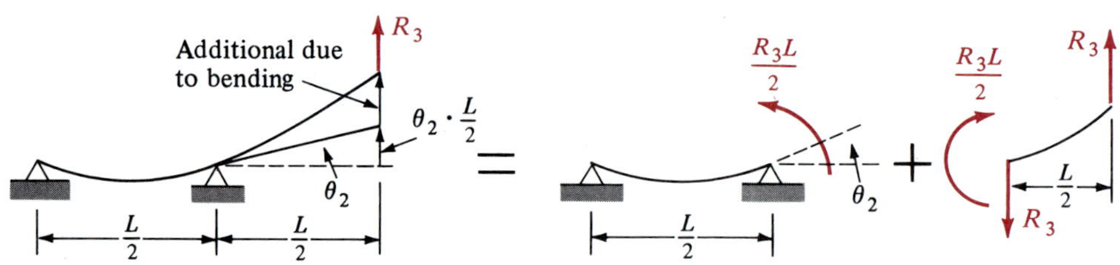

Figure 8.58 Analysis of problem 4

Using case *(f)* from Appendix D, the slope at $x = L/2$ for the problem of the end moment is given by

$$v_4'\left(\frac{L}{2}\right) = \theta_2 = \frac{ML}{3EI} = \frac{R_3L}{2}\frac{L}{2}\frac{1}{3EI} = \frac{R_3L^2}{12EI}$$

As indicated in Fig. 8.58, the corresponding displacement at $x = L$ is given by $(L/2)\theta = R_3L^3/24EI$. The additional displacement due to the bending of the portion of the beam beyond $x = L/2$ is obtained by using case *(a)* with the length of $L/2$; thus, the additional displacement is given by $R_3(L/2)^3/3EI = R_3L^3/24EI$. The total displacement at $x = L$ for problem 4 is then

$$v_4(L) = \frac{R_3L^3}{24EI} + \frac{R_3L^3}{24EI} = \frac{R_3L^3}{12EI}$$

Finally, the displacement at $x = L$ for both problems 1 and 2 is given by

$$v(L) = v_3(L) + v_4(L) = \frac{PL^3}{128EI} + \frac{R_3L^3}{12EI} = 0$$

from which

$$R_3 = -\frac{12P}{128} = -\frac{3P}{32}$$

the same as our other choice for the redundant. The redundants R_1 and R_2 can again be determined from the equilibrium equations.

In principle, the use of superposition as outlined above can be used on any statically indeterminate problem. However, if the number of redundants is large, the method becomes unwieldy and other approaches are preferable. The reader is referred to the many texts on statically indeterminate structures in this regard.

PROBLEMS

8.142 Considering the force at B as the redundant, determine the reactions for the cantilever beam supported at the free end as shown.

8.143 Considering the moment at A to be the redundant, determine the reactions for the cantilever beam supported at the free end as shown.

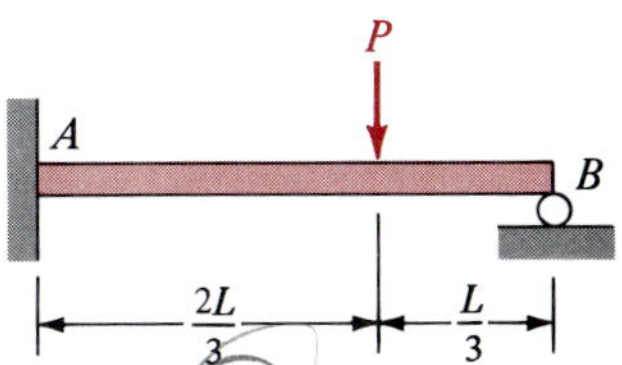

Ps. 8.142 and 8.143

8.144 Select the force at B as the redundant and determine the reactions. Note that the result will depend upon the parameter k, which can then be varied as necessary to treat the problem of the load P applied at any location between the supports.

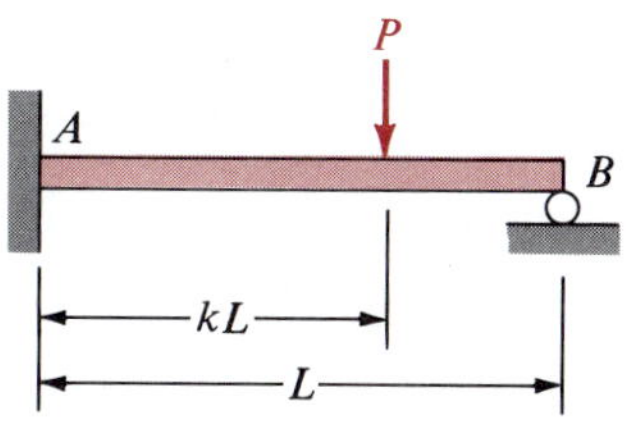

P. 8.144

8.145 Considering the force at B as the redundant, determine the reactions for the cantilever beam loaded and supported as shown.

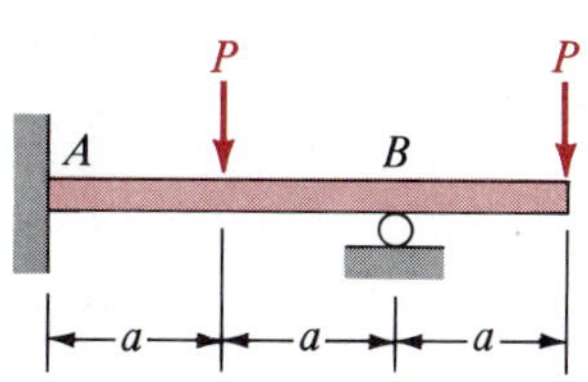

Ps. 8.145 and 8.146

8.146 Considering the moment at A to be the redundant, determine the reactions for the cantilever beam loaded and supported as shown.

8.147 Considering the force at B as the redundant, determine the reactions for the cantilever beam loaded and supported as shown.

8.148 Considering the moment at A to be the redundant, determine the reactions for the cantilever beam loaded and supported as shown.

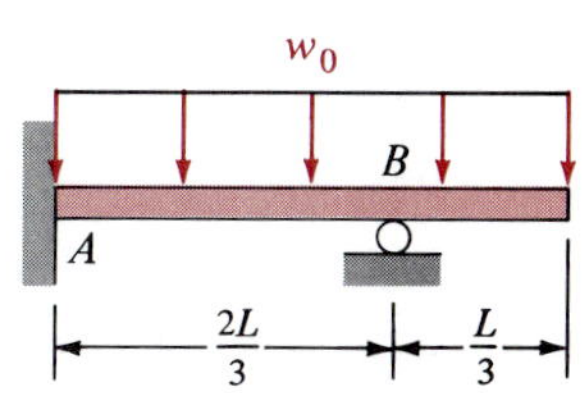

Ps. 8.147 and 8.148

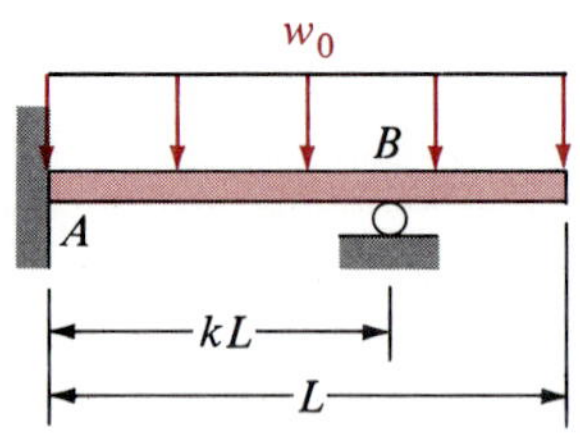

P. 8.149

8.149 Select the moment at A as the redundant and determine the reactions. Note that the result will depend upon the parameter k, which can then be varied as necessary to treat the problems of the support located at any location along the beam.

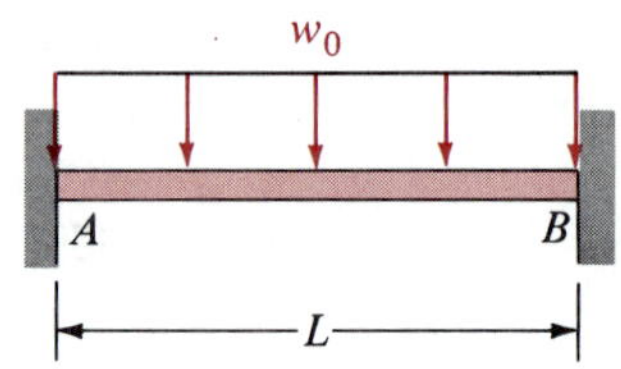

Ps. 8.150 and 8.151

8.150 For the clamped-clamped beam shown, select the moments at A and B to be the redundants and determine all the reactions.

8.151 For the clamped-clamped beam shown select the moment at A and the force at A to be the redundants and determine all the reactions.

8.152–8.161 For each of the following, determine all the reactions.

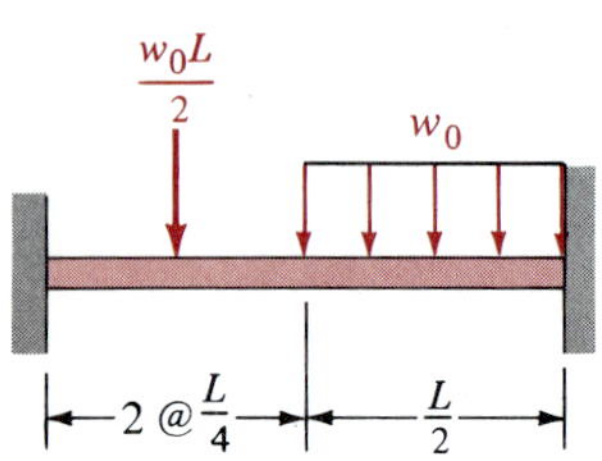

P. 8.152

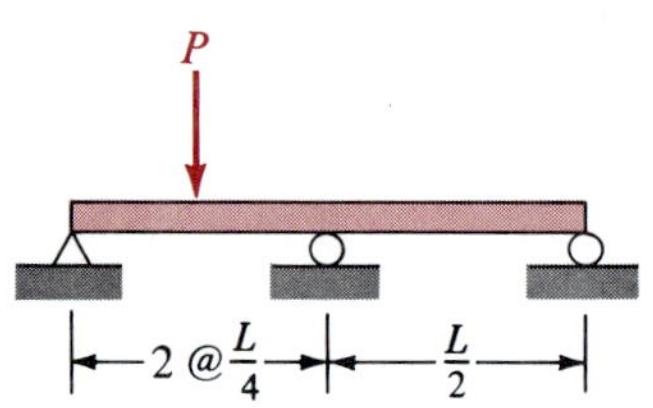

P. 8.153

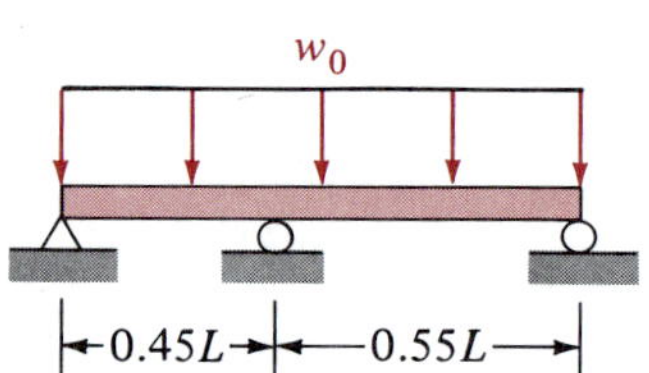

P. 8.154

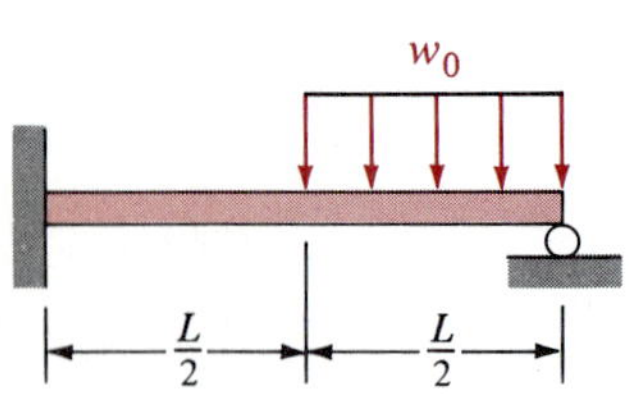

P. 8.155

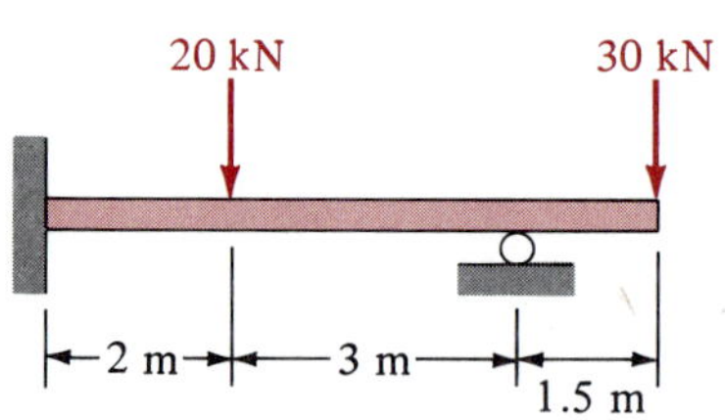

P. 8.156

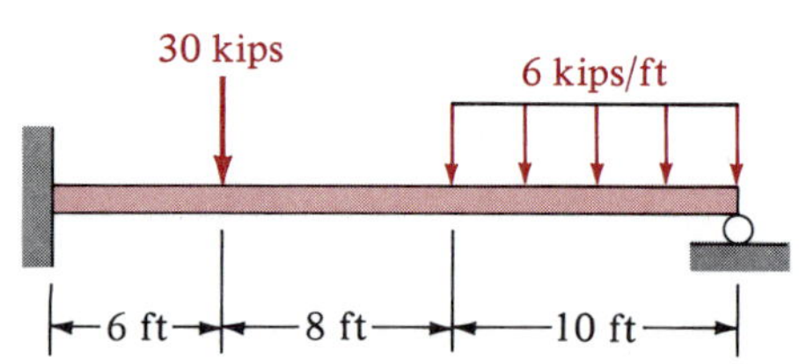

P. 8.157

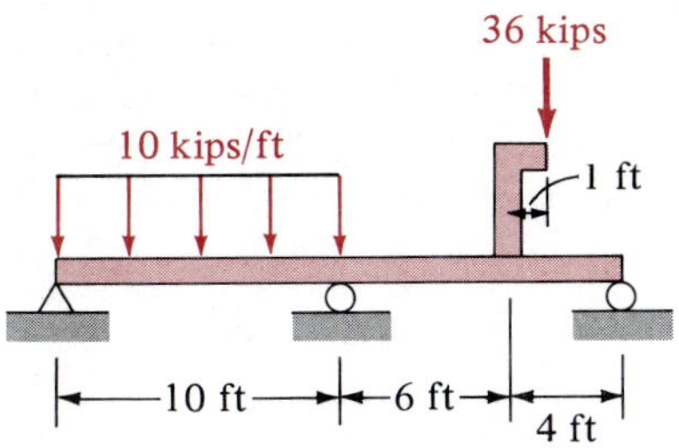

P. 8.158

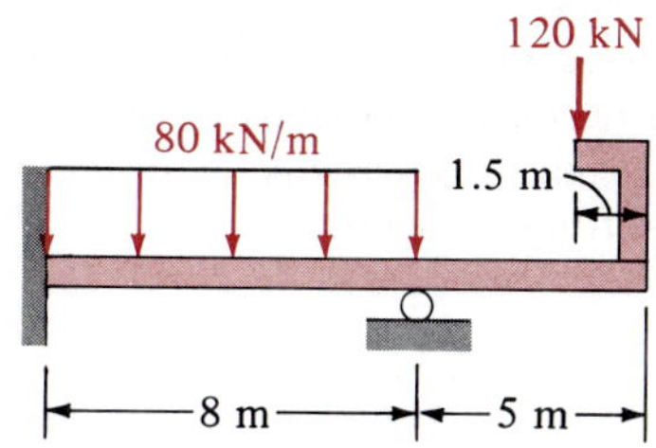

P. 8.159

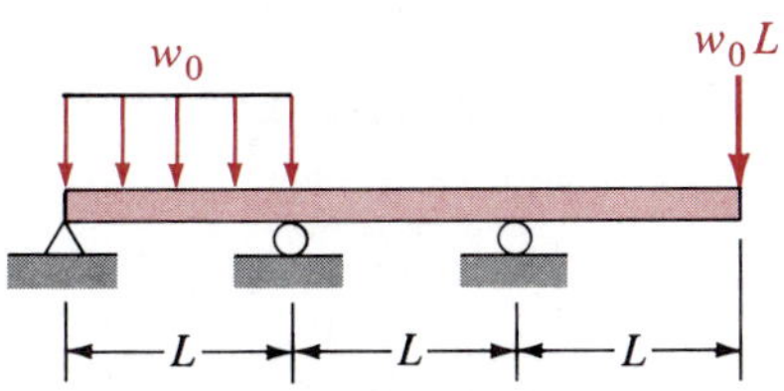

P. 8.160

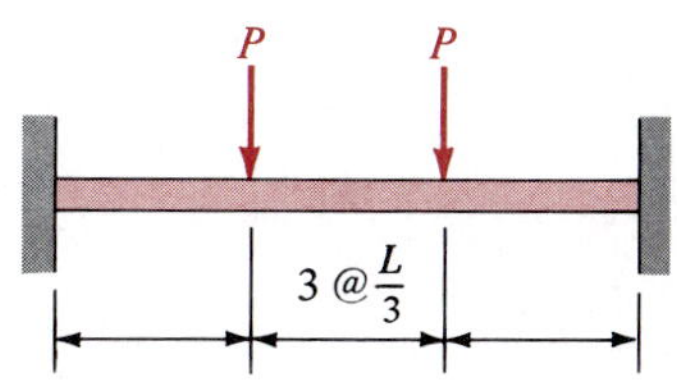

P. 8.161

8.162 Two wooden beams with identical EI are simply supported at the ends and just touch at the center when unloaded. Determine the reactions at the ends of both beams when the uniformly distributed load w_0 is applied.

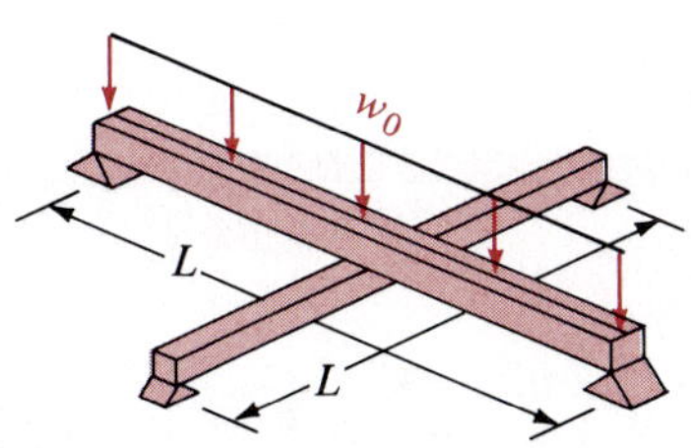

Ps. 8.162–8.164

8.163 Repeat problem 8.162 if the EI of the upper beam is twice that of the lower beam.

8.164 Repeat problem 8.162 if the EI of the two beams are the same and if the length of the upper beam is 1.5 times the length of the lower beam.

8.165 The cantilever beam shown is attached to an elastic wire that is initially unloaded. When the uniformly distributed load w_0 is applied, determine *(a)* the reactions at A, *(b)* the stress in the wire, and *(c)* the deflection at B. Take $E_1 = 70$ GPa, $I = 2 \times 10^{-5}$ m^4, $L = 2$ m, $l = 6$ m, $A = 3 \times 10^{-5}$ m^2, $E_2 = 200$ GPa, and $w_0 = 2$ kN/m.

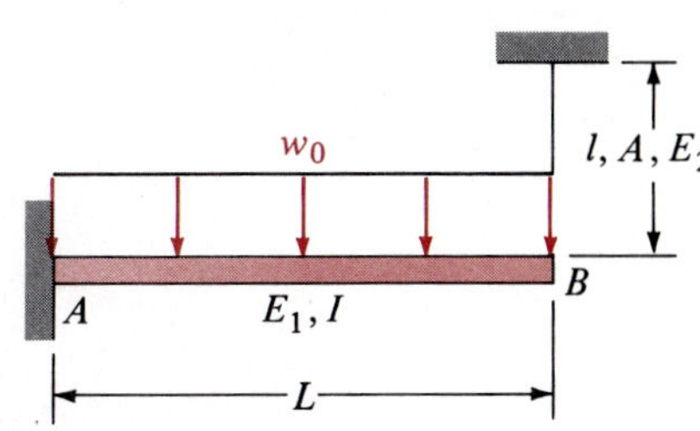

P. 8.165

8.166 The simply supported beam is attached to an unloaded elastic spring at the center and loaded uniformly as shown. Develop expressions for *(a)* the reactions at the ends, *(b)* the force in the spring, and *(c)* the shortening of the spring. Then evaluate the values of the reactions in the limit as the spring stiffness k becomes large.

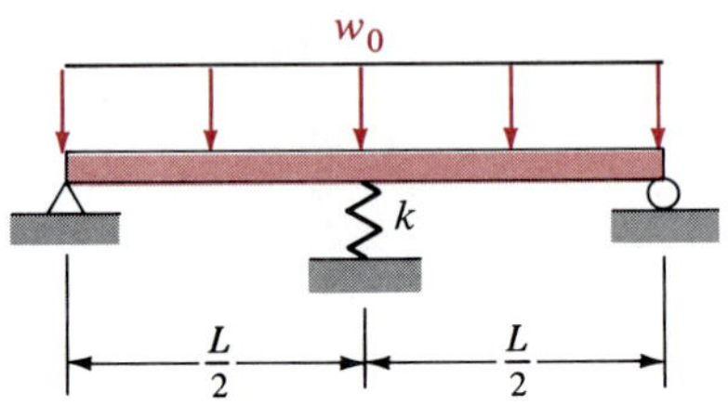

P. 8.166

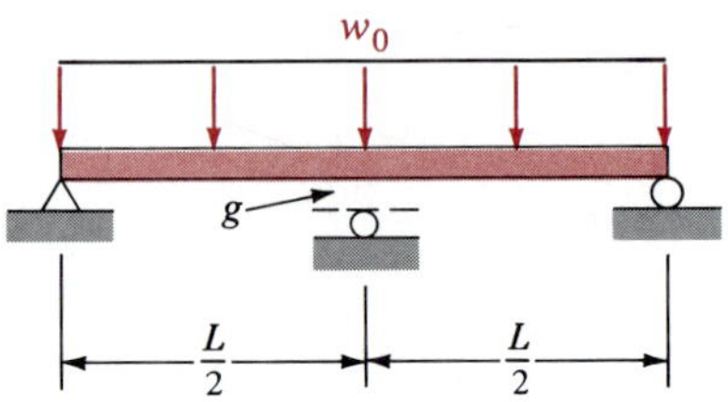

P. 8.167

8.167 The simply supported beam is loaded by the uniform load as shown. There is a gap g between the center of the beam and a rigid stop as indicated. Determine the reactions.

8.168 The support at one end of a beam that is clamped at both ends settles an amount Δ, without rotation, as shown. Determine the reactions that result at the ends of the beam.

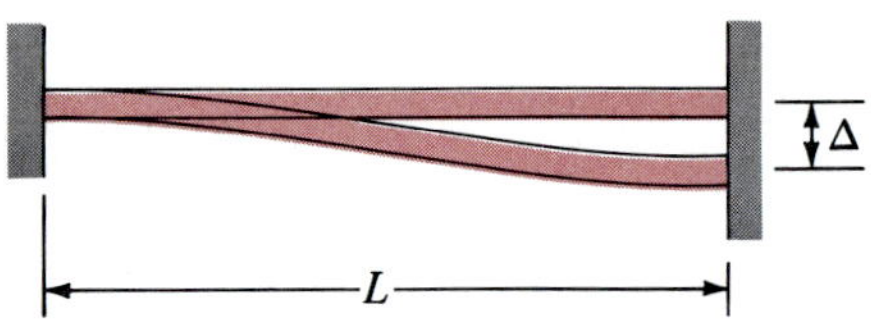

P. 8.168

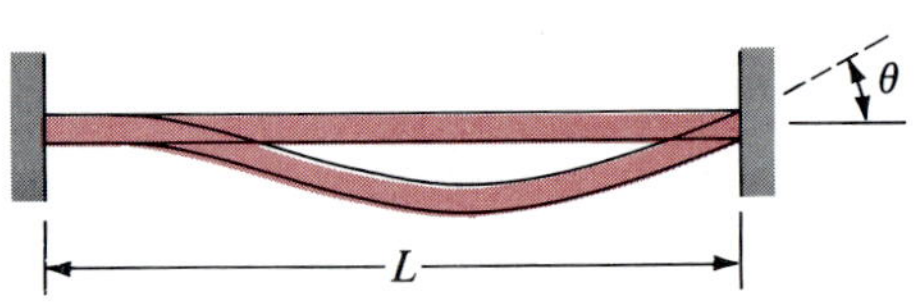

P. 8.169

8.169 The support at one end of a beam that is clamped at both ends rotates an amount θ, without vertical displacement, as shown. Determine the reactions that result at the ends of the beam.

8.170 Determine the reactions as a result of the settling of the simple support by an amount Δ as shown.

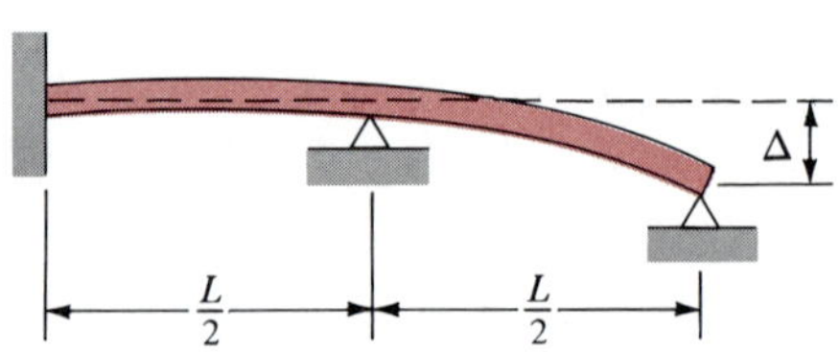

P. 8.170

8.171 For the W 530 × 92 beam shown, *(a)* determine the reactions, *(b)* draw the shear and moment diagrams, and *(c)* determine the maximum bending stress.

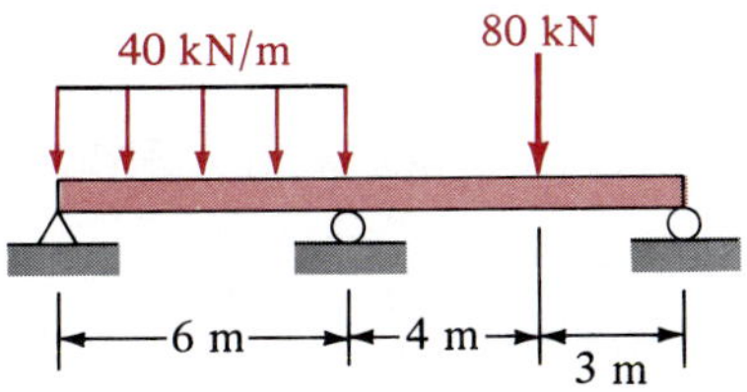

P. 8.171

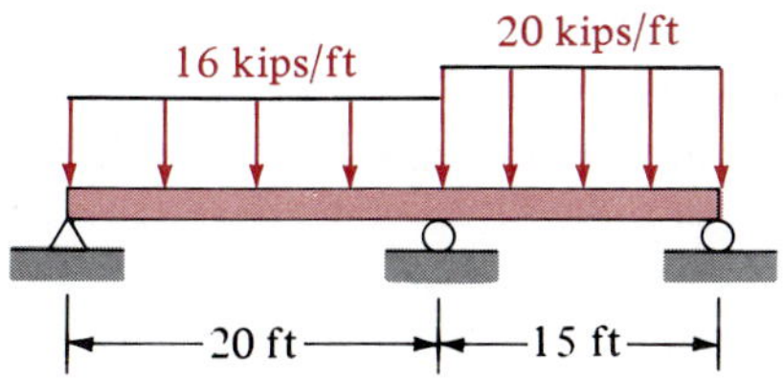

P. 8.172

8.172 For the W 24 × 104 beam shown, *(a)* determine the reactions, *(b)* draw the shear and moment diagrams and *(c)* determine the maximum bending stress.

8.173 For the W 130 × 28.1 beam supported and loaded as shown, *(a)* determine the reactions, *(b)* draw the shear and moment diagrams, and *(c)* determine the maximum bending stress.

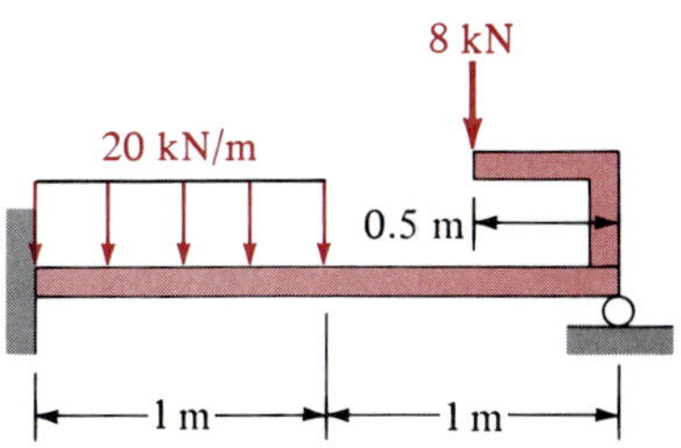

P. 8.173

8.7.2 Integration of $EIv'' = M$

When we integrate the moment-curvature relation $EIv'' = M$ in connection with a statically determinate problem, any appropriate boundary conditions on V or M will already have been satisfied in establishing the expression for M. There will be two remaining boundary conditions on v', and/or v, to determine the two constants of integration. For a statically indeterminate problem, the moment will involve unknown redundants that will be determined using the boundary conditions and/or the constraints. Let us demonstrate the process with two examples.

Example 8.19 Consider the problem of a beam clamped at one end and simply supported at the other, as indicated in Fig. 8.59. The boundary conditions are indicated at the ends of the beam. There are no constraints for this problem. Determine the reactions and the equation for the elastic curve.

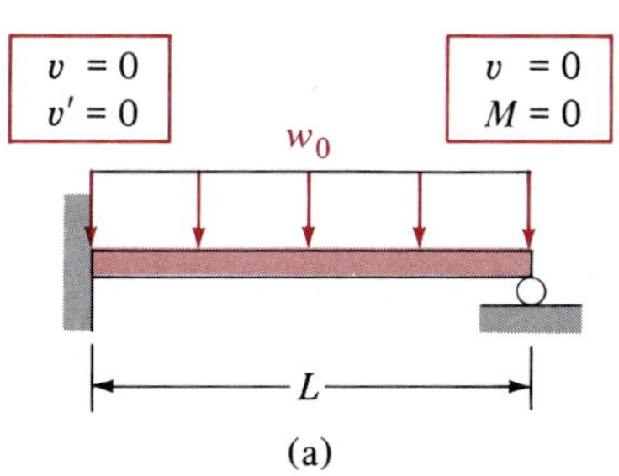

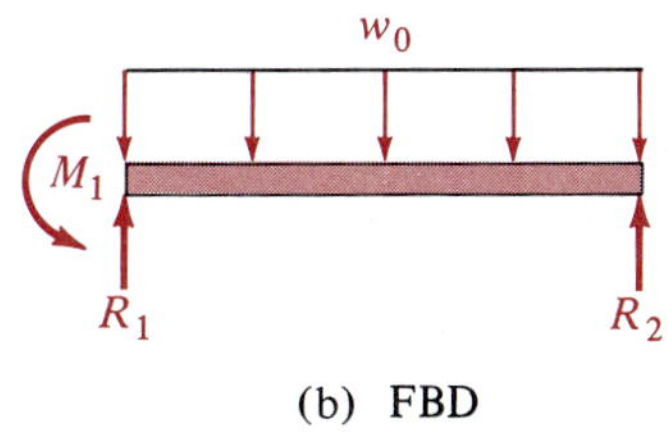

$$R_1 + R_2 = w_0 L$$

$$M_1 + R_2 L = \frac{w_0 L^2}{2}$$

Figure 8.59 Clamped–simply supported beam

Solution Consulting the free-body diagram of Fig. 8.59b, we see that the problem is statically indeterminate with one extra support (i.e., one redundant). From Fig. 8.59b, the expression for the moment can be written as

$$M(x) = R_1 x - M_1 - \frac{w_0 x^2}{2}$$

Solving the equilibrium equations for R_1 in terms of M_1, or $R_1 = M_1/L + w_0 L/2$, the moment can be expressed as

$$M(x) = M_1 \left(\frac{x}{L} - 1\right) + \frac{w_0 x(L - x)}{2}$$

where M_1, which is at this point unknown, has been selected as the redundant. Note that the boundary condition $M(L) = 0$ is satisfied. The moment-curvature relation can be written as

$$EIv'' = M_1 \left(\frac{x}{L} - 1\right) + \frac{w_0 x(L - x)}{2}$$

This equation is to be integrated subject to the three boundary conditions $v'(0) = 0$, $v(0) = 0$, and $v(L) = 0$. The integration results in

$$EIv' = M_1 \left(\frac{x^2}{2L} - x\right) + \frac{w_0(Lx^2/2 - x^3/3)}{2} + C_3$$

$$EIv = M_1 \left(\frac{x^3}{6L} - \frac{x^2}{2}\right) + \frac{w_0(Lx^3/6 - x^4/12)}{2} + C_3 x + C_4$$

Note that there are three unknowns, two constants of integration and the redundant M_1, that are to be determined from the three boundary conditions $v'(0) = 0$, $v(0) = 0$, and $v(L) = 0$. Satisfying the three boundary conditions yields

$$EIv'(0) = 0 = C_3$$

$$EIv(0) = 0 = C_4$$

$$EIv(L) = 0 = -\frac{M_1 L^2}{3} + \frac{w_0 L^4}{24}$$

from which

$$M_1 = \frac{w_0 L^2}{8}$$

We can then determine R_1 and R_2 from the equations of equilibrium as $R_1 = 5w_0 L/8$ and $R_2 = 3w_0 L/8$. The shear and moment diagrams can then be constructed, as in Fig. 8.60. Substituting for M_1, the equation for the elastic curve can be written as

$$v(x) = \left(\frac{w_0 x^2}{48EI}\right)(5Lx - 3L^2 - 2x^2)$$

as shown in Fig. 8.60. The maximum value of the transverse displacement and its location are also indicated.

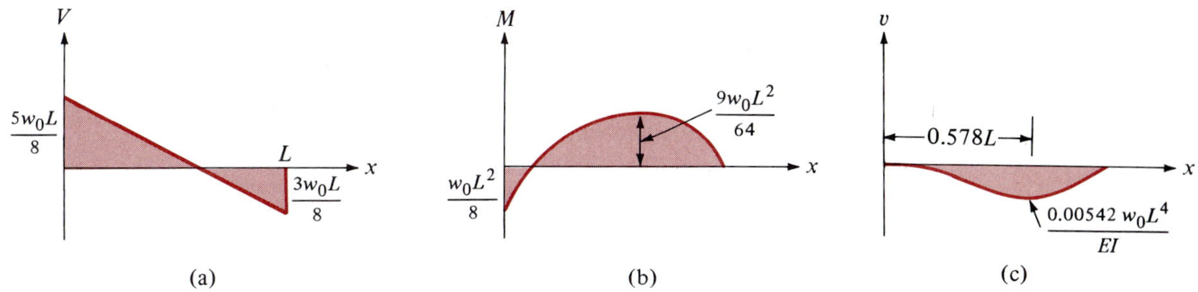

Figure 8.60 Shear, moment, and displacement curves

For this statically indeterminate example, all of the equations necessary to determine the constants of integration and the redundant are supplied by the boundary conditions. There are no constraints. It is left for the student to show that if M_1 is eliminated in favor of R_1 in the expression for the moment $M(x)$, the intermediate results are different but the final values for the reactions and the displacements are identical.

Example 8.20 For the beam shown in Fig. 8.61a, determine *(a)* the equation for the elastic curve, and *(b)* the maximum displacement between the supports. Also draw the shear and moment diagrams.

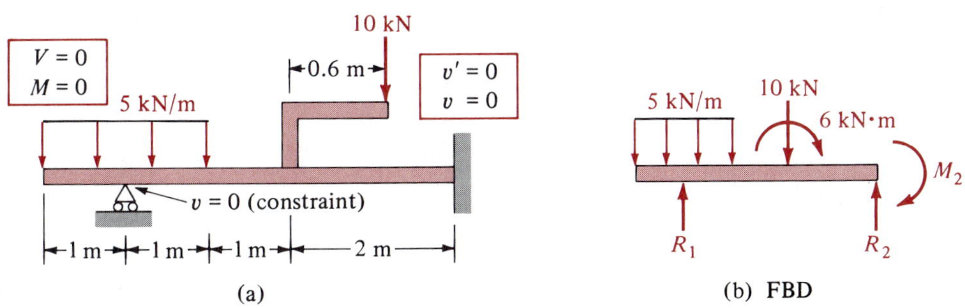

Figure 8.61 Beam under the action of a concentrated moment

Solution From the free-body diagram of Fig. 8.61b, the equilibrium equations can be written as

$$R_1 + R_2 = 20 \text{ kN}$$

$$M_2 + 4 \text{ m}(R_1) - 54 \text{ kN·m} = 0$$

so that we have a statically indeterminate problem with one redundant. Referring to the free-body diagram of Fig. 8.61b and to Fig. 8.25 for representing the moment in terms of the singularity functions, the expression for the moment can be written as

$$M(x) = \left[\frac{-5(x^2 - \langle x-2\rangle^2)}{2} + R_1\langle x-1\rangle^1 - 10\langle x-3\rangle^1 + 6\langle x-3\rangle^0 + R_2\langle x-5\rangle^1 + M_2\langle x-5\rangle^0\right] \text{kN·m}$$

where x is measured in meters. The R_2 and M_2 terms can be discarded since they will contribute nothing in the interval $0 \leq x \leq 5$ m. Note that the two boundary conditions $V(0) = M(0) = 0$ will have been satisfied in the course of correctly determining the expression for the $M(x)$. Integrating $EIv'' = M$ will produce two constants of integration C_1 and C_2, so that the expression for v will contain the three unknowns C_1, C_2, and R_1. The three conditions for determining the three unknowns will come from the two boundary conditions $v(5) = 0 = v'(5)$ and from the constraint $v(1) = 0$. The condition on the displacement at $x = 1$ m is properly referred to as a constraint, not a boundary condition.

Integrating $EIv'' = M$ gives

$$EIv' = \left[\frac{-5(x^3 - \langle x-2\rangle^3)}{6} + \frac{R_1\langle x-1\rangle^2}{2} - 5\langle x-3\rangle^2 + 6\langle x-3\rangle^1\right] \text{kN}\cdot\text{m}^2 + C_1$$

and

$$EIv = \left[\frac{-5(x^4 - \langle x-2\rangle^4)}{24} + \frac{R_1\langle x-1\rangle^3}{6} - \frac{5\langle x-3\rangle^3}{3} + 3\langle x-3\rangle^2\right] \text{kN}\cdot\text{m}^3 + C_1 x + C_2$$

The conditions $v'(5) = 0$ and $v(5)$ become, respectively,

$$0 = -\left[\frac{5(125-27)}{6} + 8\right] \text{kN}\cdot\text{m}^2 + R_1\, 8\ \text{m}^2 + C_1$$

$$0 = -\left[\frac{5(625-81)}{24} + \frac{4}{3}\right] \text{kN}\cdot\text{m}^3 + \frac{32R_1\ \text{m}^3}{3} + 5\ \text{m}\ C_1 + C_2$$

with the constraint condition $v(1)$ as

$$0 = -\frac{5}{24}\ \text{kN}\cdot\text{m}^3 + C_1\text{m} + C_2$$

Solving for C_1, C_2, and C_3 yields $C_1 = -1.911$ kN·m², $C_2 = 2.120$ kN·m³, and $R_1 = 11.447$ kN. The reactions at the right end can then be determined from the equilibrium equations as $R_2 = 8.553$ kN and $M_2 = 8.212$ kN·m. The corresponding shear and moment diagrams appear in Fig. 8.62.

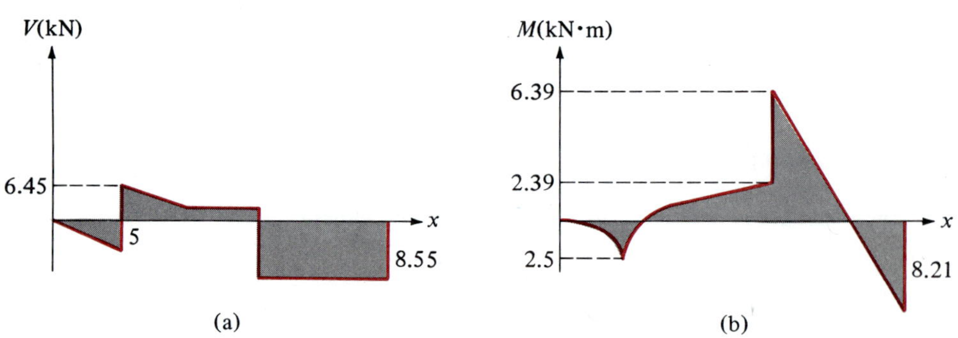

Figure 8.62 Shear and moment diagrams

The expression for the elastic curve can then be written as

$$EIv = \left[\frac{-5(x^4 - \langle x-2\rangle^4)}{24} + \frac{11.447\langle x-1\rangle^3}{6} - \frac{5\langle x-3\rangle^3}{3} + 3\langle x-3\rangle^2 - 1.911x + 2.120\right] \text{kN·m}^3$$

and is shown in Fig. 8.63

The largest value of v, which occurs at $x = 3.21$ m, has the value shown in Fig. 8.63.

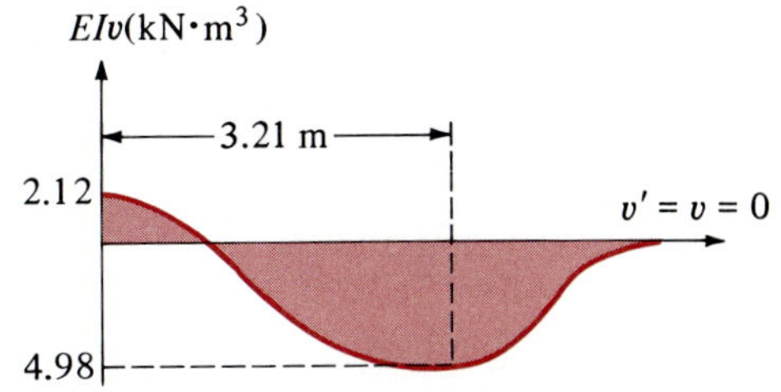

Figure 8.63 Transverse displacement

PROBLEMS

Before beginning the solution for each of the following, first draw a free-body diagram, write the equations of equilibrium, and identify and indicate all boundary conditions and constraints.

8.174 For the clamped–simply supported beam shown, choose the moment at A as the redundant and use integration of $EIv'' = M$ to determine the reactions and the expression for the displacement $v(x)$.

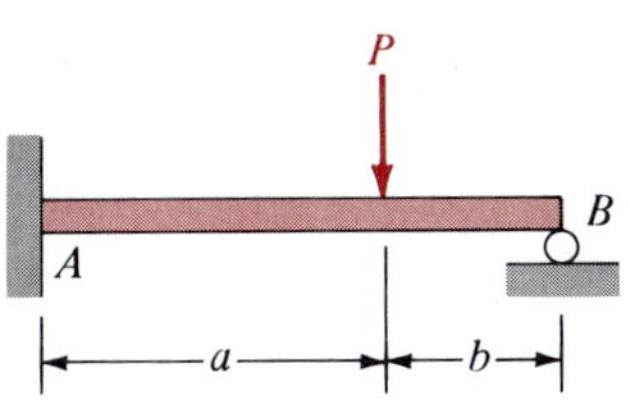

P. 8.174

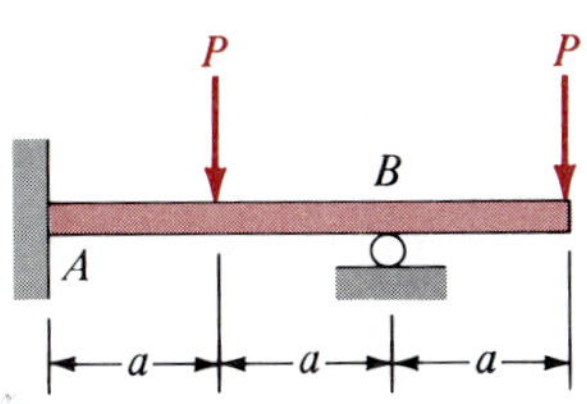

P. 8.175

8.175 For the clamped beam with the internal support shown, choose the reaction at A as the redundant and use integration of $EIv'' = M$ to determine the reactions and the expression for the displacement $v(x)$. Draw the shear and moment diagrams.

8.176 For the clamped-clamped beam loaded as shown, choose the support force and moment at the left end of the beam as redundants and use integration of $EIv'' = M$ to determine the reactions and the expression for the displacement $v(x)$.

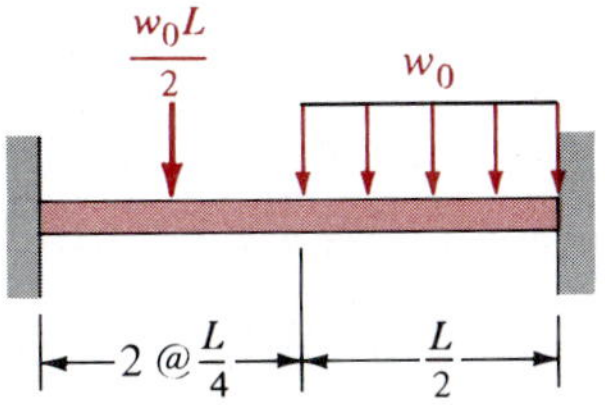

P. 8.176

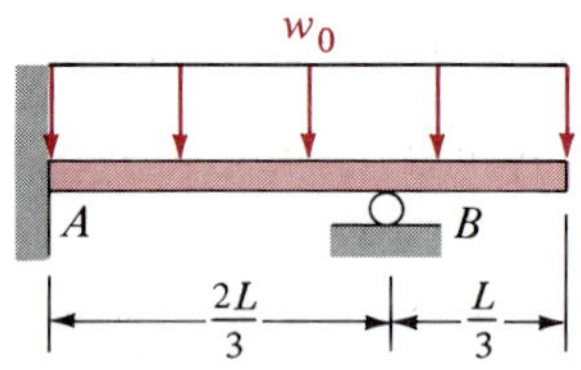

P. 8.177

8.177 For the clamped beam with the internal support shown, choose the reaction at A as the redundant and use integration of $EIv'' = M$ to determine the reactions and the expression for the displacement $v(x)$. Draw the shear and moment diagrams.

8.178 For the two-span beam loaded by a concentrated load at the middle of the right span as shown, choose the reaction force at the left end as the redundant and use integration of $EIv'' = M$ to determine the reactions and the expression for the displacement $v(x)$. Draw the shear and moment diagrams.

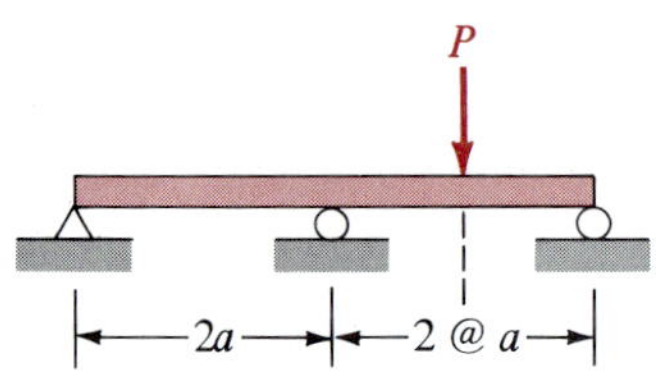

P. 8.178

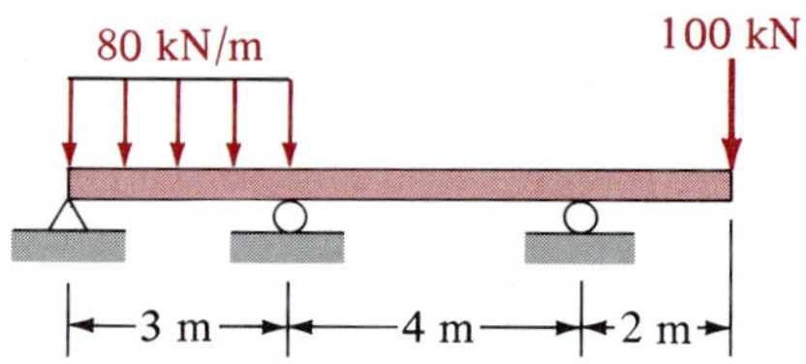

P. 8.179

8.179 For the three-span beam loaded as shown, choose the reaction at the left end as the redundant and use integration of $EIv'' = M$ to determine the reactions and the expression for the displacement $v(x)$. Draw the shear and moment diagrams.

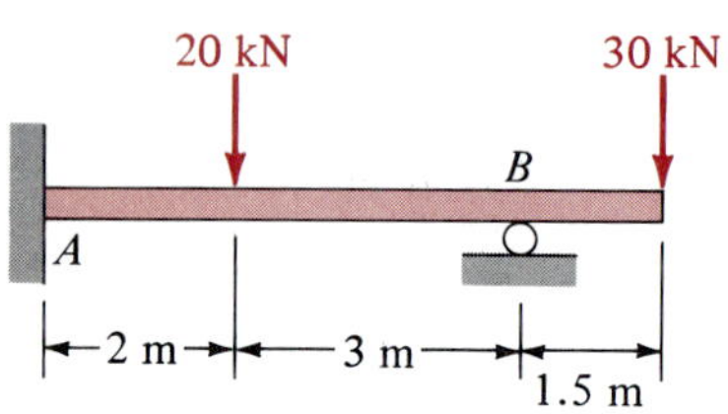

P. 8.180

8.180 In addition to the loading shown, the support at B settles an amount $\Delta = 20$ mm. Choose the moment at the left end as the redundant and use integration of $EIv'' = M$ to determine the reactions and the expression for the displacement $v(x)$. Draw the shear and moment diagrams. Take $EI = 4 \times 10^5$ N·m^2.

8.181 In addition to the loading shown, the support at B settles an amount $\Delta = 0.6$ in. Choose the moment at the left end as the redundant and use integration of $EIv'' = M$ to determine the reactions and the expression for the displacement $v(x)$. Take $EI = 2.6 \times 10^6$ kip·in^2.

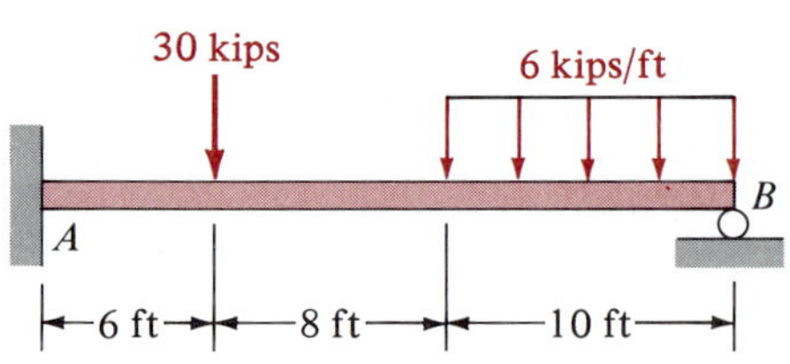

P. 8.181

8.8 SHEAR DEFORMATION IN BEAMS

As presented in Chapter 7, one of the basic assumptions leading to the equations $\sigma_x = -My/I$ for the bending stress, and $M = EI/\rho \approx EIv''$ relating the bending moment to the curvature, was that the loading consisted only of a bending moment M. It was assumed that the beam responded to the bending moment by deforming into a circular arc of radius ρ, with all lines originally perpendicular to the axis of the beam being parallel to radii, that is, plane sections remain plane and perpendicular to the neutral axis as indicated in Fig. 8.64.

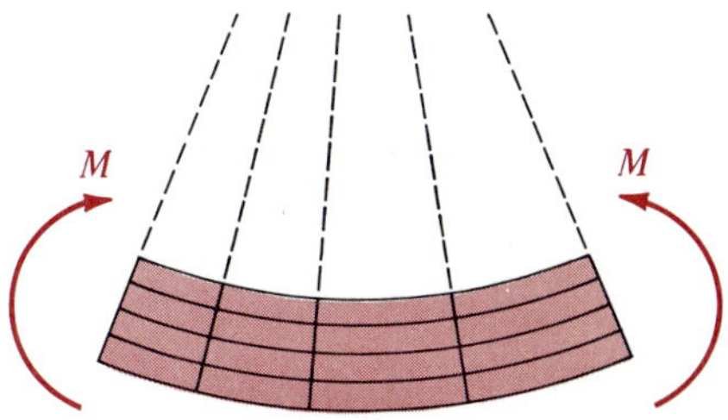

Figure 8.64 Plane sections remain plane and perpendicular deformations

In Chapter 7 and in this chapter, respectively, the equations $\sigma_x = -My/I$ and $M = EIv''$ have been used in situations where the loadings produce both moment M and a shear V throughout the beam, clearly in violation of the assumptions. In particular, the displacements determined by integrating $EIv'' = M$ must be considered as only the portion of the displacements due to transverse loading. In this section we will develop a capability for determining the additional deformations due to the shear deformation associated with transmitting a shear force V.

Consider a cantilever beam loaded as shown in Fig. 8.65a. Fig. 8.65b indicates the character, greatly exaggerated, of the deformation of the cantilever beam. The dotted lines indicate the bending deformations, whereas the curved lines correspond to the sum of the bending and shear deformations. Note that, due to the fact that the top and bottom surfaces are stress free or shear stress, the shear strain vanishes there.

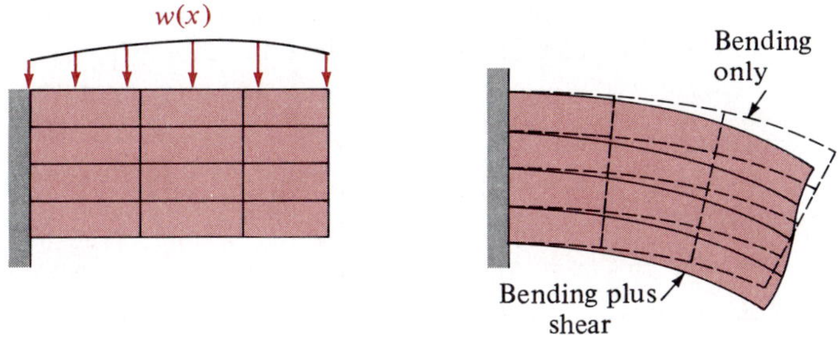

Figure 8.65 Shear deformation in a cantilever beam

The detailed character of the deformation is shown in Fig. 8.66, where the dotted lines indicate the bending deformations and the solid lines indicate the sum of the bending and shear deformations. The additional transverse displacement of the neutral axis can be seen to arise from the shear strain at the neutral axis, that is, $dv_s = \gamma_{xy}(x, 0)\, dx$, or

$$\gamma_{xy}(x, 0) = dv_s/dx$$

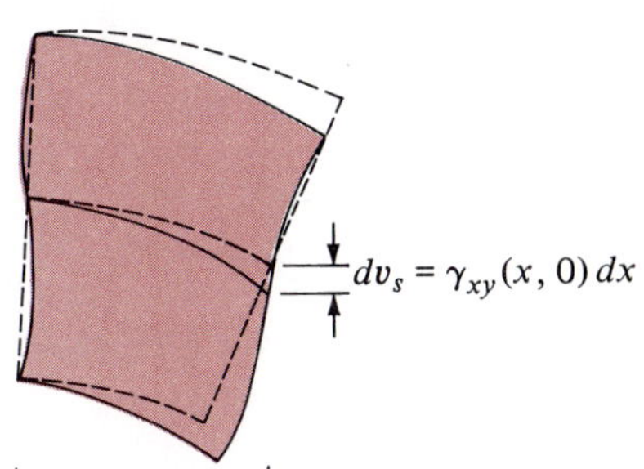

Figure 8.66 Detailed picture of the deformation of a segment of the beam

where v_s is the additional deformation due to shear. The relation between the shear force V and the shear strain γ_{xy} is developed by first recalling that the shear force V is related to the shear stress according to

$$V = -\int \tau_{xy}\, dA$$

Assuming a linearly elastic isotropic solid, the shear stress and shear strain are related according to $\gamma_{xy} = \tau_{xy}/G$, so that

$$V = -\int G\gamma_{xy}(x, y)\, dA$$

which is integrated to yield

$$V = -kAG\gamma_{xy}(x, 0)$$

where k is a factor to account for the fact that the shear strain is not constant over the depth of the cross section. The factor k depends upon the shape of the cross section and is of the order of unity for common shapes. Recalling that $\gamma_{xy}(x, 0) = v_s'$, we arrive at the relation

$$V = -kAGv_s'$$

which can be integrated to determine the additional deformation v_s' due to shear. These additional shear deformations are to be added to the corresponding bending deformations to obtain the total transverse displacement.

Example 8.21 Determine the additional transverse displacement due to shear deformation for the cantilever beam shown in Fig. 8.67a.

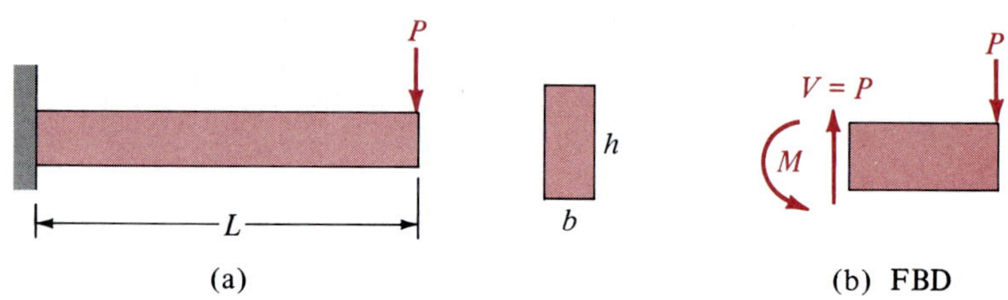

Figure 8.67 Shear deformation in a cantilever beam

Solution Referring to the free-body diagram of Fig. 8.67b, the shear V is everywhere equal to the load P, so that

$$\frac{dv_s}{dx} = \frac{-P}{kAG}$$

which can be integrated to yield

$$v_s = \frac{-Px}{kAG} + C_1$$

With $v_s = 0$ at $x = 0$, the constant of integration $C_1 = 0$ and

$$v_s = \frac{-Px}{kAG}$$

This is precisely what one would suspect, namely', that with the shear force constant along the length of the beam, the shear strain would be constant, resulting in a straight line for the displacement. This *additional* deformation due to shear is pictured in Fig. 8.68.

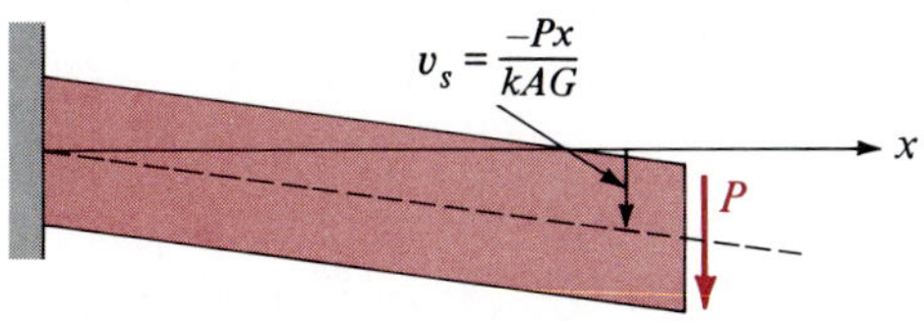

Figure 8.68 Additional deformation due to shear

Thus at $x = L$, the shear deformation results in the additional displacement

$$v_s(L) = \frac{-PL}{kAG}$$

This is to be added to the bending displacement

$$v_b(L) = \frac{-PL^3}{3EI}$$

so that the total displacement is

$$v_T(L) = -\frac{PL^3}{3EI} - \frac{PL}{kAG} = -\left(\frac{PL^3}{3EI}\right)\left(1 + \frac{3EI}{kAGL^2}\right)$$

For a rectangular section of width b and height h, the factor k is approximately equal to 5/6, so that using the relation $E = 2G(1 + \nu)$ for an isotropic solid, the total displacement can be simplified to

$$v_T(L) = -\frac{PL^3}{3EI}\left[1 + \frac{3(1 + \nu}{5}\left(\frac{h}{L}\right)^2\right]$$

For a typical value of $\nu = 0.3$ and with $h/L = 0.1$, there results

$$v_T(L) = -\left(\frac{PL^3}{3EI}\right)(1.0078)$$

showing that the additional displacement due to shear is less than 1%. For short stubby beams with h/L being large, the deformation due to shear can be a significant percentage of the total displacement. For slender beams (i.e., $h/L \leq 0.1$), shear deformation is negligible.

PROBLEMS

8.182 Given that the reactions at the ends of the clamped-clamped beam are each $w_0L/2$, develop the expression for the additional transverse displacement due to shear deformation. Compare the maximum value to the corresponding maximum value from bending only. Assume the cross section of the beam to be rectangular with $k = 5/6$.

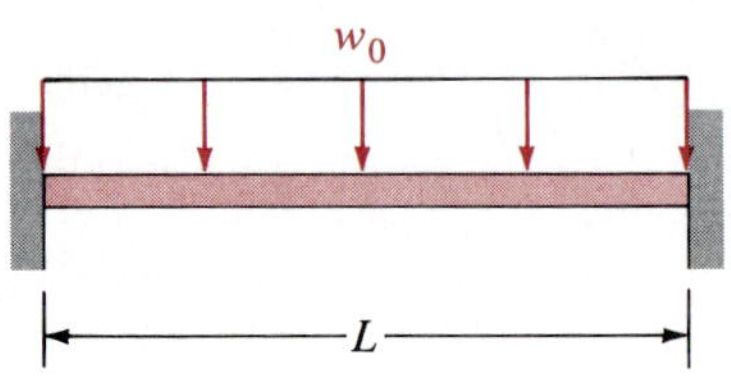

P. 8.182

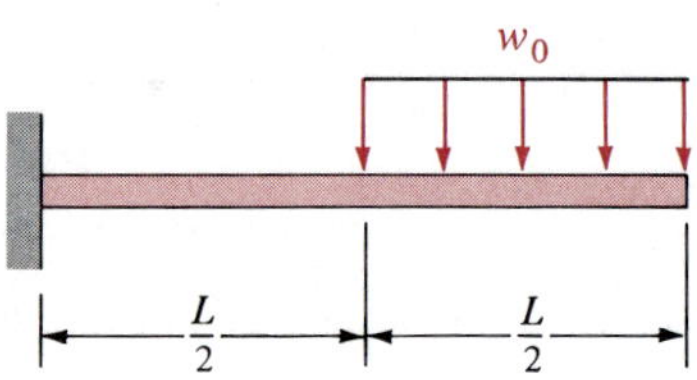

P. 8.183

8.183 For the cantilever beam loaded uniformly over a portion of its length, develop the expression for the additional transverse displacement due to shear deformation. Where does the maximum occur?

8.184 For the simply supported beam loaded uniformly over a portion of its length as shown, develop the expression for the additional transverse displacement due to shear deformation. Where does the maximum occur?

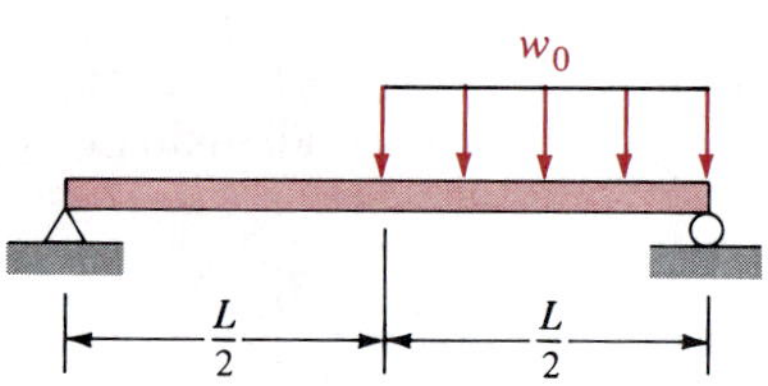

P. 8.184

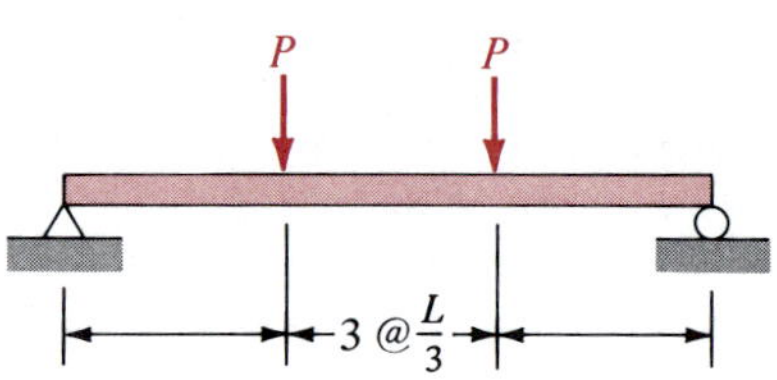

P. 8.185

8.185 For the simply supported beam under the action of two symmetrically placed concentrated loads as shown, develop the expression for the additional transverse displacement due to shear deformation. Where does the maximum occur? Take the cross section of the beam to be rectangular, with $k = 5/6$.

8.186 Develop the expression for the additional transverse displacement due to shear deformation for the beam asymmetrically loaded as shown. Where does the maximum occur?

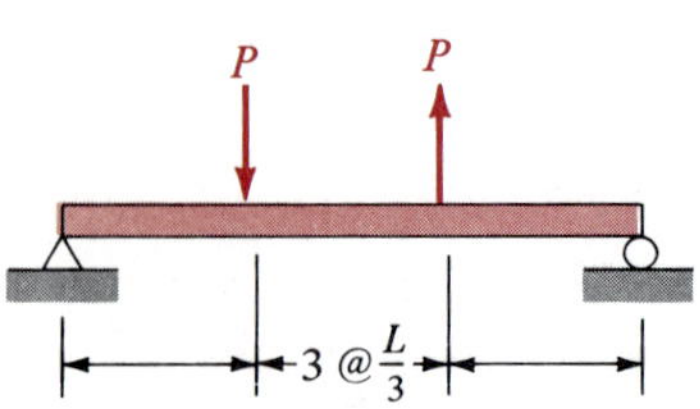

P. 8.186

8.9 DYNAMIC LOADING

In many important technical applications of beam theory, one must consider the dynamic characteristics of the situation. Critical examples include the dynamic loading on the wings of a large commercial jetliner during a nonperfect landing; the dynamic loading of an automobile bumper during impact with a stationary object; and the dynamic loading of a connecting rod in an internal combustion engine. The dynamic loading in each of these cases results in bending or additional bending that may be a major contribution to the bending and shear stresses in the beam. In this section we will present several examples to show how some of the results of our studies of beam theory can be used to obtain approximate solutions to certain types of problems involving impact, dynamic loading, and vibration.

Example 8.22 A mass M is dropped from a height h onto a uniform, simply supported beam as pictured in Fig. 8.69.

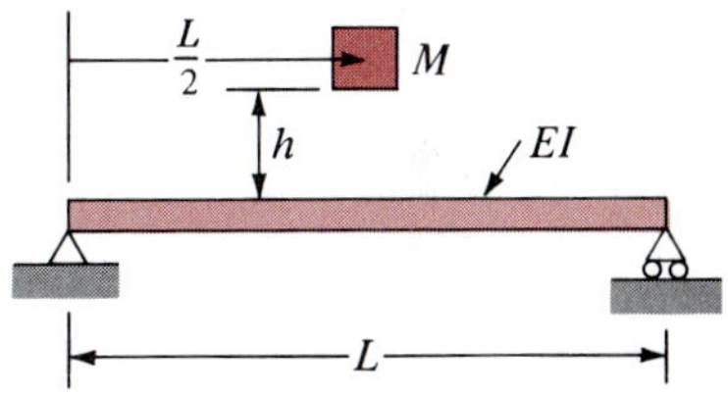

Figure. 8.69 Falling mass impacting a uniform, simply supported beam

Solution This type of problem is commonly called an impact problem, in that, after falling through the height h, the mass impacts the beam. The fundamental assumption often made in an elementary treatment of impact problems is that no energy is lost as a result of the impact. This assumption greatly simplifies the analysis, and provides results that can be quite accurate. The other assumptions involve the behavior of the beam: first, we assume that the stresses induced in the beam in decelerating the mass are in the elastic range; second, that the beam acts as a linear spring during the deceleration; and finally, that the mass m of the beam is small compared to that of the impacting mass M. With these assumptions in mind, we can outline an approximate solution to the problem posed.

The first phase of the problem is that of the mass falling through the height h. At the end of this phase, the original potential energy of the mass has been converted into kinetic energy according to

$$Mgh = \tfrac{1}{2}MV^2$$

where V is the velocity of the mass at impact.

We consider the next phase to be the very short time interval—the impact—during which the mass and the point on the beam that comes in contact with the mass reach the same velocity. This impact phase is assumed to take place in a manner that conserves energy. At the end of the impact phase, the energy of the mass and beam are given by $\tfrac{1}{2}(M + m)v^2$, so that

$$\tfrac{1}{2}(M + m)v^2 = \tfrac{1}{2}MV^2 = Mgh \qquad (a)$$

where v is the common velocity of the mass and beam at the end of the impact phase, and m is the effective mass of the beam.

The third phase is the time period during which the mass and beam come to rest at the position of maximum displacement of the center of the beam. During this phase, we again assume that energy is conserved, so that

$$-Mg\,\Delta + \tfrac{1}{2}k\,\Delta^2 = \tfrac{1}{2}(M + m)v^2 \tag{b}$$

where Δ is the distance through which the beam has displaced. The terms $-Mg\,\Delta$ and $\frac{1}{2}k\,\Delta^2$ represent the final potential energies of the mass and beam, respectively.

We can evaluate the spring constant k used in representing the potential energy of the beam as follows. For a force P applied at the center of a simply supported beam as shown in Fig. 8.70, the corresponding displacement at the point of application of the load is given by

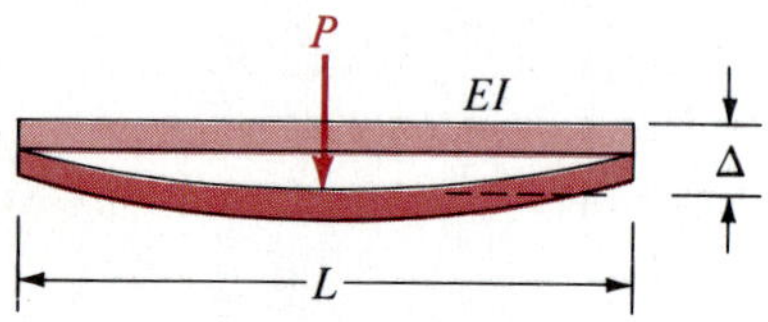

Figure 8.70 Spring constant for a centrally loaded, simply supported beam

$$\Delta = \frac{PL^3}{48EI}$$

or

$$P = \frac{48EI}{L^3}\,\Delta = k\,\Delta$$

where $k = 48EI/L^3$ is the appropriate spring constant. The essence of this result is that the beam acts as a linear spring in response to the force P between the mass and the beam, so that as a result of the assumptions made, the original problem of the mass impacting the beam, indicated in Fig. 8.71a, is equivalent to the physical problem shown in Fig. 8.71b.

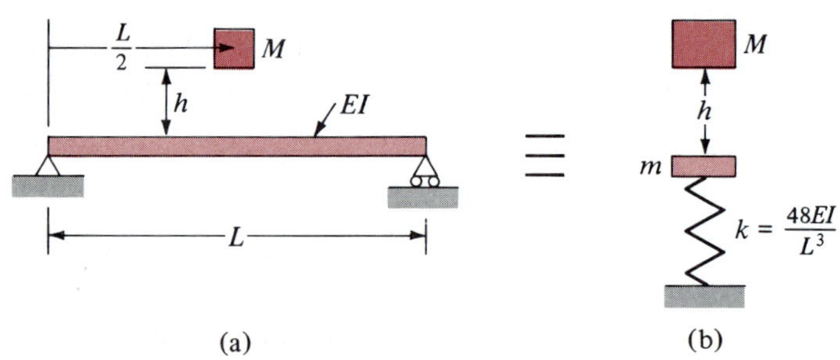

Figure 8.71 Equivalent spring problem

Combining Eqns. (a) and (b) yields

$$Mgh = -Mg\,\Delta + \tfrac{1}{2}k\,\Delta^2 \tag{c}$$

which states quite simply that the original potential energy of the mass, given by Mgh, has been entirely converted into the final potential energy of the mass and spring, given by $-Mg\,\Delta + \frac{1}{2}k\,\Delta^2$. Solving the quadratic Eqn. (c) for Δ yields

$$\Delta = \delta_{ST}\left(1 + \sqrt{1 + 2\frac{h}{\delta_{ST}}}\right) \tag{d}$$

where δ_{ST} is the corresponding static displacement given by $\delta_{ST} = Mg/k$. Note that as h approaches zero, $\Delta = 2\,\delta_{ST}$; thus, for $h = \phi$, the maximum displacement is twice the corresponding static displacement.

There are two distinct situations which must be compared and understood. First, consider the case where the mass M, acted upon by the force Mg against the corresponding gravity force, is placed in contact with the beam, with the applied force Mg then gradually reduced, thus producing a static load of Mg and its corresponding static displacement $\delta_{ST} = Mg/k$. Second, place the mass M, again acted upon by the force Mg against the corresponding gravity force, against the beam, and then suddenly release the supporting force Mg. The resulting displacement from Eqn. (d) is given by $\Delta = 2\,\delta_{ST}$, showing the dynamic effect of the sudden application of the weight. The dynamic load associated with the sudden application of the weight is $2Mg$, resulting in stresses in the beam that are twice those resulting from the corresponding static load. In the general case, where h is not zero, the dynamic load is given by

$$F = k\,\Delta = k\,\delta_{ST}\left(1 + \sqrt{1 + 2\frac{h}{\delta_{ST}}}\right) = k\,\delta_{ST}\,DLF$$

where

$$DLF = \left(1 + \sqrt{1 + 2\frac{h}{\delta_{ST}}}\right)$$

is called the **dynamic load factor.** In terms of this dynamic load factor, the bending stresses existing in the beam at the instant of maximum displacement are given by

$$\sigma = \frac{Mc}{I} = DLF\,\frac{(MgL/4)c}{I}$$

Example 8.23 A movable crane that operates in a foundry-machine shop is shown in Fig. 8.72. The operator misjudges and impacts the stops at a speed of 1 ft/sec.

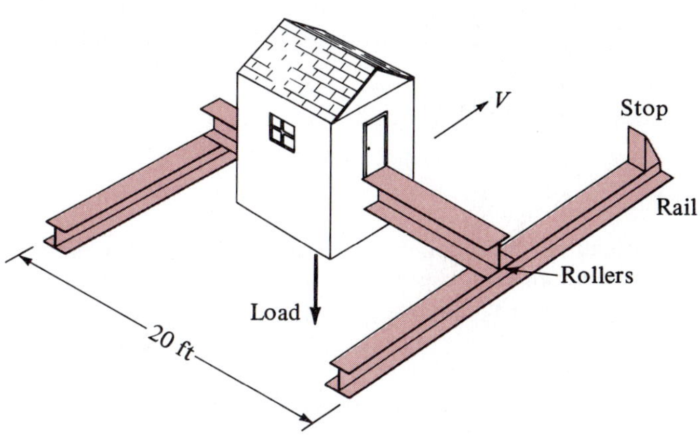

Figure 8.72 Movable crane subjected to dynamic loading

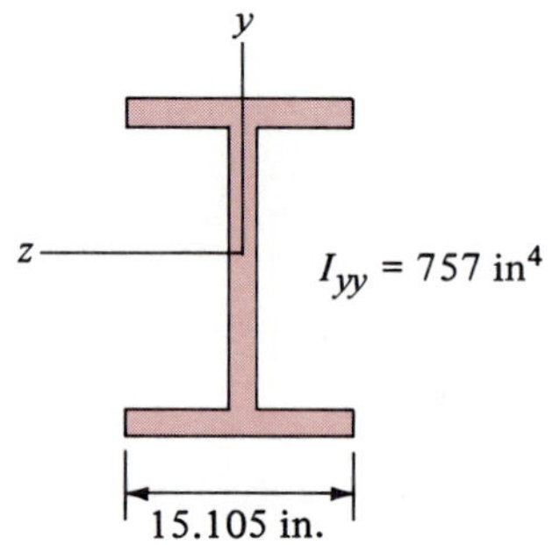

Figure 8.73 Properties of a W 30 × 211 section

Estimate the resulting additional bending stresses in the beam. The crane and its contents weigh 4 tons and the beam is a W 30 × 211 which has the properties shown in Fig. 8.73.

Solution The basic assumption is, again, that energy is conserved during impact with the stops. The kinetic energy of the crane is then converted into potential energy of the simply supported beam, assumed to act as a linear spring with stiffness $k = 48EI/L^3$, as in the previous example. For this application, the total weight of the beam is 20 ft (211 lbf/ft) = 2.11 tons, which cannot be neglected in computing the kinetic energy. For a simply supported beam it is customary to include half the mass of the beam in the kinetic energy. With M the mass of the crane, m the mass of the beam, v the initial speed, k the beam spring constant, and Δ the maximum horizontal displacement of the beam, conservation of energy equation results in

$$\tfrac{1}{2}\left(M + \frac{m}{2}\right) v^2 = \tfrac{1}{2}k\,\Delta^2$$

from which

$$\Delta = v\sqrt{\frac{M + \frac{m}{2}}{k}}$$

The corresponding dynamic load is given by

$$F = k\,\Delta = v\sqrt{k\left(M + \frac{m}{2}\right)}$$

The moment for bending about the y-axis is $M = FL/4$, resulting in

$$\sigma = \frac{vLc\sqrt{k\left(M + \frac{m}{2}\right)}}{4I_{yy}}$$

Substituting $v = 12$ in./sec, $L = 240$ in., $c = 7.553$ in., $M + m/2 = 26.14$ lbf-sec^2/in., and $E = 30 \times 10^6$ psi yields $\sigma = 9.28$ ksi, which could add significantly to the bending stresses already present due to the vertical loading. It would be wise to include the possibility of such an accidental occurrence in the design of the horizontal beam.

PROBLEMS

8.187 An automobile weighing 3500 lbf and traveling at 1 mph impacts a massive concrete abutment as shown. Assuming that the steel bumper can be modeled as simply supported between points A and B, determine the maximum bending stress in the bumper. Neglect the mass of the beam and take bending to be about the z-axis.

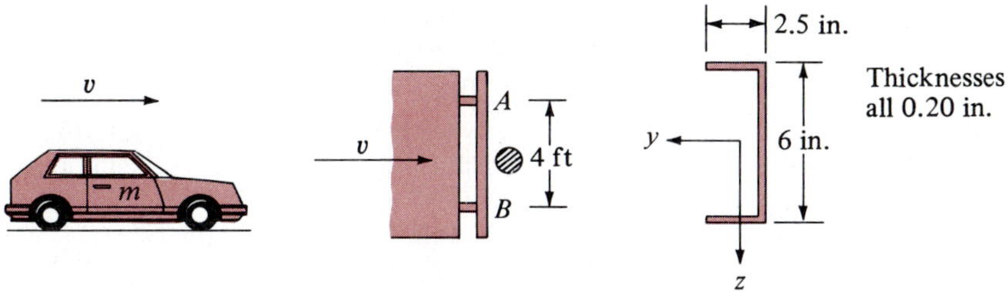

Ps. 8.187 and 8.188

8.188 Repeat the previous problem if the beam is assumed to be clamped at A and B.

8.189 A 4-kg mass is dropped from a height of 600 mm onto the simply supported steel beam as shown. Determine *(a)* the maximum deflection of the beam, and *(b)* the maximum bending stress in the beam. Neglect the mass of the beam. Take $L = 1.0$ m and $E = 70$ GPa.

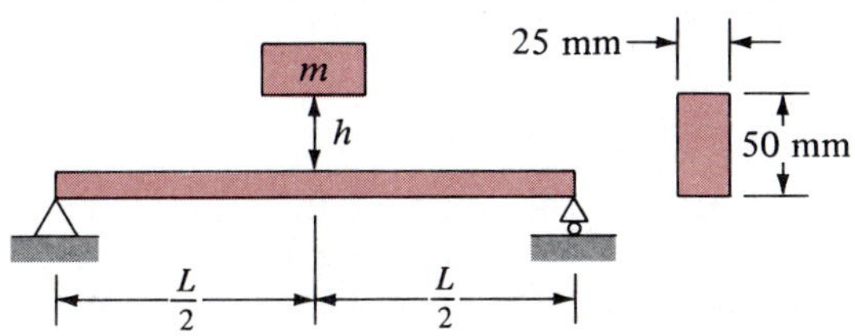

P. 8.189

8.190 A 40-lb mass moving at 5 ft/sec impacts the overhang beam as indicated. Determine *(a)* the maximum deflection of the beam, and *(b)* the maximum bending stress in the beam. Do not include any of the mass of the beam in your formulation. Take $L = 2$ ft and $E = 29 \times 10^6$ psi.

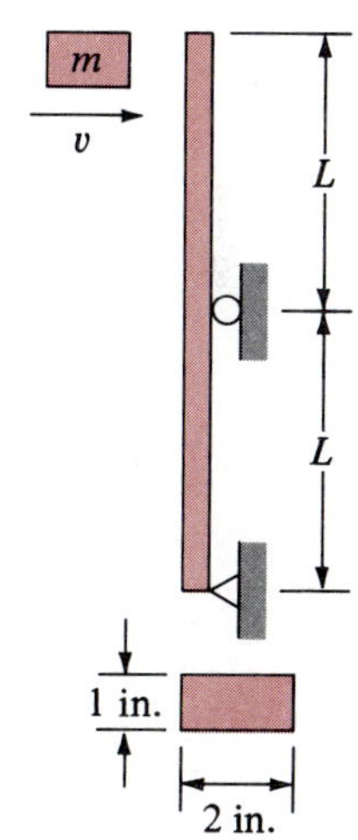

P. 8.190

8.191 A 20-kg mass is dropped from a height of 1.2 m onto the clamped beam as shown. Determine *(a)* the maximum deflection of the beam, and *(b)* the maximum bending stress in the beam. Neglect the mass of the beam. Take $L = 3$ m, $E = 70$ GPa, and the beam to be rectangular with $b = 20$ mm and $h = 50$ mm.

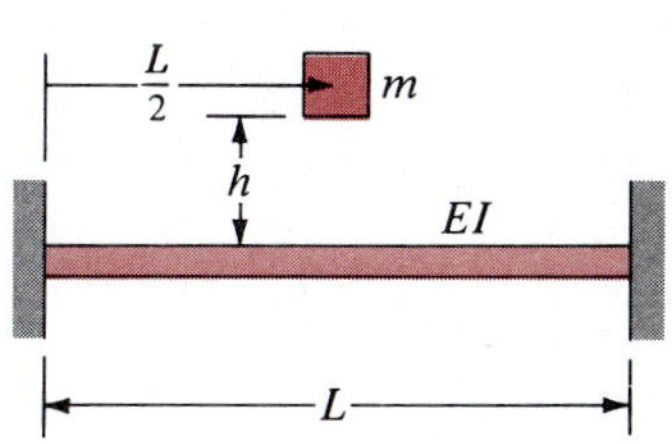

P. 8.191

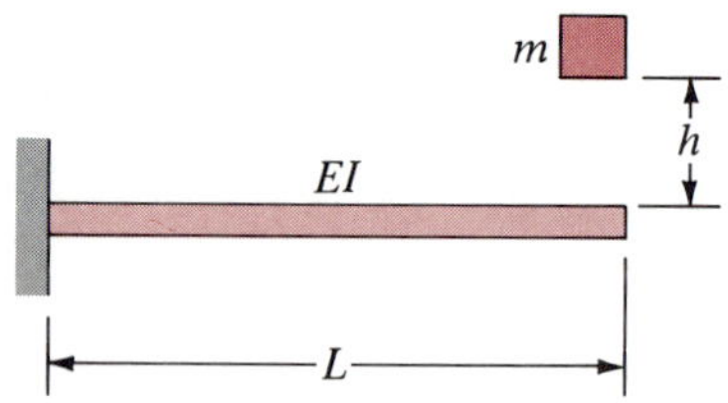

P. 8.192

8.192 A mass m is dropped onto the end of a cantilever as shown. Develop expressions for *(a)* the maximum deflection of the beam, and *(b)* the maximum bending stress in the beam. Neglect the mass of the beam.

8.193 During vibration at liftoff, a gold wire in an electric component that is located in a spacecraft experiences a constant 100-g acceleration perpendicular to its length. If the diameter of the wire is 25 μm and the wire is clamped at both ends, what is the maximum length L of the wire if the bending stress is not to exceed 70 MPa? Take the mass density of the wire to be 19,300 kg/m^3.

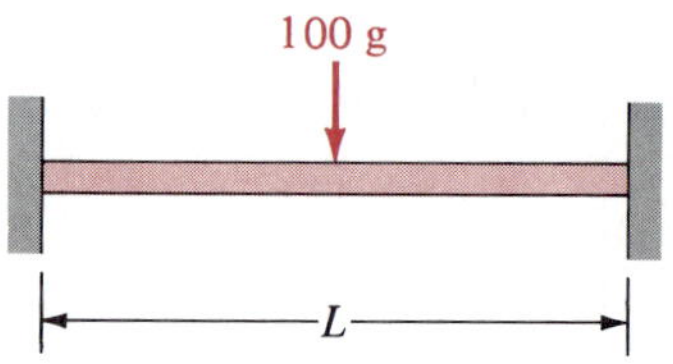

P. 8.193

8.10 CLOSURE AND OVERVIEW OF THE BENDING PROBLEM

We have demonstrated in this chapter that the elementary theory of beams can be used to assist in answering the stiffness problem. Questions of stiffness, which are answered by integrating in one form or another, either of $EIv'' = M$ or $EIv'''' = -w$, are based on these assumptions:

1. The axis of the bar, usually chosen to be the x-axis, is straight. Cross-sectional dimensions are small compared to the length of the bar and vary slowly along the axis of the bar. The shape of the cross section is such that the y-axis can be chosen as a plane of symmetry.
2. Loading consists of forces applied in the y-direction and lying in the plane of symmetry, and of moments about the z-axis.
3. Slopes associated with the transverse displacements are small.
4. The beam is composed of material behaving in a linearly elastic fashion.

If any of these assumptions is not satisfied, the elementary theory cannot be expected to predict the bending and shear stresses with any degree of accuracy. It is the responsibility of the user to ensure that, when all the necessary assumptions are not satisfied, each exception is taken into account in a rational manner.

For a statically determinate problem, is it only necessary to draw free-body diagrams and satisfy equilibrium to determine the reactions and thus the shear and moment over the length of the beam. With the moment M and the shear V known, the bending stresses can be determined using the flexure equation $\sigma_x = -My/I$, and the transverse shear stresses can be determined using $\tau = VQ/It$, thus answering the strength question. Independently, the moment can be used in $EIv'' = M$ to determine the displacements and answer the stiffness question. Thus, for a statically determinate problem, the strength and stiffness questions can be answered independently. These ideas are illustrated in the flowchart of Fig. 8.74.

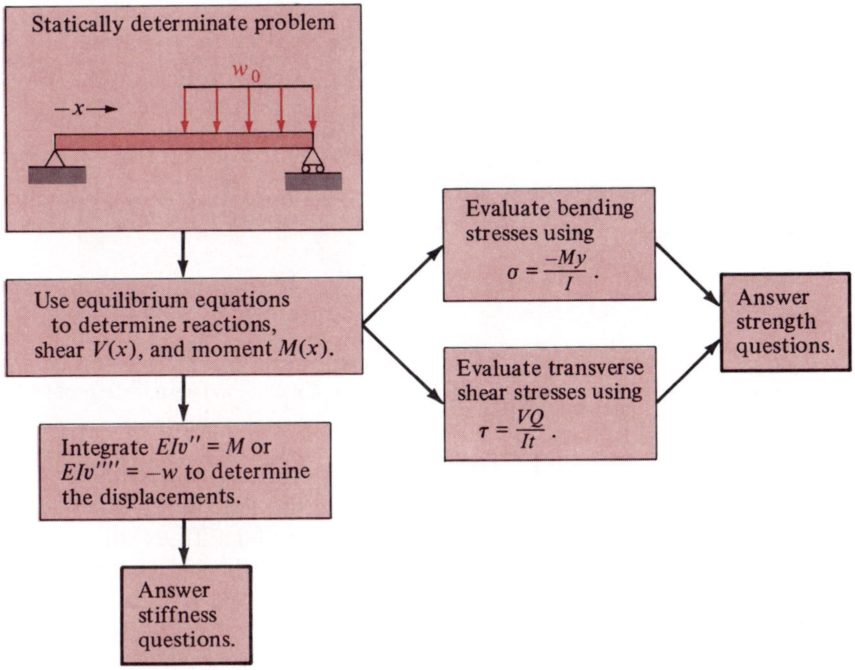

Figure 8.74 Flowchart indicating steps for a statically determinate problem

For a statically indeterminate problem, there are more unknown reactions than equations of equilibrium and it is necessary to provide the additional equations by considering the deformations. The total number of equilibrium and deformation equations must be solved for the unknowns, after which both the strength and stiffness questions can be answered. A flowchart indicating these steps is shown in Fig. 8.75.

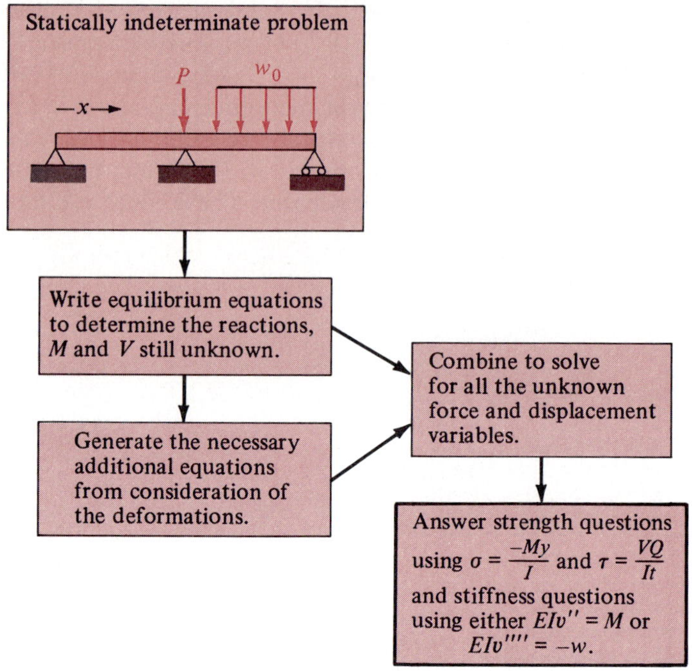

Figure 8.75 Flowchart indicating steps for a statically indeterminate problem

Although the flow of the calculations for statically determinate and indeterminate problems is different, exactly the same ideas, equations, and techniques are used to answer the strength and stiffness questions. This is helpful to keep in mind when setting up either of the two types of problems, so as to keep in focus just where one is located in the process of assessing strength and stiffness.

ANALYSIS AND DESIGN PROBLEMS

AD 8.1 The cantilever beam is to carry a distributed load of 100 lbf/in. as shown. With $\sigma_Y = 18$ ksi, determine a suitable rectangular cross section using the maximum shear stress criterion. Take as an additional constraint that the maximum transverse displacement must not exceed 20% the height of the beam.

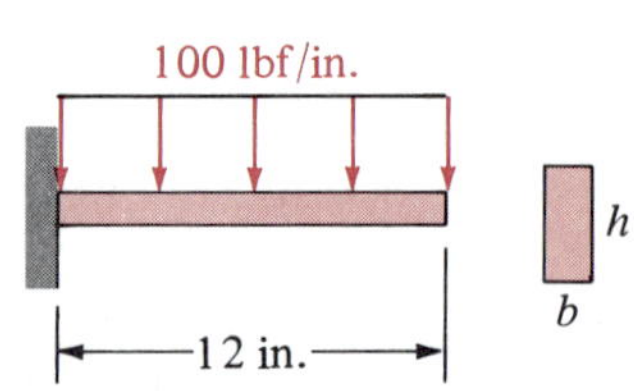

AD 8.1

AD 8.2 The overhang beam is to support the loads shown. Given that $\sigma_{ult} =$ 180 MPa, select a suitable W section for the beam using the maximum normal stress criterion with a factor of safety of 2.0. Take as an additional constraint that the maximum transverse displacement must not exceed 15% the height of the beam.

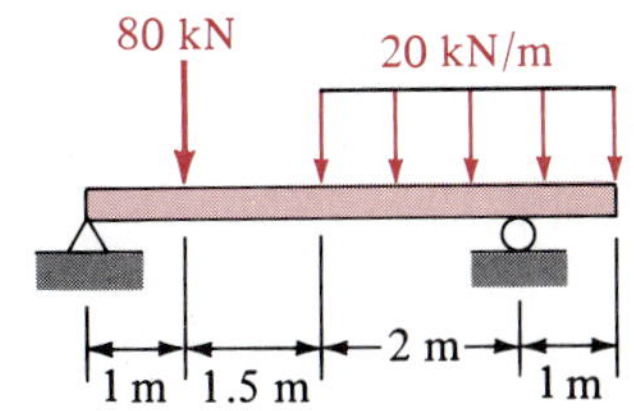

AD 8.2

AD 8.3 A circular pipe is to be used as a cantilever beam loaded as shown. If $\sigma_Y = 14$ ksi and $E = 29 \times 10^6$ psi, select a steel pipe that will suffice using the maximum shear stress criterion. Take as an additional condition that the maximum deflection is not to exceed 5% of the outside diameter of the pipe.

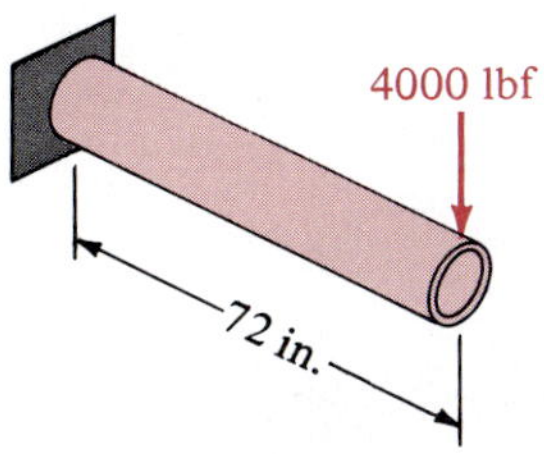

AD 8.3

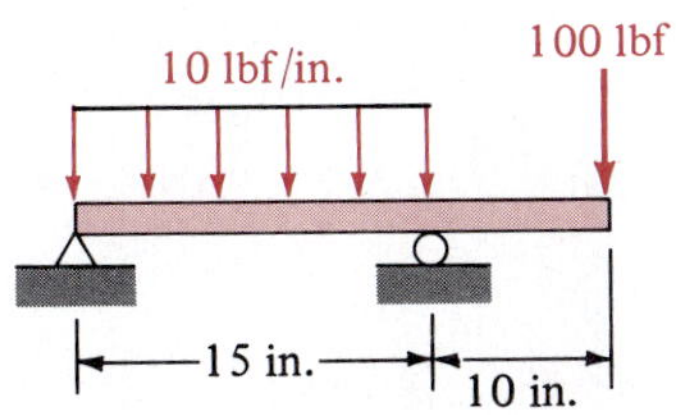

AD 8.4

AD 8.4 The beam shown is to be constructed from a hollow square section whose wall thickness is one-tenth the outside dimension and made from an aluminum with a yield stress in tension of 22 ksi and a Young's modulus of 10×10^6 psi. Using the maximum shear stress criterion, select the outside diameter of the square. Take as an additional design constraint that the transverse displacement is to be limited to 0.2 in.

AD 8.5 A beam is to support the 70 lbf/ft loading over a 40-ft span as shown and not deflect more than 1.5 in. If the beam is to be formed by gluing available 4-in. by 6-in. sections, evolve a suitable design that includes the weight of the beam. Take the allowable bending stress in the wood to be 1.5 ksi, the allowable horizontal shear stress in the wood to be 150 psi, the allowable shear stress in the glue to be 100 psi and the weight density of the wood to be 40 lbf/ft³.

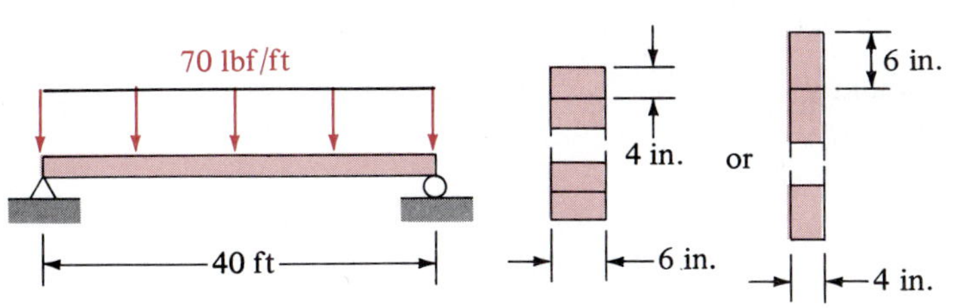

AD 8.5

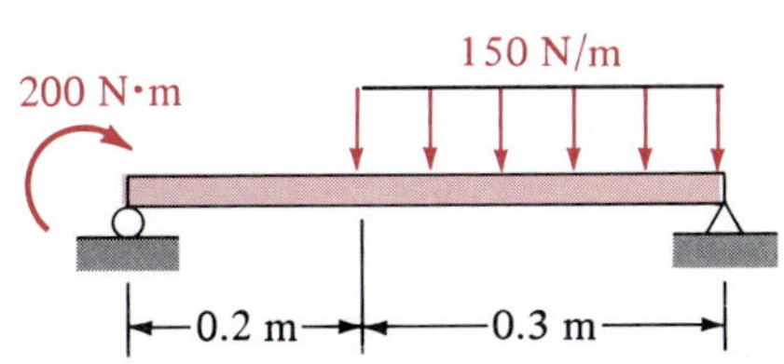

AD 8.6

AD 8.6 Using the maximum normal stress criterion, select a hollow circular section that will carry the loading shown. The material has a tensile ultimate strength of 70 MPa. Assume that the maximum deflection is limited to 10 mm.

AD 8.7 The system of beams shown is intended to support a deck that results in a 1750-lbf/ft loading on each of the 15-ft long beams. Each of the beams rests on the surrounding footings and on the main beam of length 24 ft. If the yield stress for all members is $\sigma_Y = 15$ ksi and $E = 29 \times 10^6$ psi, select W sections for *(a)* the main beam, and *(b)* each of the identical supporting beams. Also assume that the central deflection of the main beam is limited to 1 in. Use the maximum shear stress criterion.

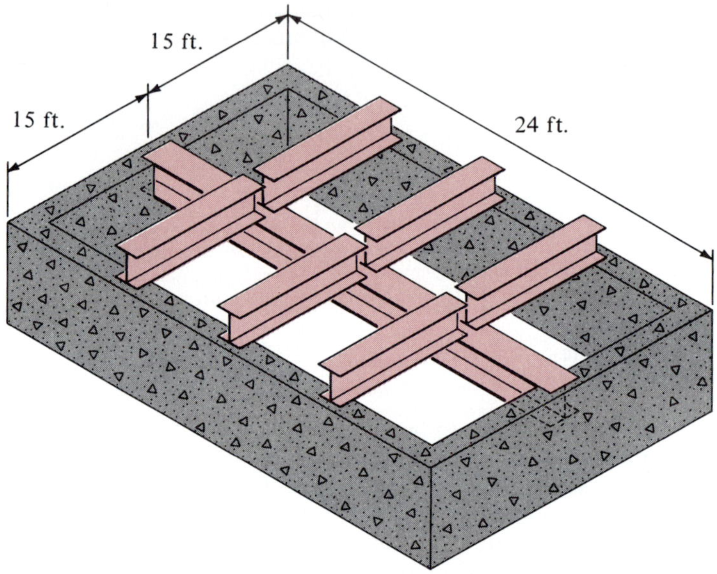

AD 8.7

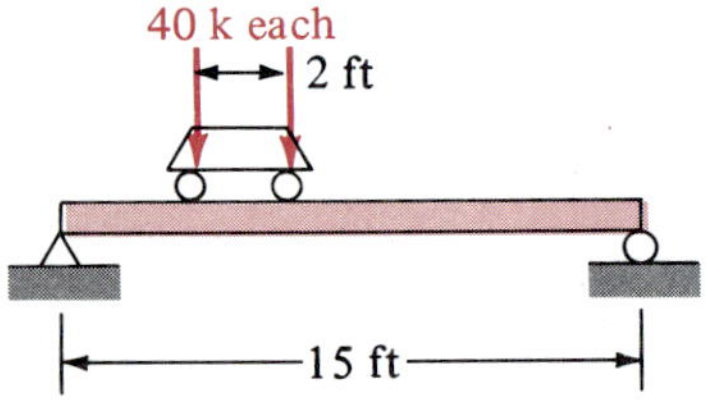

AD 8.8

AD 8.8 The simply supported beam shown is to support a variable position load consisting of two 40-kip loads separated by the 2-ft distance as shown. Given that $\sigma_Y = 9$ ksi, select a suitable W section for the beam using the Hencky-von Mises criterion and assuming that the maximum deflection is limited to 0.7 in.

AD 8.9 The 3600-kg load moves along the 5-m length of the beam as shown. If $\sigma_Y = 60$ MPa for the beam and $\sigma_Y = 70$ MPa for the tension member AB, select an S section and a required diameter for the rod of the tension member. Also require that the maximum displacement under the load is not to exceed 20 mm. Take E for both the beam and rod as 200 MPa and use the maximum shear stress criterion.

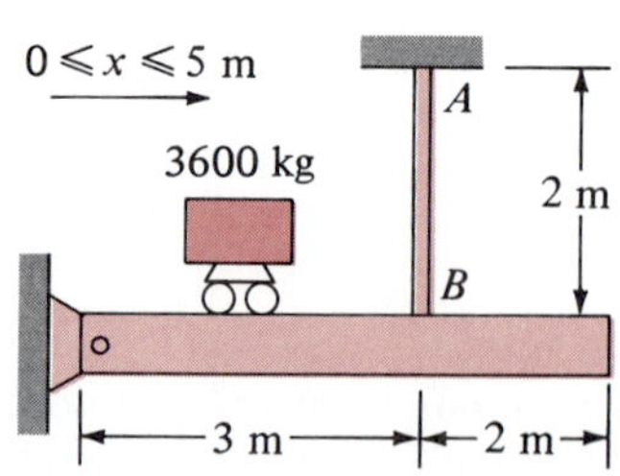

AD 8.9

AD 8.10 For the two beams shown, there is contact but no contact force at zero load. Use the maximum shear stress criterion to select S sections for each beam. For both beams, take $\sigma_Y = 110$ MPa and $E = 200$ GPa. Assume that the central deflection must not exceed 1% of the length of beam AB. Also attempt to satisfy the criterion that the maximum bending stresses in the two beams are the same.

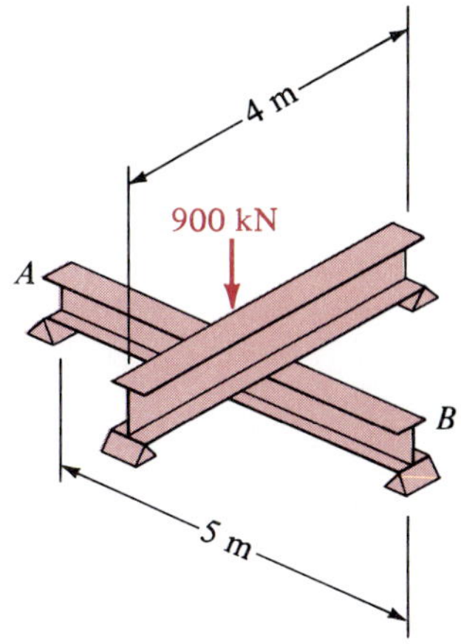

AD 8.10

AD 8.11 A beam is to be used to support the loads shown in the figure. Assuming that the material has a tensile yield strength of 22 ksi and that the Hencky-von Mises criterion is to be used, select an appropriate W section. Also assume that the displacement at the right end is not to exceed 0.025 m. Take $E = 30 \times 10^6$ psi.

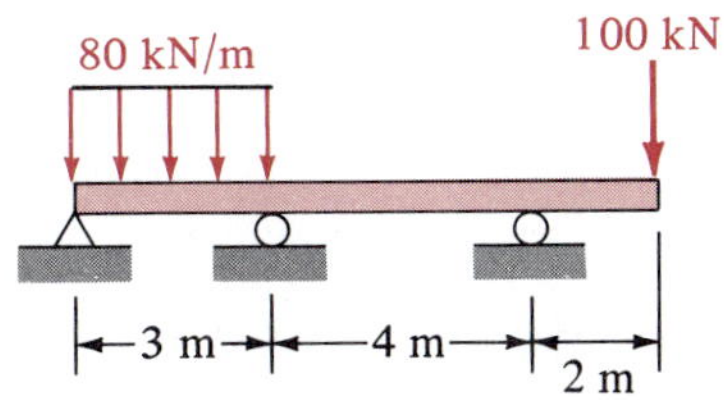

AD 8.11

AD 8.12 A hollow circular section with a wall thickness that is 20% of the outside diameter is to carry the loads shown and not deflect more than 2 in. Assuming $\sigma_Y = 19$ ksi and $E = 30 \times 10^6$ psi, determine the required outside diameter on the basis of the Hencky-von Mises criterion.

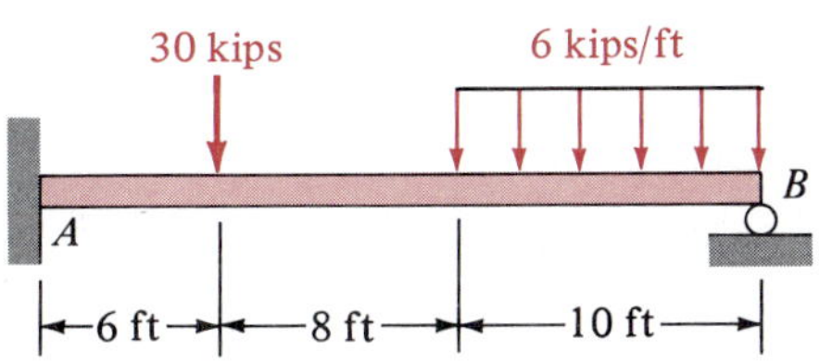

AD 8.12

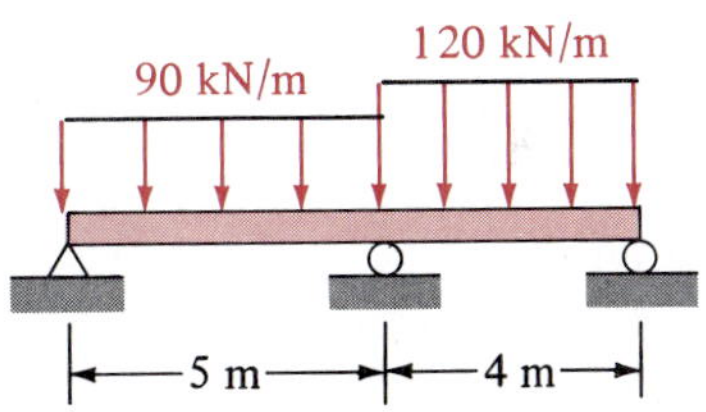

AD 8.13

AD 8.13 A beam is to be designed to support the loading shown. Given that $\sigma_Y = 110$ MPa, use the maximum shear stress criterion to select an appropriate S section. Assume that the deflection in each span is limited to 25 mm.

AD 8.14 Select a W section to carry the loading shown if the beam is composed of a material for which $\sigma_{ult} = 190$ MPa and the maximum normal stress criterion is to be used. The deflection in each span is limited to 15 mm.

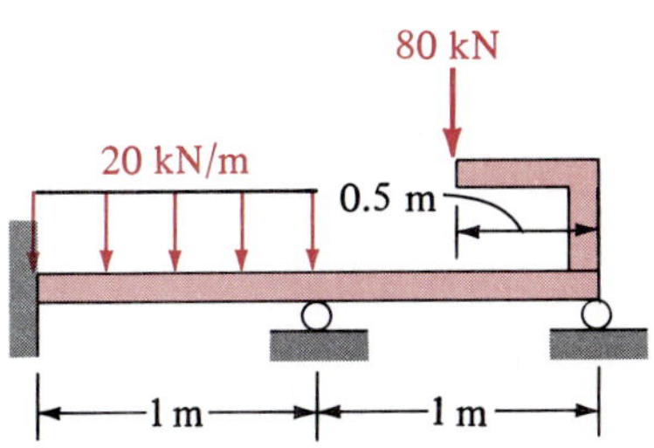

AD 8.14

CHAPTER 9 COMBINED LOADINGS

CHAPTER OUTLINE

9.1 INTRODUCTION

Chapters 5, 6, 7, and 8 were devoted to developing and applying the theories to answer strength and stiffness questions about axial, torsional, and bending deformation problems. Generally, the member we considered for each of the three types of problems was similar: a straight bar with at most a slowly varying cross section. For the axial problem, the force transmitted is axial, that is, along the x-axis of the bar. The corresponding deformations are also primarily axial, with the axis of the bar remaining straight. For the torsional problem, the force resultant transmitted is a torque parallel to the axis of the bar. The associated deformation is a twist about the axis of the bar, with the axis of the bar again remaining straight. For the bending problem, the force system transmitted is generally a transverse force and a couple perpendicular to the axis of the bar. The deformation consists of the x-axis bending into the so-called elastic curve; thus, the axis of the bar is no longer straight.

In most applications involving a straight bar, the loading is more complex than just a single force or torque parallel to the axis of the bar, or a transverse force and/or normal moment. Several typical examples are pictured in Fig. 9.1.

Fig. 9.1a shows a free-body diagram of an aircraft wheel and a portion of the landing strut during landing. The forces between the wheel and landing surface are N, the normal force, and f, the traction force. These must be balanced by the internal shear force V, the internal axial force F, and the internal moment M, resulting in combined axial and bending loads. Fig. 9.1b pictures the forces that exist in an ordinary clamp, where it is clear from the free-body diagram that the

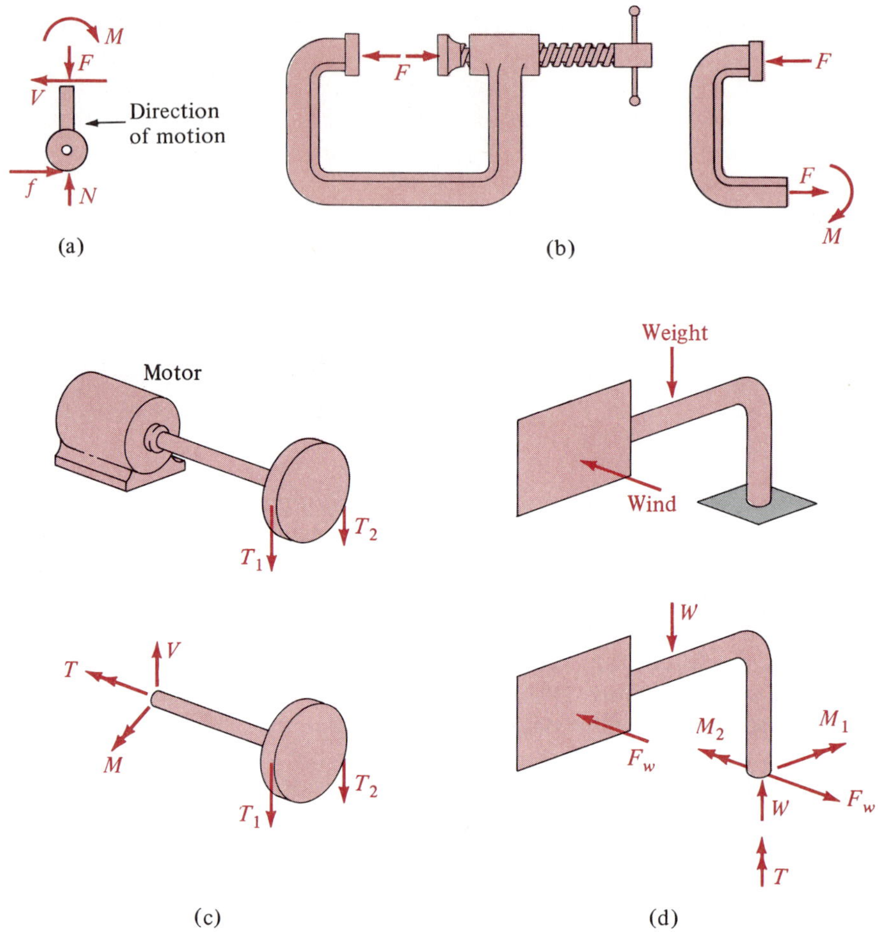

Figure 9.1 Typical examples of combined loading

internal forces consist of the axial force F and the bending moment M, that is, combined axial and bending loads. Fig. 9.1c indicates a pulley and shaft where the internal forces consist of the bending moment M, the shear force V, and the torque T. This type of loading is commonly referred to as combined bending and torsion. Last, Fig. 9.1d shows a common application where a street sign is acted upon by its own weight plus additional forces arising from the wind. The free-body diagram indicates that in the vertical portion of the sign post, there is the axial load W, the shear load F, the two moments M_1 and M_2, and the torque T: bending about two axes, shear in one direction, torsion, and axial loads. This is often referred to as combined axial, bending, and torsional loading.

In this chapter we will discuss in detail questions of strength for a variety of different combinations of loading on straight bars, using simple superposition ideas. We will also discuss the stresses that result from internal pressures in thin-walled vessels with cylindrical and spherical geometries.

9.2 COMBINED AXIAL AND BENDING LOADS

Structural members are frequently subjected to the action of loads that result in both axial and bending deformations. The manner in which the loads are applied determines whether the deformations take place in a single plane or in two planes. In this section we will develop and discuss in detail the equations necessary to handle both of these situations.

Single-Axis Bending Consider a beam subjected to a force F applied in the xy-plane and inclined to the x-axis as shown in Fig. 9.2a.

We will assume that the beam is straight, that the x-axis coincides with the centroid of the cross section, and that the cross section possesses a plane of symmetry, the xy-plane, as shown in Fig. 9.2b. The force F is resolved into the components P and Q as shown in Fig. 9.3a.

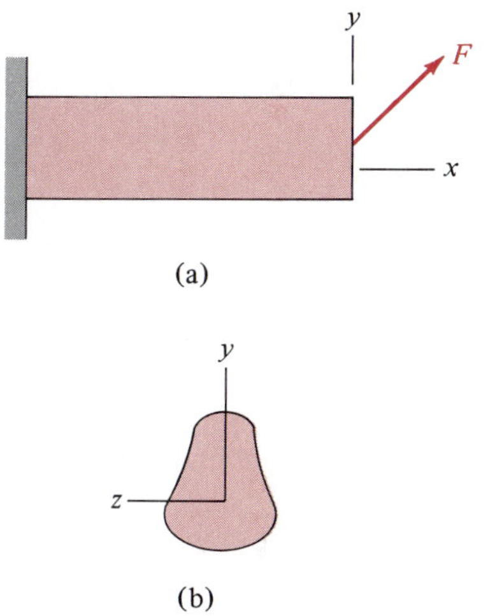

Figure 9.2 Loading that produces both axial and bending deformations

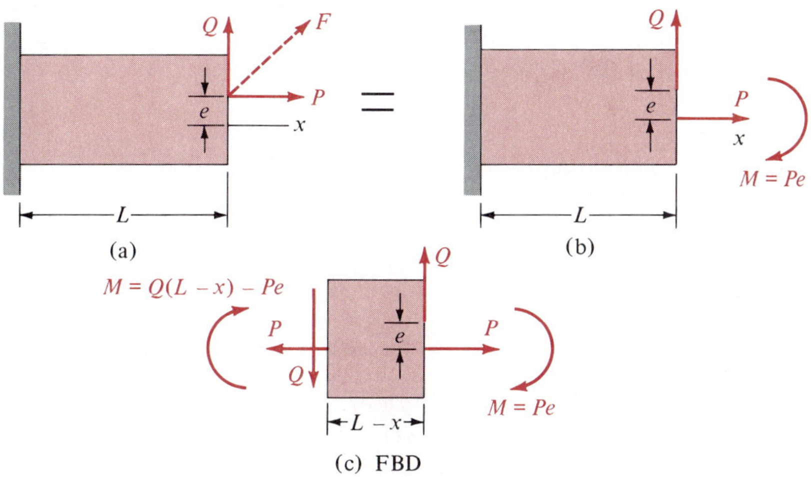

Figure 9.3 Equivalent force system

In accordance with principles of statics, the force P is then moved so that its line of action coincides with the x-axis, making it necessary to provide the compensating couple $M = Pe$ as shown in Fig. 9.3b. The two sets of forces shown in Figs. 9.3a and b are statically equivalent. Fig. 9.3c shows clearly that at any section located a distance x from the left end of the beam, equilibrium requires that an axial load P, a shear force V, and a moment $M = Q(L - x) - Pe$ be transmitted.

The basic idea we use to evaluate the stresses caused by the simultaneous action of the axial and shear forces and the bending moment is *superposition*. We recall the assumptions for geometry and loading for the axial problem: the bar must be straight, and the load must be applied through the centroid. These two assumptions are clearly satisfied in connection with the axial load P, enabling us to state that the stresses resulting from the axial load P are

$$\sigma_x = \frac{P}{A} \tag{a}$$

For the bending problem, the geometry assumptions are that the bar be straight and that the cross section possess an axis of symmetry. The loading assumptions for the bending problem are that forces are applied in the y-direction in the plane of symmetry, with moments applied about the z-axis. As these assumptions are all clearly satisfied with respect to V and M, the bending stresses can be expressed as

$$\sigma_x = -\frac{My}{I} \tag{b}$$

Superposition of these two normal stresses yields

$$\sigma_x = \frac{P}{A} - \frac{My}{I} \tag{9.1}$$

and is indicated in Fig. 9.4.

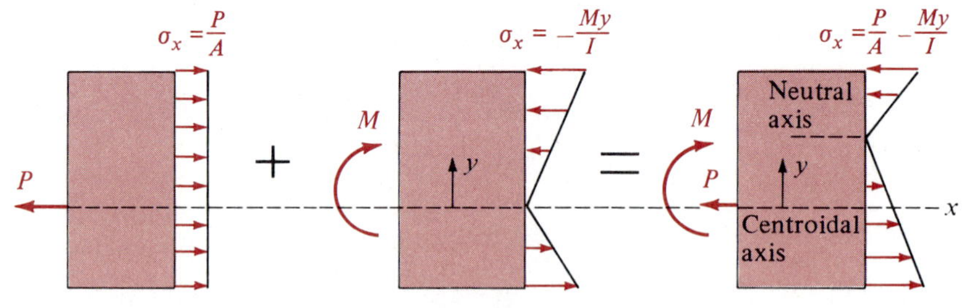

Figure 9.4 Superposition of axial and bending stresses

Note that the location of the neutral axis for combined axial and bending is offset from the centroidal axis due to the presence of the uniform stress $\sigma_x = P/A$; thus, the neutral and centroidal axes no longer coincide as is the case for pure bending. Several possibilities for the stress distribution, contingent upon the signs and magnitudes of P and M, are shown in Fig. 9.5.

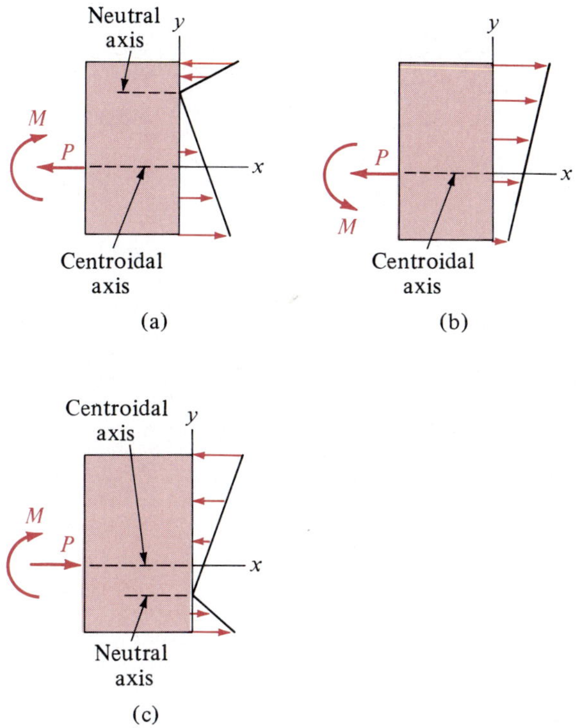

Figure 9.5 Possible axial-stress distributions

In each case the stress distribution is clearly linear, with the relative positions of the neutral and centroidal axes dependent upon P and M. Fig. 9.5a depicts a situation where the magnitude of the axial load P is sufficient to shift the neutral axis to a point substantially above the centroidal axis. Fig. 9.5b shows a case where there is no neutral axis at all due to the relatively large value of P. Fig. 9.5c indicates a case where the neutral axis has been shifted to a position below the centroidal axis due to a compressive axial load. As long as the maximum value of stress produced by the combined loading is within the elastic range of the material in question, Eqn. (9.1) will be used for computing stresses due to the combined action of axial and bending loads when the assumptions regarding geometry and loading are satisfied.

The superposition process is seen from a slightly different perspective in Fig. 9.6.

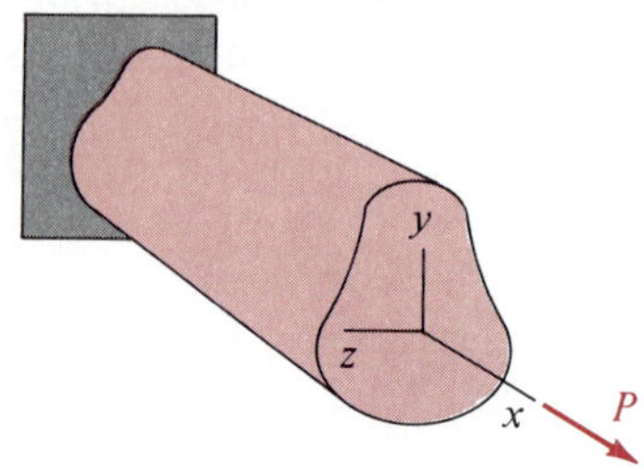

1. Straight bar.
2. No sudden changes in cross section.
3. Resultant of external forces is a single force parallel to the x-axis and passing through the centroid.

$\sigma = \frac{P}{A}$ and $AEu' = P$

+

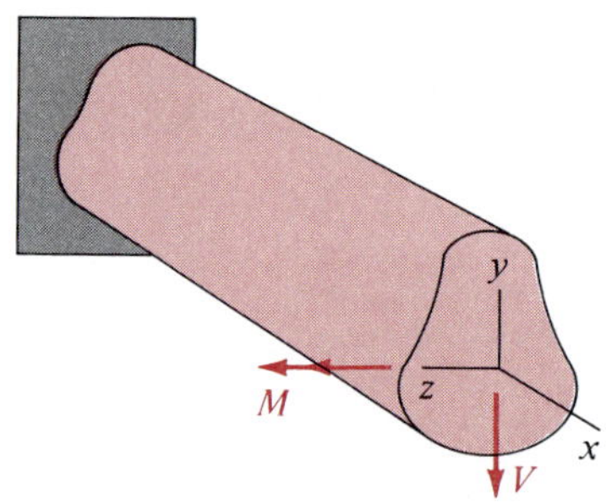

1. Straight bar, cross section symmetric about xy-plane.
2. No sudden changes in cross section.
3. Resultant of external forces is a force in the y-direction lying in the xy-plane and a moment about the z-axis.

$\sigma = -\frac{My}{I}$, $\tau = \frac{VQ}{It}$, and $EIv'' = M$

=

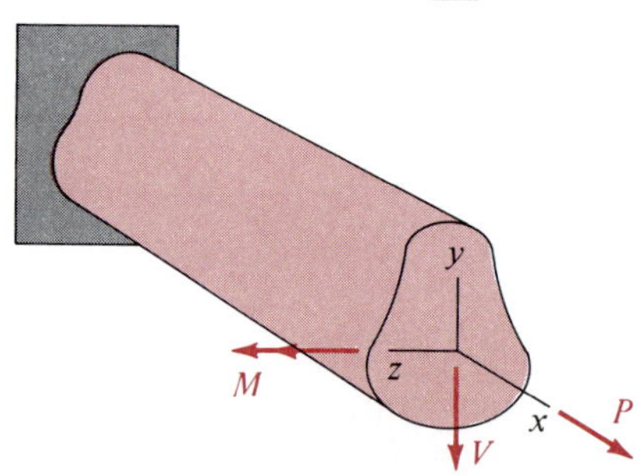

1. Straight bar, cross section symmetric about xy-plane.
2. No sudden changes in cross section.
3. Resultant of external forces is a force in the xy-plane and a moment about the z-axis.

$\sigma = \frac{P}{A} - \frac{My}{I}$, $\tau = \frac{VQ}{It}$,

$EIv'' = M$, and $P = AEu'$

Figure 9.6 Superposition of axial, bending, and transverse shear

Fig. 9.6a shows the geometry and loading of the axial deformation problem along with a list of the corresponding assumptions. Fig. 9.6b shows the corresponding information for the bending deformation problem. The assumptions shown in Fig. 9.6c include all of those present for the axial and bending problems. It is then assumed that the normal stresses can be combined as

$$\sigma_x = \frac{P}{A} - \frac{My}{I}$$

with the axial displacements determined by integrating $AEu' = P$ and the transverse displacements obtained by integrating $EIv'' = M$. The shear stresses are computed using

$$\tau = \frac{VQ}{It}$$

Example 9.1 A rectangular-cross-section beam is loaded as shown in Fig. 9.7a. Determine the normal-stress distribution.

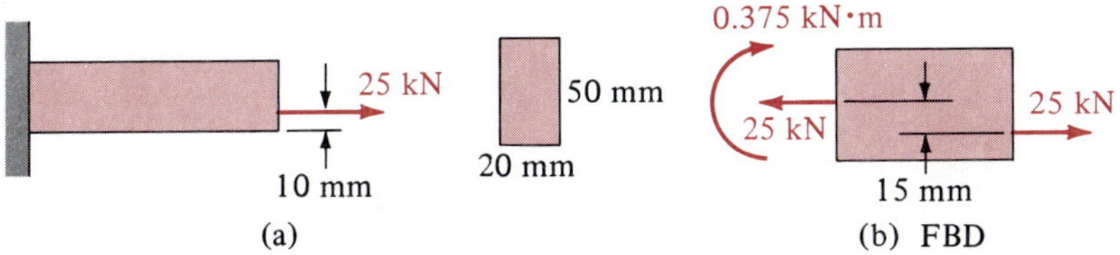

Figure 9.7 Eccentric loading of a beam

Solution From the free-body diagram drawn in Fig. 9.7b, we see that the effect of the load being applied in an eccentric manner is that there must be an internal bending moment of $M = 0.375$ kN·m for equilibrium. With $P = 25$ kN, $M = 0.375$ kN·m, $I = (0.02 \text{ m})(0.05 \text{ m})^3/12 = 2.083 \times 10^{-7} \text{ m}^4$, Eqn. (9.1) is used to compute

$$\sigma_x = \frac{25 \text{ kN}}{(0.02 \text{ m})(0.05 \text{ m})} - \frac{(0.375 \text{ kN·m})y}{2.083 \times 10^{-7} \text{ m}^4} = (25 - 1800y) \text{ MPa}$$

where y is measured in meters. The location of the neutral axis is determined by setting $\sigma_x = 0$ from which $y = 25/1800 = 0.0139$ m. The superposition of the stress is shown in Fig. 9.8.

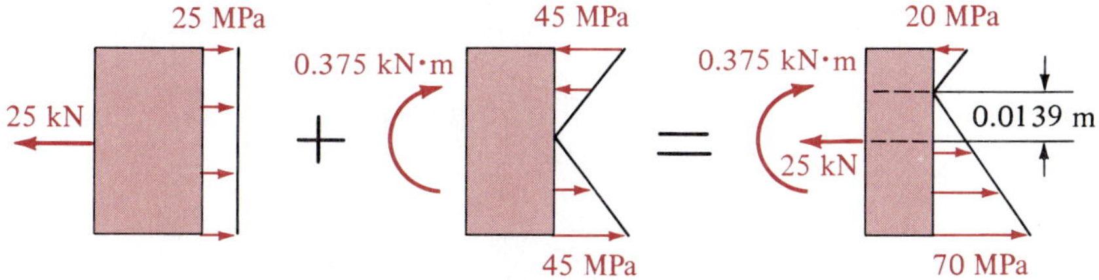

Figure 9.8 Superposition and the combined-stress distribution

Because the axial load is applied eccentrically, there results a substantial increase in the maximum stress from 25 MPa for the pure axial problem to 70 MPa for the combined loading problem.

Example 9.2 A 50-mm-thick aluminum plate hangs under its own weight as shown in Fig. 9.9a. Determine the stress distribution at the upper end of the plate. Take the mass density ρ of aluminum to be 2720 kg/m^3.

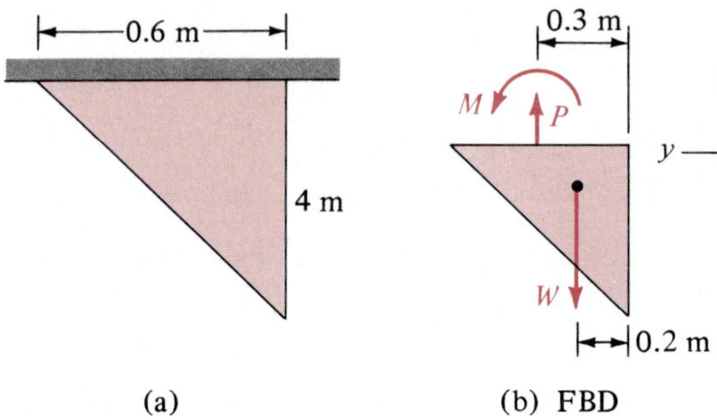

Figure 9.9 Stress distribution in a triangular steel plate

Solution The free-body diagram of the plate is shown in Fig. 9.9b. Equilibrium in the x-direction yields $P = W$, and moments about the z-axis passing through the centroid yield $M = W(0.1 \text{ m})$. The volume of the plate is 1.2 m^2 (0.05 m) = 0.06 m^3. The weight of the plate is given by ρg(volume) = (2720 kg/m^3) (9.81 m/sec^2)(0.06 m^3) = 1.60 kN. With $A = 0.03 \text{ m}^2$ and $I = (0.05 \text{ m})(0.6 \text{ m})^3/12 = 9 \times 10^{-4} \text{ m}^4$, the combined stress at the upper surface is

$$\sigma_x = \frac{1.60 \text{ kN}}{0.03 \text{ m}^2} - \frac{(0.16 \text{ kN}\cdot\text{m})y}{9 \times 10^{-4} \text{ m}^4}$$

The maximum stress will occur on the right edge of the plate where $y = -0.3$ m with the value

$$\sigma_x = 106.7 \text{ kN/m}^2 = 0.107 \text{ MPa}$$

The stress at the other edge is zero; thus, the stress distribution appears as in Fig. 9.10. The student should verify that the stress distribution is such that the weight is in equilibrium.

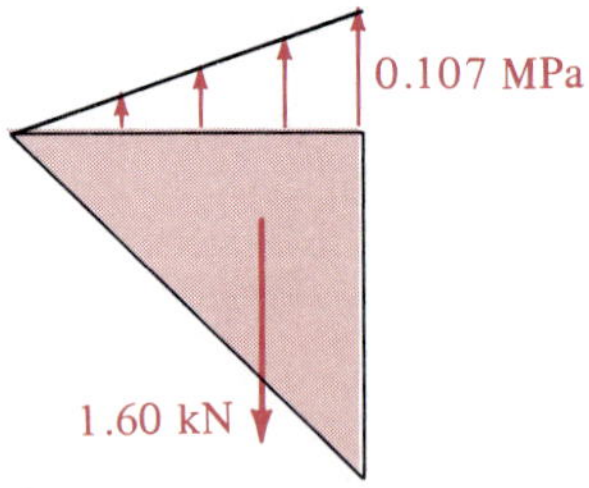

Figure 9.10 Stress distribution in a triangular plate

Example 9.3 A 2-in.-square bar is loaded as shown in Fig. 9.11a. At a section 2 in. from the support, determine *(a)* the normal-stress distribution, and *(b)* the transverse shear stress distribution.

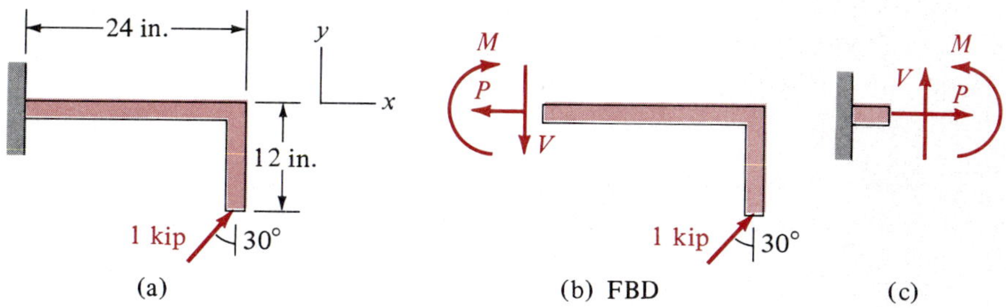

Figure 9.11 Loading of an L-shaped bar

Solution First, note that since the xy-plane is a plane of symmetry, the use of the equations $\sigma_x = P/A - My/I$ and $\tau = VQ/It$ is valid. We can use the free-body diagram of Fig. 9.11b to determine that the internal resultants, also shown in Fig. 9.11c, are $P = 1000$ lbf cos 60° = 500 lbf, $V = 1000$ lbf sin 60° = 866 lbf, and M = (866 × 21 + 500 × 11) in.-lbf = 23,686 in.-lbf. Using Eqn. (9.1), the normal stresses are

$$\sigma = \frac{500 \text{ lbf}}{4 \text{ in}^2} - \frac{23{,}686 \text{ in.-lbf}(y)}{4/3 \text{ in}^4} = (125 - 17{,}765y) \text{ psi}$$

with y measured in inches. The combined normal-stress distribution is shown in Fig. 9.12. The neutral axis is less than 0.01 in. away from the centroidal axis.

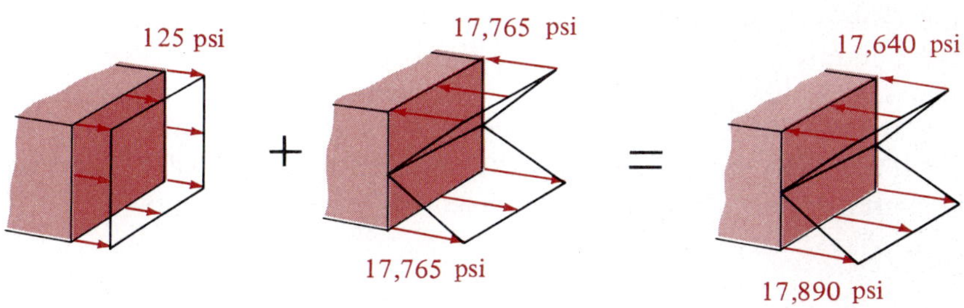

Figure 9.12 Combined normal-stress distribution

For the square cross section, the transverse shear stress distribution is parabolic, with a maximum value of

$$\tau_{\text{max}} = \frac{3V}{2A} = \frac{3(866 \text{ lbf})}{2(4 \text{ in}^2)} = 325 \text{ psi}$$

and is shown in Fig. 9.13.

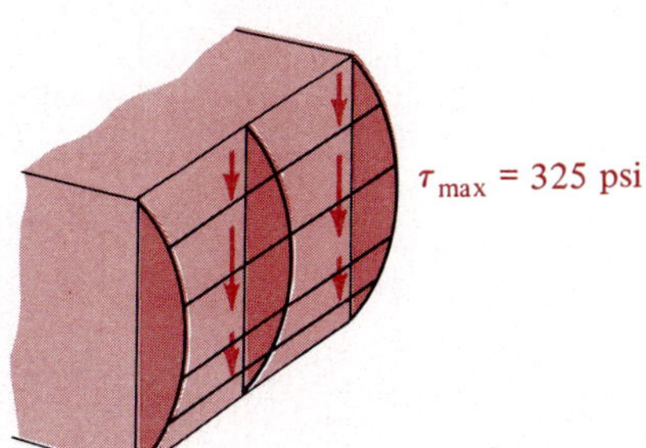

Figure 9.13 Transverse shear-stress distribution

Double-Axis Bending Consider first the case where a beam with two axes of symmetry, as in Fig. 9.14, is simultaneously acted upon by axial and transverse loads.

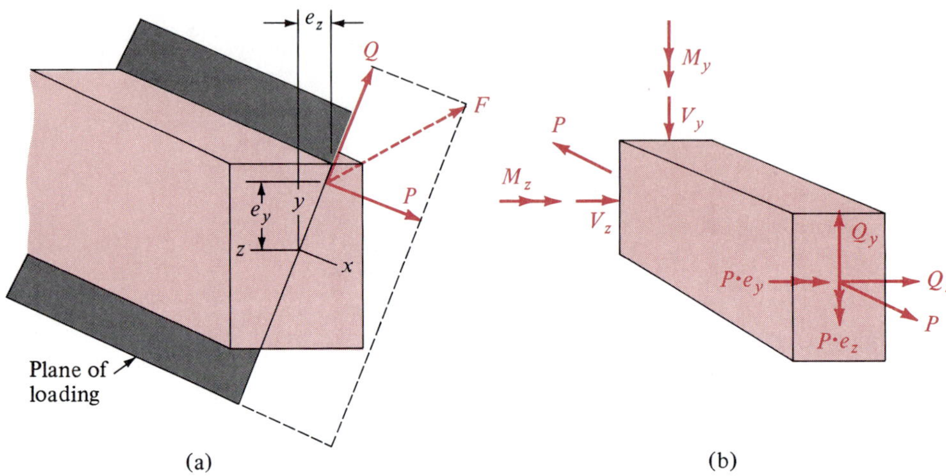

Figure 9.14 Loading for double-axis bending

As shown in Fig. 9.14a, the plane of loading containing F passes through the x-axis of the beam. This is to prevent any moment about the x-axis, with the accompanying additional shear stresses and twisting. We resolve the force F into components P in the axial direction, and Q perpendicular to the x-axis as pictured in Fig. 9.14a. We then move the axial load P to lie along the x-axis, with the necessary moments $-Pe_z$ about the y-axis and $-Pe_y$ about the z-axis, as shown in Fig. 9.14b. The resultants transmitted at a general section along the beam are then the axial force P, the shear forces V_y and V_z, and the moments M_y and M_z. The axial force will result in stresses

$$\sigma_x = \frac{P}{A}$$

the bending moment M_z about the z-axis will produce the bending stresses

$$\sigma_x = -\frac{M_z y}{I_{zz}}$$

with the bending moment M_y about the y-axis producing the bending stresses

$$\sigma_x = \frac{M_y z}{I_{yy}}$$

These individual effects are superposed to yield

$$\sigma_x = \frac{P}{A} - \frac{M_z y}{I_{zz}} + \frac{M_y z}{I_{yy}} \tag{9.2}$$

The process of evaluating the normal stresses consists of drawing the appropriate free-body diagram; determining P, M_z, and M_y; and then substituting into Eqn. (9.2). The stress distribution given by Eqn. (9.2) is linear in both y and z, and typically appears as shown in Fig. 9.15, where the superposition process is pictured.

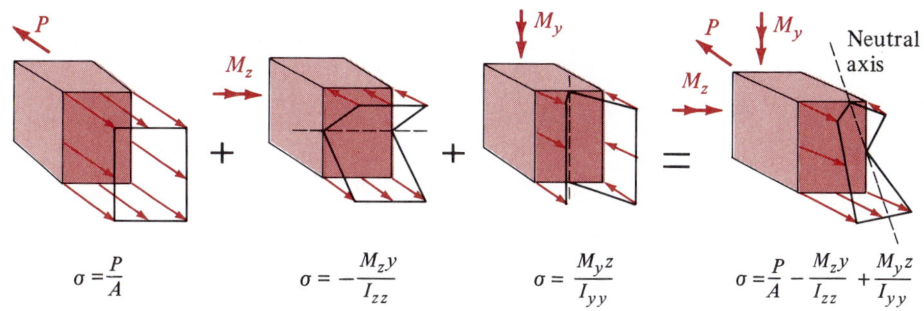

Figure 9.15 Superposition of axial and bending stresses about two axes

In this application the use of Eqn. (9.2) is justified on the basis that both the xy- and xz-planes are planes of symmetry of the rectangular cross section. As we see, the neutral axis no longer coincides with either the y- or z-axis and, depending upon all the variables involved, may not even lie in the cross section. Superposition of the axial and bending stresses in this fashion is permissible as long as the stresses lie below the proportional limit.

In general, there are shear stresses associated with both shear-force resultants V_y and V_z. The shear stresses corresponding to V_y are given by

$$\tau_{xy} = \frac{V_y Q_z}{I_{zz} t(y)}$$

where Q_z and $t(y)$ are indicated in Fig. 9.16a. The direction of the shear stress is determined by the direction of V_y.

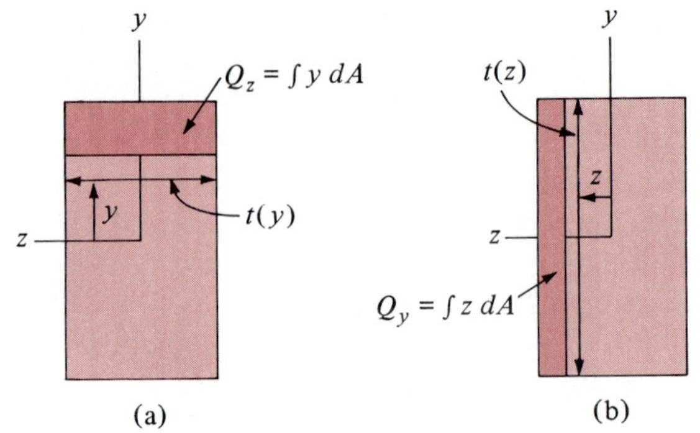

Figure 9.16 First moments and thicknesses

Similarly, the shear stresses corresponding to V_z are given by

$$\tau_{xz} = \frac{V_z Q_y}{I_{yy} t(z)}$$

where Q_y and $t(z)$ are indicated in Fig. 9.16b, and the direction of the shear stress is determined by the direction of V_y.

Example 9.4 The rectangular-cross-section block is loaded through a rigid plate as shown in Fig. 9.17a. Determine the combined stresses at a location 36 in. below the plate.

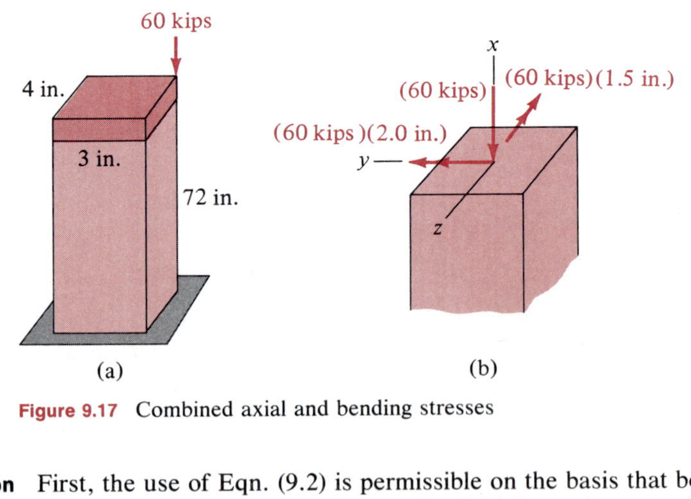

Figure 9.17 Combined axial and bending stresses

Solution First, the use of Eqn. (9.2) is permissible on the basis that both the xy- and xz-planes are planes of symmetry. The initial step is to resolve the force and determine the axial load and moments. For the choice of the coordinate shown in Fig. 9.17b, it is seen that $P = -60$ kips, $M_y = 120$ in.-kips, and

$M_z = -90$ in.-kips. The location of the cross section along the length of the beam at which the combined stresses are computed is immaterial since the axial force and moments are independent of x. With $I_{zz} = (3 \text{ in.})(4 \text{ in.})^3/12 = 16 \text{ in}^4$ and $I_{yy} = (4 \text{ in.})(3 \text{ in.})^3/12 = 9 \text{ in}^4$, Eqn. (9.2) can be used to determine

$$\sigma = \frac{-60 \text{ kips}}{12 \text{ in}^2} - \frac{(-90 \text{ in.-kips})y}{16 \text{ in}^4} + \frac{(120 \text{ in.-kips})z}{9 \text{ in}^4}$$

The equation of the neutral axis is obtained by setting $\sigma = 0$, leading to

$$5.625y + 13.33z = 5 \text{ in.}$$

the solution of which is shown in Fig. 9.18. The stresses at the four corners, labeled A, B, C, and D, are computed as

$$\sigma(A) = (-5 + 11.25 + 20) \text{ ksi} = 26.25 \text{ ksi}$$

$$\sigma(B) = (-5 + 11.25 - 20) \text{ ksi} = -13.75 \text{ ksi}$$

$$\sigma(C) = (-5 - 11.25 - 20) \text{ ksi} = -36.25 \text{ ksi}$$

$$\sigma(D) = (-5 - 11.25 + 20) \text{ ksi} = 3.75 \text{ ksi}$$

also shown in Fig. 9.18.

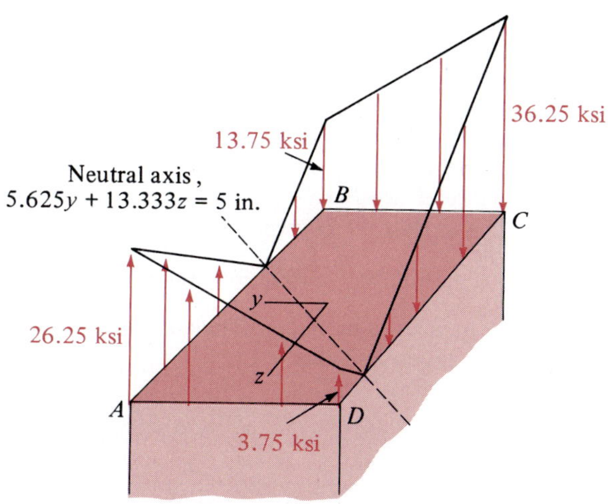

Figure 9.18 Combined stresses for double-axis bending

As we saw in the discussions of equilibrium for the cases considered above, there are (except for a case of pure bending) shear forces V_y and V_z that are also transmitted. The shear stresses associated with V_y and V_z should be included when determining the principal stresses and the maximum shear stresses that are frequently used in assessing strength. We can demonstrate this idea by means of the following examples.

Example 9.5 The rectangular member is subjected to the forces shown. Determine the principal stresses and the absolute maximum shear stress at each of the corners and midpoints of the sides at a section near the bottom.

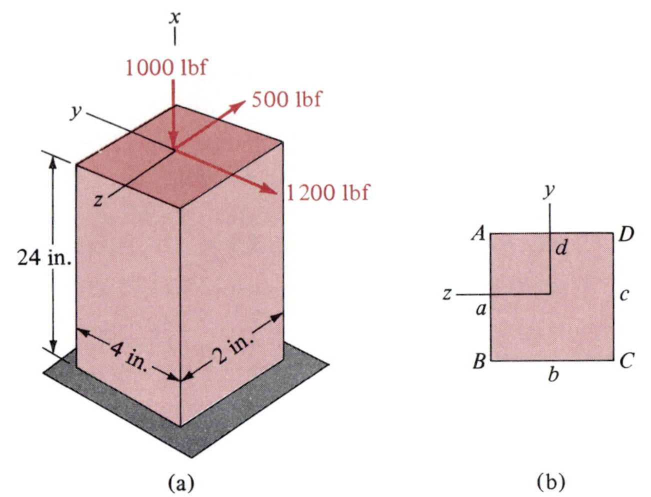

Figure 9.19 Loading of a rectangular member

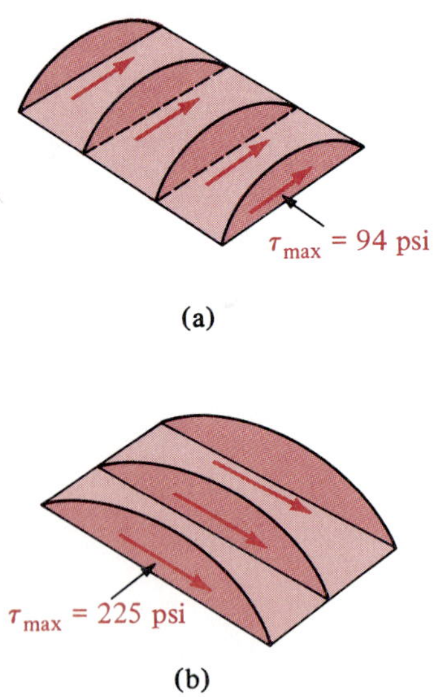

Figure 9.20 Shear-stress distributions

Solution The points at which the principal stresses and the maximum shear stress are required are shown in Fig. 9.19b. The values of the force resultants at the bottom of the member are $P = -10{,}000$ lbf, $M_z = -28{,}800$ in.-lbf, and $M_y = 12{,}000$ in.-lbf. Using Eqn. (9.2) gives

$$\sigma = (-1250 + 2700y + 4500z) \text{ psi}$$

where y and z are to be measured in inches. The values of the stress at each of the desired locations are given in Table 9.1.

Table 9.1 Combined stress values at points in the cross section

	A	*B*	*C*	*D*	*a*	*b*	*c*	*d*
σ(ksi)	8.65	−2.15	−11.15	−0.35	3.25	−6.65	−5.75	4.15

The shear stresses arising from the 500-lbf and 1200-lbf transverse loads have the parabolic distribution and the maximum midpoint values ($\tau = 3V/2A$) shown in Figs. 9.20a and b, respectively. With the normal and shear stresses known on the cross section, we can evaluate the principal stresses and maximum shear stress at each of the points in question.

Point A: A block showing the stresses acting at point *A* is shown in Fig. 9.21a. The planar block and the corresponding Mohr's circle are shown in Figs. 9.21b and c, respectively. Consulting Mohr's circle, we find the principal stresses to be $\sigma_{min} = 0$ and $\sigma_{max} = 8.65$ ksi. The maximum inplane shear stress, represented by point 3 on Mohr's circle, is $\tau_{max} = 4.33$ ksi $= \sigma_{max}/2$, and is also the absolute maximum shear stress at the point.

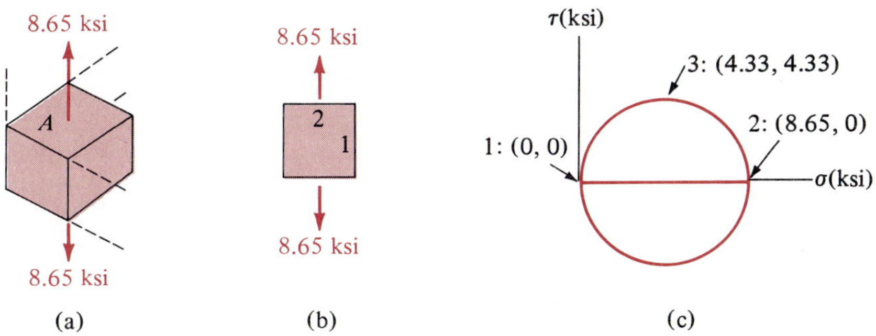

Figure 9.21 Principal stress and maximum inplane shear stress information for point A

Point a: A block showing the stresses acting at point a is shown in Fig. 9.22a. The planar block and the corresponding Mohr's circle are shown in Figs. 9.22b and c, respectively.

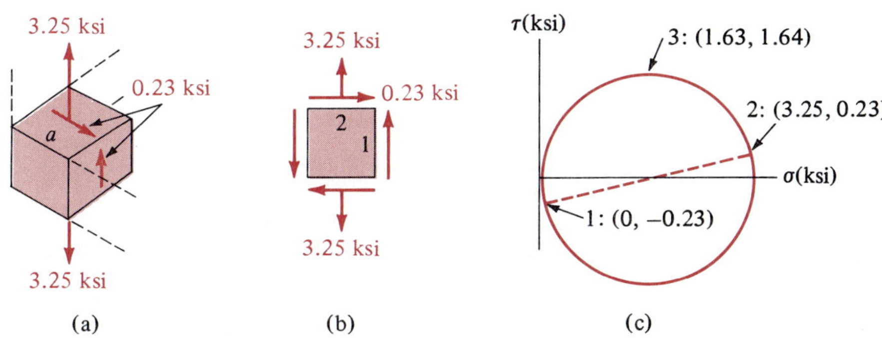

Figure 9.22 Principal stress and maximum shear stress information for point a

Again consulting Mohr's circle, we find the principal stresses to be $\sigma_{min} = -0.02$ ksi and $\sigma_{max} = 3.27$ ksi. The maximum inplane shear stress, represented by point 3 on Mohr's circle, is $\tau_{max} = (\sigma_{max} - \sigma_{min})/2 = 1.64$ ksi, and is also the absolute maximum shear stress at the point. The principal stresses and the maximum shear stresses can be evaluated at the remaining points. The results are shown in Table 9.2.

Table 9.2 Principal stresses and maximum shear stresses

	A	*B*	*C*	*D*	*a*	*b*	*c*	*d*
σ_{min}(ksi)	0.00	−2.15	−11.15	−0.35	−0.02	−6.65	−5.76	0.00
σ_{max}(ksi)	8.65	0.00	0.00	0.00	3.27	0.00	0.01	4.15
τ_{max}(ksi)	4.33	1.08	5.58	0.18	1.64	3.33	2.88	2.08

On the basis of the points checked, the principal stress of largest absolute value is the combined stress at point C, with the absolute maximum shear stress occurring at the same point. Note carefully that the maximum shear stress does not occur in connection with $\tau = VQ/It$, but rather on a 45° plane in connection with one of the principal stresses at the same point.

Example 9.6 A W 8 × 15 beam is loaded through the centroid as shown in Fig. 9.23a. Determine the maximum combined stress at the point near where the beam attaches to the wall.

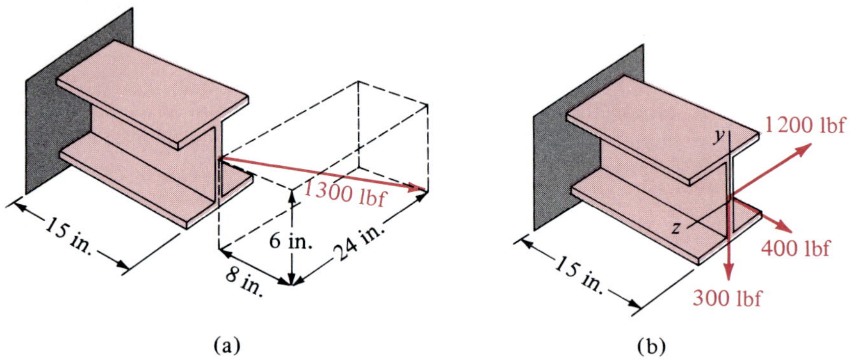

Figure 9.23 Combined stresses in a W section

Solution The 1300 lbf is resolved into components as shown in Fig. 9.23b. At the fixed end, the force resultants transmitted are $P = 400$ lbf, $M_z = -(300 \text{ lbf})(15 \text{ in.}) = -4500$ in.-lbf, $M_y = (1200 \text{ lbf})(15 \text{ in.}) = 18{,}000$ in.-lbf, $V_y = -300$ lbf, and $V_z = -1200$ lbf. Using Eqn. (9.2) and the necessary information about the W section from Appendix C yields

$$\sigma = \frac{400 \text{ lbf}}{4.44 \text{ in}^2} - \frac{(-4500 \text{ in.-lbf})y}{48 \text{ in}^4} + \frac{(18{,}000 \text{ in.-lbf})z}{3.41 \text{ in}^4}$$

$$= (90.1 + 93.75y + 5279z) \text{ psi} \qquad (a)$$

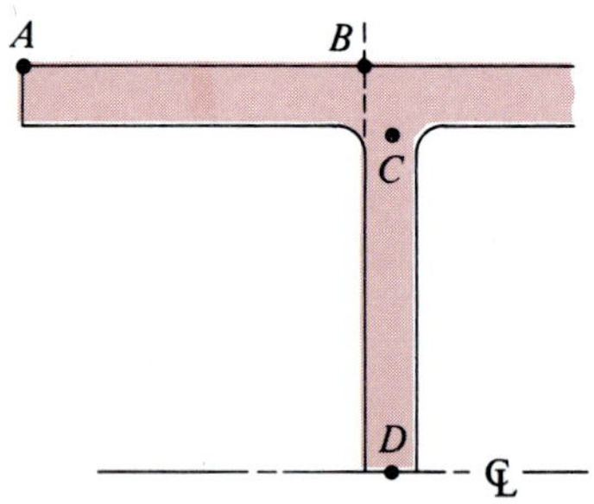

Figure 9.24 Various points in the cross section

where y and z are to be measured in inches. With all the terms in Eqn. (a) positive, it is clear that the maximum value of σ will occur when both y and z are positive, namely, $y = 4.055$ in. and $z = 2.008$ in., corresponding to point A shown in Fig. 9.24.

The value of the normal stress at point A is

$$\sigma(A) = [90.1 + 93.75(4.055) + 5279(2.008)] \text{ psi} = 11.07 \text{ ksi}$$

The normal stresses at points B, C, and D are

$$\sigma(B) = [90.1 + 93.75(4.055) + 5279(0.127)] \text{ psi} = 1.14 \text{ ksi}$$

$$\sigma(C) = [90.1 + 93.75(3.74) + 5279(0)] \text{ psi} = 0.44 \text{ ksi}$$

$$\sigma(D) = [90.1 + 93.75(0) + 5279(0)] \text{ psi} = 0.09 \text{ ksi}$$

Shear stresses: For the 300-lbf shear force acting in the negative y-direction, the shear stresses are linear in the flanges and parabolic in the web, with the values

$$\tau(A) = \frac{(300 \text{ lbf})(0)}{48 \text{ in}^4(0.315 \text{ in.})} = 0 \text{ psi}$$

$$\tau(B) = \frac{(300 \text{ lbf})(0.315 \text{ in.})(2.008 \text{ in.})(3.898 \text{ in.})}{48 \text{ in}^4(0.315 \text{ in.})} = 48.9 \text{ psi}$$

$$\tau(C) \approx \frac{(300 \text{ lbf})(0.315 \text{ in.})(4.015 \text{ in.})(3.898 \text{ in.})}{48 \text{ in}^4(0.245 \text{ in.})} = 125.8 \text{ psi}$$

$$\tau(D) = \tau(C) + \frac{(300 \text{ lbf})(3.74 \text{ in.})^2(0.245 \text{ in.})/2}{48 \text{ in}^4(0.245 \text{ in.})} = 169.5 \text{ psi}$$

all of which are shown in Fig. 9.25a. For the 1200-lbf shear force acting in the negative z-direction, the shear stresses are quadratic in the flanges with essentially no shear stresses in the web. The values at the points A, B, C, and D are

$$\tau(A) = 0$$

$$\tau(B) \approx \frac{3V_z}{2A_f} = \frac{3(1200 \text{ lbf})}{2(2.54 \text{ in}^2)} = 709 \text{ psi}$$

$$\tau(C) = \tau(D) \approx 0$$

and are indicated in Fig. 9.25b, together with the corresponding shear-stress distribution.

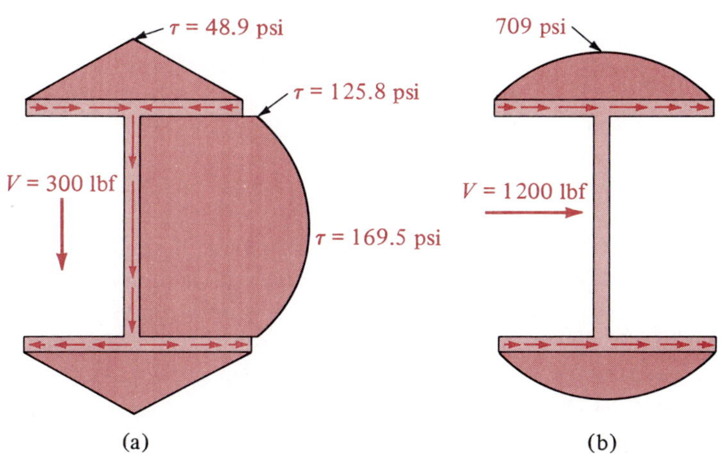

Figure 9.25 Shear-stress distributions

Principal stresses and maximum shear stress: With the normal and shear stresses known, the principal stresses and maximum shear stresses at points A, B, C, and D can be evaluated. Blocks at points A, B, C, and D are indicated in Fig. 9.26. Note that at point B, the shear stresses due to the shear forces in the y- and z-directions are in the same direction and add algebraically.

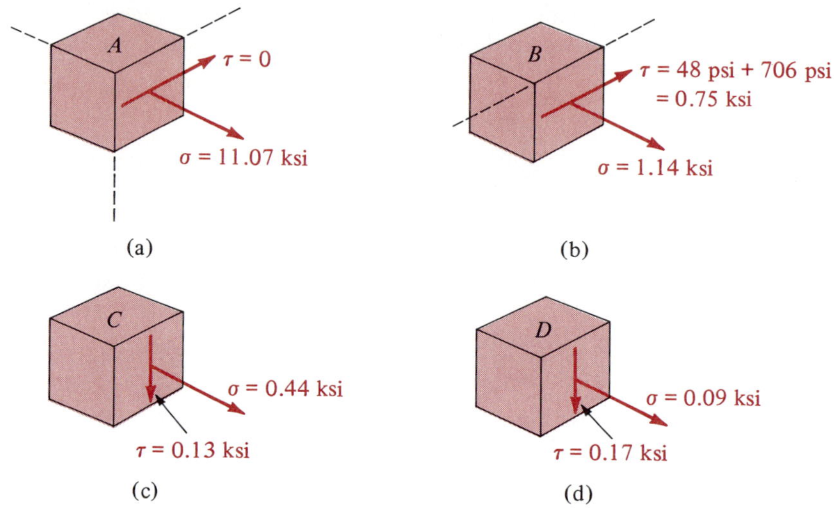

Figure 9.26 States of stress at points *A*, *B*, *C*, and *D*

The values of the corresponding principal stresses and absolute maximum shear stresses are indicated in Table 9.3.

Table 9.3 Principal stresses and maximum shear stresses

Point	σ_{max} (ksi)	σ_{min} (ksi)	τ_{abs} (ksi)
A	11.07	0	5.54
B	1.51	−0.37	0.94
C	0.48	−0.02	0.25
D	0.23	−0.14	0.18

General Axial and Bending Stresses The cases of single- and double-axis bending discussed above are valid only as long as the necessary symmetries are present. Where there is a single plane of symmetry, and the loading plane does not coincide with it, or where there is no plane of symmetry at all, the plane of loading must pass through the shear center. If the y- and z-axes are then selected to be the *principal axes*, the moment at any location along the beam is resolved into components M_y and M_z, after which the combined axial and bending stresses can be evaluated using

$$\sigma_x = \frac{P}{A} - \frac{M_z y}{I_{zz}} + \frac{M_y z}{I_{yy}}$$

where P is the axial component of the force transmitted. Alternatively, when the y- and z-axes are not principal axes,

$$\sigma = \frac{P}{A} - \frac{M_z I_{yy} + M_y I_{yz}}{I_{yy} I_{zz} - I_{yz}^2} y + \frac{M_y I_{zz} + M_z I_{yz}}{I_{yy} I_{zz} - I_{yz}^2} z$$

can be used to compute the bending stresses.

Example 9.7 The C 150 × 12.2 channel section is used as a cantilever beam and loaded as in Fig. 9.27a. Determine *(a)* the location of the shear center, and *(b)* the normal and shear stress distributions near the fixed end of the cantilever.

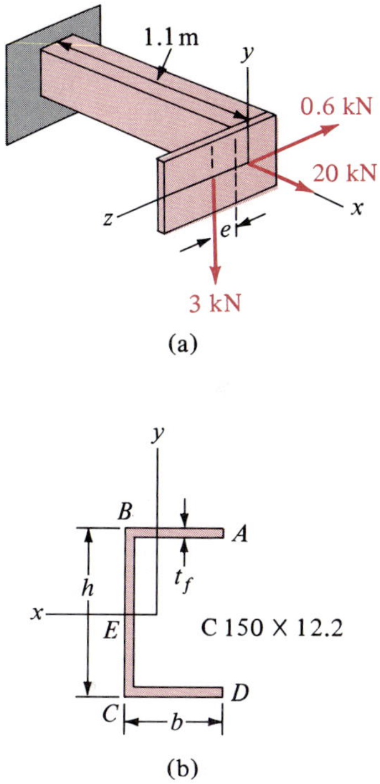

Figure 9.27 Axial and bending loading of a cantilever beam

Solution First note that all the calculations are based on the xyz-axes, which are the centroidal axes. From Section 7.10, the general expression for the distance from the web to the shear center is given by

$$e = \frac{t_f h^2 b^2}{4 I_{zz}}$$

where t_f, h, and b are shown in Fig. 9.27b. Substituting the values obtained from Appendix C yields $e = 0.075$ m. Thus the 3-kN load must be applied 75 mm to the left of the web, with the 0.6-kN and 20-kN loads applied through the centroid as shown in Fig. 9.27a. At the fixed end of the cantilever, the force resultants are $P = 20$ kN, $M_z = -3.3$ kN·m, and $M_y = 0.66$ kN·m. Again using the necessary information from Appendix C, the combined normal stresses can be expressed as

$$\sigma = \frac{20 \text{ kN}}{1.548 \times 10^{-3} \text{ m}^2} - \frac{-3.3 \text{ kN·m}}{5.45 \times 10^{-6} \text{ m}^4} y + \frac{0.66 \text{ kN·m}}{0.288 \times 10^{-6} \text{ m}^4} z$$
$$= (12.92 + 606y + 2291z) \text{ MPa}$$

with y and z measured in meters. The location of the neutral axis is determined by setting $\sigma = 0$, leading to

$$606y + 2291z = -12.92 \text{ m}$$

and is shown in Fig. 9.28. The values of the normal stresses at the points A, B, C, D, and E are

$$\sigma(A) = [12.92 + 606(0.076) + 2291(-0.035)] \text{ MPa} = -21.17 \text{ MPa}$$
$$\sigma(B) = [12.92 + 606(0.076) + 2291(\ 0.013)] \text{ MPa} = 88.75 \text{ MPa}$$
$$\sigma(E) = [12.92 + 606(0) + 2291(\ 0.013)] \text{ MPa} = 42.70 \text{ MPa}$$
$$\sigma(C) = [12.92 + 606(-0.076) + 2291(\ 0.013)] \text{ MPa} = -3.37 \text{ MPa}$$
$$\sigma(D) = [12.92 + 606(-0.076) + 2291(-0.035)] \text{ MPa} = -113.3 \text{ MPa}$$

and are also shown in Fig. 9.28. The slight variation of the normal stresses over the thicknesses of the flanges and web is also depicted in this figure.

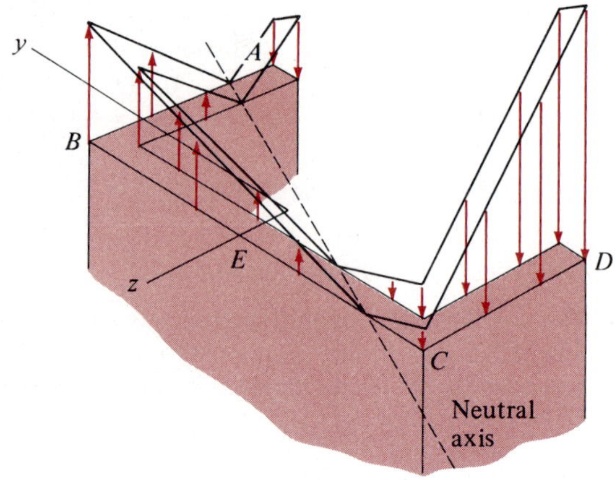

Figure 9.28 Neutral axis and normal-stress distribution

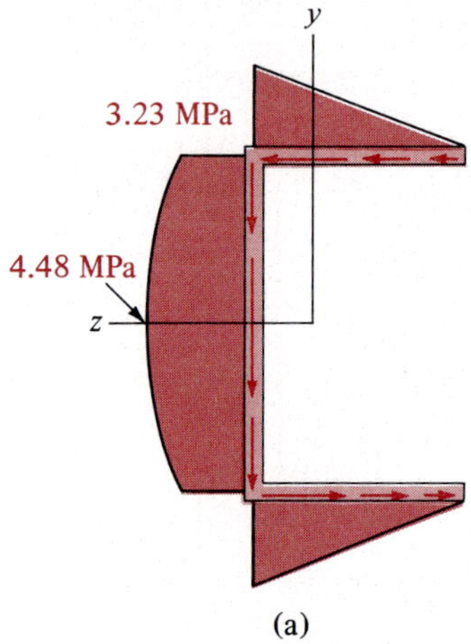

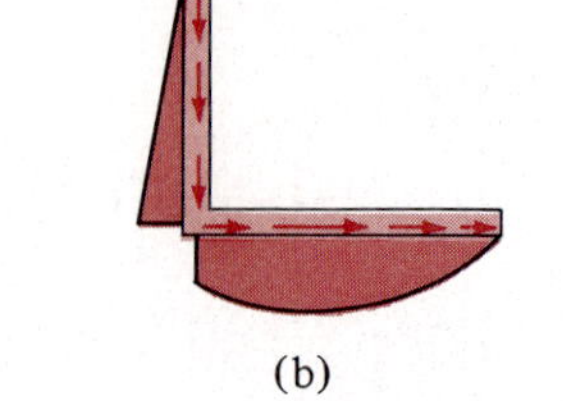

Figure 9.29 Shear-stress distributions

Shear stresses: For the 3-kN shear force in the negative y-direction, the shear stresses are linear functions of position in the flanges AB and CD, and quadratic functions of position in the web BC as shown in Fig. 9.29a. The values at A, B, E, C, and D are

$$\tau(A) = 0 \qquad (Q = 0)$$

$$\tau(B) = \frac{(3000\text{ N})(0.0087\text{ m})(0.048\text{ m})(0.0717\text{ m})}{5.45 \times 10^{-6}\text{ m}^4(0.0051\text{ m})} = 3.23\text{ MPa}$$

$$\tau(E) = \tau(B) + \frac{(3000\text{ N})(0.0673\text{ m})^2(0.0051\text{ m})/2}{5.45 \times 10^{-6}\text{ m}^4(0.0051\text{ m})} = 4.48\text{ MPa}$$

$$\tau(C) = \tau(B) \qquad \tau(D) = 0 \qquad (Q = 0)$$

For the 0.6-kN shear force in the negative z-direction, the shear stresses are quadratic functions of position in the flanges AB and CD, and linear functions of position in the web BC as indicated in Fig. 9.29b. The values at A, B, E, C, and D are:

$$\tau(A) = 0 \qquad (Q = 0)$$

$$\tau(B) = \frac{(600\text{ N})(0.048\text{ m})(0.0087\text{ m})(0.011\text{ m})}{5.45 \times 10^{-6}\text{ m}^4(0.0051\text{ m})} = 1.88\text{ MPa}$$

$$\tau(E) = 0 \qquad (Q = 0) \qquad \tau(C) = \tau(B) \qquad \tau(D) = 0$$

For this example, the values of the maximum shear stresses are generally between one and two orders of magnitude less than the maximum normal stresses. It follows that the critical point is at D where $\sigma = -113.3$ MPa and $\tau = 0$. The principal stresses are $\sigma_{\min} = -113.3$ MPa and $\sigma_{\max} = 0$. The absolute maximum shear stress is $\tau_{\max} = |-113.3\text{ MPa}|/2 = 56.65$ MPa, and acts on a 45° plane at D.

PROBLEMS

Single-Axis Axial and Bending Stresses

9.1 Determine the combined normal stresses at points A and B for the eccentrically loaded bar shown.

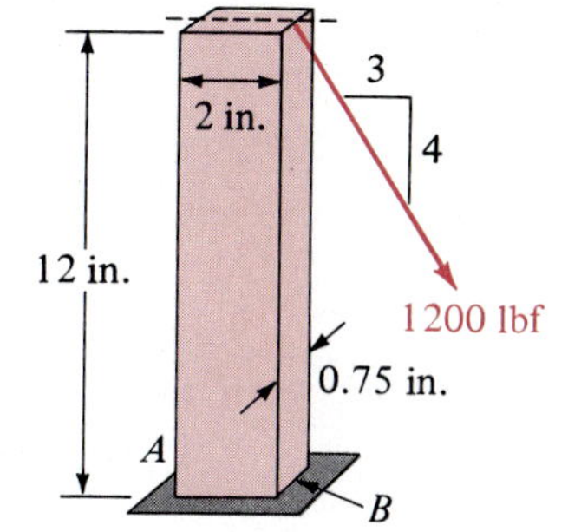

P. 9.1

9.2 Determine the combined normal-stress values at points A and B for the circular-cross-section bar shown. Take $h = 20$ mm, $d = 10$ mm, and $P = 1$ kN.

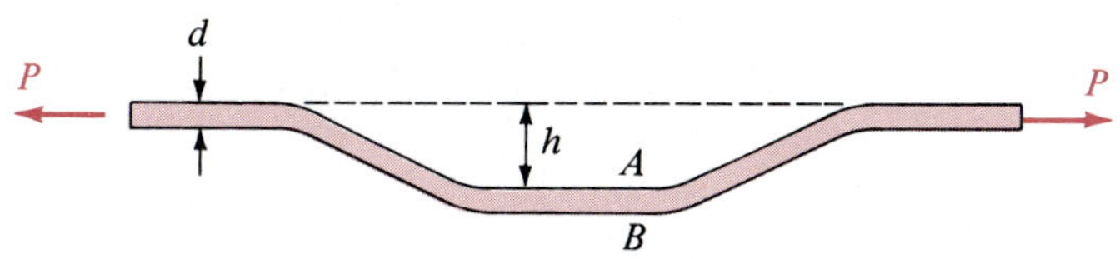

Ps. 9.2 and 9.3

9.3 Repeat problem 9.2 for $h = 1.2$ in., $d = 0.5$ in., and $P = 300$ lbf.

9.4 Determine the maximum and minimum combined normal stresses across the reduced section for the bar loaded as shown. Take the thickness to be 0.5 in., $h_1 = 1.0$ in., $h_2 = 2.0$ in., and $P = 3750$ lbf.

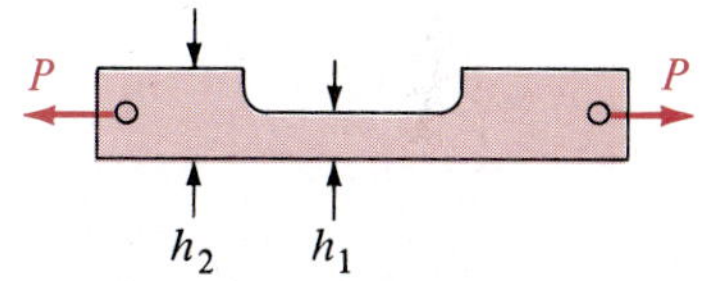

Ps. 9.4 and 9.5

9.5 Repeat problem 9.4 if $h_1 = 30$ mm, $h_2 = 70$ mm, $t = 15$ mm, and $P = 20$ kN.

9.6 Given that the allowable combined normal stresses in tension and compression are $\sigma_t = 12$ ksi and $\sigma_c = 16$ ksi, respectively, determine the allowable eccentric load P applied at the center of the flange of the C section as shown.

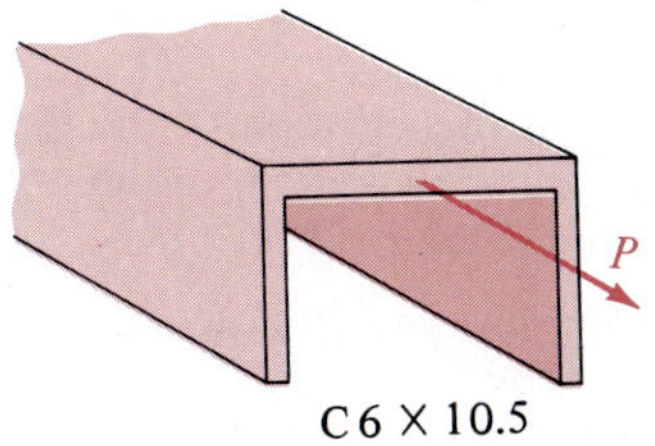

P. 9.6

9.7 A circular pipe is acted upon by an eccentric load as shown. Determine the maximum permissible value of e if the allowable combined normal stress value is 16 ksi. Take $D = 4$ in., $d = 3.5$ in., and $P = 20$ kips.

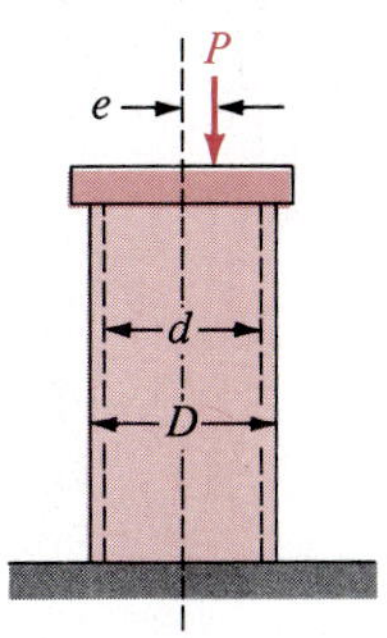

Ps. 9.7 and 9.8

9.8 Repeat problem 9.7 with $D = 75$ mm, $d = 60$ mm, $P = 80$ kN, and $\sigma_{all} = 100$ MPa.

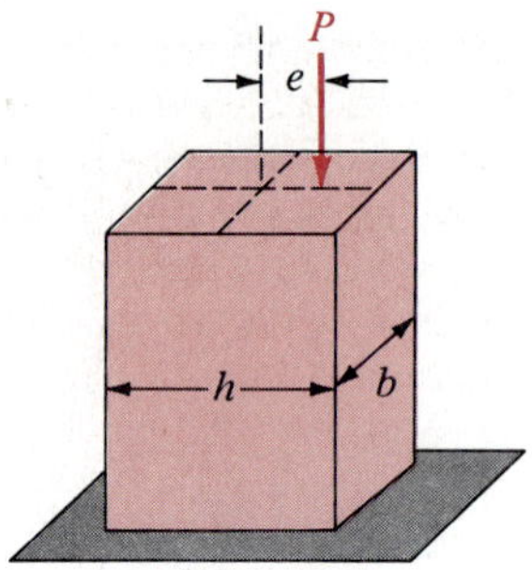

P. 9.9

9.9 A rectangular-cross-section bar is subjected to an eccentrically applied axial load as shown. If all points in the cross section are to experience a normal compressive stress, determine the maximum value of e.

9.10 The C 10 × 30 section is acted upon by the 60-kip eccentrically applied load by means of 0.5-in. plates as shown. Determine the combined normal-stress distribution.

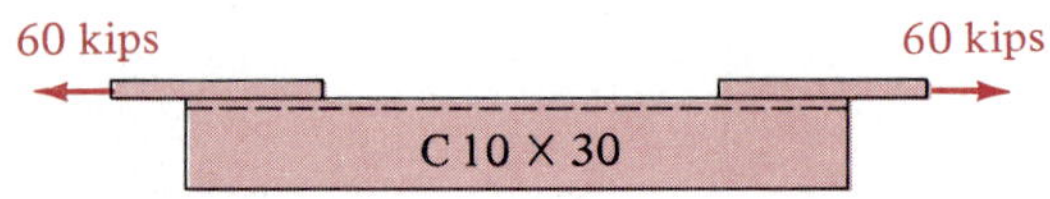

P. 9.10

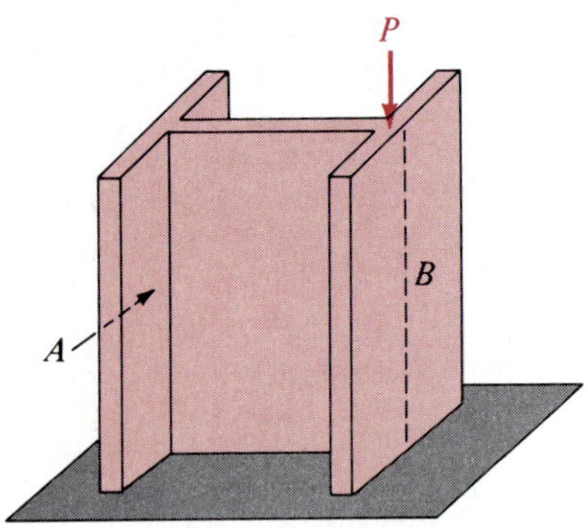

P. 9.11

9.11 The W 8 × 24 section is acted upon by the eccentric load applied at the center of one of the flanges as shown. Determine the combined normal stresses at A and B. Take $P = 80$ kips.

9.12 If the allowable combined normal stress in the steel clamp is 12 ksi, determine the largest clamp force F that can be applied by the clamp. Take $t_1 = 0.375$ in. and $t_2 = 0.1875$ in.

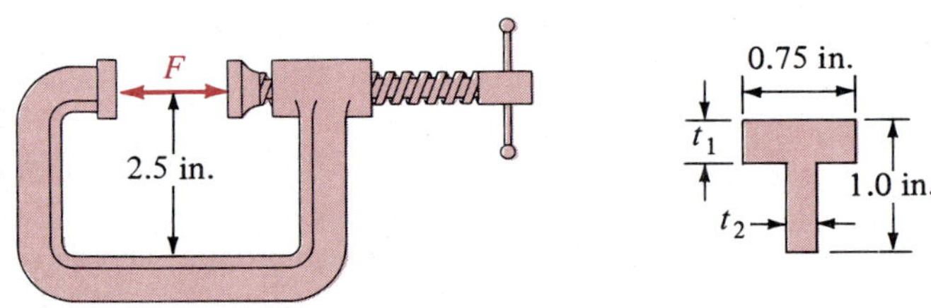

Ps. 9.12 and 9.13

9.13 Repeat problem 9.12 if the position of the flange is reversed, that is, toward the outside rather than toward the inside. What conclusion do you reach regarding the position of the flange?

9.14 The L-shaped member has a square cross section 30 mm on a side and is acted upon by the forces shown. Normal strains of 300 μ and -400 μ are measured at points A and B, respectively. Determine P and e. Take $E = 70$ GPa. Do h and L enter into the solution of the problem? Explain.

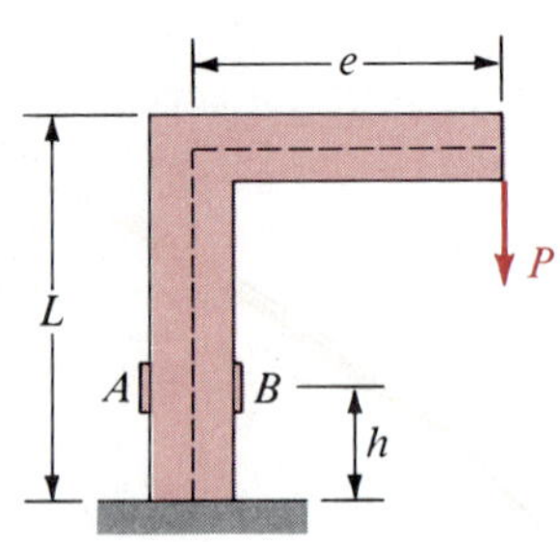

P. 9.14

9.15 The frame of a machine press is loaded as shown. If the allowable combined normal stress in the frame is 20 ksi, what is the allowable load P for a factor of safety of 2?

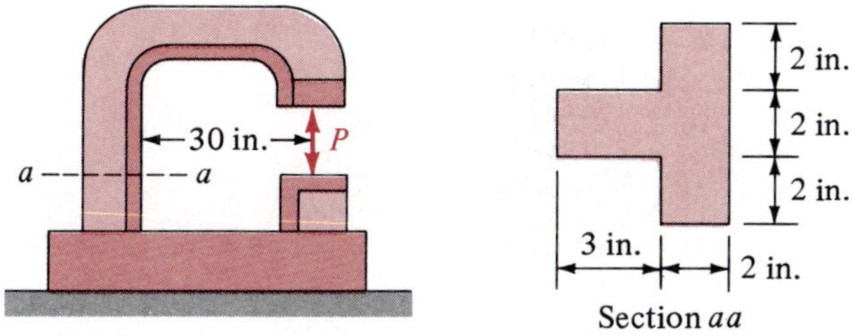

P. 9.15

9.16 For the concrete dam with the cross section shown, determine the maximum allowable height h for the water in order that the bottom corner of the dam at A not be in tension. Take the specific weights of concrete and water to be 150 lbf/ft^3 and 62.4 lbf/ft^3, respectively.

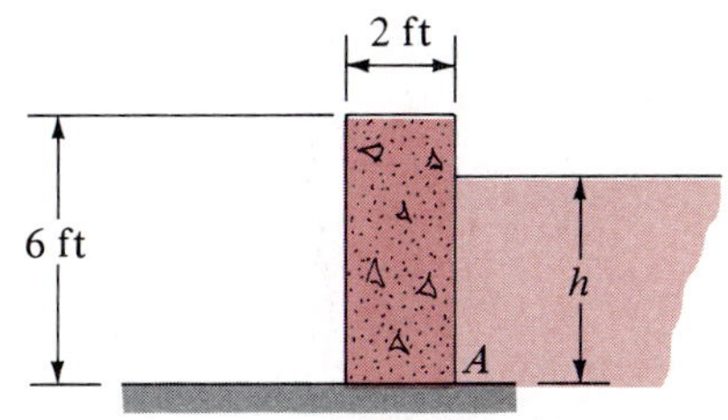

P. 9.16

9.17 The famous leaning tower of Pisa is modeled as a tube of outside diameter D and inside diameter d. How far can the tower be tipped (α = ?) before the location of the neutral axis reaches point A? Take $d/D = 0.9$ and $h = 4D$.

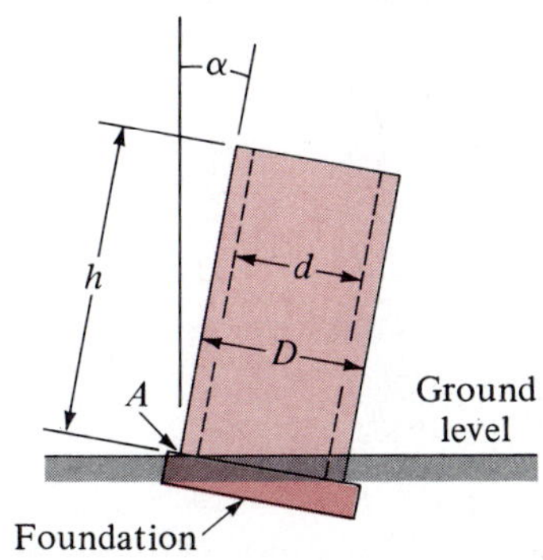

P. 9.17

9.18 If the allowable combined normal-stress level in the W 150 × 24 section is 100 MPa, determine P_{max}.

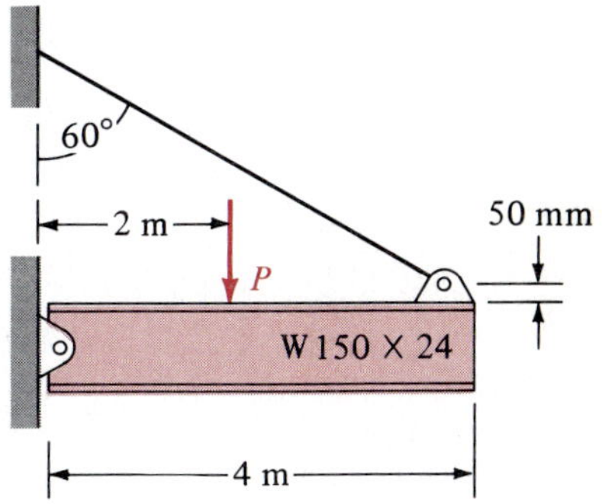

P. 9.18

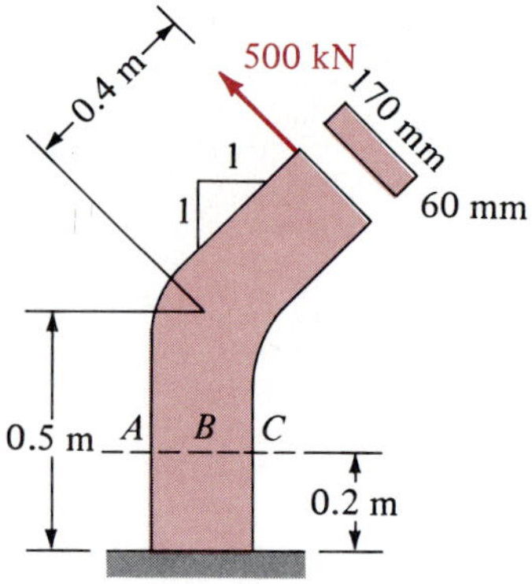

P. 9.19

9.19 Determine the combined normal stress and the transverse shear stress at points A, B, and C for the bar shown.

9.20 The rectangular-cross-section bar is acted upon by the 80-lbf force as shown. Determine the combined normal stress distribution and the transverse shear stress distribution at section AA.

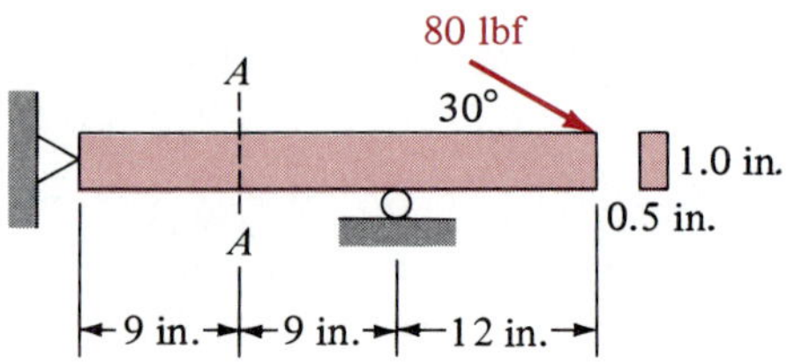

P. 9.20

9.21 The pin-jointed frame is composed of rectangular-cross-section members and loaded as shown. Determine the combined normal stress and transverse shear stress distributions at sections aa and bb. Bending of the frame members takes place about the z-axis.

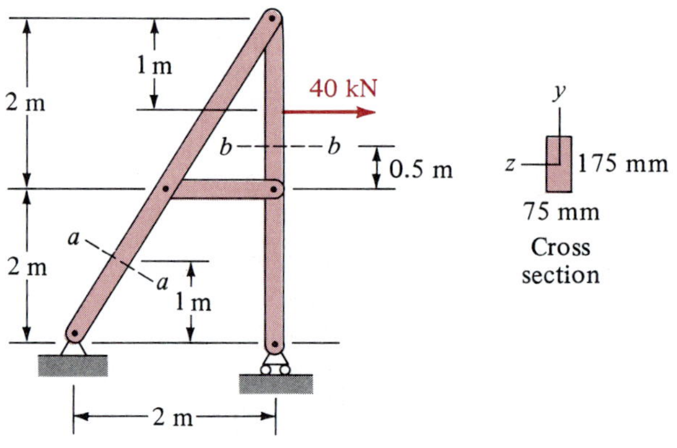

P. 9.21

9.22 Determine the principal stresses and the maximum shear stress at points A and B for the section midway between the supports for the pipe loaded as shown. Take the outside diameter of the pipe to be 3 in. and the wall thickness to be 0.25 in.

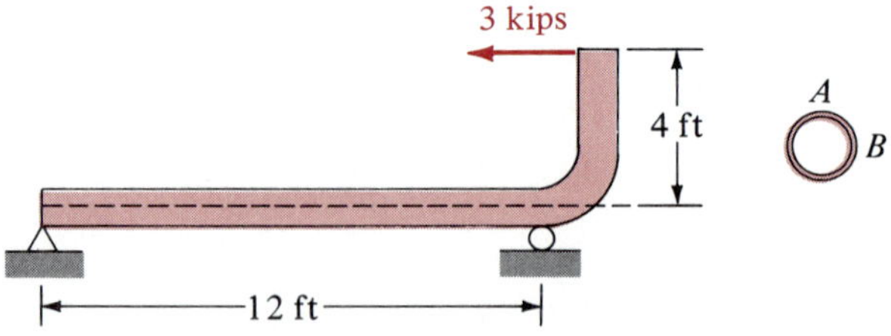

P. 9.22

Double-Axis Bending

9.23 The forces are applied through a thick plate, not shown, at the top of a cantilever formed from a W 30 × 99 section as indicated. Determine the location of the neutral axis for a section at the fixed end of the cantilever. Also determine the maximum combined normal stress. Take P = 60 kips, Q = 100 kips, and L = 48 in. Assume P is applied at the middle of the flange section.

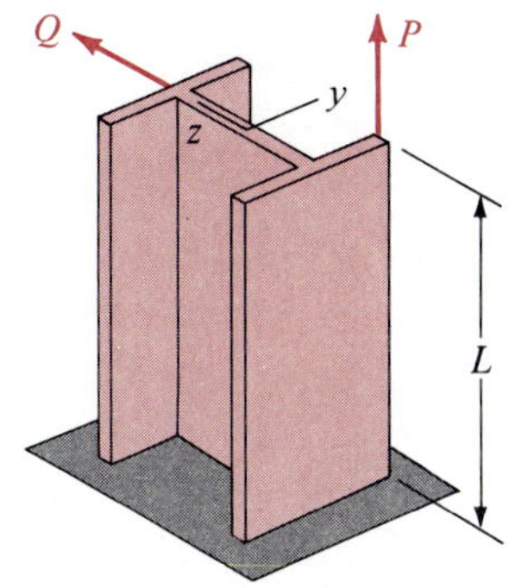

Ps. 9.23 and 9.24

9.24 Repeat problem 9.23 for a W 690 × 217 section with P = 180 kN, Q = 350 kN, and L = 1.25 m.

9.25 The rectangular member is acted upon by the loads shown. Determine *(a)* the combined normal stress at points A and B for a section at the bottom of the beam, and *(b)* the location of the neutral axis. Take F_1 = 10 kN, F_2 = 5 kN, h = 0.1 m, H = 0.15 m, b = 0.15 m, B = 0.20 m, and L = 1.4 m.

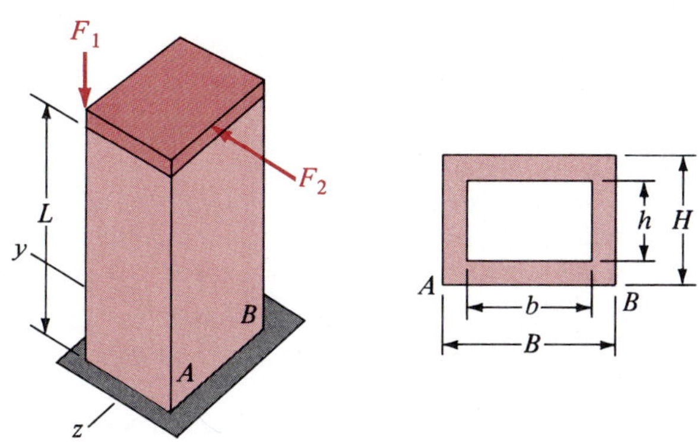

Ps. 9.25 and 9.26

9.26 Repeat problem 9.25 with F_1 = 3 kips, F_2 = 2 kips, h = 4 in., H = 6 in., b = 6 in., B = 8 in., and L = 4 ft.

9.27 The W 250 × 44.8 section is loaded through a thick plate by the forces shown. What is the allowable value of P if the allowable combined normal stress is 100 MPa?

9.28 Repeat problem 9.27 for a W 10 × 30 section with an allowable combined normal stress of 14 ksi.

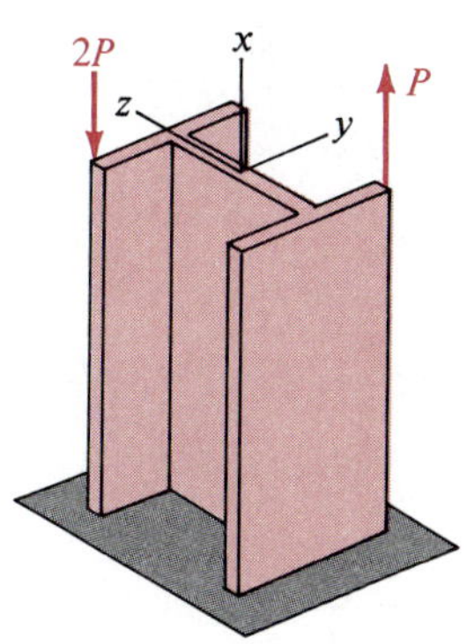

Ps. 9.27 and 9.28

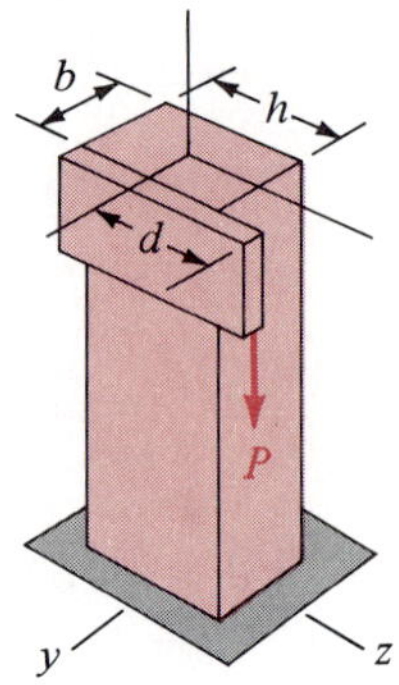

P. 9.29

9.29 The rectangular section is loaded through an extension as shown. If the allowable combined normal stress is 120 MPa, determine the largest permissible distance d. Take $h = 90$ mm, $b = 60$ mm, and $P = 75$ kN.

9.30 The beam is subjected to the forces and the uniform load shown. Determine the location and magnitude of the largest combined normal stress on a section through the point of application of P, the midpoint of the beam. The beam is simply supported in both directions at both ends. Take $L = 2$ m, $b = 0.2$ m, $h = 0.3$ m, $P = 88$ kN, $w_0 = 100$ kN/m, and $Q = 1200$ kN.

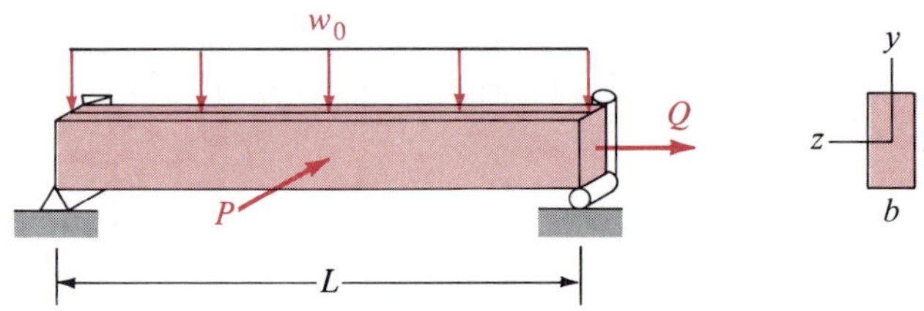

Ps. 9.30 and 9.31

9.31 For the beam of the previous problem, take $L = 2$ m, $b = 0.2$ m, $h = 0.3$ m, $w_0 = 100$ kN/m, and $Q = 800$ kN and determine the largest possible P if the largest combined normal stress is not to exceed 140 MPa.

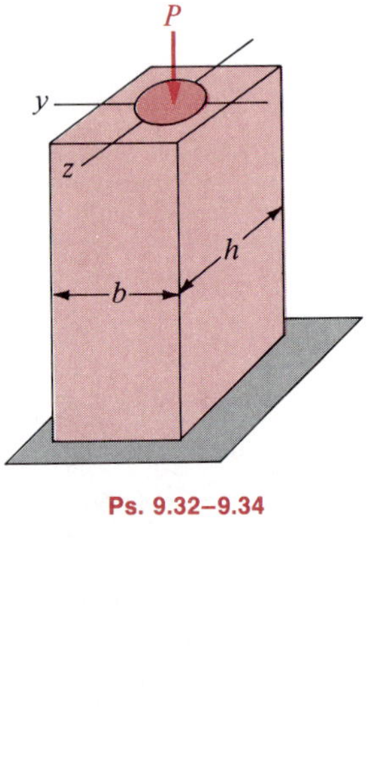

Ps. 9.32–9.34

9.32 A rectangular-cross-section column is loaded by an axial force as shown. It is desired to restrict the line of action of the force so that the combined normal stresses are compressive everywhere in the beam. Determine the equation(s) that define the region (shown shaded) over which the beam can be loaded. This region is called the **kern** of the section.

9.33 Repeat problem 9.32 for a solid circular cross section.

9.34 Repeat problem 9.32 for a hollow circular cross section with the inside radius 0.75 times the outside radius.

9.35 The cantilever beam is formed from a circular pipe with an outside diameter of 4 in. and a wall thickness of 0.375 in. What is the largest allowable value of P if the combined normal stress at the fixed end is not to exceed 16 ksi? Take $w_0 = 40$ lbf/in. and $L = 48$ in.

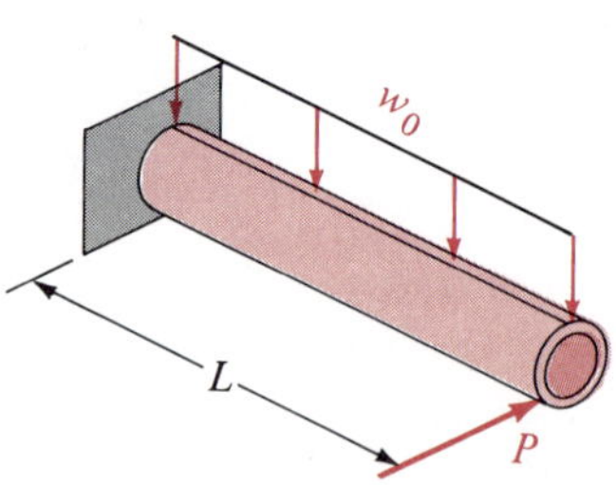

Ps. 9.35 and 9.36

9.36 Repeat problem 9.35 if the outside diameter is 100 mm and the wall thickness is 10 mm. Take $L = 1.3$ m, $P = 2$ kN, and determine the largest permissible w_0 if the maximum allowable combined normal stress is limited to 105 MPa.

9.37 A rectangular-cross-section beam is loaded by the three forces shown. At a section one quarter of the length out from the support, determine the principal stresses and absolute maximum shear stress at *(a)* point A, *(b)* point B, *(c)* point C, and *(d)* point D. Take $b = 3$ in., $h = 5$ in., $L = 20$ in., $P = 20$ kips, $Q = 5$ kips, and $R = 4$ kips.

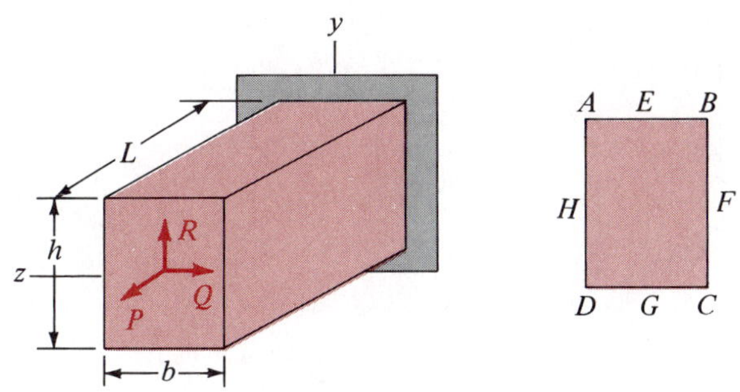

Ps. 9.37 and 9.38

9.38 Repeat problem 9.37 for points E, F, G, and H.

9.39 The W section shown is simply supported in both directions at both ends and loaded by the three forces at the center of the beam. At a section 1 in. to the left of the center of the beam, determine the principal stresses at points A, B, C, and D. Take $Q = 1000$ lbf, $R = 300$ lbf, $h = 1.0$ in., $b = 0.75$ in., $t = 0.1$ in., and $L = 20$ in.

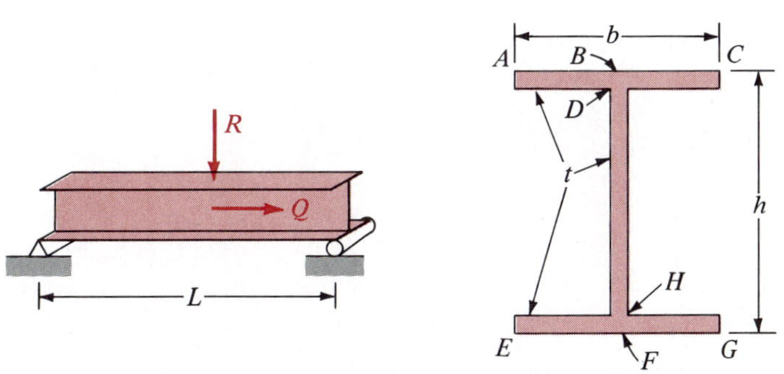

Ps. 9.39 and 9.40

9.40 Repeat problem 9.39 for points E, F, G, and H.

9.41 The hollow-box section is loaded by the forces shown. Determine the combined normal stress at points A, E, B, and F at the section indicated. Take B = 12 mm, b = 8 mm, H = 20 mm, h = 16 mm, L = 80 mm, P = 80 N, Q = 120 N, and R = 160 N.

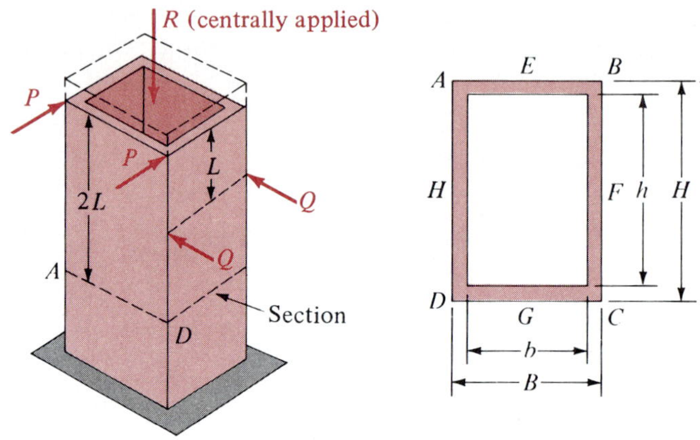

Ps. 9.41 and 9.42

9.42 Repeat problem 9.41 for points C, G, D, and H.

General Bending

9.43 The Z section is loaded through the shear center as shown. Determine *(a)* the location of the neutral axis, *(b)* the combined normal stress at points A, B, C, and D at the bottom of the beam. Take P = 8 lbf, Q = 10 lbf, R = 20 lbf, L = 6 in., b = 0.55 in., t = 0.05 in., and h = 1 in.

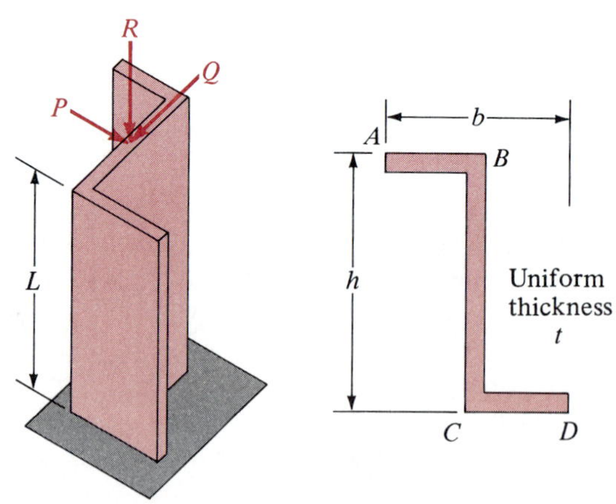

Ps. 9.43 and 9.44

9.44 Repeat problem 9.43 for P = 35 N, Q = 45 N, R = 100 N, L = 160 mm, b = 26 mm, t = 2 mm, and h = 40 mm.

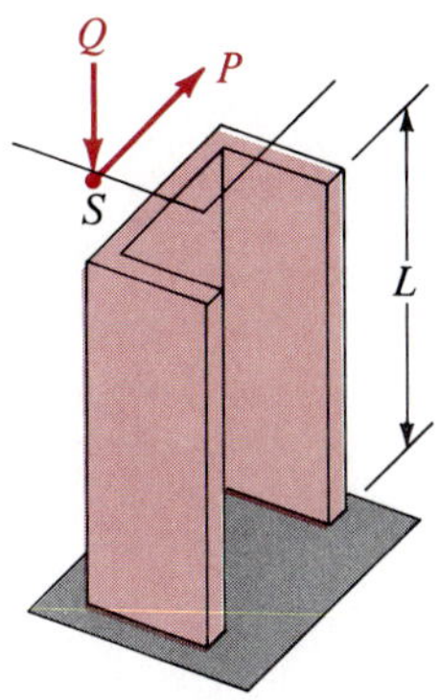

Ps. 9.45 and 9.46

9.45 For the C 75 × 6.1 section shown, determine *(a)* the location of the shear center S, *(b)* the location of the neutral axis, and *(c)* the combined stress distribution at the bottom of the column. Take $L = 0.6$ m, $P = 1$ kN, and $Q = 1.5$ kN.

9.46 For the C 3 × 6 section shown, determine *(a)* the location of the neutral axis, *(b)* the location of the shear center, and *(c)* the combined normal-stress distribution at the bottom of the column. Take $L = 30$ in., $P = 250$ lbf, and $Q = 400$ lbf.

9.47 For the L 4 × 3 × $\frac{1}{4}$ section loaded (at the shear center) as shown, determine *(a)* the location of the neutral axis, and *(b)* the maximum inplane normal stress at the support. Take $P = 150$ lbf, $Q = 100$ lbf, $R = 300$ lbf, and $L = 30$ in.

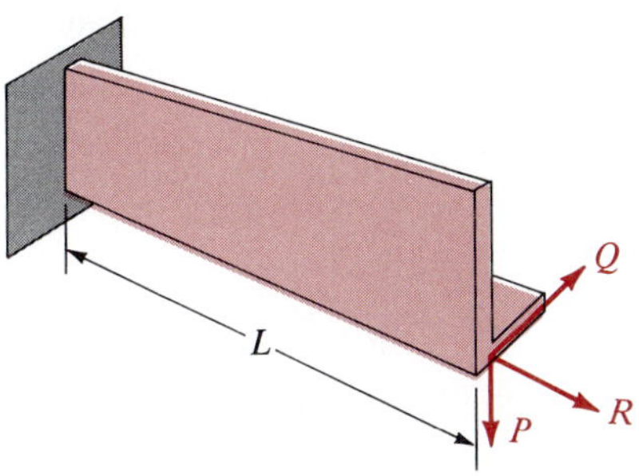

Ps. 9.47 and 9.48

9.48 For the L 102 × 76 × 12.7 section loaded (at the shear center) as shown, determine *(a)* the location of the neutral axis, and *(b)* the maximum inplane normal stress at the support. Take $P = 600$ N, $Q = 500$ N, $R = 1300$ N, and $L = 0.8$ m.

9.3 COMBINED AXIAL, BENDING, AND TORSIONAL LOADS

In the previous section, we investigated the states of stress in members simultaneously transmitting axial loads P, shear loads V, and bending moments M. On the basis of a suitable set of assumptions, we could use superposition of the stresses that arise with the transmission of the individual force resultants, to determine the resultant or combined state of stress. In this section, we will consider situations where there is also a torque being transmitted about the axis of the bar. Before discussing the combined axial, bending, and torsion problem in general, let us consider the following example.

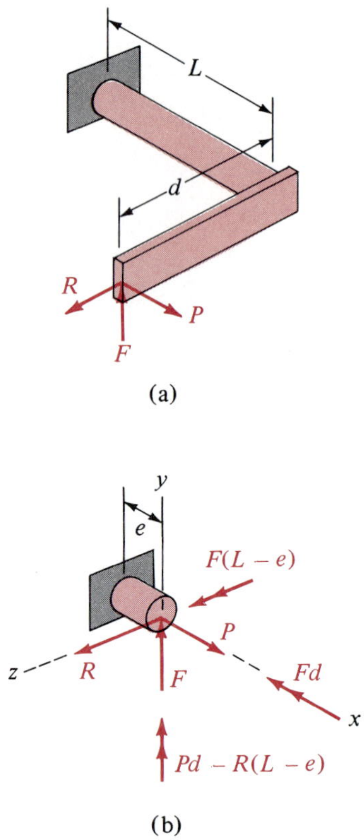

Figure 9.30 Internal forces transmitted for a typical combined axial, bending, and torsion situation

Example 9.8 A circular-cross-section bar is loaded through a rigid bar attached at the end as shown in Fig. 9.30a. Determine the force resultants and the corresponding stresses on a typical section a distance e from the fixed end.

Solution Referring to Fig. 9.30b, we see that at a typical section located a distance e from the support, all possible internal force resultants are being transmitted: an axial load P, transverse shear loads F and R in the y- and z-directions respectively, bending moments $M_y = Pd - R(L - e)$ and $M_z = F(L - e)$ about both the y- and z-axes, and a torque $T = Fd$ about the x-axis.

Using the results from the torsional theory, namely $\tau = Tr/J$ for a circular cross section, superposition is valid in terms of combining the individual results from the elementary theories for axial, torsion, bending, and transverse shear. Shown in Fig. 9.31 are the stresses associated with each of the six resultants shown in Fig. 9.30b. The equation used to compute each individual stress along with the maximum value of the stress and its location within the cross section is also given.

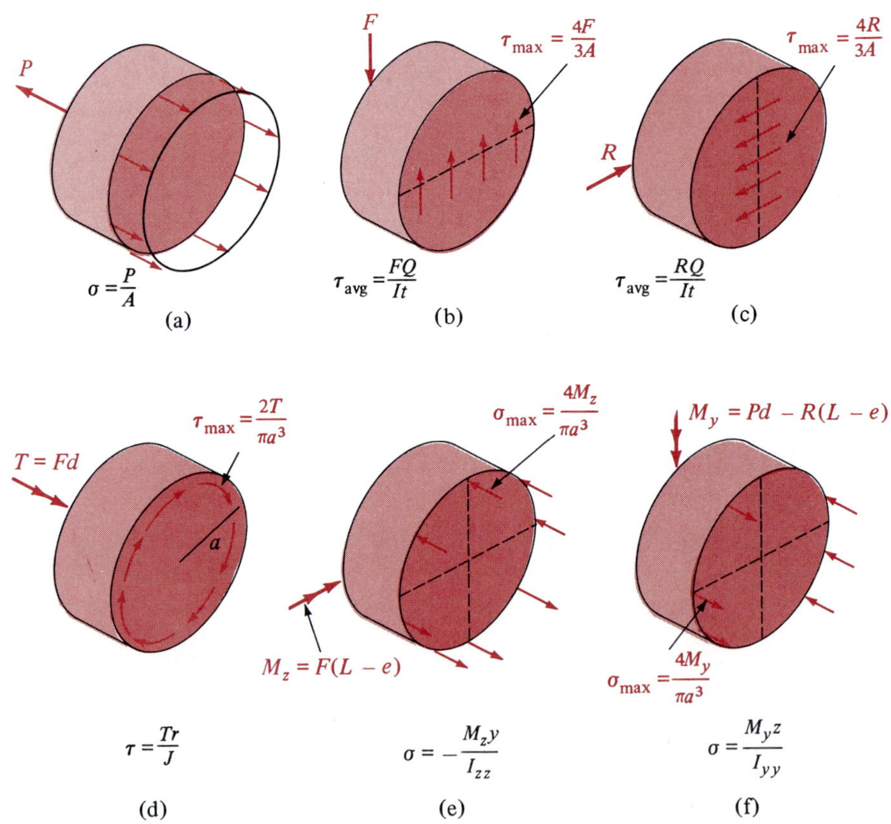

Figure 9.31 Distribution and maximum values of stresses arising from axial, shear, bending, and torsional force resultants

The individual normal stresses due to axial and bending resultants would then be added at selected points within the cross section to determine the corresponding total normal stress. Similarly, the various shear stresses would be added at selected points to determine the corresponding total shear stress. These normal and shear stress components could then be used to determine principal stresses and the maximum shear stress and to ultimately assess the strength of the bar.

It is most important to keep in mind all the assumptions that are a part of the theories represented (axial, torsional, bending, and transverse shear) when computing stresses as in Fig. 9.31. We will review the assumptions for geometry, loading, and material properties.

Geometry A review of Chapters 5, 6, and 7 shows that all of the theories are based on the assumption that *the bar is straight with no sudden changes in cross section*. In addition to satisfying this assumption, we will assume that the x-axis coincides with the centroid, and that the cross section of the bar has two planes of symmetry named the y- and z-axes. We are dealing primarily with cross sections that are circles, rectangles, thin-walled circular tubes, or thin-walled rectangular tubes. Although the elementary theories of Chapters 5, 6, and 7 can actually be combined to treat more general cross sections, we are considering only bars with doubly symmetric cross sections in this section.

Loading It is possible to resolve any force that is transmitted across a section into an axial force P coinciding with the x-axis and two transverse forces V_y and V_z, lying along the y- and z-axes. Similarly, any moment that is transmitted across a section can be resolved into a torque T about the x-axis and two bending moments M_y and M_z about the y- and z-axes.

Material Properties We assume that the material is linearly elastic, and that all stresses are below the proportional limit in tension, compression, and shear.

With the assumptions that the bar is straight and has a cross section with two planes of symmetry, it is then possible to translate the internal forces and moments referred to the xyz coordinate system into normal and shear stresses, using the elementary theories presented in Chapters 5, 6, and 7, and given in Fig. 9.31.

Example 9.9 A 2-in. solid circular bar is loaded by means of an attached rigid bar as shown in Fig. 9.32a. Determine the normal and transverse shear-stress distributions on a section 2 in. away from the support, and the worst-case principal stresses and maximum shear stress at the same location.

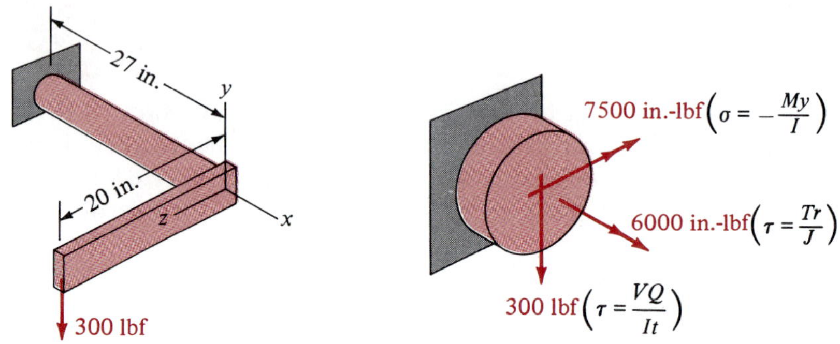

Figure 9.32 Circular bar loaded in bending, shear, and torsion

Solution Fig. 9.32b shows the internal forces and moments that must be transmitted for equilibrium. Also shown in parentheses is the type of stress that each of the resultants will produce.

7500-in.-lbf moment: The bending stress resulting from this moment is given by $\sigma = -My/I$, which evaluates to

$$\sigma = -\frac{(-7500\text{ in.-lbf})y}{0.785\text{ in}^4} = 9.55y\text{ ksi}$$

with y measured in inches. The distribution is shown in Fig. 9.33a.

6000-in.-lbf torque: The torsional stress is computed as $\tau = Tr/J$, which becomes

$$\tau = \frac{(6000\text{ in.-lbf})r}{1.571\text{ in}^4} = 3.82r\text{ ksi}$$

where r is measured in inches. The distribution is shown along a particular radius in Fig. 9.33b.

300-lbf shear: The maximum value of the shear stress due to the 300-lbf shear occurs on the z-axis and is given by

$$\tau = \frac{4V}{3A} = \frac{4(300\text{ lbf})}{3(\pi\text{ in}^2)} = 0.13\text{ ksi}$$

and is indicated in Fig. 9.33c.

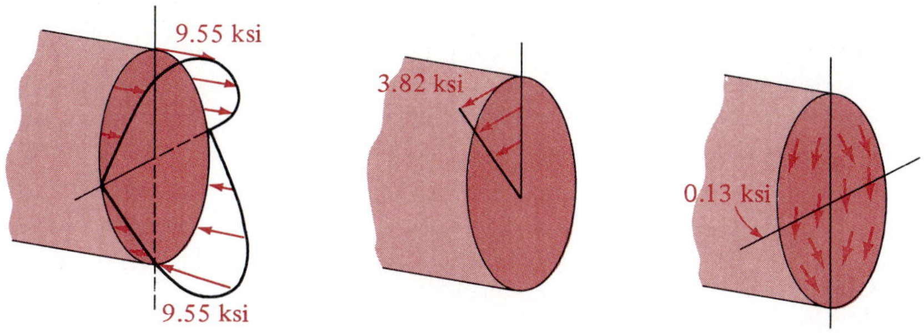

(a) $\sigma = -\frac{My}{I}$ (b) $\tau = \frac{Tr}{J}$ (c) $\tau = \frac{VQ}{It}$

Figure 9.33 Individual normal and shear stresses

At the top or bottom of the bar there is the normal bending stress of 9.55 ksi and the torsional shear stress of 3.82 ksi as shown in Fig. 9.34a.

The corresponding transverse shear stress from the 300-lbf transverse load is zero at these locations. The absolute maximum shear stress and principal stresses are then given by

$$\tau_{\text{abs}} = \sqrt{\left(\frac{9.55}{2}\right)^2 + (3.82)^2} \text{ ksi} = 6.11 \text{ ksi}$$

$$\sigma_{\min} = \left(\frac{9.55}{2} - 6.11\right) \text{ ksi} = -1.34 \text{ ksi}$$

$$\sigma_{\max} = \left(\frac{9.55}{2} + 6.11\right) \text{ ksi} = +10.89 \text{ ksi}$$

For the point located on the horizontal diameter at the outer surface in the positive z-direction, the normal stress is zero, with the shear stresses superposing to yield $\tau = (3.82 + 0.13)$ ksi $= 3.95$ ksi, as indicated in Fig. 9.34b. The absolute maximum shear stress and principal stresses are then given by

$$\tau_{\text{abs}} = \sqrt{(0)^2 + (3.82)^2} \text{ ksi} = 3.82 \text{ ksi}$$

$$\sigma_{\max,\min} = (0 \pm 3.82) \text{ ksi} = \pm 3.82 \text{ ksi}$$

Other points in the cross section could be checked, but it would turn out that for this loading, the largest values of the principal stresses and the absolute maximum shear stress occur at the top and bottom near the support.

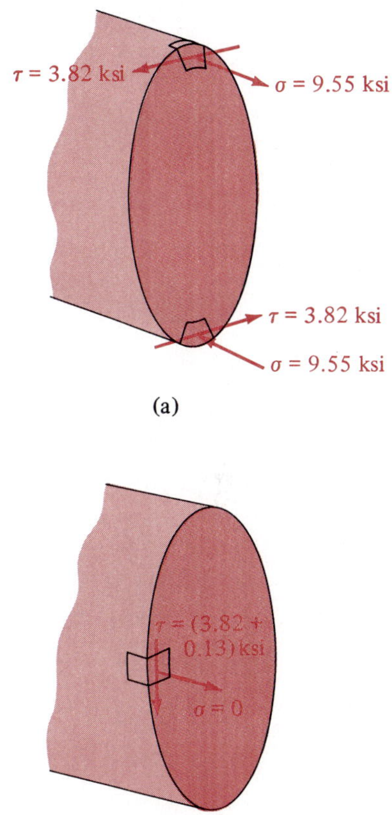

(b)

Figure 9.34 Normal and shear stresses at various locations on the cross section

Example 9.10 A square-cross-section bar is acted upon by the two forces shown. Determine the stress distributions and the point at which the maximum principal stress and the point at which the maximum shear stress occur on a section 0.1 m above the support.

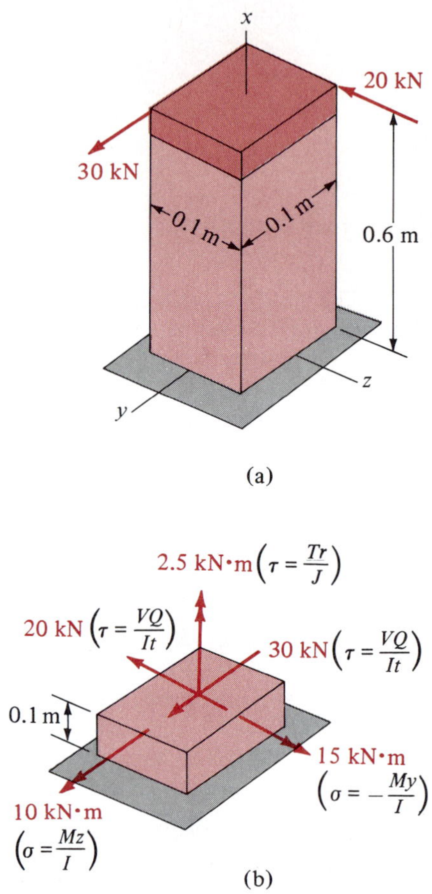

Figure 9.35 Combined loading of a square-cross-section bar

Solution Fig. 9.35b shows the internal forces that must be transmitted for equilibrium. The type of stress, normal or shear, that each of the resultants produces is indicated in parentheses. We will determine the individual stress distributions and then superpose them at a selected number of points to assess the location of the critical point in the cross section.

30-kN shear load: For the square section, the shear-stress distribution is parabolic, with a maximum value of

$$\tau_{\max} = \frac{3}{2}\tau_{\text{avg}} = \frac{3}{2}\left(\frac{30{,}000\ \text{N}}{0.01\ \text{m}^2}\right) = 4.5\ \text{MPa}$$

as indicated in Fig. 9.36a.

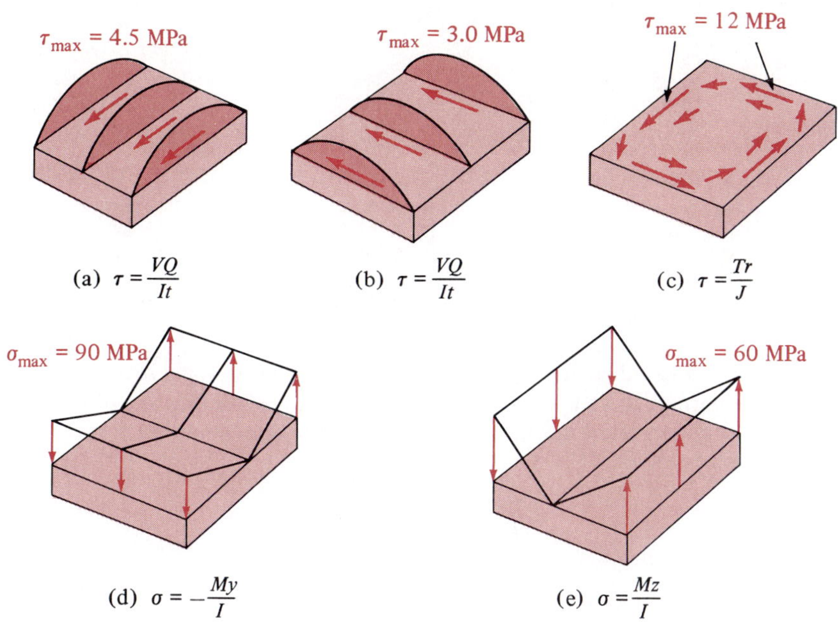

Figure 9.36 Stress distributions due to individual internal force resultants

20-kN shear load: In a completely similar fashion, the shear-stress distribution is parabolic, with a maximum value of 3.0 MPa as indicated in Fig. 9.36b.

2.5-kN·m torque: Consulting the results of Section 6.6, the maximum shear stress occurs at the midpoints of the square sides and has the value

$$\tau_{max} = \frac{2.5 \text{ kN·m}}{0.208(0.1 \text{ m})^3} = 12 \text{ MPa}$$

and is shown in Fig. 9.36c. Note also that the shear stress is zero in both directions at the corners.

10-kN·m moment: The bending stresses that result from the 10 kN·m moment about the z-axis are shown in Fig. 9.36d, and have a maximum value of

$$\sigma_{max} = \frac{6(10 \text{ kN·m})}{(0.1 \text{ m})^3} = 60 \text{ MPa}$$

15 kN·m moment: In a completely similar fashion the 15 kN·m moment about the y-axis results in a maximum bending stress of 90 MPa and is indicated in Fig. 9.36e.

At this point some judgment must be used to assess the point in the cross section that experiences the critical state of stress. We will investigate the states of stress at points A, B, and C indicated in Fig. 9.37a. Point A is the point where the normal stress is maximum; points B and C are points where the different shear contributions have the same direction.

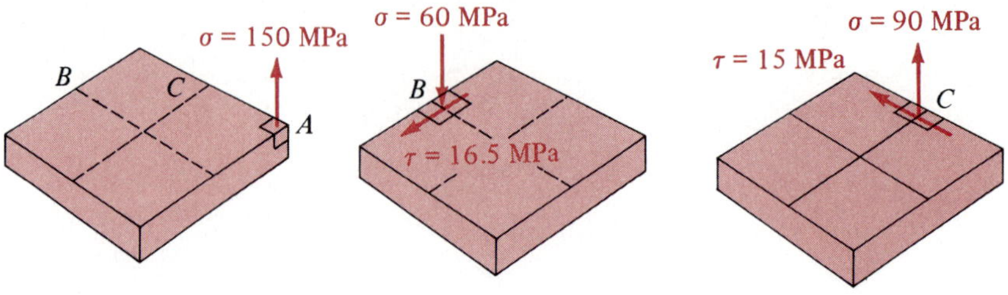

Figure 9.37 Combined normal and shear stresses at various points in the cross section

Point A: Consulting Fig. 9.36, we see that the two bending stresses add and that at the corner, there is no shear stress from either of the two shear forces or the torque, as indicated in Fig. 9.37a. The absolute maximum shear stress and principal stresses are

$$\tau_{abs} = \sqrt{\left(\frac{150}{2}\right)^2 + 0^2} \text{ MPa} = 75 \text{ MPa}$$

$$\sigma_{min} = 0 \qquad \sigma_{max} = 150 \text{ MPa}$$

Point B: Again consulting Fig. 9.36, the stresses acting at point B are as shown in Fig. 9.37b. The maximum absolute shear stress and principal stresses are

$$\tau_{abs} = \sqrt{\left(\frac{-60}{2}\right)^2 + (16.5)^2} \text{ MPa} = 34.2 \text{ MPa}$$

$$\sigma_{min} = (30.0 - 34.2) \text{ MPa} = -4.2 \text{ MPa}$$

$$\sigma_{max} = (30.0 + 34.2) \text{ MPa} = 64.2 \text{ MPa}$$

Point C: The superposition of the normal and shear stresses at point C is shown in Fig. 9.37c. With $\sigma = -90$ MPa and $\tau = 15$ MPa, the absolute maximum shear stress and principal stresses are

$$\tau_{abs} = \sqrt{\left(\frac{90}{2}\right)^2 + (15)^2} \text{ MPa} = 47.4 \text{ MPa}$$

$$\sigma_{min} = (-45.0 - 47.4) \text{ MPa} = -92.4 \text{ MPa}$$

$$\sigma_{max} = (-45.0 + 47.4) \text{ MPa} = +2.4 \text{ MPa}$$

Thus in terms of the points checked, the largest principal stress of 150 MPa and the absolute maximum shear stress of 75 MPa both occur at point A with τ_{abs} acting on a 45° plane.

PROBLEMS

In the following problems, inplane shear stress and inplane normal stress refer, respectively, to the combined shear and normal stress obtained by superposing the axial, bending, transverse shear, and torsional stresses at a particular point. The maximum shear stress and principal stresses refer to the corresponding principal values at the point based upon the use, for instance, of Mohr's circle.

9.49 A solid circular-cross-section bar is acted upon by the two loads shown. Determine the inplane shear stress and the inplane normal stress at *(a)* point *A*, the top of the bar, and *(b)* point B, on the side of the bar. Take the radius of the bar to be 20 mm, $L = 0.3$ m, and $P = 600$ N.

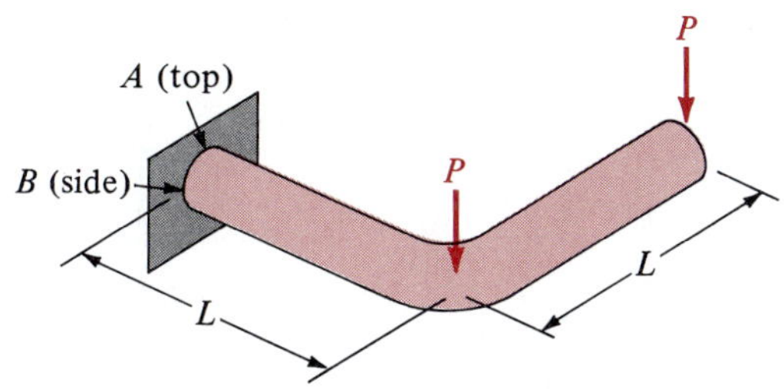

Ps. 9.49 and 9.50

9.50 For the same data as given in problem 9.49, determine the principal stresses and the absolute maximum shear stress at *(a)* point *A*, the top of the bar, and *(b)* point *B*, on the side of the bar.

9.51 A solid square-cross-section bar is acted upon by the distributed load shown. At the support, determine the inplane shear stress and the inplane normal stress at *(a)* point *A*, and *(b)* point *B*. Take $b = 40$ mm, $L = 0.6$ m, and $w_0 = 700$ N/m.

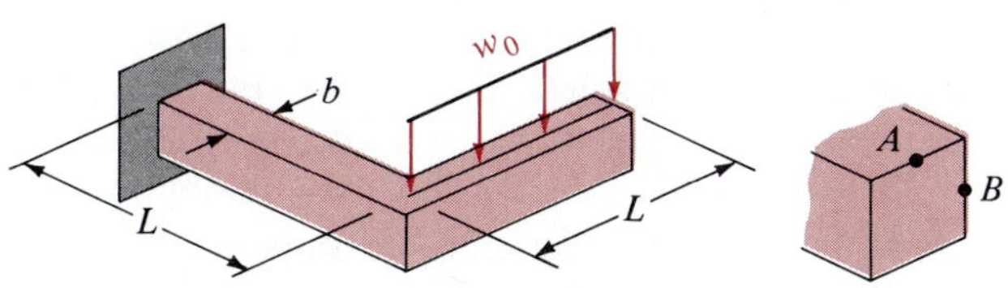

Ps. 9.51 and 9.52

9.52 For the same data given in problem 9.51, determine the principal stresses and the absolute maximum shear stress at *(a)* point *A*, and *(b)* point *B*.

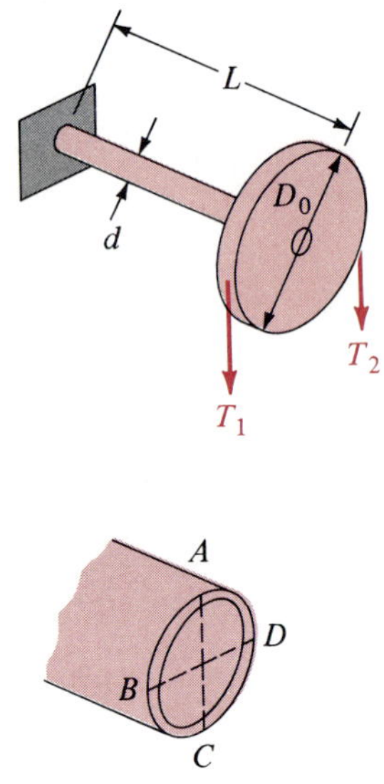

Ps. 9.53 and 9.54

9.53 A pulley attached to the end of a hollow pipe transmits the belt tensions shown. At the support, determine the inplane normal and shear stresses at points A, B, C, and D. Take $L = 1$ m, $D_0 = 0.6$ m, $T_1 = 400$ N, $T_2 = 500$ N, with the inside and outside diameters of the pipe 140 mm and 150 mm, respectively.

9.54 Repeat problem 9.53 for the principal stresses and the absolute maximum shear stress at the points A, B, C, and D.

9.55 The 1.0-in. square-cross-section bar is acted upon by the forces shown. At the support, determine the inplane shear and inplane normal stresses at *(a)* point A, *(b)* point B, *(c)* point C, on the bottom face, and *(d)* point D, opposite point B. Take $a = b = c = 8$ in., $P = 50$ lbf, $R = 100$ lbf, and $S = 70$ lbf.

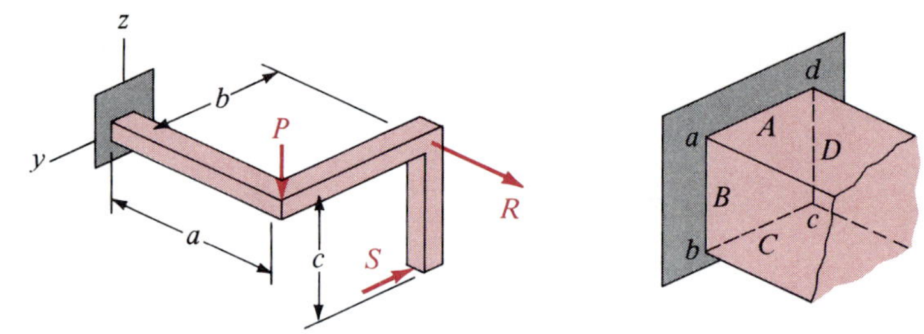

Ps. 9.55–9.58

9.56 Repeat problem 9.55 for each of the corners at the support.

9.57 Repeat problem 9.55 computing the principal stresses and the absolute maximum shear stress at points A, B, C, and D.

9.58 Repeat problem 9.55 computing the principal stresses and the absolute maximum shear stress at each of the corners.

9.59 A hollow rectangular tube is subjected to the eccentric loading shown. The distance e is measured from the centroidal axis of the tube. For a section at the support, determine the inplane shear and bending stresses at the midpoints of each of the sides. (*Hint:* For a thin-walled hollow tube the shear stress τ due to a torque T relative to the centroid of the cross section is given by $\tau = T/2At$ where A refers to the area of the tube computed on the basis of the midline of the cross section and t is the local thickness.) Take $P = 400$ lbf, $B = 1.5$ in., $b = 1.25$ in., $H = 2.0$ in., $h = 1.75$ in., $L = 14.0$ in., and $e = 6.0$ in.

9.60 Repeat problem 9.59 for the principal stresses and the absolute maximum shear stress.

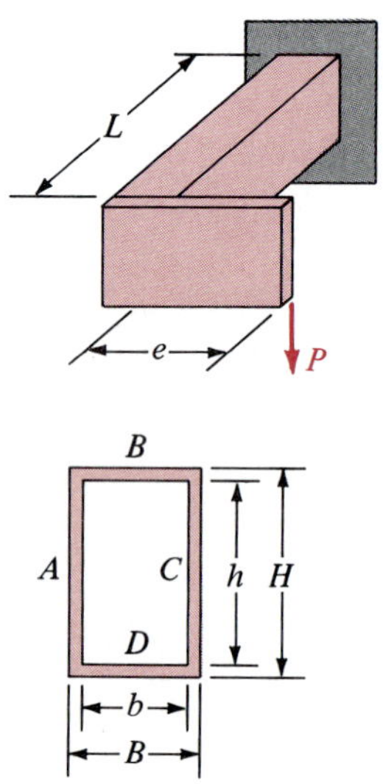

Ps. 9.59 and 9.60

9.61 A solid circular shaft is acted upon by a force applied to the rigid extension as shown. The shaft is simply supported for bending about both the y- and z-axes at both ends and clamped against torsional rotation at the left end. For a section located just to the left of the point where the rigid bar connects to the shaft, determine the inplane normal and shear stresses at points A, B, C, and D. Take the diameter of the shaft to be 4 in., $L = 6$ ft, $d = 3$ ft, $\theta = 60°$, and $P = 4000$ lbf.

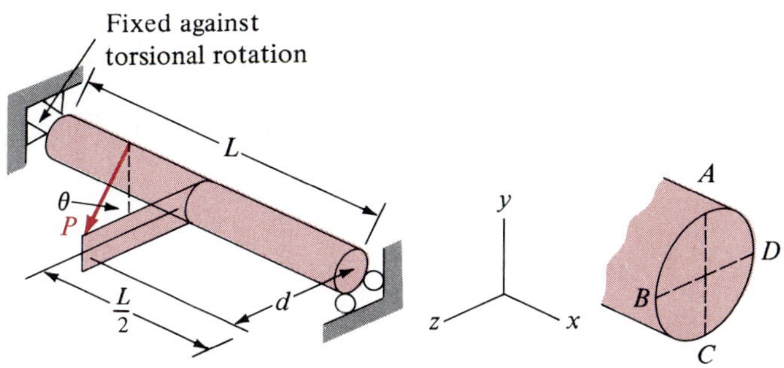

Ps. 9.61 and 9.62

9.62 Repeat problem 9.61, this time computing the absolute maximum shear stress and the principal stresses at points A, B, C, and D.

9.63 A solid circular shaft of radius a is subjected to a torque T and an axial load P as shown. If the absolute maximum shear stress is not to exceed the value τ_0, derive a relationship between T, P, a, and τ_0.

9.64 A solid circular shaft of radius a is subjected to a torque T and an axial load P as shown. If the maximum principal stress is not to exceed the value σ_0, derive a relationship between T, P, a, and σ_0.

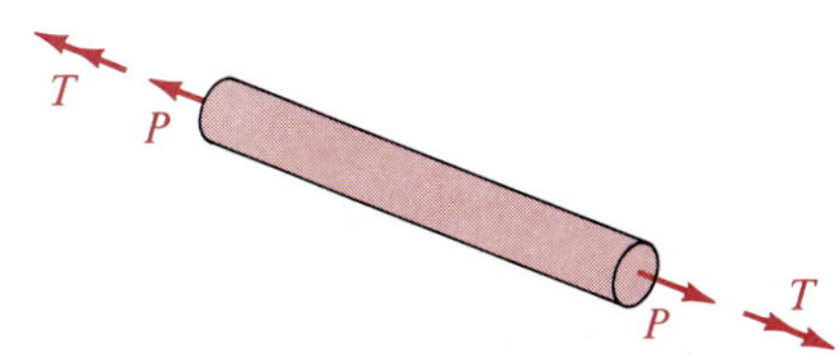

Ps. 9.63 and 9.64

9.65 A solid circular shaft of radius a is subjected to a torque T and a moment M as shown. If the absolute maximum shear stress is not to exceed the value τ_0, derive a relationship between T, M, a, and τ_0.

9.66 A solid circular shaft of radius a is subjected to a torque T and a moment M as shown. If the maximum principal stress is not to exceed the value σ_0, derive a relationship between T, M, a, and σ_0.

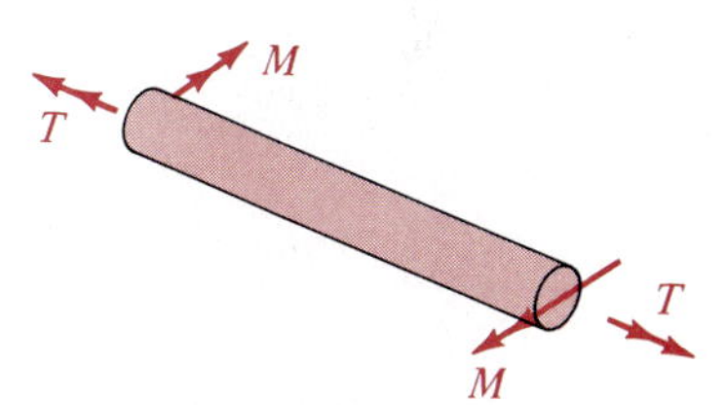

Ps. 9.65 and 9.66

9.67 A solid circular shaft of radius a is subjected to a torque T, an axial load P, and a moment M as shown. If the absolute maximum shear stress is not to exceed the value τ_0, derive a relationship between T, M, P, a, and τ_0.

9.68 A solid circular shaft of radius a is subjected to a torque T, an axial load P, and a moment M as shown. If the maximum principal stress is not to exceed the value σ_0, derive a relationship between T, M, P, a, and σ_0.

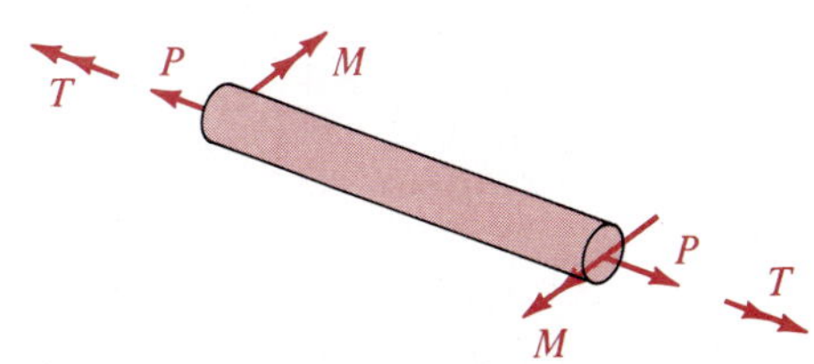

Ps. 9.67 and 9.68

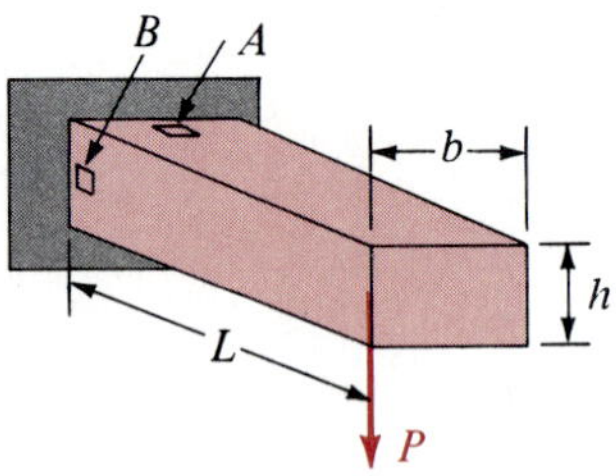

Ps. 9.69 and 9.70

9.69 The rectangular-cross-section beam is loaded by the eccentric force P as shown. Determine the inplane shear stress and the inplane normal stress at *(a)* point A, and *(b)* point B. Take $b = 40$ mm, $h = 40$ mm, $L = 200$ mm, and $P = 200$ N.

9.70 For the data of problem 9.69, determine the principal stresses and the absolute maximum shear stress at *(a)* point A, and *(b)* point B.

9.71 The traffic signal is constructed from a hollow pipe section as shown. Considering the weight W_s of the sign, the weight of the pipe, and the wind force F_w as the loads, determine the inplane shear and inplane normal stresses at the points A, B, C, and D as shown. Take the outside diameter of the steel pipe to be 8.0 in., the wall thickness to be 0.5 in., $L = 40$ ft, $h = 20$ ft, $W_s = 400$ lbf, and $F_w = 200$ lbf. The weight density of steel is 0.283 lbf/in^3.

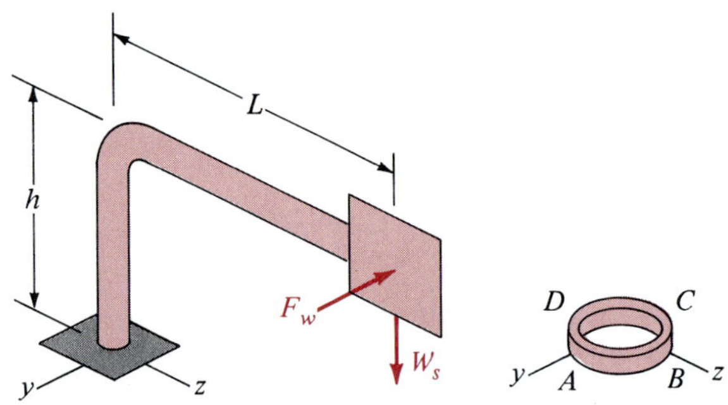

Ps. 9.71–9.74

9.72 Using the same data, repeat problem 9.71 for each of the points midway between A and B, B and C, and so forth.

9.73 Using the same data, repeat problem 9.71 determining the principal stresses and the absolute maximum shear stress at points A, B, C, and D.

9.74 Repeat problem 9.71, this time determining the principal stresses and the absolute maximum shear stress at the points midway between A, B, C, and D.

9.75 The L-shaped square-cross-section aluminum bar is loaded by its weight as shown. What is the maximum length L if the inplane normal stress at the support is not to exceed 100 MPa? Take $b = 40$ mm. Take the weight density of the aluminum to be 26.7 kN/m^3.

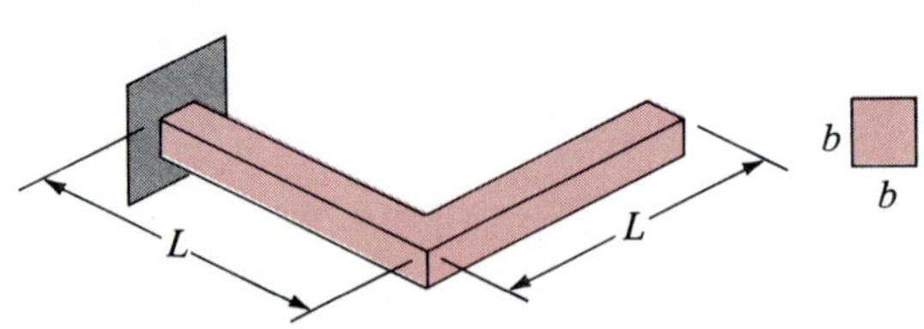

Ps. 9.75–9.82

9.76 Repeat problem 9.75 if the inplane shear stress at the support is not to exceed 50 MPa.

9.77 Repeat problem 9.75 if the allowable value of the maximum principal stress is 100 MPa.

9.78 Repeat problem 9.75 if the allowable value of the absolute maximum shear stress is 50 MPa.

9.79 Repeat problem 9.75 if the aluminum bar is circular with radius 1.0 in., and the allowable inplane bending stress at the support is not to exceed 15 ksi. Take the specific weight to be 0.1 lbf/in^3.

9.80 Repeat problem 9.79 if the inplane shear stress at the support is not to exceed 8 ksi.

9.81 Repeat problem 9.79 if the allowable value of the maximum principal stress is 15 ksi.

9.82 Repeat problem 9.79 if the allowable value of the absolute maximum shear stress is 8 ksi.

9.83 A pipe with an outside diameter D and a wall thickness of 0.5 in. is acted upon by the loads shown. If the allowable inplane normal stress on a section at the support is 16 ksi, determine D. Take $h = 8$ ft, $L = 6$ ft, $R = 1200$ lbf, $P = 2700$ lbf, and $S = 2000$ lbf.

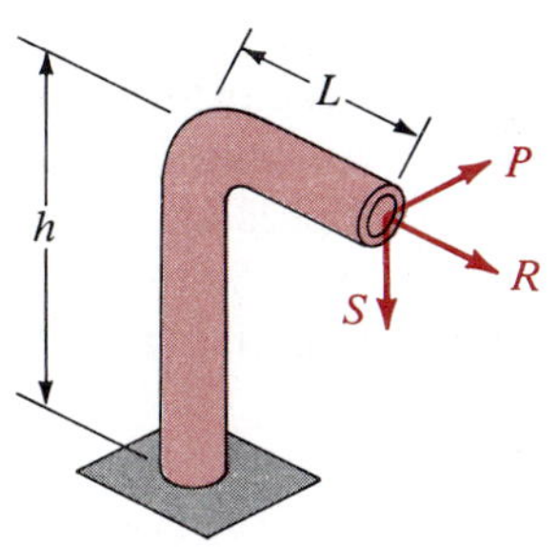

Ps. 9.83–9.86

9.84 Repeat problem 9.83 for the same data if the maximum principal stress at the section is not to exceed 16 ksi.

9.85 Repeat problem 9.83 for the same data if the inplane shear stress at the section is not to exceed 8 ksi.

9.86 Repeat problem 9.83 for the same data if the absolute maximum shear stress at the section is not to exceed 8 ksi.

9.87 The two pulleys transmit the belt tensions shown. Assume that the ends of the shaft are simply supported for bending and that the left end is restrained against torsional rotation. If the diameter of the solid shaft is 60 mm, determine *(a)* the maximum bending stress in the shaft, and *(b)* the maximum inplane shear stress in the shaft. Take the diameters of both pulleys to be 0.5 m, $T_1 = 1.3$ kN, $T_2 = 0.9$ kN, $T_3 = 0.7$ kN, $T_4 = 0.4$ kN, $L_1 = 0.9$ m, $L_2 = 1.3$ m, and $L_3 = 1.1$ m.

9.88 For the same data, repeat problem 9.87 for *(a)* the maximum principal stress in the shaft, and *(b)* the absolute maximum shear stress in the shaft.

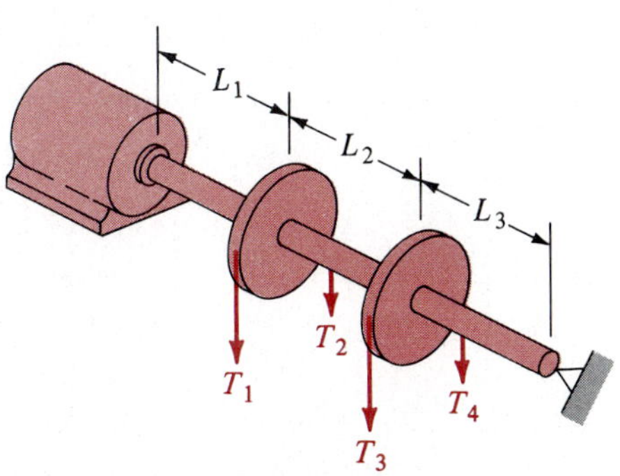

Ps. 9.87 and 9.88

9.4 THIN-WALLED PRESSURE VESSELS

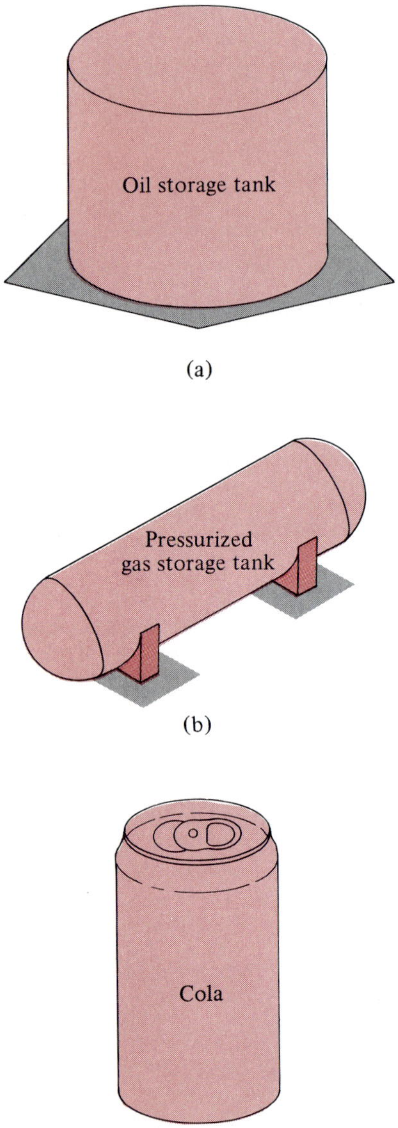

Figure 9.38 Examples of thin-walled pressure vessels

The real world abounds with examples of thin-walled pressure vessels. Several are sketched in Fig. 9.38. Fig. 9.38a shows a large-capacity cylindrical oil storage tank, where the internal pressure is due to the liquid contained within. Fig. 9.38b indicates a cylindrical vessel capped by two hemispheres, containing a gas under high pressure. Fig. 9.38c pictures an ordinary beverage container that can be idealized as a cylindrical pressure vessel capped by circular plates, which can be considered as containing a fluid under pressure. Examples also include an ordinary basketball or football, each of which contains air under pressure.

Each of these examples shares several features. First, each is intended to sustain loading in the form of an internal pressure normal to the inner surface of the vessel. Second, they are all geometrically similar from the standpoint that the wall thickness is small compared to the radius of the cylinder, sphere, and so forth. This thin-walled feature allows us to develop a simple approximate theory for determining the internal forces and stresses in the vessel. The state of stress is generally biaxial in nature, with internal forces per unit of length in both the axial and circumferential directions, that is, internal forces that are tangent to the surface of the pressure vessel. The theory will permit us to answer some of the questions regarding the strength and stiffness of thin-walled pressure vessels. (A complete accounting of the stresses and deformations in thin-walled pressure vessels is beyond the scope of this text. The interested reader should consult the various shell theories presented in advanced solid mechanics and shell theory texts.)

Cylindrical Pressure Vessels Consider the simplest geometry, namely that of a cylindrical pressure vessel closed off by two flat plates as indicated in Fig. 9.39a. We assume that the vessel contains a gas under an internal pressure p. The governing equations will be developed by considering the three basic ingredients of equilibrium, deformation, and material behavior.

Equilibrium: In order to expose the internal forces acting in the axial direction, we draw a free-body diagram associated with passing a section through the vessel perpendicular to the axis of the cylinder as indicated in Fig. 9.39b. The section is taken at a point away from the ends of the cylinder.

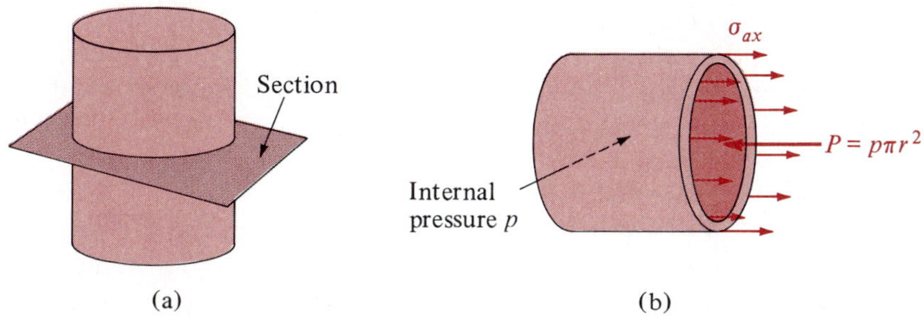

Figure 9.39 Free-body diagram exposing the internal axial forces

The free-body diagram shows the resultant force $P = p\pi r^2$ arising from the internal pressure and the induced internal forces consisting of normal stresses σ_{ax} distributed around the exposed circular section. The normal stresses σ_{ax} are assumed to be constant over the thickness of the vessel and around the circumference. With R_i and R_o the inner and outer radii of the cylinder, respectively, a sum of forces in the direction of the axis of the cylinder yields

$$\Sigma F_x = -p\pi R_i^2 + \pi(R_o^2 - R_i^2)\sigma_{ax} = 0$$

from which

$$\sigma_{ax} = \frac{pR_i^2}{R_o^2 - R_i^2} \tag{a}$$

Note that the line of action of the pressure force $P = \pi R_i^2 p$ is along the axis of the cylinder, thus we have axial loading. Write $(R_o^2 - R_i^2) = (R_o - R_i)(R_o + R_i) = t2R_{avg}$, where $t = R_o - R_i$ is the thickness and $R_{avg} = (R_o + R_i)/2$ is the average radius. The axial stress σ_{ax} can then be expressed as

$$\sigma_{ax} = pR_i^2/2R_{avg}t \tag{b}$$

Then note that $R_{avg} = R_i + t/2$, so that

$$\sigma_{ax} = \frac{pR_i^2}{2t(R_i + t/2)} = \frac{pR_i}{2t(1 + t/2R_i)}$$

Typically the geometry of a thin-walled pressure vessel is such that the ratio of wall thickness to the inside radius, namely $t/(2R_i)$, is of the order of 0.01 or less, so that $1 + t/2R_i \approx 1$, with the axial stress as

$$\sigma_{ax} = \frac{pR_i}{2t} \tag{9.3}$$

It would be possible to have nonuniform axial stresses on the exposed face that would be statically equivalent to the force $P = \pi R^2 p$ and also to have shear stresses with zero resultant force and moment acting on the exposed face. That these additional stresses are not actually present can only be determined by a more general analysis.

To expose the internal forces acting in the circumferential direction, we draw the free-body diagram in Fig. 9.40. Shown are the pressure forces, the forces associated with the axial stresses σ_{ax} determined above, and the so-called hoop stresses σ_H acting in the circumferential direction. The hoop stresses are assumed to be uniform on the exposed face. This assumption is valid as long as t/R_i is small. The sum of the forces in the axial direction is identically satisfied. A sum of forces in the circumferential direction yields

$$\Sigma F = p2R_i \, \Delta x - 2t\sigma_H \, \Delta x = 0$$

or

$$\sigma_H = \frac{pR_i}{t} \tag{9.4}$$

showing that the hoop stress σ_H is twice as large as the axial stress σ_{ax}. In what follows we will simply write R_i as R.

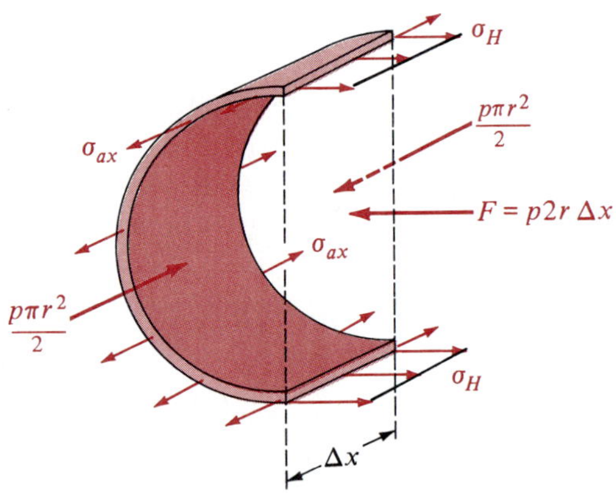

Figure 9.40 Free-body diagram exposing circumferential forces

Shear stresses on 45° planes: Consider a small differential element of the shell as indicated in Fig. 9.41.

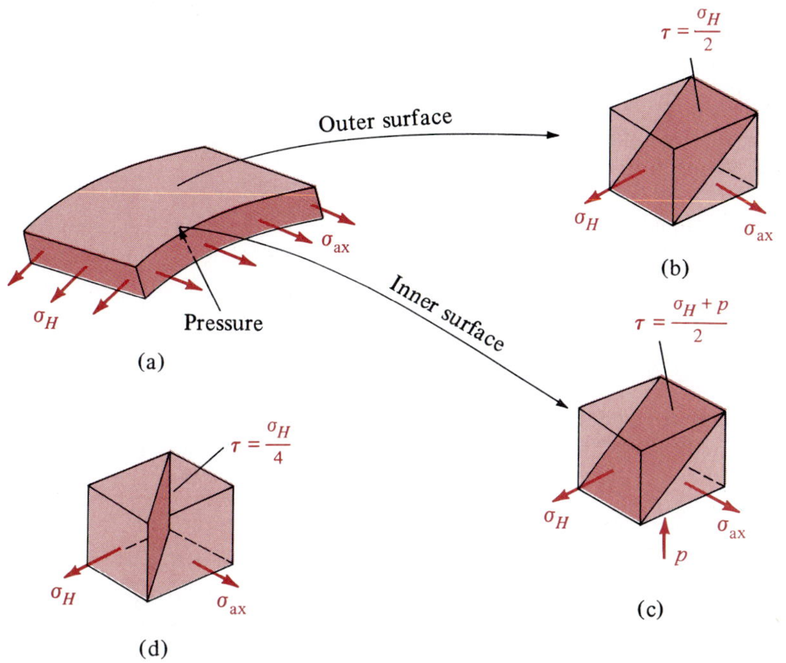

Figure 9.41 Shear stresses in the pressure vessel

At a point on the outer surface of the shell, the stresses are as indicated in Fig. 9.41b, whereas at the inner surface they are as indicated in Fig. 9.41c. In each case the stresses shown are principal stresses, since there are no shear stresses acting on any of the faces. At the outer surface, the maximum shear stress of $\tau = \sigma_H/2$ occurs on the 45° plane shown in Fig. 9.41b, and also on the perpendicular plane. At the inner surface, a shear stress of $\tau = (\sigma_H + p)/2$ occurs on the plane shown in Fig. 9.41c, as well as the perpendicular plane. For a thin shell, these two values, $\sigma_H/2$ and $(\sigma_H + p)/2$, are essentially the same since $t/R \ll 1$. A third shear stress acts on a plane bisecting the axial and hoop directions, and has the value $(\sigma_H - \sigma_{ax})/2 = \sigma_H/4$ as shown in Fig. 9.41d. Thus the absolute maximum shear stress occurs on a 45° plane at the inner surface.

Example 9.11 A cylindrical pressure vessel with a 100-in. inside diameter and 0.25-in.-thick wall is subjected to an internal pressure of 100 psi. At a point away from the ends, determine the axial and hoop stresses. Also determine the maximum shear stress at the inside and outside surfaces.

Solution For this particular geometry, the thickness-to-radius value is 0.25/50, or 1/200, a value sufficiently small to warrant the use of the approximate expression given by Eqn. (9.3) for the axial stress. Using Eqn. (9.3), the axial stress is

$$\sigma_{ax} = \frac{pR}{2t} = \frac{(100 \text{ psi})(50)}{2(0.25 \text{ in.})} = 10.0 \text{ ksi}$$

Using Eqn. (9.4) for the hoop stress there results

$$\sigma_H = \frac{pR}{t} = \frac{(100 \text{ psi})(50)}{(0.25 \text{ in.})} = 20.0 \text{ ksi}$$

The absolute maximum shear stress values at the inner and outer surfaces are $\tau_{max} = (\sigma_H + p)/2 = 10.05$ ksi and $\tau_{max} = \sigma_H/2 = 10.0$ ksi, respectively.

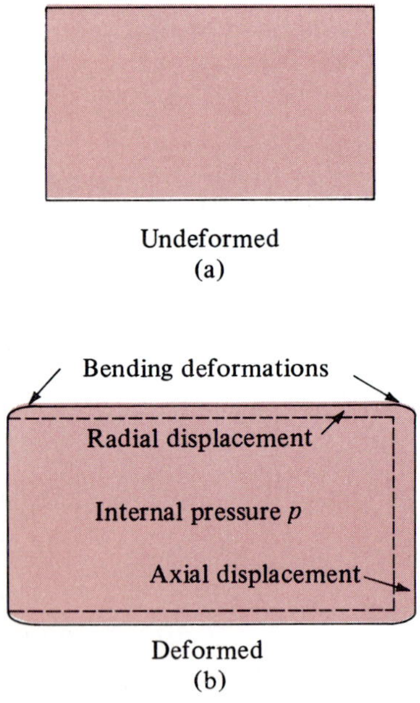

Figure 9.42 Undeformed and deformed shells

Consider a side view of a capped cylindrical shell as indicated in Fig. 9.42. The deformed shell is shown in Fig. 9.42b. We will not be able to determine the deformations in the neighborhoods of the ends of the shell on the basis of the theory presented here. As indicated in the figure, there is bending in the neighborhood of the ends of the shell. The bending is due to the constraint provided by the "cap," as it is usually called. The cap restricts the free radial expansion of the pressure vessel near the ends, thus producing local bending at the ends. The axial and hoop stresses we have computed are often called **membrane stresses,** and the corresponding theory, **membrane theory.** The membrane theory is valid only away from the ends where there are only membrane stresses. Near the ends where there are both membrane and bending stresses, the elementary theory developed here does not accurately predict the stresses or displacements.

Recall that the three basic ingredients in any solid mechanics problem are equilibrium, deformation, and material behavior. Away from the ends of the shell, equilibrium has been satisfied in computing $\sigma_H = pR/t$ and $\sigma_{ax} = pR/2t$. Let us consider the other two basic ideas necessary to relate the stresses to the strains and hence to the displacements.

Deformations: The expression for the strain in the hoop direction is developed using the definition of change in length divided by original length as follows. Consider the circular shape before and after deformation as indicated in Fig. 9.43.

With the mean radial displacement denoted by u, the strain is easily computed to be

$$\varepsilon_H = \frac{2\pi(R + u) - 2\pi R}{2\pi R} = \frac{u}{R}$$

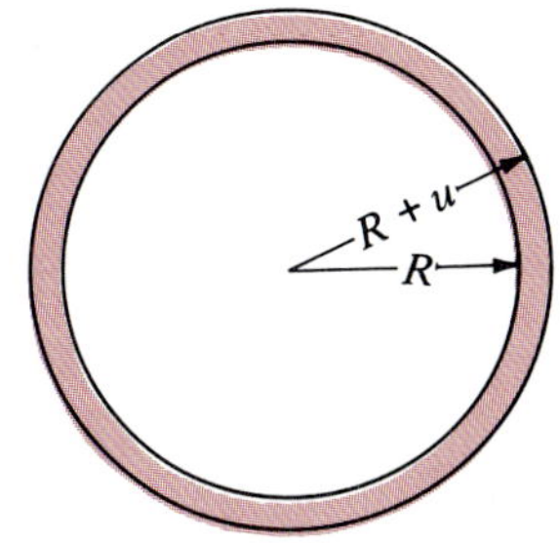

Figure 9.43 Radial deformation u of a cylinder

Assuming that all points away from the ends of the shell at a particular x-location move axially the same amount, it follows by paralleling the development carried out in Section 5.2 that the axial strain is given by

$$\varepsilon_{ax} = \frac{dw}{dx} = w'$$

where $w(x)$ is the axial component of displacement.

Stress-strain relation: We will assume that the shell is made of a linearly elastic isotropic material, so that from Eqns. 4.8

$$\varepsilon_H = \frac{\sigma_H - \nu\sigma_{ax}}{E}$$

$$\varepsilon_{ax} = \frac{\sigma_{ax} - \nu\sigma_H}{E}$$

where the effect of the stress in the radial direction, the direction of the pressure, has been ignored.

With the stresses determined from equilibrium as $\sigma_H = pR/t$ and $\sigma_{ax} = pR/2t$, the strains can be written as

$$\varepsilon_H = \frac{pR}{Et}\left(1 - \frac{\nu}{2}\right)$$

$$\varepsilon_{ax} = \frac{pR}{Et}\left(\frac{1}{2} - \nu\right)$$

Then with

$$\varepsilon_H = \frac{u}{R} = \frac{pR}{Et}\left(1 - \frac{\nu}{2}\right)$$

the radial displacement can be determined as

$$u = \frac{pR^2}{Et}\left(1 - \frac{\nu}{2}\right) \tag{9.5}$$

and is valid away from the ends of the shell. The corresponding strain in the axial direction is given by

$$w' = \varepsilon_{ax} = (pR/Et)\left(\frac{1}{2} - \nu\right)$$

The change in the length of the cylindrical shell is given approximately by integrating this equation to obtain

$$\int_0^L w'\,dx = w(L) - w(0) = (pRL/Et)\left(\frac{1}{2} - \nu\right) \tag{9.6}$$

Example 9.12 Assuming the shell in Example 9.11 to have a length of 400 in. and to be made of aluminum with $E = 10^7$ psi and $\nu = 0.3$, determine *(a)* the radial expansion away from the ends, and *(b)* the approximate change in length of the shell.

Solution The radial expansion away from the ends is given by Eqn. (9.5) as

$$u = \frac{(100\text{ psi})(50\text{ in.})^2(0.85)}{(0.25\text{ in.})(10^7\text{ psi})} = 0.085\text{ in.}$$

The corresponding increase in length is given by Eqn. (9.6) as

$$\Delta L = \frac{(100\text{ psi})(50\text{ in.})(400\text{ in.})0.2}{(0.25\text{ in.})(10^7\text{ psi})} = 0.16\text{ in.}$$

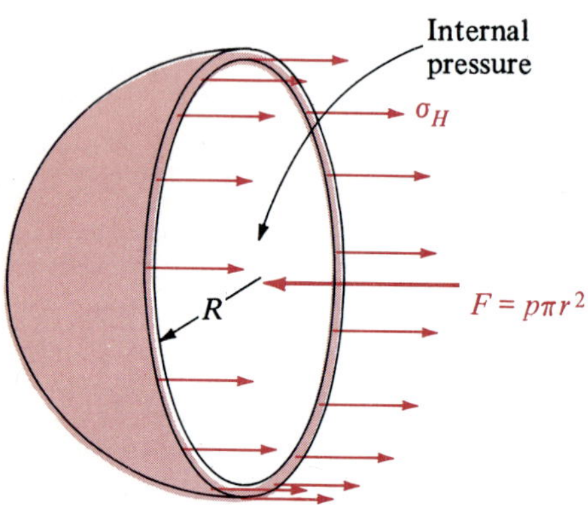

Figure 9.44 Internal forces in a pressurized sphere

Spherical Pressure Vessels Consider a spherical pressure vessel of uniform thickness t and radius R subjected to an internal pressure p. The governing equations will again be obtained by separate consideration of the three basic ingredients.

Equilibrium: A free-body diagram obtained by passing a diametral section through the sphere is shown in Fig. 9.44.

The internal forces at any point in the sphere can be exposed by such a section so that the results obtained will be valid for determining the stresses *at any point in any tangential direction* in the spherical pressure vessel. Summing forces in the direction normal to the section yields

$$\sigma_H(2\pi Rt) = p\pi R^2$$

or

$$\sigma_H = \frac{pR}{2t} \tag{a}$$

where R is the inside radius and t is the thickness of the vessel. This constant value of the normal stress acts tangentially in any direction at any point in the shell.

The values of the shear stress occurring on 45° planes at the outer and inner surfaces are $\tau = \sigma_H/2$ and $\tau = (\sigma_H + p)/2$, respectively, and represent the absolute maximum shear stress in the vessel.

Deformations: Consider a section through the shell as shown in Fig. 9.45. Following the same line of reasoning as for the cylindrical shell it follows that the strain in any direction in the spherical shell is given by

$$\varepsilon_H = \frac{u}{R} \tag{b}$$

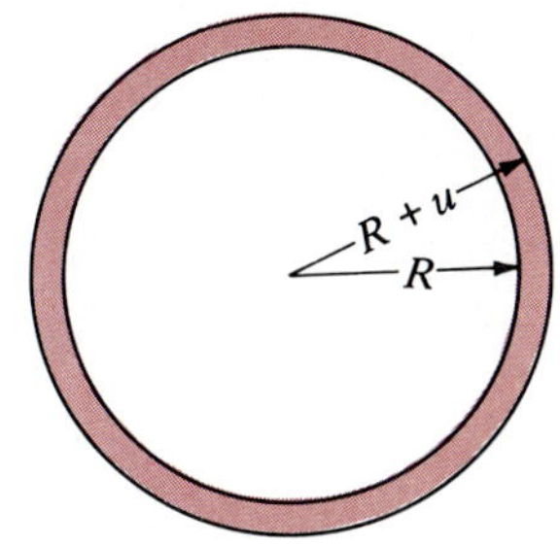

Figure 9.45 Symmetric deformations of a spherical shell

Stress-strain relation: Taking the material to be linearly elastic and isotropic, the strain in any tangential direction in the shell is given by

$$\varepsilon_H = \frac{\sigma_H - \nu\sigma_H}{E} \tag{c}$$

Eqns. (*a*), (*b*), and (*c*) are to be solved for the three unknowns ε_H, σ_H, and u. Eliminating σ_H from (*a*) and (*c*) yields

$$\varepsilon_H = \frac{pR}{2Et}(1 - \nu)$$

which with (*b*) yields

$$u = \frac{pR^2}{2Et}(1 - \nu) \tag{9.7}$$

as the radial displacement of every point in the shell.

Example 9.13 A 0.3-in.-thick steel sphere with a 10-ft diameter is subjected to an internal pressure of 40 psi. Determine the hoop stress and the radial displacement.

Solution The hoop stress in the shell, in any direction, is given by

$$\sigma_H = \frac{pR}{2t} = \frac{(40 \text{ psi})(60 \text{ in.})}{2(0.3 \text{ in.})} = 4 \text{ ksi}$$

Using Eqn. (9.7) with $E = 30 \times 10^6$ psi and $\nu = 0.3$, the radial expansion is

$$u = \frac{pR^2}{2Et}(1 - \nu) = \frac{(40 \text{ psi})(60 \text{ in.})^2}{2(30 \times 10^6 \text{ psi})(0.3 \text{ in.})} 0.7 = 0.056 \text{ in.}$$

As a matter of interest, the change in the volume is given approximately by

$$\Delta V = 4\pi R^2 u = 4\pi(60 \text{ in.})^2(0.0056 \text{ in.}) = 77 \text{ in}^3$$

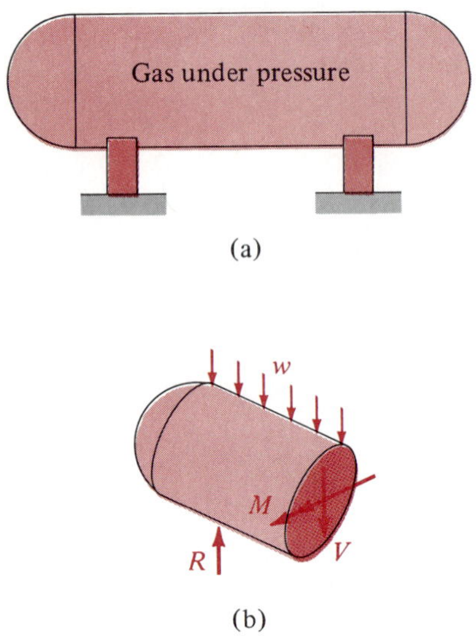

Figure 9.46 Gas storage tank

It is not uncommon for cylindrical pressure vessels to transmit other resultants in addition to those associated with the internal pressure. A typical example of a pressure vessel containing a gas such as propane is shown in Fig. 9.46a.

In addition to the axial and hoop stresses induced in the cylinder by the internal pressure, we see by referring to the free-body diagram of Fig. 9.46b that at a typical location away from the supports, there will also be a shear force and bending moment that must be transmitted due to the weight of the tank and the gas contained within. The bending and shear stresses are superposed on the circumferential and hoop stresses due to the pressure, to determine the combined stresses in the pressure vessel. In other applications it is possible to have external axial, bending, shear, and torsional loadings that result in a situation where it is necessary to superpose all the different normal and shear stresses to assess the strength of the pressure vessel. Let us examine several examples involving pressure vessels with additional loads from external sources.

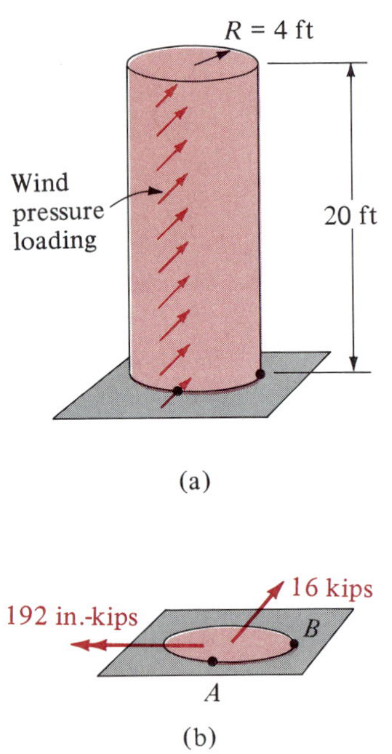

Figure 9.47 Wind-loaded pressure vessel

Example 9.14 A storage tank contains gas at an internal pressure of 20 psi. During a wind storm, the effect of the wind is judged to be a pressure of 10 lbf/ft^2 on the projected area of the tank. Determine the principal stresses and the maximum shear stress at points A and B on the surface as shown in Fig. 9.47. Take the thickness to be 0.4 in., and neglect the weight of the tank.

Solution In addition to the stresses due to the internal pressure at a section near the bottom of the tank, there will be bending stresses and transverse shear stresses from the wind loading. We will evaluate each of these in turn and then superpose them to determine the normal and shear stresses at points A and B, located a short distance above the bottom of the tank.

Bending: The second moment of the area of a thin-walled cylinder can be expressed as $I = \pi(R_o^4 - R_i^4)/4 \approx \pi R^3 t = \pi(48 \text{ in.})^3(0.4 \text{ in.}) = 1.39 \times 10^5 \text{ in}^4$. The maximum bending stress is

$$\sigma = \frac{(192{,}000 \text{ in.-lbf})(48 \text{ in.})}{1.39 \times 10^5 \text{ in}^4} = 66 \text{ psi}$$

and is indicated along with the bending stress distribution in Fig. 9.48a.

Transverse shear: Recall from Section 7.8.1 that for a thin-walled circular section, the maximum transverse shear stress is approximately $\tau_{\max} = 2\tau_{\text{avg}}$, which can be evaluated as

$$\tau_{\max} = 2\,\frac{16{,}000 \text{ lbf}}{2\pi(48 \text{ in.})(0.4 \text{ in.})} = 265 \text{ psi}$$

This maximum value is shown together with the distribution in Fig. 9.48b.

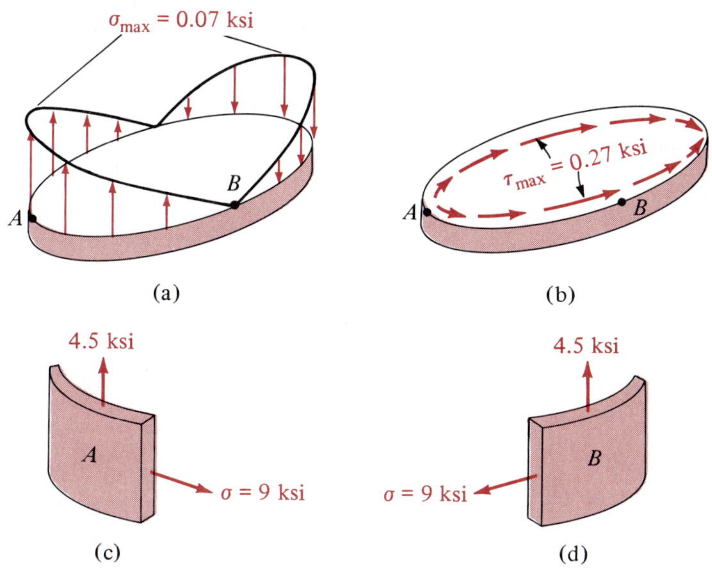

Figure 9.48 Bending, shear, and pressure stresses

Pressure stresses: The axial and hoop stresses are calculated to be

$$\sigma_{ax} = \frac{pR}{2t} = \frac{(75 \text{ psi})(48 \text{ in.})}{2(0.4 \text{ in.})} = 4.5 \text{ ksi}$$

and

$$\sigma_H = \frac{pR}{t} = \frac{(75 \text{ psi})(48 \text{ in.})}{0.4 \text{ in.}} = 9.0 \text{ ksi}$$

and are shown for points A and B in Figs. 9.48c and d, respectively.

The stresses that occur at point A are shown in Fig. 9.49a.

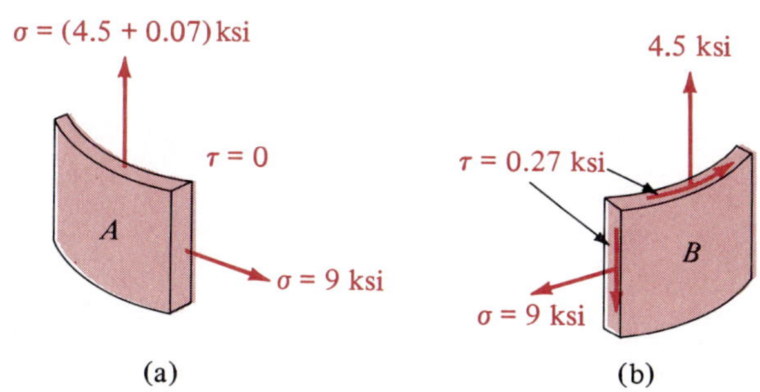

Figure 9.49 Inplane normal and shear stresses at A and B

The normal stress in the axial direction is the sum of the axial stress and the bending stress. The normal stress in the circumferential direction is the hoop stress. There is no shear stress. The principal stresses are $\sigma_{min} = 4.57$ ksi and $\sigma_{max} = 9.0$ ksi with the absolute maximum shear stress, occurring on a 45° plane, computed as

$$\tau_{abs} = \frac{(9.0 - 0)}{2} \text{ ksi} = 4.5 \text{ ksi}$$

The stresses that occur at point B are shown in Fig. 9.49b. The normal stress in the axial direction is the axial stress and the normal stress in the circumferential direction is the hoop stress. The shear stress is 0.27 ksi. The maximum inplane shear stress is computed as

$$\tau = \sqrt{\left(\frac{(9.0 - 4.5)}{2}\right)^2 + 0.27^2} \text{ ksi} = 2.27 \text{ ksi}$$

The principal stresses are then

$$\sigma_{min} = \left(\frac{(9.0 + 4.5)}{2} - 2.27\right) \text{ ksi} = 4.53 \text{ ksi}$$

and

$$\sigma_{max} = \left(\frac{(9.0 + 4.5)}{2} + 2.27\right) \text{ ksi} = 9.02 \text{ ksi}$$

with the absolute maximum shear stress again occurring on a 45° plane and having a value of $\tau_{abs} = (9.02 - 0)/2$ ksi $= 4.51$ ksi.

For the previous example, the stresses arising from the bending loads are small compared to the pressure stresses, so that the inplane stresses, the principal stresses, and the maximum shear stress differ only slightly from those associated with the pressure stresses. The next example illustrates a situation where the effects of the external loads are more significant.

Example 9.15 The cylindrical pressure vessel sustains an internal pressure of 1.5 MPa and an eccentrically applied load of 100 kN as shown in Fig. 9.50a. At a section near the support, determine the principal stresses and the maximum shear stress at points A and B. The thickness of the vessel is 30 mm.

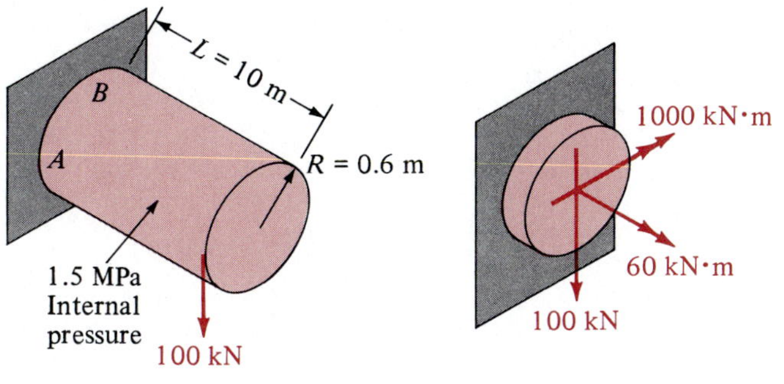

Figure 9.50 Eccentrically loaded pressure vessel

Solution The internal resultants arising from the 100-kN load are shown in Fig. 9.50b. The bending, transverse shear, and torsional shear stresses corresponding to the three internal resultants will be calculated and combined with the pressure stresses at points A and B.

Bending stress: The bending stresses with a maximum of

$$\sigma = \frac{MR}{I} = \frac{10^6\ \text{N·m}(0.6\ \text{m})}{\pi(0.6\ \text{m})^3(0.03\ \text{m})} = 29.5\ \text{MPa}$$

are indicated in Fig. 9.51a.

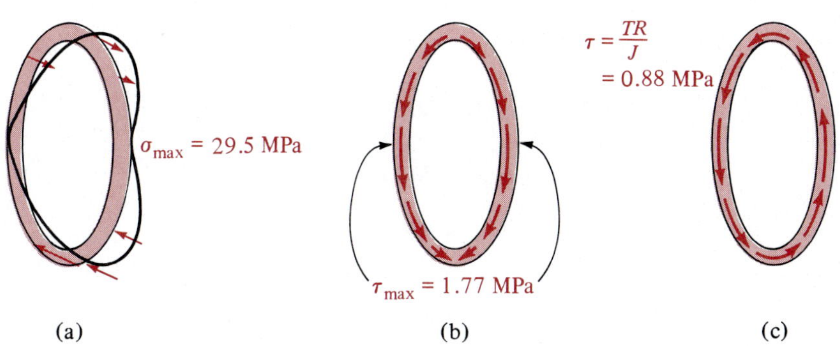

Figure 9.51 Bending, transverse shear, and torsional shear stresses

Transverse shear stresses: For the thin-walled circular section, the maximum value of the transverse shear stress is

$$\tau_{max} = \frac{2V}{A} = \frac{2(100\ \text{kN})}{2\pi(0.6\ \text{m})(0.03\ \text{m})} = 1.77\ \text{MPa}$$

and is shown along with the transverse shear-stress distribution in Fig. 9.51b.

Torsional shear stresses: The maximum value of the torsional shear stress is

$$\tau_{\max} = \frac{TR}{J} = \frac{100 \text{ kN}(0.6 \text{ m})}{2\pi(0.6 \text{ m})^3(0.03 \text{ m})} = 0.88 \text{ MPa}$$

and is shown in Fig. 9.51c.

The axial and hoop stresses associated with the internal pressure are

$$\sigma_{\text{ax}} = \frac{pR}{2t} = \frac{1.5 \text{ MPa}(0.6 \text{ m})}{2(0.03 \text{ m})} = 15 \text{ MPa}$$

and

$$\sigma_{\text{ax}} = \frac{pR}{t} = \frac{1.5 \text{ MPa}(0.6 \text{ m})}{0.03 \text{ m}} = 30 \text{ MPa}$$

These must be superposed upon the bending, shear, and torsional stresses in order to determine the combined stresses.

Point A: Since point A lies on the neutral axis of bending, the normal stresses at point A are the pressure stresses, σ_{ax} and σ_H. The transverse shear stress is maximum at point A and adds to the torsional stress as shown in Fig. 9.52a. The maximum inplane shear stress is given by

$$\tau_{\max} = \sqrt{\left(\frac{30 - 15}{2}\right)^2 + 2.65^2} \text{ MPa} = 7.95 \text{ MPa}$$

with the principal stresses as

$$\sigma_{\min} = \left(\frac{30 + 15}{2} - 7.95\right) \text{ MPa} = 14.6 \text{ MPa}$$

and

$$\sigma_{\max} = \left(\frac{30 + 15}{2} + 7.95\right) \text{ MPa} = 30.5 \text{ MPa}$$

The absolute maximum shear stress, occurring on a 45° plane, is

$$\tau_{\text{abs}} = \frac{30.5 - 0}{2} \text{ MPa} = 15.2 \text{ MPa}$$

Point B: The normal stresses at point B are the pressure stresses, σ_{ax} and σ_H, and the maximum bending stress, shown in Fig. 9.52b. The transverse shear stress is zero at point B, so that the only shear stress is the torsional stress as shown in Fig. 9.52b. The maximum inplane shear stress is given by

$$\tau = \sqrt{\left(\frac{44.5 - 30}{2}\right)^2 + 0.88^2} \text{ MPa} = 7.29 \text{ MPa}$$

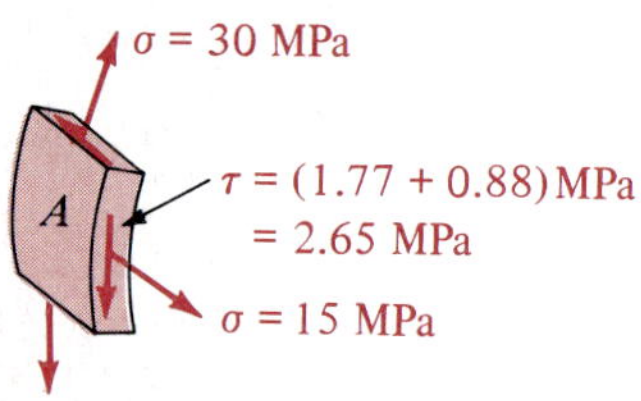

(a)

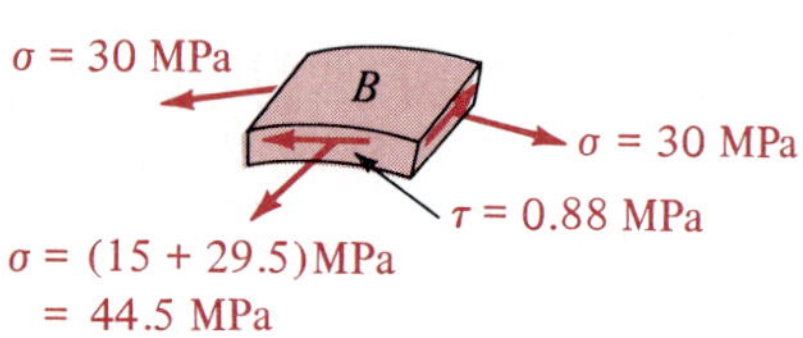

(b)

Figure 9.52 Combined stresses at points *A* and *B*

with the principal stresses as

$$\sigma_{min} = \left(\frac{44.5 + 30}{2} - 7.29\right) \text{ MPa} = 30.0 \text{ MPa}$$

and

$$\sigma_{max} = \left(\frac{44.5 + 30}{2} + 7.29\right) \text{ MPa} = 44.5 \text{ MPa}$$

The absolute maximum shear stress, occurring on a 45° plane, is

$$\tau_{abs} = \frac{44.5 - 0}{2} \text{ MPa} = 22.3 \text{ MPa}$$

Of the points checked, the maximum principal stress of 44.5 MPa and the absolute maximum shear stress of 22.3 MPa occur at point B.

PROBLEMS

9.89 A cylindrical pressure vessel with closed ends has an inside radius of 0.6 m and a wall thickness of 12 mm. If the internal pressure is 3 MPa, determine the axial and hoop stresses.

9.90 A cylindrical pressure vessel with closed ends has an inside radius of 25 in. The internal pressure is 400 psi. Determine the wall thickness if the circumferential stress is not to exceed 20 ksi.

9.91 Large pipes that serve to transport water from a reservoir to turbines in hydroelectric facilities are sometimes referred to as penstocks. For the 36-in.-diameter, 0.625-in.-wall thickness penstock shown, determine the maximum tensile stress and the absolute maximum shear stress in the penstock under static conditions.

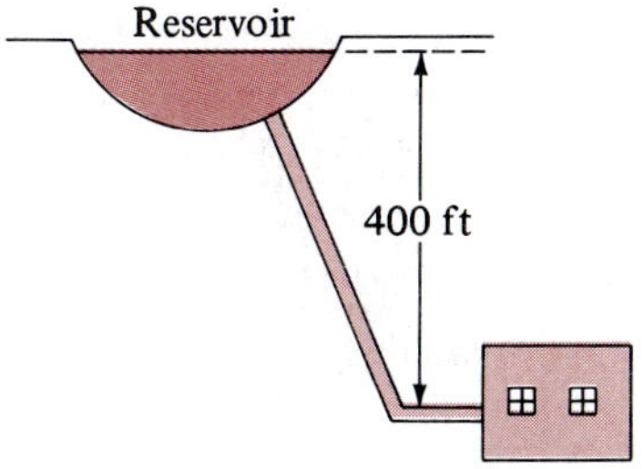

P. 9.91

9.92 A 5-m-tall cylindrical water tank is filled to a height of 4.5 m as shown. If the thickness of the tank is 15 mm, determine the maximum tensile stress and the maximum shear stress in the tank. Neglect the effect of the constraint provided at the bottom of the tank.

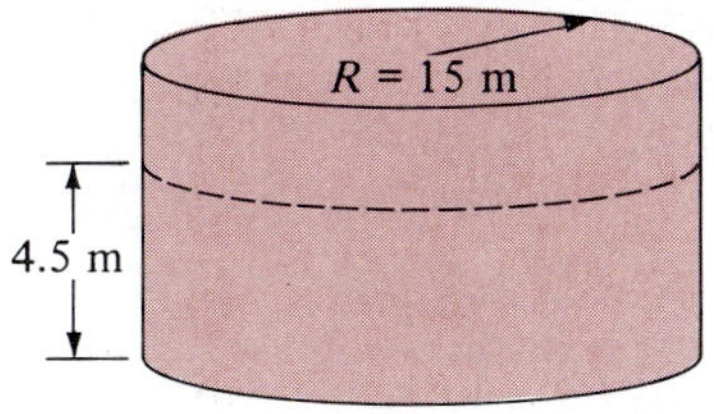

P. 9.92

9.93 A basketball with a diameter of $9\frac{5}{8}$ in. and a wall thickness of 0.125 in. is pressurized to 15 psi. Determine the tensile stress in the basketball.

9.94 A spherical pressure vessel with an outside diameter of 1.5 m is made from a steel with $\sigma_{ult} = 500$ MPa, and is pressurized to 4 MPa. If the factor of safety with respect to failure due to tensile stress is taken to be 2.5, determine the required wall thickness.

9.95 A spherical stainless steel tank is pressurized to 200 psi. If the radius is 15 in. and the thickness is 0.03 in., what is the hoop stress? What is the absolute maximum shear stress in the vessel? Where does it occur?

9.96 A steel spherical tank of inside diameter 400 mm is to contain a gas at a pressure of 4 MPa. If σ_{ult} in tension is 600 MPa, what is the required thickness for a factor of safety of 2.5 with respect to tensile failure?

9.97 A spherical shell is subjected to an internal pressure of 1000 psi. The inside diameter is 30 in. and the wall thickness is 0.625 in. What is the maximum inplane shear stress? What is the absolute maximum shear stress?

9.98 A spherical tank is subjected to an internal pressure of 6 MPa. The inside diameter is 600 mm and the wall thickness is 15 mm. What is the maximum tensile stress in the tank? What is the maximum absolute shear stress? What is the increase in volume? Take $E = 70$ GPa and $\nu = 0.3$.

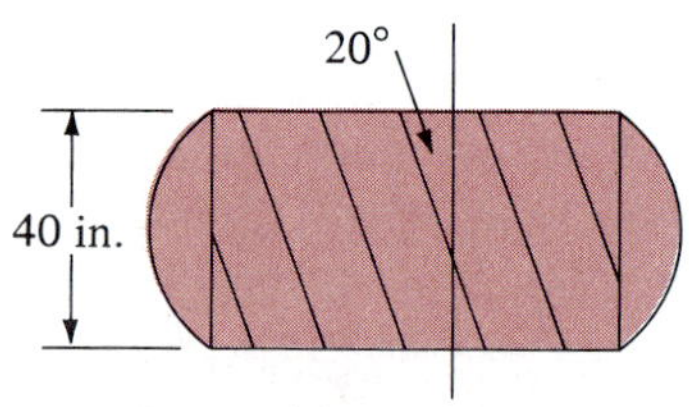

Ps. 9.99 and 9.100

9.99 The cylindrical portion of a compressed air tank is fabricated by welding a 0.3-in.-thick steel strip along a helical line as shown. If the maximum shear stress and tensile stress that the weld can withstand parallel and perpendicular to the direction of the weld are 18 ksi and 25 ksi, respectively, determine the maximum allowable internal pressure. Which governs: tensile or shear?

9.100 In the previous problem, is it possible to choose the helical angle such that the allowables on normal and shear stress are reached simultaneously?

9.101 A steel ($E = 3 \times 10^7$ psi, $\nu = 0.3$) cylindrical pressure vessel is closed by two thick circular plates bolted to flanges as shown in the figure. If the allowable tensile stress in a 1.0-in. diameter bolt is 18 ksi, determine *(a)* the number of equally spaced bolts required, *(b)* the maximum tensile and shear stresses in the vessel, *(c)* the approximate change in the length of the vessel, and *(d)* the change in the diameter near the center of the assembly.

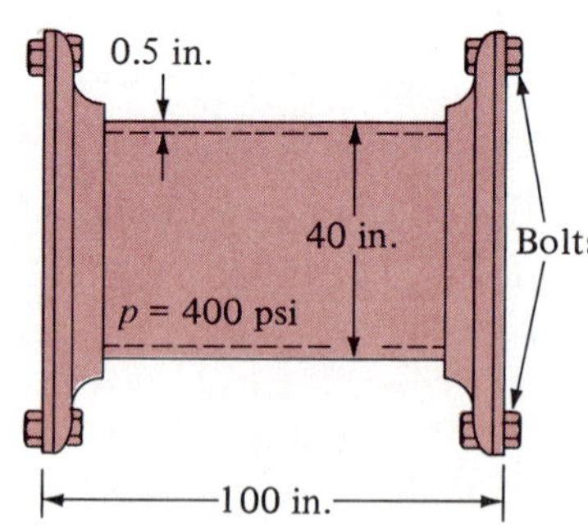

P. 9.101

9.102 A uniformly thick aluminum ($E = 10^7$ psi, $\nu = 0.3$) cylindrical tank is capped by hemispherical segments as shown in the figure. Neglecting interaction between the spherical and cylindrical portions, determine *(a)* the hoop and axial stresses in the cylindrical portion, *(b)* the hoop stress in the spherical portion, and *(c)* an approximate change in the length L of the vessel. Take $R = 15$ in., $p = 300$ psi, $L = 18$ in., and $t = 0.3$ in.

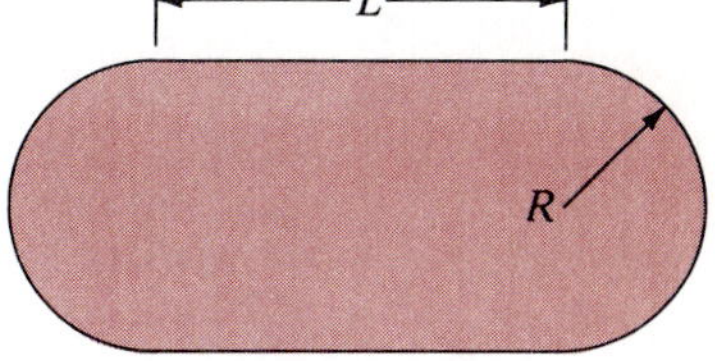

Ps. 9.102–9.104

9.103 A steel ($E = 200$ GPa, $\nu = 0.3$) cylindrical tank is capped by hemispherical segments as shown. All thicknesses are the same. The internal pressure is 2 MPa. Neglecting interaction between the spherical and cylindrical portions, determine *(a)* the maximum tensile stress in the cylindrical portion, *(b)* the maximum tensile stress in the spherical portion, *(c)* the maximum shear stress in the spherical portion, *(d)* the maximum shear stress in the cylindrical portion, and *(e)* an approximate change in the length L of the vessel. Take $R = 0.3$ m, $L = 2$ m, and $t = 8$ mm.

9.104 A cylindrical tank is capped by hemispherical segments as shown in the figure. All thicknesses are the same. Neglecting interaction between the spherical and cylindrical portions, determine the allowable internal pressure if the maximum allowable tensile stress in the shell is 150 MPa and the maximum allowable shear stress in the shell is 100 MPa. Take $R = 0.4$ m, $L = 1.8$ m, and $t = 10$ mm.

Combined Loadings on Thin-Walled Pressure Vessels

9.105 A cylindrical tank contains a gas at a pressure p. At points on the *(a)* top, *(b)* bottom, and *(c)* horizontal diameters, determine the combined normal and shear stresses at a section 2 ft away from the wall. Neglect the weight of the 0.25-in.-thick tank. Take $L = 20$ ft, $R = 3$ ft, $p = 60$ psi, and $P = 8$ kips.

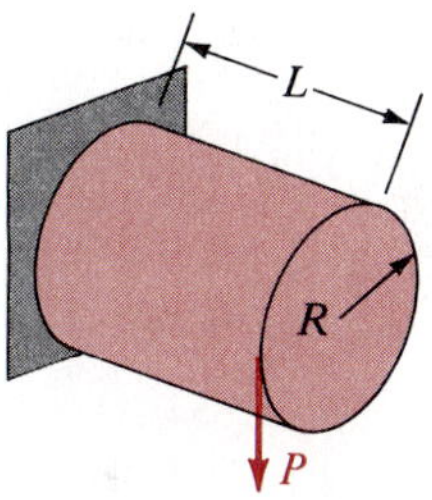

Ps. 9.105–9.108

9.106 Repeat problem 9.105 for the points midway between those for problem 9.105.

9.107 Repeat problem 9.105, computing the principal stresses and absolute maximum shear stress at the four points in question.

9.108 Repeat problem 9.106, computing the principal stresses and absolute maximum shear stress at the four points in question.

9.109 A tank is subjected to the simultaneous action of an internal pressure p and a torque T as shown. If the maximum normal stress at the center of the tank is not to exceed σ_0, develop a relationship between the pressure p, the torque T, and the stress σ_0.

9.110 For problem 9.109, develop a relationship between p, T, and τ_0 if the absolute maximum shear stress at the center of the tank is not to exceed τ_0.

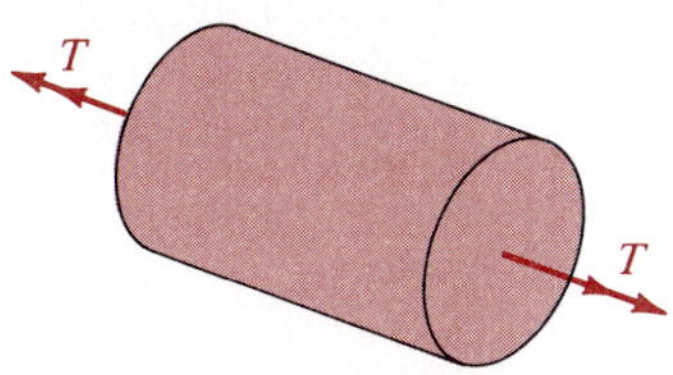

Ps. 9.109 and 9.110

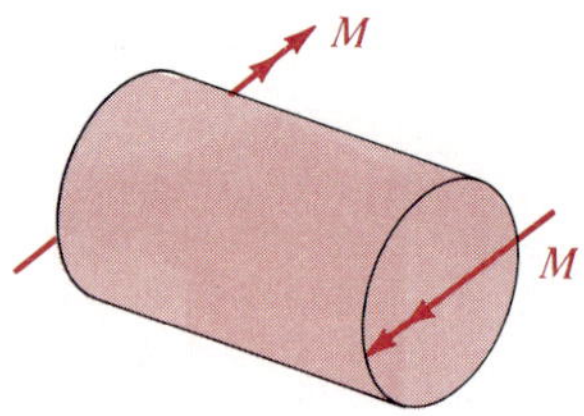

Ps. 9.111 and 9.112

9.111 A tank is subjected to the simultaneous action of an internal pressure p and a moment M as shown. If the maximum normal stress at the center of the tank is not to exceed σ_0, develop a relationship between the pressure p, the moment M, and the stress σ_0.

9.112 For problem 9.111, develop a relationship between p, M, and τ_0 if the absolute maximum shear stress at the center of the tank is not to exceed τ_0.

9.113 The cylindrical tank shown is subjected to an internal pressure p and a distributed loading w as shown. At a section through the center of the tank, determine *(a)* the combined normal and shear stresses at the top, bottom, and sides, and *(b)* the principal stresses and the absolute maximum shear stress at the top, bottom, and sides. Take $w = 1000$ lbf/ft, $L = 25$ ft, $D = 5$ ft, $p = 40$ psi, and $t = 0.125$ in.

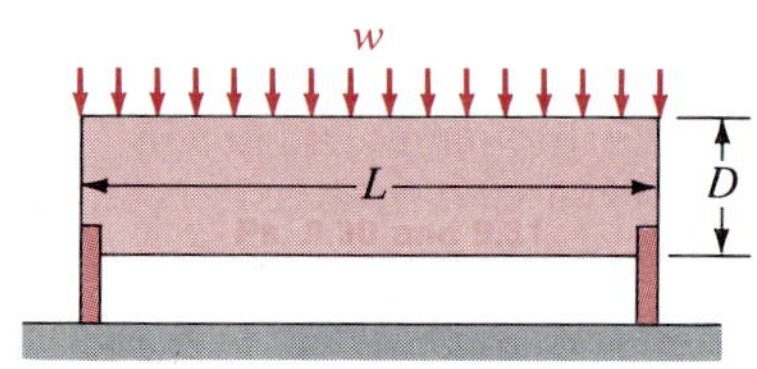

Ps. 9.113–9.116

9.114 Repeat problem 9.113 for the four points at 45° relative to the top, bottom, and sides.

9.115 Repeat problem 9.113 for $L = 8$ m, $D = 1.8$ m, $p = 0.4$ MPa, $w = 15$ kN/m, and $t = 4$ mm.

9.116 Repeat problem 9.115 for the four points at 45° relative to the top, bottom, and sides.

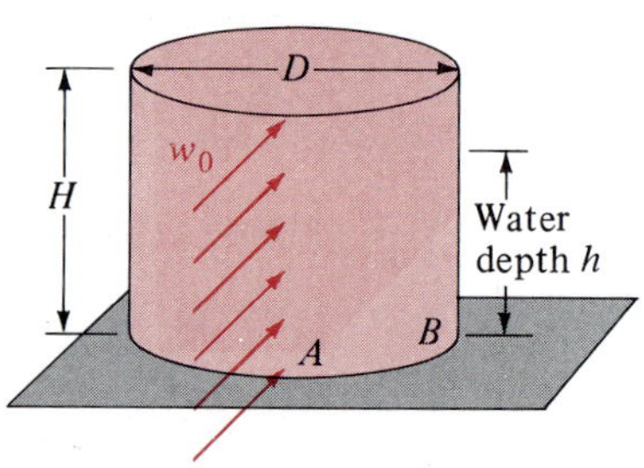

Ps. 9.117 and 9.118

9.117 A steel water tower is loaded by its own weight, by the water contained to a depth h, and by the uniform wind loading w_0 shown. Neglecting the stresses produced by the support, determine *(a)* the combined normal and shear stresses at point A, the point opposite A, and B, and *(b)* the principal stresses and the absolute maximum shear stress at point A, the point opposite A, and B. Take $H = 100$ ft, $h = 90$ ft, $w_0 = 25$ lbf/ft^2, $D = 10$ ft, $t = \frac{3}{16}$ in., and the specific weight of steel to be 0.29 lbf/in^3.

9.118 Repeat problem 9.117 for $H = 35$ m, $h = 29$ m, $w_0 = 1.2$ kPa, $D = 3$ m, $t = 5$ mm, and the specific weight of steel to be 79.6 kN/m^3.

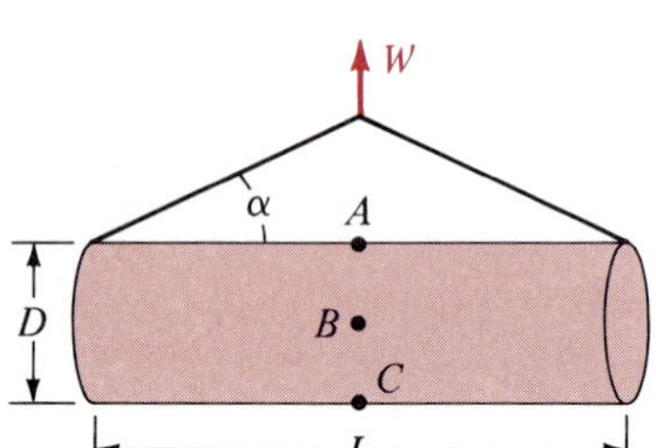

Ps. 9.119 and 9.120

9.119 A cylindrical steel tank containing an internal pressure p and having a total weight W is to be lifted by means of a cable attached centrally as shown. Determine *(a)* the principal stresses and absolute maximum shear stress at A, and *(b)* the principal stresses and absolute maximum shear stress at B. Take $p = 18$ psi, $L = 40$ ft, $D = 6$ ft, $d = 3$ ft, $W = 18{,}000$ lbf, $\alpha = 30°$, and $t = 0.2$ in.

9.120 Repeat problem 9.119 with $p = 126$ kPa, $L = 14$ m, $D = 2$ m, $d = 1$ m, $W = 80$ kN, $\alpha = 30°$, and $t = 5$ mm.

9.5 LIMITATIONS ON THE APPLICATIONS OF COMBINED LOADINGS

In previous sections of this chapter, we assumed that both the stresses and deformations resulting from a combination of different types of loadings of straight bars can be superposed. With combined axial and bending loads, for instance, it is assumed that the axial and bending problems can each be solved independently, and that the results for the stresses and displacements can then be added to determine the effect of the axial and bending loads when applied simultaneously.

Strictly speaking, superposition of axial and bending effects is not valid for the following reason. Consider the slender bar shown in Fig. 9.53.

Fig. 9.53a depicts the situation where a transverse load applied at the end of a cantilever beam produces a deflection $\Delta = PL^3/3EI$. If an axial load P is subsequently applied, as shown in Fig. 9.53b, the transverse displacements are reduced, that is, the transverse displacements and hence the bending stresses are reduced by the application of the axial load P. Considering the equilibrium requirements, the problem can be appreciated by referring to Fig. 9.54.

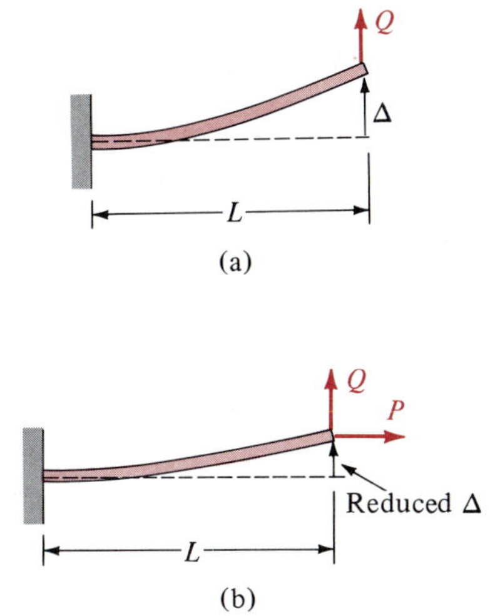

Figure 9.53 Sequentially applied axial and transverse loading

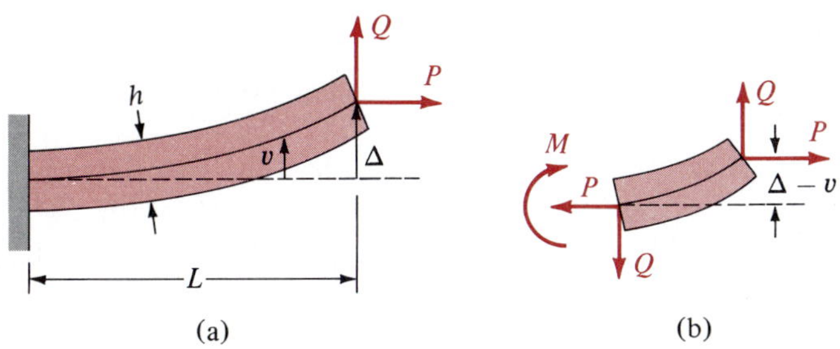

Figure 9.54 Simultaneous axial and bending effects

If we use the approximate moment-curvature relation $EIv'' = M$ in association with the free-body diagram of Fig. 9.54b, it is seen that

$$EIv'' = M = Qx - P(\Delta - v)$$

showing that the moment is decreased by the simultaneous application of the axial load P, so that the transverse displacements are affected by the inplane load. This simple demonstration shows that the effects of the separate application of the loads P and Q cannot be superposed, as symbolized in Fig. 9.55, to assess the effect of the combined loads. In effect, the two problems are actually coupled, with the inplane loads influencing the displacements and stresses of the bending problem and similarly, with the transverse loads influencing the results of the inplane problem.

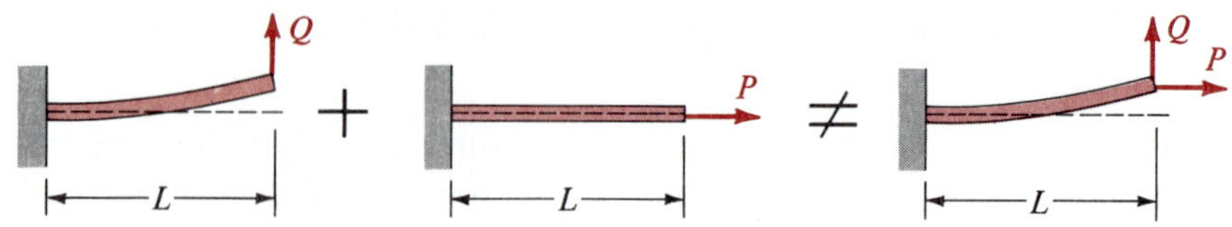

Figure 9.55 Situation where superposition is not applicable

In actuality, when the bar is "stocky," that is, when h/L is approximately 10 or less, the stresses are given quite satisfactorily by $\sigma = P/A - My/I$, and the axial load has little effect on the already small transverse displacement v. For slender bars subjected to the simultaneous action of axial, bending, and torsional loads, the axial loads can have a significant effect on the transverse displacements and the bending stresses, so that the effects of the coupling *must* be included. Generally, when computing displacements, the effect of neglecting an axial tensile load is conservative whereas the effect of neglecting an axial compressive load is nonconservative. When computing stresses, the effect of neglecting either tensile or compressive axial loads is generally nonconservative.

9.6 CLOSURE

The most important point to remember when combining the elementary theories of axial, torsional, and bending deformations is that all assumptions that are an integral part of each of the theories must also be satisfied in a combined loading application. If any of the assumptions are not satisfied, the individual theory and hence the combination will in general be invalid. For the most part, the use of superposition in determining the stresses and displacements for combined axial, torsional, and bending loads depends upon these assumptions:

1. The bar is straight with no sudden changes in section properties (i.e., area, second moment of area, and so forth).
2. Any necessary special properties of the cross section, such as planes of symmetry or circular boundary, are present.
3. The external loading produces internal force resultants consistent with those required for use in the elementary theories of axial, torsional, and bending deformation.
4. In the case of axial loads, the axial loads must be small or the bar must be short in order that the coupling between the bending and axial problems be negligible, that is, small displacements from the original straight configuration of the bar.

At a given section, we can use the results of superposing the inplane normal and the shear stresses determined from the individual theories to determine principal stresses and the absolute maximum shear stress at points within the section, and to then assess the strength. The axial, torsional, and bending displacements are determined independently by integrating the appropriate force-displacement relations for the axial, torsional, and bending problems.

As an example, the assumptions are clearly not satisfied in situations where a bar consists of several separate portions, such as in Figs. 9.56a and b, where the direction of the axis of the bar changes suddenly or there are sudden changes in geometry. The results of stresses should be viewed as valid only in the straight portions of the bars.

One must also remember that for the cylindrical pressure vessels, results obtained by combining axial and hoop stresses arising from the pressure, with the axial, torsional, and bending stresses produced by external loadings, are only valid at locations away from the ends of the shell.

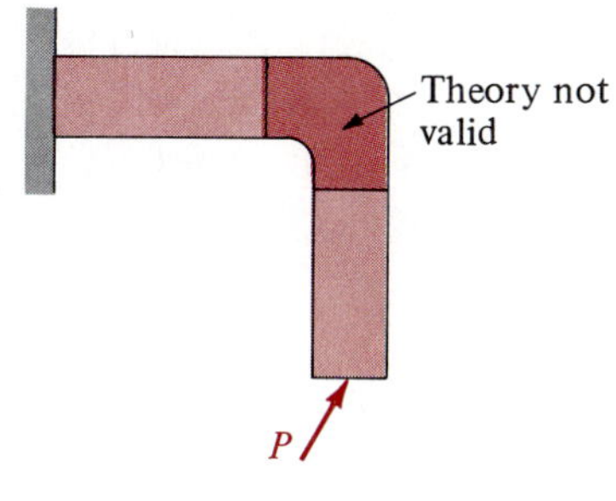

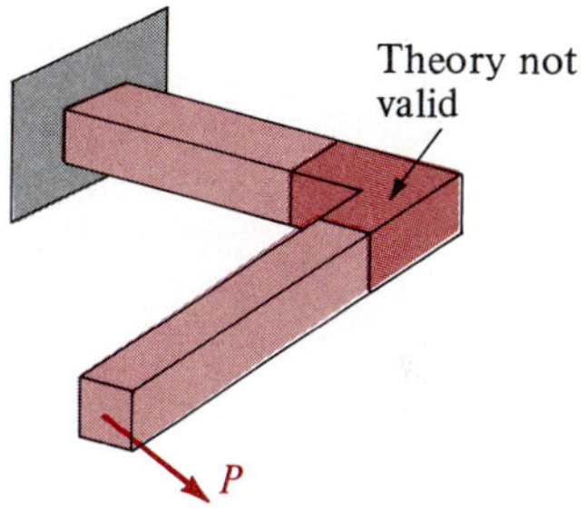

Figure 9.56 Regions in which the theory is not valid

ANALYSIS/DESIGN PROBLEMS

AD 9.1 A hollow steel pipe is loaded through a thick plate as shown. Select a suitable diameter D if $t/D = 0.05$ and the maximum principal stress is not to exceed 20 ksi. Take the factor of safety to be 2.0.

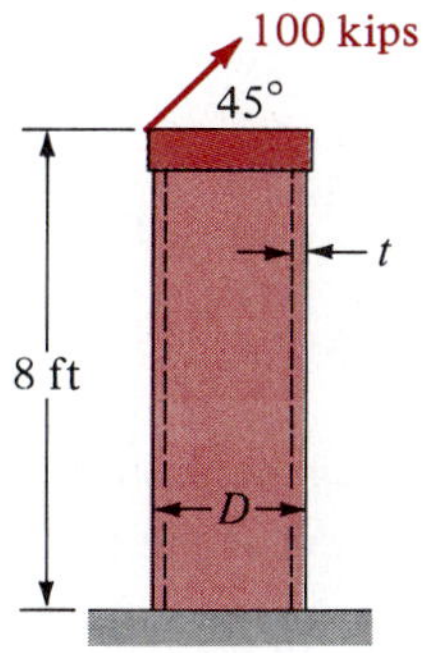

AD 9.2 Repeat AD 9.1 if the absolute maximum shear stress is not to exceed 16 ksi. Take the factor of safety to be 2.0.

AD 9.3 A hollow square is loaded as shown. Select a suitable dimension B if $t/B = 0.0625$ and the maximum principal stress is not to exceed 125 MPa. Take the factor of safety to be 2.2.

AD 9.4 Repeat AD 9.3 if the absolute maximum shear stress is not to exceed 105 MPa. Take the factor of safety to be 2.2.

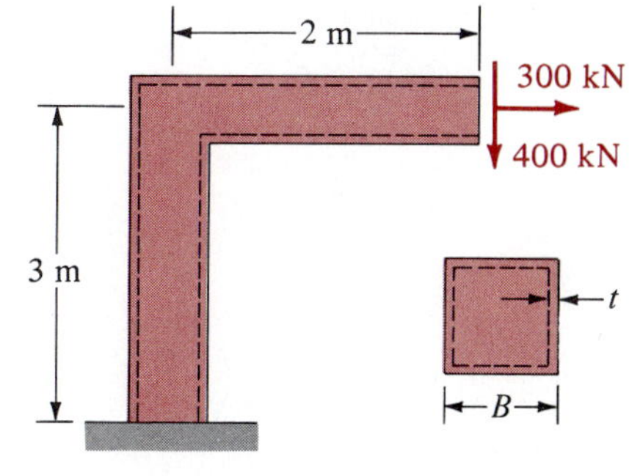

AD 9.3 and 9.4

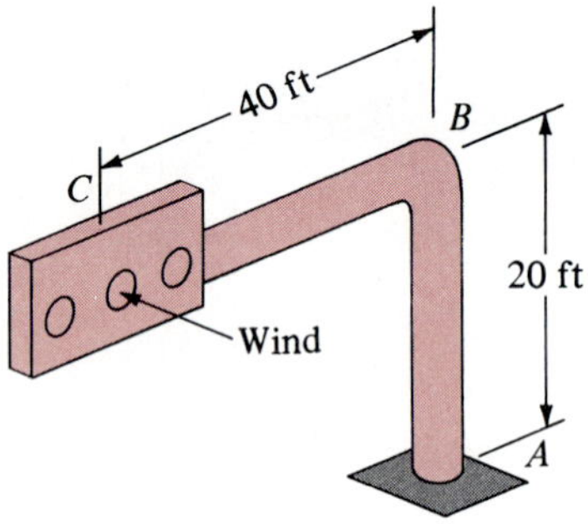

AD 9.5–9.8

AD 9.5 A traffic signal is to be constructed as shown. The signal weighs 1200 lbf and is acted upon by a wind force of 800 lbf. Assuming that the portions *AB* and *BC* are to be formed from circular pipe with $t/D = 0.1$, select a suitable value for *D* if the maximum principal stress is not to exceed 22 ksi. Neglect the weight of the pipe and take the factor of safety to be 1.6.

AD 9.6 Repeat AD 9.5 if the weight of the pipe is to be included.

AD 9.7 Repeat problem AD 9.5 if the absolute maximum shear stress is not to exceed 17 ksi and the factor of safety is taken to be 1.6.

AD 9.8 Repeat problem AD 9.5 if the weight of the pipe is to be included and if the absolute maximum shear stress is not to exceed 17 ksi. Take the factor of safety to be 1.6.

AD 9.9 The pipe hangs under its own weight as shown. If t/D is to be fixed at 0.075, what is the minimum allowable value of *D* if the absolute maximum shear stress is not to exceed 16 ksi? Take the factor of safety to be 1.5.

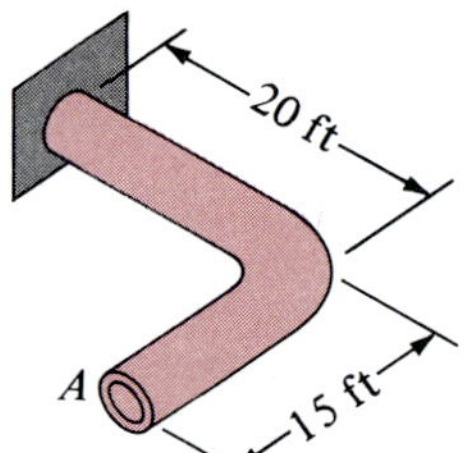

AD 9.9 and 9.10

AD 9.10 Repeat problem AD 9.9 if a weight of 2600 lbf is attached at *A*.

AD 9.11 A circular steel pipe is clamped at *C* and supported by a ball joint at *A*. If $t/D = 0.06$, select an appropriate value for *D* if the absolute maximum shear stress is not to exceed 120 MPa. Take the factor of safety on the stress to be 2.0.

AD 9.12 Repeat AD 9.11 if, in addition, the maximum vertical displacement is not to exceed 0.01 m. Assume that the factor of safety applies only to the stress.

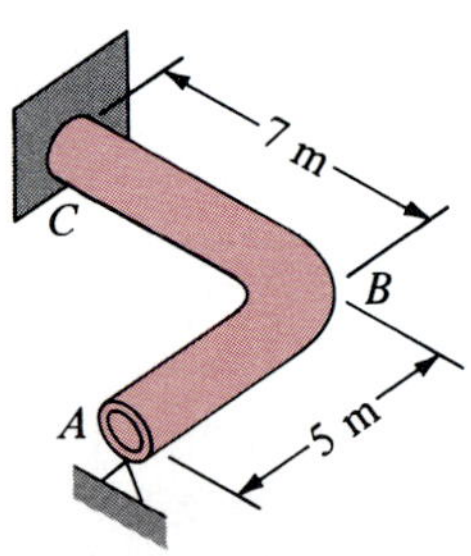

AD 9.11 and 9.12

AD 9.13 The two pulleys transmit the belt tensions shown. Assume that the ends of the shaft are simply supported for bending and that the left end is restrained against torsional rotation. Determine the required diameter of the solid shaft if the inplane shear stress is not to exceed 10 ksi. Take the diameters of both pulleys to be 18 in., $T_1 = 300$ lbf, $T_2 = 200$ lbf, $T_3 = 200$ lbf, $T_4 = 100$ lbf, $L_1 = 32.0$ in., $L_2 = 60.0$ in., and $L_3 = 56.0$ in.

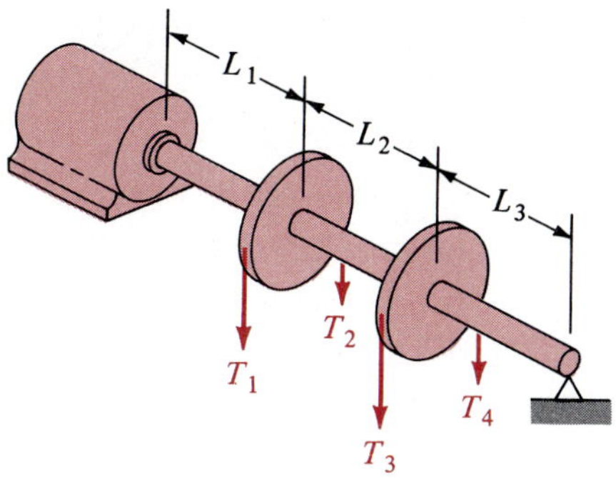

AD 9.13–9.20

AD 9.14 Repeat problem AD 9.13 for the same data if the inplane normal stress is not to exceed 17 ksi.

AD 9.15 Repeat problem AD 9.13 for the same data if the absolute maximum shear stress is not to exceed 10 ksi.

AD 9.16 Repeat problem AD 9.13 for the same data if the maximum principal stress is not to exceed 17 ksi.

AD 9.17–9.20 Repeat problems AD 9.13–9.16 if the directions of T_3 and T_4 are rotated 90° counterclockwise as viewed from the right end.

CHAPTER 10 ENERGY METHODS

CHAPTER OUTLINE

10.1 INTRODUCTION

When dealing with the statics of mechanical systems, we have two fundamental approaches for formulating the governing equations. One is the vectorial approach, wherein one or more free-body diagrams are drawn, and Newton's laws, represented as

$$\Sigma \mathbf{F} = \mathbf{0} \qquad \text{and} \qquad \Sigma \mathbf{M} = \mathbf{0}$$

are used to determine the equations of equilibrium for a deformable body. This is the approach we used exclusively to enforce the basic idea of equilibrium in the previous nine chapters.

The other approach for investigating the statics of mechanical problems is the variational approach, which uses either the principle of virtual work or the principle of stationary potential energy. When using the principle of virtual work, the equations of equilibrium for a mechanical system are derived by requiring that, during a virtual displacement of the system, the virtual work of all the external and internal forces is zero. Alternatively, using the principle of stationary potential energy for a conservative mechanical system, a function $V(x)$, representing the total potential energy of the external and internal forces, can be used to determine the equations of equilibrium by requiring that

$$\frac{dV}{dx} = 0$$

that is, that the potential energy be stationary. The resulting equilibrium equations either coincide with or are essentially the same as those determined using the vectorial approach.

When used in connection with elastic solids, the variational approaches are referred to as **energy methods** for the simple reason that, as we discussed in Chapters 5, 6, 7, and 8, a structure composed of a linearly elastic material behaves essentially as a linear spring. As an example, consider a uniform linearly elastic bar acted upon by a single axial load P as shown in Fig. 10.1a.

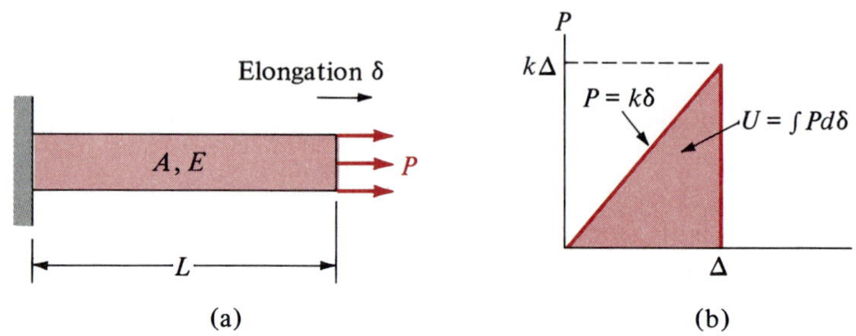

Figure 10.1 Elastic energy associated with an axial member

One of the important results of Chapter 5 was that the elongation of the bar is given by $\delta = PL/AE$, which can be restated as

$$P = \left(\frac{AE}{L}\right) \delta = k\delta$$

that is, the linearly elastic bar behaves as a linear spring with spring constant $k = (AE/L)$, as indicated in Fig. 10.1b. Thus such a bar that has been elongated an amount Δ possesses an elastic energy given by

$$U = \left(\frac{1}{2}\right) k\,\Delta^2 = \left(\frac{AE}{2L}\right) \Delta^2$$

Geometrically, U represents the area under the force-deflection curve as indicated in Fig. 10.1b.

In a completely similar fashion, results from Chapter 6 indicate that the relative angle of twist ϕ between the ends of a uniform linearly elastic circular bar transmitting a single torque T is given by $\phi = TL/JG$. This relationship can be restated as

$$T = \left(\frac{JG}{L}\right) \phi = C\phi$$

so that the uniform bar acting in torsion is essentially a torsional spring possessing an elastic energy

$$U = \left(\frac{1}{2}\right) C\phi^2 = \left(\frac{JG}{2L}\right) \phi^2$$

when the relative angle of twist between the ends is ϕ.

The primary purpose of this chapter is to demonstrate the manner in which elastic energies, such as those just mentioned for axial and torsional deformations, can be used in conjunction with several energy principles to formulate and solve problems involving elastic solids.

10.2 WORK, ENERGY, AND STATIONARY POTENTIAL ENERGY

The basis of the energy methods is contained in the definition of work. As indicated in Fig. 10.2a, the quantity of work done by a force **F** acting on a particle, or at a point on a rigid or deformable body in displacing an infinitesimal amount $d\mathbf{r}$, is

$$dW = \mathbf{F} \cdot d\mathbf{r}$$

For a displacement along a path from point A to point B the total amount of work done by the force **F** is

$$W = \int_A^B \mathbf{F} \cdot d\mathbf{r} \qquad (10.1)$$

where the line integral is to be evaluated along the path specified. The integral representing the work done can also be written as

$$W = \int_A^B F \cos \alpha \, ds$$

where α is the angle shown in Fig. 10.2a.

In Fig. 10.2b we see two distinct paths, C_1 and C_2, from A to B. Denote by W_1 the work done in traversing the curve C_1 from A to B, and by W_2 the work done by traversing the curve C_2 from A to B. In general, the work done by the force **F** depends upon the path followed in displacing the system from point A to point B, so that

$$W_1 \neq W_2$$

We describe this situation by saying that the work is *path dependent*.

As an example, consider the work done by the force of friction in moving a mass along a path on a level surface as shown in Fig. 10.3.

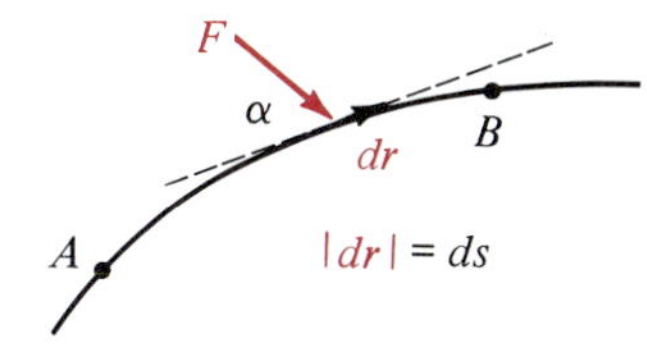

(a)

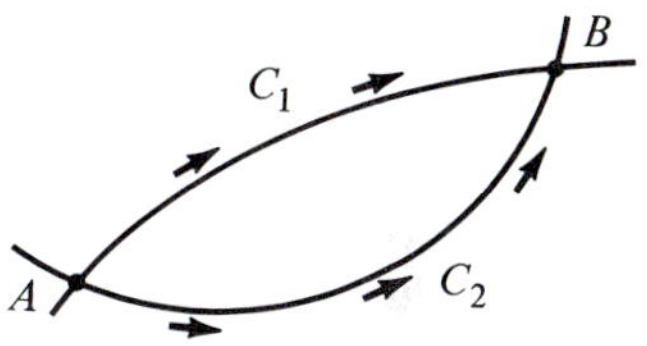

(b)

Figure 10.2 Representation of work as an integral

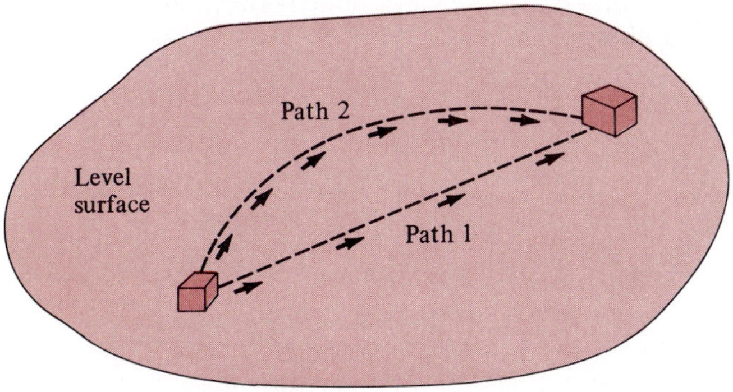

Figure 10.3 Path-dependent nature of the work done by friction

Assuming that both paths shown are traversed at constant velocity so that f, the force of friction, is the same at all points along both paths, the amount of work done by friction is

$$W_f = -f \cdot s$$

where s is the length of the path (i.e., the work done depends upon the path).

Fortunately, it turns out that many of the mechanical systems of interest are such that the force **F** is derivable from a potential according to

$$\mathbf{F} = -\nabla\phi = -\left(\mathbf{i}\frac{\partial\phi}{\partial x} + \mathbf{j}\frac{\partial\phi}{\partial y} + \mathbf{k}\frac{\partial\phi}{\partial z}\right) \tag{10.2}$$

with the result that

$$W = -\int_A^B (\nabla\phi)\, d\mathbf{r} = -\int_A^B \left(\frac{\partial\phi}{\partial x}\, dx + \frac{\partial\phi}{\partial y}\, dy + \frac{\partial\phi}{\partial z}\, dz\right)$$

or

$$W = -\int_A^B d\phi = -[\phi(B) - \phi(A)] \tag{10.3}$$

where ϕ is referred to as the potential energy associated with the force **F.**

Eqn. (10.3) states that when the force **F** is derivable from a potential according to Eqn. (10.2), the work done by the force **F** does not depend upon the path followed, but only on the values of the corresponding potential energy ϕ at the end points A and B; thus, *the work is independent of the path.*

A force for which the line integral is independent of the path is referred to as a **conservative force.** Note that for a conservative force defined according to Eqn. (10.2), the work done is the *negative* of the change in the potential energy.

Example 10.1 Consider an elastic spring fixed at one end with the other end free to move in any direction in the plane as shown in Fig. 10.4a. With e as the elongation of the spring, the characteristics of the spring are indicated in Fig. 10.4b. Determine whether there is a potential energy associated with the force in the spring, and if the answer is yes, determine the potential energy.

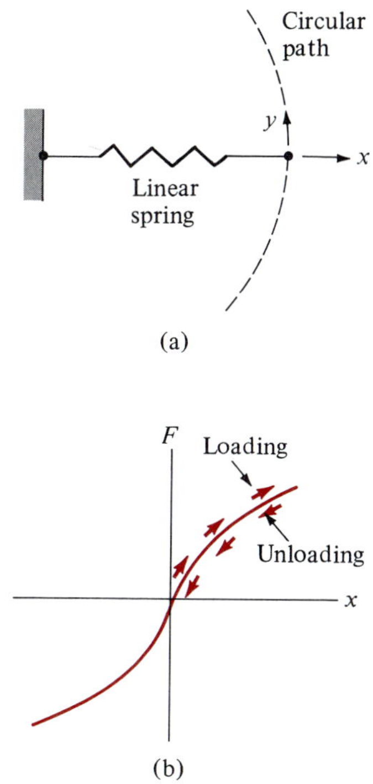

Figure 10.4 Elastic spring

Solution First we note that only those displacements that change the length of the spring result in any work being done by the spring. If for instance the end point of the spring is made to traverse a circular path of any given radius such as in Fig. 10.4a, the force in the spring is everywhere perpendicular to the path, and hence does no work. Thus we can restrict our investigation to those displacements that change the length of the spring, that is, to displacements in the x-direction. The key to answering the question about the potential energy for the elastic spring is in the assertion that the spring behaves as shown in Fig. 10.4b: that given a value of the elongation (or shortening) of the spring, there is a single value of the corresponding force in the spring. When loading and unloading take place along the same curve as shown in Fig. 10.4b, the work done by the spring is

$$W = -\int_{x_1}^{x_2} F(x)\, dx$$

and, for a given $F(x)$, depends only upon the initial and final values of the elongation x_1 and x_2, respectively. The minus sign appears because the force $F(x)$ in the spring has the direction opposite to that of the displacement dx. Thus we see that when the spring behaves as shown in Fig. 10.4b, the work done depends only upon the values x_1 and x_2; thus, the work done by the spring is independent of the path. The negative of the work is the potential energy

$$V = \int_{x_1}^{x_2} F(x)\, dx$$

and can be thought of as the area under the curve between x_1 and x_2 as shown in Fig. 10.5a.

This unique conservative behavior is distinguished from behavior in which the relation between the force and elongation depends, for instance, upon whether the spring is currently elongating or shortening, that is, path dependency, as shown in Fig. 10.5b. In this situation, the loading and unloading not only take place along different paths, but a residual elongation of the spring

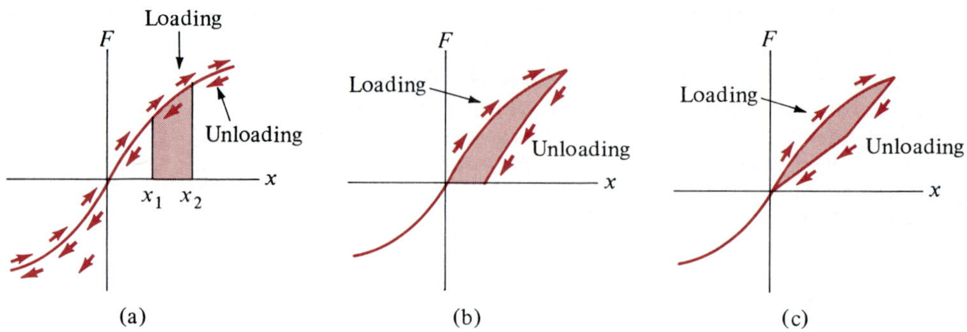

Figure 10.5 Conservative and nonconservative springs

remains when the load is completely removed. The behavior depicted in Fig. 10.5c is called **hysteresis:** even though the spring returns to its original length, the amount of work done during one cycle is equal to the area between the loading and unloading curves and is clearly not zero.

For the common special case where the value of the force in the spring is a linear function of the elongation, or $F = kx$, as pictured in Fig. 10.6, the potential energy becomes

$$V = \int_{x_1}^{x_2} kx\,dx = \frac{k(x_2^2 - x_1^2)}{2}$$

and again is commonly associated with the area under the F versus x diagram as shown in Fig. 10.6.

We will find that the potential energy associated with the linear spring, or with a component of a mechanical system that can be modeled as a linear spring, is a pivotal quantity in the energy methods we will discuss in this chapter.

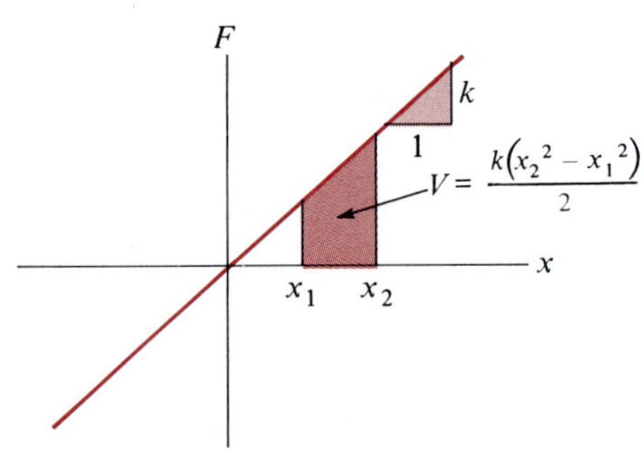

Figure 10.6 Force-elongation and energy for a linear spring

Example 10.2 Determine the potential energy of a force **f**, constant in magnitude and direction, as shown in Fig. 10.7a.

Solution First note that work will be done by the force **f** only when the force displaces in its own direction, so that the work can be represented by referring to the change in position of a coordinate x as shown in Fig. 10.7b. With $|\mathbf{f}| = f$, the work is

$$W = \int_{x_1}^{x_2} f\,dx = f(x_2 - x_1)$$

The value of W clearly depends only upon the final values of x_1 and x_2, and so is independent of the path. The corresponding potential energy is given by

$$V = -f(x_2 - x_1)$$

If the force f is applied to a point P of a body as shown in Fig. 10.7b, the work done depends only upon the displacement of P in the x-direction, and can be represented as $W = f\,x$, where x is the displacement of P in the direction of the force f. The corresponding potential energy is

$$V = -f\,x$$

A familiar example of this type of potential energy is that of a mass m positioned near the surface of the earth as sketched in Fig. 10.8.

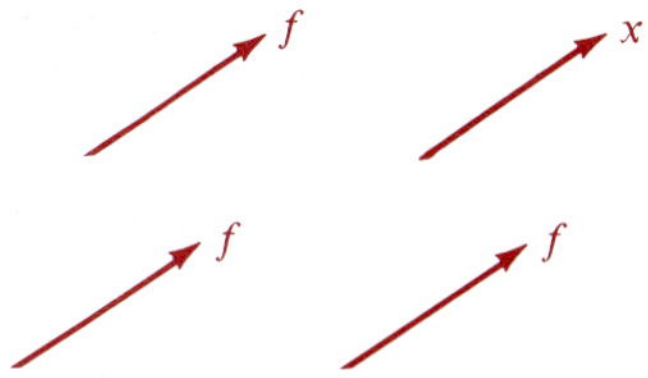

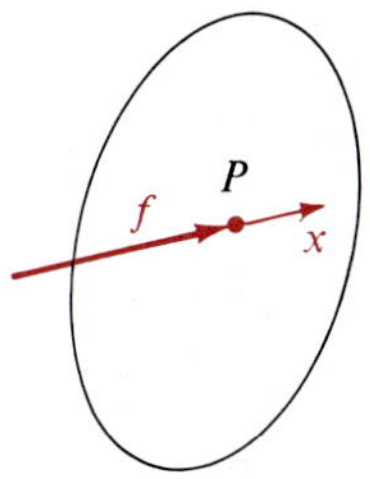

Figure 10.7 Potential energy of a force **f**

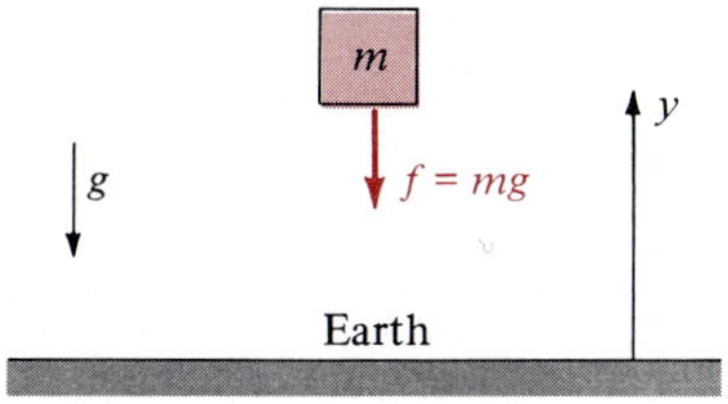

Figure 10.8 Potential energy of a mass near the surface of the earth

If the mass is displaced away from the earth an amount dy, the work done by the force mg is

$$W = -\int_{y_1}^{y_2} mg\,dy = -mg(y_2 - y_1)$$

independent of the manner in which the change in vertical position relative to the surface of the earth was accomplished. The corresponding potential energy is given by

$$V = mg(y_2 - y_1)$$

showing clearly that $\Delta V = mg(y_2 - y_1)$ is the negative of the work done. If the elevation of the mass is decreased, the work done is positive while the potential energy decreases by the same amount.

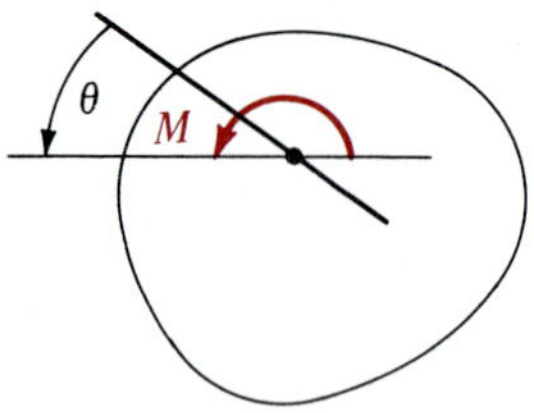

Figure 10.9 Potential energy of a constant moment M

Another very important type of potential energy is that associated with a constant moment applied to a body as shown in Fig. 10.9. The work done when the constant moment rotates through an angle θ is simply

$$W = M\,\theta$$

and depends only on the value of the angle theta. The corresponding potential energy associated with the moment is

$$V = -M\,\theta$$

Note that for either the constant force f or the constant moment M applied to a body, the definition of the potential energy has nothing whatsoever to do with the characteristics of the body to which they are applied.

For a conservative mechanical system, it is possible to derive the equations of equilibrium by invoking the **principle of stationary potential energy,** which states:

> At an equilibrium configuration of a conservative mechanical system, the total potential energy has a stationary value.

Total potential energy refers to the sum of *(a)* any internal potential energies such as elastic springs, and *(b)* the external potential energies of applied external forces.

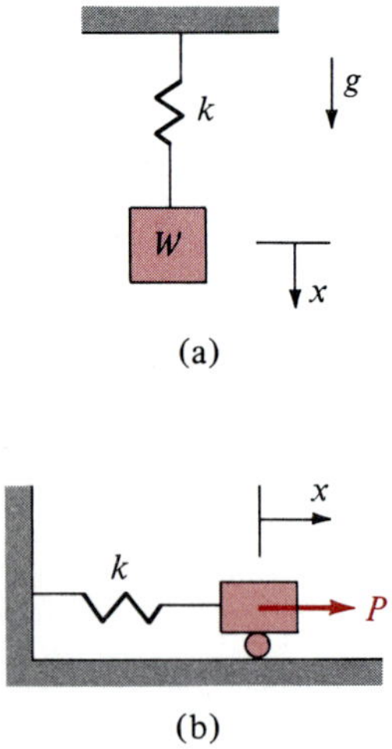

Figure 10.10 Equilibrium of a spring-mass system

Example 10.3 A linear elastic spring supports a weight W as shown in Fig. 10.10a. Using stationary potential energy, determine the equilibrium position.

Solution The potential energy of the linear spring and of the external load W can both be expressed in terms of the coordinate x, measuring the position of the weight from the *unstretched* length of the spring. The potential energy of the spring is given by

$$V_{sp} = \frac{kx^2}{2}$$

with the potential energy of the external force W as

$$V_{mg} = -Wx$$

The total potential energy is then

$$V(x) = V_{sp} + V_{mg} = \frac{kx^2}{2} - Wx$$

For this problem, x is referred to as the degree of freedom of the spring-mass system. Requiring the function $V(x)$ to be stationary leads to

$$\frac{dV}{dx} = 0 = kx - W$$

or $x = W/k$. The student should demonstrate that drawing a free-body diagram and summing forces yields the same result. Also note that if we consider the system shown in Fig. 10.10b, where again the degree of freedom x is measured from the unstretched configuration, the potential energies of the spring and the external load P are $V_{sp} = kx^2/2$ and $V_{mg} = -Px$, leading to

$$V(x) = V_{sp} + V_P = \frac{kx^2}{2} - Px$$

from which

$$V'(x) = 0 = kx - P$$

leading to $x = P/k$, which should be verified by the student by a vectorial approach.

For many applications it is necessary to represent the potential energy in terms of more than one variable, as the mechanical system in question has more than one degree of freedom. The number of degrees of freedom is the minimum number of coordinates necessary to completely specify the location of a mechanical system. Denoting the degrees of freedom for a conservative mechanical system in general as $x_1, x_2, \ldots x_n$, we can express the total potential energy as $V = V(x_1, x_2, \ldots, x_n)$, with the principle of stationary potential energy stating that each of the equations

$$\frac{\partial V}{\partial x_k} = 0 \qquad k = 1, 2, \ldots, n$$

must be satisfied.

Example 10.4 As an example of a system with two degrees of freedom, consider a rigid bar supported by springs and loaded as shown in Fig. 10.11a. Assume that before the load P is applied, the bar is in equilibrium in a horizontal configuration. Use stationary potential energy to determine the equilibrium position.

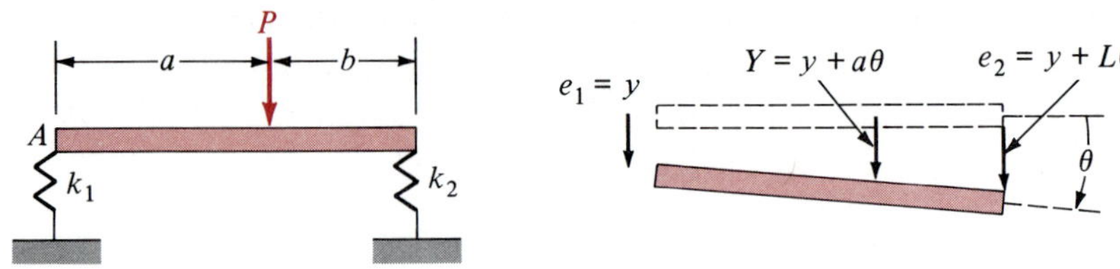

Figure 10.11 Equilibrium of a bar supported by springs

Solution In this example, the bar can move vertically and rotate about an axis perpendicular to the plane of the paper, thus there are two degrees of freedom. As shown in Fig. 10.11b, denote the vertical displacement of the left end of the bar by y, and the clockwise angle of rotation θ as shown. By specifying y and θ the location of the bar is known. The total potential energy consists of the potential energies of the two springs plus the potential energy of the external load P. As indicated in Fig. 10.11b, the elongation of spring 1 is simply $e_1 = y$, and the elongation of spring 2 is given by $e_2 = y + L\theta$, where $L = a + b$. The distance Y traveled by the load P is $Y = y + a\theta$. The total potential energy is

$$V = \frac{k_1 e_1^2}{2} + \frac{k_2 e_2^2}{2} - PY = \frac{k_1 y^2}{2} + \frac{k_2(y + L\theta)^2}{2} - P(y + a\theta) = V(y, \theta)$$

The two equilibrium equations are obtained by computing

$$\frac{\partial V}{\partial y} = 0 = k_1 y + k_2(y + L\theta) - P \qquad \text{and} \qquad \frac{\partial V}{\partial \theta} = 0 = k_2(y + L\theta)L - Pa$$

a set of linear algebraic equations to be solved for y and θ. Performing the algebra leads to $y = Pb/k_1L$ and $\theta = P(k_1 a - k_2 b)/(k_1 k_2 L^2)$. It is important to observe that the force in springs 1 and 2 can be represented, respectively, as $f_1 = k_1 e_1 = k_1 y$ and $f_2 = k_2 e_2 = k_2(y + L\theta)$, so that it is possible to rewrite the two equilibrium equations as

$$f_1 + f_2 = P \qquad \text{and} \qquad f_2 \cdot L - P \cdot a = 0$$

which clearly represent $\Sigma F_y = 0$ and $\Sigma M_A = 0$, two equilibrium equations based on the vectorial approach.

PROBLEMS

Use the principle of stationary potential to set up and solve the following problems. Neglect friction in all instances.

10.1 If the initial length of the spring is 12 in., what is the equilibrium location x of the weight? Take $k = 30$ lbf/in. and $mg = 60$ lbf.

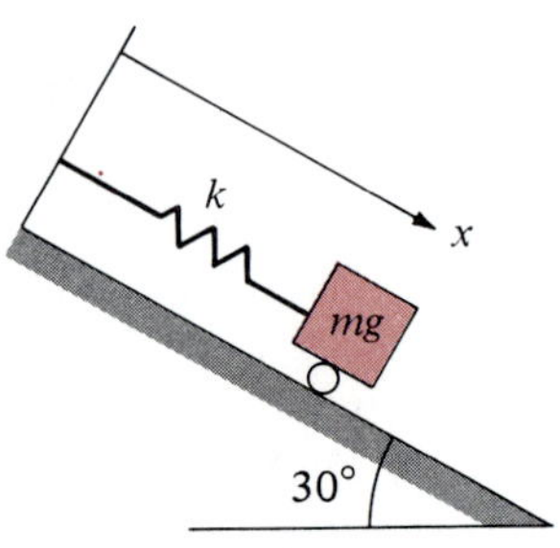

P. 10.1

10.2 Determine the equilibrium displacement u_1 of the mass m_1. Take k = 80 N/m, mg = 10 N, and ϕ = 45°.

10.3 Determine the equilibrium displacements u_1 and u_2 of the masses m_1 and m_2. Take k_1 = 60 lbf/in., k_2 = 40 lbf/in., and mg = 120 lbf.

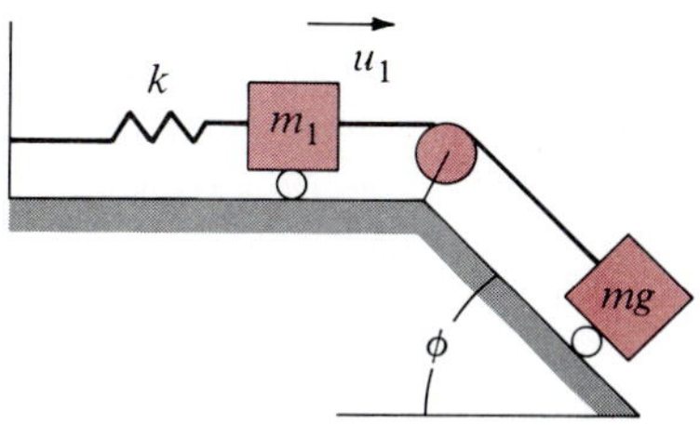

P. 10.2

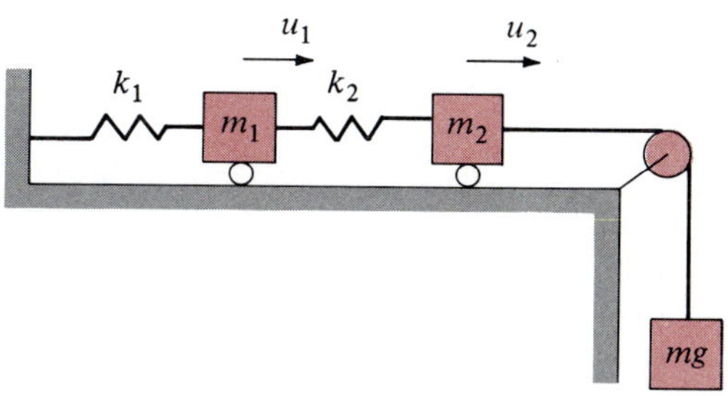

P. 10.3

10.4 Each of the bars shown has a length L and a weight W. Determine the equilibrium angle ϕ if *(a)* $P = W/2$, *(b)* $P = W$, and *(c)* $P = 2W$.

10.5 If the spring is unstretched when θ = 30°, determine the equilibrium configuration of the bar shown. Take L = 2 ft, W = 100 lbf, and k = 10 lbf/in.

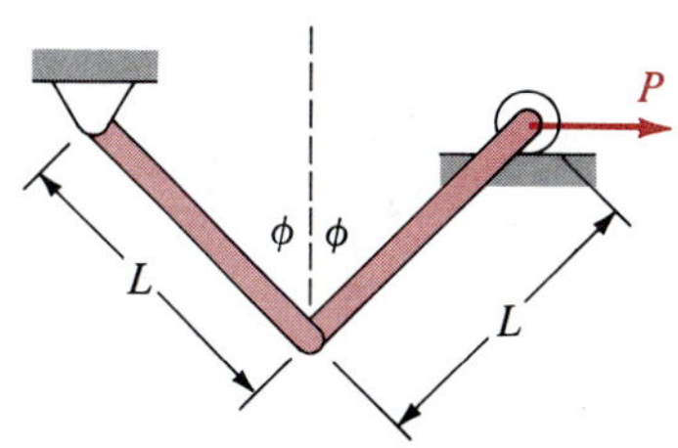

P. 10.4

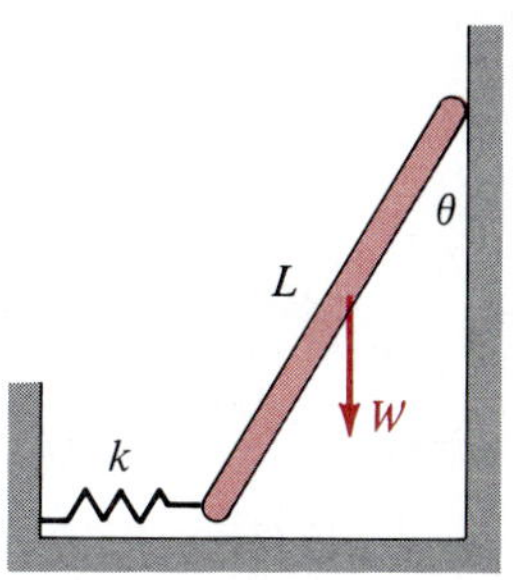

P. 10.5

10.6 Determine the equilibrium configuration of the system shown if the spring is unstretched when θ = 45°. Assume that the weight of the rigid bar is negligible. Take k = 1 kN/m, W = 400 N, a = 1.5 m, and b = 2 m.

10.7 Repeat problem 10.6 if W = 0 but the weight of the bar is 400 N.

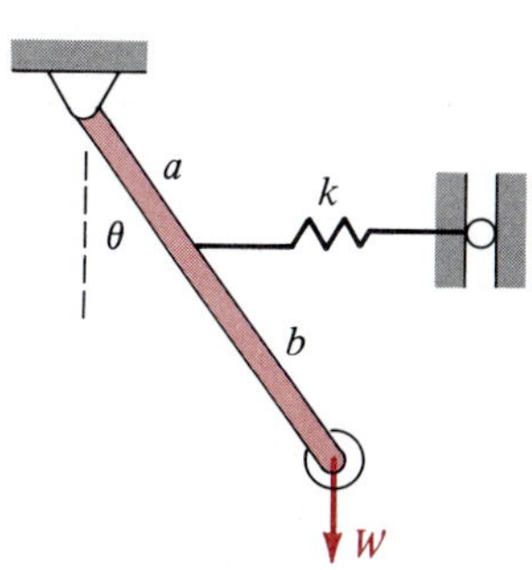

Ps. 10.6 and 10.7

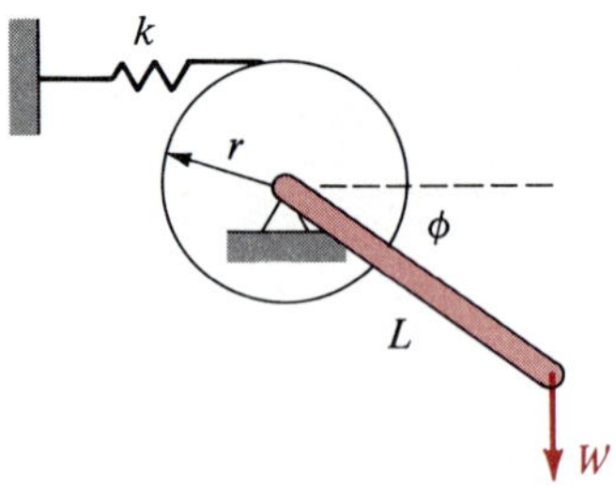

Ps. 10.8 and 10.9

10.8 If the spring is unstretched when $\phi = 20°$, determine the equilibrium position. Take $k = 10$ lbf/in., $r = 6$ in., $L = 18$ in., and $W = 10$ lbf.

10.9 Repeat problem 10.8 if $W = 0$ but the weight of the bar is 30 lbf.

10.10 The two rigid bars of negligible weight are pin-connected and supported by the linear spring as shown. If the spring is unstretched when $\theta = 53.1°$, determine the equilibrium configuration. Take $k = 1.8$ kN/m and $W = 4$ kN.

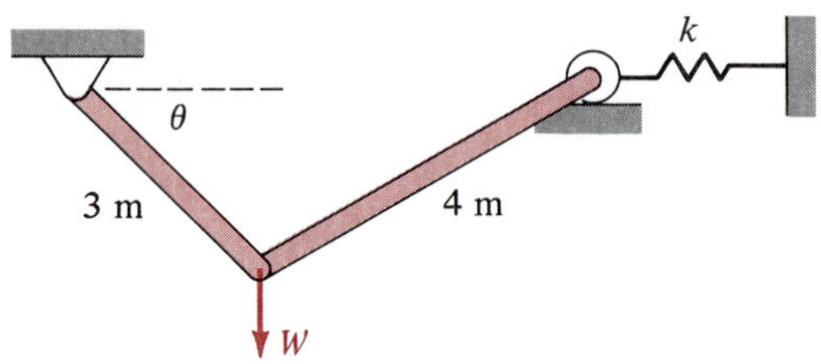

P. 10.10

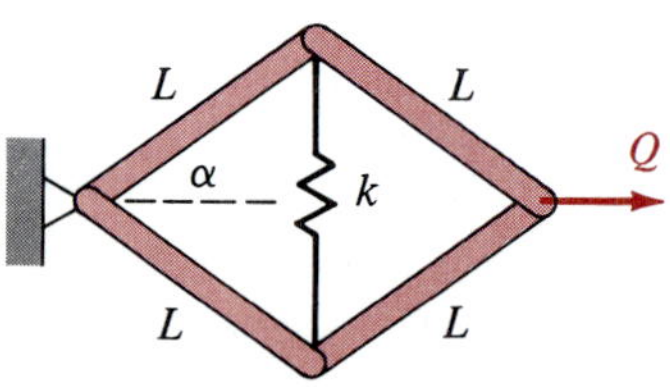

P. 10.11

10.11 Determine the equilibrium configuration of the bar if the spring is unstretched when $\alpha = \alpha_0$. Take $k = 10$ lbf/in, $Q = 20$ lbf, $\alpha_0 = 55°$, and $L = 10$ in.

10.12 For the mechanism shown, the spring is unstretched when $\alpha = 40°$. Determine the equilibrium value of α if $k = 1.8$ kN/m, $Q = 200$ N, and $L = 0.7$ m. Neglect the weights of the bars.

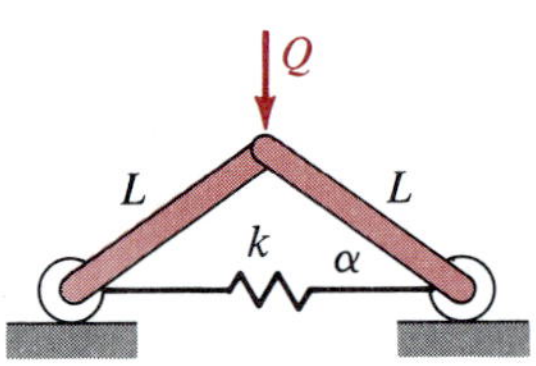

Ps. 10.12 and 10.13

10.13 Repeat problem 10.12 if $Q = 0$ but with each bar weighing 100 N.

10.14 For the mechanism shown, the spring is unstretched when $\phi = 60°$. Determine the equilibrium value of ϕ if $k = 30$ lbf/in., $L = 18$ in., $b = 9$ in., and the weight of the uniform bar is 15 lbf.

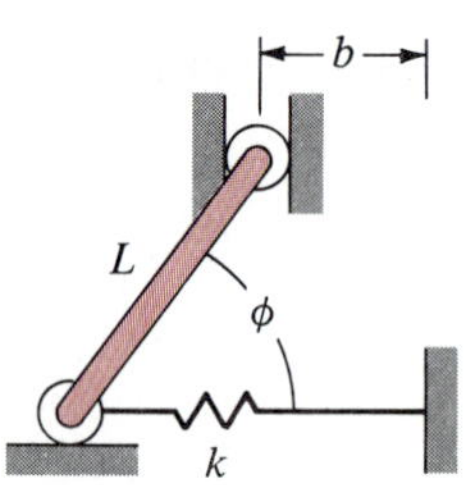

P. 10.14

10.15 Determine the equilibrium configuration(s) of the two uniform bars shown. Take $L_1 = L_2 = L$, and $W_1 = W_2 = W = P$.

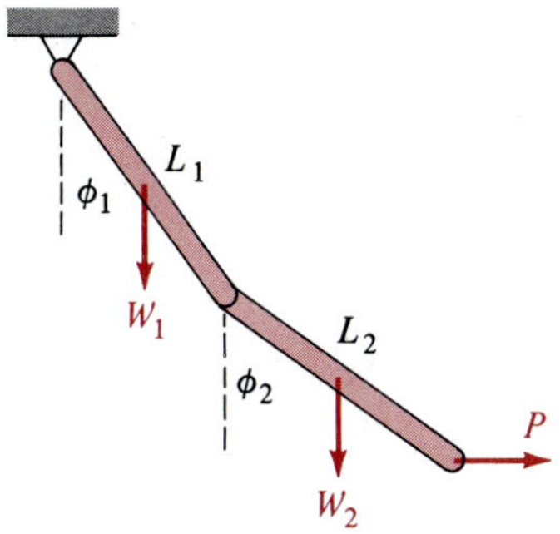

P. 10.15

10.16 Determine the equilibrium equations for the three springs supporting the weightless rigid bar as shown. Interpret these equations in terms of usual force and moment equilibrium equations. For $k_1 = k_2 = k_3 = 120$ lbf/in., $P = 90$ lbf, $x = 12$ in., and $a = 18$ in., solve the equations and determine the forces in the springs.

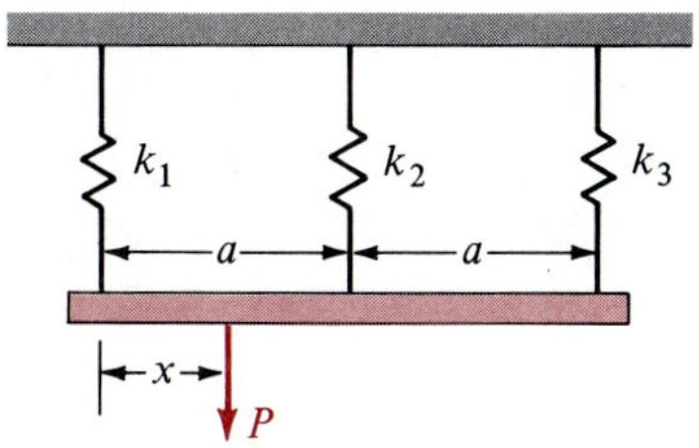

Ps. 10.16 and 10.17

10.17 Repeat problem 10.16 if the 120-lbf weight of the bar is included.

10.18 Determine the equations of equilibrium for the two rigid bars shown. Take the small angles of rotation of the two bars as the degrees of freedom. Solve the equations for $b/L = 0.5$ and $kb = 100P$.

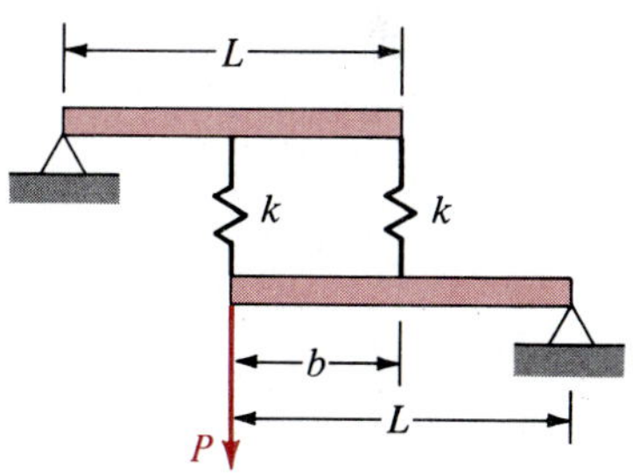

Ps. 10.18 and 10.19

10.19 Repeat problem 10.18 if the weights of the bars, each equal to $kb/100$, are the only loads acting.

10.20 Determine the equations of equilibrium for the three rigid bars shown. Take the small angles of rotations of the bars as the degrees of freedom. Solve the equations for $b/L = 0.4$ and $kb = 50P$.

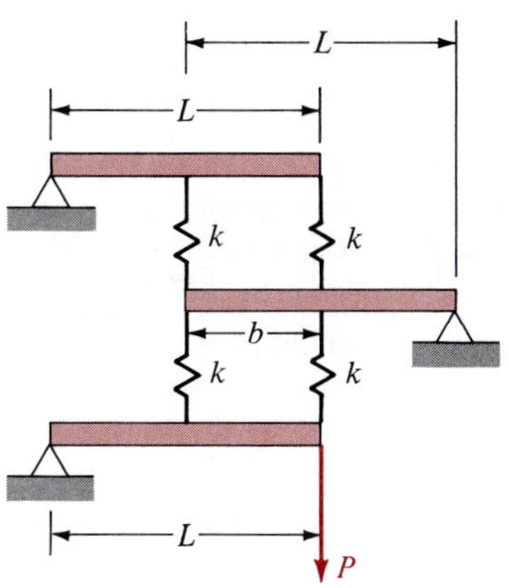

Ps. 10.20 and 10.21

10.21 Repeat problem 10.20 if the weights of the bars, each equal to $kb/50$, are the only loads acting.

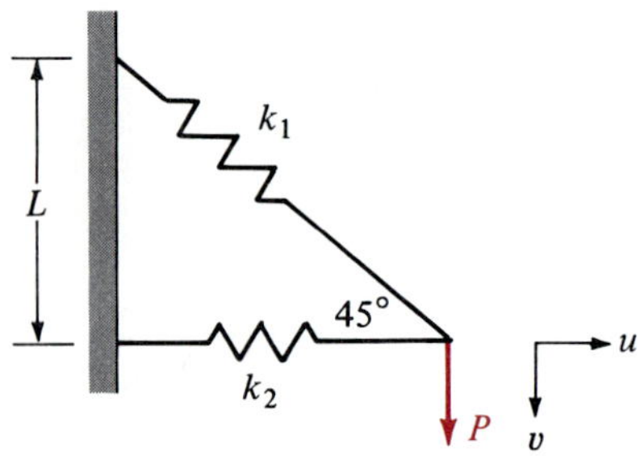

Ps. 10.22 and 10.23

10.22 Assuming u and v are small compared to the lengths of the springs, determine the equations of equilibrium.

10.23 Repeat problem 10.22 assuming that u and v are not necessarily small compared to the lengths of the springs. Show that if u/L and v/L are taken to be small, the binomial theorem can be used to reduce the equilibrium equations to those developed in problem 10.22.

10.24 Develop the equations of equilibrium for the two springs shown. Interpret them in terms of the $\Sigma \mathbf{F}$ approach.

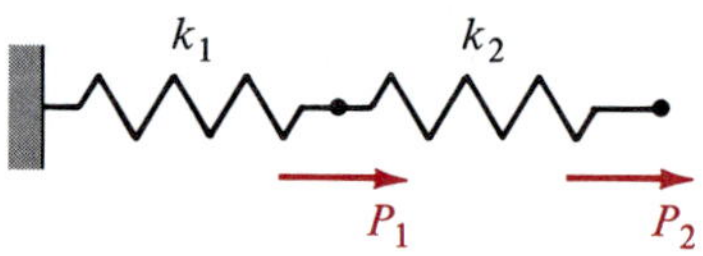

P. 10.24

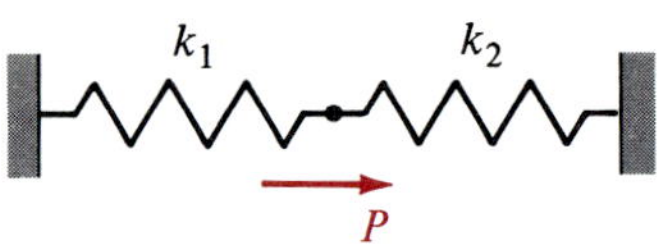

P. 10.25

10.25 Determine the equations of equilibrium for the system of springs shown. Solve the equations and determine *(a)* the displacement at the load P, and *(b)* the forces in the springs. What general conclusion can you reach regarding two springs in series loaded at the common point as shown?

10.26 Set up the equations of equilibrium for the system of springs shown. For $k_1 = k_2 = k_3 = k$ and $P_1 = P_2 = P/2$, solve the equations for *(a)* the displacements, and *(b)* the internal forces.

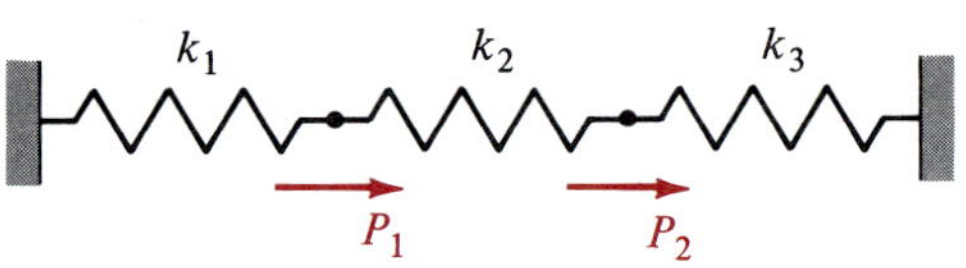

Ps. 10.26 and 10.27

10.27 Repeat problem 10.26 for *(a)* five springs of equal strength $5k/3$ and four equal loads $P/4$, *(b)* seven springs of equal strength $7k/3$ and six equal loads $P/6$, and *(c)* nine springs of equal strength $9k/3$ and eight equal loads $P/8$. Discuss the internal force distribution as the number of springs becomes large.

10.28 Set up the equations of equilibrium for the small displacements of the system of springs shown. Take $\alpha = 0.5$, $\beta = 0.6$, $k_1 = 20$ lbf/in., $k_2 = 10$ lbf/in., $k_3 = 20$ lbf/in., $P = 100$ lbf, and solve for the displacements and the forces in the springs.

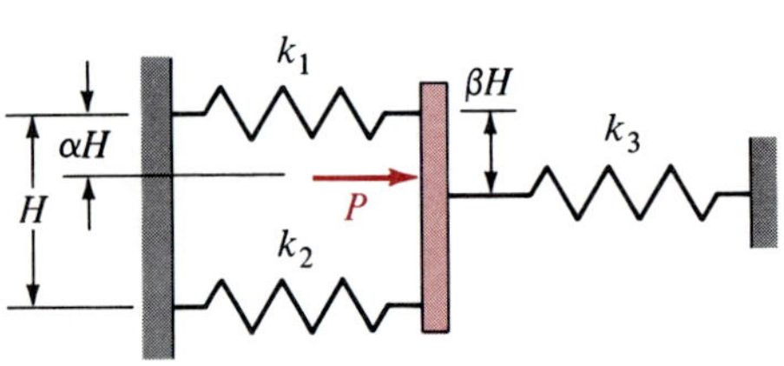

P. 10.28

10.29 The two torsional springs are subjected to the torques shown. Develop the equations of equilibrium and interpret them in terms of the ΣT approach.

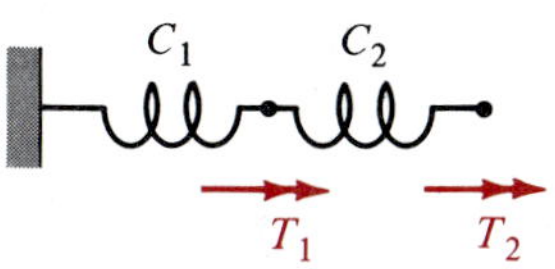

P. 10.29

10.30 Determine the equations of equilibrium for the system of springs shown. Solve the equations and determine *(a)* the rotation at the torque T, and *(b)* the torques in the springs. What general conclusion can you reach regarding two springs in series loaded at the common point as shown?

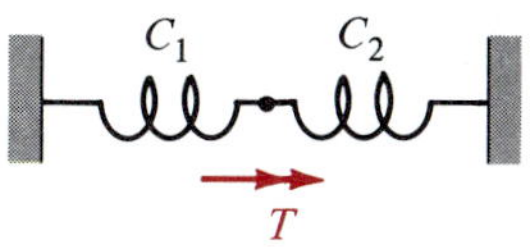

P. 10.30

10.31 Set up the equations of equilibrium for the system of springs shown. For $C_1 = C_2 = C_3 = C$ and $T_1 = T_2 = T$, solve the equations for *(a)* the rotations, and *(b)* the internal torques.

10.32 Repeat problem 10.31 for *(a)* five torsional springs of equal strength $5C/3$ and four equal torques $T/4$, *(b)* seven torsional springs of equal strength $7C/3$ and six equal torques $T/6$, and *(c)* nine torsional springs of equal strength $9C/3$ and eight equal torques $T/8$. Discuss the internal torque distribution as the number of torsional springs becomes large.

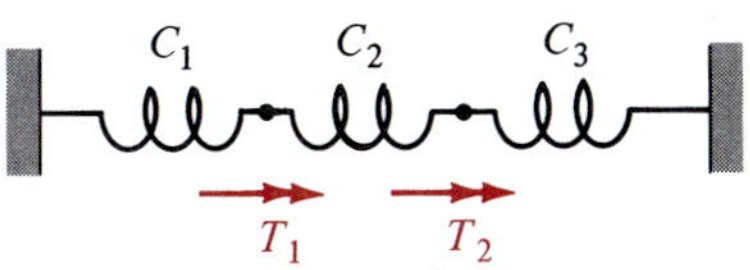

Ps. 10.31 and 10.32

10.3 STRAIN ENERGY AND STIFFNESS FORMULATIONS

As a body deforms, stresses are developed in the interior. The forces associated with these stresses do work during the deformation or straining. For an elastic body without hysteresis, the work done by the internal forces is entirely recoverable with the body returning to its original shape upon removal of the causes of the deformation. This is another way of stating that the work done is independent of the path.

Specifically, consider the relatively simple case of a bar subjected to axial loading as shown in Fig. 10.12a.

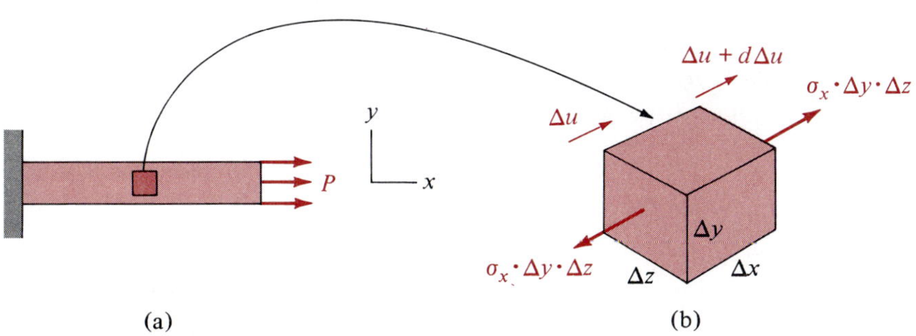

Figure 10.12 Stresses in a bar under axial loading

Isolate a volume element $\Delta \text{Vol} = \Delta x\ \Delta y\ \Delta z$ as pictured in Fig. 10.12b. During deformation, the work done by the force $\sigma_x\ \Delta y\ \Delta z$ acting on the left face will be

$$\Delta W = -(\sigma_x\ \Delta y\ \Delta z)\Delta u = -(\sigma_x\ \Delta y\ \Delta z)\varepsilon_x\ \Delta x$$

whereas the work done by the force $\sigma_x\ \Delta y\ \Delta z$ acting on the right face will be

$$\Delta W = (\sigma_x\ \Delta y\ \Delta z)(\Delta u + d\Delta u) = (\sigma_x\ \Delta y\ \Delta z)(\varepsilon_x + d\varepsilon_x)\ \Delta x$$

The total work done on the differential volume is

$$\Delta W = \sigma_x\ d\varepsilon_x\ \Delta x\ \Delta y\ \Delta z = \sigma_x\ d\varepsilon_x\ \Delta \text{Vol}$$

so that the total amount of work done by the stress σ_x per unit of volume during the deformation is

$$W = \left(\int_0^F \sigma_x\ d\varepsilon_x\right) d\text{Vol}$$

The symbols 0 and F refer to initial and final states of the body. We base the assertion that the work done is independent of the path upon the idea that for an elastic material of the type depicted in Fig. 10.13a, the relation between stress and strain is the same during loading and unloading: loading and unloading take place along the same curve.

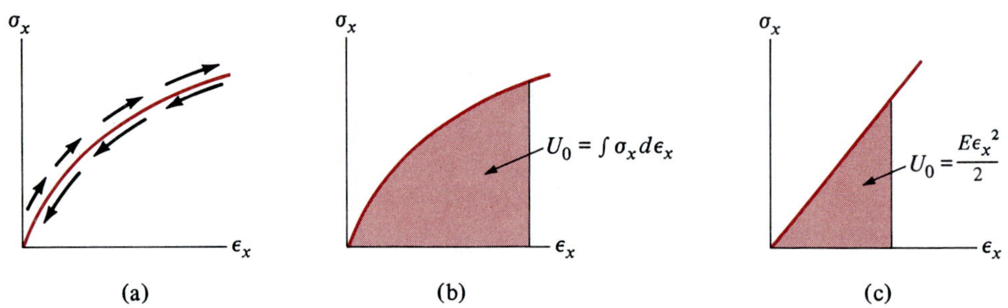

Figure 10.13 Stress-strain curve for an elastic material without hysteresis

Since the work done is independent of the path, there is an energy density U_0 such that

$$\int_0^F \sigma_x\ d\varepsilon_x = \int_0^F dU_0 = \int_0^F \frac{dU_0}{d\varepsilon_x}\ d\varepsilon_x$$

so that

$$\sigma_x = \frac{dU_0}{d\varepsilon_x}$$

Given the specific form of the stress-strain relation, $\sigma_x = \sigma_x(\varepsilon_x)$, the form of the strain-energy density can be determined by integrating

$$U_0 = \int \sigma_x(\varepsilon_x)\, d\varepsilon_x$$

Geometrically, U_0 is the area under the stress-strain curve as indicated in Fig. 10.13b.

For the special case of the linearly elastic material, the strain-energy density becomes

$$U_0 = \int E\varepsilon_x\, d\varepsilon_x = \frac{E\varepsilon_x^2}{2} \tag{10.4}$$

and corresponds to the triangular area shown in Fig. 10.13c. Note that for the case of the linearly elastic solid, $\sigma_x = E\varepsilon_x$, the strain-energy density can also be expressed in the two alternative forms

$$U_0 = \frac{E\varepsilon_x^2}{2} = \frac{\sigma_x \varepsilon_x}{2} = \frac{\sigma_x^2}{2E}$$

Consider also the state of pure shear shown in Fig. 10.14a. Loading and unloading takes place along the stress-strain curve shown in Fig. 10.14b, so that the potential for shear, called the strain-energy density for pure shear, is given by

$$U_0 = \int \tau_{xy}\, d\gamma_{xy}$$

and can be associated with the area under the shear-stress–shear-strain diagram of Fig. 10.14c.

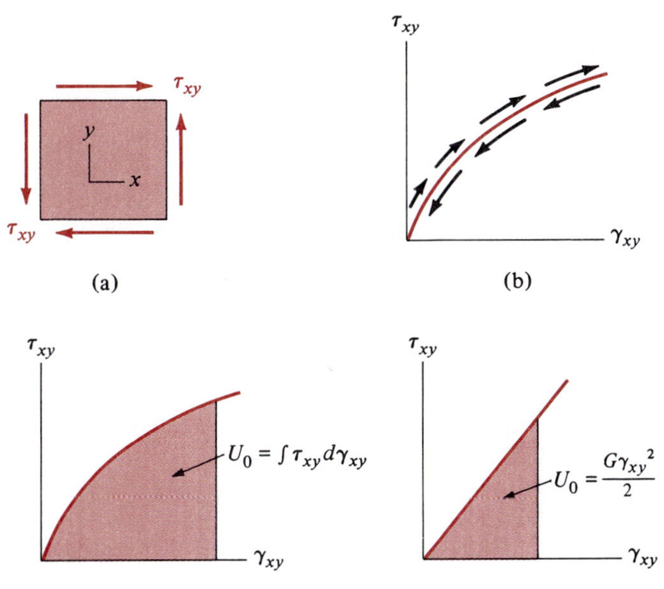

Figure 10.14 Shear-stress–shear-strain behavior for an elastic material

For a linearly elastic isotropic solid, $\tau_{xy} = G\gamma_{xy}$, and the strain-energy density for pure shear becomes

$$U_0 = \int G\gamma_{xy}\, d\gamma_{xy} = \frac{G\gamma_{xy}^2}{2} \tag{10.5}$$

corresponding to the triangular area of Fig. 10.14d. Note that for the linearly elastic solid, that is, $\tau_{xy} = G\gamma_{xy}$, the strain-energy density for pure shear can be written in the two alternative forms

$$U_0 = \frac{G\gamma_{xy}^2}{2} = \frac{\tau_{xy}\gamma_{xy}}{2} = \frac{\tau_{xy}^2}{2G}$$

both of which can be useful in specific situations.

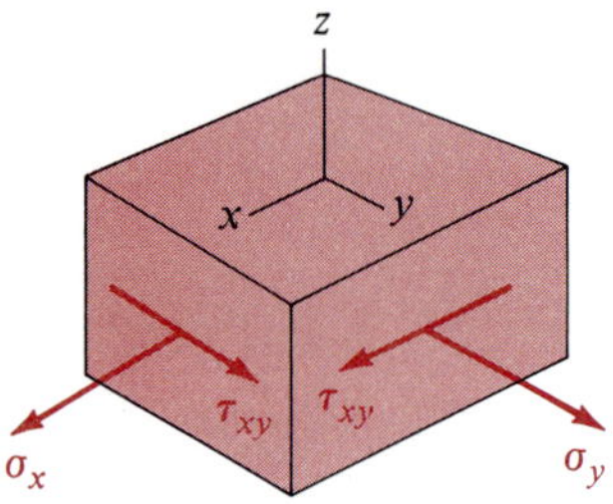

Figure 10.15 Nonzero stresses for a plane stress state

For the class of two-dimensional problems known as plane stress and shown in Fig. 10.15, all three of the stresses σ_x, σ_y, τ_{xy} are generally present simultaneously. It is left for the student to show in the problems that the strain-energy density for a linearly elastic isotropic material can be expressed as

$$U_0 = \frac{E}{2(1 - \nu^2)}(\varepsilon_x^2 + \varepsilon_y^2 + 2\nu\varepsilon_x\varepsilon_y) + \frac{G}{2}\gamma_{xy}^2 \tag{10.6}$$

Similarly, for the general three-dimensional linearly elastic isotropic solid, the strain-energy density can be represented entirely in terms of the strains as

$$\begin{aligned} U_0 = \frac{E}{2(1 + \nu)(1 - 2\nu)} &[(1 - \nu)(\varepsilon_x^2 + \varepsilon_y^2 + \varepsilon_z^2) \\ &+ 2\nu(\varepsilon_x\varepsilon_y + \varepsilon_y\varepsilon_z + \varepsilon_x\varepsilon_z)] \\ &+ \frac{G}{2}(\gamma_{xy}^2 + \gamma_{yz}^2 + \gamma_{xz}^2) \end{aligned} \tag{10.7}$$

PROBLEMS

10.33 Develop the expression for the strain-energy density U_0 for the one-dimensional linearly elastic solid, including temperature, that is, $\varepsilon_x = \sigma_x/E + \alpha T$.

10.34 Develop the expression for the strain-energy density U_0 for the one-dimensional elastic solid behaving according to $\sigma_x = E\varepsilon_x + E'(\varepsilon_x)^n$, where E and E' are constants.

10.35 For a state of plane stress the stresses, strains, and strain-energy density are related by

$$\frac{\partial U_0}{\partial \varepsilon_x} = \sigma_x \qquad \frac{\partial U_0}{\partial \varepsilon_y} = \sigma_y \qquad \frac{\partial U_0}{\partial \gamma_{xy}} = \tau_{xy}$$

Using the stress-strain relations to solve for the stresses from

$$\varepsilon_x = \frac{\sigma_x - \nu\sigma_y}{E} + \alpha T \qquad \varepsilon_y = \frac{\sigma_y - \nu\sigma_x}{E} + \alpha T \qquad \gamma_{xy} = \frac{\tau_{xy}}{G}$$

determine U_0 by integration.

10.36 When initial strain effects are included for the plane stress case, the stress-strain relations are written as

$$\varepsilon_x = \frac{\sigma_x - \nu\sigma_y}{E} + \varepsilon_{0x}(x, y) \quad \varepsilon_y = \frac{\sigma_y - \nu\sigma_x}{E} + \varepsilon_{0y}(x, y) \quad \gamma_{xy} = \frac{\tau_{xy}}{G} + \gamma_{0xy}$$

where $\varepsilon_{0x}(x, y)$, $\varepsilon_{0x}(x, y)$, and γ_{0xy} are the initial strains. Solve for the stresses and integrate the equations

$$\frac{\partial U_0}{\partial \varepsilon_x} = \sigma_x \qquad \frac{\partial U_0}{\partial \varepsilon_y} = \sigma_y \qquad \frac{\partial U_0}{\partial \gamma_{xy}} = \tau_{xy}$$

to determine the strain-energy density including initial strains.

10.37 When initial-stress effects are included for the plane stress case, the stress-strain relations can be written as

$$\sigma_x - \sigma_{0x} = \frac{E}{1 - \nu^2}(\varepsilon_x + \nu\varepsilon_y) \qquad \sigma_y - \sigma_{0y} = \frac{E}{1 - \nu^2}(\varepsilon_y + \nu\varepsilon_x)$$

and

$$\tau_{xy} - \tau_{0xy} = G\gamma_{xy}$$

Integrate the equations

$$\frac{\partial U_0}{\partial \varepsilon_x} = \sigma_x \qquad \frac{\partial U_0}{\partial \varepsilon_y} = \sigma_y \qquad \frac{\partial U_0}{\partial \gamma_{xy}} = \tau_{xy}$$

to determine the strain-energy density, including initial stresses.

10.3.1 Strain Energies for Axial, Torsional, and Bending Members

Elastic bars that transmit axial, torsional, and bending loads experience simple stress states and relatively simple deformations. In this section we will derive expressions for the strain energies contained in these specialized structural members.

Axial Deformations Consider a typical straight bar transmitting axial loads as shown in Fig. 10.16a.

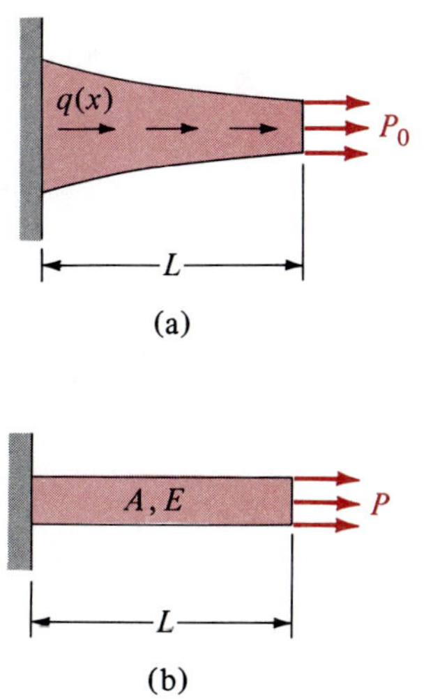

Figure 10.16 Axial deformation of a straight bar

Results from the theory developed in Chapter 5 are that the normal stress σ and internal axial load P are related by $\sigma = P/A$, and that the strain-displacement relation is $\varepsilon_x = du/dx = u'$, where u is the axial displacement of a point in the bar. Thus for a linearly elastic solid, the strain energy can be written as

$$U_0 = \frac{E\varepsilon_x^2}{2} = \frac{Eu'^2}{2}$$

with the total strain energy as

$$U = \int_{\text{Vol}} U_0 \, d\text{Vol} = \int_{\text{Vol}} \frac{Eu'^2}{2} \, d\text{Vol}$$

With $d\text{Vol} = A\,dx$, this can be expressed as

$$U = \int_0^L \frac{AEu'^2}{2} \, dx \tag{10.8}$$

For the linear material, the strain energy can be put in a form involving the internal load P by eliminating u' from Eqn. (10.8), using $u' = \varepsilon_x = \sigma_x/E = P/AE$ according to

$$U = \int_0^L \frac{AE(P/AE)^2}{2} \, dx = \int_0^L \frac{P^2}{2AE} \, dx \tag{10.9}$$

For the special case of a uniform bar under the action of a single force P at the end as pictured in Fig. 10.16b, the elongation of the bar is given by $\Delta = PL/AE$, with $u(x) = Px/AE = \Delta x/L$. The strain energy then becomes

$$U = \int_0^L \frac{AE\,\Delta^2}{2L^2} \, dx = \frac{AE\,\Delta^2}{2L} = \frac{k\,\Delta^2}{2}$$

where $k = AE/L$. Using the force-displacement relation $P = k\,\Delta$, the strain energy U can also be expressed for this special case as

$$U = \frac{P\,\Delta}{2} = \frac{P^2}{2k}$$

Torsional Deformations Consider a typical straight bar transmitting torques as shown in Fig. 10.17.

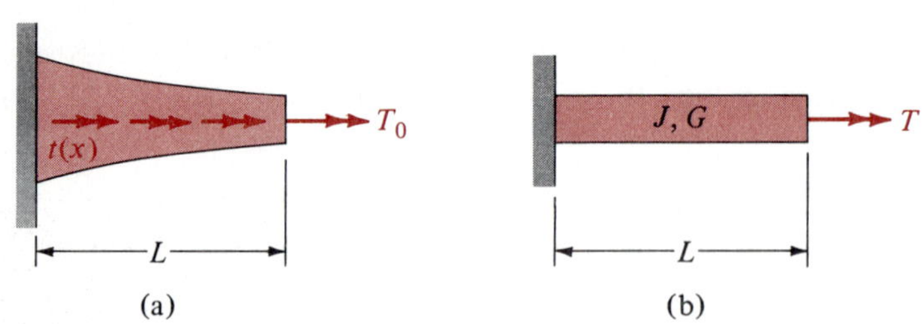

Figure 10.17 Torsional deformation of a straight bar

Results from the theory developed in Chapter 6 are that the shear stress τ and internal torque T are related by $\tau = Tr/J$ and that the strain-displacement relation is $\gamma_{x\theta} = r\phi'(x)$, where $\phi(x)$ is the angle of twist at a location along the axis of the bar. For the linearly elastic solid, the strain energy can be written as

$$U_0 = \frac{G\gamma_{x\theta}^2}{2} = \frac{G(r\phi')^2}{2}$$

with the total strain energy as

$$U = \int_{\text{Vol}} U_0 \, d\text{Vol} = \int_{\text{Vol}} \frac{G(r\phi')^2}{2} \, d\text{Vol}$$

With $d\text{Vol} = dA\,dx$, and $\int r^2\,dA = J$, this can be expressed as

$$U = \int_0^L \int_A \frac{G(r\phi')^2}{2} \, dA\,dx = \int_0^L \frac{JG\phi'^2}{2} \, dx \tag{10.10}$$

Using the torque-twist relation $T = JG\phi'$, the strain energy in torsion can be expressed in terms of the torque as

$$U = \int_0^L \frac{JG(T/JG)^2}{2} \, dx = \int_0^L \frac{T^2}{2JG} \, dx \tag{10.11}$$

For the special case of a uniform-cross-section bar under the action of a single torque T at the end as indicated in Fig. 10.17b, the relative angle of twist Φ between the ends of the bar is given by $\Phi = TL/JG$, with $\phi(x) = Tx/JG = \Phi x/L$. The strain energy then becomes

$$U = \int_0^L \frac{JG\Phi^2}{2L^2} \, dx = \frac{JG\Phi^2}{2L} = \frac{C\Phi^2}{2}$$

where $C = JG/L$. Using the torque-twist relation $T = C\Phi$, the strain energy U for this special case can also be expressed as

$$U = \frac{T\Phi}{2} = \frac{T^2}{2C}$$

Bending Deformations Consider a typical straight beam subjected to transverse loads as shown in Fig. 10.18.

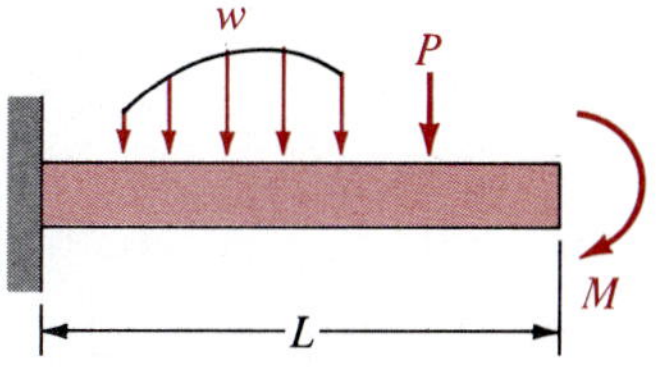

Figure 10.18 Bending deformation of a straight bar

In general the beam must transmit a shear force V and a bending moment M at every point along the axis. The bending stresses are given by the theory as $\sigma = -My/I$ and the strains as $\varepsilon = -y/\rho = -yv''$. We can express the strain energy of bending in terms of the displacements as

$$U = \int_{\text{Vol}} \frac{E\varepsilon^2}{2} \, d\text{Vol} = \int_{\text{Vol}} \frac{E(-yv'')^2}{2} \, d\text{Vol} = \int_0^L \frac{E(v'')^2}{2} \left(\int_{\text{Area}} y^2 \, dA \right) dx$$

$$= \int_0^L \frac{EI(v'')^2}{2} \, dx \tag{10.12}$$

Using the moment-curvature relation $M = EIv''$, the strain energy can also be expressed as

$$U = \int_0^L \frac{EI\left(\frac{M}{EI}\right)^2}{2}\,dx = \int_0^L \frac{M^2}{2EI}\,dx \tag{10.13}$$

Shear Energy in Bending As developed in Chapter 7, the shear-stress distribution associated with the shear force V is $\tau = VQ/It$. Expressed in terms of the stresses, the shear-strain energy is

$$U = \int_{\text{Vol}} \frac{\tau^2}{2G}\,d\text{Vol} = \int_{\text{Vol}} \frac{(VQ/It)^2}{2G}\,d\text{Vol} = \int_0^L \frac{V^2}{2GI^2}\left[\int_{\text{Area}} \left(\frac{Q}{t}\right)^2 dA\right] dx$$
$$= \int_0^L \frac{V^2}{2kAG}\,dx$$

where k is the dimensionless quantity given by

$$\frac{1}{k} = \frac{A}{I^2}\int_{\text{Area}} \left(\frac{Q}{t}\right)^2 dA \tag{10.14}$$

and accounts for the nonuniform distribution of the shear stress over the cross section. It is left for the student to show in the problems that for the rectangular cross section $k = \frac{5}{6}$, as stated in Section 8.8.

We will use the three representations

$$U = \int_0^L \frac{AEu'^2}{2}\,dx \qquad U = \int_0^L \frac{JG\phi'^2}{2}\,dx \qquad U = \int_0^L \frac{EIv''^2}{2}\,dx$$

for the strain energy associated with axial deformation, torsional deformation, and bending deformation, respectively, in the stiffness approach to formulating problems discussed in the next section.

On the other hand, we will use the three representations

$$U = \int_0^L \frac{P^2}{2AE}\,dx \qquad U = \int_0^L \frac{T^2}{2JG}\,dx \qquad U = \int_0^L \frac{M^2}{2EI}\,dx$$

for the strain energy associated with axial deformation, torsional deformation, and the bending deformation in the flexibility approach for formulating problems discussed in Section 10.4.

PROBLEMS

Axial Problems

10.38–10.43 In each of the following, consider the structure to be a collection of constant-cross-section segments transmitting a single axial force resultant, and thus express the strain energy and total potential energy in terms of the displacements indicated. Use stationary potential energy to derive the equations of equilibrium. The constants k_i represent A_iE_i/L_i.

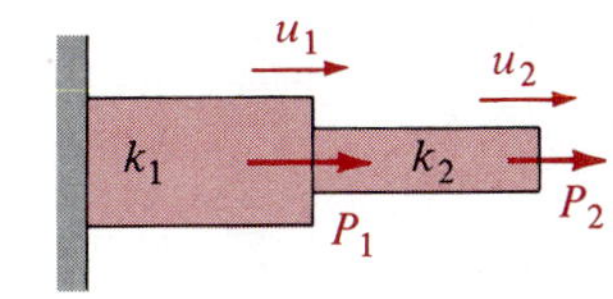

P. 10.38

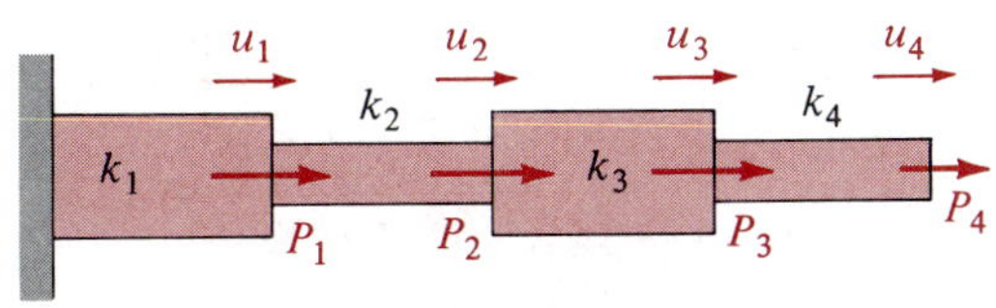

P. 10.39

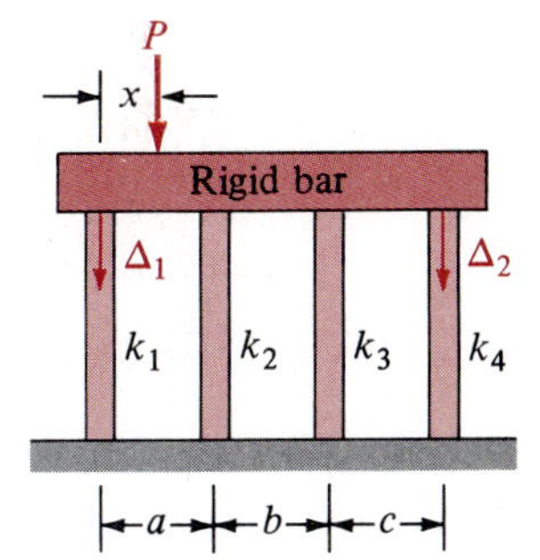

P. 10.40

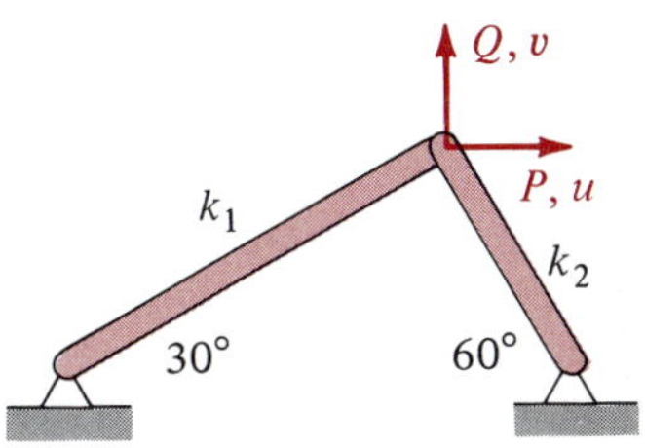

P. 10.41

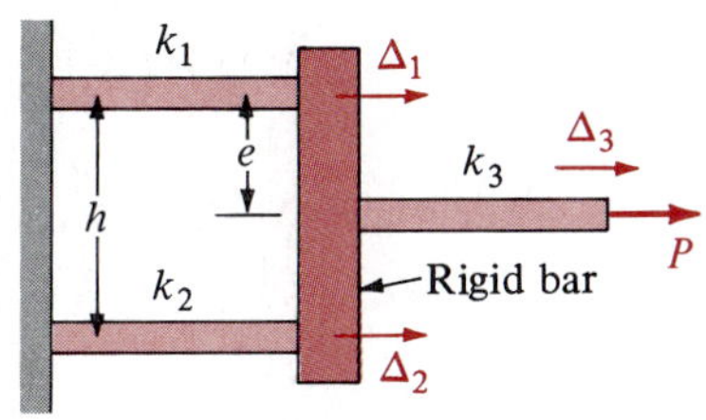

P. 10.42

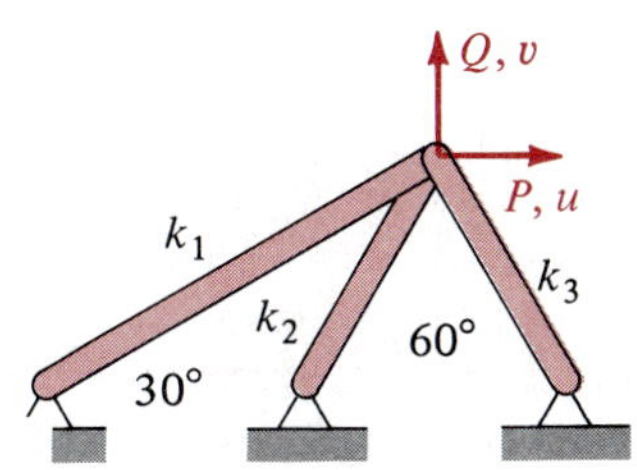

P. 10.43

Torsional Problems

10.44–10.45 In each of the following, consider the structure to be a collection of constant-cross-section segments transmitting a single axial torque, and thus represent the strain energy in terms of the rotations indicated. The constants C_i represent J_iG_i/L_i.

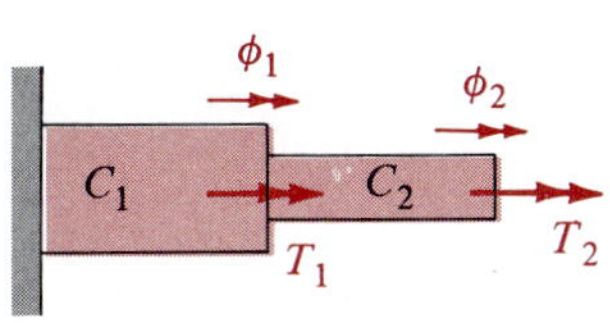

P. 10.44

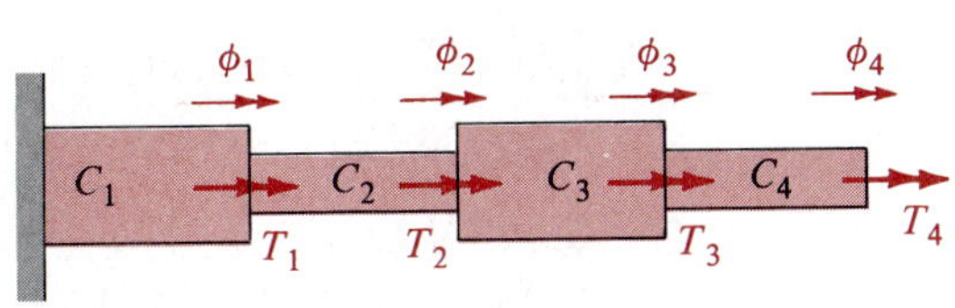

P. 10.45

10.46–10.47 In each of the following, consider the structure to be a collection of constant-cross-section segments transmitting a single axial torque, and thus represent the strain energy in terms of the rotations indicated. The constants C_i represent $J_i G_i / L_i$.

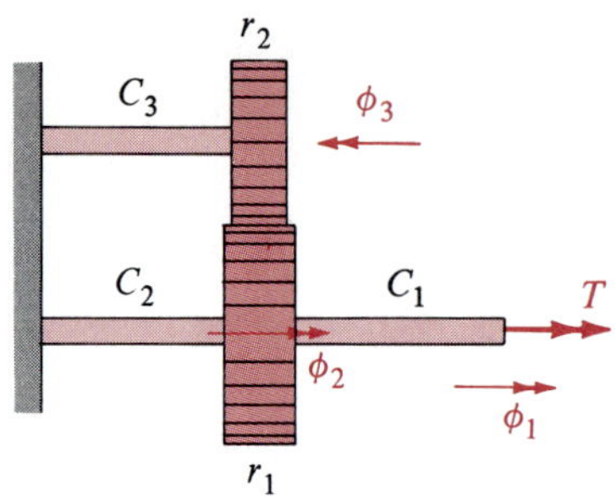

P. 10.46

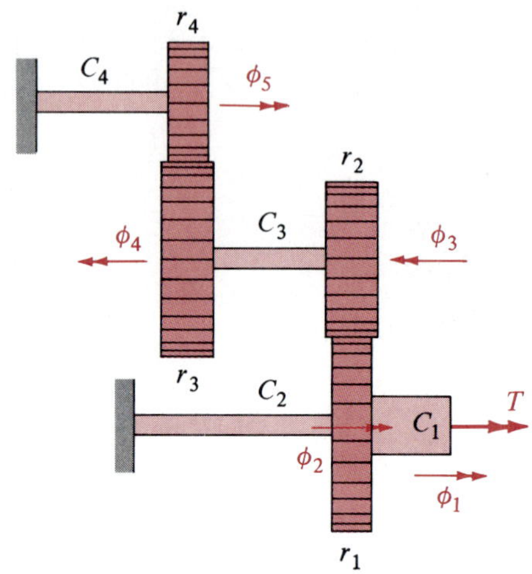

P. 10.47

Bending Problems

10.48 For the bending problem, the differential equation satisfied by the displacement $v(x)$ for an unloaded constant-cross-section segment is $EIv^{iv} = w = 0$.

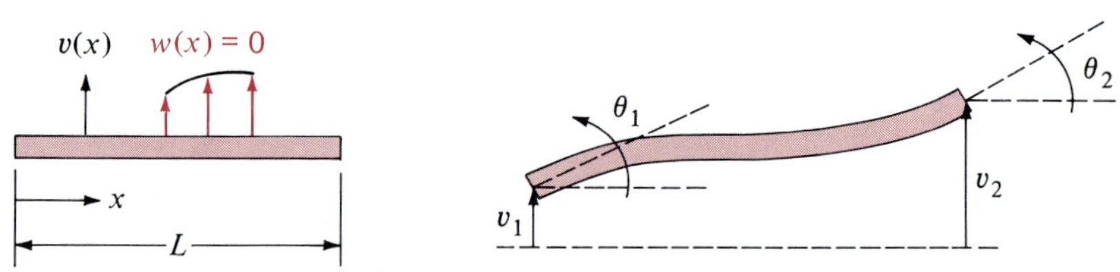

(a) (b)

P. 10.48

Show that by integrating the differential equation and satisfying the boundary conditions $v(0) = v_1$, $v'(0) = \theta_1$, $v(L) = v_2$, and $v'(L) = \theta_2$, as indicated in the figure, there results

$$v(x) = v_1\left[1 - 3\left(\frac{x}{L}\right)^2 + 2\left(\frac{x}{L}\right)^3\right] + \theta_1 x\left[1 - 2\left(\frac{x}{L}\right) + \left(\frac{x}{L}\right)^2\right]$$

$$+ v_2\left[3\left(\frac{x}{L}\right)^2 - 2\left(\frac{x}{L}\right)^3\right] + \theta_2 x\left[\left(\frac{x}{L}\right)^2 - \frac{x}{L}\right]$$

that is, the displacement $v(x)$ is represented in the segment $0 \le x \le L$ in terms of the translational degrees of freedom v_1 and v_2 and the translational degrees of freedom θ_1 and θ_2.

10.49 Using the results of problem 10.48, show that the strain energy can be evaluated in terms of the four degrees of freedom v_1, v_2, θ_1, and θ_2 as

$$U = \int_0^L \frac{EIv''^2}{2}\,dx = \frac{EI}{2L^3}(12v_1^2 + 12Lv_1\theta_1 - 24v_1v_2 + 12Lv_1\theta_2 + 4L^2\theta_1^2 - 12L\theta_1v_2 + 4L^2\theta_1\theta_2 + 12v_2^2 - 12v_2\theta_2L + 4L^2\theta_2^2)$$

10.50–10.54 For each of the following, express the strain energy and the total potential energy in terms of the degrees of freedom indicated, and determine the equations of equilibrium using stationary potential energy.

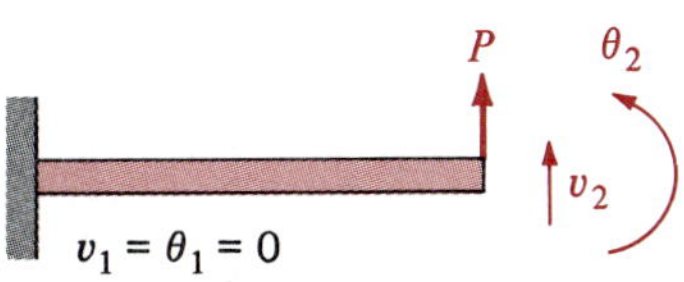

P. 10.50

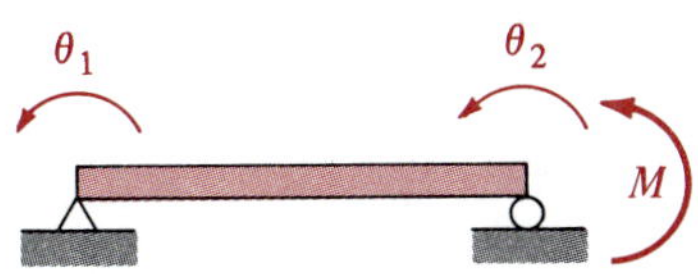

P. 10.51

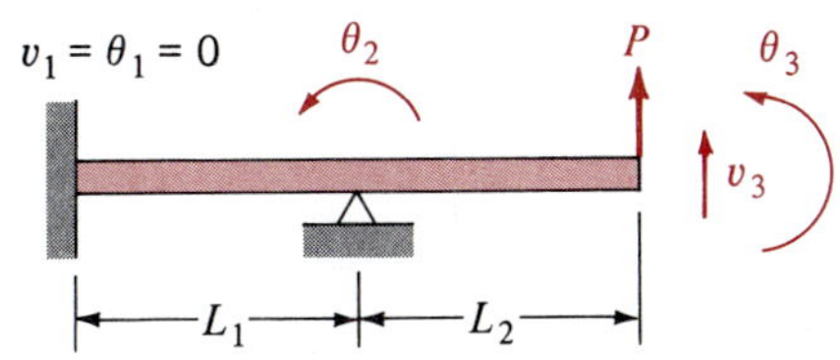

P. 10.52

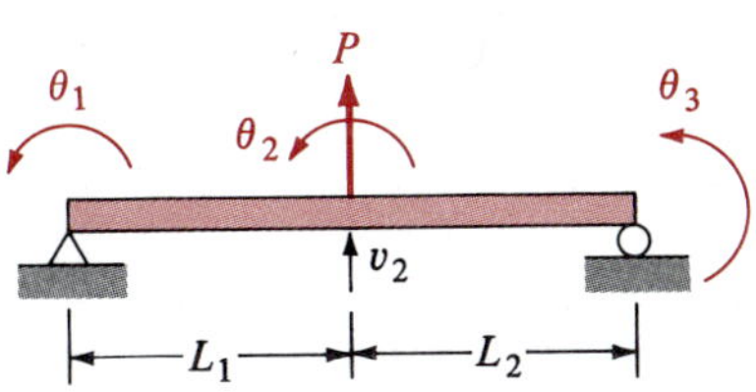

P. 10.53

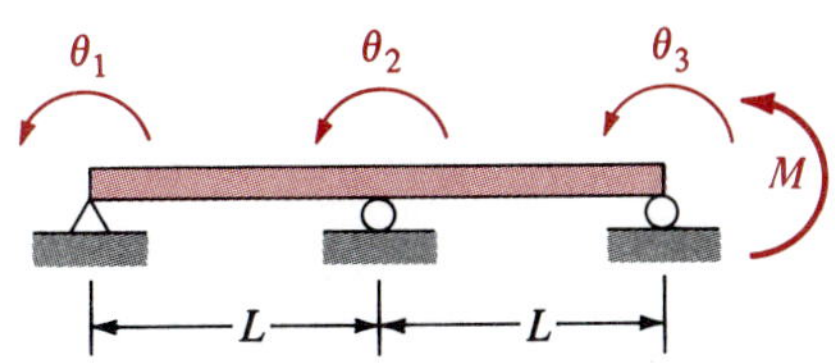

P. 10.54

10.55 Show that the expression for k given by Eqn. 10.14 evaluates to $k = \frac{5}{6}$ for the case of the rectangular cross section.

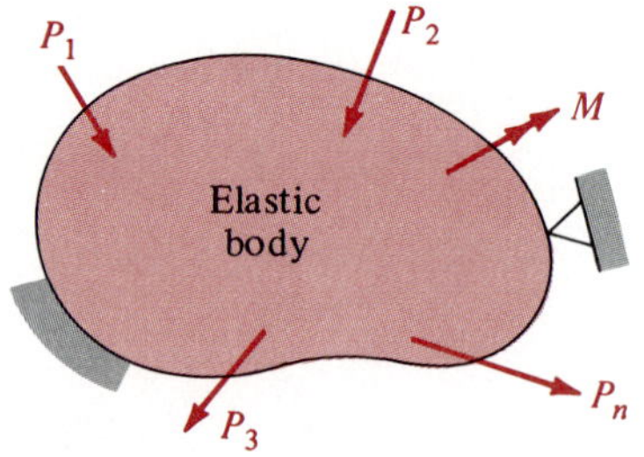

Figure 10.19 Elastic body loaded by forces and couples

10.3.2 Analysis of Structures Using Displacement Variable Formulations—Castigliano's First Theorem

Consider an elastic body or collection of elastic elements such as axial, torsional, and bending members, that is, loaded by forces and couples as shown symbolically in Fig. 10.19.

Let displacements $\Delta_1, \Delta_2, \ldots, \Delta_n$, be defined in the directions of the forces shown. If the ith load is a couple, the corresponding Δ_i is a rotation about the same axis. Recalling that for a conservative elastic system, the potential energy of the external forces is the negative of the work done by the external forces, the potential energy of the external forces can be represented as

$$\Omega = -\sum^{n} P_i \Delta_i$$

We assume that the displacements of the elastic body can be expressed in terms of the Δ_i, so that the strains and the strain energy U can be expressed in terms of the Δ_i. The total potential energy V of the system, consisting of the elastic body and the loads, can then be represented as

$$V = V(\Delta_1, \Delta_2, \ldots, \Delta_n) = U(\Delta_1, \Delta_2, \ldots, \Delta_n) - \sum^{n} P_i \Delta_i$$

We can then state the principle of stationary potential energy as

$$\frac{\partial V}{\partial \Delta_k} = \frac{\partial U}{\partial \Delta_k} - P_k = 0$$

or

$$\frac{\partial U}{\partial \Delta_k} = P_k \qquad (10.15)$$

Eqn. (10.15) is known as **Castigliano's first theorem.** It is very useful in the analysis of elastic systems whose potential energies can be expressed in terms of the displacements of the system.

If the strain energy U corresponds to a linear elastic solid, the strain energy will be a quadratic function of the displacements. If we denote the displacements as $x_1, x_2, \ldots x_n$, the strain energy will appear as the quadratic function

$$\begin{aligned} 2U = {} & k_{11}x_1^2 + k_{12}x_1x_2 + \ldots + k_{1n}x_1x_n \\ & + k_{21}x_2x_1 + k_{22}x_2^2 + \ldots + k_{2n}x_2x_n \\ & \ldots \\ & + k_{n1}x_nx_1 + k_{n2}x_nx_2 + \ldots + k_{nn}x_n^2 \end{aligned}$$

where for the linearly elastic case, $k_{ij} = k_{ji}$. Using Castigliano's first theorem leads to

$$P_1 = \frac{\partial U}{\partial x_1} = k_{11}x_1 + k_{12}x_2 + \ldots + k_{1n}x_n$$

$$P_2 = \frac{\partial U}{\partial x_2} = k_{21}x_1 + k_{22}x_2 + \ldots + k_{2n}x_n$$

$$\ldots$$

$$P_n = \frac{\partial U}{\partial x_n} = k_{n1}x_1 + k_{n2}x_2 + \ldots + k_{nn}x_n$$

which are the equilibrium equations expressed in terms of the displacements of the system. As indicated, for a linearly elastic system, these equations are linear algebraic equations, the solution of which can easily be carried out using a computer.

The coefficients $k_{ij} = k_{ji}$ are called **stiffness coefficients.** The method of solution using the displacements as the unknowns is termed the **stiffness method** or **displacement method.** The analysis of structures of all kinds is carried out almost exclusively using the displacement method. This includes the analysis of structures using the finite element method.

Example 10.5 A segmented axial member is loaded by the forces shown in Fig. 10.20. Derive (using Castigliano's first theorem) and then solve the equations of equilibrium.

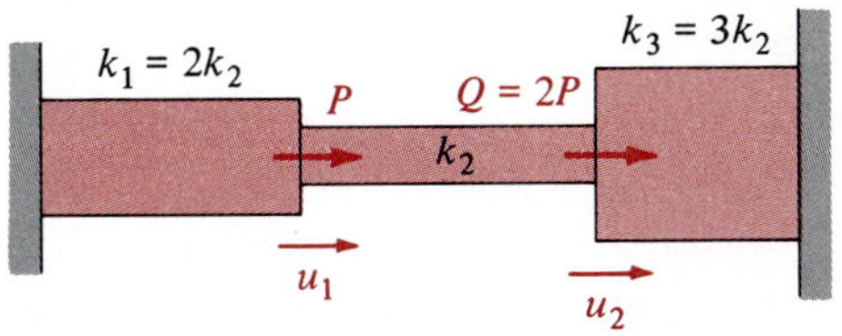

Axially loaded member

Solution For this problem it is necessary to define displacements u_1 and u_2 at the points of applications of the loads P and Q as shown in Fig. 10.20. In terms of these displacements, the elongations of the two members are $e_1 = u_1$, $e_2 = u_2 - u_1$, and $e_3 = -u_2$. The total strain energy can be written as

$$U = \frac{1}{2}k_1e_1^2 + \frac{1}{2}k_2e_2^2 + \frac{1}{2}k_3e_3^2 = \frac{1}{2}k_1u_1^2 + \frac{1}{2}k_2(u_2 - u_1)^2 + \frac{1}{2}k_3u_2^2$$

The application of Castigliano's first theorem leads to

$$\frac{\partial U}{\partial u_1} = P = k_1 u_1 - k_2(u_2 - u_1)$$

and

$$\frac{\partial U}{\partial u_2} = Q = k_2(u_2 - u_1) + k_3 u_2$$

Substituting the values for Q, k_1, and k_3, these equilibrium equations can be written as

$$3u_1 - u_2 = \frac{P}{k_2} \qquad \text{and} \qquad -u_1 + 4u_2 = \frac{2P}{k_2}$$

which can be solved to yield $u_1 = 6P/11k_2$ and $u_2 = 7P/11k_2$. The forces in the members can be computed as

$$F_1 = k_1 e_1 = 2k_2\left(\frac{6P}{11k_2}\right) = \frac{12P}{11}$$

$$F_2 = k_2 e_2 = k_2(u_2 - u_1) = k_2\left(\frac{P}{11k_2}\right) = \frac{P}{11}$$

$$F_3 = k_3 e_3 = 3k_2(-u_2) = 3k_2\left(\frac{-7P}{11k_2}\right) = -\frac{21P}{11}$$

Free-body diagrams of two parts of the structure are shown in Figs. 10.21a and b, indicating that the internal forces satisfy equilibrium.

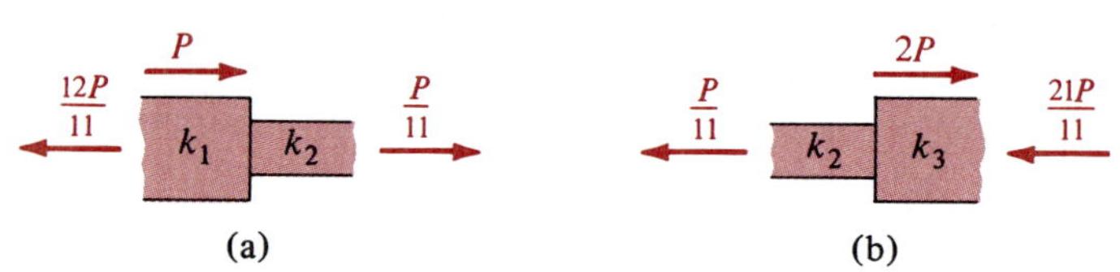

Figure 10.21 Equilibrium of internal forces

Note that although the problem is statically indeterminate, no extra effort is needed to set up and solve the equations of equilibrium.

Example 10.6 A planar truss with properties of the individual bars indicated is loaded by the vertical force as shown in Fig. 10.22a. Assuming that each member is composed of a linearly elastic material, use Castigliano's first theorem to derive the equations of equilibrium. Solve the equations and determine the internal forces in the bars.

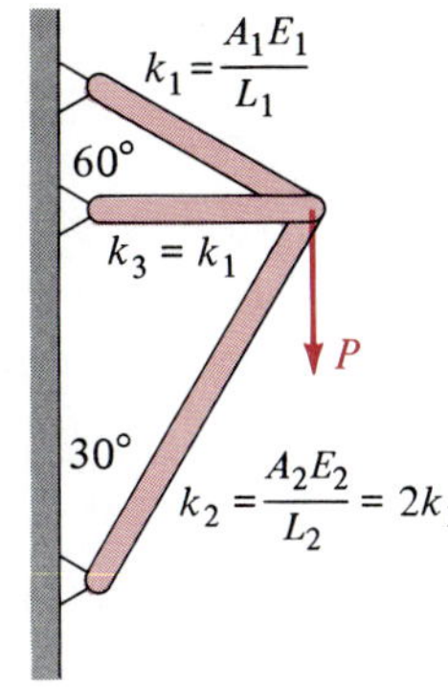

(a)

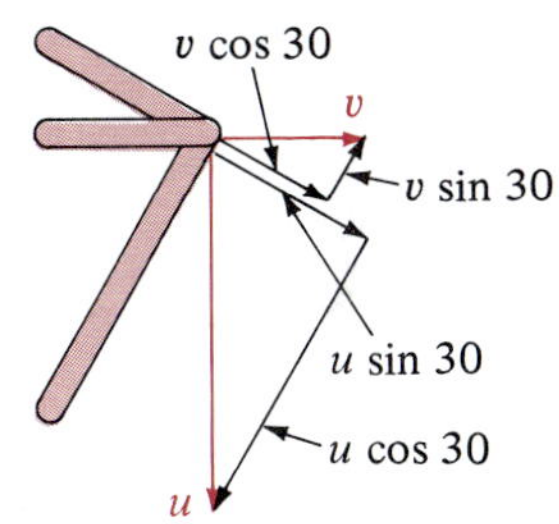

(b)

Figure 10.22 Analysis of a statically indeterminate truss

Solution In order to completely specify the deformed configuration of the two-bar truss, it is necessary to specify v and u, the horizontal and vertical displacements, respectively, of the loaded joint. In other words, the system of pinned members has two degrees of freedom. Each of the bars of the truss can be considered to be a linear spring with $k = AE/L$ and energy $U = \frac{1}{2}ke^2$, where e is the elongation. Consulting Fig. 10.22b, it is seen that the elongations of the members are

$$e_1 = v \cos 30° + u \sin 30° \qquad \text{and} \qquad e_2 = v \sin 30° - u \cos 30°$$

The total strain energy can then be written as

$$U = \frac{1}{2}k_1e_1^2 + \frac{1}{2}k_2e_2^2 = \frac{1}{2}k_1(lv + mu)^2 + \frac{1}{2}k_2(mv - lu)^2$$

where for brevity we have taken $l = \cos 30°$ and $m = \sin 30°$. Castigliano's first theorem dictates that

$$\frac{\partial U}{\partial u} = P = k_1m(lv + mu) - k_2l(mv - lu)$$

and

$$\frac{\partial U}{\partial v} = 0 = k_1l(lv + mu) + k_2m(mv - lu)$$

or

$$(k_1m^2 + k_2l^2)u + (k_1 - k_2)lmv = P$$

and

$$(k_1 - k_2)lmu + (k_1l^2 + k_2m^2)v = 0$$

Substituting for l, m, k_1, and k_2 leads to

$$1.750u - 0.433v = \frac{P}{k_1} \qquad \text{and} \qquad -0.433u + 1.250v = 0$$

with solution $u = 0.625P/k_1$ and $v = 0.217P/k_1$. The forces in the members are then given by

$$F_1 = k_1e_1 = k_1(v \cos 30° + u \sin 30°) = 0.5P$$

and

$$F_2 = k_2e_2 = 2k_1(v \sin 30° - u \cos 30°) = -0.866P$$

The student should show that these internal forces satisfy equilibrium by drawing a free-body diagram of the loaded joint.

Although the truss of Example 10.6 is statically indeterminate, we see that there was no extra effort involved in obtaining the solution. This is characteristic of the stiffness method associated with the application of Castigliano's first theorem and one of the main reasons for its extensive use. Another characteristic of the stiffness method is that the internal forces or moments are determined after solving for the displacements, whether or not the problem is statically determinate. One might think that for a statically determinate problem the internal forces could be simply determined by drawing the necessary free-body diagrams. For a complicated structure, however, it is much more efficient to use the stiffness method, determining the displacements and then the internal forces, rather than to actually draw the free-body diagrams.

Example 10.7 Use Castigliano's first theorem to derive the equilibrium equations for the statically indeterminate torsional system shown in Fig. 10.23a. Solve the equations, determine the internal torques, and show the results on properly drawn free-body diagrams. Take each of the shafts to be made from steel for which $G = 80$ GPa.

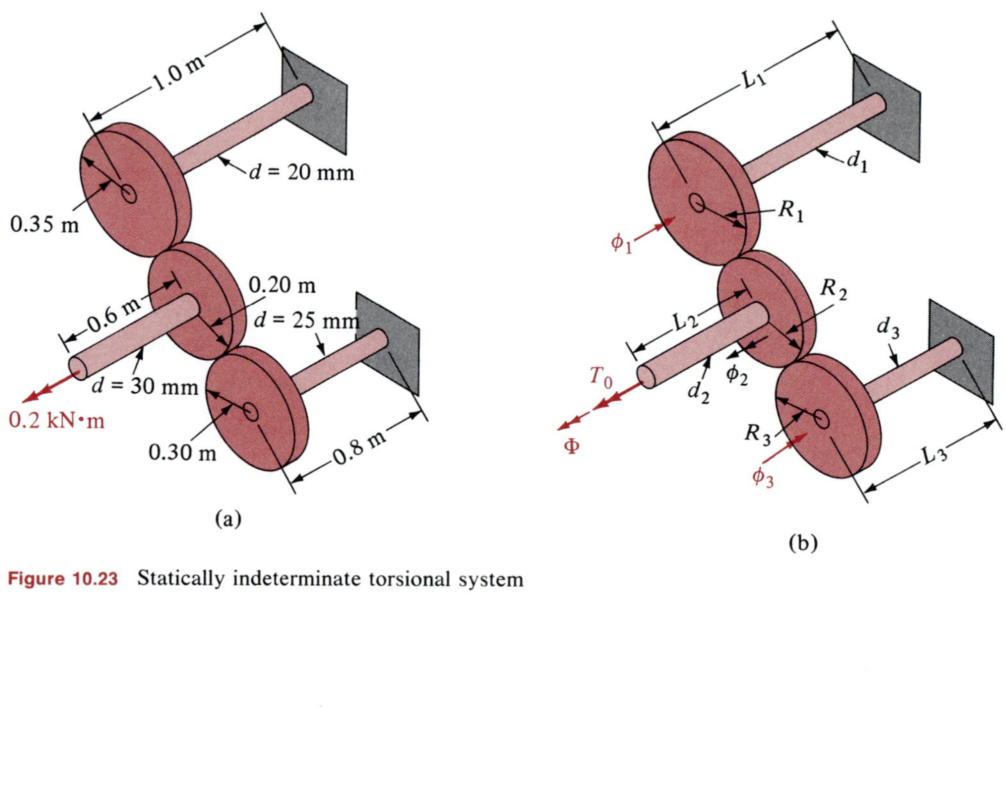

Figure 10.23 Statically indeterminate torsional system

Solution In order to facilitate setting up the problem, designate the radii of the disks and the lengths and diameters of the shafts as shown in Fig. 10.23b. Also shown in Fig. 10.23b are a set of rotations for use in representing the torsional strain energies of the shafts. These are $U_1 = \frac{1}{2}C_1\phi_1^2$, $U_2 = \frac{1}{2}C_2(\Phi - \phi_2)^2$, and $U_3 = \frac{1}{2}C_3\phi_3^2$, where the C_i are the torsional spring constants $C_i = J_iG_i/L_i$. However, if perfect contact is to be maintained between the disks, the rotations ϕ_1, ϕ_2, and ϕ_3 must satisfy the kinematical constraints

$$R_2\phi_2 = R_1\phi_1 \qquad \text{and} \qquad R_2\phi_2 = R_3\phi_3$$

between disks 1 and 2 and between disks 2 and 3, respectively, showing that only one of ϕ_1, ϕ_2, and ϕ_3 is independent. Use these constraint equations to express ϕ_1 and ϕ_3 in terms of ϕ_2 according to

$$\phi_1 = \frac{R_2}{R_1}\phi_2 = \alpha\phi_2 \qquad \text{and} \qquad \phi_3 = \frac{R_2}{R_3}\phi_2 = \beta\phi_2$$

so that the total strain energy can be represented in terms of the two degrees of freedom Φ and ϕ_2 as

$$U = U_1 + U_2 + U_3 = \frac{1}{2}C_1(\alpha\phi_2)^2 + \frac{1}{2}C_2(\Phi - \phi_2)^2 + \frac{1}{2}C_3(\beta\phi_2)^2$$

Castigliano's first theorem leads to the equilibrium equations

$$\frac{\partial U}{\partial \Phi} = T_0 = C_2(\Phi - \phi_2)$$

and

$$\frac{\partial U}{\partial \phi_2} = 0 = C_1\alpha^2\phi_2 - C_2(\Phi - \phi_2) + C_3\beta^2\phi_2$$

that can be solved for the rotations Φ and ϕ_2. Using the values shown in Fig. 10.23a, $C_1 = 1.257$ kN·m, $C_2 = 10.603$ kN·m, $C_3 = 8.835$ kN·m, $\alpha = \frac{4}{7}$, and $\beta = \frac{2}{3}$, so that the equilibrium equations become

$$10.603(\Phi - \phi_2) = 0.200 \qquad \text{and} \qquad 12.177\phi_2 - 10.603\Phi = 0$$

from which $\phi_2 = 0.0946$ rad and $\Phi = 0.1135$ rad. The torques in the shafts are computed as

$$T_1 = C_1\phi_1 = C_1\alpha\phi_2 = 1.257 \text{ kN·m}(\tfrac{4}{7})(0.0946) = 67.9 \text{ N·m}$$

$$T_2 = 200 \text{ N·m}$$

$$T_3 = C_3\phi_3 = C_3\beta\phi_2 = 3.835 \text{ kN·m}(\tfrac{2}{3})(0.0946) = 241.9 \text{ N·m}$$

A free-body diagram showing the torques in the shafts and the forces between the disks is indicated in Fig. 10.24.

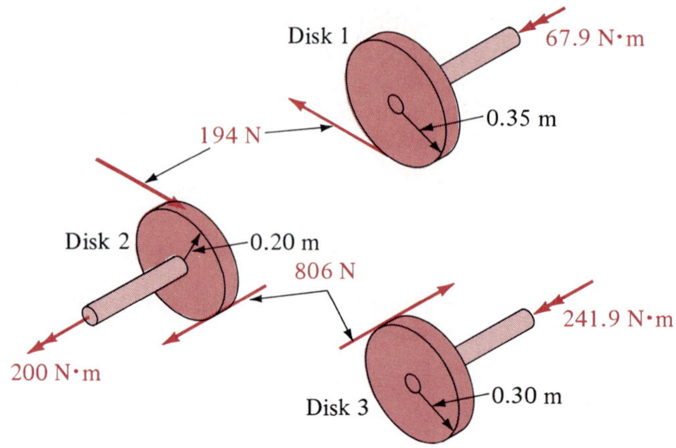

Figure 10.24 Equilibrium of the torques on individual shafts

The student should verify that the resultant torque about the axis of the shaft vanishes for each free-body diagram.

PROBLEMS

For each of the problems in this set, use strain energy and either stationary potential energy or Castigliano's first theorem to formulate the equations of equilibrium.

Axial Problems

For problems 10.56–10.69, k is used to represent the stiffness AE/L for the constant-cross-section, linearly elastic bar.

10.56 Using the angle of rotation of the rigid horizontal bar as the degree of freedom, determine *(a)* the elongations, and *(b)* the forces in the three bars. Take $A_1 = A_2 = A_3 = 1$ in^2, $E_1 = E_2 = E_3 = 10^7$ psi, $L_1 = L_2 = L_3 = 10$ ft, $a = 4$ ft, and $P = 10$ kips. Do the forces in the three bars change if all three lengths are 5 ft rather than 10 ft?

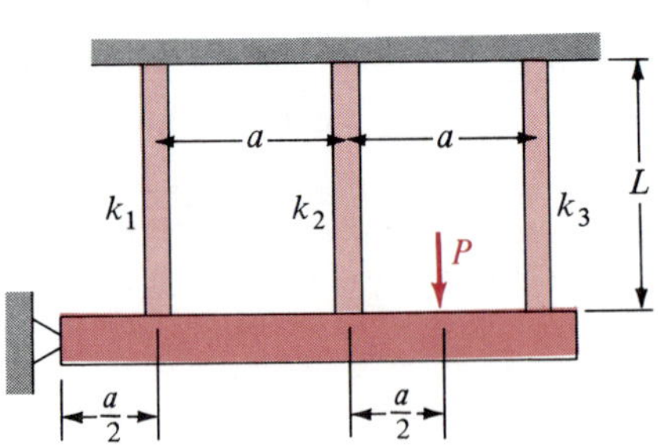

P. 10.56

10.57 Using the elongations of members 1 and 4 as the degrees of freedom, determine *(a)* the elongations in the members, and *(b)* the forces in the members. Take $k_1 = k_2 = k_3 = k_4$.

10.58 Repeat problem 10.57 taking the elongation of member 1 and the rotation of the rigid horizontal bar as the degrees of freedom.

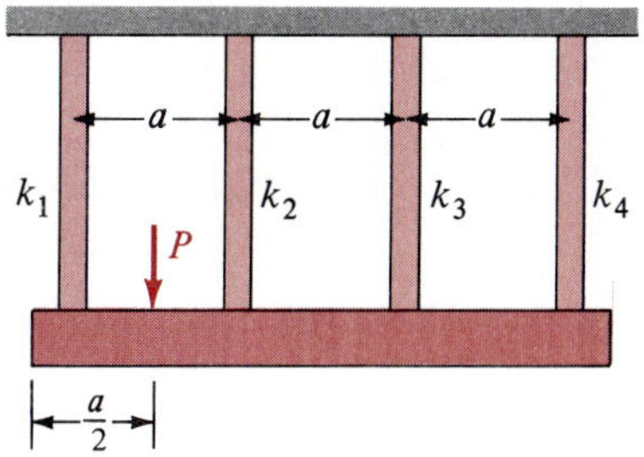

Ps. 10.57 and 10.58

10.59 The load P is applied centrally to the rigid block supported symmetrically by two aluminum members and one steel member, too short by the amount g as shown. Determine the forces in the three bars. Take $A_S = 2 \times 10^{-4}$ m^2, $A_A = 1 \times 10^{-4}$ m^2, $E_S = 200$ GPa, $E_A = 70$ GPa, $P = 20$ kN, $g = 1$ mm, and the initial length of the aluminum bars to be 1 m.

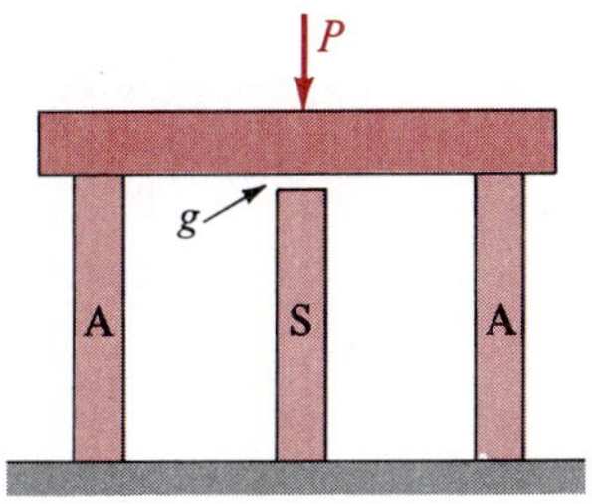

Ps. 10.59 and 10.60

10.60 Repeat problem 10.59 if the load P is applied midway between the steel bar and the aluminum bar on the left.

10.61 Determine the elongations of the two bars shown. Take $A_1 = 4 \times 10^{-4}$ m^2, $A_2 = 2 \times 10^{-4}$ m^2, $E_1 = E_2 = 70$ GPa, $L_1 = 0.7$ m, $L_2 = 0.5$ m, and $P = 30$ kN.

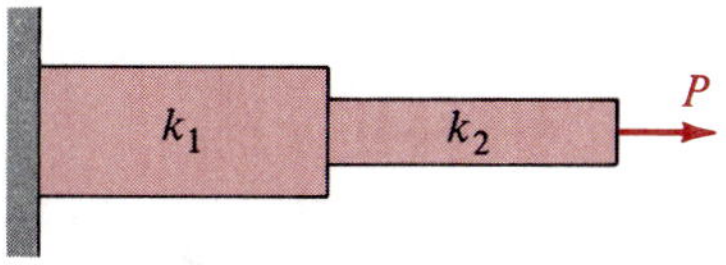

P. 10.61

10.62 Determine the displacements at the junctions of the members and the force in each of the members. Take $A_1 = 4$ in^2, $A_2 = 6$ in^2, $A_3 = 4$ in^2, $A_4 = 2$ in^2, $E_1 = E_2 = E_3 = E_4 = 18{,}000$ ksi, $L_1 = 8$ in., $L_2 = 14$ in., $L_3 = 10$ in., $L_4 = 6$ in., and $P = 12$ kips.

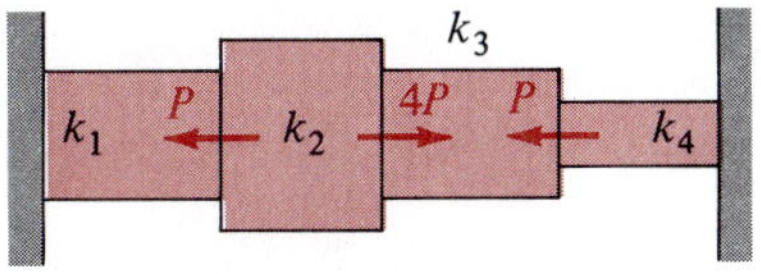

P. 10.62

10.63 For the load P applied at an interior location of the bar as shown, determine the gap g if both supports are to carry half the load.

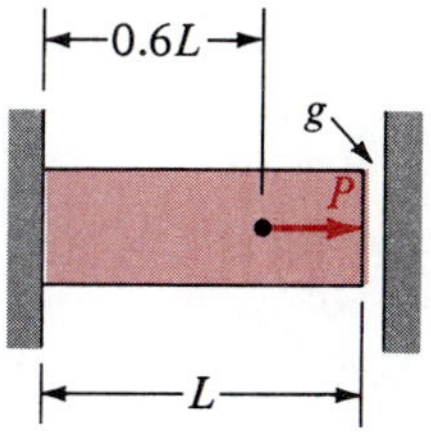

P. 10.63

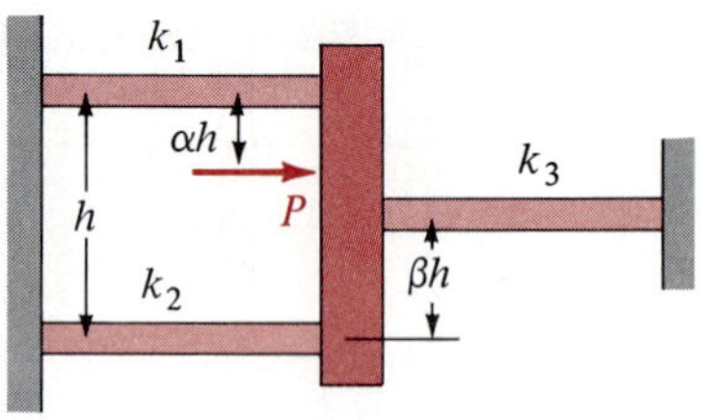

Ps. 10.64 and 10.65

10.64 Determine *(a)* the elongations of the three bars, and *(b)* the forces in the three bars. Take $A_1 = 3$ in^2, $A_2 = 3$ in^2, $A_3 = 4$ in^2, $E_1 = E_2 = E_3 = 24{,}000$ ksi, $L_1 = L_2 = 14$ in., $L_3 = 10$ in., $\alpha = \beta = 0.5$, and $P = 80$ kips. The vertical member is assumed to be rigid.

10.65 Repeat problem 10.64 if $\alpha = 0.3$ and $\beta = 0.6$.

10.66 Bar 1 is composed of a nonlinear elastic material behaving according to $\sigma = C\varepsilon^{1/3}$, where $C = 10^5$ psi. Bar 2 is linearly elastic with $E = 10^7$ psi. Determine the internal force in each bar if *(a)* $P = 3$ kips, *(b)* $P = 6$ kips, and *(c)* $P = 30$ kips. Take $L_1 = 16$ in., $L_2 = 12$ in., $A_1 = 1$ in^2, and $A_2 = 2$ in^2.

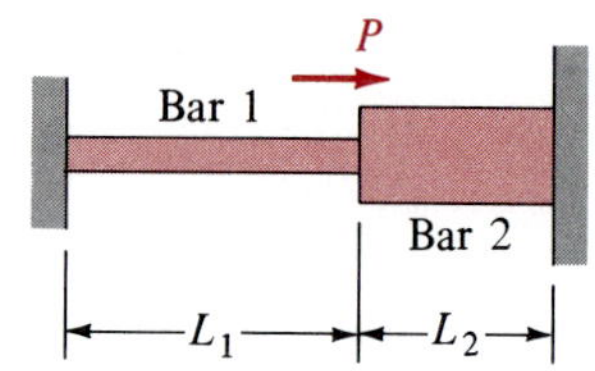

P. 10.66

10.67 Determine the horizontal and vertical displacement components at the loaded joint. Take $A_1 = 4 \times 10^{-4}$ m^2, $A_2 = 3 \times 10^{-4}$ m^2, $E_1 = E_2 = 70$ GPa, $L_2 = 0.5$ m, and $P = 30$ kN.

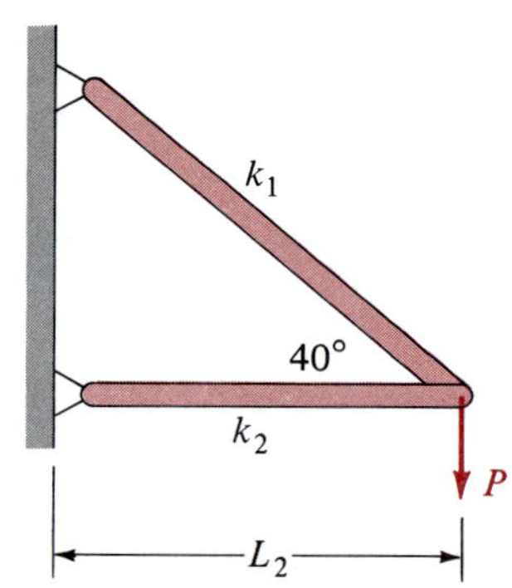

P. 10.67

10.68 Determine the horizontal and vertical displacement components at the loaded joint. Take $A_1 = 4 \times 10^{-4}$ m^2, $A_2 = 3 \times 10^{-4}$ m^2, $A_3 = 3 \times 10^{-4}$ m^2, $E_1 = E_2 = E_3 = 200$ GPa, $L_2 = 0.5$ m, $P = 40$ kN, and $Q = 30$ kN.

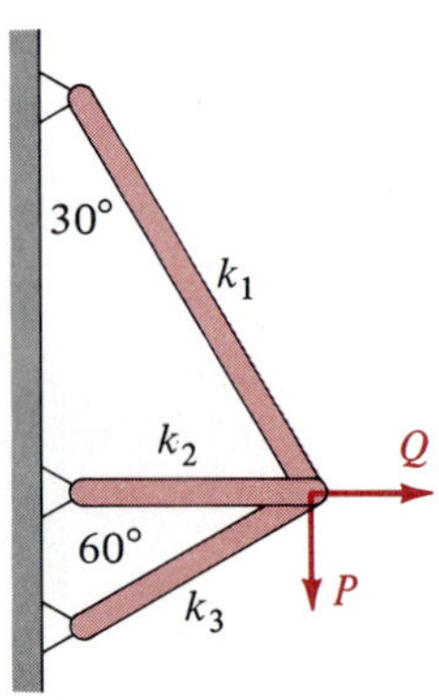

P. 10.68

10.69 Determine the horizontal and vertical displacement components at the loaded joint. Take $A_1 = A_5 = 2\ \text{in}^2$, $A_2 = A_3 = A_4 = 1.5\ \text{in}^2$, $E_1 = E_2 = E_3 = E_4 = E_5 = 10^7$ psi, $L_3 = 3$ ft, and $P = 20$ kips.

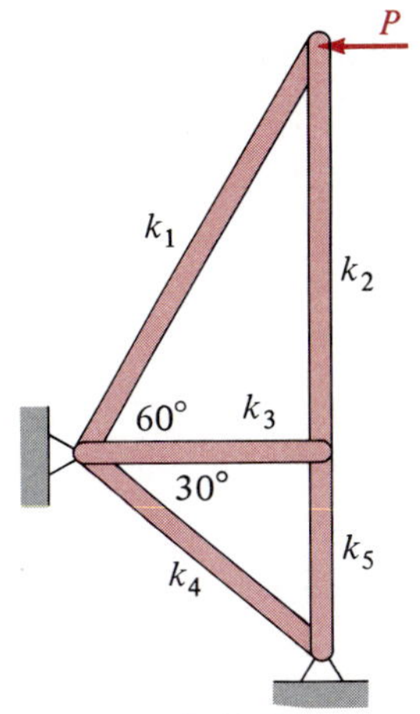

P. 10.69

Torsional Problems

For problems 10.70–10.76, C is used to represent the JG/L for the constant-radius, circular-cross-section, linearly elastic bar.

10.70 Determine the rotations at the ends of the segments of bars 1 and 2. Take $r_1 = 25$ mm, $r_2 = 20$ mm, $L_1 = 0.6$ m, $L_2 = 0.4$ m, $G_1 = G_2 = 28$ GPa, $T_1 = 0.5$ kN·m, and $T_2 = 0.3$ kN·m.

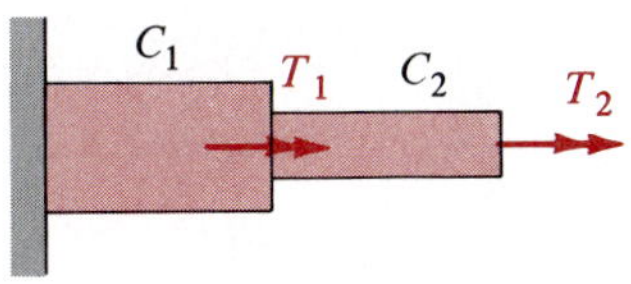

Ps. 10.70 and 10.71

10.71 Repeat problem 10.70 for $r_1 = 1.0$ in., $r_2 = 0.7$ in., $L_1 = 3$ ft, $L_2 = 2$ ft, $G_1 = G_2 = 3800$ ksi, $T_1 = 400$ ft-lbf, and $T_2 = 200$ ft-lbf.

10.72 Determine the torque transmitted by each of the torsional members. Take $r_1 = 2.0$ in., $r_2 = 1.4$ in., $r_3 = 1.8$ in., $r_4 = 0.8$ in., $L_1 = 3$ ft, $L_2 = 2$ ft, $L_3 = 1.5$ ft, $L_4 = 2.5$ ft, $G_1 = G_2 = G_3 = G_4 = 11{,}500$ ksi, $T_1 = 800$ ft-lbf, $T_2 = 600$ ft-lbf, and $T_3 = 300$ ft-lbf.

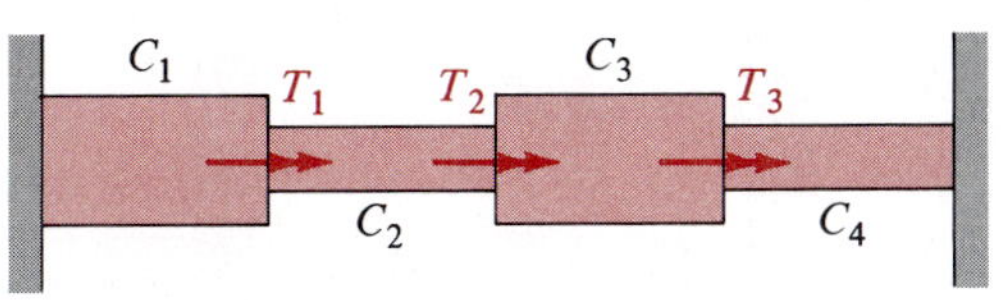

P. 10.72

10.73 Determine the angles of rotation for each of the gears. Take $r_1 = 30$ mm, $r_2 = 27$ mm, $R_1 = 0.3$ m, $R_2 = 0.25$ m, $G_1 = G_2 = 80$ MPa, $L_1 = 0.8$ m, $L_2 = 0.6$ m, and $T = 1.3$ kN·m.

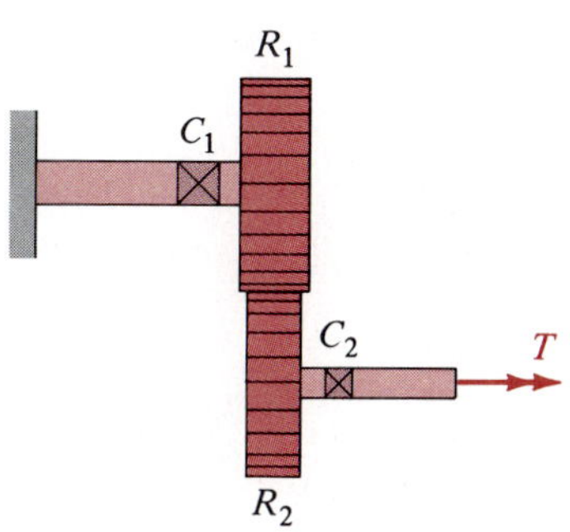

P. 10.73

10.74 Determine *(a)* the rotation of each of the gears, and *(b)* the torque in each of the shafts. Take $r_1 = 2.0$ in., $r_2 = 1.8$ in., $r_3 = 1.8$ in., $r_4 = 0.9$ in., $R_1 = 20$ in., $R_2 = 18$ in., $R_3 = 16$ in., $R_4 = 8$ in., $L_1 = 3$ ft, $L_2 = 2$ ft, $L_3 = 1.5$ ft., $L_4 = 2.5$ ft, $G_1 = G_2 = G_3 = G_4 = 11{,}500$ ksi, and $T = 3000$ ft-lbf.

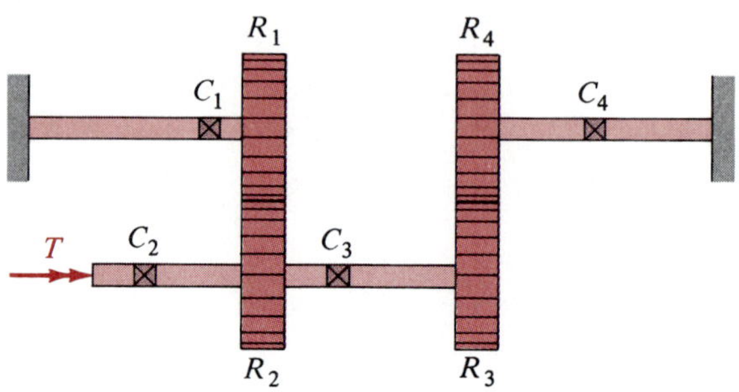

P. 10.74

10.75 Determine *(a)* the rotation of each of the gears, and *(b)* the torque in each of the shafts. Take $r_2 = r_3 = r_4 = r_5 = r_6 = 25$ mm, $r_1 = 20$ mm, $r_7 = 25$ mm, $R_2 = R_3 = R_4 = R_5 = 0.2$ m, $R_1 = R_6 = 0.28$ m, $L_2 = L_3 = L_4 = L_5 = L_6 = 2.0$ m, $L_1 = 3$ m, $L_7 = 2.4$ m, all Gs are 80 GPa, $T_1 = 0.6$ kN·m, and $T_2 = 0.4$ kN·m.

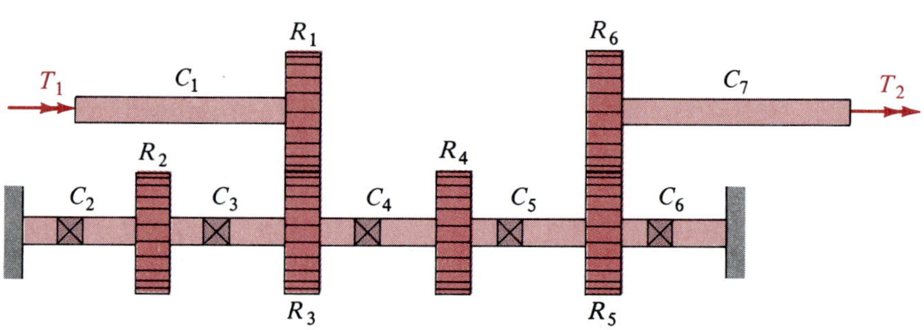

P. 10.75

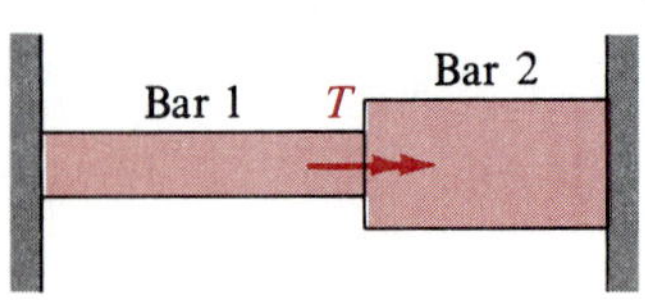

P. 10.76

10.76 Bar 1 is composed of an elastic material for which $\tau = C\gamma^{1/2}$ with $C = 3$ GPa. Bar 2 is composed of a linearly elastic material for which $G = 30$ GPa. Determine the torque in each of the bars when *(a)* $T = 1$ kN·m, *(b)* $T = 2$ kN·m, and *(c)* $T = 10$ kN·m. Take $r_1 = 20$ mm, $r_2 = 30$ mm, $L_1 = 0.4$ m, and $L_2 = 0.3$ m.

Bending Problems

10.77 Determine the deflection and rotation at the loaded end of the cantilever beam shown.

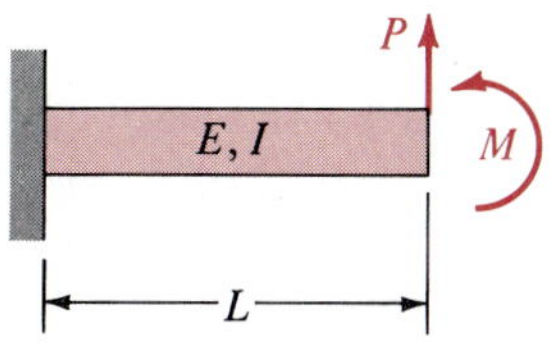

P. 10.77

10.78 Determine the rotations at the ends of the simply supported uniform beam shown.

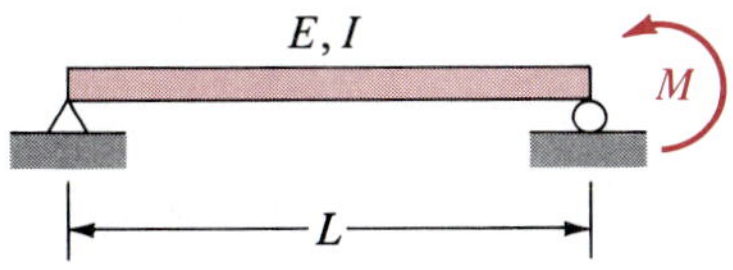

P. 10.78

10.79 Determine the rotations at the ends and the deflection at the center of the uniform beam shown.

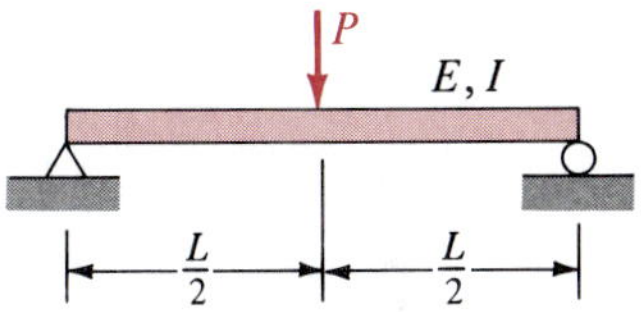

P. 10.79

10.80 Determine the rotation and deflection at the left end and the rotation at the middle support for the uniform beam shown.

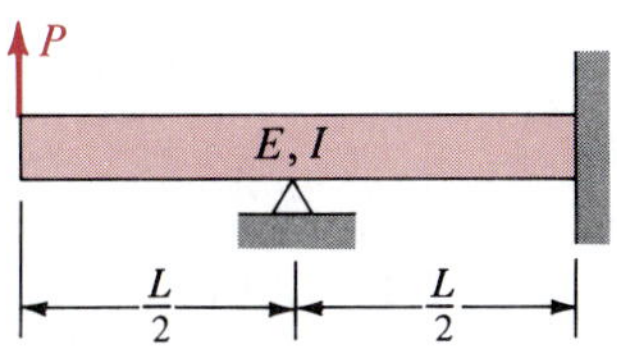

P. 10.80

10.81 Determine the vertical displacement and rotation at point 2 and the horizontal displacement and rotation at point 3. Neglect axial energy.

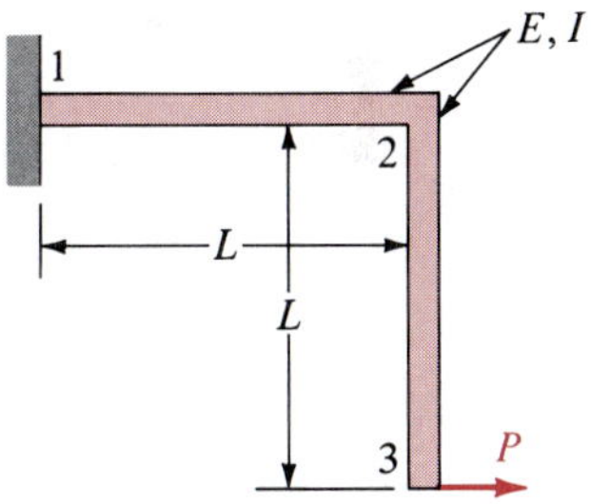

P. 10.81

10.4 COMPLEMENTARY ENERGY AND FLEXIBILITY FORMULATIONS

The stiffness method we developed in Section 10.3 consists of treating the displacements of a conservative mechanical system as the unknowns, representing the strain energy in terms of the displacements, and using stationary potential energy or Castigliano's first theorem to generate the equations of equilibrium. The actual basis of the method is the conservative nature of the material, as represented by the stress-strain diagram for a one-dimensional elastic solid as shown in Fig. 10.25a.

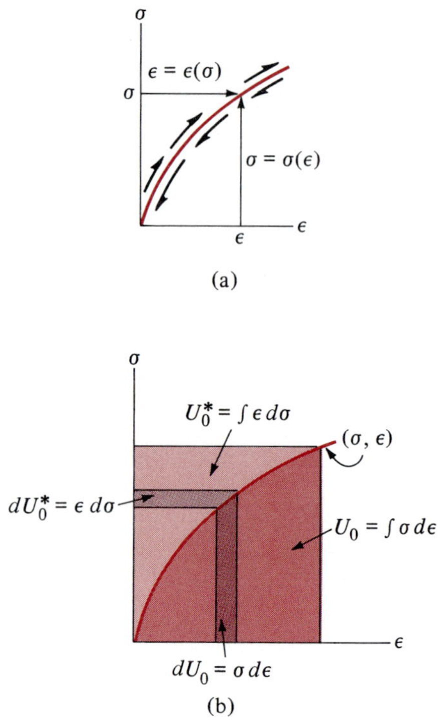

Figure 10.25 Stress-strain diagram and energy densities for a one-dimensional conservative elastic solid

As defined previously, the strain-energy density U_0 is the area under the stress-strain curve, or

$$U_0 = \int \sigma(\varepsilon)\, d\varepsilon$$

as shown in Fig. 10.25b. The total strain energy in the elastic solid is obtained by integrating the strain-energy density over the volume according to

$$U = \int_{\text{Vol}} U_0\, d\text{Vol}$$

When we apply stationary potential energy or Castigliano's first theorem, the stress is represented in terms of the strain when evaluating the strain-energy density, with the strains then represented in terms of the displacements of the elastic body when evaluating the total strain energy.

Another type of energy that leads to an alternate point of view in the analysis of elastic solids is associated with the area above the stress-strain curve as indicated in Fig. 10.25b. This energy density is called **complementary-energy density,** and is given the symbol U_0^*. The word complementary arises from the fact that

$$U_0 + U_0^* = \sigma\varepsilon$$

or, with respect to the area $\sigma\varepsilon$, U_0 and U_0^* complement each other.

We determine the properties of the complementary-energy density by referring again to Fig. 10.25b. A differential element of U_0^* can be seen to be

$$dU_0^* = \varepsilon\, d\sigma \tag{10.16}$$

or

$$U_0^* = \int \varepsilon\, d\sigma$$

representing the area above the stress-strain curve. Note carefully that as opposed to the strain-energy density, where the stress and strain were the dependent and independent variables, respectively, these roles are reversed for the complementary-energy density, that is, strain is the dependent variable and stress the independent variable. The basic differential relation given in Eqn. (10.16) between the complementary-energy density, the stress, and the strain can also be written as

$$\frac{dU_0^*}{d\sigma} = \varepsilon \tag{10.17}$$

The complementary energy U^* of an elastic body is obtained by integrating the complementary-energy density over the entire body, thus

$$U^* = \int_{\text{Vol}} U_0^*\, d\text{Vol}$$

Because the complementary-energy density is represented in terms of the stresses, the complementary energy is represented in terms of the loads on the body, rather than the displacements of the body, as was the case for the strain energy. Thus in order to evaluate the complementary energy of an elastic body, we use the equilibrium equations to determine the internal forces being transmitted by the body in terms of the external loads. We can then determine the stresses in terms of the internal force resultants, and finally, integrate the stresses over the body to obtain the complementary energy in terms of the external loads.

For this process of evaluating the complementary energy, we must be able to determine the internal forces in terms of the external loads, i.e., the problem must be statically determinate. If the problem is statically indeterminate, some of the forces, either internal forces or reactions, cannot be determined, and the complementary energy cannot be evaluated solely in terms of the known external loads. Statically indeterminate problems are discussed in Section 10.4.4.

Linearly Elastic Materials Many structures are composed of elastic bodies designed to operate so that the stresses and strains are in the elastic range of the material. In such instances, the uniaxial stress-strain diagram is linear, as in Fig. 10.26a.

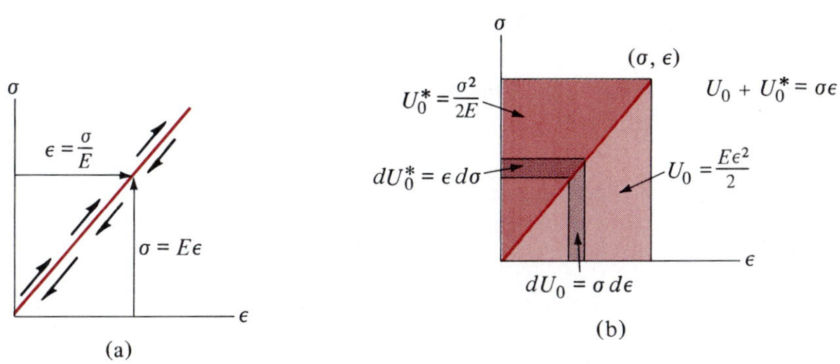

Figure 10.26 Stress-strain diagram and energy densities for a one-dimensional linearly elastic solid

The linear relationship between stress and strain is taken to be $\sigma = E\varepsilon$, where E represents Young's modulus (or the modulus of elasticity). The natures of the strain-energy and complementary-energy densities are substantially simplified as indicated in Fig. 10.26b. The complementary-energy density is then

$$U_0^* = \int \varepsilon \, d\sigma = \int \frac{\sigma}{E} \, d\sigma = \frac{\sigma^2}{2E} \tag{10.18}$$

Using the stress-strain relationship $\sigma = E\varepsilon$, we can convert U_0^* to

$$U_0^* = \frac{(E\varepsilon)^2}{2E} = \frac{E\varepsilon^2}{2} = U_0$$

showing, as is obvious from Fig. 10.26b, that the values of U_0 and U_0^* are the same for a linearly elastic solid. In this text we will keep separate the representation of the strain-energy density U_0 and complementary-energy density U_0^*, respectively, as

$$U_0 = \int \sigma(\varepsilon)\, d\varepsilon = \frac{E\varepsilon^2}{2} \qquad \text{and} \qquad U_0^* = \int \varepsilon(\sigma)\, d\sigma = \frac{\sigma^2}{2E}$$

It is left for the student to show in the problems that for a linearly elastic isotropic material, the complementary-energy density for plane stress is given by

$$U_0^* = \frac{\sigma_x^2 + \sigma_y^2 - 2\nu\sigma_x\sigma_y}{2E} + \frac{\tau_{xy}^2}{2G} \tag{10.19}$$

Similarly, for the three-dimensional isotropic solid, there results

$$U_0^* = \frac{\sigma_x^2 + \sigma_y^2 + \sigma_z^2 - 2\nu(\sigma_x\sigma_y + \sigma_y\sigma_z + \sigma_x\sigma_z)}{2E} + \frac{\tau_{xy}^2 + \tau_{xz}^2 + \tau_{yz}^2}{2G} \tag{10.20}$$

PROBLEMS

10.82 Develop the expression for the complementary-energy density U_0^* for the one-dimensional linearly elastic solid including temperature, i.e., $\varepsilon_x = \sigma_x/E + \alpha T$.

10.83 Develop the expression for the complementary-energy density U_0^* for a one-dimensional elastic solid behaving according to $\varepsilon_x = \sigma_x/E + \alpha(\sigma_x)^n$, where E and α are constants.

10.84 When initial-strain effects are included for the plane stress case, the stress-strain relations are written as

$$\varepsilon_x = \frac{\sigma_x - \nu\sigma_y}{E} + \varepsilon_{0x}(x, y) \qquad \varepsilon_y = \frac{\sigma_y - \nu\sigma_x}{E} + \varepsilon_{0y}(x, y) \qquad \gamma_{xy} = \frac{\tau_{xy}}{G} + \gamma_{0xy}$$

where $\varepsilon_{0x}(x, y)$, $\varepsilon_{0y}(x, y)$, and γ_{0xy} are the initial strains. Integrate the equations

$$\frac{\partial U_0^*}{\partial \sigma_x} = \varepsilon_x \qquad \frac{\partial U_0^*}{\partial \sigma_y} = \varepsilon_y \qquad \frac{\partial U_0^*}{\partial \tau_{xy}} = \gamma_{xy}$$

to determine the complementary-energy density including initial strains.

10.85 When initial-stress effects are included for the plane stress case, the stress-strain relations can be written as

$$\sigma_x - \sigma_{0x} = \frac{E}{1-\nu^2}(\varepsilon_x + \nu\varepsilon_y) \qquad \sigma_y - \sigma_{0y} = \frac{E}{1-\nu^2}(\varepsilon_y + \nu\varepsilon_x)$$

and

$$\tau_{xy} - \tau_{0xy} = G\gamma_{xy}$$

Integrate the equations

$$\frac{\partial U_0^*}{\partial \sigma_x} = \varepsilon_x \qquad \frac{\partial U_0^*}{\partial \sigma_y} = \varepsilon_y \qquad \frac{\partial U_0^*}{\partial \tau_{xy}} = \gamma_{xy}$$

to determine the complementary-energy density including initial stresses.

10.4.1 Complementary Energies for Axial, Torsional, and Bending Members

The useful properties of the complementary energy that we will exploit in the following sections are due to being able to represent the complementary energies of simple structural elements in terms of the loads transmitted. In this section, we will develop the expressions for the complementary energies for members composed of linearly elastic materials that transmit axial, torsional, and bending loads. Note that the use of the elementary formulae such as $\sigma = P/A$ for the axial problem, $\tau = Tr/J$ for the torsion of the bar of circular cross section, and $\sigma = -My/I$ and $\tau = VQ/It$ for the bending-shear problem assumes that all the assumptions that are an integral part of the appropriate theory are in fact satisfied.

Axial Loading Consider a straight bar transmitting axial loads as shown in Fig. 10.27a.

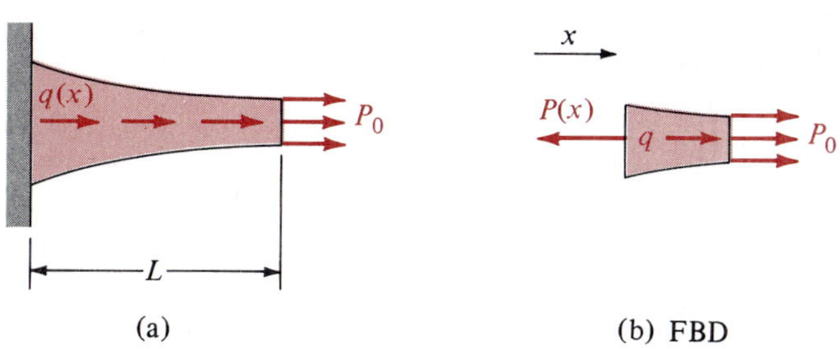

Figure 10.27 Loading and equilibrium for an axial member

When developing the complementary energy, we must be able to draw a free-body diagram and use equilibrium to determine the internal axial force $P(x)$ transmitted along the length of the bar. In the present situation, it is clear from the free-body diagram of Fig. 10.27b that requiring equilibrium leads to

$$\Sigma F_x = 0 \Rightarrow P(x) = P_0 + \int_x^L q(x)\, dx$$

expressing the internal force transmitted in terms of the external loads P_0 and $q(x)$. The results of the axial deformation theory state that $\sigma(x) = P(x)/A(x)$ is the only stress in the bar. The complementary-energy density is

$$U_0^* = \frac{\sigma^2}{2E} = \frac{P(x)^2}{2A(x)^2E}$$

with the complementary energy of the bar determined by integration as

$$U^* = \int_{\text{Vol}} \frac{P(x)^2}{2A(x)^2E}\, d\text{Vol} = \int_0^L \frac{P(x)^2}{2A(x)^2E}\left(\int_{\text{Area}} dA\right) dx = \int_0^L \frac{P(x)^2}{2A(x)E}\, dx$$

which is frequently written simply as

$$U^* = \int_0^L \frac{P^2}{2AE}\, dx \qquad (10.21)$$

with P and A considered to be functions of x when necessary. For the special case where A is constant and there is only the load P_0 applied, $P(x) = P_0$ and

$$U^* = \frac{P_0^2 L}{2AE}$$

a result that will be used frequently in the applications that follow.

Example 10.8 An axial member transmits the loads shown in Fig. 10.28a. Determine the complementary energy.

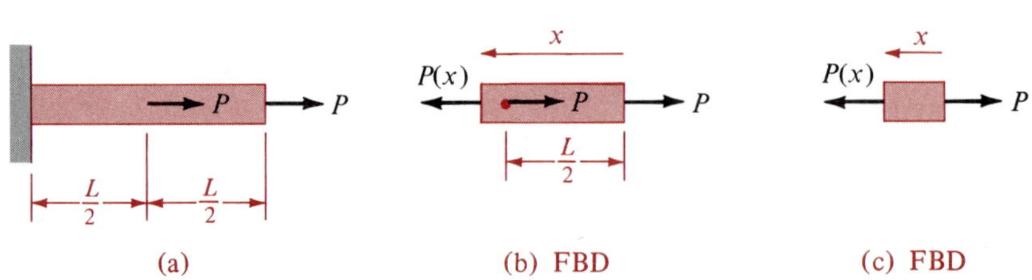

Figure 10.28 Loading and equilibrium of an axial member

Solution Referring to the free-body diagrams of Figs. 10.28b and c and requiring equilibrium, we can determine the internal axial force to be $P(x) = 2P$ for $0 \leq x \leq L/2$, and $P(x) = P$ for $L/2 \leq x \leq L$. The complementary energy is then

$$U^* = \int_0^{L/2} \frac{(2P)^2}{2EI}\,dx + \int_{L/2}^{L} \frac{P^2}{2EI}\,dx = \frac{5P^2L}{4EI}$$

Torsional Loading Consider a straight circular-cross-section bar transmitting torsional loads as shown in Fig. 10.29a. When developing the complementary energy, we must be able to draw a free-body diagram and use equilibrium to determine the internal torque $T(x)$ transmitted along the length of the bar. In the present situation, it is clear from the free-body diagram of Fig. 10.29b that requiring equilibrium leads to

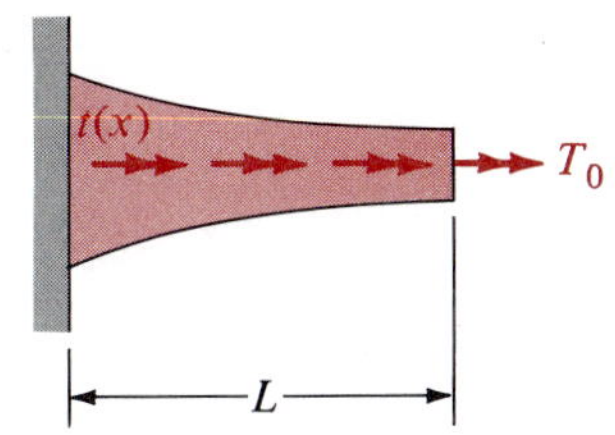

(a)

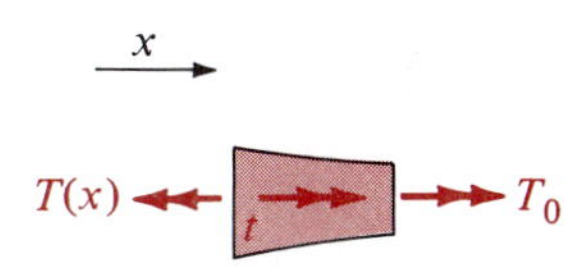

(b) FBD

Figure 10.29 Loading and equilibrium for a torsional member

$$\Sigma T_x = 0 \Rightarrow T(x) = T_0 + \int_x^L t(x)\,dx$$

expressing the internal torque transmitted in terms of T_0 and $t(x)$, the external loads on the bar. According to the torsional deformation theory, the only stress in the circular bar is $\tau(x) = T(x)r/J$. For a shear stress the complementary-energy density is

$$U_0^* = \frac{\tau^2}{2G} = \frac{T(x)^2r^2}{2J^2G}$$

with the complementary energy of the bar determined by integration as

$$U^* = \int_{\text{Vol}} \frac{T(x)^2r^2}{2J^2G}\,d\text{Vol} = \int_0^L \frac{T(x)^2}{2J^2G}\left(\int_{\text{Area}} r^2\,dA\right)dx$$

or since $(\int_{\text{Area}} r^2\,dA) = J$,

$$U^* = \int_0^L \frac{T(x)^2}{2JG}\,dx$$

frequently written simply as

$$U^* = \int_0^L \frac{T^2}{2JG}\,dx \tag{10.22}$$

where, in general, T and J can be functions of x. For the special case of a constant radius with only the torque T_0 applied, $T(x) = T_0$ and

$$U^* = \frac{T_0^2L}{2JG}$$

a result that will be used frequently in the applications that follow.

Example 10.9 The torsional member transmits the torques shown in Fig. 10.30. Determine the complementary energy of the bar.

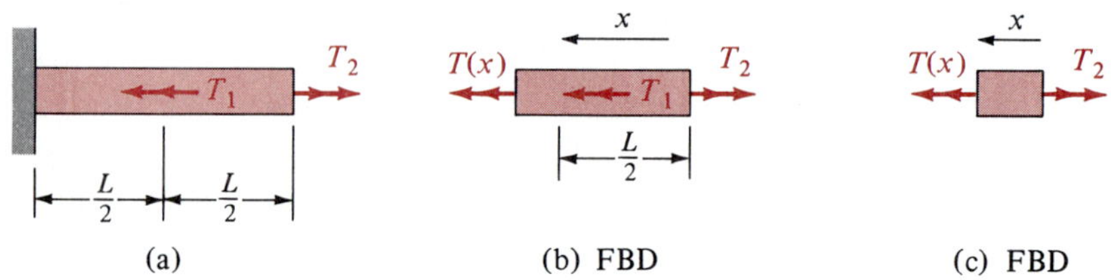

Figure 10.30 Complementary energy of a bar transmitting torques

Solution Consulting the free-body diagrams of Figs. 10.30b and c and requiring equilibrium, it is seen that $T(x) = T_2 - T_1$ for $0 \leq x \leq L/2$ and $T(x) = T_2$ for $L/2 \leq x \leq L$. The complementary energy can then be evaluated as

$$U^* = \int_0^{L/2} \frac{(T_2 - T_1)^2}{2JG}\,dx + \int_{L/2}^{L} \frac{T_2^2}{2JG}\,dx = \frac{[(T_2 - T_1)^2 + T_2^2]L}{4JG}$$

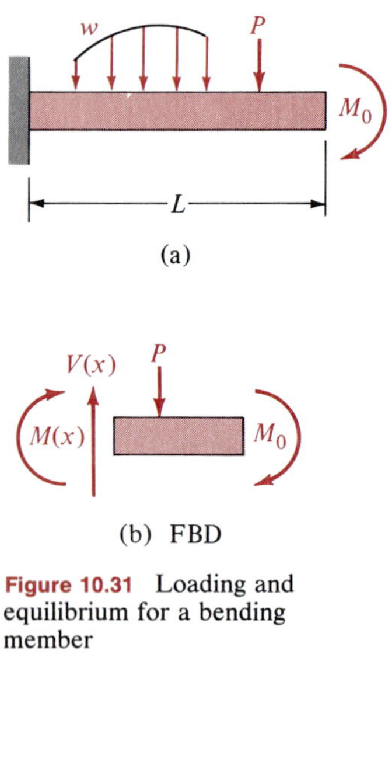

Figure 10.31 Loading and equilibrium for a bending member

Bending Deformations Consider a straight bar loaded in bending and shear as shown in Fig. 10.31a. Requiring equilibrium for the portion of the bar shown in Fig. 10.31b would determine the internal bending moment $M(x)$ and the internal shear force $V(x)$ in terms of the external loading M_0, P, and $w(x)$. The bending and shear stresses associated with these resultants are

$$\sigma = -\frac{M(x)y}{I} \qquad \text{and} \qquad \tau = \frac{V(x)Q}{It}$$

also expressed in terms of the external loads. The complementary-energy density is

$$U_0^* = \frac{\sigma^2}{2E} + \frac{\tau^2}{2G} = \frac{M(x)^2y^2}{2EI^2} + \frac{V(x)^2Q^2}{2I^2t^2G}$$

and by integrating over the volume, the complementary energy becomes

$$\begin{aligned} U^* &= \int_{\text{Vol}} \left[\frac{M(x)^2y^2}{2EI^2} + \frac{V(x)^2Q^2}{2I^2t^2G}\right] d\text{Vol} \\ &= \int_0^L \frac{M(x)^2\,dx}{2EI} + \int_0^L \frac{V(x)^2\,dx}{2kAG} \end{aligned} \tag{10.23}$$

where

$$k^{-1} = \frac{A}{I^2} \int_{\text{Area}} \frac{Q^2}{t^2}\, dA$$

is a dimensionless constant that accounts for the nonuniform distribution of the shear stress over the cross section. For a rectangular cross section, $k = \frac{5}{6}$. For most other cross sections, k ranges between about 0.8 and 2.0. The first integral in Eqn. (10.23) represents the complementary energy of bending and the second, the complementary energy of transverse shear.

Example 10.10 Consider the case of a cantilever beam acted upon by the end load P as shown in Fig. 10.32a.

Solution With x defined as shown in the free-body diagram of Fig. 10.32b, we can see by requiring equilibrium that $V(x) = P$ and $M(x) = -Px$, so that

$$U^* = \int_0^L \frac{(-Px)^2}{2EI}\, dx + \int_0^L \frac{P^2}{2kAG}\, dx = \frac{P^2L^3}{6EI}\left[1 + \frac{3(1+\nu)}{5}\left(\frac{h}{L}\right)^2\right]$$

where the second term within the brackets accounts for the shear. For a rectangular cross section with $h/L = 0.1$ and $\nu = 0.3$, this can be written as

$$U^* = \frac{P^2L^3}{6EI}\,[1.0078]$$

showing that the complementary energy of shear is less than 1% of the complementary energy due to bending. This is generally the case for thin beams, and we will ignore energy associated with transverse shear in what follows. In general, the bending complementary energy requires an integration, because the moment is invariably a function of position along the beam.

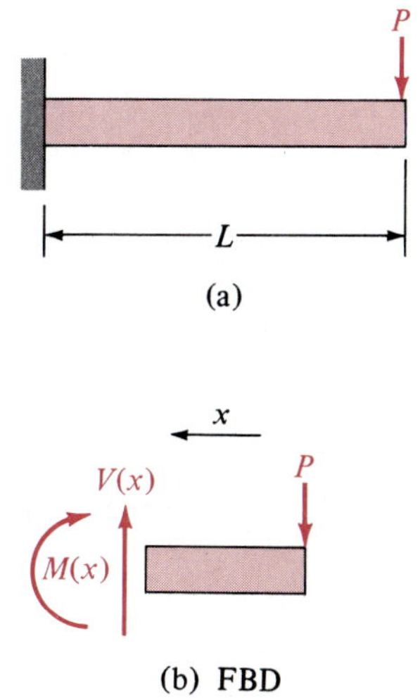

Figure 10.32 Cantilever beam acted upon by the end load P

Combined Axial and Bending Loads Fig. 10.33a shows a situation where the loading consists of combined axial and bending.

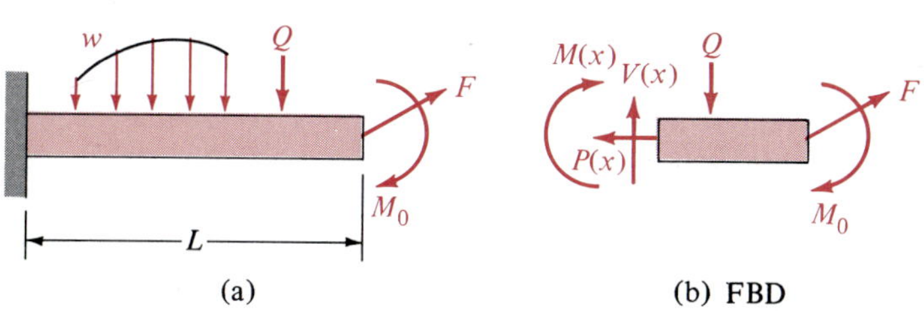

Figure 10.33 Loading and equilibrium for combined axial and bending

We give the stresses by

$$\sigma = \frac{P(x)}{A} - \frac{M(x)y}{I} \qquad \text{and} \qquad \tau = \frac{V(x)Q}{It}$$

where we assume that P, M, and V have been determined by satisfying equilibrium, and hence are expressed in terms of the external loads. We can then write the complementary energy as

$$U^* = \int_{\text{Vol}} \frac{1}{2E}\left[\frac{P(x)}{A} - \frac{M(x)y}{I}\right]^2 d\text{Vol} + \int_{\text{Vol}} \frac{1}{2G}\left(\frac{V(x)Q}{It}\right)^2 d\text{Vol}$$

which, since $\int y\, dA = 0$, can be written as

$$U^* = \int_0^L \frac{P^2}{2AE}\, dx + \int_0^L \frac{M^2\, dx}{2EI} + \int_0^L \frac{V^2\, dx}{2kAG} \tag{10.24}$$

showing that for the case of combined axial loading and single-axis bending, the individual complementary energies can simply be added to obtain the total complementary energy.

For the most general case of the simultaneous transmission of axial, torsional, bending, and shear loads, the total complementary energy can be written as

$$U^* = \int_0^L \frac{P^2\, dx}{2AE} + \int_0^L \frac{M_z^2\, dx}{2EI_z} + \int_0^L \frac{M_y^2\, dx}{2EI_y} + \int_0^L \frac{V_z^2\, dx}{2k_zAG} + \int_0^L \frac{V_y^2\, dx}{2k_yAG} + \int_0^L \frac{T^2\, dx}{2JG} \tag{10.25}$$

where y and z are the principal axes of the cross section.

PROBLEMS

Axial Problems

10.86–10.93 For each of the following, derive the expression for the complementary energy assuming a linearly elastic solid. Take k to mean AE/L.

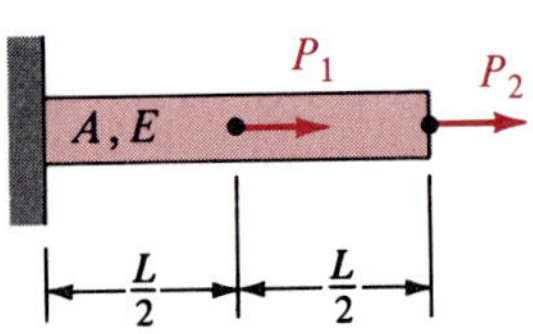

P. 10.86

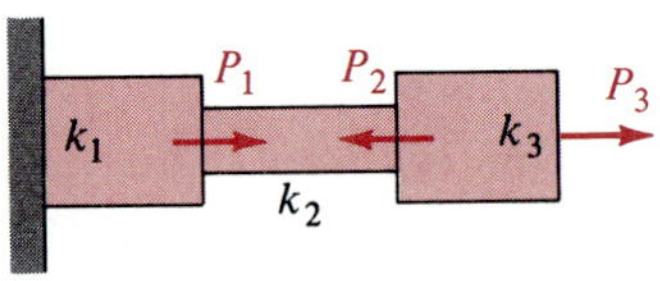

P. 10.87

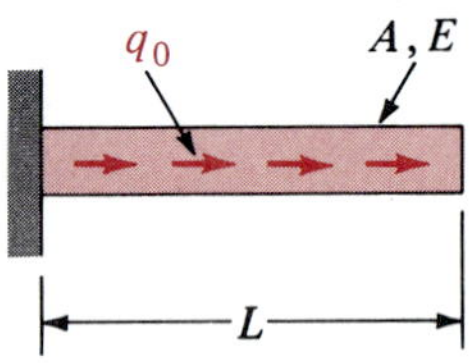

P. 10.88

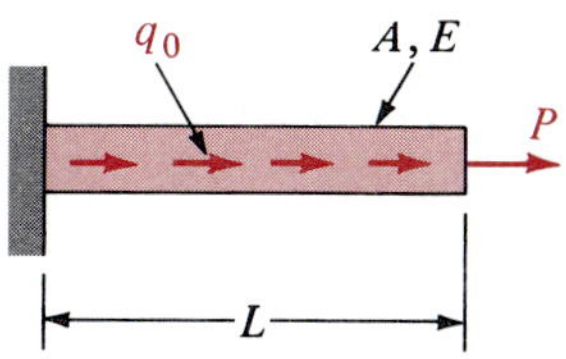

P. 10.89

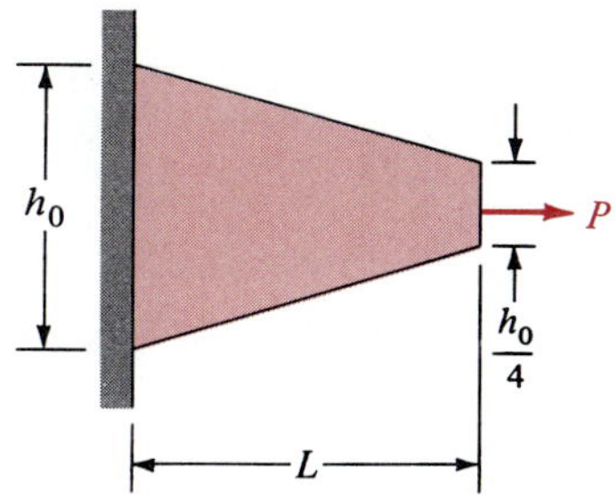

P. 10.90

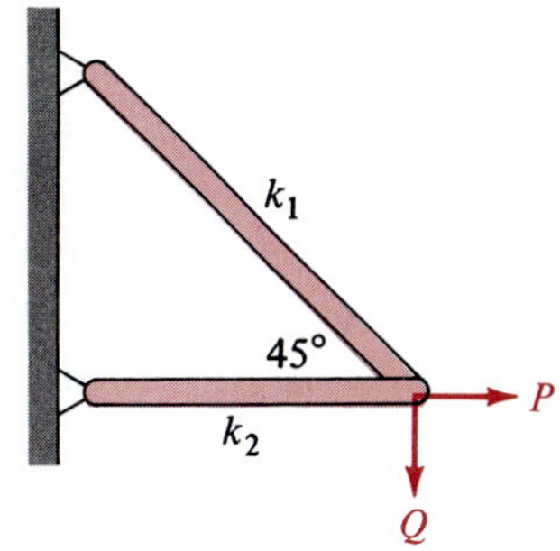

P. 10.91

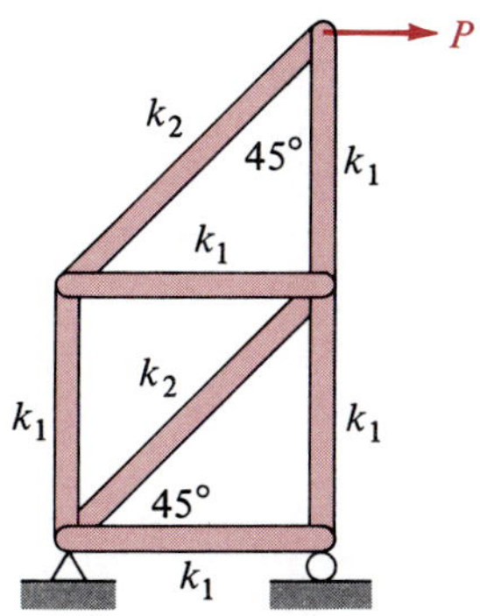

P. 10.92

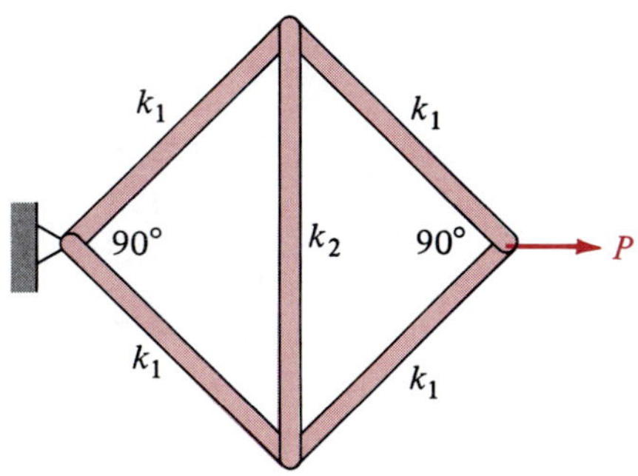

P. 10.93

Torsional Problems

10.94–10.102 For each of the following, derive the expression for the complementary energy assuming a linearly elastic solid. Take C to mean JG/L.

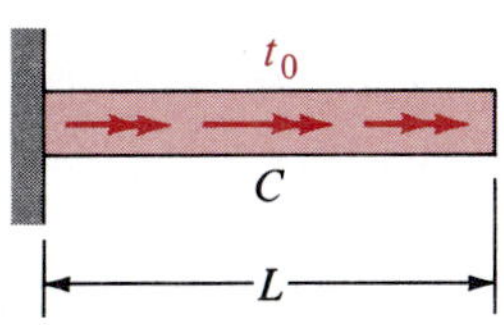

P. 10.94

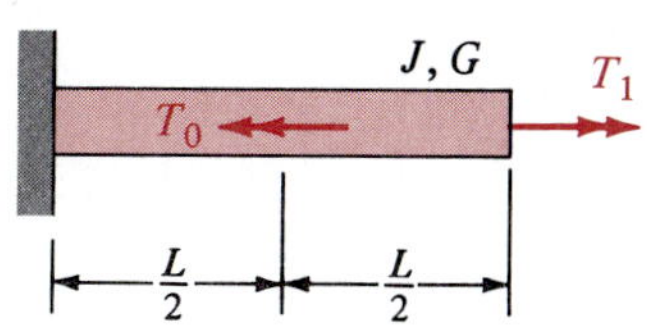

P. 10.95

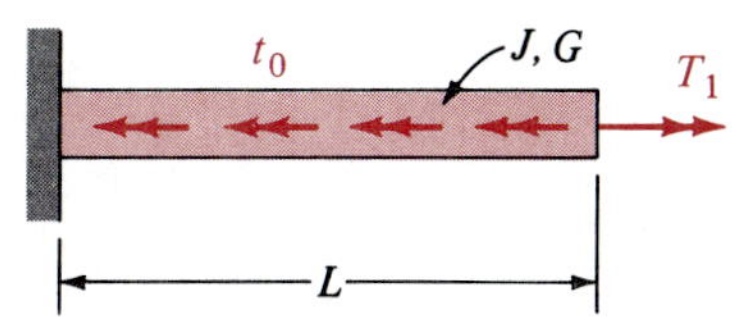

P. 10.96

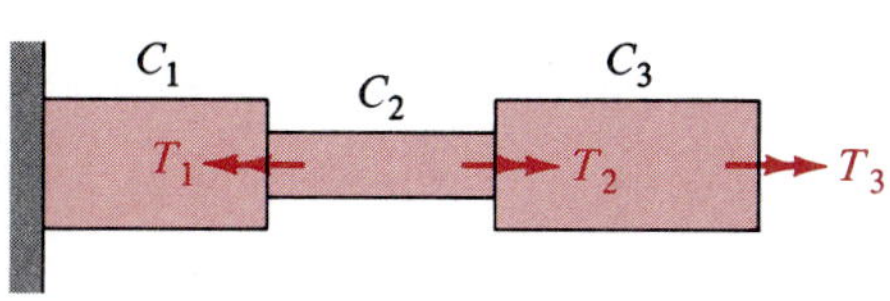

P. 10.97

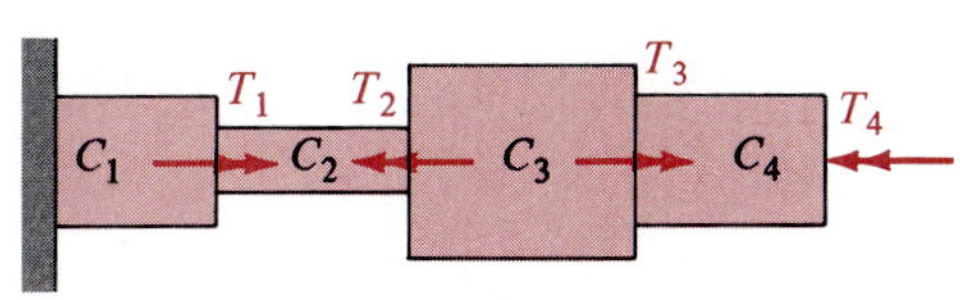

P. 10.98

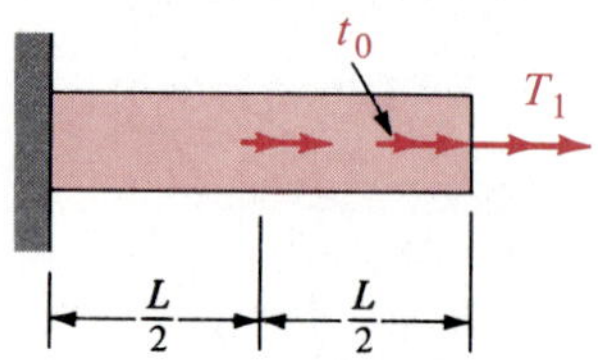

P. 10.99

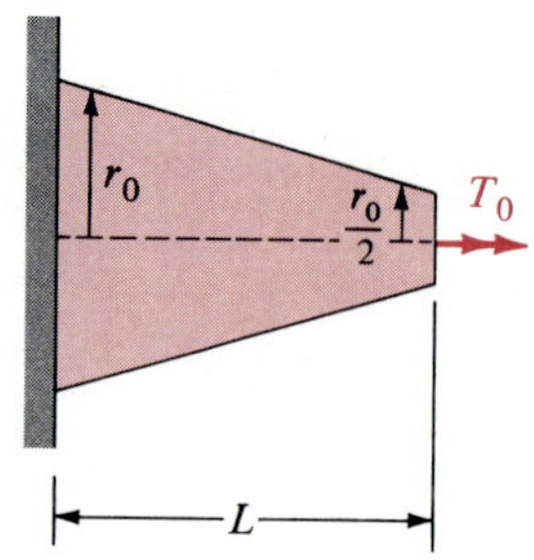

P. 10.100

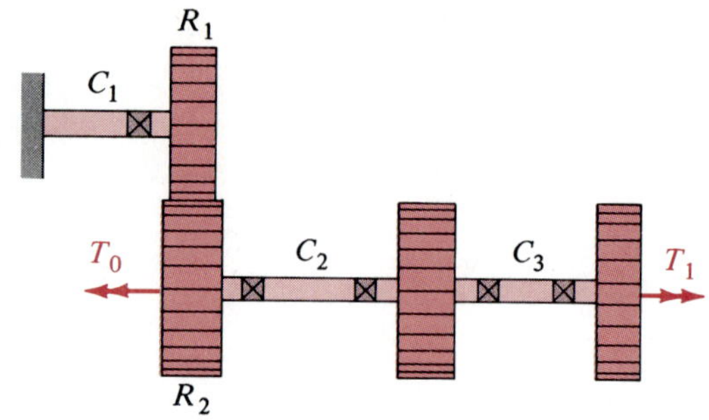

P. 10.101

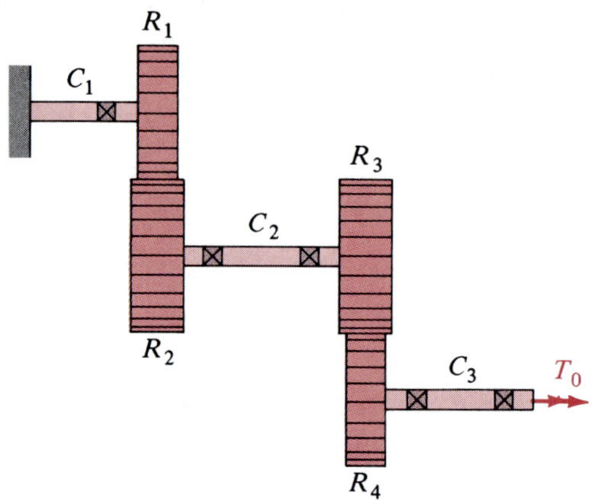

P. 10.102

Bending Problems

10.103–10.112 For each of the following, derive the expression for the complementary energy assuming a linearly elastic solid. Set up but do not evaluate the integrals.

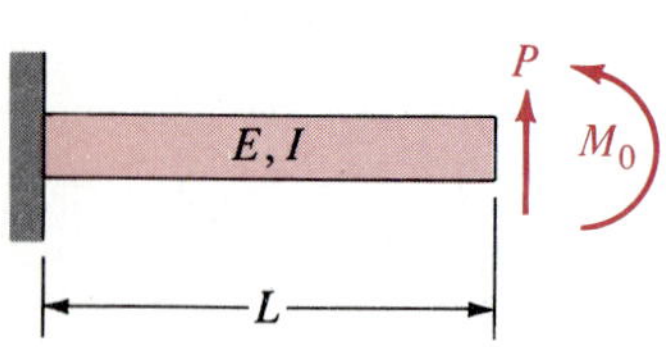

P. 10.103

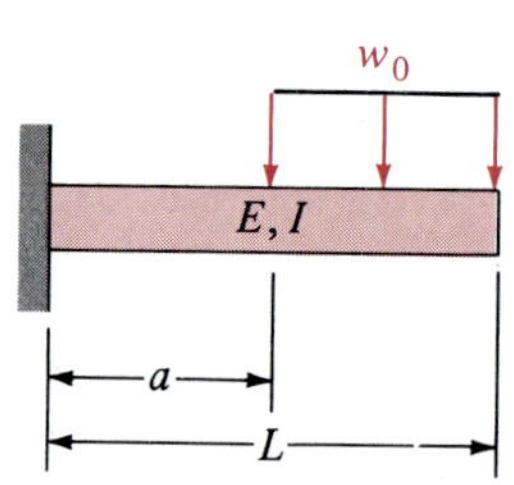

P. 10.104

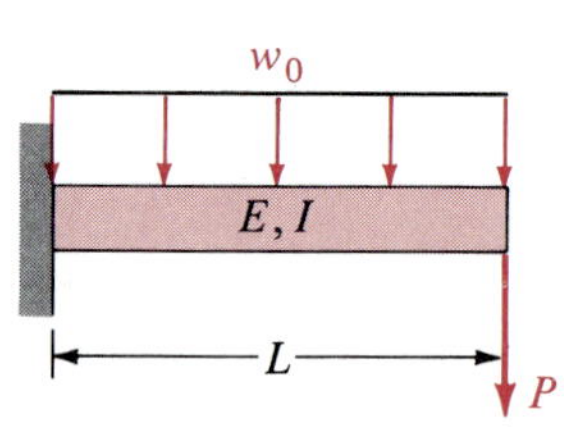

P. 10.105

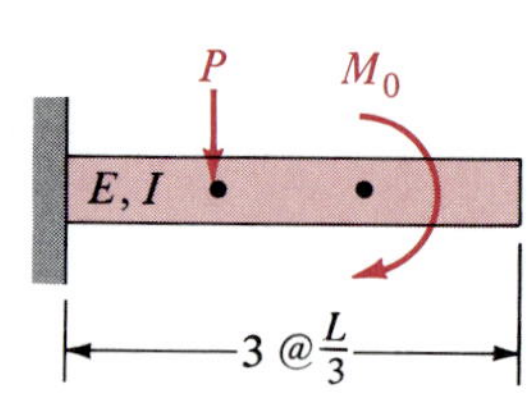

P. 10.106

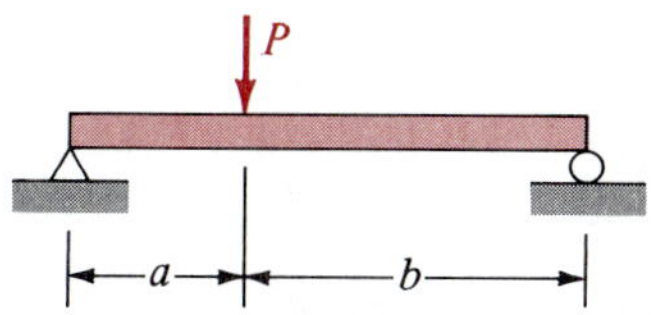

P. 10.107

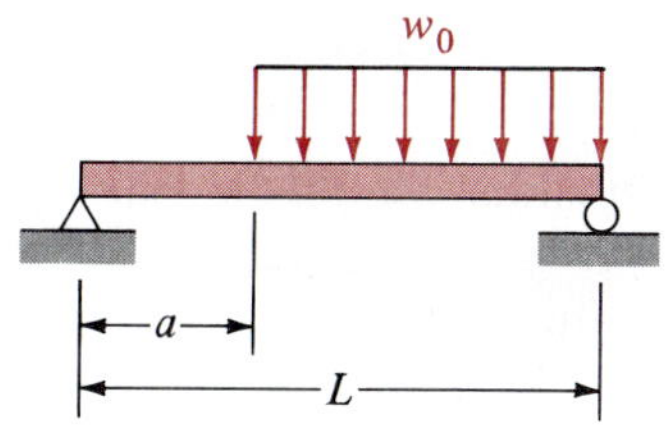

P. 10.108

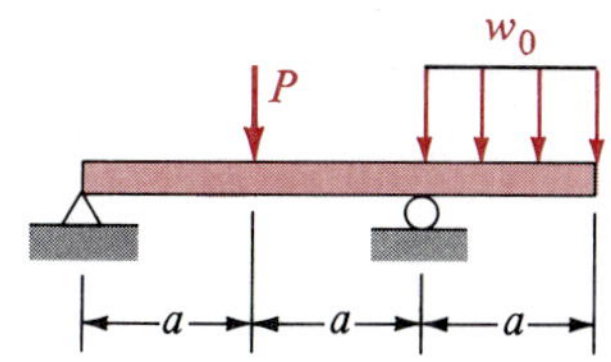

P. 10.109

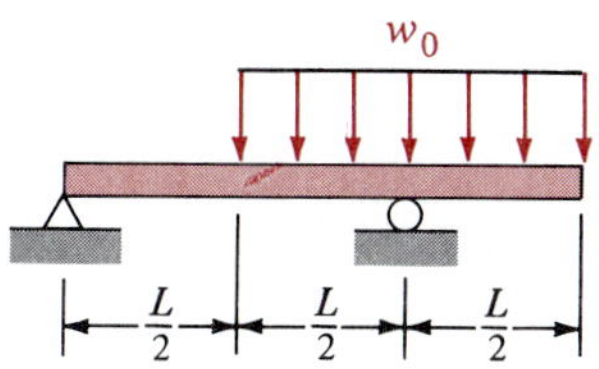

P. 10.110

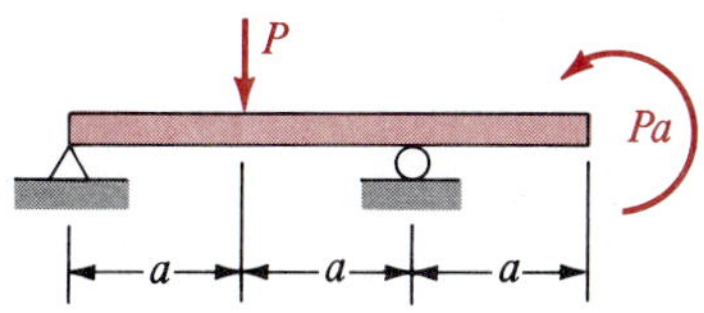

P. 10.111

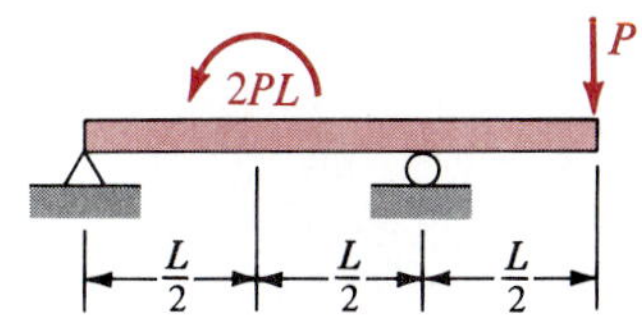

P. 10.112

Combined Loadings

10.113–10.116 For each of the following, set up all the integrals necessary to represent the complementary energy. Include axial, bending, and torsional complementary energies.

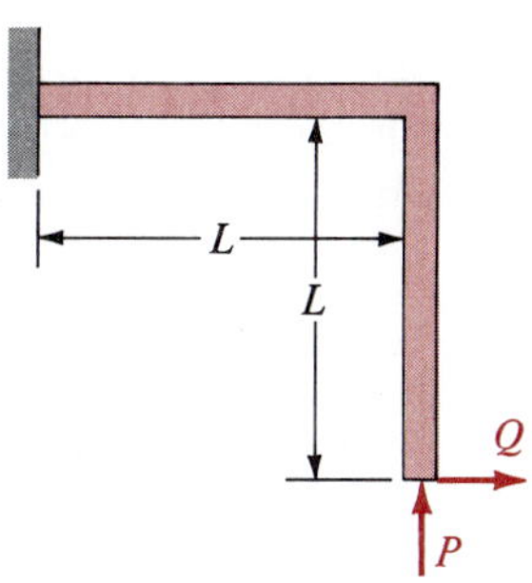

P. 10.113

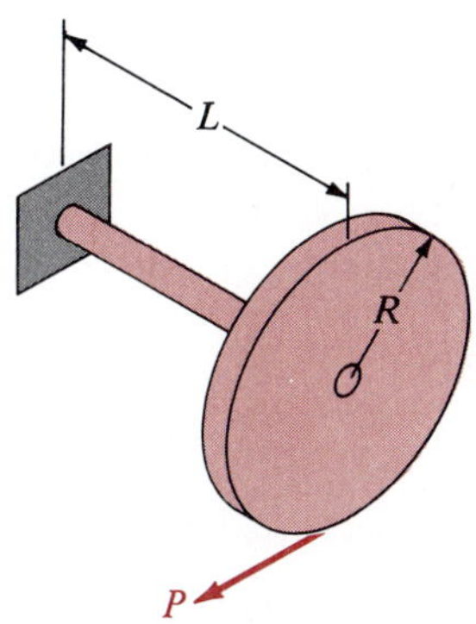

P. 10.114

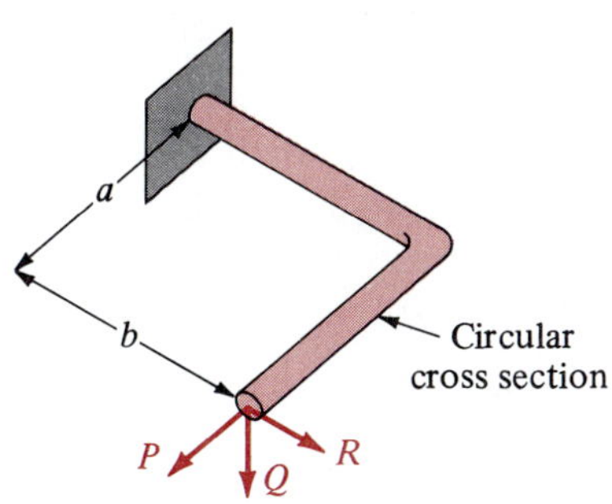

P. 10.115

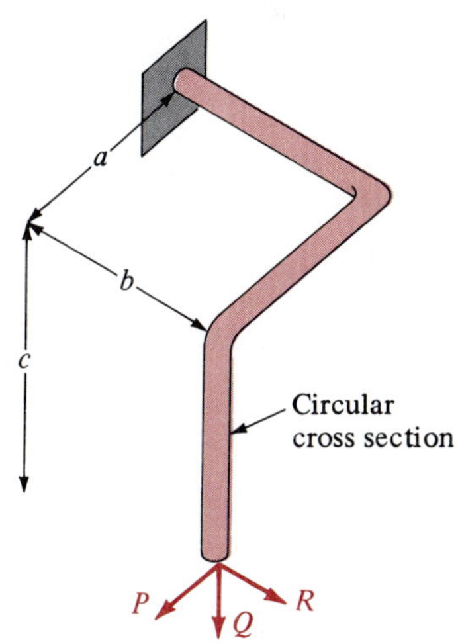

P. 10.116

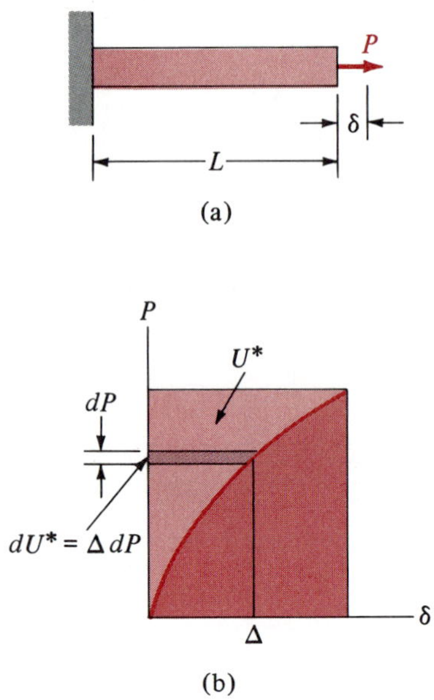

Figure 10.34 Relation between the load, elongation, and complementary energy for an axially loaded member

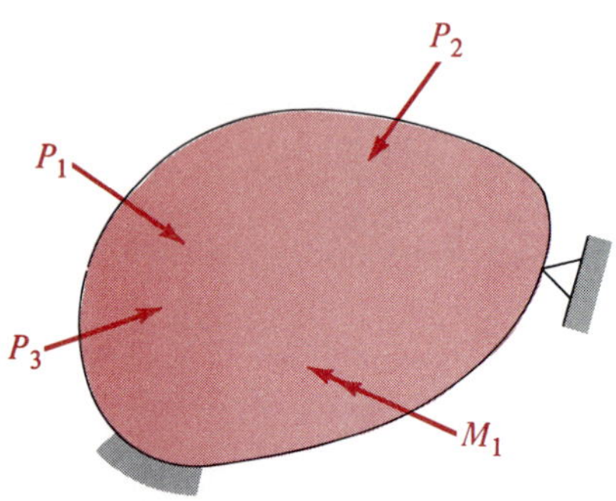

Figure 10.35 Elastic body acted upon by forces and moments

10.4.2 Analysis of Structures Using Force Variable Formulations—Engesser's First Theorem

Consider an elastic bar under the action of the load P as shown in Fig. 10.34a. Assume that when the bar is gradually loaded by the force P, the relation between the load P and the corresponding elongation δ is as indicated in Fig. 10.34b. U^*, the complementary energy for the bar, is the area above the curve. If at a particular value of the elongation, say Δ, the load is increased by an amount dP, we can see from Fig. 10.34b that the relation between the load increment dP, the elongation Δ, and the complementary energy is

$$dU^* = \Delta\, dP$$

or

$$\frac{dU^*}{dP} = \Delta \tag{10.26}$$

showing that at any value of the load P, the rate of change of the complementary energy with respect to P is the current elongation Δ.

We state without proof that for the more general situation where a statically determinate elastic body is acted on by a set of forces P_i as pictured in Fig. 10.35, the generalization of Eqn. (10.26), known as **Engesser's first theorem,** can be stated as:

> The partial derivative of the complementary energy U^* of an elastic system with respect to a load P_j equals the corresponding small displacement Δ_j of the point of application of P_j, or
>
> $$\frac{\partial U^*}{\partial P_j} = \Delta_j \tag{10.27}$$

In Engesser's first theorem, load can refer to force P_j in which case Δ_j is a displacement, or to a moment M_j in which case

$$\theta_j = \frac{\partial U^*}{\partial M_j} \tag{10.28}$$

where θ_j is the small rotation of the body at the point of application. Note that Engesser's first theorem is valid for nonlinear elastic materials. The use of the adjective ''small'' in connection with the displacement and rotation refers to the requirement that the initial and final geometries must be essentially the same, since the equilibrium equations are obviously written with respect to the initial geometry. If the displacements and rotations are large, the equilibrium equations will, in general, no longer be valid for the deformed configuration.

Example 10.11 A constant-cross-section member of area A, constructed from a material whose stress-strain behavior is as shown in Fig. 10.36b, is acted upon by the load P as shown in Fig. 10.36a. Determine the elongation of the bar.

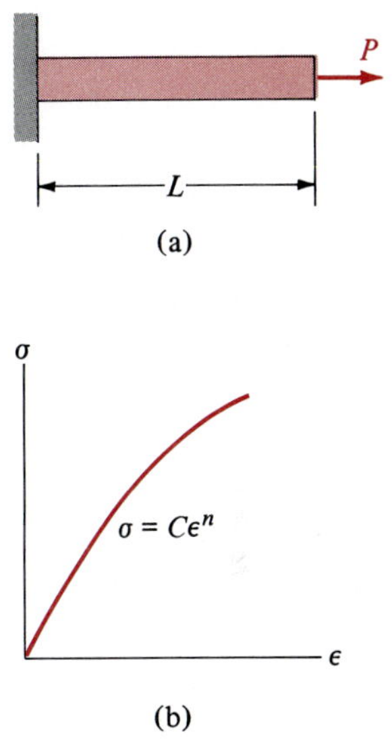

Figure 10.36 Elongation of a nonlinear elastic bar

Solution For this nonlinearly elastic bar, the complementary-energy density is given by

$$U_0^* = \int \varepsilon \, d\sigma = \int \left(\frac{\sigma}{C}\right)^m d\sigma = \frac{\sigma^{m+1}}{(m+1)C^m}$$

With $\sigma = P/A$ everywhere in the bar, U_0^* is a constant and the complementary energy can be expressed as

$$U^* = \frac{(P/A)^{m+1}}{(m+1)C^m} AL$$

Engesser's first theorem can be used to compute

$$\Delta = \frac{\partial U^*}{\partial P} = \frac{(m+1)(P/A)^m AL}{(m+1)C^m} \frac{1}{A} = \left(\frac{P}{CA}\right)^m L$$

which can be recognized as $\left(\frac{\sigma}{C}\right)^m L = \varepsilon L = \Delta$, the correct answer.

Engesser's first theorem can be used for material nonlinearities such as in Example 10.11. If the deformations are large, the changes in geometry render the equilibrium equations generally invalid, since they were written with respect to the original geometry. Thus Engesser's theorem should not be used for problems where there are geometric nonlinearities.

For the most part we will be dealing with linearly elastic structures undergoing small deformations. In these situations, the strain energy and the complementary energy are the same, and Engesser's first theorem can be written as

$$\frac{\partial U^*}{\partial P_k} = \frac{\partial U}{\partial P_k} = \Delta_k$$

The equality

$$\Delta_k = \frac{\partial U}{\partial P_k} \tag{10.29}$$

which is valid for linearly elastic structures, is known as **Castigliano's second theorem,** with the meanings of the variables P_k and Δ_k the same as in the description of Engesser's first theorem. For Castigliano's second theorem the corollary

$$\theta_k = \frac{\partial U}{\partial M_k} \tag{10.30}$$

is also valid.

In what follows we will present several examples of the use of Castigliano's second theorem. We will however continue to use the statements $\Delta_k = \partial U^*/\partial P_k$ and $\theta_k = \partial U^*/\partial M_k$ rather than $\Delta_k = \partial U/\partial P_k$ and $\theta_k = \partial U/\partial M_k$ to emphasize the fact that the complementary energy U^* is to be considered as a function of the stress or force variables, whereas the strain energy U is to be considered as a function of the strain or displacement variables.

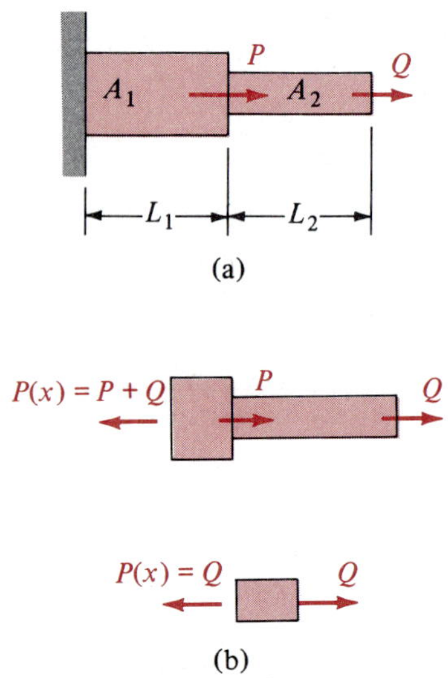

Figure 10.37 Axially loaded segmented bar

Example 10.12 A straight segmented bar composed of linearly elastic materials is loaded by the forces P and Q as shown in Fig. 10.37a. Determine the displacements at the points of application of loads P and Q.

Solution The free-body diagrams shown in Fig. 10.37b can be used to determine the loads transmitted by bars 1 and 2 as $P(x) = P + Q$ and $P(x) = Q$, respectively. As we developed in the previous section, the complementary energy for an axially loaded linearly elastic member is

$$U^* = \int_0^L \frac{P(x)^2}{2EA(x)}\,dx$$

When P and A are constant, $U^* = P^2L/2AE$, so that with U_1^* and U_2^* as the complementary energies of bars 1 and 2, respectively, the complementary energy for the segmented bar is

$$U^* = U_1^* + U_2^* = \frac{P_1^2 L_1}{2A_1E_1} + \frac{P_2^2 L_2}{2A_2E_2}$$

or

$$U^* = \frac{(P+Q)^2L_1}{2A_1E_1} + \frac{Q^2L_2}{2A_2E_2} = \frac{(P+Q)^2}{2k_1} + \frac{Q^2}{2k_2}$$

We use Castigliano's second theorem to compute

$$\frac{\partial U^*}{\partial P} = \Delta_1 = \frac{P+Q}{k_1} \qquad \text{and} \qquad \frac{\partial U^*}{\partial Q} = \Delta_2 = \frac{P+Q}{k_1} + \frac{Q}{k_2}$$

where Δ_1 and Δ_2 are the displacements of the points of application of P and Q, respectively.

Example 10.13 A planar truss is acted upon by the force Q as shown in Fig. 10.38a. Determine the vertical displacement at point B. Take A and E to be the same for all members.

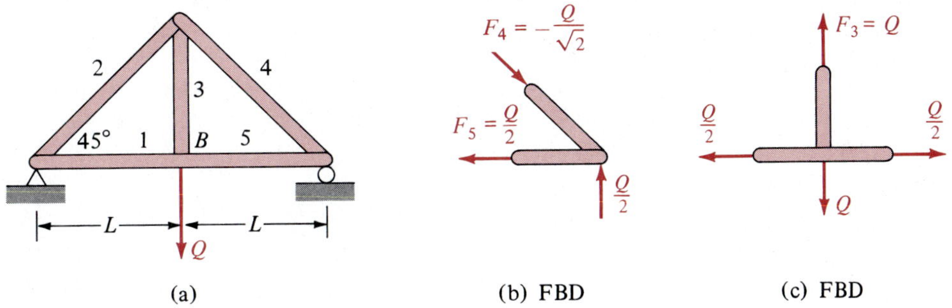

Figure 10.38 Displacements for a planar truss

Solution Using the free-body diagrams shown in Figs. 10.38b and c, we determine the internal forces P_i in each of the members. The complementary energy is the sum of the complementary energies of each of the bars and can be represented as

$$U^* = \sum_{i=1}^{n} \frac{P_i^2 L_i}{2A_i E_i}$$

The vertical displacement at B is given by

$$\Delta = \frac{\partial U^*}{\partial Q} = \sum_{i=1}^{n} \frac{P_i(\partial P_i/\partial Q)L_i}{A_i E_i} = \sum_{i=1}^{n} \frac{P_i(\partial P_i/\partial Q)}{k_i}$$

where $k_i = A_i E_i / L_i$. Our calculations are simplified if we use Table 10.1, where the first column identifies the member, the second column represents the force in the member, the third column represents the quantity $\partial P_i/\partial Q$, the fourth column is the appropriate $k_i/(AE/L) = k_i/k$, and the last column is the product of the second and third columns divided by the fourth column. The desired result is the sum of all the entries in the last column.

Table 10.1 Computation of truss displacements

Bar	P_i	$\frac{\partial P_i}{\partial Q}$	k_i/k	$P_i \frac{\partial P_i}{\partial Q}/k_i$
1	$Q/2$	$1/2$	1	$Q/4k$
2	$-Q/\sqrt{2}$	$-1/\sqrt{2}$	$1/\sqrt{2}$	$Q/(2\sqrt{2}k)$
3	Q	1	1	Q/k
4	$-Q/\sqrt{2}$	$-1/\sqrt{2}$	$1/\sqrt{2}$	$Q/(2\sqrt{2}k)$
5	$Q/2$	$1/2$	1	$Q/4k$
				$\Sigma = 2.207\,QL/AE$

This tabular approach is recommended for determining displacements in trusses when using the complementary-energy approach.

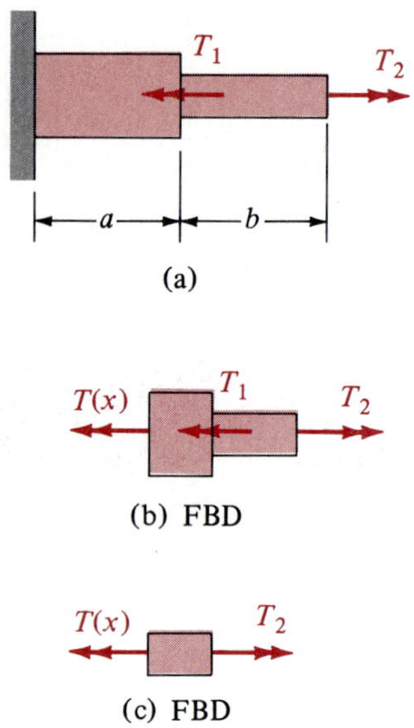

Figure 10.39 Segmented bar loaded by concentrated torques

Example 10.14 A segmented circular-cross-section bar is loaded by the torques as shown in Fig. 10.39a. Determine the rotations at the points of application of the torques T_i.

Solution Referring to the free-body diagrams of Figs. 10.39b and c, we see that the internal torques are $T(x) = T_2 - T_1$ for $0 \le x \le a$ and $T(x) = T_2$ for $a \le x \le L$, respectively, where $L = a + b$. The complementary energy of the segmented bar is the sum of the complementary energies of the two segments, that is,

$$U^* = U_1^* + U_2^* = \int_0^a \frac{(T_2 - T_1)^2}{2J_1G_1}\,dx + \int_a^L \frac{T_2^2}{2J_2G_2}\,dx$$

$$= \frac{(T_2 - T_1)^2}{2J_1G_1}\,a + \frac{T_2^2}{2J_2G_2}\,b$$

Using Castigliano's second theorem leads to

$$\frac{\partial U^*}{\partial T_1} = \phi_1 = \frac{T_1 - T_2}{J_1G_1}\,a$$

and

$$\frac{\partial U^*}{\partial T_2} = \phi_2 = \frac{T_2 - T_1}{J_1G_1}\,a + \frac{T_2 b}{J_2G_2}$$

where the positive directions of ϕ_1 and ϕ_2 are those of T_1 and T_2, respectively. Note that signs of ϕ_1 and ϕ_2 depend upon the relative values of T_1, T_2, $(T_1 - T_2)a/J_1G_1$, and T_2b/J_2G_2.

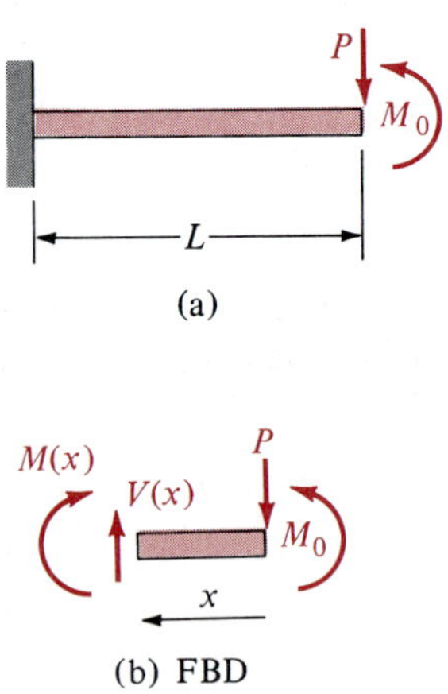

Figure 10.40 Displacements for an end-loaded cantilever beam

Example 10.15 A cantilever beam is loaded by a force and moment as shown in Fig. 10.40a. Determine the slope and deflection at the loaded end.

Solution The first step in any application of Castigliano's second theorem is to use equilibrium to determine the internal forces and moments transmitted. From the free-body diagram of Fig. 10.40b, equilibrium leads to

$$\Sigma F_y = V(x) - P = 0 \qquad \text{and} \qquad \Sigma M_z = M_0 - Px - M(x) = 0$$

so that $V(x) = P$ and $M(x) = M_0 - Px$. From our discussion of the relative values of the energies associated with bending and transverse shear in Section 10.4.1, we write that

$$U^* = \int_0^L \frac{M^2}{2EI}\,dx = \int_0^L \frac{(M_0 - Px)^2}{2EI}\,dx$$

Castigliano's second theorem is then used to calculate

$$\Delta = \frac{\partial U^*}{\partial P} = \int_0^L \frac{M(\partial M/\partial P)}{EI}\,dx = \int_0^L \frac{(M_0 - Px)(-x)}{EI}\,dx = \frac{PL^3}{3EI} = \frac{M_0L^2}{2EI}$$

and

$$\theta = \frac{\partial U^*}{\partial M_0} = \int_0^L \frac{M(\partial M/\partial M_0)}{EI}\,dx = \int_0^L \frac{(M_0 - Px)(1)}{EI}\,dx = -\frac{PL^2}{2EI} + \frac{M_0L}{EI}$$

By taking $M_0 = 0$, we see that the displacement $\Delta = PL^3/3EI$ is in the same direction as P, but that the slope or rotation $\theta = -PL^2/2EI$ is in the direction opposite to M_0—clockwise. By taking $P = 0$, the rotation $\theta = ML/EI$ has the same direction as the moment with the displacement $\Delta = -PL^2/2EI$ opposite to the direction of P. These ideas will be discussed in more detail in the next section concerning dummy loads.

Note that in the previous example, the displacement and rotation were evaluated on the basis of differentiating under the integral sign. This allows us to perform the differentiation before performing the integration. The other approach, requiring more algebra, would be to evaluate the integrals before computing the derivatives. The student should show that the two approaches give the same results.

Example 10.16 A planar L-shaped bar is loaded by the forces P and Q, and the moment M_0 as shown in Fig. 10.41a. Use Castigliano's second theorem to determine the corresponding displacements u and v and the rotation θ, also shown in the figure. Take the bars to be identical in all respects except their lengths.

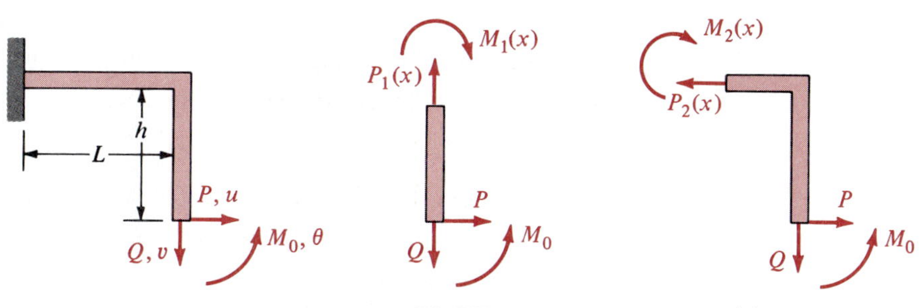

Figure 10.41 Displacements of a planar L-shaped bar

Solution The complementary energy consists of the sum of the complementary energies of the two segments of the L-shaped member. We will include axial and bending contributions to the complementary energy. Based on satisfaction of the equations of equilibrium, it is easily seen from the free-body diagram of Fig. 10.41b that $P_1(x) = Q$, and that $M_1(x) = M_0 + Px$. Similarly, from the free-body diagram of Fig. 10.41c, $P_2(x) = P$ and $M_2(x) = M_0 + Ph - Qx$. The complementary energy is then

$$U^* = U_1^* + U_2^* = \int_0^h \left(\frac{P_1^2}{2AE} + \frac{M_1^2}{2EI}\right) dx + \int_0^L \left(\frac{P_2^2}{2AE} + \frac{M_2^2}{2EI}\right) dx$$

Using Castigliano's second theorem, the displacements u, v, and the rotation θ are computed as

$$u = \frac{\partial U^*}{\partial P} = \int_0^h \left[\frac{P_1(\partial P_1/\partial P)}{AE} + \frac{M_1(\partial M_1/\partial P)}{EI}\right] dx + \int_0^L \left[\frac{P_2(\partial P_2/\partial P)}{AE} + \frac{M_2(\partial M_2/\partial P)}{EI}\right] dx$$

$$v = \frac{\partial U^*}{\partial Q} = \int_0^h \left[\frac{P_1(\partial P_1/\partial Q)}{AE} + \frac{M_1(\partial M_1/\partial Q)}{EI}\right] dx + \int_0^L \left[\frac{P_2(\partial P_2/\partial Q)}{AE} + \frac{M_2(\partial M_2/\partial Q)}{EI}\right] dx$$

$$\theta = \frac{\partial U^*}{\partial M_0} = \int_0^h \left[\frac{P_1(\partial P_1/\partial M_0)}{AE} + \frac{M_1(\partial M_1/\partial M_0)}{EI}\right] dx + \int_0^L \left[\frac{P_2(\partial P_2/\partial M_0)}{AE} + \frac{M_2(\partial M_2/\partial M_0)}{EI}\right] dx$$

Evaluating all of the integrals yields

$$u = P\left[\frac{L^3}{3EI} + \frac{h^2L}{EI} + \frac{L}{AE}\right] - \frac{QhL^2}{2EI} + \frac{M_0h(h/2 + L)}{EI}$$

$$v = -\frac{PhL^2}{2EI} + Q\left[\frac{h}{AE} + \frac{L^3}{3EI}\right] - \frac{M_0L^2}{2EI}$$

$$\theta = \frac{Ph(h/2 + L)}{EI} - \frac{QL^2}{2EI} + \frac{M_0(L + h)}{EI}$$

Note that we can write this set of equations in the form

$$\begin{aligned} u &= S_{11}P + S_{12}Q + S_{13}M_0 \\ v &= S_{21}P + S_{22}Q + S_{23}M_0 \\ \theta &= S_{31}P + S_{32}Q + S_{33}M_0 \end{aligned} \tag{10.31}$$

where the coefficients S_{ij} have the property that $S_{ij} = S_{ji}$. The coefficients S_{ij} are referred to as **flexibility coefficients.** Generally the formulation of the problem with the force variables considered to be the independent variables is called the flexibility formulation. If Eqns. (10.31) are solved for P, Q, and M_0 in terms of u, v, and θ, the results would appear as

$$\begin{aligned} P &= k_{11}u + k_{12}v + k_{13}\theta \\ Q &= k_{21}u + k_{22}v + k_{23}\theta \\ M_0 &= k_{31}u + k_{32}v + k_{33}\theta \end{aligned}$$

where $k_{ij} = k_{ji}$ are called the **stiffness coefficients,** and are associated with the stiffness method discussed in section 10.3.2.

Example 10.17 The L-shaped bar is loaded out of its plane by the force P and the torques M_0 and T_0 as shown in Fig. 10.42a. Use Castigliano's second theorem to determine the corresponding displacement u and the rotations ϕ and ψ, also shown in the figure. Take the bars to be of the same circular cross section.

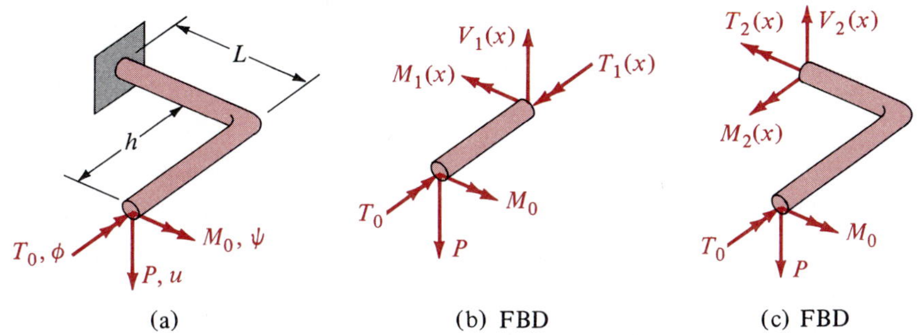

Figure 10.42 L-shaped bar subjected to out-of-plane loading

Solution Referring to the free-body diagrams shown in Figs. 10.42b and c, we see that each segment of the bar transmits a torque, a moment, and a shear. As discussed previously, we will neglect the complementary energy due to transverse shear and write

$$U^* = U_1^* + U_2^* = \int_0^h \left[\frac{T_1^2}{2JG} + \frac{M_1^2}{2EI}\right] dx + \int_0^L \left[\frac{T_2^2}{2JG} + \frac{M_2^2}{2EI}\right] dx$$

where from the free-body diagrams shown in Figs. 10.42b and c, $T_1(x) = T_0$, $M_1(x) = M_0 + Px$, $T_2(x) = M_0 + Ph$, and $M_2 = T_0 + Px$. We apply Castigliano's second theorem to compute

$$u = \frac{\partial U^*}{\partial P} = \int_0^h \left[\frac{T_1(\partial T_1/\partial P)}{JG} + \frac{M_1(\partial M_1/\partial P)}{EI}\right] dx$$
$$+ \int_0^L \left[\frac{T_2(\partial T_2/\partial P)}{JG} + \frac{M_2(\partial M_2/\partial P)}{EI}\right] dx$$

$$\phi = \frac{\partial U^*}{\partial M_0} = \int_0^h \left[\frac{T_1(\partial T_1/\partial M_0)}{JG} + \frac{M_1(\partial M_1/\partial M_0)}{EI}\right] dx$$
$$+ \int_0^L \left[\frac{T_2(\partial T_2/\partial M_0)}{JG} + \frac{M_2(\partial M_2/\partial M_0)}{EI}\right] dx$$

$$\psi = \frac{\partial U^*}{\partial T_0} = \int_0^h \left[\frac{T_1(\partial T_1/\partial T_0)}{JG} + \frac{M_1(\partial M_1/\partial T_0)}{EI}\right] dx$$
$$+ \int_0^L \left[\frac{T_2(\partial T_2/\partial T_0)}{JG} + \frac{M_2(\partial M_2/\partial T_0)}{EI}\right] dx$$

Evaluating the integrals yields

$$u = P\left(\frac{h^3}{3EI} + \frac{h^2L}{JG} + \frac{L^3}{3EI}\right) + M_0\left(\frac{h^2}{2EI} + \frac{hL}{JG}\right) + \frac{T_0L^2}{2EI}$$

$$\phi = P\left(\frac{h^2}{2EI} + \frac{hL}{JG}\right) + M_0\left(\frac{h}{EI} + \frac{L}{JG}\right) + 0\cdot T_0$$

$$\psi = \frac{PL^2}{2EI} + 0\cdot M_0 + T_0\left(\frac{h}{JG} + \frac{L}{EI}\right)$$

Note again that we can write this set of equations in the form

$$u = S_{11}P + S_{12}M_0 + S_{13}T_0$$

$$v = S_{21}P + S_{22}M_0 + S_{23}T_0$$

$$\theta = S_{31}P + S_{32}M_0 + S_{33}T_0$$

where the coefficients $S_{ij} = S_{ji}$ are flexibility coefficients for the structure of this example.

PROBLEMS

Axial Problems

Unless otherwise stated, each of the following bars is to be considered linearly elastic.

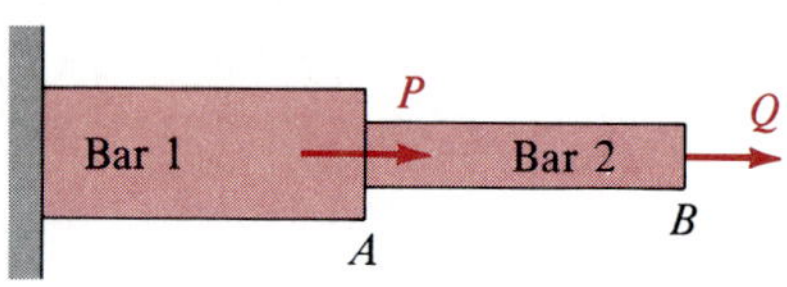

P. 10.117

10.117 Determine the horizontal displacements of points A and B for the segmented bar shown. Take $A_1 = 4$ in^2, $A_2 = 2$ in^2, $L_1 = 2$ ft, $L_2 = 3$ ft, $E_1 = E_2 = 10^7$ psi, $P = 50$ kips, and $Q = 30$ kips.

10.118 Determine the vertical displacements at points A and B. Take $E = 10^7$ psi for both members.

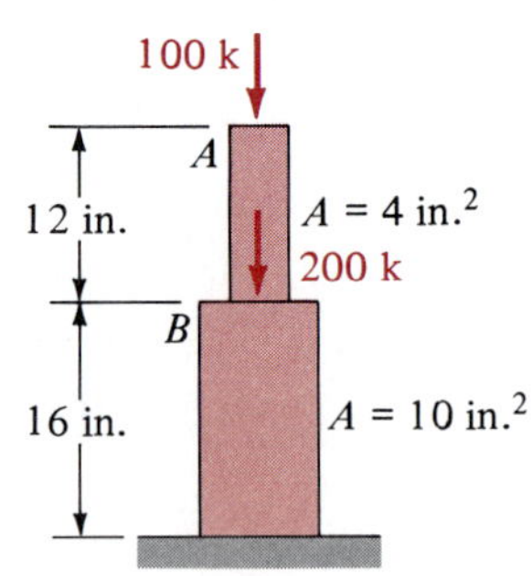

P. 10.118

10.119 Determine the horizontal displacement at point A. Take $E = 70$ GPa for all members.

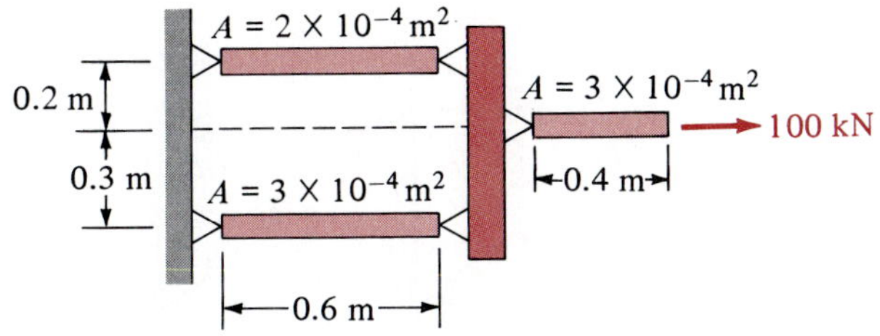

P. 10.119

10.120 A 2400-ft-long steel cable of area 2 in^2 supports a 30-kip load as shown. Determine the elongation including the weight of the cable. What percent of the total elongation is due to the weight of the cable?

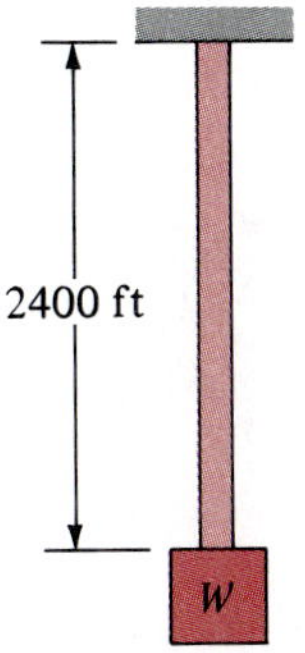

P. 10.120

10.121 Determine the elongation of the circular-cross-section bar shown. Take $E = 70$ GPa.

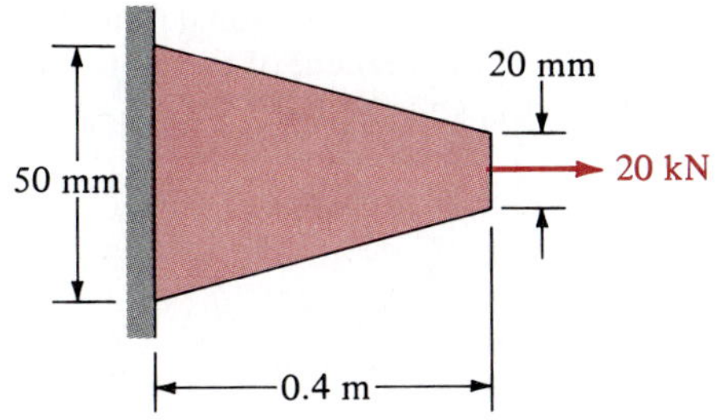

P. 10.121

10.122 Determine the vertical displacement at point C. Take $E = 10^7$ psi.

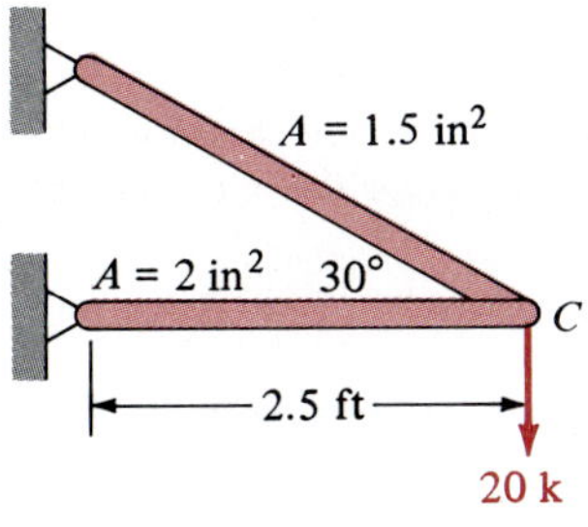

P. 10.122

10.123 Determine the horizontal and vertical displacements at point D. For all members take $A = 10^{-3}$ m^2 and $E = 70$ GPa. Take $P = 30$ kN and $Q = 20$ kN.

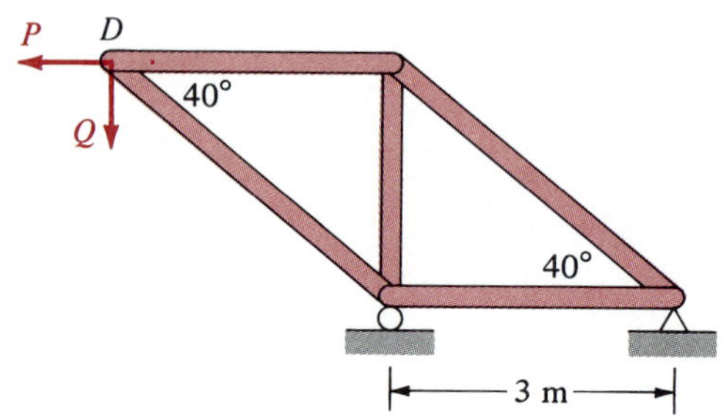

P. 10.123

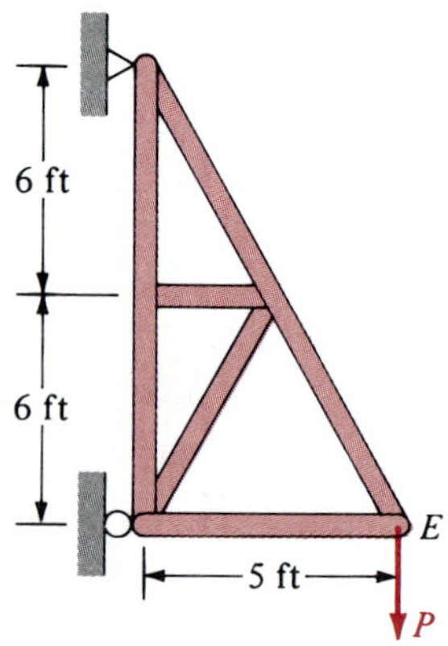

P. 10.124

10.124 Determine the vertical displacement at point E. For all members, take $A = 1.5$ in^2 and $E = 10^7$ psi. Take $P = 30$ kips.

10.125 A mass M is attached to a linearly elastic bar of area A and mass density ρ as shown. The assembly rotates about an axis through the end of the bar at an angular speed ω. Determine the elongation of the bar including the effect of the mass of the bar. What percent of the elongation is due to the mass M? What is the maximum stress in the bar? Take $\rho AL = M$.

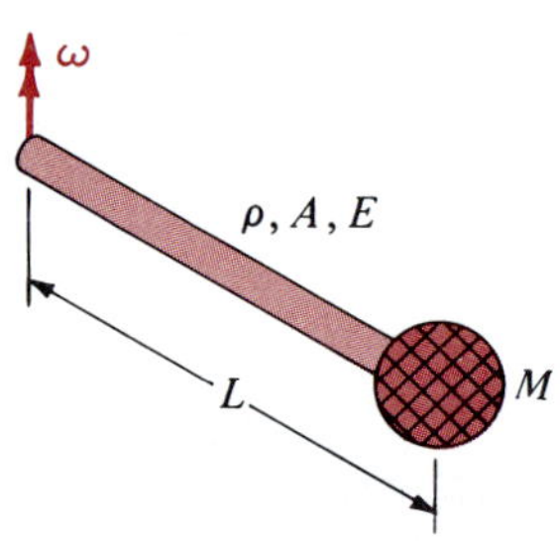

Ps. 10.125 and 10.126

10.126 For a rotating bar composed of a nonlinear elastic material behaving according to $\sigma = C\varepsilon^{1.5}$, determine the elongation neglecting the mass of the bar.

Torsional Problems

Unless otherwise stated, each of the following circular-cross-section bars is to be considered as linearly elastic.

10.127 Determine the rotations at points A, B, and C. Take $L_1 = 16$ in., $L_2 = 12$ in., $L_3 = 18$ in., $r_1 = 0.75$ in., $r_2 = 1.0$ in., $r_{3o} = 0.75$ in., $r_{3i} = 0.5$ in., $T_0 = 7$ in.-kips, $T_1 = 5$ in.-kips, $T_2 = 8$ in.-kips, and $G = 12 \times 10^6$ psi for all shafts.

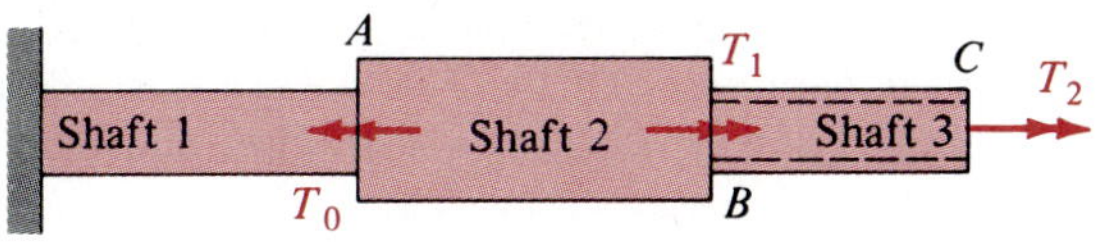

Ps. 10.127 and 10.128

10.128 Repeat problem 10.127 for $L_1 = 0.4$ m, $L_2 = 0.3$ m, $L_3 = 0.5$ m, $r_1 = 20$ mm, $r_2 = 25$ mm, $r_{3o} = 20$ mm, $r_{3i} = 15$ mm, $T_0 = 800$ N·m, $T_1 = 600$ N·m, $T_2 = 900$ N·m, and $G = 80$ GPa for all shafts.

10.129 Determine the rotation at point A. Take $G = 29$ GPa.

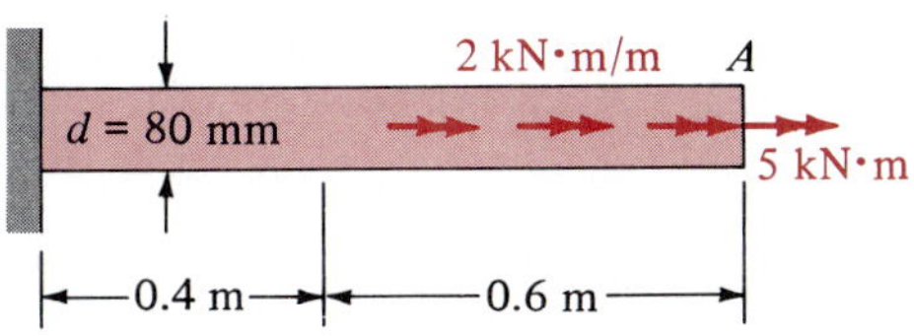

P. 10.129

10.130 Determine the rotation at point A.

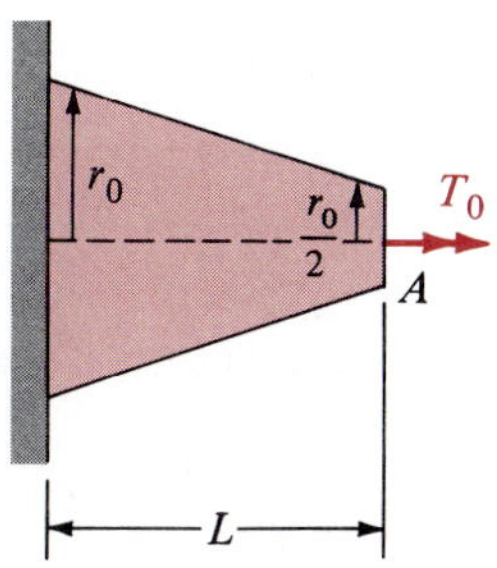

P. 10.130

10.131 Determine the rotation at point A. Take $r_1 = 0.9375r_0$, $r_2 = 0.8125r_0$, $r_3 = 0.6875r_0$, and $r_4 = 0.5625r_0$. Compare your answer with the answer to problem 10.130.

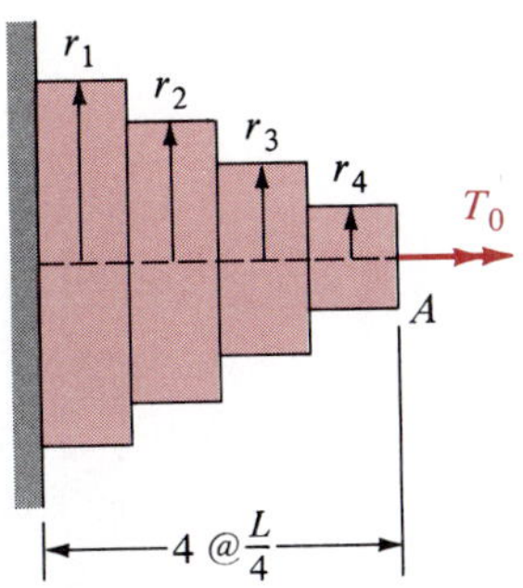

P. 10.131

10.132 Determine the rotation at point A. Take $r_0 = 4$ in., $L = 28$ in., $T_0 = 20$ in.-kips, and $G = 12 \times 10^6$ psi.

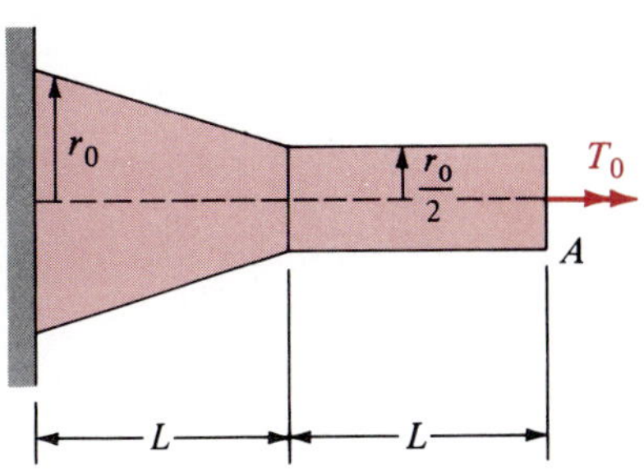

P. 10.132

10.133 Determine the rotations at points A and B. Take $L_1 = 12$ in., $L_2 = 18$ in., $L_3 = 15$ in., $L_4 = 22$ in., $r_1 = 1.2$ in., $r_2 = 0.8$ in., $r_3 = 1.0$ in., $r_4 = 1.0$ in., $R_1 = 12$ in., $R_2 = 8$ in., $T_0 = 6$ in.-kips, $T_1 = 4$ in.-kips, and $G = 4 \times 10^6$ psi for all shafts.

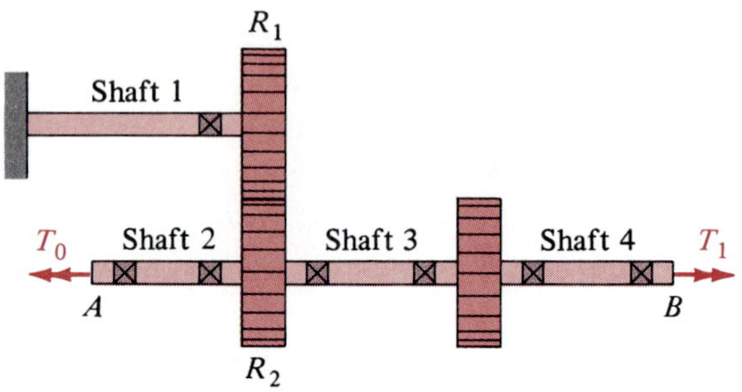

Ps. 10.133 and 10.134

10.134 Repeat problem 10.33 for $L_1 = 0.3$ m, $L_2 = 0.4$ m, $L_3 = 0.35$ m, $L_4 = 0.6$ m, $r_1 = 30$ mm, $r_2 = 20$ mm, $r_3 = 25$ mm, $r_4 = 25$ mm, $R_1 = 0.3$ m, $R_2 = 0.25$ m, $T_0 = 3$ kN·m, $T_1 = 2$ kN·m, and $G = 30$ GPa for all shafts.

10.135 Determine the rotation at point A. Take $L_1 = 0.3$ m, $L_2 = 0.4$ m, $L_3 = 0.45$ m, $r_1 = 25$ mm, $r_2 = 21$ mm, $r_3 = 22$ mm, $R_1 = 0.35$ m, $R_2 = 0.25$ m, $R_3 = 0.2$ m, $R_4 = 0.4$ m, $T_1 = 2.5$ kN·m, and $G = 30$ GPa for all shafts.

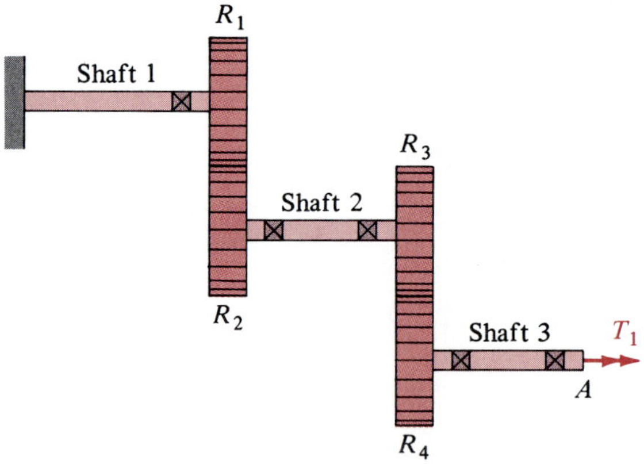

Ps. 10.135 and 10.136

10.136 Repeat problem 10.135 for $L_1 = 12$ in., $L_2 = 18$ in., $L_3 = 15$ in., $r_1 = 1.0$ in., $r_2 = 0.9$ in., $r_3 = 1.0$ in., $R_1 = 14$ in., $R_2 = 10$ in., $R_3 = 9$ in., $R_4 = 15$ in., $T_1 = 20$ in.-kips, and $G = 4 \times 10^6$ psi for all shafts.

10.137 Shaft 1 is made from a material for which $\tau = C\gamma^{0.85}$, with $C = 3.6 \times 10^5$ psi. Shaft 2 is linearly elastic with $G = 4 \times 10^6$ psi. Take $L_1 = 15$ in., $L_2 = 12$ in., $r_1 = 1.0$ in., $r_2 = 0.625$ in., $T_0 = 4$ in.-kips, $T_1 = 8$ in.-kips, and determine the rotations at A and B.

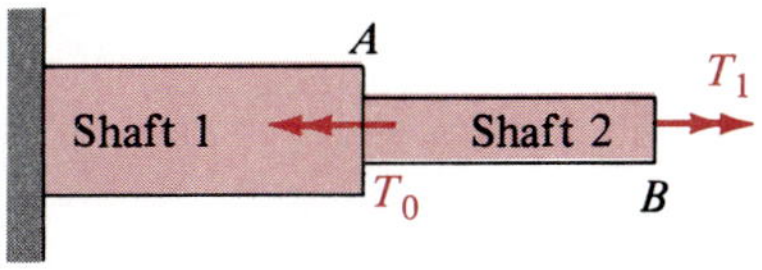

P. 10.137

Bending Problems

Unless otherwise stated, each of the following bars is to be considered as linearly elastic. Neglect energy due to transverse shear.

10.138 Determine the transverse displacement at points A and B. Take $a = 0.75L$.

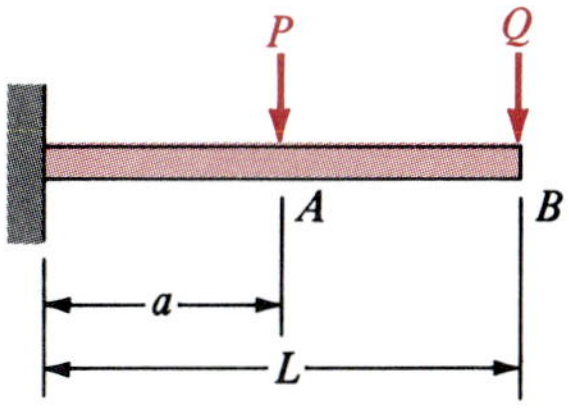

P. 10.138

10.139 Determine the transverse displacement at the right end of the cantilever beam loaded as shown. Take $E = 30 \times 10^6$ psi.

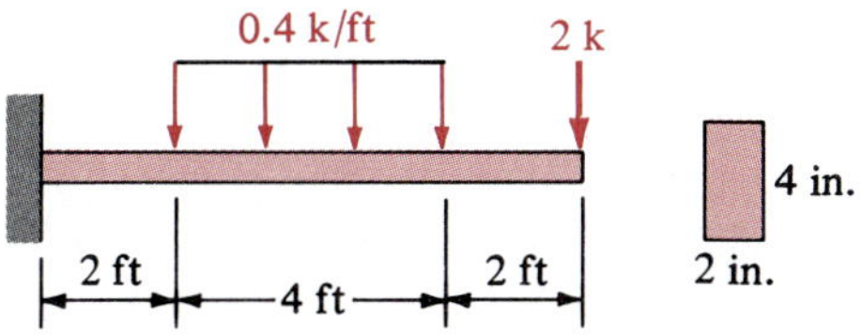

P. 10.139

10.140 Determine the transverse displacement and rotation at point B. Take $a = L/2$.

10.141 Determine the transverse displacement at point A. Take $a = 0.4L$.

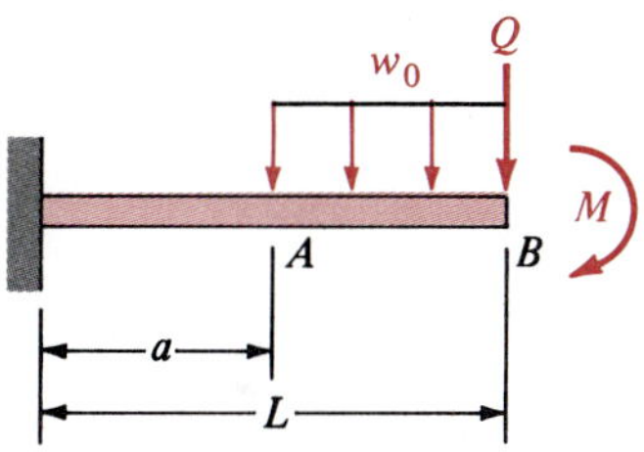

P. 10.140

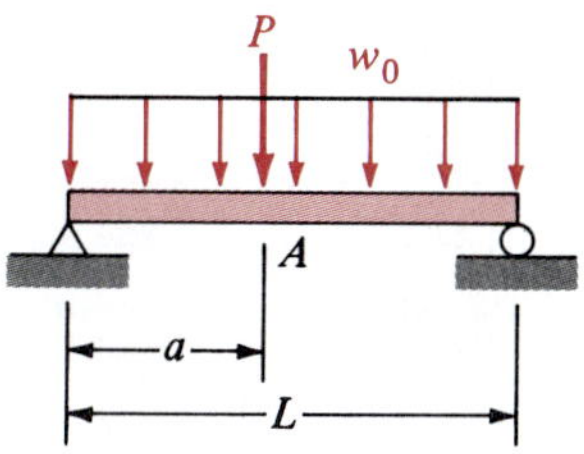

P. 10.141

10.142 Determine the transverse displacement and rotation at point A. Take $a = 0.5L$.

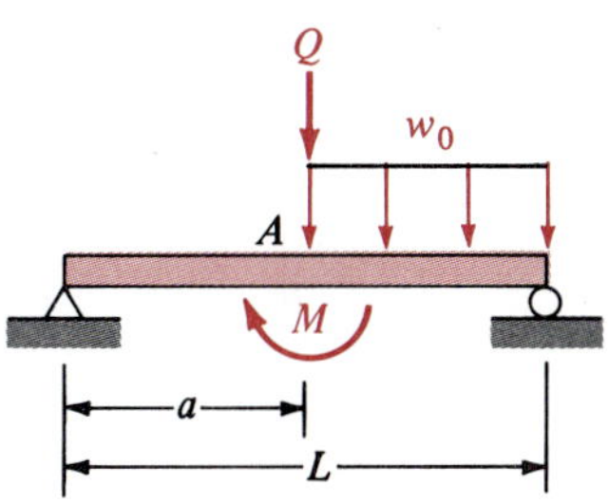

P. 10.142

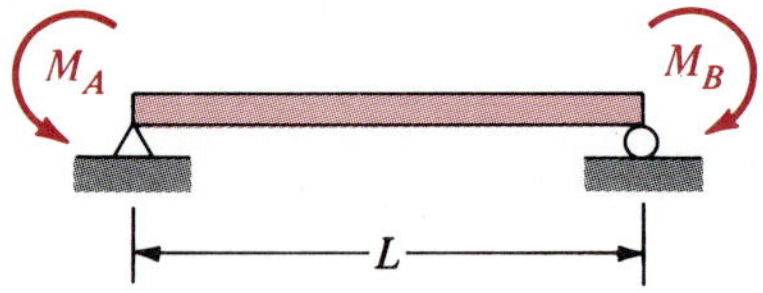

P. 10.143

10.143 Determine θ_A and θ_B.

10.144 Determine the transverse displacement at points A and B.

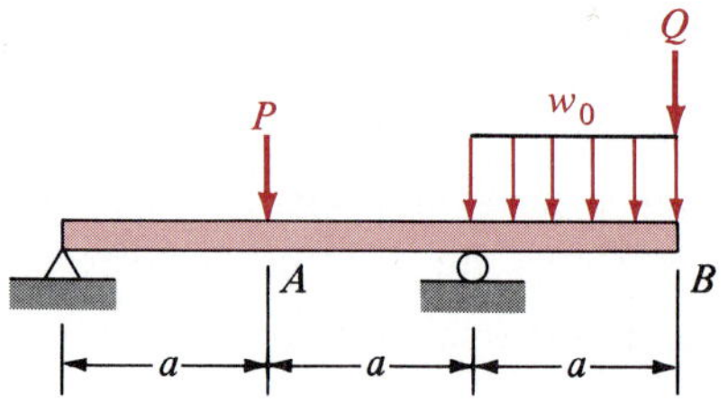

P. 10.144

10.145 Determine the displacements at points A and B. Take $E = 70$ GPa.

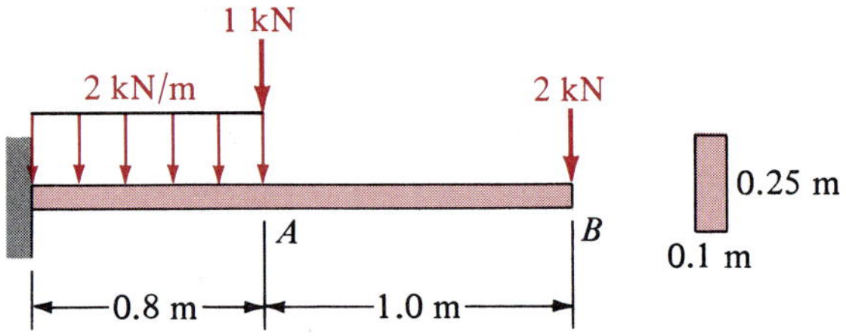

P. 10.145

10.146 Determine the displacement at point A, midway between the supports. Take $b = 2a$.

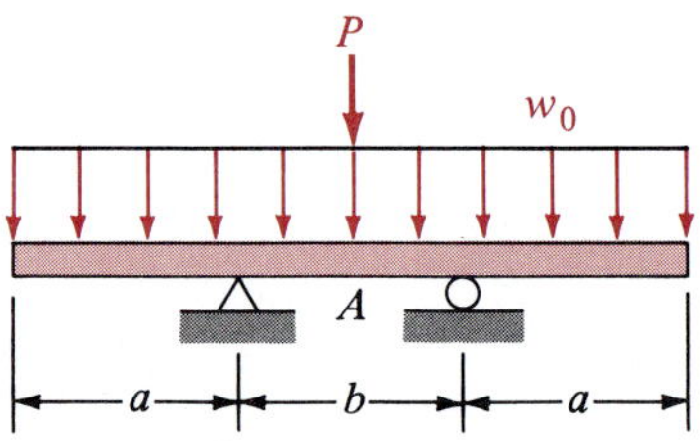

P. 10.146

10.147 Determine the displacement at point A for the rectangular-cross-section beam shown. Take the width to be the constant value b.

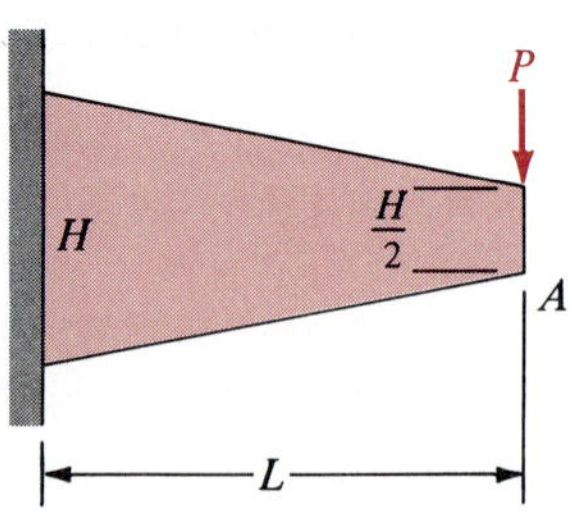

P. 10.147

The point of the next two problems is that it is possible to approximate the behavior of the displacements for a variable-cross-section beam, such as in problem 10.147, by replacing the beam with a segmented beam as indicated. In the first case, the two segments have heights that are equal to the average heights in the regions $0 \le x \le L/2$ and $L/2 \le x \le L$. In the second case, the heights are chosen using the same idea.

10.148 Determine the displacement at point A for the segmented rectangular-cross-section beam shown. Take the width to be the constant value b. Compare your answer with problem 10.147.

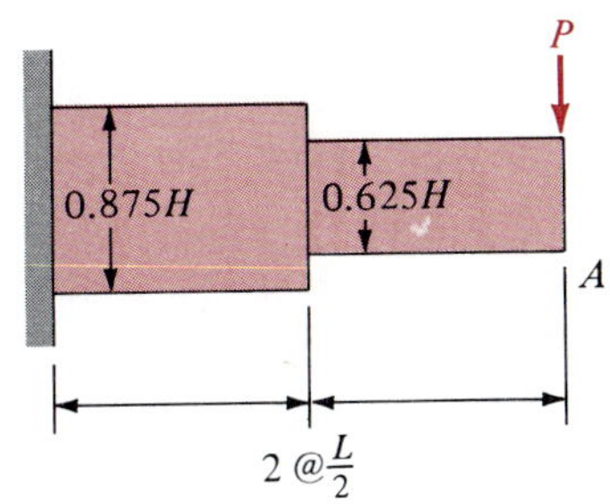

P. 10.148

10.149 Determine the displacement at point A for the segmented rectangular-cross-section beam shown. Take the width to be the constant value b. Compare your answer with the previous two problems. Take $h_1 = 0.9375H$, $h_2 = 0.8125H$, $h_3 = 0.6875H$, and $h_4 = 0.5625H$.

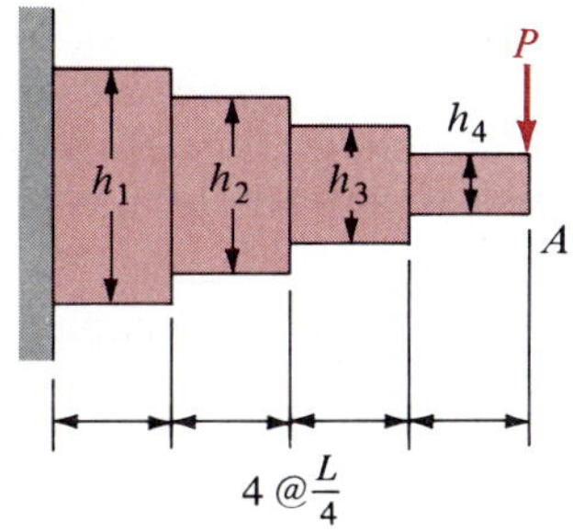

P. 10.149

10.150 Determine the displacements at points A and B for the variable-cross-section beam shown.

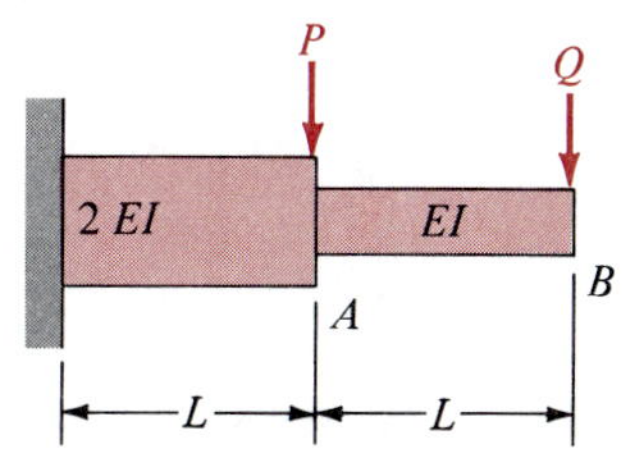

P. 10.150

10.151 Determine the displacement at point A for the variable-cross-section beam shown.

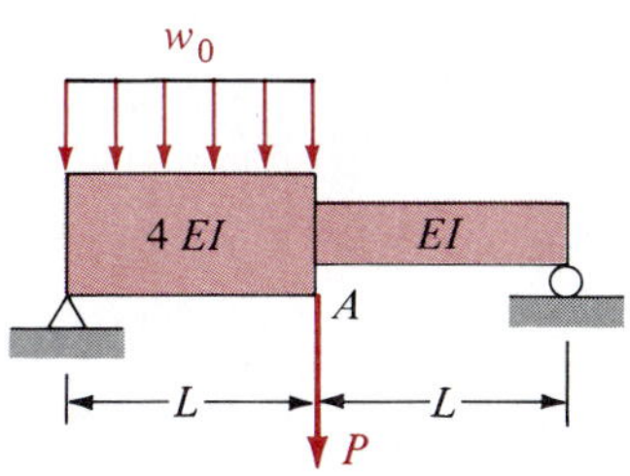

P. 10.151

10.152 The variable-cross-section beam consists of a W 6 × 25 section with two 0.5-in. by 6-in. plates attached by welding as shown. Determine the displacement at B. Take $E = 30 \times 10^6$ psi and neglect the weight of the beam.

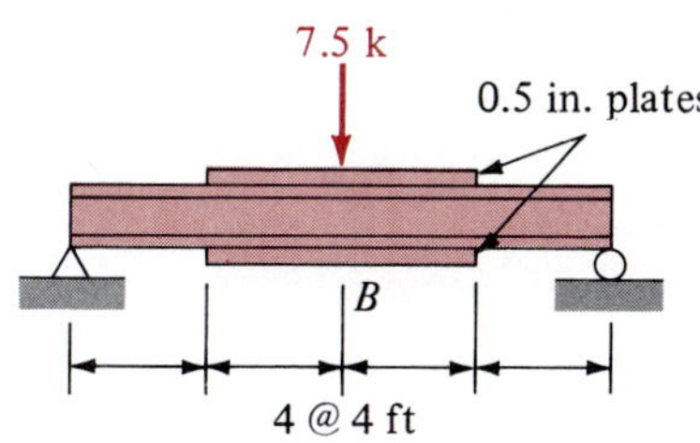

P. 10.152

10.153 Determine the displacement at A and rotation at B.

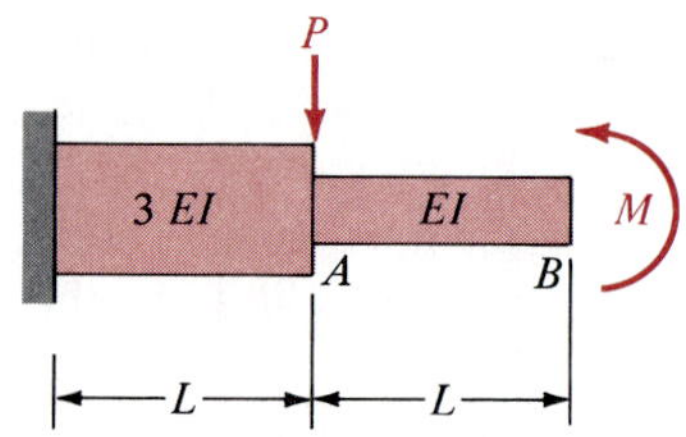

P. 10.153

Combined Loading Problems

Unless otherwise stated, take each member of the structure indicated to be linearly elastic. If there are torsional loads transmitted, assume that the cross section is circular or that a constant J, called the torsional constant, is known for the cross section so that the torsional energy can still be expressed as $U^* = \int (T^2/2JG)\, dx$. Do not include transverse shear-stress energy unless requested to do so.

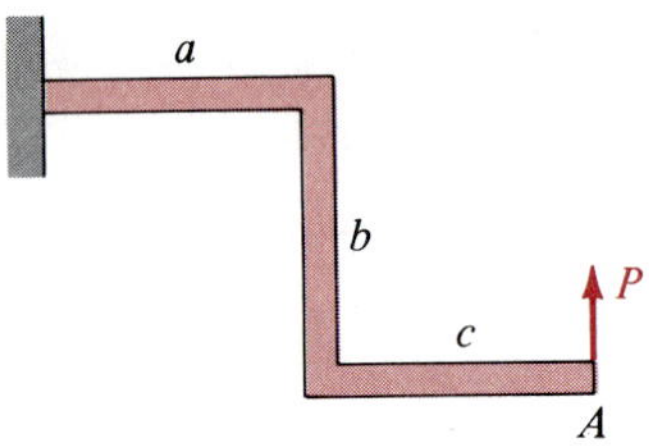

P. 10.154

10.154 Determine the vertical displacement at point A. Neglect axial effects. Specialize your results when $a = b = c = L$.

10.155 The circular-cross-section bar is supported against vertical motion at A. Determine the vertical displacement at point B. Specialize your result when $a = b = L$ and $E = 2.6G$.

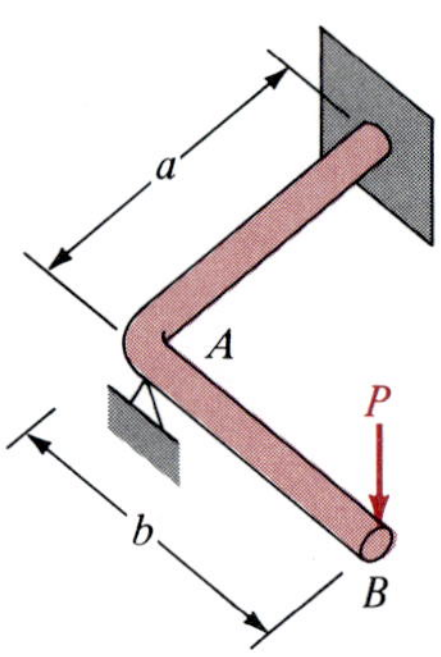

P. 10.155

10.156 Determine the vertical displacements at points A and B. The bars have identical circular cross sections.

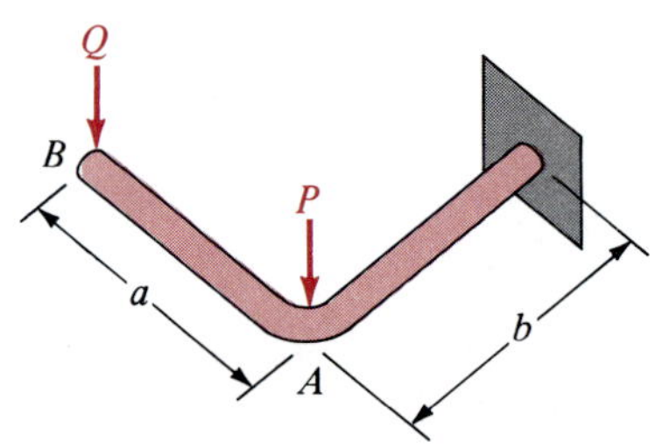

P. 10.156

10.157 Determine the vertical displacement at point A. Specialize your results when $a = b = L$. Neglect axial energy.

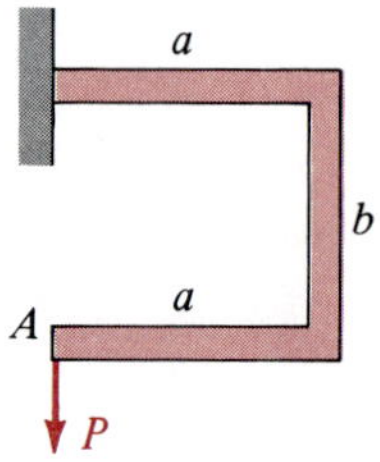

P. 10.157

10.158 Determine the horizontal displacement at point A. Specialize your results when $h = L$. Neglect axial energy.

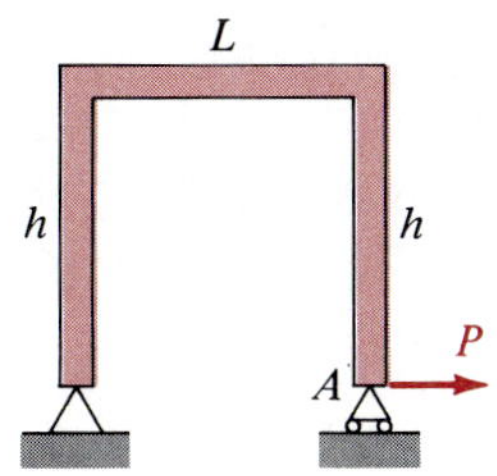

P. 10.158

10.159 For the circular-cross-section bar shown, determine the vertical displacement at point A. Specialize your results for $E = 2.6G$.

10.160 Including axial energy, determine the horizontal and vertical displacements at point A. (*Hint:* Decompose P into components $P \cos 45° = p$ and $P \sin 45° = q$.) For a circular-cross-section bar of radius $a = 0.05L$, what percentages of the two displacements are due to the axial deformation?

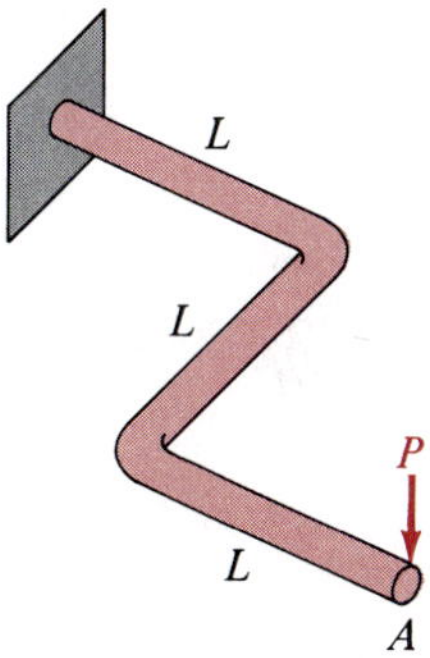

P. 10.159

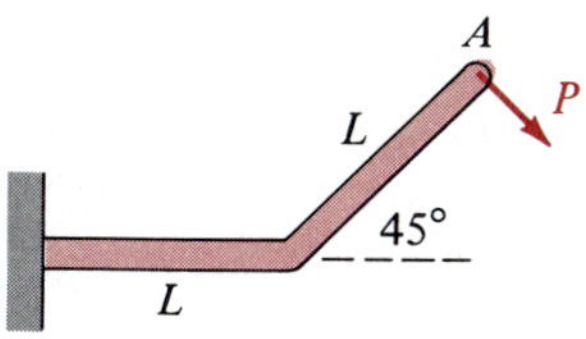

P. 10.160

10.161 Including axial energy, determine the displacement at point A. If the bar has a square cross section with $b = h = 0.1a$, how much of the displacement is due to the bending effects?

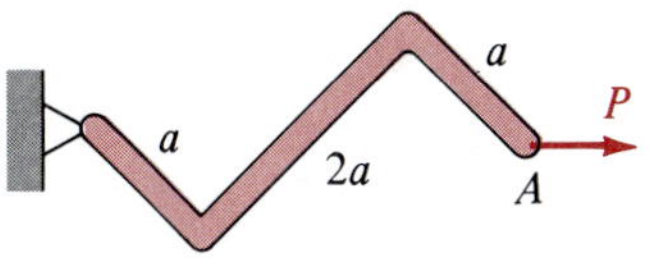

P. 10.161

10.4.3 Dummy Load Methods

Often it is necessary to be able to determine the displacement at a point in a structure at which no load is applied or in a direction other than the direction of the load at a loaded point. A technique that can accomplish this task is known as the **dummy load method.** In this section we will describe this technique and study the basis for it, namely, Castigliano's second theorem.

Recall Example 10.15, shown again in Fig. 10.43.

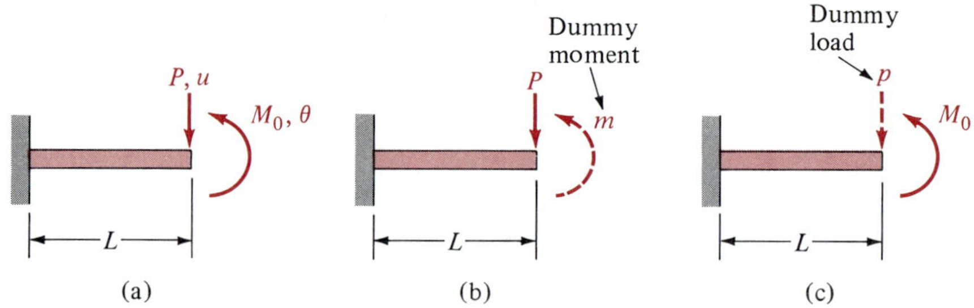

Figure 10.43 End-loaded cantilever

Using Castigliano's second theorem in connection with Fig. 10.43a, it was shown that the displacement v in the direction of P, and the rotation θ in the direction of M_0, are

$$v = \frac{PL^3}{3EI} - \frac{M_0 L^2}{2EI}$$

$$\theta = -\frac{PL^2}{2EI} + \frac{M_0 L}{EI}$$

These results are valid whether either, or both, P and M_0 are present. Note that if M_0 is taken to be zero, after calculating v and θ, it is still possible to determine both the displacement v and the slope θ due just to the action of the force P. Similarly, if P is taken to be zero, after calculating v and θ, it is again still possible to determine both the displacement v and the rotation θ due just to the action of the moment M_0.

These observations point the way to the dummy load method. In a situation where the displacement or rotation is desired in a direction at a point where there is no load, a so-called dummy load is used. We can demonstrate the idea for the end-loaded cantilever above by considering Fig. 10.43b, where the actual load P and a dummy or fictitious moment m are applied. The fact that m is fictitious is indicated by the dotted lines in Fig. 10.43b. The complementary energy can be written as

$$U^* = \int_0^L \frac{M^2}{2EI}\,dx = \int_0^L \frac{(m - Px)^2}{2EI}\,dx$$

Using Castigliano's second theorem, compute

$$\theta = \frac{\partial U^*}{\partial m} = \int_0^L \frac{(m - Px)(1)}{EI}\,dx = \frac{mL}{EI} - \frac{PL^2}{2EI}$$

and

$$v = \frac{\partial U^*}{\partial P} = \int_0^L \frac{(m - Px)(-x)}{EI}\,dx = -\frac{mL^2}{2EI} + \frac{PL^3}{3EI}$$

Allowing m to approach zero gives

$$\theta = \left.\frac{\partial U^*}{\partial m}\right|_{m=0} = -\frac{PL^2}{2EI}$$

and

$$v = \left.\frac{\partial U^*}{\partial P}\right|_{m=0} = \frac{PL^3}{3EI}$$

which are the correct results for P acting alone. Consider also Fig. 10.43c, showing the actual moment M_0 and a fictitious load p, again indicated by dotted lines. The complementary energy can be expressed as

$$U^* = \int_0^L \frac{(M - px)^2}{2EI}\,dx$$

and it follows using Castigliano's second theorem that

$$\theta = \frac{\partial U^*}{\partial M} = \int_0^L \frac{(M - px)(1)}{EI}\,dx = \frac{ML}{EI} - \frac{pL^2}{2EI}$$

and

$$v = \frac{\partial U^*}{\partial p} = \int_0^L \frac{(M - px)(-x)}{EI}\,dx = -\frac{ML^2}{2EI} + \frac{pL^3}{3EI}$$

If we now allow p to approach zero there results

$$\theta = \left.\frac{\partial U^*}{\partial M}\right|_{p=0} = \frac{ML}{EI} \qquad \text{and} \qquad v = \left.\frac{\partial U^*}{\partial p}\right|_{p=0} = -\frac{ML^2}{2EI}$$

These are the correct results for the moment M acting alone.

This is the essence of the dummy load method: A fictitious dummy load or moment is applied to the structure at the point and in the direction of the desired displacement; the complementary energy is determined, *including the effect of the dummy load(s) in writing the equilibrium equations to determine the internal forces,* after which Castigliano's second theorem is used in the usual manner. After computing the partial derivative or partial derivatives, the dummy load is set to zero, leaving intact the entire contribution of the actual loads to the displacements. It is generally much more efficient to compute the partial derivatives and then set the dummy loads to zero before evaluating the integrals. More than one dummy load can be applied at a time.

Example 10.18 The two-bar planar truss is subjected to the vertical load as shown in Fig. 10.44a. Determine the horizontal and vertical components of displacement of the loaded joint.

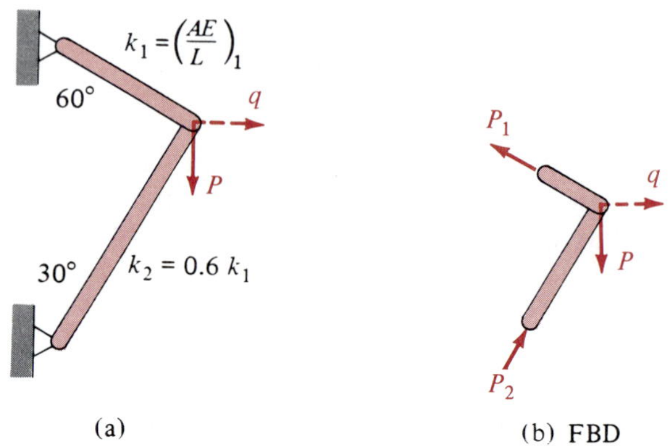

Figure 10.44 Planar truss

Solution Referring to the free-body diagram of Fig. 10.44b, the equilibrium equations are

$$\Sigma F_H = 0 = -P_1 \frac{\sqrt{3}}{2} + \frac{P_2}{2} + q$$

$$\Sigma F_V = 0 = \frac{P_1}{2} + P_2 \frac{\sqrt{3}}{2} - P$$

Solving for the internal forces P_1 and P_2 yields

$$P_1 = q \frac{\sqrt{3}}{2} + \frac{P}{2} \qquad \text{and} \qquad P_2 = P \frac{\sqrt{3}}{2} - \frac{q}{2}$$

The complementary energy can then be written as

$$U^* = \frac{P_1^2}{2k_1} + \frac{P_2^2}{2k_2} = \frac{(q\sqrt{3}/2 + P/2)^2}{2k_1} + \frac{(P\sqrt{3}/2 - q/2)^2}{2k_2}$$

Castigliano's second theorem is used to compute

$$\Delta_H = \left.\frac{\partial U^*}{\partial q}\right|_{q=0} = \frac{P\sqrt{3}}{4k_1} - \frac{P\sqrt{3}}{2.4k_1} = -\frac{0.289P}{k_1}$$

$$\Delta_V = \left.\frac{\partial U^*}{\partial P}\right|_{q=0} = \frac{P}{4k_1} + \frac{P}{2.4k_1} = \frac{2P}{3k_1}$$

The negative sign on Δ_H simply means that the displacement is to the left, opposite to the direction of q.

Example 10.19 A segmented torsional member is acted upon by the torques shown in Fig. 10.45a. The C_i represent the torsional spring constants $C_i = J_iG/L_i$. Determine the rotation at points A and B.

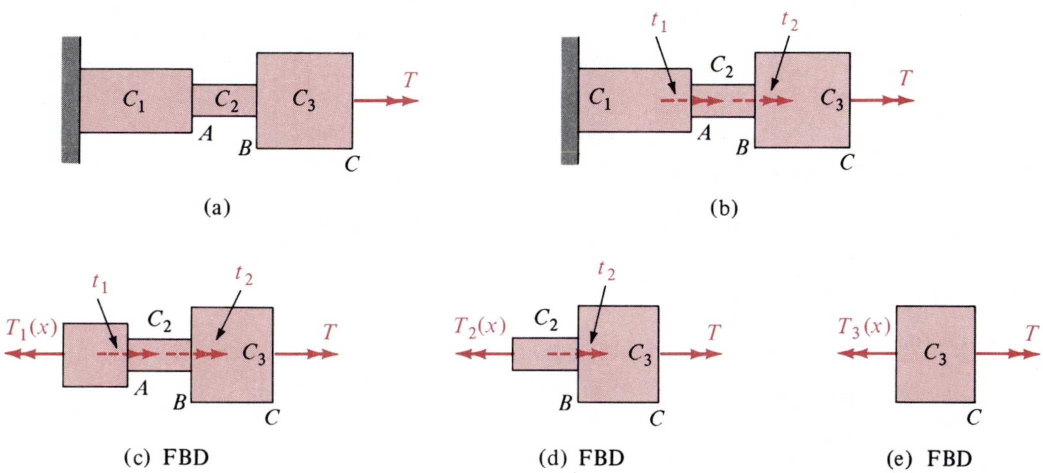

Figure 10.45 Segmented torsional member

Solution After placing dummy torques, indicated by the dotted t_1 and t_2, at points A and B as shown in Fig. 10.45b, the free-body diagrams shown in Figs. 10.45c, d, and e can be used to determine that $T_1(x) = T + t_1 + t_2$, $T_2(x) = T + t_2$, and $T_3(x) = T$, respectively. The complementary energy can be expressed as $U^* = U_1^* + U_2^* + U_3^*$, or

$$U^* = \int_0^{L_1} \frac{(T + t_1 + t_2)^2}{2J_1G_1}\,dx + \int_0^{L_2} \frac{(T + t_2)^2}{2J_2G_2}\,dx + \int_0^{L_3} \frac{T^2}{2J_3G_3}\,dx$$

Denoting ϕ_1 and ϕ_2 as the rotations at points A and B, Castigliano's second theorem is used to compute

$$\phi_1 = \left.\frac{\partial U^*}{\partial t_1}\right|_{t_1=t_2=0} = \int_0^{L_1} \frac{T}{J_1G_1}\,dx = \frac{TL_1}{J_1G_1} = \frac{T}{C_1}$$

and

$$\phi_2 = \left.\frac{\partial U^*}{\partial t_2}\right|_{t_1=t_2=0} = \int_0^{L_1} \frac{T}{J_1G_1}\,dx + \int_0^{L_2} \frac{T}{J_2G_2}\,dx$$

$$= T\left(\frac{L_1}{J_1G_1} + \frac{L_2}{J_2G_2}\right) = T\left(\frac{1}{C_1} + \frac{1}{C_2}\right)$$

both of which the student should be able to verify by independent means.

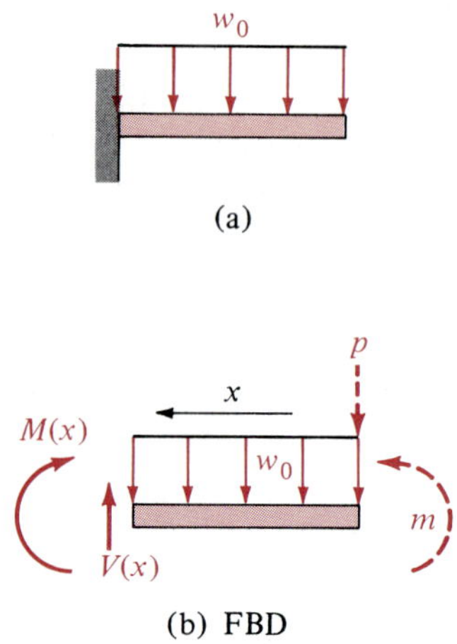

Figure 10.46 Uniformly loaded cantilever beam

Example 10.20 A cantilever beam is uniformly loaded as shown in Fig. 10.46a. Determine the deflection and rotation at the right end of the beam.

Solution Referring to the free-body diagram of Fig. 10.46b showing the dummy load p and the dummy moment m, the moment $M(x) = -m - px - w_0x^2/2$, with the complementary energy as

$$U^* = \int_0^L \left(-m - px - \frac{w_0x^2}{2}\right)^2 \frac{dx}{2EI}$$

The displacement and rotation in the directions of p and m, respectively, are given by

$$\Delta = \left.\frac{\partial U^*}{\partial p}\right|_{m=p=0} = \int_0^L \left(-\frac{w_0x^2}{2}\right)(-x)\frac{dx}{EI} = \frac{w_0L^4}{8EI}$$

$$\theta = \left.\frac{\partial U^*}{\partial m}\right|_{m=p=0} = \int_0^L \left(-\frac{w_0x^2}{2}\right)(-1)\frac{dx}{EI} = \frac{w_0L^4}{6EI}$$

showing Δ and θ to have the directions of p and m, respectively. These results can be checked by consulting Appendix D.

Example 10.21 The bent is loaded by the force P as shown in Fig. 10.47a. Determine the vertical displacement at point A. Neglect axial effects.

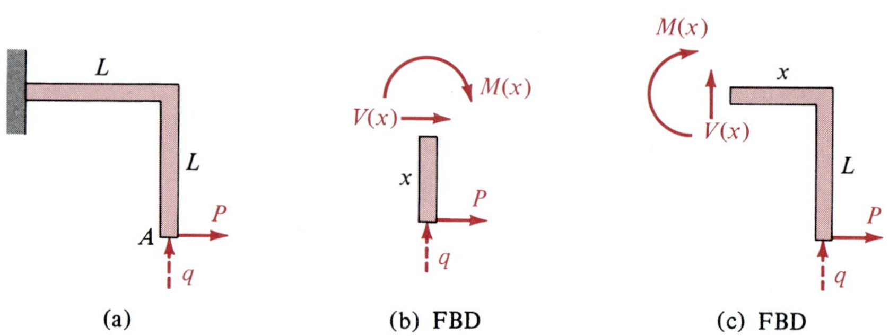

Figure 10.47 Deformations of a bent

Solution After placing the dummy load in the vertical direction at point A, the free-body diagrams of Fig. 10.47b and c are used to determine that $M_1(x) = Px$ and $M_2(x) = PL + qx$, respectively. The complementary energy is written as $U^* = U_1^* + U_2^*$ or

$$U^* = \int_0^L \frac{(Px)^2}{2EI}\,dx + \int_0^L \frac{(PL + qx)^2}{2EI}\,dx$$

from which the vertical displacement Δ can be determined as

$$\Delta = \left.\frac{\partial U^*}{\partial q}\right|_{q=0} = \int_0^L \frac{PLx}{EI} = \frac{PL^2}{2EI}$$

PROBLEMS

10.162 Determine the displacement at points A and B.

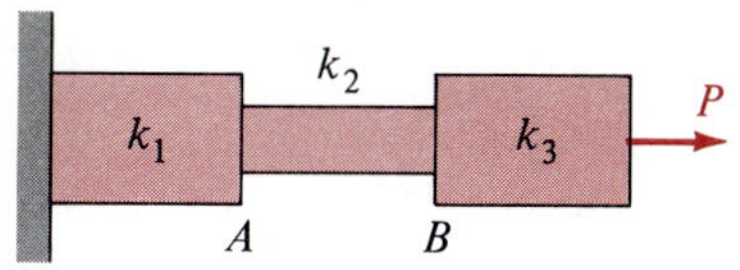

P. 10.162

10.163 Determine the displacement at point A. Take $A_1 = 1$ in^2, $A_2 = 2$ in^2, $A_3 = 0.75$ in^2, $L_1 = L_2 = 18$ in., $L_3 = 12$ in., $H = 10$ in., $\beta = 0.7$, $E = 10^7$ psi, and $P = 20$ kips.

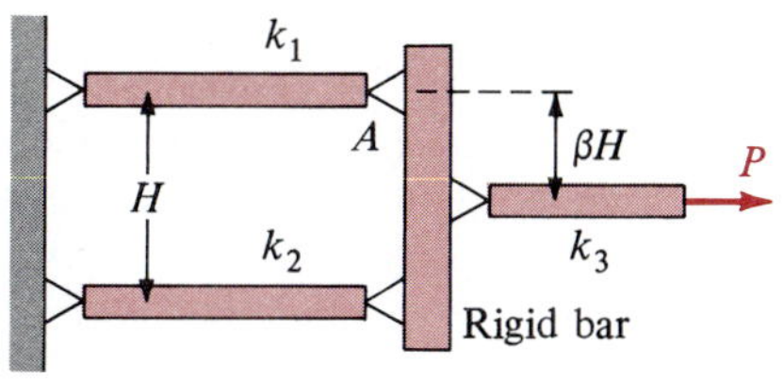

P. 10.163

10.164 Determine the horizontal and vertical components of displacement at point A. Take $A_1 = 2 \times 10^{-4}$ m^2, $A_2 = 3 \times 10^{-4}$ m^2, $L_1 = 3$ m, $L_2 = 4.24$ m, $E = 70$ GPa, and $P = 3$ kN.

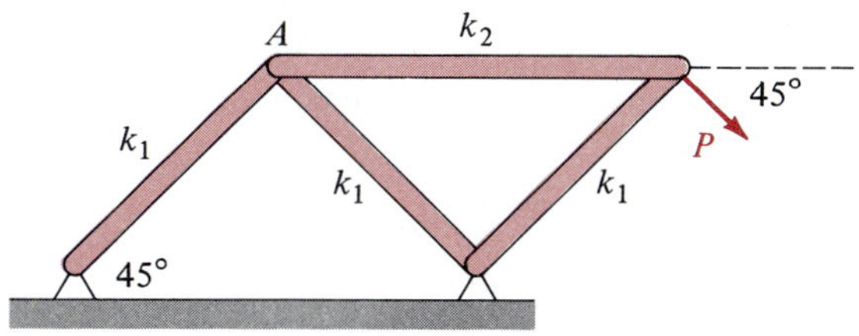

P. 10.164

10.165 The displacement at an arbitrary location ξ along the bar can be determined by placing a dummy load p at that point as shown. Show that the displacement $u(\xi)$ at ξ is exactly the same as one would obtain by integrating the governing differential equation of equilibrium $(AEu') + q(x) = 0$. This indicates the remarkable fact that the use of Castigliano's second theorem in this manner is entirely equivalent to integrating the corresponding equilibrium equation expressed in terms of the displacements.

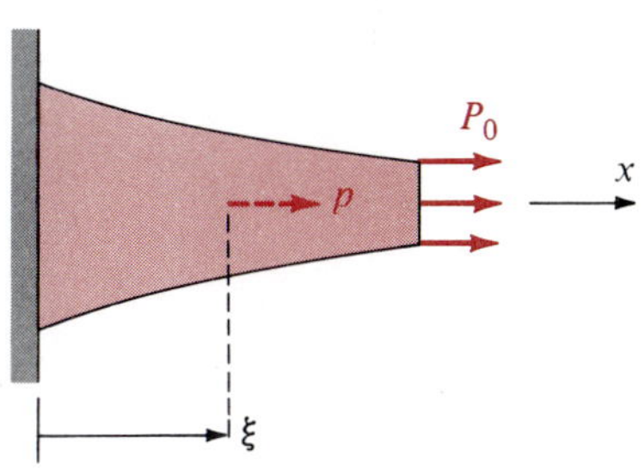

P. 10.165

10.166 For the segmented bar subjected to the torques shown, determine the rotation at A. Take $L_1 = 14$ in., $L_2 = 18$ in., $L_3 = 10$ in., $r_1 = 1.25$ in., $r_2 = 0.75$ in., $r_3 = 1.1$ in., $G_1 = G_2 = G_3 = 4 \times 10^7$ psi, $T_1 = 5$ in.-kips, and $T_2 = 7.5$ in.-kips.

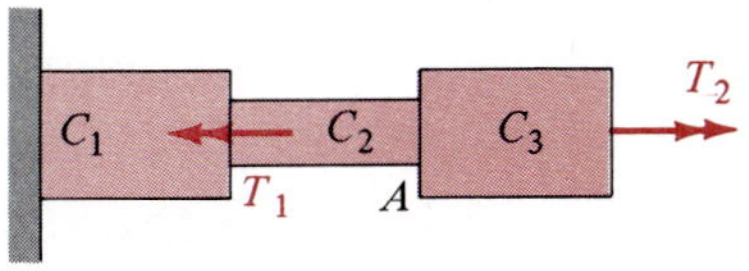

P. 10.166

10.167 Determine the rotation at point A for the variable-cross-section bar shown. Take $G = 30$ GPa.

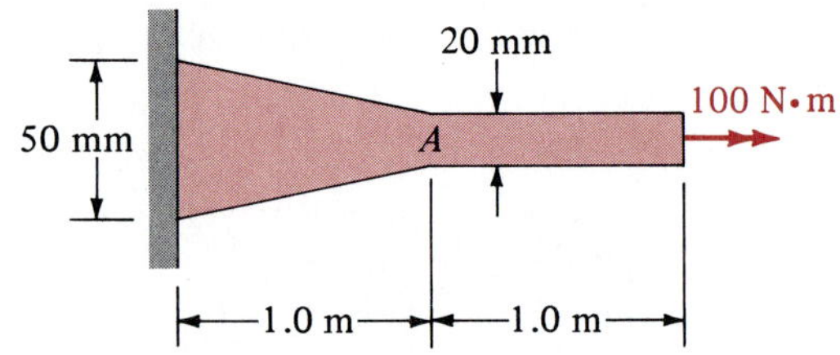

P. 10.167

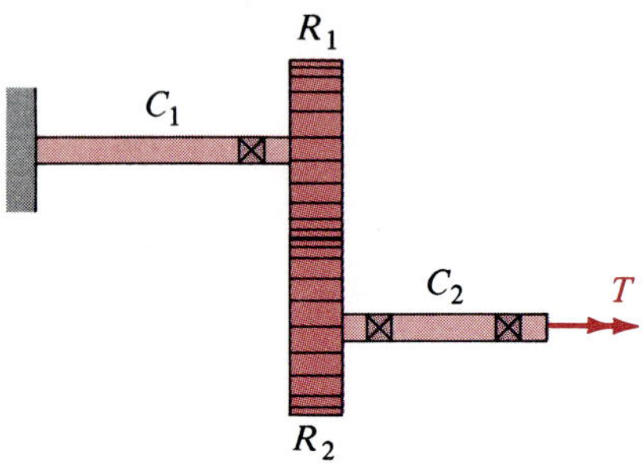

P. 10.168

10.168 Determine the rotations of disks 1 and 2. Take $L_1 = 0.4$ m, $L_2 = 0.6$ m, $r_1 = 30$ mm, $r_2 = 25$ mm, $R_1 = 0.3$ m, $R_2 = 0.24$ m, $G_1 = G_2 = 29$ GPa, and $T = 1$ kN·m.

10.169 The rotation at an arbitrary location ξ along the bar can be determined by placing a dummy torque t at that point as shown. Show that the rotation $\phi(\xi)$ at ξ is exactly the same as that obtained by integrating the governing differential equation of equilibrium, $(JG\phi')' + t(x) = 0$. This indicates the remarkable fact that the use of Castigliano's second theorem in this manner is entirely equivalent to integrating the corresponding equilibrium equation expressed in terms of the displacements.

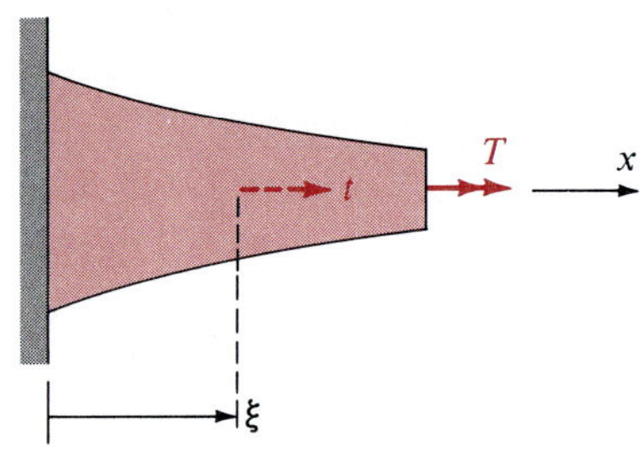

P. 10.169

10.170 Determine the deflection and rotation at point A for the W 8 × 28 section shown. Take $E = 29 \times 10^6$ psi.

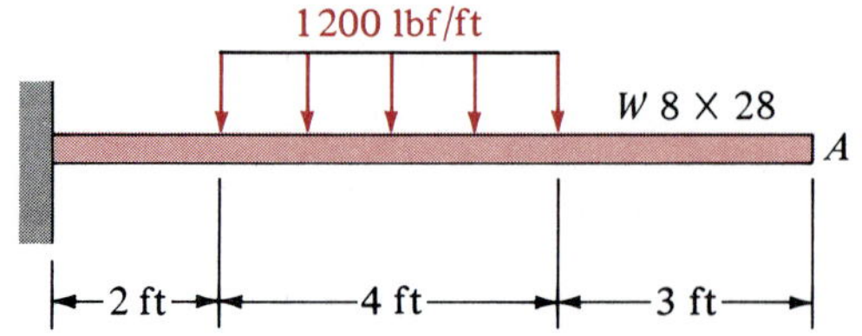

P. 10.170

10.171 Determine the rotation at A and the deflection at B for the uniform beam loaded as shown. Take $a = 0.4L$.

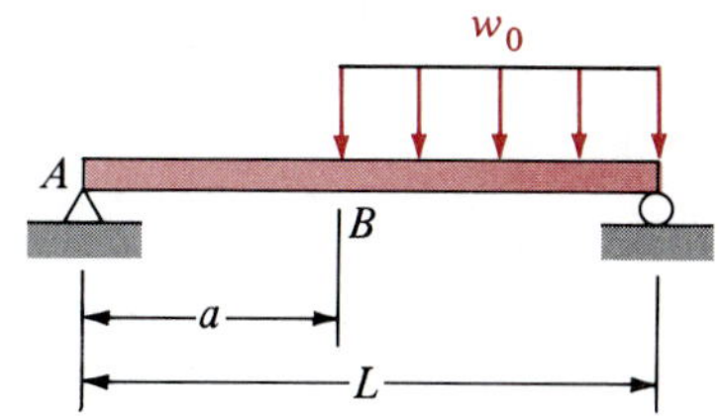

P. 10.171

10.172 Determine the rotation at A and the deflection at B for the stepped beam loaded as shown.

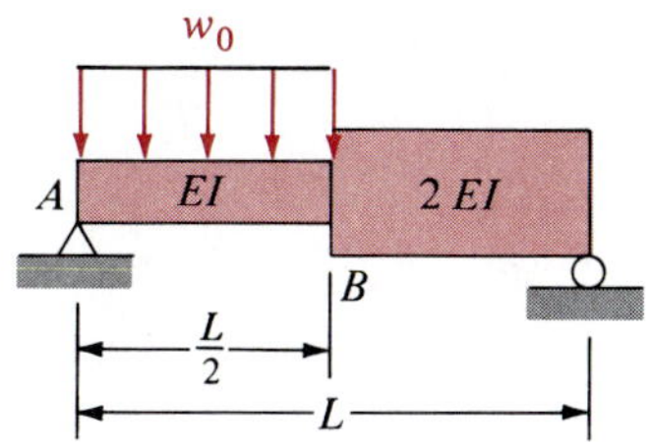

P. 10.172

10.173 The displacement at an arbitrary location ξ along the bar can be determined by placing a dummy load p at that point as shown. Show that the displacement $v(\xi)$ at ξ is exactly the same as that obtained by integrating the governing differential equation of equilibrium, $(EIv'')'' = -w(x)$. This indicates the remarkable fact that the use of Castigliano's second theorem in this manner is entirely equivalent to integrating the corresponding equilibrium equation expressed in terms of the displacements.

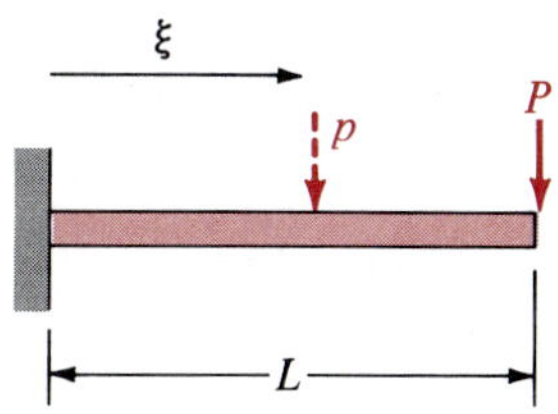

P. 10.173

10.174 The rotation or slope at an arbitrary location ξ along the bar can be determined by placing a dummy moment m at that point as shown. Show that the slope $v'(\xi)$ at ξ is exactly the same as that obtained by integrating the governing differential equation of equilibrium, $(EIv'')'' = -w(x)$.

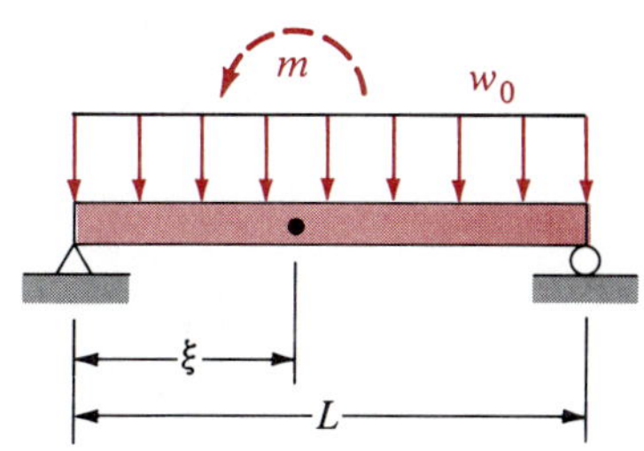

P. 10.174

10.175 Determine *(a)* the rotation at A, and *(b)* the vertical component of displacement at A for the uniform bent shown. Neglect axial energy.

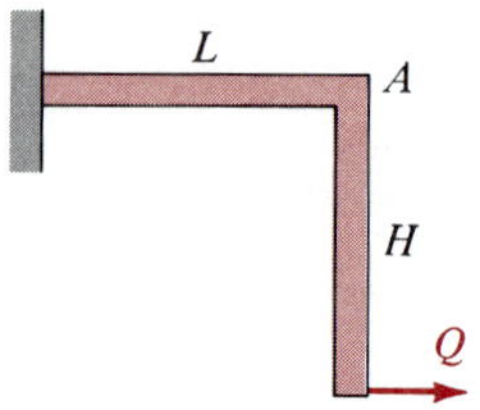

P. 10.175

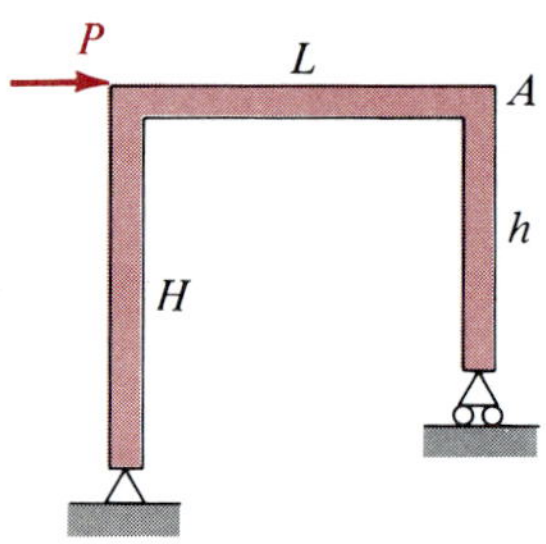

P. 10.176

10.176 For the uniform-cross-section, framed structure shown, determine *(a)* the horizontal component of displacement at A, and *(b)* the rotation at A. Neglect axial energy.

10.177 The bent shown consists of two circular-cross-section members of equal radii. Including bending and torsion, determine *(a)* the rotation about the x-axis at A, *(b)* the rotation about the y-axis at A, and *(c)* the z-component of displacement at A.

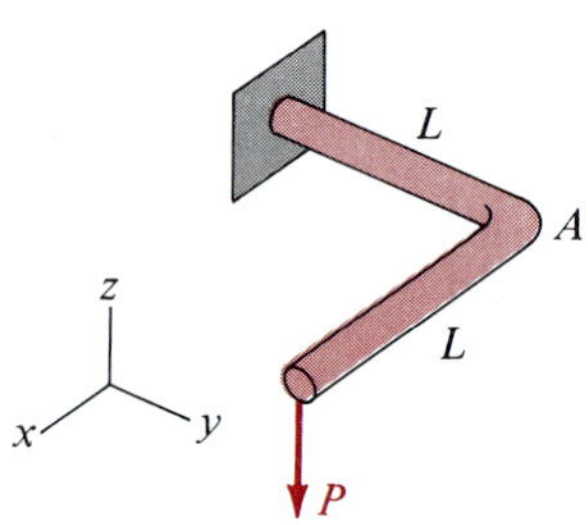

Ps. 10.177 and 10.178

10.178 The bent shown consists of two circular-cross-section members of equal radii. Including bending and torsion, determine *(a)* the rotation about the z-axis at A, *(b)* the x-component of displacement at A, and *(c)* the y-component of displacement at A.

10.179 The inplane bent is rotated at a constant rate ω about an axis as shown. Determine *(a)* the horizontal displacement at A, *(b)* the vertical displacement at A, and *(c)* the rotation at A. Include bending only.

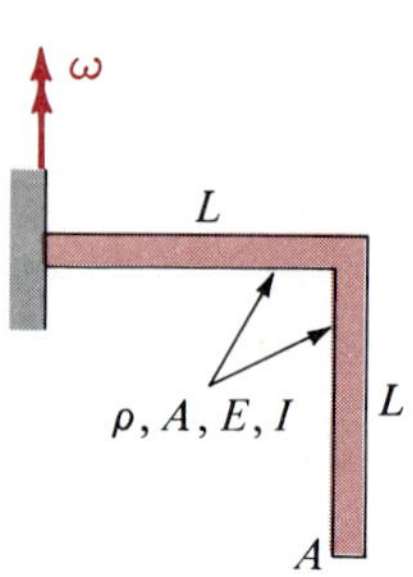

P. 10.179

10.4.4 Statically Indeterminate Structures—Principle of Least Work

The use of Engesser's first theorem or Castigliano's second theorem in connection with complementary energy is based upon the assumption that it is possible to use equilibrium to completely solve for all the reactions and subsequently to be able to represent the complementary energy in terms of the known applied loads, thus

$$U^* = U^*(P_1, P_2, \ldots, P_n)$$

The displacements can then be determined according to

$$\Delta_j = \frac{\partial U^*}{\partial P_j}$$

For a statically indeterminate problem, equilibrium still constitutes the basis of the approach used, but the number of static unknowns exceeds the available number of equilibrium equations. As discussed in Chapters 5, 6, and 8, these additional unknowns, say M in number, are termed redundants. Once the M redundants have been selected, the equations of equilibrium are solved for the remaining unknowns in terms of the redundants. Continuing to rely on equilibrium, the internal forces can then be expressed in terms of the external loads P_1, $P_2, \ldots,$ and the M redundants, which we will refer to as X_1, $X_2, \ldots, X_M$. Note that equilibrium is satisfied for any choice of the X_i.

The complementary energy can then be represented in terms of the external loads and the redundants as

$$U^* = U^*(P_1, P_2, \ldots; X_1, X_2, \ldots, X_M)$$

There remains the problem of how to select the X_i. We state without proof the **principle of least work** that provides the equations from which the M redundants can be determined:

> Of all possible choices for the redundants X_i satisfying equilibrium, the true or correct choice also satisfies compatibility, and makes the complementary energy stationary with respect to the redundants, or
>
> $$\frac{\partial U^*}{\partial X_i} = 0 \qquad i = 1, 2, \ldots, M \tag{10.32}$$

This set of M equations is then solved for the redundants. If the structure is composed of materials that are linearly elastic, the equations that result will be linear.

After solving Eqns. (10.32) for the redundants X_i in terms of the external loads P_k, the complementary energy can be expressed entirely in terms of the external loads P_k, and Engesser's first theorem or Castigliano's second theorem can then be used to determine displacements if necessary or desired. In what follows we will present several examples demonstrating the principle of least work.

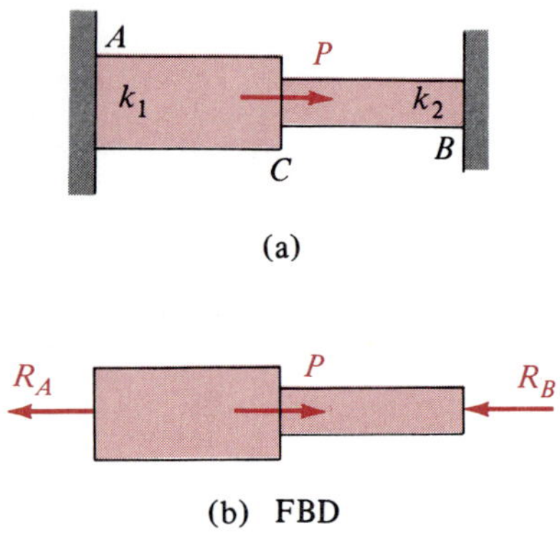

Figure 10.48 Statically indeterminate axial problem

Example 10.22 Consider the statically indeterminate problem shown in Fig. 10.48. Determine the reactions at the ends of the segmented bar and also the displacement at C.

Solution Referring to the free-body diagram of Fig. 10.48b, the single equation of equilibrium yields $R_A + R_B = P$, that is, one equation for the two unknowns R_A and R_B. Note in passing that as far as equilibrium is concerned, R_A and R_B can have any values that sum to P. Select R_A as the redundant, expressing R_B in terms of R_A as $R_B = P - R_A$. We can then write the complementary energy as

$$U^* = U_1^* + U_2^* = \frac{R_A^2}{2k_1} + \frac{R_B^2}{2k_2} = \frac{R_A^2}{2k_1} + \frac{(P - R_A)^2}{2k_2} = U^*(P; R_A)$$

that is, the complementary energy is expressed in terms of the external load P and the redundant R_A. The principle of least work is used to state that

$$\frac{\partial U^*}{\partial R_A} = 0 = \frac{R_A}{k_1} - \frac{P - R_A}{k_2}$$

from which $R_A = Pk_1/(k_1 + k_2)$. R_B can be determined from the equilibrium equation as $R_B = Pk_2/(k_1 + k_2)$. The student is asked to show in the problems that the displacement at C is given by $P/(k_1 + k_2)$.

Note that for the present example, one can interpret the principle of least work statement as

$$\Delta_A = \frac{\partial U^*}{\partial R_A} = 0$$

or, the displacement at A corresponding to the reaction R_A is zero.

Still another observation about the result

$$\frac{\partial U^*}{\partial R_A} = 0 = \frac{R_A}{k_1} - \frac{P - R_A}{k_2}$$

leads to the interpretation referred to in the principle of least work, namely, compatibility. In this regard, the term $R_A/k_1 = e_A$ represents the elongation of member 1 due to the internal force R_A. The term $e_B = (P - R_A)/k_2 = R_B/k_2$

represents the elongation of member 2 due to the internal force R_B. Thus the principle of least work equation can be written as

$$e_A - e_B = 0$$

which can be thought of as a statement of compatibility: that the elongations of the individual members must be such that the structure "fits together" after deformation. This helps to explain the part of the principle of least work statement, "also satisfies compatibility." The equations that result from enforcing the principle of least work can be interpreted, sometimes with much effort, as equations that ensure that deformations of components of the structure are compatible.

Example 10.23 As another illustration of the ideas associated with the principle of least work, consider the uniformly loaded beam shown in Fig. 10.49. Determine the reaction at the pin support.

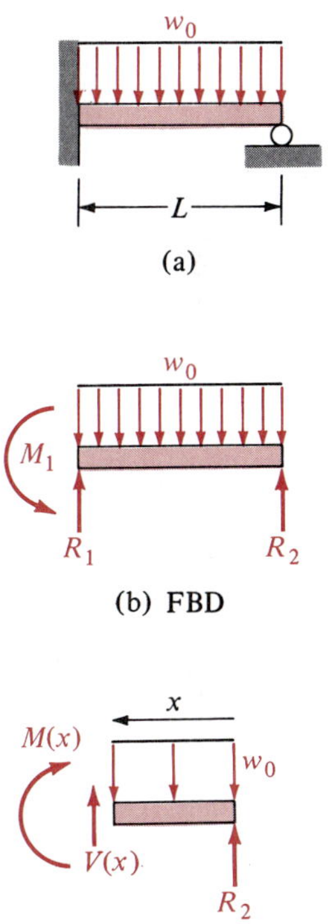

Figure 10.49 Statically indeterminate beam with one redundant

Solution From the free-body diagram of Fig. 10.49b, the equilibrium equations are $R_1 + R_2 = w_0L$ and $M_1 + R_2L = w_0L^2/2$, two equations for the three unknowns R_1, M_1, and R_2. Selecting R_2 as the redundant, for example, we can solve the equilibrium equations for the remaining unknowns R_1 and M_1 in terms of R_2 as

$$R_1 = w_0L - R_2 \qquad \text{and} \qquad M_1 = \frac{w_0L^2}{2} - R_2L$$

With x measured from the right end of the beam, the internal moment $M(x)$ can be determined using the free-body diagram of Fig. 10.49c as $M(x) = R_2x - w_0x^2/2$. The complementary energy can then be expressed as

$$U^* = \int_0^L \frac{(R_2x - w_0x^2/2)^2}{2EI}\,dx = U^*(R_2; w_0)$$

The principle of least work is applied according to

$$\frac{\partial U^*}{\partial R_2} = 0 = \int_0^L \frac{(R_2x - w_0x^2/2)(x)}{EI}\,dx = \frac{R_2L^3}{3EI} - \frac{w_0L^4}{8EI}$$

from which $R_2 = 3w_0L/8$. The equilibrium equations can then be used to determine that $R_1 = w_0L - R_2 = 5w_0L/8$ and $M_1 = w_0L^2/2 - R_2L = w_0L^2/8$. One interpretation of the equation resulting from the application of the principle of least work is simply $\Delta_A = \partial U^*/\partial R_2 = 0$, or the displacement of the point of application of the force R_2 is zero.

It is very instructive to reconsider the result

$$\frac{\partial U^*}{\partial R_2} = 0 = \frac{R_2L^3}{3EI} - \frac{w_0L^4}{8EI} \tag{a}$$

in light of Fig. 10.50, where the two problems are the cantilever loaded by the end load R_B as in Fig. 10.50a, and the cantilever loaded by the uniform load in Fig. 10.50b.

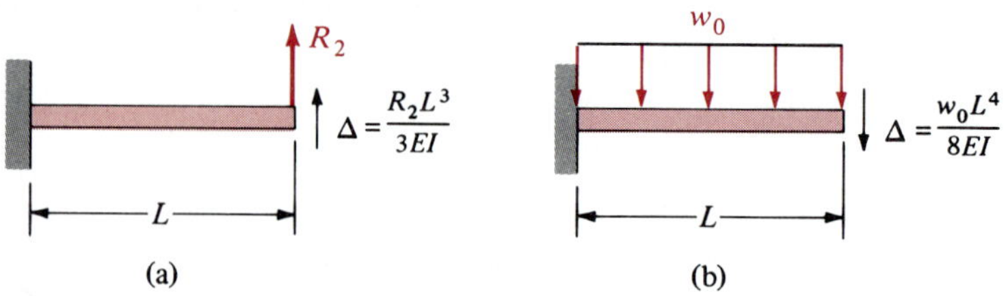

Figure 10.50 Redundant R_B determined so that $\Delta = 0$

The first term on the right side of Eqn. (a) is the displacement at $x = L$ due to the load R_B. The second term is the displacement at $x = L$ due to the uniformly distributed loading. The result of computing the partial derivative and setting it equal to zero can be interpreted as stating that of all the R_B that satisfy equilibrium, the correct value resulting in the stationary value of the complementary energy is the one that also satisfies compatibility.

Examples 10.22 and 10.23 both involve situations where the redundant is associated with a support at which the displacement is zero. It is often convenient to choose the redundant as an internal force resultant rather than a reaction. In such a situation the principle of least work continues to be valid, with the equations that result representing compatibility of the two portions of the system between which the internal force acts.

Example 10.24 To illustrate, consider the two-dimensional frame shown in Fig. 10.51a. Considering the internal moment M^* at the joint shown to be the redundant, use the principle of least work to treat the statically indeterminate problem. Neglect axial energy.

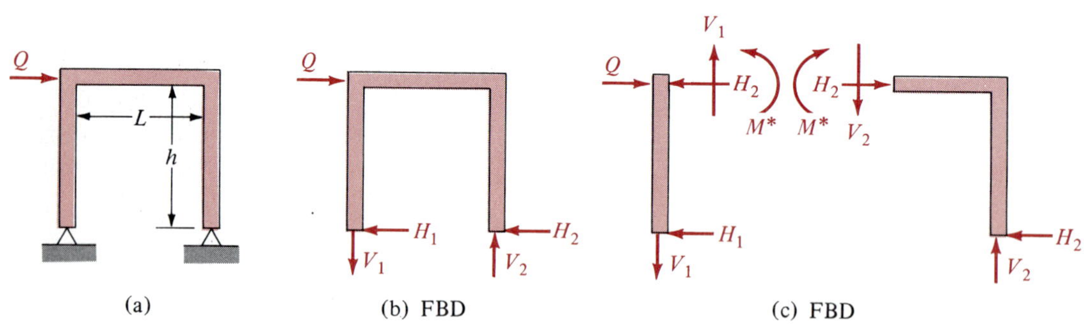

Figure 10.51 Statically indeterminate planar truss

Solution Using the free-body diagram of Fig. 10.51b, the first step is to write the three equations of equilibrium as

$$H_1 + H_2 = Q \qquad V_1 - V_2 = 0 \qquad V_2 L - hQ = 0$$

for determining the four unknowns H_1, H_2, V_1, and V_2. From the second and third equations, $V_1 = V_2 = Qh/L$. We can choose either of the reactions H_1 or H_2 as the redundant, with the other reaction determined on the basis of the first equation. The complementary energy could then be expressed in terms of whichever redundant has been chosen and the principle of least work used in the normal way, with $\partial U^*/\partial H_1 = 0$ representing the condition of zero displacement of the point of application of the reaction H_1.

In this example, however, we wish to demonstrate the idea of using an internal force as the redundant. From the free-body diagrams of Fig. 10.51c, we see the relation between the moment M^* and the reaction H_1 is $M^* = hH_1$. Intent on including only the bending energy, we can express the moment in the left vertical member in terms of the redundant M^* as $M_1(x) = H_1 x = M^* x/h$, where x is measured from the bottom. With x measured to the right, we can express the moment in the horizontal segment as $M_2(x) = M^* - V_2 x = M^* - Qhx/L$. For the right vertical member, $M_3(x) = H_2 x = (Q - H_1)x = (Q - M^*/h)x$. The complementary energy is

$$U^* = U_1^* + U_2^* + U_3^* = \int_0^h \frac{(M^* x/h)^2}{2EI}\,dx + \int_0^L \frac{(M^* - Qhx/L)^2}{2EI}\,dx$$

$$+ \int_0^h \frac{(Q - M^* x/h)^2}{2EI}\,dx = U^*(M^*; Q)$$

Using the principle of least work leads, after evaluation of the integrals, to

$$0 = \frac{\partial U^*}{\partial M^*} = \frac{M^* h}{3EI} + \frac{M^* L}{EI} - \frac{QhL}{2EI} - \frac{Qh^2}{3EI} + \frac{M^* h}{3EI}$$

from which $M^* = Qh/2$. Using the equilibrium equations, we find the values of H_1 and H_2 are each determined to be $Q/2$.

PROBLEMS

10.180 In Example 10.22, substitute R_A for U^* and show that the deflection at point C is given by $\Delta_C = dU^*/dP = P/(k_1 + k_2)$.

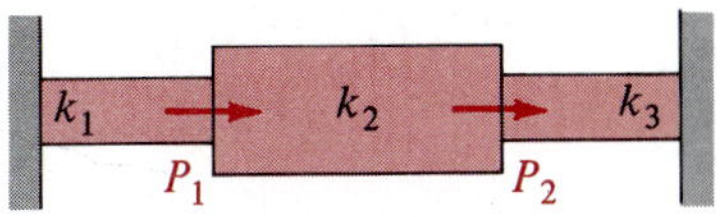

Ps. 10.181 and 10.182

10.181 Determine the internal forces in each of the members. Take the reaction at the left end to be the redundant. Use $A_1 = 2\text{ in}^2$, $A_2 = 4\text{ in}^2$, $A_3 = 3\text{ in}^2$, $L_1 = 10$ in., $L_2 = 8$ in., $L_3 = 12$ in., $E = 3 \times 10^7$ psi for all members, $P_1 = 20$ kips, and $P_2 = 30$ kips.

10.182 Repeat problem 10.181 with $A_1 = 1.2 \times 10^{-3}\text{ m}^2$, $A_2 = 2.2 \times 10^{-3}\text{ m}^2$, $A_3 = 1.8 \times 10^{-3}\text{ m}^2$, $L_1 = 0.27$ m, $L_2 = 0.2$ m, $L_3 = 0.3$ m, $E = 200$ GPa for all members, $P_1 = 80$ kN, and $P_2 = 130$ kN. Take the internal force in the middle member to be the redundant.

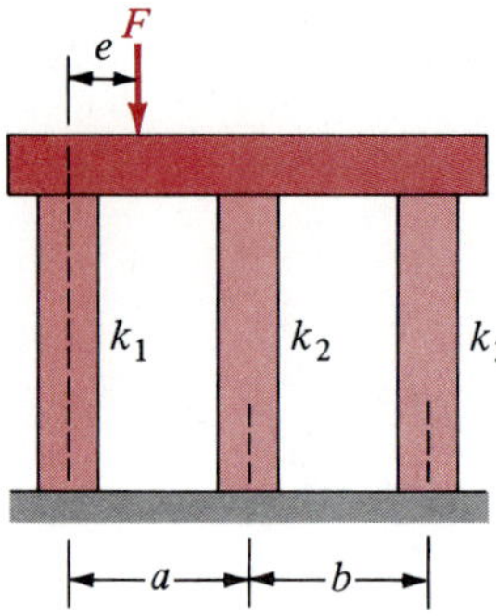

P. 10.183

10.183 The three axial members support the load F applied to the rigid plate as shown. Take $a = b$, $e = 0.6b$, $k_1 = k_2 = k_3$, and determine the forces in each of the members.

10.184 A drilling platform supports a weight W located at the coordinates $x = 0.27H$ and $y = 0.18B$. The xy coordinate system is located at the center of the platform. Determine the forces in each of the four symmetrically placed, equal-stiffness support piles. Take W = 370 tons, H = 240 ft, and B = 320 ft.

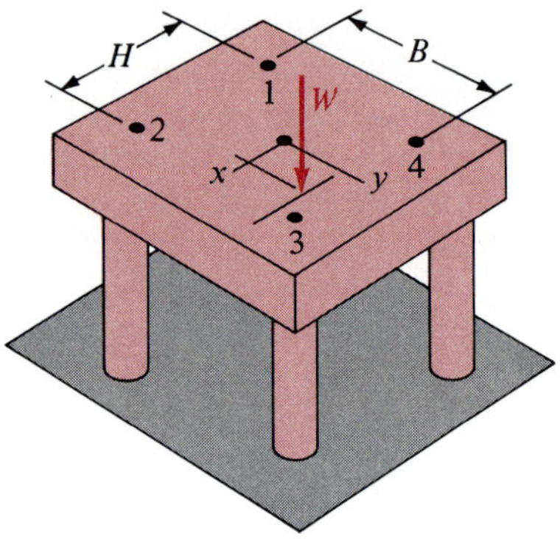

P. 10.184

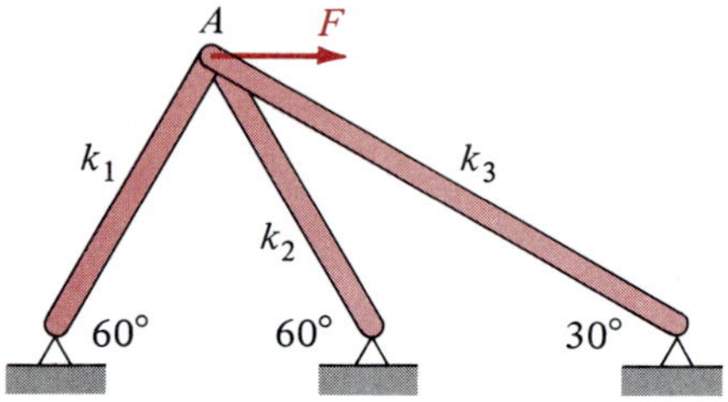

P. 10.185

10.185 Determine *(a)* the forces in the three-bar truss shown, and *(b)* the vertical and horizontal components of displacement at A. Take $A_1 = 8 \times 10^{-3}$ m^2, $A_2 = 12 \times 10^{-3}$ m^2, $A_3 = 18 \times 10^{-3}$ m^2, $L_1 = 3$ m, $E = 200$ GPa for all members, and $F = 600$ kN.

10.186 Determine the forces in the members of the truss shown. Take $A = 2$ in^2, $E = 30 \times 10^6$ psi for all members, and $Q = 80$ kips.

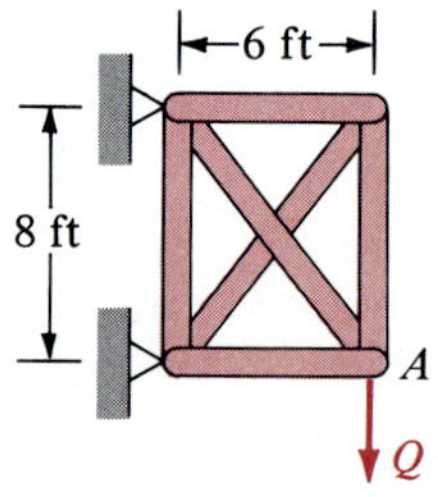

P. 10.186

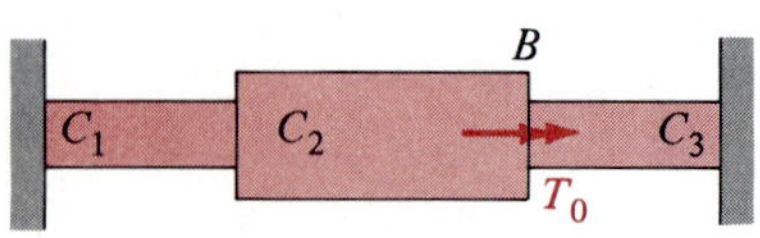

P. 10.187

10.187 Determine *(a)* the torques transmitted by each of the torsional members, and *(b)* the rotation at point B. Take $r_1 = 15$ mm, $r_2 = 24$ mm, $r_3 = 18$ mm, $L_1 = 0.3$ m, $L_2 = 0.4$ m, $L_3 = 0.25$ m, $G = 29$ GPa for all members, and $T_0 = 1.6$ kN·m.

10.188 Determine the torques carried by shafts 1 and 2. Take $C_1 = C_2 = C_3$ and $R_1 = R_2$.

10.189 Repeat problem 10.188 if $C_1 = \alpha C_2$ and $R_1 = \beta R_2$. Take $\alpha = 0.8$ and $\beta = 1.25$.

10.190 Determine what percentage of the torque T_0 is carried by each shaft. Take $C_1 = C_2 = C_3 = C_4$ and $R_1 = R_2 = R_3 = R_4$.

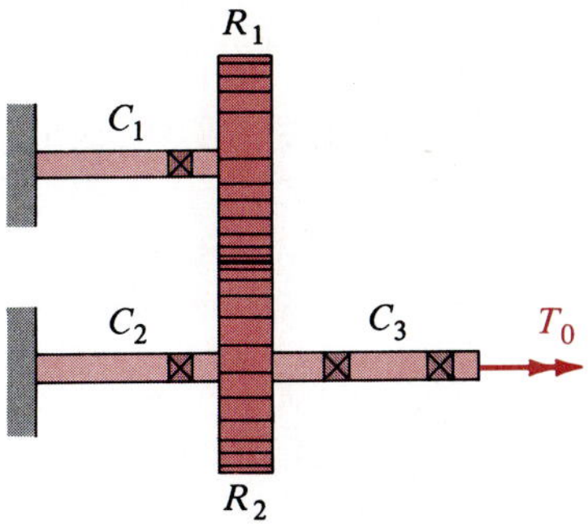

Ps. 10.188 and 10.189

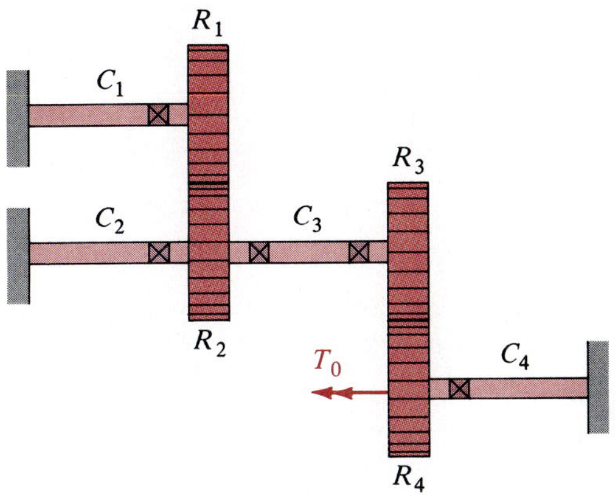

P. 10.190

10.191 Determine *(a)* the reactions at both supports, and *(b)* the displacement at the right end of the beam shown.

10.192 Determine *(a)* the reactions at both supports, and *(b)* the displacement at the load P for the beam shown. Let $a = 0.6L$.

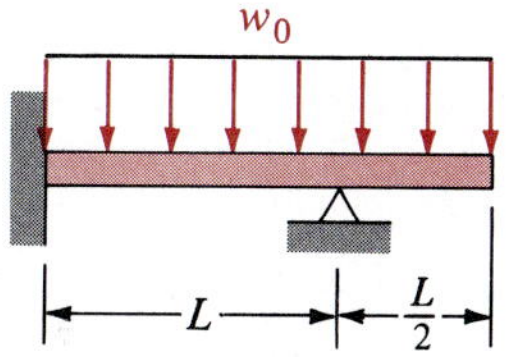

P. 10.191

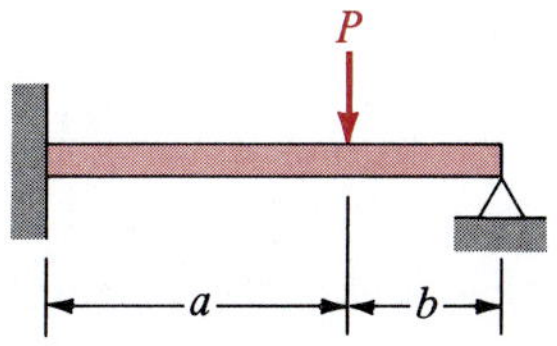

P. 10.192

10.193 Determine *(a)* the reactions at both supports, and *(b)* the displacement at the load P for the beam shown. Let $a = 0.4L$.

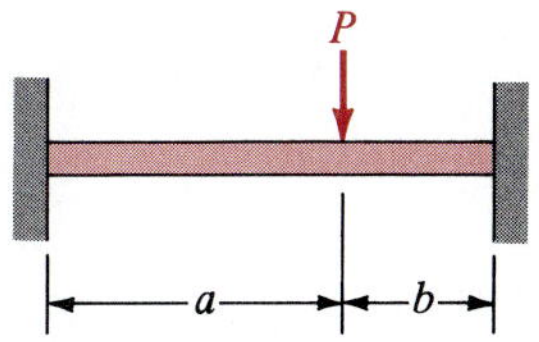

P. 10.193

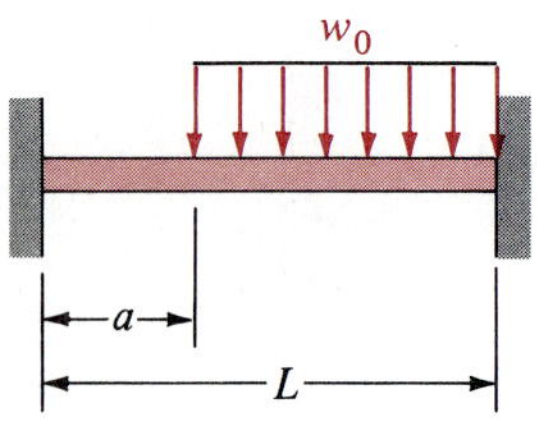

P. 10.194

10.194 Determine *(a)* the reactions at both supports, and *(b)* the displacement at the distance a from the left support. Let $a = L/3$.

10.195 Determine the force in the spring and the displacement at C.

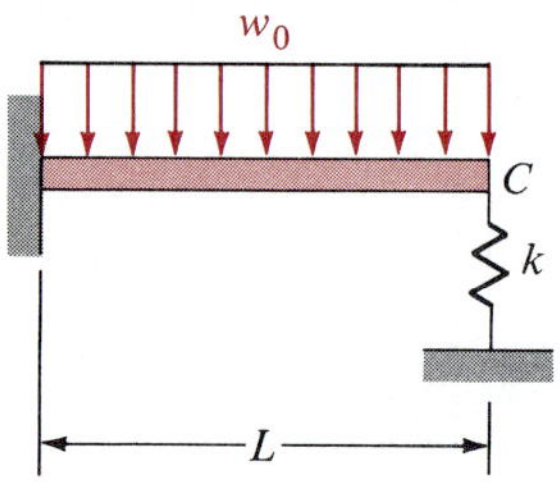

P. 10.195

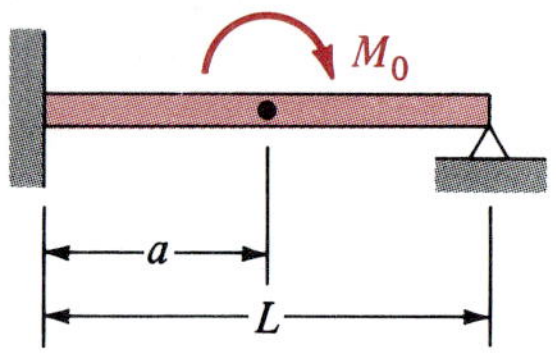

P. 10.196

10.196 Determine *(a)* the reactions at both supports, and *(b)* the rotation at the distance a from the left support. Let $a = 0.6L$.

10.197 The uniformly loaded beam is supported at the right end by a pin and a linear torsional spring; thus, the rotation at the right end is resisted by a moment that is proportional to the rotation according to $M = C\theta$. Determine *(a)* the reactions at the supports, and *(b)* the angle θ through which the spring rotates. Take $C = 2EI/L$.

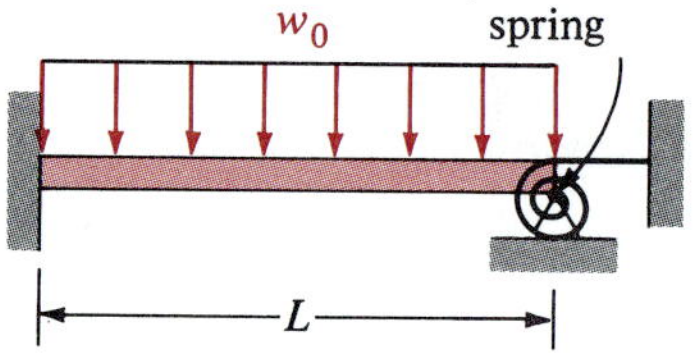

P. 10.197

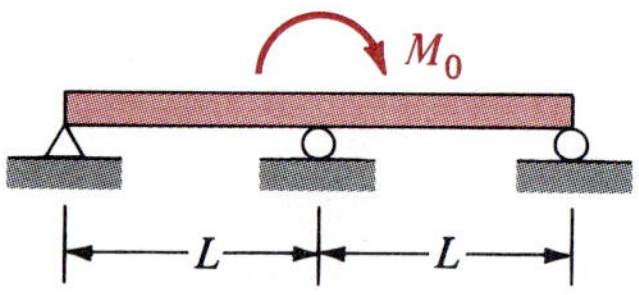

P. 10.198

10.198 Determine *(a)* the reactions at the supports, and *(b)* the rotation at the middle support.

10.199 Determine the reactions at the supports for the two-span beam shown.

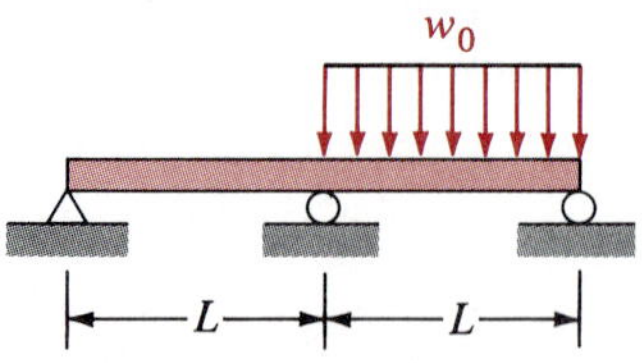

P. 10.199

10.200 Determine the reactions at the supports for the two-span beam shown.

10.201 For the variable-cross-section beam shown, determine the reactions at the supports.

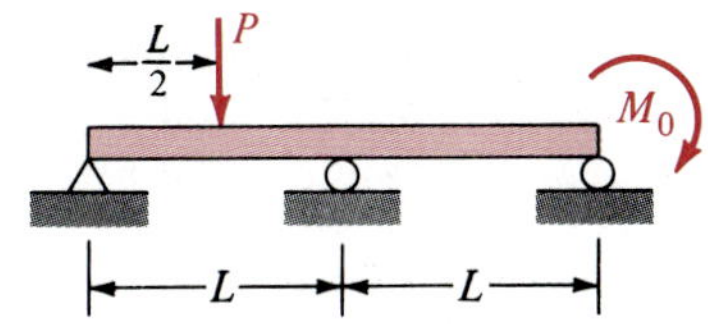

P. 10.200

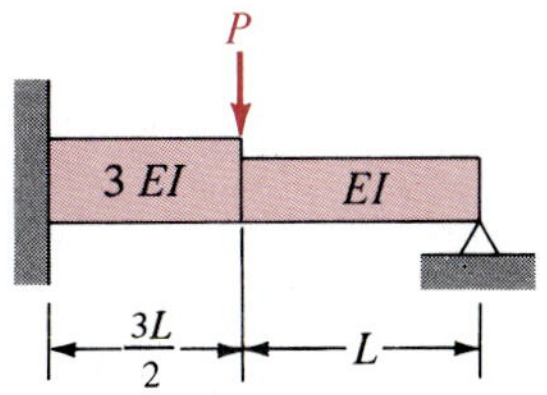

P. 10.201

10.202 Determine the reactions at the supports. Neglect axial energy. Take $H = 1.5L$, and $h = 0.5L$.

10.203 Determine the reactions at the supports. Neglect axial energy. Take $H = 2.0L$, and $h = 0.75L$.

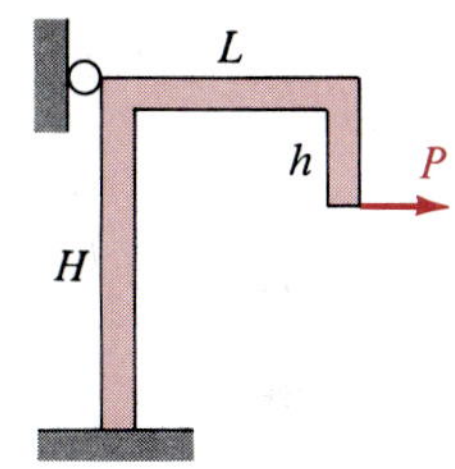

P. 10.202

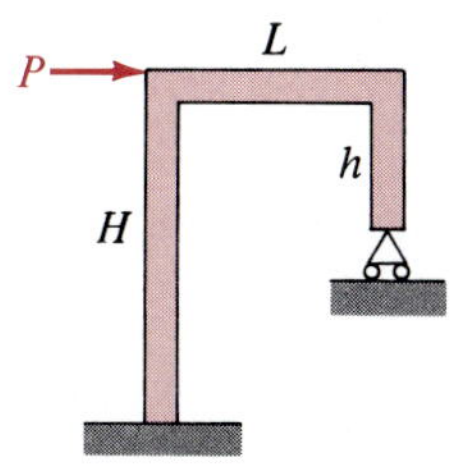

P. 10.203

10.204 Determine the reactions at the supports. Neglect axial energy. Take $H = 2.0L$, and $h = 0.75L$.

10.205 Determine *(a)* the reactions at the supports, and *(b)* the maximum internal moment for the wind-loaded frame shown. Neglect axial energy and take $H = L/3$.

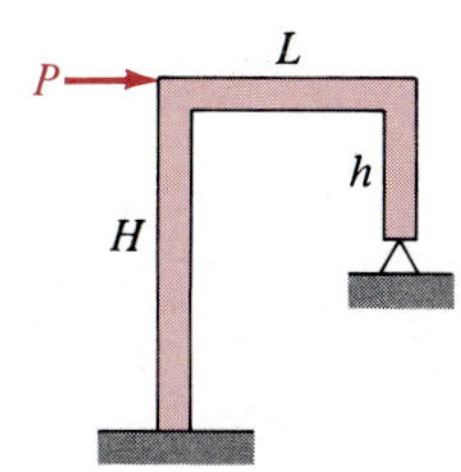

P. 10.204

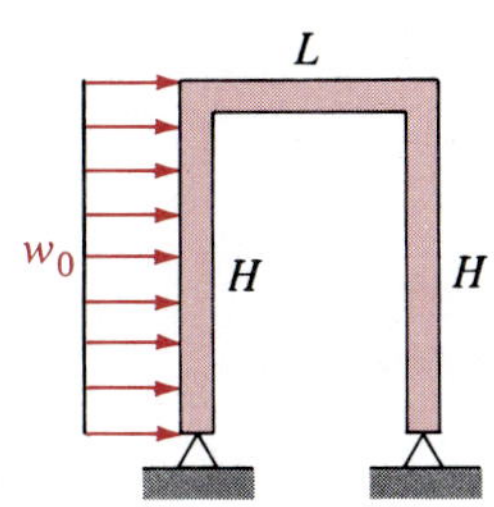

P. 10.205

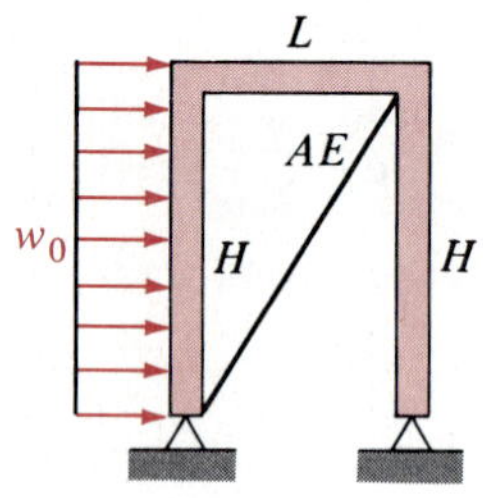

P. 10.206

10.206 Resolve problem 10.205 when a tension member is placed between two points of the structure as shown in figure P. 10.206. Take $AL^2 = 1000I$ and $H = L$. Include axial energy.

10.207 Determine *(a)* the reactions, and *(b)* the maximum internal moment for the frame shown. Take $H = L$. Neglect axial energy.

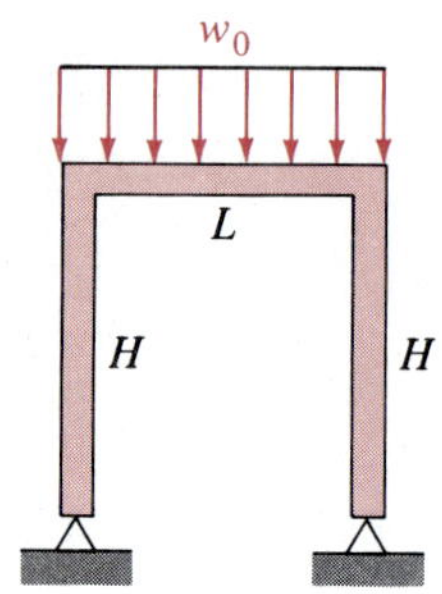

P. 10.207

10.208 Determine *(a)* the reactions and *(b)* the maximum internal moment for the frame shown. Take $H = L$ and $a = L/4$. Neglect axial energy.

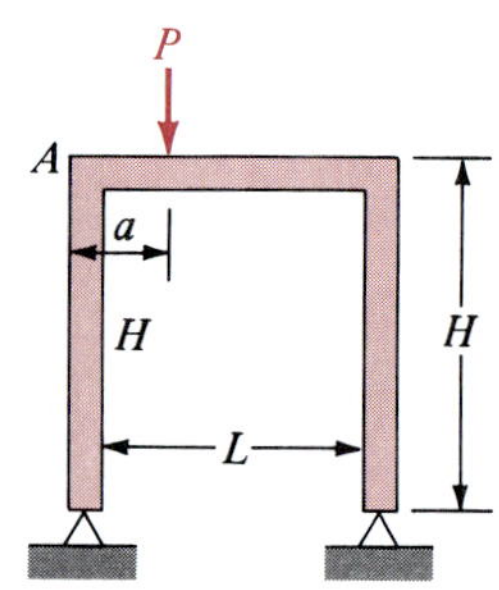

P. 10.208

10.209–10.210 Determine *(a)* the reactions at the supports and *(b)* the vertical displacement at A. Neglect axial energy.

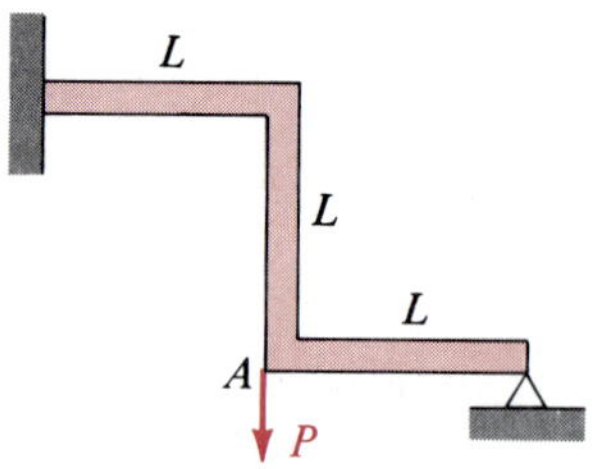

Ps. 10.209 and 10.211

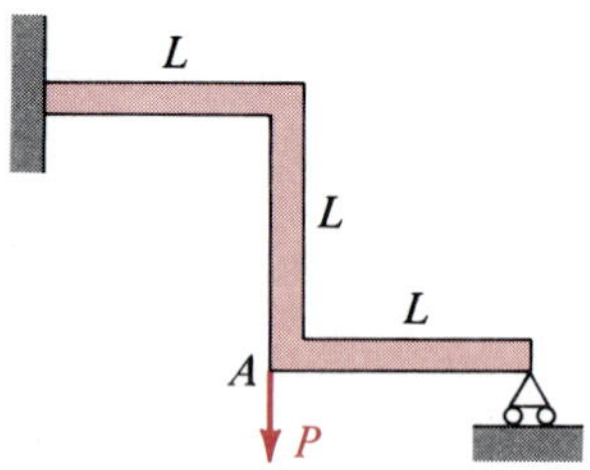

Ps. 10.210 and 10.212

10.211–10.212 Repeat problems 10.209 and 10.210 if the load P is removed and the weights of the members provide the loading.

10.213 Determine the reaction at the pin support for the structure shown. Assume each member has a circular cross section and that $E = 2.6G$.

10.214 Repeat problem 10.213 if the load P is removed and the weights of the members provide the loading.

10.215 The planar structure is restrained against all motions at A and against all translation at B. Determine the reaction at B. Assume each member has a circular cross section and that $E = 2.6G$.

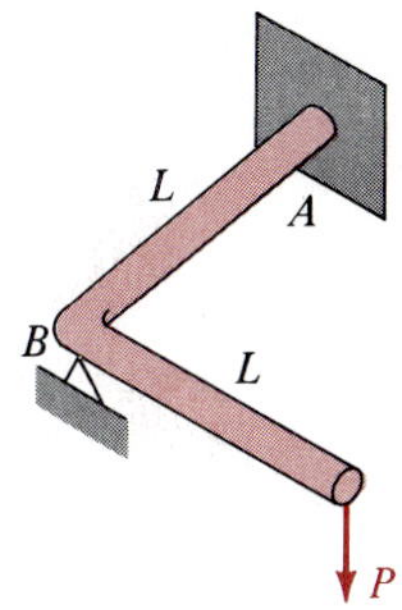

Ps. 10.213 and 10.214

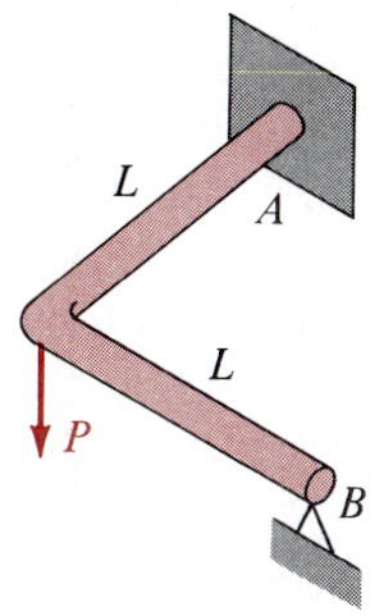

P. 10.215

10.216 The three-dimensional structure is restrained against all motion at A and against all translation at B. Determine *(a)* the reactions at the supports, and *(b)* the vertical displacement at B. Assume each member has a circular cross section and that $E = 2.6G$.

10.217 The planar structure is completely restrained against all motion at A and B. Determine *(a)* the reactions at the supports, and *(b)* the maximum internal moment. Assume each member has a circular cross section and that $E = 2.6G$.

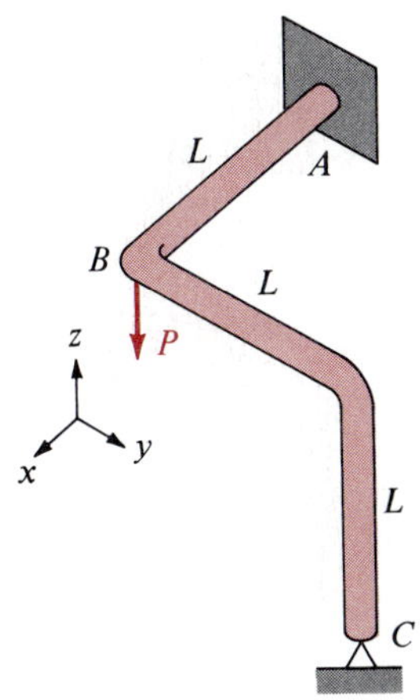

P. 10.216

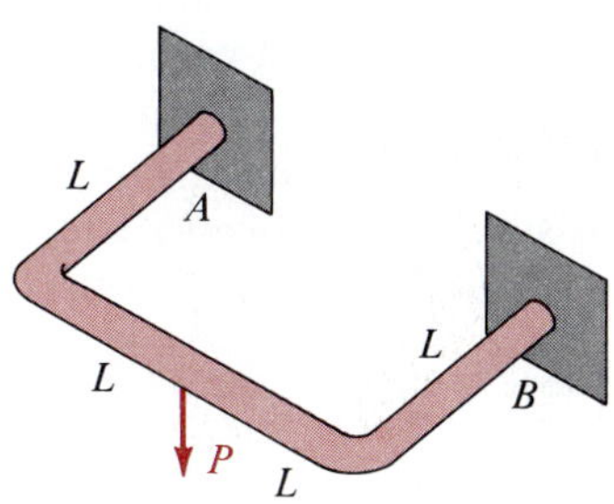

P. 10.217

10.5 CLOSURE

In this chapter we have seen that it is possible to formulate and solve problems involving elastic solids using energy concepts. As mentioned at the beginning of the chapter, the energy methods provide an alternate to the more usual vectorial approach to formulate and solve solid mechanics problems. Depending upon the information required, the energy methods for determining displacements and forces may involve less computational effort than the corresponding vectorial approach.

One basic energy approach formulates the problem beginning with the strain-energy density. The strain energy of the system is expressed in terms of a compatible set of displacements of the system, after which stationary potential energy or Castigliano's first theorem is used to generate the equations of equilibrium in terms of the displacements of the system. The equations are solved for the displacements; the internal forces can then be determined. The basic features of this approach, generally considered as a stiffness method, are indicated in the flowchart of Fig. 10.52b.

The other approach is based on the idea of the complementary-energy density. Satisfaction of equilibrium is used as the basis for determining the internal forces in the system in terms of the external loads. The complementary energy is then expressed in terms of the external loads. For statically determinate problems, either Engesser's first theorem or Castigliano's second theorem is then used to determine the displacements in terms of the external loads. The essential features of this approach, shown in Fig. 10.52a, are generally associated with ideas of flexibility.

When formulating and solving a statically indeterminate problem, the use of strain-energy ideas follows along the same path as for a statically determinate problem (i.e., Fig. 10.52b again indicates the general flow). When using complementary ideas, the principle of least work is first used to generate a set of compatibility equations that can be solved for the redundants. The displacements can then be determined in terms of the external loads using either Engesser's first theorem or Castigliano's second theorem. These steps are indicated in the flowchart of Fig. 10.53 on page 718.

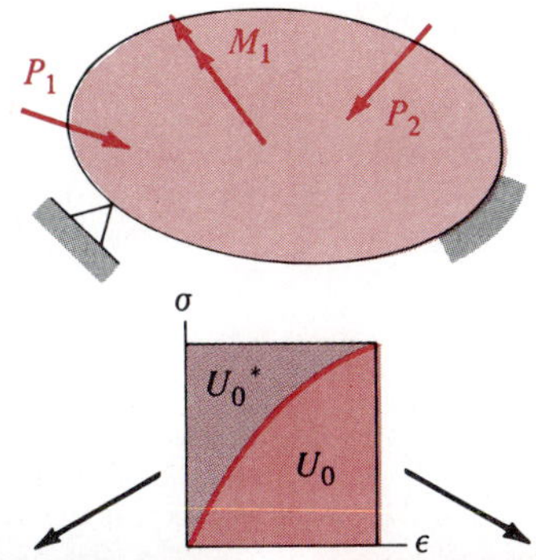

Complementary energy, flexibility formulation

Use equilibrium to solve for the internal forces in terms of the external loads P_i and hence to express the stresses in terms of the external loads P_i. Integrate the complementary energy density to get the complementary energy $U^* = \int U_0^* \, dvol = U^*(P_i)$.

↓

Use Engesser's first theorem or Castigliano's second theorem to generate the equations of compatibility according to

$$\frac{\partial U^*}{\partial P_i} = \Delta_i$$

determining the displacements in terms of the external loads.

(a)

Strain energy, stiffness formulation

Express strains in terms of a set of compatible displacements Δ_i. Integrate the strain-energy density to get the total strain energy $U = \int U_0 dvol = U(\Delta_i)$. Form total potential energy $V = U - \Sigma P_i \Delta_i$.

↓

Use the principle of stationary potential energy to generate the equations of equilibrium according to

$$\frac{\partial V}{\partial \Delta_i} = 0$$

or Castigliano's first theorem to generate the equations of equilibrium according to

$$\frac{\partial U}{\partial \Delta_i} = P_i$$

Solve these equations for the displacements.

(b)

Figure 10.52 Essential features of the strain energy and complementary energy approaches for statically determinate problems

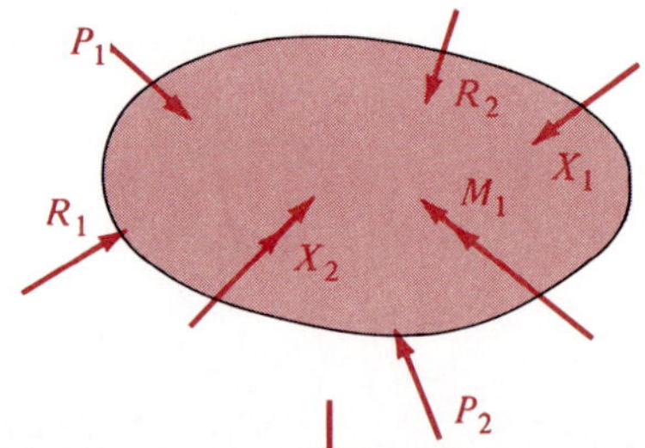

NE equations of equilibrium for NU unknowns. Solve for NE of the unknowns in terms of $M = NU - NE$ redundants X_i and the external loads P_j.

Express the complementary energy as

$$U^* = U^*(P_1, P_2, \cdots; X_1, X_2, \cdots)$$

and use the principle of least work to generate M equations of compatibility

$$\frac{\partial U^*}{\partial X_k} = 0$$

to be solved for the M redundants X_k.

Eliminate the redundants to express the complementary energy entirely in terms of the external loads and use Engesser's first theorem or Castigliano's second theorem to determine the displacements according to

$$\frac{\partial U^*}{\partial P_j} = \Delta_j$$

Figure 10.53 Essential features of the complementary energy approach for statically indeterminate problems

Observe that the strain-energy formulation begins by assuming a compatible set of displacements, and ends up with the equilibrium equations to be solved; this is known as the stiffness formulation. The complementary-energy formulation, on the other hand, begins with the satisfaction of equilibrium and ends up with the compatibility equations to be solved; this is known as the flexibility formulation. Stiffness and flexibility are, in fact, reciprocal concepts.

CHAPTER 11 STABILITY

CHAPTER OUTLINE

11.1 INTRODUCTION

In this chapter we will carry out an elementary investigation of the stability of equilibrium. In the present context, stability of equilibrium refers to what happens when an elastic system in equilibrium is disturbed slightly from that equilibrium configuration. Does the elastic system *(a)* remain in the slightly disturbed configuration, *(b)* return to the original equilibrium configuration, or *(c)* change to another nearby or faraway equilibrium configuration? These ideas are most easily discussed in terms of the classical problem of a particle which moves on a smooth surface as shown in Fig. 11.1.

For the level surface in Fig. 11.1a, the particle is in equilibrium at any point along the surface. If the particle is moved to a nearby point on the horizontal surface it will remain there since the nearby point is also an equilibrium configuration. This is referred to as **neutral equilibrium.** If the particle is displaced slightly from the equilibrium configuration at the bottom of the concave surface in Fig. 11.1b, and then released, it will eventually return to the equilibrium configuration. For the equilibrium configuration at the top of the concave surface in Fig. 11.1c, any slight disturbance will cause the particle to move permanently away from the equilibrium configuration. Cases *(b)* and *(c)* are referred to as stable and unstable equilibrium, respectively.

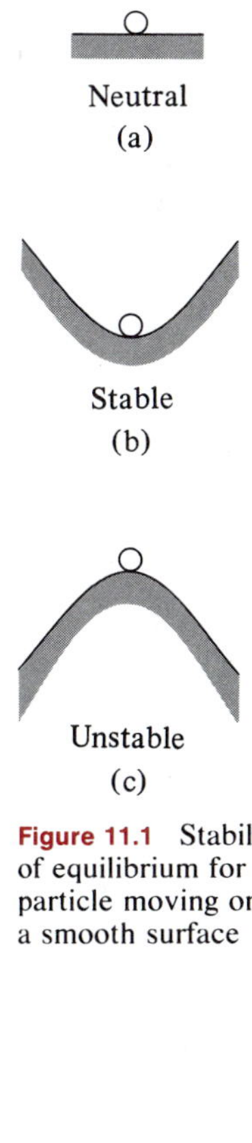

Figure 11.1 Stability of equilibrium for a particle moving on a smooth surface

Shown in Fig. 11.2 are several situations where stability of equilibrium can be important in the analysis and design of a structure. Fig. 11.2a shows a structure where two rigid bars that are initially in line, as indicated by the dotted lines, are connected by a linear torsional spring and subjected to a compressive load P. When the load reaches a certain critical level, say $P = P_{cr}$, the system may "buckle" to a nearby deflected configuration as shown. Of interest are the value of P_{cr} and whether the straight equilibrium configuration is stable under the action of the external load $P = P_{cr}$. We will determine that the value of P_{cr} depends upon the strength of the torsional spring and the lengths of the rigid bars.

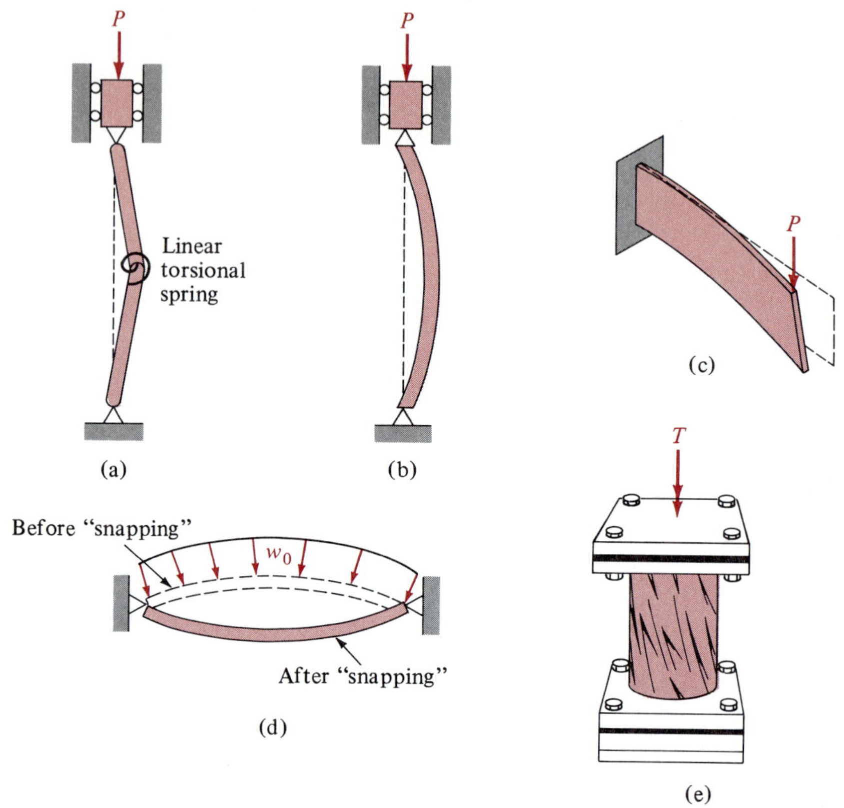

Figure 11.2 Examples of structures where stability can be a consideration

For the thin slender uniform bar of Fig. 11.2b, the phenomenon of buckling also occurs: by slowly increasing the value of the compressive load P, a value $P = P_{cr}$ is reached where the bar can also be in equilibrium in a nearby nonstraight configuration as shown. The value P_{cr} is often called the **buckling load,** in that the structure may buckle sideways as P approaches the value P_{cr}. This particular phenomenon is easily demonstrated by compressing an ordinary yardstick, an experiment the student is encouraged to perform. The value of the critical compressive force depends upon the stiffness and length of the bar.

Another instance of an instability or buckling is indicated in Fig. 11.2c, where a long uniform slender member is loaded by a shear load as shown. As the load is increased, the member may buckle sideways into a shape involving both transverse displacements and twisting about the axis of the bar. This is referred to as a lateral instability or lateral buckling. An ordinary yardstick can again be used very effectively to demonstrate this type of buckling. The value of the critical transverse force depends upon the lateral and torsional stiffnesses and upon the length of the bar.

Each of the cases shown in Figs. 11.2a, b, and c is known as a **classical** or **bifurcation instability.** The behavior of such a system is illustrated in Fig. 11.3.

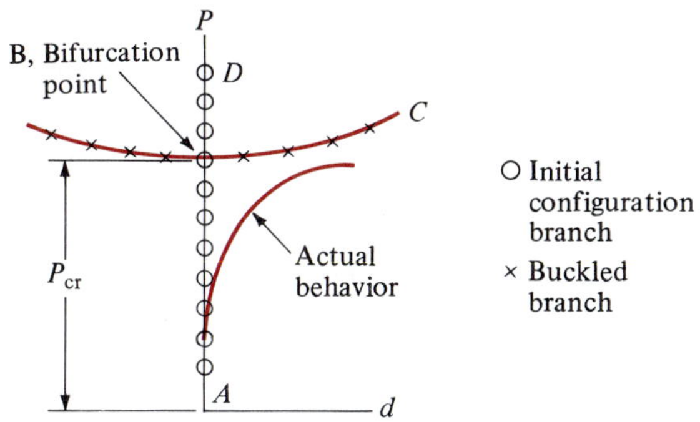

Figure 11.3 Load-displacement behavior for a bifurcation instability

The initial straight or undeformed configuration is indicated by $d = 0$, that is, the vertical axis. Nonzero values of the displacement d indicate a change in configuration from the initial or straight configuration. Upon increasing the load P gradually from zero, an *ideal* system will remain in the undeformed configuration until the critical load P_{cr} is reached, at which time buckling may occur.

This is the undeformed or initial configuration branch indicated by the circles in Fig. 11.3. According to the theory, at the bifurcation point, which is the critical load $P = P_{cr}$, the system may buckle and occupy positions on a different branch indicated by the X's, or it may be possible to increase the load beyond P_{cr} while the system remains on the undeformed branch. The point B at which the two branches meet is termed a **bifurcation point.** Emanating from the bifurcation point are two (or more in some circumstances) curves, each of which is a possible solution to the equations of equilibrium. Depending on the system under consideration, the bifurcation point B may be either stable or unstable.

Due to imperfections always present in the real world, the *actual* behavior of the system is such that there is no bifurcation or buckling, but rather, the displacements of the system from the initial or straight configuration increase gradually with an increase in the load, until there occurs a marked increase in the response as P approaches P_{cr}.

A different type of instability is demonstrated by the thin shallow arch shown in Fig. 11.2d. If the intensity of the distributed loading w_0 is gradually increased, the arch may eventually "snap through" to a completely distinct equilibrium configuration as indicated. This **snap-through** is different than the bifurcation instability of Figs. 11.2a, b, and c, in that with bifurcation buckling, all configurations in the neighbor-

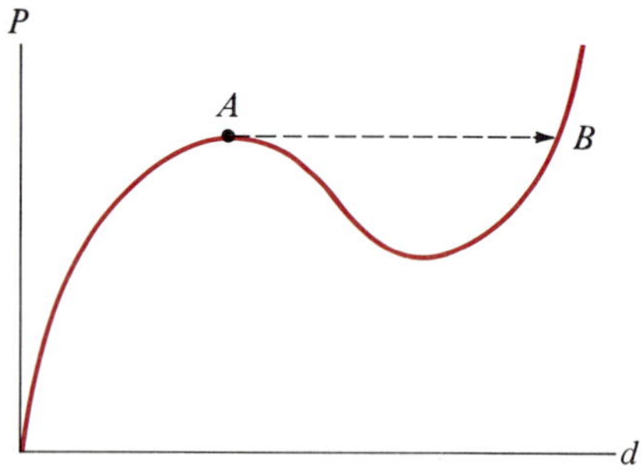

Figure 11.4 Load-deformation characteristics for snap-through buckling

hood of the initial configurations are also equilibrium configurations; whereas with the arch, the intermediate positions between the configuration at which the snap occurs and the final ''snapped configuration'' are not equilibrium configurations. The value of the critical distributed loading w_0 depends upon the stiffness and length of the bar. Another example is the snap through of the cylindrical bottom of an oil can. This particular phenomenon is frequently referred to as ''oil-canning.'' The characteristics of snap-through instability are indicated in Fig. 11.4.

As shown, there is deformation due to the loading until a critical value of the load is reached at point A. At this point the system may snap through to a completely different equilibrium configuration represented by point B. The equilibrium of the system can, for different systems, be either stable or unstable at A.

Last, consider the thin-walled cylindrical shell constructed from a linearly elastic isotropic material acted upon by a torsional moment, as shown in Fig. 11.2e. At a certain level of the torsional moment, measured by the stress τ_0, the shell will exhibit a buckling deformation consisting of a radial displacement pattern as indicated. The critical stress τ_0 depends on Young's modulus, and on both the thickness and radius of the shell. This type of instability is again a bifurcation instability.

In each of the above structures, it is most important to observe that the instability takes place in the presence of compressive stresses. This is obvious for the situations depicted in Figs. 11.2a, b, and d. The loadings in these cases result in compressive loads being transmitted throughout the structures. For the transversely loaded beam of Fig. 11.2c, the compressive stresses we refer to are the compressive bending stresses present in the lower half of the bar.

To see the presence of compressive stresses for the torsionally loaded cylindrical shell, consider the free-body diagram of a segment of the shell shown in Fig. 11.5a.

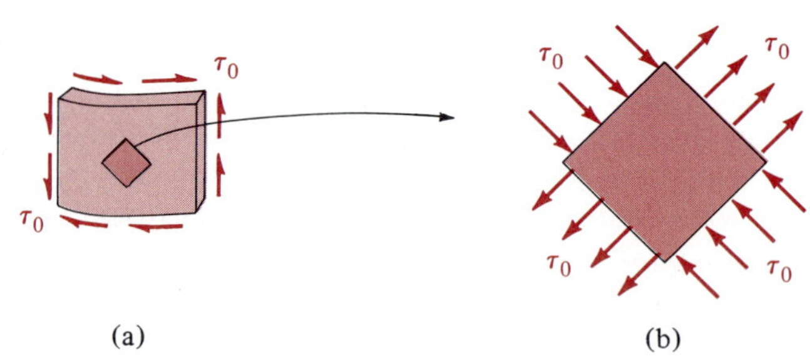

Figure 11.5 Compressive stresses on 45° planes

On a block oriented at 45° as shown in Fig. 11.5b, the stresses in one of the principal directions are clearly compressive. The directions of the waves in the buckling pattern shown in Fig. 11.2e line up nicely with the direction of the compressive stresses.

In this chapter we will be primarily concerned with the stability of straight slender bars under a variety of end conditions. We will also investigate two technically important problems: the eccentric application of a compressive load, and the problem of the bar which deviates slightly from being initially straight. These represent the realistic cases where the loading and initial geometry, respectively, are not ideal.

When analyzing a structure, the idea to keep firmly in mind is that the presence of compressive stresses should prompt the engineer to consider the *possibility* of an instability.

11.2 STABILITY OF DISCRETE SYSTEMS

The mathematical difficulties involved when we investigate the stability of any but the simplest continuous systems suggest the merit in modeling such systems as discrete rather than continuous when assessing stability. An advantage of modeling the system as discrete is that the governing equations are then algebraic rather than differential equations. In this section we will discuss the stability of discrete systems from the standpoint of an energy formulation. Fortunately, the basic stability behavior of discrete systems is very similar to that of continuous systems; thus we can gain some insight into the behavior of the corresponding continuous systems.

The starting point for our investigations is the principle of stationary potential energy. As we saw in Chapter 10, the equilibrium configurations of a conservative mechanical system can be determined by solving the equations

$$\frac{\partial V}{\partial x_i} = 0 \qquad i = 1, 2, \ldots, n$$

where V is the total potential energy of the conservative mechanical system, and x_i, $i = 1, 2, \ldots, n$, are the degrees of freedom. For a system with one degree of freedom, the stability of a particular equilibrium configuration is determined on the basis of the second derivative of the potential energy evaluated at the equilibrium configuration.

This conclusion is drawn on the basis of the expansion of the potential energy in the neighborhood of an equilibrium configuration. With x

as the single degree of freedom, and x_0 as the value of x at an equilibrium configuration, the total potential energy of a conservative mechanical system can be expressed as

$$V(x) = V(x_0) + (x - x_0)\left.\frac{dV}{dx}\right|_{x=x_0} + \frac{(x - x_0)^2}{2}\left.\frac{d^2V}{dx^2}\right|_{x=x_0} + \ldots$$

or

$$\Delta V = V(x) - V(x_0) = (x - x_0)V'(x)\Big|_{x=x_0} + \frac{(x - x_0)^2}{2} V''(x)\Big|_{x=x_0} + \ldots$$

At an equilibrium configuration $x = x_0$, the first derivative $V'(x)$ is zero, so that for small changes in position $x - x_0$, the sign of the change in the potential energy depends on

$$\Delta V = \frac{(x - x_0)^2}{2} V''(x)\Big|_{x=x_0}$$

For ΔV to be positive, meaning that for any small change in position from the equilibrium position $x = x_0$ the change in the potential energy is positive, the second derivative $V''(x_0)$ must be positive. This case is indicated in Fig. 11.6b.

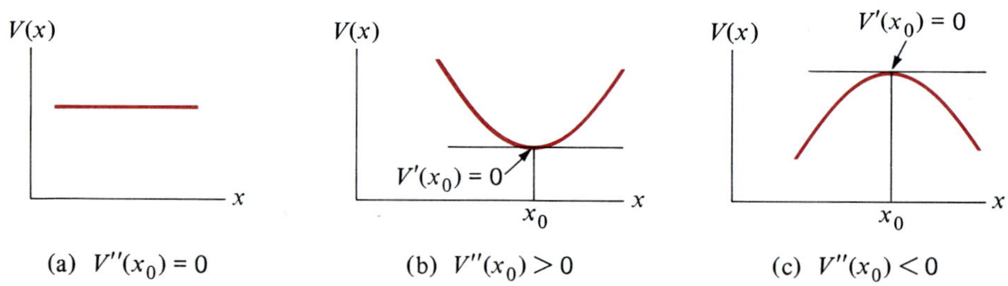

Figure 11.6 Energy criterion for stability of equilibrium

If, as indicated in Fig. 11.6c, the second derivative is negative at the equilibrium position, then the equilibrium position is unstable, since the sign of ΔV is negative for any change in position. If the second derivative is zero at the equilibrium configuration, no decision can be made without consulting the higher derivatives. This case, pictured in Fig. 11.6a is commonly called the **critical case of neutral equilibrium.**

When the second derivative vanishes at $x = x_0$, it is necessary to evaluate the higher derivatives in order to assess stability. With $V'(x_0) = V''(x_0) = 0$, write

$$\Delta V = \frac{(x - x_0)^3}{6} V'''(x_0) + \frac{(x - x_0)^4}{24} V''''(x_0) + \ldots$$

For small $x - x_0$, the sign of ΔV is governed by the term $(x - x_0)^3 V'''(x\phi)/6$. If $V'''(x_0) \neq 0$, the sign of $(x - x_0)^3 V'''(x_0)/6$ will be negative for either $x > x_0$ or $x < x_0$, that is, ΔV will be less than zero for some displacement from the equilibrium configuration. Thus if $V''(x_0) = 0$ and $V'''(x_0) \neq 0$, the equilibrium configuration is unstable. This corresponds to what is termed an inflection point of the potential energy function. If both $V''(x_0) = 0$ and $V'''(x_0) = 0$, the fourth derivative term must be evaluated. If $V''''(x_0) > 0$, the equilibrium position is stable, and if $V''''(x_0) < 0$, the equilibrium configuration is unstable. If $V''''(x_0) = 0$, higher derivatives must be considered.

It should be clear to the student that this discussion is entirely equivalent to the discussion of minima and maxima of functions of one variable, previously encountered in calculus, and that in the present class of physical problems we require that *the potential energy be a proper minimum for stable equilibrium.*

Example 11.1 Consider the problem of a rigid bar compressed by a vertical force P and connected by a horizontal spring to ground as shown in Fig. 11.7a. Determine the stability of the vertical equilibrium configuration.

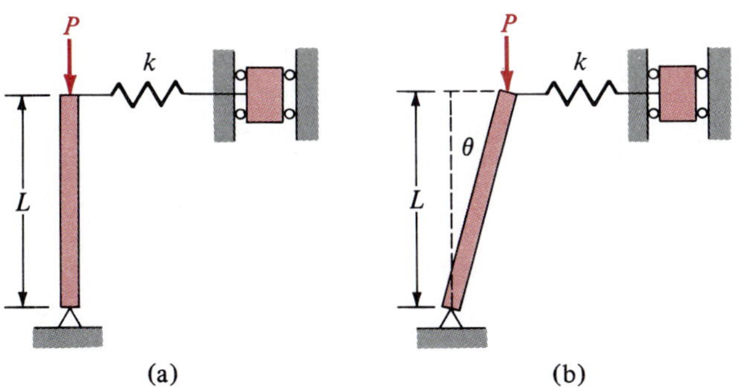

Figure 11.7 Stability of a rigid bar supported by a spring

Solution The total potential energy can be expressed in terms of the degree of freedom θ shown in Fig. 11.7b. The potential energy of the spring is $V_{sp} = (\frac{1}{2})ke^2 = (\frac{1}{2})k(L \sin \theta)^2$. With $L(1 - \cos \theta)$ as the vertical distance traveled by the load P when the bar rotates through the angle θ, the potential energy of the constant external load P is $V_P = -PL(1 - \cos \theta)$, so that

$$V = (\tfrac{1}{2})k(L \sin \theta)^2 - PL(1 - \cos \theta)$$

Setting the first derivative equal to zero yields

$$\frac{dV}{d\theta} = kL^2 \sin \theta \cos \theta - PL \sin \theta = 0$$

or

$$L \sin \theta(kL \cos \theta - P) = 0$$

The possible equilibrium configurations arise from $\sin \theta = 0$ and $kL \cos \theta - P = 0$. The root $\theta = 0$ arising from $\sin \theta = 0$ is the initially straight equilibrium configuration of interest to us. Note that $\theta = 0$ is a root irrespective of the value of P (i.e., for any value of P).

We wish to determine the range of values of P for which the equilibrium configuration $\theta = 0$ is stable. This is accomplished by computing the second derivative of the potential energy and then evaluating it at $\theta = 0$. This yields

$$V''(\theta) = kL^2(\cos^2 \theta - \sin^2 \theta) - PL \cos \theta$$

with

$$V''(0) = kL^2 - PL = L(kL - P)$$

The critical value of P, namely the value of P for which $V''(0) = 0$, is $P_{cr} = kL$. Based on the sign of $V''(0)$, the equilibrium configuration $\theta = 0$ is

stable if $P < kL$
unstable if $P > kL$

If $P = P_{cr} = kL$, $V''(0) = 0$ and the critical case of neutral equilibrium occurs. To determine whether this critical case is stable we must investigate the higher derivatives evaluated at $\theta = 0$. Computing V''' yields

$$V'''(\theta) = -2kL^2 \sin 2\theta + PL \sin \theta$$

with $V'''(0) = 0$, so that the fourth derivative must be considered. It follows that

$$V''''(\theta) = -4kL^2 \cos 2\theta + PL \cos \theta$$

with

$$V''''(0) = -4kL^2 + PL$$

At the critical load (i.e., $P = kL$), $V''''(0) = -3kL^2 < 0$; thus the straight equilibrium configuration $\theta = 0$ is unstable. Equilibrium configurations for $\theta \neq 0$ are determined from the other factor, namely, $kL \cos \theta = P - 0$. The student is asked, in the problems, to show that these equilibrium configurations are all unstable.

Information regarding the stability of the original straight equilibrium configuration and of the nearby equilibrium configurations $P = kL \cos \theta$ is shown in Fig. 11.8.

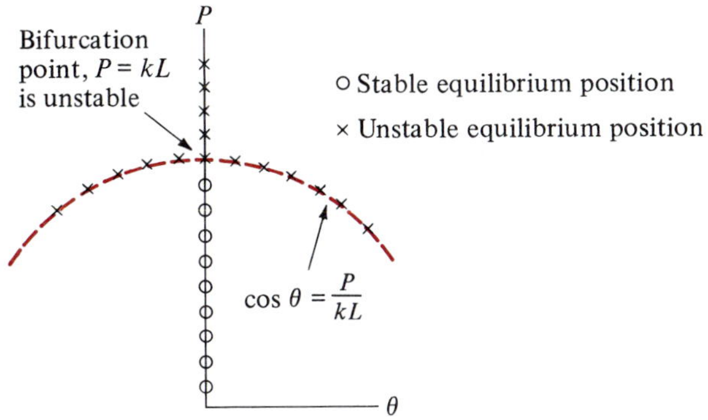

Figure 11.8 Stability of various equilibrium configurations

It is worth noting that all of the nonstraight equilibrium configurations are unstable; this means that if the bar is placed in the position $\cos \theta = P/kL < 1$ and disturbed slightly, it will not return to its original position. A thorough investigation of the problem shows that when disturbed from a nonstraight equilibrium configuration, the bar will snap through to a completely different nonstraight equilibrium configuration.

Example 11.2 A simply supported beam is to be modeled as two rigid bars connected by a torsional spring C as shown in Fig. 11.9. Determine the critical load for the column.

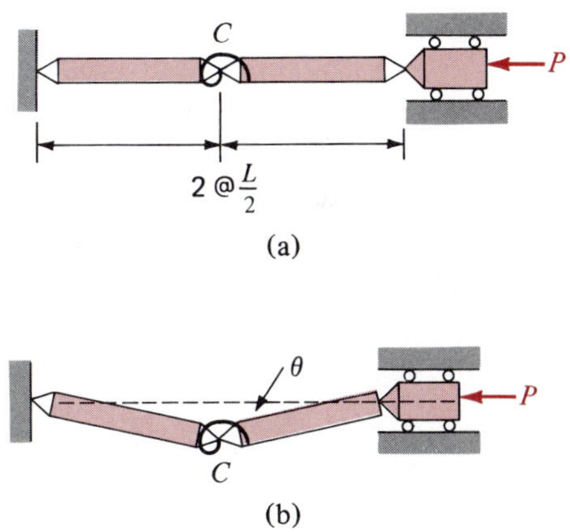

Figure 11.9 Stability of a bar-spring system

Solution Consulting Fig. 11.9b, the total potential energy consists of the energy of the spring and the energy of the load P, written as

$$V(\theta) = \frac{1}{2} C(2\theta)^2 - 2\left[\frac{PL}{2}\left(1 - \cos\theta\right)\right] = 2C\theta^2 - PL(1 - \cos\theta)$$

The equilibrium configurations are given by solutions of

$$V'(\theta) = 4C\theta - PL \sin\theta = 0 \qquad (a)$$

One root, representing the original straight configuration, is $\theta = 0$. Note that if $\theta = 0$, Eqn. (a) is satisfied for any P. The roots for nonzero θ are given by

$$\frac{\theta}{\sin\theta} = \frac{PL}{4C} \qquad (b)$$

The stability of the equilibrium configuration $\theta = 0$ is determined by considering

$$V''(\theta)|_{\theta=0} = (4C - PL\cos\theta)|_{\theta=0} = 4C - PL$$

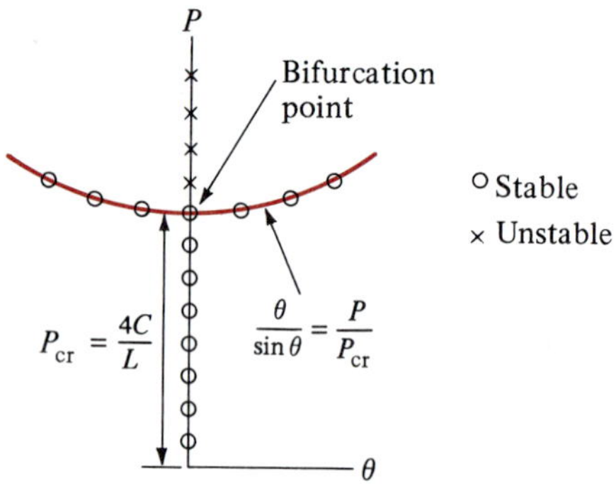

Figure 11.10 Stability information for spring-connected bars

leading to the critical value of P as $P_{cr} = 4C/L$. If $P < P_{cr}$, the straight configuration is stable, and if $P > P_{cr}$ the equilibrium is unstable. If $P = P_{cr}$, we have the critical case of neutral equilibrium and must evaluate the higher derivatives. The higher derivatives are readily evaluated at the origin as

$$V'''(\theta)|_{\theta=0} = PL \sin \theta|_{\theta=0} = 0$$

and

$$V''''(\theta)|_{\theta=0} = PL \cos \theta|_{\theta=0} = PL$$

With $V'''(0) = 0$ and $V''''(0) > 0$, the straight configuration at $\theta = 0$ with $P = 4C/L$ is stable, that is, the bifurcation point is stable. For this example the stability information is shown in Fig. 11.10.

It is left in the problems for the student to show that the nonzero equilibrium configurations given by Eqn. (b) are stable.

The energy ideas used to analyze the single degree of freedom systems considered in Examples 11.1 and 11.2 are equally applicable to systems with more than one degree of freedom. The determination of the critical loads can be carried out in a very straightforward manner by paralleling the steps outlined for the one degree of freedom system. Determining the actual stability of one of the equilibrium configurations at which an instability occurs is somewhat more detailed and will not be covered in this text.

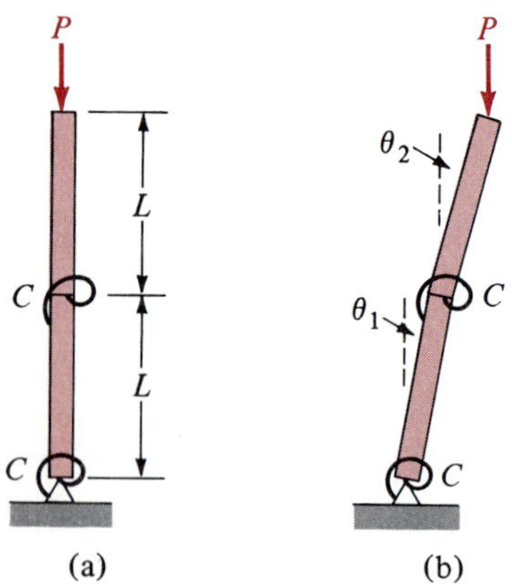

Figure 11.11 Critical loads for a system with two degrees of freedom

Example 11.3 Two bars and two torsional springs are connected as shown in Fig. 11.11a. Determine the critical loads.

Solution The system has two degrees of freedom, indicated by θ_1 and θ_2 in Fig. 11.11b. The total potential energy is composed of the potential energy of the springs and of the external load P. These are

$$V_{sp} = \frac{1}{2} C\theta_1^2 + \frac{1}{2} C(\theta_2 - \theta_1)^2$$

and

$$V_P = -PL(1 - \cos \theta_1) - PL(1 - \cos \theta_2)$$

respectively, so that

$$V(\theta_1, \theta_2) = \frac{1}{2} C\theta_1^2 + \frac{1}{2} C(\theta_2 - \theta_1)^2 - PL(1 - \cos \theta_1) - PL(1 - \cos \theta_2)$$

The two equilibrium equations are

$$\frac{\partial V}{\partial \theta_1} = C\theta_1 + C(\theta_2 - \theta_1)(-1) - PL \sin \theta_1 = 0$$

and

$$\frac{\partial V}{\partial \theta_2} = +C(\theta_2 - \theta_1) - PL \sin \theta_2 = 0$$

which can be written as

$$2C\theta_1 - C\theta_2 - PL \sin \theta_1 = 0 \qquad (a)$$

and

$$-C\theta_1 + C\theta_2 - PL \sin \theta_2 = 0 \qquad (b)$$

It is clear from these equations that $\theta_1 = \theta_2 = 0$ is a solution of the equilibrium equations for any value of P. We are interested in possible solutions for small nonzero values of θ_1 and θ_2 in the neighborhood of $\theta_1 = \theta_2 = 0$. These can be obtained by using the small-angle representations for the trigonometric functions in Eqns. (*a*) and (*b*) (i.e., by replacing $\sin \theta_1$ by θ_1 and $\sin \theta_2$ by θ_2), to obtain

$$(2C - PL)\theta_1 - C\theta_2 = 0 \qquad (c)$$

and

$$-C\theta_1 + (C - PL)\theta_2 = 0 \qquad (d)$$

This set of equilibrium equations is referred to as the **linearized equations of equilibrium.** Let us define the dimensionless term $\lambda = PL/C$ and rewrite these linearized equilibrium equations as

$$(2 - \lambda)\theta_1 - \theta_2 = 0 \qquad (e)$$

and

$$-\theta_1 + (1 - \lambda)\theta_2 = 0 \qquad (f)$$

This is a classic linear algebraic eigenvalue problem, where the nonzero solutions are forced by requiring that the determinant

$$\begin{vmatrix} (2 - \lambda) & -1 \\ -1 & (1 - \lambda) \end{vmatrix} = 0$$

from which

$$1 - 3\lambda + \lambda^2 = 0$$

The roots of this equation are $\lambda_1 = (3 - \sqrt{5})/2$ and $\lambda_2 = (3 + \sqrt{5})/2$, with the critical loads given by $P_1 = C\lambda_1/L = 0.382C/L$ and $P_2 = C\lambda_2/L = 2.618C/L$.

When P has either of these two values, it is possible for θ_1 and θ_2 to be nonzero. We determine these nonzero values by back-substituting the values for each of the critical loads into Eqns. (*c*) and (*d*). For the lower critical load, this results in

$$\left[2 - \frac{(3 - \sqrt{5})}{2}\right] C\theta_1 - C\theta_2 = 0$$

and

$$-C\theta_1 + \left[1 - \frac{(3 - \sqrt{5})}{2}\right] C\theta_2 = 0$$

or

$$\left[\frac{(1 + \sqrt{5})}{2}\right] \theta_1 - \theta_2 = 0$$

and

$$-\theta_1 + \left[\frac{(-1 + \sqrt{5})}{2}\right] \theta_2 = 0$$

These equations are of course identical in stating that $\theta_2 = 1.618\theta_1$ for the so-called mode shape into which the column will buckle at the corresponding critical load $P_1 = 0.382C/L$. The mode shape is indicated in Fig. 11.12a. The student is asked to show in the problems that when the larger critical load is substituted into the equilibrium equations, there results $\theta_2 = -0.618\theta_1$ at the critical load $P_2 = 2.618C/L$. This mode shape is indicated in Fig. 11.12b.

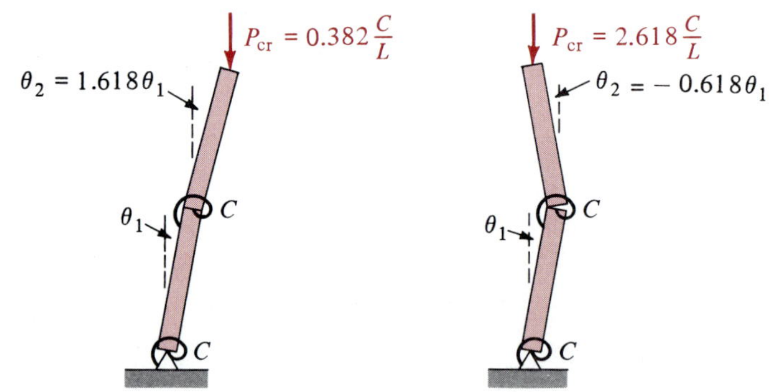

Figure 11.12 Critical loads and the corresponding mode shapes

Several comments are in order. Theoretically, if the load P is gradually increased from zero, the only solution for $P < P_1$ is $\theta_1 = \theta_2 = 0$. When P reaches P_1 the bars may remain in the straight configuration but it is also possible for the bars to buckle into the shape shown in Fig. 11.12a for arbitrary *small* values of θ_1 and θ_2. The point $P = P_1 = 0.382C/L$ for $\theta_1 = \theta_2 = 0$ is a bifurcation point. For $P_1 < P < P_2$, the equilibrium equations indicate that the only solution is $\theta_1 = \theta_2 = 0$. When P reaches the value $P_2 = 2.618C/L$, it is again theoretically possible for the bars to remain in the straight configuration or to buckle into the nonzero solution shown in Fig. 11.12b.

In practice, however, imperfections in the alignment of the bars and the line of action of the load would cause some deflection during the increase in the load P from 0 to P_1. Further, as P approaches P_1, the deflections would tend to increase at a much faster rate, making it very unlikely that the lower critical load of $P = P_1$ could actually be reached. There would be even less chance that the load could be increased beyond $P = P_1$ to ever reach $P = P_2$. The information from the linearized model and the character of the actual behavior of the system are given in Fig. 11.13.

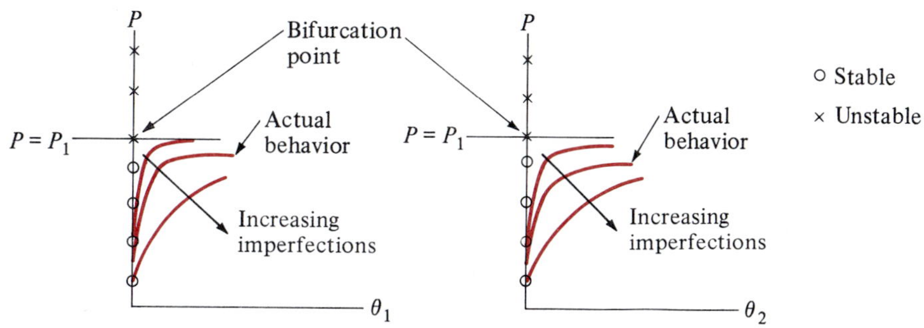

Figure 11.13 Stability information for the two-degree-freedom system

Using the linearized model, it is not possible to determine the character of the stability at the bifurcation point. It turns out for this example that the bifurcation point $P = P_1$, $\theta_1 = \theta_2 = 0$ is unstable as indicated in Fig. 11.13, and that for all $\theta_1 = \theta_2 = 0$, $P > P_1$, the system is unstable, also indicated in this figure.

PROBLEMS

11.1 Referring to Example 11.1, show that each of the equilibrium configurations given by $kL \cos \theta = P$, $\theta \neq 0$, is unstable.

11.2 Referring to Example 11.2, show that each of the equilibrium configurations given by $\theta/\sin \theta = PL/4C$, $\theta \neq 0$, is stable.

11.3–11.11 For each of the following, assume small displacements from the initial configuration (i.e., use small-angle approximations), and determine the critical load.

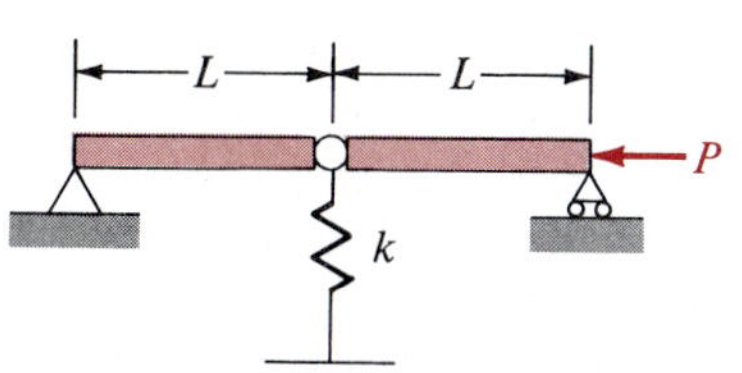

P. 11.3

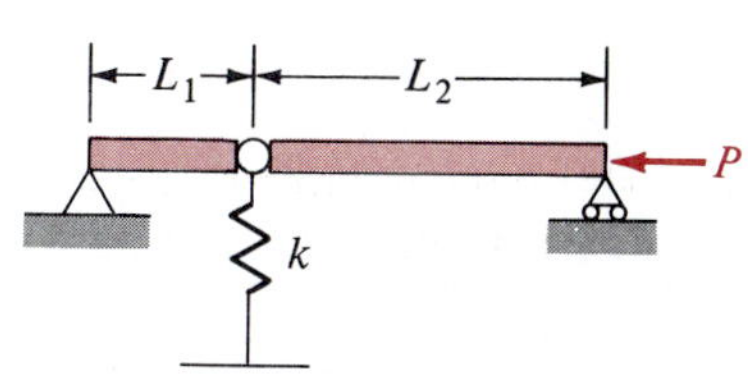

P. 11.4

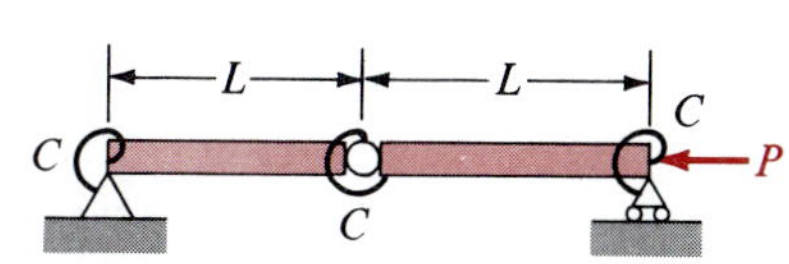

P. 11.5

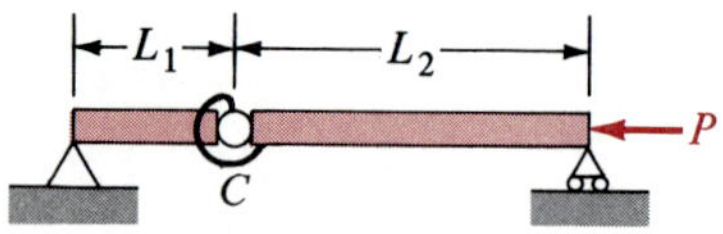

P. 11.6

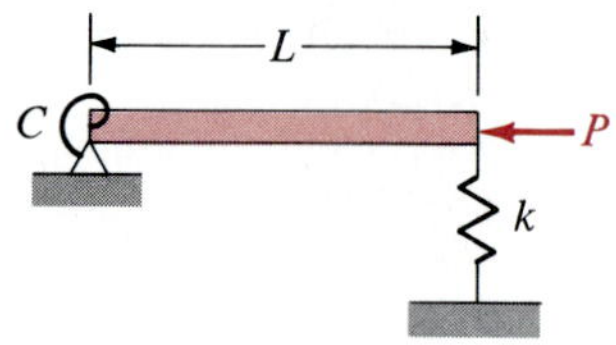

P. 11.7

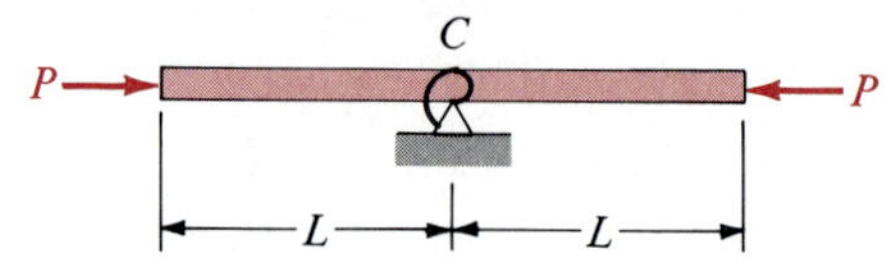

P. 11.8

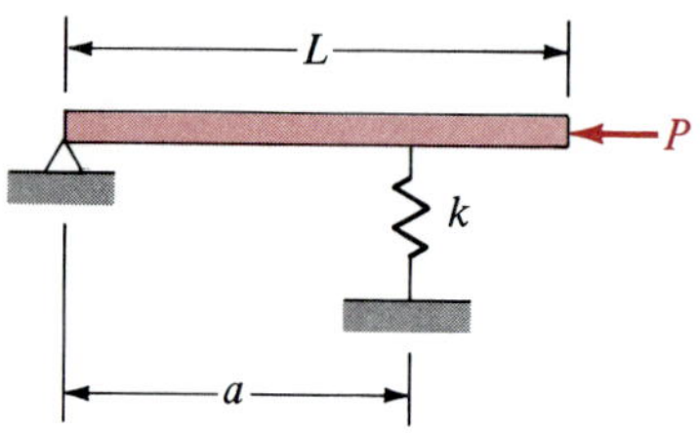

P. 11.9

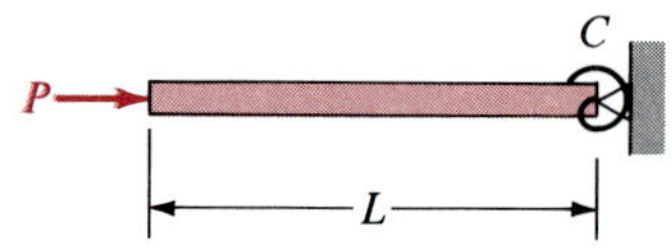

P. 11.10

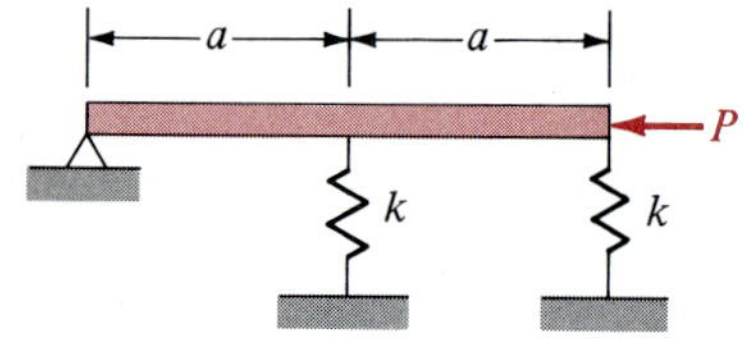

P. 11.11

11.12 Investigate the behavior of the system shown if the two bars make an angle θ_0 with the straight configuration when $P = 0$.

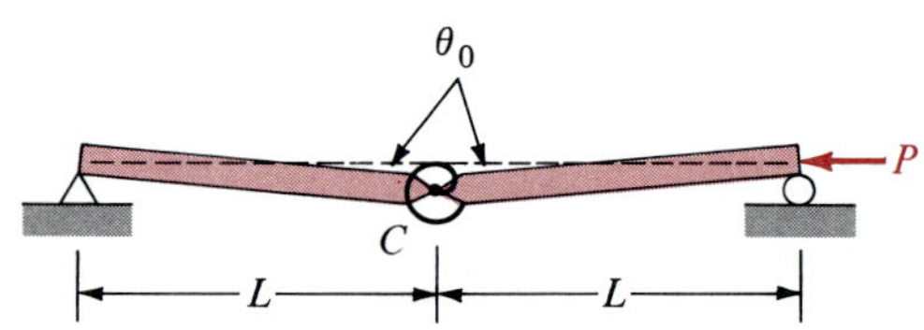

P. 11.12

11.13–11.15 For each of the following, determine the critical loads.

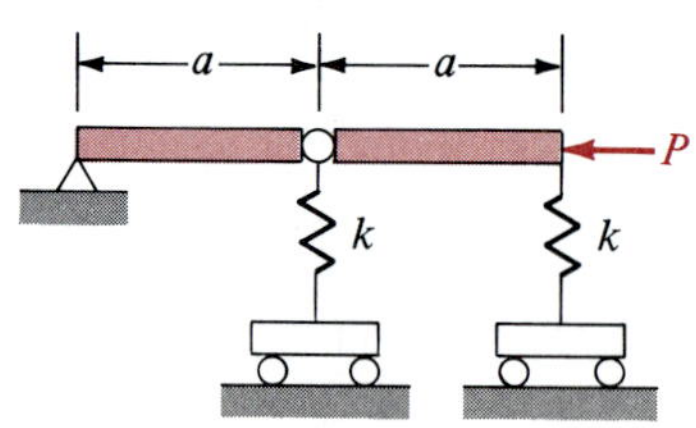

P. 11.13

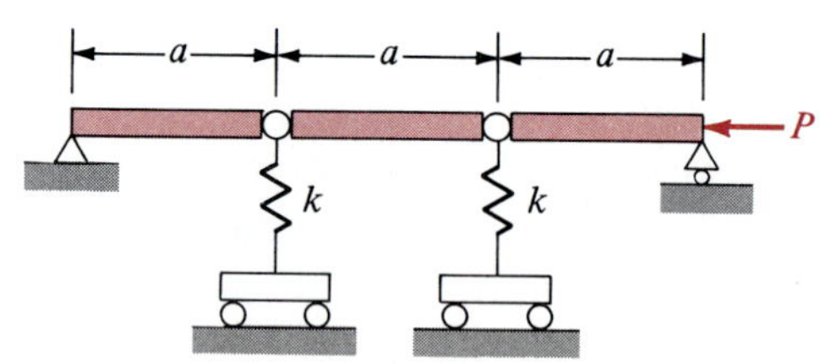

P. 11.14

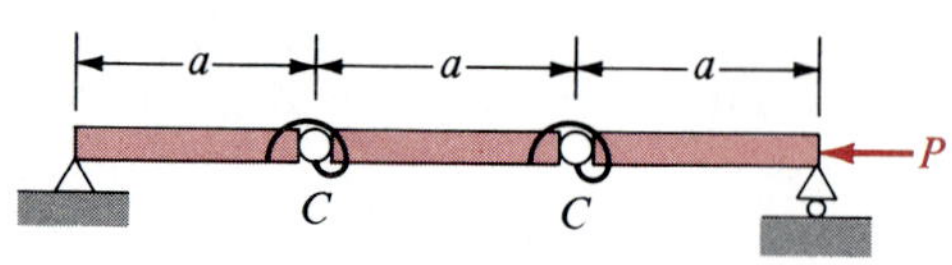

P. 11.15

11.3 COLUMNS WITH PINNED ENDS—THE EULER BUCKLING LOAD

In this section we will investigate the problem of a uniform, perfectly straight slender bar that is pinned at the ends as shown in Fig. 11.14a.

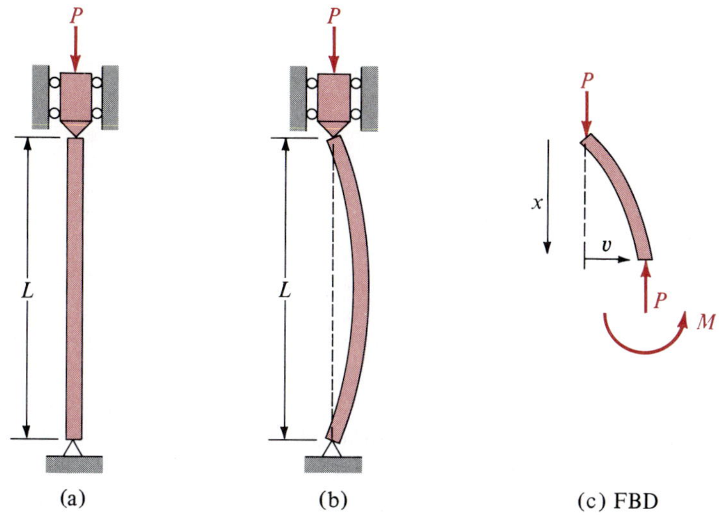

Figure 11.14 Equilibrium of a pinned-end column

The pins allow rotation about an axis perpendicular to the plane of the paper at the ends but prevent any rotation about the perpendicular direction. The cross section must be such that there is an axis of symmetry that is perpendicular to the axis of bending as shown in Fig. 11.15.

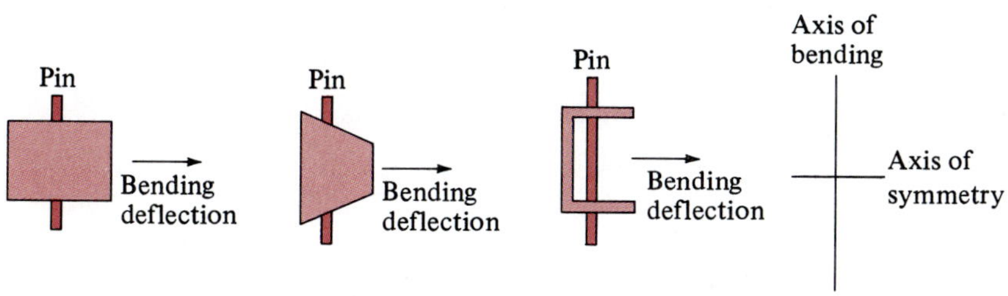

Figure 11.15 Relationship between the orientation of the cross section and the axis of bending

We will assume that the linearly elastic bar is subjected to a centrally applied compressive load P with the pin at the upper end placed on rollers to permit axial motion. The column is clearly in equilibrium in the initially straight configuration for any value of the load P. We are interested in determining whether or not the column can be in equilibrium in a nonstraight configuration under the action of the axial load P. To answer this question we must consider the equilibrium of the bar in the nonstraight configuration under the action of the load P as shown in Fig. 11.14b. A free-body diagram of a segment of the deflected bar under the action of the load P is shown in Fig. 11.14c. We use the basic moment-curvature relation $EIv'' = M$ with the free-body diagram to write

$$EIv'' = M = -Pv \tag{a}$$

or

$$EIv'' + Pv = 0 \tag{11.1}$$

where E is Young's modulus and I is the second moment of the area of the bar about an axis parallel to the axis of bending that is consistent with the rotation allowed by the pins. Eqn. (11.1) is the equilibrium equation governing the transverse displacement $v(x)$ under the action of the axial load P. The beam is prevented from displacing at the ends; thus the boundary conditions for the differential equation of equilibrium are $v(0) = v(L) = 0$. Since $v(0) = v(L) = 0$, Eqn. (a) indicates that $M(0) = M(L) = 0$, so that all four of the boundary conditions are satisfied for a beam that is simply supported or pinned at both ends.

Thus in order to determine if the bar can be in equilibrium in a nonstraight configuration under the action of an axial load, we must solve

$$v'' + k^2 v = 0$$
$$v(0) = 0$$
$$v(L) = 0$$

where $k^2 = P/EI$ is a constant parameter. The solution to the differential equation can be expressed as

$$v = C_1 \cos kx + C_2 \sin kx$$

Satisfying the two boundary conditions yields

$$v(0) = 0 = C_1(1) + C_2(0)$$

and

$$v(L) = 0 = C_1 \cos kL + C_2 \sin kL$$

From the first equation $C_1 = 0$ and from the second equation

$$C_2 \sin kL = 0$$

Taking $C_2 = 0$ leads to $v = 0$, corresponding to the straight configuration. The other possibility for satisfying the equation is to take

$$\sin kL = 0$$

which has solutions $k_n L = n\pi$ or

$$k_n^2 = \left(\frac{n\pi}{L}\right)^2 \qquad n = 1, 2, \ldots$$

The smallest root, corresponding to $n = 1$, is

$$k_1^2 = \left(\frac{\pi}{L}\right)^2 = \frac{P_1}{EI}$$

or

$$P_1 = \frac{\pi^2 EI}{L^2} = P_e$$

where P_e is known as the **Euler buckling load.** At this value of the compressive load P the corresponding nontrivial solution is

$$v(x) = C_2 \sin kx = C_2 \sin\left(\frac{\pi x}{L}\right)$$

where C_2 is arbitrary. The function $\sin(\pi x/L)$ corresponding to $n = 1$ is referred to as the **fundamental mode shape.** Note that corresponding to the values of $n > 1$, there are the solutions

$$P_n = \frac{n^2\pi^2 EI}{L^2}$$

for the buckling loads, with the corresponding mode shapes given by

$$v_n(x) = \sin(n\pi x/L)$$

The significance of these solutions will be discussed later.

The interpretation of the solution for the *ideal case* is as follows. When P is gradually increased from zero, there is no solution $v(x)$ if $P < P_e$, thus $v(x) \equiv 0$. When P reaches P_e, the straight configuration is still a solution, but any function of the form $v(x) = C_2 \sin(\pi x/L)$ is also a solution for any C_2 such that the stresses due to the axial load and the bending do not exceed the proportional limit. This means that at the Euler load P_e, the bar is in equilibrium in the straight configuration or in any nearby configuration with $v(x) = C \sin(\pi x/L)$ with C any small constant. The nonunique character of the solution is pictured in Fig. 11.16. This behavior can be approximately demonstrated by the student by compressing a yard stick as mentioned previously.

Leonhard Euler (1707–1783)

Swiss mathematician who contributed to the field of mechanics by developing the theory of columns and deriving the formula for finding the critical load in column buckling.

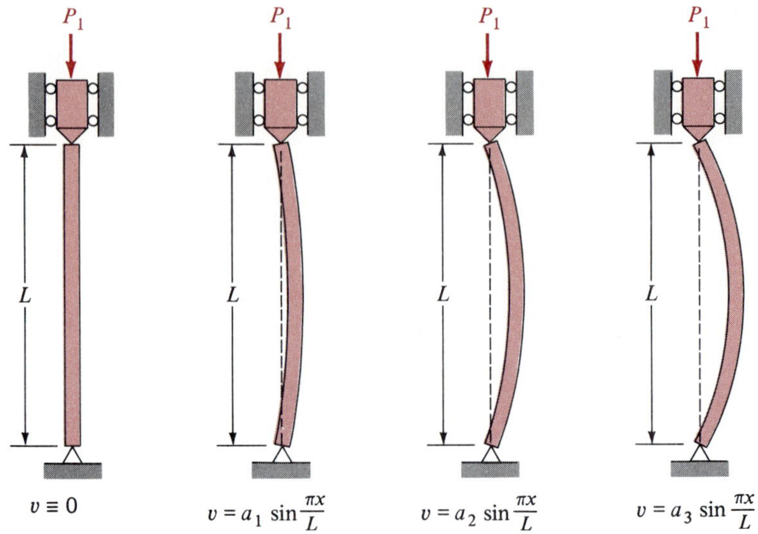

Figure 11.16 Different amplitudes of the equilibrium solution $v_1(x)$ for the same axial load

The Euler buckling load $P_e = \pi^2 EI/L^2$ is seen to be strongly dependent upon the length of the column—if the length is doubled the buckling load goes down by a factor of 4. It is also clear that the strength of the material, whether measured by yielding or fracture, does not enter into the buckling load. The material stiffness as measured by Young's Modulus does however enter into P_e, so that the buckling load for a column of a particular length and cross section can be increased by increasing E. Last, the cross-section properties enter by virtue of the second moment of the area I about the axis of rotation corresponding to the pin. Substantial increases in the buckling load can be realized by increasing the cross-sectional dimension that is perpendicular to the direction of the axis of the pin.

If, rather than being supported by pins at the ends, the bar is supported by ball joints which allow rotation about any axis, the critical load is still given by $P_e = \pi^2 EI/L^2$, where I represents the minimum of the two principal second moments of the cross section. This situation is shown for several cross sections in Fig. 11.17. As indicated for each of these cases, the buckling deformation takes place in the direction corresponding to bending about the axis for which I is minimum.

Consider again the problem of the stability of the pinned-pinned column. When P is in the range $0 < P < P_e = \pi^2 EI/L^2$ we state without proof that the straight equilibrium configuration of the column is stable.

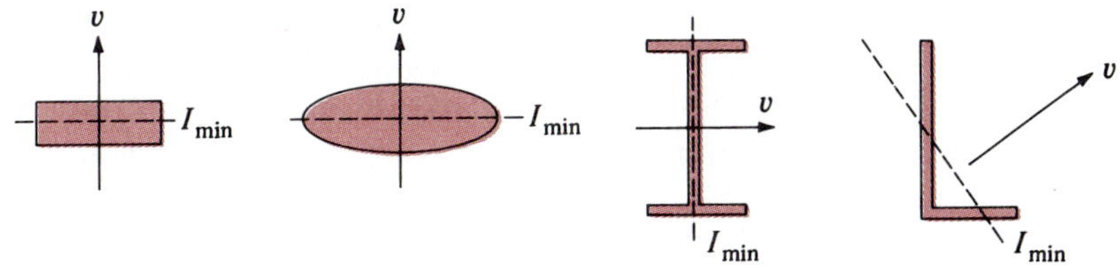

Figure 11.17 Buckling deformation perpendicular to the axis about which I is a minimum for a column with ball-joint supports at both ends

By this we mean that if the column is subjected to a small transverse disturbance, it will return to the straight configuration upon removal of the disturbance. This region of stability is shown for $P < P_e$ in Fig. 11.18. The equilibrium of the straight configuration is stable; but we cannot determine this fact on the basis of the development presented in this text.

When $P = P_e$, the elementary linear theory used in this text states that the equilibrium of the straight configuration is what is called the critical case of neutral equilibrium mentioned in Section 11.1. If the column is displaced to some nearby configuration given by $v \approx A \sin(\pi x/L)$, where A is a small arbitrary value, the equations state that the column can also be in equilibrium in that configuration when $P = P_e$. This situation is indicated by the horizontal line at $P = P_e$ in Fig. 11.18b. The result of the exact nonlinear theory is indicated by the curve that branches from the point $P = P_e$, and shows that it is actually necessary to increase P slightly as A increases. The exact theory shows that at $P = P_e$, the equilibrium of the straight configuration is stable.

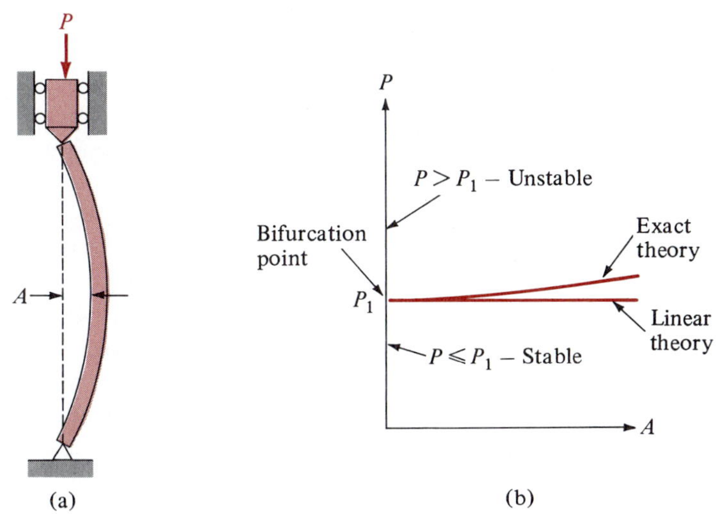

Figure 11.18 Stability of a pinned-pinned column

When $P > P_e$, the exact theory can be used to conclude that the equilibrium of the straight configuration is unstable; that is, if, under the action of $P > P_e$, the straight column is disturbed slightly the column will buckle and not return to the straight configuration. This fact is also indicated in Fig. 11.18b.

The point $P = P_e$ on the P-axis is referred to as a bifurcation point. It is a point common to two different solutions of the equilibrium equations. One equilibrium solution is the straight configuration, whereas the other equilibrium solution is the family of buckled shapes. The two families of solutions have the point $A = 0$, $P = P_e$ in common. Upon increasing P from zero up to $P = P_e$ the only solution is the straight configuration. At $P = P_e$ the perfect column may remain straight, or it may buckle into a nonstraight equilibrium configuration. The theory indicates that either state is possible at $P = P_e$.

PROBLEMS

11.16 Determine the critical load for a meter stick with cross-sectional dimensions of 25 mm by 5 mm that is pinned at both ends. Take $E = 12$ GPa. What is the corresponding compressive stress at the critical load?

11.17 Determine the critical load for a yard stick with cross-sectional dimensions of 1 in. by $\frac{3}{16}$ in. that is pinned at both ends. Take $E = 1.8 \times 10^6$ psi. What is the corresponding compressive stress at the critical load?

11.18 Determine the critical load for the rectangular-cross-section steel column with ball joints at each end. Take $b = 1$ in., $h = 2$ in., $L = 12$ ft, and $E = 29 \times 10^6$ psi. What is the corresponding compressive stress at the critical load?

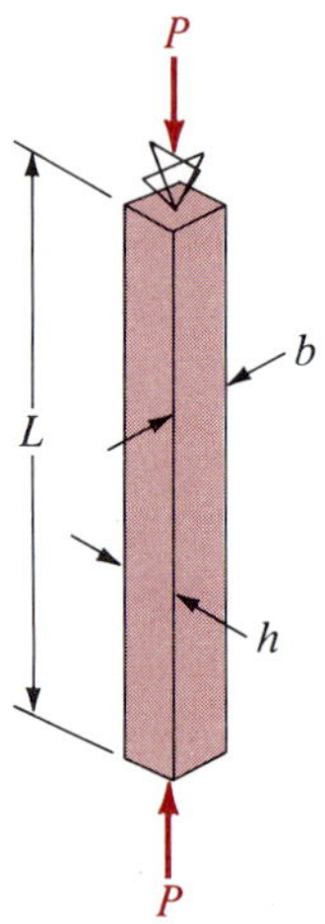

Ps. 11.18 and 11.19

11.19 Repeat problem 11.18 for an aluminum column with $b = 1.5$ in., $h = 1.0$ in., $L = 8$ ft, and $E = 10 \times 10^6$ psi.

11.20 A C 5 × 9 channel selection is supported by ball joints at both ends. Determine the critical load for a column 8 ft long. Take $E = 29 \times 10^6$ psi. What is the corresponding compressive stress at the critical load?

11.21 A W 200 × 19.3 section is supported by ball joints at both ends. With $E = 205$ GPa, determine the critical load for a column 2.5 m long. What is the corresponding compressive stress at the critical load?

11.22 A C 150 × 12.2 channel section is supported by ball joints at both ends. Determine the critical load for a column 2.7 m long. Take $E = 205$ GPa. What is the corresponding compressive stress at the critical load?

11.23 A W 6 × 12 section is supported by ball joints at both ends. With $E = 29 \times 10^6$ psi, determine the critical load for a column 16 ft long. What is the corresponding compressive stress at the critical load?

11.24 An L $2\frac{1}{2} \times 2\frac{1}{2} \times \frac{1}{2}$ section is supported by ball joints at both ends. Determine the critical load for a column 10 ft long. Take $E = 29 \times 10^6$ psi. What is the corresponding compressive stress at the critical load?

11.25 An L 76 × 76 × 9.5 section is supported by ball joints at both ends. Determine the critical load for a column 3 m long. Take $E = 205$ GPa. What is the corresponding compressive stress at the critical load?

11.26 Determine the allowable load for a 20-ft-long column made from a W 14 × 30 section. The column is pinned at both ends about an axis parallel to the web. What is the corresponding compressive stress? Take $E = 29 \times 10^6$ psi and the factor of safety to be 2.0.

11.27 For a factor of safety of 2.5, determine the allowable load for an 8.0-m-long column made from a W 200 × 86 section. The column is pinned at both ends about an axis parallel to the web. What is the corresponding compressive stress? Take $E = 205$ GPa.

11.28 For a factor of safety of 2.2, determine the allowable load for a 2.2-m-long column made from an S 200 × 34 section. The column is pinned at both ends about an axis parallel to the web. What is the corresponding compressive stress? Take $E = 205$ GPa.

11.29 Determine the allowable load for a 6-ft-long column made from an S 4 × 9.5 section. The column is pinned at both ends about an axis parallel to the web. What is the corresponding compressive stress? Take $E = 29 \times 10^6$ psi and the factor of safety to be 2.0.

11.30 For a factor of safety of 2.2, determine the allowable load for a 6-ft-long column made from a C 5 × 9 section. The column is pinned at both ends about an axis parallel to the web. What is the corresponding compressive stress? Take $E = 29 \times 10^6$ psi.

11.31 Determine the allowable load for a 2.0-m-long column made from a C 150 × 19.3 section. The column is pinned at both ends about an axis parallel to the web. Take $E = 205$ GPa and the factor of safety to be 1.92.

11.32 Two L $3\frac{1}{2} \times 3\frac{1}{2} \times \frac{3}{8}$ sections are welded along the 10-ft length to form a column pinned at both ends as shown. Determine the allowable load for a factor of safety of 2.1 if $E = 29 \times 10^6$ psi.

11.33 Repeat problem 11.32 for two L 64 × 64 × 6.4 sections 2.0 m in length. Take the factor of safety to be 2.5 and $E = 205$ GPa.

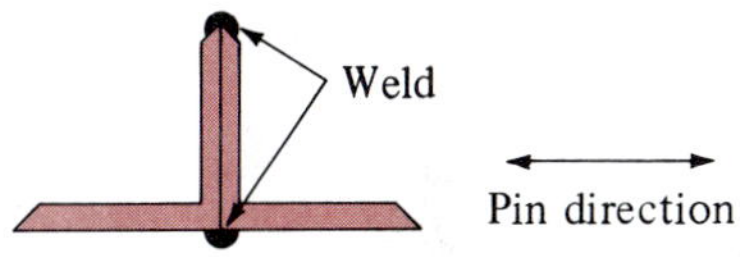

Ps. 11.32 and 11.33

11.34 Determine the allowable value of the load P with respect to buckling of member AB. AB is a circular pipe with outside diameter = 1.0 in. and wall thickness = 0.125 in., and point B is restrained against motion out of the plane. Take $E = 10 \times 10^6$ psi.

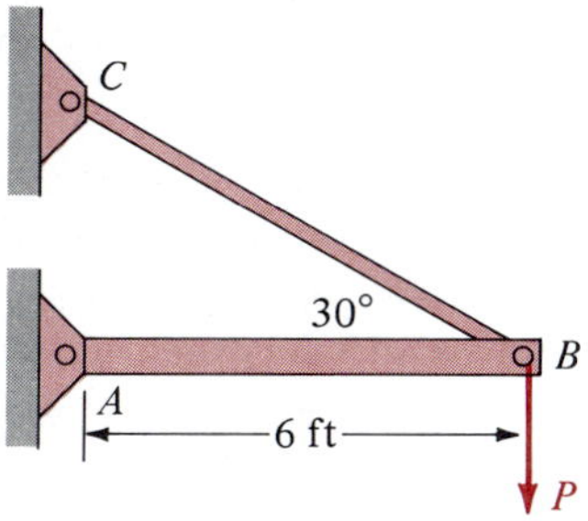

P. 11.34

11.35 For the pin-connected truss shown, determine the allowable value of the loads P. All members are composed of square box sections with outside dimension 2.0 in. and wall thickness 0.25 in., and are restrained to lie in the plane. Take $E = 10^7$ psi.

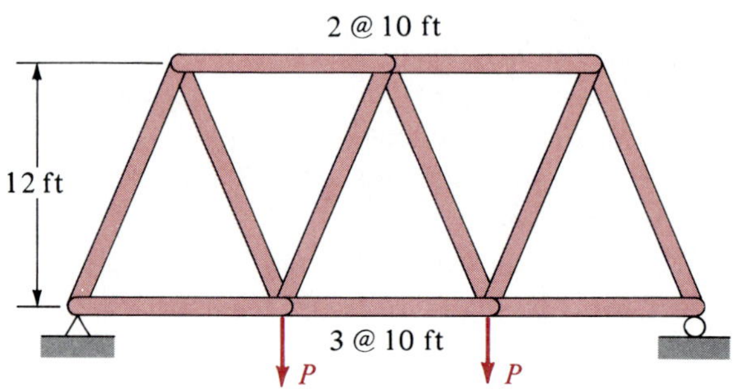

P. 11.35

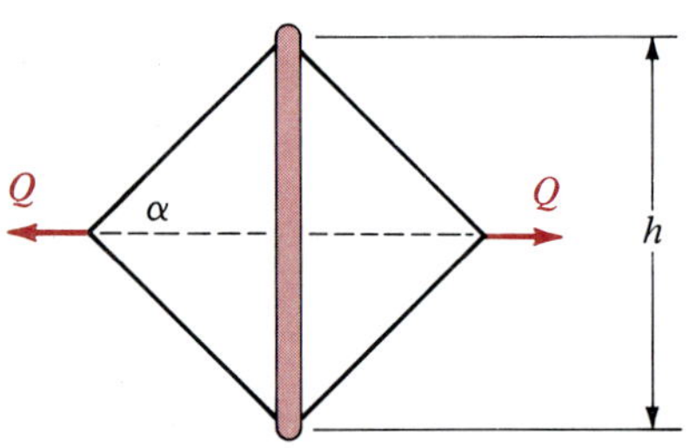

P. 11.36

11.36 The "spreader" consists of wires separated by a bar of diameter d as shown. For a factor of safety of 2.0, develop an expression for the allowable spreader force Q.

11.37 Members AC and BC, both 13 ft in length, are composed of steel pipe with an outside diameter of 2.75 in. and wall thickness of 0.1875 in. For a factor of safety of 3.0, what is the allowable value of P? The supports at A and B and the connection at C are such that the columns can be considered as pinned at each end. Take $E = 3 \times 10^7$ psi.

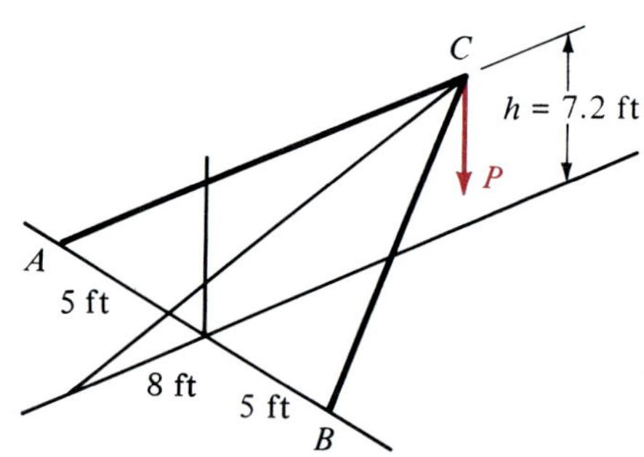

P. 11.37

11.38 The rig is used to lift heavy objects as shown. If there are stops that prevent motion of the cable CD outside of $\pm 30°$, determine the maximum allowable value of P for a factor of safety of 3.0. Assume pins at A, B, and C, and a ball joint at E. Members AC and BC are W 100×19.3 sections pinned about axes perpendicular to the web, and DE is a circular steel pipe with an outside diameter of 90 mm and a wall thickness of 6 mm. Take $E = 205$ GPa.

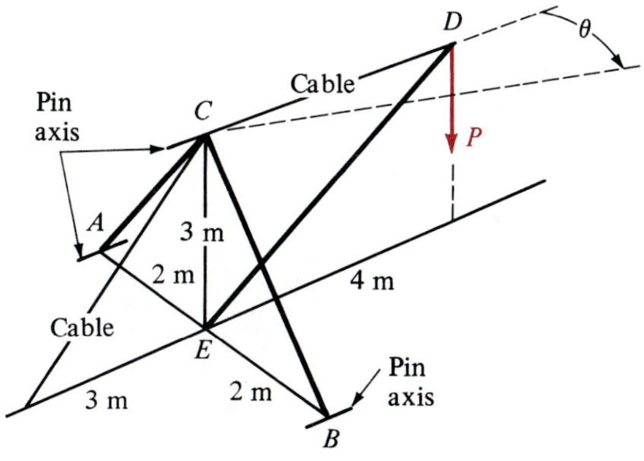

P. 11.38

11.39 Bar AC is a hollow square section of outside dimension 45 mm and wall thickness 6 mm. Determine the allowable load Q for a factor of safety of 2.75. Take $E = 70$ GPa.

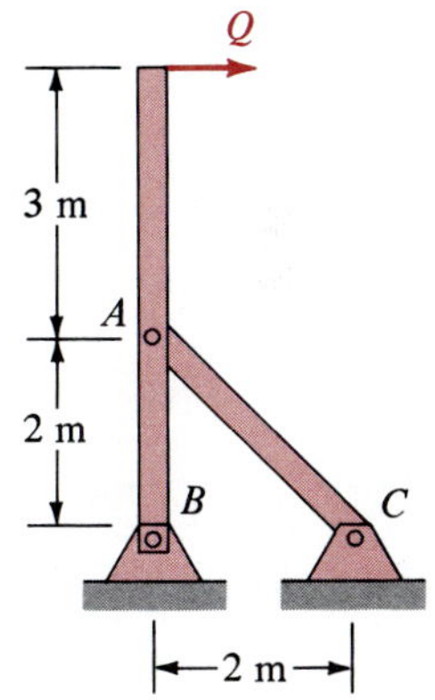

P. 11.39

11.40 Both bars shown are hollow pipes with outside diameters of 40 mm and wall thicknesses of 6 mm. Assuming that both bars are pinned at both ends and restrained against motion out of the plane of the paper, determine the allowable value of Q for a factor of safety of 2.0. Take $E = 205$ GPa.

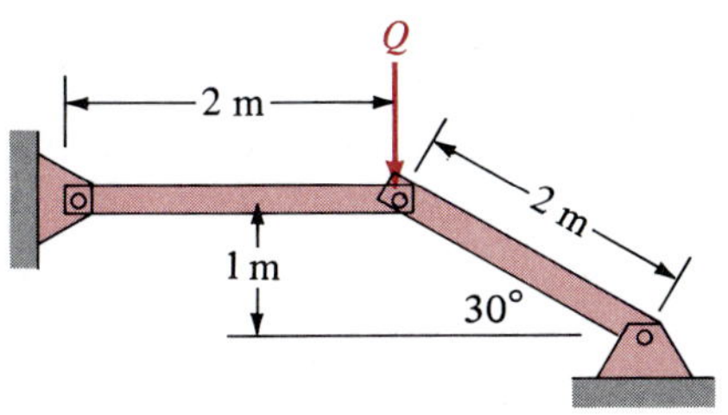

P. 11.40

11.41 Determine the critical value for the load Q applied through the rigid member to the column pinned at both ends.

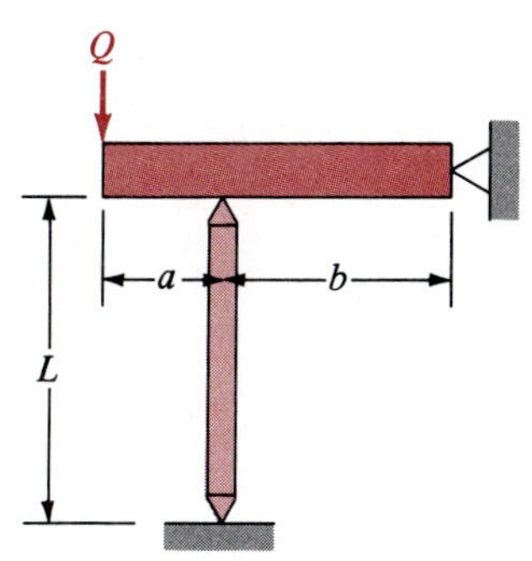

P. 11.41

11.42 Determine the critical value of P for the two pinned-end columns loaded through the rigid member.

11.43 Determine the relationship between d_1, d_2, a, and b such that the critical load P is the largest possible.

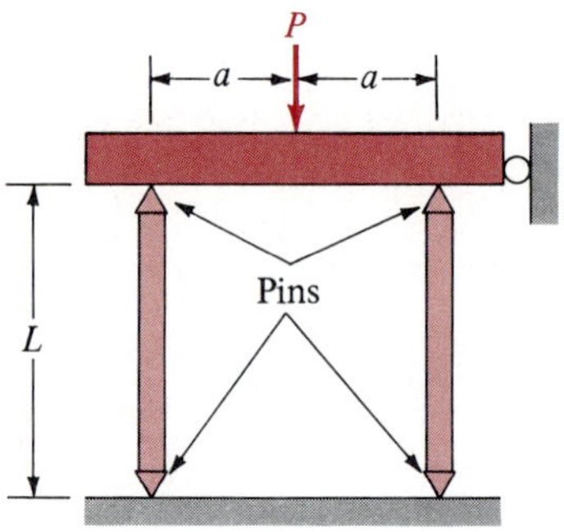

P. 11.42

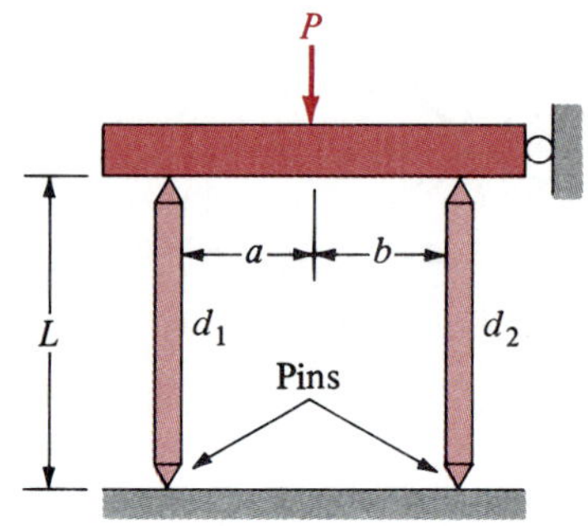

P. 11.43

11.44 Determine the largest load that can be carried by the two simply supported columns.

11.45 At what temperature increase ΔT will the uniform pinned-end bar be likely to buckle?

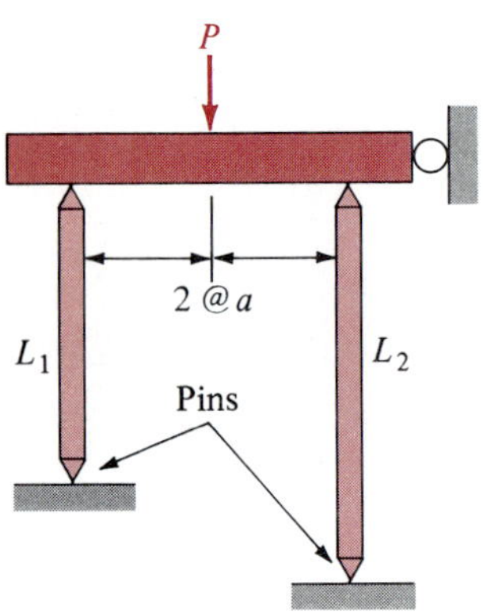

P. 11.44

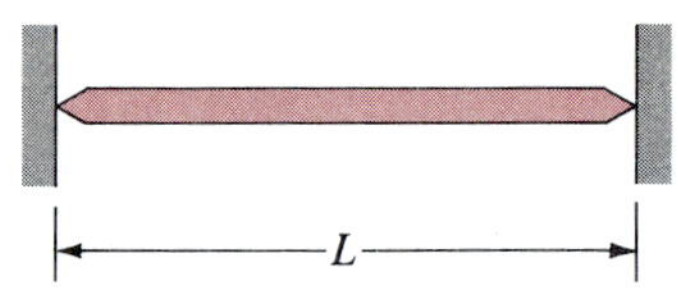

P. 11.45

11.46 What is the critical value of the load P applied indirectly to the two pinned-end columns through the rigid member?

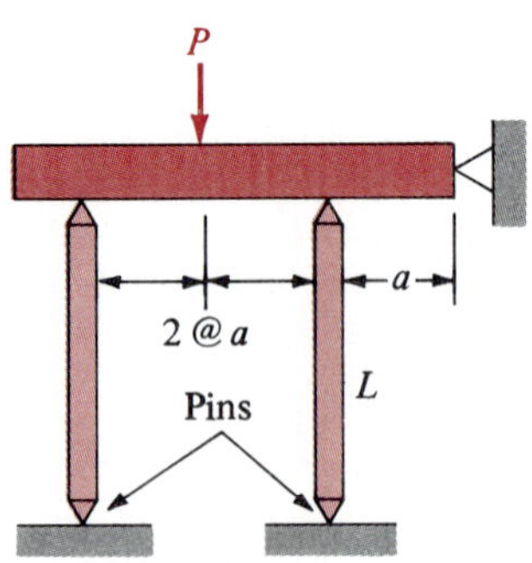

P. 11.46

11.47 What is the critical value of the load P applied through the flexible beam to the centrally located column as shown?

11.48 The uniform bar is pinned at both ends and supported by a linear spring at one end as shown. At what ΔT will the bar be likely to buckle?

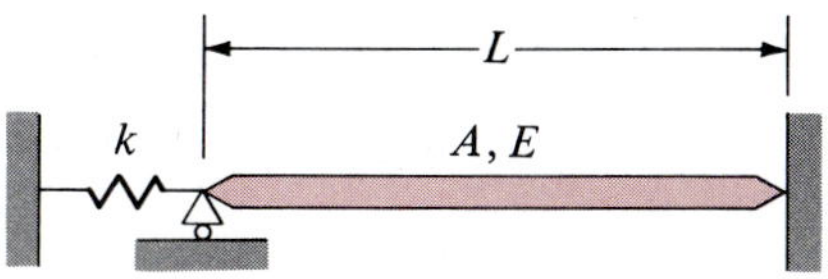

P. 11.48

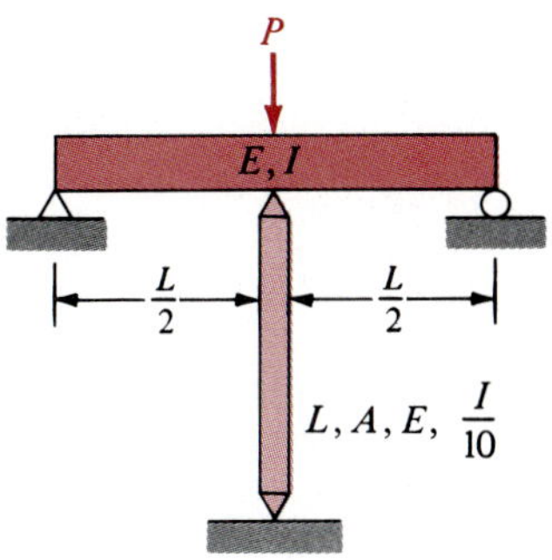

P. 11.47

11.4 OTHER BOUNDARY AND SUPPORT CONDITIONS

The case of the ideal elastic column pinned at both ends, analyzed in the previous section, is the simplest of the buckling problems associated with columns. Often, however, uniform straight members in compression are supported at the ends in other ways, as indicated in Fig. 11.19.

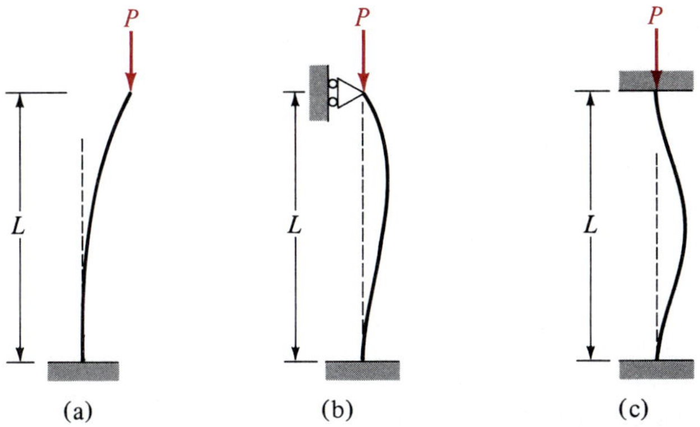

Figure 11.19 Various boundary conditions for centrally loaded ideal columns

Fig. 11.19a shows a column that is clamped at one end and free to deflect at the other end. Shown in Fig. 11.19b is a column that is clamped at one end and simply supported at the other end. The column shown in Fig. 11.19c is clamped at both ends. Other types of support conditions that are meant to model an elastic restraint against translation or rotation are also possible.

Recall from Chapter 7 that the equilibrium equation for the transverse displacement $v(x)$ was written either in the second order form as $EIv'' = M$ or in the fourth order form as $EIv'''' + w = 0$. For the case where the bar is simply supported at both ends it is possible to use the second order form as shown in the last section. In cases where one or both ends of the beam are not simply supported it is necessary to develop the fourth order form of the equation of equilibrium for a compressed slender bar, in order to properly satisfy the four boundary conditions, two at each end.

Shown in Fig. 11.20 is a free-body diagram of a segment of a bar loaded by a compressive force P.

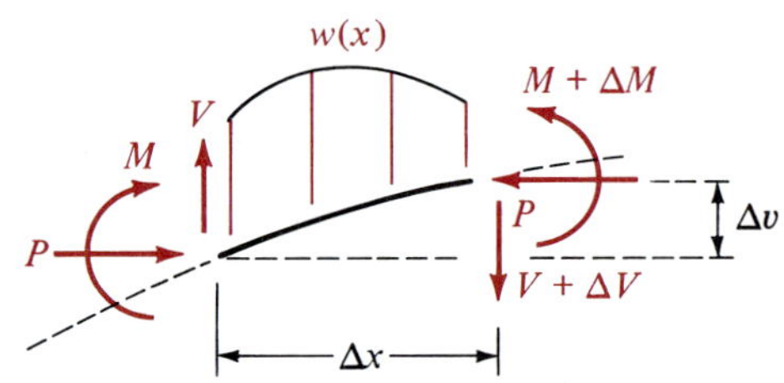

Figure 11.20 Equilibrium of a compressed bar

Requiring equilibrium of the forces yields

$$V - (V + \Delta V) - w(x^*)\,\Delta x = 0$$

or upon passing to the limit,

$$\frac{dV}{dx} + w = 0 \qquad (11.2)$$

Moment equilibrium requires that

$$-M + M + \Delta M - V\,\Delta x + [w(x^{**})\,\Delta x](\alpha\,\Delta x) + P\,\Delta v = 0$$

Again passing to the limit yields

$$\frac{dM}{dx} - V + P\frac{dv}{dx} = 0 \qquad (11.3)$$

Eliminating the shear force V between Eqns. (11.2) and (11.3) yields

$$\frac{d^2M}{dx^2} + P\frac{d^2v}{dx^2} + w(x) = 0$$

The moment-curvature relation $M = EIv''$ is then used to express the equilibrium equation entirely in terms of the displacement v as

$$EI\frac{d^4v}{dx^4} + P\frac{d^2v}{dx^2} + w(x) = 0 \qquad (11.4)$$

as the governing equation for investigating the stability of centrally compressed slender bars.

For a simple support as shown in Fig. 11.21a the boundary conditions are $v = 0$ and $M = EIv'' = 0$. For the clamped end shown in Fig. 11.21b the boundary conditions are $v = 0$ and $v' = 0$. For the free end shown in Fig. 11.21c, the boundary conditions are $M = EIv'' = 0$ and, using Eqn. (11.3), $V = EIv''' + Pv' = 0$.

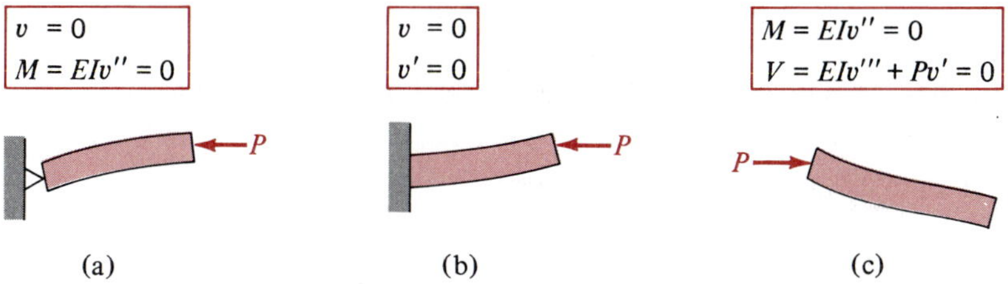

Figure 11.21 Boundary conditions for simple, clamped, and free supports in the presence of an axial load

We note that the boundary conditions for the simply supported and clamped case are the same with or without the axial load. Only the boundary condition for the free end is changed in the presence of the axial load. In what follows we will determine what effect the support conditions have on the fundamental buckling load for a column.

Clamped-Free Column The case of a centrally loaded ideal column that is free at the loaded end and clamped at the other end is shown in Fig. 11.22. The four boundary conditions, two at each end, are indicated.

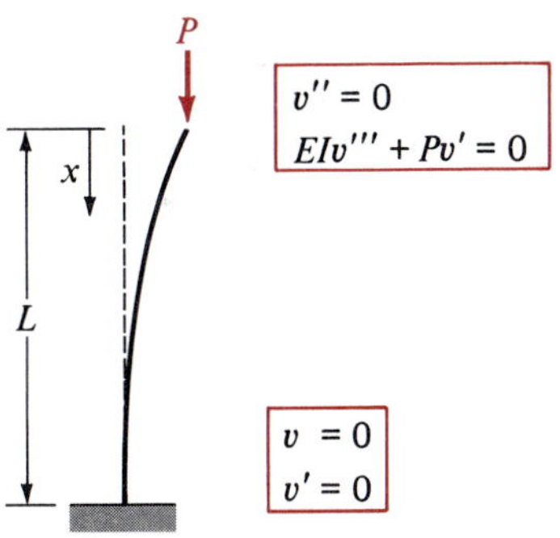

Figure 11.22 Boundary conditions for a clamped-free column

With $w = 0$, the differential equation of equilibrium and the boundary conditions are

$$EIv'''' + Pv'' = 0$$
$$v''(0) = 0$$
$$EIv'''(0) + Pv'(0) = 0$$
$$v(L) = 0$$
$$v'(L) = 0$$

The solution of the differential equation of equilibrium can be expressed as

$$v = C_1 \cos kx + C_2 \sin kx + C_3 x + C_4$$

where $k^2 = P/EI$. Satisfying the two boundary conditions at $x = 0$ yields

$$0 = C_1 k^2 \qquad \text{and} \qquad 0 = C_2 k(P - EIk^2) + PC_3$$

which, since $P = EIk^2$, yields $C_1 = C_3 = 0$. Satisfying the two boundary conditions at $x = L$ then yields

$$0 = C_2 \sin kL + C_4 \qquad \text{and} \qquad 0 = C_2 k \cos kL$$

It is left in the problems for the student to show that the lowest value of kL that satisfies these equations is $kL = \pi/2$, with the corresponding critical load as

$$P_1 = \frac{\pi^2 EI}{4L^2}$$

and that the corresponding mode or buckling shape is given by

$$v(x) = C_4\left(1 - \sin\frac{\pi x}{2L}\right)$$

shown in Fig. 11.23.

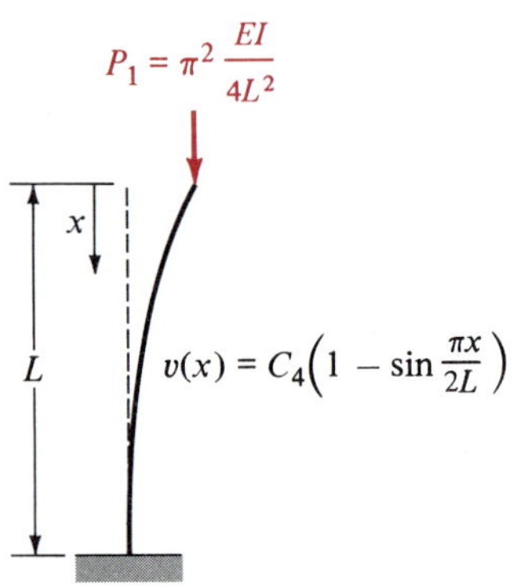

Figure 11.23 Fundamental buckling load and mode shape for the clamped-free column

Higher critical loads and mode shapes for this case will also be left for the problems.

Within the framework of the theory, the interpretation of this result is as follows. If P is slowly increased to some value less than P_1, there is no solution for v. If the column is displaced slightly and released, the column will return to the straight configuration—the equilibrium of the straight configuration is stable when $P < P_1$. When P reaches P_1 it is, according to the theory, possible for the column to remain in the straight configuration or to occupy any nearby configuration of the form $v = c[1 - \sin(\pi x/2L)]$, under the action of the load P_1. If the load is increased above P_1 there is again no solution, according to the theory, until the next critical load is reached. If the column is disturbed from the straight configuration for $P > P_1$ it will not return to the straight configuration, thus the equilibrium of the column is unstable in the straight configuration when $P > P_1$.

Pinned-Clamped Case Consider the case of a centrally loaded column that is pinned at one end and clamped at the other as shown in Fig. 11.24. The four boundary conditions, two at each end, are indicated.

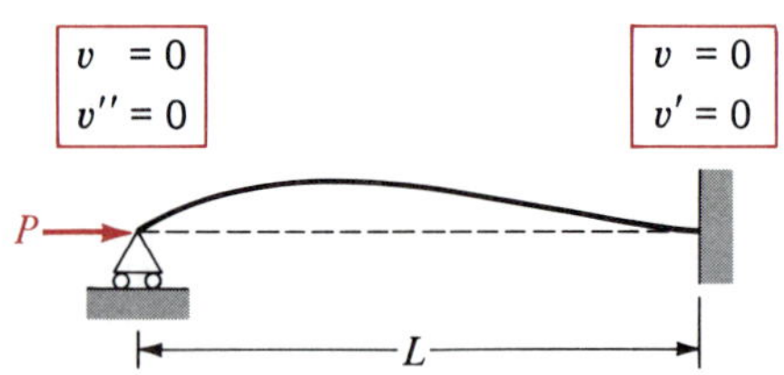

Figure 11.24 Boundary conditions for a pinned-clamped column

Then with $w = 0$ the differential equation of equilibrium and the boundary conditions are

$$EIv'''' + Pv'' = 0$$

$$v(0) = 0 \qquad v''(0) = 0$$

$$v(L) = 0 \qquad v'(L) = 0$$

The solution of the differential equation of equilibrium can be expressed as

$$v = C_1 \cos kx + C_2 \sin kx + C_3 x + C_4$$

where $k^2 = P/EI$. Satisfying the two boundary conditions at $x = 0$ yields

$$0 = C_1 + C_4 \quad \text{and} \quad 0 = -C_1 k^2$$

from which, unless we take $k = 0$, $C_1 = C_4 = 0$. Satisfying the two boundary conditions at $x = L$ then yields

$$0 = C_2 \sin kL + C_3 L \tag{a}$$

and

$$0 = C_2 k \cos kL + C_3 \tag{b}$$

Multiplying the second equation by L and subtracting from the first equation yields

$$\tan kL = kL$$

as the characteristic equation. The student should verify that the smallest root of this equation is $k_1 L = 1.43\pi$ so that

$$k_1^2 = \frac{P_1}{EI} = \left(\frac{1.43\pi}{L}\right)^2$$

or

$$P_1 = 2.045 \frac{\pi^2 EI}{L^2}$$

P_1 represents the lowest compressive load at which the column can be in equilibrium in the deflected shape shown in Fig. 11.25. Using Eqn. (a) to eliminate C_3, the equation of the deflected shape can be written as

$$v(x) = C_2 \left(\sin\left(\frac{1.43\pi x}{L}\right) + 0.976 \frac{x}{L}\right)$$

where C_2 is arbitrary. This fundamental mode shape for the pinned-clamped column is also shown in Fig. 11.25.

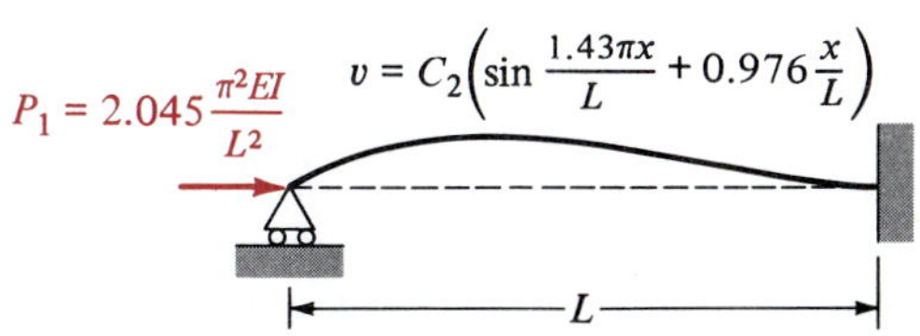

Figure 11.25 Fundamental buckling load and mode shape for the pinned-clamped column

Clamped-Clamped Column The case of a centrally loaded column that is clamped at both ends is shown in Fig. 11.26. The four boundary conditions, two at each end, are indicated.

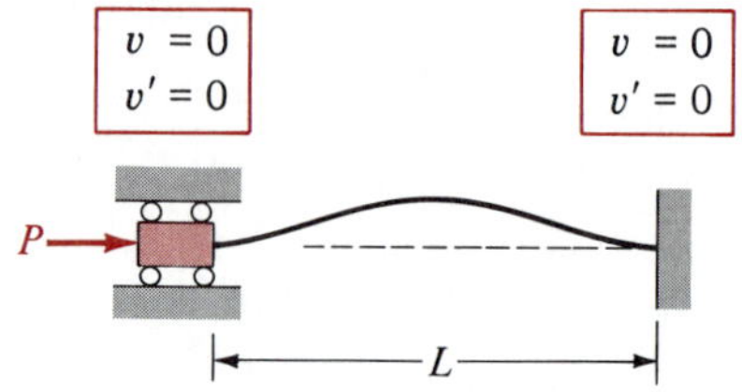

Figure 11.26 Boundary conditions for a clamped-clamped column

With $w = 0$ the differential equation of equilibrium and the boundary conditions are

$$EIv'''' + Pv'' = 0$$

$$v(0) = 0 \qquad v'(0) = 0$$

$$v(L) = 0 \qquad v'(L) = 0$$

The solution of the differential equation of equilibrium can be expressed as

$$v = C_1 \cos kx + C_2 \sin kx + C_3 x + C_4$$

where $k^2 = P/EI$. Satisfying the two boundary conditions at $x = 0$ yields

$$0 = C_1 + C_4 \qquad \text{and} \qquad 0 = C_2 k + C_3$$

Satisfying the two boundary conditions at $x = L$ then yields

$$0 = C_1 \cos kL + C_2 \sin kL + C_3 L + C_4$$

and

$$0 = -C_1 k \sin kL + C_2 k \cos kL + C_3$$

It is left in the problems for the student to perform the algebra showing that these four equations can be reduced to the single equation

$$C_1[2(1 - \cos kL) - kL \sin kL] = 0$$

from which either $C_1 = 0$ or

$$2(1 - \cos kL) - kL \sin kL = 0 \qquad (a)$$

If C_1 is taken to be zero, all of C_2, C_3, and C_4 are also zero, leading to the known straight configuration as the solution.

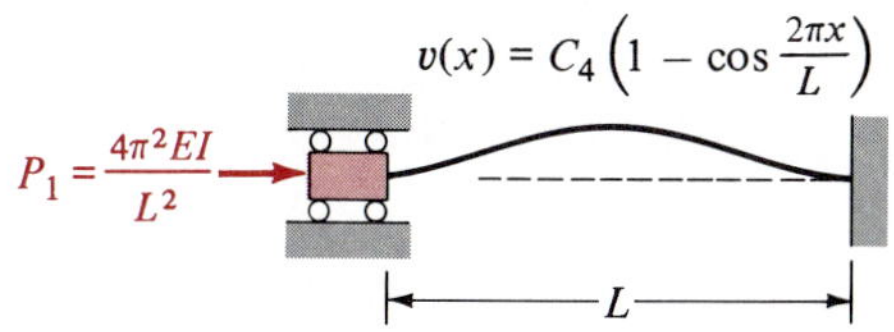

Figure 11.27 Fundamental buckling load and mode shape for the clamped-clamped column

The student is also asked to show in the problems that the lowest nonzero root of Eqn. (a) is $k_1 L = 2\pi$ and that when $k_1 L = 2\pi$ the corresponding solution can be written as

$$v(x) = C_4 \left[1 - \cos\left(\frac{2\pi x}{L}\right)\right]$$

shown in Fig. 11.27.

From $k_1 L = 2\pi$, the corresponding critical compressive force, also indicated in Fig. 11.27, is given by

$$P_1 = 4\,\frac{\pi^2 EI}{L^2}$$

Higher critical loads and mode shapes for this case will be left for the problems.

The interpretation of this result is as follows. If P is slowly increased to some value less than P_1, there is no solution for v. When P reaches P_1 it is, according to the theory, possible for the column to remain in the straight configuration or to occupy any nearby configuration of the form $v \approx 1 - \cos(2\pi x/L)$ under the action of the load P_1. If the load is increased above P_1, there is again no solution, according to the theory. Equilibrium of the straight configuration is stable for $P \leq P_1$, and unstable for $P > P_1$.

We consider also in this section the possibility of internal supports for the pinned-pinned column. Specifically, a pinned-pinned column that is also supported by a pin at its midpoint is indicated in Fig. 11.28.

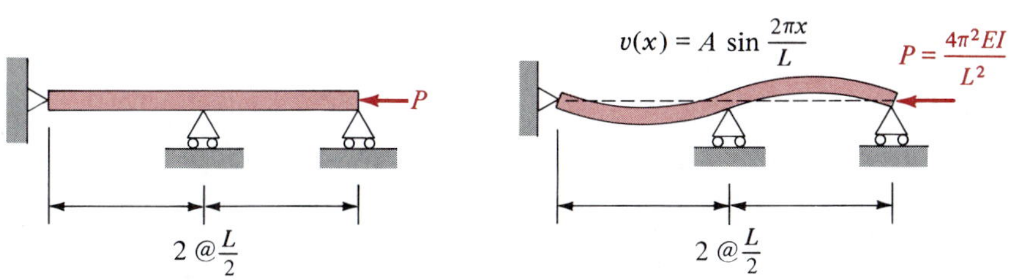

Figure 11.28 Buckling load for a pinned-pinned column with an extra pin at the center

We state without proof that the lowest buckling mode for this physical problem is $P_{cr} = 4\pi^2 EI/L^2$; this corresponds to the second buckling mode for the pinned-pinned column considered in Section 11.3. Recall from that section that there were roots $k_n L = n\pi$ for the pinned-pinned column. The buckling loads were $P_n = (n\pi/L)^2 EI$ with the buckling modes sin $(n\pi x/L)$. This is in fact the way that the second and higher modes for the pinned-pinned column can be realized in practice. In particular, the second mode is realized by placing a support at the center, forcing the column to possibly buckle into the second mode when the load reaches $P_2 = 4\pi^2EI/L^2$. Placing pins at $L/3$ and $2L/3$, as shown in Fig. 11.29, would result in the column possibly buckling at a load of $P = 9\pi^2EI/L^2$, with the shape $\sin(3\pi x/L)$.

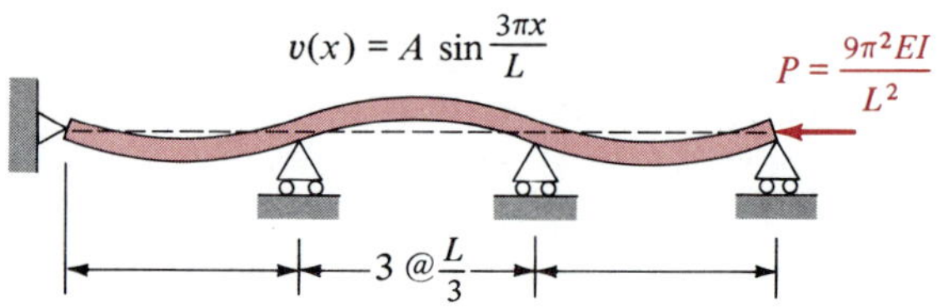

Figure 11.29 Third buckling load for a column of length L supported at the third points

The procedure for realizing the fourth and higher modes should be clear. This idea can be of enormous practical significance for increasing the buckling load of a pinned-pinned column if it is in fact possible to introduce the additional constraints at symmetrically placed internal locations. Note however that this approach cannot be used to determine buckling loads if the supports are not evenly spaced.

For the case of a clamped-clamped column the second root of the equation

$$2(1 - \cos kL) = kL \sin kL$$

is $k_2L = 2.86\pi$, with the corresponding critical load given by

$$P_2 = (2.86\pi)^2 \frac{EI}{L^2} = 8.18 \frac{\pi^2 EI}{L^2}$$

The mode shape is indicated in Fig. 11.30.

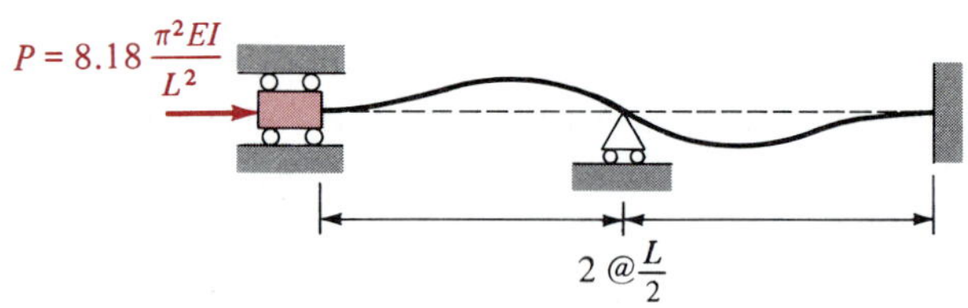

Figure 11.30 Second mode for a clamped-clamped column

It is seen that the effect of the internal symmetrically placed support is to more than double the value of the buckling load from the $4\pi^2 EI/L^2$ value for the clamped-clamped column with no internal support. Note that this case can also be considered as the pinned-clamped column with a length of $L/2$ leading to

$$P = \frac{2.045\pi^2 EI}{(L/2)^2} = 8.18\,\frac{\pi^2 EI}{L^2}$$

The third mode for the clamped-clamped column corresponds to constraining the midpoint to having zero displacement and slope as shown in Fig. 11.31.

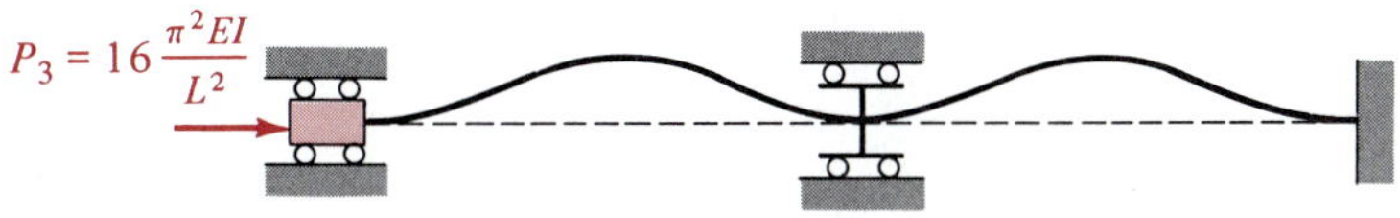

Figure 11.31 Third mode for a clamped-clamped column

The third root of

$$2(1 - \cos kL) = kL \sin kL$$

is $k_3 L = 4\pi$ with the corresponding critical load given by

$$P_3 = 16\,\frac{\pi^2 EI}{L^2}$$

The buckling load is significantly increased over either of the first two values by also forcing the slope at the middle to have the zero value. Other higher modes can be discussed for the clamped-clamped column, but these have practical significance only where it is possible to provide the appropriate constraints at the internal locations along the length of the beam.

Effective Lengths By considering the cases corresponding to the different support conditions shown in Fig. 11.32, it is possible to define an **effective length,** denoted by L_e.

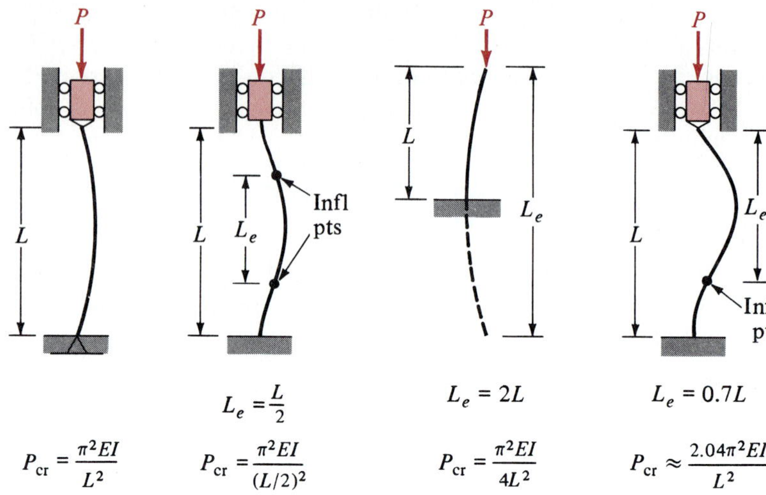

Figure 11.32 Critical loads for variously supported uniform columns

The idea is that it is possible to express the fundamental critical load for a column with supports other than pinned-pinned in terms of the critical load for the column pinned at both ends, thus

$$P_{cr} = \frac{\pi^2 EI}{(KL)^2} = \frac{\pi^2 EI}{L_e^2}$$

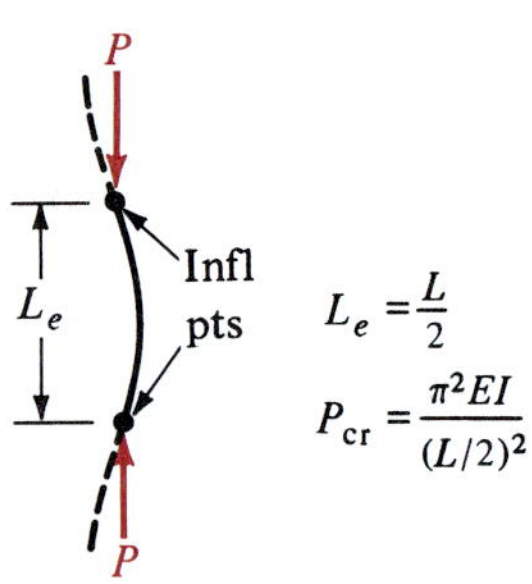

Figure 11.33 Free-body diagram of the middle half of a clamped-clamped column

where $KL = L_e$ is the effective length. As an example, consider the clamped-clamped column shown in Fig. 11.32b. The quarter points $x = L/4$ and $x = 3L/4$ are inflection points for the mode shape $v(x) = A[1 - \cos(2\pi x/L)]$, so that a free-body diagram of the inner portion of the column of length $L/2$, shown in Fig. 11.33, shows that this portion is essentially a pinned-pinned column of length $L/2$ acted upon by the compressive load P.

The critical load for such a column is

$$P_{cr} = \frac{\pi^2 EI}{(L/2)^2} = \frac{4\pi^2 EI}{L^2}$$

so that by definition, the effective length L_e is $L/2$, or $K = 0.5$. For the clamped-free case shown in Fig. 11.32c, the distance between the two inflection points is $2L$, so that L_e is $2L$ and $K = 2.0$. For the case shown in Fig. 11.32d, the effective length is approximately $0.7L$, leading to

$$P_{cr} \approx \frac{\pi^2 EI}{(0.7L)^2} = 2.041 \frac{\pi^2 EI}{L^2}$$

which is very close to the exact value of $2.045\pi^2 EI/L^2$. Thus for the case of the clamped-pinned column $L_e = 0.7L$.

The idea of an effective length can also be applied to a column supported at the midpoint as shown in Fig. 11.34, where, with an inflection point at $x = L/2$ for the mode shape of $v(x) = A \sin(2\pi x/L)$, the effective length is clearly $L_e = 0.5L$, leading to $P_{cr} = 4\pi^2 EI/L^2$. The concept of the effective length will be useful at several points in the remainder of this chapter.

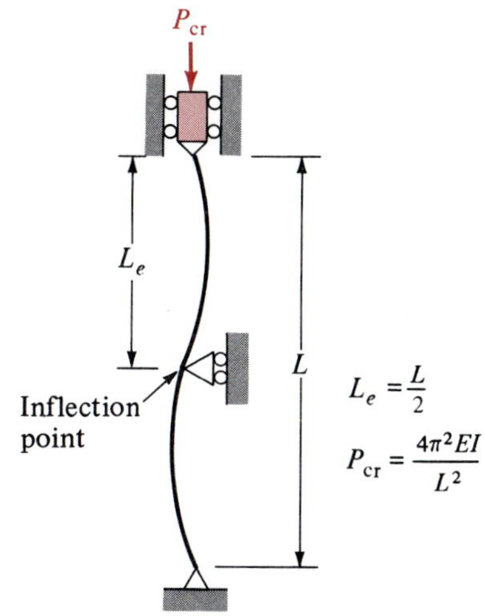

Figure 11.34 Use of effective length for a column with an internal support

PROBLEMS

11.49 A column, clamped at both ends and 15 ft in length, is constructed from a W 6 × 20 section. Determine the critical load. Take $E = 29 \times 10^6$ psi.

11.50 Repeat problem 11.49 if the column is restrained at its midpoint against deflection in the direction parallel to the flanges. Assume all motion in the perpendicular direction is prevented.

11.51 Determine the critical load for an 8-ft-long column that is clamped at one end and pinned at the other if a C 6 × 13 section is used. The axis of the pin is parallel to the web. Take $E = 29 \times 10^6$ psi.

11.52 Determine the critical load for a column, clamped at both ends, if the length is 5 m and a W 150 × 24 section is used. Take $E = 205$ MPa.

11.53 A C 180 × 22 section is to be used for a column clamped at both ends and restrained at the midpoint against deflection in the direction of the flanges. If the length of the column is 3.0 m, determine the critical load. Take $E = 205$ GPa and assume all motion in the perpendicular direction is prevented.

11.54 Repeat problem 11.53 if the restraint at the midpoint is also against rotation.

11.55 A column with a rectangular cross section with $b < h$ is supported by ball joints at both ends and restrained against deflection in the direction corresponding to the b dimension at the midpoint. There is no restraint in the direction corresponding to the h dimension. Determine the ratio of b to h in order that the critical loads for buckling in both directions are reached simultaneously.

11.56 Two columns are supported at the ends as shown. What should be the ratio of the lengths L_1 and L_2 if both columns are to buckle simultaneously?

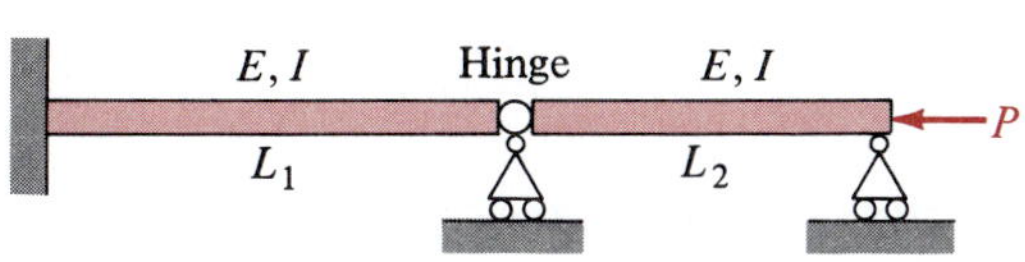

P. 11.56

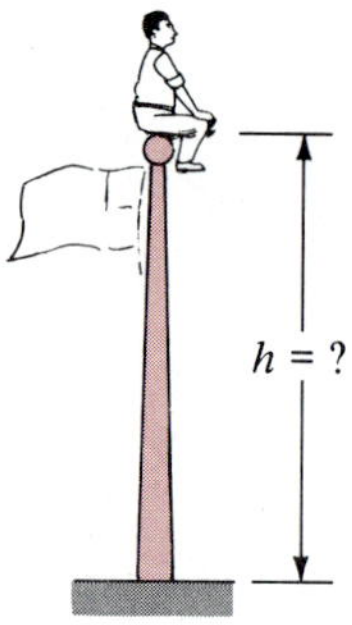

P. 11.57

11.57 A man weighing 200 lbf is to "sit a flagpole" as shown. The pole is to be constructed from a pipe with a 4.0-in. outside diameter and a 0.25-in. wall thickness. Neglecting the weight of the pole, what is the maximum height at which he should plan to sit? Assume that the man is a point at the top of the flagpole and that $E = 29 \times 10^6$ psi.

11.58 Two columns, clamped at both ends, are to support a platform of weight W which is prevented from horizontal motion as shown. Determine the critical value for W in terms of the Euler load for a single pinned-end column.

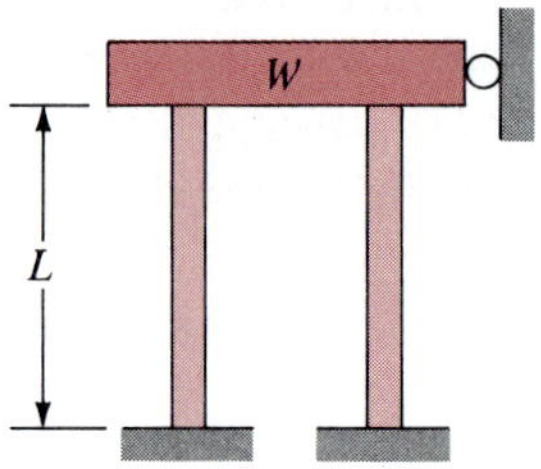

P. 11.58

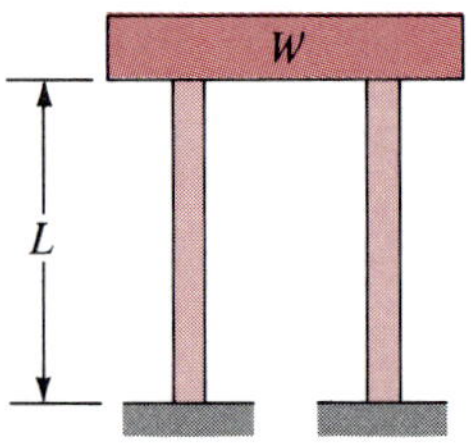

P. 11.59

11.59 Repeat problem 11.58 if the restraint against horizontal motion of the platform is removed.

11.60 Two columns, clamped at the top and pinned at the bottom, are to support a platform of weight W which is prevented from horizontal motion as shown. Given E, I, and the length of the columns, determine the critical value for W.

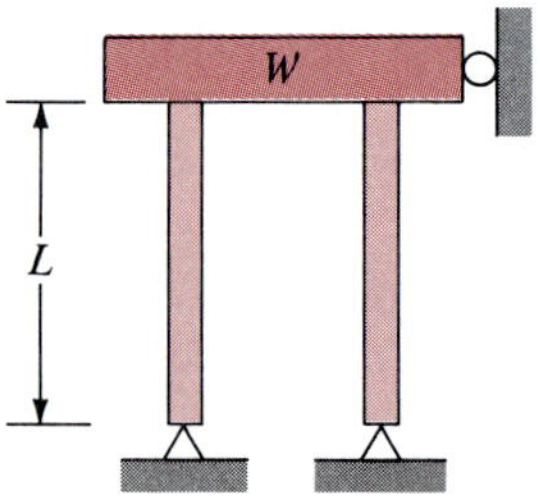

P. 11.60

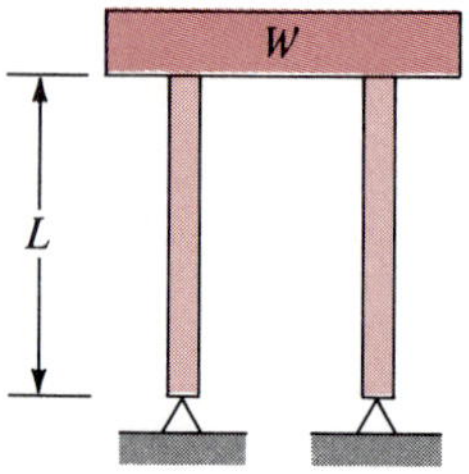

P. 11.61

11.61 Repeat problem 11.60 if the restraint against horizontal motion of the platform is removed.

11.62 A uniform platform of weight W is to be supported by four symmetrically placed clamped-clamped columns as shown. Each column is 40 ft in length and constructed from pipe that has an outside diameter of 10 in. and a wall thickness of 0.375 in. What is the allowable value of W if the factor of safety against buckling is to be 2.0?

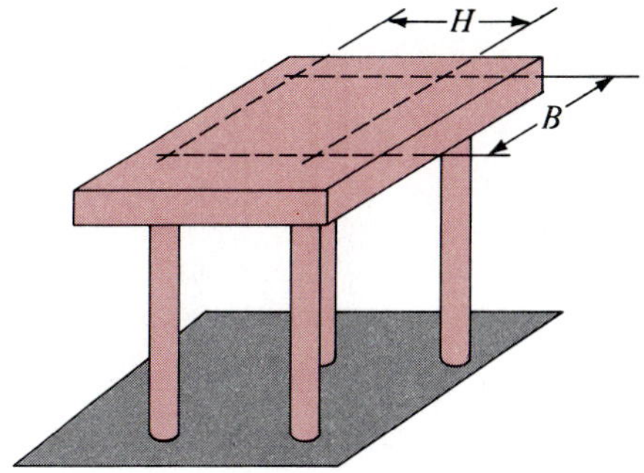

P. 11.62

11.5 IMPERFECT COLUMNS

A real column, however carefully constructed and loaded, is imperfect. These imperfections can be due to the fact that the column is initially not perfectly straight, or that the load cannot be applied in a perfectly central fashion. In this section we will investigate the qualitative and quantitative changes in the results of the previous two sections due to these two effects.

Eccentrically Applied Loads Fig. 11.35b is a schematic of the model that we will use to investigate the effects of an eccentrically applied compressive load applied to a pinned-end column shown in Fig. 11.35a.

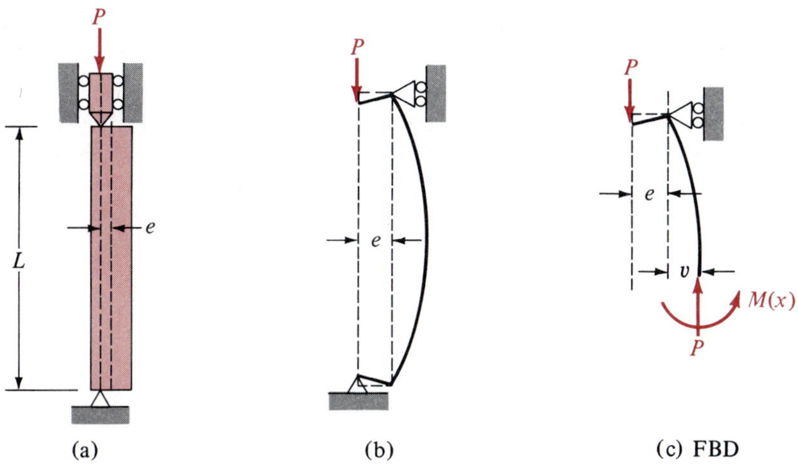

Figure 11.35 Eccentrically loaded column with pinned ends

The free-body diagram shown in Fig. 11.35c is used to set up the second order differential equation

$$EIv'' = M = -P(e + v)$$

or

$$v'' + k^2 v = -k^2 e$$

where $k^2 = P/EI$. The two boundary conditions that must be satisfied are $v(0) = v(L) = 0$. The solution of the differential equation is

$$v = C_1 \cos kx + C_2 \sin kx - e$$

Satisfying the two boundary conditions yields $C_1 = e$ and $C_2 = (1 - \cos kL)/\sin kL$ so that

$$v(x) = e\left(\tan\frac{kL}{2}\sin kx + \cos kx - 1\right) \tag{11.5}$$

In order to more readily interpret this result, consider the deflection at the center, and write

$$\frac{v(L/2)}{e} = \frac{1 - \cos(kL/2)}{\cos(kL/2)}$$

or

$$v\left(\frac{L}{2}\right) = e\left[\sec\left(\frac{kL}{2}\right) - 1\right] = e\left[\sec\left(\frac{\pi}{2}\sqrt{\frac{P}{P_{cr}}}\right) - 1\right]$$

represented in terms of the secant function. Note that as P approaches P_{cr}, the secant function, and hence the central deflection, becomes very large for any $e > 0$. The effect of the eccentricity upon the behavior is that there is no critical load as such, but rather, the transverse displacement v increases gradually as P increases and ultimately becomes unbounded as P approaches P_{cr}. This behavior is plotted in Fig. 11.36.

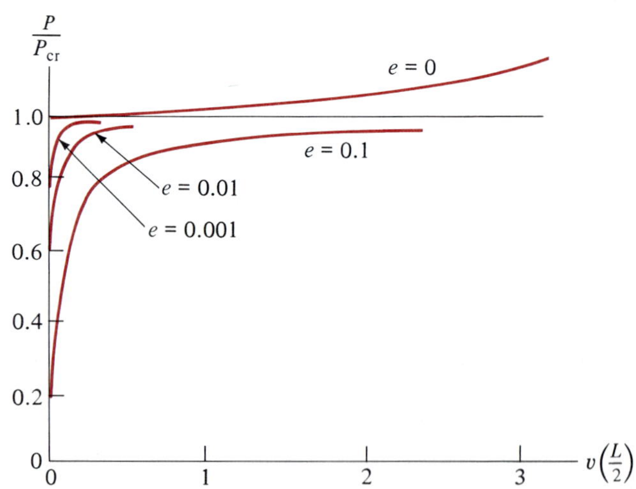

Figure 11.36 Central deflection versus axial load

Fig. 11.36 shows that when e is very small, the behavior is similar to the straight column as we would expect. The straight configuration persists until P is very near P_{cr}, at which time the deflection increases quite rapidly. For larger values of the eccentricity, the gradual increase in the deflection begins at a much earlier stage in the loading, with a more rapid increase as P approaches P_{cr}.

The case $e = 0$ represents the case of the classic buckling of the straight centrally loaded column, and indicates no deflection whatsoever until $P = P_{cr}$, at which time buckling may occur. After the column has buckled, there is a slight increase in the compressive load required to maintain equilibrium in the buckled configuration, as indicated by the corresponding nonlinear theory for the $e = 0$ curve.

All of these results are based upon the assumption that the material behaves in a linearly elastic manner, and thus that the stresses are below the proportional limit. The stress in the bar is the sum of the stresses due to the axial load and those due to bending. The equation for the stresses under combined loading of this type was developed in Chapter 8 and is

$$\sigma = \frac{P}{A} + \frac{My}{I}$$

The moment in this general expression is computed using $M = EIv''$ and, using Eqn. (11.5) for $v(x)$, can be represented as

$$M = EIv'' = -Pe\left(\tan\frac{kL}{2}\sin kx + \cos kx\right)$$

The maximum stress will be at $x = L/2$ and at the largest negative y coordinate, $y = -c$. This can be expressed as

$$\sigma = -\frac{P}{A}\left[1 + \frac{ec}{r^2}\sec\left(\frac{\pi}{2}\sqrt{\frac{P}{P_{cr}}}\right)\right] \tag{11.6}$$

showing that for $e > 0$, the maximum stress also becomes very large as P approaches P_{cr}. For our purposes it is preferable to represent Eqn. (11.6) in the form

$$\sigma = -\frac{P}{A}\left[1 + \frac{ec}{r^2}\sec\left(\frac{L}{2r}\sqrt{\frac{P}{AE}}\right)\right] \tag{11.7}$$

where we have used $I = Ar^2$, with r the **radius of gyration** of the cross section corresponding to the plane of bending. Eqn. (11.7) is known as the **secant formula.**

Given a particular material so that E is known, Eqn. (11.7) gives the maximum stress σ in the column in terms of the average stress $\sigma = -P/A$. The additional dimensionless parameters that are a part of the secant formula are L/r, the **slenderness ratio,** and ec/r^2, the **eccentricity ratio.** If we specify a limit on the maximum stress, say the value of the proportional limit, we can solve for the average stress using Eqn. (11.7). This must be done using a trial-and-error procedure or by consulting Fig. 11.37, where the solutions have been obtained numerically and represented as a set of curves for $\sigma_{max} = 36$ ksi and $E = 29 \times 10^6$ psi, typical values for structural steel. The graph can be constructed for other values of σ_{max} and for other materials.

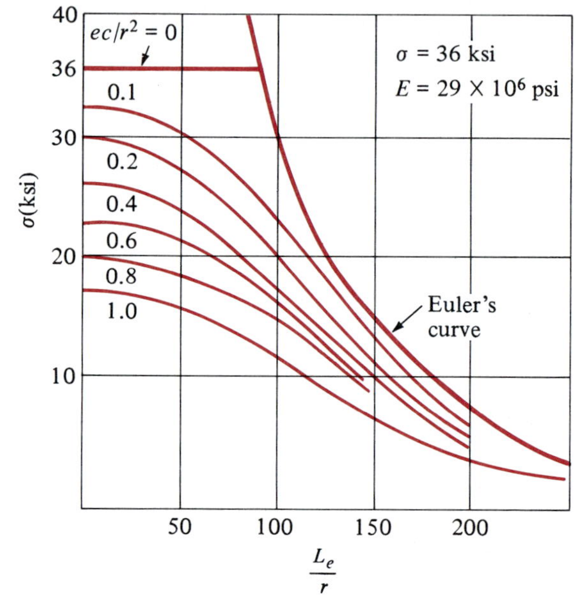

Figure 11.37 Graph of the secant equation for $\sigma_{max} = 36$ ksi and $E = 29 \times 10^6$ psi

In a general situation, the procedure for determining the maximum permissible value of P for an eccentrically loaded column would be as follows:

1. Determine values for the parameters L/r and ec/r^2.
2. Select a σ_{max} and an E and solve the secant formula for the average stress $\sigma_{avg} = P/A$.
3. Compute $P = \sigma_{avg}A$ and apply any factor of safety to this value for the load.

If a factor of safety is to be used, the secant formula should be replaced by

$$\sigma = -\frac{nP}{A}\left[1 + \frac{ec}{r^2}\sec\left(\frac{L}{2r}\sqrt{\frac{nP}{AE}}\right)\right] \tag{11.8}$$

where n is the factor of safety.

If a set of curves exists for the specific σ_{max} and E in question, the curves can of course be used to determine σ_{avg} instead of solving the secant formula.

For the case $e = 0$, the corresponding average stress at the critical load is given by $\sigma = P_{cr}/A$, or

$$\sigma_{avg} = \frac{\pi^2 EI}{AL^2} = \frac{\pi^2 E}{(L/r)^2}$$

Given a value for E, this curve, called Euler's curve, can be plotted as a function of the slenderness ratio and is shown in Fig. 11.37 for $E = 29 \times 10^6$ psi.

As e approaches zero, the curves from the secant formula approach the Euler curve for larger values of the slenderness ratio, and the horizontal curve represented by $\sigma = \sigma_{max}$ for smaller values of the slenderness ratio. Theoretically the secant formula is valid below these two curves.

The slenderness ratio is determined solely on the basis of the geometry of the column. The eccentricity ratio given by ec/r^2, on the other hand, cannot be determined theoretically: it must be determined experimentally or on the basis of experience. The engineer should consult the various structural-design codes in order to determine suitable values for this ratio in a particular setting or application. We mention in passing that when dealing with the design of structural steel pinned-end columns, it is customary to take the eccentricity ratio as 0.25.

Initially Unstraight Columns It is virtually impossible to construct a long slender column so that it is perfectly straight—there are invariably initial deviations from the ideal straight configurations. In order to assess the qualitative and quantitative effects of this type of imperfection, we will assume that $v_0(x)$ can be represented as $v_0(x) = v_0 \sin(\pi x/L)$ as shown in Fig. 11.38a. Other forms of the deviation of the column from perfectly straight could be investigated in a completely similar manner.

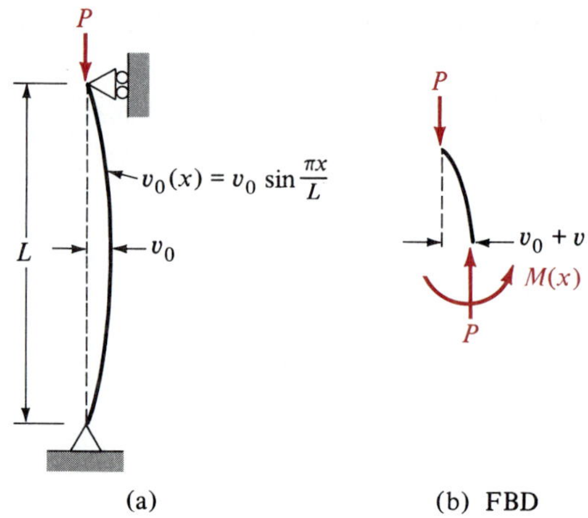

Figure 11.38 Stability of a column with initial imperfections

Referring to the free-body diagram of Fig. 11.38b, it is seen that the moment transmitted at a point along the column is due to the sum of the initial displacement plus the additional displacement due to the compressive force P. The equation of equilibrium can be written in second order form as

$$EIv'' = M = -Pv_{\text{total}} = -P(v(x) + v_0(x))$$

or

$$v'' + k^2 v = -k^2 v_0(x) = -k^2 v_0 \sin(\pi x/L)$$

where $k^2 = P/EI$. The boundary conditions are $v(0) = v(L) = 0$. The student should show that the particular solution to this differential equation is

$$v_{\text{part}} = \frac{v_0 \sin(\pi x/L)}{(\pi/kL)^2 - 1}$$

so that the general solution is

$$v(x) = C_1 \cos kx + C_2 \sin kx + v_{\text{part}}$$

Satisfying the boundary conditions leads to $C_1 = C_2 = 0$, so that

$$v(x) = v_{\text{part}} = \frac{v_0 \sin(\pi x/L)}{(\pi/kL)^2 - 1}$$

The total displacement is given by $v_{\text{total}} = v(x) + v_0(x)$ or

$$v_{\text{total}} = \frac{v_0 \sin(\pi x/L)}{(\pi/kL)^2 - 1} + v_0 \sin(\pi x/L) = \frac{v_0(\pi/kL)^2}{(\pi/kL)^2 - 1} \sin(\pi x/L)$$

$$= \frac{v_0}{1 - (kL/\pi)^2} \sin(\pi x/L)$$

At this point we observe that

$$(kL/\pi)^2 = \frac{PL^2}{\pi^2 EI} = P/P_{\text{cr}} = \beta$$

so that finally

$$v_{\text{total}}(x) = \frac{v_0}{1 - \beta} \sin(\pi x/L)$$

with a maximum at $x = L/2$ of $v_{\max} = v_0/(1 - \beta)$. This maximum value always exceeds v_0, since $P < P_{\text{cr}}$, or $\beta < 1$. A plot of the maximum central deflection as measured by v/L, versus the axial load P, is shown in Fig. 11.39 for several values of v_0/L.

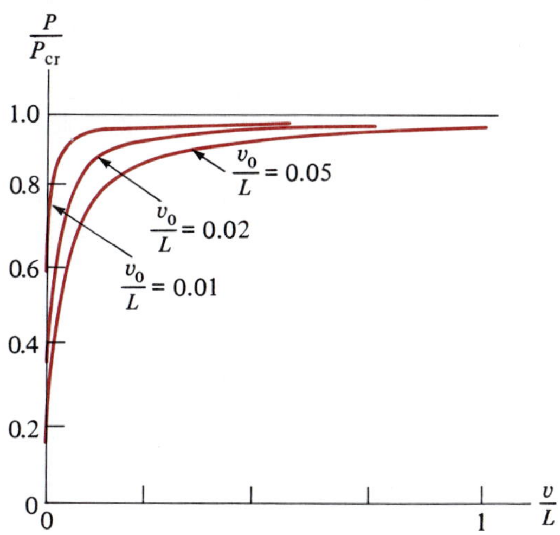

Figure 11.39 Central deflection versus load

Note that as P approaches the fundamental compressive load P_{cr}, the amplitude of the total displacement becomes very large. As was the case for the eccentrically loaded perfectly straight column, there is no buckling phenomenon, but rather the deflections approach infinity as the critical load is approached. Investigation of the stresses associated with the initially unstraight column are left to the problems.

PROBLEMS

11.63 The circular pipe is eccentrically loaded as shown. If the allowable normal stress is 12 ksi, determine *(a)* the allowable load P, and *(b)* the corresponding central deflection. Take the outside diameter of the bar to be 2 in., the wall thickness to be 0.1875 in., $E = 10^7$ psi, and $L = 8$ ft. What is the critical load if the column is loaded centrally?

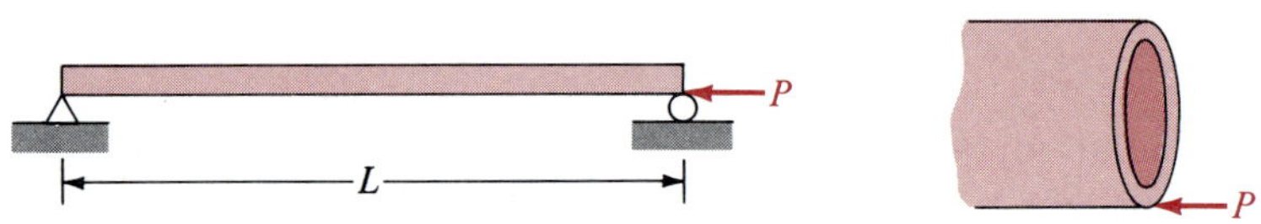

Ps. 11.63 and 11.64

11.64 If the yield stress is 90 MPa for the eccentrically loaded circular pipe shown, determine the allowable load P for a factor of safety of 2.0 with respect to permanent deformation. Determine also the corresponding central deflection. Take the outside diameter of the bar to be 40 mm, the wall thickness to be 4 mm, $E = 70$ GPa, and $L = 2.0$ m. What is the critical load if the column is loaded centrally?

11.65 The column with the section indicated is eccentrically loaded as shown. If the allowable normal stress is 100 MPa, determine *(a)* the allowable load P, and *(b)* the corresponding central deflection. Take $E = 70$ GPa, $b = 40$ mm, $h = 18$ mm, $t = 3$ mm, and $L = 1.5$ m. What is the critical load if the column is loaded centrally?

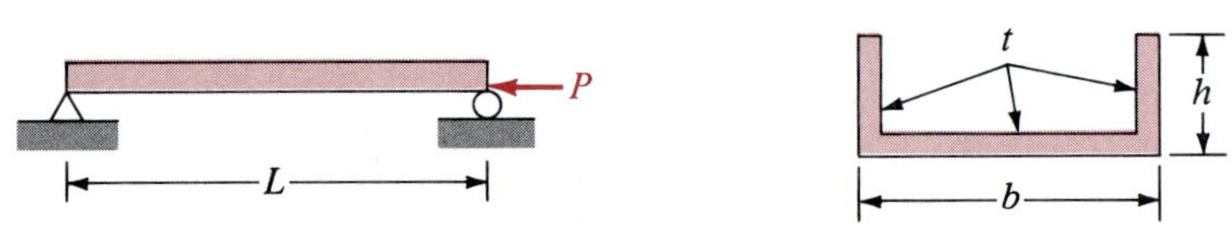

Ps. 11.65 and 11.66

11.66 Repeat problem 11.65 if the allowable normal stress is 13.5 ksi, $b = 2$ in., $h = 0.9$ in., $t = 0.20$ in., $E = 10^7$ psi, and $L = 8$ ft.

11.67 The column with the W 6 × 25 section is eccentrically loaded as shown. If the allowable normal stress is 10 ksi, determine *(a)* the allowable load P, and *(b)* the corresponding central deflection. Take $E = 3 \times 10^7$ psi, and $L = 18$ ft. What is the critical load if the column is loaded centrally?

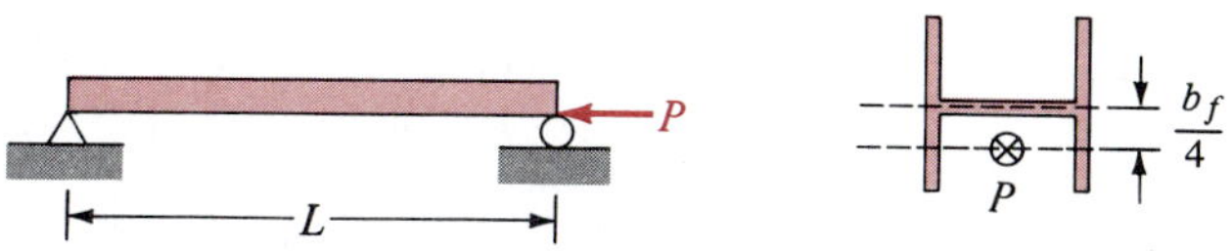

Ps. 11.67 and 11.68

11.68 Repeat problem 11.67 for a W 150 × 18 section with the allowable stress as 80 MPa, $E = 205$ GPa, and $L = 5$ m.

11.69 Given that the boundary conditions that must be enforced are those shown in the figure, show that the expressions for the displacement $v(x)$ and the maximum stress σ are, respectively,

$$v(x) = e\,\frac{1 - \cos k(L - x)}{\cos kL}$$

and

$$\sigma = \frac{-P}{A}\left[1 + \frac{ec}{r^2}\sec kL\right] = \frac{-P}{A}\left[1 + \frac{ec}{r^2}\sec\left(\frac{\pi}{2}\sqrt{\frac{P}{P_{cr}}}\right)\right]$$

where $k = \sqrt{P/EI}$, and $P_{cr} = \pi^2 EI/4L^2$.

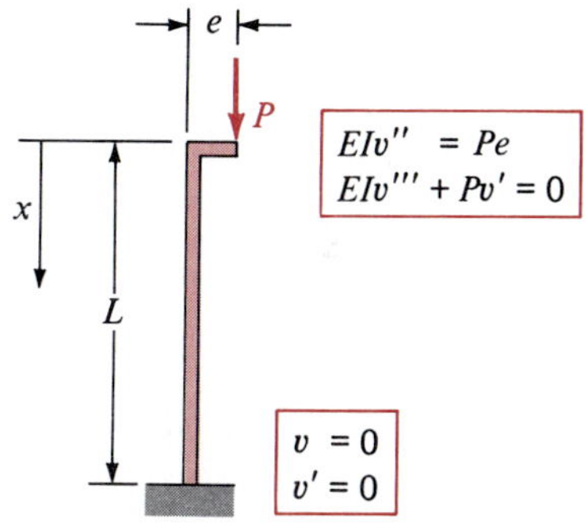

P. 11.69

11.70 The square pipe is eccentrically loaded as shown. If the allowable normal stress is 9 ksi, determine *(a)* the allowable load P, and *(b)* the corresponding central deflection. Take the outside dimension of the bar to be 2.4 in., the wall thickness to be 0.2 in., $E = 3 \times 10^7$ psi, and $L = 9$ ft. What is the critical load if the column is loaded centrally?

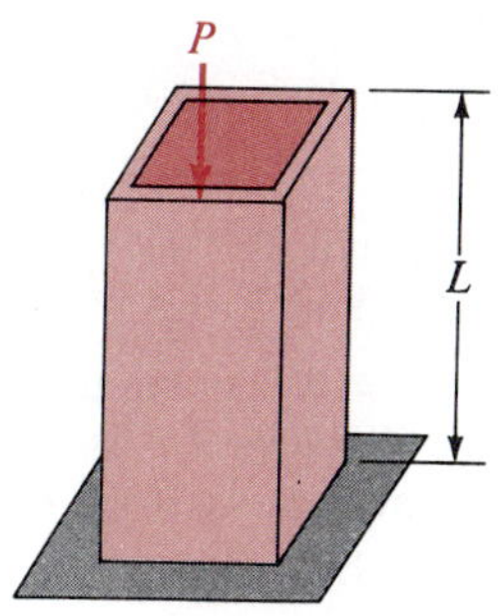

Ps. 11.70 and 11.71

11.71 If the yield stress is 90 MPa for the eccentrically loaded square pipe shown, determine the allowable load P for a factor of safety of 2.0 with respect to permanent deformation. Determine also the corresponding central deflection. Take the outside dimension of the bar to be 60 mm, the wall thickness to be 6 mm, $E = 70$ GPa, and $L = 2.8$ m. What is the critical load if the column is loaded centrally?

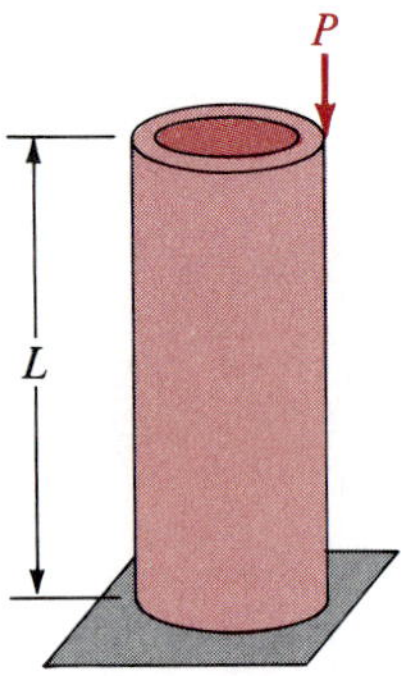

Ps. 11.72 and 11.73

11.72 The circular pipe is eccentrically loaded as shown. If the allowable normal stress is 80 MPa, determine *(a)* the allowable load P, and *(b)* the corresponding central deflection. Take the outside diameter of the bar to be 80 mm, the wall thickness to be 5 mm, $E = 205$ GPa, and $L = 4$ m. What is the critical load if the column is loaded centrally?

11.73 If the yield stress is 16 ksi for the eccentrically loaded circular pipe shown, determine the allowable load P for a factor of safety of 2.0 with respect to permanent deformation. Determine also the corresponding central deflection. Take the outside diameter of the bar to be 3.5 in., the wall thickness to be 0.1875 in., $E = 10^7$ psi, and $L = 12$ ft. What is the critical load if the column is loaded centrally?

11.74 The column with the W 8 × 18 section is eccentrically loaded as shown. If the allowable normal stress is 10 ksi, determine *(a)* the allowable load P, and *(b)* the corresponding central deflection. Take $E = 3 \times 10^7$ psi, and $L = 18$ ft. What is the critical load if the column is loaded centrally?

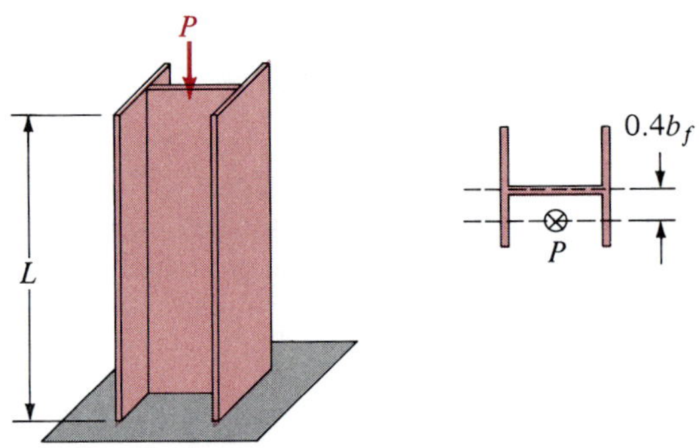

Ps. 11.74 and 11.75

11.75 Repeat problem 11.74 for a W 200 × 52 section with the allowable stress as 80 MPa, $E = 205$ GPa, and $L = 5$ m.

11.76 Develop an expression for the maximum combined stress in a column whose initial shape is given by $v_0 \sin(\pi x/L)$. Show that the result is a quadratic equation in P that can be solved without iteration.

11.6 CLOSURE

In this chapter, we studied the concept of stability of equilibrium as it pertains to simple conservative elastic systems. Most of the chapter was devoted to several aspects of the stability of equilibrium of straight slender elastic bars under the action of centrally applied compressive loads. We carried out the analysis of stability using the elementary theory of bending developed in Chapter 7; thus the theory is here subject to any restrictions or assumptions that we used in developing the corresponding elementary theory.

The type of stability demonstrated by straight ideal elastic bars is the classic bifurcation type of buckling, wherein the initial straight configuration persists until the compressive load reaches the critical value; at that time it is possible, according to the theory, for the bar to continue to be in equilibrium in the straight configuration or in any nearby deformed configuration under the action of the critical or buckling load. A real bar is not perfectly straight, nor can it be loaded in a perfectly central fashion, so that the actual behavior involves some transverse displacement of the bar prior to reaching the critical load. The realistic design of a column must take into account all of the deviations from the ideal case.

The engineer should keep in mind that whenever a thin structure, such as a bar, plate, or shell, is loaded in *compression,* it is possible for the structure to experience an instability involving buckling deformations, generally perpendicular to the thin direction. Analyses must always be made to ensure that the compressive stresses are below those that might result in buckling.

APPENDIX A PROPERTIES OF PLANE AREAS

A.1 CENTROIDS OF AREA

Consider an area as shown in Fig. A.1a.

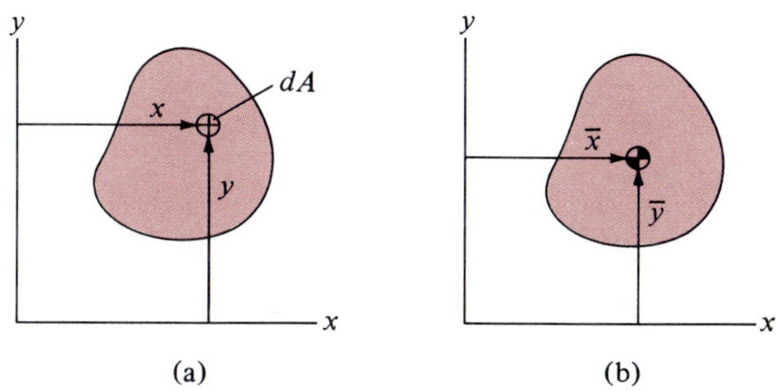

Figure A.1 Definition of first moments and centroids

The first moments of the area about the x- and y-axes are defined, respectively, as

$$Q_y = \int_{\text{Area}} x \, dA \qquad \text{and} \qquad Q_x = \int_{\text{Area}} y \, dA$$

The corresponding centroidal coordinates, that is, the distances to the centroid of the area shown in Fig. A.1b, are given by

$$\bar{x} = \frac{\int_{\text{Area}} x \, dA}{A} \qquad \text{and} \qquad \bar{y} = \frac{\int_{\text{Area}} y \, dA}{A}$$

where A is the area. In situations where the boundaries of the area can be represented by simple lines and curves, the centroidal coordinates $\bar{x}$ and $\bar{y}$ can be determined by evaluating the integrals.

Example A.1 Determine the centroidal coordinates of the area shown in Fig. A.2.

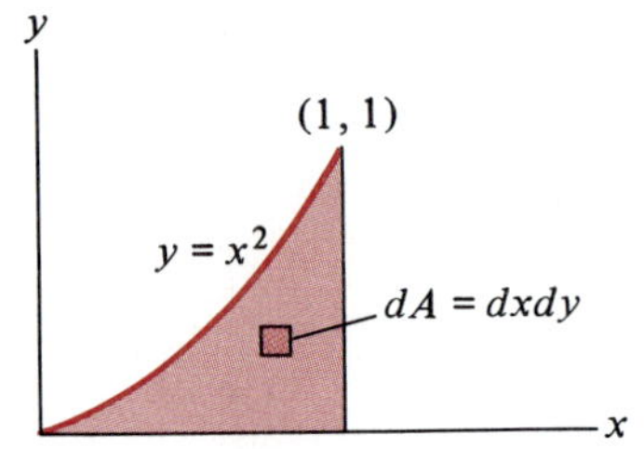

Figure A.2

Solution With $dA = dx\,dy$ as indicated, we first compute

$$A = \int_0^1 \int_0^{x^2} dy\,dx = \int_0^1 x^2\,dx = \tfrac{1}{3}$$

$$Q_y = \int_0^1 x \int_0^{x^2} dy\,dx = \int_0^1 x^3\,dx = \tfrac{1}{4}$$

from which

$$\bar{x} = \frac{\frac{1}{4}}{\frac{1}{3}} = \tfrac{3}{4}$$

Similarly

$$Q_x = \int_0^1 \int_0^{x^2} y\,dy\,dx = \int_0^1 \frac{x^4}{2}\,dx = \tfrac{1}{10}$$

so that

$$\bar{y} = \frac{\frac{1}{10}}{\frac{1}{3}} = \tfrac{3}{10}$$

In many situations the boundary of the area is such that an evaluation of the integrals is more easily carried out using a numerical approach: by dividing the area into a set of small areas ΔA_i and computing

$$A = \Sigma\,\Delta A_i \qquad Q_y = \Sigma \bar{x}_i\,\Delta A_i \qquad \text{and} \qquad Q_y = \Sigma \bar{y}_i\,\Delta A_i$$

where the $\bar{x}_i$ and $\bar{y}_i$ are the centroidal coordinates of ΔA_i. The centroidal coordinates of the area A can then be determined from

$$\bar{x} = \frac{\Sigma \bar{x}_i\,\Delta A_i}{\Sigma\,\Delta A_i} \qquad \text{and} \qquad \bar{y} = \frac{\Sigma \bar{y}_i\,\Delta A_i}{\Sigma\,\Delta A_i}$$

Example A.2 Determine the centroidal coordinates of the C-shaped region shown in Fig. A.3.

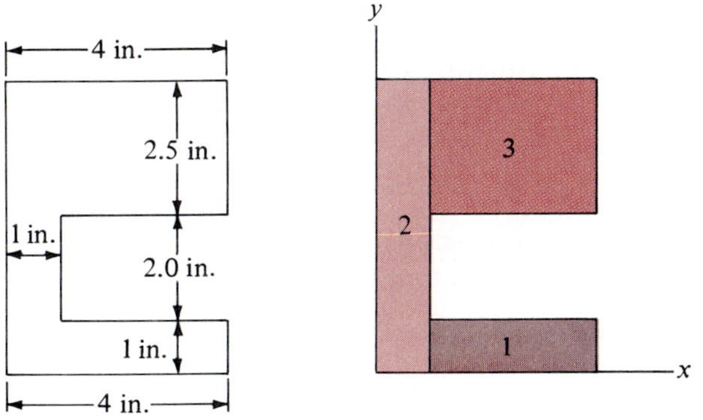

Figure A.3

Solution As indicated, subdivide the area into three pieces each having known locations for the centroids. Then construct the following table.

i	A_i(in²)	x_i(in.)	y_i(in.)	A_ix_i(in³)	A_iy_i(in³)
1	3.0	2.5	0.5	7.5	1.5
2	5.5	0.5	2.75	2.75	15.125
3	7.5	2.5	4.25	18.75	31.875
	$\Sigma =$ 16.0			$\Sigma =$ 29.0	$\Sigma =$ 48.5

from which

$$\bar{x} = \frac{29 \text{ in}^3}{16 \text{ in}^2} = 1.81 \text{ in.} \quad \text{and} \quad \bar{y} = \frac{48.5 \text{ in}^3}{16 \text{ in}^2} = 3.03 \text{ in.}$$

Note in passing that when the origin of the coordinate system is located at the centroid, it follows that $\bar{x} = \bar{y} = 0$ and that

$$\int_{\text{Area}} x \, dA = 0 \quad \text{and} \quad \int_{\text{Area}} y \, dA = 0$$

In a case where the axis is an axis of symmetry, such as the y-axis shown in Fig. A.4, it follows immediately that the centroid must lie somewhere on the axis of symmetry due to the fact that in the expression for the first moment, namely,

$$Q_y = \int_{\text{Area}} x \, dA$$

there are equal and opposite contributions to the integral from the area to the left and right of the axis of symmetry.

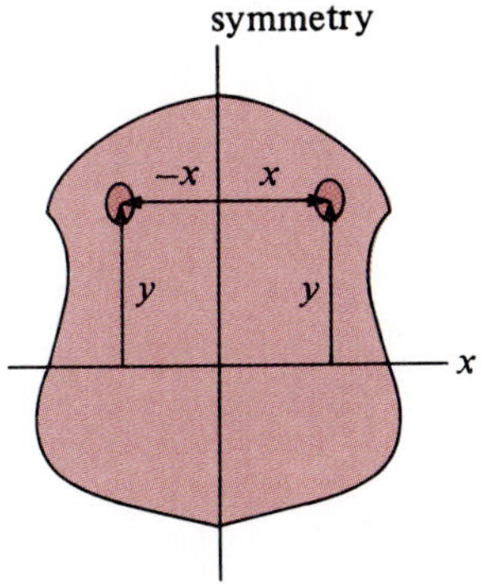

Figure A.4

A.2 SECOND MOMENTS OF AREA

With reference to Fig. A.1a, the second moments of area are defined as

$$I_x = \int_{\text{Area}} y^2 \, dA \qquad I_y = \int_{\text{Area}} x^2 \, dA \qquad \text{and} \qquad I_{xy} = \int_{\text{Area}} xy \, dA$$

When the boundaries of the area can be represented by simple lines and curves, the second moments of area can be determined by evaluating the integrals.

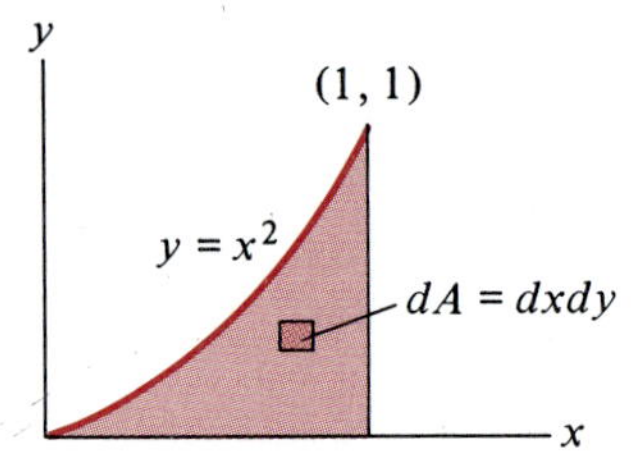

Figure A.5

Example A.3 For the area of Example A.1 determine the three second moments of the area with respect to the axes shown in Fig. A.5.

Solution With $dA = dx\,dy$, the second moment about the x-axis is

$$I_x = \int_{\text{Area}} y^2 \, dA = \int_0^1 \left(\int_{\sqrt{y}}^1 dx \right) y^2 \, dy = \int_0^1 y^2(1 - \sqrt{y}) \, dy = \tfrac{1}{21}$$

The second moment about the y-axis is

$$I_y = \int_{\text{Area}} x^2 \, dA = \int_0^1 \left(\int_0^{x^2} dy \right) x^2 \, dx = \int_0^1 x^4 \, dx = \tfrac{1}{5}$$

The product of the area is given by

$$I_{xy} = \int_{\text{Area}} xy \, dA = \int_0^1 \left(\int_0^{x^2} y \, dy \right) x \, dx = \int_0^1 \frac{x^5}{2} \, dx = \tfrac{1}{12}$$

It is frequently necessary to relate the second moments of the areas with respect to one set of axes to a second parallel set of axes as indicated in Fig. A.6.

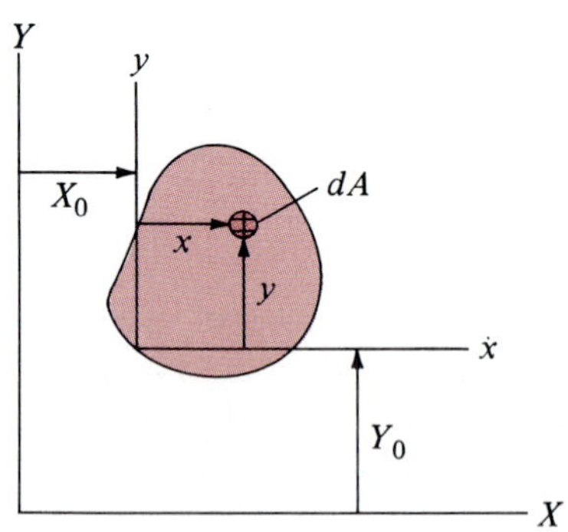

Figure A.6 Parallel axis transfer

By definition, the second moment about the X-axis is

$$I_X = \int_{\text{Area}} Y^2\, dA = \int_{\text{Area}} (Y_0 + y)^2\, dA = AY_0^2 + 2Y_0A\bar{y} + I_x$$

If the xy-axes are centroidal axes, $\bar{y} = 0$ and the transfer between the x- and X-axes becomes

$$I_X = I_x + AY_0^2$$

showing clearly that the second moment increases with the distance away from the centroidal axis. In a completely similar fashion

$$I_Y = \int_{\text{Area}} X^2\, dA = \int_{\text{Area}} (X_0 + x)^2\, dA = AX_0^2 + 2X_0A\bar{x} + I_y$$

If the xy-axes are centroidal axes, $\bar{x} = 0$ and the transfer between the y- and Y-axes becomes

$$I_Y = I_y + AX_0^2$$

Last, consider the product of the area

$$I_{XY} = \int_{\text{Area}} XY\, dA = \int_{\text{Area}} (X_0 + x)(Y_0 + y)\, dA$$

$$= AX_0Y_0 + A(Y_0\bar{x} + X_0\bar{x}) + I_{xy}$$

which, for the xy-axes as centroidal, yields

$$I_{XY} = I_{xy} + AX_0Y_0$$

The collection

$$I_X = I_x + AY_0^2$$
$$I_Y = I_y + AX_0^2$$
$$I_{XY} = I_{xy} + AX_0Y_0$$

is referred to as the **transfer formulae** between sets of parallel axes, with the very important restriction that the xy-axes are centroidal axes. Note that these transfer formulae can also be written as

$$I_x = I_X - AY_0^2$$
$$I_y = I_Y - AX_0^2$$
$$I_{xy} = I_{XY} - AX_0Y_0$$

When an area possesses an axis of symmetry such as the area shown in Fig. A.7, and when either the x- or y-axis coincides with that axis of symmetry, the product of the area $I_{xy} \equiv 0$. From observation of the figure, it follows that for every $xy\, dA$ that contributes to the integral there is an equal and opposite $-xy\, dA$, so that the sum is zero.

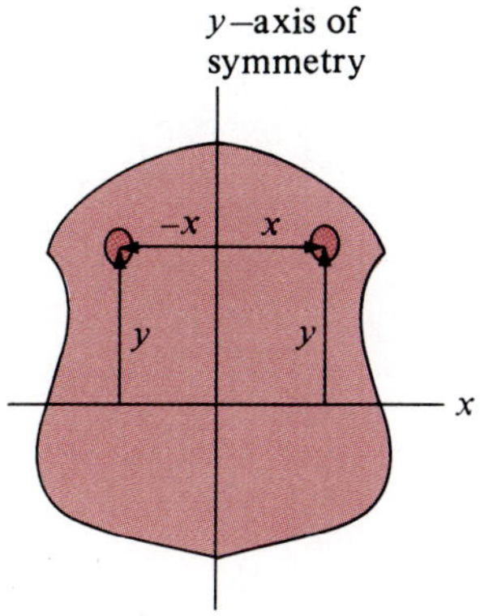

Figure A.7

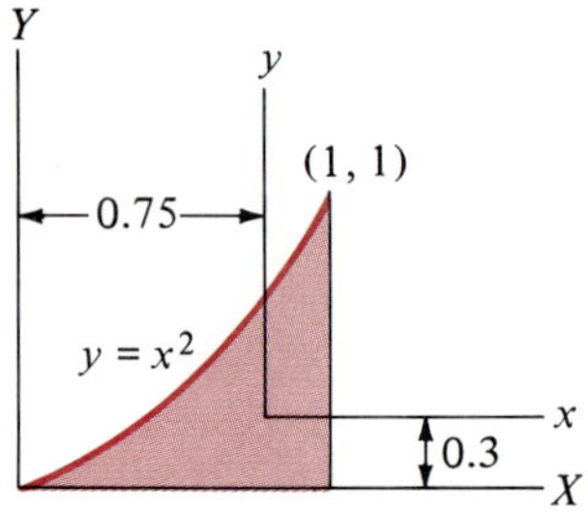

Figure A.8

Example A.4 Determine the second moments and product of the area in Examples A.1 and A.3, shown in Fig. A.8, with respect to the centroidal axes.

Solution As indicated in the figure the xy-system is the centroidal system. With $X_0 = 0.75$ and $Y_0 = 0.3$ from Example A.1 and $I_X = \frac{1}{21}$, $I_Y = \frac{1}{5}$, and $I_{XY} = \frac{1}{12}$ from Example A.3, the transfer formulae can be used according to

$$I_x = I_X - AY_0^2 = \frac{1}{21} - \frac{1}{3}\left(\frac{3}{10}\right)^2 = \frac{37}{2100}$$

$$I_y = I_Y - AX_0^2 = \frac{1}{5} - \frac{1}{3}\left(\frac{3}{4}\right)^2 = \frac{1}{80}$$

$$I_{xy} = I_{XY} - AX_0Y_0 = \frac{1}{12} - \frac{1}{3}\frac{3}{4}\frac{3}{10} = \frac{1}{120}$$

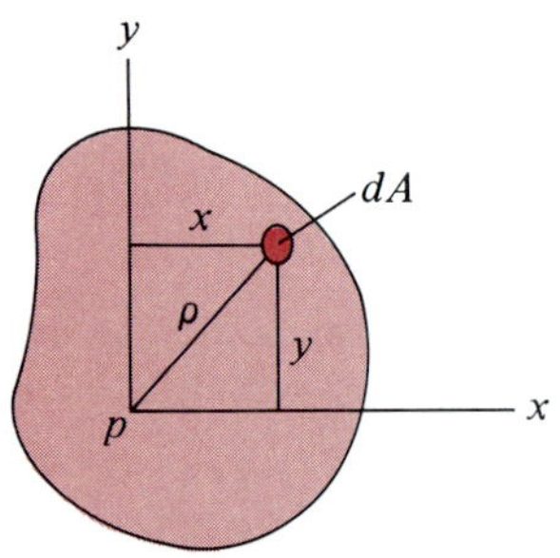

Figure A.9

Another useful property of the area is the polar moment of the area defined as

$$I_p = \int_{\text{Area}} \rho^2 \, dA$$

where as shown in Fig. A.9, ρ is the distance from the point p to the location of the element of area dA.

Recognizing that $\rho^2 = x^2 + y^2$, the polar moment of the area can be represented as

$$I_p = \int_{\text{Area}} \rho^2 \, dA = \int_{\text{Area}} (x^2 + y^2) \, dA = \int_{\text{Area}} x^2 \, dA + \int_{\text{Area}} y^2 \, dA = I_y + I_x$$

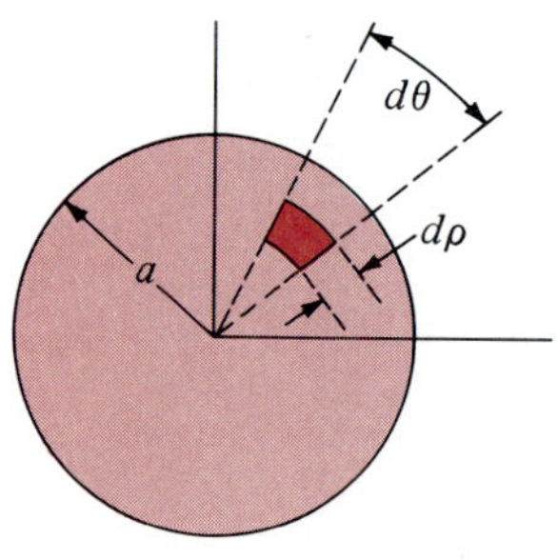

Figure A.10

Example A.5 Determine the polar moment of the area for a circle with respect to its center (see Fig. A.10).

Solution Referring to the figure, the differential area element is taken as $dA = \rho \, d\rho \, d\theta$. The integral is set up as

$$I_p = \int_0^{2\pi} \int_0^a \rho^2 \rho \, d\rho \, d\theta = \int_0^{2\pi} \frac{a^4}{4} \, d\theta = \frac{\pi a^4}{2}$$

Note that the same result is obtained using $I_p = I_y + I_x = \pi a^4/4 + \pi a^4/4$.

A.3 SECOND MOMENTS FOR COMPOSITE AREAS

A frequently encountered situation involves the calculation of the properties of areas with irregular shapes such as shown in Fig. A.11.

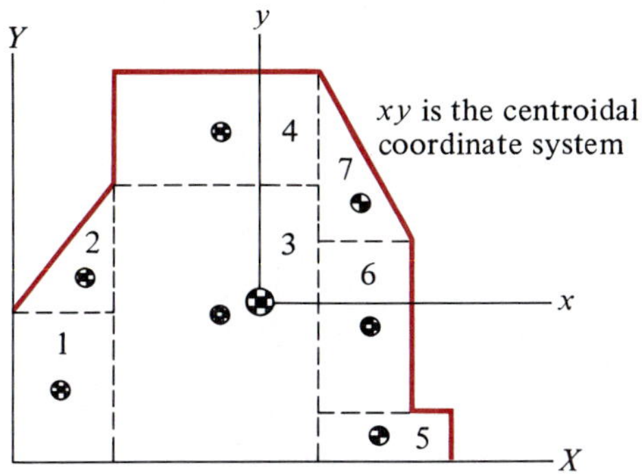

Figure A.11

Such an area is generally subdivided as shown, so that each of the subareas has known first- and second-moment properties. The location of the centroidal coordinates are then determined using

$$\bar{X} = \frac{\Sigma \bar{X}_i A_i}{\Sigma A_i} \quad \text{and} \quad \bar{Y} = \frac{\Sigma \bar{Y}_i A_i}{\Sigma A_i}$$

The second moments are then determined using

$$I_x = \Sigma I_{xoi} + \Sigma A_i \, d_{yi}^2$$
$$I_y = \Sigma I_{yoi} + \Sigma A_i \, d_{xi}^2$$
$$I_{xy} = \Sigma I_{xyoi} + \Sigma A_i \, d_{xi} d_{yi}$$

where I_{xoi}, I_{yoi}, and I_{xyoi} are the second moments of each of the individual areas with respect to their own centroidal axes. The distances d_{xi} and d_{yi} are measured from the centroid of the composite section to the centroids of the individual areas. The calculations are most easily carried out by means of a table, as indicated in the following example.

Example A.6 Determine the centroidal second moments for the area of Example A.2 (see Fig. A.12).

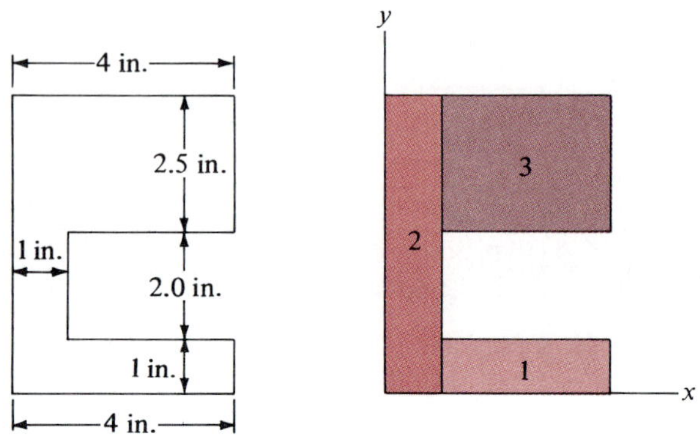

Figure A.12

Solution As indicated the area is subdivided into the areas shown. The table, part of which is repeated from Example A.2, is used as shown to carry out the calculations. Note that since each of the subareas is a rectangle with two planes of symmetry, $I_{xyoi} = 0$ for each of the subareas.

i	A_i(in^2)	x_i(in.)	y_i(in.)	A_ix_i(in^3)	A_iy_i(in^3)	dx_i(in.)	dy_i(in.)	I_{xoi}(in^4)	I_{yoi}(in^4)
1	3.0	2.5	0.5	7.5	1.5	0.69	−2.53	0.250	2.250
2	5.5	0.5	2.75	2.75	15.125	−1.31	−0.28	13.865	0.458
3	7.5	2.5	4.25	18.75	31.875	0.69	1.22	3.906	5.625
	Σ = 16.0			Σ = 29.0	Σ = 48.5			Σ = 18.021	Σ = 8.333

$$\bar{x} = \frac{29 \text{ in}^3}{16 \text{ in}^2} = 1.81 \text{ in.} \qquad \text{and} \qquad \bar{y} = \frac{48.5 \text{ in}^3}{16 \text{ in}^2} = 3.03 \text{ in.}$$

$$I_x = [18.021 + 3.0(-2.53)^2 + 5.5(-0.28)^2 + 7.5(1.22)^2]\text{in}^4 = 48.82 \text{ in}^4$$

$$I_y = [8.333 + 3.0(0.69)^2 + 5.5(-1.31)^2 + 7.5(0.69)^2]\text{in}^4 = 22.77 \text{ in}^4$$

$$I_{xy} = [0 + 3.0(-2.53)(0.69) + 5.5(-0.28)(-1.31) + 7.5(1.22)(0.69)]\text{in}^4$$
$$= 3.09 \text{ in}^4$$

A.4 ROTATION OF AXES AND PRINCIPAL SECOND MOMENTS

The second moments depend quite strongly upon the orientation of the coordinate axes. As an example, consider the thin rectangle shown in Fig. A.13.

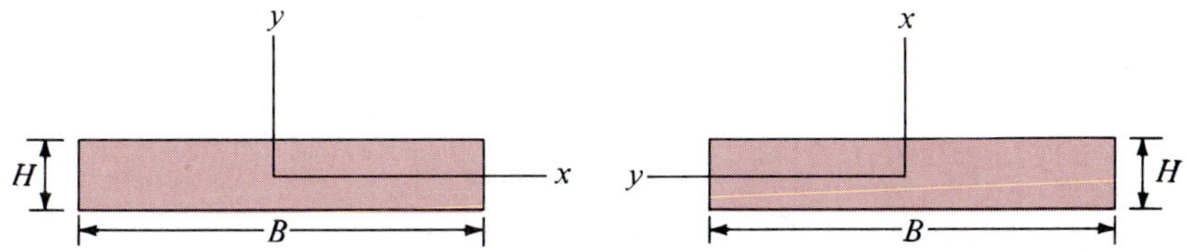

Figure A.13 Different orientations of a rectangle

In the first case $I_x = BH^3/12$ and $I_y = B^3H/12$, and in the second case $I_y = BH^3/12$ and $I_x = B^3H/12$, that is, the values have been switched as a result of having rotated the axes through an angle of 90° counterclockwise.

Consider a general case where two centroidal coordinate systems are related as shown in Fig. A.14. We will consider that the second moments are known in the xy system and that we wish to determine the corresponding second moments with respect to the $x'y'$ system obtained by rotating the xy system through a counterclockwise angle θ.

At the heart of the problem is the relation between the coordinates of a point in the region expressed in terms of each of the two sets of coordinates. In this regard consider Fig. A.15.

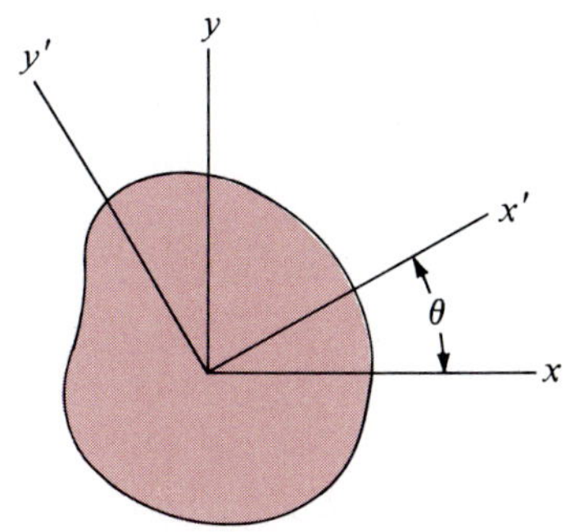

Figure A.14 Rotation of the axes

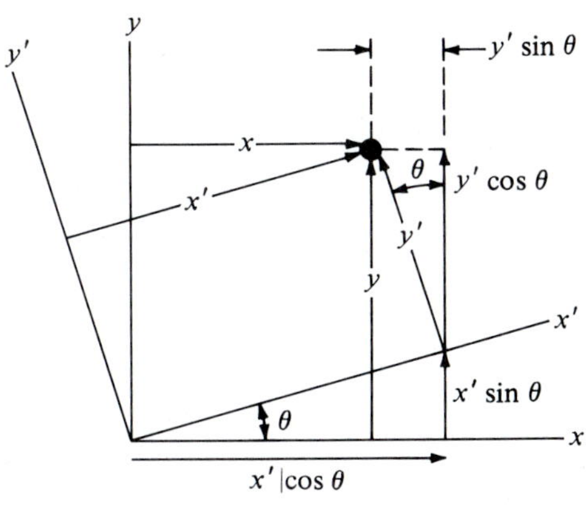

Figure A.15 Relation between xy and $x'y'$ coordinates

Consulting the figure it is seen that

$$x = x' \cos\theta - y' \sin\theta \qquad \text{and} \qquad y = x' \sin\theta + y' \cos\theta$$

or solving for the x' and y' in terms of the x and y,

$$x' = x \cos\theta + y \sin\theta \qquad \text{and} \qquad y' = -x \sin\theta + y \cos\theta$$

The second moments in the $x'y'$ system can be represented as

$$\begin{aligned} I_{x'} &= \int y'^2\, dA = \int (-x \sin\theta + y \cos\theta)^2\, dA \\ I_{y'} &= \int x'^2\, dA = \int (x \cos\theta + y \sin\theta)^2\, dA \\ I_{x'y'} &= \int x'y'\, dA = \int (-x \sin\theta + y \cos\theta)(x \cos\theta + y \sin\theta)\, dA \end{aligned}$$

Expanding and recognizing that $I_x = \int y^2\, dA$, $I_y = \int x^2\, dA$, and $I_{xy} = \int xy\, dA$ results in

$$\begin{aligned} I_{x'} &= I_x \cos^2\theta + I_y \sin^2\theta - 2I_{xy} \sin\theta \cos\theta \\ I_{y'} &= I_x \sin^2\theta + I_y \cos^2\theta + 2I_{xy} \sin\theta \cos\theta \\ I_{x'y'} &= (I_x - I_y) \sin\theta \cos\theta + I_{xy}(\cos^2\theta - \sin^2\theta) \end{aligned} \tag{A.1}$$

expressing the second moments referred to the $x'y'$ system in terms of the corresponding known values in the xy system. Eqns. (A.1) are referred to as the **transformation formulae.** Note the fact that the form is essentially identical to the corresponding stress transformation formulae in Chapter 2 and the corresponding strain transformation formulae in Chapter 3. When the transformation formulae are expressed in terms of the double-angle trigonometric functions, there results

$$\begin{aligned} I_{x'} &= \frac{I_x + I_y}{2} + \frac{I_x - I_y}{2} \cos 2\theta - I_{xy} \sin 2\theta \\ I_{y'} &= \frac{I_x + I_y}{2} - \frac{I_x - I_y}{2} \cos 2\theta + I_{xy} \sin 2\theta \\ I_{x'y'} &= \frac{I_x - I_y}{2} \sin 2\theta + I_{xy} \cos 2\theta \end{aligned}$$

By adding the first two of these equations it is easily seen that $I_{x'} + I_{y'} = I_x + I_y$, referred to as the first invariant, stating that $I_x + I_y$ has the same value in any centroidal coordinate system.

Example A.7 Determine the second moments with respect to the $x'y'$ system for the rectangle shown in Fig. A.16.

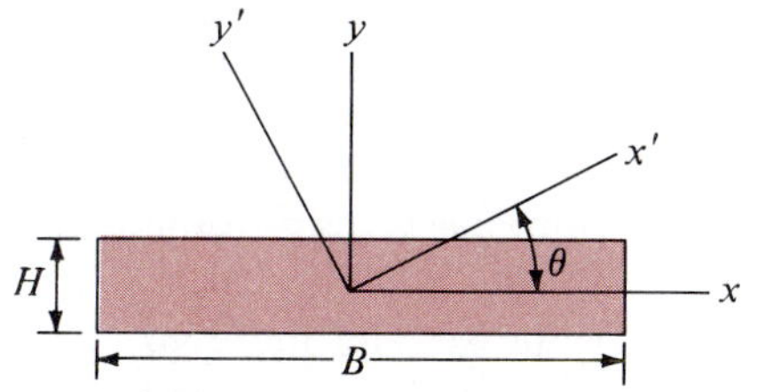

Figure A.16

Solution With $I_x = BH^3/12$ and $I_x = B^3H/12$, the use of Eqns. (A.1) yields

$$I_{x'} = (BH^3/12)\cos^2\theta + (B^3H/12)\sin^2\theta$$
$$I_{y'} = (BH^3/12)\sin^2\theta + (B^3H/12)\cos^2\theta$$
$$I_{x'y'} = (BH^3/12 - B^3H/12)\sin\theta\cos\theta$$

At $\theta = \pi/2$, $I_{x'} = (B^3H/12)$, $I_{y'} = (BH^3/12)$ and $I_{x'y'} = 0$ as should be the case since $x' = y$ and $y' = -x$. For any value of θ other than 0 or multiples of $\pi/2$, the product $I_{x'y'} \neq 0$.

Of special interest are the values of θ for which the maximum and minimum values of the second moments occur. These maximum and minimum values are called **principal values** with the corresponding values of θ called **principal axes.** To determine these values we compute the derivative of $I_{x'}$ with respect to θ, using the double-angle form of the transformation equations to obtain

$$\frac{dI_{x'}}{d\theta} = \frac{I_x - I_y}{2}(-2\sin 2\theta) - I_{xy}\, 2\cos 2\theta = 0$$

from which

$$\tan 2\theta_p = -\frac{2I_{xy}}{I_x - I_y} \tag{A.2}$$

where the roots θ_p correspond to directions of the principal axes. Solving Eqn. (A.2) yields two distinct roots for $2\theta_p$ differing by 180° in the interval $0 \le \theta \le 360°$. The two corresponding values of θ_p differ by 90° and define the principal directions or axes. One of the corresponding values of $I_{x'}$ is the maximum, the other the minimum.

For an area possessing an axis of symmetry, $I_{xy} = 0$ when either the x- or y-axis coincides with the axis of symmetry. Even if the area has no symmetry there is still an orientation of the centroidal coordinate system for which the product $I_{x'y'} = 0$. Set $I_{x'y'} = 0$ to obtain

$$\tan 2\theta = -\frac{2I_{xy}}{I_x - I_y}$$

which is exactly the same as the equation defining the orientation of the principal axes. Thus the principal axes and the axes for which $I_{x'y'} = 0$ coincide.

The principal values of the second moments can be determined by substituting the principal values θ_p into the transformation formulae. After some algebra, the results are

$$I_{\text{max}} = \frac{I_x + I_y}{2} + \sqrt{\left(\frac{I_x - I_y}{2}\right)^2 + I_{xy}^2}$$

$$I_{\text{min}} = \frac{I_x + I_y}{2} - \sqrt{\left(\frac{I_x - I_y}{2}\right)^2 + I_{xy}^2}$$

Example A.8 Determine the location of the principal axes and the values of the principal moments for the area of Example A.6 shown in Fig. A.17.

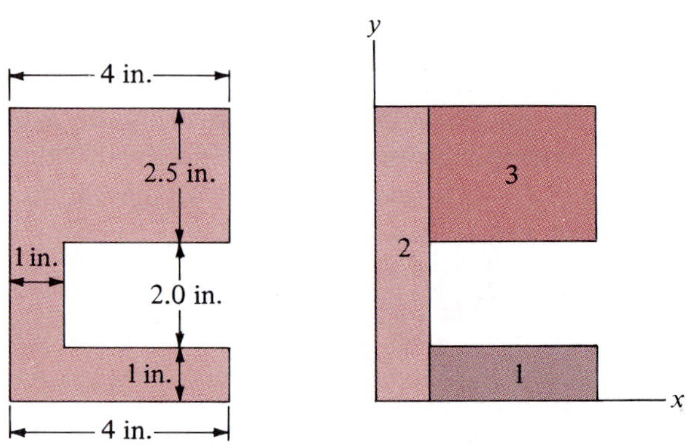

Figure A.17

Solution From Example A.6, $I_x = 48.82\ \text{in}^4$, $I_y = 22.77\ \text{in}^4$, and $I_{xy} = 3.09\ \text{in}^4$. The principal angles are given by

$$\tan 2\theta_p = -\frac{2(3.09)}{48.82 - 22.77} = -0.2376$$

from which $2\theta_p = -13.36°$ and $166.64°$, with $\theta_p = -6.68°$ and $83.32°$. The corresponding principal values for the second moments are

$$I_{\text{max}} = I_1 = \left[\frac{71.59}{2} + \sqrt{\left(\frac{26.05}{2}\right)^2 + 3.09^2}\right] \text{in}^4 = 49.18\ \text{in}^4$$

$$I_{\text{min}} = I_2 = \left[\frac{71.59}{2} - \sqrt{\left(\frac{26.05}{2}\right)^2 + 3.09^2}\right] \text{in}^4 = 22.41\ \text{in}^4$$

The orientation of the principal axes and the corresponding principal values are shown in Fig. A.18.

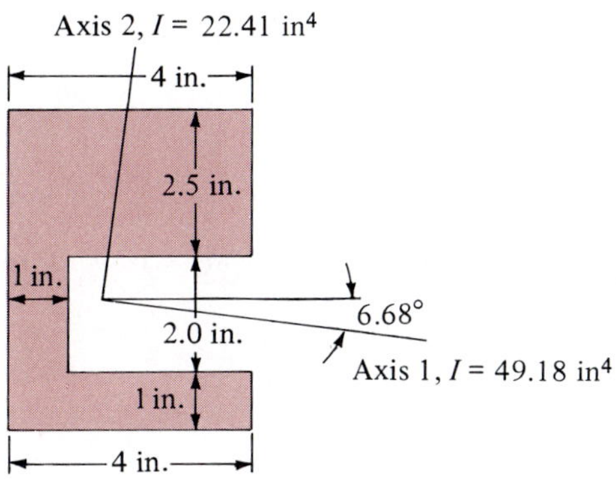

Figure A.18

Shown in Fig. A.19 are several common areas along with their centroidal locations and second moments with respect to different sets of axes.

Figure A.19 Useful area properties

Rectangle

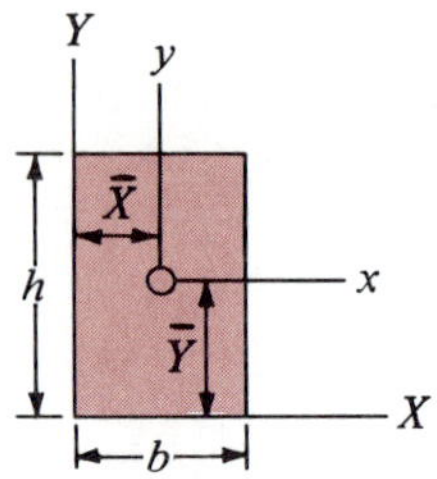

$$A = \frac{bh}{2} \qquad \bar{X} = \frac{b}{2} \qquad \bar{Y} = \frac{h}{2}$$

$$I_x = \frac{bh^3}{12} \qquad I_y = \frac{b^3h}{12} \qquad I_{xy} = 0$$

$$I_X = \frac{bh^3}{3} \qquad I_Y = \frac{b^3h}{3} \qquad I_{XY} = 0$$

Circle

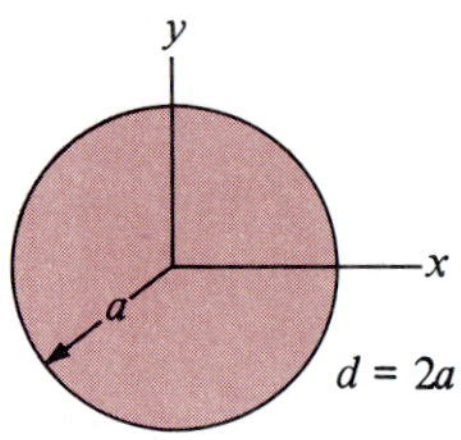

$$A = \pi a^2 = \frac{\pi d^2}{4}$$

$$I_x = I_y = \frac{\pi a^4}{4} = \frac{\pi d^4}{64} \qquad I_{xy} = 0$$

$$J = I_p = \frac{\pi a^4}{2} = \frac{\pi d^4}{32}$$

Triangle

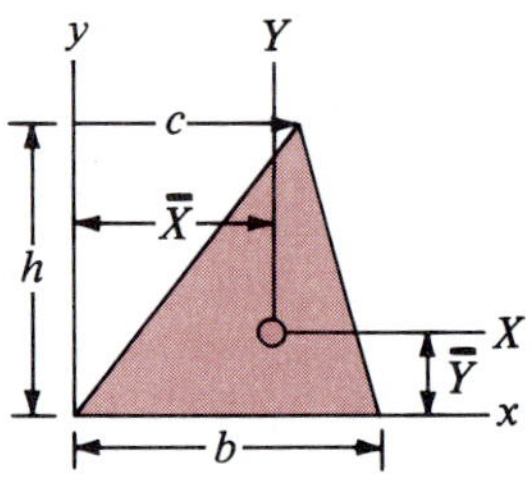

$$A = \frac{bh}{2} \qquad \bar{X} = \frac{b + c}{3} \qquad \bar{Y} = \frac{h}{3}$$

$$I_X = \frac{bh^3}{12} \qquad I_Y = \frac{bh(b^2 + bc + c^2)}{12} \qquad I_{XY} = \frac{bh^2(b + 2c)}{24}$$

Power Curve

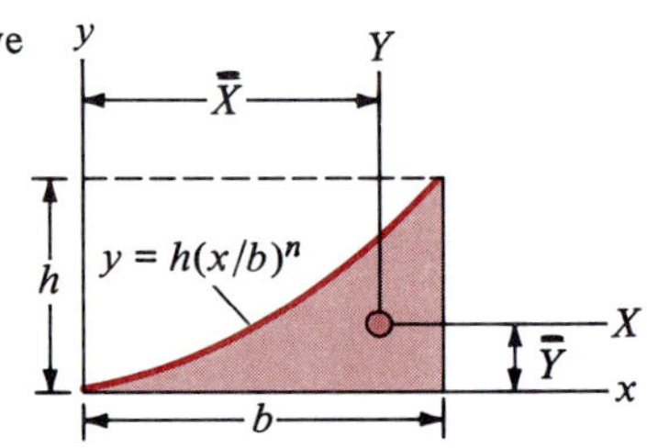

$$A = \frac{bh}{n + 1} \qquad \bar{X} = \frac{b(n + 1)}{n + 2} \qquad \bar{Y} = \frac{h(n + 1)}{2(2n + 1)}$$

$$I_X = \frac{bh^3}{3(3n + 1)} \qquad I_Y = \frac{hb^3}{n + 3} \qquad I_{XY} = \frac{b^2h^2}{4(n + 1)}$$

APPENDIX B

MECHANICAL PROPERTIES OF SELECTED ENGINEERING MATERIALS

Table B.1 Typical properties of selected materials used in engineering[1] (U.S. customary units)

Material	Specific weight, lbf/in^3	Ultimate strength: Tension, ksi	Ultimate strength: Compression,[2] ksi	Ultimate strength: Shear, ksi	Yield strength[3]: Tension, ksi	Yield strength[3]: Shear, ksi	Modulus of elasticity, 10^6 psi	Modulus of rigidity, 10^6 psi	Coefficient of thermal expansion, 10^{-6}/°F	Ductility, percent elongation in 2 in.	Poisson's ratio ν
STEEL:											
Structural (ASTM-A36)	0.284	58	—	—	36	21	29	11.5	6.5	23	0.27–0.30
High-strength–low-alloy ASTM-A242	0.284	70	—	—	50	30	29	11.5	6.5	21	0.27–0.30
Quenched and tempered alloy ASTM-A514	0.284	120	—	—	100	55	29	11.5	6.5	18	0.27–0.30
Stainless, (302) Cold-rolled	0.286	125	—	—	75	—	28	10.6	9.6	12	0.27–0.30
Annealed	0.286	90	—	—	40	22	28	10.6	9.6	50	0.27–0.30
CAST IRON:											
Gray, 4.5% C, ASTM A-48	0.260	25	95	35	—	—	10	4.1	6.7	0.5	0.2–0.3
Malleable, ASTM A-47	0.264	50	90	48	33	—	24	9.3	6.7	10	0.2–0.3
ALUMINUM:											
Alloy 2014-T6 (4.4% Cu)	0.101	70	—	42	60	32	10.6	3.9	12.8	13	0.33
Alloy 6061-T6 (1% Mg)	0.098	42	—	27	37	20	10	3.7	13.1	17	0.33
YELLOW BRASS (65% Cu, 35% Zn):											
Cold-rolled	0.306	78	—	43	63	36	15	—	11.3	8	0.34
Annealed	0.306	48	—	32	15	9	15	5.6	11.3	62	0.34
PHOSPHOR BRONZE:											
Cold-rolled (510)	0.320	81	—	—	75	40	15.9	5.9	9.9	10	0.34
Spring temper (524)	0.317	122	—	—	—	—	16	—	10.2	4	0.34
MAGNESIUM ALLOY (8.5% Al)	0.065	55	—	24	40	—	6.5	—	14.5	7	0.35
TITANIUM ALLOY (6% Al, 4% V)	0.161	130	—	—	120	—	16.5	—	5.3	10	0.33
MONEL ALLOY 400 (Ni-Cu):											
Cold-worked	0.319	98	—	—	85	50	26	—	7.7	22	0.32
Annealed	0.319	80	—	—	32	18	26	—	7.7	46	0.32
TIMBER, air dry:[4]											
Douglas fir	0.019	—	7.2	1.1	—	—	1.8	—	Varies:	—	—
Southern pine	0.022	—	7.3	1.3	—	—	1.6	—	1.7 to 2.5	—	—
CONCRETE:											
Medium strength	0.084	—	4.0	—	—	—	3.6	—	5.5	—	0.1–0.2
High strength	0.084	—	6.0	—	—	—	4.5	—	5.5	—	0.1–0.2
Nylon, molded	0.040	8	—	—	—	—	0.3	—	45	50	0.4
Polystyrene	0.038	7	13	8	—	—	0.45	—	40	4	—
Neoprene	0.045	3.5	—	—	—	—	—	—	—	500	0.45–0.5
Granite	0.100	3	35	5	—	—	10	—	4		0.2–0.3
Glass, 98% silica	0.079	—	7	—	—	—	9.6	4.1	44		0.2–0.27

[1] Properties of metals vary widely as a result of variations in composition, heat treatment, and mechanical working.
[2] For ductile metals the compression strength is generally assumed to be equal to the tension strength.
[3] Offset of 0.2 percent.
[4] Timber properties are for loading parallel to the grain.
Source: Copyright American Society of Testing Materials. Reprinted with permission.

Table B.2 Typical properties of selected materials used in engineering[1] (SI units)

Material	Density, kg/m^3	Ultimate strength			Yield strength[3]		Modulus of elasticity, GPa	Modulus of rigidity, GPa	Coefficient of thermal expansion, 10^{-6}/°C	Ductility, percent elongation in 50 mm	Poisson's ratio ν
		Tension, MPa	Compression,[2] MPa	Shear, MPa	Tension, MPa	Shear, MPa					
STEEL:											
Structural (ASTM-A36)	7860	400	—	—	250	145	200	79	11.7	23	0.27–0.30
High-strength–low-alloy ASTM-A242	7860	480	—	—	345	210	200	79	11.7	21	0.27–0.30
Quenched and tempered alloy ASTM-A514	7860	825	—	—	690	380	200	79	11.7	18	0.27–0.30
Stainless, (302)											
Cold-rolled	7920	860	—	—	520	—	190	73	17.3	12	0.27–0.30
Annealed	7920	620	—	—	275	152	190	73	17.3	50	0.27–0.30
CAST IRON:											
Gray, 4.5% C, ASTM A-48	7200	170	650	240	—	—	70	28	12.1	0.5	0.2–0.3
Malleable, ASTM A-47	7300	340	620	330	230	—	165	64	12.1	10	0.2–0.3
ALUMINUM:											
Alloy 2014-T6 (4.4% Cu)	2800	480	—	290	410	220	72	27	23.0	13	0.33
Alloy 6061-T6 (1% Mg)	2710	290	—	185	255	140	69	26	23.6	17	0.33
YELLOW BRASS (65% Cu, 35% Zn):											
Cold-rolled	8470	540	—	300	435	250	105	39	20	8	0.34
Annealed	8470	330	—	220	105	65	105	39	20	62	0.34
PHOSPHOR BRONZE:											
Cold-rolled (510)	8860	560	—	—	520	275	110	41	17.8	10	0.34
Spring temper (524)	8780	840	—	—	—	—	110	—	18.4	4	0.34
MAGNESIUM ALLOY (8.5% Al)	1800	380	—	165	275	—	45	—	26	7	0.35
TITANIUM ALLOY (6% Al, 4% V)	4460	900	—	—	825	—	114	—	9.5	10	0.33
MONEL ALLOY 400 (Ni-Cu):											
Cold-worked	8830	675	—	—	580	345	180	—	13.9	22	0.32
Annealed	8830	550	—	—	220	125	180	—	13.9	46	0.32
TIMBER, air dry:[4]											
Douglas fir	525	—	50	7.5	—	—	12.5	—	Varies:	—	—
Southern pine	610	—	50	9.2	—	—	11	—	3.0 to 4.5	—	—
CONCRETE:											
Medium strength	2320	—	28	—	—	—	25	—	10	—	0.1–0.2
High strength	2320	—	40	—	—	—	30	—	10	—	0.1–0.2
Nylon, molded	1100	55	—	—	—	—	2	—	81	50	0.4
Polystyrene	1050	48	90	55	—	—	3	—	72	4	—
Neoprene	1250	24	—	—	—	—	—	—	—	500	0.45–0.5
Granite	2770	20	240	35	—	—	70	—	7	—	0.2–0.3
Glass, 98% silica	2190	—	50	—	—	—	65	28	80	—	0.2–0.27

[1] Properties of metals vary widely as a result of variations in composition, heat treatment, and mechanical working.
[2] For ductile metals the compression strength is generally assumed to be equal to the tension strength.
[3] Offset of 0.2 percent.
[4] Timber properties are for loading parallel to the grain.

APPENDIX C

PROPERTIES OF STRUCTURAL STEEL MEMBERS

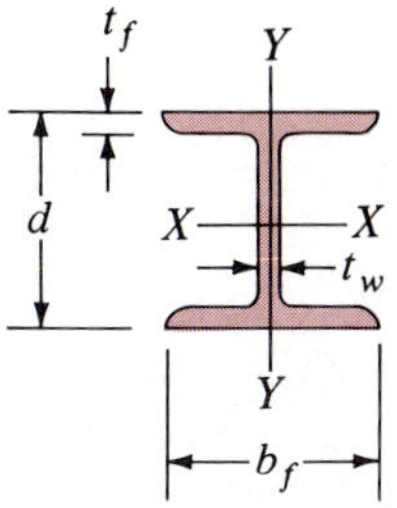

Table C.1a Properties of W shapes (wide-flange sections) (U.S. customary units)

			Flange			Axis X–X			Axis Y–Y		
Designation*	Area A, in^2	Depth d, in.	Width b_f, in.	Thickness t_f, in.	Web thickness t_w, in.	I_x, in^4	S_x, in^3	r_x, in.	I_y, in^4	S_y, in^3	r_y, in.
W36 × 300	88.3	36.74	16.655	1.680	0.945	20300	1110	15.2	1300	156	3.83
135	39.7	35.55	11.950	0.790	0.600	7800	439	14.0	225	37.7	2.38
W33 × 201	59.1	33.68	15.745	1.150	0.715	11500	684	14.0	749	95.2	3.56
118	34.7	32.86	11.480	0.740	0.550	5900	359	13.0	187	32.6	2.32
W30 × 173	50.8	30.44	14.985	1.065	0.655	8200	539	12.7	598	79.8	3.43
99	29.1	29.65	10.450	0.670	0.520	3990	269	11.7	128	24.5	2.10
W27 × 146	42.9	27.38	13.965	0.975	0.605	5630	411	11.4	443	63.5	3.21
84	24.8	26.71	9.960	0.640	0.460	2850	213	10.7	106	21.2	2.07
W24 × 104	30.6	24.06	12.750	0.750	0.500	3100	258	10.1	259	40.7	2.91
68	20.1	23.73	8.965	0.585	0.415	1830	154	9.55	70.4	15.7	1.87
W21 × 101	29.8	21.36	12.290	0.800	0.500	2420	227	9.02	248	40.3	2.89
62	18.3	20.99	8.240	0.615	0.400	1330	127	8.54	57.5	13.9	1.77
44	13.0	20.66	6.500	0.450	0.350	843	81.6	8.06	20.7	6.36	1.26
W18 × 106	31.1	18.73	11.200	0.940	0.590	1910	204	7.84	220	39.4	2.66
76	22.3	18.21	11.035	0.680	0.425	1330	146	7.73	152	27.6	2.61
50	14.7	17.99	7.495	0.570	0.355	800	88.9	7.38	40.1	10.7	1.65
35	10.3	17.70	6.000	0.425	0.300	510	57.6	7.04	15.3	5.12	1.22
W16 × 77	22.6	16.52	10.295	0.760	0.455	1110	134	7.00	138	26.9	2.47
57	16.8	16.43	7.120	0.715	0.430	758	92.2	6.72	43.1	12.1	1.60
40	11.8	16.01	6.995	0.505	0.305	518	64.7	6.63	28.9	8.25	1.57
31	9.12	15.88	5.525	0.440	0.275	375	47.2	6.41	12.4	4.49	1.17
26	7.68	15.69	5.500	0.345	0.250	301	38.4	6.26	9.59	3.49	1.12
W14 × 370	109	17.92	16.475	2.660	1.655	5440	607	7.07	1990	241	4.27
145	42.7	14.78	15.500	1.090	0.680	1710	232	6.33	677	87.3	3.98
82	24.1	14.31	10.130	0.855	0.510	882	123	6.05	148	29.3	2.48
68	20.0	14.04	10.035	0.720	0.415	723	103	6.01	121	24.2	2.46
53	15.6	13.92	8.060	0.660	0.370	541	77.8	5.89	57.7	14.3	1.92
43	12.6	13.66	7.995	0.530	0.305	428	62.7	5.82	45.2	11.3	1.89
38	11.2	14.10	6.770	0.515	0.310	385	54.6	5.88	26.7	7.88	1.55
30	8.85	13.84	6.730	0.385	0.270	291	42.0	5.73	19.6	5.82	1.49
26	7.69	13.91	5.025	0.420	0.255	245	35.3	5.65	8.91	3.54	1.08
22	6.49	13.74	5.000	0.335	0.230	199	29.0	5.54	7.00	2.80	1.04

* A wide-flange section is designated by the letter W followed by the nominal depth in inches and the weight in pounds per foot.
Source for all tables in Appendix C: *Manual of Steel Construction*, American Institute of Steel Construction. (Table continued on page C3)

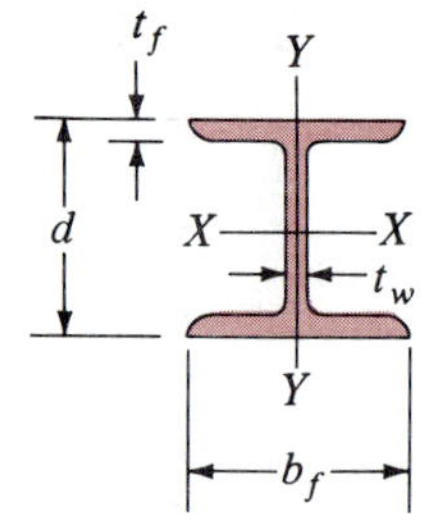

Table C.1b Properties of W shapes (wide-flange sections) (SI units)

Designation*	Area A, mm²	Depth d, mm	Flange: Width b_f, mm	Flange: Thickness t_f, mm	Web thickness t_w, mm	Axis X–X: I_x 10^6 mm⁴	Axis X–X: S_x 10^3 mm³	Axis X–X: r_x mm	Axis Y–Y: I_y 10^6 mm⁴	Axis Y–Y: S_y 10^3 mm³	Axis Y–Y: r_y mm
W920 × 446	57000	933	423	42.7	24.0	8450	18110	386	541	2560	97.3
201	25600	903	304	20.1	15.2	3250	7200	356	93.7	616	60.5
W840 × 299	38100	855	400	29.2	18.2	4790	11200	356	312	1560	90.4
176	22400	835	292	18.8	14.0	2460	5890	330	77.8	533	58.9
W760 × 257	32800	773	381	27.1	16.6	3410	8820	323	249	1307	87.1
147	18800	753	265	17.0	13.2	1660	4410	297	53.3	402	53.3
W690 × 217	27700	695	355	24.8	15.4	2340	6730	290	184.4	1039	81.5
125	16000	678	253	16.3	11.7	1186	3500	272	44.1	349	52.6
W610 × 155	19700	611	324	19.0	12.7	1290	4220	256	107.8	665	73.9
101	13000	603	228	14.9	10.5	762	2530	243	29.3	257	47.5
W530 × 150	19200	543	312	20.3	12.7	1007	3710	229	103.2	662	73.4
92	11800	533	209	15.6	10.2	554	2080	217	23.9	229	45.0
66	8390	525	165	11.4	8.9	351	1337	205	8.62	104.5	32.0
W460 × 158	20100	476	284	23.9	15.0	795	3340	199.1	91.6	645	67.6
113	14400	463	280	17.3	10.8	554	2390	196.3	63.3	452	66.3
74	9480	457	190	14.5	9.0	333	1457	187.5	16.69	175.7	41.9
52	6650	450	152	10.8	7.6	212	942	178.8	6.37	83.8	31.0
W410 × 114	14600	420	261	19.3	11.6	462	2200	177.8	57.4	440	62.7
85	10800	417	181	18.2	10.9	316	1516	170.7	17.94	198.2	40.6
60	7610	407	178	12.8	7.7	216	1061	168.4	12.03	135.2	39.9
46.1	5880	403	140	11.2	7.0	156.1	775	162.8	5.16	73.7	29.7
38.8	4950	399	140	8.8	6.4	125.3	628	159.0	3.99	57.0	28.4
W360 × 551	70300	455	418	67.6	42.0	2260	9930	179.6	828	3960	108.5
216	27500	375	394	27.7	17.3	712	3800	160.8	282	1431	101.1
122	15500	363	257	21.7	13.0	367	2020	153.7	61.6	479	63.0
101	12900	357	255	18.3	10.5	301	1686	152.7	50.4	395	62.5
79	10100	354	205	16.8	9.4	225	1271	149.6	24.0	234	48.8
64	8130	347	203	13.5	7.7	178.1	1027	147.8	18.81	185.3	48.0
57	7230	358	172	13.1	7.9	160.2	895	149.4	11.11	129.2	39.4
44.8	5710	352	171	9.8	6.9	121.1	688	145.5	8.16	95.4	37.8
39.0	4960	353	128	10.7	6.5	102.0	578	143.5	3.71	58.0	27.4
32.9	4190	349	127	8.5	5.8	82.8	474	140.7	2.91	45.8	26.4

* A wide-flange section is designated by the letter W followed by the nominal depth in millimeters and the mass in kilograms per meter.

(Table continued on page C4)

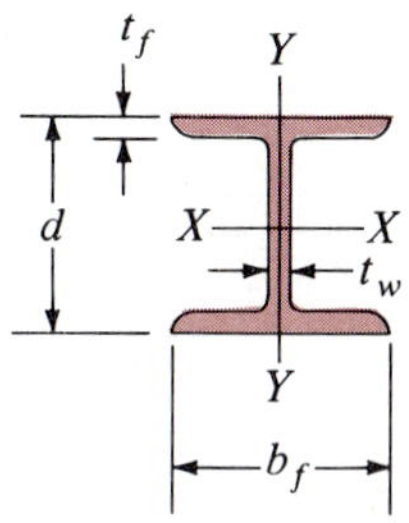

Table C.1a Properties of W shapes (wide-flange sections) (U.S. customary units)
(Continued from page C1)

			Flange		Web	Axis X–X			Axis Y–Y		
Designation*	**Area A, in²**	**Depth d, in.**	**Width b_f, in.**	**Thickness t_f, in.**	**thickness t_w, in.**	**I_x, in⁴**	**S_x, in³**	**r_x, in.**	**I_y, in⁴**	**S_y, in³**	**r_y, in.**
W12 × 96	28.2	12.71	12.160	0.900	0.550	833	131	5.44	270	44.4	3.09
72	21.1	12.25	12.040	0.670	0.430	597	97.4	5.31	195	32.4	3.04
50	14.7	12.19	8.080	0.640	0.370	394	64.7	5.18	56.3	13.9	1.96
40	11.8	11.94	8.005	0.515	0.295	310	51.9	5.13	44.1	11.0	1.93
35	10.3	12.50	6.560	0.520	0.300	285	45.6	5.25	24.5	7.47	1.54
30	8.79	12.34	6.520	0.440	0.260	238	38.6	5.21	20.3	6.24	1.52
26	7.65	12.22	6.490	0.380	0.230	204	33.4	5.17	17.3	5.34	1.51
22	6.48	12.31	4.030	0.425	0.260	156	25.4	4.91	4.66	2.31	0.848
16	4.71	11.99	3.990	0.265	0.220	103	17.1	4.67	2.82	1.41	0.773
W10 × 112	32.9	11.36	10.415	1.250	0.755	716	126	4.66	236	45.3	2.68
68	20.0	10.40	10.130	0.770	0.470	394	75.7	4.44	134	26.4	2.59
54	15.8	10.09	10.030	0.615	0.370	303	60.0	4.37	103	20.6	2.56
45	13.3	10.10	8.020	0.620	0.350	248	49.1	4.33	53.4	13.3	2.01
39	11.5	9.92	7.985	0.530	0.315	209	42.1	4.27	45.0	11.3	1.98
33	9.71	9.73	7.960	0.435	0.290	170	35.0	4.19	36.6	9.20	1.94
30	8.84	10.47	5.810	0.510	0.300	170	32.4	4.38	16.7	5.75	1.37
22	6.49	10.17	5.750	0.360	0.240	118	23.2	4.27	11.4	3.97	1.33
19	5.62	10.24	4.020	0.395	0.250	96.3	18.8	4.14	4.29	2.14	0.874
15	4.41	9.99	4.000	0.270	0.230	68.9	13.8	3.95	2.89	1.45	0.810
W8 × 58	17.1	8.75	8.220	0.810	0.510	228	52.0	3.65	75.1	18.3	2.10
48	14.1	8.50	8.110	0.685	0.400	184	43.3	3.61	60.9	15.0	2.08
40	11.7	8.25	8.070	0.560	0.360	146	35.5	3.53	49.1	12.2	2.04
35	10.3	8.12	8.020	0.495	0.310	127	31.2	3.51	42.6	10.6	2.03
31	9.13	8.00	7.995	0.435	0.285	110	27.5	3.47	37.1	9.27	2.02
28	8.25	8.06	6.535	0.465	0.285	98.0	24.3	3.45	21.7	6.63	1.62
24	7.08	7.93	6.495	0.400	0.245	82.8	20.9	3.42	18.3	5.63	1.61
21	6.16	8.28	5.270	0.400	0.250	75.3	18.2	3.49	9.77	3.71	1.26
18	5.26	8.14	5.250	0.330	0.230	61.9	15.2	3.43	7.97	3.04	1.23
15	4.44	8.11	4.015	0.315	0.245	48.0	11.8	3.29	3.41	1.70	0.876
13	3.84	7.99	4.000	0.255	0.230	39.6	9.91	3.21	2.73	1.37	0.843
W6 × 25	7.34	6.38	6.080	0.455	0.320	53.4	16.7	2.70	17.1	5.61	1.52
20	5.87	6.20	6.020	0.365	0.260	41.4	13.4	2.66	13.3	4.41	1.50
16	4.74	6.28	4.030	0.405	0.260	32.1	10.2	2.60	4.43	2.20	0.967
12	3.55	6.03	4.000	0.280	0.230	22.1	7.31	2.49	2.99	1.50	0.918
9	2.68	5.90	3.940	0.215	0.170	16.4	5.56	2.47	2.20	1.11	0.905
W5 × 19	5.54	5.15	5.030	0.430	0.270	26.2	10.2	2.17	9.13	3.63	1.28
16	4.68	5.01	5.000	0.360	0.240	21.3	8.51	2.13	7.51	3.00	1.27
W4 × 13	3.83	4.16	4.060	0.345	0.280	11.3	5.46	1.72	3.86	1.90	1.00

* A wide-flange section is designated by the letter W followed by the nominal depth in inches and the weight in pounds per foot.

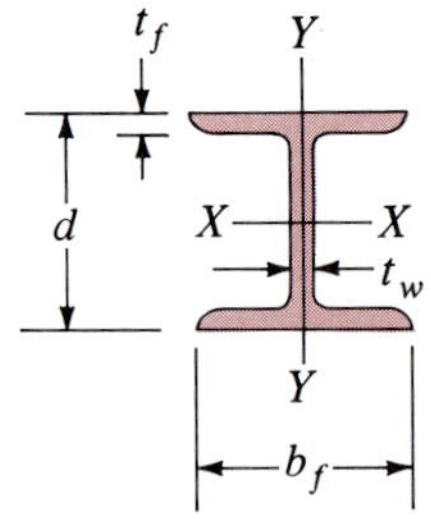

Table C.1b Properties of W shapes (wide-flange sections) (SI units)
(Continued from page C2)

			Flange		Web	Axis X–X			Axis Y–Y		
Designation*	**Area** A, **mm²**	**Depth** d, **mm**	**Width** b_f, **mm**	**Thickness** t_f, **mm**	**thickness** t_w, **mm**	I_x 10^6 **mm⁴**	S_x 10^3 **mm³**	r_x **mm**	I_y 10^6 **mm**	S_y 10^3 **mm³**	r_y **mm**
W310 × 143	18200	323	309	22.9	14.0	347	2150	138.2	112.4	728	78.5
107	13600	311	306	17.0	10.9	248	1595	134.9	81.2	531	77.2
74	9480	310	205	16.3	9.4	164.0	1058	131.6	23.4	228	49.8
60	7610	303	203	13.1	7.5	129.0	851	130.3	18.36	180.9	49.0
52	6650	317	167	13.2	7.6	118.6	748	133.4	10.20	122.2	39.1
44.5	5670	313	166	11.2	6.6	99.1	633	132.3	8.45	101.8	38.6
38.7	4940	310	165	9.7	5.8	84.9	548	131.3	7.20	87.3	38.4
32.7	4180	313	102	10.8	6.6	64.9	415	124.7	1.940	38.0	21.5
23.8	3040	305	101	6.7	5.6	42.9	281	118.6	1.174	23.2	19.63
W250 × 167	21200	289	265	31.8	19.2	298.0	2060	118.4	98.2	741	68.1
101	12900	264	257	19.6	11.9	164.0	1242	112.8	55.8	434	65.8
80	10200	256	255	15.6	9.4	126.1	985	111.0	42.8	336	65.0
67	8580	257	204	15.7	8.9	103.2	803	110.0	22.2	218	51.1
58	7420	252	203	13.5	8.0	87.0	690	108.5	18.73	184.5	50.3
49.1	6260	247	202	11.0	7.4	70.8	573	106.4	15.23	150.8	49.3
44.8	5700	266	148	13.0	7.6	70.8	532	111.3	6.95	93.9	34.8
32.7	4190	258	146	9.1	6.1	49.1	381	108.5	4.75	65.1	33.8
28.4	3630	260	102	10.0	6.4	40.1	308	105.2	1.796	35.2	22.2
22.3	2850	254	102	6.9	5.8	28.7	226	100.3	1.203	23.6	20.6
W200 × 86	11000	222	209	20.6	13.0	94.9	855	92.7	31.3	300	53.3
71	9100	216	206	17.4	10.2	76.6	709	91.7	25.3	246	52.8
59	7550	210	205	14.2	9.1	60.8	579	89.7	20.4	199.0	51.8
52	6650	206	204	12.6	7.9	52.9	514	89.2	17.73	173.8	51.6
46.1	5890	203	203	11.0	7.2	45.8	451	88.1	15.44	152.1	51.3
41.7	5320	205	166	11.8	7.2	40.8	398	87.6	9.03	108.8	41.1
35.9	4570	201	165	10.2	6.2	34.5	343	86.9	7.62	92.4	40.9
31.3	3970	210	134	10.2	6.4	31.3	298	88.6	4.07	60.7	32.0
26.6	3390	207	133	8.4	5.8	25.8	249	87.1	3.32	49.9	31.2
22.5	2860	206	102	8.0	6.2	20.0	194.2	83.6	1.419	27.8	22.3
19.3	2480	203	102	6.5	5.8	16.48	162.4	81.5	1.136	22.3	21.4
W150 × 37.1	4740	162	154	11.6	8.1	22.2	274	68.6	7.12	92.5	38.6
29.8	3790	157	153	9.3	6.6	17.23	219	67.6	5.54	72.4	38.1
24.0	3060	160	102	10.3	6.6	13.36	167.0	66.0	1.844	36.2	24.6
18.0	2290	153	102	7.1	5.8	9.20	120.3	63.2	1.245	24.4	23.3
13.5	1730	150	100	5.5	4.3	6.83	91.1	62.7	0.916	18.32	23.0
W130 × 28.1	3590	131	128	10.9	6.9	10.91	166.6	55.1	3.80	59.4	32.5
23.8	3040	127	127	9.1	6.1	8.87	139.7	54.1	3.13	49.3	32.3
W100 × 19.3	2470	106	103	8.8	7.1	4.70	88.7	43.7	1.607	31.2	25.4

* A wide-flange section is designated by the letter W followed by the nominal depth in millimeters and the mass in kilograms per meter.

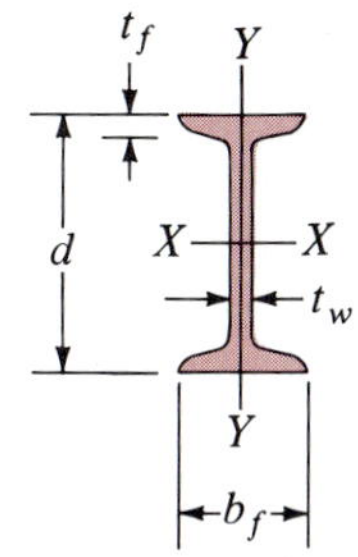

Table C.2a Properties of S shapes (American standard shapes) (U.S. customary units)

Designation*	Area A, in²	Depth d, in.	Flange Width b_f, in.	Flange Thickness t_f, in.	Web thickness t_w, in.	Axis X–X I_x, in⁴	Axis X–X S_x, in³	Axis X–X r_x, in.	Axis Y–Y I_y, in⁴	Axis Y–Y S_y, in³	Axis Y–Y r_y, in.
S24 × 100	29.4	24.00	7.247	0.871	0.747	2390	199	9.01	47.8	13.2	1.27
90	26.5	24.00	7.124	0.871	0.624	2250	187	9.22	44.9	12.6	1.30
79.9	23.5	24.00	7.001	0.871	0.501	2110	175	9.47	42.3	12.1	1.34
S20 × 95	27.9	20.00	7.200	0.916	0.800	1610	161	7.60	49.7	13.8	1.33
85	25.0	20.00	7.053	0.916	0.653	1520	152	7.79	46.2	13.1	1.36
75	22.1	20.00	6.391	0.789	0.641	1280	128	7.60	29.6	9.28	1.16
65.4	19.2	20.00	6.250	0.789	0.500	1180	118	7.84	27.4	8.77	1.19
S18 × 70	20.6	18.00	6.251	0.691	0.711	926	103	6.71	24.1	7.72	1.08
54.7	16.1	18.00	6.001	0.691	0.461	804	89.4	7.07	20.8	6.94	1.14
S15 × 50	14.7	15.00	5.640	0.622	0.550	486	64.8	5.75	15.7	5.57	1.03
42.9	12.6	15.00	5.501	0.622	0.411	447	59.6	5.95	14.4	5.23	1.07
S12 × 50	14.7	12.00	5.477	0.659	0.687	305	50.8	4.55	15.7	5.74	1.03
40.8	12.0	12.00	5.252	0.659	0.462	272	45.4	4.77	13.6	5.16	1.06
35	10.3	12.00	5.078	0.544	0.428	229	38.2	4.72	9.87	3.89	0.980
31.8	9.35	12.00	5.000	0.544	0.350	218	36.4	4.83	9.36	3.74	1.00
S10 × 35	10.3	10.00	4.944	0.491	0.594	147	29.4	3.78	8.36	3.38	0.901
25.4	7.46	10.00	4.661	0.491	0.311	124	24.7	4.07	6.79	2.91	0.954
S8 × 23	6.77	8.00	4.171	0.425	0.441	64.9	16.2	3.10	4.31	2.07	0.798
18.4	5.41	8.00	4.001	0.425	0.271	57.6	14.4	3.26	3.73	1.86	0.831
S7 × 20	5.88	7.00	3.860	0.392	0.450	42.4	12.1	2.69	3.17	1.64	0.734
15.3	4.50	7.00	3.662	0.392	0.252	36.7	10.5	2.86	2.64	1.44	0.766
S6 × 17.25	5.07	6.00	3.565	0.359	0.465	26.3	8.77	2.28	2.31	1.30	0.675
12.5	3.67	6.00	3.332	0.359	0.232	22.1	7.37	2.45	1.82	1.09	0.705
S5 × 14.75	4.34	5.00	3.284	0.326	0.494	15.2	6.09	1.87	1.67	1.01	0.620
10	2.94	5.00	3.004	0.326	0.214	12.3	4.92	2.05	1.22	0.809	0.643
S4 × 9.5	2.79	4.00	2.796	0.293	0.326	6.79	3.39	1.56	0.903	0.646	0.569
7.7	2.26	4.00	2.663	0.293	0.193	6.08	3.04	1.64	0.764	0.574	0.581
S3 × 7.5	2.21	3.00	2.509	0.260	0.349	2.93	1.95	1.15	0.586	0.468	0.516
5.7	1.67	3.00	2.330	0.260	0.170	2.52	1.68	1.23	0.455	0.390	0.522

* An American Standard Beam is designated by the letter S followed by the nominal depth in inches and the weight in pounds per foot.

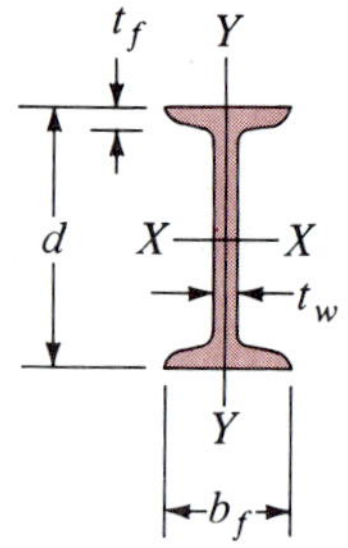

Table C.2b Properties of S shapes (American standard shapes) (SI units)

Designation*	Area A, mm²	Depth d, mm	Flange Width b_f, mm	Flange Thickness t_f, mm	Web thickness t_w, mm	Axis X–X I_x 10^6 mm⁴	Axis X–X S_x 10^3 mm³	Axis X–X r_x mm	Axis Y–Y I_y 10^6 mm	Axis Y–Y S_y 10^3 mm³	Axis Y–Y r_y mm
S610 × 149	18970	610	184	22.1	19.0	995	3260	229	19.90	216	32.3
134	17100	610	181	22.1	15.8	937	3070	234	18.69	207	33.0
118.9	15160	610	178	22.1	12.7	878	2880	241	17.61	197.9	34.0
S510 × 141	18000	508	183	23.3	20.3	670	2640	193.0	20.69	226	33.8
127	16130	508	179	23.3	16.6	633	2490	197.9	19.23	215	34.5
112	14260	508	162	20.1	16.3	533	2100	193.0	12.32	152.1	29.5
97.3	12390	508	159	20.1	12.7	491	1933	199.1	11.40	143.4	30.2
S460 × 104	13290	457	159	17.6	18.1	385	1685	170.4	10.03	126.2	27.4
81.4	10390	457	152	17.6	11.7	335	1466	179.6	8.66	113.9	29.0
S380 × 74	9480	381	143	15.8	14.0	202	1060	146.1	6.53	91.3	26.2
64	8130	381	140	15.8	10.4	186.1	977	151.1	5.99	85.6	27.2
S310 × 74	9480	305	139	16.8	17.4	127.0	833	115.6	6.53	94.0	26.2
60.7	7740	305	133	16.8	11.7	113.2	742	121.2	5.66	85.1	26.9
52	6640	305	129	13.8	10.9	95.3	625	119.9	4.11	63.7	24.9
47.3	6032	305	127	13.8	8.9	90.7	595	122.7	3.90	61.4	25.4
S250 × 52	6640	254	126	12.5	15.1	61.2	482	96.0	3.48	55.2	22.9
37.8	4806	254	118	12.5	7.9	51.6	406	103.4	2.83	48.0	24.2
S200 × 34	4368	203	106	10.8	11.2	27.0	266	78.7	1.794	33.8	20.3
27.4	3484	203	102	10.8	6.9	24.0	236	82.8	1.553	30.4	21.1
S180 × 30	3794	178	97	10.0	11.4	17.65	198.3	68.3	1.319	27.2	18.64
22.8	2890	178	92	10.0	6.4	15.28	171.7	72.6	1.099	23.9	19.45
S150 × 25.7	3271	152	90	9.1	11.8	10.95	144.1	57.9	0.961	21.4	17.15
18.6	2362	152	84	9.1	5.8	9.20	121.1	62.2	0.758	18.05	17.91
S130 × 22.0	2800	127	83	8.3	12.5	6.33	99.7	47.5	0.695	16.75	15.75
15	1884	127	76	8.3	5.3	5.12	80.6	52.1	0.508	13.37	16.33
S100 × 14.1	1800	102	70	7.4	8.3	2.83	55.5	39.6	0.376	10.74	14.45
11.5	1452	102	67	7.4	4.8	2.53	49.6	41.6	0.318	9.49	14.75
S75 × 11.2	1426	76	63	6.6	8.9	1.22	32.1	29.2	0.244	7.75	13.11
8.5	1077	76	59	6.6	4.3	1.05	27.6	31.3	0.189	6.41	13.26

* An American Standard Beam is designated by the letter S followed by the nominal depth in millimeters and the mass in kilograms per meter.

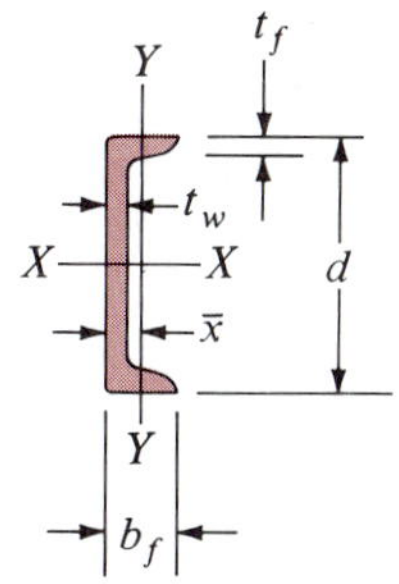

Table C.3a Properties of C shapes (American standard channels) (U.S. customary units)

Designation*	Area A, in²	Depth d, in.	Flange Width b_f, in.	Flange Thickness t_f, in.	Web thickness t_w, in.	Axis X–X: I_x, in⁴	Axis X–X: S_x, in³	Axis X–X: r_x, in.	Axis Y–Y: I_y, in⁴	Axis Y–Y: S_y, in³	Axis Y–Y: r_y, in.	Axis Y–Y: $\bar{x}$, in.
C15 × 50	14.7	15.00	3.716	0.650	0.716	404	53.8	5.24	11.0	3.78	0.867	0.799
40	11.8	15.00	3.520	0.650	0.520	349	46.5	5.44	9.23	3.36	0.886	0.778
33.9	9.96	15.00	3.400	0.650	0.400	315	42.0	5.62	8.13	3.11	0.904	0.787
C12 × 30	8.82	12.00	3.170	0.501	0.510	162	27.0	4.29	5.14	2.06	0.763	0.674
25	7.35	12.00	3.047	0.501	0.387	144	24.1	4.43	4.47	1.88	0.780	0.674
20.7	6.09	12.00	2.942	0.501	0.282	129	21.5	4.61	3.88	1.73	0.799	0.698
C10 × 30	8.82	10.00	3.033	0.436	0.673	103	20.7	3.42	3.94	1.65	0.669	0.649
25	7.35	10.00	2.886	0.436	0.526	91.2	18.2	3.52	3.36	1.48	0.676	0.617
20	5.88	10.00	2.739	0.436	0.379	78.9	15.8	3.66	2.81	1.32	0.691	0.606
15.3	4.49	10.00	2.600	0.436	0.240	67.4	13.5	3.87	2.28	1.16	0.713	0.634
C9 × 20	5.88	9.00	2.648	0.413	0.448	60.9	13.5	3.22	2.42	1.17	0.642	0.583
15	4.41	9.00	2.485	0.413	0.285	51.0	11.3	3.40	1.93	1.01	0.661	0.586
13.4	3.94	9.00	2.433	0.413	0.233	47.9	10.6	3.48	1.76	0.962	0.668	0.601
C8 × 18.75	5.51	8.00	2.527	0.390	0.487	44.0	11.0	2.82	1.98	1.01	0.599	0.565
13.75	4.04	8.00	2.343	0.390	0.303	36.1	9.03	2.99	1.53	0.853	0.615	0.553
11.5	3.38	8.00	2.260	0.390	0.220	32.6	8.14	3.11	1.32	0.781	0.625	0.571
C7 × 14.75	4.33	7.00	2.299	0.366	0.419	27.2	7.78	2.51	1.38	0.779	0.564	0.532
12.25	3.60	7.00	2.194	0.366	0.314	24.2	6.93	2.60	1.17	0.702	0.571	0.525
9.8	2.87	7.00	2.090	0.366	0.210	21.3	6.08	2.72	0.968	0.625	0.581	0.541
C6 × 13	3.83	6.00	2.157	0.343	0.437	17.4	5.80	2.13	1.05	0.642	0.525	0.514
10.5	3.09	6.00	2.034	0.343	0.314	15.2	5.06	2.22	0.865	0.564	0.529	0.500
8.2	2.40	6.00	1.920	0.343	0.200	13.1	4.38	2.34	0.692	0.492	0.537	0.512
C5 × 9	2.64	5.00	1.885	0.320	0.325	8.90	3.56	1.83	0.632	0.449	0.489	0.478
6.7	1.97	5.00	1.750	0.320	0.190	7.49	3.00	1.95	0.478	0.378	0.493	0.484
C4 × 7.25	2.13	4.00	1.721	0.296	0.321	4.59	2.29	1.47	0.432	0.343	0.450	0.459
5.4	1.59	4.00	1.584	0.296	0.184	3.85	1.93	1.56	0.319	0.283	0.449	0.458
C3 × 6	1.76	3.00	1.596	0.273	0.356	2.07	1.38	1.08	0.305	0.268	0.416	0.455
5	1.47	3.00	1.498	0.273	0.258	1.85	1.24	1.12	0.247	0.233	0.410	0.438
4.1	1.21	3.00	1.410	0.273	0.170	1.66	1.10	1.17	0.197	0.202	0.404	0.437

* An American Standard Channel is designated by the letter C followed by the nominal depth in inches and the weight in pounds per foot.

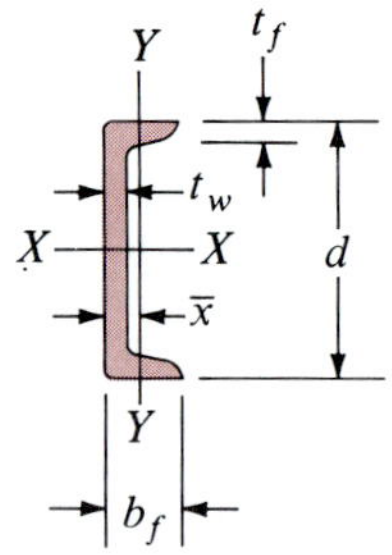

Table C.3b Properties of C shapes (American standard channels)) (SI units)

Designation*	Area A, mm²	Depth d, mm	Flange Width b_f, mm	Flange Thickness t_f, mm	Web thickness t_w, mm	Axis X–X I_x 10^6 mm⁴	Axis X–X S_x 10^3 mm³	Axis X–X r_x mm	Axis Y–Y I_y 10^6 mm	Axis Y–Y S_y 10^3 mm³	Axis Y–Y r_y mm	$\bar{x}$ mm
C380 × 74	9480	381	94	16.5	18.2	168.2	883	133.1	4.58	62.1	22.0	20.3
60	7610	381	89	16.5	13.2	145.3	763	138.2	3.84	55.5	22.5	19.76
50.4	6426	381	86	16.5	10.2	131.1	688	142.7	3.38	51.2	23.0	19.99
C310 × 45	5690	305	80	12.7	13.0	67.4	442	109.0	2.14	34.0	19.38	17.12
37	4742	305	77	12.7	9.8	59.9	393	112.5	1.861	31.1	19.81	17.12
30.8	3929	305	74	12.7	7.2	53.7	352	117.1	1.615	28.7	20.29	17.73
C250 × 45	5690	254	76	11.1	17.1	42.9	338	86.9	1.640	27.6	16.99	16.48
37	4742	254	73	11.1	13.4	38.0	299	89.4	1.399	24.4	17.17	15.67
30	3794	254	69	11.1	9.6	32.8	258	93.0	1.170	21.8	17.55	15.39
22.8	2897	254	65	11.1	6.1	28.1	221	98.3	0.949	18.29	18.11	16.10
C230 × 30	3794	229	67	10.5	11.4	25.4	222	81.8	1.007	19.29	16.31	14.81
22	2845	229	63	10.5	7.2	21.2	185.2	86.4	0.803	16.69	16.79	14.88
19.9	2542	229	61	10.5	5.9	19.94	174.2	88.4	0.733	16.03	16.97	15.27
C200 × 27.9	3555	203	64	9.9	12.4	18.31	180.4	71.6	0.824	16.60	15.21	14.35
20.5	2606	203	59	9.9	7.7	15.03	148.1	75.9	0.637	14.17	15.62	14.05
17.1	2181	203	57	9.9	5.6	13.57	133.7	79.0	0.549	12.92	15.88	14.50
C180 × 22.0	2794	178	58	9.3	10.6	11.32	127.2	63.8	0.574	12.90	14.33	13.51
18.2	2323	178	55	9.3	8.0	10.07	113.2	66.0	0.487	11.69	14.50	13.34
14.6	1852	178	53	9.3	5.3	8.86	99.6	69.1	0.403	10.26	14.76	13.74
C150 × 19.3	2471	152	54	8.7	11.1	7.24	95.3	54.1	0.437	10.67	13.34	13.06
15.6	1994	152	51	8.7	8.0	6.33	83.3	56.4	0.360	9.40	13.44	12.70
12.2	1548	152	48	8.7	5.1	5.45	71.7	59.4	0.288	8.23	13.64	13.00
C130 × 13.4	1703	127	47	8.1	8.3	3.70	58.3	46.5	0.263	7.54	12.42	12.14
10.0	1271	127	44	8.1	4.8	3.12	49.1	49.5	0.199	6.28	12.52	12.29
C100 × 10.8	1374	102	43	7.5	8.2	1.911	37.5	37.3	0.180	5.74	11.43	11.66
8.0	1026	102	40	7.5	4.7	1.602	31.4	39.6	0.133	4.69	11.40	11.63
C75 × 8.9	1135	76	40	6.9	9.0	0.862	22.7	27.4	0.127	4.47	10.57	11.56
7.4	948	76	37	6.9	6.6	0.770	20.3	28.4	0.103	3.98	10.41	11.13
6.1	781	76	35	6.9	4.3	0.691	18.18	29.7	0.082	3.43	10.26	11.10

* An American Standard Channel is designated by the letter C followed by the nominal depth in millimeters and the mass in kilograms per meter.

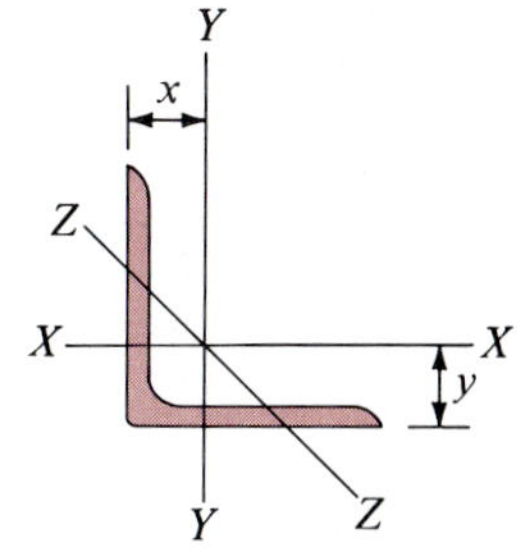

Table C.4a Properties of equal leg angle shapes (U.S. customary units)

Size and thickness, in.	Weight per foot, lb/ft	Area, in²	Axis X–X and axis Y–Y: I, in⁴	S, in³	r, in.	x or y, in.	Axis Z–Z r, in.
L8 × 8 × 1	51.0	15.0	89.0	15.8	2.44	2.37	1.56
3/4	38.9	11.4	69.7	12.2	2.47	2.28	1.58
1/2	26.4	7.75	48.6	8.36	2.50	2.19	1.59
L6 × 6 × 1	37.4	11.0	35.5	8.57	1.80	1.86	1.17
3/4	28.7	8.44	28.2	6.66	1.83	1.78	1.17
5/8	24.2	7.11	24.2	5.66	1.84	1.73	1.18
1/2	19.6	5.75	19.9	4.61	1.86	1.68	1.18
3/8	14.9	4.36	15.4	3.53	1.88	1.64	1.19
L5 × 5 × 3/4	23.6	6.94	15.7	4.53	1.51	1.52	0.975
5/8	20.0	5.86	13.6	3.86	1.52	1.48	0.978
1/2	16.2	4.75	11.3	3.16	1.54	1.43	0.983
3/8	12.3	3.61	8.74	2.42	1.56	1.39	0.990
L4 × 4 × 3/4	18.5	5.44	7.67	2.81	1.19	1.27	0.778
5/8	15.7	4.61	6.66	2.40	1.20	1.23	0.779
1/2	12.8	3.75	5.56	1.97	1.22	1.18	0.782
3/8	9.8	2.86	4.36	1.52	1.23	1.14	0.788
1/4	6.6	1.94	3.04	1.05	1.25	1.09	0.795
L3½ × 3½ × 1/2	11.1	3.25	3.64	1.49	1.06	1.06	0.683
3/8	8.5	2.48	2.87	1.15	1.07	1.01	0.687
1/4	5.8	1.69	2.01	0.794	1.09	0.968	0.694
L3 × 3 × 1/2	9.4	2.75	2.22	1.07	0.898	0.932	0.584
3/8	7.2	2.11	1.76	0.833	0.913	0.888	0.587
1/4	4.9	1.44	1.24	0.577	0.930	0.842	0.592
L2½ × 2½ × 1/2	7.7	2.25	1.23	0.724	0.739	0.806	0.487
3/8	5.9	1.73	0.984	0.566	0.753	0.762	0.487
1/4	4.1	1.19	0.703	0.394	0.769	0.717	0.491
3/16	3.07	0.902	0.547	0.303	0.778	0.694	0.495
L2 × 2 × 3/8	4.7	1.36	0.479	0.351	0.594	0.636	0.389
1/4	3.19	0.938	0.348	0.247	0.609	0.592	0.391
1/8	1.65	0.484	0.190	0.131	0.626	0.546	0.398

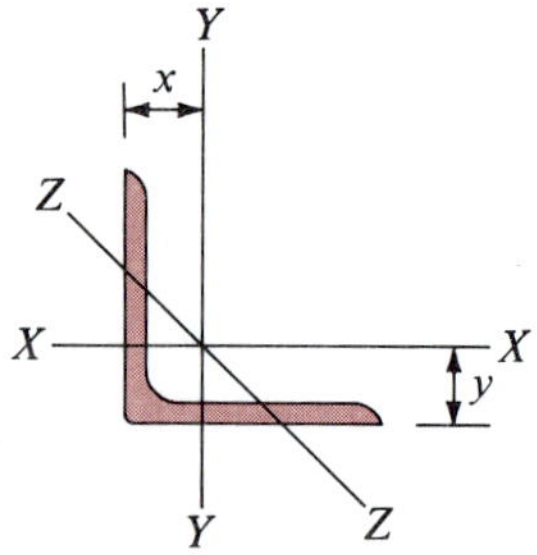

Table C.4b Properties of equal leg angle shapes (SI units)

Size and thickness, mm	Mass per meter, kg/m	Area, mm²	Axis X–X and axis Y–Y: *I* 10^6 mm	*S* 10^3 mm³	*r* mm	*x* or *y* mm	Axis Z–Z *r* mm
L203 × 203 × 25.4	75.9	9680	37.0	259	61.8	60.2	39.6
19.0	57.9	7360	29.0	200	62.8	57.9	40.1
12.7	39.3	5000	20.2	137.0	63.6	55.6	40.4
L152 × 152 × 25.4	55.7	7100	14.78	140.4	45.6	47.2	29.7
19.0	42.7	5445	11.74	109.1	46.4	45.2	29.7
15.9	36.0	4590	10.07	92.8	46.8	43.9	30.0
12.7	29.2	3710	8.28	75.5	47.2	42.7	30.0
9.5	22.2	2800	6.41	57.8	47.8	41.7	30.2
L127 × 127 × 19.0	35.1	4480	6.53	74.2	38.2	38.6	24.8
15.9	29.8	3780	5.66	63.3	38.7	37.6	24.8
12.7	24.1	3070	4.70	51.8	39.2	36.3	25.0
9.5	18.3	2330	3.64	39.7	39.5	35.3	25.1
L102 × 102 × 19.0	27.5	3510	3.19	46.0	30.1	32.3	19.76
15.9	23.4	2970	2.77	39.3	30.5	31.2	19.79
12.7	19.0	2420	2.31	32.3	30.9	30.0	19.86
9.5	14.6	1845	1.815	24.9	31.4	29.0	20.0
6.4	9.8	1252	1.265	17.21	31.8	27.7	20.2
L89 × 89 × 12.7	16.5	2100	1.515	24.4	26.9	26.9	17.35
9.5	12.6	1600	1.195	18.85	27.3	25.7	17.45
6.4	8.6	1090	0.837	13.01	27.7	24.6	17.63
L76 × 76 × 12.7	14.0	1774	0.924	17.53	22.8	23.7	14.83
9.5	10.7	1361	0.733	13.65	23.2	22.6	14.91
6.4	7.3	929	0.516	9.46	23.6	21.4	15.04
L64 × 64 × 12.7	11.4	1452	0.512	11.86	18.78	20.5	12.37
9.5	8.7	1116	0.410	9.28	19.17	19.35	12.37
6.4	6.1	768	0.293	6.46	19.53	18.21	12.47
4.8	4.6	581	0.228	4.97	19.81	17.63	12.57
L51 × 51 × 9.5	7.0	877	0.1994	5.75	15.08	16.15	9.88
6.4	4.7	605	0.1448	4.05	15.47	15.04	9.93
3.2	2.4	312	0.0791	2.15	15.92	13.87	10.11

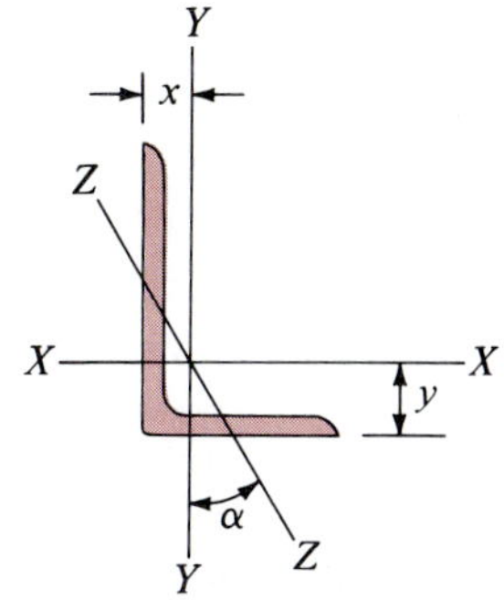

Table C.5a Properties of unequal angle leg shapes (U.S. customary units)

			Axis X–X				Axis Y–Y				Axis Z–Z	
Size and thickness, in.	**Weight per foot, lb/ft.**	**Area, in²**	I_x in⁴	S_x in³	r_x in.	y in.	I_y in⁴	S_y in³	r_y in.	x in.	r_z in.	tan α
L8 × 6 × 1	44.2	13.0	80.8	15.1	2.49	2.65	38.8	8.92	1.73	1.65	1.28	0.543
¾	33.8	9.94	63.4	11.7	2.53	2.56	30.7	6.92	1.76	1.56	1.29	0.551
½	23.0	6.75	44.3	8.02	2.56	2.47	21.7	4.79	1.79	1.47	1.30	0.558
L6 × 4 × ¾	23.6	6.94	24.5	6.25	1.88	2.08	8.68	2.97	1.12	1.08	0.860	0.428
½	16.2	4.75	17.4	4.33	1.91	1.99	6.27	2.08	1.15	0.987	0.870	0.440
⅜	12.3	3.61	13.5	3.32	1.93	1.94	4.90	1.60	1.17	0.941	0.877	0.446
L5 × 3 × ½	12.8	3.75	9.45	2.91	1.59	1.75	2.58	1.15	0.829	0.750	0.648	0.357
⅜	9.8	2.86	7.37	2.24	1.61	1.70	2.04	0.888	0.845	0.704	0.654	0.364
¼	6.6	1.94	5.11	1.53	1.62	1.66	1.44	0.614	0.861	0.657	0.663	0.371
L4 × 3 × ½	11.1	3.25	5.05	1.89	1.25	1.33	2.42	1.12	0.864	0.827	0.639	0.543
⅜	8.5	2.48	3.96	1.46	1.26	1.28	1.92	0.866	0.879	0.782	0.644	0.551
¼	5.8	1.69	2.77	1.00	1.28	1.24	1.36	0.599	0.896	0.736	0.651	0.558
L3½ × 2½ × ½	9.4	2.75	3.24	1.41	1.09	1.20	1.36	0.760	0.704	0.705	0.534	0.486
⅜	7.2	2.11	2.56	1.09	1.10	1.16	1.09	0.592	0.719	0.660	0.537	0.496
¼	4.9	1.44	1.80	0.755	1.12	1.11	0.777	0.412	0.735	0.614	0.544	0.506
L3 × 2 × ½	7.7	2.25	1.92	1.00	0.924	1.08	0.672	0.474	0.546	0.583	0.428	0.414
⅜	5.9	1.73	1.53	0.781	0.940	1.04	0.543	0.371	0.559	0.539	0.430	0.428
¼	4.1	1.19	1.09	0.542	0.957	0.993	0.392	0.260	0.574	0.493	0.435	0.440
L2½ × 2 × ⅜	5.3	1.55	0.912	0.547	0.768	0.831	0.514	0.363	0.577	0.581	0.420	0.614
¼	3.62	1.06	0.654	0.381	0.784	0.787	0.372	0.254	0.592	0.537	0.424	0.626

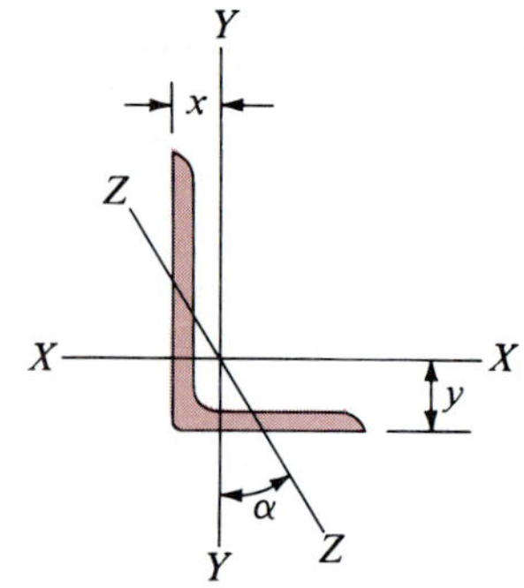

Table C.5b Properties of unequal angle leg shapes (SI units)

Size and thickness, mm	Mass per meter kg/m	Area, mm²	Axis X–X				Axis Y–Y				Axis Z–Z	
			I_x 10^6 mm⁴	S_x 10^3 mm³	r_x mm	y mm	I_y 10^6 mm⁴	S_y 10^3 mm³	r_y mm	x mm	r_z mm	tan α
L203 × 152 × 25.4	65.5	8390	33.6	247	63.3	67.3	16.15	146.2	43.9	41.9	32.5	0.543
19.0	50.1	6410	26.4	192	64.2	65.0	12.78	113.4	44.7	39.6	32.8	0.551
12.7	34.1	4350	18.44	131	65.1	62.7	9.03	78.5	45.6	37.3	33.0	0.558
L152 × 102 × 19.0	35.0	4480	10.20	102.4	47.7	52.8	3.61	48.7	28.4	27.4	21.8	0.428
12.7	24.0	3060	7.24	71.0	48.6	50.5	2.61	34.1	29.2	25.1	22.1	0.440
9.5	18.2	2330	5.62	54.4	49.1	49.3	2.04	26.2	29.6	23.9	22.3	0.446
L127 × 76 × 12.7	19.0	2420	3.93	47.7	40.3	44.5	1.074	18.85	21.1	19.05	16.46	0.357
9.5	14.5	1845	3.07	36.7	40.8	43.2	0.849	14.55	21.5	17.88	16.61	0.364
6.4	9.8	1252	2.13	25.1	41.2	42.2	0.599	10.06	21.9	16.69	16.84	0.371
L102 × 76 × 12.7	16.4	2100	2.10	31.0	31.6	33.8	1.007	18.35	21.9	21.0	16.23	0.543
9.5	12.6	1600	1.648	23.9	32.1	32.5	0.799	14.19	22.3	19.86	16.36	0.551
6.4	8.6	1090	1.153	16.39	32.5	31.5	0.566	9.82	22.8	18.69	16.54	0.558
L89 × 64 × 12.7	13.9	1774	1.349	23.1	27.6	30.5	0.566	12.45	17.88	17.91	13.56	0.486
9.5	10.7	1361	1.066	17.86	28.0	29.5	0.454	9.70	18.26	16.76	13.64	0.496
6.4	7.3	929	0.749	12.37	28.4	28.2	0.323	6.75	18.65	15.60	13.82	0.506
L76 × 51 × 12.7	11.5	1452	0.799	16.39	23.5	27.4	0.280	7.77	13.89	14.81	10.87	0.414
9.5	8.8	1116	0.637	12.80	23.9	26.4	0.226	6.08	14.20	13.69	10.92	0.428
6.4	6.1	768	0.454	8.88	24.3	25.2	0.1632	4.26	14.58	12.52	11.05	0.440
L64 × 51 × 9.5	7.9	1000	0.380	8.96	19.51	21.1	0.214	5.95	14.66	14.76	10.67	0.614
6.4	5.4	684	0.272	6.24	19.94	20.0	0.1548	4.16	15.04	13.64	10.77	0.626

APPENDIX D SELECTED BEAM DEFLECTION FORMULAE

CASE A

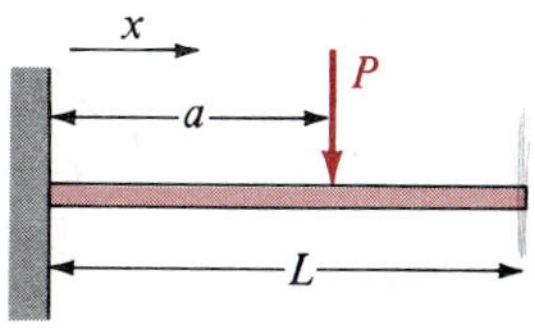

$$EIv' = -\frac{Px(2a - x)}{2} \qquad 0 \le x \le a$$

$$EIv = -\frac{Px^2(3a - x)}{6} \qquad 0 \le x \le a$$

$$EIv' = -\frac{Pa^2}{2} \qquad a \le x \le L$$

$$EIv = -\frac{Pa^2(3x - a)}{6} \qquad a \le x \le L$$

$$v'(L) = -\frac{Pa^2}{2EI}$$

$$v(L) = -\frac{Pa^2(3L - a)}{6EI}$$

CASE B

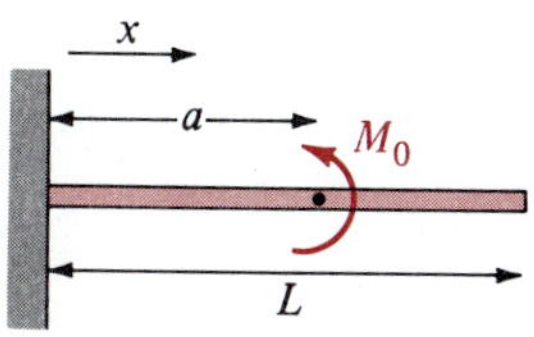

$$EIv' = M_0 x \qquad 0 \le x \le a$$

$$EIv = \frac{M_0 x^2}{2} \qquad 0 \le x \le a$$

$$EIv' = M_0 a \qquad a \le x \le L$$

$$EIv = \frac{M_0 a(2x - a)}{2} \qquad a \le x \le L$$

$$v'(L) = \frac{M_0 a}{EI}$$

$$v(L) = \frac{M_0 a(2L - a)}{2EI}$$

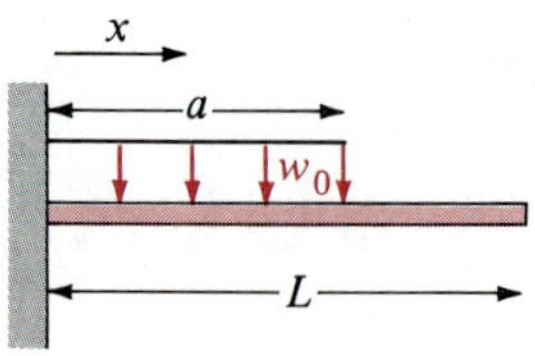

CASE C

$$EIv' = -\frac{w_0 x(3a^2 - 3ax + x^2)}{6} \qquad 0 \le x \le a$$

$$EIv = -\frac{w_0 x^2(6a^2 - 4ax + x^2)}{24} \qquad 0 \le x \le a$$

$$EIv' = -\frac{w_0 a^3}{6} \qquad a \le x \le L$$

$$EIv = -\frac{w_0 a^3(4x - a)}{24} \qquad a \le x \le L$$

$$v'(L) = -\frac{w_0 a^3}{6EI}$$

$$v(L) = -\frac{w_0 a^3(4L - a)}{24EI}$$

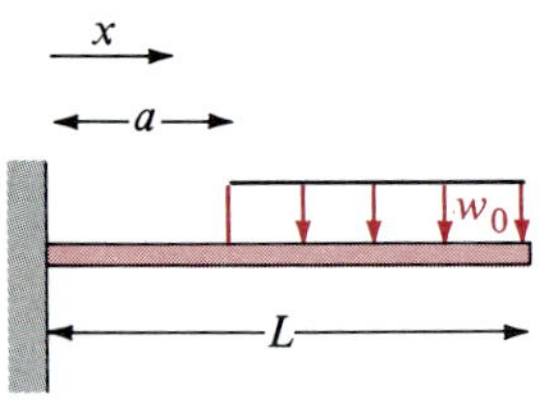

CASE D

$$EIv' = -\frac{w_0 x[L^2 - a^2 - (L - a)x]}{2} \qquad 0 \le x \le a$$

$$EIv = -\frac{w_0 x^2[6(L^2 - a^2) - 4(L - a)x]}{24} \qquad 0 \le x \le a$$

$$EIv' = -\frac{w_0[3Lx(L - x) + x^3 - a^3]}{6} \qquad a \le x \le L$$

$$EIv = -\frac{w_0[x^2(x^2 - 4Lx + 6L^2) - a^3(4x - a)]}{24} \qquad a \le x \le L$$

$$v'(L) = -\frac{w_0(L^3 - a^3)}{6EI}$$

$$v(L) = -\frac{w_0(3L^4 - 4a^3L + a^4)}{24EI}$$

CASE E

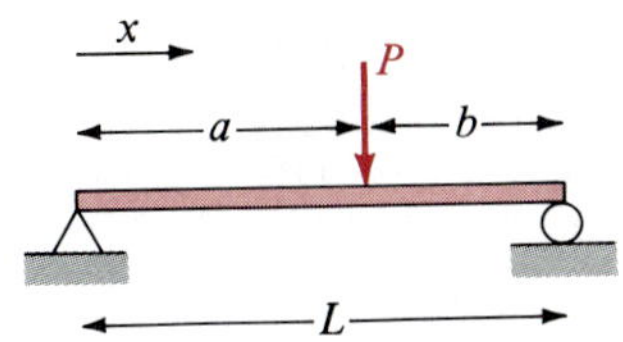

$$EIv' = -\frac{Pb[L^2 - b^2 - 3x^2]}{6L} \qquad 0 \le x \le a$$

$$EIv = -\frac{Pbx[L^2 - b^2 - x^2]}{6L} \qquad 0 \le x \le a$$

$$EIv' = -\frac{P[3L(x-a)^2 - 3bx^2 + b(L^2 - b^2)]}{6L} \qquad a \le x \le L$$

$$EIv = -\frac{P[L(x-a)^3 + bx(L^2 - b^2 - x^2)]}{6L} \qquad a \le x \le L$$

$$v'(0) = -\frac{Pab(L+b)}{6EIL}$$

$$v(L) = \frac{Pab(L+a)}{6EIL}$$

When $a \ge b$, the maximum v occurs at $x_1 = \sqrt{(L^2 - b^2)/3}$ and has the value

$$v(x_1) = v_{\max} = \frac{Pb(L^2 - b^2)^{3/2}}{9\sqrt{3}EI}$$

CASE F

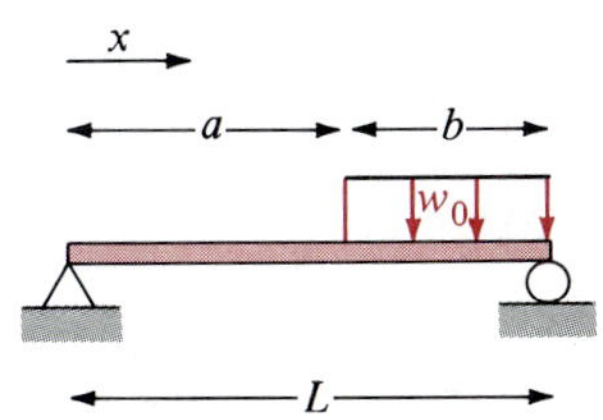

$$EIv' = -\frac{w_0 b^2(2L^2 - b^2 - 6x^2]}{24L} \qquad 0 \le x \le a$$

$$EIv = -\frac{w_0 b^2[(2L^2 - b^2)x - 2x^3]}{24L} \qquad 0 \le x \le a$$

$$EIv' = -\frac{w_0[b^2(2L^2 - b^2 - 6x^2) + 4L(x-a)^3]}{24L} \qquad a \le x \le L$$

$$EIv = -\frac{w_0[b^2x(2L^2 - b^2 - 2x^2) + L(x-a)^4]}{24L} \qquad a \le x \le L$$

$$v'(0) = -\frac{w_0 b^2[2L^2 - b^2]}{24EIL}$$

$$v'(L) = \frac{w_0 b^2(2L - b)^2}{24EIL}$$

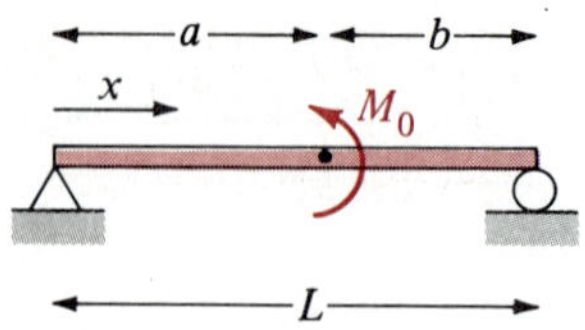

CASE G

$$EIv' = -\frac{M_0[L^2 - 3b^2 - 3x^2]}{6L} \qquad 0 \le x \le a$$

$$EIv = -\frac{M_0[x(L^2 - 3b^2) - x^3]}{6L} \qquad 0 \le x \le a$$

$$EIv' = -\frac{M_0[L^2 - 3b^2 - 3x^2 + 6L(x-a)]}{6L} \qquad a \le x \le L$$

$$EIv = -\frac{M_0[x(L^2 - 3b^2) - x^3 + 3L(x-a)^2]}{6L} \qquad a \le x \le L$$

$$v'(0) = -\frac{M_0[L^2 - 3b^2]}{6EIL}$$

$$v'(L) = \frac{M_0[2L^2 + 3b^2 - 6Lb]}{6EIL}$$

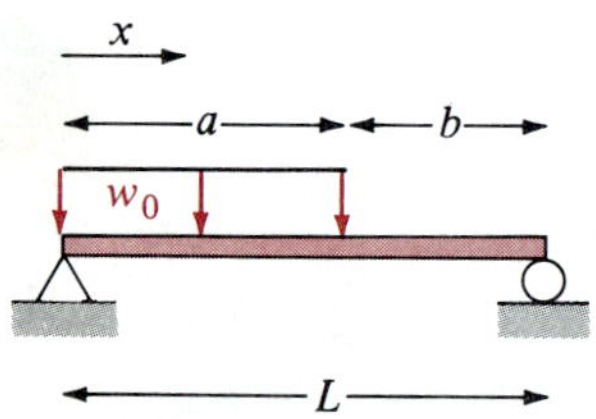

CASE H

$$EIv' = -\frac{w_0[2(L^2 - b^2)(L^2 - 3x^2) + 4Lx^3 - L^4 + b^4]}{24L} \qquad 0 \le x \le a$$

$$EIv = -\frac{w_0 x[2(L^2 - b^2)(L^2 - x^2) + Lx^3 - L^4 + b^4]}{24L} \qquad 0 \le x \le a$$

$$EIv' = -\frac{w_0\{2(L^2 - b^2)(L^2 - 3x^2) + 4L[x^3 - (x-a)^3] - L^4 + b^4\}}{24L} \qquad a \le x \le L$$

$$EIv = -\frac{w_0\{2x(L^2 - b^2)(L^2 - x^2) + L[x^4 - (x-a)^4] - x(L^4 - b^4)\}}{24L} \qquad a \le x \le L$$

$$v'(0) = -\frac{w_0(L^2 - b^2)^2}{24EIL}$$

$$v'(L) = \frac{w_0 a^2(2L^2 - a^2)}{24EIL}$$

ANSWERS TO SELECTED PROBLEMS

APPENDIX E

CHAPTER 2

2.2 $P(x) = 0, 0 < x < a; P(x) = -2P_1, a < x < a + b; P(x) = P_1, a + b < x < a + b + c$

2.4 $P(x) = q_0L/2, 0 < x < L/2; P(x) = q_0(L - x), L/2 < x < L$

2.6 $P(x) = \rho A\omega^2(L^2 - x^2)/2$

2.8 $P(x) = 4\,\pi\gamma Lr_0^2[(1 - 3x/4L)^2 - 1/64]/9$

2.10 $P(x) = -(300 + 45.45[(4 + 1.1x)^3 - 64])$ kips

2.12 $P(x) = -48.3x$ kips

2.14 $P(x) = q_0(L/2 - x + x^2/2L)$

2.16 $P(x) = q_0(L/2 - x^2/2L)$

2.18 $T(x) = T_0$

2.20 $T(x) = t_0(L - x)$

2.22 $V(x) = w_0(L/2 - x); M(x) = w_0x(L - x)/2$

2.24 $V = -M_0/L; M = M_0(1 - x/L)$

2.26 AA: $V = P; M = Pa$ BB: $V = P; M = Pa$

2.28 AA: $P_{\text{axial}} = P; V = 0; M = -Ph$ BB: $P_{\text{axial}} = 0; V = P; M = -Ph/2$

2.30 $V = P; M = PL/2; T = PL$

2.32 $F_{\text{vert}} = P; F_{\text{radial}} = 0.5Q; F_{\text{axial}} = 0.866Q; M_{\text{vert}} = 0.134QR; M_{\text{radial}} = 0.5PR; M_{\text{axial}} = 0.134QR$

2.34 $\sigma_x = 80, \sigma_y = 0, \sigma_z = -50, \tau_{xy} = -30, \tau_{xz} = 40, \tau_{yz} = 0$ MPa

2.36 $\sigma_x = 0, \sigma_y = -117, \sigma_z = 143, \tau_{xy} = 894, \tau_{xz} = -55, \tau_{yz} = 127$ MPa

2.38 $\sigma_x = -22, \sigma_y = 0, \sigma_z = 27, \tau_{xy} = 0, \tau_{xz} = -18, \tau_{yz} = 0$ ksi

2.40 $\sigma_x = 180, \sigma_y = 0, \sigma_z = -142, \tau_{xy} = 0, \tau_{xz} = 167, \tau_{yz} = 0$ MPa

2.42 1.91 ksi

2.44 @A $\sigma_{\text{avg}} = 36.6, \sigma_{\text{bear}} = 73.2, \tau_{\text{avg}} = 140$ MPa
@B $\sigma_{\text{avg}} = 0, \sigma_{\text{bear}} = 73.2, \tau_{\text{avg}} = 140$ MPa

2.46 $P = 60$ kN, 10 kN, -60 kN

2.48 $\tau_{\text{avg}} = 75$ psi

2.50 $\tau_{\text{avg}} = 19.5$ psi

2.52 *(a)* $\sigma_{\text{bear}} = 6.0$ ksi, *(b)* $\tau_{\text{avg}} = 15.3$ ksi, *(c)* $\sigma_{\text{avg}} = 2.4$ ksi

2.54 $L = 9.07$ in.

2.56 $\tau_{\text{avg}} = 26.5$ MPa, $\sigma_{\text{bear}} = 29.3$ MPa

2.58 $\tau_{\text{avg}} = 122$ psi

2.60 $\tau_{\text{avg}} = 2T/[bh(D + b/2)]$

2.62 $\sigma_{\text{avg}} = \rho\omega^2(L^2 - x^2)/2$

2.64 $p_1 = 4T/td, p_2 = 2T/td, p_1/p_{\text{avg}} = 4$

2.66 $\sigma_{\text{avg}} = 22.3$ psi, $\tau_{\text{avg}} = 16.2$ psi

2.68 $\sigma = 37.8$ MPa, $\tau = -40.7$ MPa

2.70 $\sigma = 67.2$ MPa, $\tau = 26.1$ MPa

2.72 $\sigma = 30.0$ MPa, $\tau = 87.9$ MPa

2.74 $\sigma = 111$ kPa, $\tau = 711$ kPa

2.76 $\sigma = -514$ kPa, $\tau = -387$ kPa

2.78 $\sigma = 832$ kPa, $\tau = 17.3$ kPa

2.80 $\sigma = -5.14$ ksi, $\tau = -23.4$ ksi

2.82 $\sigma = 10.2$ ksi, $\tau = 4.73$ ksi

2.84 $\sigma = 35.1$ ksi, $\tau = -3.89$ ksi

2.86 $\sigma = -13.1$ ksi, $\tau = 32.0$ ksi

2.88 $\sigma = 163$ psi, $\tau = 817$ psi

2.90 $\sigma = 605$ psi, $\tau = 11.2$ psi

2.92 $\sigma = 2.68$ ksi, $\tau = -1.06$ ksi

2.94 $\sigma = 1242$ kPa, $\tau = 232$ kPa

2.96 $\sigma = 4.68$ MPa, $\tau = 178$ Mpa

2.98 $\sigma_1 = 12.3$ ksi, $\sigma_2 = -0.32$ ksi, $\tau_{max} = 6.32$ ksi

2.100 $\sigma_1 = 35.4$ ksi, $\sigma_2 = -30.4$ ksi, $\tau_{max} = 32.9$ ksi

2.102 $\sigma_1 = 25$ ksi, $\sigma_2 = -40$ ksi, $\tau_{max} = 32.5$ ksi

2.104 $\sigma_1 = 725$ psi, $\sigma_2 = -1025$ psi, $\tau_{max} = 875$ psi

2.106 $\sigma_1 = 605$ psi, $\sigma_2 = -355$ psi, $\tau_{max} = 480$ psi

2.108 $\sigma_1 = 2.86$ ksi, $\sigma_2 = -3.36$ ksi, $\tau_{max} = 3.11$ ksi

2.110 $\sigma_1 = 1288$ kPa, $\sigma_2 = 62$ kPa, $\tau_{max} = 613$ kPa

2.112 $\sigma_1 = 123$ MPa, $\sigma_2 = -263$ MPa, $\tau_{max} = 193$ MPa

2.114 $\sigma_1 = 53.2$ MPa, $\sigma_2 = -69.2$ MPa, $\tau_{max} =$ 61.2 MPa

2.116 $\sigma_1 = 69.7$ MPa, $\sigma_2 = -200$ MPa, $\tau_{max} =$ 135 MPa

2.118 $\sigma_1 = 134.2$ MPa, $\sigma_2 = -44.2$ MPa, $\tau_{max} =$ 89.2 MPa

2.120 $\sigma_1 = 660$ kPa, $\sigma_2 = -810$ kPa, $\tau_{max} = 735$ kPa

2.122 $\sigma_1 = -256$ kPa, $\sigma_2 = -1094$ kPa, $\tau_{max} =$ 419 kPa

2.124 $\sigma_1 = 832$ kPa, $\sigma_2 = -432$ kPa, $\tau_{max} = 632$ kPa

2.126 $\sigma_1 = 26.4$ ksi, $\sigma_2 = -22.4$ ksi, $\tau_{max} = 24.4$ ksi

2.128 $\sigma = 10.2$ ksi, $\tau = 4.73$ ksi

2.130 $\sigma = 35.2$ ksi, $\tau = -3.9$ ksi

2.132 $\sigma = -13.1$ ksi, $\tau = 32$ ksi

CHAPTER 3

3.2 $u(x, y) = (B - b)x/b$, $v(x, y) = (H - h)y/h$

3.4 $u(x) = \Delta(x/L)$

3.6 $u(x, y) = U(x/b)(y/h)$, $v(x, y) = V(x/b)(y/h)$

3.8 With $\alpha = L/h$,
$\varepsilon = \sqrt{(1 + \alpha \sin \theta)^2 + \alpha^2(1 - \cos \theta)^2} - 1$,
$\varepsilon \approx \alpha\theta$

3.10 $\varepsilon = 0.571$

3.12 $\varepsilon_{AB} = 0.05$, $\varepsilon_{AB} = 0.05$

3.14
$$\varepsilon_{AB} = \sqrt{1 + 2\left[\left(\frac{a}{L}\right)\left(\frac{u}{L}\right) + \left(\frac{b}{L}\right)\left(\frac{v}{L}\right)\right] + \frac{u^2 + v^2}{L^2}} - 1 \approx \left(\frac{u}{L}\right) \sin 30° + \left(\frac{v}{L}\right) \cos 30°$$
$$\varepsilon_{BC} = \sqrt{1 - 2\left[\left(\frac{a}{L}\right)\left(\frac{u}{L}\right) - \left(\frac{b}{L}\right)\left(\frac{v}{L}\right)\right] + \frac{u^2 + v^2}{L^2}} - 1 \approx \left(\frac{u}{L}\right) \sin 30° + \left(\frac{v}{L}\right) \cos 30°$$

3.16 $\varepsilon_{DE} = 0.025$

3.18 $L^* = 1.0267$ m

3.20 $\varepsilon_{radial} = (\Delta b - \Delta a)/(b - a)$, $\varepsilon_{circ} = [\Delta a/a + \Delta b/b]/2$

3.22 $\varepsilon_{AB} = 0$, $\varepsilon_{AC} = v/h$, $\varepsilon_{AC} = \varepsilon_{AC} \sin^2 \theta = (v/h) \sin^2 \theta$ where $\tan \theta = h/L$

3.24 $\varepsilon_{AB} = \varepsilon_{AC} = 0$, $\varepsilon_{AD} = (u/h) \sin \theta \cos \theta$ where $\tan \theta = h/L$

3.28 $\varepsilon = 2(V/L)^2(x/L)^2$ with $\varepsilon_{max} = 2(V/L)^2$

3.30 $\gamma_{avg} = 0.15$

3.32 $\gamma(y) = 2(U/h)(y/h)$, $\gamma(h/2) = U/h$

3.34 $\gamma_{ACB} = 0.05$, $\gamma_{BCE} = -0.05$

3.36 $\gamma_{ACB} = -0.1$, $\gamma_{BCE} = 0.1$

3.38 $\gamma_{st} = -2[(u/b) \cos \phi \sin \phi + (v/h) \sin \phi \cos \phi]$ where $\tan \phi = h/b$

3.40 $\gamma = a\phi'$

3.42 $\varepsilon_x = e/L$, $\varepsilon_y = 0$, $\gamma_{xy} = 0$, $\varepsilon_{diag} = e/2L$, $\gamma = e/L$

3.44 $\varepsilon_x = \alpha y/bh$, $\varepsilon_y = 0$, $\gamma_{xy} = (\alpha/b)(1 + x/h)$

3.48 $\varepsilon_s = e_1 \cos^2 \theta$, $\varepsilon_t = e_1 \sin^2 \theta$, and $\gamma = 2e_1 \sin \theta \cos \theta$

3.50 $\varepsilon_s = \gamma \sin \theta \cos \theta$, $\varepsilon_t = -\gamma \sin \theta \cos \theta$, and $\gamma_{st} = \gamma(\cos^2 \theta - \sin^2 \theta)$

3.54 $\varepsilon_{min} = -62.2\ \mu$, $\varepsilon_{max} = 122.2\ \mu$, $\gamma_{max} = 184.4\ \mu$

3.56 $\varepsilon_{min} = -101.6\ \mu$, $\varepsilon_{max} = -38.4\ \mu$, $\gamma_{max} = 63.2\ \mu$

3.58 $\varepsilon_{min} = -107.7\ \mu$, $\varepsilon_{max} = 107.7\ \mu$, $\gamma_{max} = 215.4\ \mu$

3.60 $\varepsilon_{min} = -62.2\ \mu$, $\varepsilon_{max} = 122.2\ \mu$, $\gamma_{max} = 184.4\ \mu$

3.62 $\varepsilon_{min} = -101.6\ \mu$, $\varepsilon_{max} = -38.4\ \mu$, $\gamma_{max} = 63.2\ \mu$

3.64 $\varepsilon_{min} = -107.7\ \mu$, $\varepsilon_{max} = 107.7\ \mu$, $\gamma_{max} = 215.4\ \mu$

3.66 $\varepsilon_s = -0.000684$, $\varepsilon_t = -0.000916$, $\gamma_{st} = -0.000596$

3.68 $\varepsilon_s = -0.001050$, $\varepsilon_t = -0.000550$, $\gamma_{st} = -0.000400$

3.70 $\varepsilon_x = -0.00874$, $\varepsilon_y = 0.00274$, $\gamma_{xy} = 0.00405$

3.72 $\varepsilon_x = 0.00117$, $\varepsilon_y = -0.00450$, $\gamma_{xy} = 0.00127$

3.74 $\varepsilon_x = 0.00402$, $\varepsilon_y = -0.00202$, $\gamma_{xy} = 0.00097$

3.76 $\varepsilon_{min} = -0.0076$, $\varepsilon_{max} = 0.0026$, $\gamma_{max} = 0.0102$

3.78 $\varepsilon_{min} = -0.00588$, $\varepsilon_{max} = 0.00155$, $\varepsilon_{max} = 0.00736$

3.80 $\varepsilon_{min} = -0.00182$, $\varepsilon_{max} = 0.00575$, $\gamma_{max} = 0.00758$

CHAPTER 4

4.2 $E = 10^7$ psi, $\sigma_{PL} = 42$ ksi, $\sigma_{ULT} = 57$ ksi, $\sigma_Y = 44$ ksi, $E_T = 333$ ksi, $E_S = 1.7 \times 10^6$ psi, $\sigma_{FR} = 47$ ksi, $\sigma_{TRUE} = 56$ ksi, ductile

4.4 $E = 29 \times 10^6$ psi, $\sigma_{PL} = 37.8$ ksi, $\sigma_{ULT} = 57$ ksi, $\sigma_Y = 54.5$ ksi, brittle

4.6 $E = 4 \times 10^6$ psi, $\nu = 0.303$

4.8 $E = 68$ GPa, $\sigma_{PL} = 289$ MPa, $\sigma_{ULT} = 370$ MPa, $\sigma_Y = 295$ MPa, $E_T = 2.2$ GPa, $E_S = 11.3$ GPa, $\sigma_{FR} = 320$ MPa, $\sigma_{TRUE} = 413$ MPa, ductile

4.10 *(a)* $\Delta_D = 0.0438$ m, *(b)* $\Delta_D = 0.0228$ m

4.12 *(a)* $\sigma = \gamma(L - x)$, *(b)* $u(x) = (\gamma/\sigma_0)^n[(L)^{n+1} - (L - x)^{n+1}]/(n + 1)$, *(c)* $u(L) = L(\gamma L/\sigma_0)^n/(n + 1)$

4.14 elong = 0.0021 m

4.16 elong = 0.0105 m

4.18 *(a)* 0.1182 in., *(b)* 0.039 in.

4.20 $\Delta = 0.055$ in.

4.22 $\Delta = 0.0032$ m

4.24 $\Delta = 0.0029$ ft, 0.0116 ft, 0.0463 ft, and 0.185 ft

4.26 *(a)* P_1/P_2, *(b)* $\sigma_1/\sigma_2 = 0.707$

4.28 *(a)* 623 kPa, *(b)* 30.6 MPa

4.30 *(a)* $\sigma = 257$ MPA, *(b)* 91%

4.32 *(a)* 0.353 in., *(b)* 13.1×10^6 lbf

4.34 *(a)* $\sigma_1 = -(\alpha_1 b + \alpha_2 a)k_2 b E_1\, \Delta T/(k_1 a^2 + k_2 b^2)$, $\sigma_2 = -(\alpha_1 b + \alpha_2 a)\, k_1 a E_2\, \Delta T/(k_1 a^2 + k_2 b^2)$, *(b)* $\theta = (k_1\alpha_1 a - k_2\alpha_2 b)\, L\, \Delta T/(k_1 a^2 + k_2 b^2)$

4.36 *(a)* 368°F, *(b)* 49.5 ksi

4.38 *(a)* 0.04 in., *(b)* 0.085 in., *(c)* −0.0009 in., *(d)* $\Delta V = 6$ in^3

4.40 *(a)* $P = 3005$ *kN*, *(b)* $\Delta L = 0.099$ mm

4.42 *(a)* $\Delta L_x = 0.00446$ in., *(b)* $\sigma_x = \sigma_y = 12.9$ ksi

4.44 $\tau_{avg} = P(a + b)/(4\pi Lab)$, $\delta = P(b^2 - a^2)/(4\, \pi abLG)$

4.48 *(a)* 0.0428 mm, *(b)* 0.0744 mm, *(c)* 0.0480 mm

4.50 $E_L = 15$ GPa, $E_T = 4$ GPa, $\nu_{LT} = 0.09$, $\nu_{LT} = 0.24$, $G = 5$ GPa.

4.52 *(a)* 24 ksi, *(b)* 24 ksi

4.54 $-20.48 \text{ ksi} \leq \sigma_0 \leq 0.48$ ksi

4.56 *(a)* $\tau_{max} = 179.2$ MPa ⇒ no yielding, *(b)* $\tau_{max} = 192.4$ MPa ⇒ no yielding, *(c)* $\tau_{max} = 205.7$ MPa ⇒ yielding

4.58 *(a)* $\tau_{max} = 17.9$ ksi ⇒ no yielding, *(b)* $\tau_{max} = 21.4$ ksi ⇒ yielding, *(c)* $\tau_{max} = 25.2$ ksi ⇒ yielding

4.60 *(a)* $\tau_{max} = 15.1$ ksi ⇒ no yielding, *(b)* yielding

4.62 *(a)* $18.79 \text{ ksi} \geq \sigma_z \geq -20.79$ ksi, *(b)* $23.02 \text{ ksi} \geq \sigma_z \geq -25.02$ ksi

4.64 $\tau = 17.9$ ksi

4.66 $\tau = 20.3$ ksi

4.68 *(a)* $FS = 1.47$, *(b)* $FS = 1.47$

4.70 $L = 10.6$ ft

4.72 $P = 1600$ lbf based on bearing

4.74 $FS = 1.30$

CHAPTER 5

5.2 *(a)* $P = 53.3$ kips, *(b)* $\Delta = 0.0173$ in.

5.4 *(a)* $g = 18.67$ mm, *(b)* $P = 70$ kN, *(c)* $\Delta = 0.4$ mm

5.8 $P(0) = P(L) + \int_0^L q\,dx$, an equilibrium equation for the entire bar.

5.10 $\sigma = 22.9$ ksi, $\Delta = 0.0237$ *in.*

5.12 $P = 1000$ lbf

5.14 $\Delta = 0.00315$ in. & $\sigma_{max} = 2$ ksi

5.16 $\Delta = 0.00251$ in. & $\sigma_{max} = 24$ ksi

5.18 $\Delta = 1.69$ mm & $\sigma_1 = \sigma_2 = 500$ MPa

5.20 $\Delta_B = 0.00096$ in. & $\sigma_1 = 8.4$ ksi, $\sigma_2 = 11.2$ ksi

5.22 $\Delta_B = 0.00139$ in. & $\sigma_1 = 6.33$ ksi, $\sigma_2 = 8.44$ ksi

5.24 $\Delta = P(L + a)/AE$

5.26 $\Delta = -3q_0L^2/2AE$

5.28 $\Delta = q_0L^2/3AE$

5.30 As $\varepsilon \to 0$ the loading approaches that of a concentrated load

5.32 $0.4043\gamma L^2/E$

5.34 $\Delta_{approx} = 1.9821\, PL/\pi E r_0^2$, $\Delta_{exact} = 2PL/\pi E r_0^2$

5.36 $\Delta_{approx} = 0.3776\gamma L^2/E$, $\Delta_{exact} = \gamma L^2/3E$

5.38 $\Delta_{approx} = 0.3369\gamma L^2/E$, $\Delta_{exact} = \gamma L^2/3E$

5.40 0.00110 ft

5.42 0.00117 ft

5.44 0.00125 ft

5.46 $9W/8k$

5.48 $T_1 = 2P$, $T_2 = 4P$, $\Delta = 20P/k$, and $\theta = 10P/kL$

5.50 9×10^{-7} m^2, 4.58×10^{-3} m^2

5.52 $P(x) = -f_0(L - x) = -P(1 - x/L)$, $\Delta L = -PL/2AE = -f_0L^2/2AE$

5.54 $P(x) = -f_2(x^2/2L - L/2) = -2P(x^2/2L^2 - 1/2)$, $\Delta L = -f_2L^2/3AE = -2PL/3AE$

5.56 $P(x) = \rho A\omega^2[(L + r_0)^2 - x^2]/2$, $u(x) = \rho\omega^2[(L + r_0)^2 (x - r_0) - (x^3 - r_0^3)/3]/2E$

5.58 With ρ the mass density, $P(x) = \beta\rho Ax/[x_0(x + x_0)]$, $u(x) = \beta\rho\{x - L - x_0 \ln[(x + x_0)/(L + x_0)]\}$

5.60 *(a)* $T_{max} = H\sqrt{1 + 2w_0h/H}$, *(b)* $e = w_0L^3(1 + 4h^2/3L^2)hAE$

5.62 $R_2 = 37.65\ kN \leftarrow$ & $R_1 = 37.65 \leftarrow$

5.64 $R_A = 73.9\ kN \leftarrow$ & $R_D = 33.9\ kN \rightarrow$

5.66 $\Delta = 0.06$ in., $\sigma_{br} = -30$ ksi, & $\sigma_{st} = -20$ ksi

5.68 84.2 kips compression

5.70 *(a)* 570 N, *(b)* −78.1°C

5.72 $x = 47.3$ mm, $\sigma_{al} = -41.8$ MPa, & $\sigma_{st} = -122.5$ MPa

5.74 $F_1 = -k\delta/7$, $F_2 = 5k\delta/7$, $F_3 = -3k\delta/7$; $F_1 = 5P/28 - k\delta/7$, $F_2 = 10P/28 + 5k\delta/7$, $F_3 = 15P/28 - 3k\delta/7$

5.76 $\sigma_{al} = 127$ MPa & $\sigma_{st} = -113$ MPa

5.78 *(a)* $\sigma_c = -2.4$ ksi & $\sigma_{st} = -17.9$ ksi, *(b)* 0.072 in.

5.80 *(a)* $\sigma_{al} = -34$ MPa & $\sigma_{st} = -15.7$ MPa, *(b)* 0.058 mm ←

5.82 25 bars, $\sigma_{st} = -27.6$ ksi

5.84 *(a)* $\sigma_{br} = -5.84$ ksi & $\sigma_{st} = -14.2$ ksi, *(b)* 0.0467 in.

5.86 $T_1 = 0.366\ P$, $T_2 = 0.577P$, and $T_3 = 0.211P$

5.88 $R_A = 73.9\ kN \leftarrow$ & $R_D = 33.9\ kN \rightarrow$

5.90 $\sigma_c = -1.58$ ksi, $\sigma_{st} = -11.8$ ksi

5.92 $F_1 = 5P/8$, $F_2 = 10P/8$, $F_3 = 15P/8$, $\Delta_B = 15P/28k$

5.94 $\sigma_{al} = 127$ MPa, $\sigma_{st} = -113$ MPa

5.96 $\sigma_{al} = -34$ MPa, $\sigma_{st} = -15.7$ MPa, *(b)* 0.058 mm ←

5.98 *(a)* $\sigma_{st} = -14.2$ ksi, $\sigma_{br} = -5.84$ ksi, *(b)* $e = 0.047$ in.

5.100 *(a)* $3P/10K$ ↓, *(b)* $P/10$, $2P/10$, $3P/10$, and $4P/10$

5.102 Yes, $L_2 = 2L_1$, $L_3 = 3L_1$, $L_4 = 4L_1$

5.104 $\sigma_{br} = -72.4$ ksi, $\sigma_{st} = 55.0$ ksi

5.106 $\sigma_{al} = -310$ MPa, $\sigma_{st} = 318$ MPa

5.108 70 kN, 500 MPa

5.110 $F_k = P(n + 1 - 3j(n + 1 - k)/n)/[(n + 1)(2n + 1)]$

5.112 4.9 kips

5.114 50.6 kN

5.116 7.54 kips

5.118 *(a)* 0.00223 rad, *(b)* 0.00120 rad

5.120 $P = 197\ kN$, $\Delta = 0.0679$ mm

5.122 $\Delta = 0.382$ mm, $\Delta_R = 0.0062$ mm

5.124 $\Delta_H = 0.156$ in. ←, $\Delta_V = 0.282$ in. ↓; $\Delta_H = 0$, $\Delta_V = 0.052$ in. ↓

5.126 $\sigma_{bar} = -149$ MPa, $\sigma_{pipe} = -100$ MPa, $\Delta_R = 0.242$ mm

5.128 *(a)* 30.5 ksi, *(b)* 30.5 kips, *(c)* 0.110 in.

5.130 $m = 14.8$ kg, *(a)* 29.6 kg, *(b)* 7.4 kg

5.132 *(a)* 31.9 ksi, *(b)* 25.1 kips, *(c)* 0.192 in.

5.134 18.5 kg

CHAPTER 6

6.2 *(a)* 7.63 N · m, *(b)* 61.1 N · m, *(c)* 552 N · m, *(d)* 4.42 kN · m

6.4 *(a)* 18.6 N · m, *(b)* 148 N · m, *(c)* 1.19 kN · m, *(d)* 9.50 kN · m

6.6 $\tau_{r\theta}$ must be present, small if $d \approx D$

6.10 Equilibrium equation for entire bar, $t = 0$ gives torque is constant.

6.12 $\tau_{max} = t_0L(a/J)$ each segment

6.14 $\tau_{max} = (t_0L/3)(a/J)$ first and second segments, $\tau_{max} = 0$ last segment

6.16 −1.51 ksi, −0.13 ksi, −0.2 ksi, −60.4 ksi

6.18 $\tau_{max} = (t_0L/2)(a/J)$

6.20 Shaft *AD* 20 mm, shaft *EG* 27 mm

6.22 0.884 kN · m

6.24 0.496 kN · m

6.26 *(a)* & *(b)* $\{2(1 - k^4)/[(1 + k)^3(1 - k)] - 1\} \times 100$

6.28 $\tau_{max} = 40$ ksi in steel shaft, $\phi(C) = 0.0804$ rad

6.30 $d = 0.0321$ m, twist governs

6.32 $d = 1.20$ in., $A_h/A_s = 0.48$

6.34 $\tau_{max} \approx 28$ ksi, cross section doesn't vary slowly

6.36 $\tau_{max} = 170$ MPa in CD, $\phi_{max} = \phi_D = -0.0273$ rad

6.38 $\tau_{max} = 19.6$ ksi in CD, $\phi_{max} = \phi_D = 0.0275$ rad

6.40 $d_{AB} = 68$ mm, $d_{BC} = 57$ mm, $\phi_C = 0.00202$ rad

6.42 $\phi_{app}(L) = 6.32\ T_0L/\pi r_1^4G$, $\phi_{app}(L)/\phi_{ex}(L) = 0.677$

6.44 $\phi_{app}(L) = 9.03\ T_0L/\pi r_1^4G$, $\phi_{app}(L)/\phi_{ex}(L) = 0.967$

6.46 *(a)* $\phi_{app}(L) = 10.6\ T_0L/\pi r_1^4G$, $\phi_{app}(L)/\phi_{ex}(L) = 1.13$
(b) $\phi_{app}(L) = 7.62\ T_0L/\pi r_1^4G$, $\phi_{app}(L)/\phi_{ex}(L) = 0.816$

6.48 $\phi_{app}(L) = (32/27)\ T_0L/\pi r_1^3tG$, $\phi_{app}(L)/\phi_{ex}(L) = 0.790$

6.50 $\phi_{app}(L) = 1.472\ T_0L/\pi r_1^3tG$, $\phi_{app}(L)/\phi_{ex}(L) = 0.981$

6.52 *(a)* $\phi_{app}(L) = 1.73\ T_0L/\pi r_1^3tG$, $\phi_{app}(L)/\phi_{ex}(L) = 1.16$
(b) $\phi_{app}(L) = 1.30\ T_0L/\pi r_1^3tG$, $\phi_{app}(L)/\phi_{ex}(L) = 0.864$

6.54 $\tau_{max} = 163$ MPa in CD, $\phi_{max} = 0.114$ rad

6.56 $\tau_{AB} = 15.1$ ksi, $\tau_{CD} = 15.3$ ksi, $\phi(B) = 0.0941$, $\phi(C) = 0.0559$ rad

6.58 $\phi(L) = 2t_0L^2/\pi GtdD^2$

6.60 $r = 0.247$ in.

6.62 $d = 0.127$ in.

6.64 $d = 0.115$ in.

6.66 $d_{AB} = 3.7$ mm, $d_{CD} = 4.5$ mm

6.68 $d_{AB} = 0.141$ in., $d_{CD} = 0.123$ in., $d_{EF} = 0.132$ in.

6.70 $\phi_{A\text{-}C} = 0.0299$ rad, $\phi_{C\text{-}E} = 0.0317$ rad

6.72 $a = (2P/\pi\omega\tau)^{1/3}$

6.74 $T_{AB} = 1393$ ft · lbf, $T_{BC} = 607$ ft · lbf, $\phi(B) = 0.0054$ rad

6.76 $T = 344\ N \cdot m$, $\phi = 0.215$ rad

6.78 $T_{AB} = 2.53$ in · k, $T_{BC} = 0.13$ in · k, $T_{CD} = 3.47$ in · k, $\phi(B) = 0.0085$ rad, $\phi(C) = 0.00871$ rad

6.80 $d_{CD} = d_{EF} = 19$ mm, $\tau_{CD} = 110$ MPa, $\tau_{EF} = 123$ MPa

6.82 $T_2 = (d_3/d_2)^2\ (d_2/d_1)\ C_2T/[C_3 + (d_3/d_2)^2C_2]$, $T_3 = (d_3/d_1)\ C_3T/[C_3 + (d_3/d_2)^2\ C_2]$

6.84 *(a)* 1.03 ksi, *(b)* 1.45 ksi, *(c)* 12.12 in · *k*, *(d)* 2.48 in · k, *(e)* 0.00316 rad

6.86 $d_{core} = 29.8$ mm, $\tau_{st} = 128$ MPa, $\tau_{al} = 90.4$ MPa, 0.172 rad

6.88 $r = 13.6$ mm, $\tau_{st} = 285$ MPa, $\tau_{br} = 129$ MPa

6.90 $T = 44.8$ in · k, $\tau_{al} = 11.8$ ksi, $\phi = 0.0581$ rad

6.92 $r = 1.20$ in., $\tau_{al} = 17.3$ ksi, $\tau_{br} = 19.8$ ksi, $\phi = 0.0797$ rad

6.94 $T = 1.60$ kN · m, $\tau_{al} = 42$ MPa, $\phi = 0.036$ rad

6.96 *(a)* 197 MPa, *(b)* 221 MPa, *(c)* 0.309 kN · m, *(d)* 4.69 kN · m, *(e)* 0.185 rad

6.98 $T_{AB} = 1393$ ft · lbf, $T_{BC} = 607$ ft · lbf, $\phi(B) = 0.0054$ rad

6.100 $T = 344$ N · m, $\phi = 0.215$ rad

6.102 $T_{AB} = 2.53$ in · k, $T_{BC} = 0.13$ in · k, $T_{CD} = 3.47$ in · k, $\phi(B) = 0.0085$ rad, $\phi(C) = 0.00871$ rad

6.104 $d_{CD} = d_{EF} = 19$ mm, $\tau_{CD} = 110$ MPa, $\tau_{EF} = 123$ MPa

6.106 $T_2 = (d_3/d_2)^2(d_2/d_1)C_2T/[C_3 + (d_3/d_2)^2C_2]$, $T_3 = (d_3/d_1)C_3T/[C_3 + (d_3/d_2)^2C_2]$

6.108 $\tau_{max}(AA) = 1.91$ ksi, $\tau_{max}(BB) = 21.4$ ksi, $\tau_{max}(CC) = 15.3$ ksi

6.110 $\tau_{max}(AA) = 10.3$ ksi, $\tau_{max}(BB) = 28.3$ ksi

6.112 1.29

6.114 1.78

6.116 0.029 in.

6.118 0.45 in.

6.120 $T_Y = 5.50$ in · k, 7.10 in · k

6.122 *(a)* 31.4 in · k, *(b)* 41.9 in · k

6.124 $T = 2.92$ kN · m, 0.139 rad

6.126 *(a)* 85.1 in · k

6.128 *(a)* 3.68 kN · m, *(b)* 4.91 kN · m

6.130 $T = 227$ N · m, $\phi = 0.00100$ rad

6.132 $\tau_{max} = 442$ psi, $\phi = 0.00187$ rad

6.134 $d = 31.8$ mm, $d = 33.3$ mm, $d = 36.2$ mm

6.136 $T = 18.8$ in · k, $T = 13.9$ in · k, $d = 9.37$ in · k

6.138 0.736

6.140 56.7 m

6.142 Stresses and twists are the same in both

6.144 *(a)* 1.69, *(b)* 3.37

6.146 $T = 500$ N · m, $\phi = 0.111$ rad

6.148 $T = 11.1$ in · k, $\phi = 0.0766$ rad

CHAPTER 7

7.2 Bar not straight

7.4 Loading and symmetry planes don't coincide

7.6 Bar not straight

7.12 $|V_{max}| = V(0) = V(4 \text{ m}) = 5$ kN, $|M_{max}| = M(2 \text{ m}) = 7.5$ kN · m

7.14 $|V_{max}| = V(0) = V(4a) = w_0a$, $M_{max} = w_0a^2/2$ for $a \leq x \leq 3a$

7.16 $|V_{max}| = V(1 \text{ ft}) = V(5 \text{ ft}) = 2$ kips, $M_{max} = -1$ ft · kip for 1 ft $\leq x \leq$ 5 ft

7.18 $V_{max} = 5$ kN for $0 \leq x \leq 2$ m, $M_{max} = 10$ kN · m for 2 m $\leq x \leq$ 7 m

7.20 $V_{max} = -P$ for $a \leq x \leq 3a$, $|M_{max}| = M(2a^-) = M(2a^+) =$ Pa

7.22 $|V_{max}| = V(0) = V(6\Delta) = 2.5P$, $M_{max} = M(3\Delta) = 4.5P\Delta$

7.24 $V_{max} = V(15 \text{ in.}) = -142$ lbf, $M_{max} = M(15 \text{ in.}) = -1000$ in · lbf

7.26 $V_{max} = V(0.5 \text{ m}) = -71.5$ N, $M_{max} = M(0) = 20$ N · m

7.28 $V_{max} = V(40 \text{ mm}) = -175$ N, $M_{max} = M(22.5 \text{ mm}) = 1531$ N · mm

7.30 $V_{max} = -M_0/2L$ for $0 \leq x \leq 2L$, $|M_{max}| = M(L^-) = M(L^+) = M_0/2$

7.32 $V_{max} = V(0) = 25$ kN, $M_{max} = M(0) = -97.5$ kN · m

7.34 $V_{max} = V(8 \text{ in}^-) = -160$ lbf, $M_{max} = M(8 \text{ in.}) = -640$ in · lbf

7.36 $V_{max} = V(0) = w_0L/2$, $M_{max} = M(0) = -w_0L^2/6$

7.38 $|V_{max}| = V(0) = V(6 \text{ m}) = 30$ kN, $M_{max} = M(3 \text{ m}) = 60$ kN · m

7.40 $V_{max} = V(20 \text{ in}^+) = 686$ lbf, $M_{max} = M(20 \text{ in.}) = -2896$ in · lbf

7.42 $V_{max} = 400$ lbf for $0 \leq x \leq 22$ in., $M_{max} = M(0) = -7200$ in · lbf

7.44 $V_{max} = M_0/L$ for $0 \leq x \leq L$, $M_{max} = M_0$ for $L \leq x \leq 3L/2$

7.46 $V_{max} = V(a) = V(4a) = w_0a$, $M_{max} = M(2a) = 3w_0a^2/2$

7.48 $V_{max} = V(4 \text{ ft}) = -1050$ lbf, $M_{max} = M(2.69 \text{ ft}) = 689$ ft · lbf

7.50 $V_{max} = V(16 \text{ in.}) = 320$ lbf, $M_{max} = M(6 \text{ in.}) = -900$ ft · lbf

7.52 $V_{max} = V(0) = 226$ N, $M_{max} = 200$ N · m for 1.15 m $\leq x \leq$ 1.95 m

7.54 $V_{max} = -P$ for $a \leq x \leq 3a$, $|M_{max}| = M(2a^-) = M(2a^+) =$ Pa

7.56 $V_{max} = V(0) = 50.8$ kN, $M_{max} = M(2.38 \text{ m}) = 61.8$ kN · m

7.58 $V_{max} = -80$ lbf for $0 \leq x \leq 10$ ft, $M_{max} = M(3 \text{ ft}^+) = 560$ ft · lbf

7.60 $V_{max} = 400$ lbf for $0 \leq x \leq 6$ in., $M_{max} = M(0) = -7350$ in · lbf

7.62 $V_{max} = V(2a) = -2w_0a$, $M_{max} = -2w_0a^2$ for $2a \leq x \leq 3a$

7.64 $V_{max} = -15$ N for 105 mm $\leq x \leq$ 215 mm, $M_{max} = 4$ N · m for $0 \leq x \leq$ 105 mm

7.66 $V_{max} = V(2.4 \text{ m}^-) = -56.7$ kN, $M_{max} = -40$ kN · m for 2.4 m $\leq x \leq$ 4.15 m

7.68 $|V_{max}| = V(0) = V(5 \text{ m}) = 50$ kN, $M_{max} = M(2.5 \text{ m}) = 62.5$ kN · m

7.70 $|V_{max}| = V(a^-) = V(a^+) = w_0a$, $M_{max} = M(a) = -w_0a^2/2$

7.72 $V_{max} = V(40 \text{ mm}) = -175$ N, $M_{max} = M(22.5 \text{ mm}) = 1531$ N · mm

7.74 $|V_{max}| = V(0) = V(L) = w_0L/3$, $M_{max} = M(L/2) = 23w_0L^2/216$

7.78 $\varepsilon_{inner} = -0.002$, $\varepsilon_{outer} = 0.002$

7.80 $D = 1999$ in., No

7.82 $h = \varepsilon L/\pi$

7.84 *(a)* 156 in · lbf, *(b)* 1250 in · lbf, *(c)* 10^4 in · lbf, *(d)* 8×10^4 in · lbf

7.86 *(a)* 14.4 N · m, *(b)* 130 N · m, *(c)* 1.04 kN · m, *(d)* 66.7 kN · m

7.88 $v(x) = -Mx(L - x)/2EI$

7.90 *(a)* 0.00206 mm, *(b)* −0.00206 mm

7.92 *(a)* $1/\rho = M/E_2I_{NA}$, *(b)* $\sigma = -My/I_{NA}$ valid only if $E_3 = 0$

7.102 15.9 in^4

7.104 *(a)* $-192M/37bh^2$, *(b)* $168M/37bh^2$

7.106 *(a)* $-8.49M/b^3$, *(b)* $8.49M/b^3$

7.108 *(a)* $-6\sqrt{2}/[a^3(1 + 3\alpha)(1 - \alpha)^2]$, *(b)* $6\sqrt{2}/[a^3(1 + 3\alpha)(1 - \alpha)^2]$

7.110 *(a)* −238 MPa, *(b)* −179 MPa

7.112 106 MPa

7.114 1.63 kN · m

7.116 124 N

7.118 3.79 kips

7.120 6000 lbf, 1.3 in.

7.122 $F = 28.2$ kN, 47.5 mm above and 23.3 mm below the neutral axis

7.124 34.4 kips, 84.3%

7.126 (*a*) 4.13 ksi, (*b*) −4.13 ksi
7.128 (*a*) 10.9 ksi, (*b*) −10.9 ksi
7.130 (*a*) 17.1 ksi, (*b*) −17.1 ksi
7.132 (*a*) 134 MPa, (*b*) −134 MPa
7.134 (*a*) 41.8 MPa, (*b*) −27.3 MPa
7.136 (*a*) 14.1 ksi, (*b*) −14.4 ksi
7.138 (*a*) 10.7 ksi, (*b*) −10.7 ksi
7.140 (*a*) 21.8 ksi, (*b*) −21.8 ksi
7.142 (*a*) 17.2 ksi, (*b*) −17.2 ksi
7.144 (*a*) 137 MPa, (*b*) −137 MPa
7.146 (*a*) 12.3 ksi, (*b*) −12.3 ksi
7.148 753 lbf
7.150 2.54 kN
7.152 6.64 ksi
7.154 107 MPa and −64.1 MPa @ $x = 6.25$ m
7.156 49.8 kN/m
7.158 3.42 k/ft
7.160 (*a*) 13.6 ksi, (*b*) −27.3 ksi
7.162 10.5 ft
7.164 73.8 MPa @ 2 m
7.166 187 MPa
7.168 27.3 ksi
7.172 1063 N · m
7.174 (*a*) 95.8 MPa, (*b*) 178 MPa
7.176 90.9 kN · m
7.178 (*a*) 9.24 ksi, (*b*) 0.58 ksi versus 39.9 ksi and 2.75 ksi
7.180 16.0 kN
7.182 (*a*) 3.14 MPa, (*b*) 49.8 MPa
7.184 (*a*) 8.95 ksi, (*b*) 0.58 ksi
7.186 1.86 kN · m
7.188 (*a*) $-3M/bhy_c$ and $3M/bhy_t$, (*b*) $M = E^*I/\rho$, $E^* = 4E_cE_t/(\sqrt{E_c} + \sqrt{E_t})$
7.190 134 mm
7.192 4.24 in.
7.194 1600 lbf
7.198 $\tau_{\max} = 4P/3\pi a^2$, $3L/a$, $\sigma_{\max} \gg \tau_{\max}$ for normal beams
7.200 $\tau_{\max} = 2w_0L/3\pi a^2$, $3L/4a$, $\sigma_{\max} \gg \tau_{\max}$ for normal beams
7.202 1.07 m
7.204 1.80 m
7.206 $3B/bh$ @ $h/6$ below the neutral axis
7.208 $0.907V/r^2$ @ $0.488r$ above the diameter
7.210 6.79 kN, 1.13 MPa
7.212 94.3 kN, 1.12 MPa
7.214 The y-axis is not an axis of symmetry
7.216 The y-axis is not an axis of symmetry
7.218 $\tau_{\max} \approx 2V/A$
7.220 $\tau_{\max} \approx 2.7V/A$
7.222 $\tau_{\max} \approx 2.22V/A$
7.224 22.9 MPa
7.226 $\tau_{\max} = 16.4$ ksi for $0 \le x \le 6$ ft
7.228 $\tau_{\max} = 109$ MPa for $0 \le x \le 2$ m
7.230 $\tau_{\max} = 66.5$ MPa for $0 \le x \le 2$ m
7.232 $\tau_{\max} = 9.27$ ksi for $0 \le x \le 6$ ft
7.234 $\tau_{\max} = 18.8$ MPa @ $x = 0$
7.236 $\tau_{\max} = 2.72$ ksi @ $x = 0$
7.238 $\tau_{\max} = 0.57$ ksi @ $x = 8$ ft$^+$
7.240 5.05 kips
7.242 $F = (0.428\alpha + 0.0686)V/(0.451\alpha + 0.0608)$
7.244 $F = (0.214\alpha + 0.0686)V/(0.225\alpha + 0.0608)$
7.246 6.17 in.
7.248 8.73 in.
7.250 0.217 m
7.252 (*a*) 0.141 m, (*b*) 0.111 m
7.254 0.069 m
7.256 10.34 kN, 1.21 MPa
7.258 2.63 kN, 1.03 MPa
7.260 7.10 kips, 215 psi
7.262 1131 N, 0.142 MPa
7.264 1.63 in. and 2.03 in., 103 psi
7.266 75.6 mm, 157 kPa
7.268 0.122 m, 532 kPa
7.270 173 kips
7.272 $\sigma(A) = 1.20$ ksi, $\sigma(B) = -2.70$ ksi, $\sigma(C) = -1.20$ ksi, $\sigma(D) = 2.70$ ksi
7.274 8.87 MPa, 16%
7.276 6.16 ksi, 26.2%
7.278 −17.1 ksi at the top left corner
7.280 1.31 kN
7.282 0.678 in., 4.24 ksi
7.284 $(4/\pi - 1)r$ to the left of the section
7.286 $0.127h$ to the left of the section

7.290 (a) 28 mm to the left of the section, (b) A: $\sigma_{max} = 34.6$ MPa, $\sigma_{min} = 0$, $\tau_{abs} = 17.3$ MPa
B: $\sigma_{max} = 34.62$ MPa, $\sigma_{min} = -0.02$ MPa, $\tau_{abs} = 17.32$ MPa

7.292 159 lbf

7.294 23.7 ksi

7.296 (a) 16.2 ksi, (b) 20.0 ksi, (c) 31.0 ksi

7.298 (a) 58.3 N · m, (b) 98.2 N · m, (c) 149 N · m

7.300 14.2 in.

7.304 (a) $\pi a^3\sigma_Y/4$, (b) $4a^3\sigma_Y/3$, (c) 1.70

7.306 (a) $0.114\sigma_Y b^3$, (b) $0.145\sigma_Y b^3$, (c) 1.27

7.308 (a) $0.0310\sigma_Y b^3$, (b) $0.0566\sigma_Y b^3$, (c) 1.79

7.310 $M_P/M_Y = 1.27$

7.312 (a) 1.04×10^6 in · lbf, (b) 1.16×10^6 in · lbf, (c) 1.12

7.314 (a) $c = 0.0193$ m, (b) 20 m, (c) 14 MPa, (d) 13.8 mm

7.316 (a) $M = [2\beta b/(n + 2)\rho^n](h/2)^{n+2}$, (b) $2M(n + 2)/bh^2$

7.318 (b) 19.3 ksi

7.320 (a) 80.3 MPa, (b) 55.8 m

CHAPTER 8

8.2 $v(0) = 0$, $M(0) = M_0$; $v(L) = 0$, $M(L) = 0$

8.4 $V(0) = 0$, $M(0) = 40$ kN · m; $v(6\text{ m}) = 0$, $v'(6\text{ m}) = 0$

8.6 $v(0) = 0$, $M(0) = 0$; $M(1.5L) = 0$, $V(1.5L) = 0$

8.8 $v(0) = 0$, $v'(0) = 0$; $v(L) = 0$, $M(L) = M_0$

8.10 0.125 in., −0.00279 rad, 0.00279 rad

8.12 0.00297 m, −0.00264 rad, 0.00264 rad

8.14 $v(0) = v'(0) = 0$; $M(L) = 0$, $V(L) = P$, $v(x) = -P(3L^2x - x^3)/6EI$

8.16 $M(0) = V(0) = 0$; $v(L) = v'(L) = 0$, $v(x) = -w_0(x^5 + 4L^5 - 5L^4x)/120EIL$

8.18 $v(0) = M(0) = 0$; $v(L) = M(L) = 0$, $v(x) = -w_1(8L^4x - 20L^2x^3 + 15x^4L - 3x^5)/360EIL$

8.20 $v(0) = M(0) = 0$; $v(L) = M(L) = 0$, $v(x) = w_0(5L^3x^3 - 3Lx^5 + x^6 - 3L^5x)/90EIL$

8.22 $M(0) = V(0) = 0$; $v(10\text{ ft}) = v'(10\text{ ft}) = 0$
$v(x) = (-3125 + 375x)$ kip · ft³/3EI $0 \le x \le 5$ ft
$v(x) = [-3125 + 375x - 5(x - 5)^3]$ kip · ft³/3EI 5 ft $\le x \le$ 10 ft

8.24 $v(0) = v'(0) = 0$; $M(5\text{ m}) = V(5\text{ m}) = 0$
$v(x) = (40 + 40(x - 1)^3 - 5x^4 - 120x)$ kN · m³/3EI $0 \le x \le 2$ m
$v(x) = (80 - 160x)$ kN · m³/3EI 2 m $\le x \le$ 5 m

8.26 $v(0) = v'(0) = 0$; $M(7\text{ ft}) = 0$, $V(7\text{ ft}) = 5$ kips
$v(x) = (2.5x^3 - 32.5x^2)$ kip · ft³/EI $0 \le x \le 3$ ft
$v(x) = (5x^3/6 - 17.5x^2 - 45x + 45)$ kip · ft³/EI 3 ft $\le x \le$ 7 ft

8.28 $v(0) = v'(0) = 0$; $M(5\text{ m}) = 0$, $V(5\text{ m}) = 50$ kN
$v(x) = (-135x^2 + 35x^3/3 - 5x^4/12)$ kN · m³/EI $0 \le x \le 2$ m
$v(x) = (-125x^2 + 25x^3/3 - 40x/3 + 20/3)$ kN · m³/EI 2 m $\le x \le$ 5 m

8.30 (a) $\sqrt{3}PL^3/54EI$, (b) $-PL^2/6EI$, (c) $-PL^3/8EI$

8.32 (a) $0.462w_0L^4/EI$, (b) $-w_0L^3/EI$, (c) $3w_0L^4/2EI$

8.34 (a) $\sqrt{3}w_0L^4/216EI$, (b) $-w_0L^2/24EI$, (c) $-11w_0L^4/384EI$

8.36 (a) $9w_0L^4/128\,EI$, (b) $w_0L^3/3EI$, (c) $-5w_0L^4/6EI$

8.38 $v(0) = v'(0) = 0$; $M(L) = 0$, $V(L) = P$, $v(x) = -P(3L^2x - x^3)/6EI$

8.40 $M(0) = V(0) = 0$; $v(L) = v'(L) = 0$, $v(x) = -w_0(x^5 + 4L^5 - 5L^4x)/120EIL$

8.42 $v(0) = M(0) = 0$; $v(L) = M(L) = 0$, $v(x) = -w_1(8L^4x - 20L^2x^3 + 15x^4L - 3x^5)/360EIL$

8.44 $v(0) = M(0) = 0$; $v(L) = M(L) = 0$, $v(x) = w_0(5L^3x^3 - 3Lx^5 + x^6 - 3L^5x)/90EIL$

8.46 $M(0) = V(0) = 0$; $v(10\text{ ft}) = v'(10\text{ ft}) = 0$
$v(x) = (-3125 + 375x)$ kip · ft³/3EI $0 \le x \le 5$ ft
$v(x) = (75x^2 - 2500 - 5x^3)$ kip · ft³/3EI 5 ft $\le x \le$ 10 ft

8.48 $v(0) = v'(0) = 0$; $M(5\text{ m}) = V(5\text{ m}) = 0$
$v(x) = (40x^3 - 5x^4 - 120x^2)$ kN · m³/3EI $0 \le x \le 2$ m
$v(x) = (80 - 160x)$ kN · m³/3EI 2 m $\le x \le$ 5 m

8.50 $v(0) = v'(0) = 0$; $M(7\text{ ft}) = 0$, $V(7\text{ ft}) = 5$ kips
$v(x) = (2.5x^3 - 32.5x^2)$ kip · ft³/EI $0 \le x \le 3$ ft
$v(x) = (5x^3/6 - 17.5x^2 - 45x + 45)$ kip · ft³/EI 3 ft $\le x \le$ 7 ft

8.52 $v(0) = v'(0) = 0$; $M(5\text{ m}) = 0$, $V(5\text{ m}) = 50$ kN
$v(x) = (-135x^2 + 35x^3/3 - 5x^4/12)$ kN · m³/EI $0 \le x \le 2$ m
$v(x) = (-125x^2 + 25x^3/3 - 40x/3 + 20/3)$ kN · m³/EI 2 m $\le x \le$ 5 m

8.54 (a) $\sqrt{3}PL^3/54EI$, (b) $-PL^2/6EI$, (c) $-PL^3/8EI$

8.56 (a) $0.462w_0L^4/EI$, (b) $-w_0L^3/EI$, (c) $3w_0L^4/2EI$

8.58 (a) $\sqrt{3}PL^3/216EI$, (b) $-w_0L^2/24EI$, (c) $-11w_0L^4/384EI$

8.60 (a) $9w_0L^4/128EI$, (b) $w_0L^3/3EI$, (c) $-5w_0L^4/6EI$

8.62 $M(x) = [80x - 10x^2 + 10\langle x-2\rangle^2 - 240]$ kN · m

8.64 $M(x) = w_0b^2x/2L - w_0\langle x-a\rangle^2/2$

8.66 $M(x) = -P\langle x-a\rangle^1 + 2Pa\langle x-2a\rangle^0 + P\langle x-3a\rangle^1$

8.68 $M(x) = w_0Lx/2 - (w_0L/2)\langle x-L/4\rangle^1 - w_0\langle x-L/2\rangle^2$

8.70 $M(x) = M_0\langle x-a\rangle^0 - M_0\langle x-2a\rangle^0$

8.72 $M(x) = [-40 + 335x/13 - 30\langle x-4\rangle^1 - 5\langle x-10\rangle^2]$ k · ft

8.74 $M(x) = [258x/7 - 2x^2 - 70\langle x-4\rangle^1 + (568/7)\langle x-7\rangle^1]$ k · ft

8.76 $M(x) = [-65 + 15x - 10\langle x-3\rangle^1]$ k · ft

8.78 $M(x) = [400x - 2450/3 - 50x^2 + 25\langle x-3\rangle^3/3]$ kN · m

8.80 $w(x) = [20\langle x-0\rangle^{-2} - 27.5\langle x-0\rangle^{-1} + 30\langle x-4\rangle^{-1} + 40\langle x-6\rangle^0 - 82.5\langle x-8\rangle^{-1}]$ kN/m
$v(x) = [-10x^2 + 27.5x^3/6 - 5\langle x-4\rangle^3 - 5\langle x-6\rangle^4/3 - 175x]$ kN · m^3/EI

8.82 $w(x) = [-(258/7)\langle x-0\rangle^{-1} + 4 + 70\langle x-4\rangle^{-1} - (568/7)\langle x-7\rangle^{-1}]$ k/ft
$v(x) = [43x^3/7 - x^4/6 - 35\langle x-4\rangle^3/3 + 284\langle x-7\rangle^3/21 - (1193/6)x]$ k · ft^3/EI

8.84 $w(x) = [-400\langle x-0\rangle^{-1} + (2450/3)\langle x-0\rangle^{-2} + 100 - 50\langle x-3\rangle^1]$ kN/m
$v(x) = [200x^3/3 - 1225x^2/3 - 25x^4/6 + 5\langle x-3\rangle^5/12]$ kN · m^3/EI

8.86 $w(x) = P\langle x-a\rangle^{-1} - 2Pa\langle x-2a\rangle^{-2} - P\langle x-3a\rangle^{-1}$
$v(x) = P[a^2x - 2\langle x-a\rangle^3 + 12a\langle x-2a\rangle^2 + 2\langle x-3a\rangle^3]/12$

8.88 $w(x) = -(w_0L/2)\langle x-0\rangle^{-1} + (w_0L/2)\langle x-L/4\rangle^{-1} + w_0\langle x-L/2\rangle^0 - (w_0L/2)\langle x-L\rangle^{-1}$
$v(x) = w_0[Lx^3/12 - L\langle x-L/4\rangle^3/12 - \langle x-L/2\rangle^4/24 - 35L^3x/768]/EI$

8.90 $w(x) = [-840\langle x-0\rangle^{-1}/11 + 50(1-\langle x-3\rangle^0) - 1800\langle x-5.5\rangle^{-1}/11 + 90\langle x-7.5\rangle^{-1}]$ kN/m
$v(x) = [140x^3/11 - 585x/11 - 25(x^4 - \langle x-3\rangle^4)/12 + 300\langle x-5.5\rangle^3/11]$ kN · m^3/EI

8.92 $w(x) = -3w_0a\langle x-0\rangle^{-1}/4 + w_0 - 9w_0a\langle x-2a\rangle^{-1}/4$
$v(x) = w_0[ax^3/8 - a^3x/6 - x^4/24 + 3a\langle x-2a\rangle^3/8]/EI$

8.94 (a) $PL^3/48EI$ ↑, (b) $PL^2/24EI$ cw, (c) $PL^3/16EI$ ↓

8.96 (a) −0.00888 rad, (b) −0.1576 in.

8.98 (a) $-3Pa^2/2EI$, (b) $-23Pa^3/12EI$

8.100 $M = 4.375$ kN · m

8.102 $v'(A) = -59w_0L^3/648EI$, $v(B) = -w_0L^4/36EI$

8.104 $M_0 = 1.47RL$, $v(B) = -0.00504RL^3/EI$, $v'(C) = 0.0392RL^2/2EI$

8.106 Force $= (5P/16)/(1 + 3EI/kL^3)$, $\Delta = (5P/16)/(k + 3EI/L^3)$ ↓

8.108 (a) $PL^3/4EI$ ↓, (b) $PL^3/12EI$ ↓, (c) $3PL^2/8EI$ ccw, (d) $PL^3/16EI$ ←

8.110 (a) $5PL^3/6EI$ →, (b) $5PL^3/3EI$ ↓, (c) $PL^2/6EI$ ccw

8.112 (a) $PL^3/3EI$ ↓, (b) $PL^3(5/3 + \nu)/EI$ ↓, (c) $PL^3(2/3 + \nu)/EI$ ↑

8.114 $v'(A) = -81w_0L^3/6144EI$, (b) $v(B) = -19w_0L^4/4096EI$

8.116 $v'(0) = -11w_0L^3/384EI$, (b) $-19w_0L^4/2048EI$

8.118 (a) 0.00354 rad, (b) −0.130 in.

8.120 (a) $-w_0L^3/6EI$, (b) $-w_0L^4/8EI$

8.122 (a) $-7Pa^2/6EI$, (b) $-Pa^3/3EI$

8.124 (a) −0.00393 rad, (b) −0.136 in.

8.126 (a) $-Pa^2/9EI$, (b) $Pa^2/18EI$, (c) $-Pa^3/18EI$

8.128 (a) $-7Pa^2/12EI$, (b) $Pa^3/3EI$, (c) $Pa^3/24EI$

8.130 (a) −0.00100 rad, (b) −0.135 in.

8.132 (a) $-5PL^2/96EI$, (b) $PL^2/24EI$, (c) $-PL^3/64EI$

8.134 (a) $-13PL^2/288EI$, (b) $-35PL^3/2592EI$

8.136 $-0.00913wL^4/EI$ @ $x = 0.491L$

8.138 $-13PL^3/12EI$ @ $x = 0$

8.140 $-4.423PL^3/EI$ @ $x = 2.330L$

8.142 $R(B) = 14P/7$ ↑, $M(A) = 4PL/27$ ccw

8.144 $R(B) = k^2(3-k)P/2$ ↑, $M(A) = k(k-1)(k-2)PL/2$ ccw

8.146 $M(A) = Pa/8$ cw, $R(B) = 33P/16$ ↑

8.148 $M(A) = w_0L^2/36$ ccw, $R(B) = 17w_0L/24$ ↑

8.150 $M(A) = w_0L^2/12$ ccw, $M(B) = w_0L^2/12$ cw

8.152 $R(O) = 33w_0L/64$ ↑, $M(0) = 37w_0L^2/384$ ccw, $R(L) = 31w_0L/64$ ↑, $M(L) = 31\,w_0L^2/384$ cw

8.154 $R_1 = 0.217w_0L$ ↑, $R_2 = 0.630w_0L$ ↑, $R_3 = 0.153w_0L$ ↑

8.156 $M_1 = 3.3$ kN·m cw, $R_1 = 2.3$ kN ↑, $R_2 = 47.66$ kN ↑

8.158 $R_1 = 41.6$ kips ↑, $R_2 = 73.9$ kips ↑, $R_3 = 20.5$ kips ↑

8.160 $R_1 = 11w_0L/16$ ↑, $R_2 = 7w_0L/8$ ↓, $R_3 = 35w_0L/16$ ↑

8.162 Top beam $R_1 = R_2 = 11w_0L/32$; Bottom beam $R_1 = R_2 = 5w_0L/32$

8.164 Top beam $R_1 = R_2 = 0.138w_0L$; Bottom beam $R_1 = R_2 = 0.362w_0L$

8.166 *(a)* $(3 + 8\alpha)\, w_0L/16(1 + \alpha)$, *(b)* $5w_0L/8(1 + \alpha)$, *(c)* $5w_0L/8k(1 + \alpha)$ where $\alpha = 48EI/kL^3$

8.168 $R_1 = 12EI\,\Delta/L^3$ ↑, $M_1 = 6EI\,\Delta/L^2$ ccw

8.170 $R_1 = 3P/2$ ↓, $R_2 = 5P/2$ ↑, $M_1 = PL/4$ cw

8.172 $R_1 = 12.5$ kips ↑, $R_2 = 391.5$ kips ↑, $R_3 = 103.4$ kips ↑, $\sigma = 32.5$ ksi

8.174 $R_1 = Pb(3L^2 - b^2)/2L^3$, $M_1 = Pab(L + b)/2L^2$, $R_2 = Pa^2(3L - a)/2L^3$

8.176 $R_1 = 33w_0L/64$, $M_1 = 37w_0L^2/384$, $R_2 = 31w_0L/64$, $M_2 = 31w_0L^2/384$
$v(x) = w_0[33L^3x - 37L^2 x^2/2 - 32L\,\langle x - L/4\rangle^3 - 16\,\langle x - L/2\rangle^4]/384EI$

8.178 $R_1 = 3P/32$ ↓, $R_2 = 11P/16$ ↑, $R_3 = 13P/32$ ↑
$EIv(x) = Pa^2x/16 - Px^3/32 + 11P\,\langle x - 2a\rangle^3/32 - P\,\langle x - 3a\rangle^3/6$

8.180 $M_1 = 2.34$ kN · m cw, $R_1 = 2.53$ kN ↑, $R_2 = 47.47$ kN ↑
$EIv(x) = [x^3/2 - 0.078\,\langle x - 5\rangle^3 - 10\,\langle x - 2\rangle^3/3 + 7.91\,\langle x - 5\rangle^3 + 5.85x - 9.75$ kN · m^3

8.182 $v_s(x) = w_0[(L/2 - x)^2/2 - L^2/8]/kAG$

8.184 $v_s(x) = -w_0(Lx/8 - \langle x - L/2\rangle^2/2)/kAG$

8.186 $v_s(x) = -P(x/3 - \langle x - L/3\rangle^1 + \langle x - 2L/3\rangle^1)/kAG$

8.188 $\sigma = 120$ ksi

8.190 *(a)* 0.421 in., *(b)* $\sigma = 31.8$ ksi

8.192 $\Delta = (mgL^3/3EI)(1 + \sqrt{1 + 6EIh/mgL^3})$,
$\sigma = (mgLc/I)(1 + \sqrt{1 + 6EIh/mgL^3})$

CHAPTER 9

9.2 $\sigma(A) = 216$ MPa, $\sigma(B) = -191$ MPa

9.4 $\sigma_{max} = 30$ ksi, $\sigma_{min} = -15$ ksi

9.6 23.0 kips

9.8 15.2 mm

9.10 $\sigma_{top} = 15.5$ ksi, $\sigma_{bottom} = -25.8$ ksi

9.12 184 lbf

9.14 315 N, 35 mm

9.16 3.86 ft

9.18 15.3 kN

9.20 $\sigma_{top} = 13.3$ ksi, $\sigma_{bottom} = -13.0$ ksi, $\tau_{max} = 0.09$ ksi

9.22 A: $\sigma_{min} = -52.2$ ksi, $\sigma_{max} = 0$, $\tau_{max} = 26.1$ ksi
B: $\sigma_{min} = -1.88$ ksi, $\sigma_{max} = 0.48$ *ksi*, $\tau_{max} = 1.18$ ksi

9.24 $6.5 - 213.7z + 173.6y = 0$, y & z in meters, 112 MPa

9.26 $\sigma(A) = 2.43$ ksi, $\sigma(B) = 2.95$ ksi, $-0.125 - 0.938y - 0.065z = 0$

9.28 6.70 kips

9.30 $\sigma_{max} = 58.7$ MPa at $y = -0.015$ m, and $z = -0.10$ m

9.32 $\pm e_y/b \pm e_z/h < 1/6$

9.34 $e < 0.390a$

9.36 652 kN/m

9.38 E: $\sigma_{max} = 0.08$ ksi, $\sigma_{min} = -3.55$ ksi, $\tau_{abs} = 1.81$ ksi
F: $\sigma_{max} = 0.02$ ksi, $\sigma_{min} = -8.69$ ksi, $\tau_{abs} = 4.35$ ksi
G: $\sigma_{max} = 6.18$ ksi, $\sigma_{min} = -0.05$ ksi, $\tau_{abs} = 3.11$ ksi
H: $\sigma_{max} = 11.35$ ksi, $\sigma_{min} = -0.02$ ksi, $\tau_{abs} = 5.68$ ksi

9.40 E: $\sigma_{max} = 23.7$ ksi, $\sigma_{min} = 0$ ksi; F: $\sigma_{max} = 23.7$ ksi, $\sigma_{min} = -0.02$ ksi
G: $\sigma_{max} = 23.7$ ksi, $\sigma_{min} = 0$ ksi; H: $\sigma_{max} = 20.0$ ksi, $\sigma_{min} = -0.11$ ksi

9.42 -36.4 MPa, 35.0 MPa, 105.0 MPa, 68.4 MPa

9.44 *(a)* $y - 7.68z = -2.28$, *(b)* 26.5 MPa, -10.3 MPa, 8.69 MPa, -28.1 MPa

9.46 *(a)* $y + 0.374z = -0.063$, *(b)* 0.765 in., *(c)* $[-227 - 3623y - 1354z]$ psi

9.48 *(a)* $y + 2.93z = -0.059$, *(b)* 19.8 MPa

9.50 *(a)* $\sigma_{max} = 60.7$ MPa, $\sigma_{min} = -3.37$ MPa, $\tau_{abs} = 32.0$ MPa
(b) $\sigma_{max} = 13.7$ MPa, $\sigma_{min} = -13.7$ MPa, $\tau_{abs} = 13.7$ MPa

9.52 *(a)* $\sigma_{max} = 26.9$ MPa, $\sigma_{min} = -3.33$ MPa, $\tau_{abs} = 15.1$ MPa
(b) $\sigma_{max} = 9.86$ MPa, $\sigma_{min} = -9.86$ MPa, $\tau_{abs} = 9.86$ MPa

9.54 A: $\sigma_{max} = 11.3$ MPa, $\sigma_{min} = 0$, $\tau_{abs} = 5.65$ MPa
B: $\sigma_{max} = 0.60$ MPa, $\sigma_{min} = -0.60$ MPa, $\tau_{abs} =$ 0.60 MPa
C: $\sigma_{max} = 0$, $\sigma_{min} = -11.3$ MPa, $\tau_{abs} = 5.65$ MPa
D: $\sigma_{max} = 0.98$ MPa, $\sigma_{min} = -0.98$ MPa, $\tau_{abs} =$ 0.98 MPa

9.56 Point a: $\sigma = 1.06$ ksi, $\tau = 0$; point b: $\sigma = -3.74$ ksi, $\tau = 0$
Point c: $\sigma = -0.86$ ksi, $\tau = 0$; point d: $\sigma = 3.94$ ksi, $\tau = 0$

9.58 Point a: $\sigma_{max} = 1.06$ ksi, $\sigma_{min} = 0$, $\tau_{abs} =$ 0.58 ksi
Point b: $\sigma_{max} = 0$, $\sigma_{min} = -3.74$ ksi, $\tau_{abs} =$ 1.87 ksi
Point c: $\sigma_{max} = 0$, $\sigma_{min} = -0.86$ ksi, $\tau_{abs} =$ 0.43 ksi
Point d: $\sigma_{max} = 3.94$ ksi, $\sigma_{min} = 0$, $\tau_{abs} =$ 1.97 ksi

9.60 Left: $\sigma_{max} = 2.74$ ksi, $\sigma_{min} = -2.74$ ksi, $\tau_{abs} =$ 2.74 ksi
Top: $\sigma_{max} = 13.7$ ksi, $\sigma_{min} = -1.01$ ksi, $\tau_{abs} =$ 7.36 ksi
Right: $\sigma_{max} = 4.70$ ksi, $\sigma_{min} = -4.70$ ksi, $\tau_{abs} =$ 4.70 ksi
Bottom: $\sigma_{max} = -13.7$ ksi, $\sigma_{min} = 1.01$ ksi, τ_{abs} = 7.36 ksi

9.62 Point A: $\sigma_{max} = 6.23$ ksi, $\sigma_{min} = -16.2$ ksi, $\tau_{abs} = 11.2$ ksi
Point B: $\sigma_{max} = 13.4$ ksi, $\sigma_{min} = -7.64$ ksi, $\tau_{abs} = 10.5$ ksi
Point C: $\sigma_{max} = 16.0$ ksi, $\sigma_{min} = -6.04$ ksi, $\tau_{abs} = 11.0$ ksi
Point D: $\sigma_{max} = 7.29$ ksi, $\sigma_{min} = -13.0$ ksi, $\tau_{abs} = 10.2$ ksi

9.64 $P/2\pi a^2 + \sqrt{(P/2\pi a^2)^2 + (2T/\pi a^3)^2} \le \sigma_0$

9.66 $2M/\pi a^3 + \sqrt{(2M/\pi a^3)^2 + (2T/\pi a^3)^2} \le \sigma_0$

9.68 $2M/\pi a^3 + P/2\pi a^2 + \sqrt{(2M/\pi a^3 + P/2\pi a^2)^2 + (2T/\pi a^3)^2} \le \sigma_0$

9.70 A: $\sigma_{max} = 33.78$ MPa, $\sigma_{min} = -0.02$ MPa, $\tau_{abs} =$ 1.90 MPa
B: $\sigma_{max} = 0.49$ MPa, $\sigma_{min} = -0.49$ MPa, $\tau_{abs} =$ 0.49 MPa

9.72 Between A & B; $\sigma = -18.2$ ksi, $\tau = 2.33$ ksi
Between B & C; $\sigma = -21.5$ ksi, $\tau = 2.33$ ksi
Between C & D; $\sigma = 17.7$ ksi, $\tau = 2.29$ ksi
Between D & A; $\sigma = 21.0$ ksi, $\tau = 2.29$ ksi

9.74 Between A & B; $\sigma_{max} = 0.29$ ksi, $\sigma_{min} =$ -18.5 ksi, $\tau_{abs} = 9.39$ ksi
Between B & C; $\sigma_{max} = 0.25$ ksi, $\sigma_{min} =$ -21.8 ksi, $\tau_{abs} = 11.0$ ksi
Between C & D; $\sigma_{max} = 18.0$ ksi, $\sigma_{min} =$ -0.29 ksi, $\tau_{abs} = 9.14$ ksi
Between D & A; $\sigma_{max} = 20.3$ ksi, $\sigma_{min} =$ -0.25 ksi, $\tau_{abs} = 10.8$ ksi

9.76 5.56m

9.78 3.82 m

9.80 23.5 ft

9.82 13.3 ft

9.84 8.40 in.

9.86 8.92 in.

9.88 *(a)* 83.7 MPa, *(b)* 42.0 MPa

9.90 0.5 in.

9.92 $\sigma_{max} = 44.1$ MPa, $\tau_{max} = 22.1$ MPa

9.94 7.4 mm

9.96 1.67 mm

9.98 60 MPa, 33 MPa, 2.04×10^{-4} m^3

9.100 No

9.102 *(a)* $\sigma_H = 15$ ksi, $\sigma_{ax} = 7.5$ ksi, *(b)* $\sigma_H = 7.5$ ksi, *(c)* 0.0018 in.

9.104 3.75 MPa

9.106 Top left: $\sigma_H = 8.64$ ksi, $\sigma_{ax} = 5.52$ ksi, $\tau_{max} =$ 0.34 ksi
Bottom left: $\sigma_H = 8.64$ ksi, $\sigma_{ax} = 3.12$ ksi, $\tau_{max} = 0.34$ ksi
Bottom right: $\sigma_H = 8.64$ ksi, $\sigma_{ax} = 3.12$ ksi, $\tau_{max} = -0.06$ ksi
Top right: $\sigma_H = 8.64$ ksi, $\sigma_{ax} = 5.52$ ksi, $\tau_{max} =$ -0.06 ksi

9.108 Top left: $\sigma_{max} = 8.68$ ksi, $\sigma_{min} = 5.48$ ksi, $\tau_{abs} = 4.34$ ksi
Bottom left: $\sigma_{max} = 8.66$ ksi, $\sigma_{min} = 3.10$ ksi, $\tau_{abs} = 4.33$ ksi
Bottom right: $\sigma_{max} = 8.64$ ksi, $\sigma_{min} = 3.12$ ksi, $\tau_{abs} = 4.32$ ksi
Top right: $\sigma_{max} = 8.64$ ksi, $\sigma_{min} = 5.52$ ksi, $\tau_{abs} = 4.32$ ksi

9.110 $4(\tau_0 - 3pR/4t)^2 = (pR/4t)^2 + (T/2\pi R^2 t)^2$

9.112 $\tau_0 = pR/4t + M/2\pi R^2 t$

9.114 *(a)* $\pm$ 45°: $\sigma_{ax} = 2.93$ ksi, $\sigma_H = 9.60$ ksi, $\tau_{max} =$ 3.34 ksi, $\pm$ 135°: $\sigma_{ax} = 6.67$ ksi, $\sigma_H = 9.60$ ksi, $\tau_{max} = 1.47$ ksi *(b)* $\pm 45°$: $\sigma_{max} = 9.60$ ksi, $\sigma_{min} = 2.93$ ksi, $\tau_{abs} = 4.82$ ksi, $\pm 135°$: $\sigma_{max} = 9.60$ ksi, $\sigma_{min} = 6.67$ ksi, $\tau_{abs} = 4.82$ ksi

9.116 *(a)* $\pm 45°$: $\sigma_{ax} = 36.7$ MPa, $\sigma_H = 90.0$ MPa, $\tau_{max} = 26.7$ MPa, $\pm 135°$: $\sigma_{ax} = 53.3$ MPa, $\sigma_H = 90.0$ MPa, $\tau_{max} = 18.4$ MPa
(b) $\pm 45°$: $\sigma_{max} = 90.0$ *MPa*, $\sigma_{min} = 36.7$ MPa, $\tau_{abs} = 45.2$ MPa, $\pm 135°$: $\sigma_{max} = 90.0$ MPa, $\sigma_{min} = 53.3$ MPa, $\tau_{abs} = 45.2$ MPa

9.118 *(a)* A: $\sigma_{ax} = 59.6$ MPa, $\sigma_H = 85.3$ MPa, $\tau = 0$
Opposite A: $\sigma_{ax} = -65.2$ MPa, $\sigma_H = 85.3$ MPa, $\tau = 0$
B: $\sigma_{ax} = -2.79$ MPa, $\sigma_H = 85.3$ MPa, $\tau_{max} = 5.35$ MPa
(b) A: $\sigma_{max} = 85.3$ MPa, $\sigma_{min} = 59.6$ MPa, $\tau_{abs} = 42.8$ MPa
Opposite A: $\sigma_{max} = 85.3$ MPa, $\sigma_{min} = -65.2$ MPa, $\tau_{max} = 75.3$ MPa
B: $\sigma_{max} = 85.7$ MPa, $\sigma_{min} = -3.1$ MPa, $\tau_{max} = 44.4$ MPa

9.120 *(a)* $\sigma_{max} = 25.2$ MPa, $\sigma_{min} = -2.9$ MPa, $\tau_{max} = 14.1$ MPa
(b) $\sigma_{max} = 25.2$ MPa, $\sigma_{min} = 10.4$ MPa, $\tau_{max} = 1.27$ MPa

CHAPTER 10

10.2 0.88 m

10.4 *(a)* 45°, *(b)* 63.4°, *(c)* 76.0°

10.6 $\theta = 24.8°$

10.8 $\phi = 48.5°$

10.10 $\theta = 64.2°$

10.12 $\alpha = 10.3°$ and $\alpha = 34.5°$

10.14 $\phi = 6.4°$ and $\phi = 57.6°$

10.16 $F_1 = 18$ lbf, $F_2 = 30$ lbf, and $F_3 = 42$ lbf

10.18 $\theta_1 = 0.00889$ rad, $\theta_2 = 0.0111$ rad

10.20 $\theta_1 = 0.00771$ rad, $\theta_2 = 0.00888$ rad, and $\theta_3 = 0.0151$ rad

10.22 $(k_1/2 + k_2)u + k_1 v/2 = 0$ and $k_1 u/2 + k_1 v/2 = P$

10.24 $k_1 u_1 + k_2(u_1 - u_2) = P_1$ and $k_2(u_1 - u_2) = P_2$

10.26 $u_1 = u_2 = P/2k$ with $F_1 = P/2$, $F_2 = 0$, and $F_3 = -P/2$

10.28 $u_1 = 1.65$ in., $u_2 = 2.45$ in., $F_1 = 33$ lbf, $F_2 = 24.5$ lbf, and $F_3 = -42.6$ lbf

10.30 $\phi = T/(C_1 + C_2)$, $T_1 = C_1 T/(C_1 + C_2)$, and $T_2 = C_2 T/(C_1 + C_2)$

10.32 *(a)* $\boldsymbol{\phi} = (T/10C)[3 \quad 4.5 \quad 4.5 \quad 3]$
$\mathbf{T} = (T/4[2 \quad 1 \quad 0 \quad -1 \quad -2]$
(b) $\boldsymbol{\phi} = (T/14C)[3 \quad 5 \quad 6 \quad 6 \quad 5 \quad 3]$
$\mathbf{T} = (T/6)[3 \quad 2 \quad 1 \quad 0 \quad -1 \quad -2 \quad -3]$
(c) $\boldsymbol{\phi} = (T/24C)[4 \quad 7 \quad 9 \quad 10 \quad 10 \quad 9 \quad 7 \quad 4]$
$\mathbf{T} = (T/8)[4 \quad 3 \quad 2 \quad 1 \quad 0 \quad -1 \quad -2 \quad -3 \quad -4]$

10.34 $E\varepsilon_x^2/2 + E'\varepsilon_x^{n+1}/(n+1)$

10.36 $U_0 = E[(\varepsilon_x - \varepsilon_{0x})^2 + (\varepsilon_y - \varepsilon_{0y})^2 + 2\nu(\varepsilon_x - \varepsilon_{0x})(\varepsilon_y - \varepsilon_{0y})]/2(1 - \nu^2) + G(\gamma_{xy} - \gamma_{0xy})^2/2 + E(\varepsilon_{0x}^2 + \varepsilon_{0y}^2 + 2\nu\varepsilon_{0x}\varepsilon_{0y})/2(1 - \nu^2) + G\,\gamma_{0xy}^2/2$

10.38 $0 = (k_1 + k_2)u_1 - k_2 u_2 - P_1$ and $0 = -k_2 u_1 + k_2 u_2 - P_2$

10.40 $k_1\Delta_1 + k_2[\Delta_1 + a(\Delta_2 - \Delta_1)/L](1 - a/L) + k_3[\Delta_1 + (a + b)(\Delta_2 - \Delta_1)/L](1 - (a + b)/L) = P(1 - x/L)$ and $k_2[\Delta_1 + a(\Delta_2 - \Delta_1)/L](a/L) + k_3[\Delta_1 + (a + b)(\Delta_2 - \Delta_1)/L](a + b)/L = Px/L$

10.42 $k_1\Delta_1 - k_3[\Delta_3 - \Delta_1 - e(\Delta_2 - \Delta_1)/h](1 - e/h) = 0$
$k_2\Delta_2 - k_3[\Delta_3 - \Delta_1 - e(\Delta_2 - \Delta_1)/h](e/h) = 0$
$k_3[\Delta_3 - \Delta_1 - e(\Delta_2 - \Delta_1)/h] = P$

10.44 $C_1\phi_1^2/2 + C_2(\phi_2 - \phi_1)^2/2$

10.46 $C_3(r_1\phi_2/r_2)^2/2 + C_2\phi_2^2/2 + C_1(\phi_1 - \phi_2)^2/2$

10.50 $12v_2 - 6L\theta_2 = PL^3/EI$ and $-6Lv_2 + 4L^2\theta_2 = 0$

10.52 $8L^2\theta_2 - 6Lv_3 + 2L^2\theta_3 = 0$, $-6L\theta_2 + 12v_3 - 6L\theta_3 = PL^3/EI$ and $2L^2\theta_2 - 6Lv_3 + 4L^2\theta_3 = 0$

10.54 $4\theta_1 + 2\theta_2 = 0$, $2\theta_1 + 8\theta_2 + 2\theta_3 = 0$, and $2\theta_2 + 4\theta_3 = ML/EI$

10.56 $\theta = 5.71 \times 10^{-4}$ rad, $F_1 = 1143$ lbf, $F_2 = 3429$ lbf, and $F_3 = 5714$ lbf

10.58 $\Delta_1 = 11P/20k$, $F_1 = 11P/20$, $F_2 = 7P/20$, $F_3 = 3P/20$, and $F_4 = -P/20$

10.60 $F_1 = 12.8$ kN, $F_2 = 4.44$ kN, and $F_3 = 2.78$ kN

10.62 $\Delta_1 = 0.00151$ in., $\Delta_2 = 0.00484$ in., and $\Delta_3 = 0.00173$ in. with $F_1 = 13.6$ kips, $F_2 = 25.7$ kips, $F_3 = -22.4$ kips, and $F_4 = -10.4$ kips

10.64 $u_1 = u_2 = 0.00402$ in. and $F_1 = 20.7$ kips $= F_2$ and $F_3 = -38.6$ kips

10.66 *(a)* $F_1 = 2.65$ kips, $F_2 = -0.44$ kips, *(b)* $F_1 = 4.13$ kips, $F_2 = -1.87$ kips, *(c)* $F_1 = 9.21$ kips, $F_2 = -20.79$ kips

10.68 $u = 0.161$ mm and $v = 0.485$ m

10.70 $\phi_1 = 0.0280$ rad and $\phi_2 = 0.0450$ rad

10.72 $T_1 = 13.5$ in · k, $T_2 = 3.76$ in · k, $T_3 = -3.16$ in · k, and $T_4 = -6.89$ in · k

10.74 $\phi_2 = 0.00456$, $\phi_3 = 0.00396$, $\phi_1 = 0.00410$, and $\phi_4 = 0.00792$
$T_2 = 36.0$ in · k, $T_1 = 32.9$ in · k, $T_3 = -6.33$ in · k, and $T_4 = 3.17$ in · k

10.76 *(a)* $T_1 = 564$ N · m, $T_2 = 436$ N · m, *(b)* $T_1 = 897$ N · m, $T_2 = 1102$ N · m, *(c)* $T_1 = 2360$ N · m, $T_2 = 7640$ N · m

10.78 $\theta_1 = -ML/6EI$ and $\theta_2 = ML/3EI$

10.80 $v_1 = 7PL^3/96EI$, $\theta_1 = -3PL^2/16EI$, and $\theta_2 = -PL^2/16EI$

10.82 $\sigma_x^2/2E + \alpha T\sigma_x$

10.84 $(\sigma_x^2 + \sigma_y^2 - 2\nu\sigma_y\sigma_x)/2E + \tau_{xy}^2/2G + \varepsilon_{0x}\sigma_x + \varepsilon_{0y}\sigma_y + \gamma_{0xy}\tau_{xy}$

10.86 $[(P_1 + P_2)^2 + P_2^2]L/4AE$

10.88 $q_0^2L^3/6AE$

10.90 $2P^2L \ln 4/3Eth_0$

10.92 $2P^2/k_2 + 7P^2/2k_1$

10.94 $t_0^2L^3/6JG$

10.96 $(T_1^2L - t_0L^2T_1 + t_0^2L^3/3)/2JG$

10.98 $(T_1 + T_3 - T_2 - T_4)^2/2C_1 + (T_3 - T_2 - T_4)^2/2C_2 + (T_3 - T_4)^2/2C_3 + T_4^2/2C_4$

10.100 $14T_0^2L/3\pi r_0^4G$

10.102 $[1/C_3 + R_3^2/(C_2R_4^2) + R_1^2R_3^2/(C_1R_2^2R_4^2)]T_0^2/2$

10.104
$$\int_0^a [w_0(L-a)x - w_0(L^2-a^2)/2]^2\,dx/2EI + \int_0^a [w_0(L-a)x - w_0(L^2-a^2)/2 - w_0(x-a)^2/2]^2\,dx/2EI$$

10.106
$$\int_0^{L/3} [Px - M_0 - PL/3]^2\,dx/2EI + \int_{L/3}^{2L/3} M_0^2\,dx/2EI$$

10.108
$$\int_0^a [w_0(L-a)^2x/2L]^2\,dx/2EI + \int_a^L [w_0(L-a)^2x/2L - w_0(x-a)^2/2]^2\,dx/2EI$$

10.110
$$\int_0^{L/2} [0]^2\,dx/2EI + \int_{L/2}^{L}[-w_0(x-L/2)^2/2]^2\,dx/2EI + \int_L^{3L/2}[-w_0(x-L/2)^2/2 + w_0L(x-L)]^2\,dx/2EI$$

10.112
$$\int_0^{L/2} [3Px/2]^2\,dx/2EI + \int_{L/2}^{L} [3Px/2 - 2PL]^2\,dx/2EI + \int_L^{3L/2} [P(x-3L/2)]^2\,dx/2EI$$

10.114 $(PR)^2L/2JG + \int_0^L (Px)^2\,dx/2EI$

10.116
$$U_1^* = Q^2c/2AE + \int_0^c (Ps)^2\,ds/2EI + \int_0^c (Rs)^2\,ds/2EI$$
$$U_2^* = P^2a/2AE + (Rc)^2a/2JG + \int_0^a (Pc - Qs)^2\,ds/2EI + \int_0^a (Rs)^2\,ds/2EI$$
$$U_3^* = R^2b/2AE + (Pc - Qa)^2a/2JG + \int_0^a (Rc - Qs)^2\,ds/2EI + \int_0^a (Ra - Ps)^2\,ds/2EI$$
with $U^* = U_1^* + U_2^* + U_3^*$

10.118 $u_A = 0.078$ in., $u_B = 0.048$ in.

10.120 5.46 in., 74%

10.122 0.275 in.

10.124 0.387 in.

10.126 $L(ML\omega^2/AC)^{2/3}$

10.128 −0.0139 rad, 0.0231 rad, 0.0558 rad

10.130 $28T_0L/3\pi Gr_0^4$

10.132 $124T_0L/3\pi r_0^4G$

10.134 $\phi(A) = 0.171$ rad and $\phi(B) = 0.092$ rad

10.136 $\phi(A) = 0.103$ rad

10.138 $\Delta_B = QL^3/3EI + 27PL^3/128EI$
$\Delta_A = 27QL^3/128EI + 9PL^3/64EI$

10.140 $\Delta_B = QL^3/3EI + 41w_0L^4/384EI + ML^2/2EI$
$\theta_B = QL^2/2EI + 7w_0L^3/48EI + ML/EI$

10.142 $\Delta_A = QL^3/48EI + 5w_0L^4/768EI$
$\theta_A = ML/12EI + w_0L^3/384EI$

10.144 $\Delta_A = Pa^3/6EI - Qa^3/4EI - w_0a^4/8EI$
$\Delta_B = -Pa^3/4EI + Qa^3/EI + 11w_0a^4/24EI$

10.146 $\Delta_A = Pa^3/6EI - w_0a^4/24EI$

10.148 $0.606PL^3/EI_0$, 11% high

10.150 $\Delta_B = 3QL^3/2EI + 5PL^3/12EI$
$\Delta_A = 5QL^3/12EI + PL^3/6EI$

10.152 0.362 in.

10.154 $P(a + c)^3/3EI + Pc^2b/EI$, $11PL^3/3EI$

10.156 $\Delta_B = Qa^3/3EI + (Q + P)b^3/3EI + Qa^2b/JG$
$\Delta_A = (Q + P)b^3/3EI$

10.158 $5PL^3/3EI$

10.160 $\Delta_H = (PL^3/EI)[2\sqrt{2}/3 + 0.25 + (\sqrt{2}/8)(a/L)^2]$, 0.037%
$\Delta_V = (PL^3/EI)(2\sqrt{2}/3 + 3/4 + \sqrt{2}/6) = 1.93PL^3/EI$

10.162 $\Delta_B = P/k_1$, $\Delta_A = P/k_1 + P/k_2$

10.164 $\Delta_H = 0.455$ mm, $\Delta_V = 0$

10.166 $T_2(1/C_1 + 1/C_2) - T_1/C_1$

10.168 0.0130 rad, 0.0156 rad

10.170 0.189 in., 0.00211 rad

10.172 $\theta_A = w_0L^3/48EI$, $\Delta_B = w_0L^4/192EI$

10.176 $\theta_A = PHL/6EI$, $\Delta_A = (PH^2L + PH^3)/3EI$

10.178 *(a)* 0, *(b)* 0, *(c)* 0

10.182 $P_1 = 7.6$ kN, $P_2 = 87.6$ kN, and $P_3 = -122.4$ kN

10.184 $F_4 = 0.425$ W, $F_1 = 0.075$ W, $F_2 = 0.345$ W, and $F_3 = 0.155$ W

10.186 $F_{top} = 21.1$ kips, $F_{bottom} = -38.9$ kips, $F_{left} = 28.2$ kips, $F_{diag\,1} = -35.2$ kips, $F_{diag\,2} = 64.8$ kips, $F_{right} = 28.2$ kips

10.188 $T_1 = T_2 = T_0/2$

10.190 $T_1 = T_4 = T_0/3$, $T_3 = 2T_0/3$, and $T_2 = T_0/3$

10.192 $R_1 = 0.568P$, $M_1 = 0.168PL$, $R_2 = 0.432P$

10.194 $M_1 = 4w_0L^2/81$, $R_1 = 16w_0L/81$, $\Delta = 0.00152w_0L^4/EI$

10.196 $R_2 = 1.26M_0/L$, $\theta = 0.0708M_0L/EI$

10.198 $R_1 = M_0/2L \downarrow$, $R_2 = 0$, $R_3 = M_0/2L \uparrow$, $\theta = M_0L/6EI$

10.200 $R_2 = 11P/16 - 3M_0/2L$, $R_1 = 13P/32 + M_0/4L$, $R_3 = -3P/32 + 5M_0/4L$

10.202 $R_{pin} = P/2 \leftarrow$, $R_{bottom} = P/2 \leftarrow$, $M_{bottom} = PH/6$ ccw

10.204 $R_{pin} = 0.345P \leftarrow$, $R_{pin} = 0.839P \uparrow$

10.206 $T_{wire} = 0.798w_0L$, $R_{pin} = 0.0033w_0L \rightarrow$, $R_{pin} = 0.5w_0L \uparrow$

10.208 $R_{left} = 57P/640 \rightarrow$, $R_{left} = 3P/4 \uparrow$

10.210 $R_{roller} = 5P/22 \uparrow$

10.212 $R_{roller} = 10w_0L/11 \uparrow$

10.214 $R_{roller} = 11wL/8 \uparrow$

10.216 $X = 0.0616P$, $Y = -0.187P$, and $Z = 0.325P$

CHAPTER 11

11.4 $P_{cr} = kL_1/(1 + L_1/L_2)$

11.6 $P_{cr} = C(L_1 + L_2)/L_1L_2$

11.8 $P_{cr} = C/2L$

11.10 $P_{cr} = C/L$

11.12 Equilibrium given by $\theta = 2C\theta_0/(2C - PL)$

11.14 $P_{1\,cr} = ka/3$, $P_{2\,cr} = ka$

11.16 $P_{cr} = 30.8$ N, $\sigma_{cr} = 0.247$ MPa

11.18 $P_{cr} = 2.30$ kips, $\sigma_{cr} = 1.15$ ksi

11.20 $P_{cr} = 19.6$ kips, $\sigma_{cr} = 7.43$ ksi

11.22 $P_{cr} = 79.9$ kN, $\sigma_{cr} = 51.5$ MPa

11.24 $P_{cr} = 10.6$ kips, $\sigma_{cr} = 4.72$ ksi

11.26 $P_{all} = 48.7$ kips, $\sigma_{all} = 5.50$ ksi

11.28 $P_{all} = 341$ kN, $\sigma_{all} = 78$ MPa

11.30 $P_{all} = 15.9$ kips, $\sigma_{all} = 6.02$ ksi

11.32 $P_{all} = 54.3$ kips

11.34 $P_{all} = 117$ lbf

11.36 $Q_{all} = \pi^3Ed^4/(128\sqrt{3}h^2)$

11.38 $P_{all} = 22.7$ kN

11.40 $Q_{all} = 12.1$ kN

11.42 $P_{cr} = 2\pi^2EI/L^2$

11.44 $P_{cr} = 2\pi^2EI/L_2^2$

11.46 $P_{cr} = 5\pi^2EI/3L^2$

11.48 $\Delta T = (\pi^2I/AL^2\alpha)(1 + AE/kL)$

11.50 $P_{cr} = 961$ kips

11.52 $P_{cr} = 597$ kN

11.54 $P_{cr} = 2065$ kN

11.56 $L_1 = 1.43L_2$

11.58 $W_{cr} = 8P_e$

11.60 $W_{cr} = 4.09\pi^2EI/L^2$

11.62 $W_{cr} = 328$ kips

11.64 $P_{all} = 4.15$ kN, $v(L/2) = 0.012$ m, $P_{cr} = 12.82$ kN

11.66 *(a)* $P_{all} = 418$ lbf, *(b)* $v(L/2) = 2.03$ in., $P_{cr} = 493$ lbf

11.68 (*a*) $P_{\text{all}} = 36.1$ kN, (*b*) $v(L/2) = 0.018$ m, $P_{\text{cr}} =$ 101 kN

11.70 (*a*) $P_{\text{all}} = 3.74$ kips, (*b*) $v(0) = 1.06$ in., $P_{\text{cr}} =$ 9.08 kips

11.72 (*a*) $P_{\text{all}} = 10.7$ kN, (*b*) $v(0) = 0.034$ m, $P_{\text{cr}} =$ 26.3 kN

11.74 (*a*) $P_{\text{all}} = 6.00$ kips, (*b*) $v(0) = 2.38$ in., $P_{\text{cr}} =$ 12.6 kips

11.76 $\sigma = -\dfrac{P}{A}\left[1 + \dfrac{v_0 c}{r^2}\dfrac{P_{\text{cr}}}{(P_{\text{cr}} - P)}\right],$

$$A\sigma(P_{\text{cr}} - P) + P(P_{\text{cr}} - P) + P\,\frac{v_0 c}{r^2}\,P_{\text{cr}}A = 0$$

SUBJECT INDEX